MCWASP XIV: International Conference on Modelling of Casting, Welding and Advanced Solidification Processes

IOP Conference Series: Materials Science and Engineering
Volume 84

Hyogo, Japan
21-26 June 2015

ISBN: 978-1-5108-0748-8
ISSN: 1757-8981

Printed from e-media with permission by:

Curran Associates, Inc.
57 Morehouse Lane
Red Hook, NY 12571

Some format issues inherent in the e-media version may also appear in this print version.

Printed by Curran Associates, Inc. (2015)

For permission requests, please contact the Institute of Physics
at the address below.

Institute of Physics
Dirac House, Temple Back
Bristol BS1 6BE UK

Phone: 44 1 17 929 7481
Fax: 44 1 17 920 0979

techtracking@iop.org

Additional copies of this publication are available from:

Curran Associates, Inc.
57 Morehouse Lane
Red Hook, NY 12571 USA
Phone: 845-758-0400
Fax: 845-758-2634
Email: curran@proceedings.com
Web: www.proceedings.com

TABLE OF CONTENTS

IOP Conf. Series: Materials Science and Engineering **84** (2015) 011001 doi:10.1088/1757-899X/84/1/011001

PREFACE

The current volume represents contributed papers of the proceedings of the 14th international conference on "Modeling of Casting, Welding and Advanced Solidification Processes (MCWASP XIV)", Yumebutai International Conference Center, Awaji island, Hyogo, Japan on 21 – 26 June, 2016.

The first conference of the series 'Modeling of Casting, Welding and Advanced Solidification Processes (MCWASP)' was started up in 1980, and this is the 14th conference. The participants are more than 100 scientists from industry and academia, coming from 19 countries. In the conference, we have 5 invited, 70 oral and 31 poster presentations on different aspects of the modeling.

The conference deals with various casting processes (Ingot / shape casting, continuous casting, direct chill casting and welding), fundamental phenomena (nucleation and growth, dendritic growth, eutectic growth, micro-, meso- and macrostructure formation and defect formation), coupling problems (electromagnetic interactions, application of ultrasonic wave), development of experimental / computational methods and so on. This volume presents the cutting-edge research in the modeling of casting, welding and solidification processes.

I would like to thank MAGMA Giessereitechnologie GmbH, Germany and SCSK Corporation, Japan for supporting the publication of contributed papers.

Hideyuki Yasuda
Conference Chairman
Department of Materials Science and Engineering,
Kyoto University
Japan

IOP Conf. Series: Materials Science and Engineering **84** (2015) 011002 doi:10.1088/1757-899X/84/1/011002

Committee members of MCWASP XIV

Chair and Co-chairs
Hideyuki Yasuda, Kyoto University, Japan
Hervé Combeau, Ecole des Mines de Nancy, France
Charies A. Monroe, University of Alabama at Birmingham, USA

Steering Committee
Alain Jacot, ESI Group-Calcom, Lausanne, Switzerland
Andre Phillion, University of British Columbia, Canada
Andreas Ludwig, University of Leoben, Austria
Brian G. Thomas, University of Illinois, US
Charies A. Monroe, University of Alabama at Birmingham, USA
Charles-Andre Gandin, MINES- ParisTech, CEMEF, France
Christoph Beckermann, University of Iowa, US
Daan Maijer, University of British Columbia, Canada
Hani Henein, Alberta University, Canada
Hervé Combeau, Ecole des Mines de Nancy, France
Hideyuki Yasuda, Kyoto University, Japan
Ingo Steinbach, Bochum University, Germany
Ingvar L. Svensson, Jönköping University, Sweden
Itsuo Ohnaka, Osaka University, Japan
Jon Dantzig, University of Illinois, US
Koulis Pericleous, University of Greenwich, UK
Laurens Katgerman, Delft University, Netherland
Laurentiu Nastac, University of Alabama, US
Marc Schneider, MAGMA GmbH, Aachen, Germany
Mark Cross, Swansea University, UK
Mark Jolly, Cranfield University, UK
Mathew Krane, Purdue University, US
Michel Bellet, MINES- ParisTech, CEMEF, France
Michel Rappaz, Ecole Polytechnique Fédérale de Lausanne, Switzerland
Srinath Viswanathan, University of Alabama, US
Steve Cockcroft, University British Columbia, Canada
Weng-Sing Hwang, National Cheng Kung University, Taiwan

Local Organizing Committee
Akira Sugiyama, Osaka Sangyo University
Hisao Esaka, National Defense Academy
Kenichi Ohsasa, Akita University, Japan
Kohei Morishita, Kyoto University, Japan
Masato Yoshiya, Osaka University, Japan
Tomoya Nagira, Osaka University, Japan

Peer review statement

All papers published in this volume of *IOP Conference Series: Materials Science and Engineering* have been peer reviewed through processes administered by the proceedings Editors. Reviews were conducted by expert referees to the professional and scientific standards expected of a proceedings journal published by IOP Publishing.

Uncertainty Quantification in Solidification Modelling

K Fezi and M J M Krane

Purdue Center for Metal Casting Research, School of Materials Engineering, Purdue University

kfezi@purdue.edu

Abstract. Numerical models have been used to simulate solidification processes, to gain insight into physical phenomena that cannot be observed experimentally. Often validation of such models has been done through comparison to a few or single experiments, in which agreement is dependent on both model and experimental uncertainty. As a first step to quantifying the uncertainty in the models, sensitivity and uncertainty analysis were performed on a simple steady state 1D solidification model of continuous casting of weld filler rod. This model includes conduction, advection, and release of latent heat was developed for use in uncertainty quantification in the calculation of the position of the liquidus and solidus and the solidification time. Using this model, a Smolyak sparse grid algorithm constructed a response surface that fit model outputs based on the range of uncertainty in the inputs to the model. The response surface was then used to determine the probability density functions (PDF's) of the model outputs and sensitivities of the inputs. This process was done for a linear fraction solid and temperature relationship, for which there is an analytical solution, and a Scheil relationship. Similar analysis was also performed on a transient 2D model of solidification in a rectangular domain.

1. Introduction

Sophisticated numerical models have been developed to simulate solidification processes and confidence in the results is dependent on how well the input parameters are known and the practicality of comparison with limited experimental data. Some of the well-established physics incorporated into sophisticated solidification models of industrial processes include diffusion and convective transport of heat, mass, and momentum, electromagnetics, and solidification thermodynamics. When validating a model with experimental data, the discussed level of agreement is typically qualitative and frequently the assumption is made that more sophisticated physics would improve numerical results. Sensitivity analysis would be able to evaluate this question and determine which areas need more development.

Monte Carlo uncertainty quantification methods consist of sampling random values of the model inputs or system parameters that govern the model and evaluating the solution [1]. This process generates a probability distribution function (PDF) for output quantities that are of interest as a function of input uncertainties. Although Monte Carlo sampling methods are very effective, they quickly become computationally very expensive for complex numerical models. Use of polynomial chaos expansion (PCE) or global polynomial chaos (gPC), which create surrogate models, have been shown as a vast improvement in terms of computational expense without sacrificing accuracy [2]. The present study aims to quantify the sensitivities and uncertainties in a single, 1D, steady state solidification model as a template to demonstrate a method that can be used in more complex models.

2. Model Description

2.1. Description of 1D Deterministic Model

The solidification process being modeled is for continuous casting of weld filler rod consisting of primary cooling by a copper mold and secondary cooling either by water spray or natural convection. Figure 1 depicts the geometry and boundary conditions for the process.

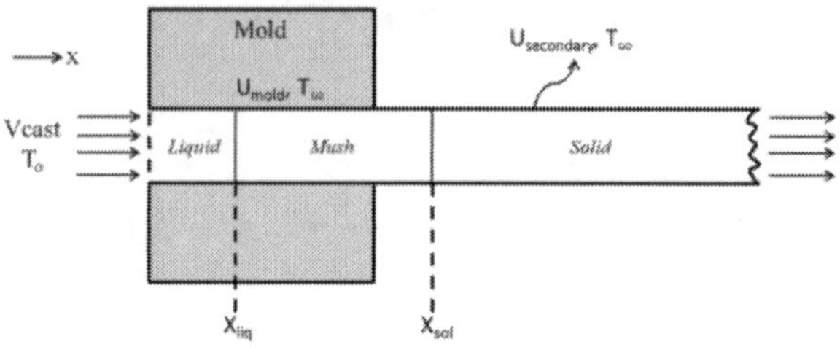

Figure 1. Schematic diagram of the simulation domain showing the boundary conditions.

This model is applied here to the prediction of weld filler rod with a radius of 1 mm. The Biot number (Bi= hR/2k) is assumed to be less than unity, so any radial temperature gradients are assumed to be very small and a one dimensional treatment of the steady state governing equation is valid:

$$kA_c \frac{\partial^2 T}{\partial x^2} - \rho c A_c V_{cast} \frac{\partial T}{\partial x} + \rho L_f A_c V_{cast} \frac{\partial f_s}{\partial x} - UP(T - T_\infty) = 0 \quad , \tag{1}$$

where A_c is the cross-sectional area, L_f is the latent heat, U is the heat transfer coefficient, c is the specific heat, k is the thermal conductivity, f_s is the fraction solid and P is the perimeter. The latent heat advection term requires a relationship between the temperature and the solid fraction. This model will use either a linear dependence for fraction solid or the Scheil model shown in equation 2, where T_m is the melting temperature of the pure material and k_p is the partition coefficient. Equation 1 was discretized using the finite volume method and upwind difference advection scheme over a uniform grid. The alloy system used in this study was Ni alloy 625 comprised of 21.5 wt. % Cr, 9.0 wt. % Mo and 3.65 wt. % Nb. The properties of alloy 625 are summarized in Table 1. The model outputs of interest are the liquidus position (X_L), solidus position (X_S), and solidification time ($t_S=[X_S-X_L]/V_{cast}$).

$$\text{Linear: } f_s = \frac{T_{liq} - T}{T_{liq} - T_{sol}} \quad \text{or Scheil: } f_s = 1 - \left(\frac{T - T_m}{T_{liq} - T_m} \right)^{1/(k_p - 1)} \tag{2}$$

Table 1. Thermophysical properties of alloy 625.

Thermal Conductivity [W/mK]	25.2	Liquidus Temperature [K]	1636
Density [kg/m^3]	8440	Solidus Temperature [K]	1430
Specific Heat [J/kgK]	670	Pure Ni Melting Temperature [K]	1728
Latent Heat [J/kg]	290,000	Partition Coefficient	0.83

Discretization of the governing equation is straightforward except when in the mushy zone, where the advection of latent heat and the fraction solid to temperature relationship becomes important. A robust discretization is achieved by substituting the fraction solid-temperature relationship into the governing equation only for the "P" control volume. For the linear relationship, this discretization is non-trivial, but for the Scheil relationship a linearization of the source term is needed, as outlined by Patankar [3]. Using the linear f_s-T dependence in equation 2 and neglecting the conduction term in equation 1 (which can be shown to be valid with a simple scaling analysis for the cases in this study) equation 1 can be solved exactly for the temperature profile in the all-liquid and mushy zones.

$$\text{Liquid zone: } \frac{T - T_\infty}{T_O - T_\infty} = \exp\left[-\left(\frac{2U}{\rho C V_{cast} R}\right)x\right] \tag{3a}$$

$$\text{Mushy zone: } \frac{T - T_\infty}{T_{liq} - T_\infty} = \exp\left[-\left(\frac{St}{St+1}\right)\left(\frac{2U}{\rho C R V_{cast}}\right)(x - X_L)\right] \tag{3b}$$

2.2. *Uncertainty Quantification Model Description*

This paper does not introduce new methods of uncertainty quantification, but uses existing methods on computational models, therefore only a brief description of the methodology will be discussed here. The uncertainty quantification model calculates a PDF for each output quantity of interest given the known uncertainty of select inputs (ρ, C, L_f, k, and U_m). This is done by assigning uncertainty to input parameters in the form of a normal distribution and then sampling outputs from the previously described 1-D model to generate a response surface, which then acts as a surrogate to running the full 1-D model. There are various sampling methods including Monte Carlo sampling (MC), Latin Hypercube sampling (LHS), etc. [4], with MC being ideal, but not all models are simplistic enough such that MC is computationally inexpensive. In this work, the Smolyak sparse grid algorithm uses the input PDFs to define the simulation conditions used to construct response surfaces of the model outputs. This algorithm requires far fewer sampling points than either MC or LHS [2,5–7]. Using the response surfaces, the uncertainty of the outputs can be gathered from the deviation of the PDFs and the relative sensitivity of the input parameters is calculated based on the Elementary Effects Method (EEM) [8,9].

The EEM calculates two sensitivity measures to determine the effect on the inputs. The mean sensitivity, μ^*, assesses the main effects on the output plus all the interaction effects of an input factor. Unlike the analytical sensitivity, which is the derivative of the output function with respect to an input factor, μ^* does not carry any sign effects and is always positive. The other parameter is the standard deviation, σ, which examines higher order effects. A high value of σ indicates the effects of one input are strongly influenced by other input factors. A low value of σ indicates a linear dependence on the input factor. This model of continuous casting of weld filler rod has been made available to the public through nanohub.org, in which users can perform uncertainty analysis for themselves [10].

3. Results and Discussion

3.1. *1D Model of Continuous Casting of Filler Rod*

The numerical casting model was compared to the analytical solution with a linear f_s-T relationship, only one heat transfer coefficient, and with conduction (Case A). The model was also run with the Scheil solidification model and two heat transfer coefficients; for these conditions there is no analytical solution (Case B). All of the calculations used a rod radius of 0.001 m, a casting speed of 0.2 m/s, a superheat of 20 K, and an ambient temperature of 300 K. Table 2 lists the input uncertainty used in both cases in the form of normal distributions, with a mean, μ, and standard deviation, σ. A level 1 Smolyak run was used to calculate each response surface in Case A for comparison to the analytical solution. Figure 2 shows the PDF of each output from the model for Case A. The uncertainty of the analytical solution (equations 4 and 5) was calculated with equation 6, where y is an output quantity of interest, x is an uncertain input, and m is the number of uncertain inputs. The comparison between the analytical and numerical uncertainties is shown in Table 3, where the analytical mean was calculated using the mean values in Table 2. The analytical and numerical solutions agree very well with each other with the most certain result being X_L followed by X_S and the most uncertain calculation being t_S.

$$X_L = -\left(\frac{\rho C R V_{cast}}{2U}\right)\ln\left(\frac{T_{liq} - T_\infty}{T_O - T_\infty}\right) \tag{4}$$

$$X_S = X_L - \frac{St+1}{St}\frac{\rho C R V_{cast}}{2U}\ln\left(\frac{T_{sol} - T_\infty}{T_{liq} - T_\infty}\right) \tag{5}$$

$$\sigma_y = \sqrt{\sum_{i=1}^{m}\left(\sigma_i \frac{\partial y_i}{\partial x_i}\right)^2} \tag{6}$$

Table 2. Uncertainty of input parameters in the form of a normal distribution.

Input Parameter	Case A	Case B
Thermal Conductivity [W/mK]	-	μ: 25 σ: 0.63
Density [kg/m^3]	μ: 8440 σ: 42.2	μ: 8440 σ: 211
Specific Heat [J/kgK]	μ: 670 σ: 3.35	μ: 670 σ: 16.5
Latent Heat [J/kg]	μ: 290,000 σ: 1450	μ: 290,000 σ: 7250
Primary Heat Transfer Coefficient [W/m^{2}K]	μ: 5,000 σ: 25	μ: 5,000 σ: 750
Secondary Heat Transfer Coefficient [W/m^{2}K]	-	μ: 10,000 σ: 1,500
Partition Coefficient	-	μ: 0.83 σ: 0.02075

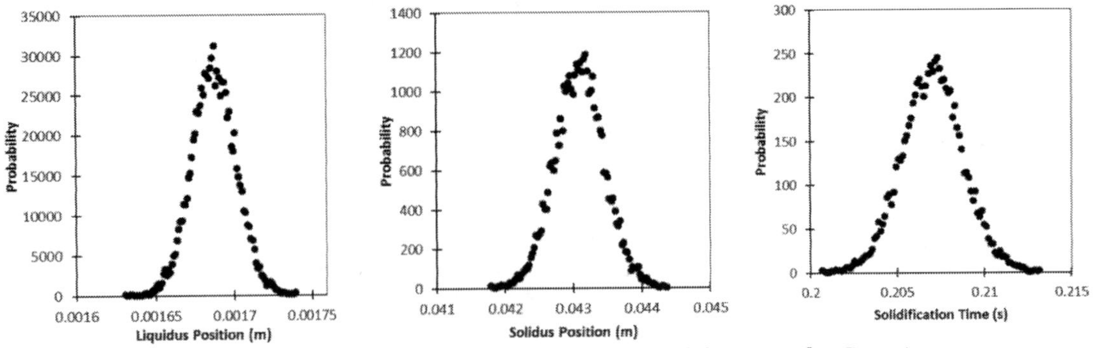

Figure 2. Probability density functions of the model outputs for Case A.

Table 3. Comparison between the analytical and model uncertainties for Case A.

Output	Analytical Solution	Case A
X_L (m)	μ: 0.00168 σ: 1.48x10^{-5}	μ: 0.00169 σ: 1.40x10^{-5}
X_S (m)	μ: 0.0431 σ: 3.59x10^{-4}	μ: 0.0431 σ: 3.58x10^{-4}
t_S (s)	μ: 0.207 σ: 1.74x10^{-3}	μ: 0.207 σ: 1.74x10^{-3}

A more complicated case (Case B) was run for which there is no analytical solution and which uses a Scheil model and has two different heat transfer coefficients. The alloy was simplified to a Ni- 9 wt. % Mo binary in which the partition coefficient was assumed to be constant. A level 3 Smolyak sparse grid algorithm was used because of the better fit over levels 1 and 2 according to the root mean square error (RMSE). The output uncertainties are given in Table 4 and the sensitivities are given in Table 5. The most uncertain output is the solidification time and the most certain is the location of the liquidus temperature. All outputs are most sensitive to the mold heat transfer coefficient, followed by material properties such as ρ, c, or L_f. To produce physically accurate results the uncertainty of the measured mold heat transfer coefficient needs to be much less than the 30% assumed here. Even though the partition coefficient determines the freezing range of the alloy, other material properties have a larger influence on the outputs.

Table 4. Output uncertainties for Case B.

X_L (m)	μ: 0.00184 σ: 2.94x10^{-4}	X_S (m)	μ: 0.054 σ: 4.31x10^{-3}	t_S (s)	μ: 0.261 σ: 1.93x10^{-2}

Table 5. Sensitivity of the uncertain inputs on the outputs for Case B.

Sensitivity Parameter	**X_L (m)**	**X_S (m)**	**t_S (s)**
Density (kg/m^3)	μ^*: 2.66x10^{-4} σ: 9.38x10^{-5}	μ^*: 4.94x10^{-3} σ: 1.42x10^{-3}	μ^*: 2.33x10^{-3} σ: 7.17x10^{-3}
Specific Heat (J/kgK)	μ^*: 2.66x10^{-4} σ: 9.39x10^{-5}	μ^*: 1.78x10^{-3} σ: 5.11x10^{-4}	μ^*: 7.57x10^{-3} σ: 2.66x10^{-3}
Latent Heat (J/kg)	μ^*: 5x10^{-7} σ: 4.97x10^{-6}	μ^*: 3.16x10^{-3} σ: 9.09x10^{-4}	μ^*: 1.58x10^{-2} σ: 4.55x10^{-3}
Thermal Conductivity (W/mK)	μ^*: 3.85x10^{-6} σ: 1.68x10^{-5}	μ^*: 5x10^{-7} σ: 4.97x10^{-6}	μ^*: 2.18x10^{-5} σ: 9.04x10^{-5}
Partition Coefficient	μ^*: 1.25x10^{-6} σ: 7.81x10^{-6}	μ^*: 1.35x10^{-4} σ: 4.77x10^{-5}	μ^*: 6.81x10-^{4} σ: 2.41x10^{-4}
Mold Heat Transfer Coefficient (W/m^{2}K)	μ^*: 2.02x10^{-3} σ: 1.09x10^{-3}	μ^*: 2.69x10^{-2} σ: 5.96x10^{-3}	μ^*: 1.24x10^{-1} σ: 3.21x10^{-2}
Secondary Heat Transfer Coefficient (W/m^{2}K)	μ^*: 0 σ: 0	μ^*: 5.74x10^{-3} σ: 6.54x10^{-3}	μ^*: 2.87x10^{-2} σ: 3.27x10^{-2}

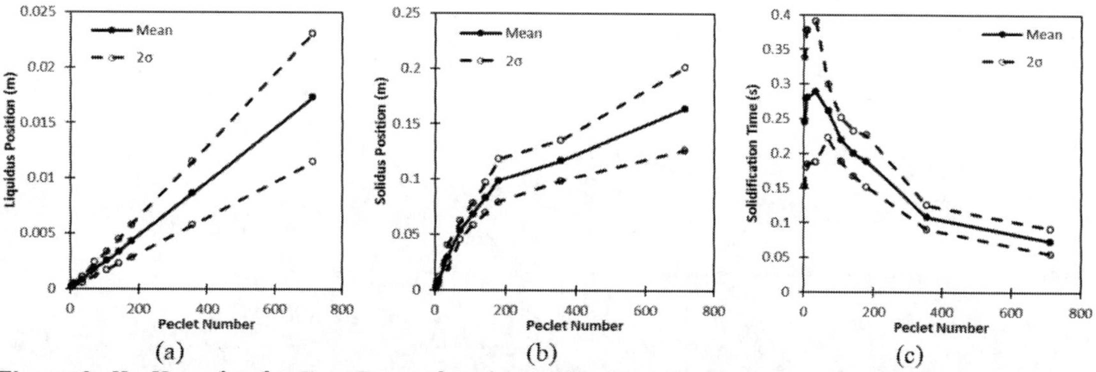

Figure 3. X_L, X_S, and t_S for Case B as a function of Pe where the black dots are the mean values and the white dots are 2σ away from the mean.

Case B was run for various Peclet numbers ($Pe=L_cV_{cast}/\alpha$) in which L_C is the characteristic length taken from the classical fin theory ($L_C = \sqrt{kA_c / U_m P}$) where A_C is the cross-sectional area [11]. Figure 3 shows the results from these simulations. Both the mean and deviation of X_L and X_S increase with Pe but at different rates. As all the heat being extracted from the liquid is through the mold, the change in X_L with Pe is linear. At low Pe, X_S has a relatively large slope as the majority of the heat extraction is through the mold and the advection of L_f is minimal. As Pe increases, the advection of L_f and the secondary heat transfer coefficient become more important causing the slope of X_S to decrease. The uncertainty in t_S is large at low Pe but decreases substantially for large Pe. At low Pe, X_L and X_S are similar in order of magnitude making small differences result in large uncertainties in t_S, while, for large Pe, X_S is orders larger than X_L so small fluctuations do not impact t_S.

3.2. 2D Model of Static Casting with Fluid Flow

The uncertainty quantification approach in Section 2.2 is now applied to a 2D continuum mixture model of transient transport phenomena during solidification of an Al-4.5 wt. % Cu alloy [12]. For this study, the permeability model of the Darcy drag in the mushy zone is adjusted to examine the effect of the uncertainty of each permeability model on average macrosegregation levels and solidification time. The 20 cm by 20 cm mold is cooled from one side using a constant heat transfer coefficient of 1000 W/m^2K. The permeability models are given in equations (7-9), where λ is the secondary dendrite arm spacing, g_l is the liquid volume fraction, and g_s is the solid volume fraction [4].

$$\text{KI:} \quad K = \frac{\lambda^2 g_l^3}{180 g_s^2} \tag{7}$$

$$\text{KII:} \quad K = \frac{\lambda^2 g_l^3}{180 g_s^{4/3}} \tag{8}$$

$$\text{KIII:} \quad K = \frac{\lambda^2 g_l^3}{80 g_s} \tag{9}$$

These models were chosen based on their behavior in the low g_s region of the mushy zone, shown in Figure 4, which is where there is flow in the mushy zone strong enough to affect heat and solute transport [5]. Table 6 gives the properties used as inputs to the model, where the mean and standard deviation are given for the secondary dendrite arm spacing. A level 3 Smolyak response surface was constructed for each permeability model and had the best fit to the simulation output. The response surfaces are shown in Figure 5 as probability density functions and plotted together to show the differences in the predicted macrosegregation number (equation 10) and solidification time.

$$M^{Cu} = \left[\frac{1}{V_{casting}} \iiint_{Volume} \left(\frac{C^{Cu}}{C_o^{Cu}} - 1 \right)^2 dV \right]^{\frac{1}{2}} \tag{10}$$

Table 6. Property values for Al-4.5 wt. % Cu alloy.

Density (kg/m^3)	2605	Thermal Expansion (1/K)	$1.17\text{x}10^{-4}$
Specific Heat (J/kgK)	1006	Melting Temperature of Pure Al (K)	933.5
Latent Heat (J/kg)	390,000	Eutectic Temperature (K)	821.2
Thermal Conductivity (W/mK)	137.5	Eutectic Composition (wt. fr. Cu)	0.33
Dynamic Viscosity (kg/ms)	0.0014	Maximum solubility of Cu in Al (wt. fr. Cu)	0.0565
Thermal Expansion (1/K)	$1.17\text{x}10^{-4}$	Secondary dendrite arm spacing (μm)	μ= 100 σ=10.5

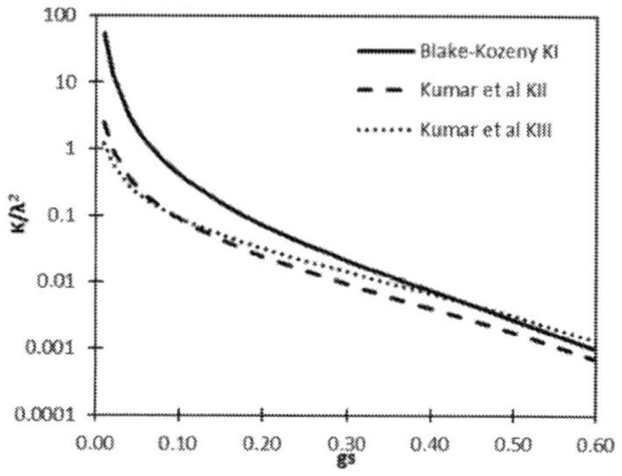

Figure 4. Normalized permeability as a function of solid fraction [4].

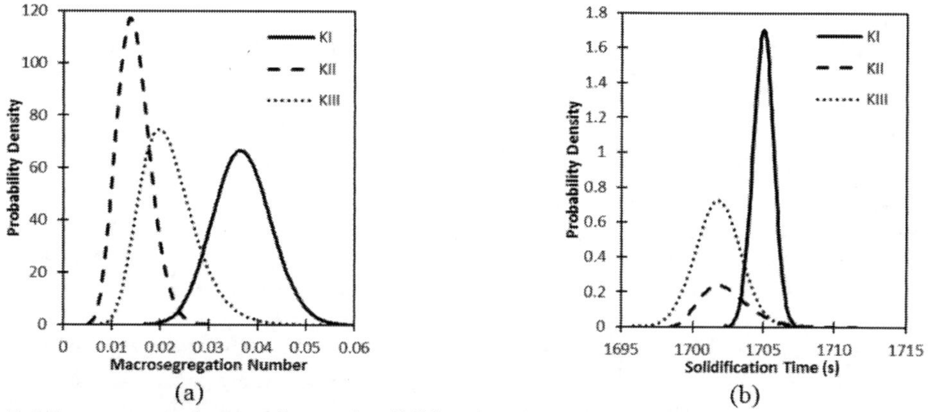

Figure 5. Macrosegregation number and solidification time probability density functions showing the effect of the permeability model chosen.

The permeability model KI has the largest amount of segregation on average and the largest uncertainty in the model prediction (Figure 5a), as expected from its larger K values at low g_s (Figure 4). The uncertainty in M^{Cu} due to KI has some overlap with KIII and even less with KII. Model KII has the narrowest range of probabilities for segregation and, while KII has a much lower mean segregation level than KI, its range is roughly the same. The mean values of solidification times for KII and KIII are the same, whereas KI has a slightly longer t_S. The higher interdendritic flow with KI causes more segregation and the enriched bulk fluid slows the overall time to freeze. Still, the mean values are within 2% of each other, indicating that the uncertainties in the K models have only a small effect on solidification time, which is expected because the flow is a secondary effect on t_S.

4. Conclusions

The procedure we used here was able to quantify uncertainties in a simplified 1D steady state solidification model of weld filler rod that agreed well with the analytical solution. A more realistic and complicated model (Case B) showed the inputs to which the model is most sensitive are the heat transfer coefficients. Material properties that had the most influence on the outputs are the specific heat, density, and, for X_S and t_S, the latent heat. In order to produce accurate results, these inputs parameters need to be measured to lesser uncertainty than the 5% assumed here. The uncertainty in t_S decreased with increasing Peclet number, while the uncertainty increased with X_L and X_S. The 2D casting models quantified the large uncertainty in segregation levels and small effect on freezing times due to uncertainty in dendritic arm spacings in three mushy zone permeability models.

References

[1] Wasserstein R L 1997 Monte Carlo: Concepts, Algorithms, and Applications *Technometrics* **39** 338–338

[2] Eldred M 2009 Recent Advances in Non-Intrusive Polynomial Chaos and Stochastic Collocation Methods for Uncertainty Analysis and Design *50th AIAA/ASME/ASCE/AHS/ASC Structures, Structural Dynamics, and Materials Conference* (Reston, Virigina: American Institute of Aeronautics and Astronautics)

[3] Patankar S V 1980 *Numerical Heat Transfer and Fluid Flow* (Washington DC: Hemisphere Publishing)

[4] Kumar A, Zaloznik M and Combeau H 2012 Study of the influence of mushy zone permeability laws on macro- and meso-segregations predictions *Int. J. Therm. Sci.* **54** 33–47

[5] Schneider M C and Beckermann C 1995 A numerical study of the combined effects of microsegregation , mushy zone permeability and flow , caused by volume contraction and thermosolutal convection , on macrosegregation and eutectic formation in binary alloy solidification *Int. J. Heat Mass Transf.* **38** 3455–73

[6] Melo M L N M, Rizzo E M S and Santos R G 2005 Predicting dendrite arm spacing and their effect on microporosity formation in directionally solidified Al-Cu alloy *J. Mater. Sci.* **40** 1599–609

[7] Wu M, Ahmadein M, Kharicha A, Ludwig A, Li J H and Schumacher P 2012 Simulation of the as-cast structure of Al-4.0wt.%Cu ingots with a 5-phase mixed columnar-equiaxed solidification model *IOP Conf. Ser. Mater. Sci. Eng.* **33** 1–8

[8] Morris M D 1991 Factorial Sampling Plans for Preliminary Computational Experiments *Technometrics* **33** 161–74

[9] Campolongo F, Cariboni J and Saltelli A 2007 An effective screening design for sensitivity analysis of large models *Environ. Model. Softw.* **22** 1509–18

[10] Kyle Fezi; Martin Hunt; Matthew John M. Krane; Benjamin P Haley; Alejandro Strachan (2014), "Continuous casting of weld filler wire," https://nanohub.org/resources/wirecasting. (DOI: 10.4231/D31N7XN67).

[11] Incropera F P, DeWitt D P, Bergman L and Lavine A S 2007 *Fundamentals of Heat and Mass Transfer* (Hoboken, New Jersey)

[12] Bennon W D and Incropera F P 1987 A continuum model for momentum , heat and species transport in binary solid-liquid phase change systems- I . Model Formulation *Int. J. Heat Mass Transf.* **30** 2161–70

A geometric algorithm for automatic riser determination and shrinkage identification in directionally solidifying castings

W Warriner[1] and C Monroe[1]

[1] Department of Materials Science and Engineering, University of Alabama at Birmingham, Birmingham, AL 35294, USA

E-mail: wwarr@uab.edu, camonroe@uab.edu

Abstract. A geometric approach for analysis of castings is outlined and contrasted with physics-based simulations. The approach uses a linear-time algorithm to create an implicit geometry representation known as a distance field. Local maxima are extracted from the distance field using heuristic search logic to determine isolated heavy sections and hotspots. Unsound sections are determined by combining hotspot results with regions of constant distance-field value. Possible uses of the algorithm as an augmentation for application of physics-based solvers is explored. The algorithm is compared against a commercially available simulation software package.

1. Introduction

Casting design engineers working with steel are familiar with the Heuver circle method for determining isolated heavy sections in 2-dimensional sections of castings. Heuver spheres are a 3-dimensional extension of Heuver circles which may be applied to fully 3-dimensional geometries. Heuver spheres of locally maximal radius indicate locations which will be last to solidify, and which may be subject to shrinkage porosity if not properly fed. Isolated heavy sections which consist of a continuum of such points, such as flat plates without taper, may be subject to centerline shrink and microporosity [1]. While it is straightforward to determine the location of locally maximal Heuver circles for a 2-dimensional casting section by inspection, it can be challenging to do the same with a fully 3-dimensional model of even modest complexity. Except at inside corners where the sand-fillet effect is significant, Heuver spheres are identical to inscribed spheres, and the latter will be used as an approximation. By using concepts related to inscribed spheres, the approximate locations of maximal Heuver spheres for polyhedral parts may be determined by a linear time algorithm, allowing for rapid insight into and automation of a portion of part rigging design.

The medial axis skeleton of a polyhedron can be thought of as the set of inscribed sphere centers for which the spheres are maximal [2] [3]. The boundary of a polyhedron is composed of 0-, 1-, and 2-dimensional boundary elements: vertices, edges, and faces. Likewise, the medial axis skeleton is composed of medial axis components: junctions, seams, and sheets [4]. Any medial axis component which does not neighbor any boundary elements will be referred to as a central component. As an example, an ell of square cross-section is shown in figure 1a, its medial axis skeleton is shown in figure 1b, and its central components are shown in figure 1c. Note that the central components consist of two linear seams connected to the sharp corners of a quarter-circle-bounded sheet.

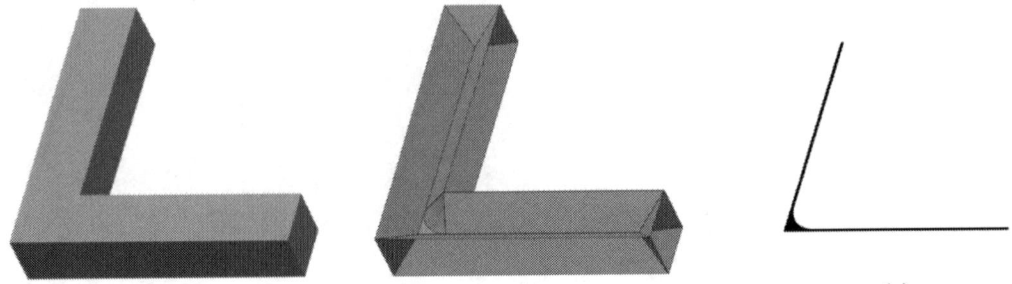

(a) (b) (c)

Figure 1. (a) Ell with square cross-section; (b) medial axis; (c) central components consisting of two linear seams and a sheet with quarter-circle boundary.

The Euclidean Distance Function (EDF) of a polyhedron is the function that gives, for each point in space, the distance between that point and the closest point on the boundary of the polyhedron. It is possible to demonstrate that the function which assigns the radii value of inscribed spheres to their center points is equivalent to the EDF. The EDF, thusly based on inscribed spheres, increases in value away from the part boundary. Likewise, solidification of a casting occurs inward from the boundary, so the EDF may be thought of as a geometric approximation to a time-to-solidification map. Central components of locally maximal value must be the last to solidify, and indicate isolated heavy sections.

The EDF is also related to the medial axis skeleton. All points on medial axis sheets are equidistant from exactly two distinct boundary elements [5]. If the two boundary elements are parallel, then their shared medial axis sheet must have constant EDF value. Similarly, all points on seams are equidistant from at least three distinct boundary elements [5]. Seams which are equidistant from three or more parallel boundary elements have constant EDF value. Such sheets and seams will be called constant sheets and seams. Because the EDF is a geometric approximation to a time-to-solidification map, it follows that two or more solidification fronts meet uniformly at constant sheets and seams. When such a meeting occurs there is a solidification instability and microporosity is expected in steel alloys [6]. Thus, from a casting perspective, constant seams and sheets may be thought of as geometric approximations to the locations of centerline shrink in steel castings. Determining the location of such sheets and seams is a major goal of our algorithm.

2. Implementation of the Euclidean Distance Function

While the medial axis is a useful approximation of centerline shrink locations, direct determination of an exact representation of the medial axis skeleton is computationally intensive and complicated [5]. Approximations to the medial axis skeleton exist [7], but unfortunately the EDF is also required to determine locations of centerline shrink. The medial axis skeleton may be approximated from the EDF by finding regions of undefined gradient. However, because the EDF is the absolute value of the solution to a non-linear hyperbolic PDE, analytic solutions even in two dimensions are often impossible to determine except in special cases. Discrete EDF approximations have been determined using a tetrahedral mesh discretization with an algorithm of linearithmic time complexity in the number of mesh elements, but the solutions require high-quality meshes which are computationally expensive, and can have significant errors [8]. Additionally, acceptable determination of the medial axis skeleton from the resulting EDF approximation may require iterative mesh refinement near the medial axis skeleton, increasing computational complexity and run-time.

Another approach is to use image analysis-based methods on a 3-dimensional image of a given geometry, whose mesh element equivalent is a volume element, or voxel. An algorithm which has linear time complexity in the number of voxels and exact computation of the discrete EDF exists for 3-dimensional images, and is used in our algorithm [9]. The discrete image-based EDF is called the Euclidean Distance Transform (EDT), and an example of the image-based EDT applied to an ell with square cross-section is shown in figure 2. The 3-dimensional image has been sectioned at half its depth into the page to give a 2-dimensional image showing the largest variation in value.

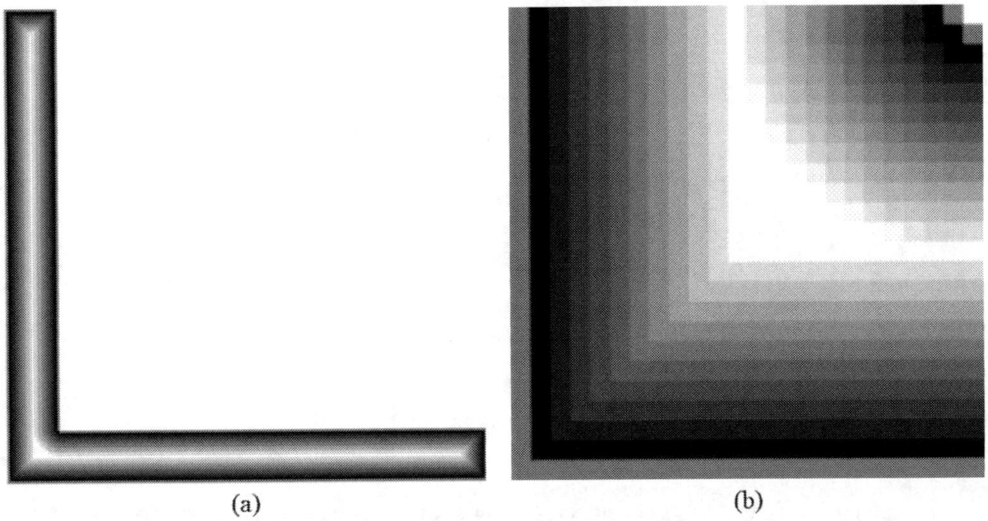

Figure 2. (a) EDT of an ell of square cross-section sectioned at half depth. Black pixels are nearest the boundary, white pixels are farthest. (b) Close-up showing constant sheet at the bend, connecting to two constant-valued seams extending into each leg, shown in white. Compare with the central components in figure 1c.

Our algorithm determines regions of constant EDT value which can be presented directly as a geometric approximation of the Niyama criterion. To determine the Niyama criterion values, and for other comparisons, a commercial simulation software MAGMA was used [10]. Using the commercial software simulation package, Niyama values were determined in a flat plate with a cylindrical feeder attached to one end [10]. Plots cut halfway through the depth of the plate are shown in figure 3. Figure 3a shows the Niyama values between 0 and 5.6 in gray with larger values in black. Figure 3b shows the central finite difference gradient magnitude of the EDT, with values near zero shown in gray and other values shown in black. Figure 3c shows the EDT with lighter values further from the plate boundary. The plate has dimensions of 19 by 5.5 by 1 inch, and the feeder has height-to-diameter ratio of 1.5 with diameter of 3 inches. The feeder's circular cross-section is tangent to one end of the plate and centered along the middle dimension. The setup of the simulation is identical to that used by Carlson and Beckermann in demonstrating that dimensionless Niyama criterion is a predictor of microporosity [11].

It is worth noting that the gray portion of the Niyama criterion in figure 3a does not match the central sheet of the gradient magnitude in figure 3b due to edge and corner effects of the physical temperature gradient. In both the physical simulation and the geometric approximation the plate shows an expansive central region where some gradient is sufficiently close to zero. In the case of the Niyama criterion the temperature gradient is close zero, and with the geometric approximation the EDT gradient magnitude is close zero. Because the Niyama criterion may be used as a predictor of microporosity [11], the EDT may be a useful geometric approximation for predicting microporosity.

(a) (b) (c)

Figure 3. (a) Niyama criterion value darkened between 0 and 2.8 in a 19 by 5.5 by 1 inch deep plate, sectioned at half depth. Plate has a 3 inch diameter, 4.5 inch deep cylindrical feeder tangent to the bottom edge centered along the width [10]. (b) Gradient magnitude of EDT of the same plate. Gray pixels have gradient value much larger than unity. (c) EDT of the same plate. Black pixels are nearest the boundary, white furthest.

3. Heuristic determination of hotspots and shrinkage

Determination of locally maximal central junctions from an EDT is relatively straightforward using a linear array search. For polyhedrons with only axially-oriented boundary elements, such as the ell of square cross section in figure 2a, constant seams and sheets are also straightforward to determine, as the central finite difference gradient magnitude values are less than unity if the corresponding voxels contain part of the medial axis skeleton. For general polyhedra it is not required that constant seams or sheets be linear or planar, nor axially oriented. Additionally, imaging discretization makes identification of exact medial axis skeleton geometry difficult and leads to error in identifying constant sheets and seams, and thus regions of centerline shrink. A non-polyhedral example geometry, a rectangular cross-section annulus, showing curved seams and sheets is shown in figure 4.

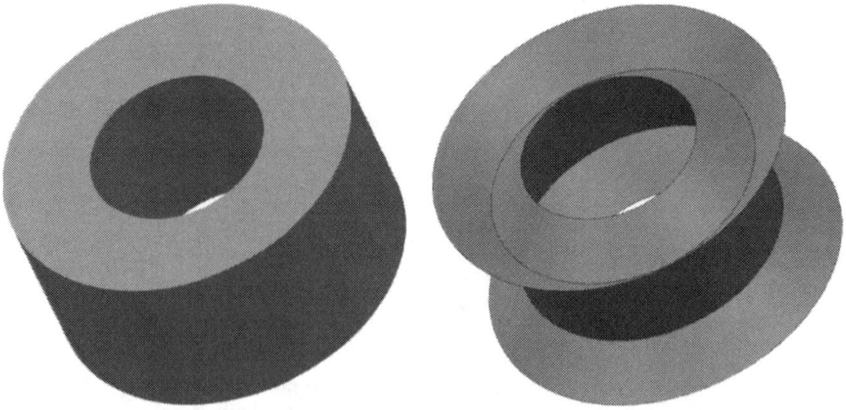

(a) (b)

Figure 4. (a) Cylindrical annulus; (b) Medial axis with non-planar central sheet darkened.

To mitigate the problems caused by discretization and oblique or curved medial axis components, heuristic searches may be used to prune a list of voxels identified as potential locally maximal components. The current implementation creates a list of voxels whose EDT values are no lower than their neighbors. The list is sorted in order of decreasing EDT value. For each, a connected basin of EDT values is created around the first element of the list down to some threshold. If the basin of a list voxel overlaps an already existing basin, it is not counted as a maximum. All surviving list voxels are declared local maxima. Extremely low EDT valued voxels are trimmed as well to remove spurious values. Once the local maxima have been located, isolated heavy sections have been determined, and with it a geometric approximation for feeder placement.

Feeder sizes may be determined by application of steel alloy feeding rules to the isolated heavy sections determined via EDT [12]. Essentially, a heuristic akin to a game is played by each of the local maxima. Starting from the largest EDT value, at each step, the local maxima voxels are assigned a feeding distance based on feeding rules from literature [12]. If a neighbor has a greater EDT value, or if it has already been assimilated, or if it is a part boundary voxel, then it may not be assimilated. If a neighboring voxel has approximately equal EDT value, it may be assimilated, but the allowed feeding distance is reduced by the distance between the voxels. If instead the neighboring voxel has lower EDT value, it may be assimilated and the allowed feeding distance is recomputed based on the new EDT value. The heuristic game must converge because either feeding fronts will have insufficient remaining feeding distance to allow further progress, or they will contact each other, or they will contact the part boundary.

When the heuristic game has converged, each isolated heavy section may be assigned a feeder volume from the feeding rules [12]. The feeding rules are a collection of non-linear functions which are dependent on several geometric measurements. Two of the measurements are characteristic length L and characteristic width W of the part, which are interpreted to mean the largest and second-largest dimensions of an equivalent plate, respectively. The third measurement used is the half-thickness, T, of the equivalent plate, and can also be interpreted to mean the maximal EDT value of the isolated heavy section [12].

The implementation currently used as of this writing uses an axially-oriented bounding box of the assimilated region to determine L and W. Associated part volume is calculated conservatively as the volume of the isolated heavy section bounding box, and T is the EDT value of the original isolated heavy section local maxima voxel or voxels. The feeding rules return a required feeder volume to achieve ideal feeding, which can be used to compute the feeder dimensions depending on desired geometry, e.g. cylindrical, prismatic polyhedral, etc. The calculation for a cylindrical feeder of aspect ratio 1.5 is shown in table 1 for the bearing block shown in figure 5. The top row of figure 5 shows the models in a 3D view as reference, while the bottom row shows a side view of each model. Figure 5a shows the original part segmented into two isolated heavy sections by our algorithm, figure 5b shows the part rigged according to the results of our algorithm as calculated in table 1, and figure 5c shows the part rigged according to a solution by Wlodawer [13]. The feeders have diameter 123 mm and have been placed without consideration for chills, feedpads, gating, or any of the other accoutrements associated with engineered rigging designs. The yield of our method is 59%, an improvement on Wlodawer's solution which has yield of 51% [13].

In Figure 6, the computation time of the same cube part with the same voxel and mesh densities and spacing are compared between the image based algorithm described here and the commercial simulation software. While the power law scaling is nearly the same, the image based algorithm is approximately 17 times faster than the commercial simulation software over the range of mesh elements used in the experiment [10]. While there are speed benefits, our algorithm does not reproduce or replace the functionality of a physics-based solver. Instead our algorithm provides a means of rapidly assessing manufacturability and estimates the locations of certain defects. Some of the components of our algorithm are parallelizable, and should show additional benefits if implemented as such.

Table 1. Computation of feeder size for part shown in figure 5 determined by image-based algorithm, assuming a cylindrical feeder of aspect ratio 1.5.

Description	Expression	Value	Units
Bounding Box Dimensions	$L \times W \times H$	225 x 160 x 130	mm
Maximum EDT	T	39.1	mm
Shape Factor	$S = \frac{L+W}{T}$	9.84	
Shape Factor Contribution	$F = S^{-0.74}$	0.184	
Bounding Box Volume	$V_B = LWH$	4.68×10^6	mm^3
Feeder Volume	$V_F = 2.51 V_B F$	2.16×10^6	mm^3
Feeder Radius	$R_F = \sqrt[3]{\frac{V_F}{3\pi}}$	61.2	mm
Feeder Height	$H_F = 3R_F$	183	mm

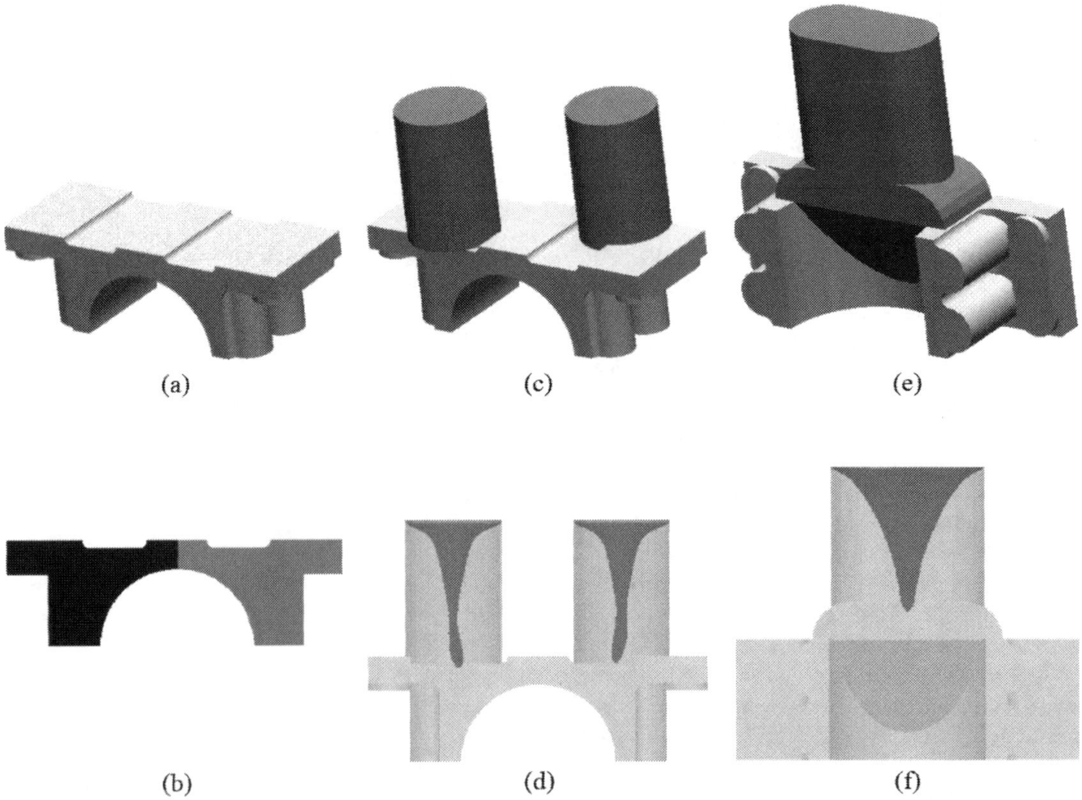

Figure 5. (a) Bearing block geometry and (b) segmentation into two isolated heavy sections by heuristic algorithm. Riser sizes and positions are calculated and attached to the original geometry to produce a (c) rigged bearing block geometry. The commercial simulation software is run on the result and (d) porosity is calculated [10]. (e) A rigging solution for the bearing block due to Wlodawer, with (f) porosity from the commercial simulation software [10] [13].

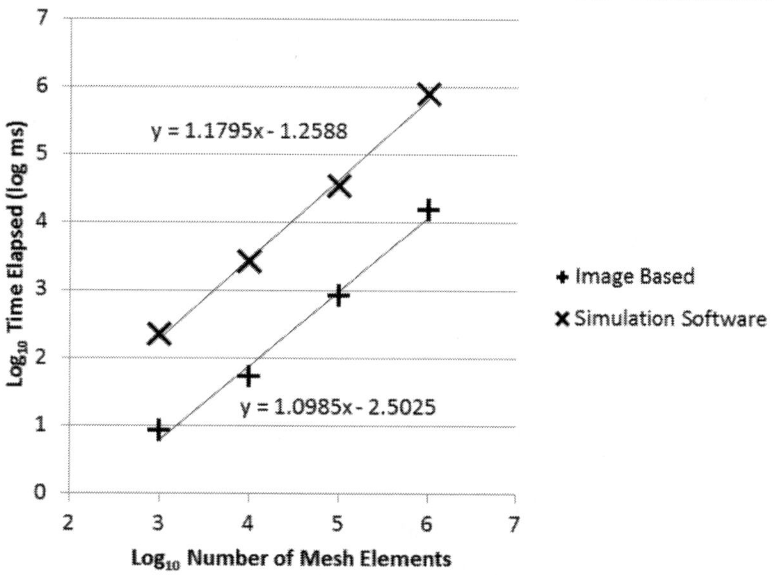

Figure 6. Timing comparison between the image-based algorithm and the commercial simulation software [10]. Error bars are smaller than is reasonably visible on the plot and are not shown.

4. Conclusions

It should be noted the authors do not believe their algorithm should serve as a replacement of physics-based casting simulation software, but instead should be used as a design tool to augment physics-based simulation software and optimization algorithms. It can be used to automate some of the design process already used by casting designers, and to create an initial seed for optimization and further design operations. Because the algorithm is fairly rapid, it is also hoped that further image-based tool development will allow for more rapid turnaround time in customer requirements versus process efficiency design iterations. Additionally, costing and other design metrics will be straightforward to plug in to the algorithm, allowing for rapid decision making among a wide variety of geometries which meet customer requirements, minimizing design time and cost, and allowing greater exploration of solution spaces to further maximize process efficiency.

References

[1] Monroe C A, Gorsky A D, Huff K R and Grandhi V R 2012 *IOP Conf. Ser.: Mater. Sci. Eng.* **33** 012004
[2] Blum H 1967 *Models Perception Speech Visual Form* **19** 362-80
[3] Pfaltz J L and Rosenfeld A 1967 *Comm. ACM* **10** 119-22
[4] Sheehy D J, Armstrong C G and Robinson D J 1996 *IEEE Trans. Vis. Comput. Graphics* **2** 62-72
[5] Culver T, Keyser J and Manocha D 2004 *Comput Aided Geom. Des.* **21** 65-98
[6] Niyama E, Uchida T, Morikawa M and Saito S 1982 *Int. J. Cast Metal Res.* **7** 52-63
[7] Miklos B, Giesen J and Pauly M 2010 *ACM Trans. Graph.* **29** 101
[8] Sethian J A and Vladimirsky A 2000 *Proc. Nat. Acad. Sci.* **97** 5699-703
[9] Maurer Jr C R, Qi R and Raghavan V 2003 *IEEE Trans. Pattern Anal. Mach. Intell.* **25** 265-70
[10] MAGMA GmbH. 2013 MAGMASOFT v5.2.0.5
[11] Carlson K D and Beckermann C 2009 *Metall. Mater. Trans. A* **40** 163-75
[12] Ou S, Carlson K D, Hardin R A and Beckermann C 2002 *Metall. Mater. Trans. B* **33** 731-40
[13] Wlodawer R 1966 *Directional Solidification of Steel Castings* Riley RV ed. Long Island City Pergamon Press

Microporosity Prediction and Validation for Ni-based Superalloy Castings

J Guo[1], C Beckermann[2], K Carlson[2], D Hirvo[3], K Bell[4], T Moreland[4], J Gu[5], J Clews[5], S Scott[1], G Couturier[1] and D Backman[6]
[1] ESI US R&D, Columbia, MD
[2] Department of Mechanical Engineering, University of Iowa,
[3] Alcoa / Howmet, Whitehall, MI,
[4] PCC Airfoils, Cleveland, OH
[5] PCC Structurals, Portland, OR
[6] Backman Materials Consulting, LLC, Saugus, MA

E-mail: jianzhengguo@gmail.com

Abstract. Microporosity in high performance aerospace castings can reduce mechanical properties and consequently degrade both component life and durability. Therefore, casting engineers must be able to both predict and reduce casting microporosity. A dimensionless Niyama model has been developed [1] that predicts local microporosity by accounting for local thermal conditions during casting as well as the properties and solidification characteristics of the cast alloy. Unlike the well-known Niyama criterion, application of the dimensionless Niyama model avoids the need to find a threshold Niyama criterion below which shrinkage porosity forms – a criterion which can be determined only via extensive alloy dependent experimentation. In the present study, the dimensionless Niyama model is integrated with a commercial finite element casting simulation software, which can now more accurately predict the location-specific shrinkage porosity volume fraction during solidification of superalloy castings. These microporosity predictions are validated by comparing modelled results against radiographically and metallographically measured porosity for several Ni-based superalloy equiaxed castings that vary in alloy chemistry with a focus on plates of changing draft angle and thickness. The simulation results agree well with experimental measurements. The simulation results also show that the dimensionless Niyama model can not only identify the location but also the average volume fraction of microporosity distribution in these equiaxed investment cast Ni-based superalloy experiments of relatively simple geometry.

1. Introduction

Porosity formed in castings leads to a decrease in the mechanical properties [2-7]. One of the most effective ways to minimize porosity defects is to design a feeding system using porosity prediction modeling. In such a way, the casting analysis can determine the location and magnitude of porosity such that the feeding system can be redesigned. This process is repeated until porosity is minimized and not likely to appear in the critical areas of the castings.

There are many models which can predict the shrinkage porosity from the pressure drop during inter- dendritic fluid flow and gas evolution [5-12]. As is the case for many casting defects observed in solidification processes, the mushy zone regime is the focus of microporosity formation. The basic mechanism of shrinkage microporosity formation is pressure drop due to shrinkage in the liquid

DISTRIBUTION STATEMENT A: Approved for public release; Distribution is Unlimited. Case Number: 88ABW-2014-5285.

[2][13][5].

The liquid densities of many alloys are lower than that of the solid phase. Hence solidification shrinkage occurs due to the metal contraction during the phase change. The pressure within the liquid decreases because of the contraction and sometimes cannot be compensated by the metallostatic pressure associated with the height of the liquid metal.

Generally speaking, there are two ways to predict the microporosity in castings. One is a parametric method by using a feeding resistance criterion function combined with macroscopic heat flow calculations [7-10]. Parametric models are easy to apply to shaped castings. Another approach is a direct simulation method [5-9][13-15]. The parametric models and/or direct methods usually derive governing equations based on a set of simplifying assumptions and solve the resulting equations numerically. By utilizing the cellular automata technique, some models can not only predict the percentage porosity but also the size, shape and distribution of the pores [10-12].

A simplified porosity model called "POROS=1" was developed by ProCAST[16] to predict piping, macroporosity, and microporosity. This model is excellent in predicting piping and macroporosity but not always effective in predicting microporosity. There are several parametric methods to predict microporosity. Among all parametric methods, the Niyama criterion[17] is the most widely used microporosity evaluation method in metal casting. This criteria proposes that shrinkage porosity will form in regions where Niyama values are below some critical threshold value. This threshold Niyama value normally is an alloy chemistry-dependent unknown determined by experimentation. The volume fraction of porosity cannot be calculated based on Niyama value. Carlson and Beckermann[1] developed a dimensionless form of the well-known Niyama criterion to directly predict the amount of shrinkage porosity with the consideration of both the local thermal conditions as well as the thermal properties and solidification characteristics of the alloy.

In this paper, the dimensionless Niyama model is integrated with the existing POROS=1 model in ProCAST to more accurately predict piping, macroporosity, and microporosity. Such an integrated model is used to predict the location-specific shrinkage porosity volume fraction during solidification of superalloy castings. The predictions are validated by comparing against radiographically and metallographically measured porosity for several Ni-based superalloy equiaxed castings that vary in alloy chemistry with a focus for the current reported effort on plates of changing draft angle and thickness.

2. Model Description

As the integrated model described in this paper is the integration of POROS=1 and dimensionless Niyama, these primary models are briefly reviewed as follows.

2.1. POROS=1 Model

There are three fundamental parameters in the POROS=1 model: *pipefs, macrofs,* and *feedlen*. The values of these parameters depend on casting process and alloy. The physical meaning of these parameters are explained in[16]. The first two parameters, *pipefs* and *macrofs* are two critical fractions of solid corresponding to the formation of piping and macroporosity. The third parameter, *feedlen* is a length parameter that determines the feeding ability in the late stages of the solidification process [16]. The detailed description of the model can be found in [16].

2.2. Dimensionless Niyama Model

The dimensionless Niyama is derived based on 1-D directionally solidifying system [1]. Darcy's law is written for such system as:

$$f_l v_l = -\frac{K}{\mu}\frac{dP}{dx} \tag{1}$$

Where f_l is the volume fraction of liquid phase, v_l is the liquids velocity in the mush, μ is the liquid dynamic viscosity, P is the pressure, x is the spatial coordinate, and K is the mushy zone permeability. The pressure drop can be calculated by:

$$\frac{dP}{dx} = \frac{\mu\beta\dot{\mathrm{T}}}{\mathrm{KG}} \tag{2}$$

Where $\beta = (\rho_s - \rho_l)/\rho_l$ is the solidification shrinkage, $\dot{\mathrm{T}}$ is the cooling rate, and G is the temperature gradient.
Porosity forms when pressure drops below a critical value. With some manipulation, the dimensionless form of Niyama can be written as:

$$Ny^* = C_\lambda \frac{G}{\dot{T}^{5/6}} \sqrt{\frac{\Delta P_{cr}}{\mu\beta\Delta T_f}} = \int_{f_{l,cr}}^{1} 180 \frac{(1-f_l)^2}{f_l^2} \frac{d\theta}{df_l} df_l \tag{3}$$

Where C_λ is a material constant, $\Delta\mathrm{T}_f$ is the freezing range, θ is the dimensionless temperature, ΔP_{cr} is the critical pressure drop, and $f_{l,cr}$ is the critical liquid fraction point when porosity forms. The critical liquid fraction can be calculated after Ny^* calculation based on the one-to-one correspondence between the integration of equation (3) and $f_{l,cr}$. Then the porosity will be calculated based on the shrinkage from the critical liquid fraction to complete solidification. Complete details of the Ny* model can be found in [1].

2.3. Integration of POROS=1 and Dimensionless Niyama Criterion (Ny)*

POROS=1 provides excellent prediction of piping and macroporosity but has a reduced level of accuracy regarding microporosity. Dimensionless Niyama (Ny*), on the other hand, exhibits good accuracy in predicting microporosity but not for piping and macroporosity. To fully take advantage of POROS=1 model and dimensionless Niyama, POROS=1 and dimensionless Niyama are now integrated such that POROS=1 is used to predict piping and macroporosity and dimensionless Niyama criterion is used to predict microporosity. The run parameter of *FEEDLEN* in the POROS=1 model is removed. Instead, the microporosity is calculated based on dimensionless Niyama value as described above.

3. Experimental Validation

In order to provide initial validation of the above developed model, several Ni-based superalloy equiaxed castings that have different alloy chemistries (IN718 and Mar-M-247) and tapered plates with different configurations were investment cast. Figure 1 shows the experiment setup. There are 24 test coupons per mold. Each coupon is 8" long and 1.8" wide with various thicknesses (0.060", 0.125", 0.250", and 0.500") and taper angles (0, 5%, 10%, and 15%). For each mold, there are 8 thermocouples, 2 inside the metal and 6 inside the shell. All coupons were X-rayed and the X-Ray images of the castings were digitized and thru thickness porosity levels were determined by detailed image analysis. Some of the test coupons were cut into halves for more detailed metallography and to confirm the X-Ray measured porosity levels.

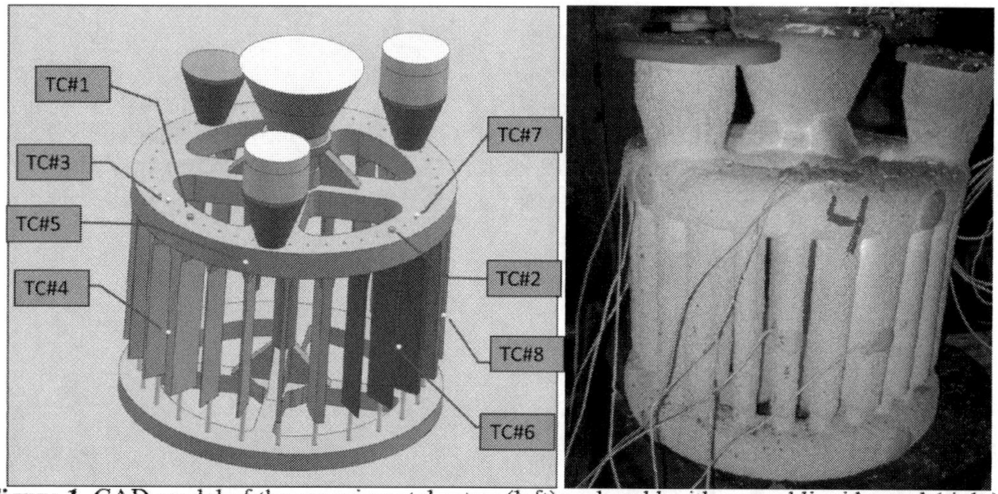

Figure 1. CAD model of the experimental setup (left) and mold with poured liquid metal (right).

Figure 2 summaries the comparison of predicted and measured microporosity for an IN718 alloy with different wall thickness and plate tapers. In the prediction, ΔP_{cr} = 4bars. In the figure, the top image of each pair shows the digitized X-Ray measured porosity from the experimental casting. The bottom images show the porosity evaluation from the casting simulation model using the integrated POROS=1 and dimensionless Niyama (Ny*) model. For a given taper angle, porosity level decreases with increasing thickness. For a given thickness, porosity decreases with increasing taper angle.

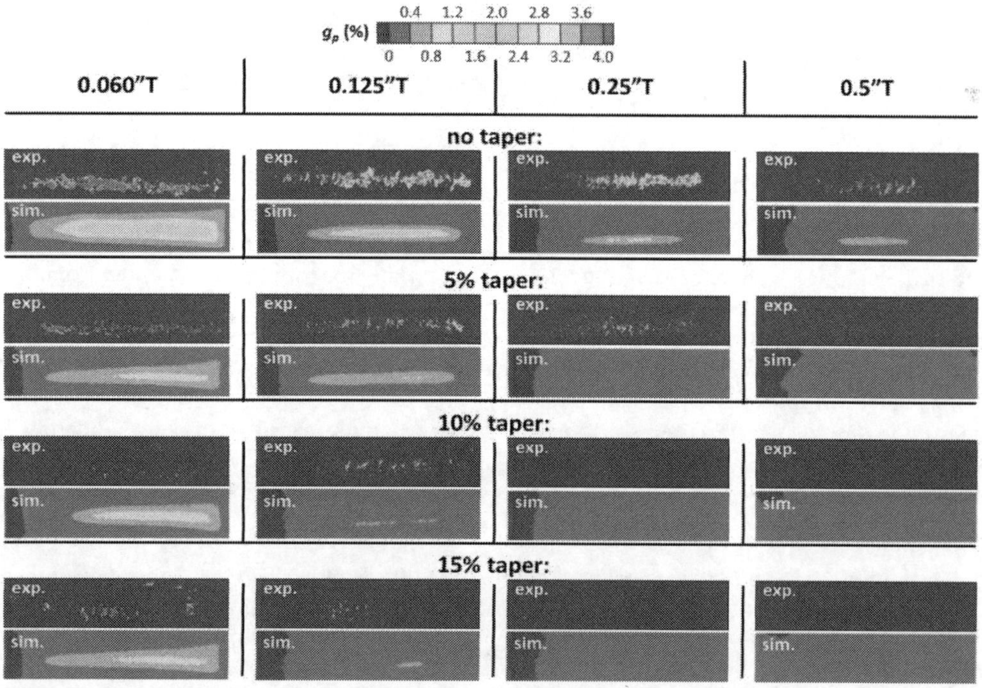

Figure 2. Comparison of prediction and measurement of porosity for different wall thickness and plate taper angles.

There are several adjustable variables in equation 3, such as the critical pressure drop ΔP_{cr} and the critical liquid fraction $f_{l,cr}$. Those variables can be adjusted to better match the experiment measurement. For example the effect of the critical pressure drop on the prediction against the experiment measurement for non-tapered plates can be found in figure 3. Error bars in the figure indicate uncertainty determined from duplicate plates. Compared to the experiment, the model can predict the trend of micro-porosity very well. For the alloy studied here, ΔP_{cr}=4bars gives the best match.

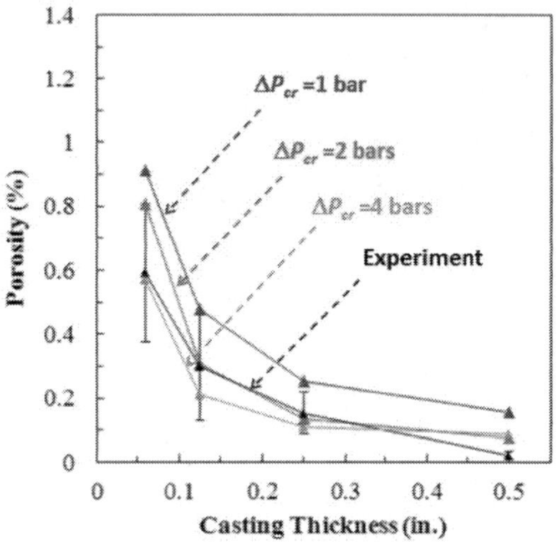

Figure 3. Comparison of porosity between simulation and experiment for non-tapered plates.

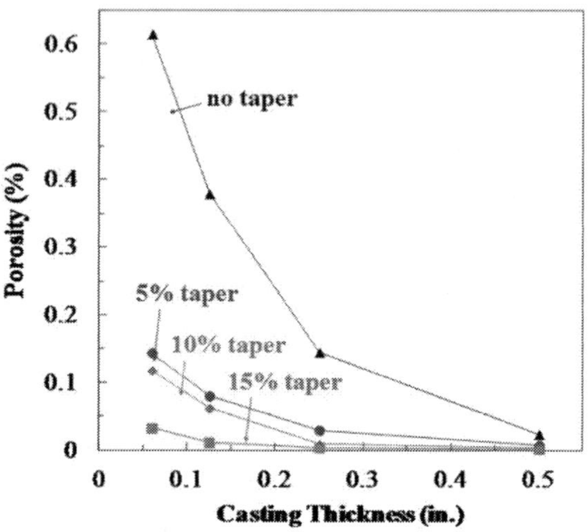

Figure 4. Effect of casting thickness and taper angle on porosity level.

Figure 4 summaries the relationship between average porosity level in a plate with different casting wall thickness and taper. Such information can be used as a reference for casting process engineers. In order to make castings with an accepted level of average plate porosity, a corresponding optimal part design is needed. For example for a plate casting with the dimension of 8" long and 1.8" wide, if the acceptable level of average plate porosity is 0.2%, the taper has to be larger than 5% for wall thickness less than 0.1 inch. On the other hand, if the wall thickness is greater than quarter inch, taper is not necessary to meet this porosity specification level.

The authors are also in the process of assessing the accuracy, verification and validation of the POROS=1 + Ny* model as applied to more complex casting geometries and processes. Results for these more detailed casting geometries will be discussed in future publications.

4. Conclusions

The dimensionless Niyama (Ny*) model is integrated with the ProCAST POROS=1 model, where this integrated model can accurately predict both macro-and micro-shrinkage porosity during solidification of superalloy castings. The integrated model is validated by comparing modeled results against measured porosity for several Ni-based superalloy equiaxed castings that have different alloy chemistries and tapered plates with different configurations. The simulation results agree well with experimental measurements for the geometries presented in this paper.

Acknowledgements

The efforts reported under this paper are a direct result of funding provided under the US Government supported Metals Affordability (MAI) Program called HOW-5 (Contract # FA8650-11-2-5229 – Agreement Order 29). Technical Leadership and Oversight of the HOW-5 project were provided by Steven Turek and Jason Blake of AFRL, Wright Patterson Air Force Base, Dayton, OH whose efforts and inputs to enhance the HOW-5 projects technical efforts are greatly appreciated. The authors also wish to thank the MAI Technical Oversight Committee (TOC) and MAI Program office for their support, guidance and review of the HOW-5 program that this porosity modeling improvement effort is a part of.

References

[1] Carlson K D and Beckermann C 2009 *Metallurgical and Materials Transaction A* **40A** 163-175
[2] Campbell J, Castings, Butterworth-Heinemann, Oxford, United Kingdom, 1991
[3] Kuznetsov A V and Vafai K 1995 *International Journal of Heat and Mass Transfer* **38** 2557-2567
[4] Sigworth G K and Wang C 1993 *Metallurgical Transactions B* **24B** 349-364
[5] Poirier D R, Yeum K and Maples A L 1987 *Metallurgical Transactions A* **18A** 1979-1987
[6] Combeau H, Carpentier D, Lacaze J and Lesoult G 1993 *Materials Science and Engineering* **A173** 155-159
[7] Sung P K, Poirier D R, Felicelli S D, Poirier E J and Ahmed A 2001 *Journal of Crystal Growth* **226** 363-377.
[8] Zhu J D and Ohnaka I 1991 *Modeling of Casting, Welding and Advanced Solidification Processes V* 435-442
[9] Sabau A and Viswanathan S 2002 *Metallurgical and Materials Transactions B* **33B** 243-255
[10] See D, Atweed R C and Lee P D 2001 *Journal of Materials Science* **36** 3423-3435
[11] Atwood R C and Lee P D 2002 *Metallurgical and Materials Transactions B* **33B** 209-221.
[12] Lee P D, Atwood R C, Dashwood R J and Nagaumi H 2002 *Materials Science and Engineering* **A328** 213-222
[13] Niyama E, Uchida T, Morikawa M and Saito S 1982 *AFS Cast Met. Res. J.* **7** 52–63
[14] Felicelli S D, Poirier D R and Sung P K 2000 *Metallurgical and Materials Transactions B* **31B** 1283-1292
[14] Anyalebechi P N 1998 *Light Metals* 827-842
[15] Kubo K and Pehlke R D 1985 *Metallurgical Transactions B* **16B** 359-366
[16] Guo J, Makinde A and Bewlay B 2009 *Modeling of Casting, Welding, and Advanced Solidification Processes – XII,* Edited b: Cockcroft S L and Maijer D M, TMS, pp. 337-344
[17] Niyama E, Uchida T, Morikawa M and Saito S 1982 *AFS Cast Met. Res. J.* **7** 52–63

IOP Conf. Series: Materials Science and Engineering **84** (2015) 012004 doi:10.1088/1757-899X/84/1/012004

A fast and efficient adaptive parallel ray tracing based model for thermally coupled surface radiation in casting and heat treatment processes.

J Fainberg and W Schaefer
MAGMA Gießereitechnologie GmbH, Aachen, Germany

E-Mail: J.Fainberg@magmasoft.de

Abstract. A new algorithm for heat exchange between thermally coupled diffusely radiating interfaces is presented, which can be applied for closed and half open transparent radiating cavities. Interfaces between opaque and transparent materials are automatically detected and subdivided into elementary radiation surfaces named tiles. Contrary to the classical view factor method, the fixed unit sphere area subdivision oriented along the normal tile direction is projected onto the surrounding radiation mesh and not vice versa. Then, the total incident radiating flux of the receiver is approximated as a direct sum of radiation intensities of representative "senders" with the same weight factor. A hierarchical scheme for the space angle subdivision is selected in order to minimize the total memory and the computational demands during thermal calculations. Direct visibility is tested by means of a voxel-based ray tracing method accelerated by means of the anisotropic Chebyshev distance method, which reuses the computational grid as a Chebyshev one. The ray tracing algorithm is fully parallelized using MPI and takes advantage of the balanced distribution of all available tiles among all CPU's. This approach allows tracing of each particular ray without any communication. The algorithm has been implemented in a commercial casting process simulation software. The accuracy and computational performance of the new radiation model for heat treatment, investment and ingot casting applications is illustrated using industrial examples.

1. Introduction

Surface-to-surface radiation is an important heat transfer mechanism in many metal casting and heat treatment processes. It needs to be considered in physical process models in order to accurately describe the heat exchange between different surfaces of the mold/casting or between the casting set-up and surrounding equipment. The numerical modelling of the radiative heat transfer in a very complex environment is a computationally expensive task due to the long range effects and the multidimensional nature of radiation. Various numerical methods can be applied for calculating thermally coupled radiation.

The type of computational model depends on the volume and radiating surface properties to be described. The most comprehensive commercially established approaches are based on different variants of the discrete ordinates model (DOM) [1, 2], where combinations of the diffusive and specular surface properties, simultaneously with volumetric emission, absorption and scattering, can be considered.

The case of grey, opaque and diffusively radiating surfaces is the most typical one for radiation processes occurring in daily foundry praxis. For applications without absorption, emission and scattering in a volume, simpler models can be applied. Usually, the radiosity method, with the calculation of view factors as a prerequisite, is applied [3, 4, 5]. Here, a net radiative flux has to be computed for the thermal balance at the radiating surface. The incident absorbed radiation flux is summed via view factors computed for pairs of radiating surfaces with direct visibility. The straight forward method for computation of view factors and thermal coupling can lead to an enormous computational effort for the complex geometries typical for industrial applications. Therefore, various accelerations based on GPU hardware and sophisticated software were recently developed. The model presented here is solely a software-based solution using the message passing protocol (MPI) parallelization technique.

Different measures for acceleration of the computational radiation model are described in detail in the paper. First, an improved space angle subdivision is introduced. The organization of the ray tracing procedure is important as it has a strong impact on the performance of the radiation model pre-processing and is described next. Finally, the method of parallelization of ray shooting is reported. The achieved computational performance is then demonstrated by means of typical foundry applications.

2. Radiating surfaces

The computational work involved in radiation modelling is linearly proportional to the number of radiation tiles used. Computational time can be saved if the fineness of the used grid surpasses the required resolution in the distribution of the integrated heat flux (=irradiance) along the radiating surface. In this case, it would be sufficient to have fewer, coarser tiles than contained in the numerical grid. The reduction can be achieved by clustering of neighboring tiles.

Here, the numerical grid currently used in the commercial casting process simulation software MAGMASOFT® was utilized directly for discretization of the radiant surfaces. The three-dimensional grid of MAGMASOFT® is made up of 3 systems of grid lines in the cartesian spatial directions, which run through the entire model and embed it in a cuboid. The individual radiant surfaces in the model are initialized by assigning as a tile a right-angled face of a grid cell with the normal vector in one of the 6 coordinate directions, if the cell face divides an opaque and a transparent material. The clustering of neighboring faces is enabled in the same plane for grid cells of the same material. Up to 3 clustered faces are permitted in each of the two lateral directions; hence a maximum of 9 faces can be joined into a tile. The list of grid cells belonging to a clustered radiation tile is assembled and stored for each radiation tile. The ray tracing starts for each composed tile from its geometric center. Radiation sources are stored for the entire tile rather than for each component. The emitted heat flux of the composed tile is calculated accounting for cell temperatures within the cluster. As a result, acceleration by a factor of 2 or slightly more is achieved on average through tile clustering in the case of complex geometries.

3. Space angle by subdivision

The spatial angle has to be discretized in order to evaluate the incident radiation flux as a sum over discrete portions of the total space angle. This sum approximates the true integral over the space angle by a numerical quadrature over a finite number of angle subdivisions. The easiest manner of subdivision is to draw the unit hemisphere around the center of the radiating surface and to project the radiating grid onto it. The individual view factor for each pair of interacting elementary radiating surfaces can be interpreted directly from the projected area on the unit hemisphere. The number of view factors to be computed and stored varies approximately quadratically with the available radiating surfaces, leading to very high computational demand in industrial applications.

Inverting the projection direction can also be used for the space angle discretization. This means that the unit sphere is subdivided in a fixed manner in advance. This subdivision is then projected onto the radiating mesh in order to map the space angle subdivision to the corresponding radiation sources, and not vice versa as in the direct view factors method. The computational and memory requirements then scale linearly with the number of surfaces.

A hemi-cube projection is the best known technique for view factor calculations of this type [6, 7]. The space angle subdivision is achieved by subdivision of the facets of the unit cube surface into a finite number of rectangles. Then, the corresponding radiation source is looked for in the direction prescribed by the middle point of the rectangle placed at the central tile position. Finally, the full subdivided hemi-cube surface is projected onto the radiating mesh with radiating sources. This method is widely used in current commercial surface radiation models.

New in this procedure is that each subdivision of the solid angle corresponds to the same view factor VF, namely $VF_J = 1/N = const$, where N is the total number of subdivisions. A homogeneous subdivision of the unit sphere of this kind in relation to view factors is not a homogeneous subdivision of its surface on account of the term $(\vec{\Omega}\vec{N})$ in the corresponding integral:

$$\int_{\Omega_i} (\vec{\Omega}\vec{\mathrm{N}})\partial\Omega = \int_{\Phi}\int_{\theta} \sin(\theta)\cos(\theta)\partial\Phi\partial\theta = \frac{\pi}{N} = const \tag{1}$$

The advantage of a subdivision of this kind is seen when calculating the incident heat flux. A view factor which remains constant can be excluded, since determination of the incident flux changes into an averaging of the outgoing flux.

$$q_{in} = \int_{(\vec{\Omega}\vec{N})>0} (\vec{\Omega}\vec{\mathrm{N}}) q_{out}\,\partial\Omega, \qquad q_{in}^{i} = \frac{1}{N}\sum_{j=1}^{N} q_{out}^{j} \tag{2}$$

A uniform distribution of the unit sphere according to the view factors is not unambiguous and can take place in innumerable ways. The method chosen here additionally requires certain symmetry in subdivision in relation to the normal vector of the radiation tile and is easy to handle.

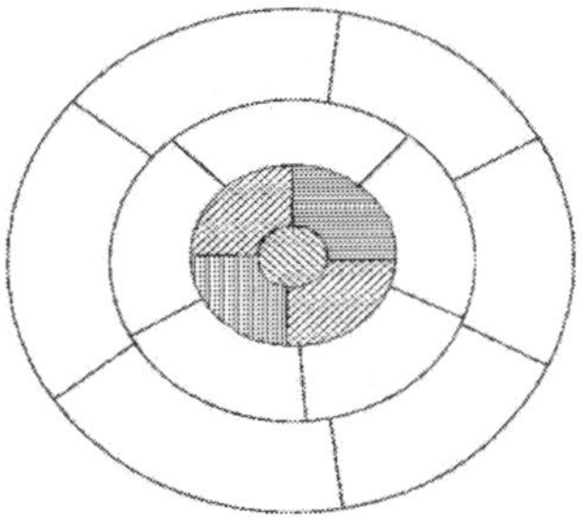

Figure 1. An illustration of the chosen principle for fixed subdivision of the unit sphere, the view "from above" along the normal direction of the radiation tile, where the subdivision resembles a target and comprises the azimuthally arranged segments, which in turn form radial rings.

The subdivision of the unit sphere starts from a circle at the north pole and is followed by a series of radially concentric rings. Each ring is then subdivided into a different number of ring segments in the azimuthal direction, as shown in Figure 1. The number of subdivisions in individual rings forms an arithmetical progression, which limits the aspect ratio of the segments.

The subdivision is parameterized by the number of rings n in the meridian direction and the number of azimuthal segments of the first ring around the north pole r. For each possible parameterization (n, r) there is a single solution for the meridian ring coordinates, so that each spherical segment defines the same view factor:

$$\theta_i = \frac{1}{2}\arccos\left(1 - \frac{(n-i+1)(2r+n+i-2)}{N}\right), \text{ with } N = \left(\frac{2r+n-1}{2}\right)n + 1 \tag{3}$$

N in equation (3) is the total number of subdivisions according to the arithmetical progression. The number of subdivisions N and therefore also the number of rays per one tile grows quadratically according to equation (3) with the number of defined meridian rings n.

The segments in a ring as shown in Figure 1 may be turned in the azimuth direction about an arbitrary free angle, so that a greater angle distance is created between segments of the adjacent azimuthal rings. Finally the system of unit vectors from the template shown in Figure 2 is rotated to match the unit vector of the north pole with the normal vector of the tile.

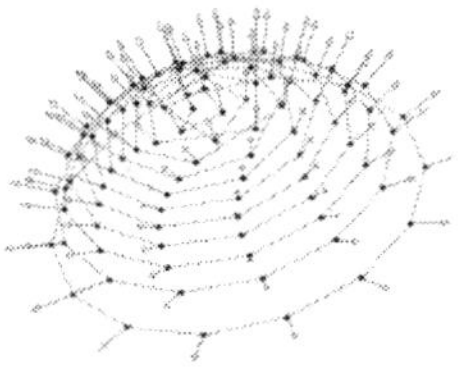

Figure 2. An example of the system of direction vectors for parameterization (r=4, n=10) resulting from the equation (3), where the total number of directions according to (3) is equal to N=86.

3.1. Hierarchical system

An excessively rough subdivision of the unit sphere leads to numerical errors in the integration of the heat flux. In order to achieve a higher angle resolution in relation to the geometry of the radiating grid, this method uses a hierarchical system of discretization levels and is similar to the various refinement stages of the numerical grid in an explicit multi-grid method.

The first level is represented by the system of spatial directions produced according to equation (3), see Figure 2. The next level is produced by subdividing each segment of the first level by halving in the azimuth and meridian direction, i.e. quartering. One exception is the first refinement of the circular region at the north pole into 4 spherical triangles. Further recursive subdivisions repeat the same procedure, and the 4 spherical triangles around the north pole are treated as abnormal spherical rectangles from the third refinement level.

This produces a geometric progression of the number of rays at each successive level. For ***k*** levels of refinement

$$N_{tot} = N\frac{4^k - 1}{3} = \left(\frac{2r + n - 1}{2}n + 1\right)\left(\frac{4^k - 1}{3}\right) \tag{4}$$

rays result overall on all refinement levels. The directional vectors according to the refinement are shown in Figure 3 A, B and C.

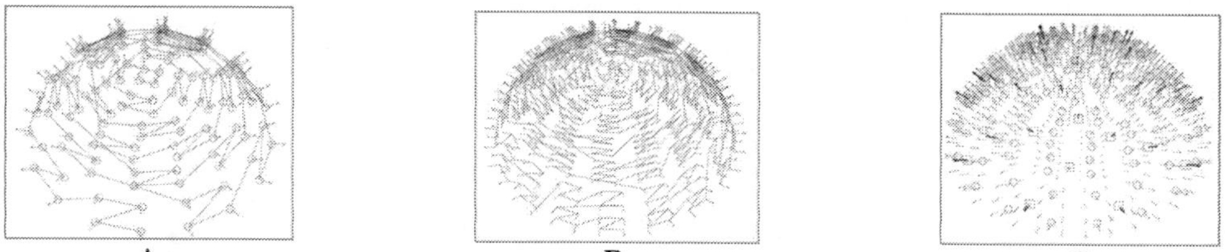

A B C

Figure 3. A: direction vectors from second refinement level created from the first level with subdivision (r, n) = (4, 10); B: third level created from the second level; C: direction vectors from all three levels, wherein vectors from different levels are characterized using different symbols at the origin of the vectors. The rays for all directions shown are sent out according to the ray tracing method.

Three refinement levels are used in the current implementation. In the hierarchical method of ray tracing, rays are initially sent out in succession from all refinement levels as shown in Figure 3C, and corresponding radiation sources are stored.

3.2. Adaptation to the geometry and temperature distribution

It would be possible to only send rays for the vectors of a single or the finest level. However, if the rays are sent by all defined refinement levels, the possibility exists for the angle resolution to be adapted locally.

The result of ray tracing for the thermal calculation without adjustment would be a complete list of radiation sources, which all result e.g. from one level of subdivision. The local adaptation of the angle discretization replaces a list of this kind with sub-lists from all discretization levels. The approach in this case is as follows:

Starting point is a discretization that contains only vectors of the most refined level and proceeds upward to the next coarser level. Each time, the radiation sources of the preceding finer level are checked. If any of the 4 quarters belongs to a different material, if the initial temperatures of the found materials differ from each other by more than a predefined threshold, or there exist both external sources and sources originated from other tiles, the radiation source of the finer level is retained. Otherwise, the 4 quarters are replaced by a single radiation source from the coarser spherical rectangle.

The procedure described above supplies k=3 lists of radiation sources for 3 levels of uniform space angle discretization. The adapted angle discretization in this case is chosen in such a manner that a concentration of representative radiation sources occurs in the angle regions where jumps of the radiation intensity with respect to the spatial direction can potentially be expected. The adapted discretization of the solid angle for the radiation tile takes place in this way through the selection of radiation sources.

One example of the adapted discretization of the solid angle is shown in Figure 4

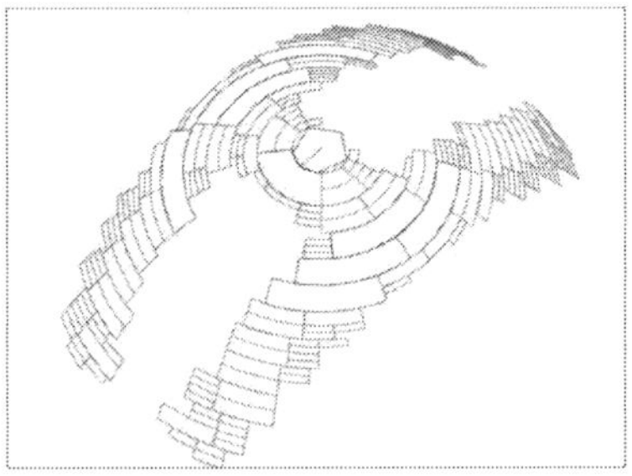

Figure 4. A view factor of an "H"-shaped face is approximated by means of an adaptive space angle subdivision. For this purpose, 3 discretization levels according to equation (3) were used, with the first level parameterized with (n=15,r=4). The number of rays is equal to 166, 664 and 2656 at the 3 discretization levels. A total of 3486 rays are sent out to scan the hemisphere. After adaptation has been carried out, 218 representative rays have been selected, which is smaller by a factor of 16 than at the finest level. The selected 218 spherical rectangles from different levels are marked in the figure.

The outgoing heat fluxes, connected to the radiation sources from different levels, must be weighted differently in the calculation of the incident heat flux. Equation (5) for the incident heat flux is supplemented by pre-exponential factors of different levels. It describes the calculation of the incident radiation flux for an arbitrary number of hierarchical levels k.

$$q_{in}^{i} = \frac{1}{N} \sum_{level=1}^{k} \left(\frac{1}{4^{k}-1} \sum_{j=1}^{N_{level}} q_{out}^{j} \right) + q_{ext} \tag{5}$$

The last term in equation (5) is the incident radiation flux from the space angle spanned on the external ambient environment in the case of half open cavities.

According to equation (5), just the outgoing total radiation fluxes of tiles have to be summed in order to obtain the total incident flux for the considered tile. The weight factors in front of the inner sum mimic the subsequent recursive subdivision of the space angle. The advantage of the developed method is that values of individual view factors are not required explicitly according to equation (5).

4. Ray tracing

A voxel-based ray tracing method with reverse ray tracing is used. All rays are initially sent out from the tile that receives the radiation and the radiation sources are traced by rays according to geometrical optics.

4.1. Voxels

The rays are sent out from all refinement levels of the space angle discretization. Ray tracing involves the search for intersections between the continuation of a ray and faces of a grid cell (voxel) in which the ray is currently located, see Figure 5.

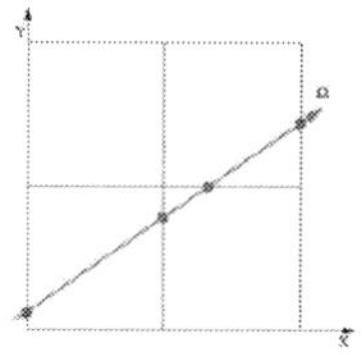

Figure 5. Ray tracing diagram on the Cartesian grid comprising the detection of a sequence of intersections with the orthogonal 2D grid.

A voxel is a cube-shaped volume where information on geometric objects is contained. In a cartesian grid, a tensor product grid is available where each grid cell becomes a voxel. Up to three tile ID's referring to three faces on the "back", "east" and "north" side of the grid cell are stored per grid cell to support ray tracing.

The three possible faces of the grid cell are checked for the next intersection with the traced ray. The faces are given by the sign of the three components of the ray direction $\vec{\Omega}$. The face of the cell with the minimum length up to the intersection is searched for, and the ray is continued as far as the located intersection by the located length ΔX^{ray}

$$X_{MIN} = \underset{l=\{i,j,k\}}{MIN}\left\{\frac{x_l - x_l^{ray}}{\Omega_l}\right\}, \quad \Delta X_i^{ray} = X_{MIN}\Omega_i \tag{6}$$

Once the next intersection has been found, the global tile ID is requested on the face of the cell, the ray tracing is stopped and the found radiation source is returned. In the case of a symmetry plane, the directional component normal to the symmetry plane is inverted and the reflected ray is traced further. Otherwise, the ray tracing procedure is repeated in the next grid cell until the ray hits a tile or leaves the bounding box of the numerical grid. In the latter case, a fixed ID of the outer space source is returned.

4.2. Chebyshev distance

The conventional ray tracing described above requires each transparent grid cell lying on the ray path to be visited, see Figure 5. The processing of empty cells is expensive for geometries with large, empty cavities and/or a fine grid there. Significant ray tracing acceleration can be achieved by ray "tunneling" over larger blocks of empty cells, but requires information about the size of empty cell blocks.

The so-called Chebyshev or chess board distance is utilized for such tunneling [8]. A two-dimensional example of Chebyshev distance usage is shown in Figure 6. The distance from the dotted grid cells is measured. Figure 6A shows the isotropic Chebyshev distance. Cells at the same distance from the dotted zero cells are arranged in a layer around the original square. The minimum distance for all Cartesian directions is used in the isotropic case. The anisotropic Chebyshev distance for the upper left quadrant is shown in Figure 6B.

Use of the Chebyshev distance allows adaptive steps in ray tracing. When the ray approaches a surface, the Chebyshev distance diminishes and with it the jump size, too. The closer a ray gets to the surface, the slower the ray tracing progress.

The anisotropic distance is measured from the zero distance cells, but only for Cartesian directions inside of the particular quadrant. It is evident from Figure 6 A and B that the anisotropic distance achieves greater values than the isotropic one. The anisotropic case allows jumps through larger empty grid blocks, as shown in Figure 6 C. In this example, 2 intersections were searched for; without using the Chebyshev distance it would be 9. In the three-dimensional case, the anisotropic Chebyshev distance is computed and stored in advance for 8 octants. For tracing of a ray in a particular direction $\vec{\Omega}$, the Chebyshev measure from the corresponding opposite octant is retrieved and used.

IOP Conf. Series: Materials Science and Engineering **84** (2015) 012004 doi:10.1088/1757-899X/84/1/012004

Despite the additional work and memory involved to support the grid jumps, the Chebyshev distance provides a substantial acceleration of the method, particularly for models with large empty / transparent spaces.

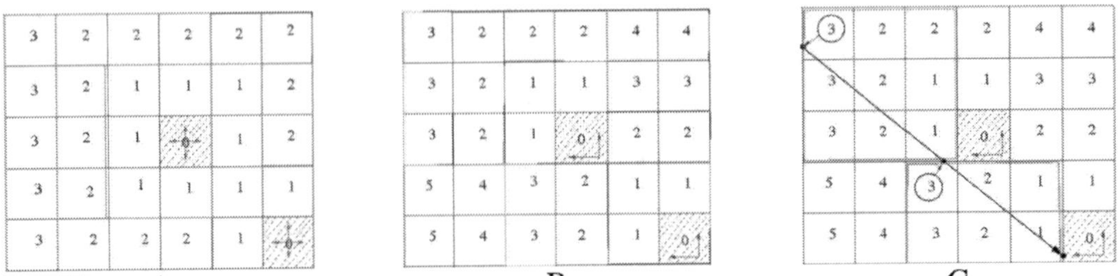

Figure 6. Calculation of the 2D Chebyshev distance. The directions observed are marked with the arrows in the zero cells in the respective cases. A: isotropic Chebyshev distance; B: anisotropic Chebyshev distance. C: an example of 2D ray tracing supported by anisotropic Chebyshev distance distribution from (B). The ray in the octant (+X,-Y) is processed assisted with distance in the octant (-X, +Y). The initial position of the ray in the intersection with a cell face at top left lies in the cell at distance 3. The next intersection with the 3x3 cube is therefore searched. Following displacement of the position to the intersection, the Chebyshev distance is taken from the next cell in the direction of the ray path. It is once again equal to 3, and the next jump leads the ray to the target at bottom right.

4.3. Parallel ray tracing

Prior to ray tracing, calculation of the anisotropic Chebyshev distance takes place on each CPU independently. Calculations and results on each CPU are identical at this point. Each CPU obtains the material distribution over the total numerical grid as the only input.

Before ray tracing starts, the number of tiles to be processed on each CPU is balanced. For this, the number of tiles located at a specific MPI partition is compared with the mean arithmetic tile number among all CPU's. The CPU's with an excessive number of tiles are treated as the tile donors, the rest as the tile acceptors. Further, the tile's data is sent from the donor to the acceptor CPU's in order to obtain a balanced tile distribution. Finally, each CPU has to process some number of own not exported and some number of imported ray bundles associated with each tile.

All rays on each CPU are dispatched independently of one another as described in Section 3.1. No MPI communication during tracing of a particular ray is required, since each CPU operates on the identical full numerical grid collected from all MPI partitions.

Radiation sources of own tiles are found, selected by means of the geometric and thermal adaptation as described in Section 2.2 and stored. Imported tiles on the CPU acceptors are processed next. The radiation sources, selected according to the adaptation procedure, are temporarily stored in a buffer.

When all tiles have been processed, the results for all imported tiles are sent backwards from the buffer by the CPU acceptors to the CPU donors. Each donor CPU may generally have several acceptors and each acceptor CPU may have several donors. Asynchronous MPI communication takes place.

5. Model applications

Three computed examples for heat treatment, investment and ingot casting are presented. They all were computed on a PC with 8 cores with default model settings. 5691 rays from 3 space angle refinement levels were traced for each tile. The used numerical grid, the numerical data of the radiation model and computational times are summarized in Table 1 below.

The results for a heat treatment of several castings in a furnace is shown in Figure 7, left, where 82,251,716 radiation sources were stored totally. A sample simulation of an investment casting is shown in Figure 7, right. A large number of radiation tiles (554,905) after agglomeration was required due to the curved, complex surfaces involved. The total number of stored radiation sources was 156,429,337.

Table 1. Statistics of three representative radiation model applications. Column descriptions: the total number of used grid cells, the total number of radiation tiles after their agglomeration, the total number of traced rays, the mean number of radiative sources after geometrical and thermal adaptation, the computational time spent for the model set-up (ray tracing), total computational time with set-up for the transient thermal modelling and the corresponding time without radiation.

	Grid size (cells)	Tiles	Total Rays	Sources /Tile	Set-up Time	Total Time	Without Radiation
Heat Treatment	9,755,200	237,702	1,352,762,082	346	147 sec.	516 sec.	53 sec.
Investment Casting	45,006,208	554,905	3,157,964,355	281	365 sec.	35 min.	21 min.
Ingot Casting	1,743,000	29,216	166,268,256	289	12.4 sec.	55 sec.	44 sec.

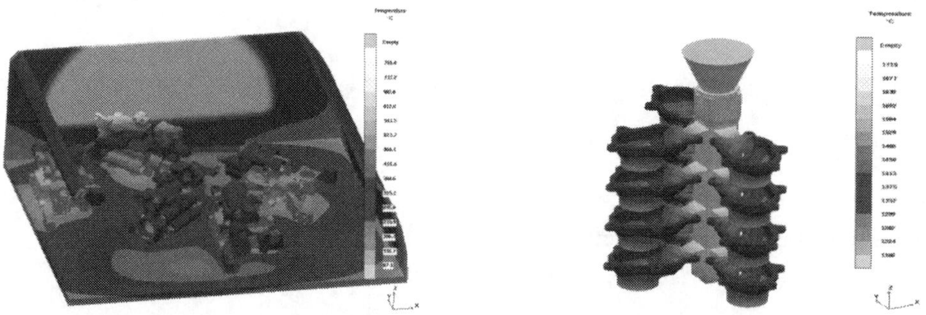

Figure 7. Left: example of surface-to-surface radiation model application, heat treatment of 5 cast pump housings in a furnace. Right: a thermal result for an investment casting project calculated using the radiation model.

A sample simulation from steel ingot production is shown in Figure 8. A symmetry boundary condition is applied for radiation and the thermal model, hence a total of 6 ingot blocks were effectively simulated. On average 2.4 radiating cell faces were joined to a single radiation tile after their agglomeration. The average number of radiation sources for the tile after thermal and geometrical adaptation was 289 – roughly 20 times less than at the finest level. The total number of stored radiation sources was 8,456,505. The parallel ray tracing performance was over 13 million traced rays per second.

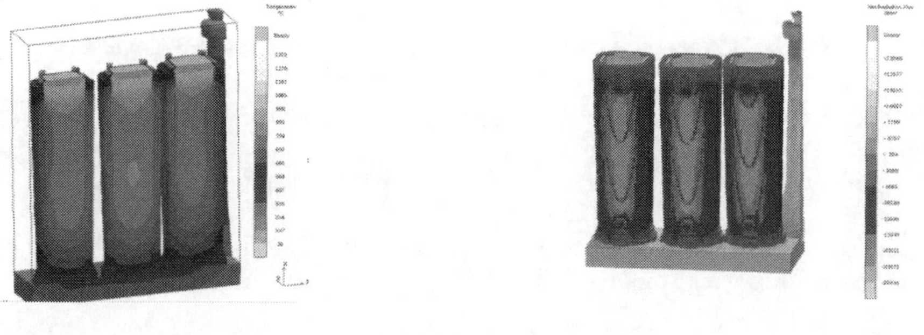

IOP Conf. Series: Materials Science and Engineering **84** (2015) 012004 doi:10.1088/1757-899X/84/1/012004

Figure 8. An example from steel ingot casting. The radiation model was applied to simulation of pouring and solidification of multiple ingots. Left: temperature at the ingots' surface during solidification. Right: computed net radiation flux density at the radiating surfaces.

The computational times from table 1 were compared with reported performance for several state of the art methods reported in [9]. The computational times for the quickest reported method *View3D-O* for 6 different numbers of radiation tiles are extrapolated by means of the power law for the number of tiles obtained in all three of the benchmark cases in table 1. The reported and extrapolated results are plotted together with the results of the current method in figure 9. A huge acceleration of 477, 1407 and 2593, respectively, is obtained with respect to the View3D-O method for the benchmark cases of ingot casting, heat treatment and investment casting.

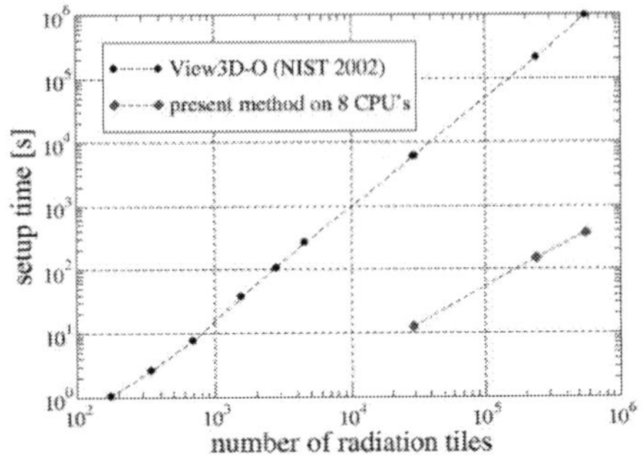

Figure 9. Computational efficiency compared with evaluation in [9]. Round symbols are times of the quickest method View3D-O, see table 5 in [9]. The symbols for number of tiles less than 10,000 are from [9]. Round symbols for three larger values are extrapolated times for the benchmark cases in Table 1. The extrapolation was done using the power law approximation time~$N^{1.7238}$. The result for the actual method in Table 1 are shown with rhomb symbols.

6. Summary

A high computational efficiency in a surface-to-surface radiation model was achieved by a complex optimization especially of the radiation model set-up. The particular computational time reductions were obtained by

- agglomeration of elementary radiating surfaces,
- adaptation of the space angle discretization by means of a hierarchical system,
- application of voxel based ray tracing with a direct utilization of a Cartesian grid,
- ray tracing "tunneling" method supported by the anisotropic Chebyshev distance.
- organization of the parallel ray tracing without any need for communication of the total grid among all CPU's.

The developed model allows robust full-scale thermal simulations including thermally coupled radiative heat exchange in very complex three-dimensional geometries using conventional multi-core hardware. The computational cost of such simulations is not significantly higher than the reference case without radiation between surfaces.

References

[1] Modest M F 1993 *Radiative heat transfer* (McGraw-Hill series) ch 15 pp 541-71, ch 20 pp 703-58
[2] Siegel R and Howell J 2002 *Thermal radiation heat transfer* (Taylor & France 4th edition) ch 16 pp 663-706
[3] Siegel R and Howell J 2002 *Thermal radiation heat transfer* (Taylor & France 4th edition) ch 6 pp 207-66
[4] Modest M F 1993 *Radiative heat transfer* (McGraw-Hill series) ch 4 pp 155-92
[5] *Fluent 6.3 user's guide* (Fluent Inc) ch 13.3
[6] Class M *CHAPARRAL 1995 - a library package for solving large enclosure radiation heat transfer problems* (Sandia National Laboratories, Albuquerque NM)

[7] Cohen M and Greenberg D 1985 *The hemi-cube: a radiosity solution for complex environments* Comp. Graphics **19** No 3 pp 31-40
[8] Abello J, Pardalos P and Mauricio G 2002 *Handbook of massive data sets* (Springer 2002)
[9] Walton G N 2002 *Calculation of obstructed view factors by adaptive integration* (NISTIR 6926)

Numerical Analysis on the Vortex Pattern and Flux Particle Dispersion in KR Method Using MPS Method

N Hirata[1], Y Xu[1] and K Anzai[1]
[1] Department of Materials Processing, Graduate School of Engineering, Tohoku University, Sendai 980-8579, Japan

E-mail: hirata@material.tohoku.ac.jp

Abstract. The mechanically-stirring vessel is widely used in many fields, such as chemical reactor, bioreactor, and metallurgy, etc. The type of vortex mode that formed during impeller stirring has great effect on stirring efficiency, chemical reacting rate and air entrapment. Many efforts have been made to numerically simulate the fluid flow in the stirring vessel with classical Eulerian method. However, it is difficult to directly investigate the vortex mode and flux particle dispersion. Therefore, moving particle semi-implicit (MPS) method, which is based on Lagrangian method, is applied to simulate the fluid flow in a KR method in this practice. Top height and bottom heights of vortex surface in a steady state under several rotation speed was taken as key parameters to compare the results of numerical and published results. Flux particle dispersion behaviour under a rotation speed range from 80 to 480 rpm was also compared with the past study. The result shows that the numerical calculation has high consistency with experimental results. It is confirmed that the calculation using MPS method well reflected the vortex mode and flux particle dispersion in a mechanically-stirring vessel.

1. Introduction

Before continuous casting process, pre-treatment of melt is carried out to control a material components. Because of its good efficiency of liquid/solid and liquid/liquid mixing, mechanically-stirring vessel has been applied to improve the efficiency of chemical reacting for a long time [1-2]. The relationship between the pattern of stirring and efficiency of mixing [3-5], mass transfer [6-8], dissolution [9], precipitation [10] etc. has been studied subsequently. The mechanically-stirring method was firstly introduced to steelmaking industry by Kanbara et al. [11], in order to improve the efficiency of desulfurization reaction rate of melt and flux. It is still the major method in steel plants of Japan even now [12]. However, the melt and the vessel is opaque, and it is difficult to observe what happens during the operation. Therefore, enough but excess desulferization agent tends to be used for the process. To reduce the use of desulferization agents, many model experimentals have been conducted using water and floating materials to understand the flux dispersion behavior in the process. Nakai et al.[13] conducted experiment to investigate the flux dispersion behaviour during the desulfurization process in steelmaking process and proposed models to evaluate the particle size of desulfurization slag. They revealed that the flux dispersion is divided into three stages, non-dispersion, transition and complete dispersion, respectively, and the desulfurization rate of each stage is strongly related to the vortex mode.

Numerical analysis of single-phase and multi-phase flow in a mechanically-stirring tank or vessel has been reported by many researchers [1-2]. Most of them are based on the conventional Eulerian algorithm. For liquid/solid flow, some of the work was carried out with Eulerian algorithm for liquid flow coupling with Lagrangian algorithm for solid particle dispersion. However, most of the researches done with Eulerian algorithm focused on the velocity or pressure field in the fluid, and a few of them report the investigation on vortex mode in a stirring vessel.

Recently, the moving particle semi-implicit (MPS) method has been popularly discussed by researchers on the field of fluid flow. MPS method was firstly proposed by Koshizuka et al [14]. And ever since then, this meshless numerical simulation method based on Lagrangian algorithm has been widely studied because of its applicability for multi-phase flow with free surface. In this practice, the MPS method was applied to investigate the vortex pattern and particle dispersion in a mechanically-stirring vessel. The vortex height and particle dispersion results of numerical calculation were discussed with published results subsequently.

2. Numerical method

2.1. The MPS method

2.1.1. Weight Function In the MPS method, governing equations consist of particle interaction equations. All interactions between particles are limited to a finite distance. The strength of interaction can be described using a weight function of distance between two particles. The gradient or other operators can be described using the weight function. Whereas numerous types of weight function have been proposed, equation (1) was used in this study [15].

$$w(r, c_k r_0) = \begin{cases} \dfrac{40}{7\pi c_k^{\,2}}\left(1 - 6\left(\dfrac{r}{c_k r_0}\right)^2 + 6\left(\dfrac{r}{c_k r_0}\right)^3\right) & (0 \le r < 0.5 c_k r_0) \\ \dfrac{10}{7\pi c_k^{\,2}}\left(2 - 2\dfrac{r}{c_k r_0}\right)^3 & (0.5 c_k r_0 \le r < c_k r_0) \\ 0 & (c_k r_0 \le r) \end{cases} \tag{1}$$

Here, r is the distance between the neighboring particles. r_0, which denotes the specific size of the particle, has the same significance as the lattice size in FDM. c_k is the kernel size coefficient and usually varies between 2 and 4. Particle number density n_i of the particle i is used in the MPS method to calculate the interaction between particle i and the surrounding particles. n_i is the sum of the weight function of the particles surrounding the particle i, and is expressed as follows.

$$n_i = \sum_{i \ne j} w\left(\left|\boldsymbol{r}_j - \boldsymbol{r}_i\right|, c_k r_0\right) \tag{2}$$

Here, $\boldsymbol{r}_i$ and $\boldsymbol{r}_j$ are the position vectors of particles i and j, respectively. n_i is equal to n_0 in the case that a particle has no surface particle around it. Assuming that the fluid is incompressible and considering that the particle number density n_i is directly related to the fluid density, we can use n_0 for the mass conservation condition in the incompressible flow analysis using the MPS method.

2.1.2. Particle interaction models Particle interaction models are used to describe differential operator in the particle method. If ϕ_i and ϕ_j are arbitrary scalars at positions $\boldsymbol{r}_i$ and $\boldsymbol{r}_j$, then the particle interaction models for the differential operators can be expressed as follows.

$$\nabla \phi_i = \frac{d}{n_i} \sum_{i \ne j} \frac{(\phi_i - \phi_j)}{\left|\boldsymbol{r}_j - \boldsymbol{r}_i\right|} \cdot \frac{(\boldsymbol{r}_j - \boldsymbol{r}_i)}{\left|\boldsymbol{r}_j - \boldsymbol{r}_i\right|} w\left(\left|\boldsymbol{r}_j - \boldsymbol{r}_i\right|, c_k r_0\right) \tag{3}$$

$$\nabla^2 \phi_i = \frac{2d}{n_i} \sum_{i \ne j} \frac{(\phi_i - \phi_j)}{\left|\boldsymbol{r}_j - \boldsymbol{r}_i\right|^2} w\left(\left|\boldsymbol{r}_j - \boldsymbol{r}_i\right|, c_k r_0\right) \tag{4}$$

Equation (3) is a gradient model and equation (4) is a Laplacian model. Suffixes *i* and *j* represent the assigned numbers of particles. *d* is the number of space dimensions.

2.2. Fluid flow Analysis

The governing equations for incompressible flow consist of the continuity equation and Navier-Stokes equation as follows.

$$\frac{D\rho}{Dt} = 0 \tag{5}$$

$$\frac{D\boldsymbol{u}}{Dt} = -\frac{1}{\rho}\nabla p + \nu\nabla^2\boldsymbol{u} + \boldsymbol{f} \tag{6}$$

Here, ρ indicates density (kg/m^3); $\boldsymbol{u}$, velocity vector (m/s); p, pressure (Pa); v, kinematic viscosity (m^2/s); and $\boldsymbol{f}$, the body force vector including gravity. Here, *D/Dt* is the symbol for Lagrangeian differential equation without convection terms.

Flow calculation by the MPS method is based on the predictor-corrector method similar to that in the case of the FDM or other conventional methods. The major differences between the MPS method and the FDM are the formulation of the Poisson equation of pressure and the calculation of particle position. In the MPS method, the tentative particle position is calculated using tentative velocity in the prediction phase; therefore, the correction in position is calculated along with the correction in velocity in the correction phase. The Poisson equation for pressure in the MPS method is described using the tentative particle number density n^* and n_0 as follows. n^* is the particle number density calculated using the tentative particle positions.

$$\nabla^2 p^{k+1} = -\frac{\rho}{\Delta t^2}\frac{n^* - n_0}{n_0} \tag{7}$$

3. Hydrodynamic Experiment in Past Research

Hydrodynamic experiment and results published by Nakai et al. [13] are used to verify the numerical simulation by MPS method. As illustrated in Fig.1, a four-blade Rushton turbine is set in the center of the vessel with the whole impeller immersed in a certain depth *L*. ΔH_1 and ΔH_2 are the differential value of vortex bottom height and top height from the initial water height H_0, respectively. Domain A and Domain B are defined as the left side in the sight of the space between the vessel wall and impeller blades, so that the particle dispersion could be studied in these two domains. The boundary to separate these two domains is the central line of the initial water height. The numerical calculation at *L*=214mm and *L*=338mm are carried out. The results at different rotation speed (*RS*), varying from 80 to 480 rpm, are obtained numerically to compare with past experimental results by Nakai et al. Nakai et al. counted the number of particles in domain A and domain B after reaching steady state by evaluating the interfacial area between the particles and water using image processing method.

Table1 gives the vessel size and stirring information in the research by Nakai et al. and numerical analysis in this study. The calculation conditions are same as published research. Only flux density is different because of the calculation stability.

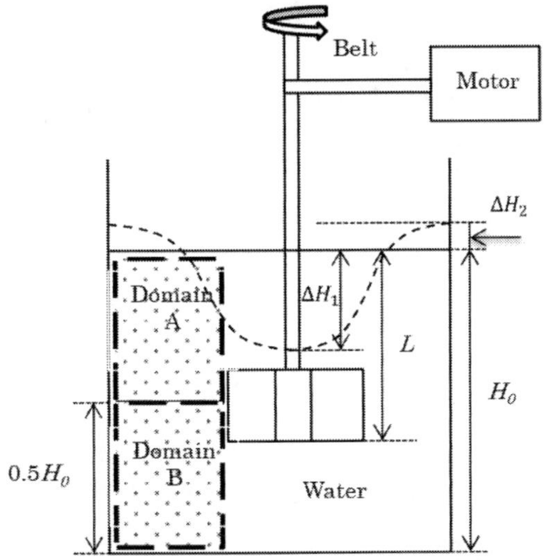

Figure 1. Schematic of hydrodynamic experiment vessel.

Table 1. Hydrodynamic experiment conditions.

Item		Water model by Nakai et al. [13]	Numerical simulation (this study)
Vessel	Height	695mm	695mm
	Diameter	523mm	523mm
Water	Depth (H_0)	414mm	414mm
Rotation Speed (RS)		80~480rpm	80~480rpm
Impeller	Height	110mm	110mm
	Diameter	203mm	203mm
	Width	52mm	52mm
Impeller Immersion Depth (L)		214mm, 338mm	214mm, 338mm
Particle Density		0.03g/m^3	0.5g/m^3

4. Results and discussions

4.1. Vortex Height

There are two critical heights for the vortex, ΔH_1 and ΔH_2 as shown in figure 1. Figure 2 shows numerical results of ΔH_1 and ΔH_2 variation with rotation speed and compares that with reported data by Nakai et al[13]. ΔH_1 and ΔH_2 are measured after the flow behavior reaches to the steady state. As is illustrated in figure 2(a) and figure 2(b), numerical results of ΔH_1 and ΔH_2 variation for L=214mm case with rotation speed match good with that reported by Nakai et al. ΔH_1 is minor when RS is lower than 80 rpm, and increased fast to a steady level around 250mm when RS grows up to 200 rpm. Numerical ΔH_2 variation also shows good agreement with the past study. It presents quasi-linear feature with RS. Figure 2(c) and figure 2(d) are the results for the case of L=338mm. The calculated results coincide with the results in the past study quite well. The calculated ΔH_1 increases from about 10 mm at RS=80 rpm to 414mm (Vessel bottom) at RS=360 rpm continuously. In the past study, ΔH_1 rises up to 414mm at RS=320 rpm. The comparison of numerical and reported experimental results on ΔH_1 and ΔH_2 variation shows that the numerical method well reflects the vortex mode in a mechanically-stirring vessel.

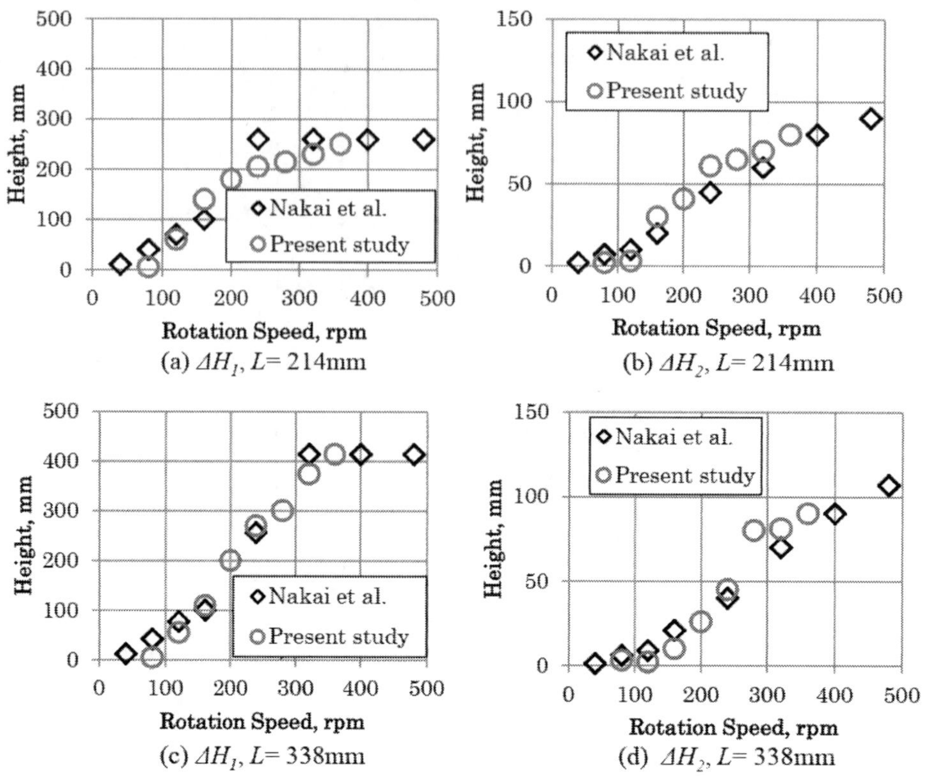

Figure 2. ΔH_1 and ΔH_2 variation with rotation speed.

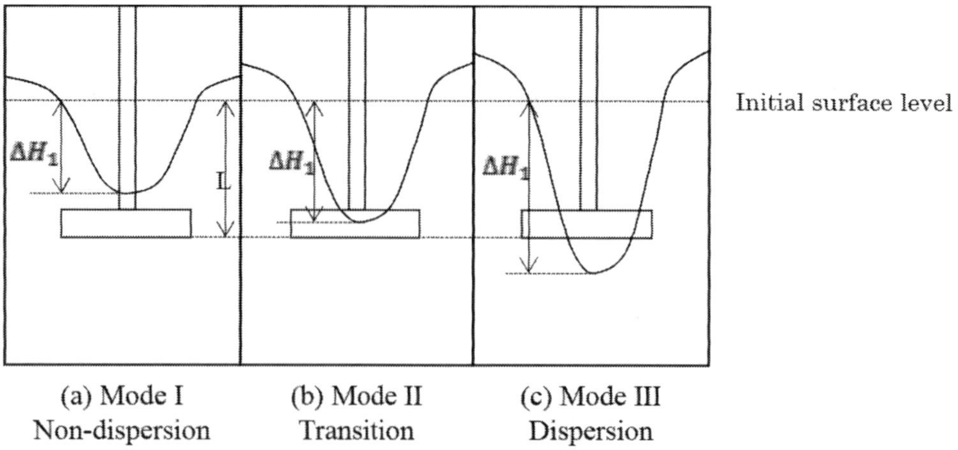

Figure 3. The three types of the vortex mode in the stirring vessel.

4.2. Relationship between the vortex mode and particle dispersion

There are three types of vortex mode in a mechanically-stirring vessel according to the relative position of the vortex bottom and the impeller, and the vortex mode correlates the rotation speed of the impeller [13]. Figure 3 is the schematics of vortex mode and the figure 4 shows the particle dispersion information of numerical results at steady state. As shown in Figure 3 and 4, the flux particle dispersion behavior is strongly related to the vortex mode. Mode I : Non-dispersion stage, the vortex bottom is higher than the impeller, in this stage, there is no particles dispersed in the water. Mode II : Transition stage, the vortex bottom is right on the height range of impeller and the particle dispersion emerges. Mode III : Complete dispersion stage, the vortex bottom in under the impeller and much air is entrapped; the particles disperse in the whole scale of the vessel.

In figure 4(a), which illustrates the particle dispersion for L=214mm case, very few particles enter the water at RS=160rpm. When the rotation speed goes up to 240rpm, a great amount of particles begin entering the water, as shown in figure 4(b). Finally, all of the particles enter the water as shown in figure 4(c), where RS=320rpm. Figure 4(d) to figure 4(f) shows the results in the case of L=338mm. No particle enters the water at 160 rpm rotation speed and a small number enters when the rotation speed grows up to 240 rpm. Even in 320 rpm, not all of the particles can enter the water. The mixing efficiency of 214mm impeller immersion depth is better than that of 338 mm impeller immersion depth according to the numerical results.

4.3. Particle dispersion in the domain A and B

In the study by Nakai et al.[13], the amount of particle dispersion in the domain A and B are counted by the method of evaluating the particle occupying area percentage in the picture. Figure 5 shows the counted particle numbers of numerical simulation in this study and past research done by Nakai et al. in domain A and B when the stirring status becomes steady state. The immersion depths of impeller are L=214mm and L=338mm. The particle size in the report by Nakai et al. was 2mm, in contrast to 13mm in numerical simulation. As a result, the total particle number in numerical calculation in this study is fewer than that in the past research. This is also the reason why the vertical axises in figure 4 have different magnitude. In this discussion, the tendency of particle dispersion variation is compared.

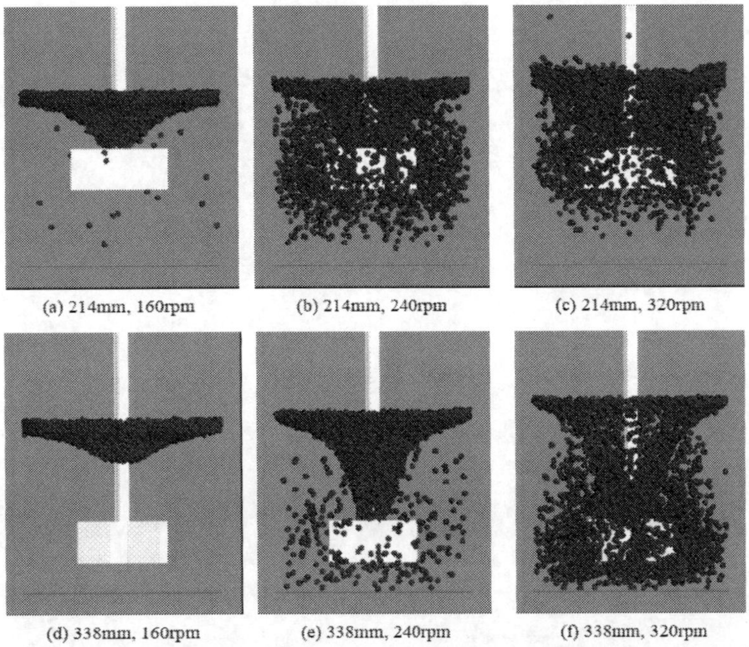

(a) 214mm, 160rpm (b) 214mm, 240rpm (c) 214mm, 320rpm

(d) 338mm, 160rpm (e) 338mm, 240rpm (f) 338mm, 320rpm

Figure 4. Numerical simulation result of particle dispersion.

Figure 5(a) and figure 5(b) is the calculated results of the number of flux particle counted in the domain A and B. When the impeller rotation speed is low and the vortex type is mode I, few particles are entrapped into the liquid. With the rotation speeding up, vortex type turns to mode II, massive particles are drawn into domain A both in L=214mm and L=338mm cases. However, there is no obvious difference for that in domain B. If the rotation speed keeps growing up, particles are found to decrease for *L*=214mm case and increase for *L*=338mm case in domain A. The particle amount dramatically rises in domain B for *L*=214mm case while it decreased to almost none for *L*=338mm case.

Figure 5(c) and figure 5(d) show the water model experimental results by Nakai et al. [13]. In domain A, the particle amounts changing curve is quite similar to the numerical results for *L*=214mm case while the *L*=338mm case has some difference in mode III. Numerical results do not grow up as dramatically as experimental results do. In domain B, the amount of particles for *L*=214mm case shows almost no change after the stirring turns to mode II in the experiment whereas the amount decreased to almost zero when the stirring turns to mode III in numerical results. Another difference is that the number of particles dispersed in domain B in the experimental results keeps a steady level when the rotation speed rises up to certain strength while that for numerical results keeps growing up.

One of the reason for the differences of particle dispersion behaviour in the domain A and B is that the air entrapment has not been considered in numerical analysis. In the actual experiment, much air entrapment emerged over *RS*=320rpm according to the report by Nakai et al. The bubbles generate at high rotation speed are drawn into the deep area, and emerge the particles from the water.

Particle methods is suitable for the multi-phase and multi-physics calculation, therefore the proposed method can be extend to calculate not only the flux dispersion behavior, but chemical reation and efficiency of desulferization. As a result, significant contribution is expected in the reduction of the desulferization agent consumption. In addition, the results in this study show that the proposed method can capture the dispersion behavior of flux without numerical diffusion. This advantage is also expected in capturing the behavior of inclusions or bubbles in the continuous casting process, by which considerable defects occurs.

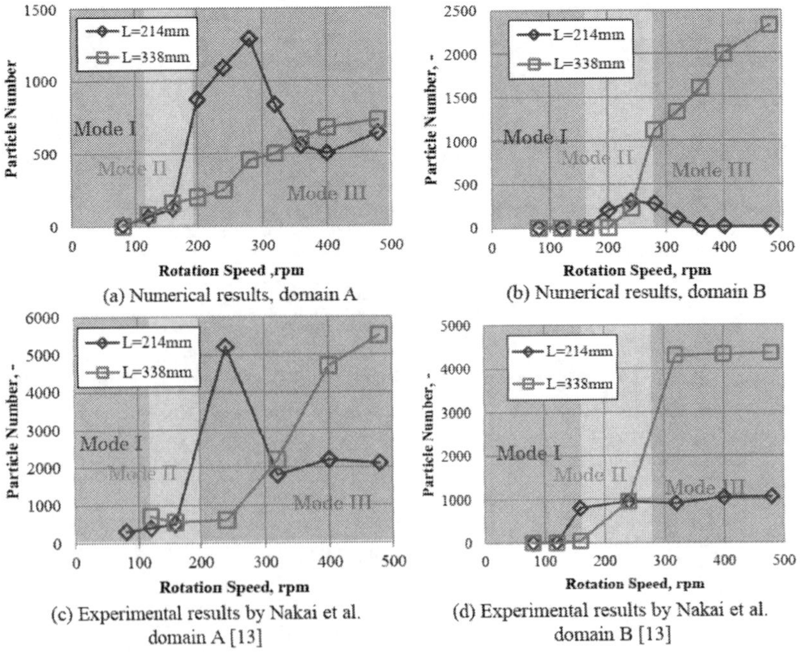

Figure 5. Relationship between number of particles and rotation speed.

5. Conclusions

MPS method was applied to analyze fluid flow in a mechanically-stirring vessel in this paper. Numerical results of particle dispersion in domain A and B and the differential of vortex top height and bottom height with initial water height, named as ΔH_1 and ΔH_2, were compared with past report by Nakai et al. The comparison shows that the numerical method reflected the vortex pattern well.

The numerical results of particle dispersion in the water present consistency with Nakai et al.'s report. In the first stage of stirring, there is no or few particles enter the water. The amount of particles entrapped into the water increase with the rotation speed growing up and totally be entrapped finally. Particle amounts in certain domains are also counted to compare with past research; the numerical results of particle amount variation have relative consistency with that of past research. There are some differences between numerical particle dispersion results in domain A and domain B. It is attributed to the neglecting of air entrapment in numerical analysis.

References

[1] Joshi J B, Nere N K, Rane C V, Murthy B N, Mathpati C S, Patwardhan A W, and Ranade V V 2011 *The Canadian Journal of Chemical Engineering*, **89** 23
[2] Joshi J B, Nere N K, Rane C V, Murthy B N, Mathpati C S, Patwardhan A W, and Ranade V V 2011 *The Canadian Journal of Chemical Engineering*, **89** 754
[3] Ochieng A, Onyango M S, Kumar A, Kiriamiti K, Musonge P 2008 *Chemical Engineering and Processing* **47** 842
[4] Guillard F, Tragardh C 2003 *Chemical Engineering and Processing* **42** 373
[5] Kasat G R, Khopkar A R, Ranade V V, Pandit A B 2008 *Chemical Engineering Science*, **63** 3877
[6] Martín M, Montes F J, Galán M A 2010 *Chemical Engineering Science* **65** 3814
[7] Kerdouss F, Bannari A, Proulx P, Bannari R, Skrga M, Labrecque Y 2008 *Computers and Chemical Engineering* **32** 1943
[8] Brüning S, Weuster-Botz D 2014 *Chemical Engineering Research and Design* **92** 240
[9] Hartmann H, Derksen J J, Akker H.E.A. van den 2006 *Chemical Engineering Science* **61** 3025
[10] Jaworskic Z, Nienow A W 2003 *Chemical Engineering Journal* **91** 167
[11] Kanbara K, Nisugi T, Shiraishi O, Hatakeyama T 1972 *Tetsu-to- Hagané* **58** 34
[12] Ogawa Y, Maruoka N 2014 *Tetsu-to- Hagané* **100** 434
[13] Nakai Y, Sumi I, Matusno H, Kikuchi N, Kishimoto Y 2010 *ISIJ International* **50** 403-410
[14] Koshizuka S, Oka Y 1996 *Nuclear Science and Engineering* **123** 421
[15] Hirata N, Anzai K 2011 *Journal of Japan Foundry Society* **83** 259

IOP Conf. Series: Materials Science and Engineering **84** (2015) 012006 doi:10.1088/1757-899X/84/1/012006

Using four-phase Eulerian volume averaging approach to model macrosegregation and shrinkage cavity

M Wu[1,2], **A Kharicha**[1,2], **A Ludwig**[2]
[1] Christian-Doppler Lab for Advanced Process Simulation of Solidification & Melting,
[2] Chair for Simulation and Modelling of Metallurgical Processes,
Dept. of Metallurgy, Univ. of Leoben, A-8700 Leoben, Austria

E-mail: menghuai.wu@unileoben.ac.at

Abstract. This work is to extend a previous 3-phase mixed columnar-equiaxed solidification model to treat the formation of shrinkage cavity by including an additional phase. In the previous model the mixed columnar and equiaxed solidification with consideration of multiphase transport phenomena (mass, momentum, species and enthalpy) is proposed to calculate the as-cast structure including columnar-to-equiaxed transition (CET) and formation of macrosegregation. In order to incorporate the formation of shrinkage cavity, an additional phase, i.e. gas phase or covering liquid slag phase, must be considered in addition to the previously introduced 3 phases (parent melt, solidifying columnar dendrite trunks and equiaxed grains). No mass and species transfer between the new and other 3 phases is necessary, but the treatment of the momentum and energy exchanges between them is crucially important for the formation of free surface and shrinkage cavity, which in turn influences the flow field and formation of segregation. A steel ingot is preliminarily calculated to exam the functionalities of the model.

1. Introduction

Typical steel ingots solidify with a consequence of shrinkage cavity in the hot top, microscopic shrinkage porosity and macrosegregation located somewhere in the cross section [1], as illustrated in Fig. 1. Those undesired casting defects result from the volume change during solidification, and the resulting feeding flow, thermal-solutal convection and crystal sedimentation. The formation mechanisms of them are complex, and they interact with each other. People have tried different approaches to model or predict the formation of them [2-9], but no model is available to calculate the shrinkage cavity/porosity together with macrosegregation of the art as shown in Fig. 1 in a coupled manner. The formation process of the above defects is multiphasic in nature.

This work is to extend a previous 3-phase mixed columnar-equiaxed solidification model [10] to treat the formation of shrinkage cavity by

Figure 1. Example of the shrinkage cavity and macrosegregation (sulphur print) in a steel ingot (reproduced from literature [1]).

including an additional gas (or covering slag) phase [11]. Four phases are necessarily considered: primary liquid melt, equiaxed crystals, columnar dendrite trunks, and gas. The final goal is to develop an Eulerian multiphase solidification model to calculate the as-cast columnar-equiaxed structure including CET (columnar-to-equiaxed transition), shrinkage cavity/porosity, and macrosegregation at the process scale for industry castings.

2. The numerical model

2.1 Brief model description

- Phase definition: primary liquid (ℓ), solidifying equiaxed phase (e), columnar phase (c), and gas phase (g). Their volume fractions (f_ℓ, f_e, f_c, f_g) sum up to unit. The mass transport equations considering the mass transfer due to solidification are solved.
- There is no mass transfer between gas and other phases, and gas phase (or covering slag) is supposed to be immiscible with other metal phases, and be sucked into the casting domain to feed the shrinkage cavity. The interface between gas and other phases is explicitly solved. The microscopic porosity forms in the deep inter-dendritic region is treated differently (Section 2.3).
- The primary liquid, equiaxed and gas phases are moving phases, for which the corresponding Navier-Stokes equations are solved to get $\vec{u}_\ell$, $\vec{u}_e$ and $\vec{u}_g$, but $\vec{u}_c \equiv 0$.
- Enthalpy equations for all 4 phases are solved. Due to the fact of relatively large thermal diffusivity, we assume that only one temperature (T) represents each volume element. Therefore, a large interphasic volume heat exchange coefficient is applied to balance the temperatures among the phases.
- Three volume-averaged concentration fields (c_ℓ, c_e, c_c) are solved for the three metal phases. Thermodynamic equilibrium condition always applies at the liquid-solid interface, and solute partition occurs during solidification at the liquid-solid interface.
- A diffusion-governed growth kinetic is considered to calculate the growth of crystals.
- Ideal morphology for columnar phase is assumed: cylindrical (cellular) tree trunks. A simplified dendritic morphology for equiaxed phase is assumed: the growth kinetics is modelled according ideal spheres; while for the hydrodynamics a concept of "grain envelope" is employed.
- The columnar dendrite trunks are assumed to originate from the mould wall. Neither nucleation of columnar trunks nor equiaxed-to-columnar transition (ECT) is taken into account.
- Heterogeneous nucleation and transport of the equiaxed crystals are considered. Grain fragments brought into the mould during filling, further fragmentation of dendrites during solidification and the attachment of equiaxed grains into columnar tips (as a part of the columnar phase) are ignored.
- The size (d_e) and number density (n_e) of equiaxed grains and the diameter (d_c) of the columnar trunks are explicitly calculated, while a constant value for the primary arm spacing (λ_1) of columnar dendrites is assumed.

Details of the three-phase mixed columnar-equiaxed solidification model were described elsewhere [10]. Treatment of the gas phase, the formation of shrinkage cavity and porosity is described below.

2.2 Treatment of interactions between gas and other phases

As no mass/species transfer between the gas and other metal phases is considered, only enthalpy and momentum conservation equations for the gas phase are necessarily solved.

$$\frac{\partial}{\partial t}\left(f_g \rho_g h_g\right) + \nabla\cdot\left(f_g \rho_g \vec{u}_g h_g\right) = \nabla\cdot\left(f_g k_g \nabla\cdot T_g\right) + Q_{\ell g}^{ex} + Q_{eg}^{ex} + Q_{cg}^{ex} \quad (1)$$

$$\frac{\partial}{\partial t}\left(f_g \rho_g \vec{u}_g\right) + \nabla\cdot\left(f_g \rho_g \vec{u}_g \otimes \vec{u}_g\right) = -f_g \nabla p + \nabla\cdot\bar{\bar{\tau}}_g + f_g \rho_g \vec{g} + \vec{U}_{\ell g}^{ex} + \vec{U}_{eg}^{ex} + \vec{U}_{cg}^{ex} \quad (2)$$

Key feature by introducing a new gas phase in the mixed columnar-equiaxed solidification model is to treat the exchange terms, as superscripted with 'ex'. They are summarized in Table 1.

Table 1. Momentum and energy exchange terms between gas and other metal phases

$Q^{ex}_{\ell g} = H^*_{\ell g} f_\ell f_g (T_\ell - T_g)$ $Q^{ex}_{eg} = H^*_{eg} f_e f_g (T_e - T_g)$ $Q^{ex}_{cg} = H^*_{cg} f_c f_g (T_c - T_g)$	$H^*_{\ell g}$ [W·m^{-3}·K^{-1}]: the volume heat exchange coefficient is modeled according to Ranz-Marshall [11, 12]. H^*_{eg} and H^*_{cg} [W·m^{-3}·K^{-1}]: a constant volume heat exchange coefficient (500) is assumed.	
$\vec{U}^{ex}_{\ell g} = K_{\ell g} \cdot (\vec{u}_\ell - \vec{u}_g)$ $\vec{U}^{ex}_{eg} = K_{eg} \cdot (\vec{u}_e - \vec{u}_g)$ $\vec{U}^{ex}_{cg} = K_{cg} \cdot (\vec{u}_c - \vec{u}_g)$	$K_{pg} = \frac{\rho_{pg} R_e C_D}{144 \tau_{pg}} d_{pg} A_{pg}$ Symmetric model [13] with minor modifications is applied. Subscript 'p' indicates different metal phases: 'ℓ', 'e' or 'c'.	$R_e = \rho_{pg} \lvert \vec{u}_p - \vec{u}_g \rvert d_{pg} / \mu_{pg}$ $C_D = \begin{cases} 27.6 & R_e \le 1 \\ 24(1+0.15 R_e^{0.687})/R_e & 1 < R_e \le 10^3 \\ 0.44 & R_e > 10^3 \end{cases}$ $\tau_{pg} = \rho_{pg} d^2_{pg} / 18 \mu_{pg}$ $A_{pg} = 6 f_p f_g / d_{pg}$ $\mu_{pg} = \min(\mu_p, \mu_g)$; $\rho_{pg} = \min(\rho_p, \rho_g)$; d_{pg}: modeling parameter (diameters) as function of volume fractions of involving phases.

2.3 Simplified microscopic shrinkage porosity model

Microscopic shrinkage porosity is not explicitly modelled, but the Niyama criterion modified by Carlson and Beckermann [2, 3] is implemented. We name it as CBN (Carlson-Beckermann-Niyama) criterion, Eq. (3). The possibility of occurrence of shrinkage porosity and the amount of pores (volume fraction of voids) decrease with the local CBN criterion. In other words, the smaller the CBN criterion, the more probably the shrinkage porosity would occur, and the more amount of pores might be.

$$\mathrm{CBN} = C_\lambda \frac{G}{\dot{T}^{5/6}} \sqrt{\frac{\Delta P_{cr}}{\mu_\ell \beta (T_{Liquidus} - T_{Eutectic})}} \quad (3)$$

Where C_λ is a material constant, 1.44 x 10^{-4} [m·(K/s)$^{1/3}$], being used for determining the secondary dendrite arm spacing; G and $\dot{T}$ are the temperature gradient and cooling rate, which are evaluated at a critical temperature of T_{cr} $(=0.1 \cdot T_{Liquidus} + 0.9 \cdot T_{Eutectic})$, assuming that $T_{Eutectic}$ corresponds to the end solidification; β is the solidification shrinkage, $(\rho_s - \rho_\ell)/\rho_\ell$; ΔP_{cr} is the critical pressure drop when a pore-nucleus with a critical radius of r in the deep mushy zone can overcome capillary force $(2\sigma/r)$ and metastatic pressure (P_{static}) to grow. ΔP_{cr} is 1.01 x 10^5 Pa for steel [3]. As no threshold of CBN for the occurrence of pores can be determined in advance, this criterion is used only for qualitative evaluation of the occurrence possibility of the microscopic shrinkage porosity.

Although the current model offers the possibility to calculate the interdendritic flow, it is extremely hard to solve the feeding in the deep mush near the end of solidification due to the convergence difficulty, therefore a so-called 'simplified porosity model' is introduced [14, 15]. When the temperature drops below the critical temperature of T_{cr}, volume shrinkage by solidification is ignored, i.e. the solidification is assumed to occur without volume change. This treatment is made by assuming that the rest melt in the deep mush solidifies as a solid-pore mixture with the mixture density equal to the liquid density.

3. Simulation settings

A 2.45 ton ingot was preliminarily simulated in full 3D (average mesh size of 10 mm) and 2D (average mesh size of 5 mm). The ingot had a square cross-section; however, for the 2D calculation an axis-symmetry is assumed. Casting conditions, process parameters, experimentally measured macrosegregation and shrinkage cavity information were reported previously [16, 17]. The alloy is multicomponent, but we take a simplified binary alloy (Fe-0.45 wt.%C) currently. The material properties and some important boundary conditions are summarized in Table 2 and Fig. 2. No mould filling is calculated, and the mould is assumed to be initially filled with liquid melt of 1770 K (above the liquidus 1768.95 K). As solidification starts, the casting shrinks and it sucks the gas phase from the top 'pressure inlet'. A zero-gradient boundary condition is applied at the 'pressure inlet' for other quantities: temperature, concentrations, equiaxed number density. Nucleation parameters of equiaxed crystals are: $n_{\max} = 5.0\text{x}10^{-3}$ m^{-3}, $\Delta T_N = 9.0$ K (different from 5 K as used in [9, 17]), $\Delta T_\sigma = 2.0$ K. For simplicity, the growth of the equiaxed crystals is treated with globular (sphere) morphology, but for modelling the hydrodynamics a concept of "grain envelope" is considered by assuming a certain amount of interdendritic liquid (volume fraction of 0.3). Calculation with 4-phase model including gas phase is extremely time consuming (demanding tiny small time step), especially when a 3D calculation is performed. Therefore, a covering slag (liquid) which is also mostly used in praxis is considered instead of gas phase.

Table 2. Thermodynamic & physical properties of materials

Steel		
melting point of pure iron, T_f	K	1805.15
liquidus slope, m	K (wt.%)$^{-1}$	-80.45
equilibrium partition coefficient k	-	0.36
melt density, ρ_ℓ	kg·m^{-3}	7027
solid density, ρ_e , ρ_c	kg·m^{-3}	7324
specific heat, c_p^ℓ , c_p^c , c_p^e	J·kg^{-1}·K^{-1}	500
thermal conductivity, k_ℓ , k_e , k_c	W·m^{-1}·K^{-1}	34.0
latent heat, L	J·kg^{-1}	2.71×10^5
viscosity, μ_ℓ	kg·m^{-1}·s^{-1}	4.2×10^{-3}
thermal expansion coefficient, β_T	K^{-1}	1.07×10^{-4}
solutal expansion coefficient, β_c	wt.%$^{-1}$	1.4×10^{-2}
dendritic arm spacing, λ_1	m	5×10^{-4}
diffusion coefficient (liquid), D_ℓ	m^2·s^{-1}	2.0×10^{-8}
diffusion coefficient (solid), D_e , D_c	m^2·s^{-1}	1.0×10^{-9}
Covering slag		
viscosity, μ_{slag}	kg·m^{-1}·s^{-1}	0.01
density, ρ_{slag}	kg·m^{-3}	2500
specific heat, c_{slag}^g	J·kg^{-1}·K^{-1}	1815
thermal conductivity, k_{slag}	W·m^{-1}·K^{-1}	4.0

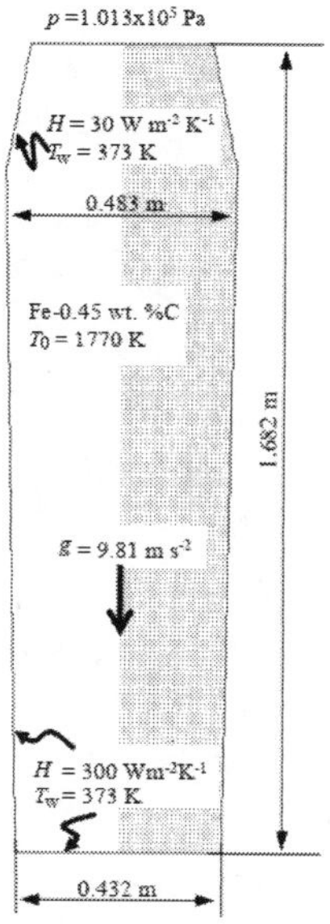

Figure 2. Configuration of the 2.45-ton industry steel ingot.

4. Numerical results

Example of the solidification sequence at 600 s is shown in Fig. 3. The overall solidification process is mainly governed by heat transfer. The cooling and solidification starts from the mould wall. Columnar phase develops from mould wall and grows towards casting centre. In the meantime equiaxed grains nucleate and grow in front of columnar tips, and those equiaxed grains start to sink and try to settle in the bottom region. The melt is dragged downwards along the columnar tip front by the sinking grains, which in turn induces a rising melt flow in the middle of the ingot. Thermal-solutal convection contributes to the interdendritic flow and global melt flow in the bulk. Both $\vec{u}_\ell$ and $\vec{u}_e$ fields are instable. As the flow of the melt and motion of the equiaxed grains are fully coupled with other transport phenomena (energy, species and mass), the instability of the flow patterns will directly influence the solidification sequence. Sedimentation of crystals in the bottom region will cause the volume fraction of the equiaxed phase to reach a quite high level. As f_e in the lower part is high enough (larger than the so-called mechanical blocking limit of 0.49, Fig. 3(b)), the columnar-to-equiaxed transition (CET) occurs. In the upper part of the ingot, the columnar tips can continue to grow, as no sufficient equiaxed crystals exist in front of them. The flow and crystal sedimentation are key mechanisms for the formation of macrosegregation in ingot casting, and they were discussed previously [9, 10, 17]. The final segregation result is shown in Fig. 4. A large positive segregation area just below the top shrinkage cavity is predicted. A layer of negative segregation zone is found in the bottom region, which coincides with the equiaxed sedimentation zone. The volume of this sedimentation zone depends strongly on the nucleation (parameters) of the equiaxed crystals. Note that the long calculation time (2 weeks in SGI cluster with 12 cores of 3.5 GHz) does not allow performing parameter study in 3D.

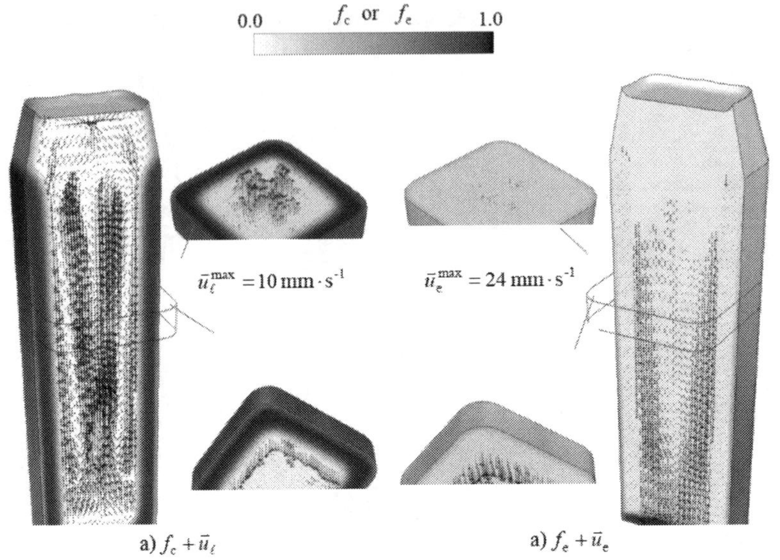

Figure 3. Solidification sequence at 600 s. Both f_c and f_e in the centre vertical section are shown in grey scale with light for lower and dark for high volume fraction. The velocity of the melt, $\vec{u}_\ell$ is shown together with f_c, while the velocity of the equiaxed crystals, $\vec{u}_e$ is with f_e. Additionally, the phase distributions and velocity fields in 2 horizontal sections are also shown.

Formation of the shrinkage cavity is shown in Fig. 5. The free surface, i.e. the interface between the liquid slag and metal phase, is represented by the iso-surface of $f_g = 0.5$ with 'g' standing for the slag phase. The numerical thickness of the interface is confined in 2~3 neighbouring elements. With the progress of solidification, the forming cavity is continuously filled by the slag from the inlet. The final volume of the cavity is predicted to be 0.0136 m^3, i.e. 4.04% of the volume of the entire ingot (V_0= 0.3362 m^3), which agrees with the solidification shrinkage rate of the alloy, $(\rho_s - \rho_\ell)/\rho_s$ = 4.05%. The formation of the microscopic shrinkage porosity is calculated indirectly by the CBN criterion, i.e. a post-processing of the modelling result of the thermal field. The final distribution of the CBN criterion is shown in Fig. 6. The smallest CBN criterion is located in the middle of the casting centreline, indicating the most possible position with the highest amount of shrinkage pores there.

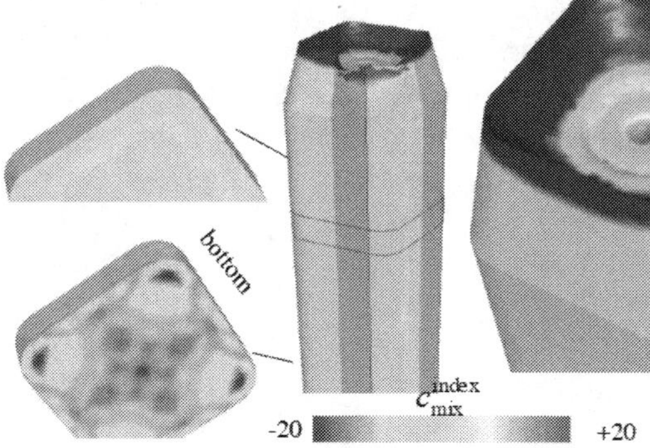

Figure 4. Final segregation distribution. The segregation index, $c_{mix}^{index} = 100 \times (c_{mix} - c_0)/c_0$, is used to evaluate the macrosegregation, where c_{mix} is the local mixture concentration among the existing metal phases (equiaxed, columnar and rest eutectic).

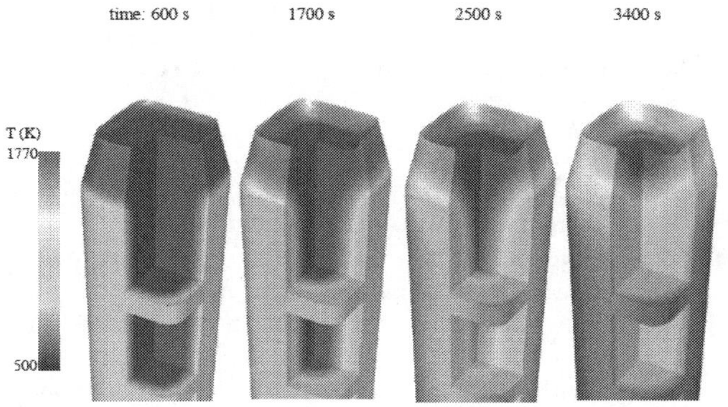

Figure 5. Cavity formation sequence in the hot top. Temperature field is shown in colour scale.

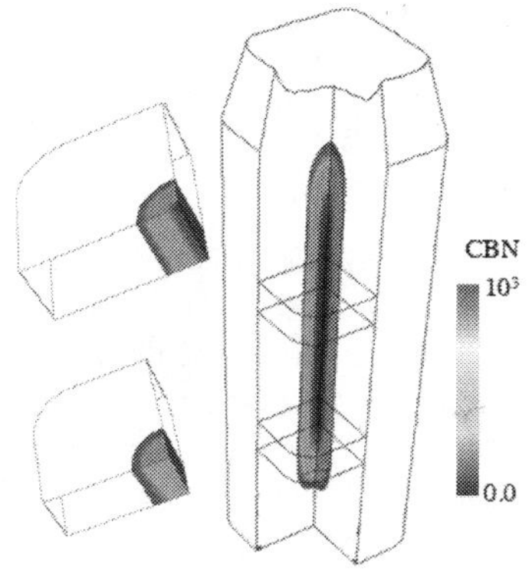

Figure 6. CBN criterion for microscopic shrinkage porosity.

The same casting was calculated in 2D axis symmetry (but the average mesh size is smaller, half of that used for 3D). The solidification sequence, the shrinkage cavity and CBN criterion of the 2D result are quite similar to those of 3D. The global macrosegregation pattern is also predicted similar: the layer of negative segregation in the equiaxed sedimentation zone and the large positive segregation below the top shrinkage cavity. The final macrosegregation pattern, and the predicted shrinkage cavity and CBN criterion are shown in Fig. 7, which are quite comparable with the experiment. Similar profile of shrinkage cavity is predicted, but the volume of cavity is smaller than that of the experiment. The macrosegregation along the casting centreline is also plotted and compared with the experiment, as shown in Fig. 8. Both 2D and 3D calculations have predicted similar variation features of the macrosegregation along the centreline, which agree qualitatively with the experiment. Between 0.2 and 1.0 meter from the bottom the predicted segregation profiles agree quantitatively with the experiment. As comparison, one more calculation was made with the previous 3-phase model without considering the formation of shrinkage cavity. The segregation profile is also plotted (green curve) in Fig. 8, which shows larger discrepancy from the experiment. The agreement between the new 4-phase model and the experiment is obviously improved.

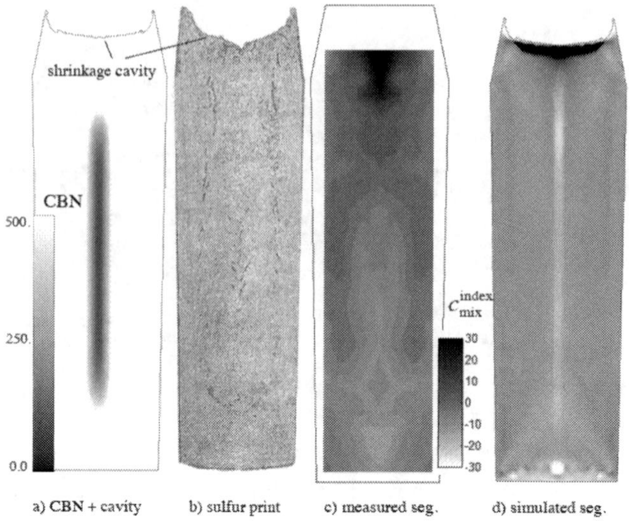

Figure 7. Comparison of the numerical simulation (2D axis symmetry with grid size of 5 mm) with the experiment as reported in [16]: a) numerical simulation of the shrinkage cavity and position of the microscopic shrinkage porosity by CBN criterion; b) sulphur print; c) reconstructed segregation map in grey scale out of chemical analysis of 54 drilling samples; d) predicted segregation map in grey scale.

5. Conclusion and discussion

The calculation of a 2.45 ton ingot has demonstrated the functionalities of the newly proposed 4-phase model. The goal to use an Eulerian volume averaging approach to model the as-cast columnar-equiaxed structure including CET (columnar-to-equiaxed transition), shrinkage cavity/porosity, and macrosegregation at the process scale for industry castings seems to be achievable. Most significance of the model lies in the consideration of the interaction between the solidification shrinkage and the formation of macrosegregation.

The current model is an extension of the previous mixed columnar-equiaxed solidification model [10] by including an additional gas (or covering liquid slag) phase. The new phase is supposed to be immiscible with other phases. It has no mass and species transfer with other phases, but the treatment of momentum and energy exchanges with other phases becomes crucially important to get a sharp interface between the liquid slag and other metal phases. Both 3D and 2D calculations show that the thickness of the interface between the slag and metal phases is confined in the 2~3 neighbouring elements. It means that the refinement of grid size improves the sharpness of the interface, i.e. the accuracy of the shape of the cavity. It is here verified that the shrinkage cavity can be dealt with Eulerian approach. The volume of the formed cavity in the hot top coincides well with the total volume shrinkage by solidification of the ingot.

The implemented CBN criterion is a further indicator for the possible occurrence of microscopic porosity. As the reported experimental data of this ingot [16] is not sufficient to verify the microscopic shrinkage result, no comment can be made here. However, many valuable works were done by other researchers [2, 3].

As seen from the segregation profile along the casting centreline (Fig. 8), the improvement of the current 4-phase model is significant in comparison with the previous 3-phase model which ignores the formation of shrinkage cavity. Detailed comparison of the two models will be presented in the late publication. One remark worth mentioning is that the modelling results presented previously [9, 17] were based on different nucleation parameters for the equiaxed crystals, with which the amount of the equiaxed phase, i.e. the bottom cone-shape equiaxed zone, seems to be overestimated for this casting. An additional finding based on the comparison study (3-phase model vs. 4-phase model) is that the solidification shrinkage adapts the multiphase flow in/near the columnar tip front slightly, and this minor adaptation influences the formation of the so-called quasi A-segregates. This finding needs further confirmation as well.

Few other points about the current macrosegregation model and modelling results should also be discussed. Although the key segregation features of industry ingot can be 'reproduced' numerically, the quantitative discrepancy from the reality still exists. This might be attributed to (i) the simplified dendritic morphology which exaggerates sedimentation induced negative segregation at the bottom; (ii) the grid size which is still too large to resolve the possible channel segregates; (iii) the shortage of adequate process parameters and materials properties; (iv) the simplification of multicomponent alloy system with a binary alloy. Finally, it is aware that the calculation cost increases intensively with the increase of the number of phases.

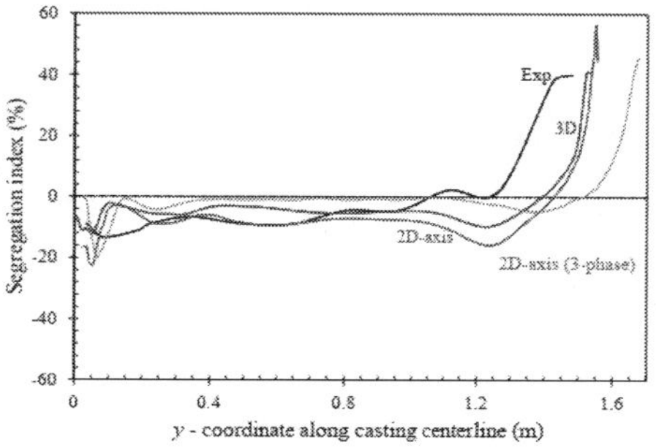

Figure 8. Comparison of centreline segregation (c_{mix}^{index}) between simulations (2D in red and 3D in blue) and experiment in black. The result of the previous 3-phase model (2D) was also plotted in green.

Acknowledgement
This work was financially supported by the FWF Austrian Science Fund (P23155-N24), FFG Bridge Early Stage (No. 3893791), and the Austrian Federal Ministry of Economy, Family and Youth and the National Foundation for Research, Technology and Development within the framework of the Christian Doppler Laboratory for Advanced Process Simulation of Solidification and Melting.

References

[1] Moore J and Shah N 1983 *Int. Metals Rev.* **28** 338-356
[2] Niyama E, Uchida T, Morikawa M and Saito S 1982 *AFS Cast Met. Res. J.* **7** 52-63
[3] Carlson K D, Beckermann C 2009 *Metall. Mater. Trans.* **40A** 163-175
[4] Nastac L and Marsden K 2013 *Int. J. Cast Metal Res.* **26** 374-382
[5] Wu M, Schädlich-Stubenrauch J, Augthun M, Sahm P and Spiekermann H 1998 *Dent. Mater.* **14** 321-328
[6] Ludwig A, Gruber-Pretzler M, Mayer F, Ishmurzin A and Wu M 2005 *Mater. Sci. Eng. A* **413-414** 485-489
[7] Wu M, Könözsy L, Ludwig A, Schützenhöfer W and Tanzer R 2008 *Steel Res. Int.* **79** 637-644
[8] Wu M, Li J, Ludwig A and Kharicha A 2013 *Comp. Mater. Sci.* **79** 830-840
[9] Wu M, Li J, Ludwig A and Kharicha A 2014 *Comp. Mater. Sci.* **92** 267-285
[10] Wu M and Ludwig A 2006 *Metall. Mater. Trans. A* **37A** 1613-1624
[11] Wang T, Yao S, Zhang X, Jin J, Wu M, Ludwig A, Pustal B and Bührig-Polaczek A 2006 *Jinshu Xuebao/Acta Metallurgica Sinica* **42** 584-590
[12] Ranz W and Marshall W 1952 *Chem. Eng. Prog.* **48** 141-146
[13] ANSYS FLUENT 12.0 User's Guide, Copyright © 2009 by ANSYS, Inc, April 2009
[14] Mayer F, Wu M and Ludwig A 2010 *Steel Res. Int.* **81** 660-667
[15] Wu M, Domitner J and Ludwig A 2012 *Metall. Mater. Trans. A* **43** 945-963
[16] Iron Steel Inst. 1926 *Report on the heterogeneity of steel ingots, J. Iron Steel Inst.* **103** 39-151
[17] Li J, Wu M, Kharicha A and Ludwig A 2014 *Int. J. Heat and Mass Transfer* **72** 668-679

Multi-scale finite element modelling of solidification structures by a splitting method taking into account the transport of equiaxed grains

T T M Nguyen[1], **H Combeau**[2], **M Založnik**[2], **M Bellet**[1], **Ch-A Gandin**[1]

[1] MINES ParisTech, CEMEF UMR CNRS 7635, 06904 Sophia Antipolis, France
[2] Université de Lorraine, IJL UMR CNRS 7198, 54011 Nancy, France

E-mail: thi-thuy-my.nguyen@mines-paristech.fr

Abstract. In solidification processes of large industrial castings and ingots, the transport of solid in the liquid has an important effect on the final grain structure and macrosegregation. Modeling is still challenging as complex interactions between heat and mass transfers at microscopic and macroscopic scales are highly coupled. This paper first presents a multi-scale numerical solidification model coupling nucleation, grain growth and solute diffusion at microscopic scales with heat and mass transfer, including transport of liquid and solid phases at macroscopic scales. The resolution consists of a splitting method, which considers the evolution and interaction of quantities during the process with a transport stage and a growth stage. This splitting reduces the nonlinear complexity of the set of considered equations and provides an efficient numerical implementation. It is inspired by the work of Založnik *et al.* [1,2], which used a finite volume method (FVM). The present work develops the solution based on the finite element method (FEM). Numerical results obtained with this model are presented and simulations without and with grain transport are compared to study the impact of solid-phase transport on the solidification process and on the formation of macrosegregation.

1. Introduction

Solidification modelling accounting for melt convection and solid movement has been limited. The first important works were proposed with a volume-averaged model consistently connecting microscopic phenomena to macroscopic transports [3-6]. The transport of the solid structure was based on a population balance written for the equiaxed grains. Similar solidification modeling was later further developed [1,2,7-10]. The numerical resolution algorithm proposed by Založnik and Combeau [1] efficiently dealt with the complexity of the strongly coupled problem by using a splitting method. This splitting method was successfully applied to large industrial castings [11]. It is worth noticing that these models were based on the finite volume method (FVM), while the use of the finite element method (FEM) has been rarely considered for volume-averaged multiphase multiscale modelling of solidification. Besides ongoing advancements with FVM, the present study aims at developing FEM solvers when solid transport is to be considered. In this paper, a numerical FEM solidification model is presented accounting for microscopic phenomena as well as for the motion of solid and liquid. First, the macroscopic conservation equations and the constitutive relations describing interfacial interactions and exchanges are summarized. The resolution method and the numerical implementation for the set of non-linear equations are then detailed. Finally, numerical simulations using the current model are performed to validate the numerical implementation. Results demonstrate an efficient FEM resolution scheme implemented for the purely convective transport problem, which is difficult to solve numerically with the FEM in the absence of diffusive effects.

2. Model description

2.1. Macroscopic conservation equations and constitutive relations

The macroscopic conservation equations of mass, momentum, species and energy for each phase α, (1)-(4) respectively, are obtained by averaging the microscopic equations over a representative elementary volume. Details on this averaging method can be found elsewhere [12].

Mass
$$\frac{\partial(g^\alpha \rho^\alpha)}{\partial t} + \nabla\cdot(g^\alpha \rho^\alpha \langle \mathbf{v}^\alpha\rangle^\alpha) = \Gamma^\alpha + \Phi^\alpha \quad (1)$$

Momentum
$$\frac{\partial(g^\alpha \rho^\alpha \langle \mathbf{v}^\alpha\rangle^\alpha)}{\partial t} + \nabla\cdot(g^\alpha \rho^\alpha \langle \mathbf{v}^\alpha\rangle^\alpha \langle \mathbf{v}^\alpha\rangle^\alpha) = -g^\alpha \nabla\langle p^\alpha\rangle^\alpha + \nabla\cdot(g^\alpha\langle \boldsymbol{\tau}^\alpha\rangle^\alpha) + g^\alpha\langle \mathbf{f}^\alpha\rangle^\alpha + \mathbf{M}^{\alpha,d} + \mathbf{M}^{\alpha,\Gamma} + \mathbf{M}^{\alpha,\Phi} \quad (2)$$

Species
$$\frac{\partial(g^\alpha \rho^\alpha \langle w^\alpha\rangle^\alpha)}{\partial t} + \nabla\cdot(g^\alpha \rho^\alpha \langle w^\alpha\rangle^\alpha \langle \mathbf{v}^\alpha\rangle^\alpha) = -\nabla\cdot(g^\alpha\langle \mathbf{j}^\alpha\rangle^\alpha) + J^{\alpha,J} + J^{\alpha,\Gamma} + J^{\alpha,\Phi} \quad (3)$$

Energy
$$\frac{\partial(g^\alpha \rho^\alpha \langle h^\alpha\rangle^\alpha)}{\partial t} + \nabla\cdot(g^\alpha \rho^\alpha \langle h^\alpha\rangle^\alpha \langle \mathbf{v}^\alpha\rangle^\alpha) = -\nabla\cdot(g^\alpha\langle \mathbf{q}^\alpha\rangle^\alpha) + Q^{\alpha,J} + Q^{\alpha,\Gamma} + Q^{\alpha,\Phi} \quad (4)$$

where the superscript α represents phase α; the notation $\langle\,\rangle$ indicates the average value; g is the volume fraction; ρ is the mass density; $\mathbf{v}$ is the velocity vector; w is the concentration; h is the enthalpy; p is the pressure; $\boldsymbol{\tau}$ is the viscous stress; $\mathbf{f}$ is the body force per unit volume; $\mathbf{j}$ is the species flux; $\mathbf{q}$ is the heat flux; Γ, Φ are the change rates of mass due to the phase change and the nucleation.

Except Φ, the other terms related to the nucleation, noted with the superscript Φ, are neglected. The change rates of momentum, solute mass and energy due to the phase change are expressed respectively as

$$\mathbf{M}^{\alpha,\Gamma} = \bar{\mathbf{v}}^{\alpha*}\Gamma^\alpha \quad\Big|\quad J_i^{\alpha,\Gamma} = \bar{w}^{\alpha*}\Gamma^\alpha \quad\Big|\quad Q^{\alpha,\Gamma} = \bar{h}^{\alpha*}\Gamma^\alpha \quad (5)$$

The solute flux at the solid/liquid interface (6) is defined as a function of the difference of the variation between the average solute concentration in the phase $\langle w^\alpha\rangle^\alpha$ and at the interface, $w^{\alpha*}$. Over a representative elementary volume, the temperature is considered to be uniform, therefore the heat flux $Q^{\alpha,J}$ is equal to zero.

$$J^{\alpha,J} = \frac{\rho^\alpha S_V D_w^\alpha}{\delta_w^\alpha}(w^{\alpha*} - \langle w^\alpha\rangle^\alpha) \quad (6)$$

The interfacial terms for mechanical interactions between the solid and liquid phases are distinguished considering two regimes: the slurry regime in which solid grains are small and freely move in the liquid phase; and the porous regime in which they become larger, heavier and aggregate to fixed regions (e.g., packed beds of equiaxed grains or mold boundaries).

slurry regime
$$\mathbf{M}^{l,d} = -\mathbf{M}^{s,d} = \frac{1}{2V_e}\rho^l A_e C_D \left|\langle \mathbf{v}^s\rangle^s - \langle \mathbf{v}^l\rangle^l\right|\left(\langle \mathbf{v}^s\rangle^s - \langle \mathbf{v}^l\rangle^l\right) \quad (7)$$

porous regime
$$\mathbf{M}^{l,d} = -\mathbf{M}^{s,d} = -\frac{(g^l)^2 \mu^l}{\mathrm{K}}\langle \mathbf{v}^l\rangle^l \quad (8)$$

where $\bar{\mathbf{v}}^{\alpha*}$, $\bar{w}^{\alpha*}$, $\bar{h}^{\alpha*}$ are respectively the averages of velocity, concentration and enthalpy of phase α over the interfacial area; S_V is the interfacial area concentration; D_w, D_T represent the solutal and thermal diffusivities; δ_w^α, δ_T^α are the characteristic solutal and thermal diffusion lengths [13] ; A_e, V_e are the cross section area and the volume of grains respectively; C_D is the drag coefficient [14]; the permeability K is modelled by the Kozeny-Carman equation.

The distribution of solid grains is modelled by the population balance equation as

$$\frac{\partial N}{\partial t} + \nabla \cdot (N \langle \mathbf{v}^s \rangle^s) = S_{nucl} \tag{9}$$

where N is the grain density per unit volume and S_{nucl} represents the nucleation source.

2.2. Resolution method and numerical implementation

Strong coupling exists between the quantities in the above equations, describing at the same time macroscopic transports and microscopic phenomena. It requires dedicated algorithms to find the solution of the set of non-linear equations. In this section, the implementation of a splitting method proposed earlier [1,2] is presented. The general resolution algorithm is shown in Figure 1. In the first step, the mixture energy conservation obtained by adding up equations (4), written for the solid and liquid phases with equal intrinsic average temperatures for the phases, i.e. equation (10) in Figure 1, is solved to determine the average specific enthalpy $\langle h \rangle$. Note that the phase densities are assumed constant and equal for all phases in the present model. In the next step, the evolution of the total mass, of the grain density and of the mass of species are computed in the transport stage. For that purpose, Eqs. (2), (3) and (9) are solved by keeping only the transport terms and neglecting those related to nucleation and growth of the equiaxed grains, which leads to Eqs. (11-13) in Figure 1. The values deduced from the transport stage for the solid fraction, g_{tr}^s, and the compositions $\langle w^s \rangle_{tr}^s$ and $\langle w^l \rangle_{tr}^l$, are used to initialize the subsequent nucleation and growth stage. The number of nucleated grains is determined by Eq. (14). The growth rate of solid grains, $\Gamma^s = -\Gamma^l$, and the composition at the interface, w^{s*} and w^{l*}, are determined by the energy balance and the interfacial solute mass balance, Eqs. (15) and (16) in Figure 1 respectively. Knowing these quantities it is possible to determine the variations of the solid fraction as well as those of the intrinsic average compositions for each phase by Eqs. (17) and (18) and hence the actual solid fraction, g_{gr}^s and compositions, $\langle w^s \rangle_{gr}^s$ and $\langle w^l \rangle_{gr}^l$. The final evolution of quantities includes contributions of these two stages. The solid and liquid velocities as well as the pressure are determined by solving the momentum conservation Eqs. (2) and the continuity Eq. (19). At each time-step, iterations are performed until all variables are converged, as schematized by the CV diamond and the feedback arrow on the left hand side of Figure 1. This ensures coherency between the solution of the complete set of equations with the FEM at the macroscopic scale (Eqs. (10-13), (19) and (2) in Figure 1) with the local solutions for the microscopic processes described by Eqs. (14-18) in Figure 1.

In the work presented here, a particular treatment is carried out by adding the artificial diffusion terms to Eqs. (11-13) in order to stabilize the FEM resolution of the purely advective transport equations. The artificial diffusion coefficient D_M used in this study depends on the characteristic mesh sizes, $he_{\mathbf{v}^\alpha}^2$, and on the spatial variation of velocities, as seen in Figure 1. This treatment is inspired by the treatments used for supersonic reacting flows [15] but it is modified to adapt to solidification modelling. In order to respect the mass conservation and to ensure consistence of all transported quantities, the same artificial coefficient is applied to the energy, Eq. (10). All terms related to the artificial diffusion are marked with yellow squares in Figure 1.

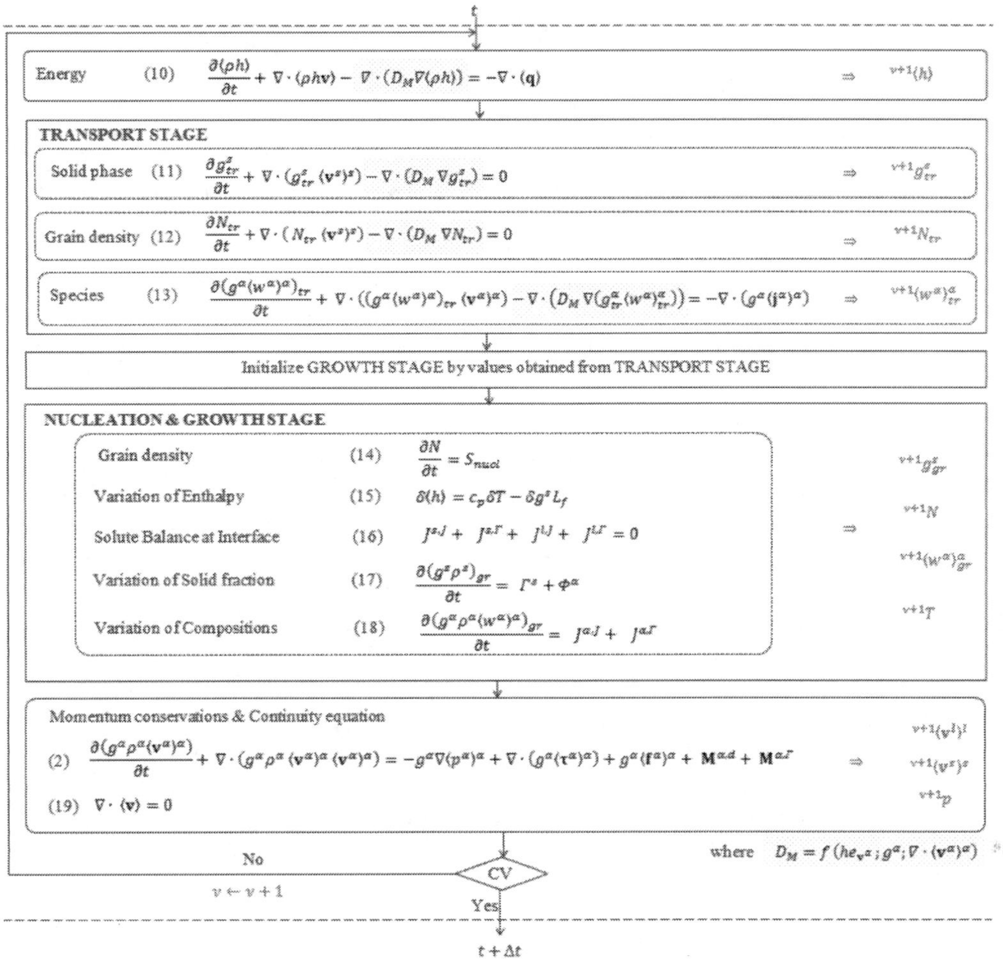

Figure 1. Resolution algorithm

3. Simulation results and discussion

3.1. 1D simulation – Study of the growth stage

For verification purposes the model is first tested for a case of pure diffusion. This tests its capacity to account for the grain growth kinetics at the microscopic scale. Both liquid convection and solid transport are neglected. Solidification occurs in a one-dimensional manner. This is achieved by defining a 1 mm thick and 100 mm wide domain cooled from one of its narrow sides. Six sensors are defined at along the sample, two at both ends and four between them with a regular spacing of 2 cm. The thermophysical properties of the material are taken from reference [16] and are representative of a binary Al-4wt%Cu alloy, except for the solute diffusivity coefficients and the grain density. These two parameters are control parameters. They are used to switch between a case with a very high solute diffusion, so-called "infinite" diffusion and the other case for a lesser diffusion, so-called "finite" diffusion, as indicated in Figure 2.

Figure 2 shows a comparison of the cooling curves (left) and solid fraction evolutions (right) at six sensors between the "infinite" solute diffusion case (green curves) and the "finite" diffusion case (red curves): a) with high diffusion in the solid phase; b) with no solute diffusion in the solid phase. In the case with "finite" diffusion, it can be noticed that the first red curve on the cold side presents a recalescence close to the liquidus temperature, which is not computed when liquid composition is well mixed, i.e. for the green curves. This phenomenon is captured by the present model because it takes into account nucleation and grain growth kinetics. It depends on both the values of the grain density and the liquid diffusion coefficient. Yet the red curves are found to depart very little from the green curves that were verified to exactly retrieve the lever rule approximation. With high values of the Lewis number (the ratio between heat diffusivity and mass diffusivity), solidification at microscopic scale is dominated by solute diffusion. A limited solute diffusion results in a slow phase transformation at the beginning of solidification which is illustrated by a discrepancy of solid fraction evolutions between the "infinite" and "finite" diffusion cases. The discrepancy decreases as solidification progresses. This can be explained by the fact that during solidification only a part of the liquid solute content is transferred to the solid due to solute partitioning. The rest that is rejected from the solid builds up in the liquid between grains. The increase of solutal gradients around the solidification front and the extension of solid-liquid surfaces accelerate solute mixing. In the case of no solute diffusion in the solid phase, a formation of eutectic structure is predicted toward the end of solidification when the eutectic temperature is reached, as shown in Figure 2b. The amount of eutectic is proportional to the final vertical jump of the solid fraction curves to reach unity. It was also checked that the green curves in Figure 2b tends toward the solution with the Gulliver-Scheil approximation, further validating the implementation of the microscopic growth stage.

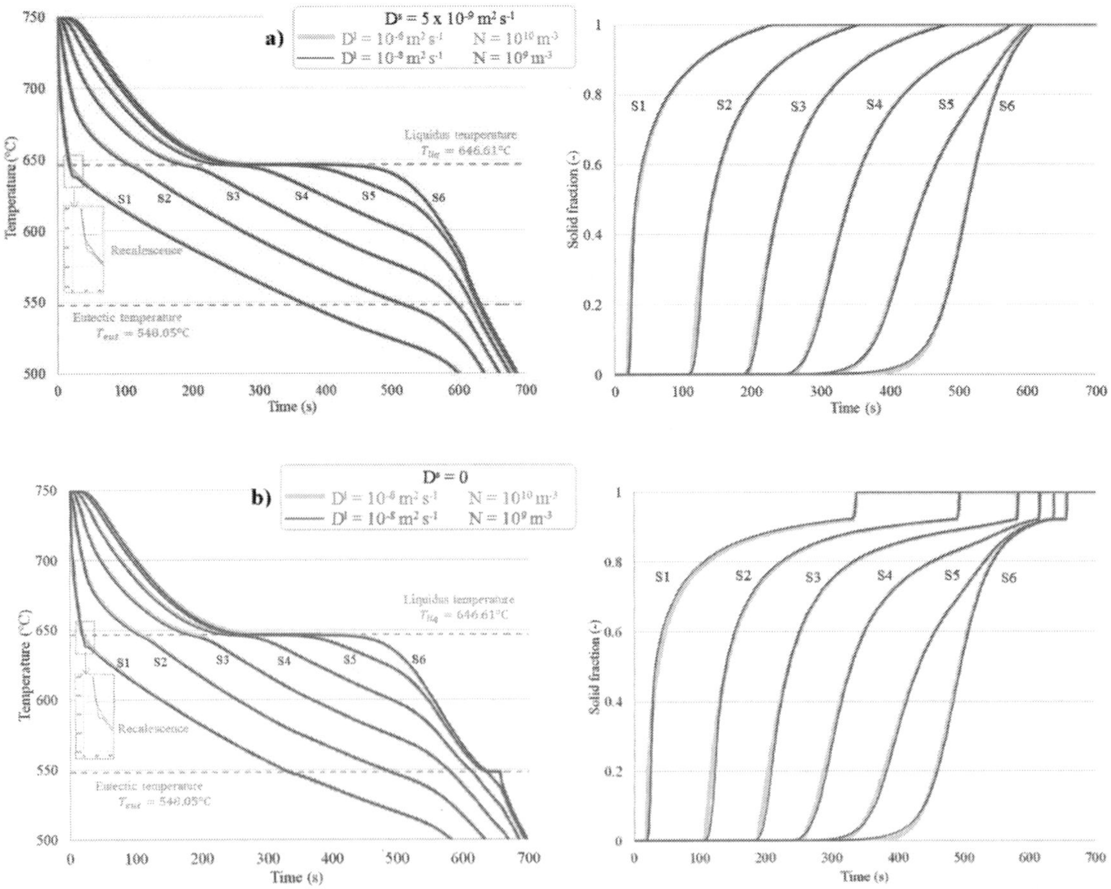

Figure 2. Cooling curves (left); Temporal evolutions of the solid fraction (right)

3.2. 1D simulation – Study of the transport stage

In the previous section, the growth stage combined with the energy resolution showed the ability of the present model to model solidification without motion. In the present section, the transport stage is carried out by studying the sedimentation of solid grains. This is first done neglecting the nucleation and growth stages, i.e. with a given number of grains of constant size. The properties chosen for this test come from a binary Sn-5wt%Pb [2]. At the beginning, the grains are distributed with a fixed density, 10^9 grains m^{-3}, within a subdomain of 20 mm to 80 mm of a 10 mm × 100 mm sample, as shown in Figure 3a at time t = 0 s. The fraction of solid in this subdomain, shown in Figure 3c-top, is 0.1, thus fixing the grain radius to 288 μm. The intrinsic solute composition of the liquid is 5 wt%Pb in the entire sample and the intrinsic composition of the solid is 0.328 wt%Pb where the solid is present and zero elsewhere. The average composition is thus not homogeneous at 5 wt%Pb. The initial temperature, 498.057 K, is computed by using the lever rule so that it directly corresponds to the solid fraction and the compositions. An unstructured triangular mesh with a size of 0.5 mm and a constant time step 0.01 s are used.

The sedimentation process is illustrated through the maps of solid fraction and its profiles at 2 s, 5 s and 10 s, as shown in Figures 3a and 3c-top. From the initial state, the solid grains fall with a velocity of around 7 mm s^{-1}, computed through a balance of forces acting on the grains (Figure 3b). The first grains arrive at the sample bottom after 3 s and begin to accumulate until the characteristic packing fraction is reached (taken as 0.3). During the computation, the velocity of the grains moving towards the zone near the packing area is adjusted so that these grains can land smoothly in the packed bed, i.e. reaching $\langle \mathbf{v}^s \rangle^s = 0$ at the packing limit. To be more specific, on the one hand, the solid velocity is reduced by viscous effect, as the viscosity strongly increases when the solid fraction is close to the packing value [17]. On the other hand, if the grains go so fast, their velocity is multiplied by a factor calculated by dividing their current distance to the packing limit by the distance which they can go due to their velocity projected on the gradient of solid fraction. Without these numerical remedies, the solid fraction of the packed zone can exceed the packing fraction, which is caused by a brutal stop of grains arriving in the packed area.

As the grain motion involves the transport of other quantities, it is important to verify the consistency of their transports. Figure 3c-center presents the profiles of the average composition at 0 s, 2 s (before the grains have reached the bottom of the domain), 5 s (accumulation below the packing limit) and 10 s (end of packing). They show consistency with those of the solid fraction. When the solid grains sediment, solute-rich liquid with 5wt%Pb rises and fills the region left by the grains. The average composition of 3.6 wt% Pb in the negative segregation zone at the bottom corresponds precisely to the final state of 30% of solid at 0.328 wt%Pb and 70% of liquid at 5 wt% Pb. In addition, as expected for a pure transport simulation, the temperature does not change during this process. This is seen in Figure 3c-bottom. A good overall conservation of all quantities is also verified, the maximum relative errors of solute mass and energy being about 10^{-6} and 10^{-5}, respectively.

This section deals with difficulties in FEM modelling of purely advective transport, as its solution presents numerical instabilities. To remedy these difficulties an addition of adequate artificial diffusion is necessary along with the aforementioned modifications of the solid-velocity calculation.

These remedies are necessary to be able to solve the equations of the present model with the FEM. In this work the addition of artificial diffusion was done and examined very carefully, in order to ensure consistency between all the quantities and the conservation of mass and energy.

IOP Conf. Series: Materials Science and Engineering **84** (2015) 012007 doi:10.1088/1757-899X/84/1/012007

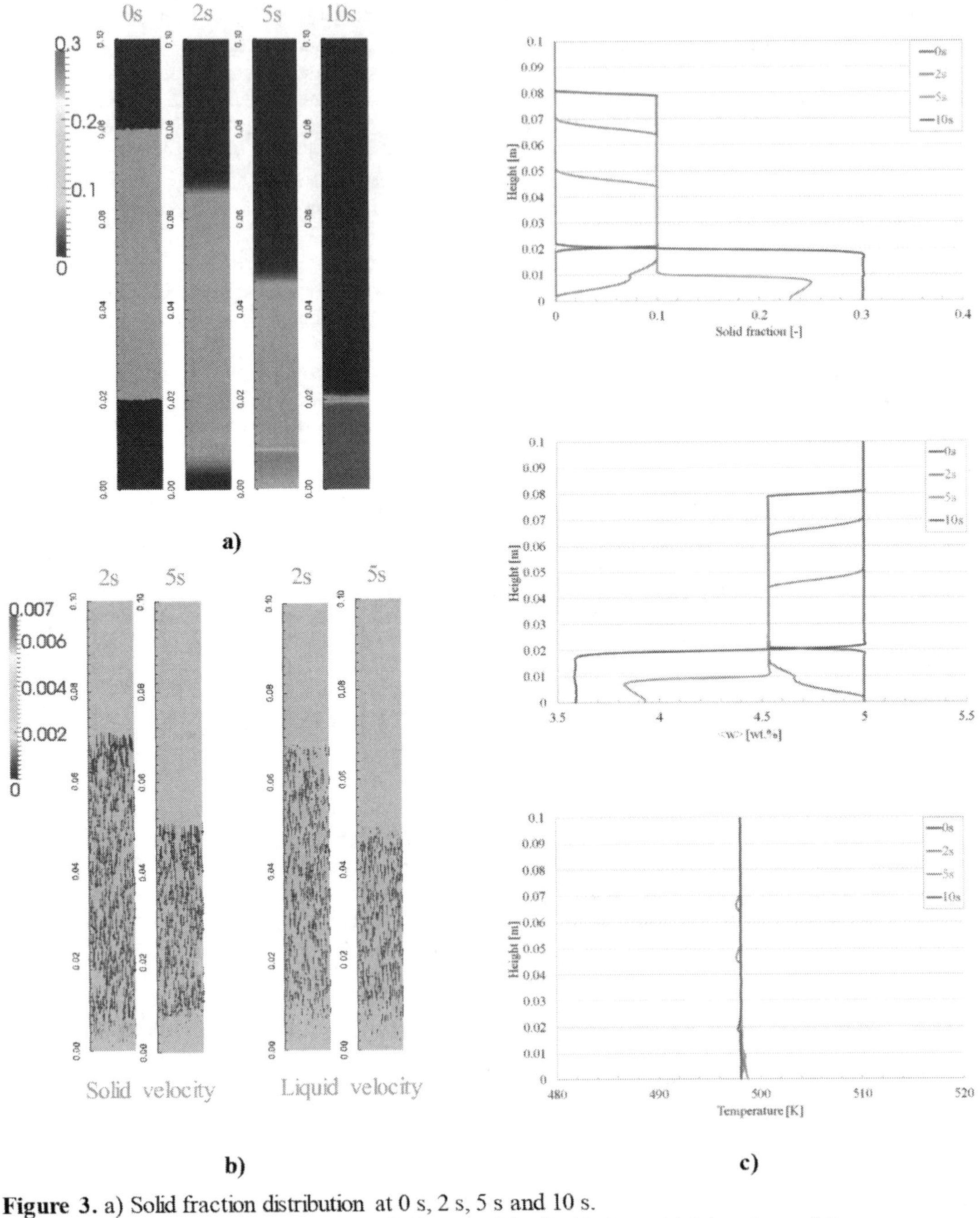

Figure 3. a) Solid fraction distribution at 0 s, 2 s, 5 s and 10 s.
b) Vectors of the solid phase (left) and of the liquid phase (right) at 2 s and 5 s.
c) Profiles of the solid fraction (top), the average composition (center) and the temperature (bottom) at 0 s, 2 s, 5 s and 10 s.

3.3. 2D solidification simulations

Following the 1D simulations, two 2D solidification cases are simulated and compared. Both cases involve a Sn-5wt%Pb alloy. Case 1 accounts for thermo-solutal liquid convection but considers a fixed solid phase and Case 2 includes both liquid convection and solid movement due to liquid advection and sedimentation. The sample of 0.06 m x 0.1 m size is cooled from the left hand side. The initial liquid metal temperature is 260°C. An unstructured triangular mesh with constant size, 1 mm, and a time step of 0.01 s are used.

In Case 1 there is no solid transport, therefore the terms relating to the solid motion in the equations presented in section 2 are set equal to zero; the packing fraction is considered as equal to zero. In Case 2, the full model is applied accounting for grain growth and transport phenomena, in which the solid phase is supposed to be stationary when its fraction reaches the packing value.

The simulation results of Case 1 and Case 2 are compared in Figures 4 and 5. Figure 4 displays the solid velocity vectors, the liquid velocity streamlines and the solid fraction distribution. The black line represents the limit between the packed and the slurry zones. The solidification sequence in these two cases is quite different. In Case 1, initial solid grains nucleated on the cooled side are fixed there and when solidification proceeds, newly formed solid grains attach to the existing solid layers. The global solidification front thus advances from left to right. This is close to a columnar growth situation. In Case 2, after nucleation due to the gravity force and the liquid advection, solid grains sink downwards to the bottom as well as are transported far away by liquid flow. During this process, some grains travel in the slurry region and strongly affect the movement of the liquid phase, while the largest grains settle and accumulate as an increasing solid layer forming from at the bottom.

In addition, solidification progresses faster in Case 2 compared to Case 1. This can be explained by the fact that the solid grains that settle in the vicinity of the cold mold are replaced by a hotter liquid phase. Consequently, the global heat exchange between the metal and the mold takes place more efficiently and results in faster solidification, as shown at 360 s.

The final macrosegregation maps are shown in Figure 5. Although higher solute content is found at the bottom area for both cases, its formation mechanism is quite different. In Case 1, it is primarily governed by natural convection in the zone with relatively high liquid fraction. The solid zone extends from left to right and the rejected solute is progressively accumulated in the bottom-right region due to the higher density of the solute-enriched liquid. A high solute concentration is therefore found at the bottom-right corner.

For Case 2, at the initial stage, sedimentation of solute-depleted grains leads to a rigid solid layer and a negative segregation in the lower part. As the process evolves, an intergranular downward flow builds up in the packed porous zone, which brings the solute from the upper to the lower region. At the very bottom of the cavity, the intensity of the flow through the mushy zone is sufficient to lead to remelting and finally forms a positive segregation.

Through this section, we study the effect of the solid transport on the solidification sequence as well as the formation of macrosegregation. The results of the solid fixed case obtained from the present formulation with high values of solute diffusion coefficients retrieve those simulated with the lever rule model from the literature using FVM [2], and our FEM simulation. For the solid transport modeling, further evaluation that would be necessary will be carried out in the future.

IOP Conf. Series: Materials Science and Engineering **84** (2015) 012007 doi:10.1088/1757-899X/84/1/012007

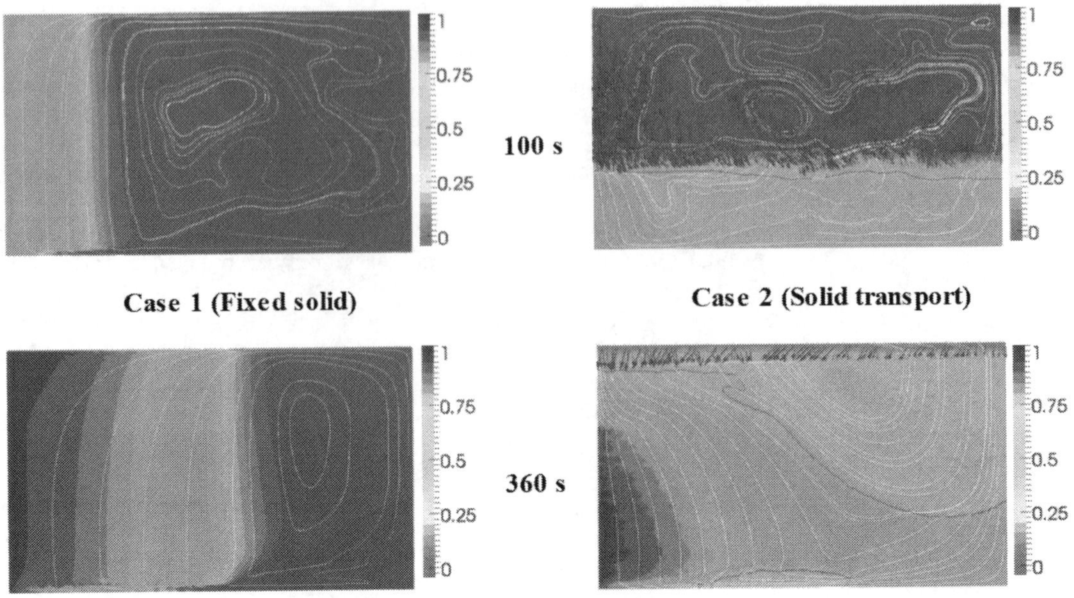

Figure 4. Solid fraction distribution displayed with (violet arrows) solid velocity vectors and (white contours) liquid velocity streamlines.

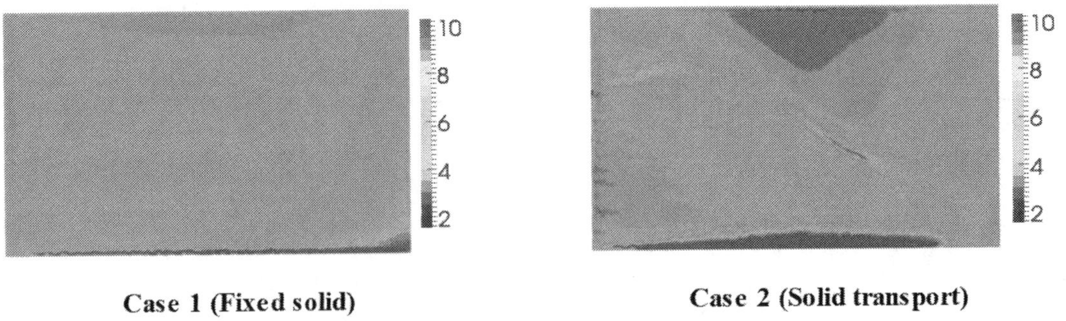

Figure 5. Final macrosegregation maps.

4. Conclusion

This paper presents a FE numerical model taking into account microscopic processes as well as the macroscopic transport phenomena. The resolution is based on splitting scheme, in which the transport stage is solved independently from the nucleation and growth stages. The coupling between the two stages is achieved through an iterative procedure. Simulations were first carried out to test the capabilities of this model to deal with pure growth and pure transport situations. Then the model is applied to solidification with a fixed solid phase and with the transport of solid grains. The main results can be summarized as follows.

- The present model incorporates a microsegregation approach able to simulate solidification paths with finite diffusion in solid and liquid phases, including the recalescence phenomenon.

- The vertical sedimentation of particles has been simulated by the FE model and consistent time and space evolutions of all quantities during transport have been verified.
- Compared to the fixed solid case, results obtained with this new FE model for solid transport show an important influence of solid motion on solidification progression and on formation of macrosegregation.

Simulation results would be further assessed by comparing with other works and experimental results. Combined with mesh adaptation techniques, the present model should provide industry with a promising tool for modeling castings. Further development should encompass the extension to multicomponent alloys and the consideration of various morphologies of the solid phase.

Acknowledgments
The financial support from ArcelorMittal, Industeel, Aubert & Duval, Ascometal and Aperam is gratefully acknowledged.

References

[1] Založnik M and Combeau H 2010 *Comput. Mater. Sci.* **48** 1-10
[2] Založnik M, Kumar A and Combeau H 2010 *Comput. Mater. Sci.* **48** 11-21
[3] Ni J and Beckermann C 1991 *Metall. Trans.* B. **22** 1991-349
[4] Beckermann C and Viskanta R 1993 *Appl. Mech. Rev.* **46** 1-27
[5] Wang C Y and Beckermann C 1996 *Metall. Mater. Trans.* A. **27** 2754-64
[6] Wang C Y and Beckermann C 1996 *Metall. Mater. Trans.* A. **27** 2765-83
[7] Ludwig A and Wu M 2002 *Metall. Mater. Trans. A* **33** 3673-83
[8] Wu M, Ludwig A, Buhrig-Polaczek A, Fehlbier M and Sahm P R 2003 *Int. J. Heat. Mass. Tran.* **46** 2819-32
[9] Wu M, Li J, Ludwig A and Kharicha A 2013 *Comput. Mater. Sci.* **79** 830-40
[10] Wu M, Li J, Ludwig A and Kharicha A 2014 *Comput. Mater. Sci.* **79** 267-85
[11] Combeau H, Založnik M, Hans S and Richy P E 2009 *Metall. Mater. Trans.* B. **40** 289-304
[12] Rappaz M, Bellet M and Deville M 2003 *Numerical Modeling in Materials Science and Engineering* (Verlag Berlin Heidelberg New York : Springer)
[13] Bedel M 2014 *Etude de la formation des structures de solidification et des macroségrégations en coulée semi-continue d'aluminium* chapter 2 p 55-66
[14] Guillemot G, Gandin Ch A and Combeau H 2006 *ISIJ Int.* **46** 880-95
[15] Cook A W and Cabot W H 2004 *J. Comput. Phys.* **195** 594-601
[16] Ni J and Beckermann C 1993 *J. Mater. Process. Manuf. Sci.* **2** 217-231
[17] Graham A L 1981 *Appl. Sci. Res.* **37** 275-86

IOP Conf. Series: Materials Science and Engineering **84** (2015) 012008 doi:10.1088/1757-899X/84/1/012008

Transient permeability in macrosegregation of static casting in binary alloys: Use of CDF statistical model for analysis

I. Vušanović

Faculty of Mechanical Engineering, University of Montenegro, 81000 Podgorica, Montenegro

E-mail: igor.vusanovic@ac.me

Abstract. Recently, a statistical method for analyzing macrosegregation predicted by casting simulations or measured in experimental systems has been proposed [1]. The basis of this approach is to collect and order the distribution of nodal concentration values in the final solidified product and thereby create a cumulative distribution function (CDF), presented in a log-log or semi-log format. By considering changes in the shape of the CDF, this statistical tool has been used to study grid convergence and the effect of key physical parameters in 2D casting simulations of the Al-4.5%Cu alloy [2]. These simulations assumed fixed constant representative values for the parameters, in particular a fixed arm dendrite arm spacing in the permeability model for the solid/liquid mushy region. Here the intention is to see how consideration of an arm spacing that coarsens with time (Voller [3]) might influence the shape of the CDF curves. To achieve this, calculations are performed for the recent benchmark experimental study on the Sn-3%Pb alloy presented by Hachani et al [5]. The comparison of the CDF from this experimental benchmark with the one from the simulation, is used to determine the effect of transiently adjusting the permeability during the simulation. Results indicate that an appropriate coarsening model in the permeability calculation of the simulation may improve the comparisons between the measured and simulated CDF. Any level of improvement, however, is on the order of a reasonable expectancy for the resolution of the experimental measurement.

1. Introduction

Macrosegregation, as key defect phenomenon which occurs in many metal casting systems has been widely investigated, both with experiments[4-6], and numerical simulations[7-15].. Recently, Voller and Vušanović [1] proposed a simple statistical tool for comparing experimental and model results. The key is to collect and order the distribution of nodal concentration values in the final solidified product and thereby create a cumulative distribution function (CDF). When it is presented, in a log-log or semi-log form, this CDF curve can be used to determine the volume fraction of the solidified casting above a given level of segregation. In its first application the utility of the method was demonstrated using the experimental measurement data presented in Quillet et al. [6]. The CDF statistical method was subsequently used to study the sensitivity of casting simulation model parameters [2]. In this later work it was shown that the shape and spreading of the CDF curves is, not surprisingly, sensitive to change in the parameters that control the fluid flow in the mushy region. Flow in the mushy region is dominantly driven by thermal and solutal expansions of the liquid phase, while the main resistance to the flow comes from the rigid network of solid dendrites. In most current simulation technologies the permeability constant (κ_0) in the Carman Kozeny equation [8] controls the resistance of flow in the mushy region. Typically this is calculated in terms of an assumed fixed secondary arm space (DAS):

$$\kappa_0 = \frac{DAS^2}{180} \quad (1)$$

The objective of this paper is to use the CDF tool to investigate the influence of a transient permeability due to the coarsening effects, and thus provide advice for improvements of the mathematical modeling of flow in the mushy region.

2. Physical and mathematical model of benchmark experiment

Recently, researchers [4-6] have performed sets of controlled benchmark casting experiments in a rectangular cavity using binary Sn-Pb and Sn-Bi alloys. These experiments report temperatures, compositions and phase fractions on grid of positions in the final cast. To demonstrate the CDF statistical method we have previously used the Sn-10%Bi alloy measurements [1]. By contrast in this study we use the physical model and casting conditions for benchmark solidification of the Sn–3%Pb alloy performed by Hachani et al. [5]. In this benchmark experiment the alloy was solidified in a rectangular cavity of 100x60x10mm size, using two controlled water cooled heat exchangers. In the simulation model presented here, the material properties of Sn-3%Pb, given in Table 1, were taken from Table E.1 in ref [5]. In addition the imposed boundary temperature conditions used in the simulation matched those reported in [5] for the case of a cooling rate of $0.03Ks^{-1}$. The simulation of this system is based on the standard macrosegregation mixture model of mass, momentum, solute and heat conservation introduced by Bennon and Incropera [8]:

$$\frac{\partial \rho}{\partial \tau} + \nabla \cdot (\rho u) = 0 , \quad (2)$$

$$\frac{\partial}{\partial t}(\rho u) + \nabla \cdot (\rho u u) = \nabla \cdot (\mu \nabla u) - \frac{\partial p}{\partial x} - \mu K^{-1} u , \quad (3)$$

$$\frac{\partial}{\partial t}(\rho v) + \nabla \cdot (\rho u v) = \nabla \cdot (\mu \nabla v) - \frac{\partial p}{\partial y} - \rho g [\beta_T (T - T^{rerf}) + \beta_C (C_L - C_L^{ref})] - \mu K^{-1} v , \quad (4)$$

$$\frac{\partial}{\partial t}(\rho C) + \nabla \cdot [(\rho C_L) u] = 0 , \quad (5)$$

$$\frac{\partial}{\partial t}(\rho H) + \nabla \cdot [(\rho c T + \rho L) u] = \nabla \cdot (k \nabla T) , \quad (6)$$

$$K = \kappa_0 \frac{f^3}{(1-f)^2} . \quad (7)$$

Where $u_L=(u_L, v_L)$ is the liquid velocity, $u=fu_L$ is the mixture velocity assuming no solid movement $(u_s=0)$, f is the liquid fraction, $H=cT+fL$ is the enthalpy, c the specific heat, k thermal conductivity, and the temperature T is related to the liquid solute concentration C_L through $T=T_m+m_L C_L$, where T_m is the fusion temperature of Sn and m_L is the slope of the liquidus line. Numerical solution of equation (2)-(7) is obtained using a square grid of control volumes. The fluid flow equations are solved implicitly (with a SIMPLER method for pressure calculation [18]), while the thermal and solutal equations are solved with the time explicit scheme suggested by Voller et al. [16]. This combination of time schemes may be computationally inefficient, but does not induce numerical noise as we observed in using of fully implicit schemes [19].

Table 1. Thermophysical properties of Sn-3wt%Pb alloy

Fusion temperature T_m (C)	232	Latent heat of fusion L (kJ/kg)	57.512
Slope of the liqudus line (K/wt%)	-1.274	Dynamic viscosity μ (kg/m s)	1.737e-3
Density ρ (kg/m^3)	6965	Thermal expansion coefficient $\beta_{T,L}$(K^{-1})	0.95e-4
Specific heat of liquid c_L (J/kgK)	243	Liquid solutal expansion coefficient $\beta_{S,L}$	-0.53
Specific heat of solid c_S (J/kgK)	209	Dendrite arm spacing DAS (μm)	90
Thermal conductivity k_L (W/mK)	33	Partition coefficient k_{par} (-)	0.06562
Thermal conductivity k_S (W/mK)	64		

3. Results and discussions

Fig.1 shows the CDF's for the 50 measured solute concentration values reported for the CR=0.03 Ks^{-1} experiment in ref [5]. Creating such a CDF from either experimental or numerical simulation results is as follows [1]. Assuming that both experimental data and numerical results of casting simulation are given in a form of a grid with *(N)* uniform sized control volumes, the nodal segregation value predictions $M_j=C_j/C_0$, on complete solidification, are arranged in descending order; the entry at $j=1$ being the smallest nodal value predicted and the entry at $j=N$ being the largest. Each entry is then associated with an inverse Weibul "plotting position" determined as:

$$F_j = 100\frac{j}{N+1} \quad j = 1,2..N \; . \tag{8}$$

The y axis of the CDF (in log scale) is the value $100\times(1-F_j)$ which represents percentage (%) volume fraction greater that a given normalized segregation factor $M=(C/C_0)$ where C is the value of computed nodal mixture composition (equation [5]), or measured solute, while C_0 is the nominal composition.

The x axis is the value of M in a linear scale. Also to indicate that the measurements will not be perfect we have added as dot/dashed lines +/- 5% of the reported measurement. It is emphasized that these are not meant to represent the actual error in the experimental measurements but rather to indicate that errors are present.

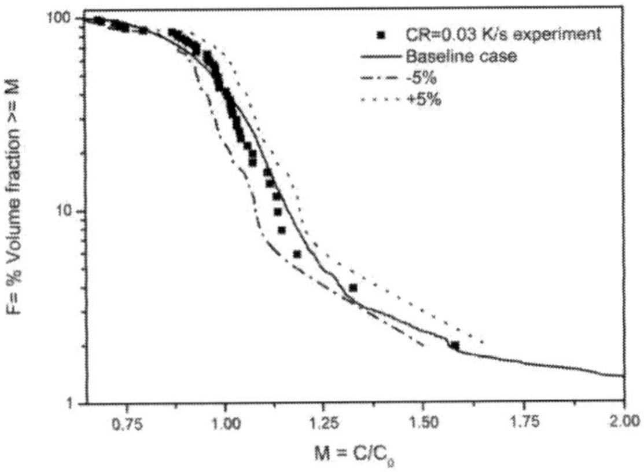

Figure 1. CDF profiles of experimental results [5] and baseline case

Variation of model parameters is shown in Table 2. The first line in this table is our baseline case which matches the case with cooling rate of CR=0.03Ks^{-1} reported in [5]. In the first two cases (1 and 2) we vary the permeability constant (κ_0) through the dendrite arm spacing (DAS) as is given in equation (1). The consistency of this baseline case is established by solving at a finer grid (see case 3 in Table 2). In our previous research [2] we conclude that the liquid flow (driven by thermal solutal contractions) in the mushy region is a key control on the formation of macrosegregation. To

IOP Conf. Series: Materials Science and Engineering **84** (2015) 012008 doi:10.1088/1757-899X/84/1/012008

demonstrate this here we conduct a number of cases that differ from out baseline, see Table 2. In tests Case 1 and Case 3 we vary the permeability of the mushy region through changing the fixed value of the dendrite arm spacing (DAS) (see equation (1). In case 4 we investigate the impact of increasing the thermal solutal expansion coefficients. As we conclude in previous research [2], flow in the liquid and mush is a key parameter that mostly influence on macrosegregation, and it was driven with thermal and solutal contractions of liquid phase. The last case in Table 2 is an attempt to investigate how the use of a variable DAS in the permeability equation (5) might alter the shape of the CDF.
As a model for the coarsening of the secondary arm space here we use:

$$DAS = a + b \times (1 - f)^C , \qquad (9)$$

This is based on coarsening model originally proposed by Kirkwood [21], where the secondary arm space scales as time $t^{1/3}$, i.e, their coarsen as time $t^{2/3}$. Assuming that solid fraction ($f_S = 1 - f$) is square root of time [21], this would require exponent C in equation (8) to be $C=2/3$.

Table 2. Variations of process parameter for CDF analysis of macrosegregation results

Case	Permeability	Micro model	Grid size	$\beta_{T,S}/\beta_{T,Snom}$ (K^{-1})
Baseline ———	DAS=90 μm	Scheil	100x60	1.0
1 ———	DAS=60 μm	Scheil	100x60	1.0
2 ———	DAS=150μm	Scheil	100x60	1.0
3 ———	DAS=90 μm	Scheil	200x120	1.0
4 ———	DAS=90 μm	Scheil	100x60	1.2
5 ———	DAS=f(f_S)	Scheil	100x60	1.0

Macrosegregation pattern for baseline case with increased grid size is shown on Fig. 2. Cooling is imposed from the right vertical side (see L10 – L30 in ref. [5]) and flow in the cavity has a clockwise direction, with a first formed solid on the right side. Enriched liquid accumulates on the left while the solidification proceeds in time, and the solidification on the left side starts when the temperature of the left side of cavity falls below the liquidus temperature which is now decreased due to the segregation from the right side.

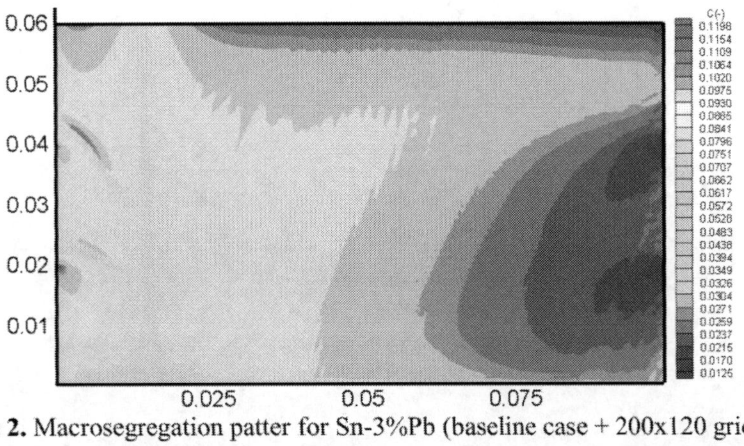

Figure 2. Macrosegregation patter for Sn-3%Pb (baseline case + 200x120 grid size)

The last part of domain to solidify has a tree shape 2cm from the left boundary wall and with some channels on the same side, due to the enhanced heat flux (see Fig. 4 in Ref. [5]) around $t=6300$ *sec*, close to the latest stage of solidification. The wide light blue bay near the right top side of casting is

caused by flow reversing and strong re-melting occurring in the early stage of solidification, while the bottom part on the right side remains negative, with less influence of re-melting. It was noted both in similar numerical study of this benchmark experiment performed by Carrozzani et al. [20] that segregation channels are observed in experiment but not in their numerical simulations. As a reason for that they suspect on anisotropy of the mushy zone regarding permeability, which is good driving point to explore this issues in some future study. Our experience with channels is that they can be formed in zone with low heat flux and with increased mixture concentration [19], but the hypothesis of anisotropy is worthwhile and deserve attention.
The CDF curve covering 99% of the casting for the baseline case (DAS=90 μm) together with experimental data taken from Ref. [5] is shown in Fig. 3a. We note that the CDF of the baseline case predictions is a reasonable match to the CDF from the experimental measurement. Our only concern with the results in Fig. 3 is that, in the range $1<M<1.3$, the simulation CDF has a slightly flatter slope when compared with the experimental CDF. In Fig. 3b predicted CDF curves for different and constant permeability values are shown. These results clearly indicate that changes in the permeability have a significant effect on the shape of the CDF, in particular this parameter controls the slope of the CDF in the mid range, $1<M<1.3$. This motivates us to examine if use of the transient arm spacing model of equation (8) in the permeability model, equation (1), might increase the slope of the CDF in the range$1<M<1.3$. The grid convergence of our predictions is confirmed in Fig. 4a which shows that for 98% of the casting the CDF from our chosen grid and one with double the resolution are essentially the same. As we believe that intensity of fluid flow has a key role in macrosegregation formation, we increase for 20% intensity through the increasing of thermal and solutal expansion coefficient (β_T, β_S) to test their influence on CDF curve, and results are shown on Fig. 4b. As can be seen, this doesn't help much in negative ($M<1$) and central zone ($M\sim1$) where we expect a flatter line in the negative zone, followed by narrow gradient in central part. Instead of that we have further extension in positive zone ($M>1.13$) due to the increased flow and therefore macrosegregation.

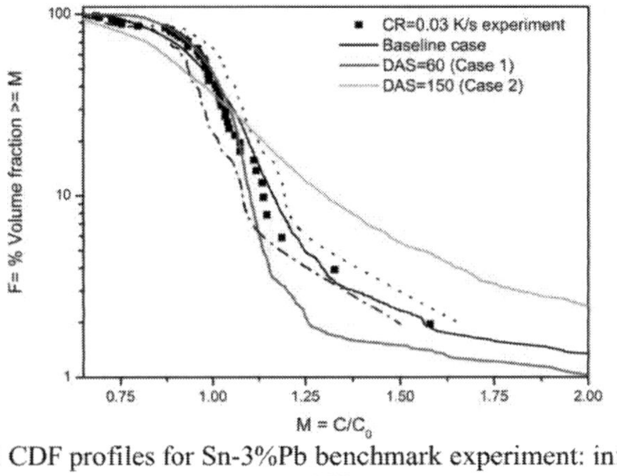

Figure 3. Statistical CDF profiles for Sn-3%Pb benchmark experiment: influence of different permeability

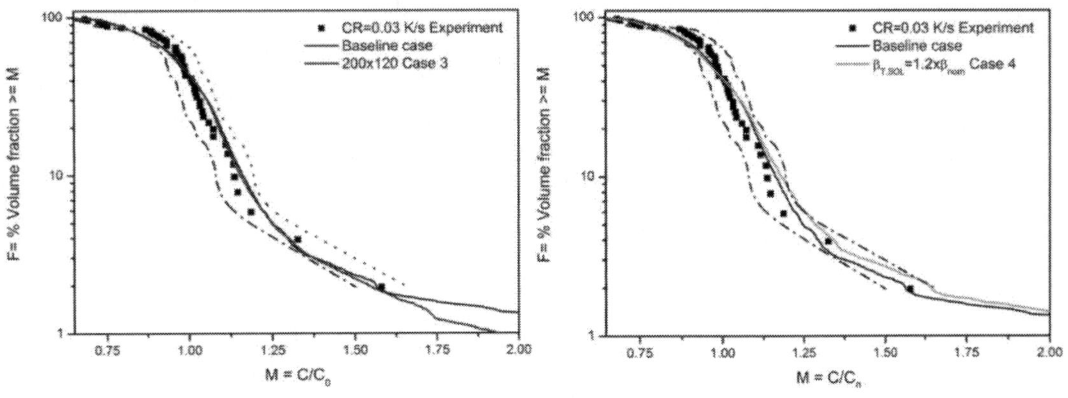

a) b)

Figure 4. Statistical CDF profiles for Sn-3%Pb benchmark experiment

a) refined grid b) influence of enhanced fluid flow

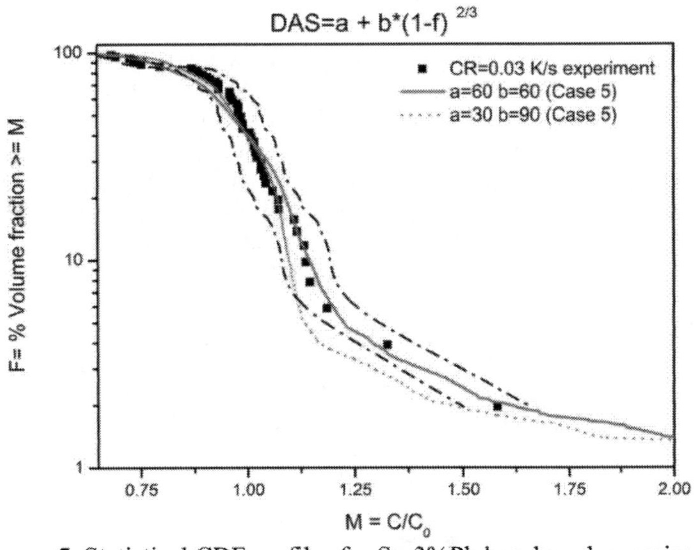

Figure 5. Statistical CDF profiles for Sn-3%Pb benchmark experiment with coarsening model proposed with equation (8)

The results in Figure 5 confirm this, indicating the use of a transient DAS size, consistent with the expected $t^{-2/3}$ coarsening rate, does indeed increase the slope of the CDF in the mid range$1<M<1.3$, Indeed it can be argued that the CDF generated by the choice $DAS=60+60(1-f)^{2/3}$ microns is a better match to the experimental CDF.

4. Conclusions

Our intention in this study was to use the recently proposed CDF analysis with the recent reported experimental measurements on an Sn-3%Pb alloy is to investigate how changing the two-dimensional model parameters that control flow in mushy affect the shape and the spread of predicted CDF's. Our particular interest was to try and understand how modification of the permeability model through control of the secondary arm space would influence simulation predictions. Beyond this we also had an interest in investigating other flow controlling features such as anisotropic permeability and solid phase motion. In the first place the analysis clearly demonstrates that the chosen size of the secondary arm space in the two-dimensional permeability model can have quite a dramatic influence on the shape and spread of the predicted CDF. And that if an appropriate transient secondary arm spacing is used in the permeability model the predicted CDF more closely follow the CDF of the experimental measurements. It is important to note that these findings are based on a 2-D model and there is some reason to speculate, due to the higher number of degrees of freedom for flow in the mush, that a 3-D model (e.g., see ref [20]) may match experimental observations without the need of a transient arm spacing. In this context, however, it is also important to recognize that, for the particular Sn-3%Pb experimental casting, a basic macrosegregation model, using a constant arm space in a fixed isotropic mushy region, also produces acceptable agreement between simulated and measured CDFs. This has two significant outcomes. First it suggests that the CDF of the Sn-3%Pb macrosegregation measurements represent a first order benchmark against which basic macro segregation simulations codes can be tested. Second it suggests that additional experimental measurements are needed if the CDF tool can produce a signal that will be able to fully test simulation models that incorporate anisotropic moving solid mushy regions.

Acknowledgment:
The author thanks Prof. V.R. Voller for reviewing and commenting on drafts of this manuscript.

References

[1] Voller V R and Vušanović I 2014 *Int. J. Heat Mass Transfer* **79** 468.
[2] Vušanović I and Voller V R 2014 *4th Int. Conf. on Adv. in Solidification Processes, Beaumont Estates*, Old Windsor, UK
[3] Voller V R 2000 *Int. J. Heat Mass Transfer* **43** 2047.
[4] Wang X and Fautrelle Y 2009 *Int. J. Heat Mass Transfer* **52** 5624.
[5] Hachani L, Saadi B, Wang X D, Nouri A, Zaidat K, Bouzida A B, Derouiche L A, Raimondi G and Fautrelle Y 2012 *Int. J. Heat Mass Transfer* **55** 1986.
[6] Quillet G, Ciobanas A, Lehmann P and Fautrelle Y 2007 *Int. J. Heat Mass Transfer* **50** 654.
[7] Mehrabian R and Keane M and Flemings M C 1970 *Metall. Trans.* **1** 1209.
[8] Bennon W D and Incropera F P 1987 *Int. J. Heat Mass Transfer* **30** 2161
[9] Beckermann C and Viskanta R 1988 *PhysicoChem. Hydrodyn.* **10** 195.
[10] Voller V R , Brent A D and Prakash C 1989 *Int. J. Heat Mass Transfer* **32** 1719.
[11] Ni J and Beckermann C 1991 *Metall. Trans. B* **22** 349.
[12] Beckermann C and Wang C Y 1995 *Annual Review of Heat Transfer* **6** 115.
[13] Schneider M C and Beckermann C 1995 *Metall. Trans. A* 1995 **26** 2373.
[14] Krane M J M and Incropera F P 1997 *Int. J. Heat Mass Transfer* **40** 3827.
[15] Krane M J M 2004 *App. Math. Model* **28** 95.
[16] Pardeshi R, Voller V R, Singh A K and Dutta P 2008 *Int. J. Heat and Mass Transfer* **51** 3399.
[17] Voller V R, Mouchmov A and Cross M, 2004 *App. Math. Model.* **28** 79.
[18] Patankar S 1980 *Numerical Heat Transfer and Fluid Flow* (New York: Hemisphere).
[19] Vušanović I and Voller V R 2014 *Mat. Sci. Forum* **790-791** 74.
[20] Carozzani T, Gandin C A, Digonnet H, Bellet M, Zaidat K and Fautrelle Y 2013 *Metall. Trans. A* **44A** 873
[21] Kirkwood D H 1985 *Material Science and Engineering* **73** L1.

Numerical simulation of multi-mini-pot pouring process of a 13-ton steel ingot

Jun Li[1], Honghao Ge[1], Bin Wang[1], Menghuai Wu[2], Jianguo Li[1]

[1]School of materials science & engineering, Shanghai Jiaotong University, Shanghai 200240, China
[2]Chair of Simulation and Modeling of Metallurgy Processes, University of Leoben, 8700 Leoben, Austria

E-mail: li.jun@sjtu.edu.cn

Abstract. Heavy ingots up to several hundred tons for power plant forgings exhibit large scale grain size and morphology differences, which are harmful to the design of forging parameters, and severe macrosegregation, which could not be eliminated during the subsequent forging and heat treatment processing. To cast these ingots with more homogeneity, a multi-mini-pot (MMP) poring technique is proposed, in which liquid metal is poured by multi-mini-pot (MMP) with intermission between each pot and solidification occurs step by step. In this paper, we are focus on the prediction of macrosegregation in MMP pouring process. A three-phase model is employed to study the MMP pouring process for a 13-ton heavy ingot. The main features of this three-phases model in such a heavy ingot can be quantitatively modelled: growth of columnar dendrite trunks; nucleation, growth and sedimentation of equiaxed crystals; thermosolutal convection of the melt; solute transport by both convection and crystal sedimentation; and the columnar-to-equiaxed transition (CET). The results shown that the MMP pouring technique tend to decrease the macrosegregation significantly by compared with the conventional method.

1. Introduction

Ingot casting is a common process in foundry. More alloyed metallic materials are now cast in ingot for further processing with increased quality to meet with the higher service windows, such as machine tools, forgings for power plant, etc [1]. Unfortunately, it seems that foundry man could not cast satisfactory ingot to meet the need of increasing desire from final products. Heavy ingots up to several hundred tons for power plant forgings exhibit large scale grain size and morphology differences, which are harmful to the design of forging parameters, and severe macrosegregation, which could not be eliminated during the subsequent forging and heat treatment processing. Therefore, to suppress macrosegregation during casting process has always been the focus of casting research [2-3]. Flemings [4] investigated the effect of the hot top on macrosegregation. He proposed that an ingot with ideal hot top can significantly reduce the macrosegregation by (1) eliminating formation and settling of equiaxed grains from the hot top and (2) promoting a positive temperature gradient from the ingot and hot top region. He also proposed a new method that rotated the ingot during solidification [4]. Sang et al. [5] present a new technique for reducing macrosegregation in steel ingots. They added solid steel balls to the melt during the pouring process. With this technique the macrosegregation was reduced, the formation of A-type segregation was prevented. However, this technique is difficult to control, and some other defects will happen. Tanaka presented a multi concentration pouring technique to restrain the formation of macrosegregation during casting process [6-7]. The serious macrosegregation defects which were unavoidable for conventional casting were improved significantly. However, when the ingot size is larger than 180-ton the improvement is unapparent.

Authors proposed a novel technique: multi-mini-pot (MMP) pouring for the casting of heavy ingot [1]. The liquid metal is poured by multi-mini-pot with intermission between each pot and solidification occurs step by step. The composition and amount of each pot are designated depending on the controls grain differences and segregation level of the target ingot. The model for multi-mini-pot pouring process is in consistent with the classical solidification theory, where the segregation will tend to the minimum as the amount of each pot goes to the minimal.

The principal intention for this paper is to validate the probability of MMP technique by numerical simulation. Therefore, a three-phase model [8-10] is employed to simulate MMP pouring process and to predict the formation of macrosegregation.

2. Process model of multi-mini-pot pouring

Generally, in the production of ingots, alloy, as large amount as needed, is melt in one (for small ingot) or more (for heavy ingot) furnaces, and then poured by one or more pots into a mould continuously. Solidification occurs after and progress freely to the end. Pouring time is only a little part compared with solidification time, during which intensive melt flow results in untraceable solute profile and cooling rate decreases gradually as solidification proceeds. As foundryman can cast small ingots with a quality as needed in heavy ingot, heavy ingot casting can be divided into several small ingots. As schematically shown in figure 1, an ingot with weight W is divided into N parts, each time t_i, $W_i(=W/N)$ melt is poured into the mould. Next W_i melt is then added at the time when (t_{i-1})'s pouring become solid with 20%-30% left. We choose this value based on numerical parameters study and also the earlier laboratorial experiment study [1]. New pouring melt will mix with the former 20%-30% left and next solidification begins. This process repeats till the moult is full.

3. General description of the model

With MMP pouring technique the solidification process of a 13-ton big-end-up steel ingot (Fe-0.2 wt.% C) was simulated. Here a 2-D axis-symmetrical simulation was performed to approximate the solidification behaviour of this ingot. The configuration of this ingot, labelled with the necessary boundary and initial conditions, is described in figure 2. All the thermodynamic and physical properties used in this study are the same as reference [8,11], as shown in table 1.

The columnar-equiaxed mixed three-phase model [8, 12-13] was employed to simulation the solidification process. Details of the numerical model are described elsewhere [8, 12-13]. A brief outline of the model and simulation settings are given below:

- Mould filling for each pot is ignored. Solidification starts with an initial concentration Fe-0.2 wt. %C and an initial temperature of 1840 K.
- Three phases are defined: the primary liquid phase (ℓ), the equiaxed phase (e), and the columnar phase (c). The corresponding phase fraction is f_ℓ, f_e and f_c, with $f_\ell + f_e + f_c = 1$. Both the liquid and equiaxed phases are moving phases, for which the corresponding Navier-Stokes equations are solved. The columnar phase is assumed to stick to the wall, and solidifies from the wall towards the bulk melt. Thus, no momentum equation for the columnar phase is considered.
- Columnar dendrites are approximated by growing cylinders starting from the mold wall towards the casting centre. The advance of the columnar tip front was tracked during the solidification.

Table 1. Thermodynamic & physical properties

Property	Symbol	Units	Quantity
Melting of pure iron	T_f	K	1805 [8]
Liquidus slope	m	K (wt.%)$^{-1}$	-8000 [11]
Equilibrium partition coefficient	k	-	0.18 [11]
Reference density	ρ_ℓ , ρ_e , ρ_c	kg·m^3	7060 [11]
Solid-liquid density difference	$\Delta\rho$	kg·m^3	150 [8]
Specific heat	c_p^ℓ , c_p^c , c_p^e	J·kg^{-1}·K^{-1}	500 [8]
Thermal conductivity	k_ℓ , k_e , k_c	W·m^{-1}·K^{-1}	34.0 [8]
Latent heat	L	J·kg^{-1}	2.71×10^5 [8]
Viscosity	μ	Kg·m^{-1}·s^{-1}	4.2×10^{-3} [8]
Thermal expansion coefficient	β_T	K^{-1}	8.9×10^{-5} [11]
Solutal expansion coefficient	β_c	wt.%$^{-1}$	1.416×10^{-2} [11]
Dendritic arm spacing	λ_1	m	5×10^{-4} [8]
Diffusion coefficient (liquid)	D_ℓ	m^2·s^{-1}	2.0×10^{-8} [8]
Diffusion coefficient (solid)	D_e , D_c	m^2·s^{-1}	1.0×10^{-9} [8]

- For equiaxed grain a simple method is employed to handle the dendritic morphology [10].
- A three-parameter heterogeneous nucleation law is used for the nucleation of equiaxed grains. No fragmentation and grain attachment are currently considered. Growth of the columnar trunk and equiaxed grain is governed by diffusion.
- The permeability in the columnar mush zone is modelled according to Blake-Kozeny, while the drag law between the melt and equiaxed phase is modelled according to Wang's approach [14].
- Growth of columnar primary tips is stopped when the volume fraction of equiaxed phase in front of them reaches 0.49 (hard blocking criterion) [15].
- Packing limit for the equiaxed phase is set as $f_e + f_c = 0.637$ [16], and the equiaxed crystals are trapped when the trapping limit of f_c 0.2 is reached [17].

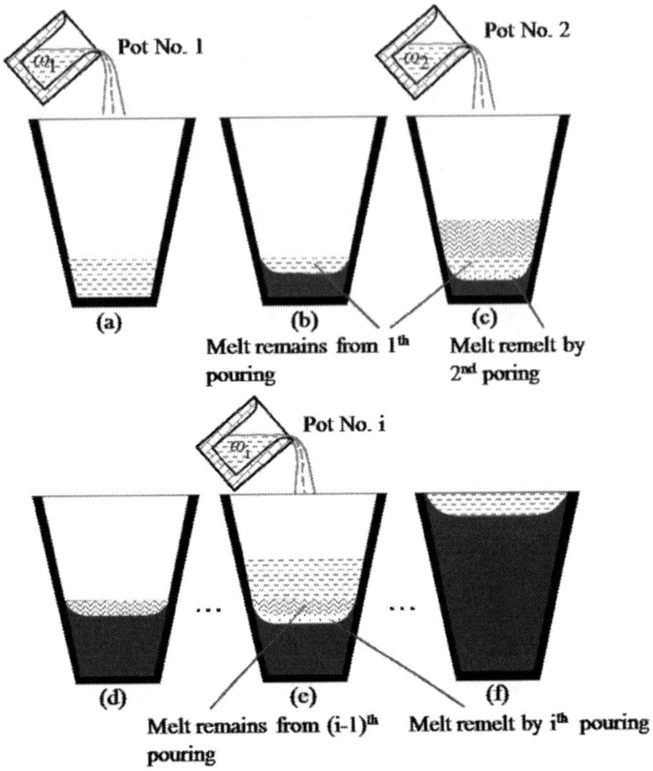

Figure 1. Process model of MMP pouring

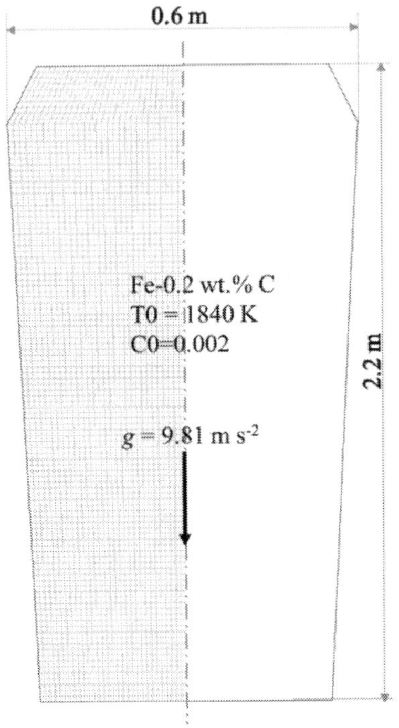

Figure 2. Configuration of 13-ton steel ingot

4. Simulation result and discussion

The solidification sequence of MMP pouring process was presented in figure 3. As for MMP pouring technique, the ingot was divided into 10 mini-ingot (each pot can be considered as one mini-ingot), each mini-ingot solidified step by step. The current pot be poured into the model when the previous pot has only 20-30% liquid melt remained, therefore, the solidification of each mini-ingot behaved approximatively individually. For the solidification process of each pot the melt was in a shallow-broad domain. The molten steel can only flow and develop within this small domain. Therefore the melt velocity is quite small, as seen from figure 3, after 15 minutes of pouring the maximum melt velocity for each pot is around 0.002 m s^{-1}. The solidification front, as well as the mush zone, behave as a shallow U-type other than deep V-type as in the conventional pouring technique [8]. With shallow flow domain, the sedimentary equiaxed grain did not have sufficient time to accumulate in the bottom of the ingot which result in the significant cone shape negative segregation for the conventional casting [8].

Figure 4 shows the comparison between conventional and MMP casting after 15 minute of pouring. The left part shown the melt velocity vector and the right part indicated the liquid volume fraction. As for conventional casting, since at the early stage of solidification the molten steel domain is very large, due to the thermalsolutal convection and the sedimentation of equiaxed crystal the liquid flow can develop to a big value, as large as 0.018 m s^{-1}, in this large domain as seen in figure 4(a). For MMP solidification the ingot was divided into 10 mini-ingots, and they solidified gradually approximately one by one. During the solidification process of each pot the melt domain is about one tenth of the conventional solidification. Which determined that the melt flow did not have sufficient space to get full developing. Subsequently, the liquid velocity was significantly decreased by compared with the conventional casting, which is about one eighteenth smaller than the conventional one.

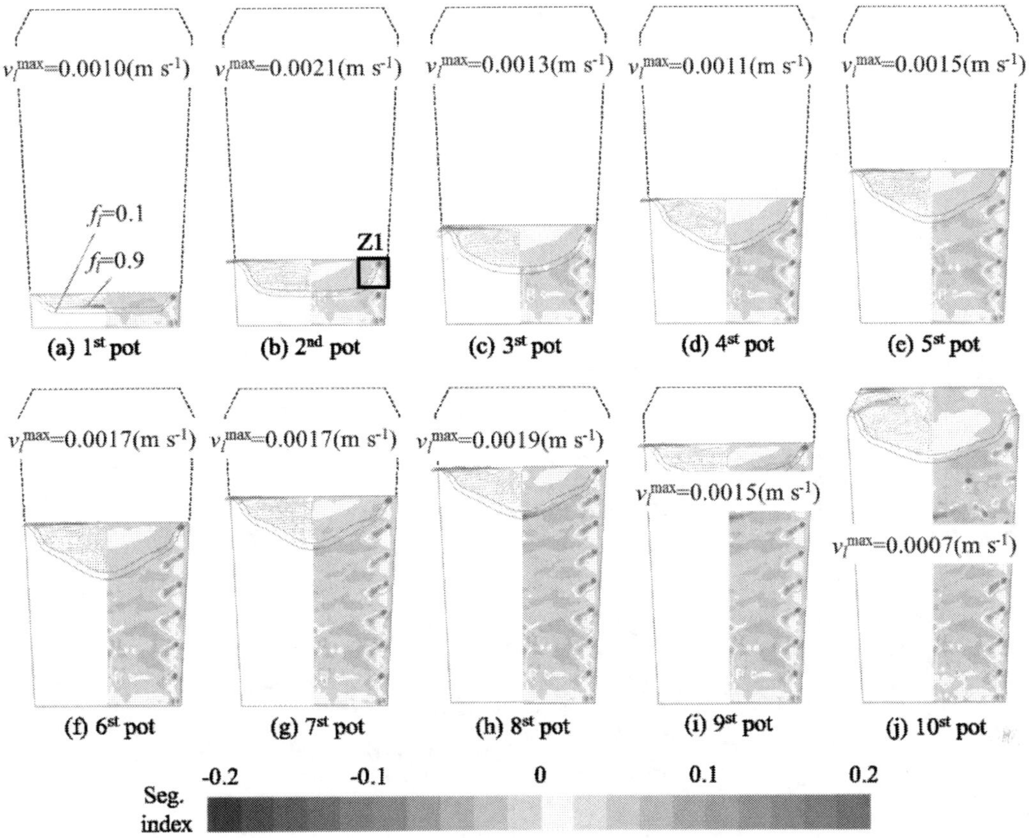

Figure 3. Solidification sequence of MMP pouring processing. All the figures were selected after 15 minutes after the pouring of each pot. The left part shown the melt velocity vector. The right part shown the macrosegregation index ($(c_{mix}-c_0)/c_0$). The liquid volume fraction iso-lines were also shown as the red line.

Figure 5 shows the final macro segregation pattern. As for conventional casting, the core of negative segregation in the ingot bottom and positive segregation in the top region were predicted which is correspond to the classic solidification theory and other researches [8, 12]. The reason for the core of negative segregation in the bottom region could be in two aspect: on one hand, the sedimentation of equiaxed crystal from upper region to the bottom; on the other hand, the equiaxed crystal moved from the side mushy zone to the bottom region by sedimentation and molten steel flow. Since the molten steel flow is very unstable and chaos in the domain (see figure 4(a)), the macrosegregation distribution is also uneven. There is some small positive or negative segregation patches near the ingot edge.

For MMP technique, by contrast, there is neither core of negative segregation zone in the bottom nor obvious positive segregation zone in the hot top region, however, a periodic distribution of negative-positive segregation layer is easily be found in the central of the ingot. Since the melt domain for MMP is small and shallow there is fewer equiaxed crystal settling from the upper to the bottom. Moreover, since the liquid velocity in this case is very small it is difficult to drive the equiaxed crystal from the side mushy zone to the centre bottom region. The equiaxed crystal become more scatter than the conventional casting. Therefore, although there is a negative segregation region in the bottom of each mini-ingot, the severity of negative segregation was significantly decreased.

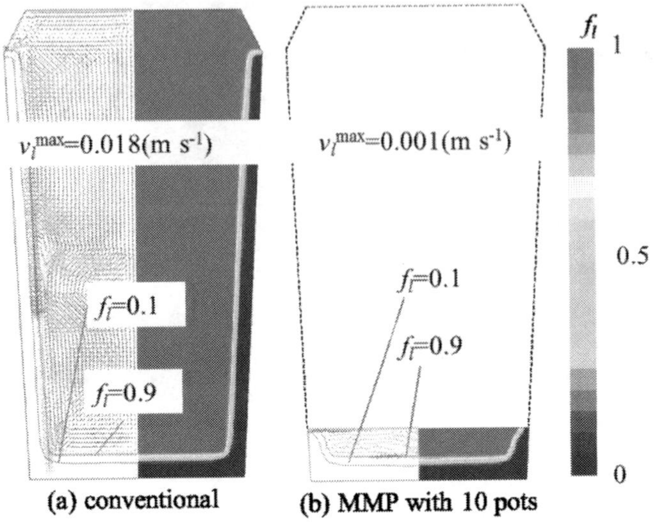

Figure 4. The comparison between conventional and MMP solidification after 15 minute of pouring: (a) conventional with one pot; (b) MMP with 10 pots. The left part shown the melt velocity vector. The right part indicated the liquid volume fraction.

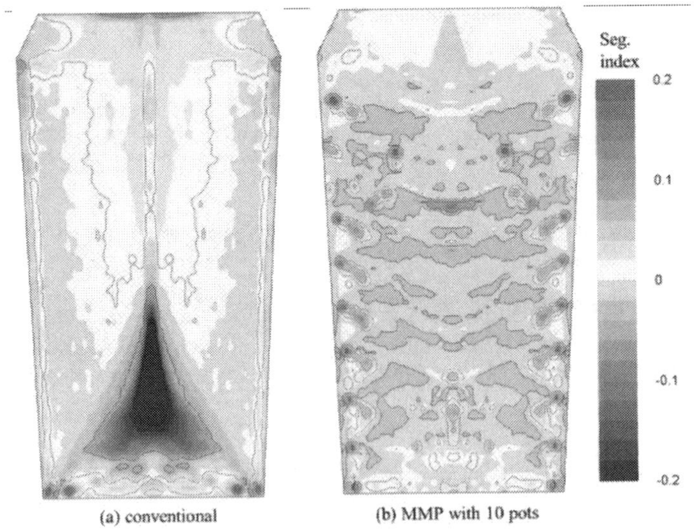

Figure 5. The predicted final macrosegregation pattern: (a) conventional with one pot; (b) MMP with 10 pots.

In addition, there are some positive segregation patches which are periodic distribution near the ingot wall. The formation of this positive segregation can be understand as following: under the melt-air interface about two or three cells distance there are horizontal melt flow from the solidification front (melt-rich region) to the centre of the ingot (relative melt deplete region), as seen from figure 3; this flow bring more solute into the volume and bring out less solute out of the volume; subsequently, result in a positive segregation patch. Figure 6 elaborates this mechanism more distinct. As discussed by Wu [18], in the mush zone in the region where the melt velocity is in the opposite direction with the concentration gradient may tend to form a positive segregation zone. The formation of this positive segregation patch in coincide with their opinion. This kind of positive segregation, to some extent, indicated the shortcoming of the current processing parameter for the MMP technique. Therefore, more effort should be implement to overcome this defect, and increasing the final homogeneity more effectively.

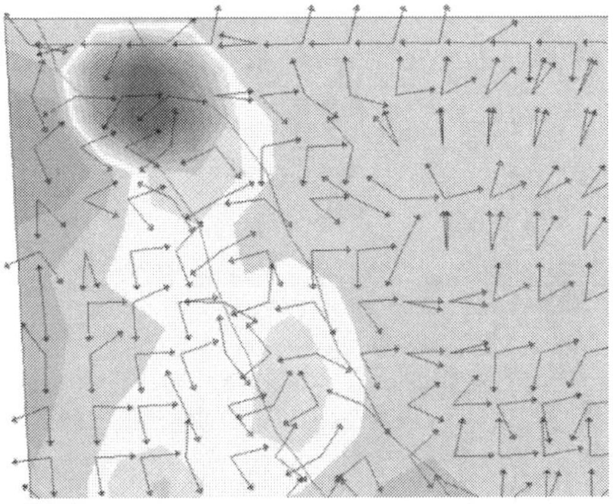

Figure 6. The relationship between positive segregation patch and the melt velocity and concentration gradient. The color map indicate the concentration field. The blue vectors show the melt velocity. The black vectors indicate the concentration gradient. The result was zoom in from **Z1**, as shown in figure 3(b).

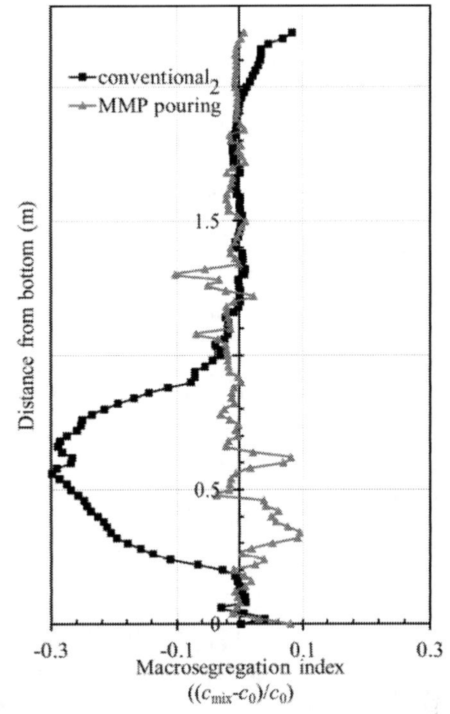

Figure 7. The comparison of central line macrosegregation distribution between conventional and MMP pouring techniques.

In order to obtain more detail information about the macrosegregation, the centreline segregation were compared and shown in figure 7. As we can see, the severity of macrosegregation was significantly decreased by MMP casting. The maximum (magnitude) negative segregations were predicted as one third of the conventional casting.

5. Summary

A novel technique, Multi-Mini-Pot (MMP) pouring, for cast heavy ingot was present to diminish the macrosegregation. In order to validate this novel technique, a mixed columnar-equiaxed three-phase model was employed to simulate both the MMP pouring solidification and conventional solidification process. The result shown, for the solidification of a 13-ton heavy ingot, that by compare with the conventional technique the MMP pouring technique: 1) decreased the melt velocity dramatically, the maximum melt velocity was decreased from 0.018 m s^{-1} to 0.001 m s^{-1}; the core of negative segregation in the bottom region was eliminated; the intensity of negative segregation was decreased significantly which is from -0.3 to -0.1. In brief the solidification of heavy ingots may be divided into several steps by a multi-mini-pot pouring to increase homogeneity. However, there are some positive patch in near the ingot wall which was the shortcoming of the current casting parameters for MMP pouring technique. Further parameter optimum and technology improvement should be implement to overcome this shortcoming.

Acknowledgments
This work is sponsored by National Natural Science Foundation of China (Grant No. 51404152), Shanghai Pujiang Program (Grant No. 14PJ1404800), Shanghai International Cooperation Project (Grant No. 14140711000), and National Basic Research Program of China under contract No.2011CB012900.

Reference

[1] Li J, Wang B and Li J 2014 in: *The 4rd International Conference on Advances in Solidification Processes*, 2 (Old Windsor, UK)
[2] Gu J P and Beckermann C 1999 *Metallurgical and Materials Transactions A* **30** 1357-66,
[3] Flemings M C 2000 *ISIJ International* **40** 833-41
[4] Flemings M C 1976 *Scand. J. Metall.* **5** 1-15
[5] Sang B G, Kang X H and Li D Z 2010 *J. Mater. Processing Technology* **210** 703-711
[6] Tanaka Y and Ishiguro T 1997 *Phys. Stat. sol.* **160** 305-320
[7] Tanaka Y and Sato I 2011 *J Nuclear Mater.* **417** 854-859
[8] Li J, Wu M H, Ludwig A and Kharicha A 2014 *Int. J. Heat Mass Transfer* **72** 668-679
[9] Wu M H, Li J, Ludwig A and Kharicha A 2014 *Comp. Mater. Sci.* **92** 267-285
[10] Li J, Wu M, Ludwig A, Kharicha A and Schumacher P 2014 *Mater. Sci. Forum* **790-791** 121-126
[11] Gouttebroze S, Liu W, Bellet M and Combeau H in "Proceeding s MCWASP XI, 11th Int. Conf. on Modeling of Casting, Welding and Advanced Solidification Processes, Opio: France 2006"
[12] Wu M H and Ludwig A 2006 *Metall. Mater. Trans. A* **37A** 1613-1631.
[13] Wu M H and Ludwig A 2007 *Metall. Mater. Trans. A* **38A** 1465-1475.
[14] Wang C and Beckermann C 1995 *Metall. Meter. Trans. B* **26B** 111-9
[15] Hunt J D 1984 *Mater. Sci. Eng.* **65** 75-83
[16] Wu M and Ludwig A 2009 *Acta Mater.* **57** 5621-31.
[17] Ludwig A and Wu A 2005 *Mater. Sci. Eng.* A **413-414** 109-114
[18] Wu M H, Könözsy L, Ludwig A, Schützenhöfer W and Tanzer R 2008 *Steel Res. Int.* **79** 56–63.

Effect of liquid volume on the evolution of solidified structure in horizontal centrifugal casting

H Esaka[1,3], Y Kataoka[2,3] and K Shinozuka[1,3]
[1] Department of Materials Science and Engineering, National Defense Academy
[2] Graduate student, National Defense Academy
[3] 1-10-20 Hashirimizu, Yokosuka 239-8686 Japan

E-mail: hesaka@nda.ac.jp

Abstract. Horizontal centrifugal casting process is widely applied for long products. However, it often has macro-segregation problems, such as band segregation. In order to understand solidification process in this process, in-situ observation using transparent substances has been performed. Here, the effect of liquid volume in the mold was investigated. Succinonitrile-water alloy was introduced in the glass cell and filling ratio of organic alloy was varied from 70 to 100 vol. % and the rotation rate was set to be 250 rpm. Equiaxed grains formed in case of low filling ratio. On the other hand, no equiaxed grain formed when the filling ratio was 100 vol. %. Strong fluid flow may be induced by the fluctuation of free surface and this may form equiaxed grains. However, when the area of free surface decreased with increasing filling ratio, the influence of free surface diminished and fluid flow was stabilized. Therefore, well-developed columnar zone grew and no equiaxed grains formed.

1. Introduction

Horizontal centrifugal casting is widely used for production of pipes, rolls and so on. Advantage of this process is to press molten metal against the permanent mold wall due to centrifugal force. Owing to this force, fine solidified structure and sound castings without porosity and/or non-metallic inclusion can be obtained [1]. On the other hand, there are sometimes problems of macroscopic segregation. The countermeasure to the segregation problems has not been established because the formation mechanism of macroscopic segregation is still unclear. Further, solidification process is not well understood [2-4] because the centrifugal casting itself is essentially dynamic process.

Thus, an experimental technique of in-situ observation using succinonitrile-water alloy system for horizontal centrifugal casting process using an organic substance has been developed by the present authors [5-7]. Here, an important result was unveiled, such as the oscillation of equiaxed grains synchronizing to rotation of mold and violent fluid flow due to fluctuation of free surface. Especially, the vigorous and unsteady fluctuation of free surface may form equiaxed grains and make them oscillate back and forth. Therefore, in this study, in order to understand the effect of free surface on the evolution of solidified structure, in-situ observations have been made changing the liquid volume in the glass cells.

2. Experimental Procedure

In order to make an in-situ observation of horizontal centrifugal casting, basically the same equipment as the previous study was used [5,6]. A schematic view of experimental apparatus is shown in figure 1. A glass cell, which contained succinonitrile-water alloy, was vertically set along the rotating axis. The glass cell, 100 mm in outer diameter and 1.3 mm gap, was constructed from two pieces of round glass and some small pieces of glass with epoxy resin. Here, the filling ratios were varied from 70 to 100 vol.%. A high vision video camera was installed on the rotating shaft to make an in-situ observation. Since the frame rate was 30 frames per second, rotation speed was limited to 250 rpm. One of the biggest advantageous points was that the camera for in-situ observation rotates with the glass cell. Thus, what is rotating with the mold, it is stationary in the movie. On the other hand, what is moving relative to the mold, it is moving in the movie.

– Using this system, one could observe the evolution of solidified structure. The observed area was

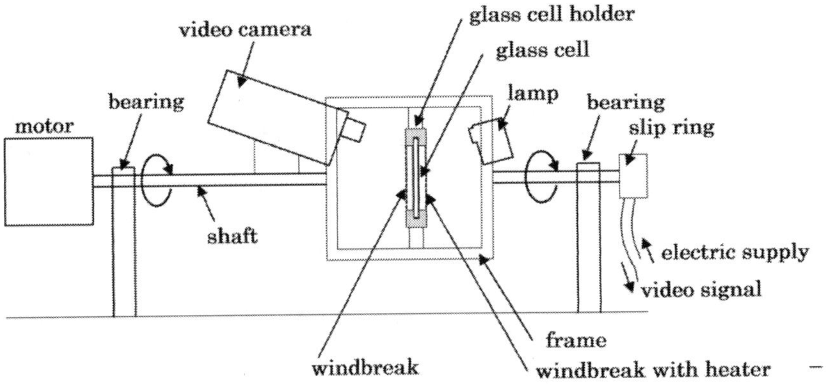

Figure 1. A schematic drawing of experimental apparatus for horizontal centrifugal casting.

approximately 40 mm x 50 mm. Basically, the solidification proceeded inwards from the outer surface, since a positive temperature gradient was imposed. In this study, rotation rate was fixed to be 250 rpm, which corresponded to 3.5 G at the outer surface of the glass cell.

3. Experimental Results and Discussions

3.1. Solidification Process

When the filling ratio was 70 vol.%, the snap shots of in-situ observation are shown in figure 2. The interval of each snap shot was 7 seconds. Since figure 2 a) is the initial stage of solidification, the major part in the glass cell was liquid phase. The glass cell holder (the left hand side) was formerly cooled by water, the solidification started from the outside of the glass cell. In this snap shot, 'white part' is the solid phase and transparent part is the liquid phase. Here, the columnar dendrite, which grew from the outside of glass cell, was observed and some solid particles (equiaxed grains) are visible near the solid/liquid interface. Furthermore, the free surface was also observed on the right hand side.

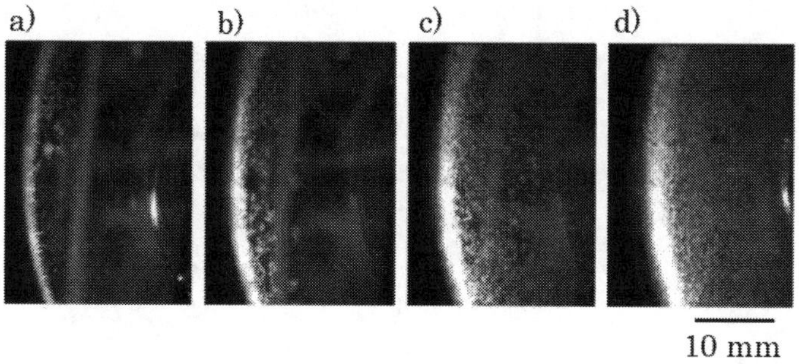

Figure 2. Snap shots of in-situ observation of horizontal centrifugal casting at a rotation rate was 250 rpm. The filling ratio was 70 vol.%.

Equiaxed grains formed much and they stopped the growth of columnar dendrites. Equiaxed grains are observed near the columnar zone and in this snap shot there is no equiaxed grain near the free surface, because it is still initial stage of solidification (figure 2 b)). Then, in all region of the liquid phase, so many equiaxed grains were formed (figure 2 c)). Here, almost all equiaxed grains were independent and moved up and down synchronously in the observed area. Free surface fluctuated or violently moved according to the rotation. Figure 2 d) shows a further elapsed snap shot. Here, the volume fraction in the mushy zone increased and the movement of equiaxed grains calmed down because the apparent viscosity of solid/liquid mixture increased. Thus, the shape of free surface gradually became a circle. There are some dark spots, where the solid fraction was lower than the other area. They may result in the segregation spots.

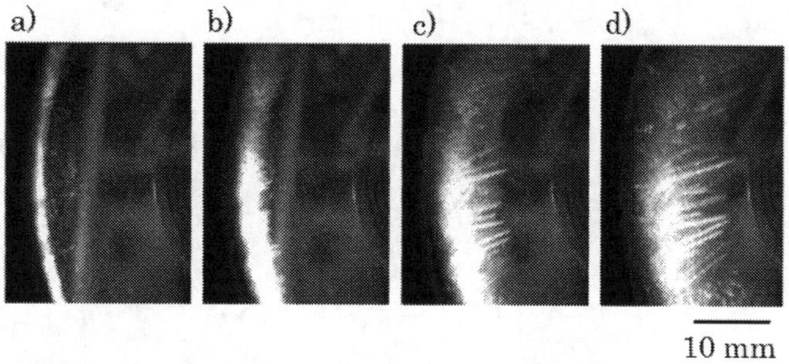

Figure 3. Snap shots of in-situ observation of horizontal centrifugal casting at a rotation rate was 250 rpm. The filling ratio was 100 vol.%.

When the filling ratio was 100 vol.%, the snap shots of in-situ observation are shown in figure 3. The interval of each snap shot was 15 seconds. The columnar dendrites started to grow from the outside of the glass cell, which is the same as shown in figure 2a). The columnar dendrites continued to grow inward and no equiaxed grains formed (figure 3 b) - d)). Columnar dendrites with favored growth direction grew faster and they were selected via grain selection.

3.2. Equiaxed zone ratio

The areal fraction of equiaxed zone was measured in the observation area. The relationship between the filling ratio and the areal fraction of equiaxed zone is shown in figure 4. When the filling ratio is less than 80%, the equiaxed zone ratio is constant and approximately 80%. The area of the columnar dendrite was very small. Almost all inner regions consist of equiaxed grains. The areal fraction of equiaxed zone decreased steeply with increasing the filling ratio. When the filling ratio was 100%, solidified structure consists of columnar dendrites and no equiaxed grains are formed.

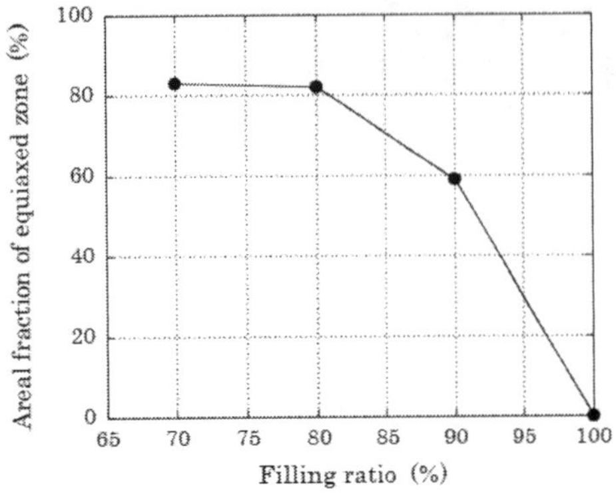

Figure 4. Relationship between the filling ratio and areal fraction of equiaxed zone.

3.3. Liquid flow pattern affected by free surface

In order to understand the liquid flow pattern, model experiments have been performed. The glass cell was filled with silicone oil and small plastic particles. The filling ratio of silicone oil was 70 and 100 vol.%. The kinematic viscosity of silicone oil was 5 cSt and the density of plastic particles is slightly larger than the silicone oil. These glass cells were set in the glass cell holder and in-situ observation were carried out at a rotation rate of 250 rpm. The locus of a particle was analyzed to characterize the fluid flow in the horizontal centrifugal casting.

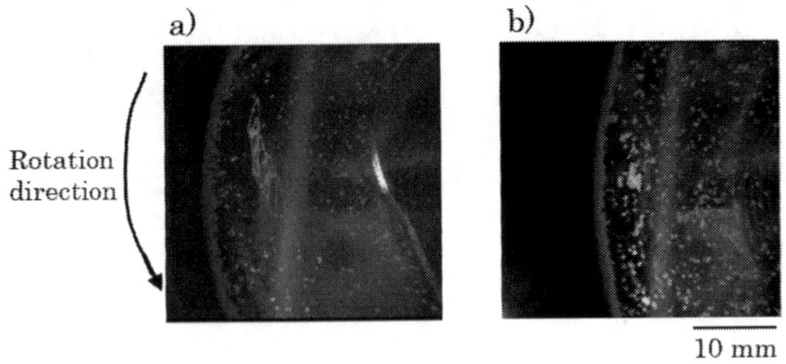

Figure 5. Loci of a plastic particle in the model experiment for horizontal centrifugal casting process.
a) the filling ratio was 70 vol.%
b) the filling ratio was 100 vol.%

The experimental results are shown in figure 5. Figure 5 a) indicates the locus of the particle in case of the filling rate of 70 vol.%. The particle moved up and down, indicating elongated circles and traveled toward the anti-rotating direction. Furthermore, it moved little by little towards the radial direction due to centrifugal force. On the contrary, in case of the filling ratio of 100 vol.%, the locus of the particle is shown in figure 5 b). The plastic particle moved outwards, indicating small circles. No traveling towards the anti-rotating direction was observed. Small vibration is due to gravitational acceleration. The amplitude of vibration in case of the filling ratio of 70 vol.% is larger than that in case of the filling ratio of 100 vol.%. This may be attributed the effect of free surface. The fluctuation of free surface may enhance the movement of plastic particles or equiaxed grains. When the filling ratio is 70 vol.%, traveling toward the anti-rotating direction was observed. On the other hand, when 100 vol.%, this movement was not observed. The difference between these two conditions is presence of free surface. Therefore, travelling towards the anti-rotating direction may be also caused by the fluctuation of free surface.

In the model experiment using silicone oil and small plastic particles, the particles clearly moved gradually outward. The reason of this is acceleration in radial direction due to centrifugal force. In the in-situ observation using succinonitrile-water alloy, it was difficult to observe the outward movement of equiaxed grains. Since the solidification proceeds from outside to inside, there is no room for equiaxed grain to move outward.

4. Conclusions

The experimental study for in-situ observation of horizontal centrifugal casting process has been performed changing the filling ratio in the glass cell. Following conclusions have been derived.

1) When the filling ratio is low, liquid phase or solid/liquid mixture not only vibrates but flows because of the fluctuation of free surface. In this case, equiaxed grains form because of vigorous flow.

2) When the filling ratio is 100%, liquid phase vibrates but does not flow. Equiaxed grains do not form because there is no macroscopic and vigorous flow in the liquid phase.

References

[1] Royer S and Vasseur S 1988 *Metals Handbook* vol15 9th ed.(Metals Park Ohio: TMS) p 296

[2] Kang I and Ohnaka I 1997 *J. Japan Foundry Eng. Soc.* **69** 119

[3] Kang I and Ohnaka I 1997 *J. Japan Foundry Eng. Soc.* **69** 240

[4] Kang I and Ohnaka I 1997 *J. Japan Foundry Eng. Soc.* **69** 815

[5] Esaka H, Hiramoto J, Miyahara S and Shinozuka K 2010 *Materials Science Forum* **649** 325

[6] Esaka H, Hiramoto J, Miyahara S, Tokusashi K and Shinozuka K 2011 *Tetsu-to-Hagané* **97** 369

[7] Esaka H, Tokusashi K and Shinozuka K 2013 *Tetsu-to-Hagané* **99** 149

Application of a macroscopic model to predict the band segregation induced by shear deformation of semisolid

S Morita[1], H Yasuda[1], T Nagira[1], C M Gourlay[3], K Morishita, M Yoshiya[1], A Sugiyama[4]

[1] Department of Materials Science and Engineering, Kyoto University, Sakyo, Kyoto 606-8501, Japan

[2] Department of Adaptive Machine Systems, Osaka University, Suita, Osaka 565-0871, Japan

[3] Department of Materials, Imperial College, London, SW7 2AZ, U.K.

[4] Osaka Sangyo University, Daito, Osaka 574-8530, Japan

E-mail: yasuda.hideyuki.6s@kyoto-u.ac.jp

Abstract. Semisolid deformation during solicitation can cause some casting defects such as the band segregation. Since the defect formation originates in nature of semisolid, it is of interest to build a model including the nature and to predict the defect formation. In-situ and time-resolve X-ray imaging has proved that rearrangement of solid grains in semisolid dominantly controls deformation and localization of shear strain leads to the band segregation. On the basis of the observations, a macroscopic model, which explicitly includes the rearrangement, is proposed. The model uses two-phase flow model and introduces hydrostatic stresses to express the rearrangement. The model was applied to simple shear of semisolid to confirm instability against shear. Fluctuation of solid fraction gradually increased and consequently shear deformation was localized. The model is also applied to pseudo 2D deformation in a shear cell, of which dimension was the same as that used in the X-ray imaging. The calculation result qualitatively agreed with the experimental results. It was concluded that the model has a potential to simulate the localization of shear and the band segregation. For further improvement, to measure some physical properties, which are closely related to the rearrangement, is required.

Introduction

It has been known that deformation of semisolid during solidification could cause some casting defects. For example, segregation band observed in centrifugal casting [1] and die casting [2,3], centerline segregation in twin-roll casting [4,5] and even V-type segregation [6] in continuous casting of steel are closely related to semisolid deformation. Semisolids of metallic alloys shows various mechanical phenomena. Thixotropic properties [7,8], agglomeration / disagglomeration [7], deformation of solid globules [9], and dilatancy during solid-grain rearrangement [10-12] have all been reported. It is of interest to know how the semisolid deforms and consequently the band segregation forms during solidification. However, the model that represents the formation of band segregation induced by deformation has not been established well, comparing to the models for predicting macrosegregation induced by melt flow in the mushy region. The aim of this study is to build a model, which explicitly includes nature of semisolid deformation, and to represent the formation of segregation band.

IOP Conf. Series: Materials Science and Engineering **84** (2015) 012011 doi:10.1088/1757-899X/84/1/012011

In the semisolid deformation, interactions between solid grains take place and strongly influence dynamics of semisolid [13]. Apparent viscosity depending on shear rate [14], thixotropy [15] and normal stress on the shear plane against shear [16] are typical examples for semisolid. A previous research pointed out that the band segregation formed due to the rearrangement of solid grains during the deformation [10]. Recently, in-situ and time-resolve X-ray imaging has been performed to observe semisolid deformation [12,17-22]. The deformation of semisolid of Al-15mass%Cu alloys with a solid fraction of 48% was characterized [12,18,19]. The rearrangement of solid grains induced dilation where shear deformation occurred and localization of the shear strain was observed in the shear-deformed domain [18]. The qualitative and quantitative data obtained by the observations are useful to build and to validate a semisolid-deformation model. This paper demonstrates a model, which is build on the basis of the observation, and some calculation results. The validity is also discussed, comparing to the observations.

Macroscopic model

1.1. 2.1 Governing equations

The governing equations were proposed in the previous study [23]. This study slightly modifies the equations to be consistent with the observation. The governing equations of the liquid phase and the solid phase (grains) are given using the Navier-Stokes equations.

$$\rho\frac{\partial}{\partial t}\left(f_l u_i\right)+\rho\frac{\partial}{\partial x_j}\left(f_l u_i u_j\right)=-f_l\frac{\partial P}{\partial x_i}+\frac{\partial}{\partial x_j}\tau^l_{ij}-F^{sl}_i \tag{1a}$$

$$\rho\frac{\partial}{\partial t}\left(f_s v_i\right)+\rho\frac{\partial}{\partial x_j}\left(f_s v_i v_j\right)=-f_s\frac{\partial P}{\partial x_i}+\frac{\partial}{\partial x_j}\tau^s_{ij}+\frac{\partial}{\partial x_i}\left(\sigma_s+\sigma_r\right)+F^{sl}_i \tag{1b}$$

$$\frac{\partial f_l}{\partial t}+\frac{\partial}{\partial x_i}\left(f_l u_i\right)=0 ,\quad \frac{\partial f_s}{\partial t}+\frac{\partial}{\partial x_i}\left(f_s v_i\right)=0 \tag{2a, b}$$

$$f_l+f_s=1 \tag{3}$$

The stresses σ_s and σ_r express the rearrangement of the solid grains. The force of F^{sl}_i is an interaction between the liquid and the solid phases. Since the solid particles suspended in the liquid phase are expected to have lower Reynolds number, Stokes' law is used as the interaction. Thus, the force per unit volume is simply given by

$$F^{sl}_i=n_s f^{sl}_i=-\frac{18\mu}{d_s^2}f_s\left(v_i-u_i\right) \tag{4}$$

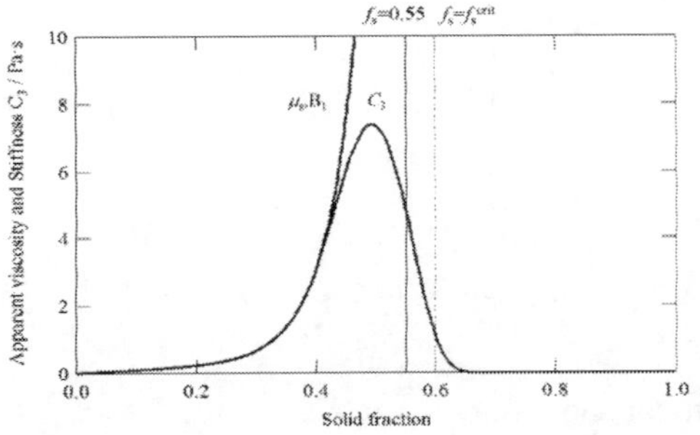

Figure 1 Example of apparent viscosity μ_s, stiffnesses C_3 and B_1 as a function of solid fraction.

1.2. Rearrangement of solid grains

In the model [23], the solid phase was treated as a viscoplastic fluid. The in-situ observation showed that collisions and rearrangement of solid particles dilate the liquid-filled spaces during shear [12,17]. As discussed in the previous paper [23], a hydrostatic stress, σ_s, operating only the solid particles, is introduced to represent the dilation. In addition, another hydrostatic stress, σ_r, is also introduced. The full stress tensor of the solid phase is given by

$$\begin{bmatrix} \sigma^s_{11} & \sigma^s_{12} & \sigma^s_{13} \\ \sigma^s_{21} & \sigma^s_{22} & \sigma^s_{23} \\ \sigma^s_{31} & \sigma^s_{32} & \sigma s_{33} \end{bmatrix} = \begin{bmatrix} \sigma_s + \sigma_r + \tau^s_{11} & \tau^s_{12} & \tau^s_{13} \\ \tau^s_{21} & \sigma_s + \sigma_r + \tau^s_{22} & \tau^s_{23} \\ \tau^s_{31} & \tau^s_{32} & \sigma_s + \sigma_r + \tau^s_{33} \end{bmatrix}. \quad (5)$$

The deviator stress tensor of the solid phase is simply expressed by the following equation.

$$\tau^s_{ij} = 2\mu_s \dot{\varepsilon}^s_{ij} \quad (6)$$

Here μ_s is the apparent viscosity of solid particles. Since the viscosity of the liquid phase is much smaller than that of the semisolid at solid fractions more than 0.3, the apparent viscosity of the solid particles should be nearly equal to the apparent viscosity of semisolid. Thus, the apparent viscosity of the solid phase is estimated from the experimentally measured values.

The stress σ_s is proportional to normal / shear strain rates.

$$\sigma_s = -\left[C_3 \dot{\varepsilon}^{oct}_{shear} - B_1 \dot{\varepsilon}^{oct}_{norm}\right] \quad (7)$$

Here, C_3 is defined as a stiffness to shear strain rate and B_1 to normal strain rate. The octahedral shear and normal strain rates are used in this study, because the hydrostatic stresses σ_s and σ_r should not depend on coordination. Figure 1 shows an example of the apparent viscosity μ_s, the stiffness C_3 and the stiffness B_1 as a function of solid fraction. The apparent viscosity μ_s is estimated from the viscosity measured in Sn-Bi semisolid [13]. For simplicity, the stiffness B_1 is assumed to be equal to the apparent viscosity. At lower solid fractions, the stiffness C_3 is relatively low because solid grains hardly interact each other. Thus, it is considered that C_3 is equal to the apparent viscosity μ_s in the lower solid fractions. At a certain solid fraction, C_3 should exhibit a maximum and then decreases due to build-up of solid network. In this study, C_3 have a maximum at solid fraction of 0.48. It is note that the solid-fraction-dependence of C_3 strongly depends on morphology of solid grains. The value of 0.48

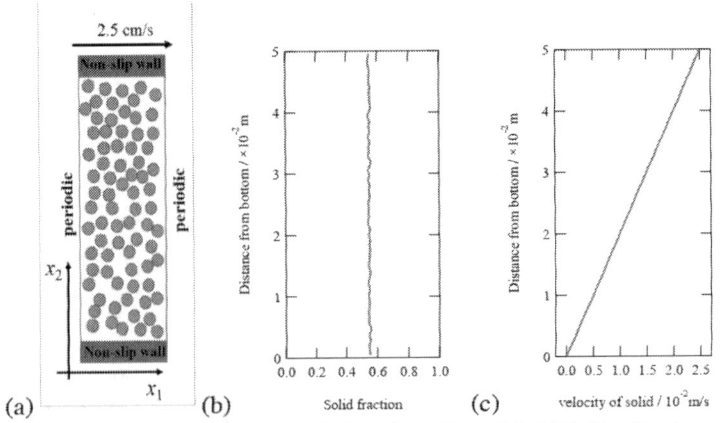

Figure 2 (a) Schematic illustration of calculation domain and initial distribution of (b) solid fraction and (c) velocity of solid in the x_1 direction.

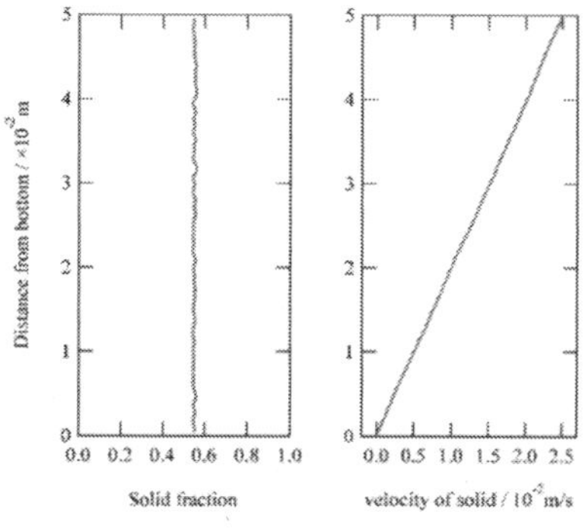

Figure 3 Distribution of solid fraction and velocity of the solid phase at 30s. The rearrangement was not included in the calculation.

Table 1 Calculation condition of simple shear

Calculation domain / mm	Mesh size / mm	Boundary condition
x:0.5, y:50, z:0.5	0.5 x 0.5 x 0.5	x_1: periodic, x_2: non-slip wall

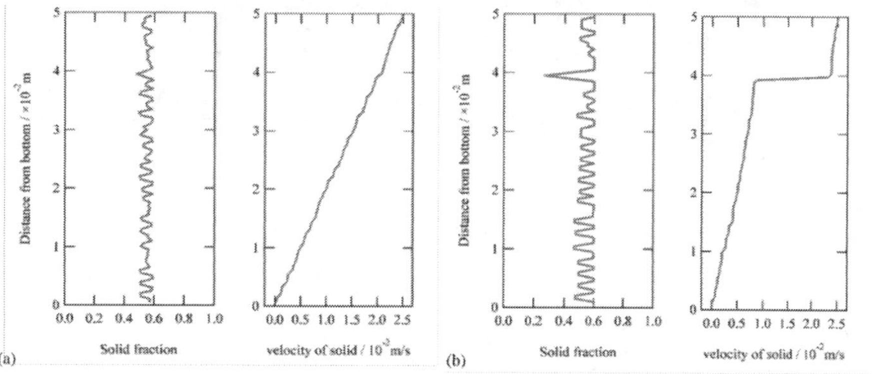

Figure 4 Distribution of solid fraction and velocity of the solid phase in the x_1 direction at (a) 7s and (b) 30s. The rearrangement was included in the calculation.

does not have a particular meaning and is set for descriptive purpose.

The stress σ_r in eq. (5) is an additional term to express a repulsive force between the solid grains. When the solid fraction exceeds a critical solid fraction f_{scrit} where the solid grains are closely packed, the solid grains physically contact each other and further increase in solid fraction hardly occurs. Namely, the stress σ_r acts like pressure in the fluid flow. In this study, the critical solid fraction f_{scrit} is set to be 0.6.

Application of the macroscopic model

1.3. Simple shear

A linear stability analysis in the previous study [23] showed that instability against simple shear appeared in semisolid when the following condition was satisfied.

$$\frac{\partial C_3}{\partial f_s} < 0 \tag{8}$$

However, it is not clear how the shear band is developed. Numerical calculation was performed to confirm the instability and to know how the shear band forms. Figure 2 shows the initial condition of

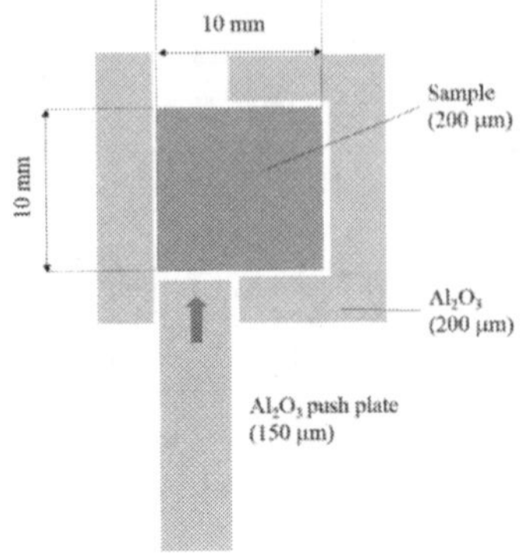

Figure 5 Configuration of semisolid deforrmation observed by the in-situ X-ray imaging.

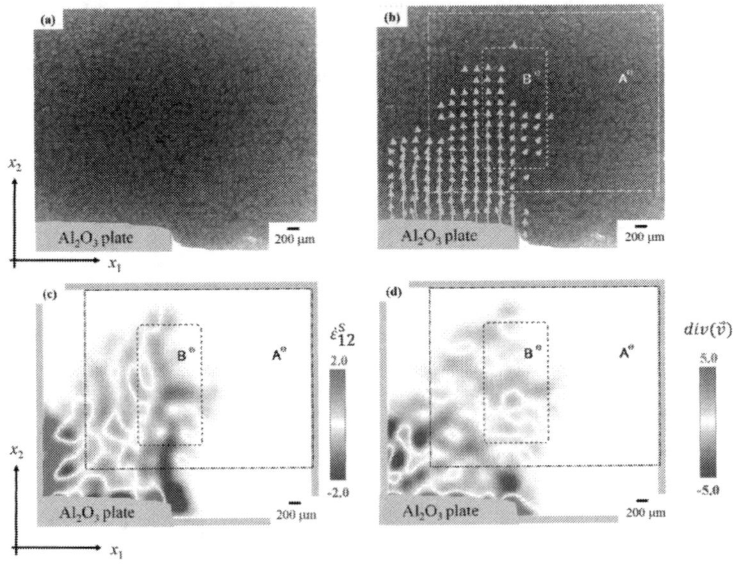

Figure 6 (a) Radiograph of the semisolid deformation after 3.1d increment of the push plate motion (d: mean grain diameter). (b) Solid displacement rate, (c) shear strain rate, (d) divergence of solid velocity in 0.5d increment of the push plate motion.

calculation. The solid fraction between 0.54 and 0.56 is randomly given at each element and average solid fraction in the casting is 0.55. The velocities of the solid and the liquid phases in the x_1 direction are the same at the beginning. The velocities at the bottom (x_2=0mm) and the top (x_2=50mm) are 0 mm/s and 25mm/s, respectively. The initial velocities in the x_2 direction are zero. Calculation domain size, mesh size and boundary condition are listed in Table 1.

Figure 3 shows the distribution of solid fraction (color) and velocity (arrows) of the solid phase after 30s, when the hydrostatic stress σ_s is not introduced (C_3=0) in the calculation. It means that the rearrangement of the solid grains is considered. Solid fraction and velocity do not obviously change during the calculation. Even the solid fraction is slightly fluctuated, the semisolid is stable against the shear and nearly steady state is achieved.

In contrast, the shear band forms when the hydrostatic stress σ_s is included in the calculation, as shown in Figure 4. The fluctuation of solid fraction gradually increases at the initial stage as shown in Figure 4(a). Since the apparent viscosity depends on solid fraction, the velocity in the x_1 direction

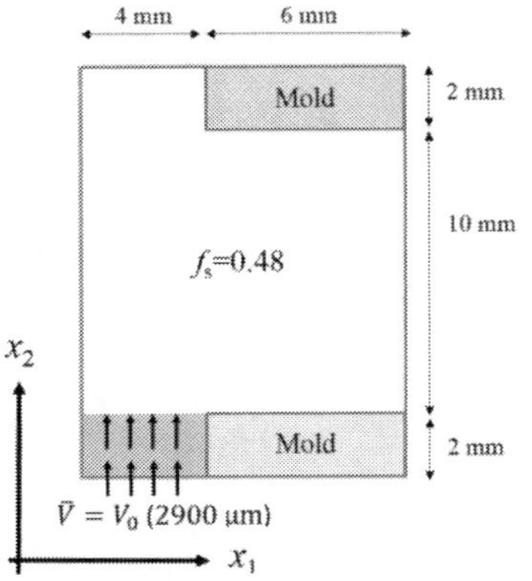

Figure 7 Calculation domain to compare with the in-situ observation.

fluctuates approximately 10%. The solid fraction has the lowest value at 4.9 mm from the bottom. Localization of shear strain occurs the shear band clearly forms at the position, as shown in Fig. 4(b). Once the shear band forms, the shear strain rates except the localization position are significantly reduced. As a result, the hydrostatic stress σ_s becomes sufficiently low to achieve the nearly steady state.

1.4. Deformation in Shear cell

Figure 5 shows the configuration of a shear cell used in the in-situ observation [24]. Alumina plate was inserted into the cell at 2900 μm/s and the semisolid (mean diameter of grains: 200μm) with 200 μm in thickness was deformed. The localization of shear deformation occurred at the beginning of

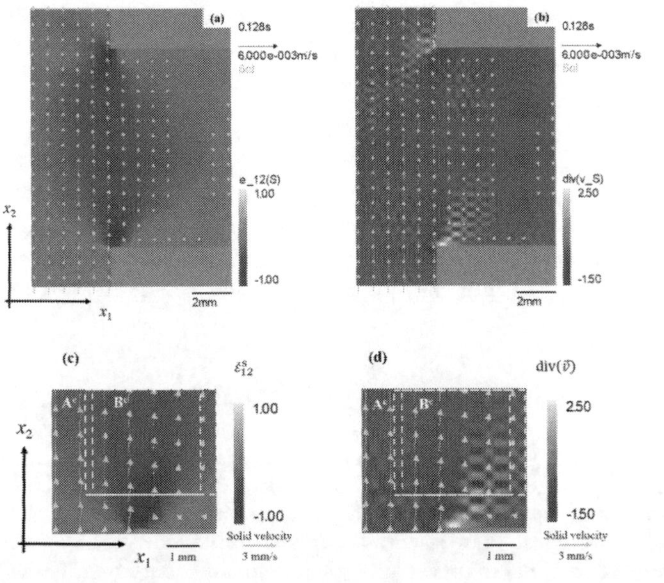

Figure 8 Calculated results. (a) Strain rate $\dot{\varepsilon}_{12}$ and (b) divergence of solid velocity $\mathrm{div}(\bar{v})$ with flow pattern of solid at 0.128 s. (c) and (d) are enlarged views of (a) and (b), respectively.

deformation. For example, only 3.1d motion of the solid grains was sufficient to cause the localization. Figure 6(a) shows a transmission image of the semisolid during deformation at 0.21s (3.1d). The velocity of solid grains was evaluated by image processing. The green arrows indicate motion of the solid grains. Shear strain rate and divergence of velocity of the solid grains are also shown in Figures 6 (c) and (d). Shear strain rate in the region B^e (within shear band) was 1.92 s^{-1} while shear strain rate in the region A^e (outside shear band) was 0.19 s^{-1}. The shear strain was clearly localized at the right-top of the alumina plate (region B^e). In addition, values of the divergence in the region B^e were mostly positive, indicating decrease in solid fraction. It should also be noted that the values of the divergence fluctuated from time to time even though time-average values were positive.

A model calculation for the shear cell is done to verify validity of the model. Figure 7 shows the configuration of calculation. The solid and the liquid phases are inserted into the cell from the inlet (left-bottom) and are ejected from the outlet (left-top) at 2900 μm/s. At the beginning, velocities of the solid and the liquid phases are zero. Physical properties used in the calculation are the same as those used in the simple shear.

Figure 8(a) shows the distribution of shear strain rate and the flow patter of the solid phase at 0.128s. As indicated by the blue color, the shear strain is localized at the right-top of inlet. When the calculation is done for Newtonian fluid, the localization is much weak. Namely, the introduction of the hydrostatic stress causes the localization. As shown in Fig. 8(b), values of the divergence are positive, although they fluctuate significantly. The position of the localization obtained by the calculation coincides with that observed in the experiments. Thus, the proposed model qualitatively simulates the coupling of the shear deformation and the decrease in solid fraction.

There is still quantitative discrepancy between the observation and the calculation. Shear strain rates in the regions A^c and B^c are 0.29 s^{-1} and 0.59 s^{-1}, respectively. The localization in the calculation is relatively weak, comparing to the observation. Several possibilities to cause the discrepancy are considered. One is the accuracy of stiffnesses C_3 and B_1, and the other is looseness of the semisolid at the beginning.

The stiffness C_3 and B_1 should depend on not only solid fraction but also morphology of solid grain. We assumed that the value of C_3 is equal to the apparent viscosity μ_s at low solid fraction, because a water-model experiment shows that C_3 is in the same order of μ_s [24]. However, no experimental data for the stiffnesses are available for metallic systems. Stiffness C_3 can influence degree of the localization. To measure the physical properties of the semisolid qualitatively is required for further development of the macroscopic model.

The looseness of solid grains is not included in the present model. In the semisolid before deformation, the solid grains are isolated by the liquid phase and most of them do not physically contact each other. Namely, the liquid phase can exist in the gap between the solid grains. Thus, the external force is not transmitted by contact of the solidi grain at the beginning. The force starts to be transmitted between the solid grains after the solid grains physically contact. This looseness can enhance the localization at the beginning. To include the looseness is a problem to be solved in the next step.

Summary

The macroscopic model, which includes the rearrangement of the solid grains during deformation, is examined.

(1) In the 1D simple shear of semisolid, the model well reproduces the localization of shear strain and the decrease in solid fraction. Consequently the band segregation is reproduced in the calculation.
(2) Semisolid deformation in the shear shell used in the in-situ observation is calculated. The calculated result qualitatively agrees with the observation results. The agreement indicates that the proposed model has potential to predict the shear band and consequently the band segregation.
(3) There are still quantitative discrepancies between the calculation and the observation. Further understanding of some physical properties such as the stiffnesses C_3 and B_1 is needed. In addition, the looseness of the semisolid is also included in the model to simulate the initial stage of deformation.

Symbols

u_i : flow velocity of liquid. i is direction
v_i : flow velocity of solid particles
f_s : fraction of solid
$\varepsilon^{s,l}_{ij}$: Strain rate tensor of solid (s) / liquid (l)
$\sigma^{s,l}_{ij}$: Full stress tensor of solid (s) / liquid (l), including the hydrostatic stress
$\tau^{s,l}_{ij}$: Deviator stress tensor of solid (s) / liquid (l)
σ_s, σ_r : Normal stress induced by collisions/rearrangement of solid particles
P : Pressure in liquid phase
F_i^{sl} : Interaction between liquid and solid particles
μ_s: Apparent viscosity of solid particles

Acknowledgements

The synchrotron experiments were performed at beamlines BL-20XU and BL-20B2 of SPring-8 (Proposal Nos. 2011A1209. 2011B1096, 2012A1110, 2012B1173, 2013B1242) with approval of the Japan Synchrotron Radiation Research Institute (JASRI). This work was supported by Grants-in-Aid for Scientific Research (KAKENHI) (Grant No 24226018) from JSPS.

References

[1] Kang I, Ohnaka I, 1997 *J Japan Foundry Engineering Society* **69** 240
[2] Gourlay C M, Laukli H I, Dahle A K, Metall 2007 *Mater Trans A* 38 1833
[3] Otarawanna S, Gourlay C M, Laukli H I, Dahle A K 2009 *Mater Charact* **60** 1432
[4] Yun M, Lockyer S A, Hunt J D 2000 *Mater Sci Eng A* **280** 116
[5] Lockyer S A, Yun M, Hunt J D, Edmonds V D 1996 *Mater Charact* **37** 301
[6] Suzuki K, Miyamoto T 1974 *ISIJ international* **14** 296
[7] Young KP, Riek R G, Flemings M C 1979 *Metals Technology* **6** 130
[8] Flemings M C 1991 *Metall Trans* **22A** 957
[9] Tzimas E, Zavaliangos A 1999 *Acta Materialia* **47** 517
[10] Gourlay C M, Dahle A K 2007 *Nature* **445** 70
[11] Gourlay C M, Meylan B, Dahle A K 2008 *Acta Mater* **56** 3403
[12] Gourlay C M, Dahle A K, Nagira T, Nakatsuka N, Nogita K, Uesugi K and Yasuda H 2011 *Acta Mater* **59** 4933
[13] Flemings M C 1991 *Metall Trans B*, **22B** 269
[14] McLelland A R A, Hendaerson H G, Atkinson H V, Kirkwood D H 1997 *Mater Sci Eng A*, **A232** 110
[15] Modigell M, Koke J 2001 *J Mat Proc Technol* **111** 53
[16] Hunt M L, Zenit R, Campbell C S, Brennen C E 2002 *J Fluid Mech* **442** 1
[17] Nagira T, Gourlay C M, Sugiyama A, Uesugi M, Kanazawa Y, Yoshiya M, Uesugi K, Umetani K and Yasuda H 2011 *Scripta Mater* **64** 1129
[18] Nagira T, Yokota H, Morita S, Yasuda H, Yoshiya M, Gourlay C M, Sugiyama A, Uesugi K and Umetani K 2013 *ISIJ International* **53** 1195
[19] Fonseca J, O'Sullivan C, Nagira T, Yasuda H and Gourlay C M 2013 *Acta Mater* **61** 4169
[20] Nagira T and Yasuda H 2014 In-situ Studies with Photons, *Neutrons and Electrons Scattering II*, (Switzerland: Springer) p 231
[21] Gourlay M, O'Sullivan C, Fonseca J, Yuan L, Kareh K M, Nagira T and Yasuda H 2014 *JOM* **66** 1415
[22] Nagira T, Morita S, Yokota H, Yasuda H, Gourlay C M, Yoshiya M, Sugiyama A, Uesugi K, Takeuchi A and Suzuki Y 2014 *Metall Mater Trans A* **45** 5613
[23] Morita S, Yasuda H, Nagira T, Gourlay C M, Yoshiya M, Sugiyama A 2012 *IOP Conf Ser: Mater Sci Eng* **33** 012053
[24] Nagira T, Morita S, Yokota H, Yasuda H, Gourlay C M, Yoshiya M, Sugiyama A, in preparation.

IOP Conf. Series: Materials Science and Engineering **84** (2015) 012012 doi:10.1088/1757-899X/84/1/012012

Probability of twin boundary formation associated with the nucleation of equiaxed grains on icosahedral quasicrystal templates

Güven Kurtuldu[1], **Michel Rappaz**[1]

[1] Computational Materials Laboratory, Institute of Materials, Ecole Polytechnique Fédérale de Lausanne, Station 12, CH-1015 Lausanne, Switzerland

E-mail: guven.kurtuldu@epfl.ch

Abstract. Recently, we have shown that minute Cr additions (typically 0.1 wt%) to Al-Zn alloys solidified in a uniform temperature field lead to the formation of fine equiaxed fcc Al grains [1]. Furthermore, these grains exhibit an unexpectedly large number of twin, or near-twin, relationships with their nearest neighbors and some of them even show a 5-fold symmetry multi-twin relationship with a common ⟨110⟩ direction. Similar observations have been made for yellow gold alloys (Au-12.5wt%Cu-12.5wt%Ag) inoculated with very small amounts of Ir (5-200 ppm) [2]. These results become fully consistent when one considers that the primary fcc phase forms on facets of icosahedral quasicrystals (iQC's), either by heteroepitaxy solidification or by peritectic transformation, with the following relationship: $\langle 111\rangle_{fcc} \parallel$ 3-fold symmetry iQC axes, $\langle 110\rangle_{fcc} \perp$ 2-fold symmetry iQC axes. The present study contributes to a better understanding of the frequency of twin boundary formation by the nucleation of fcc phase from an iQC template. A simple stereological model for the formation of equiaxed grains on such iQC templates has been derived. It is based on a 3D Voronoi tessellation of randomly distributed and oriented iQC template centers. Each iQC nucleation template site is the origin of 20 fcc grains with the heteroepitaxy relationships mentioned above on the 20 facets of the iQC. Therefore, the neighboring grains having a common iQC nucleation site contributes to the twin boundary frequency, while those coming from different iQC sites have random grain boundary orientations. The twin boundary frequency of the grains nucleated from iQC templates seen in 2D metallographic sections is compared with that measured in EBSD reconstructed grain structures.

1. Introduction

A new nucleation mechanism has recently been evidenced in which icosahedral quasicrystals (iQC's) are responsible for the formation of the fcc Al phase during solidification of an Al-20wt.%Zn alloy when a small mount of Cr is added [1]. The density of twin boundaries and the grain size in Al-Zn with and without Cr is shown in the EBSD reconstructed grain structures of Fig. 1. While the average grain size of Al-20wt.%Zn sample is greater than 1 mm (Fig. 1(a)), that of Al-20wt.%Zn-0.1wt.%Cr sample is only 135 µm (Fig. 1(b)). Furthermore, a twin boundary frequency of 2.3% was present in Al-Zn-Cr alloys. These boundaries are highlighted with white segments in Fig. 1(b).

Several multiple-twinned (MT) nearest neighbor grain configurations were also identified with a symmetry relationship which could be explained only if they were assumed to form on

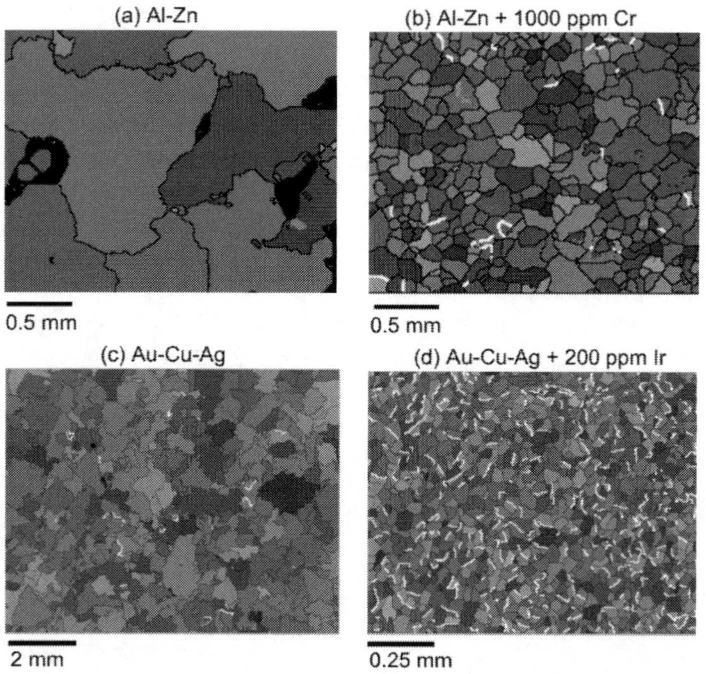

Figure 1. EBSD maps of different as-solidified alloys show the grain refinement effect of minute Cr and Ir additions to Al and Au alloys, respectively. a) Al-20wt.%Zn, b) Al-20wt.%Zn-1000 ppm Cr, c) Au-12.5wt.%Cu-12.5wt.%Ag and d)Au-12.5wt.%Cu-12.5wt.%Ag-200 ppm Ir (after References [1, 2]). Twin boundaries are outlined with white lines.

a single regular, or interlocked, icosahedron with the epitaxial relationship: $\langle 111\rangle_{fcc} \parallel$ 3-fold symmetry iQC axes, and $\langle 110\rangle_{fcc} \perp$ 2-fold symmetry iQC axes. These experimental findings are a fairly clear indication that fcc grains nucleate from an icosahedral template, known to be present as iQC's in the supercooled liquid or as building blocks of the crystalline structure of the approximant $Al_{45}Cr_7$ phase [3, 4]. The phase initiated from a melt can be the one that is closest in free energy to that of the liquid, and not necessarily the most thermodynamically stable one [5]. Metastable iQC in supercooled metallic liquid is the most probable candidate to form initially due to its low interfacial energy [6, 7]. Upon further solidification and cooling, the iQC's disappear due to peritectic reaction between liquid, fcc Al and approximant phase or iQC [8], while the fcc phase grows with multiple twin or near-twin relationships between the various grains. Once the fcc phase reaches a critical radius, the solid-liquid interface destabilizes, thus leading to the formation of twinned dendrites [9, 10, 1].

This iQC enhanced nucleation mechanism evidenced in Al-Zn-Cr alloys has been shown to operate also in yellow gold alloys (Au-12.5wt.%Cu-12.5wt.%Ag) when minute amounts of Ir ($\leq$ 200 ppm) are added to the melt [2]. Inoculation of such alloys with Ir is a grain refining technique currently used in industry and known for nearly half a century [11, 12, 13]. As can be seen in Fig. 1(c-d), it has also a drastic effect on the grain size in yellow gold alloys: it is about 250 µm in the specimen without Ir addition, and only 30 µm in the specimen with 200 ppm Ir (Fig. 1(d)). The grain refinement is accompanied by a very large increase of the fraction of twin boundaries, from less than 1% without Ir to 11% with 200 ppm Ir. It also promotes the formation of MT grains in a way comparable to minute Cr additions to Al-Zn alloys. Several grains are shown to have orientation relationships compatible with the icosahedron or interlocked icosahedron geometry. All these findings lead to the same conclusion that icosahedral symmetry is responsible for this nucleation mechanism, although no iQC has been reported so far in such gold alloys.

In the present study, a model has been developed to calculate the frequency of twin boundaries formed by this iQC enhanced nucleation mechanism. It is based on the hypothesis of randomly distributed iQC's centers in 3D with a random orientation. Each iQC leads to the formation of 20 MT fcc grains with the heteroepitaxy relationship mentioned before. The twin boundary frequency of the fcc grains nucleated from these iQC templates and the grain structures are then deduced in 2D metallographic sections for both Al-Zn:Cr and Au-Cu-Ag:Ir alloys.

2. Model

A stereological model for the formation of equiaxed grains has been derived based on the 3D Voronoi tessellation of random iQC nucleation centers of random crystallographic orientation. Each nucleation site produces 20 fcc grains with the heteroepitaxy relationships compatible with the 20 facets of an icosahedron, and are thus multiple twinned. Therefore, the neighboring fcc MT grains having a common iQC nucleation center contribute to the twin boundary frequency, while boundaries between grains formed on different iQC's do not.

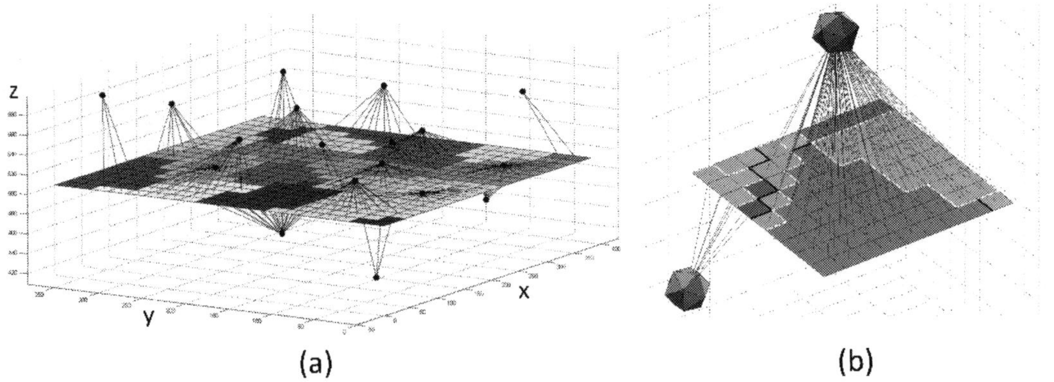

Figure 2. Schematic view of the model used for the nucleation of equiaxed grains from randomly distributed and oriented iQC templates: (a) Determination of orientations and grain boundaries in a 2D metallographic section; (b) Twin boundaries formed between the grains nucleated on the adjacent facets of a single iQC template. Only the iQC centers giving a contribution to the 2D section are shown in (a). While grain boundaries are shown by black lines, twin boundaries are highlighted by white lines.

In a first step, iQC nucleation centers are positioned pseudo-randomly within a given volume: the volume is subdivided into a certain number of regular cubic volume elements, within which a single iQC with a random orientation is positioned (i.e., the size of a cubic element is directly related to the mean volume occupied by the 20 MT fcc grains). Assuming that all the iQC's and fcc MT grains form at the same time and grow at the same speed in the uniform temperature field, the random boundaries correspond to the Voronoi tessellation of the iQC's centers. The grain structure is then calculated in a 2D metallographic section by a scan along its $x-$ and $y-$directions with a regular grid spacing h corresponding to the pixel size of an EBSD reconstructured map. For each position (x_i, y_i), the model first identifies the iQC center I which is the closest (see Fig. 2(a)). Then, the facet ν of this iQC on which the direction connecting pixel i to the center I falls is searched (see Fig. 2(b)). The crystallographic orientation attributed to this pixel i is based on the orientation of facet $I\nu$ and using the heteroepitaxy relationship described before. Please note that a perfect icosahedral symmetry is kept during the growth of the fcc grains, i.e., each individual grain has the exact epitaxial relationship of $\langle 111 \rangle_{fcc} \parallel$ 3-fold iQC symmetry axes, and $\langle 110 \rangle_{fcc} \perp$ 2-fold iQC symmetry axes with facet $I\nu$. Therefore, the misorientation of 1.47 deg. from a perfect twinning relationship between two neighboring fcc MT grains is kept during growth for the sake of simplicity. This is not the case in Al-Zn:Cr and yellow gold:Ir alloys investigated by EBSD analysis, in which five grains issued from 5 iQC facets sharing a common 5-fold symmetry axis exhibit at the end 4 twins and one near-twin relationship with a 7-8 deg. misorientation.

Once the whole surface has been scanned, the frequency of twin grain boundaries is determined among this 2D simulated section by checking the crystal orientations of adjacent grains for each grain boundary within a 5 deg. tolerance, as in the EBSD analysis of real grain structures (Fig. 1).

3. Results and Discussion

Figure 3(a) shows the simulated structure of randomly oriented grains on a 2D plane in the case of regular nucleation centers which are not iQC's. Taking a random orientation of a grain with respect to a reference grain, the probability to be in a twin relationship within the 5 deg. tolerance can be calculated as 0.3% according to Reference [14]: This is confirmed by our calculations: 3336 twin boundaries with a 5 deg. tolerance have been identified from 10^6 grain boundaries. There is only one grain boundary with a twin relationship in Fig. 3(a,b) for 159 grains and 430 grain boundaries. If now 20 grains nucleate on the 20 facets of icosahedral templates having the same nucleation centers in the volume, the average grain size decrease (factor of about $20^{3/2}$) is accompanied by an increase in twin density due to grains nucleated on the adjacent facets of the same iQC (Fig. 3(c,d)). Each iQC template generates 1 to 15 MT fcc grains visible in a 2D section.

In a volume of $25{\times}25{\times}6$ mm^3 and on 2D section of $25{\times}25$ mm^2 having a pixel size of 10 µm, 950 regular grains have been simulated with 915 µm average grain diameter, 10 twin boundaries and 2722 grain boundaries (Fig. 3(a,b)). The average number of neighboring grains is 5.7, which is similar to what was observed in EBSD measurements of Al- and Au-based alloys. In the case of iQC enhanced nucleation (Fig. 3(c,d)) with the same nucleation centers, 7475 grains have been calculated with an average grain diameter of 326 µm. The total number of grain boundaries and the number of twin boundaries are 20634 and 8546, respectively. This means that the frequency of twin boundaries increases from 0.3% to 41% if all the grains are nucleated on iQC templates. This frequency of twin boundaries value contains also the error associated with the introduction of pixels to scan the plane which produces artificial boundaries at the meeting point of 5 MT fcc grains, as for EBSD measurements having a given step size. However, it has been checked that this error does not depend on the step size, if it is smaller than 5% of average grain diameter. The average number of grain neighbors for MT fcc grains is 5.5 and does not change significantly, although many triangular MT fcc grains formed by this mechanism as shown in Fig. 3(c,d).

In [2], the effect of different Ir amounts on the grain size and twin frequency has been investigated in yellow gold alloys and is summarized in Fig. 4(a). The error bars for the grain size correspond to the standard deviation of the distribution and the ones for twin density correspond to a 99% confidence interval for the corresponding total number of grain boundaries. With only 5 ppm addition of Ir, the grain size drops from 250 µm to 90 µm. Further addition of Ir continues to decrease the grain size, but to a lower extent. The grain size drops to about 60-70 µm for 10-50 ppm Ir addition and it further decreases to 30-40 µm with 100-200 ppm Ir addition. As mentioned above, the decrease in grain size is accompanied by an increase in twin

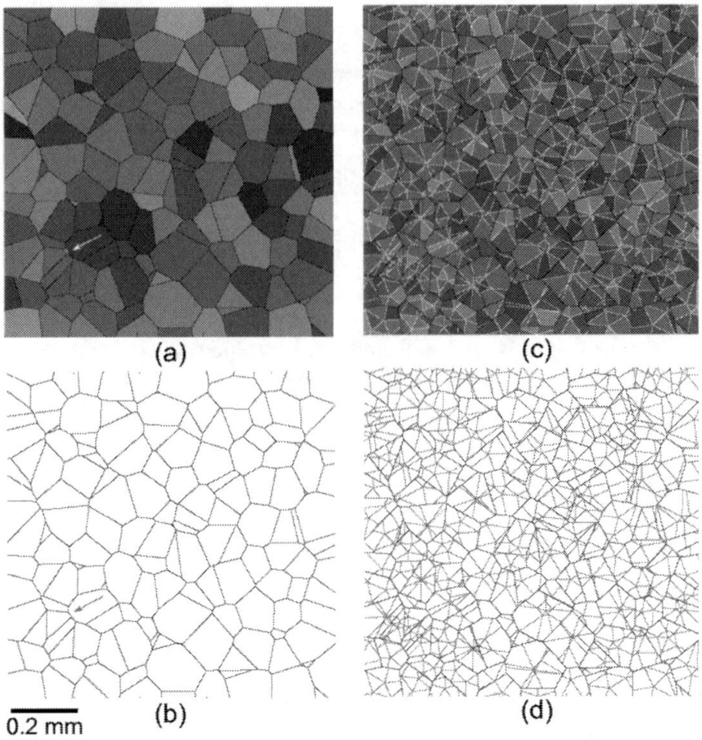

Figure 3. (a,b) Simulated grain structure of equiaxed grains in a 2D section. (c,d) Grain structure for MT fcc grains formed by the iQC nucleation mechanism with the same density of nucleation centers in the volume. While grain boundaries are shown with black lines, twin boundaries are identified with red or white lines.

frequency from less than 1% without Ir to 11% with 200 ppm Ir.

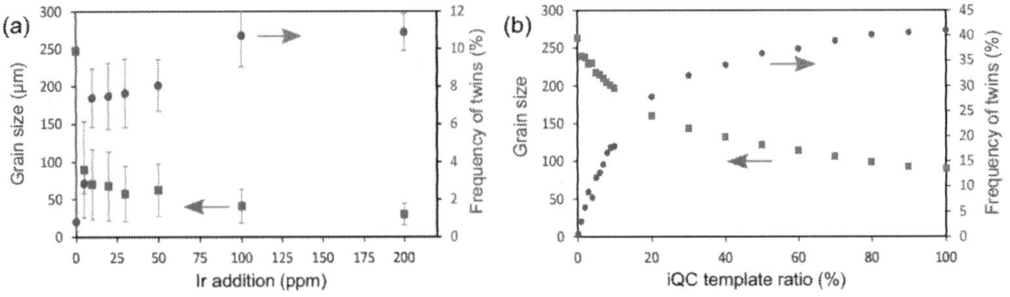

Figure 4. (a) Grain size and frequency of twins as a function of Ir addition in Au-Cu-Ag alloys [2]. While error bars show the standard deviation for grain size, they show the error limits for twin frequency with a confidence interval of 99%. (b) Average grain size and frequency of twin boundaries for different icosahedral template ratios by taking into account two different nucleation mechanism: (i) one grain nucleates from one site, (ii) 20 grains nucleate from one iQC template.

For the twin boundary densities investigated in Al and gold alloys [1, 2], there can be a competition between a classical nucleation mechanism (each grain nucleated on a single site) and the iQC enhanced nucleation mechanism. In order to analyze the effect of this competition on grain size and twin density, different calculations have been performed and the equiaxed grain structure characterized with various percentages of iQC nucleation sites for a fixed total density of nucleation sites. Fig. 4(b) shows the effect of the percentage of iQC templates on the average grain size and twin frequency. For a low iQC template ratio, up to 20%, the twin frequency increases from 0.3% to about 30%. Further increase in the iQC template ratio from 20 to 100% generates only 11% additional twin boundary frequency. Similar tendency is present for the grain size as a function of the iQC template ratio. The reason for this behavior is that, since each additional iQC template generates 20 new grains in the volume, the ratio of 20 grains over total number of grains decreases with the increase in total number of grains, which is itself proportional to the fraction of iQC template.

Figure 4(b) demonstrates that, although the iQC enhanced nucleation mechanism generates twin boundaries, the decrease in grain size for a certain twin frequency is very limited in the simulations as compared with the values measured in gold alloys (Fig. 4(a)). For a twin frequency of 11% in gold alloy with 200 ppm Ir, the grain density is multiplied by a factor of 600, which cannot be more than 20 in the simulations for a fixed number of nucleation sites. Therefore, the addition of Ir does not only promotes the formation of iQC templates, it also significantly increases the density of nucleation sites as could be expected.

Addition of minute amounts of Cr and Ir also promotes the formation of MT fcc grains in Al-Zn and yellow gold alloys, respectively. However, the number of MT grains observed in EBSD maps is very limited as compared with simulation which produces always MT grains for the iQC enhanced nucleation mechanism. This might be due to a growth mechanism which is not accounted for in the simulations. When 20 MT fcc grains form, they are all assumed to grow with a spherical symmetry. However, having a twin between two MT fcc grains creates a cusp in order to satisfy mechanical equilibrium at the triple line between the two twinned fcc grains and the liquid. Salgado et al. [15] have shown that this naturally leads to a twinned dendrite with a liquid groove in between, during directional solidification. In equiaxed growth with a uniform temperature field, one of the MT fcc grains can overgrow its neighbor by solute rejection. This mechanism would explain why we have less MT occurrences in metallographic sections and also why we have a lower fraction of twin boundaries in reality.

Figure 5(a) shows the EBSD reconstructed map of 9 grains in a 20 ppm Ir-gold alloy [2]. The six 5-fold symmetry axes of an icosahedron have been revealed by their twin or near-twin orientation relationships. Figure 5(b) shows the same occurrence of 9 grains nucleated on a single iQC template. Both measured and simulated grains are numbered from 1 to 9 in order to demonstrate that the grains having the same number in Fig. 5(a) and (b) have similar orientation relationships. 6 groups of MT fcc grains sharing a common $\langle 110 \rangle$ direction, i.e., formed on 5 facets of an iQC around a 5-fold symmetry axis, have been identified. They reproduce the 6 5-fold symmetry axes of an iQC. These grain groups are (1/2/3/4/5), (5/4/6/7), (3/4/6/9/8), (7/2/3/8), (1/2/7/6/9) and (5/1/9/8). The detailed orientations of experimental grains can be found in Ref. [2]. Adjacent grains in the same group have MT relationship as shown in Fig. 5(b) by white lines. While grains on EBSD map have complex morphologies, simulated ones form convex polygons due to the limitations of the model.

Figure 5(c) shows another example for the simulated grain morphology. If an iQC template is very close to the cutting plane and one of the 5-fold symmetry axis is perpendicular to the plane, 10 MT fcc grains with 5 different crystal orientations (indicated as A, B, C, D and E in Fig. 5(c)) can be observed in the section. It looks as if we had a 10-fold MT relationship.

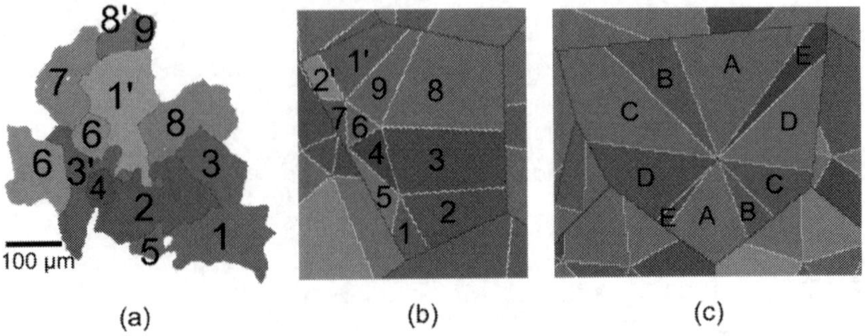

Figure 5. (a) EBSD reconstructed map of 9 MT fcc grains in Au-Cu-Ag alloy with 20 ppm Ir [2]. These MT grains share 6 common ⟨110⟩ directions by groups which reproduce the 6 5-fold symmetry axes of iQC. These 6 grain groups are (1/2/3/4/5), (5/4/6/7), (3/4/6/9/8), (7/2/3/8), (1/2/7/6/9) and (5/1/9/8). Adjacent grains in the same group have MT relationships (After Ref. [2]). (b) Simulated 9 MT fcc grains nucleated on a common iQC site having the same crystal symmetry as in (a). (c) Simulation of 10 grains with 5 different crystal orientations (A, B, C, D and E) when the iQC template is close to the 2D section and one of the 5-fold symmetry axis is perpendicular to the section. Twin bounadries are highlighted by white lines.

4. Conclusion

A simple stereological model for the formation of equiaxed grains by the iQC enhanced nucleation mechanism has been developed. It is based on randomly distributed and oriented iQC template centers. Each iQC template generates 20 MT fcc grains with the heteroepitaxy relationships on the 20 triangular facets of the iQC. Therefore, neighboring grains having a common nucleation site contributes to the twin boundary frequency, while those coming from different iQC sites have random grain boundary orientations. The morphology, occurrences of multiple twins and twin boundary frequency of the grains nucleated from iQC templates seen in 2D metallographic sections is compared with those obtained by EBSD reconstructed maps in yellow gold alloys. It has been found that the twin boundary frequency increases from 0.3% for a randomly oriented regular fcc grain to 41% when all the fcc grains are formed on iQC's.

References

[1] Kurtuldu G, Jarry P and Rappaz M 2013 *Acta Mater* **61** 7098–108
[2] Kurtuldu G, Sicco A and Rappaz M 2014 *Acta Mater* **70** 240–8
[3] Cooper M J 1960 *Acta Crystallogr* **13** 257–63
[4] He Z B, Zou B S and Kuo K H 2006 *J Alloy Compd* **417** L4–8
[5] Ostwald W 1897 *Z. Phys. Chem* **22** 289–330
[6] D Herlach, P Galenko and D Holland-Moritz 2007 *Metastable solids from undercooled melts* (Pergamon)
[7] Lee G W, Gangopadhyay A K, Croat T K, Rathz T J, Hyers R W, Rogers J R and Kelton K F 2005 *Phys Rev B* **72** 174107
[8] Kurtuldu G, Jessner P and Rappaz M 2015 *J Alloy Compd* **621** 283–6
[9] Henry S, Rappaz M and Jarry P 1998 *Metall Mater Trans A* **29** 2807–17
[10] Salgado-Ordorica M A and Rappaz M 2008 *Acta Mater* **56** 5708–18
[11] Nielsen J P and Tuccillo J J 1966 *J Dent Res* **45** 964–9
[12] Ott D and Raub C J 1981 *Gold Bull* **14** 69–74
[13] Renner H, Schlamp G, Hollmann D, Lüschow H M, Tews P, Rothaut J, Dermann K, Knödler A, Hecht C, Schlott M and Others 2000 *Ullmann's Encyclopedia of Industrial Chemistry*
[14] Warrington D H and Boon M 1975 *Acta Metall Mater* **23** 599–607
[15] Salgado-Ordorica M A, Desbiolles J L and Rappaz M 2011 *Acta Mater* **59** 5074–84

IOP Conf. Series: Materials Science and Engineering **84** (2015) 012013 doi:10.1088/1757-899X/84/1/012013

Heterogeneous nucleation of the primary phase in the rapid solidification of Al-4.5wt%Cu alloy droplet

A Maitre[1], A-A Bogno[1], M Bedel[2], G Reinhart[2] and H Henein[1]

1Department of Chemical and Materials Engineering, University of Alberta, Edmonton, AB, Canada
2Aix-Marseille University & CNRS, IM2NP UMR 7334, Campus Saint-Jérôme, Cas142, 13397 Marseille Cedex 20, France

Email: maitre@ualberta.ca

Abstract. This paper reports on rapid solidification of Al-Cu alloys. A heterogeneous nucleation/growth model coupled with a thermal model of a falling droplet through a stagnant gas was developed. The primary undercooling as well as the number of nucleation points was compared with Al-Cu alloy droplets produced by Impulse Atomization (IA). Based on experimental results from Neutron Diffraction, secondary (eutectic) phases were obtained. Then, primary and secondary undercoolings were estimated using the metastable extensions of solidus and liquidus lines calculated by Thermo-Calc. Moreover, Synchrotron X-ray micro-tomography has been performed on Al-4.5wt%Cu droplets. The undercoolings are in good agreement. Results also evidence the presence of one nucleation point and are in agreement with the experimental observations.

1. Introduction

Manufacturing of most metallic alloy products involves solidification at some stage. Mechanical properties of these products are generally related to their solidification microstructures. Depending on the final application of a product, a certain type of microstructure is more appropriate compared to another. For a product that requires directional properties, a microstructure of columnar grains is needed while isotropic properties are satisfied with an equiaxed structure. Generally, post-processing of the solidified materials is required to obtain the final product with desired properties. These post-solidification treatments are generally time-consuming and therefore increase the production cost without fully eliminating solidification related defects such as segregation. Therefore, it is important to understand all the dynamics involved in the formation of solidification microstructures in order to control the properties of the final products. As dendrites growth from an undercooled melt depends a great deal on the nucleation undercooling. Therefore, determination of undercooling and the resulting growth rate, recalescence, microsegregation/phase fraction and grain size is very important.

Al-Cu alloys (4.5, 5, 10 and 17 wt% Cu) have been produced by IA and the last three compositions were analysed in our previous papers [1, 2]. IA is a single fluid atomization technique that is capable of producing droplets of controlled size having a relatively narrow distribution and a predictable cooling rate. The alloys (350 to 450g) were melted in a graphite crucible by means of an induction furnace and atomized at 850°C in an almost oxygen free chamber (10ppm) under Nitrogen, Helium or Argon atmospheres. The atomized droplets rapidly solidify during their fall by losing heat to the stagnant atmosphere and are quenched in a collecting beaker filled with oil to avoid any solid-state reaction that could alter the microstructures. Subsequently, the solid droplets are washed, dried and sieved into different sizes. A detailed description of the IA process is given in [3]. A solidification model of a droplet falling through a stagnant gas, based on heterogeneous nucleation/growth was developed. The primary undercooling as well as the number of nuclei is considered and the results were compared to experimental results of a rapidly solidified Al-4.5 wt%Cu droplet, generated by IA.

2. Determination of the number of nuclei

2.1. Thermal model

For a hot spherical droplet moving in a cooler surrounding fluid, thermal energy is transferred within the droplet by conduction and between the surface of the droplet and the surrounding gas by forced convection, conduction and radiation. An equation describing the rate of heat energy lost at the droplet surface is given by:

$$q = h_{eff}\, A\left(T_m - T_\infty\right) \tag{1}$$

where h_{eff} is the effective heat transfer coefficient and consists of the additive contribution of convection, conduction and radiation heat transfer mechanisms. A is the surface area of the droplet, T_m is the droplet surface temperature and T_∞ is the free stream gas temperature (300 K). For alloys such as aluminum, the effective heat transfer for a moving droplet is dominated by convection. Radiation heat transfer can be significant for higher temperature alloy systems (iron based alloys).

One approach to quantify the convective component (h_c) of h_{eff} has been through the use of semi-empirical equations [4-10] in which the Nusselt number (Nu) is averaged over the entire droplet surface. The general form of these semi-empirical equations for calculating the effective heat transfer coefficient is

$$\mathrm{Nu} = \frac{H_C.D}{k_g} = A + B.\mathrm{Pr}^m.\mathrm{Re}^n \tag{2}$$

where D is the droplet diameter, $\mathrm{Re} = \dfrac{\rho_g.v.D}{\mu_g}$, $\mathrm{Pr} = \dfrac{C_p^g.\mu_g}{k_g}$, k_g, C_p^g, ϱ_g and μ_g are the conductivity, heat capacity, density and viscosity of the gas respectively and A, B, m and n are constants.
Considering the correlation proposed by Whitaker [3] and a correction term (due to the definition of Nu at Re=0) [11], equation (2) can be rewritten as

$$\mathrm{Nu} = \frac{H_C.D}{k_g} = 2.0\left(\frac{B}{k_s(\alpha+1)}\frac{T_s^{\alpha+1} - T_\infty^{\alpha+1}}{T_s - T_\infty}\right) + \left(0.4\,\mathrm{Re}^{1/2} + 0.06\,\mathrm{Re}^{2/3}\right)\mathrm{Pr}^{0.4}\left(\frac{\mu_\infty}{\mu_S}\right)^{1/4} \tag{3}$$

where k_s is the gas conductivity at the droplet surface temperature (T_s), B is the pre-power coefficient from the variation of the gas conductivity with temperature, α is the power coefficient in this same equation (Table 1) and $\dfrac{\mu_\infty}{\mu_s}$ is the ratio between the viscosity at the free stream gas temperature and the viscosity at the droplet surface temperature.

Table 1 – Thermophysical properties of nitrogen, helium and argon [12-14]

Property (all T values in K)	**Nitrogen**	**Helium**	**Argon**
Density (kg/m^3)	$335.55\ T^{-0.9971}$	$37.303\ T^{-0.9559}$	$539.23\ T^{-1.0205}$
Specific heat (J/kg.K)	$978 + 0.182T$	5197	520
Dynamic Viscosity (Pa.s)	$3.79\times10^{-7}T^{0.6766}$	$4.3679\times10^{-7}T^{0.6702}$	$2.38\times10^{-7}T^{0.6702}$
Conductivity (W/m.K)	$0.0344\times10^{-2}\ T^{0.7609}$	$0.2778\times10^{-2}\ T^{0.7025}$	$0.0186\times10^{-2}\ T^{0.7915}$

Several assumptions were made in the model formulation:
1 - Internal temperature gradients in droplets are negligible (Biot < 0.1 in the droplet size studied);
2 - The time for stream break up and spheroidization of ligaments emanating from the orifice plate is very small compared to the solidification time [15];
3 - The initial velocity of the droplet exiting the orifice is 0.5 m/s [16, 17];
4 - The ambient gas temperature remains constant during atomization;
5 - For radiation heat transfer, a droplet emissivity $\varepsilon = 0.1$ was used [18];
6 - Thermal interaction between droplets is negligible;
7 - Droplet diameter decreases during solidification.

Based on these assumptions, the governing heat transfer equation for droplet cooling is:

$$\frac{dT_m}{dt} = -\frac{6h_{eff}}{\rho_m . Cp_m . D}(T_m - T_\infty) \tag{4}$$

where $\frac{dT_m}{dt}$ is the change in droplet temperature with time, ϱ_m and Cp_m are the droplet density and specific heat, respectively. The h_{eff} term used in Equation (4) was calculated using a linearized radiation term and h_c calculated with Equation (3). During solidification, an effective Cp_m (in Equation (4)) was calculated as

$$Cp_m = \frac{dH}{dT_m} = \frac{df_s}{dT_m} . \Delta H_f + Cp \tag{5}$$

where fs is the fraction solid at temperature T_m and ΔH_f is the latent heat of fusion. The equation (4) is solved using a fourth order Runge-Kutta method. More details can be found in [11].

2.2. *Nucleation model*

This model considers a heterogeneous nucleation with growth in the frame of a cellular method. The start temperature is taken just below the liquidus temperature. At each step the temperature is determined using equation (4) and the critical radius r^* as well as the corresponding energy ΔG^* related to the formation of a nucleus at this temperature are calculated using equation (6) to (8)

$$r^* = \frac{2\gamma T_L}{\Delta H_f \Delta T} \tag{6}$$

$$\Delta G^{*} = \frac{16\pi\gamma^{3}T_{L}^{2}}{3\Delta H_{f}^{2}\Delta T^{2}} f(\theta) \tag{7}$$

$$f(\theta) = \frac{(1-\cos(\theta))^{2}.(\cos(\theta)+2)}{4} \tag{8}$$

where γ is the surface energy, T_L is the liquidus temperature, ΔT is the primary phase undercooling and θ is the wetting angle.

It is then possible to determine the nucleation rate $I_F(T)$ (9) and the nuclei density $n(T)$ (10). The probability density function $P(T)$ (11) for the nucleation of the droplet and the probability $P_m(T)$ for a mesh to become a cluster can be determined as well

$$I_F(T) = 10^{30} \times e^{\left(\frac{-\Delta G^{*}}{kT}\right)} \tag{9}$$

$$n(T) = \int_{T(t)}^{T(t+\delta t)} \frac{I_F(T)}{Q} dT \tag{10}$$

$$P(T) = \frac{I_F(T)V_{droplet}}{Q} \exp(-n(T)V_{droplet}) \tag{11}$$

$$P_m(T) = \frac{[n(T(t+\delta t)) - n(T(t))]V_{droplet}}{x_{mesh}} \tag{12}$$

x_{mesh} is the number of mesh in the system and $V_{droplet}$ is the volume of the considered droplet. At each step, r^*, ΔG^* and so $P(T)$ and $P_m(T)$ are calculated. The model used is partially stochastic as a random number is taken and compared to the probabilities and can be divided in two parts ($P(T)$ for the nucleation start and $P_m(T)$ for the number of clusters). Firstly, a random number is compared to $P(T)$ at each step, if the number is lower than $P(T)$ then nucleation starts. This provides us the nucleation temperature (T_L-ΔT) and so the undercooling. In a second time and only if the nucleation is allowed, a run is done on the meshes contained in the droplet (due to memory issue, only a box of $10\times10\times10\ \mu m^3$ is considered for the number of clusters and not for the entire droplet). All the meshes are investigated. Each time, a different random number is selected and compared to $P_m(T)$. If this number is higher than $P_m(T)$, the mesh will stay liquid otherwise it will become a cluster. It should be noted that a homogeneous nucleation model has been attempted. Due to a too small value of I_F (i.e. too "important" value of ΔG^*) no nucleation was observed.

2.3. Growth model

During the growth process, the radius of each cluster is updated for each step by using

$$r(t+\delta t)=r(t)+v.\delta t \qquad (13)$$
$$r(t=0)=r^*$$

where v is the growth velocity determined by the LKT model [19] (Figure 1). It was assumed that the first introduced cluster will be the first to grow and other clusters will follow. Each time the size of a cluster is updated using eq (12), the temperature of the whole droplet, using eq (4), is determined. If the temperature of the droplet becomes higher than the temperature required for the stability of some clusters, these clusters are then re-melted. Thus a distinction exists between the number of determined clusters and the number of nuclei (clusters which do not re-melt).

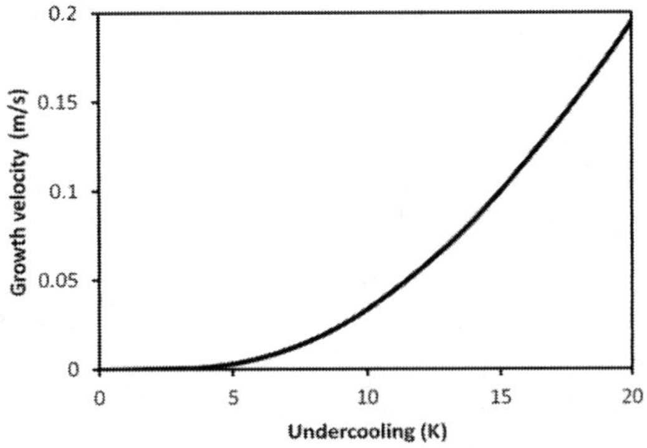

Figure 1: Evolution of the growth velocity for Al-4.5wt%Cu with the primary undercooling.

3. Results and discussions

Al-4.5wt Cu droplets of different sizes were obtained by IA under Ar and He. This work focuses on a droplet with average size (diameter) of 196 μm atomized in He (data used for the model are summarize in Table 2).

Table 2 – Properties of the Al-4.5wt%Cu alloy under investigation

gas	C_p^l (J.m^{-3}K^{-1})	ΔH_f (J.m^{-3})	T_m (K)	θ (°)	Diameter (μm)	γ (J/m^2)	Q (K/s)	Growth velocity (m/s)
He	3.25×10^6	1.07×10^9*	920	23	196	0.093*	5650	0.044

* A. Roosz, E.Halder and H. E. Exner, Mater. Sci. Tech., 1986, vol. 2, pp. 1149-1155

Using the model described in the previous section, the primary dendritic nucleation undercooling is predicted to be 11K for the investigated droplet, a value that is in good agreement with the experimentally obtained value of 10.9K±3(error inherent to experimental measurements). Indeed, experimentally the primary dendritic nucleation undercooling is obtained by combining semi-empirical coarsening models of secondary dendrite arms spacing with the eutectic nucleation undercooling estimated through the measurement of eutectic fraction within the investigated droplet microstructure by Rietveld refinement analysis of neutrons diffraction data. The eutectic fraction is estimated by extending the solidus and liquidus lines of the Al-rich corner of Al-Cu phase diagram using Thermo-Calc. A detailed description of this analysis is given in [1, 2]. However, it is worth

noting that the time/temperature variation of the coarsening parameter was not taken into account in [1, 2] whereas this variation is considered in the present case.
Figure 2 shows the number of clusters distribution (modelling was performed for 30 different seeds). The number of clusters introduced in the 10×10×10 μm^3 box varies between two and ten (depending on the seed number of the random number generator) with an average of six clusters.

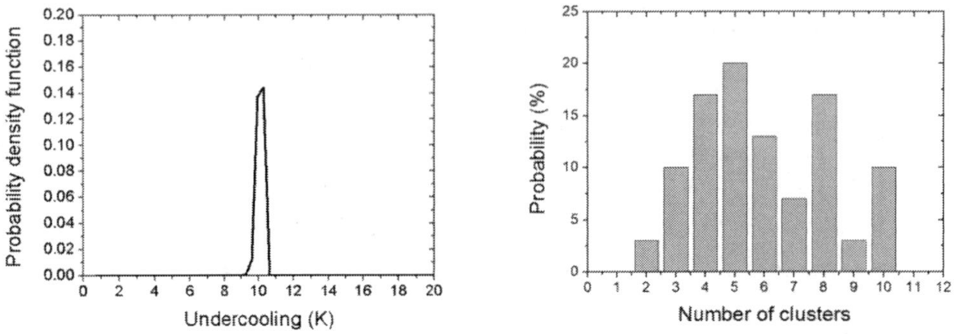

Figure 2: Probability density function for the considered droplet (a) and distribution of the number of clusters (b) (performed for 30 seeds of the random number generator).

When the first step of growth occurred, the solid fraction formed during the growth of the first cluster leads to a sufficient increase in the temperature of the whole droplet (recalescence) that the other clusters become unstable. As the temperature is calculated for the entire droplet and not only for the 10×10×10 μm^3 in size box, this means that the other clusters being introduced in the droplet will re-melt. This results in the presence of only one nucleus. The model was run for a droplet with average of 925 μm as well as for a droplet atomized in N_2 and the same trend was observed.

Synchrotron X-ray micro-tomography has been performed on several (99) Al-4.5wt%Cu alloy droplets with sizes ranging from 100 to 355 μm atomized under He [20]. The image acquisition and the 3D reconstructions were carried-out on the ID19 beamline of the ESRF (European Synchrotron Radiation Facility). A series of X-ray images with contrast due to the density and composition difference between the primary α-Al phase which appears in dark grey and the eutectic (α-Al+θ-Al_2Cu) which appears in light grey and the porosity appearing as black areas within the droplet. These images obtained at different view angles were then reconstructed into 3D droplet microstructures. Such 3D reconstructions of the inner structure of the droplets enabled us to locate the nucleation position by determining the intersection of the primary arms of the dendritic structure. A typical image obtained with synchrotron X-ray micro-tomography is given in Figure 3. The slice in the droplet is chosen in this figure in such a way that the nucleation centre can be clearly visualized. This work highlighted the presence of only one nucleus in 95 droplets out of the 99 droplets visualized. In the 4 other droplets, only 2 nuclei were observed. These experimental observations are in agreement with our model prediction.

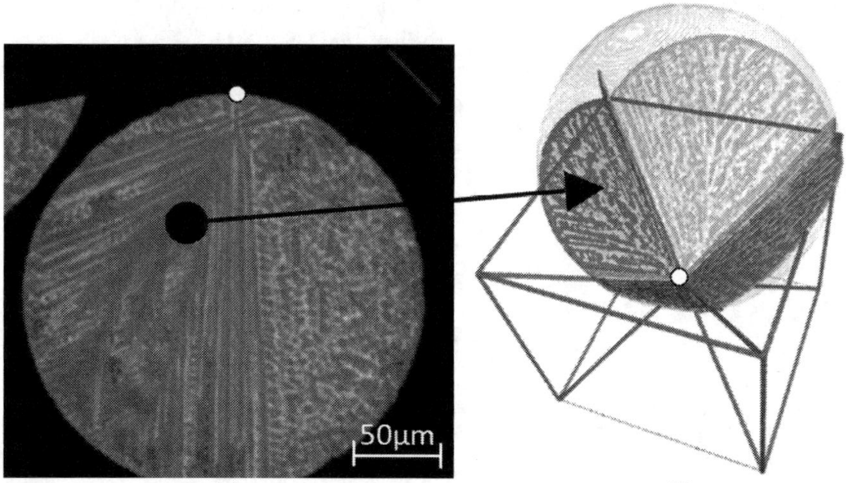

(a) (b)

Figure 3: Synchrotron X-ray micro-tomography result showing (a) a droplet cross-section and (b) the corresponding 3D-reconstruction with characteristic planes and the position of the nucleation centre (white dot) in a 290 μm diameter Al-4.5wt%Cu droplet atomized in helium.

4. Conclusions

Primary dendritic nucleation undercooling of an IA Al-4.5wt%Cu droplet is investigated. A numerical prediction of the primary dendritic nucleation undercooling was achieved based on a thermal model coupled with a heterogeneous nucleation/growth model of a droplet falling through a stagnant gas. The model prediction is compared with experimental results of primary dendritic nucleation undercooling estimated using a metastable extension of solidus and liquidus lines of the phase diagram by Thermo-Calc and a semi-empirical coarsening model of secondary dendrite arms spacing. The comparison yields a good agreement between the experimental result and the model prediction of primary dendritic nucleation undercooling of the investigated droplet. Furthermore, the number of nuclei in the droplet was numerically analysed. It appears that only one nucleus is expected to form, a results that is confirmed by the evidence of a single nucleation point on 3D images of IA droplets obtained by synchrotron X-ray micro-tomography analysis

5. Acknowledgements

The authors are grateful to the Natural Sciences and Engineering Research Council of Canada (NSERC), the Canadian Space Agency (CSA), the European Space Agency (ESA) and the Agency National de Recherche (ANR-France) for their financial support. Dr. Jonas Valloton is also acknowledged for fruitful discussions.

References

[1] Herlach D M Galenko P and Holland-Moritz D 2007 *Metastable Solids from Undercooled Melts* (Oxford Pergamon Materials Series Elsevier)
[2] Bogno A, Delshad Khatibi P, Henein H and Gandin Ch-A 2014 *Under review in Met. Trans.*
[3] Bogno A, Khatibi P D, Henein H and Gandin Ch-A 2013 *Light Metals for Transportation*, (MS&T 2013)
[4] Ranz W E and Marshall W R Jr 1952 *Chem. Eng. Progress* **48**(3) 141-146
[5] Whitaker S 1972 *AIChE Journal* **18**(2) 361-371
[6] White F M 1974 Viscous Fluid Flow *McGraw-Hill Book Co., New York, NY* 213
[7] Vilet G C and Leppert G 1961 *Journal of Heat Transfer* **5** 163-175

[8] Jackson J D and Hatchman J C 1999 *Proc. Instn. Mech. Engrs.* **213A** 45-56
[9] Sayegh N N and Gauvin W H 1979 *AIChE Journal* **25**(3) 522-534
[10] Kramer H 1946 *Physica* XII **2-3** 61-80
[11] Wiskel J B, Henein H and Maire E 2002 *Can. Metall. Quart* **41**(1) 97-110
[12] Yule A J and Dunkley J J 1994 *Atomization of Melts: For Powder Production and Spray Deposition* (Claredon Press, Oxford, England) 28432
[13] Welty J R, Wicks C E and Wilson R E 1984 *Fundamentals of Momentum, Heat and mass Transfer* (3rd ed) (New York: John Wiley and Sons) pp 759
[14] Weast 1980 *CRC Handbook of Chemistry and Physics* (61th ed) (CRC Press inc. Boca Raton) pp D62, E2-E3
[15] Rabinovich V A Nedostup V I, Vasserman A A and Veksler L S 1988 *Thermophysical Properties of Neon, Argon, Krypton and Xenon* (10th ed) (USSR) pp 384-385, 417-418, 423-424
[16] Yuan D 1998 *Ph.D. Thesis, University of Alberta*
[17] Wiskel J B and Henein H 1999 *Proceedings of Fluid Flow Phenomena in Metals Processing* eds. El-Kaddah N, Robertson D G C, Johansen S T and Voller V R, TMS, (Warrendale) 517-524
[18] Kreith F 1973 *Principles of Heat Transfer* (3rd ed) (New York: Harper and Row) pp 236, 316-317, 339
[19] Galenko P K and Danilov D A 1997 *Physics Letters* A **235** 271-280
[20] Bedel M Reinhart G Bogno A Nguyen-Thi H Boller E Gandin Ch-A and Henein H 2014 being published *IOP Conf. Ser. Mater. Sci. Eng.*

Real-time synchrotron x-ray observations of equiaxed solidification of aluminium alloys and implications for modelling

A Prasad[1], E Liotti[2], S D McDonald[1,3], K Nogita[1,3], H Yasuda[4], P S Grant[2] and D H StJohn[1]

[1] Centre for Advanced Materials Processing and Manufacturing (AMPAM), School of Mechanical and Mining Engineering, The University of Queensland, Brisbane, QLD 4072, Australia

[2] Department of Materials, University of Oxford, Oxford, OX1 3PH, UK

[3] Nihon Superior Centre for the Manufacture of Electronic Materials (NS CMEM), School of Mechanical and Mining Engineering, The University of Queensland, Brisbane, QLD 4072, Australia

[4] Department of Materials Science and Engineering, Kyoto University, Kyoto 606-8501, Japan

Corresponding author's e-mail address: d.stjohn@uq.edu.au

Abstract. Recently, in-situ observations were carried out by synchrotron X-ray radiography to observe the nucleation and growth in Al alloys during solidification. The nucleation and grain formation of a range of Al-Si and Al-Cu binary alloys were studied. When grain refiner was added to the alloys, the location of the nucleation events was readily observed. Once nucleation began it continued to occur in a wave of events with the movement of the temperature gradient across the field of view due to cooling. Other features observed were the settling of the primary phase grains in the Al-Si alloys and floating in the Al-Cu alloys, the effects of convection with marked fluctuation of the growth rate of the solid-liquid interface in the Al-Si alloys, and an absence of fragmentation. The microstructures are typical of those produced in the equiaxed zone of actual castings. These observations are compared with predictions arising from the Interdependence model. The results from this comparison have implications for further refinement of the model and simulation and modelling approaches in general. These implications will be discussed.

1. Introduction

The results of two studies of equiaxed solidification observed by synchrotron real-time X-ray experiments are presented. One study was undertaken on a range of Al-Si alloys at the Spring-8 (Japan) synchrotron and the other on Al-Cu alloys at the Diamond Light Source (UK) synchrotron. Although experiments were undertaken on both unrefined and refined alloys, this paper only focuses on alloys that have been refined by an AlTiB master alloy. The purpose of these studies was to observe the nucleation events and the grain formation processes occurring during the formation of the equiaxed zone in a casting environment similar to that of gravity and low pressure die casting methods. As expected the solidification processes at this early stage of solidification are complex and dynamic. However, a number of features are clearly observed that provide a picture of the key factors affecting the final as-cast grain size.

The following sections describe the main observations and present images taken from the X-ray videos to illustrate the important features occurring during nucleation and grain formation including the effect of gravity and convection on grain growth and movement. The final section discusses the implications of these observations on the development of models for the prediction of as-cast grain size and the simulation of the nucleation stage and subsequent development of the equiaxed grains.

2. Experimental Methods

2.1. Real-time x-ray of Al-Si alloys

The Spring-8 experiments consisted of performing real time in-situ solidification experiments on Al - 1, 4, 7 and 9 wt%Si refined with 0.1 wt%Ti from the Al3Ti1B master alloy. Samples of a given alloy sandwiched between Al_2O_3 plates 100 μm apart were inserted into a furnace and underwent a heating and cooling cycle designed to capture melting and solidification of the sample while being exposed to the X-ray beam. The experiments were performed in low-resolution as well as high-resolution beam lines. Detailed procedures for selecting radiation energy, image processing algorithm, experimental set-up etc., have been described previously in a number of articles [1-4]. A low thermal gradient was present, although not deliberately controlled in all experiments [2, 5]. The cooling rate was set to 20 K/min for all cases. Note that there is a lack of contrast between the liquid and solid phases due to there being little difference in X-ray absorption and, therefore, the X-ray energy was set as low as 16 keV to enhance the contrast between solid and liquid phases. The energy of 16keV enabled the detection of solidifying dendrites in the liquid phase by absorption contrast. As discussed in previous work [1], phase contrast also enhances the contrast at the solid - liquid interface. As a result the contrast is not as good as that usually observed in Al-Cu alloys and some of the images can be difficult to see clearly.

2.2. Real-time x-ray of Al-Cu alloys

In-situ X-ray radiography studies of the solidification of Al-15 wt%Cu refined by 0.1 wt%Ti from a Al5Ti1B master alloy were carried out at the B16 beamline at the Diamond Light Source using a specially designed solidification rig. The rig consisted of a small Bridgman furnace for directional solidification composed of two independently controlled plate heaters separated by a small gap. Thin foil samples 200 μm thick were encapsulated within 100 μm quartz coverslips and their solidification was imaged by recording the transmitted X-ray signal. Details about the furnace and the experimental set up can be found in [6]. Experiments were carried out with and without the presence of a controlled temperature gradient parallel to gravity. Initially the temperature of both heaters was increased above the melting point of the alloy until the sample was fully melted. When a thermal gradient was needed during solidification it was imposed by decreasing the temperature of one of the two heaters while remaining above the melting point. When thermal stability was reached the cooling of the sample was initiated by lowering the temperature of both heaters at a constant rate until the entire field of view was solid.

3. Nucleation

Figures 1 and 2 show the solidification sequence of an Al-7wt.%Si alloy and an Al-15wt.%Cu alloy, respectively. From the X-ray videos the progress of nucleation can be readily observed. Figures 1 and 2 show selected images to illustrate the features of nucleation observed in the videos. In Figure 1 the temperature gradient is positive from the bottom left to the top right of the field of view. New grains are progressively nucleated as the temperature gradient moves through the melt due to cooling. The arrows marked on Figure 1 highlight the position of the most recent wave of nucleation. In Figure 2 the temperature gradient is very low and close to isothermal in the field of view and nucleation progresses from the top of the field of view. The reason for this observation is that Cu-rich liquid segregates to the bottom due to having a higher density. The liquid of lower Cu content has a higher liquidus temperature and thus begins to solidify first. It can be clearly seen that nucleation occurs in waves of events similar to the Al-Si alloy. The wave-like nature is particularly obvious in the 12.5s image in Figure 2. What cannot be observed easily in Figure 2 but is clearly visible in the videos is that once nucleation of grains occurs they then begin to float upwards. The effect of gravity and density differences between phases is discussed in section 5.

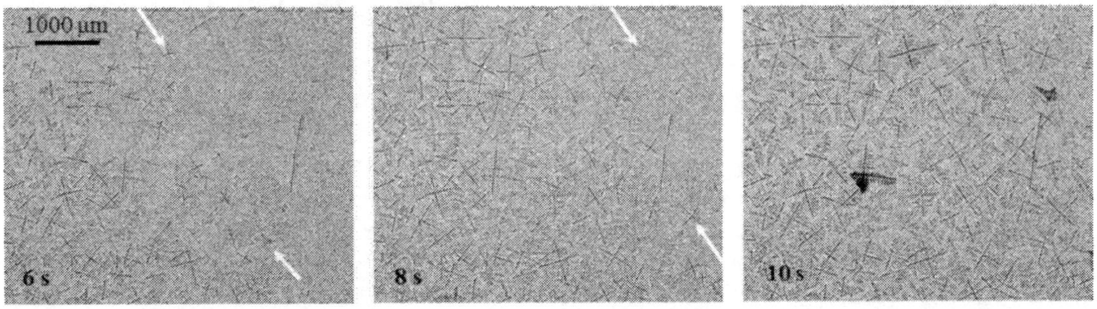

Figure 1. Images showing the time-sequence of wave-like grain nucleation in an Al-7wt%Si alloy with 0.1% Ti from an Al3Ti1B master alloy. The formation of grains starts at the bottom left hand corner and proceeds towards the top right hand corner of the field of view. The arrows show the demarcation between nucleated and non-nucleated areas. The time of 0 s is taken from the image prior to the image where the first grains are observed.

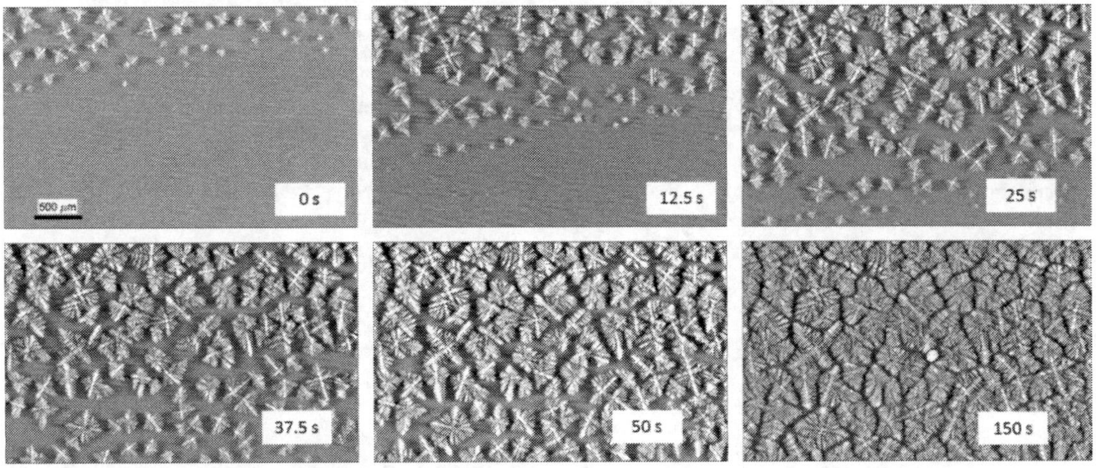

Figure 2. Image sequence of the solidification of an Al-15wt%Cu containing Al5Ti1B master alloy in close to isothermal conditions (G < 1.5 K/mm) with a cooling rate of 20 K/min. Grains are seen appearing in a wave from the top of the sample towards the bottom. Nucleation began at the top of the field of view due to solute (Cu) segregation in the bottom of the sample. The final frame (t = 150 s) shows the final microstructure of the fully solid sample.

4. Grain Formation

Once nucleation occurs the grains grow with a dendritic morphology. The growth of the Al-Si primary dendrite arms was measured for the Al-Si alloys and it was found that the growth rate fluctuates with a periodicity between 0.2 and 0.4 s where the growth rate changes from a low rate to several 100 µm/s and then back again as apparent in Figure 3. Fluctuations in the rate of growth have previously been observed by Mathiesen et. al. in an Al-15wt.%Cu alloy [7].

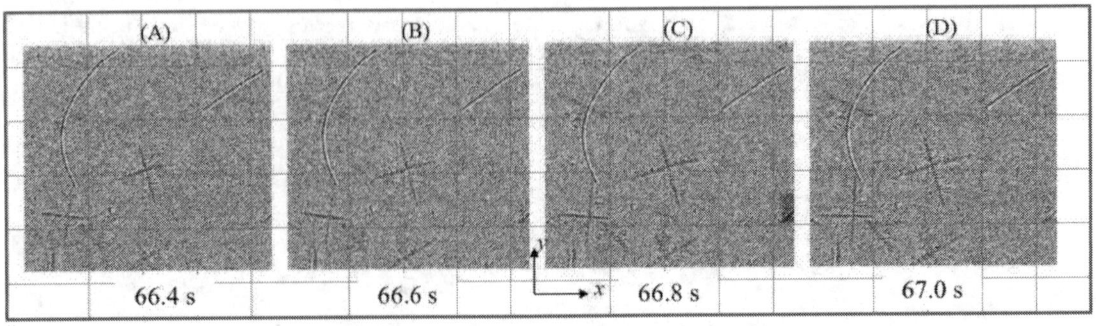

Figure 3. The fluctuation of the growth rate of dendrite tips in the Ti refined Al-7wt%Si alloy. A grid has been superimposed to help view the relative dendrite tip movements. The images are taken at 0.2 s intervals beginning at 66.4 s after the start of recording. In the centre grain, along the +x direction, there is negligible tip growth between (A) and (B), which accelerates in (C), and continues to grow in (D). Along the +y direction there seems to be no growth between (A)-(C), but rapid growth in (D). In the –y direction there appears to be some growth between (A)-(C), which seems to decelerate in (D). These differences in the amount of growth manifest as tip growth rate fluctuations.

Images from the synchrotron video were analyzed for the rate of increase of the number of grains. Individual grains were manually counted on each frame as a function of time for refined alloys of all compositions (Al-1, 4, 7 and 9 wt%Si). The grain count is presented as a function of time in Figure 4(a). The maximum number of grains was observed for 1 wt%Si, steadily decreasing with the increasing Si composition. The maximum grain count was converted to grain number density by dividing the grain count by the area of the field of view. As expected, the grain density decreases with increasing composition. There is more than an order of magnitude difference in grain number density between the 1 and 9wt%Si alloys. This observation is important when analyzing the effect of solute concentration in grain refined alloys. Clearly, in Al-Si alloys, there is a marked difference in inoculant efficiency with change in Si concentration. This information is often obscured in grain size data (see figure 5) which is the more widely used parameter.

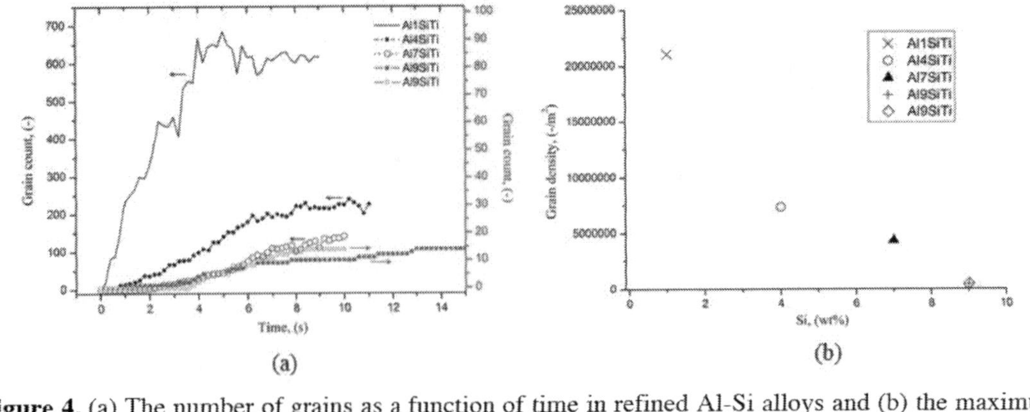

Figure 4. (a) The number of grains as a function of time in refined Al-Si alloys and (b) the maximum grain count from (a) converted to grain number density per unit area (m^2).

The grain size of the Al-Si alloys is plotted as a function of Si concentration in Figure 5. For the synchrotron experiments the maximum grain number (Figure 4) for a given alloy composition was converted into grain size by assuming a perfect spherical shape for the grains occupying the sample volume. The data for refined and unrefined alloys show a clear difference with data for the unrefined alloys showing larger grain sizes as expected. The data obtained from the synchrotron experiments fit well with those from the casting experiments of previous researchers. Irrespective of the presence of refining agents, the data shows an increase in grain size beyond ~3 wt%Si. Below this concentration, unrefined alloys show a decrease in grain size as the Si composition increases. The refined alloys tend to show the same grain size for Si compositions lower than 3%. The higher growth restriction factor with increasing solute content is expected to result in smaller grain sizes at higher compositions. However, Al-Si alloys show an opposite trend and the increase in grain size with Si content is known as Silicon poisoning [8].

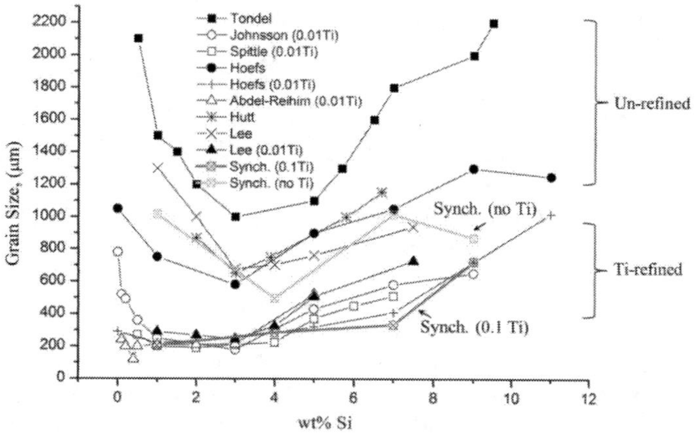

Figure 5. Grain size as a function of Si composition from casting and synchrotron experiments for both unrefined and refined alloys.

5. Effect of Gravity and Convection

It was observed that the relative movement of grains in Al-Cu alloys compared to Al-Si alloys was reversed. In the Al-Cu alloys the upward movement of grains is illustrated in Figure 6 over a time interval of 3 seconds. This upward drift of grains, also observed in other research on similar Al-Cu compositions [9], can have a significant effect with the newly nucleated grains moving from their nucleation sites up into the semisolid region assisting the packing of grains.

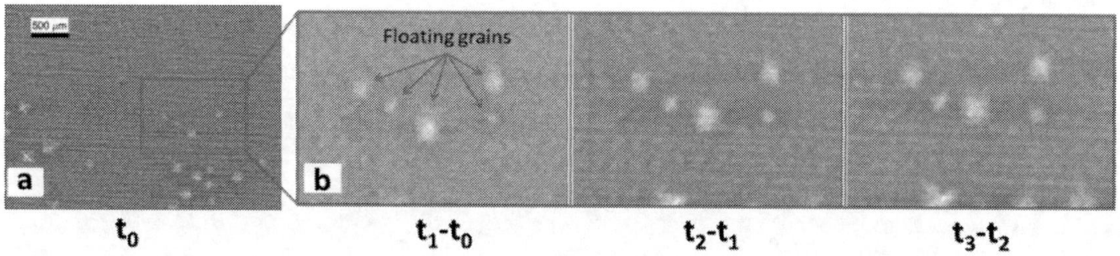

Figure 6. (a) The location of the grains at time $t_0 = 0$ s, (b) differential images between four frames taken at 1 s intervals ($t_1 = 1$ s, $t_2 = 2$ s, $t_3 = 3$ s) showing the floating of the four dendrites in the center of the field of view (highlighted in magenta).

On the other hand, the primary Al dendritic grains in the Al-Si alloys have a tendency to sink. This is more noticeable when the grains are small. As both the Al-Si and Al-Cu primary phases grow they begin to interact with each other preventing further sinking or floating. Figure 7 illustrates the degree of movement of the primary grains in the refined Al-4 wt%Si alloy over a period of 5 seconds. Note that some grains do not move over this time period (identified by dashed lines) while others sink (solid lines). This maybe because these grains remain stuck between the Al_2O_3 plates as the grains are already the same size as the thickness of the sample.

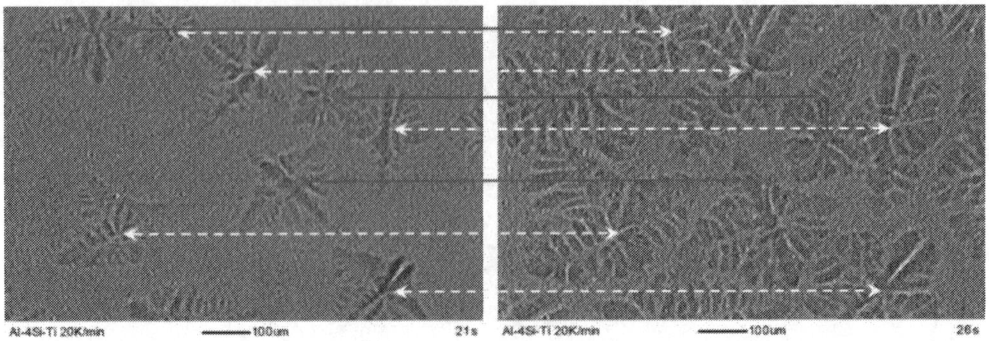

Figure 7. Primary Al grains in a refined Al-4 wt%Si alloy growing and sinking simultaneously with time. The change in the location of the center of the grain is marked by the ends of the red solid lines, with the movement shown as a downward arrow. Not all grains move as indicated by the dotted white lines. Images are from the high-resolution synchrotron experiments.

The relative difference in grain motion between the Al-Si and Al-Cu alloys can be understood with reference to Figure 8. This figure shows the density difference between the solid and liquid phases at the point of initial nucleation (fraction solid approaching zero) for a range of hypoeutectic Al-Cu and Al-Si compositions. Data for the Al-Cu alloys are taken directly from [10] while those for the Al-Si alloys come from the equations provided in [11] with the required temperatures taken from the equilibrium liquidus temperature for that composition. For all hypoeutectic Al-Si compositions the density of the newly nucleated solid grains is greater than that of the liquid. For the Al-Cu alloys a similar condition holds when the Cu concentration is less than around 10.5 wt%. However, as the concentration of Cu exceeds this level (as in the experimental alloys of this paper) the density of the newly nucleated grains is less than that of the liquid. As discussed in [9] the velocity of the grains will depend on the relative contribution of the buoyancy forces and the drag due to interaction with the sample walls, which is a function of viscosity, grain size and mould dimensions. The effect of the grain motion on microstructure development will also depend on the direction of travel relative to the thermal gradient.

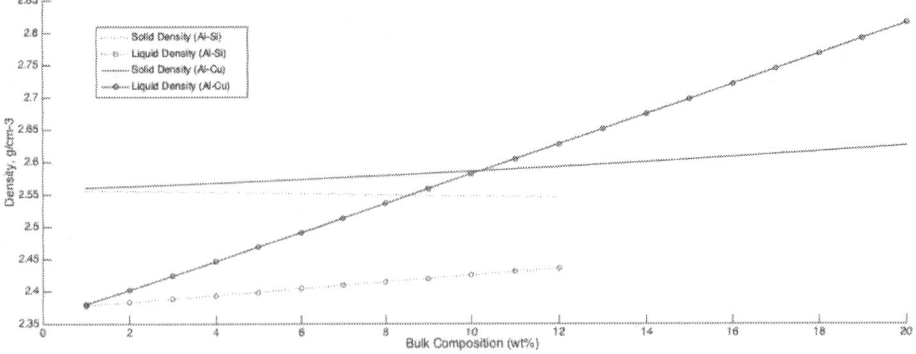

Figure 8. The density of solid and liquid phases early in solidification for a range of Al-Si and Al-Cu alloys. Data from [10, 11].

Of interest for the study of casting quality is the ability of real-time synchrotron studies to observe how the growing grains pack together to form the final as-cast macrostructure. In Figure 2 and more readily observed in the video, the grains appear to rotate to fit into the spaces formed by the previous layer of grains and sometimes vertical channels of liquid between the grains seem to form. The channels most probably form as the intergranular liquid is displaced by the rising lighter Al grains. This movement of the liquid would in turn assist the rotation of the grains. This process may affect the formation of casting defects such as hot tearing and porosity formation.

6. Other observations

In addition to the wave-like nature of nucleation shown in Figures 1 and 2 it was noted that once nucleation occurs no further nucleation is observed to occur between adjacent grains. This process of nucleation supports the predictions of the Interdependence model which was further validated by numerical modelling [12]. It can also be observed that once the grains nucleate they do not remelt even when they move considerable distances from the point of nucleation particularly in the Al-Cu alloys. This survival is due to the low applied temperature gradients allowing the formation of constitutional supercooling throughout much of the liquid which protects the grains from remelting. A significant observation from viewing the videos is that there is no evidence of fragmentation in either of the alloy systems studied; implying fragmentation may not be favoured during equiaxed solidification compared with directional solidification where much evidence of fragmentation has been reported in the literature.

7. Implications for Modelling and Simulation

The modelling of grain nucleation is difficult and many models simply set a matrix of nucleant locations which can trigger nucleation when the liquidus is reached during cooling. However, the nucleant particles, in this case TiB_2 particles, usually have a broad range of particle sizes and thus potencies that is difficult to model [13]. Because the observations of this research indicate that nucleation occurs on the most potent particles, most of the smaller particles can be ignored as they are unlikely to ever nucleate a grain [13]. The observation of fluctuating growth rates in the Al-Si alloys show just how complex the casting environment is. To be able to predict fluctuations thermal and solutal convection needs to be modeled. Recent modelling research is attempting to deal with the effects of convection during solidification [14, 15]. The thermo-solutal convection may also affect the local undercooling ahead of the dendrite tips which in turn may affect the size of the nucleation free zone [2, 13] and the subsequent nucleation rates. Add the effect of relative changes in density within the semisolid melt and the physics becomes complicated. Thus, analytical models would be inadequate for the accurate prediction of as-cast grain size and its variation throughout the macrostructure compared to sophisticated numerical models such as reported in [16]. However, the Interdependence model which is an analytical model, is remarkably effective as an indicator of relative, if not actual, grains size [13].

8. Concluding remarks

As shown in this work the grain size outcomes of these real-time synchrotron studies correspond to that of actual castings. Therefore, from observations of the solidification of the alloys studied, information on nucleation and grain formation provide useful data that improves our understanding of the solidification mechanisms in actual commercial castings. Not only are we now able to quantify the rate and progress of nucleation events but also gain information on the role of density differences and convection in affecting the final as-cast grain structure. These studies allow us to test the established theories of nucleation and grain growth and also analytical and numerical models with regard to their validity and weaknesses providing avenues for their improvement. The complexity of the modelling task in providing accurate simulations has been highlighted by the observed interaction between alloy chemistry and the changing thermal, relative phase density and fluid states in facilitating solidification and the arrangement of grains within the final as-cast macrostructure. Further real-time synchrotron studies will be undertaken to gain quantitative information to improve our knowledge and validate the analytical and numerical models used by the authors.

Acknowledgements

This work is supported in part by Australian Research Council (ARC) Discovery projects (DP120101672 and DP140100702), and a Grant-in-Aid for Scientific Research (S) (24226018) from JSPS, Japan. Synchrotron X-ray radiography experiments were performed at SPring-8 synchrotron (2012B1430, 2014A1540, 2014A1541), funded by the Australian Synchrotron international synchrotron access program (AS/IA123/5664, AS/IA142/8198) and at ESRF (France) and the Diamond Light Source (UK).

References

[1] Nogita K, Yasuda H, Prasad A, MacDonald SD, Nagira T, Nakatsuka N, Uesugi K, StJohn DH. 2013 *Materials Characterization* 134.
[2] Prasad A, McDonald SD, Yasuda H, Nogita K, StJohn DH. *Acta Materialia*;Submitted.
[3] Terada Y, Yumoto H, Takeuchi A, Suzuki Y, Yamauchi K, Uruga T. 2010 *Nuclear Instruments and Methods* **616** 270.
[4] Yasuda H, Nagira T, Yoshiya M, Nakatsuka N, Sugiyama A, Uesugi K, Umetani K. 2011 *ISIJ International* **51** 402.
[5] Zeng G, MacDonald SD, Gourlay CM, Uesegi K, Terada Y, Yasuda H, Nogita K. 2013 *Metallurgical and Materials Transactions A*.
[6] Liotti E, Lui A, Vincent R, Kumar S, Guo Z, Connolley T, Dolbnya IP, Hart M, Arnberg L, Mathiesen RH, Grant PS. 2014 *Acta Materialia* **70** 228.
[7] Mirihanage WU, Arnberg L, Mathiesen RH. In-situ observation of transient columnar dendrite growth in the presence of themo-solutal convection. In: Ludwig A, Wu M, Kharicha A, 2012 editors. IOP Conference Series: *Materials Science and Engineering*, **33**. Schladming, Austria: IOP.
[8] Easton M, Prasad A, StJohn D. 2014 *Materials Science Forum* **794-796** 161.
[9] Delaleau P, Mathiesen RH, Schaffer PL, Arnberg L, Beckermann C. 2009 Modeling and In-Situ X-Ray Video Microscopy of Confined Equiaxed Grain Growth and Buoyant Motion in Al-Cu, .
[10] Ganesan S, Poirier DR. 1987 *Metallurgical and Materials Transactions A* **18** 721.
[11] Magnusson T, Arnberg L. 2001 *Metallurgical and Materials Transactions A* **32** 2605.
[12] Prasad A, Yuan L, Lee PD, StJohn DH. 2013 *Acta Materialia* **61** 5914.
[13] D.H.StJohn, M.Qian, M.A.Easton, P.Cao. 2011 *Acta Materialia* **59** 4907.
[14] Du L, Zhong R. 2014 *Metallurgical and Materials Transactions B* **45** 2504.
[15] Yuan L, Lee PD. 2010 *Modelling and simulation in Materials Science and Engineering* **18** 13.
[16] Acharya R, Bansal R, Gambone JJ, Das S. 2014 *Metallurgical and Materials Transactions B* **45** 2247.

Prediction of as-cast grain size of inoculated aluminum alloys melt solidified under non-isothermal conditions

Qiang Du[1], **Yanjun Li**[2]
1 SINTEF Materials and Chemistry, Trondheim, Norway
2 Department of Materials Science and Engineering, NTNU, 7491 Trondheim, Norway
E-mail: qiang.du@sintef.no

Abstract. In this paper, a multi-scale as-cast grain size prediction model is proposed to predict as-cast grain size of inoculated aluminum alloys melt solidified under non-isothermal condition, i.e., the existence of temperature gradient. Given melt composition, inoculation and heat extraction boundary conditions, the model is able to predict maximum nucleation undercooling, cooling curve, primary phase solidification path and final as-cast grain size of binary alloys. The proposed model has been applied to two Al-Mg alloys, and comparison with laboratory and industrial solidification experimental results have been carried out. The preliminary conclusion is that the proposed model is a promising suitable microscopic model used within the multi-scale casting simulation modelling framework.

1. Introduction

Prediction of as-cast grain size of inoculated aluminum alloys have been an active research topic in the past 15 years since the development of the "Free Growth" nucleation model by Greer et al [1-7]. Most of the recent numerical grain size prediction models have followed the approach pioneered by Maxwell and Hellawell (hereafter referred to as MHN model[2]) [8], in which the continuous solid grain size distribution is described by discrete size classes. The influences of solute content, the type of grain refiners, inoculant particle size distribution and cooling rate have been taken into account. While it is clear that with given alloy composition, cooling and inoculation conditions quantitative predictions of the final as-cast grain size could be made even for multicomponent alloys, a gap still exists to apply the MHN-type model to predict grain structure at the scale of a typical casting ingot.

The prediction of grain structure at the scale of casting component would require a multi-scale model to address both of macroscopic heat and solute transportation, melt flow, solid grain floating and microscopic grain nucleation and growth. Cellular Automaton Finite Element (CAFE) method, originally proposed by Gandin and Rappaz in [9] with the most recent development reported in [10], is a popular approach for this purpose. Indeed CAFE had been applied by Greer and his co-workers [11, 12] to predict as-cast grain size of inoculated aluminium alloys. Although these application successfully illustrated the trends in the observed variation of grain diameter with refiner addition level and cooling conditions [11, 12], the limitations of CAFE modelling in predicting the

[1] To whom any correspondence should be addressed.
[2] It should be noted that the size class subdivision concept in the MHN model is identical to the Kampmann-Wagner Numerical (KWN) model used in precipitation kinetics modelling.

constitutional undercooling was also revealed. As stated by Quested in [12], the CAFE model's limitations originate from its assumption that dendritic growth occurs under steady state and dendrite growth rates are equal to those in isolated dendritic arrays. This assumption means that the interactions among the growing grains (i.e., soft impingement) have not been taken into account. To overcome the limitations, it requires the determinations of both solute and temperature fields and their interdependence in the melt. It should be noted that the temperature field vary at the scale of the casting component (~0.1-1 m) while the solute at the scale of grain size (~10-100 μm) due to the huge disparity between thermal and solutal diffusion coefficients.

In this work, we are aiming to develop such a multi-scale model by integrating a MHN-type microscopic model (solute field) into macroscopic heat diffusion equation (temperature field). To do so the most recent CALPHAD-coupled multi-component as-cast grain size prediction model [1] is extended to cover the full primary phase solidification range (instead of the initial solidification stage). Then this microscopic model is applied to compute the release of the latent heat in all the finite volume elements used in solving the macroscopic heat transfer equation. Meanwhile the numerical solution of the heat diffusion equation gives a thermal history of each volume element, which is a required input to the microscopic model to compute the volume element's solidification path. To achieve the cross-scale coupling numerically the splitting technique, borrowed from combustion modelling [13] and successfully applied for homogenization heat treatment modelling [14], is adopted.

The paper is organized as follows: Section 2 is devoted to model description and Section 3 to results and discussions. Conclusions are given in Section 4.

2. Model description

The proposed multi-scale grain size prediction model consists of a macroscopic heat transfer equation with a latent heat release source term and a microscopic model describing grain nucleation and growth.

$$\rho C_p \frac{\partial T}{\partial t} = \nabla \cdot (D \nabla T) + L \frac{df_s}{dt} \qquad \text{(Eq.1)}$$

T is temperature, L is the volumetric latent heat, f_s is solid volume fraction, ρ, C_p and D are the weighted density, specific heat capacity and diffusivity, which are expressed by

$$\begin{aligned} \rho &= f_l \rho^L + f_s \rho^S \\ C_p &= f_l C_p^L + f_s C_p^S \\ D_T &= f_l D_T^L + f_s D_T^S \end{aligned} \qquad \text{(Eq.2)}$$

ρ^L ρ^S are the density of liquid and solid phase. C_p^L C_p^S are the specific heat of liquid and solid phase. D_T^L D_T^S are the thermal conductivity of liquid and solid phase.

An explicit Finite Volume Method is employed to solve the heat transfer equation at the scale of the whole ingot. The calculation of the changing rate of solid fraction is done by the microscopic MHN-type model. This model is applied to each volume element, and able to track grain nucleation and growth. It is an extension of the one reported in [1] to the full solidification range. The next subsection is devoted to the microscopic model.

2.1. The microscopic as-cast grain size prediction model

The microscopic model adopted in this multi-scale framework is based on the as-cast grain size prediction model reported in [1]. The main assumptions adopted in the extended model include:

I. Solid grains are of spherical shape and their growth/dissolution is solely controlled by diffusion in the melt.
II. Local equilibrium modified by the Gibbs-Thomson effect prevails at the solid-liquid interface; thermal and kinetic undercoolings are neglected.
III. The diffusion field surrounding each growing solid grain is at quasi-steady state.

The model consists of a nucleation model and a growth model. Below a brief description of the nucleation model is given, which is followed by the growth model with a focus on their modification to cover the full solidification range.

Greer's free growth model [1] is adopted to describe nucleation on inoculant particles. According to this model, the number of growing grains is determined by a free-growth condition. For an undercooling of ΔT, the critical diameter, d_c, of refiner particle to initiate a free growth grain is given by:

$$d_c = \frac{4\sigma}{\Delta T \Delta S_V} \qquad \text{(Eq.3)}$$

where σ interfacial energy, ΔS_V entropy of fusion. All the particles with diameter above this critical diameter are active. One of the key input parameters to the free growth model is inoculation particles size distribution density function, which has been obtained by fitting to the experimentally measured distribution [15] as:

$$PSD(d) = N_V \frac{\exp(-\frac{d}{d_0})}{d_0} \qquad \text{(Eq.4)}$$

where d is the diameter of disc-shaped inoculant particle , d_0 is the characteristic width of the distribution ($d_0 = 0.72\ \mu m$) and N_v is the total number of TiB_2 particles per unit volume of aluminum melt when Al-5Ti-1B grain refiners are added[1]. N_v could be calculated according to the following empirical rule [15]: for 1 Part Per Thousand (PPT) by weight Al-5Ti-1B addition, N_v is equal to $1.15\times10^{12} m^{-3}$. Via Eq. (3) and Eq.(4) nucleation rate could be calculated for a given undercooling.

The growth model consists of four group of equations as described in [1]. The first set of equations is derived from solute conservation law accounting for the melt enrichment/depletion resulted from the growth of solid grains. The second equation is derived from heat balance calculating the local temperature change induced by heat extraction and latent heat release. The third and fourth sets of equations are applicable to each discrete size class; the former comes from the interfacial local equilibrium assumption (modified by the Gibbs-Thomson effect) and the latter are multi-component growth rate equations. The growth rate equations still follow the ones used in [1], i.e.:

$$\dot{R}_j = \frac{D_i \Omega_i}{R_j} \qquad \text{(Eq.5)}$$

D_i and Ω_i are the diffusivity and dimensionless super-saturation of alloying component i, respectively. However to cover the full solidification range, the concept in the standard Avrami analysis that the fraction of the real volume increment during each time increment is proportional to the volume fraction of untransformed is employed. Therefore

$$df_s^{Real} = df_s(1 - f_s) \qquad \text{(Eq. 6)}$$

The treatment of secondary phase formation is still in progress.

2.2. The direct coupling of the microscopic microstructure evolution with macroscopic heat transfer

While the FVM-based algorithm is used to solve the heat transfer equation (Eq.1) across the casting component scale, the microscopic model introduced in Section 2.1 is applied to each volume element of the 1D domain to capture the solidification kinetics resulting from the variation in local thermal history. These two models are tightly coupled and involve two length scales. This coupling is mathematically analogous to the one in homogenization heat treatment modeling reported in [16]. The splitting method adopted in [16] is employed in this paper to deal with the coupling. The interested readers are referred to Ref. [16] and the references therein for the complete details.

3. Results and discussions

In this section, the proposed model will be applied to both of laboratory scale and industrial casting experiments. The model predictions are going to be examined in the light of the experimental results. The model's input parameters are listed in Table 1, and they are taken from Ref. [12]. The simulation domain is in 1D with adiabatic heat boundary condition on one side. On the other side, the two-step heat-transfer coefficients identified in [12], i.e. 707 W m^{-2} K^{-1} for the first 25 seconds and 9000 W m^{-2} K^{-1} for the rest, is employed to mimic DC casting heat extraction condition. Grain refiner addition level, alloy composition and the simulated domain geometry are illustrated within in the text below.

Table 1: Thermo-physical properties and nucleation input parameters in the simulations

Property	**Value**		
Interfacial energy, σ	0.158 J/m^2		
Molar volume of liquid and solid phase	1.0×10^5 m^3		
Entropy of fusion, ΔS_V	1.11×10^6 J K^{-1} m^{-3}		
Latent heat per unit volume, ΔH_V	9.5×10^8 J m^{-3}		
Heat capacity of melt	2.58×10^6 J K^{-1} m^{-3}		
Heat capacity of solid	3.1×10^6 J K^{-1} m^{-3}		
Thermal conductivity of melt	95 W K^{-1} m^{-1}		
Thermal conductivity of solid	215 W K^{-1} m^{-1}		
Diffusivity of Mg	Diffusion constant (m^2/s)	activation energy (kJ/mole)	Evaluated at 650 °C
	9.90×10^{-5}	71.6	8.78×10^{-9}
Linear binary phase diagram data of Mg	Liquidus slope (K/at%)		Partition coefficient
	-6.2		0.3

3.1. The application of the model to DC casting simulator experiments

The DC casting simulator, described by Quested in [12], is able to reproduce the variation in solidification conditions that occur through the thickness of a full-scale DC ingot. The simulator is composed of a tube furnace containing a slightly tapering steel crucible (with internal diameter of 43.8 mm at top and 40.8 mm at the base) that fits closely to the surrounding insulation of the tube furnace. At the base of the crucible is a removable base plug, which sits above a water jet. During solidification, the top of the crucible is covered with an insulating blanket.

The proposed model has been applied to the casting simulator experiment of Al-3.5Mg-0.2Fe-0.05Si alloy. The 1D simulation domain is 21 cm in length consisting of 105 finite volumes. The multi-component alloy is approximated as binary Al-3.8at%Mg to decrease the computational time. The model's grain size predictions are plotted together with the measured and predicted (with CAFE model) ones by Quested [12] in Fig. 1. It is clear the proposed model has a better agreement with the measured grain size than the one obtained by CAEF model performed by Quested et al [12].

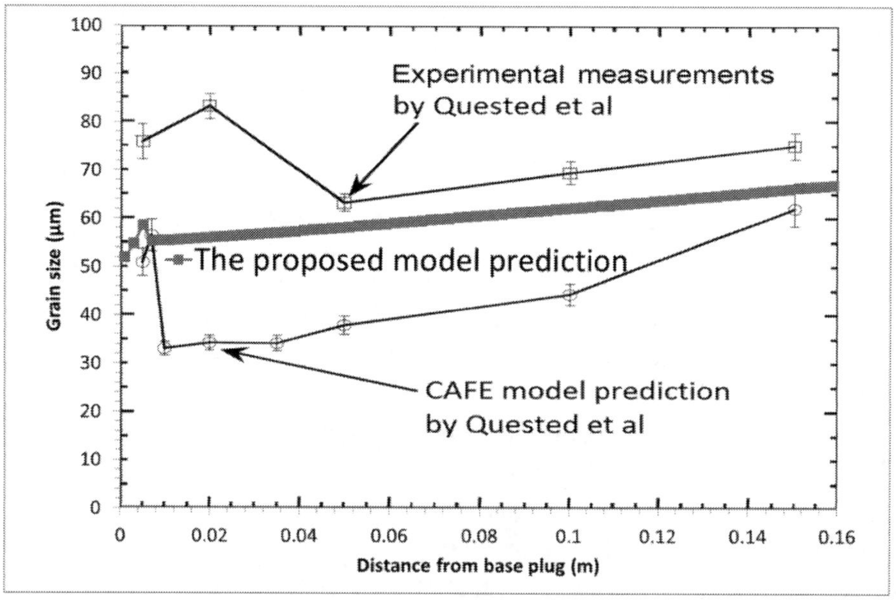

Fig. 1 The proposed model predictions for Al-3.5Mg-0.2Fe-0.05Si DC-simulator samples; the predicted and measured grain size by Quested are also reproduced for comparison.

3.2. The application of the model to industrial DC casting trial of an AA5182 alloy

In the industrial DC casting trial reported in [17], the melt, with the nominal composition of Al-0.36wt%Fe-4.12wt%Mg-0.22 wt%Mn-0.19wt%Si and grain refiner addition level of 4 PPT, was DC-cast into an ingot of dimensions of 510 mm in thickness and 1880 mm in width. The variation of grain size through the thickness of the ingot was also reported, which is reproduced in Fig. 2. The measurement shows initially grain size increases with increasing distance from the skin of the ingot. At about the distance of 150 mm to the ingot surface (or 105 mm to the ingot center) it reaches a maximum value of 207 μm. Then after it drops down to a local minimum value of 167 μm, it finally rises to 195 μm at the ingot center. It is interesting to note the largest grain sizes do not appear at the ingot center where the distance for heat to diffuse to the ingot skin (where heat is extracted) is the longest. It is at the location about 105 mm away from the center. This observation was believed to be due to "the lack of a heat reservoir in the melt" in [17].

The proposed model is able to shed some lights on the mechanisms behind this complex grain size variation with the distance. The 1D simulation domain is 25.4 cm in length consisting of 127 finite volumes. The multi-component alloy is approximated as a binary Al-5.8at%Mg to decrease the computational time. The predicted grain size variation is also plotted in Fig. 2. It is in consistence with the experimental measurement that the largest grains do not appear at the center but at a location about 40 mm away from the center. To explain this observation the predicted cooling curves during the whole nucleation stage for these two positions are plotted in Fig.3. It is surprising to see the

IOP Conf. Series: Materials Science and Engineering **84** (2015) 012015 doi:10.1088/1757-899X/84/1/012015

correlation between the predicted grain size and the average cooling rate at the center and the one 40 mm away from the center. Although the former has smaller grain size, its average cooling rate is lower by 0.07K/s. Even the average cooling rates do not correlate with the predicted grain size in the usual manner, the cooling rates at the latter nucleation stage do. Therefore maybe instant cooling rate should be used in grain size-cooling rate correlation relation. Indeed as shown in Fig. 4, at the same temperature, the one with the slower instant cooling rate has a lower grain number density due to the longer time to reach the same temperature, therefore the more solid grain growth solutally stifling more nucleation event.

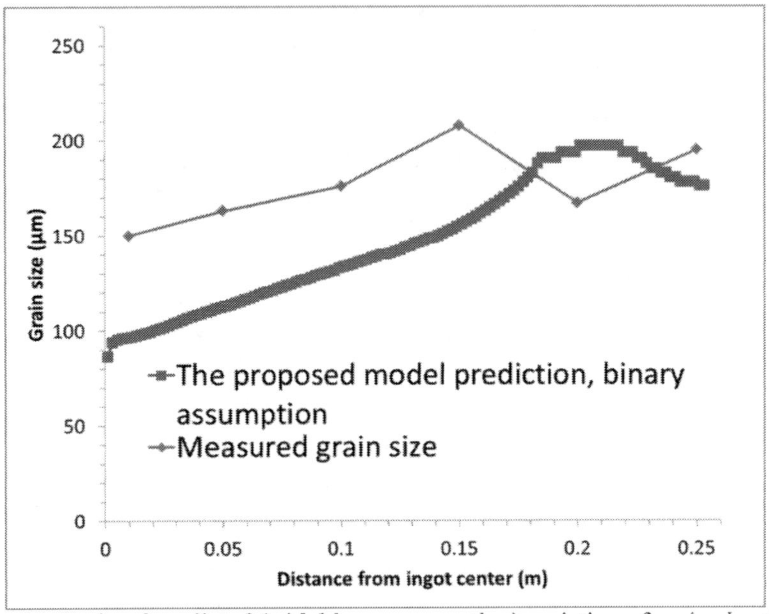

Fig. 2 The measured and predicted (with binary assumption) variation of grain size through the thickness of the ingot.

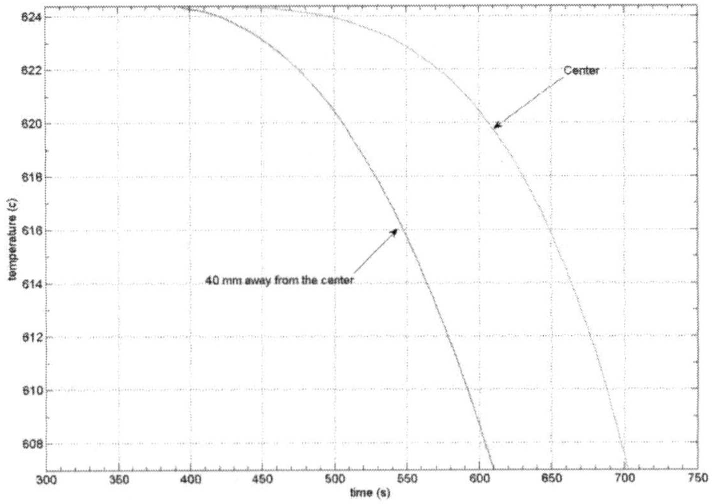

Fig. 3 The predicted cooling curves at two selected volume cells in the 1D simulation domain

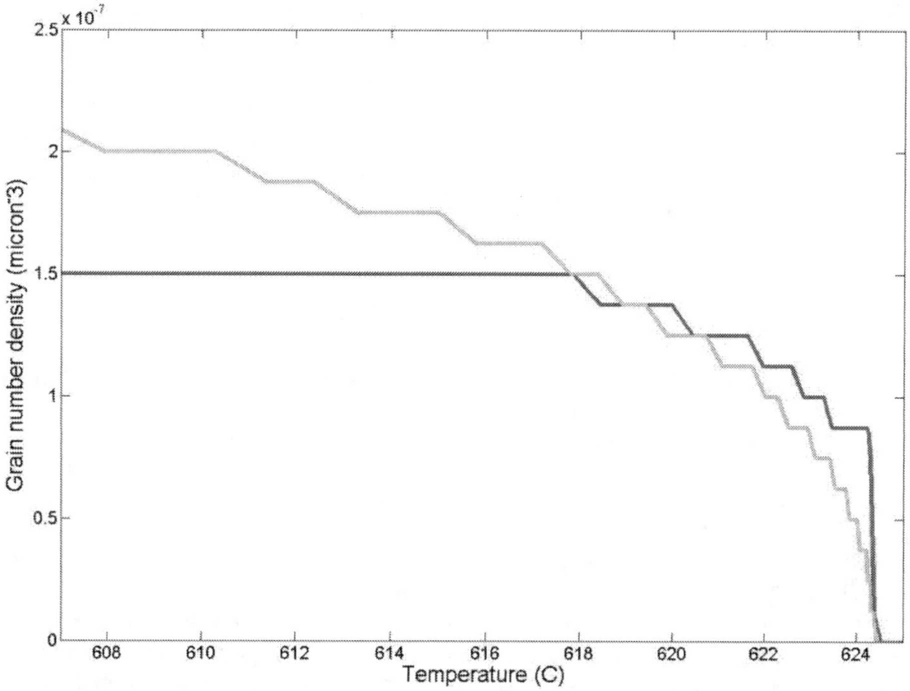

Fig. 4 The predicted grain number density evolution with temperature at two selected volume cells in the 1D simulation domain

The discrepancy between the experimental measurement and model predictions could be due to the settling of "travelling" crystals, which has not been considered in the performed simulations.

4. Conclusions

A multi-scale modeling framework has been established to predict grain structure at the scale of casting component. The model has been applied to a DC casting simulator experiment and industrial DC casting trials, and the model predictions have been compared with the reported experimental results. The preliminary conclusion is that the MHN-type model is a promising suitable microscopic model used within the multi-scale casting simulation modelling framework. Simulations without adopting the binary assumption are in progress.

5. References

[1] Du Q, Li Y. An extension of the Kampmann–Wagner numerical model towards as-cast grain size prediction of multicomponent aluminum alloys. 2014 *Acta Materialia* 71(0):380-9.
[2] Men H, Fan Z. Effects of solute content on grain refinement in an isothermal melt. 2011 *Acta Materialia* 59(7):2704-12.

[3] Quested TE, Greer AL. Grain refinement of Al alloys: Mechanisms determining as-cast grain size in directional solidification. 2005 *Acta Materialia* 53(17):4643-53.
[4] Greer AL, Bunn AM, Tronche A, Evans PV, Bristow DJ. Modelling of inoculation of metallic melts: Application to grain refinement of aluminium by Al-Ti-B. 2000 *Acta Materialia* 48(11):2823-35.
[5] StJohn DH, Qian M, Easton MA, Cao P. The Interdependence Theory: The relationship between grain formation and nucleant selection. 2011 *Acta Materialia* 59(12):4907-21.
[6] Shu D, Sun B, Mi J, Grant PS. A quantitative study of solute diffusion field effects on heterogeneous nucleation and the grain size of alloys. 2011 *Acta Materialia* 59(5):2135-44.
[7] Quested TE, Greer AL. The effect of the size distribution of inoculant particles on as-cast grain size in aluminium alloys. 2004 *Acta Materialia* 52(13):3859-68.
[8] Maxwell I, Hellawell A. A simple model for grain refinement during solidification. 1975 *Acta Metallurgica* 23(2):229-37.
[9] Gandin CA, Rappaz M. A 3D Cellular Automaton algorithm for the prediction of dendritic grain growth. 1997 *Acta Materialia* 45(5):2187-95.
[10] Carozzani T, Gandin C-A, Digonnet H, Bellet M, Zaidat K, Fautrelle Y. Direct Simulation of a Solidification Benchmark Experiment. 2013 *Metall and Mat Trans A* 44(2):873-87.
[11] Vandyoussefi M, Greer AL. Application of cellular automaton-finite element model to the grain refinement of directionally solidified Al-4.15 wt% Mg alloys. 2002 *Acta Materialia* 50(7):1693-705.
[12] Quested T. Solidification of Inoculated Aluminium Alloys. Cambridge University of Cambridge; 2004.
[13] Pope SB. Computationally efficient implementation of combustion chemistry using in situ adaptive tabulation. 1997 *Combustion Theory and Modelling* 1(1):41-63.
[14] Du Q, Poole WJ, Wells MA, Parson NC. Microstructure evolution during homogenization of Al–Mn–Fe–Si alloys: Modeling and experimental results. 2013 *Acta Materialia* 61(13):4961-73.
[15] Greer AL, Quested TE, Spalding JE. Modelling of grain refinement in directional solidification. In: Schneider WA, editor. Light Metals 20022002. p. 687-94.
[16] Du Q, Poole WJ, Wells MA, Parson NC. Microstructure evolution during homogenization of Al-Mn-Fe-Si alloys: Modeling and experimental results. 2013 *Acta Materialia* 61(13):4961-73.
[17] Li YJ, Arnberg L. Solidification structures and phase selection of iron-bearing eutectic particles in a DC-cast AA5182 alloy. 2004 *Acta Materialia* 52(9):2673-81.

IOP Conf. Series: Materials Science and Engineering **84** (2015) 012016 doi:10.1088/1757-899X/84/1/012016

Evolution of the dendritic morphology with the solidification velocity in rapidly solidified Al-4.5wt.%Cu droplets

M Bedel[1], G Reinhart[1], Ch-A Gandin[2], A-A Bogno[3], H Nguyen-Thi[1], H Henein[3]
[1] Aix-Marseille University & CNRS, IM2NP UMR 7334, Campus Saint-Jérôme, Case 142, 13397 Marseille Cedex 20, France
[2] MINES ParisTech & CNRS, CEMEF UMR 7635, 06904 Sophia Antipolis, France
[3] AMPL, Department of Chemical and Materials Engineering, University of Alberta, Canada

E-mail: marie.bedel@im2np.fr

Abstract. The microstructure morphology of Al-4.5wt.%Cu droplets formed by the Impulse Atomization technique is investigated. Three-dimensional reconstructions by synchrotron X-ray micro-tomography of several droplets reveal different morphologies in droplets of similar diameter and produced in the same batch. Moreover, microstructural features also indicate that the development of the dendrite arms occurs in some droplets along <111> crystallographic axes instead of the usual <100> directions observed in conventional casting for the same alloy. It has been observed that such an unusual growth direction of the dendrites is directly related to the solidification velocity. We underpin these results by carrying out comparisons with a solidification model. Predictions are used to discuss the change of dendrite growth direction, as well as the existence of a dendrite growth direction range for a given type of droplets. In addition, the effect of the droplet size and the cooling gas on the dendrite growth direction range observed experimentally is also investigated by using the model.

1. Introduction

Rapid solidification techniques have been developed as they enable to obtain a wide variety of structures which cannot be formed under conventional solidification processes [1]. They differ by the way to form the liquid as a strip or a droplet and by the method of heat extraction. Atomization techniques are used to make metallic powders which are used for making a desired object by pressing or by sintering [2]. The liquid metal generated as a stream breaks up into droplets by Rayleigh-Plateau instability, which subsequently solidify in a much colder medium. In the Impulse Atomization (IA) technique the liquid is pushed through a nozzle plate to form the liquid streams [3].
In order to deepen the understanding of the microstructure formation in the droplets, synchrotron X-ray micro-tomography was carried out at the European Synchrotron Radiation Facility (ESRF, Grenoble, France). Three-dimensional reconstructions of a large number of droplets were obtained, enabling the inner microstructure of the droplets to be statistically analysed for the first time. In a previous paper, we showed that four distinct morphologies could be identified in droplets of the same size and from the same batch [4]. Such a range of morphologies can be linked to a range of solidification velocities for the droplets. Indeed, while Rappaz and co-workers highlighted the <100>

[1] marie.bedel@im2np.fr

IOP Conf. Series: Materials Science and Engineering **84** (2015) 012016 doi:10.1088/1757-899X/84/1/012016

to <110> continuous transition induced by the solute composition variation in Al-Zn alloys [5], the work of Chan and co-workers [6] and Gudgel et al. [7] on NH_4-Cl and more recently the work of Castle et al. on Cu-Ni [8] and Chen et al. on Al-Fe [9] showed that the growth orientation transition from <100> to <111> occurs when the solidification velocity increases.

The purpose of the present study is to use a solidification model [10] [11] to get a better understanding of the experimental observations. Although the model does not provide direct information on the growth orientation of the grain microstructure, the final morphologies of the droplet can be linked to the evolution of the predicted growth velocity. Moreover, the impact of the droplet size and the cooling gas on the predicted solidification velocity is also compared to experimental observations.

2. Experimental and modelling details

This study focuses on Al-4.5wt.%Cu droplets, as this alloy has already been considered in previous works on powder atomization [11][12]. The droplets are formed by the IA technique [3]. Its principle is as follows: the alloy, melted in a furnace, is then pushed through a nozzle plate by mechanical impulses. The resulting liquid jets break into small droplets due to the Rayleigh-Plateau instability. While falling through a stagnant atmosphere of argon or helium gas, the droplets lose their heat and solidify. When they reach the oil quench bath placed 4 meters below, they are already fully solid. This technique leads to a size distribution of the droplets, whose diameter can vary from less than 200 μm to more than 1 mm in the same batch. The droplets are then sieved into several size classes by the technique described in [13].

Once the droplets are formed and sieved, their microstructure morphology is investigated for different size ranges and for the two cooling gases. For this purpose, synchrotron X-ray micro-tomography is used post-mortem [14]. This technique provides a three-dimensional reconstruction of the droplets where the grey level depends on the X-ray transmission of the phase. The primary phase (Al) being less absorbing than the eutectic ($Al+Al_2Cu$), the last formed solid appears in lighter grey in the tomography reconstructions. The image acquisition and the 3-D reconstruction were performed on the ID19 beamline at the ESRF. The resolution is 0.56 μm/pixel (field of view of 1146 μm), which enables us to study several small droplets at once. The statistical analysis of the droplet morphology has be carried out using ImageJ software [15]. The final droplet microstructure is the result of a complex three dimensional competition between dendrite arms. Therefore we only consider for this study the cross-sections showing characteristic morphologies, as shown in Figure 1, where the four distinct morphologies observed in the hundred-plus studied droplets are represented.

Droplet solidification is predicted by a solidification model detailed in [10]. The volume-averaged mass conservation equations are written and solved with limited diffusion in all phases as in standard microsegregation analyses. Even though coupling with a heat balance is made, a uniform temperature is assumed in the droplet. The heat extraction rate is computed thanks to an atomization model [12]. The solidification model also accounts for nucleation undercooling of the phases and the growth kinetics of the dendritic and eutectic microstructures [11][12]. It predicts the time evolution of temperature, volume fraction and solute composition of the phases. In the model, a single nucleation event per phase is assumed to take place at the centre of a spherical droplet. Also reaction for each successive microstructure is approximated by radial growth in the droplet.

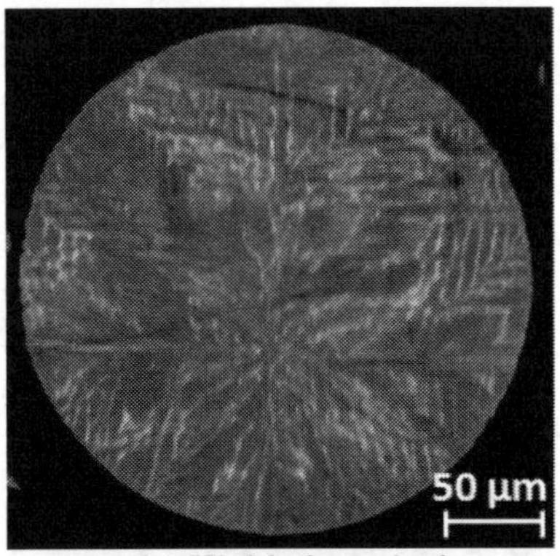

(a) Highly branched

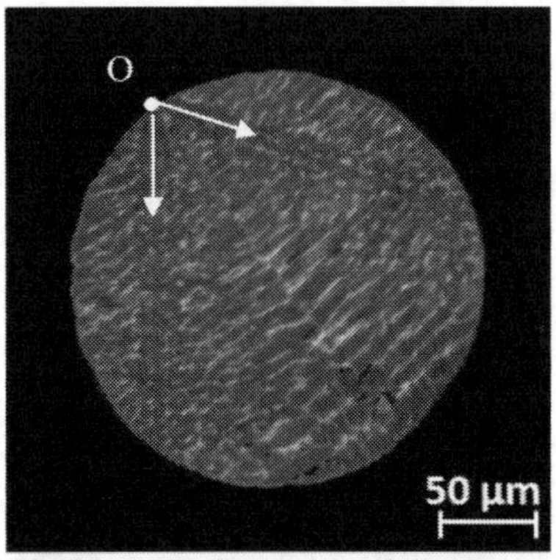

(b) Highly branched with <111> primaries

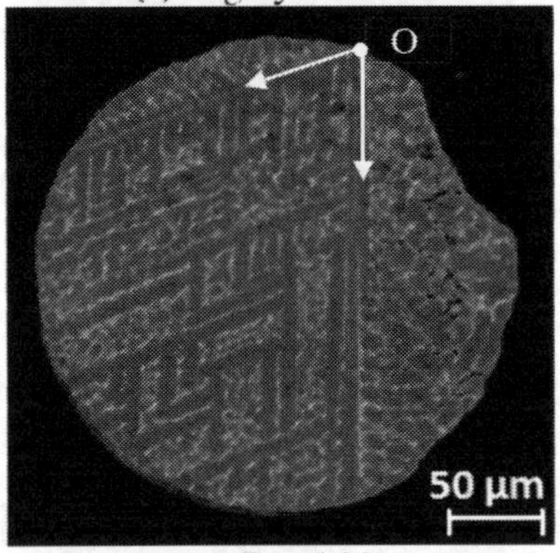

(c) Dendritic

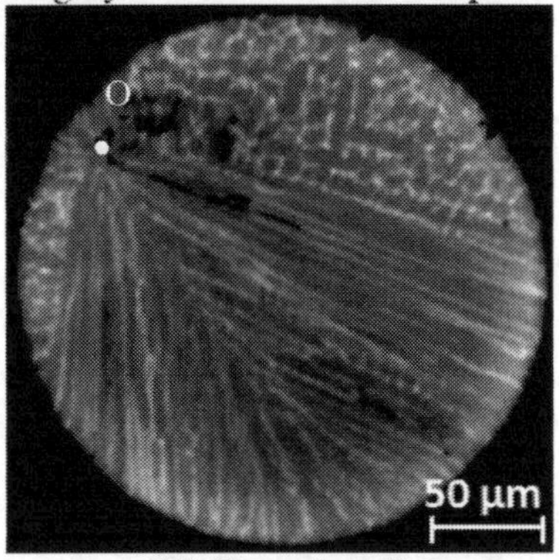

(d) Finger bundle

Figure 1. Examples of the four morphologies identified in the Al-4.5wt.%Cu droplets of diameter between 250 and 300 μm: (a) highly branched morphology, (b) highly branched morphology with visible primary arms oriented along <111> directions, (c) dendritic morphology and (d) finger bundle morphology. The nucleation position noted O is shown by a white dot and the primary arms by white arrows.

For all calculations, the Al-Cu phase diagram and the phase enthalpy and equilibrium composition are taken from the PBIN thermodynamic database [16]. Properties such as the liquidus slope and the partition coefficient thus vary with temperature and composition. The elements densities are taken from [17] and the solute diffusion coefficients in the solid and liquid phases from [18] and [19], respectively. They depend on the temperature as $D=D_0.\exp(-Q/RT)$ with R the molar gas constant. The Gibbs-Thomson coefficients for the primary Al-rich phase and the Al_2Cu phase are given in [20] and the Al-rich/liq and Al_2Cu/liq contact angles in [21]. The secondary dendrite arm spacing is obtained by linear interpolation of the experimental measurements from tomography analyses. For droplets solidified in argon, $\lambda_2 = 4.5017 + 0.0149 \times d$ and in helium, $\lambda_2 = -0.4902 + 0.0224 \times d$, where d is the droplet diameter in microns. The process parameters used by default in the present calculations are given in Table 1.

Table 1: Default parameters values used in the simulations

Alloy parameters	Value	Unit
Phase diagram	PBIN [16]	
Density	2982	$kg.m^{-3}$
Diffusion coefficient of Cu		
solid Al-rich phase D_0^s	$6.47x10^{-5}$	$m^2.s^{-1}$
liquid D_0^l	$1.05x10^{-7}$	$m^2.s^{-1}$
solid Al-rich phase Q^s	135000	$J.mol^{-1}$
liquid Q^l	23804	$J.mol^{-1}$
Gibbs-Thomson coefficients		
solid Al-rich/liquid interface	$2.41x10^{-7}$	m.K
solid Al_2Cu/liquid interface	$0.55x10^{-7}$	m.K

Process parameters	Value	Unit
Alloy composition in copper	4.5	wt.%
Gas temperature	293	K
Initial droplet velocity	0.5	$m.s^{-1}$
Initial droplet temperature	933	K
Nucleation undercooling	5	K
Droplet diameter	300	μm

3. Results and discussion

3.1. Dendrite growth direction evolution

A major result of the synchrotron X-ray micro-tomography analysis is the variety of dendrite morphologies for droplets solidifying under the same process conditions. These morphologies have been described in details in [4] by combining electron backscatter diffraction (EBSD) and X-ray micro-tomography analysis. We only remind here their main characteristics. Some of the droplets grow in the usual <100> directions and present a highly branched microstructure, as illustrated in Figure 1-a. A structure growing first in <111> directions and then in <100> directions can also be observed as in Figure 1-b. The two other types of morphologies are fully growing along <111> directions, with a dendritic (Figure 1-c) or a finger-bundle morphology (Figure 1-d).

In order to explain the growth direction change from <111> to <100> observed in some droplets (Figure 1-b), we consider the growth kinetics predicted by the model for a 300 μm diameter droplet solidifying in argon. The time-evolution of the temperature T and the dendrite tip growth velocity v_{tip} are shown in Figure 2-a. The simulation enables the distinction of different growth regimes. Firstly, the temperature continuously decreases after reaching the nucleation temperature (imposed through the nucleation undercooling ΔT_n). After nucleation, the dendrite tip growth velocity increases very rapidly until the beginning of recalescence (t=0.0154s indicated by the vertical dashed line in Figure 2-a). Then, the heat released by the solidification more than compensates the heat extracted from the droplet and the temperature increases while the overall solid fraction still increases in the droplet. As the calculated average liquid composition does not increase yet, the liquidus temperature does not vary significantly. Thus the undercooling, which is the driving force for the growth, decreases as the temperature increases and the dendrite tip velocity decreases accordingly. In a second step, an almost permanent regime is reached where the temperature and the growth velocity tend to stabilize. Finally, when the grain envelope approaches the droplet periphery, the solute enrichment of the extra-dendritic liquid is not negligible anymore and growth is slowed down. From this time (t≈0.075s), we observe

IOP Conf. Series: Materials Science and Engineering **84** (2015) 012016 doi:10.1088/1757-899X/84/1/012016

that the solid fraction evolution approaches the Gulliver-Scheil approximation as diffusion in the solid phase is limited and solute is well mixed in the liquid phase.
These different growth regimes can explain the dendrite growth direction change observed in droplets by considering that the growth orientation is linked to the growth velocity. Indeed, the first solid that forms very fast can grow along the <111> directions if its growth velocity reaches values beyond the growth orientation transition. Then the growth velocity decreases and the last part of the droplet would grow in the <100> directions, as observed Figure 1-b.

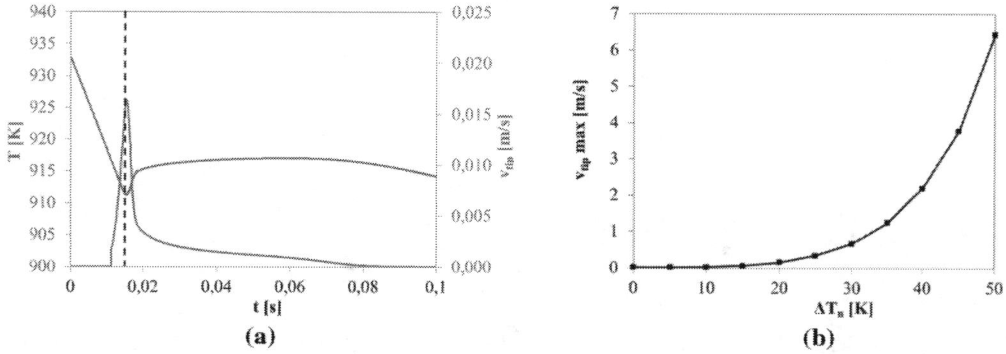

Figure 2. (a) Predicted time evolution of temperature and tip velocity during the solidification beginning of a droplet cooled in argon whose diameter is equal to 300μm, for a nucleation undercooling fixed at ΔT_n=5°C. The dashed line corresponds to both the maximal tip velocity and the local minimum temperature and (b) Predicted maximal tip velocity reached during solidification of a droplet cooled in argon whose diameter is equal to 300μm, depending on the imposed nucleation undercooling value.

3.2. Dendrite growth direction distribution

The previous analysis of the droplet growth kinetics is not sufficient to explain why droplets solidified under the same conditions in the same batch reveal the diverse oriented microstructures shown in Figure 1. The four morphologies are indeed found in similar proportions for 40 droplets of the same batch and whose diameter is between 250 and 300 μm, as can be seen further in Figure 3. While their cooling history can be considered as similar in a first approximation, we actually have no precise information on the temperature of the nucleation event in each droplet. An average nucleation temperature can be estimated through the secondary dendrite arm spacing and the eutectic fraction for a range of droplets by the technique developed in [22]. However, a range of nucleation temperature can exist. Therefore, we varied the nucleation undercooling ΔT_n in the model from 0°C to 50°C to look at its impact on the growth velocity during the droplet solidification. When ΔT_n is increased, the start of the growth is delayed but the initial driving force for growth is higher. This explains why the maximum velocity is multiplied by almost a factor 400 when ΔT_n is increased from 5°C to 50°C, as can be seen in Figure 2-b. It is also worth noting that varying ΔT_n only impacts the solidification start as the solidification tends to follow very rapidly the Gulliver-Scheil model afterwards. Although we do not know a priori the growth velocity limit inducing the growth orientation change, we can see that the higher ΔT_n is, the faster the first solid is formed and the larger the fraction of rapidly formed solid would be. Thus, the temperature of the nucleation event can significantly alter the kinetics of the droplet solidification and as a consequence the growth orientation of the microstructure. A range of nucleation undercooling could therefore help explain the variety of dendrite growth directions observed in droplets solidified under the same process conditions.

IOP Conf. Series: Materials Science and Engineering **84** (2015) 012016 doi:10.1088/1757-899X/84/1/012016

3.3. Effect of droplet size

The previous comparison between experimental observations and the numerical simulations was made for a given size of droplet. We also observe from the synchrotron X-ray micro-tomography data analysis that the droplet size has a strong impact on the microstructure distribution. Indeed, the 4 dendrite morphologies are almost identically represented in droplets of diameter 250 μm < d < 300 μm, while <111> oriented microstructures are more present in smaller droplets of diameter 0 μm < d < 212 μm, as shown in Figure 3.

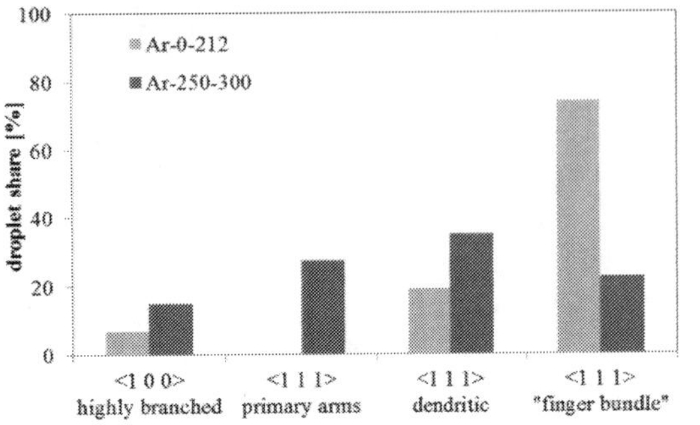

Figure 3: Droplet share of the four morphologies for the droplets solidified in argon which diameter is between 0 and 212 μm (73 droplets) and between 250 and 300 μm (40 droplets).

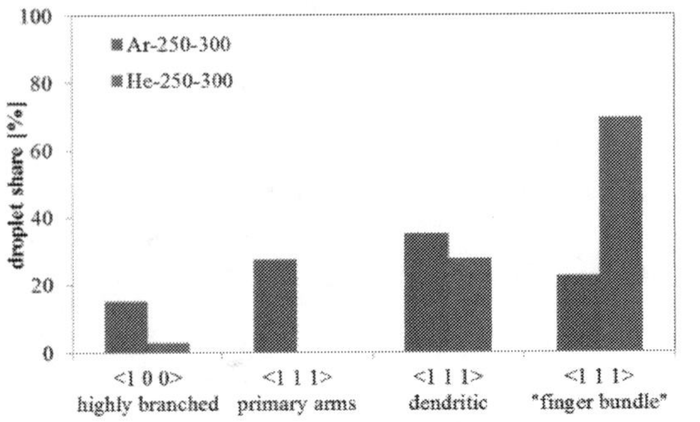

Figure 4: Droplet share of the four morphologies for droplets which diameter is between 250 and 300 μm, solidified in argon (40 droplets) and in helium (36 droplets).

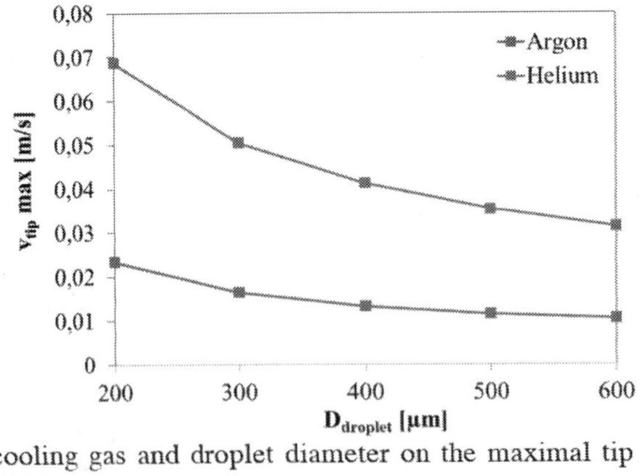

Figure 5: Impact of cooling gas and droplet diameter on the maximal tip velocity, for a nucleation undercooling equal to 5°C.

The impact of the droplet size on the maximum predicted velocity is illustrated in Figure 5. The predicted maximum dendrite velocity is almost doubled when the droplet diameter decreases from 600 μm to 200 μm, independently of the cooling gas, as the tendency is observed for droplets solidified in argon (red curve) as for droplets solidified in helium (blue curve). Figure 5 compiles results for an imposed nucleation undercooling ΔT_n=5°C but the tendency was found identical for a range of nucleation undercoolings from 0°C to 50°C. Thus, according to the model, the intensity of the rapid solidification step increases significantly when the droplet size decreases, for the two tested cooling gases and for a range of nucleation undercoolings. This can explain the promotion of the fully <111> growth orientation for smaller droplets, observed experimentally (Figure 3). In addition, we can note that mostly <111> growth orientation is obtained for the 0-212 μm diameter droplets

3.4. Effect of cooling gas

The impact of the cooling gas has also been investigated by considering droplets cooled in argon and helium atmospheres. The micro-tomography analysis shows that, all others parameters being similar, the <111> finger bundle morphology is promoted in helium in comparison with argon, as illustrated in Figure 4. As shown in Figure 5, much larger maximum growth velocities during the rapid solidification step are predicted by the model for droplets solidified in helium, all other parameters being equal. This trend is actually the same for the range of nucleation undercooling and droplet diameter investigated in the present study. Since the thermal conductivity of helium is more than 8 times larger than the one of argon, the heat extraction of the droplet is much faster when cooled in helium atmosphere. The intensity of the rapid solidification step is then larger for droplets solidified in helium, which is in agreement with the fact that the <111> finger bundle morphology is more represented in this case.

4. Conclusion

We have studied the microstructure and more precisely the dendrite growth direction of droplets formed by the IA technique using synchrotron X-ray micro-tomography. Droplets formed under the same process conditions and with the same size have different morphologies. By statistical analysis, we could estimate the morphology distribution of the droplets for different droplets sizes and for two cooling gases. The impact of the nucleation undercooling, droplet size and cooling gas are discussed based on simulations with a solidification model. We have showed that the observed experimental trend can be retrieved using the model assuming that the dendrite growth direction can change from <100> to <111> if the solidification velocity becomes large enough.

However, a more precise model accounting for the anisotropy of interfacial free energy and the anisotropy of attachment kinetics is needed to directly link the dendrite growth direction in a droplet to its solidification velocity. To this, a phase field model is also currently under development.

5. Acknowledgements

This study has been performed in the framework of MIMOSA project funded by the French National Research Agency (ANR) in collaboration with the Natural Sciences and Engineering Research Council of Canada. The authors would like to thank Gildas Guillemot from CEMEF for fruitful discussions.

References

[1] Jones H 1984 *J. Mater. Sci.* **19** 1043–76
[2] Prasad A 2006 *PhD Thesis* 70
[3] Henein H 2002 *Mater. Sci. Eng. A* **326** 92–100
[4] Bedel M, Reinhart G, Bogno A-A, Gandin C-A, Jacomet S, Boller E, Nguyen-Thi H and Henein H 2015 *Acta Mater.* **89** 234–46
[5] Haxhimali T, Karma A, Gonzales F and Rappaz M 2006 *Nat. Mater.* **5** 660–4
[6] Chan S-K, Reimer H-H and Kahlweit M 1976 *J. Cryst. Growth* **32** 303–15
[7] Gudgel K A and Jackson K A 2001 *J. Cryst. Growth* **225** 264–7
[8] Castle E G, Mullis A M and Cochrane R F 2014 *Acta Mater.* **66** 378–87
[9] Chen J, Dahlborg U, Bao C M, Calvo-Dahlborg M and Henein H 2011 *Metall. Mater. Trans. B* 1–11
[10] Tourret D and Gandin C-A 2009 *Acta Mater.* **57** 2066–79
[11] Prasad A, Mosbah S, Henein H and Gandin C-A 2009 *ISIJ Int.* **49** 992–9
[12] Wiskel J B, Henein H and Maire E 2002 *Can. Metall. Q.* **41** 97–110
[13] Federation M P I 2012 (S.l.: Metal Powder Industry)
[14] Nguyen-Thi H, Salvo L, Mathiesen R H, Arnberg L, Billia B, Suery M and Reinhart G 2012 *Comptes Rendus Phys.* **13** 237–45
[15] Abramoff M D, Magalhães P J and Ram S J 2004 *Biophotonics Int.* **11** 36–42

[16] PBIN, TCS thermodynamic database v1.2, Thermo-Calc Software AB, 2008 (Stockholm, SE)
[17] Totemeier W F G C 2004 *Smithells Metals Reference Book (Eighth Edition)* (Oxford: Butterworth-Heinemann) pp 1–27
[18] Heringer R, Gandin C-A, Lesoult G and Henein H 2006 *Acta Mater.* **54** 4427–40
[19] Du Y, Chang Y A, Huang B, Gong W, Jin Z, Xu H, Yuan Z, Liu Y, He Y and Xie F-Y 2003 *Mater. Sci. Eng. A* **363** 140–51
[20] Kurz W and Fisher D 1986 (4711, Aedermannsdorf, Switzerland: Trans Tech Publications Ltd, Trans Tech House)
[21] Gandin C-A, Mosbah S, Volkmann T and Herlach D M 2008 *Acta Mater.* **56** 3023–35
[22] Bogno A-A, Delshad Khatibi P, Henein H and Gandin C-A 2013 *Materials Science & Technology Conference and Exhibition 2013* vol 2 (Montreal, QC, Canada) pp 1153–60

Contactless ultrasonic treatment of melts using EM induction

V Bojarevics[1], G Djambazov[1], GSB Lebon[1] and KA Pericleous[1],
[1]Centre of Numerical Modelling and Process Analysis, University of Greenwich, Park Row, London, SE10 9LS, UK

E-mail: k.pericleous@gre.ac.uk

Abstract. Ultrasound Treatment (UT) is commonly used in light alloys during solidification to refine microstructure, or disperse immersed particles. A sonotrode probe introduced into the melt generates sound waves that are strong enough to produce cavitation of dissolved gases. The same method cannot be used in high temperature melts, or for highly reactive alloys, due to probe erosion and melt contamination. An alternative, contactless method of generating sound waves is proposed and investigated theoretically in this paper, using electromagnetic (EM) induction. In addition to strong vibration, the EM induction currents generate strong stirring in the melt that aids distribution of the UT effect to large volumes of material. In a typical application, the same induction coil surrounding the crucible used to melt the alloy may be adopted for UT with suitable frequency tuning. Alternatively - or in addition - a top coil may be used. For industrial use, instead of multiple sonotrodes as has been the practice in scaling up, modelling shows that one simply has to alter the coil geometry and current to suit. To reach sinusoidal pressure fluctuations suitable for cavitation it may be necessary to tune the induction coil frequency for resonance, given the crucible dimensions.

1. Introduction

Intense vibration of the liquid metal during solidification is known to refine microstructure, remove trapped gas and reduce porosity [1-4]. Vibration can be achieved in various ways: shaking the crucible, inserting a mechanical stirrer in the melt or, more commonly, using an ultrasonic horn e.g. in the DC casting of aluminium ingots [3]. During ultrasonic treatment (UT), frequencies around 20 kHz were found to be the most effective [3], a fact attributed to the onset of cavitation. Gas micro-bubbles exist in the liquid either due to dissolved gases coming out of solution as the melt cools (e.g. H_2 in Al), air trapped with oxides during mould filling, or gas attached to the surface of immersed particles. Bubbles, subjected to an imposed pressure sine-wave, first expand to many times their original size at the pressure minimum and then collapse violently as the pressure rises, generating high speed micro-jets and shock waves. These events fragment emerging dendrites, or facilitate the breakup and dispersion of particle clusters to populate nucleation sites. Evidence of cavitation has been seen in situ recently [5], found in post-solidification analyses and further shown to depend on a critical pressure threshold [3]. Of relevance to the present study, ultrasonic processing is one of the candidate techniques for the production of metal-matrix nano-composites (MMNCs).

There are however, several disadvantages in immersive stirring/vibration techniques: contact with the liquid metal leads to contamination of the melt and conversely erosion of the immersed probe surface, requiring frequent replacement. These problems are limiting when the technology is to be applied to high temperature alloys (ODS steel, nickel superalloys) or to highly reactive metals (Ti, Zr,…). Another limiting factor of the immersed sonotrode approach is the localised effect of the cavitation region, which leads to long processing times, high cost/energy usage and therefore it

IOP Conf. Series: Materials Science and Engineering **84** (2015) 012017 doi:10.1088/1757-899X/84/1/012017

remains most suitable for small treatment volumes. EM induction is an attractive alternative. It provides melt stirring and heat. What is overlooked is the ability of the AC component of the induced Lorentz force to generate a strong sound field within the melt, equivalent to that produced by an immersed sonotrode, but without contact. Vives [6] was the first to investigate experimentally such a non-contact technique, using a combination of static and AC magnetic fields. Other investigators followed this lead, using different configurations; for AlSi hypereutectic alloys [3,7], for grey iron melts [8] and aluminium alloys, etc. In each case, different thresholds of pressure amplitude were found to be necessary and frequencies ranging from 50 Hz to 50 kHz were used. There has been no systematic study of the sound field generated in these situations, which led us in a previous publication [9] to examine the sound field generated by the cylindrical induction coil in an induction crucible, coupling the compressible sound wave equations with the pressure source generated by the electromagnetic field and the Rayleigh-Plesset equations for cavitating bubbles. This simulation demonstrated that pressure amplitudes likely to cause cavitation of dissolved gas can be generated, provided the AC frequency is tuned to approach wave resonant conditions – a similar conclusion was reached by Vives, who considered the design of a resonant EM cavity for this purpose (figure 1,[6]). In the present contribution, we extend this idea via the introduction of a *tuned top coil* (VibroEM patent #PCT/EP2013/067896) inserted into the liquid metal volume, but nevertheless still using EM repulsion to prevent contact with the metal. In this very close proximity to the surface, intense vibrations and melt stirring can be produced which can be controlled via the depth of immersion.

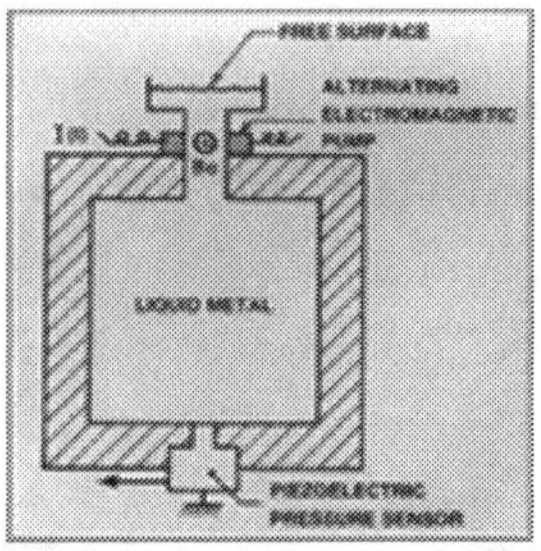

Figure 1: A resonant EM cavity idea from Vives [6]

The remainder of the paper is arranged as follows: In section 2, the problem to be addressed is described, followed by the set of equations for the magnetic and sound fields and bubble evolution. Results of simulations using aluminium and silicon melts are given, followed by discussion and conclusions.

2. Problem setup

A typical ceramic induction crucible is shown in figure 2, together with the two alternative excitation arrangements. The cylindrical induction coil and top coil can be operated independently of each other and at different frequencies. The function of the cylindrical coil remains that of heating/melting and stirring, whilst the top coil is there to provide controlled vibration. The top coil is allowed to move vertically so that it can gradually enter the melt, forming a depression in the free surface. In reality, both coils contribute to the soundfield within the melt. The ceramic crucible is enclosed in a graphite container that is itself surrounded by the coil. The crucible is filled with aluminium.

The flow and thermal fields containing the EM force and Joule heating are solved using an incompressible spectral collocation technique, assuming harmonic averaging for mean values as described in previous publications, e.g. [10]. Time accurate information is extracted within each AC cycle once a pseudo-steady condition is reached, to enable the driving sound source evaluation for the compressible acoustic simulation.

In addition to the aluminium crucible where the main consideration of the study is the uniform distribution of refining particles, a second industrial case has also been simulated, where silicon melt is subjected to AC EM field to expel impurities such as SiC.

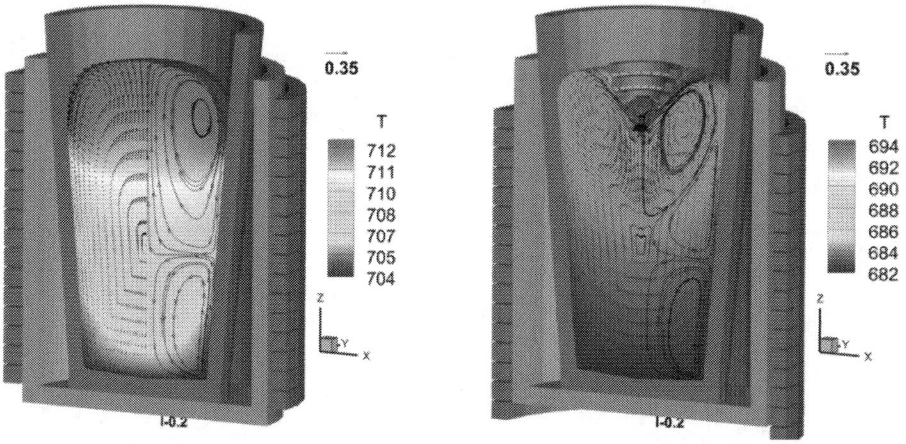

Figure 2: Two alternative arrangements for melt stirring and vibration of the melt. The computed RMS velocity and temperature field are shown, with the figure on the left [9] representing a traditional arrangement with a cylindrical coil (2 kHz, 1.6 kA), the one on the right the new concept of the immersed top coil.

3. Theoretical analysis for the EM force in a cylindrical crucible

To demonstrate the approach followed, we consider the case of an infinitely long cylindrical crucible coaxially surrounded by a cylindrical coil:

EM field: $$\boldsymbol{B} = B_R \cos \omega t + B_I \sin \omega t \quad (1)$$

where, $$B_R = B_o e^{-\frac{x}{\delta}} \cos\frac{x}{\delta}\ ;\ B_I = -B_o e^{-\frac{x}{\delta}} \sin\frac{x}{\delta} \quad (2)$$

AC current: $$\boldsymbol{J} = J_R \cos \omega t + J_I \sin \omega t \quad (3)$$

where, $$J_R = \sigma \frac{\omega}{2} \delta(B_R + B_I);\ J_I = \sigma \frac{\omega}{2} \delta(-B_R + B_I) \quad (4)$$

(Where x is the cylindrical r-coordinate measured from the external wall inwards and δ is the skin layer thickness)

Then, Lorentz force on liquid : $$\boldsymbol{F} = \boldsymbol{J}\text{x}\boldsymbol{B} \quad (5)$$

Time averaged force: $$\bar{F} = \frac{1}{2}(J_R B_R + J_I B_I) = \frac{1}{2\mu\delta} B_o^2 e^{-2\frac{x}{\delta}} \quad (6)$$

Time-dependent component: $$\tilde{F} = \frac{1}{2\mu\delta} B_o^2 e^{-2\frac{x}{\delta}}[1 + \sqrt{2}\cos(2\omega t - 2\frac{x}{\delta} + \frac{\pi}{4}] \quad (7)$$

Note, the Lorentz force doubles the supply frequency.

The resulting sinusoidal force is shown graphically in figure 3. It is evident that the force is only significant in the outermost part of cylinder, in the skin layer, decaying rapidly away from the wall at a frequency of 10 kHz. Amplitude decay is accompanied by a gradual change of phase.

The observations gleaned from this idealised 1D problem are valid for a realistic crucible simulation as shown in figures 4 and 5. Figure 4 shows on the left a schematic representation of an experimental silicon refining crucible [11], together with the resulting flow and temperature field on the right. The quartz crucible is surrounded by graphite, necessary to melt the solid charge using resistance heating. As the silicon melts and its conductivity increases a vigorous flow results due to the induced RMS Lorentz force. The sinusoidal component of this force is shown in figure 5. On the left,

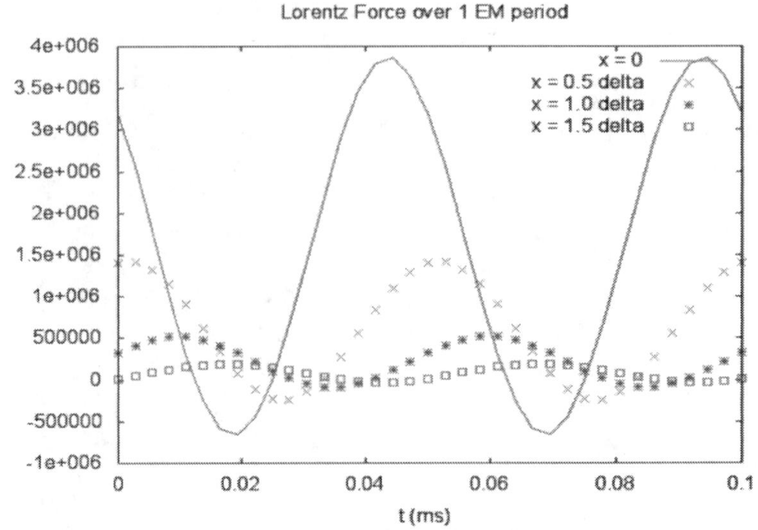

Figure 3: EM force in liquid Al at 20 kHz, at four radial positions 'x', inwards from the surface, showing a rapid decay. The solid line represents the surface value (skin friction depth – δ= 2.38 mm for aluminium).

the instantaneous force variation diminishes rapidly in amplitude outside the skin layer (at a frequency of 1 kHz) and although the mean force is always compressive, the resulting pressure troughs reach a significant negative gauge value.

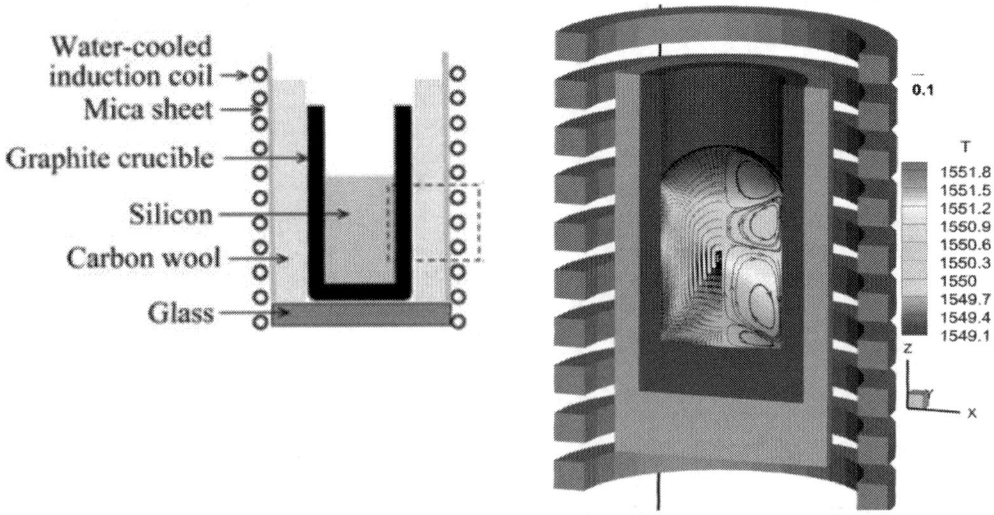

Figure 4: Experimental Graphite induction crucible [11] used to remove SiC impurities from recycled PV silicon. The melt acquires a domed shape due to the induction force and a complex mixing pattern develops. The small temperature variation is due to radiation loss at the top surface.

4. Electromagnetic sound generation

The sinusoidal part of the Lorentz force is the source of vibration, hence source of sound in the liquid. A physical manifestation of this can be seen in the mercury experiment depicted in figure 6.

4.1 Linearised Euler Equations in Liquids

The generation of sound waves by the alternating magnetic force was simulated using a Computational Acoustics approach. In 1D, the equation of mass continuity and the simplified momentum equations (no convection or viscous terms), yield the following system:

$$\frac{\partial p}{\partial t} + \rho c^2 \frac{\partial \tilde{u}}{\partial x} = 0 \qquad \rho \frac{\partial \tilde{u}}{\partial t} + \frac{\partial p}{\partial x} = f;\ f = |\mathbf{F}| - \bar{F} \tag{8}$$

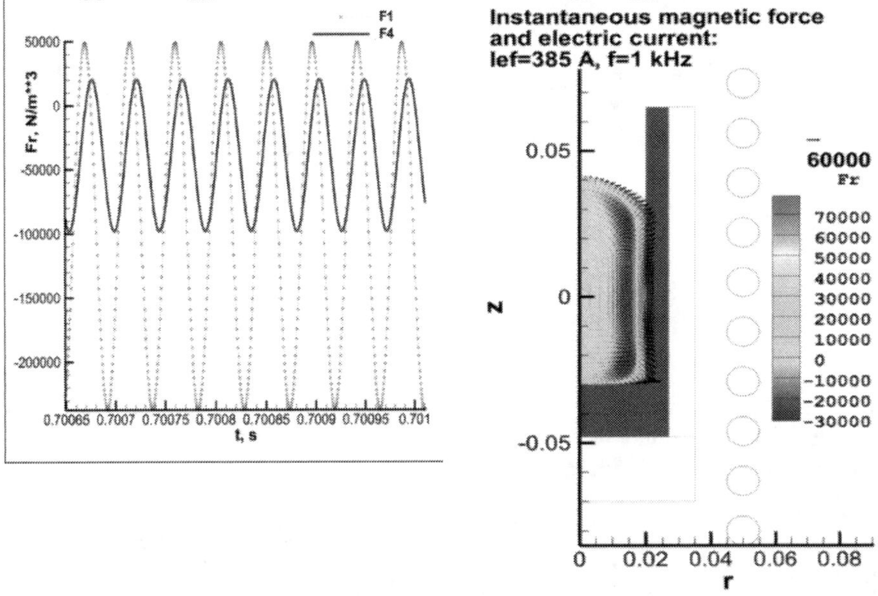

(a) (b)

Figure 5: Instantaneous radial force variation (a) in the skin layer, (b) in the bulk of a silicon-melting crucible.

This system of 1st order partial differential equations was discretised onto a *fully-staggered* mesh formed of the spatial axis x and the temporal axis t. The acoustic pressure values p were stored on the cell faces of this mesh and the acoustic velocity perturbations $\tilde{u}$ were in the cell centres in the middle of each time step – this leads to the 'leap frog' numerical scheme [12]. The time-dependent component of the force f was evaluated in the 'u' positions and added to the right-hand side of the momentum equation.

Figure 6: Waves on surface of mercury generated by AC EM

4.2 Bubble evolution and cavitation

The Keller-Miksis equation governs the dynamics of a spherical bubble, including the effects of acoustic radiation and yields satisfactory results with large forcing amplitudes [13]:

$$\left(1 - \frac{1}{c_\infty}\dot{R}\right) R\ddot{R} + \frac{3}{2}\left(1 - \frac{1}{3}\frac{1}{c_\infty}\dot{R}\right)\dot{R}^2 = \left(1 - \frac{1}{c_\infty}\dot{R}\right)\frac{1}{\rho}\left[p_B(t) - p_\infty(t) - p_c\left(t - \frac{R}{c_\infty}\right)\right] + \frac{R}{\rho_\infty c_\infty}\frac{dp_B(t)}{dt} \tag{9}$$

Figure 7 shows the rapid expansion and then collapse of a single bubble placed in the aluminium melt subjected to a driving EM frequency of 10 kHz. The soundfield generated by the bubble collapse, responsible for the breakup of particle agglomerations, is not included in the simulations shown here.

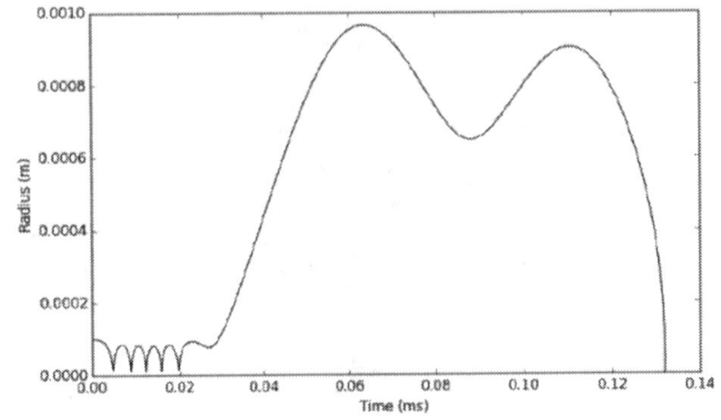

Figure 7: Hydrogen bubble collapse in aluminium at 700°C predicted using equation (9)

5. Sound field results

In the following figures, samples of computed soundfields using side and top coils are shown, together with the equivalent values for the traditional immersed sonotrode operating at 20 kHz.

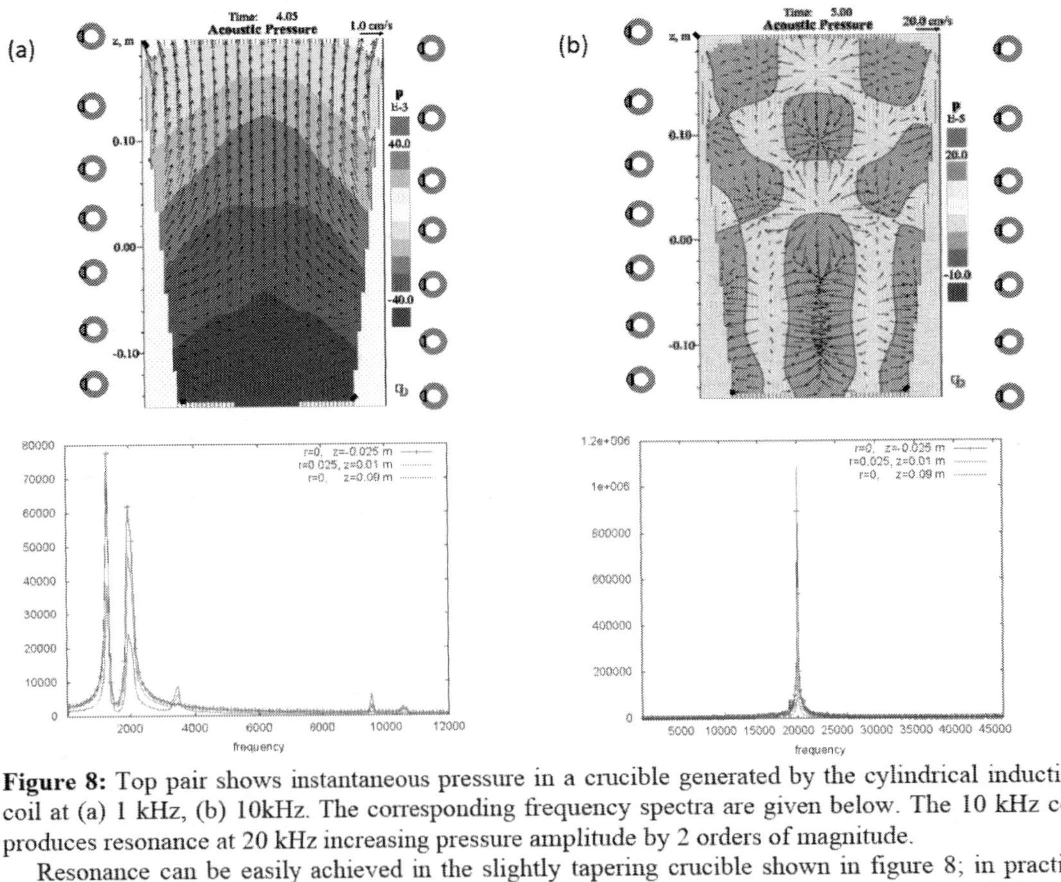

Figure 8: Top pair shows instantaneous pressure in a crucible generated by the cylindrical induction coil at (a) 1 kHz, (b) 10kHz. The corresponding frequency spectra are given below. The 10 kHz coil produces resonance at 20 kHz increasing pressure amplitude by 2 orders of magnitude.

Resonance can be easily achieved in the slightly tapering crucible shown in figure 8; in practice and for safety a frequency approaching resonance may be preferable. The figure also highlights the substantial difference in sound pressure patterns between 1 kHz – good for mixing, and 20 kHz, good for cavitation. Since resonance depends on the relative values of wavelength and crucible dimensions, the frequency will need to be adjusted on a case-by-case basis. Figure 9, depicts the soundfield generated by the immersed vibration coil (RHS) as well as the surface depression and stirring caused by its presence (LHS). The middle pictures show the sound 'beat' caused by the near-resonant condition at two different locations in the core of the crucible (top) with the bottom figure showing how the driving frequency excites resonance at three different peaks.

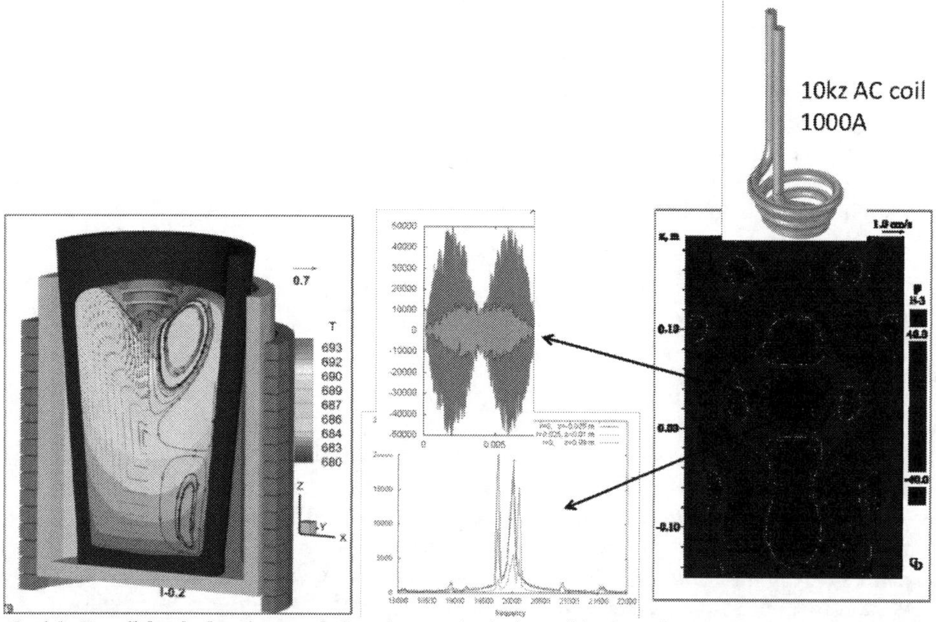

Figure 9: (a) Crucible, inductive melt heating and stirring, (b) Soundwave 'beat' as forcing frequency approaches resonance, (c) Acoustic pressure spectrum – multiple peaks due to conical crucible design, (d) Instantaneous sound field

It is instructive to see what the algorithm makes of the sonotrode sound field in the same geometry, and that is given in figure 10. The general patterns of peaks and troughs are similar, but one notes that at the settings given, the vibrating sonotrode produces relatively higher pressure amplitudes (by a factor of 3).

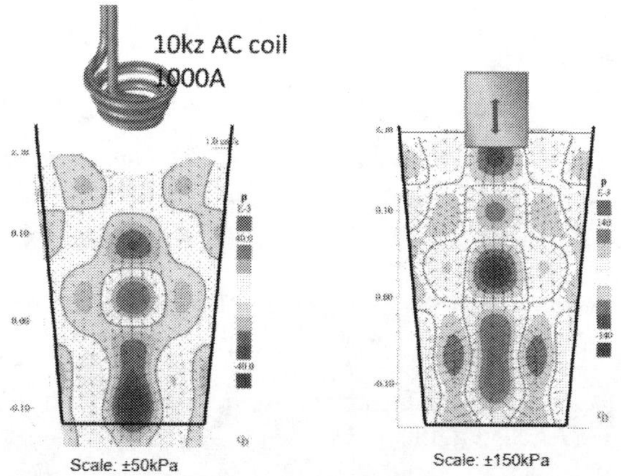

Figure 10: Comparative sound fields generated by the top coil and sonotrode for the same geometrical configuration

6. Conclusions

It has been shown through a combination of MHD and Computational Acoustic simulations that EM induction can be used as a contactless replacement of the mechanical sonotrode. Then,

- The mean EM force acts in a thin skin layer and promotes strong bulk mixing (which a mechanical sonotrode does not!)
- The sinusoidal part of the force generates sound waves
- The AC frequency can be adjusted to produce near resonance pressure amplitudes for cavitation (e.g. so as to disperse particle clusters)

The EM "sonotrode" is contactless, therefore

- It can be used for high temperature or reactive metals (Ni, Ti, Zr, ODS steel, etc.)
- Avoids contamination of the melt
- Can fit in with existing induction coil arrangements

Work is in progress to test this setup experimentally in the near future.

7. References

[1] Campbell J, 1981, Effect of vibration during solidification, *Int Metals Rev*, **vol. 26**, 2, pp.71-108
[2] Abramov O, 1994, Ultrasound in Liquid and Solid Metals (Boca Raton, FL: CRC Press), p. 289
[3] Eskin GI, Eskin DG, 2003, *Ultrasonics Sonochemistry*, **vol. 10**, 4–5, p. 297
[4] Wang G, Dargusch MS, Qian M, Eskin DG, StJohn DH, 2014, *J Crystal Growth* **vol 408,** p119
[5] Xu WW, Tzanakis I, Srirangam P, Terzi S, Mirihanage WU, Eskin DG, Lee PD, In situ Synchrotron Radiography of Ultrasound Cavitation in a Molten Al-10Cu Alloy; TMS2015 144th Annual Meeting & Exhibition; 15–19 March, 2015 – to appear
[6] Vives C., J. , 1996, *J Crystal Growth*, **vol. 158**, 1–2, pp 118–127
[7] Verma A, Tewari SP, Prakash J, 2011, *Int. J. Engg Sci and Techn*, **vol. 3**, no. 6, p. 5215 – 5220.
[8] Dong J, Cui J, Zeng X, Ding W, 2005, Effect of Low-Frequency electromagnetic vibration on cast-ability, microstructure and segregation of large-scale DC ingots of high-alloyed Al, *Materials Transactions*, **vol. 46**, no.1, p. 94 – 99.
[9] Djambazov G, Bojarevics V, Lebon B and Pericleous K, 2014, Contactless Acoustic Wave Generation in a Melt by Electromagnetic Induction, *Light Metals 2014*, TMS, Grandfield J (Ed.)
[10] Bojarevics V, Harding RA, Pericleous KA and Wickins M, 2004, The development and experimental validation of a numerical model of an induction skull melting furnace. *Metall. and Mat. Trans. B*, **35** (4), pp. 785-803.
[11] Kadkhodabeigi M, Safarian J, Tveit H, Tangstad M, Johansen ST, 2012, Trans. *Nonferrous Met. Soc. China* **22**, pp 2813–2821
[12] Djambazov GS, Lai CH and Pericleous KA, 2000, Staggered-mesh computation for aerodynamic sound, *AIAA Journal*, **38** (1), pp. 16-21.
[13] Alehossein H, and Qin Z , 2007, Numerical analysis of Rayleigh-Plesset equation for cavitating water jets. *Int. J. Numer. Meth. Engng.* **72**:780_807.

Acknowledgments

The authors acknowledge financial support from the ExoMet Project (co-funded by the European Commission (contract FP7-NMP3-LA-2012-280421), by the European Space Agency and by the individual partner organizations).
Dr Bruno Lebon, acknowledges the financial support of the EPSRC (grant EP/K00588X/1) in developing the cavitation model used in this work

The effects of natural, forced and thermoelectric magnetohydrodynamic convection during the solidification of thin sample alloys

A. Kao[1], N. Shevchenko[2], O. Roshchupinka[2], S. Eckert[2], K. Pericleous[1]
[1]Centre of Numerical Modelling and Process Analysis, University of Greenwich, Park Row, London, SE10 9LS
[2]Helmholtz-Zentrum Dresden-Rossendorf, Institue of Fluid Dynamics, P.O. Box 510119, 01314 Dresden, Germany

E-mail: a.kao@greenwich.ac.uk

Abstract. Using a fully coupled transient 3-dimensional numerical model, the effects of convection on the microstructural evolution of a thin sample of Ga-In25%wt. was predicted. The effects of natural convection, forced convection and thermoelectric magnetohydrodynamics were investigated numerically. A comparison of the numerical results is made to experimental results for natural convection and forced convection. In the case of natural convection, density variations within the liquid cause plumes of solute to be ejected into the bulk. When forced convection is applied observed effects include the suppression of solute plumes, preferential secondary arm growth and an increase in primary arm spacing. These effects were observed both numerically and experimentally. By applying an external magnetic field inter-dendritic flow is generated by thermoelectrically induced Lorentz forces, while bulk flow experiences an electromagnetic damping force. The former causes preferential secondary growth, while the latter slows the formation of solute plumes. This work highlights that the application of external forces can be a valuable tool for tailoring the microstructure and ultimately the macroscopic material properties.

1. Introduction

Convective transport of solute is known to have a dramatic effect on the evolution of the underlying microstructure during solidification. In many cases the fluid flow is an inherent part of the process, for example, natural convection is significant in alloys that have relatively large density differences between the various components causing plumes of solute to be ejected back into the melt. These can lead to preferential flow channels (or chimneys) and ultimately the formation of freckles. This phenomenon is common in Nickel based super alloys which are widely used in the manufacturing of gas turbine blades [1-3]. Performing direct *in situ* experimental studies on Nickel based super alloys is very difficult due to the high solidification temperature. An alternative system to study the underlying mechanism is the Gallium-Indium system, which is liquid at room temperature and easy to handle. Furthermore, due to a relatively large density difference between Gallium and Indium, the system exhibits similar flow behaviour to Nickel based super alloys. An *in situ* radiographic technique has been developed by Eckert and co-workers [4-6]. In this work the same experimental technique is used to validate the numerical studies.

Numerical models have been used to investigate the formation of freckles in castings [7-9], however they have primarily focused on the effects of natural convection. This work focuses on the introduction of external forces in the liquid to modify the fluid dynamics. The first is the application of a magnetic wheel, which generates an alternating magnetic field providing conditions analogous to forced convection close to the solidification front. The second is the application of a static external magnetic field, which has two effects, the first is electromagnetic damping and the second is thermoelectric magnetohydrodynamics (TEMHD). The latter is caused by the interaction with thermoelectric currents localised to the solidification front giving a resultant Lorentz force, hence driving fluid flow. In this work a fully coupled transient 3-dimensional numerical model was applied to each of these cases. The model uses an Enthalpy based method originally developed by Voller [10]. The Enthalpy based method is a level set method that uses an iterative finite difference scheme to solve the Gibbs-Thompson condition and conservation of enthalpy. This method has been extended by Kao *et al.* [11,12] into 3-dimensions and to include fluid flow and electromagnetics. It has been used to investigate TEMHD to control microstructure evolution [13], the role of secondary arms in TEMHD [14] and the effect of slow rotating fields [15].

Understanding the effect of external forces on microstructure is fundamentally important for both the generation of new materials and improving current techniques. The capability to predict the outcome of microstructural modifications due to externally applied forces will allow for macroscopic material properties to be tailored by selective design of these forces.

2. Governing Equations

The evolution of the microstructure is modelled using an enthalpy-based method. The volumetric enthalpy H is defined as the sum of sensible latent heats

$$H = c_p T + fL \tag{1}$$

where, c_p is the volumetric specific heat capacity and L is the volumetric latent heat of fusion. For binary alloys the concentration potential V in terms of the solute concentration C and partitioning coefficient k is defined as

$$V = \frac{C}{f(1-k)+k}. \tag{2}$$

The interfacial temperature (T^i) is given by the Gibbs-Thompson condition

$$T^i = T_m - \frac{\gamma(\theta,\phi)}{L} T_m \kappa - m\left(C_0 - C_l^i\right) \tag{3}$$

where T_m is the melting temperature, $\gamma(\theta,\phi)$ is the surface energy, κ the curvature, m the liquidus slope, C_0 the bulk solute concentration and C_l^i the liquid solute concentration at the interface. Due to the low thermal Péclet number (Pe $<10^{-2}$) and large time scales involved, the thermal field can be assumed to be at a steady state and diffusion dominated. Consequently the thermal field at a given time is known explicitly from the imposed thermal gradient, G and nucleation temperature, T_0. However, due to a large Lewis number, Le = 10000, mass transport needs to be solved by

$$\frac{dC}{dt} = \nabla \cdot (D_l \nabla V) - \nabla \cdot (\mathbf{u} V) \tag{4}$$

where D_l is the liquid mass diffusivity. In the solid the Scheil assumption, $D_s = 0$ is used.
Due to the low Reynolds number encountered on the scale of microstructures and thin sample experiments, a simplified form of the Navier-Stokes equation is used, representing Stokes flow

$$\rho \frac{d\mathbf{u}}{dt} = -\nabla p + \mu \nabla^2 \mathbf{u} + \mathbf{F} \tag{5}$$

where ρ is the density, $\mathbf{u}$ the velocity, p the pressure, μ the dynamic viscosity and $\mathbf{F}$ is the sum of all the body forces. Conservation of mass for incompressible flow gives

$$\nabla \cdot \mathbf{u} = 0. \tag{7}$$

For natural convection, the body force $\mathbf{F}$ in (5) is given by

$$\mathbf{F}_{NC} = \rho \mathbf{g} f \beta (C - C_0) \tag{8}$$

where β is the compositional expansion coefficient and $\mathbf{g}$ is the gravitational acceleration. In the case of forced convection, $\mathbf{F}_{FC}$ is assumed to be a constant body force acting over the entire domain. Furthermore, due to the relatively small thickness, W of thin samples and low Reynolds number the flow between the plates can be assumed to be Poiseuille flow. By selecting a characteristic bulk mean velocity $\bar{\mathbf{u}}$ the body force can be calculated from

$$\mathbf{F}_{FC} = \frac{\bar{\mathbf{u}}\mu}{2W^2}. \tag{9}$$

For TEMHD, the current density, $\mathbf{J}$ is given by a generalised form of Ohm's law including an extra term accounting for thermoelectric currents [16]

$$\frac{\mathbf{J}}{\sigma} = \mathbf{E} - S\nabla T + \mathbf{u} \times \mathbf{B} \tag{10}$$

where σ is the electrical conductivity, $\mathbf{E}$ the electric field, S the Seebeck coefficient and $\mathbf{B}$ the DC magnetic field. Conservation of charge gives

$$\nabla \cdot \mathbf{J} = 0. \tag{11}$$

The thermoelectric currents interact with the magnetic field generating a Lorentz force given by

$$\mathbf{F}_{TE} = \mathbf{J} \times \mathbf{B}. \tag{12}$$

3. Problem Description

Solidification of Ga-In25wt%. alloy in a Hele-Shaw cell with dimensions 25 X 35 X 0.15mm^3 was observed by X-ray radiography. The Hele-Shaw cell was cooled at the bottom by a Peltier cooler at a rate of 0.01K/s. An electric heater at the top with variable heating power was applied to maintain a thermal gradient of 1K/mm over the entire cell. A magnetic wheel was applied at the top of the cell to provide forced convection conditions. The sample is cooled from a molten state and at time t = 0s nucleation occurs and solidification of the microstructure propagates from the base of the sample to the top. A schematic of the experimental setup is given in figure 1.

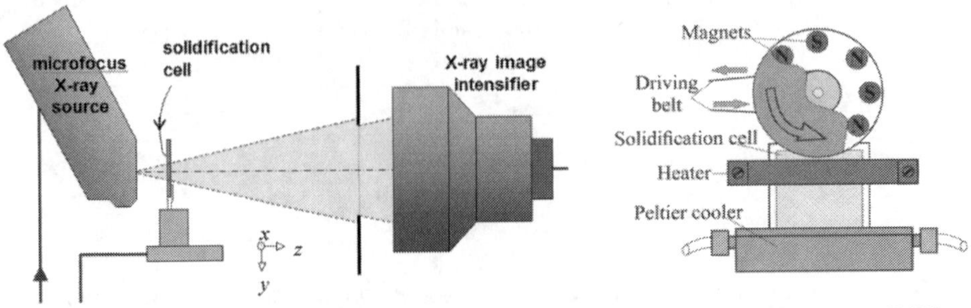

Figure 1. Experimental setup. Left: X-ray radiography system. Right: Hele-Shaw solidification cell with electric heater, Peltier cooler and magnetic wheel.

The numerical model simulates a 5 X 5 X 0.1mm^3 region of the solidification cell. The computational grid comprises cubes with a length of 5μm. In the latter stages of solidification, when the interface reaches 5/8ths of the length of the domain, the domain begins to move at the interface velocity tracking the tips. Table 1 describes the boundary conditions for concentration, temperature, pressure, velocity and current. The location of the faces is shown in figure 2 which gives the relative size of the numerical domain. Periodic boundary conditions are applied on the *x* faces normal to the thermal gradient representing an infinitely wide sample.

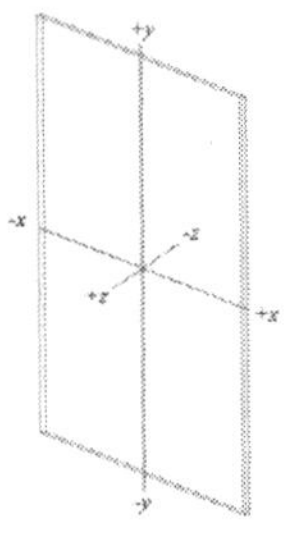

Figure 2. Computational domain.

Table 1. Boundary conditions for numerical model

Face	C	T	p	**u**	**J**
$-x$	Periodic				
$+x$	Periodic				
$-y$	$\frac{dC}{dy}=0$	$T=T_C$	$\frac{dp}{dy}=0$	$\frac{du_x}{dy}=u_y=\frac{du_z}{dy}=0$	$\nabla\cdot\mathbf{J}=0$
$+y$	$C=C_0$	$T=T_H$	$p=0$	$\frac{du_x}{dy}=\frac{du_y}{dy}=\frac{du_z}{dy}=0$	$\mathbf{J}=0$
$-z$	$\frac{dC}{dz}=0$	$\frac{dT}{dz}=0$	$\frac{dp}{dz}=0$	$u_x=u_y=u_z=0$	$\frac{d\Psi}{dz}=0$
$+z$	$\frac{dC}{dz}=0$	$\frac{dT}{dz}=0$	$\frac{dp}{dz}=0$	$u_x=u_y=u_z=0$	$\frac{d\Psi}{dz}=0$

The numerical model assumes that the crystallographic orientation is parallel to the thermal gradient. However, the orientation cannot be controlled experimentally. The experimental results presented below are for cases were nucleation was approximately parallel to the thermal gradient. The material properties used in the numerical model are given in Table 2.

Table 2. Material properties used in the numerical model

c_p	2.38 X 10^6	J/m^3/K	ρ	6250	Kg/m^3
L	4.2 X 10^8	J/m^3	μ	1.25 X 10^{-3}	Pas
k	0.5	-	β	1.66 X 10^{-3}	Wt%$^{-1}$
m	1.8	K/wt%	S	2.5 X 10^{-7}	V/K
D_l	1.2 X 10^{-9}	m^2/s	σ	3.0 X 10^6	S/m

4. Results

4.1 Natural convection

This section investigates the evolution of the microstructure from natural convection only. The absence of forced convection or an external magnetic field provides a control solution that is used to compare against cases when additional external forcing is applied.

Numerical and experimental results of the time evolution of the microstructure are given in figure 3. The figure shows numerical results for the velocity and both numerical and experimental results for the solute concentration. The numerical results show that during the early stages of growth convective rolls form around the nucleation site. Upward flow directed toward the bulk melt, forms plumes of enriched liquid stunting local growth, while downward flow directed onto the solidification front, transports bulk material encouraging growth. The plumes are sustained by drawing enriched solute from the inter-dendritic network. However, as the primary arms grow into the downward flow and the inter-dendritic porosity decreases the convective rolls destabilise and the plumes migrate along the solidification front to the furthest extent of solidification. The upward flow stunts growth at this new location, but the downward flow, now directed toward the previously stunted regions, encourages growth. This transient process then repeats itself. The experimental results show a similar transient behaviour of the plume and there is a reasonable agreement between the predicted primary arm spacing and concentration profiles.

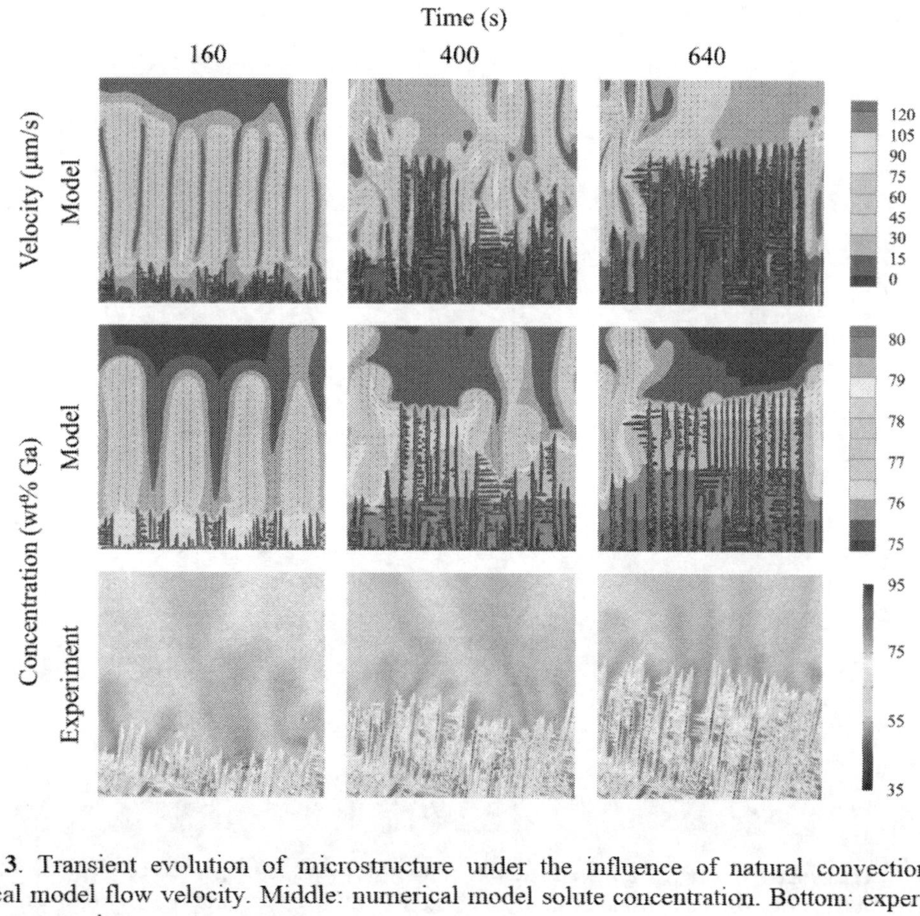

Figure 3. Transient evolution of microstructure under the influence of natural convection. Top: numerical model flow velocity. Middle: numerical model solute concentration. Bottom: experimental solute concentration

4.2 Forced convection

This section introduces an additional body force in the liquid directed perpendicular to the direction of solidification representing forced convection. Experimentally, the rotation of the magnetic wheel introduces electromagnetic forces with greatest intensity at the top of the sample. Further down the sample a pressure gradient forms across the solidification front. The numerical model is calibrated to the experimental results, by selecting a characteristic bulk mean flow. In this case, experimental observations have shown the velocity ahead of the solidification front to be around 200μm/s. This velocity is comparable but larger than those observed and predicted in natural convection.

Figure 4 shows the transient evolution of the microstructure under the influence of forced convection. The numerical results show that as flow passes around and over a dendrite, solute is transported downstream forming an extended enriched region. The incident flow on a dendrite can have two effects which depend on the solute concentration transported. Incident flow that has been enriched after passing an upstream dendrite will stunt downstream dendrites. However, if the incident flow transports bulk material, then growth of both the primary tip and secondary arms will be promoted. This flow-dependent competition between neighbouring grains causes a significant increase in the primary arm spacing compared to the control case. Natural convection is still present and plumes of enriched solute are released into the bulk. However, due to bulk mean flow the plumes are transported downstream of the sample. These mechanisms are also observed experimentally with preferential growth of the secondary arms and an increase in the primary arm spacing clearly shown in figure 4.

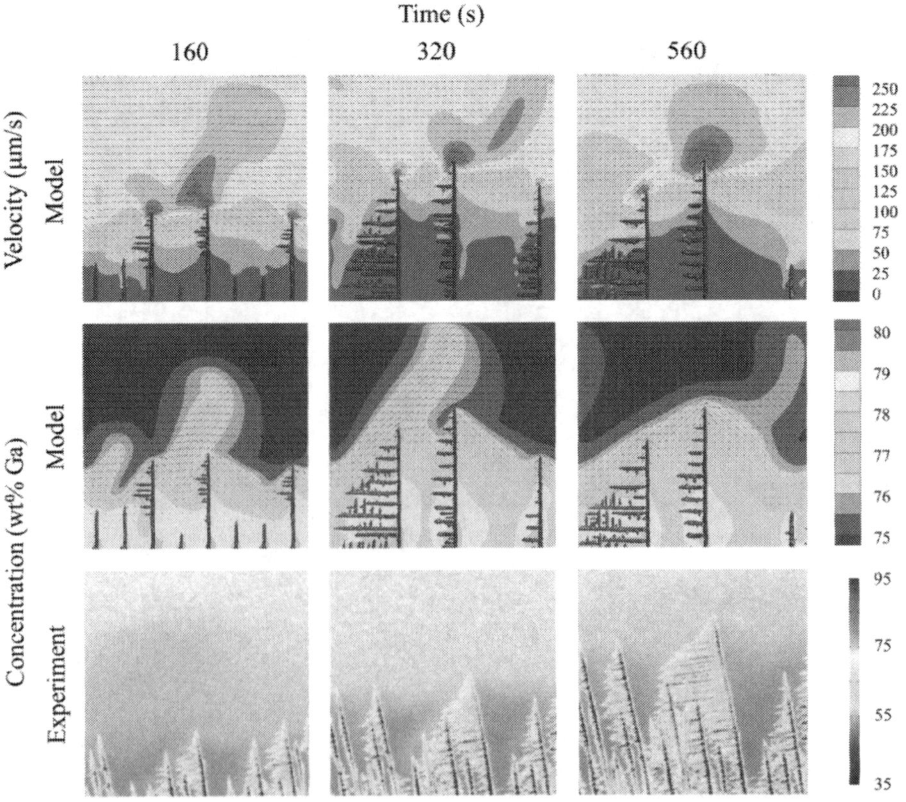

Figure 4. Transient evolution of microstructure under the influence of forced convection. Top: numerical model flow velocity. Middle: numerical model solute concentration. Bottom: experimental solute concentration

IOP Conf. Series: Materials Science and Engineering **84** (2015) 012018 doi:10.1088/1757-899X/84/1/012018

4.3 Thermoelectric magnetohydrodynamics

Figure 5 shows transient numerical results for the velocity, concentration and current density. The current passes up through the dendrites, emanating from the tips and then back down through the inter-dendritic liquid. Application of a 1T external magnetic field generates a Lorentz force, driving flow in inter-dendritic regions where there is a significant current. However as the porosity decreases the inter-dendritic flow becomes restricted. This creates a region of relatively high velocity that tracks the solidification front. The velocity is slower than the natural and forced convection cases, about half and a quarter respectively. However, the underlying mechanism is similar to forced convection, as this region of velocity transports liquid downstream. Depending on the concentration of the liquid, growth is either promoted or stunted. However, in contrast to the case of forced convection, the absence of currents in the bulk combined with electromagnetic damping gives no bulk mean flow. Moreover, electromagnetic damping slows the formation of plumes of solute, where after 360s the first plumes are only just emerging. In comparison to the case of only natural convection several plumes had formed and destabilised within this time scale. The inability to effectively eject solute into the bulk causes the inter-dendritic region to become more enriched compared to the other cases.

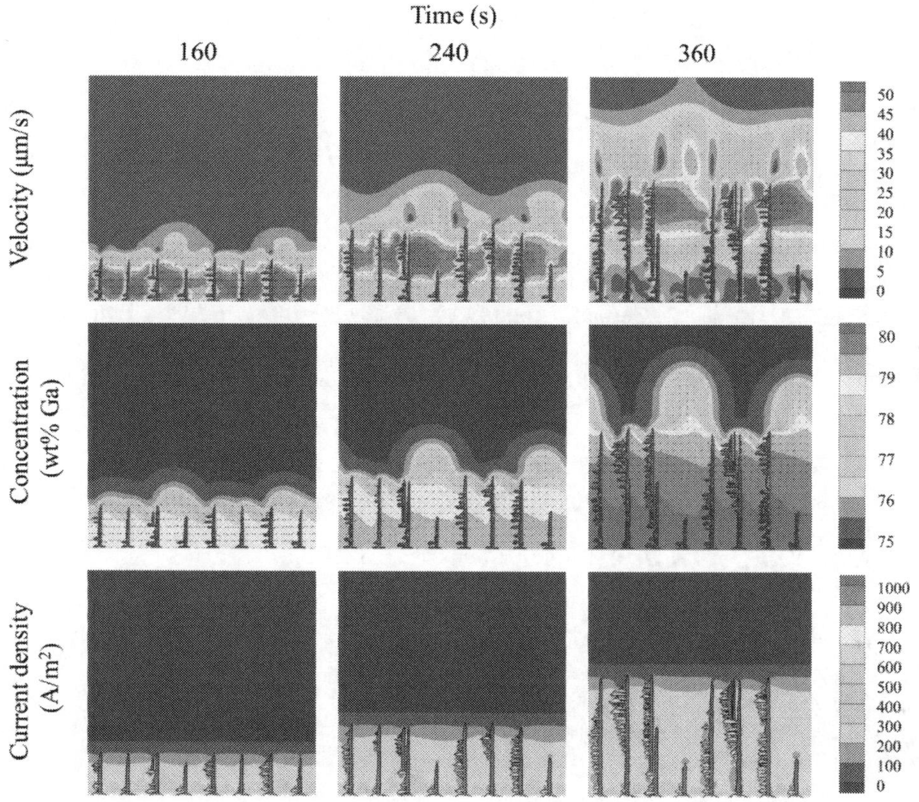

Figure 5. Numerical results for the transient evolution of microstructure under the influence of an external 1T DC magnetic field. Top: flow velocity. Middle: solute concentration. Bottom: current density.

5. Conclusion

The introduction of convection during solidification has been shown to have a significant impact on the evolution of the microstructure. The effects of natural, forced and thermoelectric magnetohydrodynamic convection during solidification of a Ga-In25%wt. thin sample were investigated. Natural convection is present in this system due to significant variations in density between the two components causing plumes of solute to be ejected into the melt. This mechanism was observed both experimentally and numerically.

Using a magnetic wheel, forced convection conditions were generated close to the solidification front. By using the observed bulk mean velocity an equivalent body force was calibrated for the numerical model. Both the numerical and experimental results of forced convection show that the microstructure exhibits preferential growth of secondary branches and an increase in primary arm spacing. These changes are caused by the downstream transport of solute. The final case investigated the effects of thermoelectric magnetohydrodynamics. Numerical simulations show similarities to the forced convection case, with inter-dendritic flow causing preferential growth of secondary arms. In the melt where there are no thermoelectric currents, the flow is dominated by electromagnetic damping. Consequently the formation of solute plumes is significantly slowed compared to the other cases.

Acknowledgments

The research is supported by the EPSRC (EPK011413/1) and the German Helmholtz Association in form of the Helmholtz-Alliance "LIMTECH".

References

[1] Reed R C, Tao T and Warnken N 2009 *Acta Mater.* **57** pp 5898-5913
[2] Madison J, Spowart J, Rowenhorst D, Aagesen L K, Thornton K and Pollock T M 2010 *Acta Mater.* **58** pp 2864-2875
[3] Beckermann C, Gu J P and Boettinger W J 2000 *Metall. and Mater. Trans. A* **31A** pp 2545-2557
[4] Boden S, Eckert S, Gerbeth G 2010 *Mater Lett.* **64** pp 1340-1343
[5] Shevchenko N, Boden S, Gerbeth G and Eckert S 2013 *Metall. And Mat. Trans. A* **44** pp 3797-3808
[6] Boden S, Eckert S, Willers B and Gerbeth G 2008 *Metall. Mat. Trans. A* **39** pp 613-623
[7] Pericleous K A, Djambazov G Ward M, Yuan L and Lee P D (2013) *Metall and Mat Trans A* **44** (12) pp 5365-5376
[8] Guo, J, and Beckermann, C 2003 *Num Heat Transfer, A* **44**, pp 559-576
[9] Karagaddea S, Yuan L, Shevchenko N, Eckert S and Lee P D 2014 *Acta Mater.* **79** pp 168-180
[10] Voller V 2007 *Int J. Heat Mass Transfer* **51** pp 823-834
[11] Kao A and Pericleous K 2012 *J. Algor. Comp. Tech.* **(6)1** pp 173-201
[12] Kao A and Pericleous K 2012 *J. of Iron and Steel Research International* **(19)1** pp 317-321
[13] Kao A and Pericleous K 2012 *Modeling of Casting, Welding, and Advanced Solidification Processes - XIII (MCWASP XIII). IOP Conf. Series: Mat. Sci. Eng.* **33** 012045
[14] Kao A and Pericleous K 2012 *Magnetohydrodynamics* **(48)2** pp 361-370
[15] Kao A, Lee P D and Pericleous K 2014 *ISIJ International* **54(6)** pp 1283-1287
[16] Shercliff J A 1979 *J. Fluid Mech.* **(91)2** pp 231-251

Modeling of the effect of a thermoelectric magnetic force onto conducting particles immersed in the liquid metal

Y Du Terrail Couvat[1,2], O Budenkova [1,2,6], A Gagnoud [1,2], G. Salloum Abou Jaoude[3,4], H Nguyen-Thi [3,4], G Reinhart [3,4], J Wang [5], Z-M Ren [5] and Y Fautrelle[1,2]
[1] CNRS, SIMAP, UMR 5266 Grenoble, France
[2] SIMAP, Grenoble Institute of Technology, France
[3] IM2NP, Aix-Marseille and Toulon University, France
[4] CNRS, IM2NP, UMR 7334, Aix-Marseille, France
[5] Department of Materials Sciences, Shanghai University, China

Abstract. Simulation of a thermo-electromagnetic force which acts on a conducting particle immersed into liquid metal is performed using multi-gird multi-physics software AEQUATIO. To verify numerical solutions a model thermoelectric problem is solved using two methods. In the first one a phase function is used to indicate the phase transition whereas in the second the solid particle is described with a real frontier of a simplified shape. Numerical and analytical solutions for a model problem qualitatively agree but strong oscillations are observed in a numerical solution with a phase function. Further AEQUQTIO is applied for calculation of the velocity of a dendrite fragment observed in-situ in experiment of solidification of AlCu alloy. Numerical solution gives a good agreement with the experimental observation.

1. Introduction

Electromagnetic fields are used in metallurgical processes since several decades. Traditionally, static magnetic fields are applied to suppress the liquid motion in the bulk liquid due to the braking effect of the Lorentz force. The latter is a result of interaction between the electric current, induced by a moving liquid, and the imposed magnetic field. Yet, during solidification processes bulk materials are subjected to temperature gradients and thermo-physical properties of materials can change abruptly because of phase transition. This leads to appearance of a thermoelectric current near the liquid-solid interface and further to the thermoelectric magnetic force (TEMF) which is a result of interaction between the thermoelectric current and the applied magnetic field. Theoretical studies of the effect of TEMF force on the convective flows were made several decades ago [1] and experimental evidences were first found in crystal growth [2-3]. Post-mortem analysis of samples of alloys solidified under the action of static magnetic field showed remarkable results [4]. Nowadays, advanced in-situ observations made with X-rays allow one to identify the effect of external fields in real time [5-6]. However, despite diverse experimental results, very limited number of numerical simulations can be cited and most of them are applied either for a very small calculation domain [7] or for simple geometries [8]. This is related to the fact that an intensive thermoelectric current appears in a very thin layer in the bulk liquid near the solid-liquid interface and a fine calculation mesh is required to capture this. On the other hand, the effect of the thermoelectric magnetic force spreads over large distance in a

[6] To whom any correspondence should be addressed at olga.budenkova@simap.grenoble-inp.fr

form of convection and transport of the solid phase in the liquid, as strains and deformation of the solidified structure. Therefore, to study numerically the effect of the TEMF in solidification of alloys one needs either to use an average approach which would account lower-scale information at macro-scale or to employ a multi-grid technique in order to link processes at microscale and at macroscale in a direct way.

A numerical toolbox AEQUATIO is intended for use of multiple grids in a calculation domain and couples solutions obtained at different scales. The present work was inspired by the idea of modeling of a particular phenomenon observed in the experiment on solidification of Al-4wt%Cu alloy which is briefly described in the next section. Mathematical statement of the problem and numerical approach are presented in sections 3 and 4. Then comparisons between the analytical and numerical results for a model problem are made and results of simulation are compared with experimental observations.

2. Experimental observation

In the experiment performed with in-situ X-ray observation at ESRF a liquid Al-4wt%Cu alloy was solidified from the bottom in a crucible 6mm×40mm×0.2mm. A scheme of the set-up is shown in figure 1. A static magnetic field was generated by a permanent magnet fixed close to the crucible.

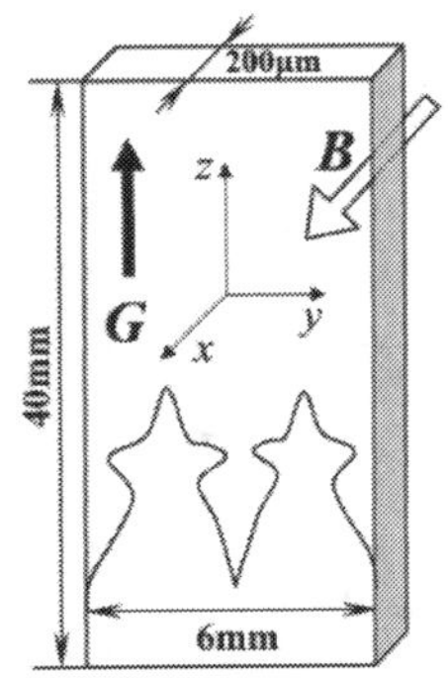

Figure 1. Geometry for the experiment on solidification of Al-4wt%Cu performed at ESRF, G=3000 K/m and B=0.08T [9].

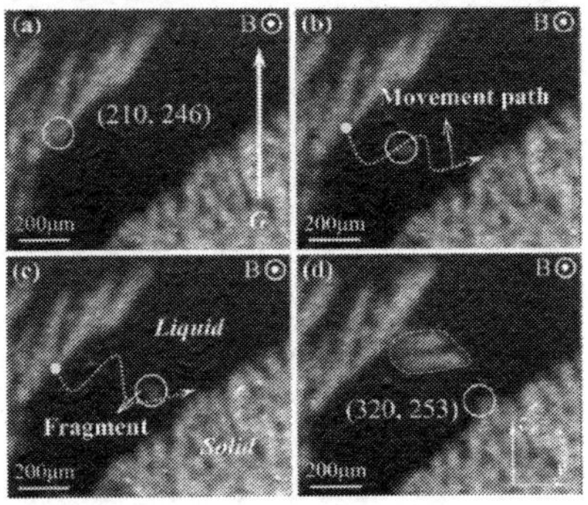

Figure 2. Registered detachment of dendrite pieces and their motion in horizontal direction, time interval between the frames (a)-(b) and (b)-(c) is 1.4s. Frame (d) shows detachment of a larger fragment [9].

In the experiment several events of detachment of dendrite pieces were observed. A sequence of frames which shows motion of detached pieces in the horizontal direction is given in figure 2. Since this effect was not observed in experiments without magnetic field, there are strong grounds to believe that thermoelectric magnetic force acting on the solid phase is at the origin of this phenomenon. Image treatment showed that the average velocity of the motion of a dendrite fragment depends on its size and is about 130-200µm/s [9].

3. Statement of the problem

Below mathematical formulations for two problems are given. The first one is related to the case of the experiment whereas the second concerns a model thermoelectric problem which we used to verify numerical results obtained with AEQUATIO.

3.1. Statement of the problem for the experimental case

The system of equations to be solved for each phase includes stationary equation for the energy transfer, continuity equation for the electric current and relation between the density of the electric current and temperature gradient:

$$\nabla(\lambda_i \nabla T_i) = 0 \tag{1}$$

$$\nabla \vec{j}_i = 0 \tag{2}$$

$$\vec{j}_i = -\sigma_i \nabla U_i - \sigma_i S_i \nabla T_i \tag{3}$$

In (1-3) the index $i = S, L$ is related either to the solid or to the liquid phase and U is an electric potential, which, in fact, should be found. Notations for other variables and their values for the Al-4wt%Cu alloy are given in the table 1. The boundary conditions for the temperature are fixed temperature values at the top and bottom of the calculation domain and no heat flux through the lateral boundaries:

$$T\big|_{z=z\min} = T_1 \tag{4a}$$

$$T\big|_{z=z\max} = T_2 \tag{4b}$$

$$\nabla T \cdot \vec{n}\big|_{wall} = 0 \tag{4c}$$

For the current we presume insulating conditions for all surfaces bounded the calculation domain:

$$\vec{j}_i \cdot \vec{n}\big|_{wall} = 0 \tag{5}$$

At the solid-liquid interface conditions of continuity of heat fluxes and normal component should be satisfied that gives following conditions for the temperature and electric potential, respectively:

$$\lambda_S \nabla T_S \cdot \vec{n}\big|_{SL} = \lambda_L \nabla T_L \cdot \vec{n}\big|_{SL} \tag{6}$$

$$(\sigma_S \nabla U_S + \sigma_S S_S \nabla T_S) \cdot \vec{n}\big|_{SL} = (\sigma_L \nabla U_L + \sigma_L S_L \nabla T_L) \cdot \vec{n}\big|_{SL} \tag{7}$$

The thermoelectric magnetic force (per unit volume) acting in each phase is given as

$$\vec{f}_{TEM,i} = \vec{j}_i \times \vec{B} \tag{8}$$

The velocity and the trajectory of the particle can be found from Newton's equation:

$$\rho\tau \frac{d\vec{V}_S}{dt} = \vec{F}_{TEM,S} - \vec{F}_{Stokes} \tag{9}$$

where τ is the volume of the particle and the TEMF acting onto the whole particle is

$$\vec{F}_{TEM,S} = \int_\tau \vec{f}_{TEM,S} d\tau \tag{10}$$

Stokes' drag force is taken proportional to the velocity of the particle and to the viscosity of the solid:

$$\vec{F}_{Stokes} = A \cdot \vec{V}_S \tag{11}$$

where coefficient A depends on the particle geometry. If the thermoelectric magnetic force does not change with particle motion, then the analytical solution of (9) is given as

$$\vec{V}_S = \frac{\vec{F}_{TEM,S}}{A}\left[1 - \exp\left(-\frac{A}{\rho\tau}t\right)\right] \tag{12}$$

3.2. Statement of a model problem

A simple thermoelectric problem which allows analytical solution for the system of equation (1)-(3) is a conducting solid sphere immersed into a conducting *infinite* media. The boundary conditions (4a)-

(4c) and (5) are replaced with a constant one-dimensional temperature gradient and zero electric current at infinity. Let us suppose that the thermal gradient G in infinity is directed along the z-axis. Then the analytical solution for the problem shows that the thermoelectric current in the whole domain circulates in the plane (z,y) with the maximal density in the solid particle and decays with a third power of a distance in the bulk liquid [10]. If a uniform static magnetic field acts in the x direction as $\vec{B} = B_0\vec{i}_x$, then the analytical solution for the thermoelectric force acting in the liquid and in the solid calculated using (8) and presented in Cartesian coordinates system with the origin at the centre of the sphere has the following form:

$$\vec{f}_{TEM,S} = -B_0\frac{\sigma_S\sigma_L}{\sigma_S + 2\sigma_L}\cdot(S_S - S_L)\cdot G\frac{3}{2+\lambda_S/\lambda_L}\cdot\vec{i}_y \quad (13)$$

$$\vec{f}_{TEM,L} = -B_0\frac{\sigma_S\sigma_L}{\sigma_S + 2\sigma_L}\cdot(S_S - S_L)\cdot G\frac{3}{2+\lambda_S/\lambda_L}R^3\cdot\left(\frac{3xz}{r^5}\vec{i}_y + \frac{2z^2 - x^2 - y^2}{r^5}\vec{i}_z\right) \quad (14)$$

Note that the force which acts onto particle $\vec{f}_{TEM,S}$ is constant and is directed along the y-axis in a negative direction if a difference of Seebeck coefficient is positive and vice versa.

4. Numerical approach

AEQUATIO is a toolbox based on the finite elements methods [11]. The spatial discretization for partial differential equations is obtained with Galerkin method and one or several calculation meshes can be used in the domain. The solution is sought in the elements' nodes and interpolation polynomials of Lagrange are of the second degree. In the present problem coupled equations for the energy and electric current (1)-(3) are projected over two meshes: the one which is constructed for the whole domain and which is coarse and a significantly finer mesh for a sub-domain around the particle (figure 3). The boundary conditions for the sub-domain are obtained via interpolation of data from the coarse mesh. We construct only one system of linear equations taking into account the two meshes.

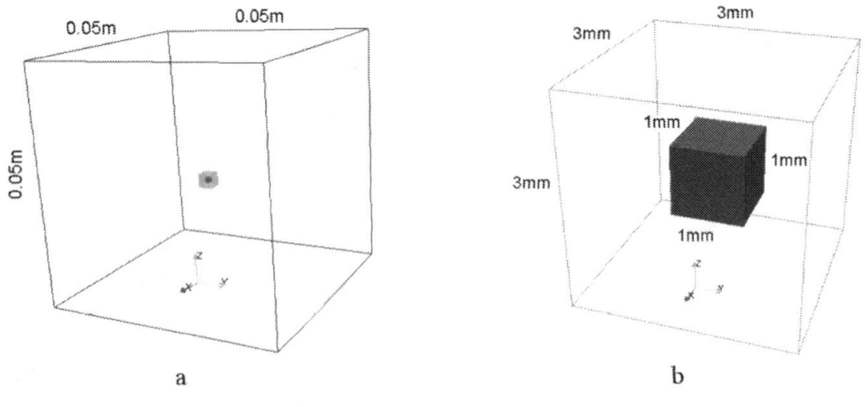

Figure 3. Illustration to the domain and sub-domain division in AEQUATIO toolbox. a: complete geometry of the problem with a calculation domain and sub-domain in its centre, particle inside the sub-domain seen as a point and b: sub-domain and the particle

The aim of the present work was to implement the thermoelectric magnetic equations into AEQUATIO toolbox using two different methods described below and to apply one of them for simulation of the experimental observation. The first method implied a straightforward utilization of a "phase function" $\phi(\vec{r})$ which defines the properties of materials thus indicating transition between the two phases but without using a sharp boundary between them. The second method is to describe a real frontier for a particle of simplified from, like, for example, cubic. Motivation for the use of phase function is driven by methods like volume of fluids or phase field where it is employed to calculate the evolution of the solid phase. One of advantages of implementation of phase function is that we can use a regular mesh in the sub-domain even if the particle has irregular form. Another advantage is that governing equations (1)-(5) can be solved in a unified manner without special treatment of the conditions at the solid-liquid interface.

5. Results and discussion

Results of numerical simulations presented below are given for points within the sub-domain containing the particle. Local values for the density of the electric current and for the thermoelectric magnetic force are calculated on a grid with obtained values in the mesh nodes handled with interpolations and derivation techniques used in finite elements method.

5.1. Comparison of the analytical solution and numerical results with phase function for a model problem

At the first step a comparison between the analytical solution and the ones obtained using AEQUTIO toolbox were made. In an analytic model a sphere of a diameter of 1 mm was taken, i.e.R=50μm is used in (11)-(14). To satisfy the condition of infinite media the size of the 3D computational domain in AEQUATIO simulations was taken 50 times larger than the sphere: 50×50×50 mm and was meshed with 4 elements in each direction. The phase function corresponded to a sphere, i.e. had a form

$$\phi(\vec{r}) = \begin{cases} 1, & if \quad \vec{r} \le R \\ 0 & if \quad \vec{r} > R \end{cases} \tag{15}$$

The size of a cubic sub-domain contained a spherical particle was four times larger than the particle: 4×4×4 mm. Every property γ for materials used in the calculations was presented as linear combination of the value for the solid γ_S and the liquid γ_L phase:

$$\gamma(\vec{r}) = \gamma_S \cdot \phi(\vec{r}) + \gamma_L \cdot (1 - \phi(\vec{r})) \tag{16}$$

We considered two combinations for the values of Seebeck coefficients in the liquid and in the solid phase. The first case with $S_S = -1\cdot10^{-6}$ V/K and $S_L = -2\cdot10^{-6}$ V/K corresponds to values measured for Sn-Pb alloy [12]. The difference $S_S - S_L = 1\cdot10^{-6}$ V/K is positive, and the TEMF acts along the y-axis in negative direction. In figure 4a we present distributions of the force along the line parallel to the z-axis which passes through the center of the sphere and lies within the sub-domain used in numerical simulations. The maximal value of the force is inside the particle for $-0.5\cdot10^{-3}\text{m} \le z \le 0.5\cdot10^{-3}$ m. Then the TEMF decreases rapidly and almost disappears toward boundaries of the sub-domain which are at $\pm 1\cdot 10^{-3}\, m$. Surprisingly, in numerical simulations qualitatively correct results are obtained already if a sub-domain is meshed only with 5 elements in each direction, but the value of force is significantly higher than the one predicted theoretically. With increasing the number of cells in a sub-domain, numerical and analytical results become closer and for the sub-domain mesh containing 30 cells in each direction the maximal difference between the numerical and analytical solution is about 8%, yet oscillations appear in numerical solution.

Situation changes dramatically for other values of Seebeck coefficients which were measured for an Al-Cu alloy and are given in the table 1. In this case the difference $S_S - S_L = -1.6\cdot10^{-6}$ V/K is negative and TEMF is co-directed with y-axis. The analytical model still predicts monotone behavior for the force with transition from the liquid to the solid phase, but in the numerical solution strong oscillations of the force is observed near the solid-liquid phase transition (figure 4b). Nevertheless, apart from two jumps the qualitative behavior of the numerical solution is not close to the analytical one. We suppose that oscillations in numerical solution are related to a discontinuity of the chosen phase function and can be minimized if a smooth variation of physical properties with transition between the two phases is assured.

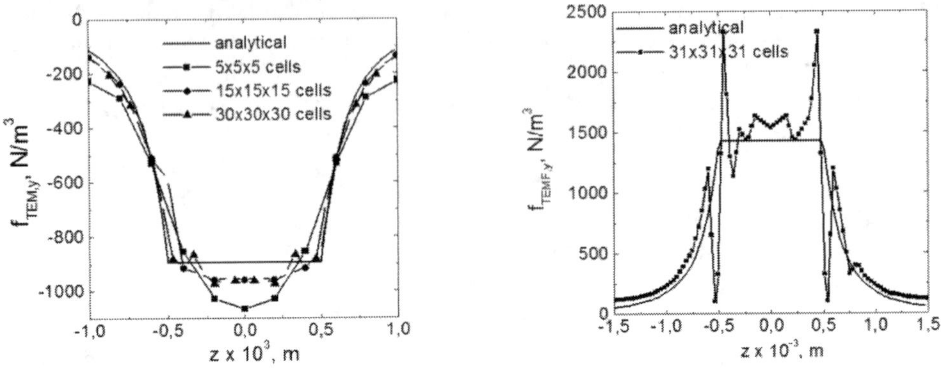

a b

Figure 4. Analytical and numerical solutions obtained in a model problem for the y-component of the thermoelectric force along the z-axis, origin of the coordinate system is at the centre of the sphere. The sphere in numerical solution is given with the phase function (15). a. $S_S{-}S_L{=}1{\cdot}10^{-6}$V/K and b: $S_S{-}S_L{=}{-}1.6{\cdot}10^{-6}$V/K, other properties are as in table 1.

5.2. Comparison of the analytical solution for a sphere and numerical results for a cubic particle with a real frontier

To verify that solutions with a real frontier do not contain oscillations, calculations were made for a cubic particle of a size 1mm×1mm×1mm with explicit indication of its boundary. A sub-domain containing the particle was 4mm×4mm×4mm whereas the size of the calculation domain was kept as in previous case 5cm×5m×5cm. Quantitative comparison between numerical results obtained for a cubic particle and analytical solution for a sphere is, of course, not possible, but one can expect similarity of these solutions and this is observed in figure 5a-b.

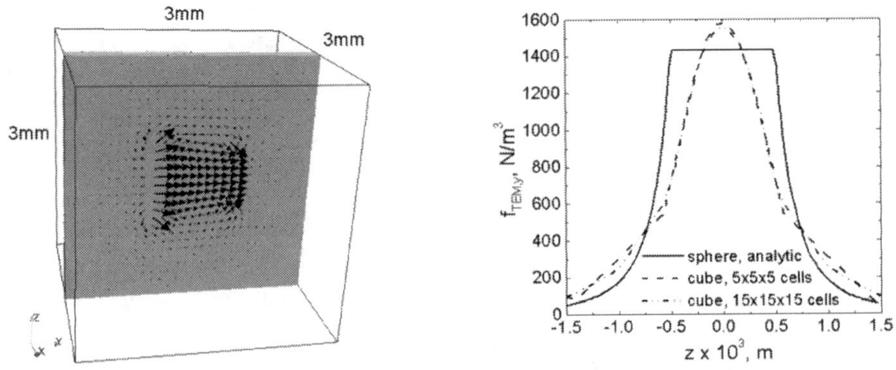

a b

Figure 5. Numerical results for a cubic particle 1mm×1mm×1mm inside the infinite media. a: Vector field of the thermoelectric magnetic force presented in a sub-domain in a plane (yz) which passes through the centre of the particle. b: TEMF distribution along the z-axis with the origin in centre of the particle, analytical solution for a sphere is given as a reference.

5.3. Comparison of the results of numerical simulation and experimental observation

Because of the size of the volume where solidification of Al-4wt%Cu was performed some effects related to the constraints of the thermoelectric current compared with the analytical problems could be expected. Properties of the solid and liquid phases were similar to that given in the table 1. Geometry of the calculation domain for modelling with AEQUATIO was taken as the one presented in figure 1 but with the thickness of 300μm. A sub-domain containing a cubic particle 100μm×100μm×100μm was placed near the centre of the calculation domain. We considered sub-domains of two sizes: 200×400μm×400μm and 200×800μm×800μm with the same space discretization. One can observe that the size of the sub-domain affects results of calculation drastically both in terms of the smoothness of the solution and in values as shown in figure 6.

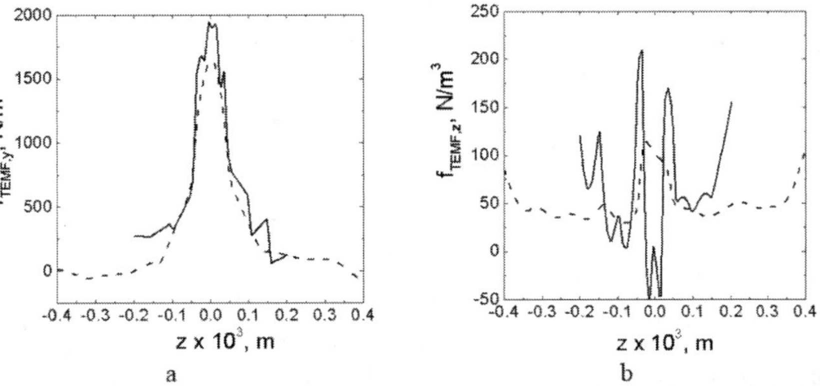

Figure 6. Calculated distribution of the y-component (a) and z-component (b) of the TEMF within sub-domains along the z-axis which passes through the centre of each sub-domain, parallel to the thermal gradient. Solid line and dashed lines are for sub-domain 200×400μm×400μm and 200×800μm×800μm, respectively. A particle is between $-0.05\cdot10^{-3}$m and $0.05\cdot10^{-3}$m.

The second sub-domain provides smooth variation of the y-component of the thermoelectric magnetic force in the liquid surrounding the particle (dashed line in figure 6a) and smaller values for its z-component (dashed line in figure 6b) which appear due to numerical errors. Furthermore, the maximal value calculated for the $f_{TEMF,y}$ in this case is about 1600N/m^3, i.e. the same value as obtained for a cubic particle in the infinite media in previous section (figure 5b). The values of the force at the boundary of the particle at $z=\pm0.05\cdot10^{-3}$m are also similar to the values obtained in previous section. In other words, the thermoelectric current is not affected by the presence of the walls of the domain.

For further calculation of the particle velocity and trajectory a larger sub-domain was used. Due to stable thermal gradient there is no thermo-convection. According to experimental observation, particle is moving preferentially in an horizontal direction with some oscillations in vertical direction. These oscillations may be attributed to solute convection. In the liquid phase, Lorentz force decrease sharply. It is expressed as a function of r^3, with r the distance from the center of the particle. We suppose that the liquid flow around the particle will change its shape but will not affect the motion.
The components of the thermoelectric magnetic force integrated over the volume of the particle according to the (10) are $F_{TEM}=(0.0,\ 1.7037\cdot10^{-9},\ 9.39\cdot10^{-11})$. The z-component of the TEMF, as said above, appeared because of the numerical errors and was 18 times smaller than the y-component. According to (9) a viscosity drag force is required to calculate the evolution of the particle velocity. Navier-Stokes equations are not implemented yet into AEQUATIO toolbox and because of this an analytical approximation was used in calculations. For a sphere with the radius R, the coefficient in (11) is known to be $A = 6\pi\mu_L R$. As approximation, the equivalent radius for a cubic particle can be taken $R_{eq} = D/\sqrt{\pi}$ where D is the size of the cube side to keep the same cross section of the particle that in our case gives R_{eq}=56.4 μm. Since the particle moves in the horizontal direction the temperature distribution does not change around it and therefore thermoelectric magnetic force does not change either. The stationary situation when TEMF and viscous drag force are equal occurs less than in a half of the second as shown in eq.(12) and the value of the stationary velocity is given as

$$\vec{V}_S = \frac{\vec{F}_{TEM,S}}{A} = \frac{\vec{F}_{TEM,S}}{6\pi\mu R_{eq}}$$

For our case this gives a velocity of 670µm/s. This value is about 3 times larger than the one estimated from the experimental results and theoretically in [9]. This difference can be explained by following factors. The first one is related to uncertainty in the physical properties of the materials, especially for the Seebeck coefficients. Thus in [9] other values for S_S and S_L were used and their difference was two times smaller than in the present work, i.e. the resulting TEMF was also two times smaller. The second reason is that the dendrite fragment does not have a regular shape: even if it has a spherical envelope its solid fraction is smaller that decreases the total TEMF and increases viscous drag force. For a cube integrated TEMF is approximately 2 times higher than the one of equivalent sphere. Finally, a drag force may be higher than we used in our approximation due to a small thickness of the crucible.

Table 1. Properties used in simulations.

Property	Units	Solid	Liquid
Thermal conductivity, λ	W/(m·K)	150	95
Thermoelectric power, S	V/K	$-5.4\cdot10^{-6}$	$-3.6\cdot10^{-6}$
Electrical conductivity, σ	$(\Omega\cdot m)^{-1}$	$1.3\cdot10^{7}$	$3.8\cdot10^{6}$
Dynamic viscosity, μ	Pa·s	-	$2.4\cdot10^{-3}$
Density, ρ	kg m^{-3}	2400	2400

6. Conclusion

Numerical simulations made with multi-physics multi-grid numerical toolbox AEQUATIO showed good agreements with analytical solutions. Oscillations of numerical solutions observed in the present work are probably related to a straightforward implementation of the phase function and can be suppressed via correct choice of the latter. In simulations of the experimental problem, constraints effect for the electric current related to the size of the simulation domain were not found. However, strong sensitivity of the results to the space discretization was revealed.

References

[1] Shercliff J A 1979 *J. Fluid Mech.* **91** 231
[2] Karklin J and Mikelson A J 1981 *J. Cryst. Growth* **52** 524
[3] Gelfgat Yu and Gorbunov L A 1989 *Dokl. Acad. Nauk SSSR* **306** 667
[4] Li Xi 2007 Thèse de docteur de l'INP Grenoble
[5] Yasuda H., Kato S, Shinba T, Nagira T, Yoshiya M, Sugiyama A, Umetani K and Uesugi K 2010 *Mat. Sci. Forum* **649** 131
[6] Nguyen-Thi H, Wang J, Salloum Abou Jaoude G, Reinhart G, Kaldre I, Mangelinck N, Ren ZM, Buligins L, Bojarevics A, Fautrelle Y, Budenkova O and Lafford T 2014 *Mat. Sci. Forum* **790-791** 420
[7] Kao A and Pericleous K, 2012 *IOP Conf. Ser.: Mater. Sci. Eng.* **33** 012045
[8] Li Xi, Gagnoud A, Wang J, Li Xialong, Fautrelle Y, Ren Z M, Lu X, Reinhart G and Nguyen-Thi H 2014 *Acta Mater.* **73** 83
[9] Wang J, Fautrelle Y, Ren Z M, Li X, Nguyen-Thi H, Mangelinck-Noel N, Salloum Abou Jaoude G, Zhong Y B, Kaldre I, Bojarevics A and Buligins L, 2012 *Appl. Phys.Lett.* **101** 251904
[10] Humbert E, Wang J, Budenkova O, Ren Z M and Fautrelle Y 2014 *Proc. Int. Sci. Colloquium Modelling for Electromagnetic Processing, MEP 2014, Hannover* 43
[11] Du Terrail Couvat Y, Gagnoud A and Triwong P **2008** *Proc. NUMELEC* 2008, Belgique,
[12] Kaldre I, Fautrelle Y, Etay J and Buligins L 2011 *Mod. Phys. Lett.* **25** 731

Advanced Numerical Modeling of the Dispersion of Ceramic Nanoparticles during Ultrasonic Cavitation Processing and Solidification of 6061-based Nanocomposites

D Zhang and L Nastac

The University of Alabama, Tuscaloosa, AL, 35487, USA

email: lnastac@eng.ua.edu

Abstract. The metal-matrix-nano-composites (MMNCs) in this study consist of a 6061 alloy matrix reinforced with 1.0 wt.% SiC 50 nm diameter nanoparticles that are dispersed uniformly within the matrix in large volume using an ultrasonic cavitation dispersion technique (UCDS) available in the Solidification Laboratory at UA. The required ultrasonic parameters to achieve the required cavitation for adequate degassing and refining of the aluminium alloy as well as the fluid flow characteristics for uniform dispersion of the nanoparticles into the 6061 matrix are being investigated in this study by using an in-house developed CFD ultrasonic cavitation model. The multiphase CFD model accounts for turbulent fluid flow, heat transfer and solidification as well as the complex interaction between the solidifying alloy and nanoparticles by using the Ansys's Fluent Dense Discrete Phase Model (DDPM) and a particle engulfment and pushing (PEP) model. The PEP model accounts for the Brownian motion. SEM analysis was performed on the as-cast MMNC coupons processed via UCDS and confirmed the distribution of the nanoparticles predicted by the current CFD model. A parametric study was performed using the validated CFD model. The study includes the effects of magnitude of the fluid flow and ultrasonic probe location (gravity direction).

1. Introduction

Aluminum-based metal matrix composites (MMCs) have been extensively studied and widely used in the aerospace, automotive and military industries due to their high strength-to-weight ratios and enhanced mechanical and thermal properties including specific modulus, superior strength, stiffness, good wear resistance, fatigue resistance and improved thermal stability [1-3]. However, the particles commonly used are micron-sized which has a counterpart that the ductility of the MMCs deteriorates with high ceramic particle concentration [4]. Consequently, more attention has been drawn to metal matrix nanocomposites (MMNCs), since the properties of metallic alloys reinforced by ceramic nanoparticles (with dimensions less than 100 nm) would be enhanced considerably while the ductility of the matrix is retained [5-11], because the incorporation of nano-sized reinforcements in metals and alloys will result in an Orowan-type strengthening mechanism where line defects are pinned by a uniform dispersion of particles within the grains, causing the dislocations to bow, and experimental evidence also shows that a significant degree of grain refinement (i.e., Hall-Petch strengthening) often occurs with additions of nanoparticles [12]. However, it is extremely difficult to obtain uniform dispersion of nano-sized ceramic particles in liquid metals due to high viscosity, poor wettability in the metal matrix, and a large surface-to-volume ratio, which results in agglomeration and clustering [4].

Currently, several fabrication technologies including high-energy ball milling [8, 11], in-situ synthesis [7], electroplating [13], and ultrasonic technology (UST) [4-5] are most commonly used, among which UST is supposed to be more reliable and cost effective.

During the solidification process, the particles will either be pushed or engulfed by the solidification front, among which particle pushing will always lead to segregation even clustering of the particulate reinforcement, which is undesirable as it results in non-homogeneous response and lower macroscopic mechanical properties. In general, it is considered that whether particles are pushed or engulfed during solidification depends on the velocity of the particle relative to the solidification front according to several previous models describing such particle engulfment and pushing phenomena [14-18]. However, these models only predict the behavior in the coarse (>> 1 μm) and fine particle (~ 1 μm) systems, and they don't accurately describe the ultrafine particle (<< 1 μm) system, presumably because the models rely on continuum mechanics. But these models cannot explain the evidence in MMNCs that nanoparticles can indeed be engulfed and distributed throughout the material and are not necessarily concentrated in grain boundary or interdendritic regions. As proposed by Ferguson [12], for sufficiently small particles, "Brownian Motion" can partially or completely counteract forces such as viscous drag, gravity and thermal/concentration gradients, thus leading to engulfment rather than pushing.

In this study, Ansys's Fluent Dense Discrete Phase Model (DDPM) [19] was adapted. The DDPM accounts for turbulent fluid flow, heat transfer, and the complex interaction between the molten alloy and nanoparticles during the melting and unidirectional solidification processes. Based on the theory proposed by Ferguson, all of the nanoparticles are assumed to be engulfed by the solidification front and no entrapment will occur. The dispersion of SiC nanoparticles with different injection positions, fluid flow intensities, and probe locations have been investigated in detail.

2. Model Description

The geometry of the model is shown in Figure 1. The ultrasonic probe has a diameter of 40 mm. The liquid aluminium alloy is 6061. It has a density of 2685 kg/m^3. The SiC nanoparticles with an average particle size of 55 nm and density of 3216 kg/m^3 are treated as inert-particles. The mass flow rate of the SiC nanoparticles is 0.014 kg/s. Thus, 1.0 wt.% of SiC nanoparticles can be injected at about 15 mm beneath the ultrasonic probe for 1 sec. The multiphase computational fluid dynamics (CFD) model accounts for turbulent fluid flow, heat transfer, and the complex interaction between the molten alloy and nanoparticles by using the ANSYS Fluent DDPM and k-ω turbulence model [19]. The CFD model is described in detail below.

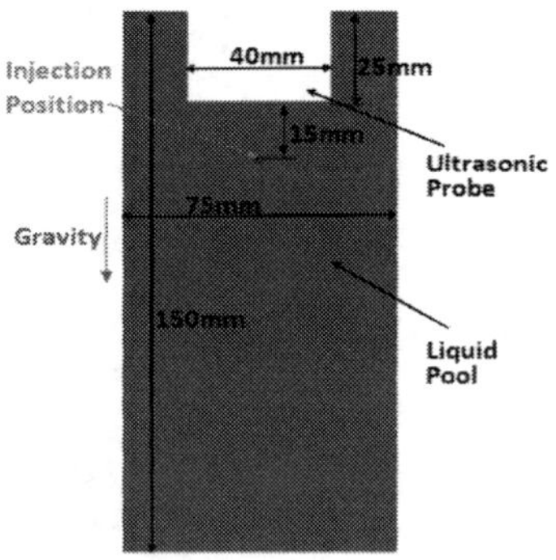

Figure 1. Geometry model

2.1. Fluid Flow Model

In the Eulerian DDPM multiphase model an Eulerian treatment is used for each phase, and the discrete phase (nanoparticles) is designated as a granular phase. The volume fraction of the particulate phase is accounted for in the conservation equations.
The continuity equation for the phase q is

$$\frac{\partial}{\partial t}\left(a_q \rho_q\right) + \nabla \cdot \left(a_q \rho_q \mathbf{u}_q\right) = \dot{m}_{pq} - \dot{m}_{qp} \tag{1}$$

The momentum balance for the phase q yields

$$\frac{\partial}{\partial t}\left(a_q\rho_q\mathbf{u}_q\right)+\nabla\cdot\left(a_q\rho_q\mathbf{u}_q\mathbf{u}_q\right)=-a_q\nabla P+\nabla\cdot\left[a_q\mu_q\left(\nabla\mathbf{u}_q+\nabla\mathbf{u}_q^{\ T}\right)\right]+a_q\rho_q g+\mathbf{f}_{DPM}+\mathbf{f}_{other} \quad (2)$$

where a_q is the phasic volume fraction, ρ_q is the density, $\mathbf{u}_q$ is the velocity, μ_q is the molecular viscosity, and P is the pressure shared by all phases. $\dot{m}_{pq}$ characterizes the mass transfer from the pth to q^{th} phase, and $\dot{m}_{qp}$ characterizes the mass transfer from phase q to phase p. The momentum exchange term, $\mathbf{f}_{DPM}$, is considered only in the primary phase equations. The source term, $\mathbf{f}_{other}$, includes the virtual mass force, lift force, and turbulent dispersion force etc.
Equations (1) and (2) do not solve for the velocity field and volume fraction of the discrete phase. Their values are obtained from the Lagrangian tracking solution.

2.2. Solidification Model

The enthalpy method is used in the solidification model. The energy conservation equation of the enthalpy-formulation is

$$\rho\frac{\partial h}{\partial T}+\rho\nabla\cdot(\mathbf{u}h)=\nabla\cdot(k\nabla T)+Q_L \qquad \text{with} \qquad h=h_{ref}+\int_{T_{ref}}^{T}c_p dT \quad (3)$$

where h is the sensitive enthalpy, h_{ref} is the reference enthalpy at the reference temperature T_{ref} ; c_p is specific heat; ρ and k are the density and thermal conductivity of the melting aluminium alloy, respectively, and $\mathbf{u}$ is velocity of the fluid.

The source term Q_L concerning the latent heat in a single phase solidification model can be written as

$$Q_L=\rho L\frac{\partial f_s}{\partial T} \quad (4)$$

where L is the latent heat and f_s is the solid fraction, which is assumed to vary linearly with the temperature in the mushy zone.

The mushy region is treated as a porous medium, and the porosity in each cell is set equal to the liquid fraction f_l in that cell. The momentum sink due to the reduced porosity in the mushy zone is:

$$\mathbf{S}=\frac{f_s^2}{(f_l^3+\varepsilon)}A_{mush}\mathbf{u} \quad (5)$$

where ε is a small number (0.001) to prevent division by zero; A_{mush} is the mushy zone constant, which measures the amplitude of the damping. The dynamic viscosity of the fluid also depends on the solid fraction which is related to temperature. It varies from 0.009 kg/(m·s) at the liquidus temperature to 1.0 kg/(m·s) below the coherency solid fraction (0.5).

2.3. Particle Tracking Model

The trajectory of a discrete phase particle is predicted by integrating the force balance on the particle. The force balance equates the particle inertia with the forces acting on the particle, and it is written as

$$\frac{d\mathbf{u}_p}{dt} = \mathbf{F}_D + \mathbf{F}_G + \mathbf{F}_B + \mathbf{F}_{virtual-mass} + \mathbf{F}_{pressure-gradient} + \mathbf{F}_{lift} + \mathbf{F}_{interaction} \tag{6}$$

where $\mathbf{u}_p$ is the particle velocity, and all the terms at the right-hand are with a unit of force/unit particle mass. The detailed description of the force terms shown in Equation (6) are presented in [19].

The chaotic effect of turbulence on the particle trajectories is accounted for using the stochastic tracking approach, i.e., the discrete random walk (DRW) model:

$$\mathbf{u} = \bar{\mathbf{u}} + \zeta\sqrt{2k/3} \tag{7}$$

where $\bar{\mathbf{u}}$ is the mean fluid velocity in the trajectory equation (9), ζ is a normally distributed random number, and k is the local turbulent kinetic energy.

Equation (6) can be cast into the following general form:

$$\frac{d\mathbf{u}_p}{dt} = \frac{1}{\tau_p}(\mathbf{u} - \mathbf{u}_p) + \mathbf{a} \tag{8}$$

where the term $\mathbf{a}$ includes accelerations due to other forces except drag force. Integrating the transport equation (8) for the path of each particle yields

$$\frac{d\mathbf{x}_p}{dt} = \mathbf{u} \tag{9}$$

where $\mathbf{x}_p$ is the particle position.

2.4. Boundary Conditions

The ultrasonic cavitation model is presented in Refs. [20, 21]. The ultrasonic cavitation model is uncoupled from the CFD nano-dispersion model because of significantly different time scales. The frequency used in the UST experiments is 18 kHz. However, modelling of ultrasonic cavitation takes place at a very small time scale (of the order of 10^{-5} to 10^{-7} s) and the fluid flow time scale is of the order of 10^{-2} to 10^{-3} s in this case. In order to account for the effects of ultrasonic stirring on the fluid flow-nanoparticle interaction, the ultrasonic probe surface is set as a velocity inlet sinusoidal profile with a frequency of 100 Hz (see Fig. 2), and therefore the time step in the nanodispersion simulations can be much higher (*i.e.*, 0.001s), which makes it possible to model the fluid flow with nanoparticles for much longer times with a reasonable computational time. After the initial transient, which is about 1s after adding the nanoparticles, and for the velocity profile shown in Fig. 2, the characteristics of the fluid flow and nanodispersion will remain unchanged. A similar trend was found at high frequencies.

All the walls are considered to be adiabatic before the solidification starts when the particles are well dispersed in the furnace. After that, the heat transfer boundary condition for the bottom wall of the furnace is set as constant temperature (500 °C) and all the other walls are still treated as adiabatic. All of the discrete phase boundaries are set as reflect. The interface between liquid 6061 alloy and air is pressure outlet and the other boundaries are set as wall.

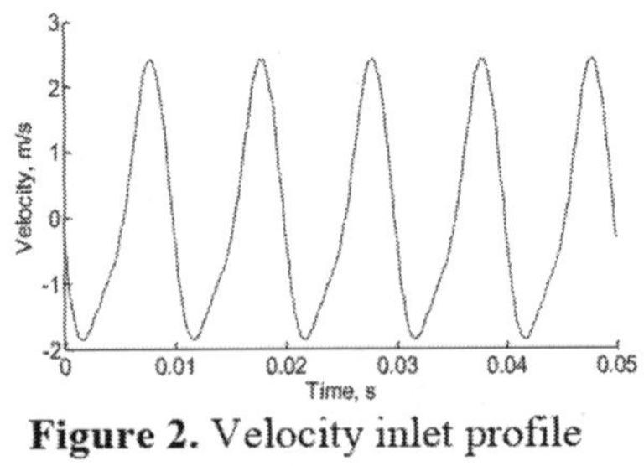

Figure 2. Velocity inlet profile

2.5. Solution Procedure

The SiC nanoparticles are injected at every fluid flow time step with a mass flow rate of 0.014 kg/s in the first sec. The distribution of the particle diameters varying from 45 nm to 65 nm follows the Rosin-Rammler expression. Particles are tracked at every time step after the fluid velocity field is solved. Because of the low volume fraction of the discrete phase, one-way coupling is employed, which neglects the effect of the discrete phase on the fluid turbulence.

3. Simulation Results and Discussion

Figure 3 shows the fluid flow and particle distributions after 1.0s, 3.0s, 10s and 30s, respectively. It can be seen that the flow is much stronger at the center of the furnace. It demonstrates that the particles have little effect on the fluid flow because of the one-way coupling, and the gravity direction is the same as the direction of the stronger flow in the center, so it will help to develop the fluid flow. Therefore the fluid flow after 3.0s is almost the same as that after 1.0s because both of them are fully developed (see Figs. 3(a) and 3(b)). The fluid flow is damped in the mushy zone, where the particles will move relatively slow and stop moving in the solid (see Figs. 3(c) and 3(d)). At 1.0s when the injection is stopped, newly injected particles are still at the center of the furnace where they are injected, since more time is needed for all of them to be dispersed into the liquid.

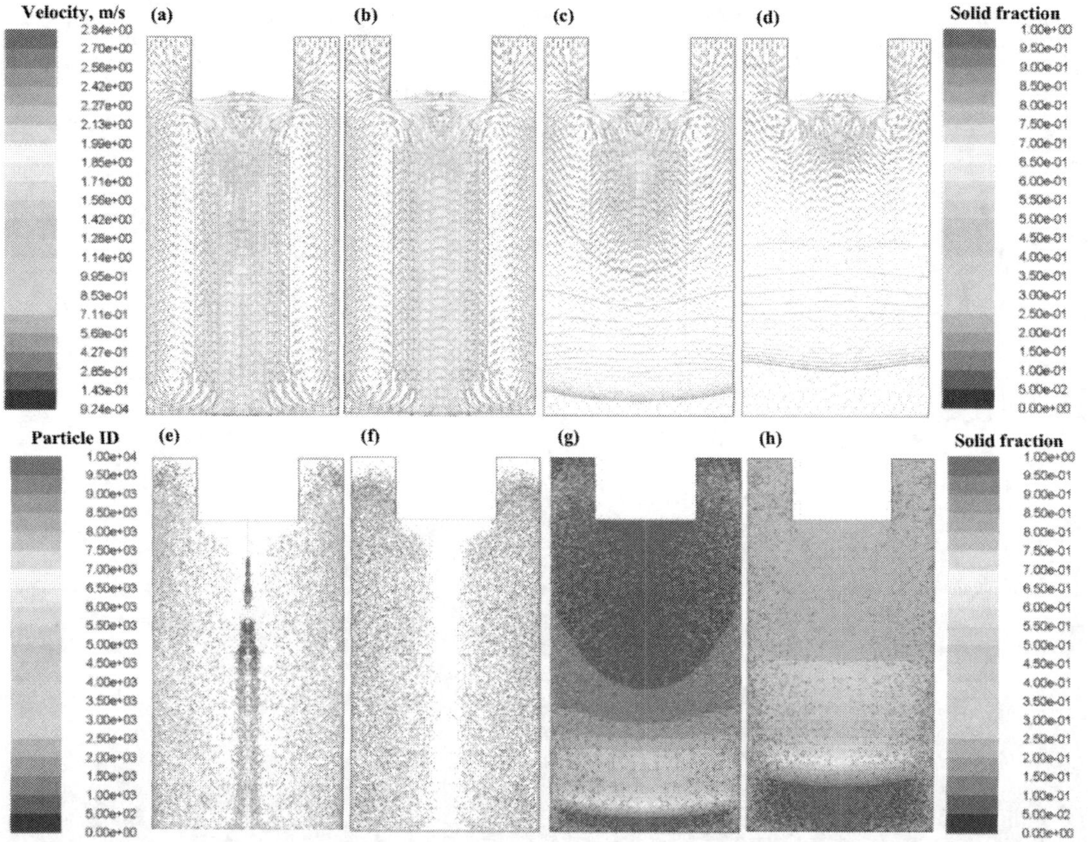

Figure 3. Fluid Flow and particle distributions (gravity direction downward, strong flow): (a) and (e) after 1.0s; (b) and (f) after 3.0s; (c) and (g) after 10s; and (d) and (h) after 30s.

IOP Conf. Series: Materials Science and Engineering **84** (2015) 012020 doi:10.1088/1757-899X/84/1/012020

Two seconds later, the particles have been dispersed pretty well from the bottom to the top, but a little more particles tend to stay near the wall, and fewer nanoparticles at the center where the flow is stronger, which indicates the nanoparticles couldn't disperse well in strong flows (see Fig. 3(f)). Additionally, the particle distribution stays almost the same henceforth which means the distribution becomes stable. Then, the solidification process starts at 3.0s, and in order to model the unidirectional solidification process, all the walls except the bottom one are treated as adiabatic. After 10s, we can see that there are some agglomerations in the solid (see Fig. 3(g)) because the nanoparticles approaching the solid/liquid interface will be engulfed based on the assumption that the Brownian motion will dominate during the particle engulfment/pushing process. It can be seen from Fig. 3(h) that after 30s the nanoparticles are dispersed quite well throughout the metal matrix, which is also validated by SEM analysis [22, 23]. To investigate the effect of the magnitude of the fluid flow on the nanodispersion, the velocity magnitude is changed to 1/10th of the original one shown in Fig. 2.

Figure 4 shows the fluid flow and particle distributions after 1.0s, 3.0s, 10s and 30s, respectively. As the fluid flow is weaker, the particle dispersion is slower. However, the flow pattern is similar as before (see Figs. 4(a) and 4(b)). But the flow fields after 1.0s and 3.0s are not exactly the same, because the flow after 1.0s is not fully developed. By the time when the injection is stopped (1.0s), most of the particles are at the top of the furnace where they are injected, because more time is needed to disperse them well in the liquid metal (see Fig. 4(e)).

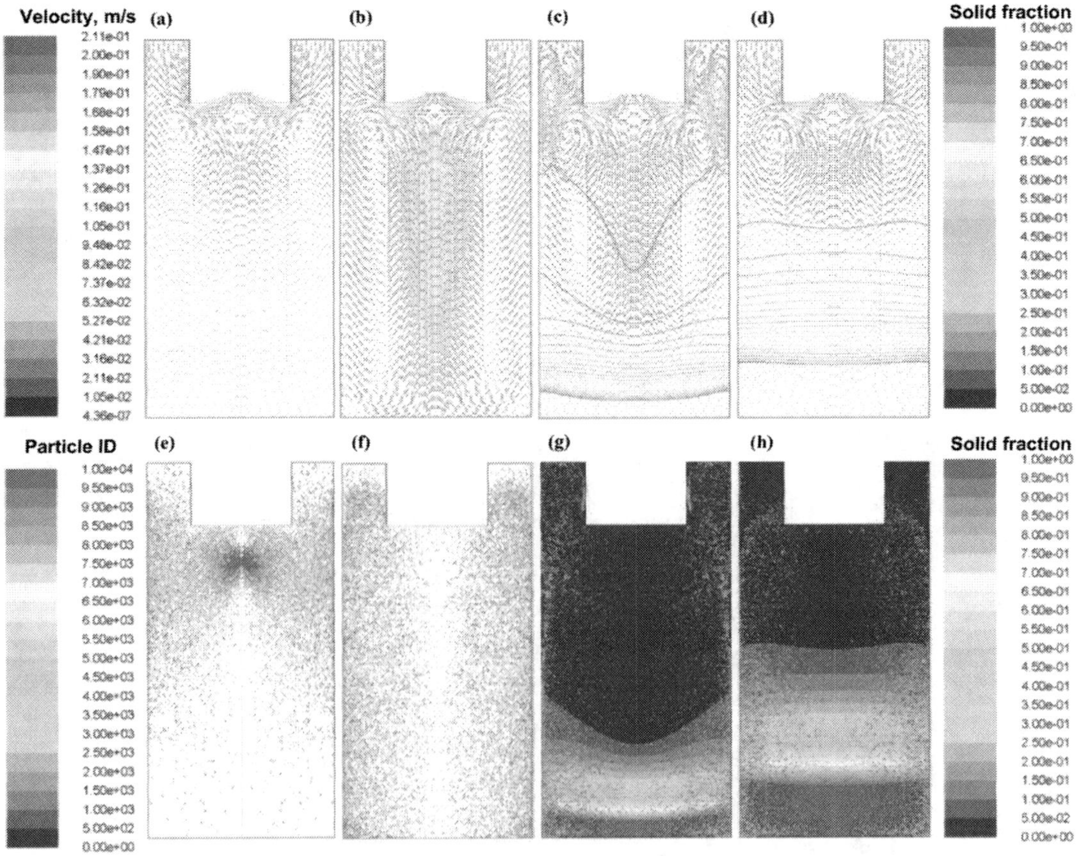

Figure 4. Fluid Flow and particle distributions (gravity direction downward, weak flow): (a) and (e) after 1.0s; (b) and (f) after 20s; (c) and (g) after 30s and (d) and (h) after 50s.

After 20s, the distribution of the nanoparticles becomes uniform and stable, and even more uniform than the case with a strong flow, because the difference between the fluid flow at the center of the furnace and that near the wall is not as significant as in the former case (strong flow). Then when the solidification starts 10s later, the overall distribution of the nanoparticles stays almost the same (see Fig. 4(g)). But after 50s, it is obvious that more particles are captured by the mushy zone, and almost no particles are at the top of the furnace, which indicates that the uniformity of the particles deteriorates (see Fig. 4(h)). Therefore, we can conclude that weak flow cannot disperse the nanoparticles well during the solidification process. To investigate the effect of the gravity direction, the ultrasonic probe is placed at the bottom of the furnace instead of the top as shown in Figure 5(a). Figure 5 presents the fluid flow and particle distributions after 1.0s, 3.0s, 10s and 20s, respectively.

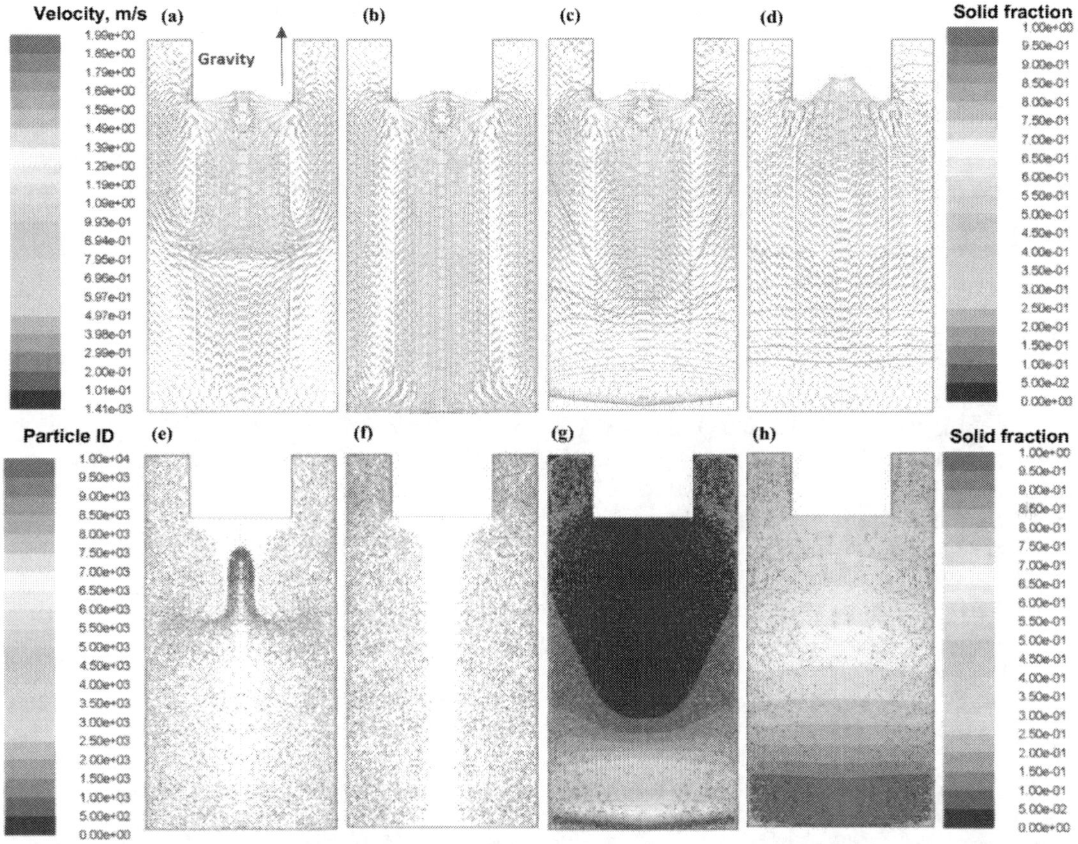

Figure 5. Fluid Flow and particle distributions (gravity direction upward, strong flow)): (a) and (e) after 1.0s; (b) and (f) after 3.0s; (c) and (g) after 10s and (d) and (h) after 20s.

It can be seen from Figs. 5(a) and 5(b) that the flow fields are not the same though a one-way coupling is used. This is because the gravity direction is opposite to the direction of the stronger flow in the center of the furnace, thus retarding the development of the fluid flow and resulting in a different flow pattern at 1.0s. But after 3.0s, the fluid flow is fully developed and becomes almost the same as that of the case with a different gravity direction, because the gravitational force is relatively weak compared to the drag force, if the flow is very strong. When the solidification starts after 3.0s, the fluid flow is damped in the mushy zone and becomes weaker, and then the gravity begins to play an import role in the flow. After 20s, the flow in the center even has an opposite direction as before, which is the same as the gravity direction (see Fig. 5(d)), and accelerate the solidification process. As the solidification continues, the entire domain is fully solidified after only around 30s (this result is not shown in Fig. 5). The gravity effect on the dispersion of the nanoparticles is not significant in the fluid metal and the particle distribution is pretty uniform as before (see Fig. 5(f)). But as explained above, the gravity will make a big difference during the solidification process, reducing the solidification time and producing well dispersed nano-composites quickly.

4. Concluding Remarks and Future Work

The DDPM model coupled with the k-ω turbulence model is used to investigate the distribution of SiC nanoparticles into a 6061 matrix with different flow magnitudes and gravity directions under ultrasonic stirring and unidirectional solidification processing conditions. It was determined that a strong flow can disperse the nanoparticles relatively well, especially when compared with a weak flow condition. Also, it was demonstrated that the gravity direction can play a major role in the nano-dispersion during the solidification process. When the gravity direction is upward with respect with the location of the ultrasonic probe, the solidification time is decreased and thus well dispersed nano-composites can be produced faster than in the case when the gravity direction is reversed. The CFD nanodispersion model will be fully coupled with the ultrasonic cavitation model. The simulations will then be performed via a high performance computing cluster available at the University of Alabama.

References

[1] Kaczmar J W, Pietrzak K, Wlosinski W 2000 *J. Mater. Process. Technol.* **106** 58-67
[2] Durisinova K, Durisin J, and Orolinova M et al. 2012 *J. Alloys Compd.* **525** 137-142
[3] William C H 1998 *Mater. Sci. Eng., A* **244** 75–79.
[4] Yang Y, Lan J, Li X 2004 *Mater. Sci. Eng., A* **380** 378-383
[5] Cao G, Konishi H, Li X 2008 *Mater. Sci. Eng., A* **486** 357-362
[6] Shin J H, Choi H J, and Cho M K et al. 2014 *J. Compos. Mater.* **48** 99-106
[7] Dikici B, Gavgali M, Bedir 2010 *J. Compos. Mater.* **45** 895-900
[8] El-Daly A A, Abdelhameed M, and Hashish M et al. 2013 *Mater. Sci. Eng., A* **559** 384-393
[9] Jiang X, Galano M, Audebert F 2014 *Mater. Charact.* **88** 111-118
[10] Shin J H, Bae D H 2014 *Mater. Chem. Phys.* **143** 1423-1430
[11] Borgohain C, Acharyya K, and Sarma S et al. 2013 *J. Mater. Sci.* **48** 162-171
[12] Ferguson J B, Schultz B F, and Rohatgi P K et al. 2014 *Light Metals TMS* 1383-1388.
[13] Sautter F K 1963 *J. Electrochem. Soc.* **110** 557
[14] Kim J K, Rohatgi P K 1998 *Metall. Mater. Trans. A* **29** 351-358
[15] Uhlmann D R, Chalmers B, and Jackson K A 1964 *J. Appl. Phys.* **35** 2986-2993
[16] Stefanescu D M, Moitra A, Kacar A S, and Dhindaw B K 1990 *Metall. Trans. A* **21** 231-239
[17] Shangguan D, Ahuja S, and Stefanescu D M 1992 *Metall. Trans. A* **23** 669-680
[18] Kaptay G 2001 *MetalL. Mater. Trans. A* **32** 993-1005
[19] Fluent 6.3: User's Guide Manual Fluent Inc. and Ansys's Fluent 2006 http://ansys.com/
[20] Nastac L 2011 *Metallur. and Mater. Trans.* B **42**(6) 1297-1305.
[21] Nastac L 2014 *Numerical Modeling of Fluid Flow and Solidification Characteristics of Ultrasonically Processed A356 Alloys. ISIJ Intern.* **54**(8) 1830-1835.
[22] Liu X, Jia S, Nastac L 2014 *Intern. Journal of Metal Casting (AFS)* **8** 1 1-12.
[23] Jia S, Allison P G, Rushing T W, Nastac L, Book Proceedings of *Advances in the Science and Engineering of Casting Solidification: An MPMD Symposium Honoring Doru Michael Stefanescu*, Eds. Nastac L *et al.*, TMS Conference, Orlando, FL, March 15-19, 2015 1-8.

Effect of Electromagnetic Brake on Decreasing Unbalanced Flow in Mold

Yuji Miki and Kohei Furumai

JFE Steel Corporation, Steelmaking Research Dept., Steel Research Laboratory, JFE Steel Corporation, 1 Kokan-cho, Fukuyama, Hiroshima 721-8510

E-mail: y-miki@jfe-steel.co.jp

Abstract. In order to clarify the mechanism of suppression of unbalanced flow in the mold under high throughput conditions in continuous casting, the effect of using an electromagnetic brake on decreasing molten steel momentum and suppressing unbalanced flow in the mold was investigated. (1) The measured value of the effect of the electromagnetic brake on decreasing molten steel momentum was consistent with the calculated value. Molten steel momentum could be reduced by more than 50% when the Stuart number was more than 3. 5. (2) Increased mold level fluctuations caused by unbalanced flow in the mold are suppressed by applying the optimum magnetic flux density with an electromagnetic brake, even under high throughput conditions.

1. Introduction

Inclusions and bubbles which are trapped by the solidified shell during continuous casting and remain in the slab can cause defects during cold or hot rolling. In recent years, higher surface quality has been required under high throughput conditions in continuous casting.

High throughput conditions in continuous casting induce penetration of inclusions into the deep region[1)] and entrainment of mold flux caused by increased mold level fluctuations[2)], and thereby degrade the quality of slabs. However, it is possible to reduce penetration of inclusions into the deep region and suppress mold level fluctuations by applying an electromagnetic brake in the mold[3)]. It is also known that an unbalanced flow is induced by nozzle clogging or sliding nozzle gate movement, and this unbalanced flow increases under high throughput conditions[4-7)]. When a large unbalanced flow occurs in the mold, areas of large or small molten steel flow velocity in the mold occur (Fig.1)[8)].

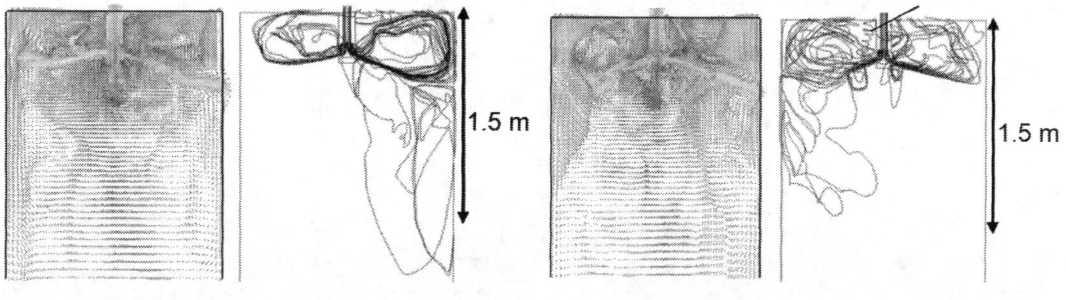

Fig. 1 Velocity vector and 1 mm bubble trajectories of time-dependent unbalanced flow calculated by LES model[8)].

In a large upward flow velocity area at near the narrow side of the mold, mold flux entrainment is caused by the large mold level fluctuations at the meniscus of the mold[9)]. In the deeper region of the mold, penetration of inclusions is caused by the large downward flow velocity[10)]. In a small molten steel flow velocity area at the inner shell, entrapment of inclusions and/or bubbles at the solid–liquid interface increases due to the decrease of washing effects[11)]. If the unbalanced flow in the mold is large, it may also cause breakout due to remelting of the solidified shell at the narrow side area by the large discharge flow rate from the submerged entry nozzle (SEN)[12)]. As techniques for suppressing unbalanced flow in the mold, study of the shape of the immersion nozzle, electromagnetic stirring in the mold, and application of an electromagnetic brake to the immersion nozzle have been reported[13-15)]. However, the relationship between unbalanced flow in the mold and the distribution of defects entrapped on the solidified shell in the mold, and the effect of an electromagnetic brake in the mold on unbalanced flow had not been investigated. In this paper, we measured the molten steel momentum in a mold that has two static magnetic fields in the upper and lower areas of the mold and the distribution of defects entrapped on the solidified shell in order to evaluate the effects of the electromagnetic brake on the distribution of defects entrapped on the solidified shell, unbalanced flow and the molten steel momentum in the mold.

2. Experimental

Experiments were performed at No. 3 continuous caster (No. 3 CC) at JFE Steel East Japan Works(Chiba District), where the strand is equipped with the electromagnetic FC (Flow Control) mold[3)]. A schematic diagram of the FC mold is shown in Fig. 2.

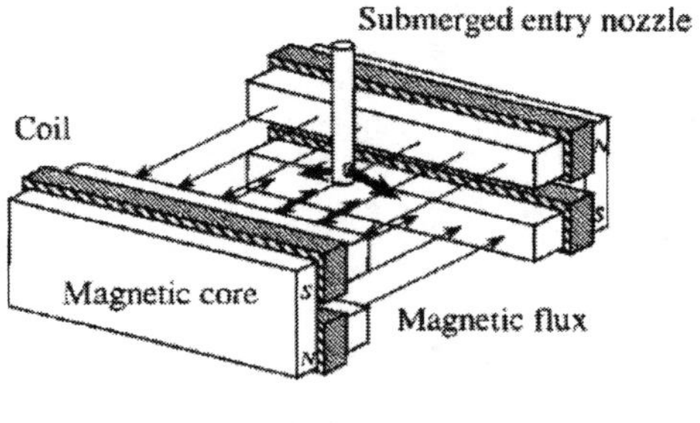

Fig. 2 FC mold[3)].

The FC mold uses two static magnetic fields, an upper field at the slab meniscus level to control the meniscus metal flow velocity, and a lower field to control the lower part of the mold in order to minimize the penetration depth of the steel jets from the SEN(Fig.3)[8)].

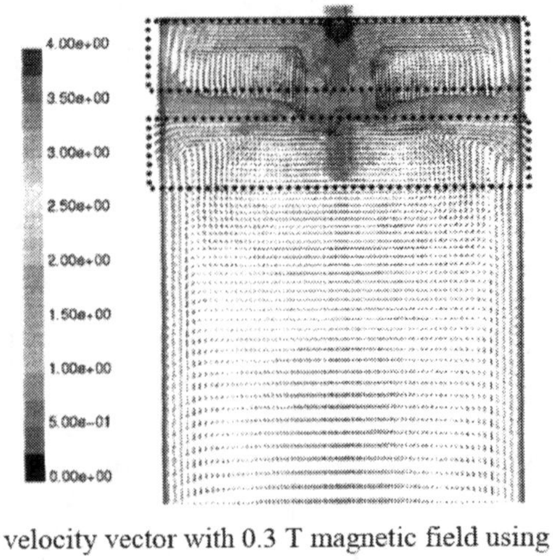

Fig. 3 Flow velocity vector with 0.3 T magnetic field using FC mold[8)].

In the FC mold, the discharge holes of the SEN are located between the center of upper magnetic field and the center of the lower magnetic field. In this experiment, a three-layer type sliding gate was used. The opening and closing direction is the mold width direction. Casting was performed while blowing Ar gas through the SEN (two hole-type pool bottom nozzle). The experimental conditions are listed in Table 1. The casting speed was from 1.2 m/min to 1.7 m/min, and the maximum throughput was 5.3 ton/min. During the test, the magnetic flux density of the electromagnetic brake in the FC mold was varied under various casting conditions. The concentration of C was varied from 0.0015wt% to 0.0018wt%. Ar gas was blown into the mold through the SEN. To evaluate the molten steel momentum reduction effect of the electromagnetic brake, the angles of dendrites in the slabs were measured, and the flow velocity of the molten steel in the mold was calculated using Eq. (1)[16].

To evaluate the solidification constant (2.58×10^{-2} m/s$^{1/2}$) calculated by Eq. (2), iron-sulfur alloys were added to the molten steel during casting, and the solidified shell thickness in the slabs was measured with the sulfur concentrations. The solidification rates in Eq. (1) were calculated by Eq. (3). To measure the upward molten steel velocity, samples were cut from the short side of the slabs. After the samples were polished and etched with picric acid, the angles of dendrites were measured at 1mm intervals by using a microscope.

$$\theta = \left(\frac{0.35 \cdot C_0^{\ 2}}{C_0^{\ 2} + 0.0005} + 0.65 \right) \cdot 11.5 \cdot V_F^{\ -0.177} \cdot \log\left(\frac{5.38 \cdot 10^{-1} \cdot V_F^{\ 2.08}}{V} \right) \tag{1}$$

$$D = K\sqrt{t} = K\sqrt{Z / V_c} \tag{2}$$

$$V = \frac{dD}{dt} \tag{3}$$

Where, θ: angles of dendrites [degree], V_F: molten steel flow velocity [m/s], V: growth rate [m/s], C_0: carbon content in molten steel [mass%], D: solidified shell thickness [m], K: solidification constant [m/s$^{1/2}$], t: time [s], Z: distance from meniscus [m] and Vc: casting velocity [m/min].

3. Calculation Method

The molten steel flows in the mold were calculated in order to evaluate the effect of the electromagnetic brake[11]. The standard k-ε model in Fluent was employed to calculate the molten steel flows in the mold[17-18]. The equation of continuity, Navier-Stokes equation and law of the conservation of energy in Fluent are expressed by Eqs. (4)-(6).

$$\frac{\partial v_x}{\partial x} + \frac{\partial v_y}{\partial y} + \frac{\partial v_z}{\partial z} = 0 \tag{4}$$

$$\rho \left[\frac{\partial \vec{v}}{\partial t} + (\vec{v} \cdot \nabla)\vec{v} \right] = -\nabla p + \eta \nabla^2 \vec{v} + \vec{F} \tag{5}$$

$$\rho \left[\frac{\partial \vec{T}}{\partial t} + (\vec{v} \cdot \nabla) T \right] = -\frac{k}{C_p} \nabla^2 T + \frac{Q}{C_p} \tag{6}$$

Where, v: velocity [m/s], ρ: molten steel density, p: pressure [Pa], η: viscosity coefficient [0.0057Pa·s][19], F: external force [N/m^3], T: temperature [K], k: thermal conductivity [34W/K/m][20], Cp: specific heat at constant pressure [753J/kg/K][21], Q: latent heat flux of solidification [J/m^3/s], x: width direction in mold, y: thickness direction in mold and z: continuous casting direction.

We calculated the solid phase ratio was calculated from the temperature field. Q was calculated by coupling the flow calculations with a subroutine of a changing coefficient of viscosity. The electromagnetic braking forces were calculated by using Eqs. (7)-(8) while determining external magnetic field B.

$$\vec{J} = \sigma(\vec{E} + \vec{V} + \vec{B}) = \frac{1}{\mu}\nabla \times \vec{B} \tag{7}$$

$$\frac{\partial \vec{B}}{\partial t} + (\vec{v}\cdot\nabla)\vec{B} = \frac{1}{\sigma\mu}\nabla^2 \vec{B} + (\vec{B}\cdot\nabla)\vec{v} \tag{8}$$

Where, J: electric current density [A/m^2], E: electric field [V/m], magnetic flux density [T], σ: electric conductivity [7.14×10^{6}S/m][22) and μ: magnetic permeability [1.26×10^{-6}H/m][23).

4. Experimental Results and Discussion

4.1. Effect of electromagnetic brake on molten steel momentum

Fig. 4 shows the relationship between magnetic flux density and decreasing ratio of molten steel momentum in the mold. The vertical axis is the ratio of measured or calculated molten steel momentum to momentum without magnetic field. The measured molten steel momentum ρV_m was determined from the angles of dendrites(Eq.(1)) and the calculated molten steel momentum ρV_c was determined by calculation by Eqs. (4)-(6). ρV_e is the momentum in absence of the magnetic field and was calculated by Eqs. (9)-(13)[24). γ, h_{TD}, l_1, l_2, l_3, q, a, B', ϕ, and d (Eqs. (9)-(13)) depend on the experimental conditions.

$$V_d = \gamma\sqrt{(1-\zeta_4)(1-\zeta_3)}\cdot\sqrt{2g[c^2(h_{TD}+l_1)+l_2+l_3)]} \tag{9}$$

$$c = 0.364q^{0.65} \tag{10}$$

$$\zeta_3 = 1.1\left(1-\frac{a}{B'}\right)^2 \tag{11}$$

$$\zeta_4 = 1.16 - 0.015\phi \tag{12}$$

$$\rho V_e = \rho V_d\left(\frac{X}{6.3d}\right)^{-1} \tag{13}$$

Where, V_d: discharge flow rate from submerged nozzle [m/s], γ: ratio of average flow velocity and maximum flow velocity from discharge portion of submerged nozzle[24), g: acceleration of gravity [m/s^2], h_{TD}: molten steel depth in tundish [m], l_1: length of upper nozzle [m], l_2: length of under nozzle [m], l_3: distance between under nozzle and molten steel surface level in submerged nozzle [m], c: coefficient of discharge of free fall flow in submerged nozzle [-], q: molten steel throughput [ton/min], a: cross-sectional area of falling flow in submerged nozzle [m^2], B': cross-sectional area in submerged nozzle [m^2], ϕ: discharge angle of submerged nozzle [degree], V_e: molten steel collision velocity on narrow side of mold [m/s], X: horizontal distance between discharge port of submerged nozzle and mold wall [m] and d: diameter of discharge port in submerged nozzle [m].

When the magnetic flux density is small, the actual measured momentum value is equivalent to that in the absence of the influence of a magnetic field obtained by Eq. (13).

On the other hand, when the magnetic flux density is large, the actual measured momentum value is smaller, and it was confirmed to be about half of the maximum momentum compared with the case of not applying the magnetic brake. The calculation results of the molten steel momentum were in good agreement with the actual measured values at the molten steel throughput values of 4.7 and 5.4 ton/min. Electromagnetic braking force F in a mold can be expressed by Eq. (14).

$$\vec{F} = \vec{J} \times \vec{B} \tag{14}$$

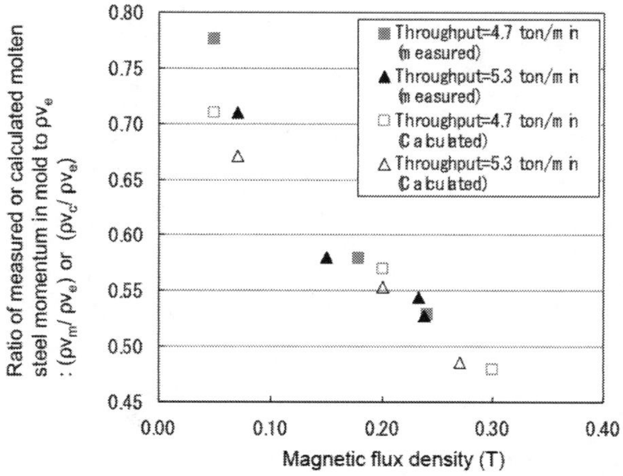

Fig.4 Relationship between magnetic flux density and ($\varrho v_m / \varrho v_e$) or ($\varrho v_c / \varrho v_e$).

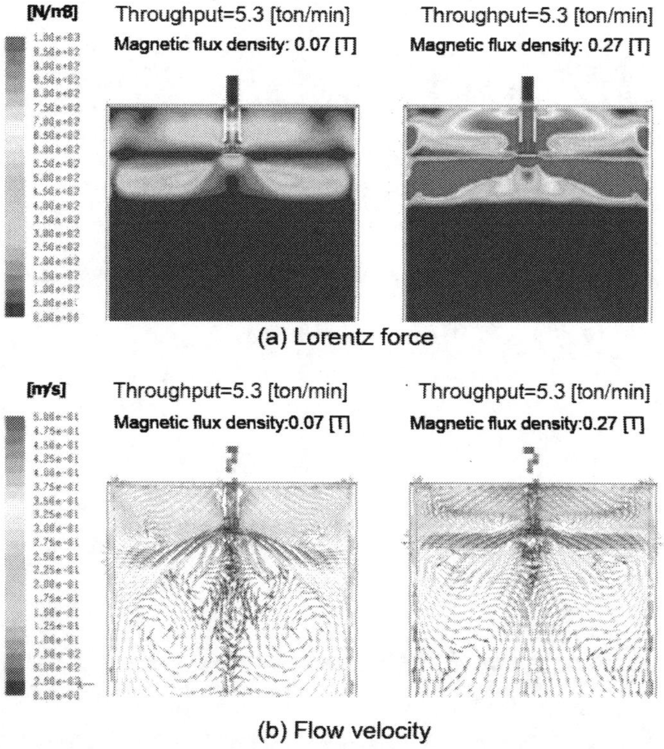

Fig.5 Calculation results of Lorentz force and flow velocity.

Fig. 5 shows the results of the calculated Lorenz force and the flow velocity in the mold under the magnetic flux density conditions of 0.07 and 0.27 T. The effect of the Lorenz force on the molten steel was increased by increasing the magnetic flux density. As a result, the momentum of the molten steel collision on the narrow sides of the mold was reduced. The Stuart number N was evaluated in order to investigate the effect of electromagnetic force. The Stuart number N is the ratio of inertial force and external force represented by Eq. (15)[23].

$$N = \frac{|\sigma(\vec{v} \times \vec{B}) \times \vec{B}|}{|\rho(\vec{v} \cdot \nabla)\vec{v}|} = \frac{B^2 \sigma L}{\rho v} \tag{15}$$

Where, L: characteristic length [m] ≒ mold width [m]

Fig. 6 shows the relationship between N and the magnetic flux density under different throughput conditions. In order to increase the effect of the electromagnetic brake force and overcome inertial force under large throughput conditions, it is necessary to apply a larger magnetic flux density compared with the case of low throughput conditions. From this, it is important to apply an adequate electromagnetic brake corresponding to increased throughput conditions. Fig. 9 shows the relationship between N and the ratio of measured or calculated molten steel momentum in the mold to the momentum calculated by Eq. (13).

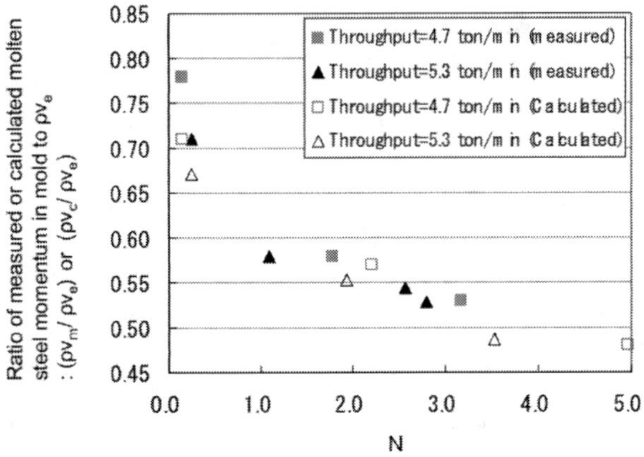

Fig.6 Relationship between N and ($\varrho v_m/\varrho v_e$) or ($\varrho v_c/\varrho v_e$).

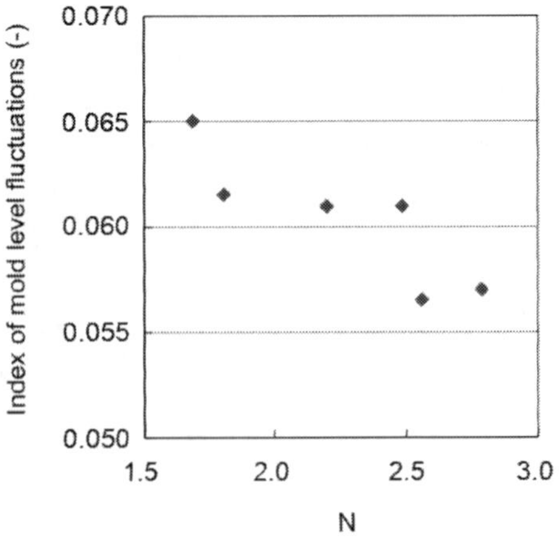

Fig.7 Relationship between Stuart number N and mold level fluctuations.

A correlation can be observed between the ratio of the measured or calculated molten steel momentum in the mold and the momentum calculated by Eq. (13) and N. The momentum reduction effect of the electromagnetic brake in the mold is large with large N. Under the different throughput and magnetic flux density conditions, it is considered that the molten steel momentum in the mold can be reduced to about 50%, compared to that in case an electromagnetic brake is not applied, when N is more than 3.5.

4.2. Effect of electromagnetic brake on mold level fluctuations

Fig. 7 shows the relationship between N and increases in mold level fluctuations due to unbalanced flow in the mold under different magnetic flux density conditions. Mold level fluctuations decreases with increasing N. These results are the same as the results for the relationship between the difference of molten steel momentum and molten steel throughput in the mold. Based on these results, it can be stated that the unbalanced distribution of defects and increased mold level fluctuations caused by an unbalanced flow in the mold are suppressed by applying the optimum magnetic flux density with an electromagnetic brake, even under high throughput conditions.

5. Conclusion

The effect of using an electromagnetic brake on decreasing molten steel momentum and suppressing unbalanced flow in the mold was investigated. The following knowledge was obtained.
(1) The measured value of the effect of the electromagnetic brake on decreasing molten steel momentum was consistent with the calculated value. Molten steel momentum could be reduced by more than 50% when the Stuart number N was more than 3.5.
(2) Increased mold level fluctuations caused by unbalanced flow in the mold are suppressed by applying the optimum magnetic flux density with an electromagnetic brake, even under high throughput conditions..

References

[1] Miyamura K, Kaneko N and Kanamaru K **1979**: Tetsu-to-Hagané, 65 S229.
[2] Kosaka S, Yokoya S, Iwata K, Tsukaguchi Y, Yoshihara M and Hayashi H **2003**: CAMP-ISIJ, 16 949.
[3] Idogawa A, Kitano Y and Tozawa H **1996** : Kawasaki Steel Giho, 28 46.
[4] Kubo N, Kubota J and Ishii T **2001**: ISIJ Int., 41 1221.
[5] Gupta D, Chakraborty S and Lahiri A.K **1997** : ISIJ Int., 37 654.
[6] Wang Y.H. **1990** : Steelmaking Conf. Proc., 73 473.
[7] Sasai K, Yamamura H, Tsutsumi N, Mizukami Y and Esaka H **1992** : CAMP-ISIJ, 5 223.
[8] Miki Yand Takeuchi S **2003** : ISIJ Int., 43 10 1548.
[9) Teshima T, Kubota J, Suzuki M, Ozawa O, Masaoka T and Miyahara S **1992** : Tetsu-to-Hagané, 79 576.
[10] Tanaka H, Imamura A and Nishihara R **1992** : Tetsu-to-Hagané, 78 1464.
[11] Miki Y, Ohno H, Kishimoto Y and Tanaka S **2011** : Tetsu-to-Hagané, 97 423.
[12] Nakai D, Yamamoto Y, Miyake T and Nakata H **2008**: Proceedings of the 4th International Congress on Science Technology of Steelmaking, Gifu, Japan, 97.
[13] Tsukaguchi Y, Hayashi H, Kurimoto H, Yokoya S, Marukawa K and Tanaka T **2009** : Tetsu-to-Hagané, 95 33.
[14] Morishita M, Tai Y, Ayata K and Katsuta J **1997**: CAMP-ISIJ, 10 832.
[15] Morishita M, Terauchi M, Nakao M, Koga Y and Nakaoka T **2007** : CAMP-ISIJ, 20 867.
[16] Esaka H, Suter F and Ogibayashi S **1996** : ISIJ Int., 36 1264.
[17] Fluent 6.3 Manual **2007**, ANSYS Inc., 10 Cavendish Court, Lebanon, NH, U.S.A.
[18] Launder B.E. and Spalding D.B **1972** : Lectures in Mathematical Models of Turbulence. Academic Press, London, England.
[19] Nakanishi K, Saito T and Shiraishi Y **1967** : J. Jpn. Inst. Met. 31 881.
[20] Touloukian Y.S. **1967** : Thermophysical Properties of High Temperature Solid Materials, Vol. 1, Macmillan Co. New York.
[21] Elliott J.F. and Gleiser M **1963** : Thermochemistry for Steelmaking, AIME.
[22] Hirayama K, Kuwano R, Ono Y and Yagi S **1970** : Tetsu-to-Hagané, 56 S91.
[23] Iwai K **2004** : Bull. Iron Steel Inst. Jpn., 9 705.
[24] Imamura A, Kusano A and Moritama N **1992** : Tetsu-to-Hagané, 78 439.

Effect of sand dilation on core expansion during steel casting

D Galles and C Beckermann[1]
Department of Mechanical and Industrial Engineering, University of Iowa

E-mail: becker@engineering.uiowa.edu

Abstract. The thermo-mechanical behavior of the bonded sand used for molds and cores has a strong effect on dimensions of steel castings. Experiments are conducted in which a thick-walled hollow carbon steel cylinder is cast using a silica sand core. The temporal evolution of the inner diameter of the cylinder is measured in-situ during solidification and cooling by utilizing quartz rods connected to LVDTs (Linear Variable Differential Transformers). It is found that the inner diameter increases significantly during the initial stages of solidification when the steel offers little restraint to core expansion. Without accurately modeling this initial core expansion, the final cylinder dimensions at room temperature cannot be predicted. Preliminary simulations using the measured linear thermal expansion coefficient of the core considerably under-predict the measurements, which suggests that shear induced sand dilation also contributes to core expansion. The Drucker-Prager Cap model, which can predict dilative behavior, is used to simulate the mechanical behavior of the core. Utilizing this model in conjunction with an elasto-visco-plastic constitutive law for the steel, the stress simulations successfully predict the observed dimensional changes in the casting during solidification.

Introduction

During pattern design for steel sand castings, dimensional changes are estimated by specifying pattern allowances (PA [%] = {(pattern feature size – casting feature size)/ casting feature size} × 100). Several physical phenomena contribute to pattern allowances. Thermal contraction of the steel reduces unconstrained linear dimensions by approximately 2.2%. The sand mold and, specifically, cores impact dimensions in two ways. First, cores restrain the casting from free shrink to generate distortions (i.e., plastic strains). Second, cores expand into the mold cavity to reduce casting dimensions while the solidifying steel is weak and offers little restraint.

While the aforementioned mold effects have been previously investigated [1], an additional phenomenon that can also affect pattern allowances is sand dilation, i.e., volumetric expansion of a granular material due to a shear force. This behavior is illustrated in figure 1; in the undisturbed state, the sand grains are tightly-packed and contain small air voids between the grains (figure 1(a)). After a shear force is applied, however, the irregularly-shaped sand grains translate and/or rotate, causing the voids to grow, resulting in expansion of the sand aggregate (figure 1(b)).

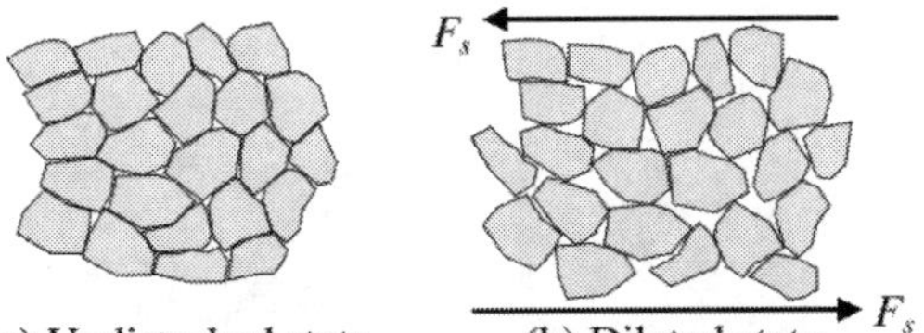

(a) Undisturbed state (b) Dilated state

Figure 1. Sand dilation. (a) The undisturbed state of densely-packed sand. After a shear force, F_s, is applied, the voids between sand grains increase, resulting in dilation (i.e., volumetric expansion of the sand aggregate), as shown by the (b) dilated state.

Dilation is well-known within the geological community but has received little attention in the foundry industry and therefore, has (to the authors' best knowledge) not been considered as a significant contributor to pattern allowances. The purpose of this study is to measure the evolution of a casting feature during solidification in order to determine the effect of sand dilation. The results will then be used to validate a stress model employing the Drucker-Prager Cap (DPC) constitutive law, a pressure-dependent model capable of predicting dilative behavior.

For the experiments, a hollow carbon steel cylinder with a core is produced; the inner diameter of the cylinder is continuously measured in-situ with LVDTs. Thermocouples are utilized to record the temperature in the casting and at several locations in the core. For the simulations, a one-way temperature-displacement coupling is adopted; the transient temperature fields are solved first using commercial casting simulation software (MAGMASOFT® [2]). The results are then exported to a general purpose finite element stress analysis code (ABAQUS® [3]) to predict dimensional changes throughout the casting. By comparing the in-situ measurements to the predicted strains in the core and casting, sand dilation is quantified and predicted.

Description of experiments

1.1. Experimental setup and casting procedure

The casting geometry consisted of a hollow thick-walled cylinder with dimensions (in mm) shown in figure 2(a). The experimental setup is depicted in figure 2(b). The outer dimensions of the cope (280 mm length × 280 mm width × 75 mm height) and drag (280×280×200) are not drawn to scale. The change in the inner diameter at the mid-height of the cylinder was continuously measured by utilizing two identical assemblies consisting of a quartz rod, quartz tube, and LVDT. One end of the quartz rod was flattened into a disc (7 mm in diameter) using an oxy-acetylene torch and inserted through pre-drilled holes in the drag and core. The disc was butted to the outer diameter of the core, as shown in figure 2(b). In order to transmit displacement, the quartz rod passed through a quartz tube, which traversed the mold cavity. The other end of the quartz rod was attached to an LVDT, which continuously measured the displacement from one side of the inner diameter. The other assembly measured displacement on the opposite side of the cylinder. The LVDT measurements were added together to calculate the temporal evolution of the inner diameter. It is obvious from figure 2(b) that both LVDT measurements could not be taken at the same height. Therefore, one measurement was taken approximately 5 mm above the cylinder mid-height, while the other was taken 5 mm below.

Temperatures were measured at several locations; type K thermocouples were inserted through the bottom of the drag and into the core at radial distances of 6, 9, 15, and 25 mm from the vertical core-casting interface. Also, the thermocouples were staggered circumferentially to minimize the influence from other thermocouples. Finally, a type B thermocouple was encased in a quartz tube and inserted into the mold cavity to measure the temperature of the steel.

To build the molds, Unimin® IC55 silica lake sand was bonded with a phenolic urethane no-bake (PUNB) binder system. The binder (1.25% of mold weight) was mixed using a 55:45 ratio of part 1 (PEPSET® 1000) to part 2 (Techniset® 6435). The cope and drag were hand packed, whereas the cores were manually rammed. The core weights varied less than 2% among all experiments.

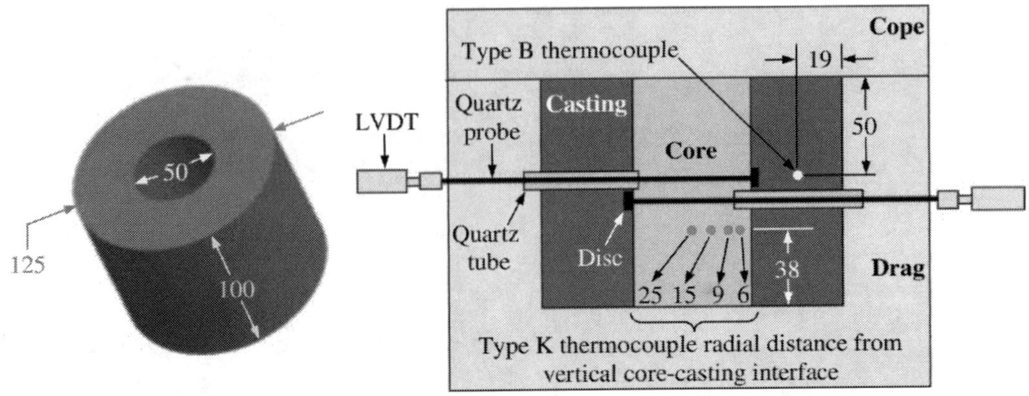

(a) Casting geometry (b) Experimental setup

Figure 2. (a) The casting geometry and (b) experimental setup at the casting mid-plane. Units in mm.

The steel was melted in a 300 lb induction furnace at the University of Northern Iowa Metal Casting Center. The target chemistry was ASTM A216 grade WCB carbon steel. Instead of using a pouring cup, liquid metal was poured directly into the mold cavity, after which the cope was placed on top of the drag. This methodology was chosen to eliminate mechanical interactions between the casting and the outer mold. In total, 5 castings were produced. For the first 4 castings, LVDT measurements and temperatures in the steel and the "25" location in the core (see figure 2(b)) were taken. For the 5th casting, no LVDT measurements were taken; only temperatures were recorded in the steel and at all core locations in figure 2(b).

1.2. Experimental Results

The changes in the inner diameter measured by the LVDTs (termed "LVDT measured") are plotted in figure 3 on (a) complete (40000 s) and (b) 600 s time scales. Figure 3(b) shows that the inner diameter expands to a maximum value (ranging from 1.15 mm in Cylinder 2 to 1.3 mm in Cylinder 1) at 200 s. This expansion occurs during solidification while the steel was mostly liquid and offered little restraint. After 200 s, the inner diameter remained constant until approximately 300 s, at which time solidification neared completion. After 325 s, the inner diameter decreases due to thermal contraction of the solid steel. The "wiggle" in the displacement curves seen at 4000 s in figure 3(a) can be attributed to a volumetric expansion associated with a solid-state phase transformation in the steel.

In addition to the LVDT measurements, pattern allowances were measured after shakeout at the three inner diameter locations indicated in figure 4(a). The results are shown in figure 4(b). The actual core dimensions (rather than pattern dimensions) were used to calculate pattern allowances to remove any variability due to the molding process. Measurements on the casting and core were taken with digital calipers. The dashed horizontal line in figure 4(b) denotes the free shrink of the steel (2.2%) and serves as a reference; any deviation from this line represents the core's contribution to dimensional changes. The largest deviation can be observed at the mid-height, while smaller deviations can be seen at the top and bottom. As a result, the inner surface of the cylinder evolved into a barrel-shaped profile. The variation in the pattern allowances with height can be explained by the local solidification times. When the steel solidifies more quickly, the impact of the core on dimensional changes is less and the resulting pattern allowance is closer to the free shrink line. This is the case at the top and bottom of the cylinder. At mid-height, the steel solidifies more slowly, core expansion occurs over a longer time, and the pattern allowance is farther away from the free shrink line. Another observation from figure 4(b) is that ID_{top} contained more scatter than the pattern allowances at the other locations, which can be explained by the manner in which the castings were

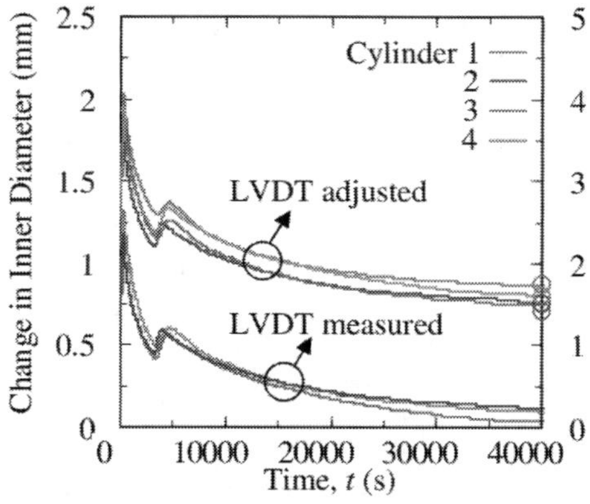

(a) Complete time scale

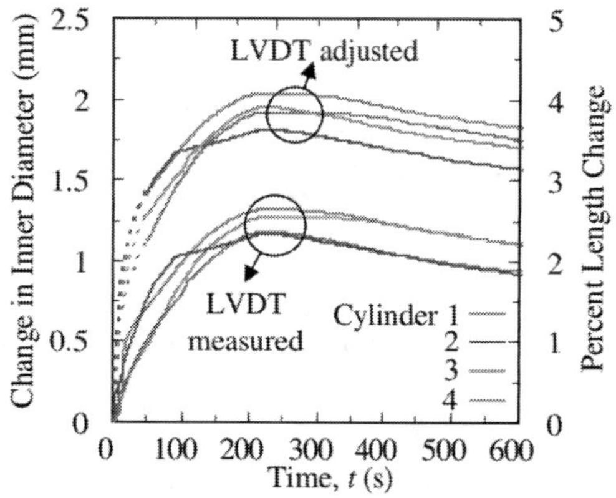

(b) Initial 600 s

Figure 3. Measured change in inner diameter plotted on (a) complete and (b) 600 s time scales. The symbols on the secondary vertical axis on the complete time scale correspond to the pattern allowances at ID_{mid} in figure 4 and serve as the basis for shifting the measured curves upward.

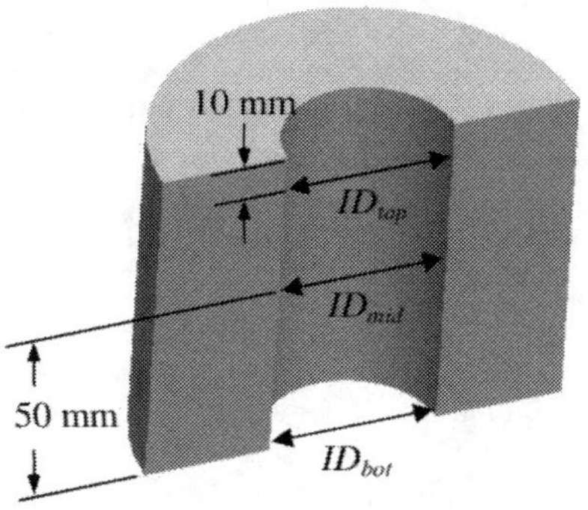

(a) Feature locations

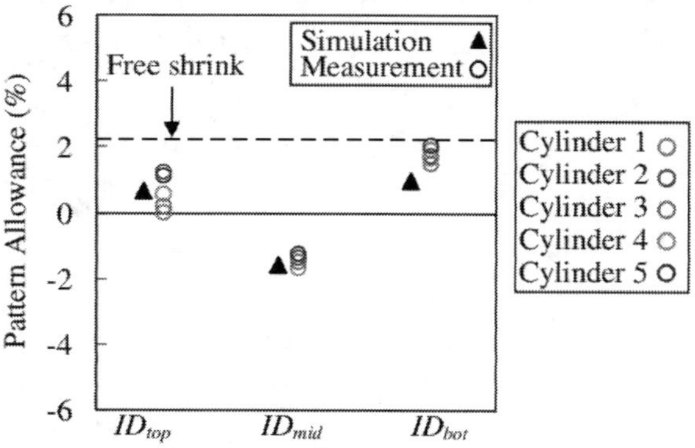

(b) Pattern allowances

Figure 4. Pattern Allowances. (a) Feature locations. (b) Measured and simulated pattern allowances.

poured. To avoid spilling steel onto the foundry floor, the mold was filled within a few mm from the top, leaving an air gap between the casting and cope. The variation in air gap thicknesses between the experiments likely affected local solidification times, which in turn influenced the pattern allowances at the top.

From figures 2(b) and 4(a), it can be seen that ID_{mid} is the same feature that was measured by the LVDTs. Therefore, the percent change of the inner diameter at room temperature in figure 3(a) should match the negative pattern allowance of ID_{mid}. Unfortunately, this is not the case, as a maximum expansion of 0.27% at room temperature was measured by the LVDTs, whereas the negative pattern allowances at ID_{mid} were about 1.75%. Hence, approximately 1.5% (0.75 mm) additional core expansion was measured by the calipers over the LVDTs. This discrepancy can be attributed to the LVDT measurements. As confirmed below by the stress simulations, the thermal contraction of the steel after 325 s is accurately measured by the LVDTs. Therefore, the discrepancy between the LVDT measurements and pattern allowances originated from the LVDTs during the initial 325 s. During these early times, the steel adjacent to the core at mid-height existed in a liquid or semi-solid state, which resulted in slippage between the quartz probes and the steel. In order to account for this

IOP Conf. Series: Materials Science and Engineering **84** (2015) 012022 doi:10.1088/1757-899X/84/1/012022

slippage, the entire LVDT curves in figure 3 were shifted upward to match the measured pattern allowance at room temperature (40,000 s). The negative pattern allowances are shown as open circles on the secondary (right) vertical axis in figure 3(a). Then, the LVDT curves were modified during the initial 50 s to smoothly increase from zero to the shifted measurements. These modifications are shown as dashed lines in figure 3(b). The resulting curves are termed "LVDT adjusted" and will be used below to validate the stress model predictions.

Thermal simulations

Thermal simulations of the casting process were performed using MAGMASOFT®, a detailed description of which can be found elsewhere [4]. Excellent agreement between measured and predicted temperatures was achieved at all times, as shown by the temperature vs. time plots on complete and 600 s time scales in figures 5(a) and 5(b), respectively. The predicted transient temperature fields were subsequently transferred to the finite element mesh used in the stress simulations.

Stress Simulations

Stress simulations were performed with ABAQUS®. Only the casting and core were simulated, as the outer mold had minimal impact on the inner diameter evolution. For preliminary simulations, only thermal strains were calculated in the core and cylinder. These simulations are important for two reasons. First, if the predicted thermal expansion of the core under-predicts the measured inner diameter expansion (see figure 3(b)), then dilation of the core must indeed be significant. Second, if the predicted thermal strains in the cylinder match the LVDT results after 325 s, the method used to shift the LVDT curves in figure 3 is justified. Thermal strains were predicted using the linear thermal expansion (LTE) curves shown in figure 6. The steel LTE was calibrated in a previous study [4], with the onset of thermal contraction of the steel (1410°C) occurring at a solid fraction of 0.97. The LTE for bonded silica sand was measured with a dilatometer at the University of Northern Iowa. The strong expansion at 1470°C represents the transformation to cristobalite.

The predicted variations in the inner diameter of the steel cylinder at mid-height from the purely thermal strain simulation are plotted in figure 7. The inner diameter is predicted to begin contraction after 250 s. The predicted and measured curves are exactly parallel after 325 s, which confirms that the dimensional changes measured by the LVDTs after 325 s were indeed solely due to thermal contraction of the steel. Hence, the method used to shift the measured LVDT curves in figure 3 is justified.

The outer diameter of the core at mid-height is predicted by the purely thermal strain simulations to increase to a maximum value of 0.5 mm at about 150 s (see figure 7(b)). Subsequently, the diameter increases marginally until 6,000 s (due to the decreasing sand LTE between 600°C and 1470°C in figure 6), after which it decreases as the core cools to room temperature. The maximum expansion of the core at 150 s (0.5 mm) is far less than the measured maximum "LVDT adjusted" value of 1.9 mm. Therefore, 1.4 mm of expansion cannot be explained by thermal strains in the core. This suggests that dilation of the core indeed plays an important role in determining the dimensional changes of the casting. One caveat that would create additional thermal expansion is cristobalite formation, which was not predicted by the simulations. This is because the maximum predicted core temperature (i.e., the interface value at 200 s in figure 5(b)) was lower than the transformation temperature of 1470°C. However, impurities in the sand mold are known to reduce the temperature at which cristobalite forms. Hence, an additional experiment was performed, in which layers of sand were extracted in 1 mm increments from the mold-metal interface from a steel casting after it cooled to room temperature. Because the cristobalite transformation process is essentially irreversible [5], any cristobalite formed at high temperatures also remains at room temperature. The sand samples were tested using x-ray diffraction, from which minimal amounts of cristobalite were detected. Therefore, expansion due to cristobalite formation can be excluded as a reason for the observed core expansion.

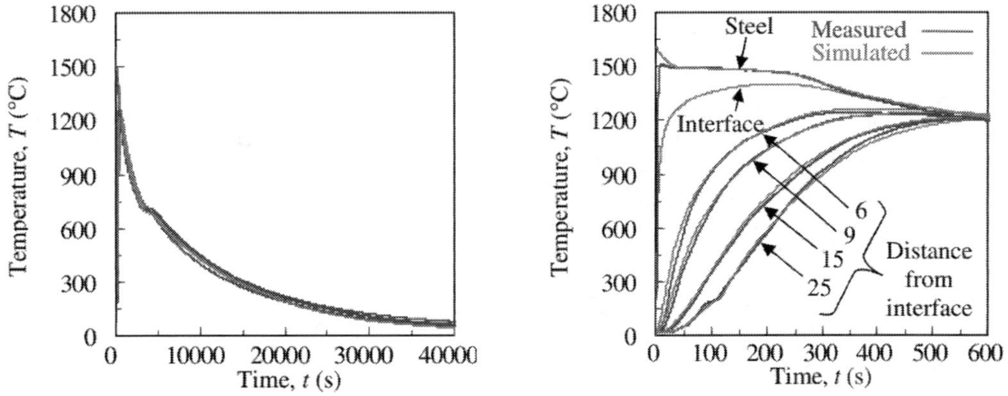

(a) Complete time scale (b) Initial 600 s

Figure 5. Comparison between measured and predicted temperatures in the steel and throughout the core on (a) complete and (b) 600 s time scales.

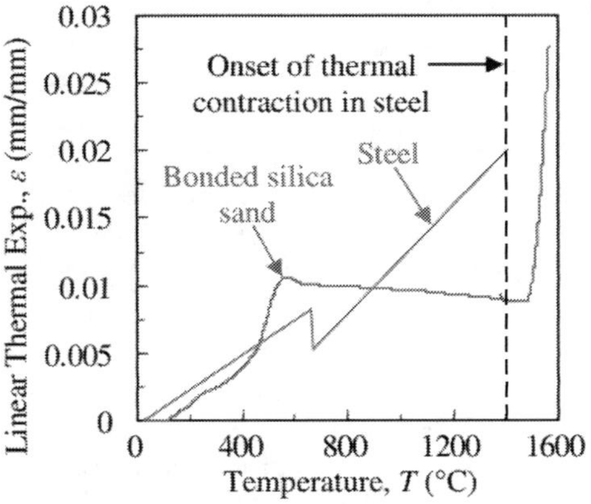

Figure 6. Linear thermal expansion (LTE) for the steel and bonded silica sand. The onset of thermal contraction in the steel (1410°C) corresponds to a solid volume fraction of 0.97.

After the preliminary simulations, full stress simulations were performed using an elasto-visco-plastic constitutive law (implemented in a user-defined UMAT subroutine) for the steel and the Drucker-Prager Cap (DPC) model for the core. Features of the elasto-visco-plastic model include rate, hardening, and temperature effects, as well as pressure-dependent deformation during solidification. The complete model, including the calibration of its parameters, is described elsewhere [4]. The preliminary simulations showed that negligible viscoplastic strains are generated after solidification, which implies the simulations are insensitive to changes in the viscoplastic parameters. However, one parameter that requires consideration is the coherency solid fraction, g_{coh}, which defines the solid fraction at which the steel can transmit stresses. For the current study, $g_{coh} = 0.75$ was specified. Any g_{coh} lower than this value generated insufficient core expansion.

Figure 8 shows the DPC yield surface in q (shear stress) vs. p (pressure) space. The critical feature of the DPC model for the present study is the ability to predict dilation, which occurs during yielding on the shear failure surface, F_s. Other features of the model include a pressure-dependent yield surface containing a cap, F_c, which prevents dilation at high pressures. The transition yield surface, F_t (defined by parameter α), provides a smooth transition between F_s and F_c to avoid numerical issues. The compressive strength of the material, p_b, defines the position of the cap on the pressure axis. A detailed explanation of the model can be found in a suitable plasticity textbook.

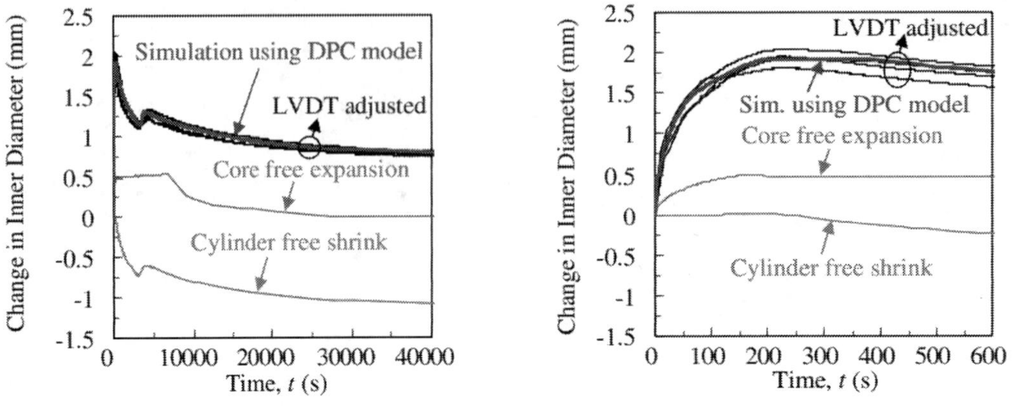

Figure 7. Comparison between measured and predicted inner diameter temporal evolutions on (a) complete and (b) 600 s time scales.

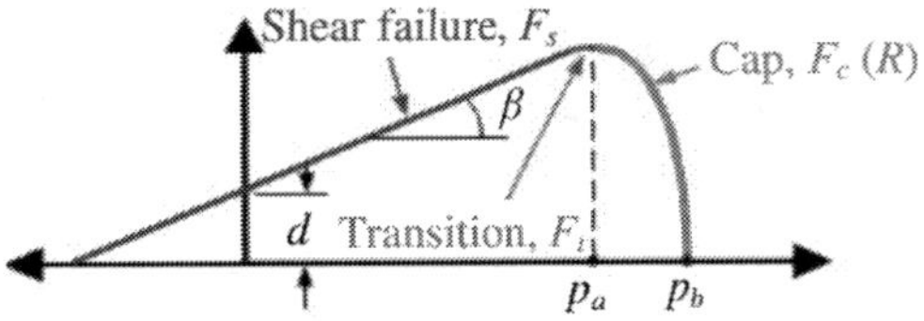

Figure 8. The Drucker-Prager Cap (DPC) yield surface is defined by the shear failure (F_s), cap (F_c), and transition (F_t) surfaces. Dilation is predicted when yielding occurs on F_s.

The mechanical properties of the sand core were estimated as follows. At temperatures less than 180°C, the modulus of elasticity (*E*) and cohesion (*d*) were taken from tests on PUNB bonded sand [6]. After the binder pyrolyzes (≈ 200°C), experimental stress-strain data for dense sand at room temperature [7-9] were used as a first estimate to determine the high-temperature DPC model parameters. Elastic properties (*E* and *ν*) were taken from [7]. Hardening measurements of Lade et al. [8] were used to determine p_b. The cohesion (*d*), friction angle (*β*), and cap eccentricity (*R*) were estimated from data in Hettler and Vardoulakis [9]. At 25°C and 180°C, the larger cohesion values required either an increase in p_b or decrease in *R*. A parametric study found neither to have an impact on predicted dimensions for this study. Therefore, *R* was decreased to a small value at these temperatures. The final DPC model parameters are summarized in table 1. Initially, the stress simulations predicted excessive core expansion. To match the measured and predicted inner diameter evolutions (see below), the cohesion was increased from 0.025 to 0.05 MPa. This increase is not unexpected, as chemical reactions during binder decomposition will likely increase the strength [6].

Table 1. Parameters used in Drucker-Prager Cap Model

T (°C)	*β* (degrees)	*d* (MPa)	*R*	*α*	p_b (MPa)	*E* (MPa)	*ν*
25	37.6	1.8	0.01	0.01	5	3403	0.3
180	37.6	0.2	0.01	0.01	5	282	0.3
> 300	37.6	0.05	0.85	0.01	5	61	0.3

Figure 7 shows that using the DPC model excellent agreement between the measured and predicted time evolutions of the inner diameter at mid-height is obtained. In particular, the early expansion of the core by 1.9 mm is predicted correctly. As shown in figure 4(b), the pattern allowances are predicted well not only at mid-height, but also at the top and bottom of the cylinder. The predicted barrel-shaped profile of the core can be clearly observed in figure 9. The von Mises stresses and pressures at 150 s are also plotted in figure 9. The von Mises stresses in the core range from 50 kPa to 400 kPa. They are induced by the uneven heating of the core and the non-uniform solidification and contraction of the steel around it. Recall that the existence of these shear stresses is essential for the prediction of dilation.

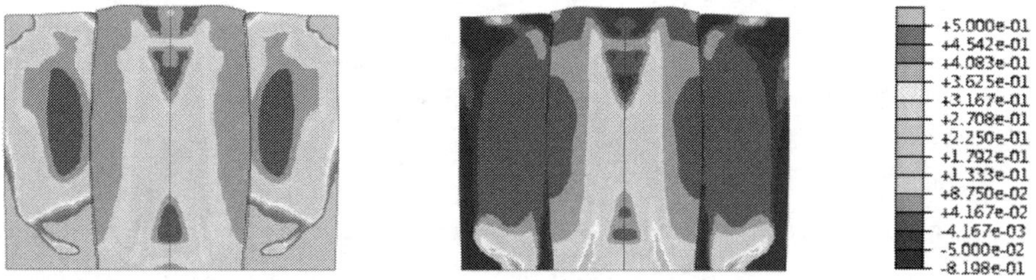

(a) von Mises stress (b) Pressure

Figure 9. (a) von Mises stress and (b) pressure at 150 s. Units in MPa. Deformation factor = 5.

Conclusion

In this study, the effect of sand dilation on core expansion during steel casting was studied through experiments and stress simulations. The inner diameter evolution of a thick-walled hollow steel cylinder was measured in-situ with LVDTs during casting. It was found that the measured expansion of the inner diameter far exceeded the predicted thermal expansion of the core. To account for this additional expansion, core dilation was considered. In order to model dilation, the Drucker-Prager Cap (DPC) model was utilized in the stress simulations. The simulation successfully predicted the measured evolution of the inner diameter. The measured pattern allowances revealed a barrel-shaped profile of the inner diameter, which was also predicted by the simulations. This study sheds considerable light on the physical phenomena responsible for dimensional changes during casting of steel and the resulting pattern allowances. Dilation of the mold or core sand, in addition to thermal expansion, can play a significant role during early times, when the steel offers little resistance to deformation. The stress simulations in this study are preliminary in that the material parameters for the DPC model, in particular at high temperatures, were only estimated from room temperature sand data.

References

[1] Peters F, Voight R, Ou S Z and Beckermann C 2007 *Int. J. Cast Metals Res.* **20** 275-87
[2] ABAQUS®, Abaqus, Inc., Providence, RI
[3] MAGMASOFT® v4.6, Magma GmbH, Aachen, Germany
[4] Galles D and Beckermann C 2013 *67th SFSA Technical and Operating Conf.* (Chicago, IL)
[5] Carniglia S C 1992 *Handbook of Industrial Refractories Technology* (Park Ridge, NJ) Noyes Pub
[6] Thole J and Beckermann C 2010 *Int J Metalcasting* **4** 7-18
[7] Brown P T 1995 *Geotechnical Testing J.* **18** 259-70
[8] Lade P V, Yamamuro J A and Bopp P A 2006 Geomechanics II: Testing, Modeling, and Simulation ed P V Lade and T Nakai pp 87-102
[9] Hettler A and Vardoulakis I 1984 *Geotechnique* **34** 183-98

Acknowledgements

This research is sponsored through the Defense Logistics Agency through the American Metalcasting Consortium and the Steel Founders' Society of America.

Flow Dynamics of green sand in the DISAMATIC moulding process using Discrete element method (DEM)

E Hovad[1,3], P Larsen[3], J H Walther[1], J Thorborg[1,2] and J H Hattel[1]
1 Department of Mechanical Engineering, Technical University of Denmark (DTU)
2 MAGMA Giessereitechnologie GmbH, Kackertstr. 11, 52072 Aachen, Germany
3 DISA Industries A/S, Højager 8, 2630 Taastrup, Denmark

E-mail: emilh@mek.dtu.dk

Abstract. The DISAMATIC casting process production of sand moulds is simulated with DEM (discrete element method). The main purpose is to simulate the dynamics of the flow of green sand, during the production of the sand mould with DEM. The sand shot is simulated, which is the first stage of the DISAMATIC casting process. Depending on the actual casting geometry the mould can be geometrically quite complex involving e.g. shadowing effects and this is directly reflected in the sand flow during the moulding process. In the present work a mould chamber with "ribs" at the walls is chosen as a baseline geometry to emulate some of these important conditions found in the real moulding process. The sand flow is simulated with the DEM and compared with corresponding video footages from the interior of the chamber during the moulding process. The effect of the rolling resistance and the static friction coefficient is analysed and discussed in relation to the experimental findings.

1. Introduction

The DISAMATIC process is extensively used in casting of metal parts for the automotive industry for making breaking disks, crank shafts, engine blocks etc. In order to ensure a high quality of the components, it is important to control the manufacturing process of the mould, so that it is homogeneous and stable. A short explanation of the DISAMATIC process for manufacturing of the mould, is given in [1]. The following figure 1, is an illustration of the sand shot in the mould chamber. The flow dynamics of the sand shot inside the mould chamber is investigated with DEM and compared to the video footage, capturing the flow dynamics in the real mould chamber. The sand shot is driven by air pressure, initializing the flow of sand from the hopper above, into the chamber below. The sand fills out the mould geometry in the chamber at around the time 1 s.

Hence, the focus of the present work is on the sand shot and how this first part of the moulding process can be modelled numerically with the DEM, in this study only the sand phase is investigated. The first period of the sand shot 0.6 s is investigated, where the space underneath the three "ribs" is filled with green sand (see figure 3). Earlier research has been done, using DEM to simulate the sand flow [2] as well as the subsequent squeezing process [3] in the application of green sand moulding. Obtaining the right DEM parameters to realistically simulate the process is important, but at the same time also quite challenging [4]. This especially goes for the green sand material properties due to the small particle size around 0.2 mm as well as for the description of adhesion. The latter is often neglected in DEM simulations and this is also the case for the present work.

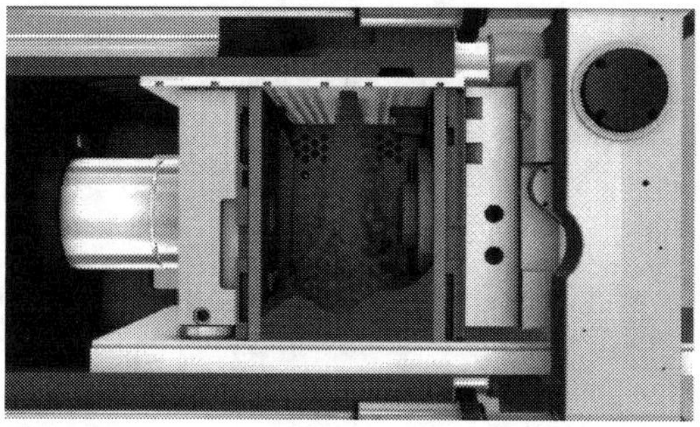

Figure 1. The sand shot, the sand coming from the hopper down into the chamber. The simulation of the sand shot is the topic of the present work.

DEM has also been used to simulate the lost foam process [5], where it was suggested that the rolling resistance and the Coulomb sliding friction are the most important parameters for the flow behaviour in this process. The fitting of rolling resistance and sliding friction parameters was studied in [4], in which different experimental tests were used for calibration. Particle scaling is frequently needed in the discrete element method [6], in this study two different particle sizes are used for the simulations.

2. Discrete element method (DEM).

DEM has received increased attention the last decade and the general application areas of the method can be found in e.g. [7]. A general review of the method's theoretical foundation is also given by [8] and a comparison of different frequently used DEM models was made by Di Renzo et. Al [9]. With the advancement in computational power and the introduction of parallel computing in DEM, the method also seems to be convenient for granular flow [10] and [11]. The commercial software STAR-CCM+ has been used for the 2-D simulations.

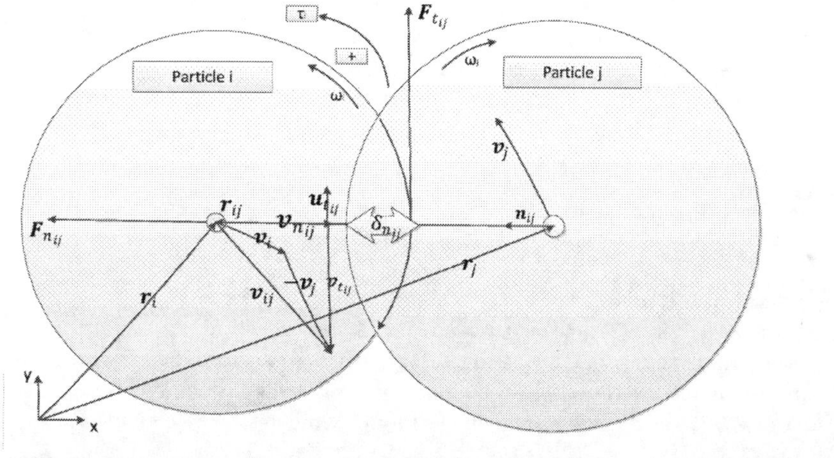

Figure 2. Particle i-th impact with particle j-th, the force exerted on the i-th particle in the normal direction is $\boldsymbol{F}_{n_{ij}}$ and in the tangential direction is $\boldsymbol{F}_{t_{ij}}$.

In figure 2, the impact distance $r_{ij} = \left\|\boldsymbol{r}_{ij}\right\|_2$ of the spherical i-th particle center of mass ($\boldsymbol{r}_i$) is found by the spherical particle j-th center of mass ($\boldsymbol{r}_j$) as

$$\boldsymbol{r}_{ij} = \boldsymbol{r}_i - \boldsymbol{r}_j \tag{1}$$

In DEM there is a normal direction ($\boldsymbol{n}_{ij} = \frac{\boldsymbol{r}_{ij}}{r_{ij}}$). The relative velocity of the two particles is

$$\boldsymbol{v}_{ij} = \boldsymbol{v}_i - \boldsymbol{v}_j$$

decomposed to a velocity in the normal direction ($\boldsymbol{v}_{n_{ij}}$) and a velocity in the tangential ($\boldsymbol{v}_{t_{ij}}$) to find the elastic force from the "spring" and the viscous damping force from the "dashpot" [11][12]. The normal velocity is given by

$$\boldsymbol{v}_{n_{ij}} = (\,\boldsymbol{n}_{ij} \cdot \boldsymbol{v}_{ij})\boldsymbol{n}_{ij} \tag{2}$$

The tangential velocity vector is defined as

$$\boldsymbol{v}_{t_{ij}} = \boldsymbol{v}_{ij} - \boldsymbol{v}_{n_{ij}} - (\,\boldsymbol{\omega}_i R_i + \boldsymbol{\omega}_j R_j) \times \boldsymbol{n}_{ij} \tag{3}$$

The normal displacement is $\delta_{n_{ij}} = \left\|\boldsymbol{v}_{n_{ij}}\right\|_2 \Delta t$ and the tangential displacement vector $\boldsymbol{u}_{t_{ij}}$ is found by integration of the tangential velocity $\frac{d\boldsymbol{u}_{t_{ij}}}{dt} = -\boldsymbol{v}_{t_{ij}}$

The simplified Hertz-Mindlin (H-MDns) force model with non-linear damping is used [13] [14] [15] and now that the displacements has been found the normal interaction force can be found

$$\boldsymbol{F}_{n_{ij}} = \boldsymbol{n}_{ij} K_n \delta_{n_{ij}}{}^{3/2} - N_n \boldsymbol{v}_{n_{ij}} \tag{4}$$

The tangential interaction force

$$\boldsymbol{F}_{t_{ij}} = \boldsymbol{u}_{t_{ij}} K_t - N_t \boldsymbol{v}_{t_{ij}} \tag{5}$$

K_n is the normal stiffness and K_t is the tangential stiffness, note that N_n is the normal and N_t the tangential non-linear damping model [13].

There is a max tangential force due to the Coulomb's law,

$$\left\|\boldsymbol{F}_{t_{ij}}\right\|_2 < \left\|\mu_s \boldsymbol{F}_{n_{ij}}\right\|_2 \tag{6}$$

μ_s is the static friction coefficient. Finally the total force on the particle is,

$$\boldsymbol{F}_i^{tot} = m_i \boldsymbol{g} + \textstyle\sum_j (\boldsymbol{F}_{n_{ij}} + \boldsymbol{F}_{t_{ij}}) \tag{7}$$

The tangential forces give a final torque on the i-th particle,

$$\boldsymbol{T}_i = -R_i \textstyle\sum_j \left(\boldsymbol{n}_{ij} \times \boldsymbol{F}_{t_{ij}}\right) \tag{8}$$

From this the acceleration, velocity and position is calculated by Newton second law, incrementally for each time step. The rolling resistance chosen is the Constant Torque Method first used by Zhou et. Al [16]. The relative rotation between the two particles is defined as $\boldsymbol{\omega}_{rel} = \boldsymbol{\omega}_i - \boldsymbol{\omega}_j$ and the Constant Torque Method is used to calculate the rolling resistance, defined as

$$\boldsymbol{T}_{\text{rol}} = -\frac{\omega_{rel}}{|\omega_{rel}|} \mu_r R_{eq} \boldsymbol{F}_{n_{ij}} \tag{9}$$

So the tangential force $\boldsymbol{T}_i$ is counteracted by a torque from the rolling resistance $\boldsymbol{T}_{\text{rol}}$ and the rolling resistance is μ_r.

3. Settings for the simulation

In the following figure 3, a simulation of the sand shot is shown with the green sand coming from top of the chamber with the flow rate and the initial velocity calculated from the experimental footage.

The flow rate is calculated by the circle diameter e.g. d=0.004 m, filling time 1.05 s, the assumed packing fraction of 0.75 and the total area of the mould which is 0.25 m^2 and finally this gives the flow rate $\frac{0.75 \cdot 0.25 m^2}{(0.002m)^2 \pi \cdot 1.05\ s} = 14210\ particles/s$. The diameter of 0.002 m gives the 56840 particles/s.

The particle velocity at the inlet is found by a simple 2-D bulk flow calculation based on the areas filling time of sand and orifice length from the actual chamber and experimental video footage.

This area has the height going to the third rib, $h = 0.3\ m$ (see figure 3) and the width of the chamber is $w = 0.48\ m$, then this area is $0.3\ m \cdot 0.48\ m = 0.144\ m^2$.

The orifice width is 0.04 m and the filling time of the area is around 0.6 s, so the average vertical velocity is estimated to be $v_y = \frac{0.144m^2}{0.04m \cdot 0.6\,s} = 6.0\,m/s$. For a more realistic velocity distribution a normal distribution is assumed around the calculated mean value and the standard deviations is given in table 1.

Table 1. The velocity distribution.

Direction	Mean value (m/s)	Standard deviation (m/s)	Velocity range Min (m/s)	Max (m/s)
v_x	0	0.1	-0.1	0.2
v_y	-6.0	0.1	-5.8	-6.2

The time step is chosen to be Δt=0.00001 s, plots of the velocities are made for every 500 time step, corresponding to the time 0.005 s, 0.01 s, 0.015 s and 0.02 s etc.

Table 2. Material values for the simulation.

Material properties	Value
Solid density – green sand	1600 kg/m3
Solid density – chamber wall	7500 kg/m3
Young's modulus – green sand	17000 MPa
Young's modulus – chamber wall	200000 MPa
Poisson ratio – green sand	0.3
Poisson ratio – chamber side	0.3
Coefficient of restitution particle-particle	0.01
Coefficient of restitution particle-wall	0.01
Gravity	9.82 m/s2

In table 2 typical values have been chosen for the Poisson ratio and Young modulus for the steel in the chamber wall. For the sand it is chosen to be the same value as a similar material, brick in STAR-CCM+. The sand in the mould has a density of 1200 kg/m^3 after the sand shot and before squeezing. The density of "representative" sand particles is $\frac{1200\,\mathrm{kg/m^3}}{0.75} = 1600\frac{\mathrm{kg}}{\mathrm{m^3}}$, due to the packing fraction.

The coefficient of restitution is chosen to be very small and very close to critical damping. This is due to the high damping properties of the bentonite coated green sand and Young modulus have been suggested to be of less importance compared to the rolling resistance (equation (9)) and static friction coefficient (equation (6)), [5].

Table 3. Simulation parameters, where the rolling resistance is μ_r and the static friction coefficient is μ_s.

Diameter	μ_s	μ_r	Injection rate
4 mm	0.8	0.6	14210 particles/s
4 mm	0.8	0.9	14210 particles/s
4 mm	1	0.6	14210 particles/s
4 mm	1	0.9	14210 particles/s
2 mm	0.8	0.6	56841 particles/s
2 mm	0.8	0.9	56841 particles/s
2 mm	1	0.6	56841 particles/s
2 mm	1	0.9	56841 particles/s

In table 3, choosing different values for the rolling resistance and static friction coefficient simulates changing the flow ability of the particles. This way, the effect of changing the moisture and bentonite content in the "real" green sand can be simulated as this changes the flow ability.

4. Results of simulation and experiment

The result that will be presented show when the flow front passes underneath the three "ribs" into the three "cavities" at the right side wall, seen in the following figure 3. The flow front is hard to define and track exactly in DEM, so six times intervals t_1-t_6 are constructed for the arrival of the particles at the six positions. These six times intervals t_1-t_6 will be measured for the flow front for different parameter values of the rolling resistance, static friction coefficient and particle radius and will then be compared to the experimental results.

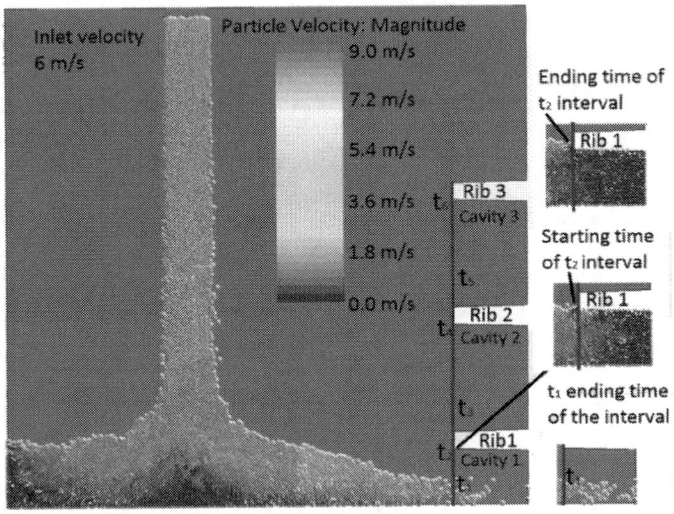

Figure 3. Tracking the front of the flow profile at the 6 different positions (t_1-t_6). An example of interval start and end time is shown in the bottom right corner of the figure for the time interval denoted t_1. An example of interval start and ending is shown in the right top of the figure for the time interval denoted t_2. The magnitude of the velocity is plotted with a scaling of 0-9 m/s the scale going from minimum dark blue, 0 m/s to the maximum velocity red 9 m/s.

In figure 3, the starting time for the interval, is when the flow front reaches the brown line and the ending time of the interval is when the bulk flow of the flow front has reached the brown line. The flow reaches the "cavities" underneath the three "ribs", when the particles cross the brown line into cavity 1 (t_1), cavity 2 (t_3) and cavity 3 (t_5). The starting time of the interval is when the "loosely packed particles" reach into the cavities and ending time of the interval is when the bulk flow or "closely packed particles" reach the cavities. The starting times of t_2, t_4 and t_6 intervals are when the flow front reaches the bottom edge of rib 1, 2 and 3 subsequently. Here the starting times of the time intervals are again when the "loosely packed" particles hit the bottom edge of the rib in the time step. The ending time of t_2, t_4 and t_6 intervals are again defined as when the "closely packed particles" reaches the bottom edge rib in the time step. The following table 4 will show all the intervals t_1-t_6 for the different simulations and selected plots will be made showing the flow profile passing the different positions and compared to the video footage.

Table 4. The six time intervals t_1-t_6, where each interval (start-end) indicates the period, when the flow front reaches the selected positions at the brown line. The experimental values are shown in the last row.

Simulations values	t_1 [s]	t_2 [s]	t_3 [s]	t_4 [s]	t_5 [s]	t_6 [s]
μ_r=0.6, μ_r=0.8, 4 mm	0.155-0.170	0.215-0.225	0.255-0.270	0.380-0.390	0.435-0.445	0.585-0.590
μ_r=0.9, μ_r=0.8, 4 mm	0.175-0.205	0.225-0.245	0.275-0.290	0.380-0.390	0.450-0.460	0.585-0.600
μ_r=0.6, μ_r=1, 4 mm	0.160-0.165	0.225-0.230	0.250-0.260	0.385-0.395	0.440-0.450	0.590-0.605
μ_r=0.9, μ_r=1, 4 mm	0.175-0.185	0.235-0.240	0.255-0.270	0.410-0.420	0.475-0.480	0.625-0.665
μ_r=0.6, μ_r=0.8, 2 mm	0.145-0.155	0.215 -0.220	0.245-0.250	0.370-0.380	0.415-0.425	0.550-0.560
μ_r=0.9, μ_r=0.8, 2 mm	0.155-0.165	0.220-0.225	0.240-0.245	0.380-0.390	0.435-0.440	0.585-0.595
μ_r=0.6, μ_r=1, 2 mm	0.150-0.165	0.210-0.215	0.235-0.240	0.365-0.370	0.440-0.450	0.540-0.550
μ_r=0.9, μ_r=1, 2 mm	0.155-0.160	0.215-0.225	0.240-0.250	0.380-0.385	0.430-0.440	0.590-0.595
Experiment	0.254-0.262	0.304-0.313	0.325-0.342	0.429-0.441	0.467-0.492	0.575-0.591

In table 4, the different values for the six time intervals t_1-t_6 are listed for the different simulations values and the experiment. A mean is calculated for each of the two diameters overall simulation times, this includes all the 4 times interval for each of the six columns with the times intervals t_1-t_6. Maximum and minimum times are also found from the 4 times interval for each of the six columns with the times intervals t_1-t_6. So for all the simulations with the diameter e.g d=4, a mean, maximum and minimum is calculated for each of the six time intervals t_1-t_6. The results from table 4 are depicted in figure 4, with the mean, maximum and minimum values plotted for the two diameters and the experiment.

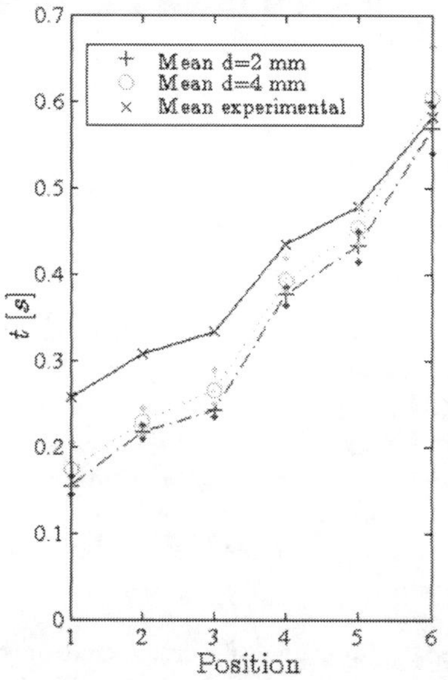

Figure 4. The mean, maximum and minimum times for t_1-t_6 for all the simulations compared to the experiment. The solid black line is linear interpolation of the experimental mean; the cross is the actual value of the experimental mean. The circle is the mean of the diameter d=4 mm and the plus is the diameter d=2 mm. The green dots are the intervals minimum and maximum of the diameter 4 mm and the blue dots represents the diameter 2 mm.

IOP Conf. Series: Materials Science and Engineering **84** (2015) 012023 doi:10.1088/1757-899X/84/1/012023

In figure 4, the simulation of diameter 4 mm mean is occurring around 0.06 s earlier than the first times t_1-t_3 from the experimental mean's and the difference narrows as the flow passes at the times t_4–t_5, at t_6 the simulations reached this point later than the experiment. Similarly dynamics occurs for the simulation of diameter 2 mm, where the mean is occurring around 0.08 s earlier compared to the experimental mean, for the first times t_1-t_3 and the difference narrows as the flow passes at the times t_4–t_5 and at the time t_6 the simulations are very close to the experimental value (difference 0.01). The larger diameters of 4 mm, have longer t_1-t_6 times than the diameter of 2 mm. The specific simulation chosen is the diameter of 2 mm, rolling resistance μ_r=0.9 and static friction coefficient μ_s=1, because this simulation's qualitative flow dynamics is very similar to the experiment. In figure 5 (simulation) and figure 6 (experiment) the progression of the flow front at t_2 is presented and the results are compared. The flow profile of the experiment figure 6, shows a quite conical pile shape (free surface shape) and the simulation exhibits a very similar conical shaped flow profile, although the actual time of the occurrence (t_2) is somewhat different i.e. 0.1 s, as earlier mentioned.

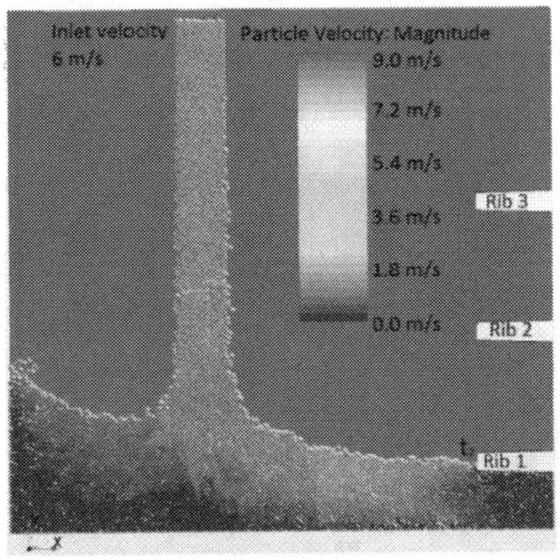

Figure 5. The simulation in the interval of t_2 plot taken at the time 0.225 s (μ_r=0.9, μ_s=1, 2 mm).

Figure 6. The experiment: t_2 at the time of 0.32 s, the contour of the flow profile can be seen on the rear wall.

5. Conclusion

The sand shot in the DISAMATIC process has been investigated with experimental video footage of the chamber and also simulated with DEM "representing" the sand particles. The dynamical flow properties of the green sand have been quantified by tracking the flow fronts arrival in six positions for the "ribbed" geometry. The Experimental video footage flow front and the DEM simulations flow front are compared for these six filling positions in the times (t_1-t_6) and plotted together with mean, minimum and maximum values. Selected plot for a specific chosen simulation is compared with the experimental footage, when the flow fronts has reached the rib 1 bottom edge position (t_2) and these flow fronts profiles similarities is discussed.

The main findings of the discrete element method for modelling green sand flow during production of DISA moulds is presented below.

- The dynamic flow behaviour of the particles in the simulation is similar to the experiment
- The behaviour of the time intervals (t_1-t_6) for the simulation is also similar to the experiment

- The flow ability changes to a more viscous flow with larger rolling resistance and static friction coefficient, resembling the behaviour of moisture and bentonite content in green sand
- With the correct particle velocity, flow rate, damping coefficient, rolling resistance and static friction coefficient it is possible to simulate the experimental video footage very well

The experimental data t_1 is slower with a time delay of 0.1 s for the diameter 2 mm and 0.8 s for the diameter 4 mm compared to these simulations, properly due to a non-constant flow rate in the inlet, especially overestimating the initial flow rate. In later studies the flow rate will be fitted more precisely with the experimental video footage and selected material values will also be tested experimentally. The density will also be measured from experiments and compared to the simulations. Overall it is possible to represent the flow quite well with a DEM model especially qualitative behaviors of the flow and for the different times t_1-t_6.

References

[1] DISA Industries A/S 2013 DISA 231/DISA 231 Var. Sand Moulding System Instructions for Use
[2] Makino H, Maeda Y and Nomura H 2002 Computer Simulation of Various Methods for Green Sand Filling *Trans. Am. Foundry Soc.* **110** 1–9
[3] Maeda Y, Maruoka Y, Makino H and Nomura H 2003 Squeeze molding simulation using the distinct element method considering green sand properties *J. Mater. Process. Technol.* **135** 172–8
[4] Grima A P and Wypych P W 2011 Discrete element simulations of granular pile formation: Method for calibrating discrete element models *Eng. Comput.* **28** 314–39
[5] Rojek J, Zarate F, de Saracibar C A, Gilbourne C and Verdot P 2005 Discrete element modelling and simulation of sand mould manufacture for the lost foam process *Int. J. Numer. Methods Eng.* **62** 1421–41
[6] Sakai M, Takahashi H, Pain C C, Latham J-P and Xiang J 2012 Study on a large-scale discrete element model for fine particles in a fluidized bed *Adv. Powder Technol.* **23** 673–81
[7] Zhu H P, Zhou Z Y, Yang R Y and Yu A B 2008 Discrete particle simulation of particulate systems: A review of major applications and findings *Chem. Eng. Sci.* **63** 5728–70
[8] Zhu H P, Zhou Z Y, Yang R Y and Yu a. B 2007 Discrete particle simulation of particulate systems: Theoretical developments *Chem. Eng. Sci.* **62** 3378–96
[9] Di Renzo A and Di Maio F P 2004 Comparison of contact-force models for the simulation of collisions in DEM-based granular flow codes *Chem. Eng. Sci.* **59** 525–41
[10] Cleary P W 2009 Industrial particle flow modelling using discrete element method *Eng. Comput.* **26** 698–743
[11] Walther J H and Sbalzarini I F 2009 Large-scale parallel discrete element simulations of granular flow *Eng. Comput.* **26** 688–97
[12] Cundall P A and Strack O D L 1979 A discrete numerical model for granular assemblies *Géotechnique* **29** 47–65
[13] Tsuji Y, Tanaka T and Ishida T 1992 Lagrangian numerical simulation of plug flow of cohesionless particle in a horizontal pipe *Powder Technol.* **71** 239–50
[14] Hertz H 1881 Über die Berührung fester elastischer Körper *J. für die reine und Angew. Math.* **171** 156–71
[15] Mindlin R D and Deresiewicz H 1953 Elastic spheres in contact under varying oblique forces *Am. Soc. Mech. Eng. -- Trans. -- J. Appl. Mech.* **20** 327–44
[16] Zhou Y C, Wright B D, Yang R Y, Xu B H and Yu a. B 1999 Rolling friction in the dynamic simulation of sandpile formation *Phys. A Stat. Mech. its Appl.* **269** 536–53

Combination of microscopic model and VoF-multiphase approach for numerical simulation of nodular cast iron solidification

E Subasic[1], C Huang[1], J Jakumeit[1] and F Hediger[1]

[1] Access e.V., Aachen, Germany

E-mail: e.subasic@access.rwth-aachen.de

Abstract. The ongoing increase in the size and capacity of state-of-the-art wind power plants is highlighting the need to reduce the weight of critical components, such as hubs, main shaft bearing housings, gear box housings and support bases. These components are manufactured as nodular iron castings (spheroid graphite iron, or SGI). A weight reduction of up to 20% is achievable by optimizing the geometry to minimize volume, thus enabling significant downsizing of wind power plants. One method for enhancing quality control in the production of thick-walled SGI castings, and thus reducing tolerances and, consequently, enabling castings of smaller volume is via a casting simulation of mould filling and solidification based on a combination of microscopic model and VoF-multiphase approach. Coupled fluid flow with heat transport and phase transformation kinetics during solidification is described by partial differential equations and solved using the finite volume method. The flow of multiple phases is described using a volume of fluid approach. Mass conservation equations are solved separately for both liquid and solid phases. At the micro-level, the diffusion-controlled growth model for grey iron eutectic grains by Wetterfall et al. is combined with a growth model for white iron eutectic grains. The micro-solidification model is coupled with macro-transport equations via source terms in the energy and continuity equations. As a first step the methodology was applied to a simple geometry to investigate the impact of mould-filling on the grey-to-white transition prediction in nodular cast iron.

1. Introduction

The development of renewable energies is central to current energy policies worldwide. As an example, building new power supply systems and increasing the share of renewable energy in the total electricity requirement by 2050 by at least 80% is a goal of the Renewable Energy Sources Act of the German Ministry for Economic Affairs and Energy [1]. Such initiatives strongly encourage the wind energy sector to produce even larger wind power plants, alongside upgrading already existing facilities. As key suppliers to wind power plant manufacturers, foundries are also involved in the process: larger windmills demand larger castings. In this situation, both foundries and windmill manufacturers are confronted with a range of restrictions when transporting castings and installing windmills. Consequently, there is a need to reduce the weight of critical parts (hubs, main shaft bearing housings, gear box housings and support bases). These components are manufactured by casting nodular cast iron (spheroidal graphite iron, SGI). Current estimations by foundry experts are that optimization of the geometry aimed at minimizing volume can achieve a weight reduction of up to 20%, enabling significant reduction in the weight of wind power plant.

Over the last couple of decades, casting simulation of mould filling and solidification has been recognized as a powerful tool in ensuring the quality and robustness of casting processes. However, the above-described challenges from industrial practice allow additional improvements to existing software solutions. The solidification process of large castings is characterized by convection-driven phenomena, which could be taken account of through application of the principles of computational fluid dynamics. At the same time, local development of the graphite phase in cast iron needs to be taken into account. A combination of microscopic model and VoF-multiphase approach is proposed here as a method of enhancing quality control during the production of thick-walled SGI castings. Thus, the possibility of reducing tolerances and, consequently, enabling castings of smaller volume is to be expected.

Coupled fluid flow with heat transport and phase transformation kinetics during solidification is described by partial differential equations and solved using the finite volume method. The flow of multiple phases is described using the volume of fluid approach [2, 3]. Mass conservation equations are solved separately for both the liquid and solid phase. At the micro-level the diffusion-controlled growth model for grey iron eutectic grains by Wetterfall et al. [4] is combined with a growth model for white iron eutectic grains by Nastac and Stefanescu [5]. The micro-solidification model is coupled with macro-transport equations via source terms in the energy and continuity equation.

As a first step the methodology was applied to a simple geometry to investigate the impact of variation of the relevant micro-scale parameters on grey-to-white transition prediction in nodular cast iron.

2. Model description

The kinetics of solidification have been modelled by many authors. Within the scope of this work, fundamental microscopic phenomena such as nucleation, undercooling, primary and eutectic solidification are described and incorporated into a macroscopic heat and fluid flow calculation. One option of enhancing the predictive capability of this model with a fully-developed macroscopic model was explored.

2.1. General multiscale/multiphase modeling frame and VoF model

To model the casting and solidification process for a cast iron alloy precisely, the macroscopic transport phenomena and the microscopic solidification kinetics have to be considered at different scales at the same time. The general multiscale/multiphase modelling framework has been introduced by Beckermann and Viskanta [6] and adopted in the present research. The macroscopic conservation equations for each phase are obtained by averaging the microscopic (exact) equations over the volume V_0. Definition of the volume averaging of a quantity Ψ in phase k is given by:

$$\langle\Psi\rangle = \frac{1}{V_0}\int_{V_0} X_k\,\Psi_k\,dV \tag{1}$$

where X_k is a phase function equal to unity in phase k and zero elsewhere. The volume fraction of phase k is defined as

$$\alpha_k = \frac{1}{V_0}\int_{V_0} X_k\,dV = \frac{V_k}{V_0} \tag{2}$$

The averaged macroscopic conservation equation of mass is given as

$$\frac{\partial}{\partial t}(\alpha_k\,\rho_k) + \nabla\cdot(\alpha_k\rho_k\boldsymbol{v}) = \sum_{j,j\neq k}\Gamma_{kj} \tag{3}$$

where Γ_{kj} is the interfacial mass transfer due to phase change from phase j to phase k. The energy equation is analogous:

$$\frac{\partial}{\partial t}\left(\alpha_k\, \rho_k\, \langle \mathrm{h_k}\rangle^k\right) + \nabla\cdot\left(\alpha_k \rho_k\, \langle \mathrm{h_k}\rangle^k\, \boldsymbol{v}\right) = \sum_{j,j\neq k} \mathrm{Q}_{kj}^{\Gamma} + \mathrm{Q}_{kj}^{\mathrm{d}} \tag{4}$$

where Q_{kj}^{Γ} and Q_{kj}^{d} are the interfacial energy transfers due to phase change and diffusion respectively. The source terms in equations (3) and (4) represent the connection between the macroscopic transport equations and the microscopic solidification model. These source terms are described in subsection 2.2.

The VoF model description assumes that all immiscible fluid phases contained in a control volume are of the same velocity, pressure and temperature field. Adding the volume-averaged equations, such as (3) and (4) will provide a set of basic governing equations describing momentum, mass and energy transport in a single-phase flow. The equations are solved for an equivalent fluid whose physical properties are calculated as functions of the physical properties of its constituent phases and their volume fractions:

$$\rho = \textstyle\sum_i \alpha_i\, \rho_i,\quad \mu = \sum_i \alpha_i\, \mu_i \;\text{ and }\; c_p = \sum_i \frac{(c_p)_i \rho_i}{\rho}\, \alpha_i \tag{5}$$

The conservation equation of volume fraction α_i is

$$\frac{d}{dt}\int_{v} \alpha_i\, dV + \int_{S} \alpha_i\, v\, \cdot d\boldsymbol{a} = \int_{V} S_{\alpha_i}\, dV \tag{6}$$

where S_{α_i} is the source or sink of the i-th phase

2.2. The eutectic growth model

The first analytical model for describing growth of the graphite/austenite eutectic was proposed by Wetterfall et al. [4], who derived the equation of the advancing velocity of the graphite/austenite interface by taking account of mass conservation at the interface, under the assumption that diffusion in austenite is in a steady state. Stefanescu [7] also derived the austenite shell growth velocity

$$\frac{dR_\gamma}{dt} = D_C^{\gamma}\, \frac{R_G}{(R_G - R)r_\gamma}\, \frac{C^{G/\gamma} - C^{L/\gamma}}{C^{\gamma/L} - C^{L/\gamma}} \tag{7}$$

where D_C^{γ} is the diffusivity of carbon in austenite. $C^{G/\gamma}, C^{\gamma/L}$ and $C^{L/\gamma}$ are the carbon concentration at the graphite/austenite, austenite/liquid and liquid/austenite interface respectively. They are functions of temperature and can be derived from the Fe–C phase diagram. The local fraction of solid is

$$\alpha_s(\boldsymbol{x},t) = N(\boldsymbol{x},t)\, \frac{4}{3}\, \pi R^3(\boldsymbol{x},t) \tag{8}$$

where the volumetric density of nucleation $N(\boldsymbol{x},t)$ is a function of time and space. To describe nucleation, many authors use the model from Oldfield [8]:

$$N = \mu_N \Delta T^2 \tag{9}$$

In this work, the number of activated nuclei is assumed to be constant from the beginning of the calculation. From equation (8), the volume fraction transfer rate is calculated as

$$\left(\frac{\partial \alpha_s}{\partial t}\right)_{ext} = 4\,\pi N\, R^2\, \frac{dR_\gamma}{dt} \tag{10}$$

To account for grain impingement, the Johnson-Mehl approximation is used. The volume transfer rate in equation (10) is called extended volume fraction transfer rate, which is obtained under the assumption that there is still sufficient liquid melt present to allow the grains to grow freely without impingement. Only its fraction is real; if we assume that nucleation occurs randomly, this fraction is approximated as volume fraction of melt, and thus the real volume fraction transfer rate of grey eutectic is calculated as

$$\frac{\partial \alpha_s}{\partial t} = 4\,\pi N_{ge} R^2 \frac{dR_\gamma}{dt}\,\alpha_l \tag{11}$$

To cover the complete physical model during a eutectic reaction, it is also necessary to model the stable-to-metastable transition. Growth of white eutectic can be modelled based on a nucleation model and velocity at which the solid/liquid interface advances. However, there is some difficulty in handling microscopic morphology since white eutectics have lamellar and rod shapes [5]. In the present work, the emphasis is on implementation of growth competition between grey and white eutectics. A kinetic model that assumes spherical grain growth from Nastac and Stefanescu [5] is adopted.

$$\frac{\partial \alpha_s}{\partial t} = 4\,\pi N_{we} R^2 \mu_{we}\,\Delta T^2\,\alpha_l \tag{5}$$

where N_{we} is the grain density (m^{-3}) and μ_{we} is the growth coefficient of white eutectic. Many authors have proposed a value of μ_{we} in the range 6.0×10^{-7} to 2.4×10^{-6} m · s^{-1} · K^{-2}.

3. Model parameter study

3.1. Simulation settings, case description

The gray and white eutectic solidification kinetics model has been implemented in STAR-CCM+. The newly introduced parameters are tested and their impacts on phase transition kinetics observed. When applying this model, special attention must be paid to very sensitive parameters. Some parameters may be very difficult to obtain. If it can be shown that a parameter is not sensitive, the use of an approximated value is acceptable. In test cases material parameters used by Wu and Ludwig [9] are shown in Table 1.

Table 1. Material parameters used in test case (from Wu and Ludwig [9]).

Density	$\rho^l = \rho^{ge} = \rho^{we} = 7027.25$ kg · m^{-3}
Heat	$c_p^l = c_p^{ge} = c_p^{we} = 808.25$ J · kg^{-1} · K^{-1}
Thermal conductivity	$K^l = K^{ge} = K^{we} = 33.94$ W · m^{-1} · K^{-1}
Mass latent heat of phase change	$\Delta h = 256476$ J · kg^{-1}

Table 2. Values of microscopic solidification parameters to be tested.

Value	Lowest	Lower	Default	Higher	Highest
D_C^A (m^2 · s^{-1})	–	9.0e–11	4.0e–10	1.0e–9	–
R_A^0 (m)	–	1.0e–9	1.0e–8	1.0e–7	1.7e–6
μ_{we} (m · s^{-1} · K^{-2})	–	0.8e–6	2.4e–6	7.2e–6	–
N_{ge} (m^{-3})	–	5.0e+9	5.0e+10	1.0e+12	5.0e+12
N_{we} (m^{-3})	–	5.0e+8	1.1e+9	2.2e+9	–
ΔT_E (°C)	–	10	30	60	–
dT/dt (K· s^{-1})	–	0.02	0.2	2.0	–

In equation (7) the diffusion coefficient of carbon in austenite D_C^γ was reported by Liu and Elliott [10] to have a value in the range of 9×10^{-7}~1.0×10^{-5} cm^2 · s^{-1}. The initial radius of the austenite shell R_A^0 is also required, although this parameter is difficult to measure and is not found in the literature. To estimate an approximately reasonable range, the critical nucleus radius for homogeneous nucleation of copper 1.28×10^{-9} m proposed by Kurz and Fisher [11] is taken as the lower limit of the initial radius of the austenite shell. The upper limit is approximated to be 1% of the final grain radius.

For the growth kinetic of white eutectic, the key parameter is the growth coefficient μ_{we}. In this study, 2.4×10^{-6} m · s^{-1} · K^{-2} by Nastac and Stefanescu [5] it is applied as default value. The volumetric nucleus density for grey and white eutectics, N_{ge} and N_{we}, used in equations (11) and (12) are estimated according to the values given by Liu and Elliott [10] and Pedersen and Tiedje [12]. The metastable eutectic temperature should be dependent on local composition. The microsegregation model of the third element is not currently included in this implementation; according to Nastac and Stefanescu [5], it is reason to estimate the range to be 10~60 K lower than stable eutectic temperature.

3.2. Discussion of results

The impact of microscopic solidification parameters on the solidification results (temperature T, growth velocity of grey and white eutectic grains V, fraction of solid f_s, fractions of grey and white eutectic f_{ge} and f_{we}, grey-to-white transition G/W, radius of grey and white eutectic grains R and mass transfer rate for grey and white eutectic M_{ls}) are summarized in Table 3.

Table 3. Statistics showing the impact of parameters used in the microscopic solidification model on solidification prediction. 0 means no impact, 0+ means weak positive impact, + means significant positive impact, and ++ means strong positive impact, similar for negative impact 0–, –, ––.

	T	V	f_s	f_{ge}	f_{we}	G/W	R	M_{ls}
D_C^A	0	+	0	+	–	+	+	0
R_A^0	0	0	0	0	0	0	0	0
μ_{we}	+	+	0	0	0	0	0	0
N_{ge}	+	+	0+	++	––	++	0	0+
N_{we}	+	–	0	0	0	0	–	0
ΔT_E	–	0–	–	–	+	+	0+	0–
dT/dt	––	++	–	––	++	––	–	++

The diffusion coefficient of carbon in austenite D_C^A will not impact the cooling curve much, since the latent heat values of both grey and white eutectic are assumed to be the same. The volume fraction and total mass transfer rate is also not greatly impacted. The grain growth velocity of grey eutectic is proportional to D_C^A, so that the volume fraction of grey eutectic also increases significantly. At the same time, the amount of white eutectic is impacted by the increased amount of grey eutectic. The final ratio of grey and white eutectic phases is significantly impacted by the D_C^A. The initial radius of the austenite shell R_A^0 has no impact on predicted solidification results. The growth coefficient of white eutectic μ_{we} was shown to have significant impact on cooling curve and grain growth velocity, but no impact on fractions of grey and white eutectic, grain eutectic radius and mass transfer rate. The volumetric nucleus density of grey eutectic N_{ge} (Figure 1) strongly impacts the predicted fractions of grey and white eutectic and shows a significant impact on cooling curve and solidification velocity. The volumetric nucleus density of white eutectic N_{we} has almost no impact on the predicted solidification results. The difference between stable and metastable eutectic temperature ΔT_E showed a weak impact on solidification velocity, eutectic grain radius and mass transfer rate, as well as significant impact on cooling curve and fractions of grey and white eutectic. The cooling rate applied dT/dt is a very important parameter, since its impact on solidification results is very marked.

As an example, results from the parameter test for the volumetric nucleus density of grey eutectic N_{ge} are given here. The impact of N_{ge} on the predicted cooling curve, growth velocity of grey and white eutectic grains, fractions of grey and white eutectic, radius of grey and white eutectic grains, latent heat release rate and mass transfer rate for grey and white eutectic is shown in Figure 1.

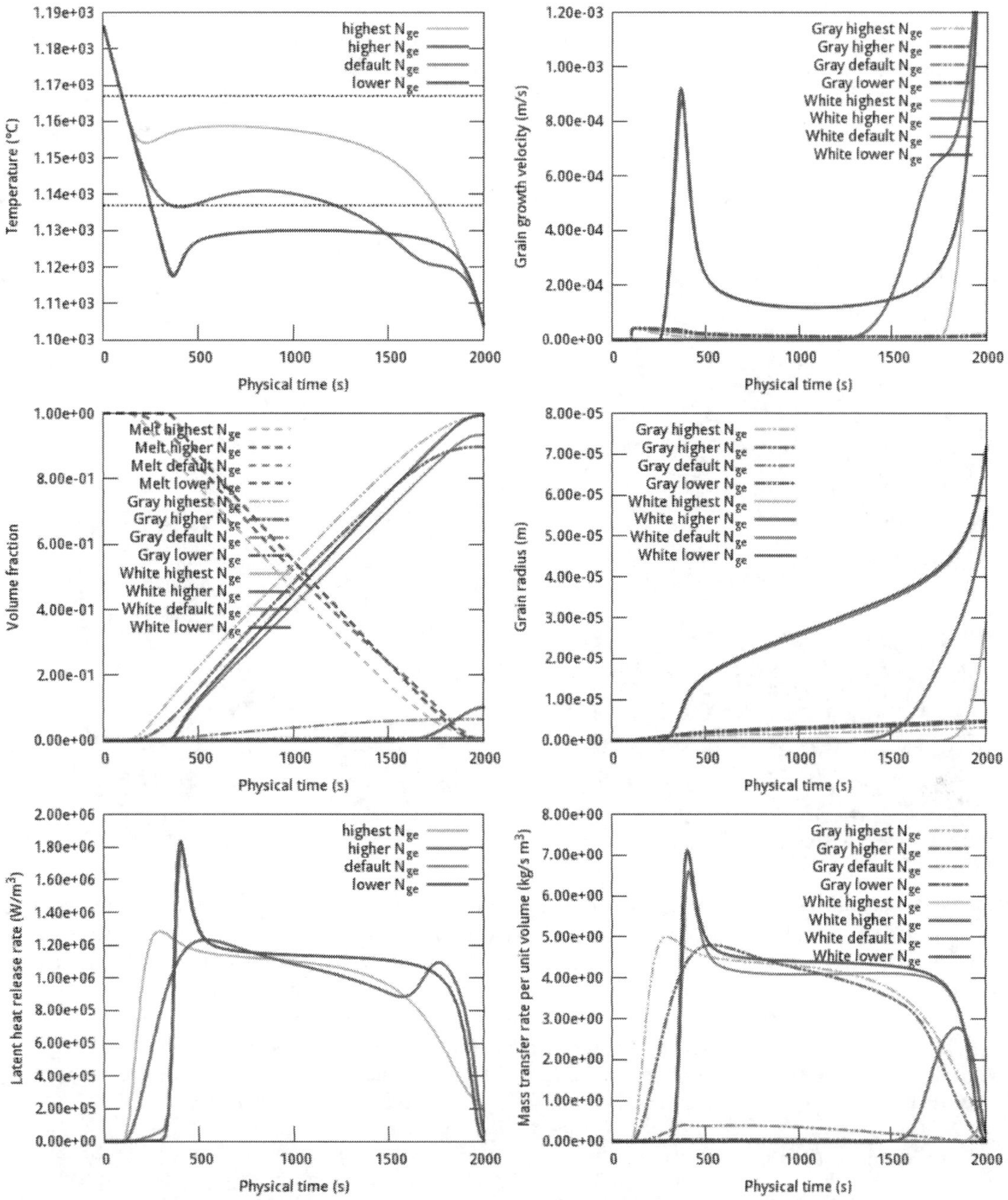

Figure 1. Parameter test shows the impact of the volumetric nucleus density of grey eutectic N_{ge} on the solidification results: predicted cooling curve (top left), growth velocity of grey and white eutectic grains (top right), fractions of grey and white eutectic (middle left), radius of grey and white eutectic grains (middle right), latent heat release rate (bottom left) and mass transfer rate for grey and white eutectic (bottom right). The various N_{ge} (m^{-3}) values are shown in colour: 5.0e+9 (lower), 5.0e+10 (default), 1.0e+12 (higher) and 5.0e+12 (highest); the line style represents different phases: melt, grey eutectic and white eutectic.

4. Casting trial simulation

The enmeshment of the simulated casting trial geometry is shown in Figure 2. Alongside cast blocks, the gating system and the sand mould, the flasks are also modelled. Results of the solidification simulation are shown in Figure 3: the temperature distribution and fraction of liquid, grey eutectic and white eutectic in a vertical cross-section of a 500 mm thick block. For cast alloy, a ferritic SGI grade EN-GJS-400-18-LT is used and the mould is made of furan resin sand. The initial melt temperature is 1305°C.

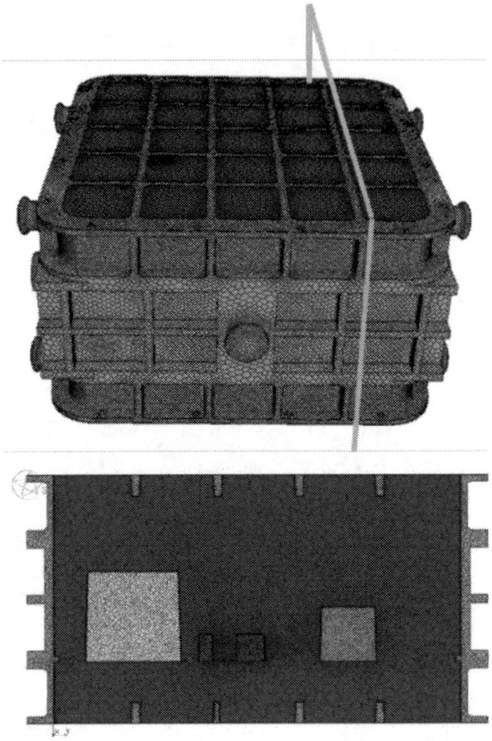

Figure 2. Geometry enmeshment of five blocks, gating system, sand mould and flasks; polyhedral mesh with prism layers in 3D-view (top) and vertical section (bottom)

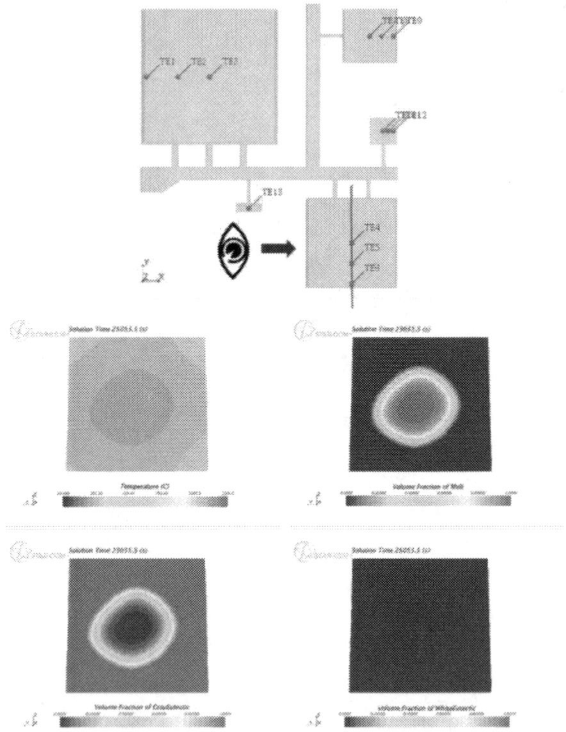

Figure 3. Temperature distribution (middle left) and fractions of liquid (middle right), grey eutectic (bottom left) and white eutectic (bottom right) in cross-section of a 500 mm thick block (top) at approx. 7 h solidification time

A first comparison between measured and simulated temperature curves in a 500 mm thick block is given in Figure 4. The difference in predicted solidification time (approx. 10%) indicates a necessity for further optimization of the heat transfer conditions on the metal/mould interface.

5. Conclusion, Outlook

This paper describes coupling the general multiscale/multiphase modelling framework with the VoF model. An overview of the parameters used in the microscopic model is given, followed by the test case description. Finally, the simulation results of the parameter test are presented. As example, the impact of volumetric nucleus density of grey eutectic on the solidification simulation results (cooling curve, solidification velocity, fractions of grey and white eutectic, eutectic grain radius and mass transfer rate) is given. In the parameter study presented it is shown that variation of the initial radius of the austenite shell R_A^0 has no impact on the afore-mentioned solidification results. At the same time, the importance of selecting a correct value for cooling rate, volumetric nucleus density of grey eutectic and the difference between stable and metastable eutectic temperature is highlighted. Ongoing

IOP Conf. Series: Materials Science and Engineering **84** (2015) 012024 doi:10.1088/1757-899X/84/1/012024

research covers the eutectoid reaction, phase transformations in the solid state and prediction of local mechanical properties.

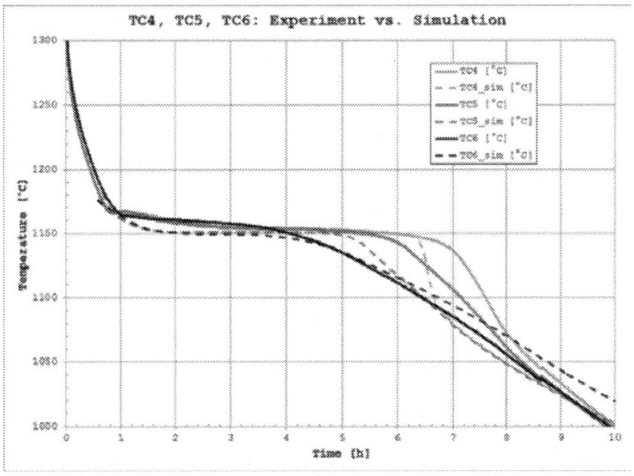

Figure 4. Comparison of measured and calculated temperature curves in a 500 mm thick block

Acknowledgments

Research work presented in this paper is a part of the research project "Numerical and experimental investigation of the solidification process in thick-walled castings made of nodular cast iron for wind power plants - DiWaGussGJS". Authors would like to thank the German Federal Ministry for Economic Affairs and Energy (Bundesministerium für Wirtschaft und Energie) for its financial support of this work.

References

[1] The Federal Ministry for Economic Affairs and Energy, Germany 2014 Renewable Energy Sources Act §1 6
[2] Jakumeit J, Subasic E and Bünck M 2014 Shape Casting: 5th International Symposium, ISBN: 978-1-118-88818-6
[3] Jakumeit J, Laqua R, Peric M and Schreck E 2006 Proceedings of Modeling of Casting, Welding and Advanced Solidification Processes XI 63
[4] Wetterfall S-E, Fredriksson H and Hillert M 1972 Solidification process of nodular cast iron *J. Iron Steel Inst.* **210** 323–33
[5] Nastac L and Stefanescu D M 1997 Modeling of the stable-to-metastable structural transition in cast iron *Adv. Mat. Research* **4-5** 469–78
[6] Beckermann C and Viskanta R 1993 Mathematical modelling of transport phenomena during alloy solidification *Appl. Mech. Rev.* **46** 1–27
[7] Stefanescu D M 2007 Modelling of cast iron solidification – The defining moments *Met. Mat. Trans.* A **38** 1433–47
[8] Oldfield W 1966 A quantitative approach to casting solidification: freezing of cast iron *Trans. ASM* **59** 945–61
[9] Wu M and Ludwig A 2006 A three-phase model for mixed columnar-equiaxed solidification *Met. Mat. Trans.* A **37** 1613–31
[10] Liu J and Elliott R 1998 Numerical modelling of the solidification of ductile iron *J. Cryst. Growth* **191** 261–7
[11] Kurz W and Fisher D J 1986 *Fundamentals of solidification* Trans Tech Publications Ltd 244
[12] Pedersen K M and Tiedje N S 2008 Graphite nodule count and size distribution in thin-walled ductile cast iron *Mat. Characterization* **59** 1111–21

Comparison of residual stresses in sand- and chill casting of ductile cast iron wind turbine main shafts

M R Sonne[1*], J O Frandsen[2] and J H Hattel[1]

[1]Department of Mechanical Engineering, Technical University of Denmark
[2]Global Castings A/S, Denmark

[*]Corresponding author: mrso@mek.dtu.dk

Abstract. In this work, simulations of pouring, solidification and cooling, and residual stress evolution of sand and chill cast wind turbine main shafts is performed. The models are made in the commercial software MAGMAsoft. As expected, the cooling rate of the sand casting is shown to be much lower than for the chill casting, resulting in a very course microstructure. From the simulations the nodule count is found to be 17 nodules per mm^2 and 159 nodules per mm^2 for the sand and chill casting, respectively, in the critical region of the main bearing seat. This is verified from nodule counts performed on the real cast main shafts. Residual stress evaluations show an overall increase of the maximum principal stress field for the chill casting, which is expected. However, the stresses are found to be in compression on the surface of the chill cast main shaft, which is unforeseen.

1. Introduction

The work presented in this paper is a part of a larger project called REWIND. The aim of this project is to gain knowledge for improvement of the reliability of critical wind turbine components, specifically aiming at the metallic parts in the rotor hub and the drivetrain, see Figure 1.

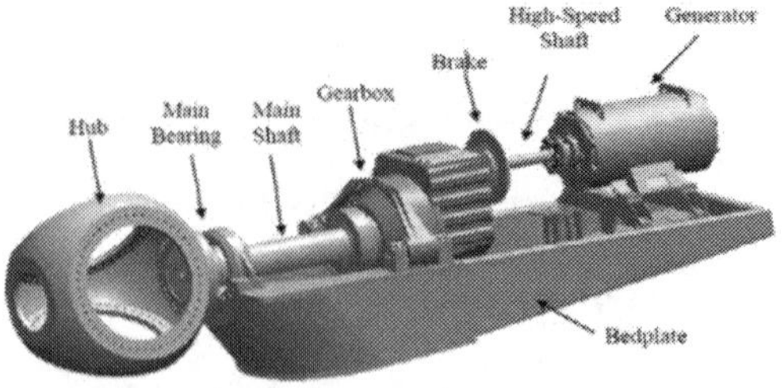

Figure 1. Schematic illustration of the main components in the nacelle of a wind turbine, showing the placement of the main shaft considered in this work.

One of the very large and critical components is the main shaft, which transfers the torque from the rotor hub into the gearbox. The main shaft is a huge cast component with a weight of approximately 12 tons. The material used for this part is ductile cast iron, which due to its spheroidal graphite has an increased ductility compared to standard grey cast iron. The growth of these nodules is influenced by many different parameters in the manufacturing process. Of course, the ability for the inoculants, such as magnesium, to react with the oxygen and sulphur, creating MgO and MgS which floats to the surface, allows more isotropic growth. Another important factor is the cooling rate, which controls the size of the final nodules. Here, the ability of the mould to absorb the heat during solidification and cooling plays an important role. Until just recently, the main shafts for wind turbines have in industry been casted in sand moulds. But due to the low cooling rates, this process sometimes resulted in unfavourable microstructures, leading to unwanted mechanical properties and reduced reliability of the components. In order to increase the cooling rate, and thereby hopefully improve the microstructures, the main shafts can as an alternative be chill cast in permanent moulds made of ductile cast iron, which have a much higher thermal conductivity than the sand moulds. The disadvantage of using a chill for the castings is its more rigid construction and the increased thermal gradients in the casting, leading to possible higher and undesirable residual stresses in the main shafts. The aim of this paper is to address this problem by numerical simulations of the casting process. Both sand and chill casting of a specific wind turbine main shaft component will be modelled, and the residual stresses in critical regions will be compared and related to the resulting microstructures of the part. The simulations are performed in the commercial software MAGMAsoft in three subsequent steps: (i) mould filling, (ii) solidification and cooling, (iii) stress calculations. The numerical simulations are validated by comparison with real castings of the specific main shaft, which also have been performed in both sand moulds and chills.

2. Theory

The simulation of the casting processes in this work is carried out in the finite volume based commercial software MAGMAsoft [1]. The simulation consists in essence of three subsequent steps: (i) mould filling, (ii) solidification and solid state cooling, (iii) stress calculation. During mould filling, the flow field is found from solving the momentum equations and the continuity equation under the assumptions of a Newtonian fluid and incompressibility, i.e.

$$\rho \dot{u}_i + \rho\left(u_j u_{i,j}\right) = \sigma_{ij,j} + p_i + \rho g \tag{1}$$

$$\sigma_{ij} = -\delta_{ij} p + \mu(u_{i,j} + u_{j,i}) \tag{2}$$

$$u_{i,i} = 0 \qquad (1)$$

which is combined with the energy equation including the convective terms, in order to obtain the temperatures during mould filling

$$\rho c_p \dot{T} + u_i\left(\rho c_p T_{,i}\right) \;\; = \left(k T_{,i}\right)_{,i} + \dot{Q}''' \tag{3}$$

This is followed by the solidification analysis in which the energy equation eq (3) is solved in combination with the release of latent heat. It should be emphasized that during solidification the convective terms in eq (3) account for the natural convection taking place, and might have a considerable effect on the thermal field during solidification. However, this natural convection is neglected in the present paper in order to simplify the problem, hence reduce the calculation time. This reduces eq (3) to the heat conduction equation

$$\rho c_p \dot{T} = \left(k T_{,i}\right)_{,i} + \dot{Q}''' \tag{4}$$

Kotas and Hattel [2] investigated the influence of neglecting the natural convection during solidification in large steel castings, and found that the results without convection was very similar and showed the right tendencies compared so simulations including the convective terms during solidification.

For calculation of the transient as well as the residual stress field in the workpiece, a standard mechanical model based on the solution of the three static force equilibrium equations is used, i.e.

$$\sigma_{ij,i} + p_{,i} = 0 \tag{5}$$

Where p_i are the body forces at any point within the plate and $\sigma_{ij,i}$ is the stress tensor. The well-known Hooke's law and linear decomposition of the strain tensor as well as small strain theory are applied together with the expression for the thermal strain

$$\varepsilon_{ij}^{tot} = \varepsilon_{ij}^{el} + \varepsilon_{ij}^{pl} + \delta_{ij}\varepsilon^{th} \tag{6}$$

$$\sigma_{ij} = C_{ijkl}^{el}\varepsilon_{kl}^{el} = C_{ijkl}^{el}\left(\varepsilon_{ij}^{tot} - \varepsilon_{kl}^{pl} - \delta_{kl}\varepsilon^{th}\right) \tag{7}$$

$$C_{ijkl}^{el} = \frac{E}{1+\nu}\left[\frac{1}{2}\left(\delta_{ik}\delta_{jl} + \delta_{il}\delta_{jk}\right) + \frac{\nu}{1-2\nu}\delta_{ij}\delta_{kl}\right] \tag{8}$$

$$\varepsilon_{ij}^{tot} = \frac{1}{2}\left(u_{i,j} + u_{j,i}\right) \tag{9}$$

$$\varepsilon^{th}\left(T_1 \rightarrow T_2\right) = \int_{T_1}^{T_2} \alpha dT \tag{10}$$

The plastic strain is based on the standard J_2 flow theory with a temperature dependent von Mises yield surface. For isotropic hardening, the elastic range expands from the initial yield stress, when a load above the yield stress is applied.

$$f(\sigma_{ij}) = \frac{3}{2}s_{ij}s_{ij} = (\sigma_e)_{\max}^2 \tag{11}$$

2.1. *Microstructural evolution*

The microstructure of the ductile cast iron is characterized by the presence of spherical graphite particles or nodules throughout the material. These graphite nodules nucleate on small inclusions during the solidification. During eutectic solidification the graphite nodules will be surrounded by austenite, and growth of the nodules will then occur by diffusion of carbon from the liquid through the austenite shell [3]. For the final material, the graphite nodule count and size distribution are very important parameters as it influences the mechanical properties of the materials. The count of the nodules is in MAGMAsoft performed as a function of the thermal conditions according to the formulation by Oldfield [4]

$$N_A = A \cdot \Delta T^B \tag{12}$$

Where N_A is the nodule count per unit area, A is a nucleation constant and B is an exponent dependent on the effectiveness of the inoculation. ΔT is the super cooling, which is defined by the phase diagram of the considered materials composition. It is observed that the growth of the eutectic spheres is not taken into account in the microstructural model. This is a limitation of the model as both nodule count and size distribution are important for the properties of the final material [5].

3. Application

As mentioned in the introduction, the application of the numerical model is in this case the main shaft of a 3 MW wind turbine, which connects the forces from the blades on the hub, to a torque as input to the gearbox, see Figure 1. The main shafts have until now been cast in sand moulds, which as mentioned, in some cases gave unwanted microstructures (low nodule count some places in the geometry). The new development in the foundry technology is to cast the shafts in chills, which due to higher cooling rates, increases the nodule count, and hence improves the microstructural properties. The disadvantage of this process though, is an increase in residual stresses due to the higher constrains coming from the permanent mould compared to a sand mould as also mentioned in the introduction. These residual stresses can have a negative effect on the subsequent loading of the component in service (bending and torsion), but the residual stress state could also be in favour of the performance, i. e. with compressive stresses in the surface to prevent crack propagation. In the following, simulations of both sand and chill casting will be performed, and the results in terms of temperature fields, microstructure and residual stress states are compared.

3.1. Geometry and material properties

The overall dimensions of the main shaft geometry are given in Figure 2.

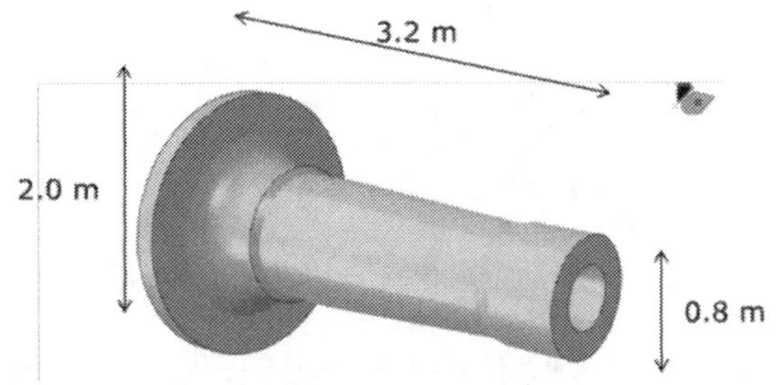

Figure 2. The overall geometry of the main shaft.

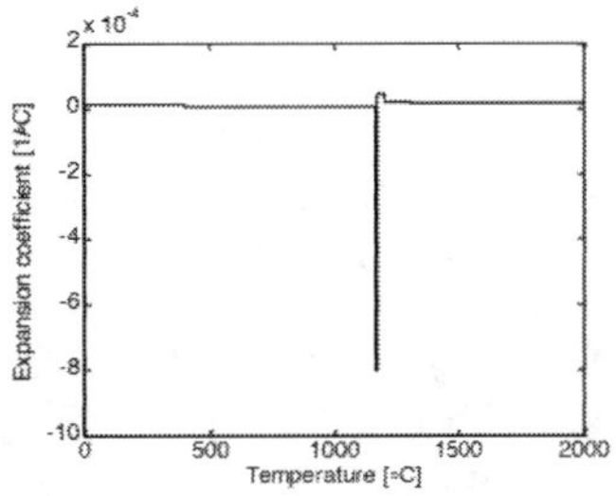

Figure 3. Thermal expansion coefficient as a function of temperature for GJS-400.

The main shafts are cast in the ductile cast iron GJS-400 with the composition as shown in Table 1.

Table 1. GJS-400 composition

C [wt.%]	Si [wt.%]	Mn [wt.%]	Mg [wt.%]	Cu [wt.%]	S [wt.%]
3.6	2.65	0.15	0.036	0.1	0.02

One of the major challenges for modelling the residual stresses in the cast component is the description of the thermal expansion component. For ductile cast irons this is special because the metal will expand during cool down due to the formation of the graphite nodules. This expansion is very much dependent on the initial inoculation, and for that reason, it can in MAGMAsoft be changed to fit if the inoculation is poor or good, and serves in this way as a fitting parameter for the simulation. It was in this work found, that with a middle inoculation, the simulations gave the best results compared to the experiment. For this setting, the temperature dependent thermal expansion is prescribed as shown in Figure 3. It is observed that the expansion coefficient is negative in the range of graphite expansion. The reason for this is how the software calculates the thermal strain, which is the change in temperature from old to new times the expansion coefficient. And as this change is negative, the thermal expansion coefficient must also be negative in order to induce a positive thermal strain.

3.2. The numerical model

The simulations are for both sand and chill casting performed according to the five steps shown in Figure 4. First pouring of melt into the mould cavity is performed, where the flow is calculated based on the Navier-Stokes equations. Then the resulting temperature field is used as initial conditions for the solidification state, where the solidification of the GJS-400 material is simulated. After solidification, the casting is further cooled down to a temperature of 400 °C, where the casting then is shaken out of the mould, which in the numerical model is performed by removing the mould cells and prescribing a film radiation on the surface of the casting. After further cooling to a maximum temperature in the cast component of 50 °C, the machine allowance is removed, resulting in the final geometry of the main shaft. In the numerical model this is done by "killing" the cells of the machine allowance, and recalculate the equilibrium state of stresses in the main shaft. During solidification, cooling, shake out and machining, the stress-strain analysis is performed.

Figure 4. The five steps in the simulations.

The total number of cells was chosen to be 4.000.000 in both simulation cases, with approximately 400.000 cavity cells for the fluid flow calculations. With this discretization, the calculation time was around two days on an eight cores work station. In Figure 5 and 6, the overall geometry of the two castings is shown. For the sand casting simulation (Figure 5), the filling cavity (casting and machine allowance) is surrounded by the sand drag and cope boxes modelled as cylinders, with the properties of Furan sand. For the chill casting (Figure 6), the permanent mould is modelled as one piece with the same mechanical properties as the cast material (this was also the case for the real casting). The gating system was in this case not modelled as in the real situation, since this is confidential, but the filling behavior and the overall filling time of 200 s is identical, which means that the overall temperature field after mould filling should be comparable with the temperatures in the real castings.

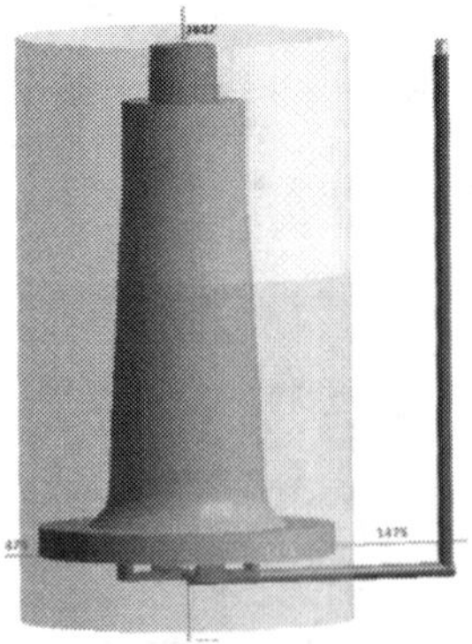

Figure 5. The overall geometry of the sand casting including the gating system.

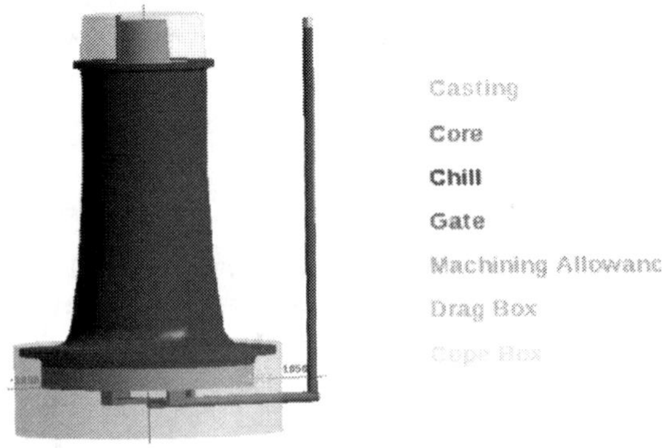

Figure 6. The overall geometry of the chill casting including the gating system

IOP Conf. Series: Materials Science and Engineering **84** (2015) 012025 doi:10.1088/1757-899X/84/1/012025

4. Results

Sand and chill casting simulations are now compared in terms of temperature history, nodule count and final residual stresses. The nodule counts were furthermore verified by comparison with the experiment, where nodule count at a critical point of the main shafts was performed by tomography.

4.1. Temperatures

In Figure 7, a contour plot of the final temperature distribution in the sand cast main shaft is shown on the cross section. The minimum temperature (blue) is 50 °C, where as it is observed that some points in the bulk material is still around 200 °C (yellow), when the calculations are stopped. In this point, the temperature history of the sand and chill castings are compared (see Figure 8 and 9). First of all it is observed, that the cooling of the sand casting takes around eight days (200 hours) compared to the one day (26 hours) of the chill casting. This was also observed for the casting experiments. In Figure 9, the temperature history during solidification in the bulk part of the main shaft is compared. It is here observed that the solidification time for the sand casting is almost five times larger than the chill casting, which of course, will have a huge effect on the resulting microstructure.

4.2. Nodule count

The modelling of the microstructure was compared with actual nodule counts performed at the main bearing seat (see Figure 10 and 11), as this is known to be a critical place for the shafts.

The nodule count was performed by tomography of which some SEM images of the sand and chill casting samples are shown in Figure 12 and 13. The volumes of the tomographic cylinders were 6.28 mm^3 and 50.24 mm^3 for the sand and chill casting, respectively, with 5 μm cut-off value for the tomography. With a sample diameter of 2 mm for the chill casting, the nodule count was calculated to be 228 nodules per mm^2. For the sand casting, the sample diameter was 4 mm and the nodule count was calculated to be 13 nodules per mm^2. This agrees very well with the values obtained in the models on Figure 10 and 11, where the nodule count is around 17 for the sand casting and 159 for the chill casting. Still, there are some deviation in the chill casting, which suggests that the cooling rate should be higher in the simulations, which could be due to a too low heat transfer coefficient between casting and mould.

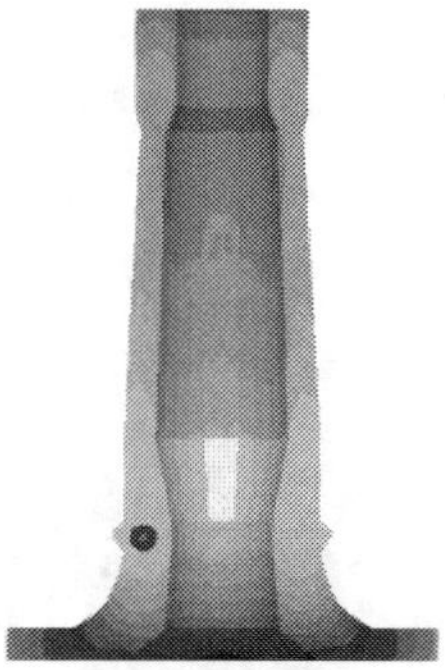

Figure 7. Contour plot of the temperatures on the cross section of the main shaft, and the point where the temperature curves are extracted.

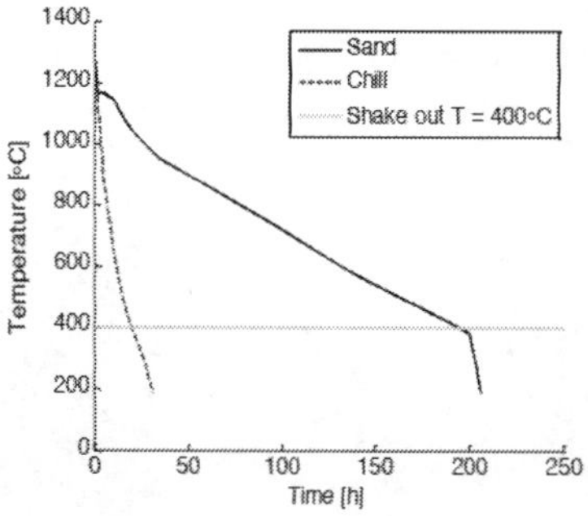

Figure 8. The overall temperature history at the point in Figure 7, for sand and chill casting, respectively.

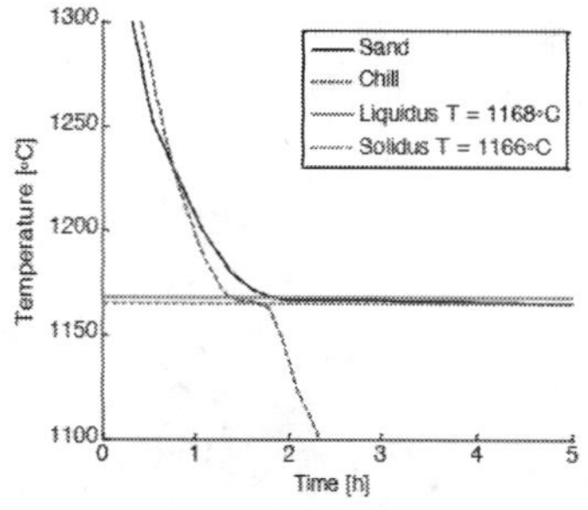

Figure 9. The temperatures during solidification at the point in Figure 7, for sand and chill casting, respectively.

IOP Conf. Series: Materials Science and Engineering **84** (2015) 012025 doi:10.1088/1757-899X/84/1/012025

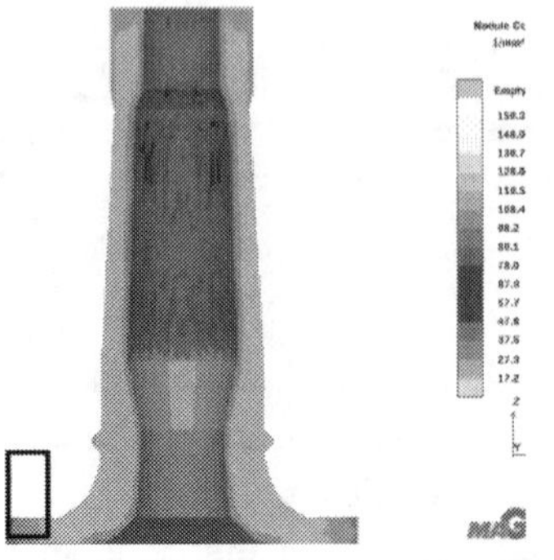

Figure 10. Nodule count per mm^2 found in the sand casting simulation. The square indicates where the nodule count was performed on the actual shaft.

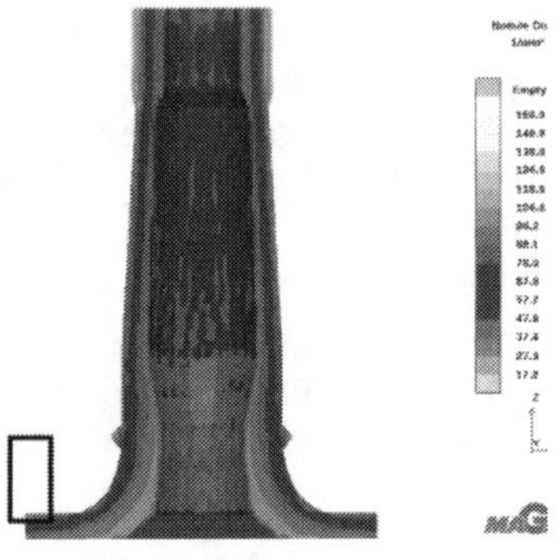

Figure 11. Nodule count per mm^2 found in the chill casting simulation. The square indicates where the nodule count was performed on the actual shaft.

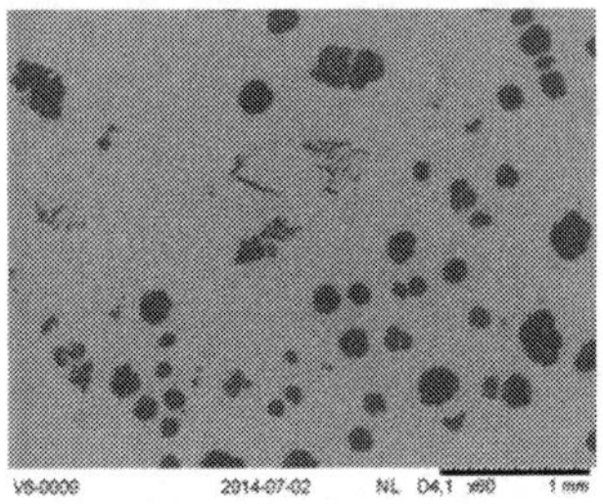

Figure 12. SEM image for the nodule count performed on the sand casting in the area shown in Figure 10.

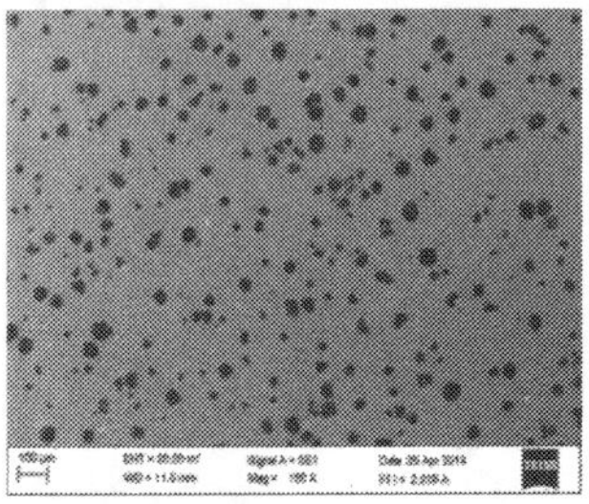

Figure 13. SEM image for the nodule count performed on the chill casting in the area shown in Figure 11.

4.3. Residual stresses

What now is interesting is to see if these enhanced micrustructures for the chill cast main shaft, has a negative effect on the corresponding residual stresses. In Figure 14 and 15, the maximum principal stress is plotted on the cross section of the sand and chill cast main shafts respectively. The stresses are normalized with respect to the yield stress due as it is confidential.

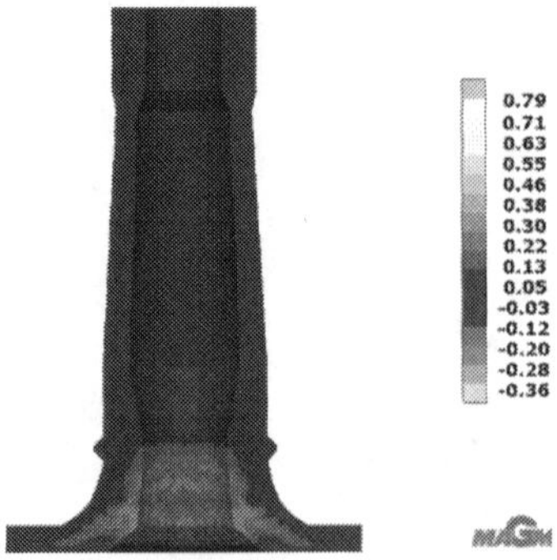

Figure 14. Maximum principal stress normalised with respect to the yield stress for the sand casting at room temperature.

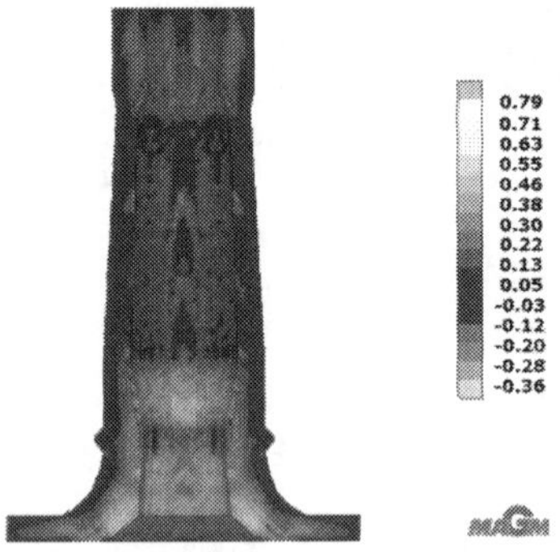

Figure 15. Maximum principal stress normalised with respect to the yield stress for the sand casting at room temperature.

It is observed that the residual stresses in the sand casted shaft are very low, with small stresses in tension on the inside of the shaft geometry. For the chill cast shaft, it is observed that the stress state in general is higher and higher tensional stresses are dominating the inside geometry of the shaft geometry. But what is an interesting finding is that the residual stresses on the outside of the shaft, which are known to be for more critical for the main shaft in-service, are in compression, which actually is beneficial for the reliability of the shaft.

5. Conclusion

Simulations of pouring, solidification and cooling and residual stress evolution of sand and chill cast wind turbine main shafts has been performed. The models were developed in the commercial software MAGMAsoft. As expected, the cooling rate of the sand casting was shown to be much lower than for the chill casting, resulting in a very course microstructure for the sand cast component. From the simulations the nodule count was found to be 17 nodules per mm^2 and 159 nodules per mm^2 for the sand and chill casting, respectively, in the critical region of the main bearing seat. This was verified from nodule counts performed on the real cast main shafts. Residual stress evaluations showed an overall increase of the maximum principal stress field for the chill casting, which was expected. But, the stresses were found to be in compression on the surface of the chill cast main shaft, which was unforeseen.

6. Acknowledgment

This work is financially supported by the Danish council for Strategic Research, which is highly acknowledged.

References

[1] 'MAGMAsoft: v.5,2, "reference manual", MAGMA GmbH, Aachen, Germany
[2] Kotas P and Hattel J H 2012 Modelling and simulation of A segregates: Part I, *Materials Science and Technology* **28** 872
[3] Wetterfall S-E, Fredriksson H and Hillert M. 1972 Solidification process of nodular cast iron. *J Iron Steel Inst* **210** 323
[4] Oldfield W 1966 *Trans. of the ASM* **59** 945
[5] Tanaka Y, Yang Z and Miyamoto K 1995 Evaluation of fatigue limit of spheroidal graphite cast iron *Mater Trans* **36** 749

Nomenclature

α	Thermal expansion coefficient
μ	Viscosity
ϱ	Density
v	Poisson's ratio
A	Nucleation constant
B	Inoculation exponent
c_p	Specific heat
E	Young's modulus
g	Gravitational acceleration
k	Thermal conductivity
N_A	Nodule count pr unit area
T	Temperature
ΔT	Super cooling
$\dot{Q}'''$	Volumetric heat source term
p_i	Pressure/external force vector
u_i	Velocity/displacement vector
ε_{ij}^{el}	Elastic strain tensor
ε_{ij}^{pl}	Plastic strain tensor
ε^{th}	Thermal strain
ε_{ij}^{tot}	Total strain tensor
σ_{ij}	Stress tensor
σ_e	Von-Mises equivalent stress
s_{ij}	Deviatoric stress tensor
C_{ijkl}^{el}	Elastic constitutive fourth-order tensor
$f(\sigma_{ij})$	Yield function

Modelling and simulations of ductile iron solidification-induced variations in mechanical behaviour on component and microstructural level

J Olofsson[1], K Salomonsson[2] and I L Svensson[1]

[1]Department of Materials and Manufacturing, School of Engineering, Jönköping University, Box 1026, SE-551 11 Jönköping, Sweden
[2]Department of Product Development, School of Engineering, Jönköping University, Box 1026, SE-551 11 Jönköping, Sweden

E-mail: jakob.olofsson@jth.hj.se

Abstract. The mechanical behaviour and performance of a ductile iron component is highly dependent on the local variations in solidification conditions during the casting process. Here we show a framework which combine a previously developed closed chain of simulations for cast components with a micro-scale Finite Element Method (FEM) simulation of the behaviour and performance of the microstructure. A casting process simulation, including modelling of solidification and mechanical material characterization, provides the basis for a macro-scale FEM analysis of the component. A critical region is identified to which the micro-scale FEM simulation of a representative microstructure, generated using X-ray tomography, is applied. The mechanical behaviour of the different microstructural phases are determined using a surrogate model based optimisation routine and experimental data. It is discussed that the approach enables a link between solidification- and microstructure-models and simulations of as well component as microstructural behaviour, and can contribute with new understanding regarding the behaviour and performance of different microstructural phases and morphologies in industrial ductile iron components in service.

1. Introduction

Cast iron is a group of materials which exhibit highly interesting combinations of material properties for applications in e.g. the automotive and transportation areas. The materials are usually classified in three wide groups according to the morphology of the graphite particles found within the material; flake graphite iron (FGI), compacted graphite iron (CGI) or spheroidal graphite iron (SGI). In FGI the graphite particles are found as thin flakes with sharp edges. In CGI the edges are somewhat rounded (vermicular), while in SGI the graphite has formed more or less round graphite nodules. The difference in graphite morphology between these three materials are illustrated in figure 1. The graphite morphology is very important since it controls as well the mechanical as physical material behaviour. In general terms, FGI exhibit low strength but high thermal conductivity, SGI high strength and lower thermal conductivity, while CGI has intermediate strength and conductivity [1]. In this work the SGI material, also known as ductile iron, is studied. The local variations in microstructure formed during solidification are addressed, and the effects of these variations on the mechanical behaviour of a cast SGI component are modelled, both on the macro- and micro-scale.

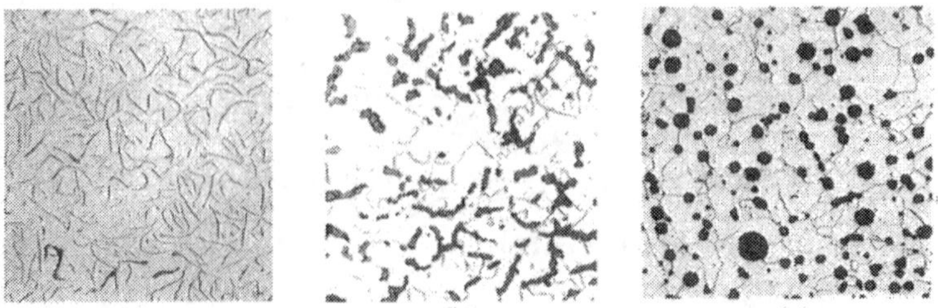

(a) (b) (c)

Figure 1. Typical graphite morphology of (a) FGI (b) CGI and (c) SGI [1]

1.1. Macro-scale behaviour induced by micro-scale variations during solidification

Even for a specific class of cast iron, the graphite on different locations of a cast geometry exhibit local variations in morphology. Measures as roundness, nodularity and nodule count are typically used to further describe the character of the graphite precipitates. Though the precipitates are distributed in a 3D volume, these measures are typically identified from micrographs of a polished flat 2D section of the material.

The graphite morphology is not constant for a specific alloy or component geometry. The graphite morphology is affected by several local parameters which vary throughout the geometry during solidification. One important factor is the local cooling rate, which is highly related to the geometry of the component (e.g. section thickness) and the extraction of heat from the component during solidification. Another important factor is the local chemical composition. Graphite morphology is affected e.g. by the magnesium (Mg) content when the graphite forms, which in turn depends on e.g. the chemical composition of the alloy, the effectiveness of the Mg-treatment, Mg-fading during pouring and solidification and chemical segregation. The graphite morphology thus display local variations throughout the volume of a cast component [2]. Since graphite morphology affects material behaviour, the material behaviour exhibits local variations throughout a cast component.

Not only graphite morphology but also the matrix between the graphite nodules exhibits local variations throughout a component due to local variations in solidification conditions. The matrix typically consists of ferrite, pearlite or combinations of both. In a ferritic-pearlitic matrix, the ferrite is typically formed around the graphite nodules during the solid-state transformation stage of the solidification process. Since ferrite and pearlite have significantly different material behaviour, the local ferrite-pearlite fraction have a great influence on the local material behaviour throughout a component [2]. As for graphite morphology, the area fractions ferrite-pearlite fractions provides information about a 2D cross section, while these microstructural constituents are in fact dispersed through the 3D volume of the casting.

1.2. The closed chain of simulations

The material behaviour throughout a cast iron component thus exhibit significant local variations, which are dependent on complex interactions between component design, casting process parameters and several chemical and physical phenomena, during solidification. To address this problem, two of the current authors have previously presented a simulation strategy called *the closed chain of simulations for cast components* [3]. The strategy is schematically outlined in figure 2. The strategy is based on numerical solidification models implemented into a casting process simulation, which include the entire casting process including e.g. chemical composition, mould filling and solidification. The casting process simulations are performed using MAGMASOFT™ [4], which is based on the Finite Difference Method (FDM) for solving the governing equations. Local microstructure evolution during solidification and solid-state transformation is predicted using previously developed microstructure formation models, e.g. [5]. Material characterization models which relate microstructure and mechanical behaviour [6] are applied on the local level, i.e. in every element of the FDM mesh. The predicted material behaviour is then incorporated into a macro-scale Finite Element Method (FEM) simulation of the behaviour of the component in service, using an in-house developed software to generate element specific microstructure-based local material definitions [3]. The simulation strategy has also been applied to cast aluminium components [7]. It has been verified, using Digital Image Correlation (DIC), that by taking these local variations into account, a more accurate prediction of the distribution of stresses and strain is obtained compared to using traditional homogeneous or tabulated material behaviour [8].

The macro-scale simulation is thus able to predict the effect of the local variations in mechanical behaviour onto the global macro-scale stress distribution of the component. It is however unable to predict the effect of the macro-scale stress on micro-scale stress distributions around microstructural features in the loaded region.

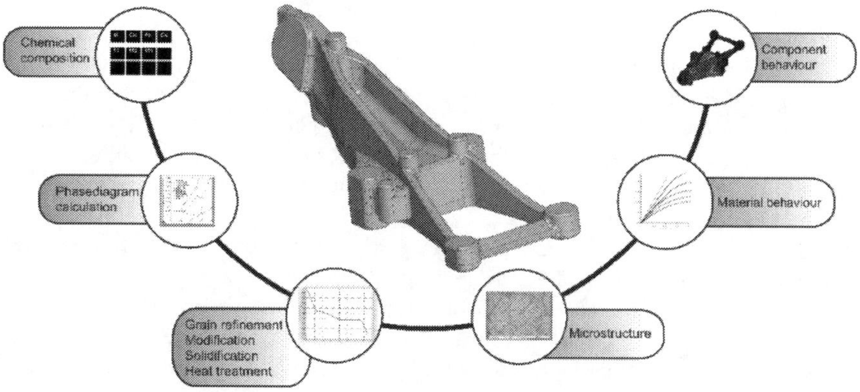

Figure 2. Illustration of the closed chain of simulations for cast components [3].
Figure reprinted with permission of Elsevier.

The current contribution aims to present ongoing work on an extension of *the closed chain of simulations*. Here the macro-scale FEM simulation results serves as input for a FEM simulation of the micro-scale stress distribution using a representative model of the microstructure. 3D X-ray tomography is applied to generate a representative geometrical model of a real micro-structure, including ferrite, pearlite and graphite nodules. The material properties of each phase is determined using surrogate model based optimization. From the model, as well volume fractions of ferrite, pearlite and graphite as graphite morphology characteristics can be determined in 3D. By predicting the micro-scale distribution of stresses and strains, an increased understanding of the performance of a cast component can be obtained. This will provide important information to aid engineers within mechanical design, solid mechanics and materials science to develop more reliable and robust components.

2. Macro-scale characterization and FEM simulation

In this work, an engine support for a truck was used as a demonstrator. The size of the component is approximately 0.5 m in length, and 0.3 m in height and width. The same geometry and FEM setup has been used in previous work [2, 6, 9], where also the setup of the casting process simulation and the macro-scale FEM simulation is described [9]. In previous work, the Hollomon equation was applied to describe the elasto-plastic material behaviour in the macro-scale simulation as

$$\sigma = K_H \cdot \varepsilon_{pl}^{\ n_H} \tag{1}$$

where the strength coefficient K_H and the strain hardening exponent n_H are introduced to relate true stress σ and plastic strain ε_{pl}. Following findings in recent work [6], the Ludwigson equation was applied in the current work. The Ludwigson equation provides an extension of the Hollomon equation for small plastic strains and is formulated as

$$\sigma = K_H \cdot \varepsilon_{pl}{}^{n_H} + e^{(k_L + n_L \cdot \varepsilon_{pl})} \tag{2}$$

The Ludwigson equation has been found to describe the the two-slope plastic behaviour found in SGI [10] with higher accuracy than the Hollomon equation [6]. Material characterization models relating the parameters K_H, n_H, k_L and n_L to fraction pearlite and nodularity have previously been derived [6], and implemented into the casting process simulation. The local predictions of microstructural features and mechanical behaviour has been shown to be in good agreement with experimental data [2, 6].

Figure 3 shows the stress distribution predicted by the macro-scale FEM simulation at a certain load using three different types of material data; tabulated homogeneous material data and local material data using the Hollomon and the Ludwigson equations. A comparison of the results obtained shows that the implementation of the Ludwigson equation affects the predicted results for stresses and strains already before the tabulated 0.2% offset yield stress of 350 MPa is reached. As also found in previous work [9], the local variations in microstructure causes a redistribution of stresses and strains, an effect which will be even more significant at higher loads. The results show the importance of providing accurate and local material data to the FEM simulation. The difference between using tabulated material data and accurate material modelling using the Ludwigson equation is in this case around 20% difference in predicted stress level. This difference is of utmost importance for the accuracy of predictions of as well static as fatigue performance of the component, which are crucial aspects when designing the component. This highlights the relevance of taking the effect of local microstructure into consideration when analysing the stress distribution in cast components.

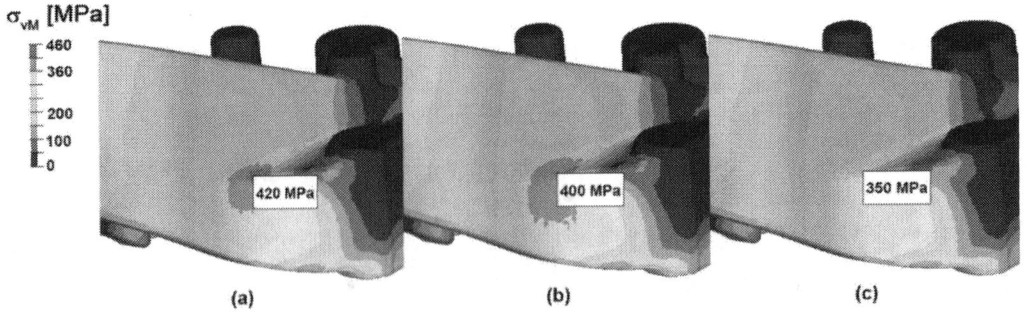

Figure 3: Predicted von Mises stress in a region of the component subjected to load using (a) homogeneous tabulated material data, and local data using (b) the Hollomon equation and (c) the Ludwigson equation.

Based on the macro-scale FEM simulation, a critical region of the component subjected to high stresses and strains can be identified, and selected for further studies of the response of the microstructure in this region. In the current semi-multiscale approach, the stresses and strains predicted by the microstructure-based macro simulation are used as input for the subsequent micro-scale simulation. For these micro-scale studies, it is important that a relevant and representative stress and strain distribution is used in order to provide relevant results, thus the macro-scale FEM simulation with local Ludwigson data is selected as the basis for the micro-scale FEM analyses.

3. Micro-scale simulation

Studies of microstructures in 3D can be done in various ways. Focused Ion Beam (FIB) is commonly used to generate 3D microstructures, see e.g. [11-12]. Another method which also falls under the same scope is to study the 3D microstructures by deep-etching [13]. However, both of these techniques are rarely used for the purpose of generating FEM-models, but merely for quantitative and/or qualitative analyses. Instead, X-ray tomography (XT) is the prominent method for generating 3D FEM-models of microstructures [14-16].

In the present work, X-ray tomography was used to generate a model of the actual microstructure of the SGI, see figure 4a. From this model, information about the 3D graphite morphology and volume fractions of the different matrix constituents can easily be extracted. Results from these measurements will be further evaluated and presented in upcoming work.

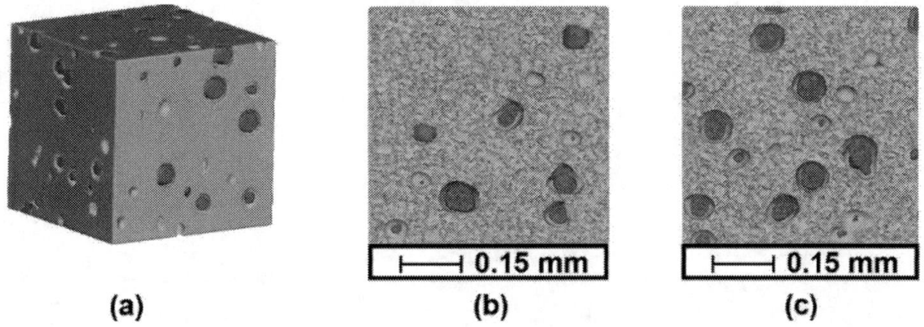

Figure 4. (a) 3D microstructure of the SGI (b-c) Examples of 2D microstructure representations.

If the material curve describing the elasto-plastic behaviour of the separate phases is known, this 3D model can be directly used for FEM simulations of the behaviour and performance of the microstructure in 3D. These curves are however not easily determined. In this work the material curve has been determined using a reversed engineering strategy, where surrogate modelling is used to determine material curve for each individual phase based on experimental tensile testing data.

In order to save computational power and time, the 3D microstructure was sliced to provide 2D cross sections which were used in an optimization algorithm to reverse engineer the material properties for the different phases, see figure 4. The 2D representation, i.e. the representative volume element (RVE), of the microstructure was analysed by applying boundary conditions that originated from critical regions in the macro-scale simulation according to figure 5.

The reversed engineering scheme was based on an assumed material behaviour of the individual metallic phases of the microstructure. In the present study a frequently used relation was used for describing the behaviour of the different phases, namely the Ramberg-Osgood relation

$$E\varepsilon = \sigma + \alpha\left(\frac{|\sigma|}{\sigma_Y}\right)^{n-1} \tag{3}$$

where σ is the Cauchy stress, σ_Y the yield stress, E the Young's modulus, ε the strain and α and n are material specific constants. The material behaviour of the graphite is considered to be elastic.

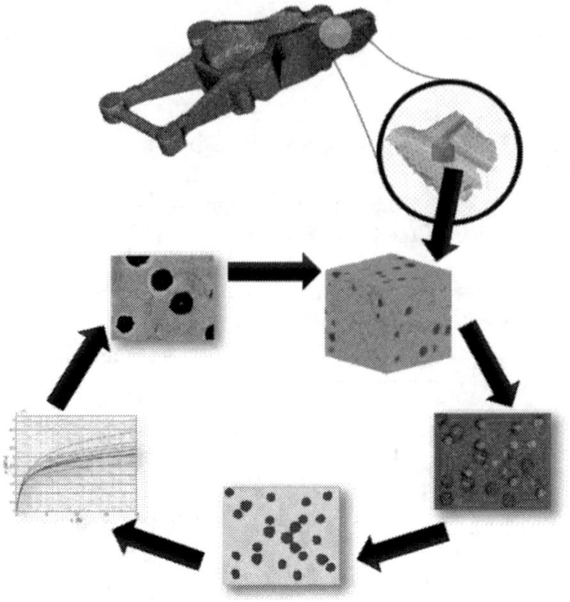

Figure 5. Schematic illustration of the reversed engineering strategy applied on a microstructural region identified from the macro-scale FEM simulation of the component.

An objective function in the form of a fitness functional to be minimized was then formulated as

$$f_j\left(\sigma_{Y,},\alpha_j,n_j\right)=\sum_i^k W_i\left\|\sigma_i^{sim}-\sigma_i^{\exp}\right\| \tag{4}$$

Here σ^{sim} is the simulated average stress of the RVE, and the corresponding stress measured from physical experiments is given by σ^{exp}. The L_2-norm is denoted by ‖...‖ and W is a weight factor. The objective function contains three variables per phase, namely σ_Y, α and n. When considering two phases, the objective function will contain six variables by keeping Young's modulus and Poisson's ratio fixed. Since the simulations are fairly time consuming it was appropriate to generate a so-called surrogate model. This model was established by generating a set of simulation models by providing combinations of variables. The fitness values for the resulting simulations were used to generate the surrogate model by use of a radial basis function (RBF), see e.g. [17]. The RBF is in the form of

$$f\left(\tilde{x}\right)=\sum_{i=1}^n \lambda_i A_{ij}+\sum_{i=1}^n \beta_i B_{ij} \tag{3}$$

where the first term is the RBF and the second term is the augmented bias. The minimization of the objective function is done using the Hooke and Jeeves pattern search method. An example of the generated surrogate model for variables σ_Y and n is shown in figure 6.

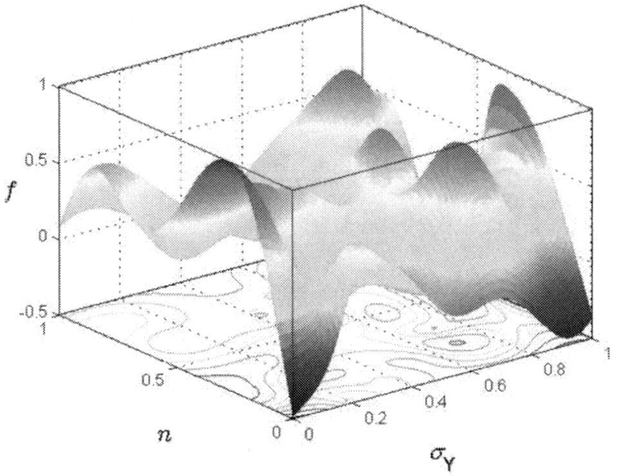

Figure 6. Example of the surrogate model using RBF with *a priori* bias.

Using the surrogate model, the combination of parameters for each phase that best fits the experimental data was determined. By using this data for the micro-scale FEM simulations, the behaviour and performance of the microstructure subjected to the stresses and strains provided by the macro-scale simulation can be studied. Future work will be aimed at e.g. studying the distribution of stresses and strains in different phases when the microstructure model is subjected to load, and the effect of different graphite morphologies on the components performance. The micro-scale FEM simulation can also be applied to study the initiation and propagation of cracks at cyclic loading and fatigue. The current framework thus enables a link from solidification- and microstructure-modelling to component simulations of as well macro-scale as micro-scale behaviour and performance. By integrating these simulations into the design process for cast components, new possibilities to design cast components with high performance, robustness and quality are enabled.

4. Conclusions

The local solidification conditions in a cast component are critical for the microstructure formation and thus for the mechanical behaviour of a cast iron component. A framework for studying the effect of variations in microstructure due to variations in solidification conditions throughout a cast component has been outlined. The framework is a semi-multiscale approach to predict as well the macro-scale distribution of stresses and strains in the component as well as the response and performance of the microstructure within a critical region. X-ray tomography has been used to generate a representative model of the microstructure in a critical region. Surrogate modelling has been successfully applied to generate the material data needed for the micro-scale FEM simulation based on experimental data. The framework enables new possibilities to include solidification and microstructure modelling into the design process for industrial cast components.

5. Acknowledgements

The work has been performed in the CCSIM2 project within the research profile CompCAST at the School of Engineering, Jönköping University. The Swedish Knowledge Foundation is gratefully acknowledged for financial support of the research profile. The authors would also like to thank Professor Ragnvald Mathiesen and Professor Arne Dahle for providing the X-ray tomography, and Kaveh Amouzgar for assistance with the optimization routine.

References

[1] Davis J R 1996 *Cast Irons* Materials Park, Ohio. ASM International

[2] Svensson I L and Sjögren T 2009 On modeling and simulation of mechanical properties of cast irons with different morphologies of graphite *Int. J. Metalcast.* **3** 67-77

[3] Olofsson J and Svensson I L 2012 Incorporating predicted local mechanical behaviour of cast components into finite element simulations *Mater. Des.* **34** 494–500

[4] MAGMA Gießereitechnologie GmbH 2011 *MAGMASOFT v. 5.2* http://www.magmasoft.com

[5] Wessén M and Svensson I L 1996 Modeling of ferrite growth in nodular cast iron *Met. Mater. Trans. A* **27** 2209-20

[6] Svensson I L and Olofsson J 2014 On microstructure-based mechanical behaviour of a ductile iron component *Proc. 10th Int. Symp. on the Science and Processing of Cast Iron – SPCI10* Mar del Plata, Argentina

[7] Olofsson J and Svensson I L 2014 Closed chain simulations of a cast aluminium component – Incorporating casting process simulation and local material characterization into stress-strain simulations *ISIJ Int.* **54** 259-65

[8] Olofsson J, Svensson I L, Lava P and Debruyne D 2014 Characterisation and investigation of local variations in mechanical behaviour in cast aluminium using gradient solidification, Digital Image Correlation and finite element simulation *Mater. Des.* **56** 755–62

[9] Olofsson J and Svensson I L 2013 The effects of local variations in mechanical behaviour – Numerical investigation of a ductile iron component *Mater. Des.* **43** 7264–71

[10] Larsson D, Sjögren T and Svensson I L 2008 Modelling the plastic deformation behaviour of cast irons *Giessereiforschung* **60** 8-16

[11] Velichko A, Holzapfel C, Siefers A, Schladitz K and Mücklich F 2008 Unambiguous classification of complex microstructures by their three-dimensional parameters applied to graphite in cast iron *Acta Mater.* **56** 1981-90

[12] Hatton A, Engstler M, Leibenguth P and Mücklich F 2011 Characterization of graphite crystal structure and growth mechanisms using FIB and 3D image analysis *Adv. Eng. Mater.* **13** 136-44

[13] Imasogie B I and Wendt U 2004 Characterization of graphite particle shape in spheroidal graphite iron using a computer-based image analyzer *J. Min. Mater. Char. Eng.* **3** 1-12

[14] Toda H, Sinclair I, Buffiére J Y, Maire E, Khor K H, Gregson P and Kobayashi T 2004 A 3D measurement procedure for internal local crack driving forces via synchrotron X-ray microtomography *Acta Mater.* **52** 1305-17

[15] Fischer G, Nellesen J, Anar N B, Ehrig K, Riesemeier H and Tillmann W 2013 3D analysis of micro-deformation in VHCF-loaded nodular cast iron by mCT *Mater. Sci. Eng.* **577** 202-9

[16] Seo D, Tomizato F, Toda H, Uesugi K, Takeuchi A, Suzuki Y and Kobayashi M 2012 Spatial resolution of synchrotron X-ray microtomography in high energy range: Effect of X-ray energy and sample-to-detector distance *Appl. Phys. Lett.* **101** 1-5

[17] Amouzgar K and Stromberg N 2014 An approach towards generating surrogate models by using RBFN with a priori bias. *Proc. ASME 2014 International Design Engineering Technical Conference & Computers and Information in Engineering Conference IDETC/CIE* Buffalo, NY, USA

Modeling of damage in ductile cast iron – The effect of including plasticity in the graphite nodules

T Andriollo[1], J Thorborg[2], N S Tiedje[1] and J Hattel[1]
[1]Department of Mechanical Engineering, Technical University of Denmark, DK-2800 Kgs. Lyngby, Denmark
[2]MAGMA GmbH, D-52072 Aachen, Germany

E-mail: titoan@mek.dtu.dk

Abstract. In the present paper a micro-mechanical model for investigating the stress-strain relation of ductile cast iron subjected to simple loading conditions is presented. The model is based on a unit cell containing a single spherical graphite nodule embedded in a uniform ferritic matrix, under the assumption of infinitesimal strains and plane-stress conditions. Despite the latter being a limitation with respect to full 3D models, it allows a direct comparison with experimental investigations of damage evolution on the surface of ductile cast iron components, where the stress state is biaxial in nature. In contrast to previous works on the subject, the material behaviour in both matrix and nodule is assumed to be elasto-plastic, described by the classical J2-flow theory of plasticity, and damage evolution in the matrix is taken into account via Lemaitre's isotropic model. The effects of residual stresses due to the cooling process during manufacturing are also considered. Numerical solutions are obtained using an in-house developed finite element code; proper comparison with literature in the field is given.

1. Introduction

Since its commercial introduction in 1948, ductile cast iron (DCI) has constantly found new fields of application, ranging from the automotive sector to the wind power industry. Castings made of DCI are especially attractive as they combine high ductility and strength together with lower price compared to traditional low carbon steel. Surprisingly, despite extensive experimental work carried out in the last 60 years [1], little has been done to cast light on the microscopic features which determine the constitutive behavior of the material at the macroscopic scale.

In some papers published in literature in the 90's, DCI has been assumed to behave as a porous material [2][3][4], in consideration to the "soft" nature of the graphite and the weak strength of the nodule-matrix interface, for which the Gurson-Tvergaard-Needleman model reasonably applies [5].

On the other side, in the last decade several researchers have pointed out that some effects observed during deformation of DCI cannot be explained by a simple voided-matrix material model. First of all, it has been proved that fatigue in DCI cannot be analyzed using a linear elastic fracture mechanics framework [2]; this might be related to the fact that, according to the imposed stress intensity factor, different competing damage mechanisms seem to be active in the matrix and/or in the nodule [6]. Moreover, tensile and compression tests of DCI samples conducted at different temperatures have highlighted large differences in the deformed nodule shapes [7]: if the nodule stiffness and strength were negligible compared to the matrix in the entire range of temperatures considered, the graphite

IOP Conf. Series: Materials Science and Engineering **84** (2015) 012027 doi:10.1088/1757-899X/84/1/012027

Nomenclature

$\bar{\varepsilon}_{ij}$	Volume average of total strain tensor	D	Damage variable
$\bar{\varepsilon}_{ij}$	Volume average of total strain tensor	S, s	Lemaitre's damage evolution parameters
$\bar{\sigma}_{ij}$	Volume average of the stress tensor	Y	Energy release rate
R_v	Triaxiality function	f	Yield function
s_{ij}	Deviatoric part of the stress tensor	k, n	Isotropic hardening parameters
δ_{ij}	Kronecker delta	p	Equivalent Von Mises plastic strain
ε_{ij}^{tot}, ε_{ij}^{e}, ε_{ij}^{p}	Total/elastic /plastic strain tensor	p_{crit}	Critical effective plastic strain for damage evolution
σ_{ij}	Stress tensor		
σ_e	Equivalent Von Mises stress	r	Hardening variable
σ_y, σ_y^0	Actual / initial yield stress	α	Thermal expansion coefficient
E	Young's modulus	λ	Plastic multiplier
E^T	Tangent modulus	ν	Poisson's ratio

should always deform in the same manner. Finally, the almost continuous variation of the DCI Young's modulus with respect to strain in the very early deformation range [8] can hardly be explained by plasticity induced only by the presence of spherical voids.

Recently, Bonora & Ruggiero [9] have drawn attention to the role played by residual stresses which develop during the cooling phase of the manufacturing process as a result of the thermal expansion coefficient mismatch between the matrix and the graphite nodules. In their work, the constitutive response of a ferritic DCI has been numerically simulated using a micromechanical approach based on a 2D axisymmetric unit cell, assuming linear elastic behavior of the graphite and allowing plasticity together with damage evolution in the metallic matrix.

The present work extends the findings of Bonora & Ruggiero by investigating the influence of graphite stiffness and yield strength on the elastic parameters of ferritic DCI during tensile uniaxial testing. Special focus is put on explaining how inelastic deformation induced by the combined action of residual stresses and external loading affects the material response in the early deformation range.

2. Micromechanical model

Microscopic observations of local deformation and damage evolution in cast irons are taken, in most of the cases, on the surface of the component under examination [10][11][12], where the stress state is biaxial in nature. In order to produce results which are closely comparable with experimental findings, a 2D micromechanical model subjected to plane-stress conditions is developed. Graphite nodules are assumed to be spheres of equal size homogeneously dispersed in the ferritic matrix; the DCI microstructure is therefore schematized as a periodic square unit cell with a central circular graphite nodule (figure 1). The ratio between nodule radius and cell side is taken to be 0.39; this gives a graphite area fraction of approximately 12%, corresponding to the graphite volume fraction typical of GJS 400-18 according to EN 1563 ferritic DCI [4], which will be considered as reference material throughout the analysis.

Because of symmetry considerations, only ¼ of the unit cell is analyzed (figure 2). Periodic boundary conditions are applied in order to generate axial normal loading, yet fulfilling continuity of displacement and surface tractions along the cell boundaries, as discussed in [13]. Following Bonora & Ruggiero, the nodule-matrix transition is modelled as a frictionless contact interface, without any tensile strength in the normal direction, i.e. only compressive normal stresses can be transmitted across the interface. The mesoscopic Young's modulus E_{DCI} for the periodic unit cell is defined as:

$$E_{DCI} = \lim_{\bar{\varepsilon}_{11} \to 0} \left(\frac{\partial \bar{\sigma}_{11}}{\partial \bar{\varepsilon}_{11}} \right) \tag{1}$$

where $\bar{\sigma}_{11}$, $\bar{\varepsilon}_{11}$ and $\bar{\varepsilon}_{22}$ denote normal components of the mesoscopic stress and strain, given by the average over the unit cell volume of the corresponding microscopic quantities.

IOP Conf. Series: Materials Science and Engineering **84** (2015) 012027 doi:10.1088/1757-899X/84/1/012027

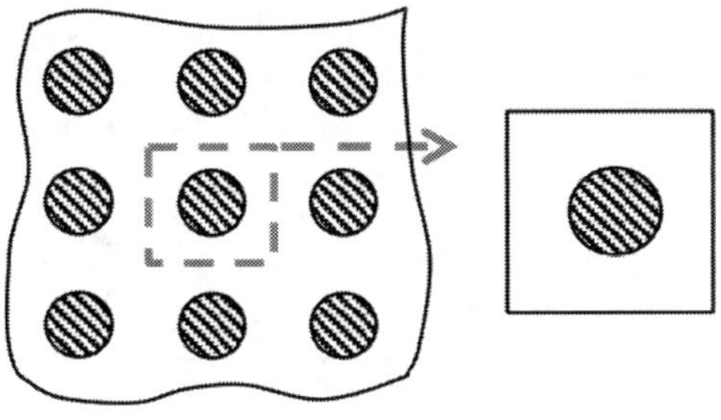

Figure 1. Assumed DCI microstructure, with identification of the periodic unit cell.

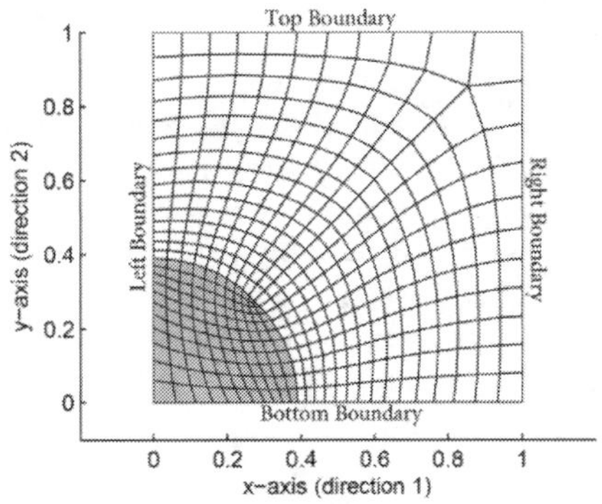

Figure 2. View of 2D mesh of ¼ of the periodic unit cell.

3. Constituents material behavior

3.1. Ferritic matrix

In light of its soft ductile nature, the constitutive behavior of the ferritic matrix is assumed to be described by the isotropic Lemaitre's damage model [14], whose basic equations in Cartesian components are summarized as follows:

- additive strain decomposition:

$$\varepsilon_{ij}^{tot} = \varepsilon_{ij}^{e} + \varepsilon_{ij}^{p} + \delta_{ij}\alpha\Delta T \tag{2}$$

- elastic constitutive law:

$$\frac{\sigma_{ij}}{1\text{-}D} = \frac{E}{1+v}\left[\varepsilon_{ij}^{e} + \frac{v}{1\text{-}2v}\delta_{ij}\varepsilon_{kk}^{e}\right] \tag{3}$$

- flow rule:

$$\dot{\varepsilon}_{ij}^{p} = \frac{3s_{ij}}{2\sigma_e}\frac{\dot{\lambda}}{1\text{-}D} \tag{4}$$

- yield function:

$$\mathrm{f} = \frac{\sigma_e}{1\text{-}D} - \sigma_y(\mathrm{r}) \leq 0, \qquad \sigma_e = \left(\frac{3}{2}s_{ij}s_{ij}\right)^{1/2} \tag{5}$$

- isotropic hardening rule:

$$\sigma_y = k(r+r_0)^n, \qquad r_0 = \left(\frac{\sigma_y^0}{k}\right)^{1/n} \tag{6}$$

- effective plastic strain increment and hardening parameter increment:

$$\dot{p} = \frac{\dot{r}}{1\text{-}D}, \qquad \dot{r} = \dot{\lambda} \tag{7}$$

- damage evolution law:

$$\dot{D} = \left(\frac{Y}{S}\right)^s \dot{p}, \qquad \text{if } p > p_{crit} \tag{8}$$

- energy release rate:

$$\mathrm{Y} = \frac{\sigma_e^2 R_v}{2E(1\text{-}D)^2}, \qquad R_v = \frac{2}{3}(1+v) + 3(1\text{-}2v)\left(\frac{\sigma_{kk}}{3\sigma_e}\right)^2 \tag{9}$$

- consistency condition:

$$\mathrm{f} \leq 0, \qquad \dot{\lambda} \geq 0, \qquad \mathrm{f}\dot{\lambda} = 0 \tag{10}$$

It may be noticed that knowledge of 9 material parameters is required: 3 thermo-elastic (E,v,α), 3 related to plastic flow (σ_y^0,k,n) and finally 3 related to damage (p_{crit},S,s). In principle, an additional parameter specifying the conditions at which crack nucleation occurs would be necessary: however, in the present analyses damage never exceeds 0.1, which is well below the critical fracture initiation threshold for common metals and alloys.

Values for Young's modulus, linear thermal expansion coefficient and initial yield stress, together with their temperature dependence, are reported in table 1; regarding Poisson's ratio, a constant value of 0.3 is assumed [15]. The remaining five parameters to be entered in the model have been determined on the basis of a data-fitting optimization procedure [16], minimizing the error between the predicted numerical results and the reference uniaxial stress-strain curve at room temperature for ferrite reported in [15]. Calculated values are given in table 2.

As no information is available for the post-yielding behavior of the ferritic matrix at higher temperatures, plastic flow and damage evolution parameters are assumed to be constant, except for the temperature dependence of the initial yield stress previously mentioned. Time-dependent deformation mechanisms are also neglected.

Table 1. Material properties for ferritic matrix (after Bonora & Ruggiero [9]).

Temperature (°C)	Young's modulus (GPa)	Thermal exp. coefficient ($\times10^{-5}$ °C^{-1})	Initial yield stress (MPa)
25	210.0	1.25	297
250	153.8	1.50	194
500	102.5	1.60	137
750	41.4	-	96
900	20.0	-	70
1000	0.1	2.40	60

Table 2. Plastic flow and damage evolution parameters for ferritic matrix.

Plastic flow factor k (MPa)	Plastic flow exponent n	Damage factor S (MPa)	Damage exponent s	Critical eff. plastic strain p_{crit} (mm/mm)
818.0	0.245	0.357	0.167	5.33x10^{-3}

3.2. Graphite nodules

As correctly pointed out by Bonora & Ruggiero, very little has been published in literature regarding the mechanical behavior of DCI nodules. At present, available mechanical data come mainly from nano-indentation measurements, which have provided values for Young's modulus in the range 15 GPa [10] to 28 GPa [17]; however, such numbers are not in agreement with Bonora & Ruggiero's numerical findings, who have proposed much higher stiffness values, in the order of 300-375 GPa.

In the present work, the graphite nodules are assumed to follow a linear elastic-perfectly plastic behavior described by the classical J2-flow theory of plasticity, and different values of Young's modulus and yield strength are considered in order to assess the related effects on the constitutive response of the entire unit cell. Poisson's ratio and linear thermal expansion coefficient are assumed to be 0.15 and 2.5x10^{-6} °C^{-1} respectively; no temperature dependence of any material parameter is taken into consideration.

4. Finite element modeling and simulations setup

The mechanical behavior of the unit cell is investigated by means of an "in-house" developed implicit FE C-code based on a small strains formulation. The mesh, shown in figure 2, is composed by 4-node isoparametric plane-stress quadrilateral elements with full integration. The nodule-matrix contact interface is modeled using a penalty-based method in association with an isoparametric discretization based on shape functions [18]. Non-linearity, arising from the constituent's material behavior and the contact condition, is numerically handled by means of a full Newton-Raphson scheme in combination with an implicit elastic predictor-return mapping algorithm [19]. For the purpose of improving convergence, a suitable expression for the consistent tangent modulus in the special case of plane stress and isotropic Lemaitre's damage model is derived and implemented in the code.

Numerical analyses are performed by initially applying a uniform temperature decrease to the entire cell, from 1000 °C down to room temperature, in order to simulate the cooling stage during manufacturing of ferritic DCI, in agreement with the considerations reported by Bonora & Ruggiero. Subsequently, a mesoscopic strain $\bar{\varepsilon}_{11}$ of $5x10^{-3}$ is progressively imposed to investigate the unit cell behavior during uniaxial tension.

A first set of simulations is run by varying the graphite Young's modulus in the range 1 to 500 GPa, while keeping its yield strength very high, which is equivalent to assuming linear elastic behavior of the nodules. Then, the graphite Young's modulus is kept fixed at selected values and tests are performed for graphite yield strength $\sigma_{y,g}$ in the interval 10 to 500 MPa. An additional simulation with "zero" graphite stiffness is also carried out, in order to have a direct comparison with the voided-matrix model mentioned in the introduction.

5. Results

5.1. Effect of graphite Young's modulus

In this section, the outcome of the numerical analyses focusing on the influence exerted by the nodules stiffness on the unit cell constitutive response in the early deformation range is presented, under the assumption of infinite graphite yield strength.

Figure 3 shows the obtained mesoscopic stress-strain tensile curve for selected values of E_g. The variation of the curve initial slope, which corresponds to the definition of the mesoscopic Young's modulus given in equation (1), is instead reported in figure 4. Starting from the voided matrix value corresponding to zero graphite stiffness, 3 different ranges may be identified: a first one in which E_{DCI} exhibits a small linear growth with E_g, a second one in which E_{DCI} drops abruptly to approximately 55 % of its initial value, and finally a third one in which E_{DCI} returns to increase, but not linearly. This particular behavior is related to the twofold effect played by an increase in the graphite stiffness: from one side, it makes the entire unit cell stiffer, as the nodule offers greater resistance to be deformed to a horizontal "oval" shape by the surrounding matrix during tensile loading; from the other side, it drives higher residual stresses at the end of the cooling stage, increasing the risk of promoting plastic deformation in the matrix, with consequent loss of stiffness.

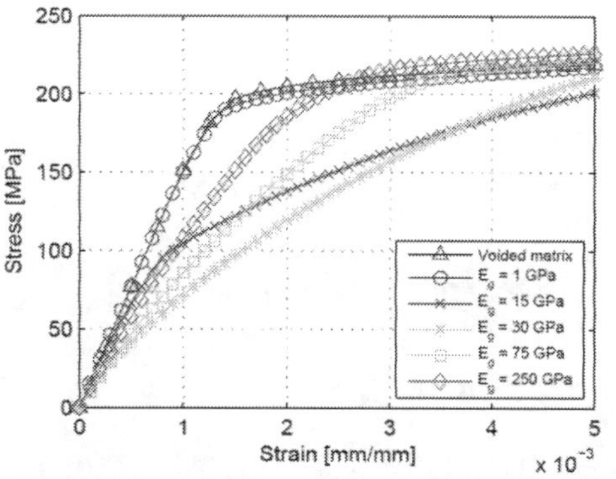

Figure 3. Mesoscopic tensile curves for selected values of the graphite Young's modulus.

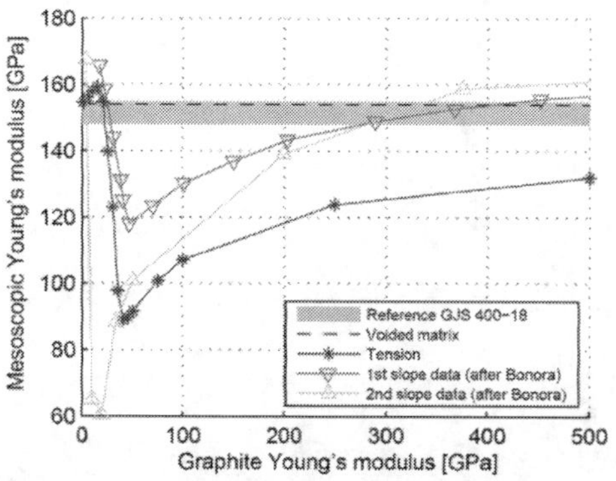

Figure 4. Calculated mesoscopic Young's modulus as a function of the graphite stiffness. Data are compared with findings from [9].

IOP Conf. Series: Materials Science and Engineering **84** (2015) 012027 doi:10.1088/1757-899X/84/1/012027

It is worth noticing that the shape of the calculated mesoscopic tensile stress-strain curves depicted in figure 3 closely matches the results reported by Bonora & Ruggiero in [9], figure 9; this is particularly evident when comparing the curves obtained for E_g = 15 GPa, which show exactly the same type of slope "break point" in the very early deformation range. This is indicative of the capacity of the two models to simulate the same physical mechanisms, despite Bonora & Ruggiero's analysis being based on a 2D axisymmetric cell coupled with a different damage formulation. Moreover, the mesoscopic Young's modulus evolution with E_g describes a pattern very similar to that of the "first slope" recorded by the two abovementioned authors, as shown in figure 4. Regarding this point, however, no comment is made in their paper about the initial linear elastic regime at very low E_g values, where an increment in the graphite Young's modulus produces a proportional increase in the global cell stiffness, the voided matrix value representing a lower bound.

5.2. Effect of graphite yield strength

The previous results have highlighted the fundamental role played by the residual stresses in determining the mechanical response of the unit cell in the early deformation range. However, it is likely that all analyses presented so far have somehow been biased by overestimating the amount of plasticity induced in the matrix by assuming infinitely high graphite yield strength $\sigma_{y,g}$. In order to investigate this point, the influence of nodule yielding on the predicted mesoscopic quantities is studied for two values of graphite Young's modulus: 15 and 300 GPa, which correspond to the literature values mentioned in section 3.2.

From a qualitative point of view, it is seen from figure 5 that very low graphite yield strength values provide mesoscopic tensile curves which are similar to that of the voided matrix. The reason is the following. During initial cooling the stress state in the nodule is homogeneous; therefore, the yield point is reached simultaneously on the whole area occupied by graphite. As perfect plasticity does not allow any hardening of the graphite, the stress state in the entire cell does not vary anymore during the remaining uniform temperature decrease; shrinkage of the matrix around the nodule is simply accommodated by plastic flow of the latter in the 3rd dimension. As a consequence, if the graphite yield strength is very low, the residual stress pattern in the matrix becomes "frozen" at the very beginning of the cooling process, when the magnitude of the stresses is still negligible. In addition, during subsequent tensile testing, the nodule opposes with very little resistance to deformation, again because of its very low yield strength.

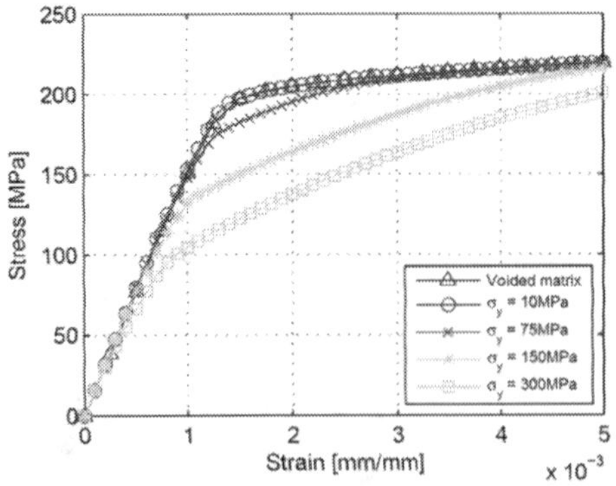

Figure 5. Mesoscopic tensile curves for selected values of the graphite yield strength, obtained using a graphite stiffness of 15 GPa.

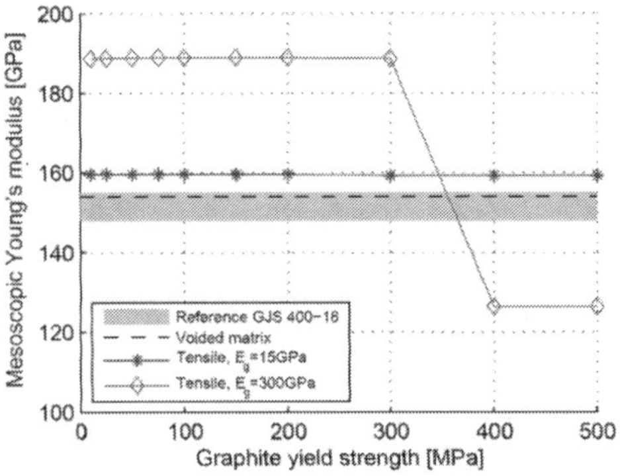

Figure 6. Calculated mesoscopic Young's modulus as a function of the graphite yield strength.

The above discussion leads to the conclusion that when the graphite yield strength is decreased, the unit cell tends to behave according to the voided matrix model, irrespective of the graphite stiffness.

Conversely, for sufficiently high values of $\sigma_{y,g}$, the nodule never yields during testing, and the mesoscopic tensile curves become the same as those reported in Figure 3: for instance, in figure 5 this happens for $\sigma_{y,g}$ equal to 300 MPa.

The effect of graphite yield strength on the mesoscopic Young's modulus is reported in figure 6. It may be observed that in the case with E_g = 15 GPa, the cell stiffness is almost independent of $\sigma_{y,g}$, and its value is slightly higher than that of the reference voided matrix. This is due to two things. First, as pointed out in section 5.1, such graphite stiffness is too low to produce any yielding in the matrix at the end of the cooling stage, even if plastic flow in the nodule is neglected. Moreover, when subsequent tensile loading begins, unloading takes place in the nodule, implying that "almost" the entire unit cell is the elastic regime, no matter the value of $\sigma_{y,g}$. In this situation, the presence of the nodule simply hinders lateral contraction of the matrix during tensile testing, increasing the mesoscopic cell stiffness.

It is interesting to notice that the same reasoning applies when E_g = 300 GPa, but only for values of the graphite yield strength below 300 MPa. In this case, yielding of the nodule prevents yielding of the matrix during cooling, which would otherwise take place. At the same time, the high elastic graphite stiffness leads to an exceptionally high value of the mesoscopic Young's modulus, which would never be achieved if plastic flow in the nodule was inhibited, as comparison between figure 6 and figure 4 demonstrates. In fact, as soon as $\sigma_{y,g}$ is increased beyond the critical threshold of approx. 400 MPa, matrix yielding at the end of initial cooling occurs and E_{DCI} reduces dramatically. Evidence of the fundamental role that graphite strength plays in determining the cell mesoscopic stiffness via affecting the plastic yielding distribution is given in figure 7 and figure 8.

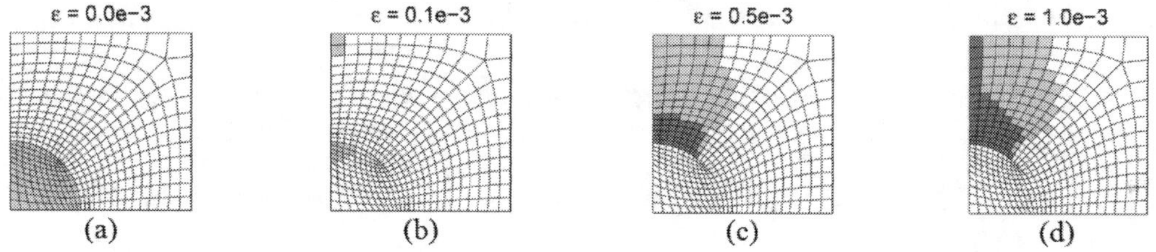

Figure 7. Plastic yielding (green) and damage (red) evolution at the beginning of tensile testing, for $\sigma_{y,g}$ = 300 MPa and E_g = 300 GPa. The mesoscopic strain level is indicated above each contour.

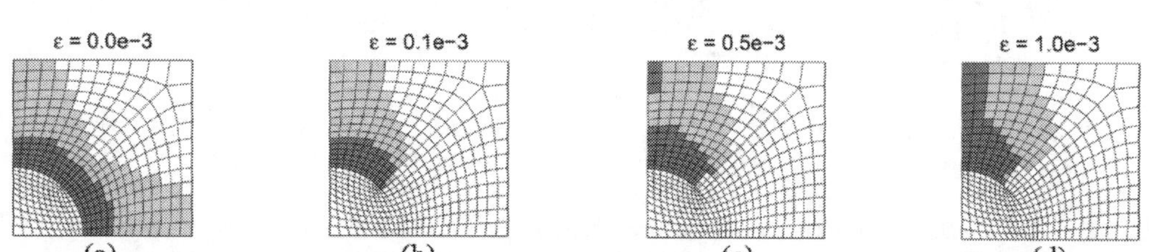

Figure 8. Plastic yielding (green) and damage (red) evolution at the beginning of tensile testing, for $\sigma_{y,g}$ = 400 MPa and E_g = 300 GPa. The mesoscopic strain level is indicated above each contour.

6. Conclusions

In the present paper a micromechanical model based on a 2D periodic unit cell subjected to plane-stress conditions has been used to investigate the ductile cast iron behavior in the early deformation range. The influence of graphite stiffness and strength on the mesoscopic "elastic" parameters have been analysed and discussed, extending and clarifying the findings obtained by Bonora & Ruggiero in a previous work. In particular, the mechanical response of the unit cell has proved to be strongly affected by the imposition of a threshold on the maximum load carrying capacity of the nodule.

According to the present results, very high graphite Young's modulus values, in the range of 300 to 400 GPa, are probably to be excluded. In fact, if the graphite yield strength is assumed to be finite and of reasonable values, yielding of the nodule would prevent matrix plastification during the initial cooling stage, thus leading to values of the mesoscopic stiffness above the admissible range. As a consequence, graphite Young's modulus values in the order of a few tens of GPa appear more sensible, in agreement with nano-indentation data reported in literature.

References

[1] Labrecque C and Gagne M 1998 Review ductile iron: 50 years of continuous development *Can. Metall. Q.* **37** 343–78
[2] Dong M J, Tie B, Béranger A S, Prioul C and François D 1997 Damage Effect on the Fracture Toughness of Nodular Cast Iron *Adv. Mater. Res.* **4-5** 181–8
[3] Berdin C, Dong M J and Prioul C 2001 Local approach of damage and fracture toughness for nodular cast iron *Eng. Fract. Mech.* **68** 1107–17
[4] Steglich D 1998 Micromechanical modeling of damage and fracture of ductile materials *Fatigue Fract. Eng. Mater. Struct.* **21** 1175–88
[5] Gurson A 1977 Continuum theory of ductile rupture by void nucleation and growth: Part 1 - Yield criteria and flow rules for porous ductile media *J. Eng. Mater. Technol. ASME* **99** 2 – 15
[6] Di Cocco V, Iacoviello F and Cavallini M 2010 Damaging micromechanisms characterization of a ferritic ductile cast iron *Eng. Fract. Mech.* **77** 2016–23
[7] Hervas I, Bettaieb M B and Hug E 2013 Damage mechanisms evolution of ductile cast irons under thermomechanical loadings *Int. J. Mater. Prod. Technol.* **47** 23
[8] Kohout J 2001 A simple relation for deviation of grey and nodular cast irons from Hooke's law *Mater. Sci. Eng. a* **A313** 16 – 23
[9] Bonora N and Ruggiero A 2005 Micromechanical modeling of ductile cast iron incorporating damage. Part I: Ferritic ductile cast iron *Int. J. Solids Struct.* **42** 1401–24
[10] Dierickx P, Verdu C, Reynaud A and Fougeres R 1996 A study of physico-chemical mechanisms responsible for damage of heat treated and as-cast ferritic spheroïdal graphite cast irons *Scr. Mater.* **34** 261–8
[11] Di Cocco V, Iacoviello F, Rossi A and Iacoviello D 2014 Macro and microscopical approach to the damaging micromechanisms analysis in a ferritic ductile cast iron *Theor. Appl. Fract. Mech.* **69** 26–33
[12] Sjögren T, Persson P E and Vomacka P 2010 Analysing the Deformation Behaviour of Compacted Graphite Cast Irons Using Digital Image Correlation Techniques *Key Eng. Mater.* **457** 470–5
[13] Drago A and Pindera M 2007 Micro-macromechanical analysis of heterogeneous materials: Macroscopically homogeneous vs periodic microstructures *Compos. Sci. Technol.* **67** 1243–63
[14] Lemaitre J 1985 A continuous damage mechanics model for ductile fracture *J. Eng. Mater. Technol. ASME* **107** 83 – 89
[15] Zhang K S, Bai J B and François D 1999 Ductile fracture of materials with high void volume fraction *Int. J. Solids Struct.* **36** 3407–25
[16] Lemaitre J and Desmorat R 2005 *Engineering damage mechanics : ductile, creep, fatigue and brittle failures* (Springer)
[17] Pradhan S K, Nayak B B, Sahay S S and Mishra B K 2009 Mechanical properties of graphite flakes and spherulites measured by nanoindentation *Carbon N. Y.* **47** 2290–2
[18] Wriggers P 2006 *Computational Contact Mechanics* (Springer)
[19] Neto E de S, Peric D and Owens D R J 2008 *Computational methods for plasticity : theory and applications* (Wiley)

IOP Conf. Series: Materials Science and Engineering **84** (2015) 012028 doi:10.1088/1757-899X/84/1/012028

In-situ observation and modelling of solidification and fluid flow on GTAW process

A Chiocca, F Soulié, F Deschaux-Beaume and C Bordreuil

LMGC - UMR 5508, Université Montpellier 2, CC048 Place Eugène Bataillon 34095 Montpellier Cedex 5, France

E-mail: alexis.chiocca@univ-montp2.fr

Abstract.

An experimental setup is presented in order to obtain experimental data during solidification of a static weld pool after arc extinction with a GTAW process. Several devices have been set up to extract three kinds of measurements: *(i)* solidification front velocity *(ii)* fluid flow velocity at the vicinity of the front *(iii)* temperature field in the solid part. A high-speed camera is used to film the interface during welding at microscopic level and an infra-red in order to take the temperature field around the weld pool in the solid part. After processing and calibration of the videos, the experimental results are compared to theoritical results founded on an adapted model from the KGT [1] and from the one of Gandin et al. [2]. All the tests are done thin plate of Cu-30wt.%Ni.

1. Introduction

The microstructure of welded joints mainly determines their mechanical properties. The weld bead microstructure results from the solidification process occurring in the weld pool at the solid/liquid interface. This process of solidification includes mechanisms that are driven by heat and fluid flow. The understanding of solidification mechanisms in relation with weld pool behaviour is then a key point to analyse microstructure formation and to control joint strength.

Compared to the common case of solidification such as casting, welding processes induce high thermal gradients and high solutal gradients at the solid/liquid interface and the front velocity is much faster. Due to these solidification conditions, columnar and equiaxed dendritic microstructures are broadly observed. The characteristic length of the weld pool is around 10mm. In the weld pool, fluid flow and heat transfer can be complex and can vary from point to point along the interface [3]. Depending on the position on the solid/liquid interface, the thermal gradient and the solidification speed change. At the trailing edge, the solidification speed is maximal and the thermal gradient is minimal but at the fusion line, it is the opposite. To understand the microstructural evolution along the interface, information from two scales are necessary: *(i)* macroscopic fields around the interface and *(ii)* microscopic observations. Macroscopic fields (temperature gradient, for example) gives driven force for the solidification. On in-situ observations, Zhao et al. [4] proves that the fluid flows have complex shapes that can influence solidification. Microscopic observations tells us the behaviour of the interface (stability and/or fragmentation). Delapp et al. [5] based on microscopic observations showed that fluid flow could produce fluctuation of the front speed giving ripples on the weld bead.

Table 1. Parameter used for the analytical model.

Parameters	Values
Liquidus slope, m $(K/wt.\%Ni)$	3.89
Partition coefficient, k	1.40
Nominal alloy composition C_0 $(wt.\%Ni)$	30
Liquid density, ρ (g/cm^3)[8]	8.1
Gibbs-Thomson coefficient, Γ $(K.m)$[2]	1.96e-7
Stability constant, $\sigma*$[2]	0.0253
Dynamic viscosity, μ $(Pa.s)$[9]	3.1e-3
Liquidus temperature, T_l (K)[8]	1520
Diffusion coefficient D_l (m^2/s)[9]	1e-9
Solidification range ΔT_r (K)[10]	1472-1520

Modelling dendritic growth under weld solidification conditions is still a major issue. Different works introduced important phenomena observed during solidification. Hunziker et al. [6] gives, for the case of welding, an exponential relation between the solidification speed and the total undercooling. Gandin et al. [2] have simulated a dendritic growth with fluid flow showing the interactions between the growth velocity and the fluid flow ahead the interface.

The present work wants to develop a solidification model in welding conditions with the help of different experiments. Experiments are done on Cu30wt.%Ni plate and the front velocity, the fluid flow velocity and the thermal field are analysed under different thermal conditions. Then the comparison is done with the model which is adapted from two different models [2, 6].

2. In situ Observations

In this part, experimental setup is described with the two scale devices to investigate solidification under welding conditions. One purpose is to compare microstructure evolution with different solidification conditions induced by different process parameters. The main process parameters are the intensity of welding and the down-slope. The intensity has a direct control on the power input and the down-slope allow the solidification to be slower by decreasing the intensity before the end of welding with a negative slope. In order to estimate the kinetic of solidification, front velocity, the fluid flow velocity and the thermal fields will be measured in-situ.

Details of the experimental setup and errors made in the measurement can be found in an other article [7].

2.1. Material

For an efficient comparison with the theoritical models, the simple binary alloy Cu30Ni has been chosen. This alloy is composed of 70% of copper and 30% of nickel and its phase diagram has only a solid solution in the solid part. The physical parameters useful for the model are listed on table 1. The partition coefficient, the diffusion coefficient and the liquidus slope are supposed constant. The liquid density is taken at the liquidus temperature.

2.2. Experimental setup

The experimental setup design is illustrated on figure 1. The observations are done on a static weld pool after or during the arc extinction. The weld pool generated by a GTAW torch is

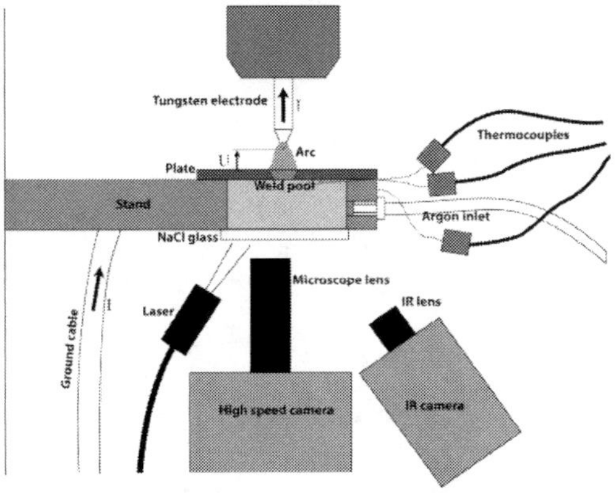

Figure 1. Drawing of the set-up showing the different measurements done around the plate.

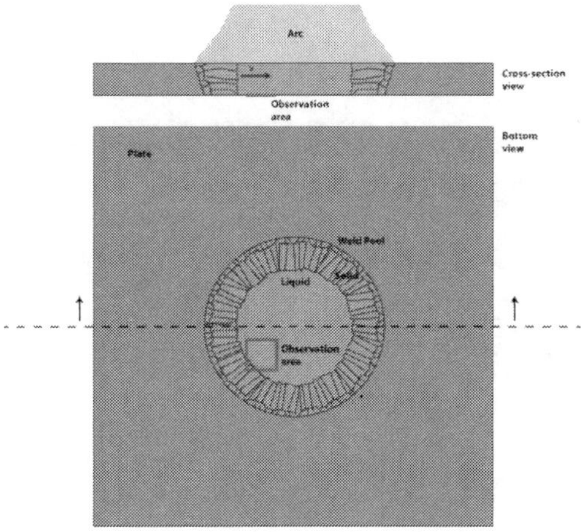

Figure 2. Geometrical description of the observed system, at the top a cross-section view and at the bottom a bottom view.

full-penetrated to allow the observation on the back side. On this side, The liquid metal is protected from oxidation by a chamber full of argon. The coupon is 1.6 mm thick and 70 mm of diameter.

Observations of the weld pool for solidification front and fluid flow measurements are obtained thanks to an high speed camera. This camera is placed under the plate and observe the area represented by a square on figure 2. The observation area is lighted up by a LASER in order to insure the segmentation of the liquid and solid phase. The back side of the plate is filmed with the IR camera and two thermocouples are placed below the plate for the calibration of the temperature field. The observation area represent the whole weld pool and a large solid part.

2.3. Field measurements

2.3.1. Thermal fields IR camera gives access to temperatures in the solid phase and near the interface on the back surface, after a calibration step with the thermocouples.

The solid part temperature of the sample along a the dashed line on figure 3 is plotted on figure 4. The temperature increases from the edge to the centre of the sample. Due to the calibration procedure, the measured temperature is only available in the solid phase. Nevertheless, the temperature gradient in the solid phase can be approximated. With this setup configuration, the gradients measured are about $2 * 10^5\ K/m$ and the temperatures at the interface are about $1300\ K$ (Fig. 3 and 4).

2.3.2. Velocities fields Two kind of velocities are extracted from the high speed films: the velocities of the solidification front and the velocities of the fluid flow. The velocities of the solidification front are obtained by detecting contours of solidification front in images. The field of velocities in the liquid are computed by Particle Image Velocimetry (PIV) processing [11]. Velocities of particles at the vicinity of the front are estimated around 50 mm/s (Fig. 5 and 6) and front speed are between 1.5 mm/s and 4 mm/s (Fig. 7 and 8) depending on the adjustments of the power source.

Fluid flow curves (Fig. 6) show a peak at the vicinity of the front. The projection of the velocity of the fluid particle is high when it goes out the front and decreases while approaching the weld pool centre. Only few eddies are observed on videos telling us that the flow is mainly

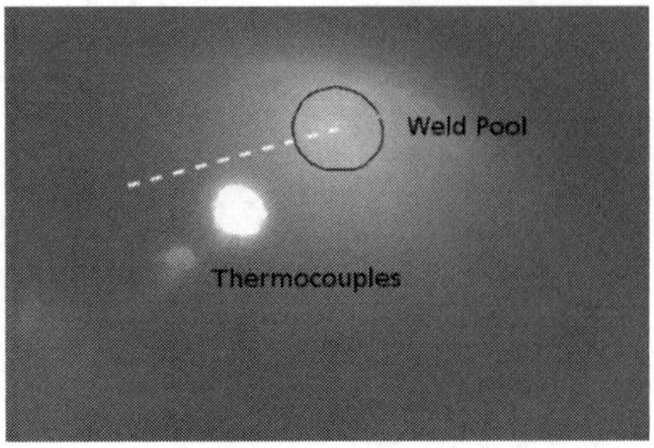

Figure 3. Thermal Field under the plate. The plain contour ——circle the weld pool.

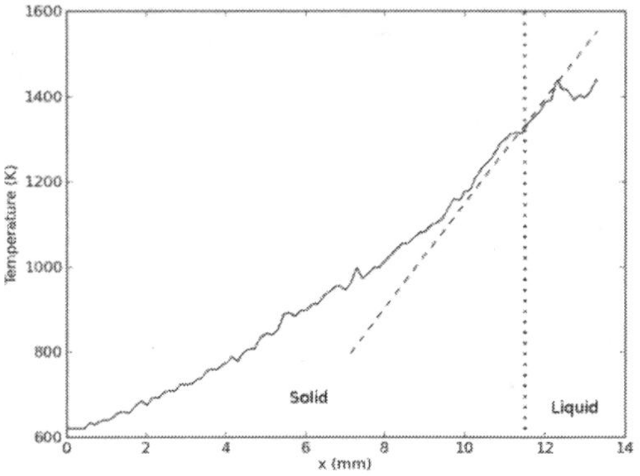

Figure 4. The ——curve correspond to temperature along the dashed line - - - -on figure 3. The dashed line - - - -is the temperature gradient at the interface. The approximative limit between solid and liquid is represented by the dotted line · · · · · ·.

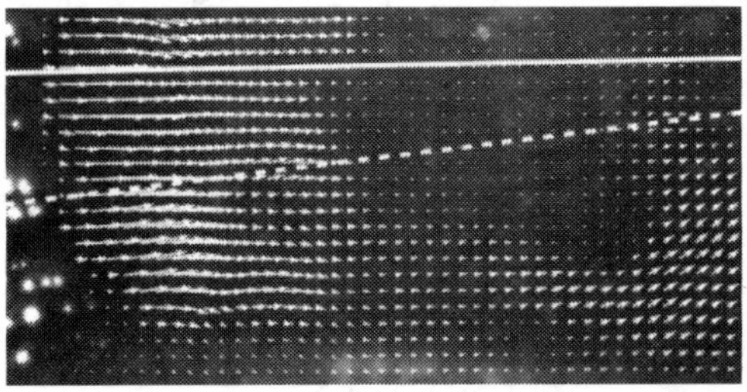

Figure 5. Velocity field ahead the front. On the left the dendrite tips growing from the left to the right and on the left the liquid pool.

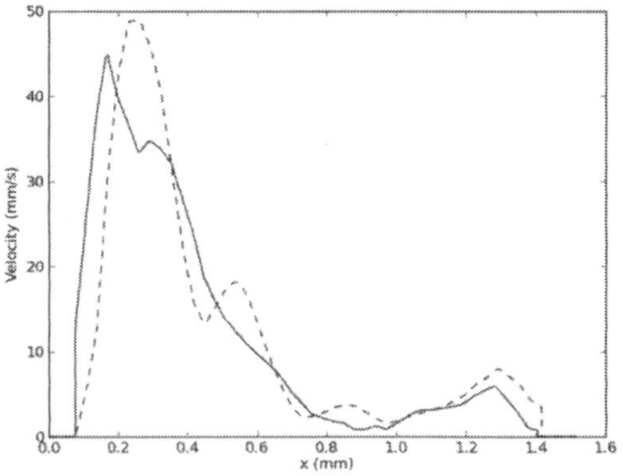

Figure 6. Evolution of the velocity vectors from the left to the right along the straight lines on figure 5.

laminar. For the back side, a vortex is observed from the interface to the center of the weld pool. Fluid velocity are more important at the extinction of the arc and then tends to decrease.

The decreasing of the radius of the weld pool (Fig. 8) is almost linear with a decrease of the slope at the end of the curve. Irregularity all along the curve might be caused by oscillation of the weld pool.

2.4. Observations

A complete observation test is illustrated on figures 3, 5 and 7. For this test the intensity applied on the electrode is 70 A during 5.5 seconds without down-slope. After extracting and processing data, the solidification speed is 3.43 mm/s (Fig. 8) and the velocity of the fluid flow at the vicinity of the interface is 39 mm/s (Fig. 6). The field of temperature allow us to estimate the

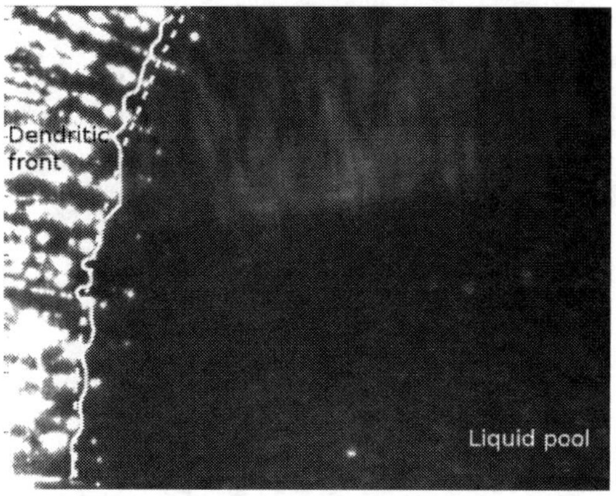

Figure 7. Contour detection of the front with in plain line ——the raw contour and in dashed line - - - -the estimated circle.

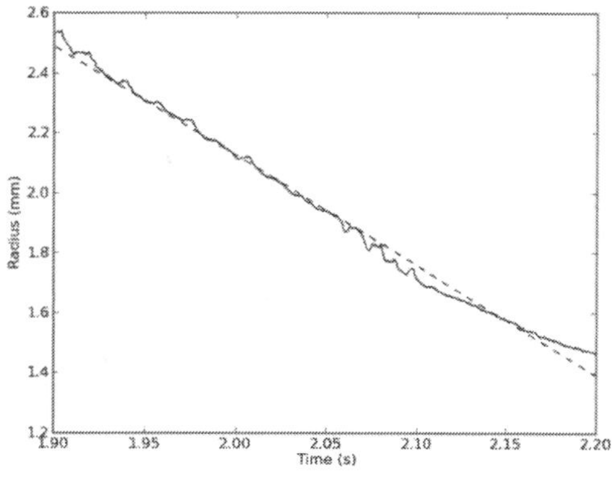

Figure 8. In plain line ——the decreasing of the radius of the weld pool across time and in dashed line - - - -the mean slope.

gradient near the interface at $1.82 * 10^5$ K/m and the total undercooling at 191 K (Fig. 4). The undercooling is equal to the difference between the temperature at the interface and the liquidus temperature.

During high speed imaging, small particles are ejected from the interface. The videos tell us that the solidification is mainly columnar and that at the very end some equiaxed grains seems to be observed. All the dendrite tips are submerged during growth. The tip grows below the liquid surface, the dendrite is then visible when the primary branches grow, becoming fatter and joining with its neighbour dendrites.

With measurements, dimensionless numbers can be estimated in order to appreciate the heat and mass transfer in this configuration. In particular, the Péclet numbers show the dominance of convection over diffusion $P_v = \frac{R*v}{D_l}$ and $P_u = \frac{R*u}{D_l}$ with v the solidification speed, R the radius of the dendrite tip and $u = u_{piv} - v$ the velocity of the liquid with respect to the front. These Péclet numbers are respectively estimated at 10^{-1} and 10^0 in order of magnitude, demonstrating that the liquid is well homogenized. The value of the Reynolds number $Re = \frac{\rho u L}{\mu}$ with the experimental velocity is estimated at 1300 confirming the laminar state.

3. Welding solidification modeling

It exists a lot of theoretical models on solidification. The variations in the models depend on the conditions of solidification. For example, with slow or rapid solidification. In the case of slow solidification or low Péclet number [12] assumptions can accelerate computations of the model. For welding solidification, the conditions taken for this study are the one found in the KGT model [1]. Kurz et al. have developed a model for rapid solidification with assumptions that can be applied to welding. This model results in two equations with two unknowns:

$$\Delta T = mC_0\left(1 - \frac{1}{1-(1-k)\Omega}\right) + \frac{2\Gamma}{R} \qquad R = 2\pi\sqrt{\frac{2\Gamma}{mG_c\xi - G}} \tag{1}$$

with R the radius of the dendrite tip, ξ a factor present in the model [1] and $\Omega = Iv()$ the supersaturation equal to the Ivantsov function.

Even if it is suitable for welding, it doesn't take into account the fluid flows. It is only a diffusive model.

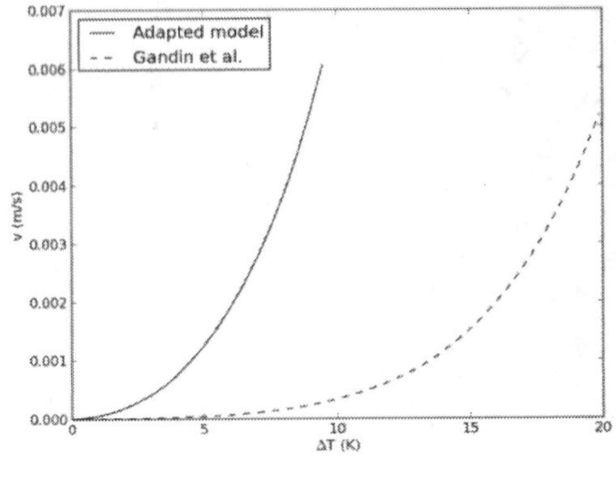

Figure 9. Solidification speed with respect to the undercooling.

3.1. Fluid Velocities

To implement the fluid flow without using a fully numerical model, one solution is to add a boundary layer δ ahead the front. It is the solution presented in the article of Gandin et al. [2]. It takes into account the intensity and the direction of the flow by changing the thickness of the layer. Changes are made on the supersaturation:

$$\Omega = P_v exp(P_v)(E_1(P_v) - E_1(P_v(1+\frac{2\delta}{R}))) \qquad \frac{\delta}{R} = \frac{2}{ARe_{2r}^B Sc^C sin(\theta/2)} \tag{2}$$

$$E_1(P_v) = \int_{P_v}^{\infty} \frac{exp(-\tau)}{\tau} d\tau \tag{3}$$

with A, B and C parameters defined in the article of Gandin et al.[2], Re number of Reynolds and Sc number of Shmidt. However this model, as it is described, is made for low Péclet numbers corresponding to slow solidification.

3.2. Welding growth

To take into account high thermal gradients and velocities measured during the experiment, a model is developed. The system to solve is the same one as in equation 1 but the definition of the supersaturation is the one of equation 2. This model shows an exponential relation between the undercooling and the front velocity (Fig. 9). Fluid flow velocity has the strongest influence when it is in the opposite direction of the front growth. But when it is in the same direction, the influence on solidification is weakest and is found null as it is shown in the study of Gandin et al. [2]. In this model, the temperature gradient has no influence on front velocity but a big influence on the stability limit of the planar front. As the gradient increase, the stability limit velocity increases too [1].

4. Results and discussions

The models presented previously show relations between intensity and direction of the fluid flow, velocity of the front velocity and thermal parameters. In this section, the purpose is to know if the results coming from the models can be observed experimentally. Different tests are summarized on table 2.

Table 2. Results on four tests.

Test number	Intensity	Down-Slope	v	u_{piv}	G	ΔT
1	70 A	0 s	3.46 mm/s	39 mm/s	182.4 K/mm	191 K
2	80 A	0 s	3.03 mm/s	49 mm/s	184.6 K/mm	185 K
3	90 A	0 s	2.90 mm/s	50 mm/s	187.9 K/mm	215 K
4	90 A	2 s	1.38 mm/s	50 mm/s	123.8 K/mm	226 K

4.1. Macroscopic results

An increase of power input (70 to 90A) leads to larger and hotter weld pool. This can explain the two main differences in the tests : a lower front velocity for 90A and lower fluid velocity for 70A. With higher temperature in the weld pool, it is expected that the front needs to transform more energy to change phase. With higher temperature on the arc surface due to more energetic plasma, the gradient in temperature can be more important leading to higher Marangoni forces inducing higher fluid velocities. The undercooling at the interface, corresponding to the difference between liquidus temperature and the temperature measured at the interface, does not seem to be influenced by the power input of the welding process. Looking at the energy increasing from test 1 to test 4, the undercooling values between test 1 and test 4 have not big variations.

According to the model, the solidification front velocity increase with the level of undercooling at the interface. Between the different test, the same level of undercooling gives different value of solidification speed. With the experimental values applied on the adapted model (Fig. 9), the undercooling found vary between 5 K and 8 K. It is very few compared to the experimental results about 200 K. The values might be wrong because the limit between liquid and solid is blur. The temperatures measured at the interface are underestimated because it is not the temperature of dendrite tip. The dendrites tip might be submerged in the liquid while the measurement is done at the surface, corresponding to a point before the tip where the dendrite has already cooled.

4.2. Interface behaviour

At the microscopic level, several variations have to be highlighted. The direction of solidification and the direction of the fluid flow are mainly in the same direction. In the presented model, the fluid flow has in that case no influence on dendritic growth velocity.

From a test to an other, the morphology is dendritic columnar in the window of observation of the high speed camera. If the observation area is placed on the centre of the weld pool, at the end of the solidification, grains which do not come from the front grow inside the liquid bulk. These grains are either equiaxed grains that have nucleated in the bulk or columnar grains coming from the upper side of the plate. The first assumption could be qualitatively explained. At the end of the solidification, the gradient can be lower, the solidification speed accelerate and impurities driven by the flow might be blocked in the centre. If the top of the weld is solidified at first, columnar grains can nucleate and grow downward until the bottom of the weld.

The dendrite spacing measured on all the tests do not show great variations over the dendrites velocity. The spacing between dendrites is more sensitive to the temperature gradient. The changes of spacing between the different measurements have been found coherent with the law of David et al. [13]: $d = a(G^2 v)^{-0.25}$

4.3. Growth modeling during solidification

During the solidifcation after arc extinction, it is found that the undercooling keep a quasi constant value while the gradient decrease. These results are coherent with the theory. A constant undercooling verify a constant value of velocity according to the model. The decreasing of the gradient show the flattening of the temperature field and may cause the growth of the supposed equiaxed grains in the centre. It has already been found that discrepancy are observed for fluid flow direction in the range of $[\frac{7\pi}{4}; \frac{\pi}{4}]$ [2]. The model presented in this case is finally found not pertinent for the direction of fluid flow. However the fluid flow does not seems to have an influence on front velocity speed. For three equal velocities of fluid, the front velocities are different.

5. Conclusion

Experimental results have been presented on solidification of a binary alloy (Cu-wt.%30Ni) processed by GTAW. Several welding conditions have been carried out to vary solidification conditions. Assumption have been made to explain the results and to compare them to the modelling data. A lot of data, that was not measured yet, are determined in this study: the solidification front velocity, the fluid flow velocity and the thermal field.

The macroscopic observations explain the solidification behaviour. The velocities are driven by the thermal conditions, the undercooling only plays a role for the position of the dendrite front between solidus and liquidus.

The lack of accuracy of the thermal measurements are the main problem of the difficulty to set experiments against modelling. It is not possible to measure the temperatures of dendrite tip inside the liquid. A way of improvement could be a prediction of the dendrite temperature tip with the available informations on the gradient and the temperatures.

References

[1] Kurz W, Giovanola B and Trivedi R 1986 *Acta Metallurgica* **34** 823–30
[2] Gandin C A, Guillemot G, Appolaire B and Niane N 2003 *Materials Science and Engineering: A* **342** 44–50
[3] Kou S 2003 *Welding Metallurgy* (John Wiley & Sons, Inc.)
[4] Zhao C X, Van Steijn V, Richardson I M, Saldi Z and Kleijn C R 2009 *ASM Proceedings of the International Conference: Trends in Welding Research* (Pine Mountain, GA, United states) pp 201–10
[5] DeLapp D, Cook G, Strauss A and Hofmeister W 2005 *Proceedings of the 7th Internatioanl Conference on Trends in Welding Research* 97–102
[6] Hunziker O, Dye D and Reed R 2000 *Acta Materialia* **48** 4191–201
[7] Chiocca A, Bordreuil C, Deschaux-Beaume F and Soulié F *Science and Technology of Welding and Joining*
[8] Brillo J and Egry I 2003 *International journal of thermophysics* **24** 1155–70
[9] Chakraborty N 2009 *Applied Thermal Engineering* **29** 3618–31
[10] Hüpf T, Cagran C, Kaschnitz E and Pottlacher G 2010 *International Journal of Thermophysics* **31** 966–74
[11] Prasad A K 2000 *Curr. Sci. (India)* **79** 51–60
[12] Dantzig J A and Rappaz M 2009 *Solidification* 1st ed (EPFL Press)
[13] David S and Vitek J 1989 *International Materials Reviews* **34** 213–45

Ultra-fast in-situ X-ray studies of evolving columnar dendrites in solidifying steel weld pools.

W U Mirihanage[1,2], M Di Michiel[3], R H Mathiesen[1]
[1]Departmet of Physics, Norwegain University of Science and Technology, Trondheim, Norway
[2]School of Materials,The University of Manchester, Manchester, UK
[3]ESRF – Europian Synchtrotron , Grenoble, France

wajira.mirihanage@manchester.ac.uk

Abstract. High-brilliance polychromatic synchrotron radiation has been used to conduct in-situ studies of the solidification microstructure evolution during simulated welding. The welding simulations were realized by rapidly fusing ~ 5 mm spot in Fe-Cr-Ni steel. During the solid-liquid-solid phase transformations, a section of the weld pool was placed in an incident 50-150 keV polychromatic synchrotron X-ray beam, in a near-horizontal position at a very low inclination angle. Multiple high-resolution 2D detectors with very high frame rates were utilized to capture time resolved X-ray diffraction data from suitably oriented solid dendrites evolving in the weld pool. Comprehensive analysis of the diffraction data revealed individual and overall dendritic growth characteristics and relevant melt and solid flow dynamics during weld pool solidification, which was completed within 1.5 s. Columnar dendrite tip velocities were estimated from the experimental data and during early stages of solidification were exceeded 4 mm/s. The most remarkable observation revealed through the time-resolved reciprocal space observations are correlated to significant tilting of columnar type dendrites at their root during solidification, presumably caused by convective currents in the weld pool. When the columnar dendrite tilting are transformed to respective metric linear tilting velocities at the dendrite tip; tilting velocities are found to be in the same order of magnitude as the columnar tip growth velocities, suggesting a highly transient nature of growth conditions.

1. Introduction

In a fusion welding process, the ability to weld a metal to itself or to another strongly depends on dendritic growth dynamics, which determine the morphological evolution of the advancing solidification interfaces and the spatiotemporal pattern building at the advancing front. Thus microstructure at the weld center, where the final growth fronts meet and solidification terminates, is dictated by the full history of the crystal growth process. Weld center boundaries form critical morphological features and are the regions where weld failures often occur. In most instances weld fusion zone (FZ) is deemed as a mini casting and solidification behavior of the FZ is derived from the extrapolation of the scientific knowledge gained from the investigations of ordinary solidification processes, which usually take place at orders of magnitude lower cooling rates [1,2] compared to the real welds.

With the advancement of the computational material science, many solidification mechanisms were hypothesized via computer simulations, eg. ref. [3,4] establishing self-consistent theories. Synchrotron X-ray or neutrons studies [5-8] emerged as a powerful tool during last two decades to investigate microstructure evolution in solidifying metallic alloys in-situ. X-ray and neutron experimental parameters are generally limited by the mutual compromise between the material properties, beam brilliance, detector capability and the required time resolution. Although a few experimental attempts were focused on welding related phase transformations eg. Ref. [8-10] employing X-ray diffraction (XRD) techniques, however a critical allusion of the solidification features in the weld pool, especially related to the crystal growth were not reported until recently [11]. In this contribution, we provide further analysis of columnar growth and related dynamic effects found in ultra-fast synchrotron XRD experiments [11].

2. Experimental configuration

The experiments were carried out at the European Synchrotron Radiation Facility's ID15A beamline. Fe-Cr-17.3%-Ni-11.1%-Mo-2.1%-C<0.1% austenitic steel was used as the sample material and 2 mm thick and 30 mm x 20 mm rectangular samples were used for the experiments. The primary heat source employed to simulate a real spot-welding process comprised seven 1 kW Halogen lamps equipped with elliptical reflectors. The lamps were mounted in a custom-built rig that allowed for individual adjustments to superpose infrared and near infrared energy emitted from all the lamps into a common focal spot of appox. 5 mm diameter at the center of the sample Argon flow was brought in at the sample surface just adjacent to the weld spot to reduce oxidation rates at the pool surface and weld pool depths in the experiments were varied between 0.6 – 1.5 mm. Generally melting initiated after a few seconds from switching on the Halogen-lamps to full power, followed by stabilization of a molten pool within 1-3 seconds. As soon as complete melting in the part of the pool was verified through the illuminated incident X-ray beam, pool freezing was initiated by powering off the halogen heaters and the power-off time is referred hereafter as t = 0 s . Typically, the observed region of the weld pool re-solidified completely within 1.0 - 1.5 s. For each welding experiment, a sample was placed on top of a resistive heater with the surface normal tilted 0.035 rad towards the incident beam, as outlined in Fig.1. The incident X-rays were 50 - 150 keV polychromatic. The incident beam was positioned concentric to the pool surface center, and with the sample tilt and the incident beam geometry, the region monitored throughout the experiments corresponded to a 1 mm wide region stretching diametrically across the pool surface. Diffraction data were collected with a Pco.dimax® CMOS equipped with an image intensifier, The CMOS spans a 2016 x 2016 pixel array, and was read at 1.0 kHz frame. Ref.[11] provides further details on this experimental configuration.

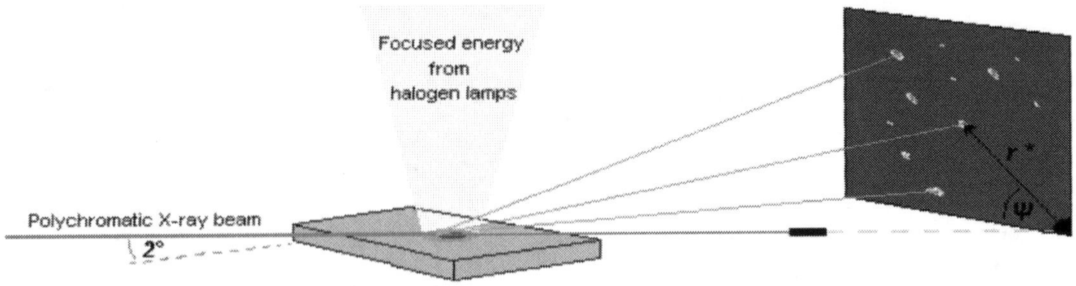

Fig 1. Schematic configuration of time resolved X-ray diffraction experiments.

3. Experimental observations and data analysis

During the experimental observations, a remarkable motion of individual discontinued XRD peaks in the early stage of solidification was appeared. At the beginning, these discontinued XRD peaks moved in very fast radially and each peak only appeared for few milliseconds only. After a while, lasting XRD peaks were appeared with reducing angular movement rates, the average angular motion rate for one experiment is shown in Figure 2.a. It was ascertained that the combination of the polychromatic beam and the ultra-high frame rate is essential to realize a traceable evolution of XRD peaks, especially at < 400 ms. Subsequent to the discontinued XRD peaks, continued XRD peaks began to appear with, still substantially high moving rates, but the their moving rates drastically reduced to a minimum during next 500 ms. A considerable grain re-orientation is attributed via such XRD peak movements and if a monochromatic X-ray beam or a fewer time resolutions was employed, only a mist like diffraction pattern could be detected e.g. ref.[10]. Initial discontinued XRD peaks were attributed to equiaxed type free moving grains in the melt pool and, subsequently appeared continued XRD peaks were attributed for columnar grains [11].

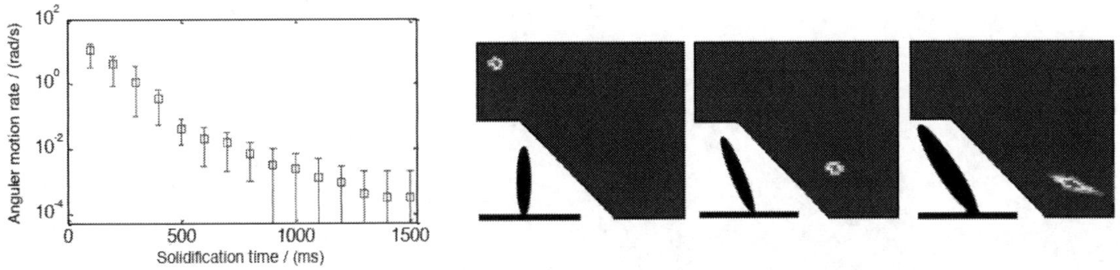

Figure 2. **a**) Average angular motion rate for XRD peaks binned for every 100ms, error bars indicate the reported minimum and maximum during the period. **b**) Schematic illustration of XRD observations for corresponding columnar crystal tilting, and growth. The tilting angle is exaggerated here for better illustration.

In order to analyze the data related to the columnar type grains quantitatively, time evolving integrated diffracted intensity, I_h reciprocal lattice vector, r^* length along the major axis(angular direction), d_a and azimuth angle, ψ for each XRD peak was extracted from the background corrected image sequence(s). While ψ was observed to remain nearly at a constant value, prominent variations in I_h , r^* and d_a were recognized, in relation to the growth and orientation of a crystal. The observed XRD peak and potentially attributed crystal characteristics are schematically illustrated in Figure 2.b. As mentioned in the previous section above, XRD data were collected frame by frame at a fixed rate of 1.0 kHz and in order to analyze each frame, raw pixel intensities are required to undergo few processing steps for necessary quantification. Background intensity caused by diffuse scattering from air or the liquid metal was assessed from frames captured with the weld pool completely molten, and found to be weak and fairly constant across the solid angle covered by the detector. Accordingly, basic corrections were performed simply by subtracting the mean background intensity value from all pixels to arrive at a set of net pixel intensities for each frame. After background intensity subtraction, individual XRD peaks emerged as islands of interconnected pixels with net intensities above the pre-defined threshold in each frame. For complete mathematical details of the image processing procedure readers are referred to ref.[11].
According to the theory of kinematical diffraction I_h from a crystal fully bathed in the incident beam, and for a XRD peak associated with the reciprocal lattice vector h= ha* + kb* + lc*, where h, k, l are the Miller indices, and a*, b*, c* reciprocal unit cell vectors, can be expressed as,

$$I_{\mathbf{h}} = V_c^{-2} K_{\mathbf{h}} \, |\mathbf{F}_{\mathbf{h}}|^2 \cdot V_g \quad (1)$$

where V_c the volume of the unit cell, and V_g the X-ray illuminated grain volume. K_h is a factor that collects reflection dependent experimental corrections and universal physical constants, and the former may be neglected here as all required corrections have been done on the pixel intensities. Finally, $|\mathbf{F_h}|$ is the amplitude of the structure factor. As long as crystal growth is analysed on a relative scale, and under the assumption that the effect of thermal contraction can be neglected, the time evolution of the crystal volume can be expressed in terms of a relative fraction,

$$f_g(t) = \frac{V_g(t)}{V_\infty} = \frac{I_h(t)}{I_\infty} \tag{2}$$

where V_∞ and I_∞ represent the final X-ray illuminated grain volume and the corresponding intensity scattered in direction $\mathbf{k_h}$ at sequence termination when the weld pool is completely solidified.

4. Discussion

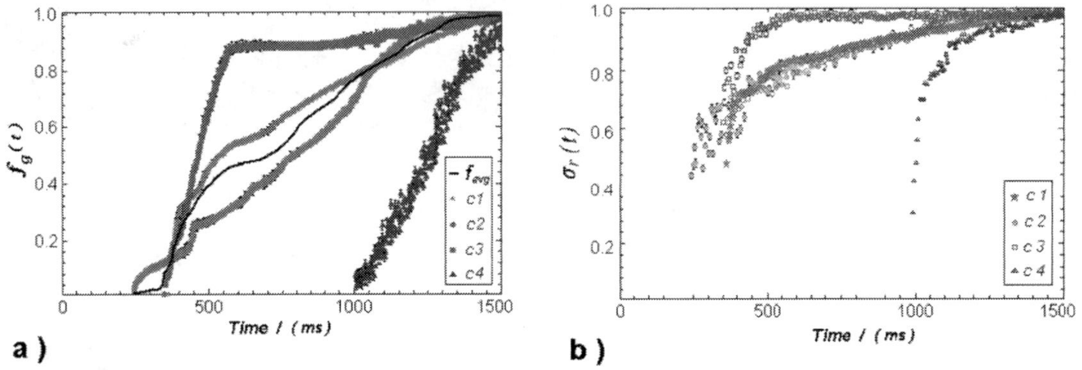

Figure 3. **a)** Overall solid fraction, $f_{avg}(t)$ of the weld pool and volume fraction evolution, $f_g(t)$ of four selected solidifying columnar crystals. **b)** Time evolution of relative XRD peak width, d_a (with respect to the highest measurable) related for the same crystals. Values are binned for every 5 ms to enhance data readability in the figure.

As the lasting (continued) XRD peaks observed in the experiments were attributed to columnar crystals [11]. Angular motion of such a XRD peak can be characterized as re-orientation via tilting of the crystal, as illustrated in Figure 2.b. Angular motion of XRD peaks could also occurred due to the thermal contraction, but observed dr^*/dt were ~3-4 orders of magnitude above the resulting dr^*/dt from the lattice parameter changes [12] at ~ 300 K/s cooling rate. In fact, thermal contraction effects are distinguishable in the latter part of the XRD image sequence(s). By successful indexing of a few reflection pairs belonging to the same columnar crystals with respect to *<100>* growth direction [2,10] allowed a number of the columnar crystal re-orientations and growth to be analyzed in more detail with $I_h(t)$ of the respective XRD peaks. The time evolution of a set of individual columnar type grain volume fraction, $f_g(t)$ evolutions were extracted and one example grain volume fraction curve is presented in Figure 3.a. Most of the analyzed columnar crystals that grow with primary tip directions at angles < 15° with respect to the incident beam. The average columnar grain length measured from the corresponding micrographs was taken as an input to the calculations, in the relevant experiments. The derived $V_g(t)$ evolution for columnar crystals were computationally fitted to an evolving parabolic shape envelope. The initial growth rates found to be up to 4 mm/s initially, but were dropped by two orders of magnitude or more within 200 ms. In addition to the increase in intensity, the expansion of the XRD peak spot size, mainly along the radial direction (measured as d_a) was also observed later. However, the observation was identified as an effect of the crystal volume expansion, as suggested via a strong correlation between d_a and V_g rather than an evidence of a deformation (bending along the crystal trunk), through the similarity between the XRD peaks I_h increase (which is represented by the crystal volume fraction) and the relevant XRD peak width change presented in Figures 3.a and 3.b.

Experimental evidence on columnar crystal tilting from the root as explained here is novel in the case of welding or similar rapid solidification processing, and is to our knowledge, not incorporated in any simulation models, although its presence should be well-recognized. In the processes such as what we have studied, where high cooling rates employed, advection and convection is extremely strong and we suggest the flow driven solid-liquid momentum as the driving force for the tilting of solidifying columnar crystals. Very low yield strength of the alloy near melting point together with the relatively slender crystal roots could also be critically contributed for such as crystal tilting. As established via computational studies [1,13], even the absence of the arc pressure or Lorentz forces; surface tension driven forces can impose a reasonably strong flow in the weld pool, as similar to the conditions exists in the experiment(s). The maximum fluid flow value calculated for the experimental conditions via an established analytical relationship [1] is over 400 mm/s. The presence of a strong convection was evident in the post-welded samples which typically showed pitting in the center of the weld together with a continuous protruded ring of solidified material outside its edges.

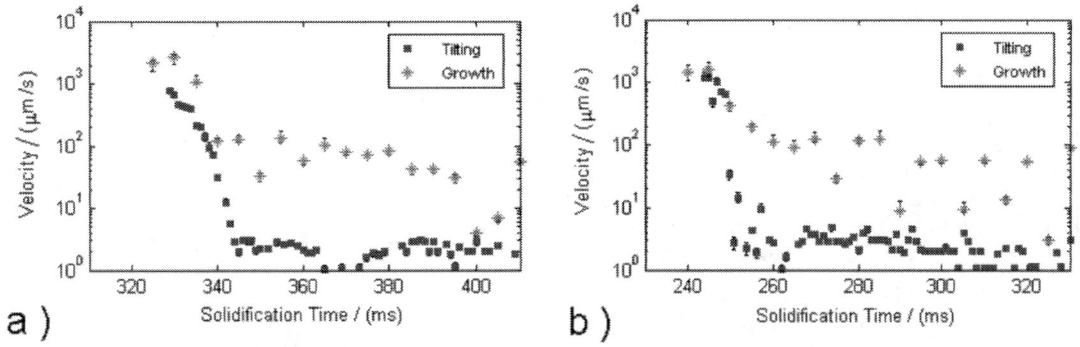

Figure 4. (**a** and **b**)Evolution of growth and tilting for two representative columnar crystals. Average growth velocities are binned for every 5 ms and both the crystals grow with their major component parallel to the incident X-ray beam direction.

The reason for not incorporating such crystal dynamics presented here to the current solidification science may only be justified by a complete lack of relevant experimental data. Owing to the high temperatures tilting take place, no evidence from the post characterization could be anticipated. At the preliminary spell of the solidification, the columnar dendrite growth rates and the bending velocities are found to be nearly at the same order of magnitude as illustrated in Figure 4. As a implication of simultaneous growth and rotating/tilting and prevailing flow, the solute boundary layer around the growing crystals would be nearly wiped out, and may leave the anisotropy and diffusion control effects [1,3,4,14] nontrivial for microstructure formation in the solidification conditions presented here and could expect generally in fusion welding or any other rapid solidification process.

Our experimental evidence emphasize sever inadequacy to affiliate conventional crystal growth theories on weld or similar rapid solidification and also serves to underline the role of liquid flow as a controlling mechanism for the crystal growth, at least in the early to intermediate stages of such rapid solidification. Thus, the experimental results serves to stress that melt flow and liquid to solid momentum transfer is an obligatory aspect to address and account in future theoretical work and simulation models for weld solidification or any other similar solidification conditions which involve strong flow, rather than relying on direct extrapolation of conditions and exchange terms used for ordinary transient solidification processes.

5. Summary

Synchrotron radiation have been used to carry out in-situ time resolved studies of evolving microstructures in steel alloy weld solidification, with ultra-fast 2D X-ray diffraction pattern capture rate. Experimental results with subsequent data analysis brings a quantitative insight to the highly dynamic physical phenomena existing in weld solidification processes or any other similar dendritic crystal growth taking place under very high melt flow environments. The study has shown that substantial crystal rotation, tilting and motion is present, especially in the earlier stages of solidification, which is likely to have a strong influence on the transient behavior of growing crystals in welds. The experimental findings presented here will influence and emphasize on the critical factors to be considered in the progression of the solidification and subsequent defect formation mechanisms theories related to the weld process. Furthermore, the experiment and the results set an example for future extremely fast in-situ XRD.

References

[1] DebRoy T and David SA 1995 *Rev. Modern Physics* **67** 85
[2] David S A, Babu S S and Vitek J M 2003 *JOM* **55** 14
[3] Jeong J H, Goldenfeld N and Dantzig J A 2001 *Phys. Rev. E* **64** 041602
[4] Granasy L *et al.* 2004 *Nature Mater.* **2** 645
[5] Mathiesen R H *et al.* 1999 *Phys Rev Lett.* **83** 5062
[6] Buffet A *et al.* 2007 *Physica Status Solidi: Applications and Materials Science* **204** 2721
[7] Yasuda H *et al.* 2011 *ISIJ International* **51** 402
[8] Elmer J W, Wong J and Ressler T 2000 *Scripta Mater.* **43**, 751
[9] Babu S S *et al.* 2002 *Acta Mater.* **50** 4763
[10] Yonamura M *et al.* 2010 *J. App. Physics* **107** 013523
[11] Mirihanage W U *et al.* 2014 *Acta Mater.***68** 159
[12] Seki I and Nagata K 2005 *ISIJ International* **45**, 1789
[13] Tanaka M and Lowke J J 2006 *J. Phys. D: Appl. Phys.* 40, R1
[14] Tomorr H *et al.* 2006 *Nature Mater.* **5** 660

On the hot cracking susceptibility of a semisolid aluminium 6061 weld: Application of a coupled solidification-thermomechanical model

HR Zareie Rajani[1] and AB Phillion[1]

[1] School of Engineering, The University of British Columbia, Kelowna, BC, Canada

E-mail: hamid.zareie.rajani@ubc.ca

Abstract. A coupled solidification-thermomechanical model is presented that investigates the hot tearing susceptibility of an aluminium 6061 semisolid weld. Two key phenomena are considered: excessive deformation of the semisolid weld, initiating a hot tear, and the ability of the semisolid weld to heal the hot tear by circulation of the molten metal. The model consists of two major modules: weld solidification and thermomechanical analysis. 1) By means of a multi-scale model of solidification, the microstructural evolution of the semisolid weld is simulated in 3D. The semisolid structure, which varies as a function of welding parameters, is composed of solidifying grains and a network of micro liquid channels. The weld solidification module is utilized to obtain the solidification shrinkage. The size of the micro liquid channels is used as an indicator to assess the healing ability of the semisolid weld. 2) Using the finite element method, the mechanical interaction between the weld pool and the base metal is simulated to capture the transient force field deforming the semisolid weld. Thermomechanical stresses and shrinkage stresses are both considered in the analysis; the solidification contractions are extracted from the weld solidification module and applied to the deformation simulation as boundary conditions. Such an analysis enables characterization of the potential for excessive deformation of the weld. The outputs of the model are used to study the effect of welding parameters including welding current and speed, and also welding constraint on the hot cracking susceptibility of an aluminium alloy 6061 semisolid weld.

1. Introduction

During fusion welding of aluminum alloys, cracks sometimes form and grow behind the weld pool, within the two-phase mushy zone, rupturing liquid films that are present at grain boundaries [1]. This phenomenon, known as hot tearing or hot cracking, is prevalent in both industrial casting and welding processes.

The mechanisms of hot tearing have been widely studied in casting processes (e.g. [2-11]). Two factors are frequently reported: 1) initiation of hot cracks by deformation of the mushy zone, and 2) healing of hot cracks by circulation of the molten metal within the mushy zone. Early studies by Pellini [4] and later Magnin et al. [5] showed that excessive deformation of the mushy zone in semisolid material arising from solidification shrinkage and thermal contraction serves to develop hot cracks. In addition, the formation of a hot crack is strongly linked to the grain morphology. The initiated mushy zone micro-cracks will only survive if the microstructural configuration does not allow for liquid feeding to heal the resulting defect [8, 9]. Vernède [2], and Sistaninia [12] have shown that the width of the micro liquid channels lying along the grain boundaries controls the healing ability of the mushy zone, where wider channels improve the healing process. The theory and knowledge of hot cracking in casting is also applicable to welding. As reviewed by Coniglio et al. [6], excessive tensile strain within a semisolid weld will lead to the formation of a hot crack. The role of the microstructure configuration on liquid feeding, and consequently hot cracking, is also extendable to welding [3]. Therefore, analyzing a semisolid weld in terms of deformation and microstructure can reveal insight into the susceptibility of the weld to this defect.

Due to the short lifetime of the semisolid weld and also the high temperature condition during welding, real-time experiments cannot be utilized to study the microstructural configuration of the mushy zone weld region and its healing ability, nor the deformations within the mushy zone. On the other hand, numerical techniques appear to be more feasible. Recently, some studies have been conducted to numerically reconstruct the microstructure of the semisolid weld at the meso scale. Zhong et al. [13] and Bordreuil et al. [14] have used cellular automaton (CA) method to reconstruct the 2-D microstructure of the semisolid weld. Zareie Rajani et al. [15] have also reconstructed the 3-D meso-scale microstructure of the semisolid weld composed of solidifying grains and micro liquid channels through a granular model of solidification. Unlike microstructural configuration, the deformation of the mushy zone during welding has not yet been examined using mathematical modeling at the meso scale. This is because of the challenges relating to simulating the complex force field deforming the semisolid weld [16, 17]. This transient force field acts on the interface between the base metal and the semisolid weld, *i.e.* the fusion surface, and evolves due to the thermomechanical response of the base metal to the solidification of the semisolid weld.

In this paper, a coupled thermomechanical-solidification model is presented to investigate the hot cracking susceptibility at the scale of the weld microstructure. The model consists of 1) a thermomechanical analysis module that determines the transient force field on the weld and then evaluates deformation of the mushy zone perpendicular to the weld line; and 2) a solidification module that reconstructs the microstructure of the mushy zone and yields part of the boundary conditions used in the thermomechanical analysis module. The model is applied to various welding scenarios to study the effect of welding parameters on formation of hot cracks during fusion welding of AA 6061.

2. Model description

2.1. Thermomechanical analysis module

The transient forces that act on the fusion surface and deform the semisolid during welding occur due to mechanical coupling between the base metal and the weld pool. The thermomechanical analysis module numerically solves the equation of conservation of momentum on the base metal of the weld to obtain the material flow within this domain caused by thermal stresses and solidification contraction during welding, and consequently to determine the reaction forces that evolve on the fusion surface, $\vec{f}^{\text{reaction}}_{\partial\Omega_{\text{fusion|base metal}}}$. Utilizing Newton's third law of motion, the external force field acting on the semisolid weld during welding, *i.e.* $\vec{f}_{\partial\Omega_{\text{fusion|semisolid weld}}}$, is then calculated as:

$$\vec{f}_{\partial\Omega_{\text{fusion|semisolid weld}}} = -\vec{f}^{\text{reaction}}_{\partial\Omega_{\text{fusion|base metal}}} \tag{1}$$

In order to numerically solve the equation of conservation of momentum, the thermomechanical analysis module requires definition of a simulation domain with boundary conditions. The numerical solution to the transient stress analysis is then obtained using the commercial FEM solver ABAQUS.

2.1.1. Simulation domain

The simulation domain for the thermomechanical analysis consists of the base metal of the weld, and is shown in Figure 1. Due to symmetry, only one half of the base metal is simulated. The base metal is assumed to be a thin plate, and separated from its surroundings via a boundary ($\partial\Omega$) composed of seven different surfaces: one parabolic fusion surface ($\partial\Omega_{\text{fusion|base metal}}$) representing the contact surface of the base metal and the weld line, one symmetry surface ($\partial\Omega_{\text{symmetry}}$) lying on the symmetry plane of the weld, four free surfaces ($\partial\Omega_{\text{free}}$) that are not in contact with anything but the air, and the back surface ($\partial\Omega_{\text{back}}$) that is assumed to be either free or constrained. Based on the defined Cartesian coordinate system, the depth and width of the plate respectively lie along the X and Y directions, and the Z direction follows the length of the plate and consequently the line of welding. The width and thickness of the base metal are assumed to be 50 mm and 3 mm, while the length of the domain is set at 20 mm. The simulation domain is then discretized through a nonuniform mesh composed of 8-node brick elements. The material properties are assumed to be elastic-thermoplastic with strain hardening, obeying the von Mises plasticity criterion. The relevant temperature-dependent flow stress curves for AA 6061 are shown in Figure 2 [18, 19]. ABAQUS obtains the temperature variation of flow stress of the base metal, σ_o, through linear interpolation of the given flow stress functions for various temperatures.

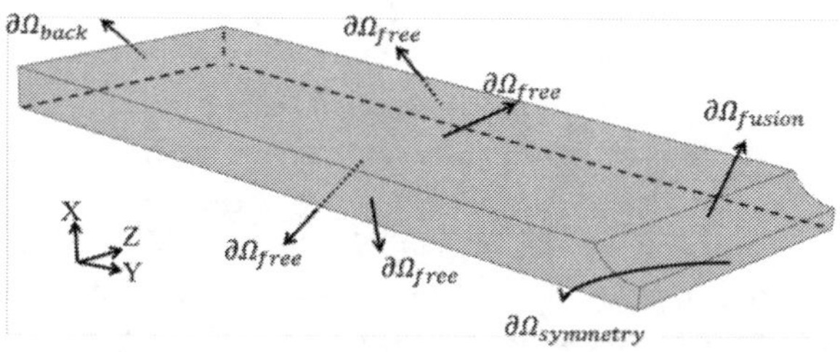

Figure 1. The simulation domain representing the base metal.

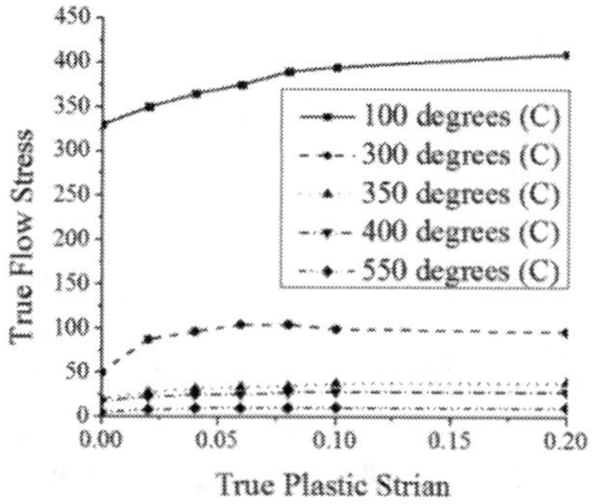

Figure 2. Temperature-dependent flow stress curves for AA 6061.

2.1.2. Boundary conditions

For the boundary conditions, it is assumed that the four surfaces in contact with air have a homogenous boundary condition, $\vec{f}|_{\partial\Omega_{\text{free}}} = 0$, where $\vec{f}$ is the applied force. Dirichlet boundary conditions are applied to the remaining three surfaces. The symmetry surface is constrained in the Y direction due to the symmetry as well as the X and Z directions since the weld line is normally clamped during welding to prevent misalignment defects. On the back surface, different displacement Dirichlet boundary conditions can be applied, *i.e.* $\vec{U}|_{\partial\Omega_{\text{back}}} = \vec{U}_o$ where $\vec{U}_o$ is the imposed displacement, to model various welding constraints including clamping ($\vec{U}_o = 0$), and application of tension/compression ($\vec{U}_o \neq 0$). It is also possible to assume that the base metal is not clamped during welding by setting a homogenous boundary condition for the back surface. The Dirichlet boundary condition applied to the fusion surface, *i.e.* $\vec{U}|_{\partial\Omega_{\text{fusion|base metal}}} = \vec{D}_{(\vec{x},t)}$, is the most complex since the displacement field on the fusion surface, $\vec{D}_{(\vec{x},t)}$, is imposed by nonuniform solidification of the semisolid weld rather than a mechanical tool. This boundary condition can be obtained by performing a meso-scale solidification simulation within the weld pool.

2.2. Solidification module

The solidification module determines the displacement field on the fusion surface, and completes the boundary conditions needed by the thermomechanical analysis module. It has previously been shown [3] that the displacement field on the fusion surface is mainly affected by the lateral solidification shrinkage that is perpendicular to the weld line. Based on conservation of mass, the lateral solidification contraction Y_{Sh} for a confined semisolid domain can be expressed as:

$$Y_{\text{Sh}} = Y_{\text{o}}\sqrt[3]{g_{\text{s}}}(1-\alpha) \tag{2}$$

where $\alpha = \sqrt[3]{\rho_{\text{l}}/\rho_{\text{s}}}$ is the linear shrinkage coefficient, ρ_{l} and ρ_{s} are the liquid and solid densities respectively, and Y_{o} and g_{s} denote the lateral length and volumetric solid fraction. For a small solidification increment, equation (2) can be rewritten in differential form:

$$dY_{\text{Sh}} = \frac{1}{3}Y_{\text{o}}(1-\alpha)g_{\text{s}}^{-\frac{2}{3}}dg_{\text{s}} \tag{3}$$

Integration of equation (3) yields the gradual, accumulation of the lateral solidification contraction within the confined space as a function of solid fraction. Since the semisolid is not able to transfer stress before a continuous solid bridge forms across the confined domain, *i.e.* mechanical coalescence, the integration domain is limited to values of solid fraction at which the semisolid is percolated, *i.e.*:

$$Y_{\text{Sh}}|_t = \int_{g_{\text{s}}|_{\text{p}}}^{g_{\text{s}}|_t}\frac{1}{3}Y_{\text{o}}(1-\alpha)g_{\text{s}}^{-\frac{2}{3}}dg_{\text{s}} = Y_{\text{o}}(1-\alpha)\left(\sqrt[3]{g_{\text{s}}|_t} - \sqrt[3]{g_{\text{s}}|_{\text{p}}}\right) \tag{4}$$

In Equation (4), $Y_{\text{Sh}}|_t$ represents the evolution in linear solidification contraction with time, $g_{\text{s}}|_t$ denotes the average solid fraction of the confined space as a function of time, and $g_{\text{s}}|_{\text{p}}$ represents the average solid fraction at percolation, assumed to occur at a value of g_{s} =0.95. The confined elements used in this study are shown in Figure 3. These elements, given the name *bar elements*, lie along the Y direction connecting the weld center to the fusion surface. The ensemble of the bar elements and their corresponding displacement functions ($Y_{\text{sh}}|_t{}^i$), where the index i represents an individual bar element, yields the Dirichlet boundary condition acting on the fusion surface of the thermomechanical module:

$$\vec{U}|_{\partial\Omega_{\text{fusion|base metal}}} = \vec{D}_{(\vec{x},t)} = \begin{cases} D_Y|_t = Y_{\text{sh}}|_t{}^i \quad , \quad \vec{x} \in A_Y{}^i \\ D_X|_t = D_Z|_t = 0 \end{cases} \tag{5}$$

where the term $A_Y{}^i$ refers to the projection of bar element i onto the portion of the fusion surface containing the given point in $\vec{D}_{(\vec{x},t)}$.

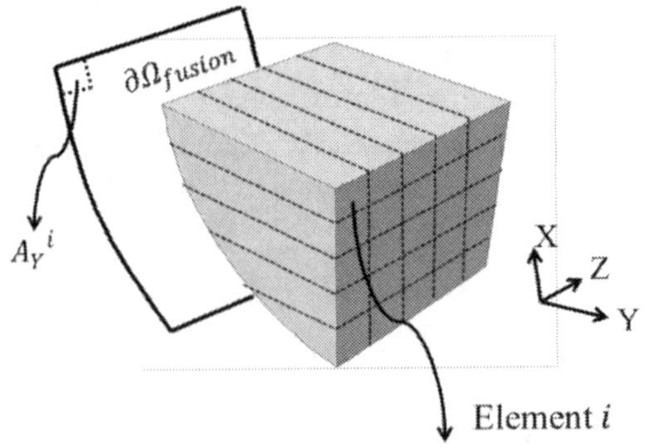

Figure 3. The bar element used to obtain the local solid fraction within the semisolid weld.

It should be noted that the bar elements are conceptual only; no simulation is performed directly on them. The displacement field given by Equation (5) is obtained by determining the average evolution in fraction solid with time within each bar element, *i.e.* $g_{\text{s}}|_t{}^i$, and then applying Equation (4) to calculate the correct displacement value. The evolution in fraction solid is generated from the meso-scale solidification model of Zareie Rajani and Phillion [15] for AA6061 welds to reconstruct the 3D microstructure of the semisolid weld. The salient features of this model are presented below.

Through a 3D unstructured grid, the meso-scale solidification model first generates an ensemble of elements representing the microstructure of the weld composed of both columnar and equiaxed grains. As Figure 4a shows, the columnar grains are extended from the fusion boundary towards the center of the weld, while the equiaxed grains are located at the center of the weld. Only one half of the weld is modeled due to symmetry. Weld microstructure varies as a function of welding parameters. The model contains a lookup table linking welding parameters to equiaxed and columnar grain sizes, allowing different microstructures (unstructured grids) to be created based on the choice of welding conditions.

After discretization, solidification within the developed grains is simulated through tracking of the solidification fronts splitting the grains into separate solid and liquid regions [9, 12, 15]. Tracking of the solidification front in each grain is carried out by combining an evolving temperature field within the weld pool generated by the Rosenthal equation [15, 20, 21] with a fraction solid curve for AA 6061 generated by Thermocalc for Scheil solidification. As Figures 4b-c indicate, the meso-scale solidification model yields a microstructure weld composed of solid grains and a network of micro liquid channels at different times during the welding process; the empty space represents the liquid channels spreading along the grain boundaries. The solidification module predicts the evolving semisolid microstructure morphology during the welding process. The corresponding evolution in solid fraction can be averaged over any arbitrary confined space. The solidification module performs this averaging to obtain the transient solid fraction within each bar element to allow for calculation of the displacement field on the fusion surface. The simulated microstructure can also be used to assess the healing ability of the semisolid weld. The average width of the micro liquid channels is assumed as the relevant criterion, where wider channels improve the liquid feedability [2, 12].

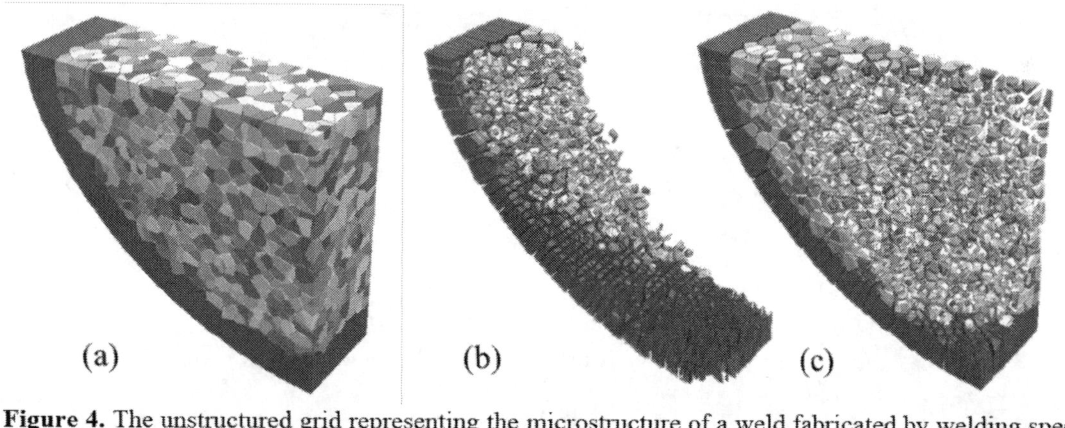

Figure 4. The unstructured grid representing the microstructure of a weld fabricated by welding speed of 3 mm/s and welding current of 120 A (a). The microstructure of the same weld during welding at t=0.2 s (b) and t=0.6 s (c).

2.3. Solution technique

Once the solidification module computes the displacement boundary condition acting on the fusion surface of the base metal, the thermomechanical analysis module simulates the evolution of the reaction forces on the base metal during welding. Equation (1) is then applied to extract the transient force field that deforms the semisolid weld. Since the lateral component of tensile force plays the key role in welding hot cracking [3], the tensile force components in Y that apply on bar elements with $g_s > g_s|_p$ are averaged and then divided by the percolated area of the fusion surface to obtain the average tensile stress acting on the semisolid weld. In order to gain insight into hot crack susceptibility, the average tensile stress values are then compared against known critical tensile stress values from the literature. A recent experimental study by Giraud and et al. [22] showed that the percolated semisolid AA 6061 cannot bear tensile stresses greater than 5 MPa. This value is considered as the critical criterion for defect formation and thus average tensile stresses greater than 5 MPa are assumed to cause excessive deformation and consequently initiate defects.

3. Results and discussion

The calculated average tensile stresses acting on the fusion surface of unclamped AA 6061 semisolid welds as a function of time fabricated by various welding parameters are shown in Figure 5 for different welding speeds between 2 and 5 mm/s. By comparing the results within individual figures it can be seen that increasing the welding current at constant welding speed generates larger average tensile stresses on the fusion surface, increasing the risk of severe deformation within the semisolid weld. Such an observation can be explained through the shrinkage displacement. By increasing the welding current, larger welds form, and as demonstrated in Equation (2), lead to larger amounts of solidification contraction. Higher magnitudes of shrinkage contraction will then create larger reaction forces on the fusion surface. This result matches prior experimental studies [3] where it is shown that high welding current and large weld pools increase the risk of hot cracking.

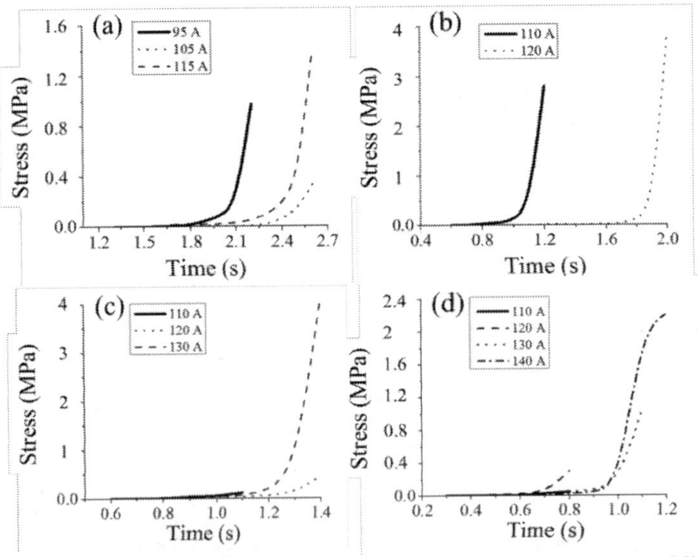

Figure 5. The calculated average tensile stress for various welding currents at welding speeds of: (a) 2 mm/s; (b) 3mm/s; (c) 4 mm/s; (d) 5 mm/s.

In terms of welding speed, Figure 5 indicates that this parameter greatly affects the average tensile stress deforming the semisolid weld. At a low welding speed of 2 mm/s, Figure 5a, the maximum average tensile stress is relatively low. With increasing welding speed to 3 and 4 mm/s, Figures 5b and 5c, the maximum average tensile stress significantly rises, approaching the critical condition for defect formation. However, a further increase in welding speed to 5 mm/s, Figure 5d, appears to result in a decrease in the maximum average tensile stress. In other words, the simulations predict that the maximum average tensile stress and consequently the susceptibility to hot cracking are highest at an intermediate range of welding speeds. This result can be explained through the interaction between two main factors that control the material flow within the base metal: 1) solidification shrinkage causing material flow within the base metal since larger welds are accompanied by more lateral solidification contraction and consequently bigger driving forces for the materials flow, and 2) the amount of resistance against material flow that strongly depends on the temperature of the base metal since colder base metals are accompanied by higher elastic moduli and larger yield strengths. Therefore, according to the Newton's third law of motion, cold base metals and large welds generate high magnitudes of reaction forces, while hot base metals and smaller welds create lower magnitudes of reaction forces. As a result, the small average tensile stresses observed at low welding speeds can be linked to the excessive softness of the base metal that stems from high temperature conditions during slow welding. By increasing the welding speed, the heat input decreases, leading to a colder base metal. Consequently, the reaction force and the average tensile stress deforming the semisolid weld increase. Although the base metal is coldest at high welding velocities, the reaction forces do not keep growing since there is a significant drop in the size of the weld. In this simulation, by increasing the welding speed from 4 mm/s to 5 m/s at a constant welding current of 120 A, the depth of weld penetration decreased from 2 mm to 1 mm. It is hypothesized that very small welds do not generate enough linear solidification contraction to maintain the ascending trend of the reaction forces.

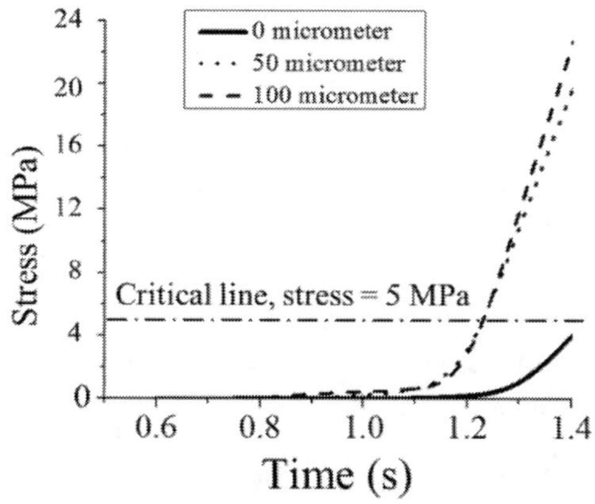

Figure 6. The average tensile stresses on the weld fusion surface under different constraining conditions.

Table 1. The average thickness of the micro liquid channels for various welding parameters.

Welding speed (mms^{-1})	Welding current (A)	Average thickness of micro liquid channels (μm)
2	110	10±2
3	105	11±2
3	120	13±3
4	110	10±3
4	130	6±1
5	105	9±2
5	125	8±1
5	140	8±2

Another welding parameter that is well known to affect the susceptibility of the weld to hot cracking is the constraining condition of the weld [3]. In Figure 6, three different constraining conditions are compared for a weld produced at a speed of 3 mm/s using a current of 120 A. In the first, the back surface of the plate was clamped ($\vec{U}_0 = 0$). In the second and third, the back surface was displaced in tension along the Y direction. The displacement was initiated at the time corresponding to the time that at least one of the bar elements within the semisolid weld had percolated. The displacement reached the maximum value, 50 μm and 100 μm, when the weld had fully solidified. As can be seen in the figure, the application of tensile displacement on the back surface of the plate during welding results in the average tensile stresses along the fusion surface exceeding the critical value, making the weld severely susceptible to defect formation.

Initiated defects within the semisolid weld will lead to hot cracking if the network of micro liquid channels fails to heal the defects, i.e. feed the molten aluminum adequately. Hence, in addition to the stress applied to the semisolid weld, the feeding ability of the microstructure has to also be known in order to assess hot cracking susceptibility. Table 1 lists the average thickness of the micro liquid channels within a percolated microstructure for various welding parameters. The results are extracted from the solidification module. Based on the obtained results, the liquid films present at a welding speed of 4 mm/s and current of 120 A are the narrowest, and thus, have the lowest healing ability. Further discussion of the micro liquid channel widths can be found in [15].

4. Conclusion

The present model proposes a method to couple simulations of solidification in a weld pool at the meso-scale with thermomechanical deformation of the base metal in order to assess hot cracking susceptibility. The solidification module reconstructs the 3D microstructure of semisolid AA 6061 welds, enabling determination of the average thickness of the micro liquid channels within the semisolid weld, and also the linear solidification contraction. The thermomechanical module utilizes the displacements imparted by the weld shrinkage along with various boundary conditions to extract the tractions on the fusion surface that deforms the semisolid weld during solidification.

Based on the obtained results, the semisolid weld undergoes the maximum deformations and also the lowest healing ability at a moderate welding speed of 4 mm/s and a high level of welding current. The results also indicate that any external tensile stress during welding will further increase the semisolid weld's susceptibility to hot cracking. Future work will focus on development of a simulation to localize the deformations within the semisolid weld based on the tractions on the fusion surface.

Acknowledgments

The authors wish to thank the American Welding Society (AWS), Rio Tinto AlCan, and the Natural Sciences and Engineering Research Council of Canada (NSERC) for financial support.

References

[1] Boettinger WJ, Coriell SR, Greer AL, Karma A, Kurz W, Rappaz M, Trivedi R 2000 *Acta Mater.* **48** 43-70.
[2] Vernède S, Jarry P, Rappaz M 2006 *Acta Mater.* **54** 4023-4034.
[3] Kou S 2003 *Welding Metallurgy* (New Jersey: John Wiley & Sons Inc.).
[4] Pellini WS 1952 *J. Foundry* **80** 124-133.
[5] Magnin B, Maenner L, Katgerman L, Engler S 1996 *Mater. Sci. Forum* **217-222** 1209-1214.
[6] Coniglio N, Cross CE 2013 *Int. Mater. Rev.* **58** 375-397.
[7] Prokhorov NN 1962 *Russ. Castings Production* **2** 172-175.
[8] Rappaz M, Drezet JM, Gremaud M 1999 *Metall. Mater. Trans.* A **30A** 449-455.
[9] Sistaninia M, Phillion AB, Drezet JM, Rappaz M 2012 *Acta Mater.* **60** 3902-3911.
[10] Mitchell JB, Cockcroft SL, Viano D, Davidson C, StJohn D 2007 *Metall. Mater. Trans.* A **38A** 2503-2512.
[11] Hao H, Maijer DM, Wells MA, Phillion AB, Cockcroft SL 2010 *Metall. Mater. Trans.* **41** 2067-2077.
[12] Sistaninia M, Phillion AB, Drezet JM, Rappaz M 2012 *Acta Mater.* **60** 6793-6803.
[13] Zhong XH, Dong ZB, Wei YH, Ma R 2009 *J. Cryst. Growth.* **311** 4778-4783.
[14] Bordreuil C, Niel A 2014 *Comp. Mater. Sci.* **82** 442-450.
[15] Zareie Rajani HR, Phillion AB 2014 *Acta Mater.* **77** 162-172.
[16] Slużalec A 2005 *Theory of Thermomechanical Processes in Welding* (Dordrecht, The Netherlands: Springer).
[17] Lindgren LE 2007 *Computational Welding Mechanics: Thermomechanical and Microstructural Simulations* (New York: CRC Press).
[18] Prassad YVRK, Sasidhara S, Gegel HL, Malas JC 1997 *Hot Working Guide: A Compendium of Processing Maps* (Ohio: ASM International).
[19] Mahabunphachai S, Koç M 2010 *Mater. Design* **31** 2422-2434.
[20] Zareie Rajani HR, Torkamani H, Sharbati M, Rygan S 2012 *Mater. Design* **34** 51-57.
[21] Yadaiah N, Bag S 2014 *Int. J. Ther. Sci.* **86** 125-138.
[22] Giraud E, Suery M, Coret M 2010 *Metall. Mater. Trans. A* **41A** 2257-2268.

Detached Melt Nucleation during Diffusion Brazing of a Technical Ni-based Superalloy: A Phase-Field Study

B Böttger[1], M Apel[1], B Laux[2] and S Piegert[2]

1) Access, Aachen, Germany
2) Siemens AG, Power and Gas, Large Gas Turbines, Engineering, Berlin, Germany

E-mail: b.boettger@access.rwth-aachen.de

Abstract. Advanced solidification processes like welding, soldering, and brazing are often characterized by their specific solidification conditions. But they also may include different types of melting processes which themselves are strongly influenced by the initial microstructures and compositions of the applied materials and therefore are decisive for the final quality and mechanical properties of the joint. Such melting processes are often not well-understood because - compared to other fields of solidification science – relatively little research has been done on melting by now. Also, regarding microstructure simulation, melting has been strongly neglected in the past, although this process is substantially different from solidification due to the reversed diffusivities of the involved phases. In this paper we present phase-field simulations showing melting, solidification and precipitation of intermetallic phases during diffusion brazing of directionally solidified and heat-treated high-alloyed Ni-based gas turbine blade material using different boron containing braze alloys. Contrary to the common belief, melting of the base material is not always planar and can be further accompanied by detached nucleation and growth of a second liquid phase inside the base material leading to polycrystalline morphologies of the joint after solidification. These findings are consistent with results from brazed laboratory samples, which were characterized by EDX and optical microscopy, and can be explained in terms of specific alloy thermodynamics and inter-diffusion kinetics. Consequences of the gained new understanding for brazing of high-alloyed materials are discussed.

1. Introduction

Diffusion brazing is a process which has gained industrial importance, e.g. as build-up or repair brazing of gas and aircraft turbine components made of Ni-/Co-based high temperature materials [1]. To achieve an isothermal solidification process and hence to avoid formation of brittle eutectic phases, a filler material similar to the base metal is applied which has been modified by melting point depressing elements (typically boron) [2-4]. Figure 1 [5] shows the commonly assumed mechanism of the brazing process based on a simple binary Ni-B phase diagram. At brazing temperature, the filler melts due to its high content of boron (①). Then the base metal is rapidly melting until the liquid phase reaches equilibrium composition (②). But as boron rapidly diffuses into the base metal, isothermal solidification is possible (③). If the brazing time is long enough, complete solidification occurs (④), and the material can reach mechanical properties of the level of the base material (⑤).

For identifying suitable filler materials and process parameters, a deeper understanding of the microstructural changes during the brazing process including melting is needed, which can be achieved by computer simulation. The task is not only to minimize the brazing time, but also to avoid formation of new grain boundaries (epitaxial brazing).

For reaching this goal, among the different available simulation techniques, the phase-field approach appears to be most promising. Phase-field models have become very popular for the simulation of microstructure evolution during solidification [6-8]. The software MICRESS® [9] is based on the phase-field concept for multiphase systems which has been extended to multicomponent systems by direct coupling to thermodynamic databases [8,10]. Since then the software has been continuously developed not only for advanced scientific applications but also as a simulation tool for microstructure formation in technical processes [11-13].

In this paper, MICRESS® simulation results for brazing of CM247 using a Ni-15% Cr-4% B braze alloy are presented and compared to experimental results for various brazing times. A complex melting process with detached melt nucleation is suggested which accounts for the observed melting width and rows of new grains which typically emerge in the microstructure during brazing.

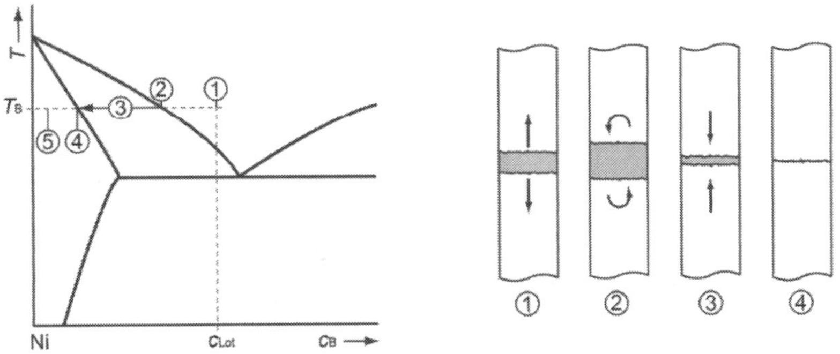

Figure 1: Principle of diffusion brazing [5]

2. Experimental

Brazing experiments were performed using small plate samples of solution heat treated equiaxed CM247 base material and melt-spun amorphous foils of the braze alloy with thickness of 50 μm. SEM images of three samples with different brazing times (½Δt, Δt, 2Δt, exact times are confidential), which have been cut perpendicular to the braze gap, are shown in Figure 2. Characteristic microstructural features are the polycrystalline nature of the molten zone with two or more rows of newly formed grains as well as boride precipitates on both sides of the gap which reach further into the base material with increasing brazing time.

Furthermore, the area covered by boride precipitates was evaluated using image analysis. For slices of 5 μm width parallel to the braze gap, the average area of all borides was measured and compared to simulation results (Figure 8). Individual types of borides were not distinguished.

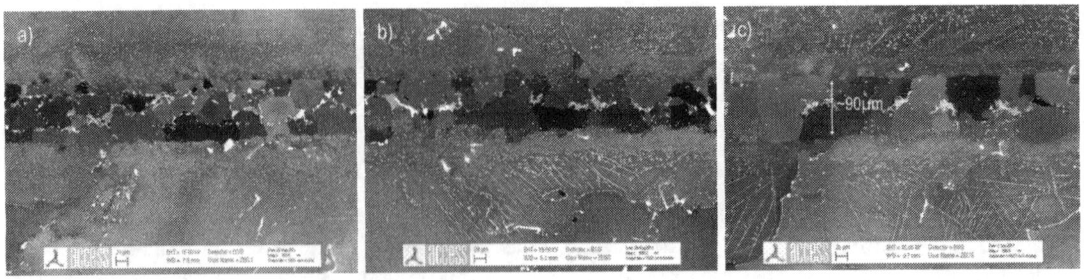

Figure 2: SEM images of three samples with different brazing times: a) ½Δt, b) Δt, c) 2Δt. The width of the molten zone was about 90 μm in all cases.

3. Phase-Field Modelling

3.1. Multiphase-field model and MICRESS® software

The multiphase-field theory describes the evolution of multiple phase-field parameters $\phi_\alpha(\vec{x},t)$ in time and space. They reflect the spatial distribution of multiple grains with different orientation and/or of multiple phases with different thermodynamic properties. At the interfaces, the phase-field variables change continuously over an interface thickness η which can be defined as being large compared to the atomic interface thickness but small compared to the microstructure length scale. Their time evolution is calculated by a set of phase-field equations deduced by the minimization of the free energy functional [8]:

$$\dot{\phi}_\alpha = \sum_\beta M_{\alpha\beta}(\vec{n})\left(\sigma^*_{\alpha\beta}(\vec{n})\,K_{\alpha\beta} + \frac{\pi}{\eta}\sqrt{\phi_\alpha\phi_\beta}\;\Delta G_{\alpha\beta}(\vec{c},T)\right) \quad (1)$$

$$K_{\alpha\beta} = \phi_\beta\nabla^2\phi_\alpha - \phi_\alpha\nabla^2\phi_\beta + (\pi/\eta)^2(\phi_\alpha - \phi_\beta) \quad (2)$$

In equation (1), $M_{\alpha\beta}$ is the mobility of the interface as a function of the interface orientation, described by the normal vector $\vec{n}$. $\sigma^*_{\alpha\beta}$ is the anisotropic surface stiffness, and $K_{\alpha\beta}$ is related to the local curvature of the interface. The interface, on the one hand, is driven by the curvature contribution $\sigma^*_{\alpha\beta}$ $K_{\alpha\beta}$ and on the other hand by the thermodynamic driving force $\Delta G_{\alpha\beta}$. This driving force which is a function of temperature T and local composition $\vec{c}$ couples the phase-field equations to the multi-phase diffusion equations

$$\dot{\vec{c}} = \nabla\sum_{\alpha=1}^{N}\phi_\alpha\vec{D}_\alpha\nabla\vec{c}_\alpha \qquad \text{with } \vec{c} \text{ defined by} \qquad \vec{c} = \sum_{\alpha=1}^{N}\phi_\alpha\vec{c}_\alpha \quad (3)$$

and $\vec{D}_\alpha$ being the multicomponent diffusion coefficient matrix for phase α. $\vec{D}_\alpha$ and $\Delta G_{\alpha\beta}$ are calculated online from databases for the given concentration and temperature.

The above equations are implemented in the software package MICRESS® [9] being used for the simulations throughout this paper. Direct coupling to thermodynamic and mobility databases is accomplished via the TQ-interface of Thermo-Calc Software [14]. The thermodynamic driving force ΔG and the solute partitioning are calculated separately using the quasi-equilibrium approach [8], and are introduced into the equation for the multiple phase-fields (equation (1)).

3.2. Model parameters and data

Thermodynamic data were provided by online coupling to the database TTNI7 [15] using the elements given in Table 1 and the phases listed in Table 2. In the same way, diffusion data for γ including cross-terms were taken from the MOBNI1 mobility database [14]. As MOBNI1 has no corresponding data for γ' and MOBNI2 is not compatible to the coupled TTNI7 thermodynamic database, the diffusion matrix of γ' was tabulated as function of temperature along equilibrium for the base metal.

Compositions of both materials are shown in Table 1. Phase-related data like nucleation conditions, interface mobility μ and interface energy σ between all phases and melt or γ are given in Table 2.

3.3. Initial Microstructure

Creating the initial microstructure of the base metal from metallographic images was not the preferred method in view of the complexity of the observed microstructures, the small size of precipitates, and the difficulty to obtain concentration distributions which are consistent with the nominal alloy composition. Therefore, the initial microstructure was created by 2D phase-field simulation of solidification with subsequent solution heat treatment according to the Siemens internal material specification. Solidification was simulated at the scale of two ¼ equiaxed grains including a grain boundary and essentially using the parameter sets given in Table 1 and 2. A self-consistent homoenthalpic approach [16] was applied to assure that microstructure formation was consistent with the macroscopic temperature boundary condition which leads to a realistic initial microstructure.

Table 1: Chemical compositions (in wt %) used for simulation

	C	Cr	Co	Mo	W	Ti	Al	B	Ta	Hf
CM247	0.07	8.0	9.0	0.4	9.3	0.6	5.45	0.01	3.1	1.2
braze foil	0.06	15.0	-	-	-	-	-	4.0	-	-

Table 2: Phases and phase related parameters

No	name in database [15]	identity/ constitution	diffusion data	nucleation in phase	$\mu_{0/x}$ (cm^4/Js)	$\mu_{1/x}$ (cm^4/Js)	$\sigma_{0/x}$ / (J/cm^2)	$\sigma_{1/x}$ / (J/cm^2)
0	LIQUID	melt	10^{-6} cm^2/s	1 (region)	-	1×10^{-7}	-	7×10^{-5}
1	FCC_A1#1	γ	MOBNI1	0	1×10^{-7}	1×10^{-5}	7×10^{-5}	1×10^{-5}
2	FCC_A1#2	(Hf,Ta)C	-	0,1	1×10^{-7}	8.2×10^{-9}	1×10^{-4}	1×10^{-4}
3	MB2_C32	$(Ti,Hf)B_2$	-	0,1	1×10^{-7}	2.4×10^{-8}	1×10^{-4}	1×10^{-4}
4	FCC_L12	γ'	MOBNI2	0,1	1×10^{-7}	6.1×10^{-8}	7×10^{-5}	4×10^{-5}
5	M3B2_TETR	$(W,Cr,Mo)_3B_2$	-	0,1	1×10^{-7}	2.4×10^{-8}	1×10^{-4}	1×10^{-4}
6	M2B_ORTH	$(Cr,Mo,W)_2B$	-	0,1	1×10^{-7}	2.4×10^{-8}	1×10^{-4}	1×10^{-4}
7	MB_ORTH	(Cr, Mo,W)B	-	0,1	1×10^{-7}	2.4×10^{-8}	1×10^{-4}	1×10^{-4}
8	CR5B3	$(Cr, Mo)_5B_3$	-	0,1	1×10^{-7}	2.4×10^{-8}	1×10^{-4}	1×10^{-4}
9	M2B_TETR	$(W,Cr,Mo)_2B$	-	0,1	1×10^{-7}	2.4×10^{-8}	1×10^{-4}	1×10^{-4}
10	M3B	$(Cr,W,Co)_3(B,C)$	-	0,1	1×10^{-7}	2.4×10^{-8}	1×10^{-4}	1×10^{-4}

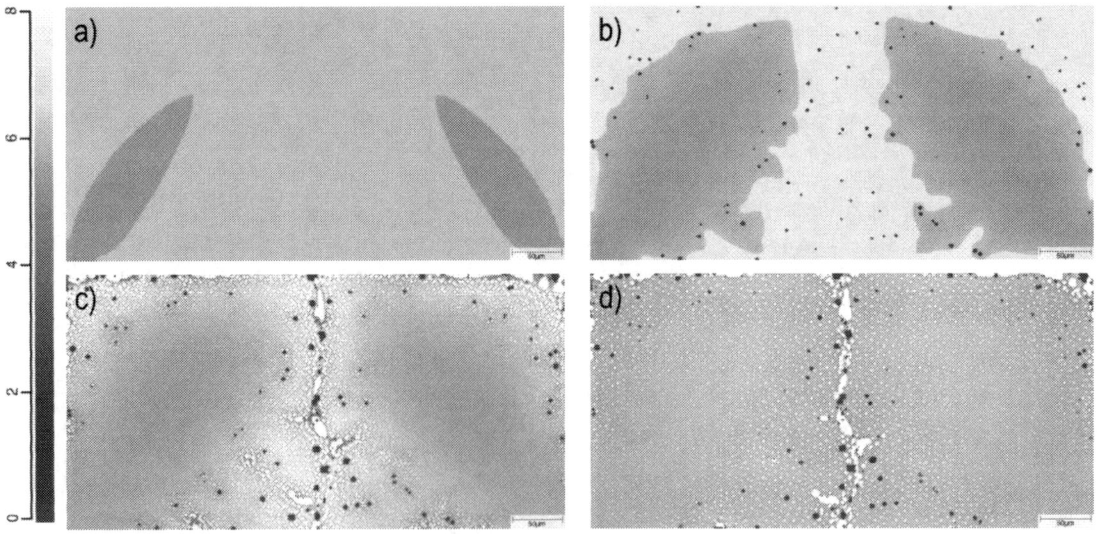

Figure 3: Simulated Al-distribution (wt%) during solidification at a) 1365 °C, b) 1345 °C, after solidification at c) 800 °C and d) after homogenisation and subsequent cooling to 800 °C.

Figure 3a-c shows the Al-distribution during the solidification process. Dark particles are carbides (FCC_A1#2) which nucleate in the melt modelled by a seed density model [13,17]. The γ' phase appears as bright regions along the grain boundary. Although grid resolution was rather low (500x200 cells with Δx=1μm), γ' precipitation from the solid was simulated as individual small particles represented by a small phase fraction in a single grid cell (analytical curvature model [13]).

Solution heat treatment was simulated for the specified industrial conditions including heating from 800 °C. As shown in Figure 3d, the concentration gradients of Al and small γ' particles were dissolved while larger eutectic γ' remained nearly unchanged. Few smaller γ' precipitates formed during cooling down to 800 °C afterwards. The resulting microstructure was stored for use in the brazing simulations.

3.4. Simulation of Diffusion Brazing

The simulation domain for brazing had a size of 500x550 grid cells (Δx=1 μm) and consisted of three grains representing half of the filler (grain 1), one side of the base metal (grain 2) and a connection zone for a one-dimensional extension for far-field diffusion of boron (grain 3, see Figure 4a). While grain 1 and 3 were defined to have a constant composition corresponding to the braze alloy and base metal composition in Table 1, respectively, grain 2 was replaced by the initial microstructure in Figure 3d. This was achieved using a specific restart algorithm which reads the detailed microstructure information from the previous simulation and further allows complete filling of the space originally covered by grain 2 using mirror symmetry (Figure 4b). The zones where melting and boride precipitation is expected and mirror symmetry at the bottom are also indicated in Figure 4b).

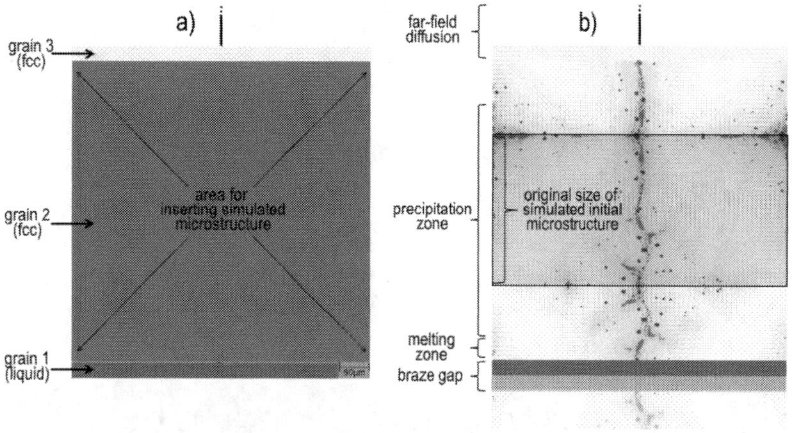

Figure 4: Schematic representation of the simulation domain for brazing: a) before, b) after replacing grain 2 with the symmetrically expanded results of the simulated initial microstructure of Figure 3d. Note that due to mirror symmetry at the bottom only half of the braze gap is simulated.

During brazing a homogeneous temperature is assumed. In the simulation scenario, heating starts from 800 °C with a rate of 25 K/min up to the brazing temperature. Then, temperature is kept constant for ½Δt, Δt, or 2Δt followed by cooling to 800 ° C with -25 K/min. Melting of the filler was assumed to occur instantaneously at 1061 °C (according to DSC experiments). This "Melting" was implemented as a sudden raise of the diffusivity in grain 1 to D_{liq} and of the interface mobilities $\mu_{0/x}$ from 0 to the values specified in Table 2. In order to obtain a realistic solidification time, the diffusivities of B in γ from MOBNI1 had to be reduced by a factor of 5.

The progression of the liquid-solid interface during brazing is shown in Figure 5. If the simulation is performed according to the simple mechanism of Figure 1, melting of the base metal is observed, with a maximum shortly after brazing temperature is reached (Figure 5a). Due to the inhomogeneous microstructure of the base metal non-planar melting is observed. But there is no indication that formation of new grains during solidification should occur. And the average width of the (total) molten zone does not exceed 65 μm in contrary to the experimentally found ~90 μm (Figure 2). Nevertheless, this smaller simulated value is supported by thermodynamic calculations: a mixture of 50 parts braze alloy and 40 parts base metal (corresponding to 90 μm maximum melting width) according to Thermo-Calc equilibrium calculations using the TTNI7 database should consist only to 73.3% of liquid at brazing temperature!

Therefore, an alternative mechanism including detached nucleation of liquid inside the base metal was taken into consideration. In a first simple approach, nucleation of liquid was allowed along a fixed row at 15 μm behind the initial solid-liquid interface. This relatively large distance was chosen to prevent early mixing of the new liquid with the liquid braze alloy. Interestingly, the simulations clearly showed that detached liquid nucleation leads to rapid melting inside the base metal which is accompanied by γ' precipitation and formation of a semisolid region inside the base material. In Figure 5b we can observe that this semi-solid region finally breaks up and gets connected to the main liquid in the more segregated central region around the grain boundary, while accelerated solidification is observed laterally. Later (Figure 5b, right) the remaining semi-solid regions have vanished and the liquid gap starts closing.

Thus, the new mechanism which includes detached melt nucleation, with the simple approach of a nucleation line, seems to explain as well the formation of new grains as the increased width of the brazing gap in the experiments (Figure 6).

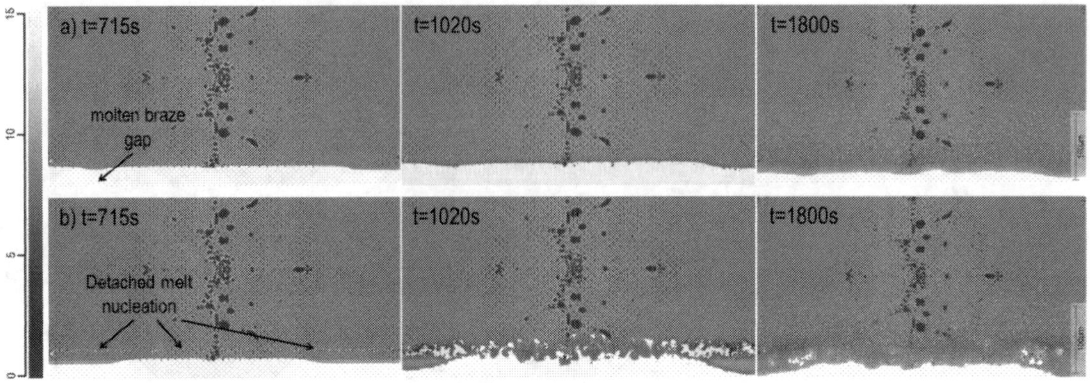

Figure 5: Cr composition profile (wt%) in the lower part of the simulation domain (Figure 4) during the initial stage of the brazing process, showing melting and re-solidification during brazing: a) without, and b) with detached melt nucleation. The bright zones indicate liquid regions.

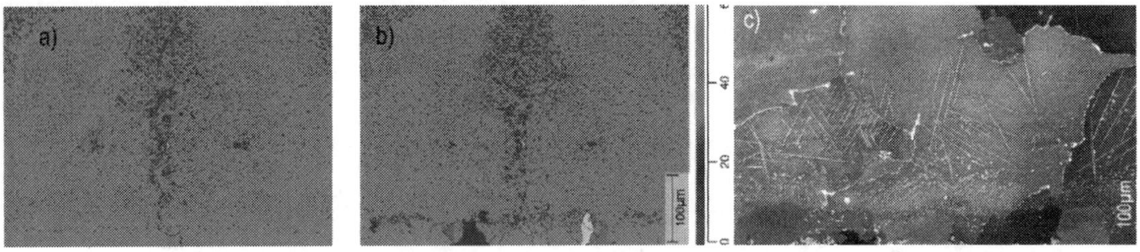

Figure 6: a) Formation of new grains during brazing. Shown is the grain orientation (in degrees) after 2Δt brazing for simulation a) without and b) with detached melt nucleation. c): SEM image of a brazed sample after 2Δt.

Besides of the melting and re-solidification process, precipitation of borides was observed. Nucleation of the different boride types listed in Table 2 was allowed as well at the interfaces to the melt as inside the γ phase. Figure 7 shows the location of the borides after Δt. Most abundant are the Ti-rich MB2_C32 borides. They form bigger precipitates close to the gap, which have grown from the melt, and smaller particles which nucleate inside the base metal. The size of the latter increases with the distance from the gap, which is due to their decreasing growth rate and the fact that growth is limited by the available amount of Ti. This observation is consistent with experimental microstructures (Figure 2), though the morphology change to needle-like structures could not be correctly modelled owing to the low resolution of the simulation. The lower amounts of W-rich M3B3_TETR borides visible in Figure 7 left could not be directly identified in the experimental micrographs.

Finally, Figure 8 compares the simulation results with the measured distribution of the total boride area for three different brazing times. There is a good qualitative agreement with respect to the maximum and width of the distribution as well as to the shift and broadening with increasing brazing time. The plot of the simulated borides over time (Figure 8b) reveals that during brazing MB2_C32 is the most abundant boride type. All others appear intermediately but vanish at longer times, with the exception of M3B3_TETR. Note that this phase slightly grows again at the end due to cooling.

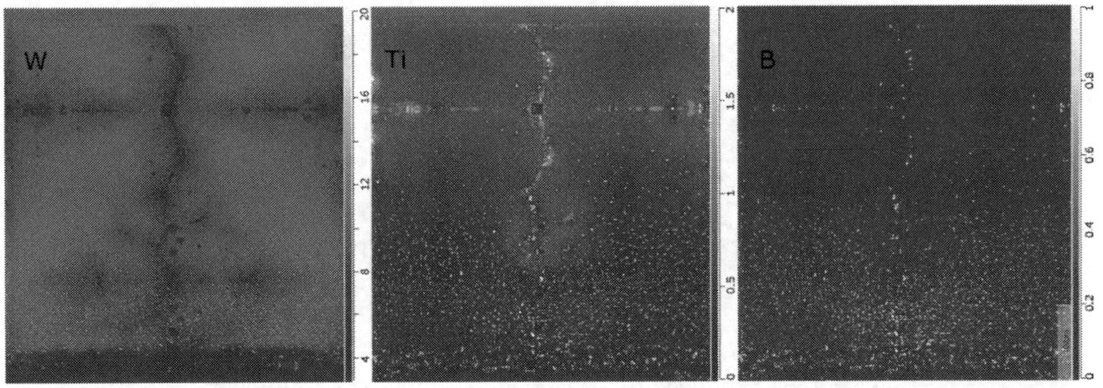

Figure 7: Concentration distribution of W, Ti and B indicating the location of borides after Δt.

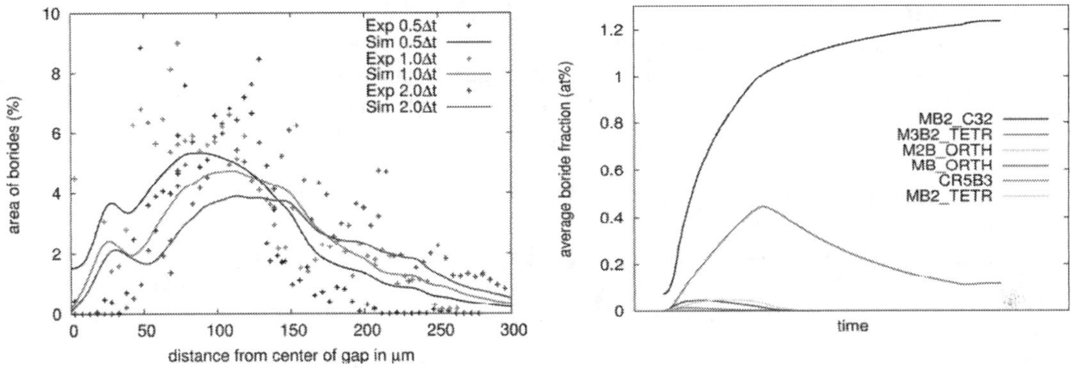

Figure 8: Area of total borides after brazing as a function of the distance from the gap center (left) and simulated average molar fraction of individual boride types during the $2\Delta t$ brazing cycle.

4. Discussion

Although this has not yet been discussed in literature, it is obvious that detached melting during brazing may occur if the base metal plus the additional high boron activity provided by fast B diffusion results in a lower melting point compared to the braze alloy itself. This is the case for the high-alloyed CM247 when combined with a braze alloy which has only 15% Cr.

The reason why we believe the base metal is not melting from the surface is shown in Figure 9. After equilibrium is reached at the solid-liquid interface, there is a depletion of substitutional elements close to the solid-liquid interface which prevents further melting (①). Instead, there is a region closely behind this zone with unchanged compositions where the B activity is high and may lead to melting. Therefore, a thin detached liquid film can form below the surface (②) which has a composition different to the primary melt and which rapidly grows due to fast B diffusion (③). In case of CM247 simultaneous γ' precipitation occurs in this zone. As the original solid layer which separates the two liquids can be very thin, material inhomogeneity or other effects may easily lead to its rupture (④). The remnants of the solid layer form a row of new grains (⑤). Alternatively, if the layer melts completely, nucleation may also occur on the γ' particles which start dissolving once the liquid composition is altered by mixing. After complete mixing of the two liquids, the same process ①-⑤ may be repeated deeper into the base metal while the inner part of the gap is already partially solidified. The fact that detached melting and solidification of the primary liquid occur simultaneously can explain the observed width of the molten zone of 90 μm without contradicting the equilibrium calculations mentioned in Section 3.4.

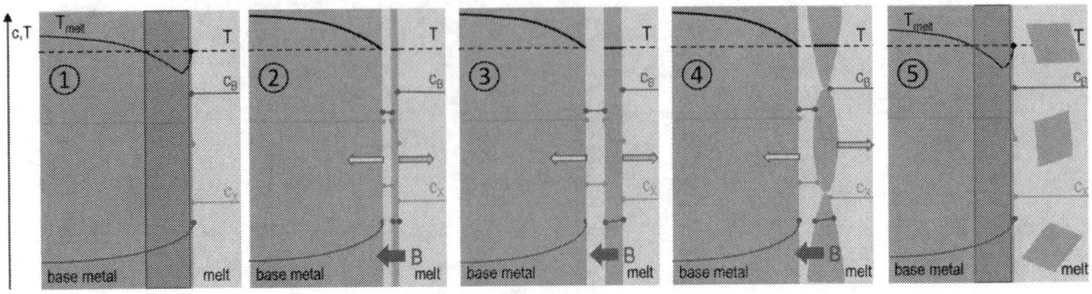

Figure 9: Schematic representation of detached melting. Shown are the solid-liquid interfaces with the temperature and concentration profiles of B and an arbitrary substitutional element X.

5. Conclusions and Outlook

It has been shown that, despite of the high complexity of the alloys and the process, simulation of brazing can be achieved using the phase-field method. Good agreement with experiments was reached with respect to morphology and boride distribution. Furthermore, a new mechanism of detached melting has been postulated and discussed, which allows explaining not only the formation of new grains, but also of an increased width of the molten zone compared to thermodynamic estimations. Although, due to the low grid resolution, including this mechanism into modelling was rather rough, the effect of detached melting was clearly confirmed and turned out to be of utmost importance for understanding the experimental findings.

The results of this paper also shed new light on the criteria for choosing braze alloys, especially for the case of epitaxial brazing. The question arises whether new high-alloyed filler materials or improved process parameters can prevent formation of new grains in case of high-alloyed materials.

References

[1] Demo W A and Ferrigno S J. 1992 *Adv. Mater. Process.* **141** 43–5
[2] Duvall D S, Owczarski W A and Paulonis D F 1974 *Weld. J.* **53** 203–14
[3] Ekrami A, Moeinifar S and Kokabi A H 2007 *Mat. Sci. Eng.* A **456** 93–8
[4] MacDonald W D and Eagar T W 1998 *Metall. Mater. Trans.* A, **29** 315–25
[5] Hoppe B 2003 Dissertation Technische Universität Braunschweig/Germany
[6] Wheeler A A, Boettinger W J and Mc Fadden G B 1993 *Phys. Rev.* E **47** 1893-909
[7] Kim S G, Kim W T and Suzuki T 1999 *Phys. Rev.* E **60** 7186-97
[8] Eiken J, Böttger B and Steinbach I 2006 *Phys. Rev.* E **73** 066122
[9] MICRESS software, http://www.micress.de
[10] Böttger B, Grafe U, Ma D and Fries S G 2000 *Mater. Sci. Technol.* **16** 1425-28
[11] Schmitz G J et al 2012 *Int. J. Adv. Eng. Sci. Appl. Math.* **2** 4 126-39
[12] Böttger B, Apel M, Eiken J, Schaffnit P and Steinbach I 2008 *Steel Res. Int.* **79** 608-16
[13] Böttger B, Eiken J and Steinbach I 2006 *Acta Mater.* **54** 2697-704
[14] Themo-Calc Software, http://www.thermocalc.se
[15] Thermotech Ltd., http://www.thermotech.co.uk
[16] Böttger B, Eiken J and Apel M 2009 *J. Comput. Phys.* **228** 6784-95
[17] Böttger B, Apel M, Santillana B and Eskin D G 2013 *Metall. Mater. Trans.* A **44A** 3765-77

IOP Conf. Series: Materials Science and Engineering **84** (2015) 012032 doi:10.1088/1757-899X/84/1/012032

A scale adaptive dendritic envelope model of solidification at mesoscopic scales

A Kharicha[1], M Stefan-Kharicha[1*], A Ludwig[1] and M Wu[1,2]
[1] Chair for Modeling and Simulation of Metallurgical Processes, University of Leoben, Austria
[2] Christian-Doppler Laboratory for Advanced Process Simulation of Solidification & Melting, University of Leoben, Austria

E-mail: [*] abdellah.kharicha@unileoben.ac.at

Abstract. We present the Scale Adaptive Dendritic Envelope (SADE) model of solidification at mesoscopic scales. The new approach is based on the rescaling of the microscopic laws on the desired resolution scale of simulation. The diffusivity, the Gibbs Thomson coefficient, and the fraction which solidifies with the solid front are modified to create large fictitious dendrites (envelope) whose tips growth at the same speed as the microscopic tips. The model is inspired from the methodology of the turbulence models that filter the scales that are smaller than the simulation grid size, such as the Large Eddy Simulation (LES) approach. Here, the solidified structure scales which are smaller than the grid size, such as the tip radius and the smallest arms, are modelled with sub-grid scale models. The envelope growth is coupled with a sub-grid model to account for the contribution of the unresolved secondary arms to the phase transformation. The model is applied to constitutionally undercooled domains with different grid sizes that are larger than the initial secondary arm spacing and much larger than the microscopic tip radius. A similar primary arm spacing is predicted regardless of the grid resolution. However the result obtained with smaller meshes resolve more dendrites branches than with coarser meshes.

1. Introduction

Dendrites are common solidification morphology in both pure materials and alloys. Complex interplay between macroscopic heat and solute transport and microscopic interfacial phenomena results in various dendritic solidification structures. The interaction between dendritic branches and tips are responsible for the features of the grain inner structure. These features are responsible for the macroscopic properties, such as the mush permeability, as well as for the development of freckles and channel segregations. In term of modelling a huge scale gap exists between microscopic phase-field simulations of the dendrites shape, and the volume-averaged simulations dedicated to the prediction of the macrosegregation and freckles. The phase-field method necessitates extremely small grid size to correctly catch the diffusion of solute through the liquid/solid interface. Typically the grid size is several orders of magnitude smaller than the tip radius. Great progress in numerical implementation were recently achieved by using adaptive meshing techniques, multigrid methods, and GPU parallelization. However, even with those improvements the simulations of dendritic arrays still requires considerable computing resources. To fill the gap between the phase-field scale and the full volume-averaged simulation, Steinbach and Beckerman [1-2] have introduced a model where the complex dendrite morphology is approximated with its envelope. The envelope growth model is deduced from the velocity of dendrite tips which are calculated with an LGK model. The melt on the envelope surface as well as inside is assumed to be well mixed and in equilibrium with the solid field. The actual structure of the dendritic solid is not resolved, but a volume-averaged solid fraction is used. The model was successfully applied to equiaxed dendritic interactions, as well as for columnar growth

(Delaleau et al.[3]). More recently, Tourret et al. [4] presented the multiscale dendritic needle network approach to model well ramified dendritic microstructures. The primary, secondary and higher order branches are represented by a network of needles that interact through the solutal field. The range of stable primary arm spacing for an Al-7% Si alloy was calculated.

We present here an alternative approach based on the rescaling of the microscopic growing laws to desired scale of simulation. The aim is to be able to simulate the generation and the growth of primary arms for meshes where the grid size is much larger than the tip radius but still smaller than the expected primary arm spacing, λ_1. A previous work was presented in 2012 [5], where this approach was applied to the macroscale solidification of an NH_4Cl-H_2O alloy. In order to catch the smallest solute plumes scale the solidification model simulated directly the envelope of the columnar dendrites. The mushy interior of the dendrites made of secondary and tertiary arms is modelled with a volume-averaged method. The predicted solidification front and the primary arms spacing fitted with the observed ones. The envelope front started from all the walls with a planar front. The growth morphology changed quickly into cellular, and finally bifurcated into a columnar dendritic type [5]. The model was able to predict the occurrence of several flow regimes resulting from the interplay between thermal and solutal natural convections. Flow regimes such as turbulent, stratified and meandering flows were predicted [6-8]. In the present analysis we further develop the theoretical foundation of the model. It will be shown that it is possible to build a dendrite envelope where the tip radius is larger than the real dendrite tip by a specific factor. The factor is directly related to the solid fraction being solidified with the growth of the envelope. The model self-adapts to resolve only the scales that are larger than the chosen mesh size, smaller dendrite arms spacing are modelled with a sub-grid model.

2. The Model

2.1. Rescaling of the microscopic dendritic growth

Let us consider a microscopic dendrite with tip velocity, V_t , radius, and its secondary arms (Figure 1a). The secondary arms that are generated at the dendrite trunk grow and compete, resulting in progressive increase of the secondary arm spacing with the distance from the trunk. Let us consider further a fictitious dendrite that has the same tip velocity but much larger tip radius. The microscopic tips are located inside the volume of the fictitious dendrite renamed as "dendrite envelope". Similarly to the microscopic dendrite, the dendrite envelope must develop new secondary and tertiary arms. These envelope arms are assumed to surround tightly the major microscopic arms surviving the competition between smaller arms. The envelope contains an undefined volume of liquid, and a volume of solid equal to the solid contained within the real dendrite.

The aforementioned quality of the dendrite envelope can be achieved by assuming that the dendrite envelope solidifies a fraction of solid smaller than 1. We will show later the solid fraction must be equal to the ratio between the envelope tip radius, R_E, and the microscopic real tip radius, R_t (Figure 1b).

Let us first define the ratio $f_s^0 = R_t / R_E$. If the diffusivity is scaled with this factor, $D_E = D_l / f_s^0$, the tip envelope Péclet number becomes strictly equal to the microscopic tip Péclet number:

$$\mathrm{Pe}_E = \frac{V_t R_E}{2D_E} = \mathrm{Pe} = \frac{V_t R_t}{2D_l}. \tag{1}$$

This Péclet number is directly related to the melt super-saturation, Ω , at a distance, δ , by the exact steady state 3D solution derived by Cantor-Vogel [9]:

$$\Omega = \mathrm{Pe}\exp(\mathrm{Pe})(E_1(\mathrm{Pe}) - E_1(\mathrm{Pe}(1+\frac{\alpha}{\mathrm{Pe}}))) \tag{2}$$

Where E1 is the exponential integral function. Although the simulations are performed in 2D Hele-Shaw cell, the microscopic tip is small enough so that it develops a 3D paraboloid of revolution between two plates. The parameter α is the ratio between the distance δ and solute diffusion layer $\alpha = \delta/(D_l / V_t)$, in the present study this parameter is chosen to be in order of 1 ($\alpha \sim 1$).

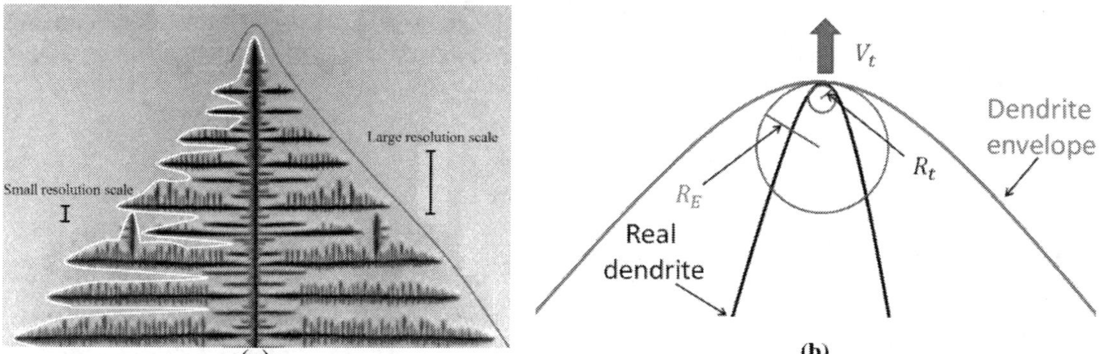

Figure 1. Schematic illustration of the envelope model used in the SADE approach. (a) A tight (white) envelope and a loose (red) envelope of an ammonium-bromide crystal in a supersaturated aqueous solution [10]. The SADE model targets to predict a tight envelopes with meshes of sufficiently small resolution scale. If larger grid size are used looser envelopes are predicted. (b) Concept of envelope used in the SADE model. The envelope tip has a much larger tip than the real microscopic tip, but shares the same tip velocity. With the interface advection the real dendrite solidifies totally the volume passed, the envelope solidifies only a fraction of solid equal to ratio between the real to the envelope tip radius.

With the choice of scaling, the original microscopic Ivantsov equation is still valid for the envelope. The tip radius and the solute diffusivity are both modified to conserve the magnitude of the Péclet number. To complete the determination of the tip radius and velocity, the tip selection criterion must also be rescaled. It can be achieved by again rescaling the diffusivity and the Gibbs-Thomson coefficient $\Gamma_E = \Gamma / f_s^0$:

$$R_t^2 V_t = \frac{\Gamma D_l}{|m_L|(C_S^* - C_l^*)\sigma^*} = \beta \Rightarrow R_E^2 V_t = \frac{\Gamma_E D_E}{|m_L|(C_S^* - C_l^*)\sigma^*} = \frac{\beta}{(f_S^0)^2}, \tag{3}$$

where m_l is the liquidus slope, σ^* is the selection parameters, C_S^* and C_l^* are the equilibrium concentrations at the solid and liquid interface. For isothermal conditions β is constant.

By rescaling the molecular diffusivity and the Gibbs Thomson coefficient with the factor $1/f_s^0$ we can build an object which develops a tip radius R_E larger than the real tip radius R_t, and growth at the same speed as the microscopic tip, V_t. In other words, the Stefan condition at the envelope surface must fulfil

$$V_n (C_S^* - C_l^*) = D_E \frac{\partial C}{\partial n}, \tag{4}$$

where V_n is the normal velocity of the envelope-liquid interface, and $\partial C / \partial n$ denotes the macroscopic normal solute gradient. Since the interface curvature and the Gibbs-Thomson coefficient are both rescaled, the curvature undercooling of the envelope is equal to microscopic one. Therefore, the equilibrium concentration at the envelope interface C_l^* is not modified. By simply transferring f_s^0 to the left hand side of Eq. (4), we can re-interpret the scaling factor f_s^0 as a solid fraction:

$$f_s^0 V_n (C_S^* - C_l^*) = D_l \frac{\partial C}{\partial n}. \tag{5}$$

The desired dendrite envelope can be built by using the normal solute diffusivity D_l, a rescaled Gibbs-Thomson coefficient, and by solidifying the fraction f_s^0 as the envelope passes.

IOP Conf. Series: Materials Science and Engineering **84** (2015) 012032 doi:10.1088/1757-899X/84/1/012032

The applicability of the model will depend mainly on the chosen grid size. The model can be applied in two ways. The first way is to choose f_s^0, then the envelope velocity is computed either with Eqs. (2) and (3) for highly curved regions, and with Eq. (5) when the envelope curvature is well resolved. In this case f_s^0 must be properly chosen so as not to build an envelope which is larger than the expected primary arm spacing. The second way is to shift the envelope with the microscopic tip velocity, Eqs. (2) and (5), and compute the solid fraction that would fulfill Eq. (3). After some simple algebra, Eq. (3) can be rewritten in the following form:

$$f_s^0 = \frac{\Gamma}{2|m_L|(C_S^* - C_l^*)\sigma^* R_E \mathrm{Pe}}. \tag{6}$$

Since the Péclet number is fixed, the required solid fraction to be solidified with the envelope is proportional to the tip curvature. This fraction is calculated every time step at the tips, and kept constant for further lateral growth where Eq. (5) is applied.

While the main tip envelope growth considers only one real tip is considered, the lateral growth represents the growth of numerous unresolved secondary arm tips. This situation is more difficult because the competition between secondary arm tips cannot be directly resolved. In order to keep the model simple, we consider the number of arms fixed. The grid size is assumed to be sufficiently large so that only one tip survives when crossing the next volume elements. These secondary arms will then be considered as a new lateral "primary" tip. New branches (primary, secondary and tertiary) are emitted by the inherently unstable growth of the dendrite envelope surface.

2.2. *Volume averaged equations and sub-grid model for the unresolved secondary arms*

The 2D isothermal equations governing conservation of species and phase fraction are written as follows

$$\frac{\partial(\rho f_l C_l)}{\partial t} = \nabla\cdot(f_l \rho D_l \nabla C_l) - \rho C_s^* \frac{\partial f_s}{\partial t} \tag{7}$$

$$\rho \frac{\partial f_s}{\partial t} = f_s^0 \int V dl + M_{la} \tag{8}$$

where f_l, f_s, C_l, D_l and ρ are the liquid and solid fractions, the concentration of the solute (NH_4Cl), the solute diffusivity and the density. The terms on the right hand side of Eq. (8) stand for the mass transfer through an element of the envelope surface dl, and M_{la} the sub-grid mass transfer due to the growth of unresolved secondary arms (see Eq. (10)). Depending on whether the envelope represents a tip or a lateral growth, the front velocity V represents V_t or/and V_n. The mush topology can be represented as parallel or crossed cylindrical sticks. We use here the multiphase volume-averaged approaches developed by Wu et al. [11]. This model was developed for periodic alignment of primary arm columnar trunks. Two arms are considered as unresolved if the distance that separates them is smaller than one grid size. The resolved arms are generated by instabilities of the dendrite envelope surface. Thus, the model does not expressively separate primary and secondary arms in the classical meaning, the number of resolved arms depends on the choice of mesh resolution. By "resolved" we mean that the presence of arms are reflected in the variation in volume fraction. The model of Wu et al. [11] is used in the present analysis with an average secondary arms spacing, λ_2. The number of secondary arms that germinates along the trunk can be estimated from the observation that the ratio $\lambda_2^0 / R_t \sim 5$ [12] is approximatively constant, λ_2^0 represents the initial spacing. The initial number of sub-grid arms that grow in lateral directions is $\Delta x / \lambda_2^0$. This number is assumed constant within the volume element where they are generated. For the volume elements that has been passed by the envelop front, a diffusion-controlled growth model around cylindrical trunks is assumed. The growth velocity in the radial direction of a cylindrical trunk is assumed very slow, so that a quasi-steady growth can be used,

$$V_a = \frac{D_l}{R_a}\Omega \ln^{-1}(\frac{R_f}{R_a}) \tag{9}$$

where R_a is the average radius of the unresolved arm trunks, $R_f = \lambda_2^0/2$ is the maximum packing radius of a secondary dendrite arm trunk. For our 2D dendrites, the trunks are assumed to be parallel and lying on the same plane, so that the surface area of the secondary arm trunks per volume is $S_a = 2\pi R_a/(\lambda_2^0)^2$. The net volume-averaged mass-transfer rate for elements within the inner dendrite envelope mushy region is:

$$M_{la} = V_a\left(\frac{2\pi R_a}{(\lambda_2^0)^2}\right)\rho\ f_l \tag{10}$$

Table 1: Boundary conditions and effective parameters used for simulations.

Property/Symbol	Value	Property/Symbol	Value
C_0	0.3	C_s^*	1
C_l^*	0.266	α (Eq 1).	1.0
Temperature	17 ^{0}C	ρ_l	1078 kg/m^3
β (Ref.[13])	12.6e-18	m_L	4.8 K
Γ_i	8.05e-5 K m	D_l	2.5e-9 m^2/s

3. Application of the model

The dendritic envelope model is applied to the solidification of a H_20-30wt.%NH_4Cl alloy at 290 K (Ω = 0.1), with the thermophysical properties given in Table 1. To illustrate the capability of the model to self-adapt to different mesh sizes, the simulations are performed with the same grid, but with two different scales. The first scale is 20 mm x 200 mm with a grid size of 200 µm (Fig. 2) and the second 0.5 mm x 5 mm with a grid size of 50 µm (Fig. 3). In order to prevent the interface from remaining flat, a random factor is applied to the computed front velocity at a level of 1%. Initially one primary tip is introduced per volume element. Table 2 summarises some quantitative results. The initial interface (mush/liquid) is initially assumed flat on the right half of the bottom boundary. The second half on the left is perturbed with a sinusoidal. In Fig. 2b the interface develops kind of lamella structures which very quickly transforms into columnar. The "secondary" arms are generated as soon as the distance between primary vertical arms is large enough (Fig. 2d). These arms represent the resolved part of the secondary horizontal arms. The unresolved arms are invisible, hidden in the solid fraction predicted in each volume element. The solid fraction reaches its maximum on the arm axis at about 7 %. It can be noticed that few volume elements are almost free of solid.

During the growth (Fig. 2b-f) the number of primary arms decreases by a mechanism of competition. In the very beginning $\lambda_1 \sim$ 0.4-0.8 mm, in the end of the calculation λ_1 is in average much larger 0.6-3 mm. As could be expected different spacing developed on the perturbed left side compared to the right side. It is interesting to notice that the "regularly" perturbed side develops a more irregular interface (Fig. 2b-d). This irregularity leads to a faster selection of the primary arm spacing compared to the right side. However even in the right hand side, λ_1 seems to tend towards ~ 3mm. At the later stage (Fig. 2e-f), a relatively steady tip velocity (~ 450 µm/s) is reached which correspond to an effective tip super-saturation of 0.06 smaller by 40% from the global super-saturation (Ω = 0.1).

The simulation is now performed on a mesh which is 4 times finer. The initial state assumes a perfectly plane interface. The very first developments shows an irregular growth of two islands (Fig. 3a). Later columnar structures with side branches become visible. The dendrite envelope is now able to resolve tertiary arms (Fig. 3d-f). These tertiary arms transform in new primary arms when the conditions allow it (Fig. 3d-f). It must be emphasized that even in this higher resolution case many secondary arms are still unresolved. In each volume element located on an arm axis, 7 to 20 small unresolved arms are predicted according to the ratio λ_2^0 / R_t . These arms are considered within the sub-grid model and are assumed to not develop more than one grid size in length. As in the previous case the competition between the numerous initial primary arms result in the progressive increase of λ_1, from 0.15-0.5 mm in Fig. 3a to 0.4-5 mm in Fig. 3f. Although not yet stable the primary spacing predicted with the finer mesh is very similar to the one predicted with the coarser mesh. Due to the smaller mesh size, the maximum solid fraction reached in the dendrite trunks is naturally higher 13 % (compared to 7% previous calculation). Due to the higher resolution of the dendrite envelope surface, the volume elements which are almost free of solid (in blue) are more numerous than in the coarser case.

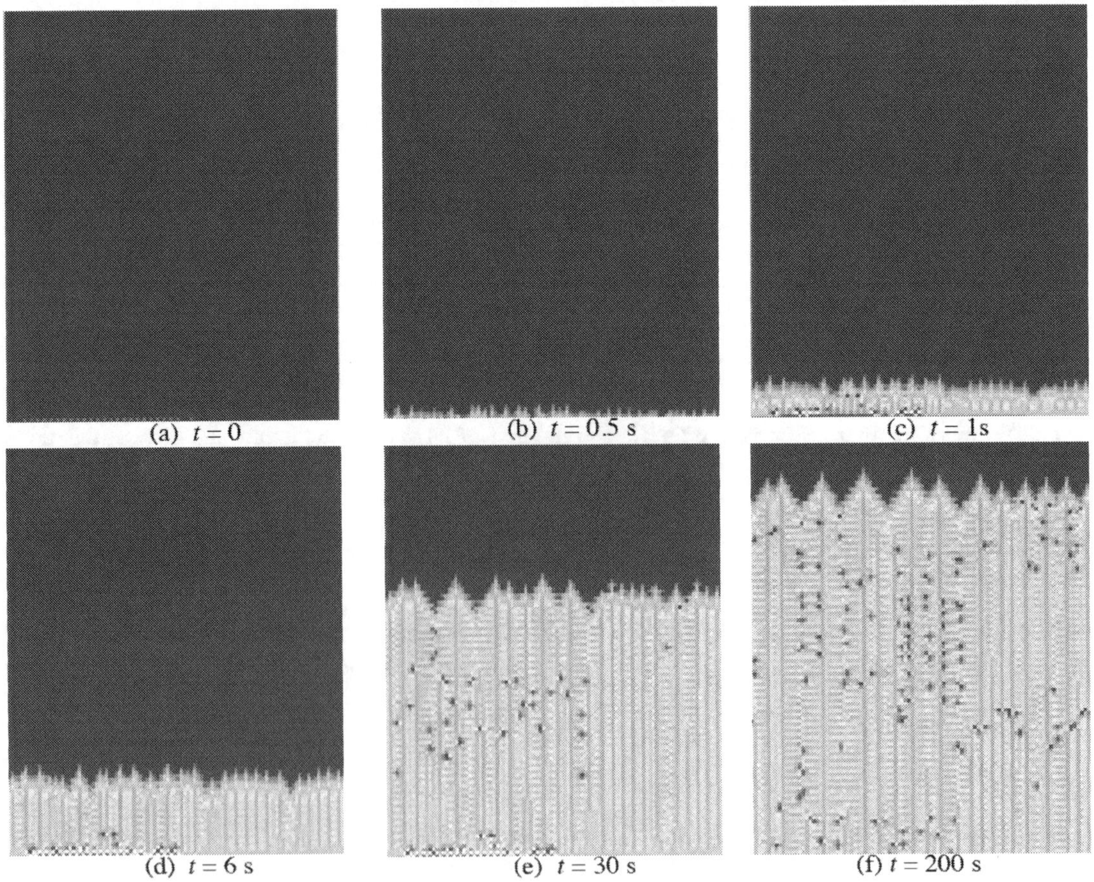

Figure 2. Solid fraction (0-0.1(red)) of columnar growth simulated with the dendrite envelope model on a low resolution mesh (domain width: 2 cm wide, grid size: 0.2 mm).

The concentration field near the dendrite tips is shown in Fig. 4. In the inter-dendritic region the equilibrium concentration (~0.266) is reached at a distance of about 3 mm from the liquid/envelope surface, at larger distance solidification is thermodynamically stopped. The model predicts the same decreasing length for both high and low resolution simulations.

After an initial transient regime, the front made of envelope tips reaches a relatively stable velocity (~360 μm/s). Some strong fluctuations exist especially during the elimination of trunks due to competition between the tips. Around 20% difference exists in the predicted steady tip velocity between the coarser and finer mesh size simulations. The main reason lies in the fact that the domain used in the simulation with the finer mesh is probably not long enough to reach a final steady state. Another problem is the correct estimation of the concentration field ahead of the tip, which in the present model depends on the solidification rate within the first cell containing the envelope surface. The concentration of interest is at a distance $\delta \approx D_l / V_t$ of the tip, which depending on the mesh accuracy is close but not strictly equal to the concentration given by the volume averaged equation (Eq. 7). The ratio of the envelope to the microscopic tip radius R_E/R_T is typically in the range of 120-250 for the coarser mesh and in the range of 40-60 for the finer mesh.

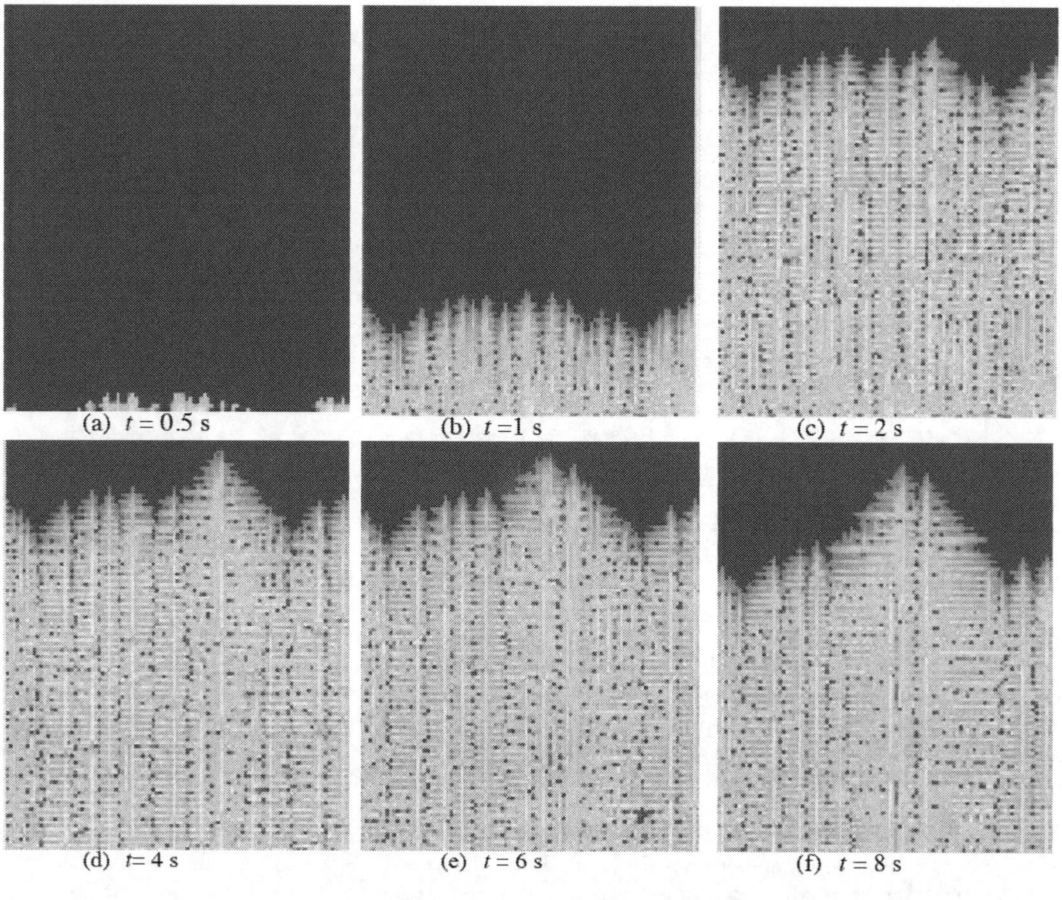

Figure 3. Solid fraction (0-0.2 (red)) of columnar growth simulated with the dendrite envelope model on the higher resolution mesh (domain width: 5 mm cm wide, grid size: 0.05 mm).

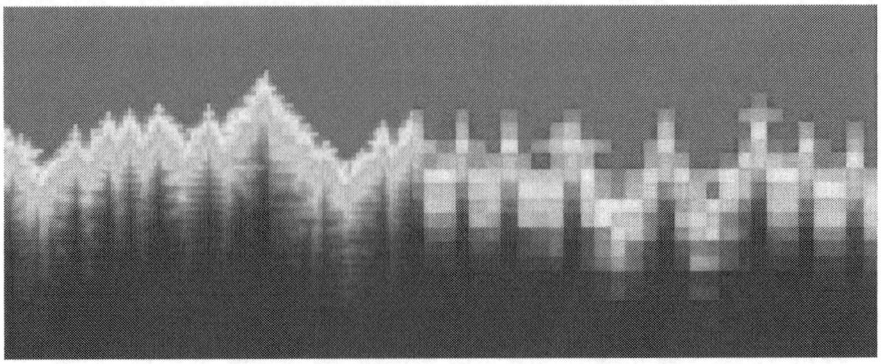

Figure 4. Zoom on a typical concentration field near the tips of the dendrite envelopes predicted with high (left) and low resolution (right) simulations (0.266 (blue), 0.3 (red), domain height: 4 mm).

Table 2. Summary of some quantitative results.

	Coarse mesh	Fine mesh
Mesh size	0.2 mm	0.05 mm
Calculation domain	20x100 mm^2	5x20 mm^2
Dendrites trunk solid fraction	7 %	13%
Steady tip velocity	460 µm/s	360 µm/s
Final λ_1	0.6-3 mm	0.4-5 mm

4. Final remarks and conclusions

We presented a dendrite envelope approach based on the rescaling of the microscopic growing laws to desired scale of simulation. This is an alternative approach based on the rescaling of the microscopic growing laws to desired scale of simulation. The aim is to be able to simulate the generation and the growth of primary arms for meshes where the grid size is much larger than the tip radius but still smaller than the expected primary arm spacing, λ_1. This model is inspired from the methodology of the turbulence models that filter the scales that are smaller than the simulation grid size, such the Large Eddy Simulation (LES) model. The elimination of the small scales from the Navier–Stokes equations reduces the computational cost of the simulation with respect to the Direct Numerical Simulation (DNS). The latter resolves every scale of the solution, is prohibitively expensive for almost all geometry or flow configurations. In the LES model the equations are transformed by using a Sub-Grid Scale (SGS) model which implicitly accounts for the eddies smaller than the filter width (grid size). A large number of specific SGS grid-dependant models was built according to the turbulence theory and the available computational resources. Only the large scales of the flow field are resolved allowing better fidelity than alternative approaches based on the Reynolds-Averaged Navier-Stokes (RANS) methods. It is noteworthy that most of the results of LES are grid dependent, however, depending on the capability of the sub-grid scale model, statistical grid independence can be achieved.

If a parallel can be made between the field of dendritic solidification and turbulence, the contrast between the phase-field/sharp interface and the volume-averaged methods can be represented by the contrast between the DNS and RANS methods. Just as for the LES with turbulent eddies, the SADE approach models the smallest (and most expensive) scales of the solution, rather than resolving them as phase-field does. The simulated dendrites are grid-dependant. However if simulations on two different meshes are performed, only the mean and higher order statistics of results must be compared. In the present work we have focused on the largest scale of the variations of the solid fraction (i.e. λ_1) and the concentration field. If the mesh is sufficiently small, the ratio $f_s^0 = R_t / R_E$ will tend towards 1, the SADE model transforms into a classical sharp-interface method for dendritic growth.

If we persist in the comparison between LES and the present model, a great effort should be put on the development of two sub-grid models, the first is for the growth and competition of the unresolved arms, and the second for an accurate estimation of the thermo-solutal conditions ahead of the dendrite tips.

Acknowledgement

This work was supported by FWF Austrian Science Foundation (P 22614-N22) for which the authors kindly acknowledge.

References

[1] Steinbach I, Beckermann C, Kauerauf B, Li Q and Guo J 1999 *Acta Metall.*, **47** 971
[2] Steinbach I, Diepers H J and Beckermann C 2005 *J. Cryst. Growth* **275** 624
[3] Delaleau P, Beckermann C, Mathiesen R H and Arnberg L 2010 *ISIJ International* **50** 1886
[4] Tourret D and Karma A 2013 *Acta mater.* **61** 6474
[5] Kharicha A, Stefan-Kharicha M, Wu M and Ludwig A 2012 *IOP Conf. Series: Mater. Sci. Eng.* **33** 012115
[6] Kharicha A, Stefan-Kharicha M, Ludwig A and Wu M 2013 *Metall. Mater. Trans. A* **44** 650
[7] Kharicha A, Stefan-Kharicha M, Ludwig A and Wu M, 2013 *Metall. Mater. Trans. A* **44** 661
[8] Stefan-Kharicha M, Kharicha A, Wu M and Ludwig A 2014 *Fluid Dyn. Res.* **46** 041424
[9] Cantor B and Vogel A 1977 *J. Cryst. Growth* **41** 109
[10] Couder Y, Maurer J, González-Cinca R and Hernández-Machado A 2005 *Phys. Rev. E* **71** 031602
[11] Wu M and Ludwig A 2007 *Metall Mater Trans A* **38** 1465
[12] Hansen G, Liu S Z, Lu S and Hellawell A 2002 *J. Cryst. Growth* **234** 731
[13] Liu S, Lu S-Z and Hellawell A 2002 *J. Cryst. Growth* **234** 751

Evaluation of dendrite morphology using fractal dimension and dimensionless perimeter in unidirectionally solidified Al-Si Alloys

K Ohsasa, Y Natsume, T Sekiya and T Hatayama

Akita University, Tegata Gakuenmachi, Akita, 010-8502, Japan

Email:sasa@gipc.akita-u.ac.jp

Abstract. The dendrite morphology of unidirectionally solidified Al-Si alloys was evaluated by measuring the fractal dimension and dimensionless perimeter of dendrites. In an unidirectional solidification experiment, columnar crystals grew from a bottom chill and columnar to equiaxed transition (CET) occurred at the upper part of an ingot. Then, equiaxed crystals were formed at the top of the ingot. Different dendrite morphology was observed in longitudinal, transverse and oblique sections, however, the fractal dimension or dimensionless perimiter of the dendrites in the sections with same local solidification time showed same values, and continuously decreased with increase in the local solidification time through columnar, CET and equiaxed regions. It can be considered that the fractal dimension and dimensionless perimiter of dendrites are controlled by local solidification time and irrespective of dendrite morphology. This result demonstrated the potential of the fractal dimension and dimensionless perimiter as a parameter for estimating local solidification time of an ingot in which the measurement of SDAS is difficult.

1. Introduction

A dendrite structure is the most frequently observed structure during the solidification of an alloy. Dendrite pattern characterize the micro segregation pattern and mechanical properties of materials. Hence, quantitative evaluation of dendrite morphology is very important for predicting the properties of mushy zone composed of dendrite arrays such as interdendritic fluidity, i.e. the permeability. To evaluate the dendrite morphology, secondary dendrite arm spacing (SDAS) is commonly used for a columnar structure, however, the measurement of SDAS is relatively difficult because a specimen must be cut parallel to the growth direction of columnar dendrites. For equiaxed crystals, the measurement of SDAS is usually difficult and the grain size is often used to evaluate the morphology of the equiaxed structure. Development of a unified method is desirable to evaluate the dendrite morphology of both columnar and equiaxed crystals.

It is known that complex dendrite pattern exhibits self-similar structure, i.e. the fractal structure. Several attempts have been carried out for applying the fractal theory to describe the morphology of materials structure[1-4)]. Yang *et al.*[4)] calculated the fractal characteristic of dendrite and cellular structures of a Ni-based superalloy under various cooling conditions. Sanyal *et al.*[5)] described the mushy zone of an alloy as a network of continuous fractal structure, and they calculated the permeability of the mushy zone based on the fractal dimensions of the dendrite network. Fractal dimension may be an effective parameter for representing the complexity of dendrite array of an alloy in comparison with the SDAS. The authors also tried to use fractal dimension for evaluating the morphology of dendrite structure[6-8]. In addition, the authors also tried to use a new parameter "dimensionless perimeter" to evaluate the dendrite morphology and estimated the permeability of the mushy zone based on a phase-field simulation[6-8].

The first aim of this study is to measure the fractal dimension and dimensionless perimeter of experimentally observed real dendrites and second aim is to examine the characteristics and possibility of the fractal dimension and dimensionless perimeter for evaluating the dendrite morphology of solidified alloys based on a unidirectional solidification experiment.

2. Experiment

Al-3, 5 and 7mass%Si alloys were used for the unidirectional solidification experiment. A mold made of an adiabatic material with 30mm in inner diameter and 200mm in height was preheated to 1000K, then the mold was placed on a water-cooled chill box. A molten Al-Si alloy with super heat of 100K was poured into the mold. Temperatures were measured by six thermocouples set at the center of the mold along with the distance from the water-cooled chill surface. Local solidification time at each position were obtained from recorded cooling curves. The local solidification time deceased with increase in the distance from the chill surface. After the completion of the solidification, an ingot was cut at transverse section and a half of the ingot was etched by an acid reagent and macro grainstructure was revealed. Other half ingot was cut into several pieces and polished, etched and dendrite structures at longitudinal, transverse and oblique sections were revealed. Experimental apparatus was shown in figure.1.

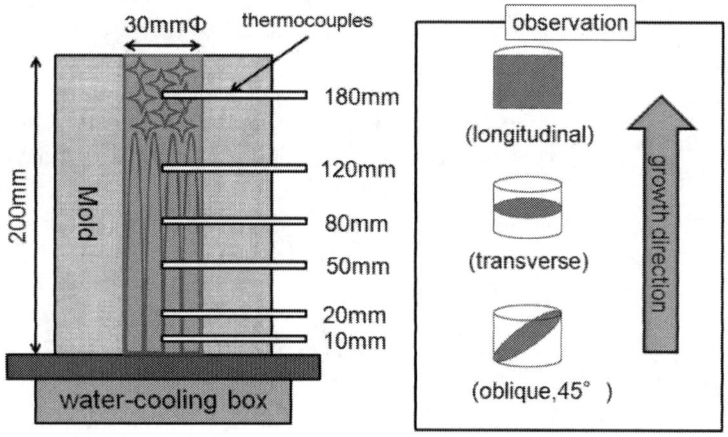

Figure 1. Experimental apparatus and microstructure observation procedure.

The fractal dimensions of observed dendrites at each section were measured by using the box counting method[9]. Procedure of the box counting method is as follows; an area including dendrites is divided into square boxes with the size of r. Then, the number of boxes, N, in which the S/L interface is included, is counted. Then, the size of the box, r, is changed and the same procedure is repeated. If following relationship holds true between the number of the boxes, N, and the size of the box, r, the geometry of dendrite shows fractal characteristic.

$$N(r) = r^D \quad (1)$$

The exponent, D, in the equation (1) is called "fractal dimension" and its value can be obtained from the slope of the plot line of log r vs. log N. In the measurement of the fractal dimension, 8 box sizes were used and total number of boxes was changed from 32 to 42658 during the measurement.

In previous works by the authors, a new index "dimensionless perimeter" of dendrites was introduced for evaluating the morphology of dendrites based on a 2D phase-field simulation[6-8].

"dimensionless perimeter" was calculated as follows; first, the perimeter of a simulated dendrite L_1 is measured. Then the perimeter L_2 of a circle, whose area is same as the simulated dendrite, is measured. The "dimensionless perimeter" of the dendrite was defined as the ratio of L_1/L_2. The dimensionless perimeter increases when dendrite morphology becomes complex. Therefore, dimensionless perimeter is another index for showing the complexity of dendrite shape. The measurement of the perimeter is easy for a simulated dendrite, however, the measurement of the perimeter is difficult for a experimentally observed dendrite at the cutting surface of a sample.

In this study, we extended the definition of "dimensionless perimeter" to real structures observed in the cutting surface of an sample as shown in figure.2.

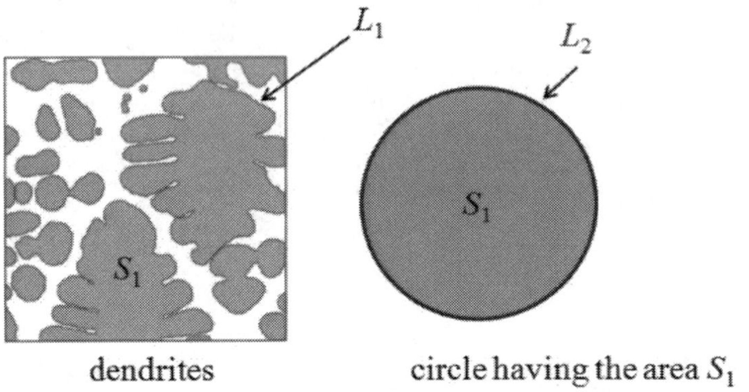

$$L = \frac{L_1}{L_2}$$

S_1 : whole area of primary dendrites
L_1 : whole perimeters of primary dendrites
L_2 : perimeter of the circle having the area S_1

Figure 2. Procedure for measuring dimensionless perimeter of dendrites in an experimentally observed structure.

Whole perimeters of dendrites observed in the section of a sample were measured and summed up to L_1. Whole areas of primary dendrites were measured and summed up to S_1. Then the perimeter L_2 of the circle, whose area is same as S_1, is measured. The "dimensionless perimeter", L, of dendrites is defined as the ratio of L_1/L_2.

3. Results and discussion

3.1 Solidified structures

Figure 3 shows solidified structures of a unidirectionally solidified Al-5mass%Si alloy ingot. Grain macrostructure is shown in the left side of the figure 3. Columnar crystals grew from a bottom chill and columnar to equiaxed transition (CET) occurred at the upper part of the ingot. Dendrite structures at lomgitudinal and transverse sections are shown in the right side of the figure 3. In the longitudinal section of a columnar structure region, columnar dendrites and secondary dendrite arms can be clearly seen. On the other hand, in the transverse section, dendrite morphology seems to be equiaxed structure. In the CET region, both columnar and equiaxed dendrites are observed at the longitudinal section, however, only equiaxed morphology is observed at the transverse section. In the equiaxed region, only equiaxed morphology is observed both at the longitudinal and transverse sections. It is clear that different morphology is observed depending on the observed section in a sample.

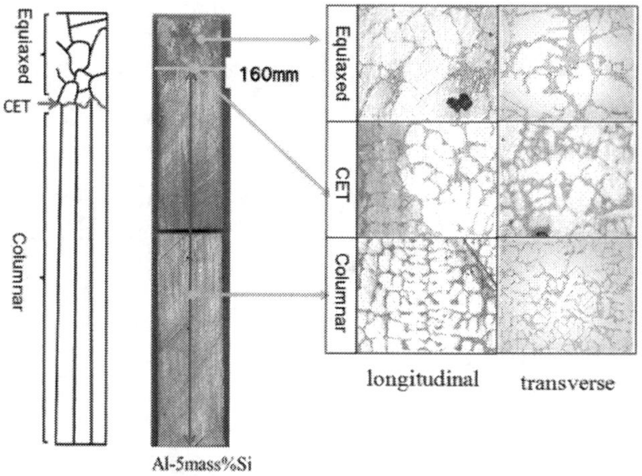

Figure 3. Structures of a unidirectionally solidified Al-5mass%Si alloy ingot.

3.2 Fractal dimension and dimensionless perimeter of dendrites

Figure 4 shows changes in the fractal dimension (Fig.4 (a)) and dimensionless perimeter (Fig.4 (b)) along with local solidification time (corresponds to the distance from the chill). Dendrite structures at longitudinal, transverse and oblique sections at the local solidification time of about 55s (positions enclosed by a circle in upper figures (a) and (b)) are shown in figure.4(c).

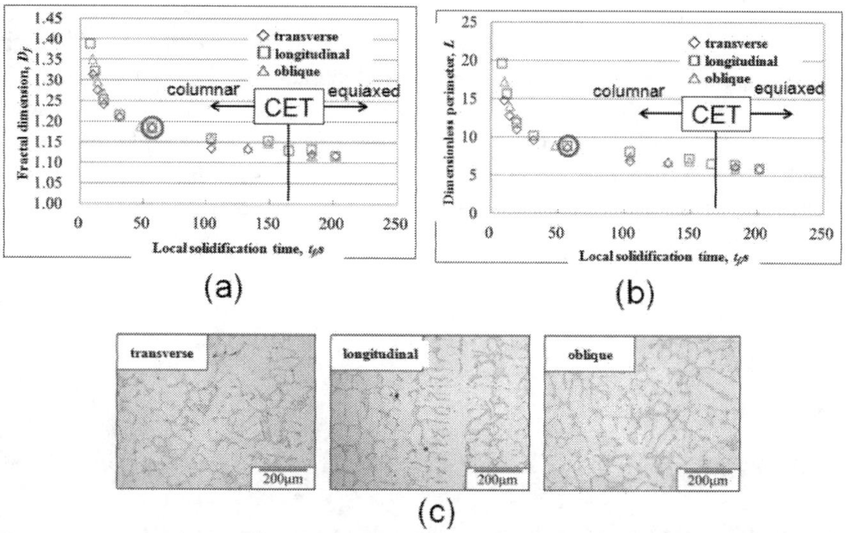

Figure 4. Changes in the fractal dimension (a) and dimensionless perimeter (b) of dendrites along with local solidification time in Al-5mass%Si alloy. Lower photographs show the dendrite structures of longitudinal, transverse and oblique sections at the local solidification time of about 55s (positions enclosed by a circle in upper figures (a) and (b))

Fractal dimension and dimensionless perimeter continuously decreased with increase in local solidification time through columnar, CET and equiaxed regions. In the figure (c), different dendrite morphology is observed in each section. However, in spite of different morphology, it is clear that the

IOP Conf. Series: Materials Science and Engineering **84** (2015) 012033 doi:10.1088/1757-899X/84/1/012033

fractal dimensions or dimensionless perimeters of each section at the same local solidification time show same values. This result demonstrates that local solidification time can be estimated by measuring fractal dimension or dimensionless perimiter in an arbitrary section, even in a case that the measurement of SDAS is difficult.

3.3 Relation between fractal dimension and SDAS

Secondary dendrite arm spacing (SDAS), λ_2, can be measured only at the longitudinal section in a columnar dendrite region. Figure 5 shows relationship between fractal dimension, D_f, and SDAS, λ_2, measured in the columnar dendrite region in the unidirectionally solidified Al-5mass%Si alloy ingot.

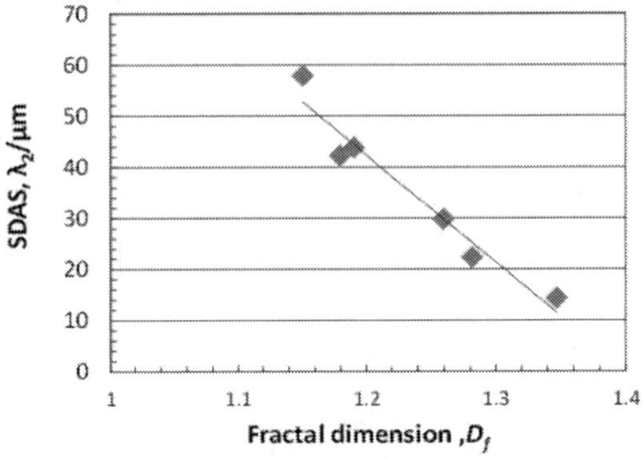

Figure 5. Relationship between fractal dimension, D_f, and secondary dendrite arm spacing (SDAS), λ_2, measured at the columnar region of the longitudinal section in the unidirectionally solidified Al-5mass%Si alloy ingot.

Negative-correlation is shown between the fractal dimension, D_f, and secondary dendrite arm spacing (SDAS), λ_2. It seems that the measurement of the fractal dimension of dendrites is relatively easy in comparison with the measurement of the SDAS. By obtaining a relationship between fractal dimension and SDAS, SDAS may be presumed from the fractal dimension even in a case that the measurement of SDAS is difficult.

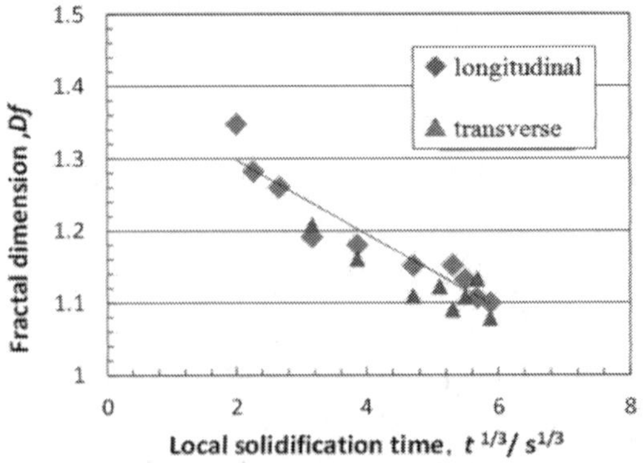

Figure 6. Relationship between cube root of local solidification time, $t^{1/3}$ and fractal dimension, D_f in the Al-5mass%Si alloy ingot

Figure 6 shows the relationship between the cube root of local solidification time, $t^{1/3}$, and fractal dimension, D_f, in the unidirectionally solidified Al-5mass%Si alloy ingot. Negative-correlation is realized between cube root of local solidification time, $t^{1/3}$, and fractal dimension, D_f. It is well known that the secondary dendrite arm spacing SDAS, λ_2, is proportional to the cube root of local solidification time[10-11)]. Negative-correlation is shown between fractal dimension, D_f, and SDAS, λ_2, as shown in the figure.5. Hence, the relationship between the cube root of local solidification time, $t^{1/3}$, and fractal dimension, D_f, shown in the figure 6 can be understood. The results shown in the figures 5 and 6 suggest that there a close relationship between fractal dimension, D_f, and SDAS, λ_2. Further study will be desired.

3.4 Effect of initial solute content

Figure 7 shows changes in the fractal dimension (Fig.7 (a)) and dimensionless perimeter (Fig.7 (b)) along with local solidification time in unidirectionally solidified Al-3, 5 and 7mass%Si alloys. Dendrite structures in the longitudinal section at the solidification time of about 100s (positions enclosed by an ellipse in upper figures (a) and (b)) are shown in figure.7 (c).

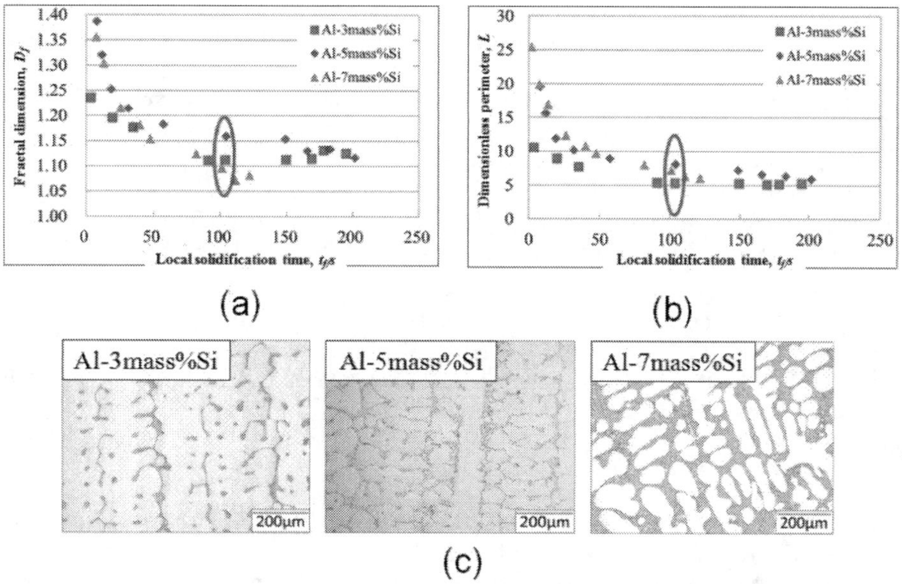

Figure 7. Changes in the fractal dimension (a) and dimensionless perimeter (b) of dendrites along with local solidification time in Al-3, 5 and 7mass%Si alloys. Lower photographs show the dendrite structures in the longitudinal section at solidification time of about 100s (positions enclosed by an ellipse in upper figures (a) and (b)).

Fractal dimension and dimensionless perimeter basically increased with increase in initial solute content. This result means that the dendrite morphology of an alloy with higher solute content is more complex than that of an alloy with lower solute content in the case if solidification time is same. This result can be explained based on the "constitutional undercooling". The degree of the "constitutional undercooling" of an alloy with higher solute content is larger than that of an alloy with lower solute content. High degree of the "constitutional undercooling" makes *S/L* interface unstable and produces more complex dendrite morphology. However, the fractal dimension of Al-7mass%Si alloy decreased quickly along with increase in local solidification time and shows lowest value when local solidification time becomes longer. This result may be attributed to following two reasons. First reason is that the range of fractal dimension in two dimensional is relatively narrow from 1 to 2. Hence, the effect of error arising from the measurement of fractal dimension is relatively high in comparison with that of the dimensionless perimeter. The range of the dimensionless perimeter is relatively wide from 5 to 30 as shown in the figure.7. Therefore, it is expected that the effect of the error arising from the measuring process of the dimensionless perimeter is relatively small. Second reason may be considered as follows. The dendrite morphology of Al-7mass%Si alloy shown in the figure 7 (c) is relatively simple. The amount of eutectic of Al-7mass%Si alloy is largest of about 49% in the three alloys. The volume of liquid surrounding primary crystals may affect primary dendrite morphology. Further study about this will be desired.

It should be noted that the permeability of solidifying alloys was estimated[6-8] by regarding the dimensionless perimeter as the tortuousity factor[12] of interdendritic liquid channels. Relatively high value of the dimensionless perimeter made such calculations possible. From the point of view for engineering applications, it seems that dimensionless perimeter has more advantage than fractal dimension due to its relatively higher value in comparison with that of fractal dimension.

4. Conclusions

Unidorectional solidification experiments were carried out for Al-3, 5, and 7mass%Si alloys. Fractal dimension and dimensionless perimeter of dendrites observed at longitudinal, transverse and oblique sections in the solidified ingot were measured along with the distance from the bottom chill. Obtained results are summarized as follows;

1) Different dendrite morphology was observed at longitudinal, transverse and oblique sections, however, the fractal dimension or dimensionless perimiter of the dendrites in each section with same local solidification time showed same values

2) The fractal dimension and dimensionless perimiter of the dendrites in each section continuously decreased with increase in local solidification time through columnar, CET and equiaxed regions.

3) Negative-correlation is shown between the fractal dimension, D_f, and secondary dendrite arm spacing (SDAS), λ_2.

4) Negative-correlation is realized between cube root of local solidification time, $t^{1/3}$, and fractal dimension, D_f.

5) If local solidification time is same, the dendrite morphology of an alloy with higher solute content is more complex than that of an alloy with lower solute content. However, the fractal dimension of Al-7mass%Si alloy decreased quickly along with increase in local solidification time and showed lowest value when local solidification time becomes longer.

References

[1] Kleiser T and Bocek M 1986 *Z. Metallkd.* **77** 582
[2] Hornbogen E 1987 *Z. Metallkd.* **78** 622
[3] Tanaka M J 1992 *Mater. Sci.* **27** 4717
[4] Yang Y, Xiong Y and Liu L 2001 *Sci. Tech. Adv. Mat.* **2** 101
[5] Sanyal D, Ramachandrarao P and Gupta O P 2006 *Chem. Eng. Sci.* **61** 307
[6] Ishida H, Natsume Y and Ohsasa K 2009 *ISIJ International* **49** 37
[7] Ohsasa K, Katsumi T, Sugawara R and Natsume Y 2013 MCWASP XIII IOP Publishing IOP Conf. Series: *Materials Science and Engineering* **33** 012108
[8] Sugawara R, Itoh T, Natsume Y and Ohsasa K 2013 *Tetsu-to-Hagane*, **99** 126
[9] Matsushita N 2002 "*Furakutaru no Buturi*" Shokabou 36
[10] Bower TF, Brody HD and Flemings MC 1966 *Trans. Met. Soc. AIME* **236** 624
[11] Kurz W and Fisher DJ 1998 "*Fundamentals of Solidification*" 4th Ed. Trans. Tec. Publication Swizerland 85
[12] Piwonka T S and Flemings M C 1966 *Trans. metall. Soc.A.I.M.E.* **236** 1157

IOP Conf. Series: Materials Science and Engineering **84** (2015) 012034 doi:10.1088/1757-899X/84/1/012034

How to solve complex problems in foundry plants - future of casting simulation -

I Ohnaka
Prof. Emeritus of Osaka University, ie-Solution Corporation,
Higashitoyonaka-cho, Toyonaka-shi, Osaka 560-0003, JAPAN

E-mail: ohnaka@ams.eng.osaka-u.ac.jp

Abstract. Although the computer simulation of casting has progressed dramatically over the last decades, there are still many challenges and problems. This paper discusses how to solve complex engineering problems in foundry plants and what we should do in the future, in particular, for casting simulation. First, problem solving procedures including application of computer simulation are demonstrated and various difficulties are pointed-out exemplifying mainly porosity defects in sand castings of spheroidal graphite cast irons. Next, looking back conventional scientific and engineering research to understand casting phenomena, challenges and problems are discussed from problem solving view point, followed by discussion on the issues we should challenge such as how to integrate huge amount of dispersed knowledge in various disciplines, differentiation of science-oriented and engineering-oriented models, professional ethics, how to handle fluctuating materials, initial and boundary conditions, error accumulation, simulation codes as black-box, etc. Finally some suggestions are made on how to challenge the issues such as promotion of research on the simulation based on the science-oriented model and publication of reliable data of casting phenomena in complicated-shaped castings including reconsideration of the evaluation system.

1. Introduction

Recent rapid development of computer aided engineering(CAE) has been transforming the heavily experience-dependent foundry technology into a modern and science-based one. Modelling and simulation such as those presented in the MCWASP are crucial elements in the CAE in the foundry industry and they have been progressing toward ICME(Integrated Computational Materials Engineering), by adding multi-scale-analysis for estimation of solidification structure and performances, and by combining various process simulations such as molding and heat treatment, etc.

Although we have now practical commercial casting simulation codes which have been widely used in the foundry and its related industries, foundry engineers still have many difficulties to solve their engineering problems. This paper discusses those problems including the application of casting simulation. Further, the conventional research works are looked back and what we should do in order to solve the challenges and problems are discussed.

2. Procedures to solve the problems and application of computer simulation

First, let us look back how we are now solving real problems in plants and extract issues on the problem solving. As an example of real problems, porosity defects in already produced sand castings of spheroidal graphite cast irons are selected in order to get a clear picture.

When occurrence of porosity defects is reported from the inspection section or customers, the engineers should solve the problem as quickly as possible. The procedures could be as follows and should be carried out with imagining various causes avoiding specific preconception. They include 5W-2H, namely Which casting, When, Where, What kind, Who involved, How and How many. Note that the procedures after finding suitable measures to solve the problem are omitted here because of the limited space.

Step 1- Investigate the appearance and morphology of defects, defect sites, and frequency of the occurrence

The size, morphology(spherical, irregular, inside smoothness and colour, etc), remains in the defects and others are observed and measured by naked eyes or a stereomicroscope, scanning electron-micrometer, etc, depending on time, cost and resources. If possible, it is better to measure gas components in the defects. The dye penetrant testing, ultrasonic testing, etc can be used to find the defect sites. Further, identify when the defects occurred, if possible, the date and time.

Step 2 - Investigate the change of manufacturing during the period while the defects appeared

It is helpful to use the effect-cause diagram as shown in figure.1. In actuality it is better to use more detailed effect-cause diagrams specialized for various porosity defects based on engineer's own experiences and information obtained from others. Although the following includes many items to check, they should be limited to important ones by considering the results of step 1.

a) Check charts and numerical values of control items

This includes grade and amount of base metals, chemical composition of inoculants, chemical composition, gas content and carbon equivalent of the melt, consumption rate of new sand and binders, total clay or dust content, moisture content in the mixture, permeability, etc.,

b) Check the behaviour of workers or operators and try to see the change in operation, particularly pouring. Careless slag removing, discontinuous pouring, taking no care to keep constant pouring temperature, etc. cause the defect problems.

c) Check machine or equipment change, maintenance situation such as part-change frequency, and wear and deformation of pattern and molding flask, precision control of measuring instruments, etc.

d) Check ambient change such as weather, temperature and humidity.

Step 3 - Identify the defect type and causes

Identifying the defect type is very important, because it is closely related to their causes and measures. The porosity defects are classified into shrinkage porosity, gas porosity, hot tearing and cracks. Gas defects are sub-divided into blows or gas holes, pinholes, blisters and others. There are many handbooks[1] explaining their appearance, morphology and causes. We should use such conventional knowledge and experiences to differentiate those defects and find causes.

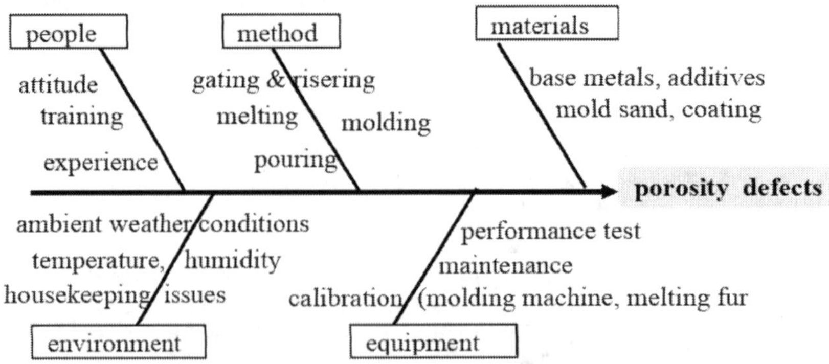

Figure 1. An example of effect - cause diagram for porosity defects

IOP Conf. Series: Materials Science and Engineering **84** (2015) 012034 doi:10.1088/1757-899X/84/1/012034

For example, if they are jagged holes or spongy with dendrites, they are usually shrinkage porosity. Although cavities with smooth, clear and bright inner surface are usually considered as gas defects, sometimes it could be shrinkage cavity generated due to melt or solidification shrinkage at early stage of solidification. However, if the gas content in the melt is higher than a critical value and the shape is spherical or elongated, they are probably gas defects. It should be noted that the gas porosity tend to be located in heavy section or the change of section, causing confusion between shrinkage and gas porosities.

Anyway the identification of detailed gas porosity type except the hot tearing and cracks is usually difficult, and requires integration of existing knowledge and experiences. Further, we should aware that the text books or handbooks are not always correct or do not include the latest research results.

The causes are speculated form consideration on the defect type and the changes found in step 2.

If clear changes are found for a few factors, it is rather easy to solve this problem. For example, if moisture content in the sand mixture was higher than the normal value and the defects look like gas porosity, the cause could be the entrapment of gas generated from the mold. However, we often cannot find noticeable changes for the defects. One of the reasons is that continuously recorded data are limited even for daily operation.

Step 4 - Application of casting simulation

It is useful to apply casting simulation not only for identifying the causes of the defects but also for finding solutions or measures.

1) The case that the simulation has already performed.

1.1) Compare the simulation results with observed defects sites.

If the agreement is not bad, repeat the simulation to find proper conditions which could eliminate the defects. For example, if the defects site is in a heavy section and the defect seems shrinkage porosity, re-run the simulation with changing the risers, cavity dimensions, chills or pouring temperature depending of situation considering available time and cost, etc. The modulus method [2] is useful in this case but the other cases require much more experiences.

If the agreement is bad, re-simulate with revised input data as follows:

a) Check the change of casting condition, gating and risering systems and cavity dimensions.

They could have been changed from the original ones due to various reasons such as defect measures, wear, repair and customer's request.

b) Check what kinds of functions in the simulation code were used.

This requires to imagine the cause of the defects. If the defects look like gas defects and melt gas content is a bit larger than usual, we should select the function which can consider the gas content.

c) Check the change of pouring method, pouring temperature, pouring speed and mold including mold coating.

The real pouring temperatures usually vary with time and date even for the same casting. In particular, if a manual pouring is used, it is often necessary to observe the pouring operation such as pouring interval and if there is a possibility of pouring temperature change, it is better to measure it.

d) Check other input data.

These include physical properties of melt and mold, heat resistance between the casting and the mold, shrinkage coefficients, critical solid fraction for melt flow and criteria used to estimate defects such as Niyama criterion[3], etc.

One of the practical ways to evaluate the properness of the initial and boundary conditions and physical properties is to compare the simulated and measured temperatures.

e) Re-run the simulation

f) Evaluate the results and consider next steps

These require experiences and a comprehensive knowledge about casting including the defect formation mechanisms. If the causes of the defects and measures are found, then proceed to trial casting with the revised casting conditions.

The following is one example of practical ways to evaluate the simulation results:

i) The case where the defect sites are near the observed ones

Proceed to find proper conditions which eliminate the defects as mentioned above.

ii) The case where the defect sites are far from the observed ones.

Check the input data and re-run the simulation. In particular, attentions must be paid to the thermal conductivity of the sand mold and boundary conditions, because they often change the defect sites.

If the difference can be explained by using imagination, then it is often useful to go further, namely, to find solutions by using the simulation supplementing with imagination. For example, if the real defects are located above the estimated one, it is possible that the difference was caused by floating-up of gas bubbles or thermal convection during solidification. If such phenomena were considered in the simulation, check the input data and re-run the simulation.

iii) The case where the simulation shows no clear defects.

If defect criteria were used, re-run the simulation with revised ones.

iv) The case where the simulation codes cannot predict the defects.

Even in this case, some defects can be estimated from the simulation. For example, gas or dross entrapment, which are related to the gas porosity defects, could be judged from melt flow behaviour during the mold filling.

2) The case that the simulation has not yet been performed.

Perform the simulation with input data proper to the situation where the defects occurred, following the procedures such as shown below:

2-1) Select proper functions in the simulation code available and run it.

2-2) Evaluate the results and proceed to next steps

These are the same as those in the item b and f in *1.1)* mentioned above.

3. Difficulties in the application of simulation

More detailed discussions on some of the following are available in [4].

3.1 Getting proper input data

3.1.1 Physical properties.

Although rather good data bases exist for metallic materials, usually slightly or considerably different materials are used in plants. Therefore, it is often difficult to correctly judge the input property data, in particular, which are sensitive to minor elements, even when data on materials with similar composition exist. Further, there are limited reliable data on the relationship between temperature and fraction solid. In particular, in the case of the porosity defects problem in spheroidal graphite cast irons, it is often not good to use a constant relationship even when the chemical compositions are almost the same, because the graphite precipitation is strongly affected by melt treatment including inoculation. The solidification and liquid shrinkage data of practical alloys are also not enough, particularly for cast irons. There are many physical properties which cannot be determined only by chemical composition. These include surface tension of melt, wetting angle between the melt and mold surface, etc.

The properties of sand mold vary from plant to plant and with time, because the density and chemical composition of sand mold are different. For example, because the thermal conductivity is affected by apparent density, it varies with molding condition and position in mold. Further, gas evolution in mold and condensation also affect the thermal and mechanical properties.

3.1.2 Initial and boundary conditions.

The initial conditions such as pouring temperature and speed, mold surface conditions vary with time and affect the mold filling and solidification.

The heat resistance between casting and mold varies with melt flow, gas evolution, deformation of solidified region, etc. Although heat resistance data have been reported, it is not sure if the data can be generally used, because the casting condition and the cavity shape, etc. are often different. Therefore usually we use the values with better fitting with measured temperatures or those provided by the data base of simulation code. However, it is not sure if the value used are correct or not.

IOP Conf. Series: Materials Science and Engineering **84** (2015) 012034 doi:10.1088/1757-899X/84/1/012034

Further, although estimation methods of the thermal resistance have been reported[5, 6], it is not easy to apply them to real casting, in particular, for sand molds.

3.1.3 Input parameters.

Parameters such as critical solid fractions for the liquid flow through dendrites or mass feeding vary with the shape and distribution of the solid phase which are affected not only by the chemical composition but also by fluid flow and solidification condition and inoculation. In particular, defect estimation criteria such as Niyama criterion vary not only the chemical composition but also with the shape and size of castings, etc. Therefore, the selection of such parameters often requires experiences or research.

In particular, the porosity estimation in spheroidal graphite castings by such criteria is not easy, because of the expansion of graphite precipitation. Although we have developed a simulation code where the defect is estimated by tensile stress in mushy region[7], it requires an empirical critical stress depending on cast alloy and mold strength.

3.2 Selection of proper functions in simulation code

Selecting proper functions requires a comprehensive knowledge about the identification of the defects and their formation mechanisms. Unfortunately or fortunately there are currently not many functions available. For example, there exist few codes having functions which can consider the effects of the change in mold coating, mold additives or gas contents in melt. Therefore currently it is not a big issue but it would be a mater in the future when various functions become available.

Further, it is often not easy to judge the degree of properness of the functions from user's manual of simulation codes. This is because the details of functions are usually not disclosed. For example, the "function of gas entrapment" is greatly affected by how to treat the gas phase at the mold / melt interface including estimation of contact site of melt on the mold, and formation of gas bubbles and their size, and coalescence of bubbles, etc. Since such details are usually not disclosed, it is difficult to theoretically evaluate it. Even if they have been disclosed, the accuracy cannot be evaluated without comparing the results with reliable data. Unfortunately reliable data are difficult to get for many cases such as the gas entrapment. Further, entrapment of gas generated from mold requires the data on gas generation which vary with chemical composition.

3.3 Evaluation of simulation results and modelling

Although simulation codes have to be verifies and validated, the verification is almost impossible, because the real phenomena which the simulation aims to realize haven't been known. Therefore, we usually just compare the simulation results with limited measured data and say the code has been validated if they agree.

However, it does not guarantee that the accuracy of the simulation code is good enough or not for the problems to be solved. This is because that the data used for the validation have some errors caused by measurement, errors of input data and fluctuation of real phenomena. Further, it is not sure if the code has the equivalent accuracy for other cases which are different from those used in the validation.

It is much more difficult to evaluate the accuracy of the through-process simulation or the ICME, because the impact of error accumulation on the final results hasn't been well known. One possible way to evaluate the accuracy might be a sensitivity analysis but it becomes more difficult if the number of input data increases.

Although the error due to modelling may become smaller in the future because of its improvement, the error due to input data such as the initial and boundary conditions may not decease as shown in figure 2.

3.4 There are many cases where the application of the simulation is not enough to solve problems

In reality there are many cases where the application of simulation is not so useful to find proper measures against the problems, though we cannot say it is useless, because at least the simulated mold filling and solidification sequences are more accurate than those estimated from experiences.

The reasons why the simulation fails to find the causes comes mainly from so many things related to the problems. For example, the scrap ratio due to casting defects usually varies with time as shown in figure 3. Namely it increases gradually or abruptly and decreases with time by various measures taken. It is rather easy to identify the real cause if dominated changes occurred in a few related matters such as mentioned in *Step 3* of section 1. Even in this case, it is still difficult to solve the problem with simulation codes because they often don't have enough functions. In many cases, various possible measures are taken at the same time, because dominated change cannot be found or the time to solve the problems are very limited, resulting in fail of identifying clear causes of the problems. Such situation is repeated and similar defects repeatedly occur. Particularly, if defects suddenly appeared and disappear in a short period, it is very difficult to identify the cause because the same condition cannot be recreated.

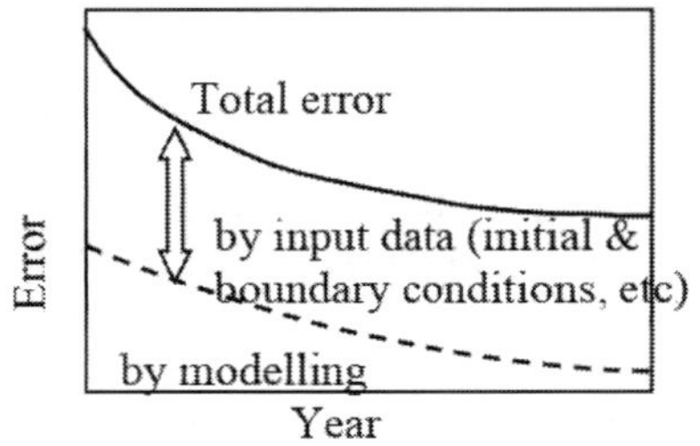

Figure 2. Expected simulation error

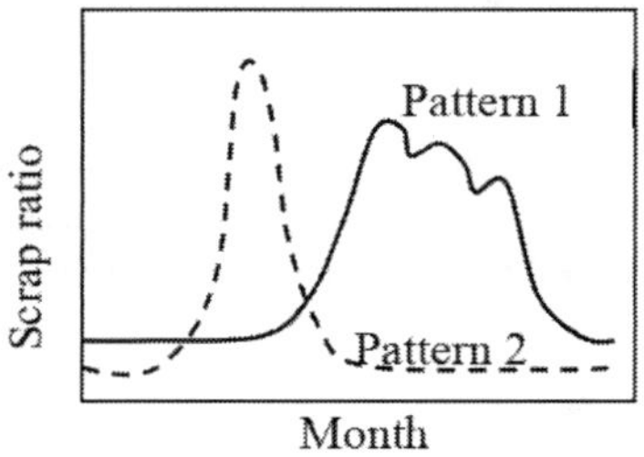

Figure 3. Scrap ratio change with time

4. Reflection of conventional research on casting

A great deal of effort has been put into solving engineering problems in casting by many people. Now let us look back on what we have done in the past and consider what we have to improve in the future from the point of view of problem solving.

4.1 *Flow of research*

Over the past several thousand years people have conducted engineering works, in particular, how to produce good castings mainly by trial-and-error approaches. Because the casting process consists of melting, molding, casting and post processing, and each process requires its own special technologies and skills, most of engineering works have been done individually in these processes.

From about the beginning of 20th century the scientific method has been gradually introduced in the casting field. It was natural for people to think that it is necessary to understand the casting phenomena and to be able to predict the consequences of actions taken, because it is very difficult to solve complex casting problems just by trial and error.

From around mid-20th century scientific research has been greatly increased, in particular, for solidification of metals. The scientific research means the research conducted by the scientific method, namely making hypotheses or conjectures, deriving predictions from them as logical consequences, and then carrying out experiments based on those predictions, followed by testing of those hypotheses with experimental data or observations.

The reason why solidification of metal was selected as a major scientific research subject would be that it is a fundamental of metallic materials and related to many processes including casting. Further, the testing of hypotheses is rather easy under the condition where the shape of specimen is limited to simple-shape. This is because that complicated-shape of specimen causes not only a risk of unexpected phenomena but also difficulties in realization of repeatable initial and boundary conditions, therefore, repeatability of observed phenomena or measured data, resulting in difficulty of proving the correctness of the hypotheses.

On the other hand, it is not easy to conduct such scientific research for processing and process-dependent phenomena such as casting defects, because it is usually difficult to realize clear and repeatable initial and boundary conditions. Therefore, rigorous scientific works delayed for those subjects. However, engineering science-based research has been also greatly developed from about mid-20th century by introducing physical and engineering science such as heat transfer and chemical engineering but without rigorous testing.

From about 1960 together with rapid development of computer, quantitative analysis of phenomena have been developed. Namely, first numerical calculation of temperature and fraction solid has been developed with solving energy balance equation considering the latent heat of solidification, followed by development of mold filling analysis considering moving melt front. Because the accurate estimation requires accurate input data such as thermal properties of alloys and molds, relationship between solid fraction and temperature, thermal resistance between mold and melt, measuring methods of those data have been developed and various useful data have been accumulated. Further, the understanding various phenomena such as the boundary phenomena between mold and casting, and heat and mass transfer in green sand molds have progressed.

From a little later or similar time computer simulation has been developed. The computer simulation is defined here as the reproduction of physical phenomena by using computer and models. It is not just solving established differential equations such as Navier-Stokes equation. Most of casting simulation codes are engineering products based on engineering-oriented models explained later in section 4.2.1.

Together with such rapid progress in research, various casting process and casting alloys have been developed and knowledge on casting have tremendously increased. Further, in order to get deeper understanding, research subjects have been divided into various subjects as shown in figure 4(a) and various equipments necessary to the individual research have been developed. These situations have made it difficult for a single person to handle many subjects. As a result, many disciplines, specialization and specialists have appeared.

4.2 Difficulties and barriers

Although the knowledge obtained by those engineering and scientific activities described above are now very useful to solve various problems, there are still a lot of challenges. Some of them are discussed below:

4.2.1 Science, engineering, science- and engineering-oriented models, and artistic model.

Because scientific research is now so popular that people often confuse the research as science and engineering research. However, the goal of the research as science is to establish a theory or understanding, while the goal of the engineering research is to solve real problems. Therefore as shown in figure 4(b) engineers specify the problems to solve from complex phenomena and create solutions integrating explicit and tacit knowledge obtained from research and experiences. They should

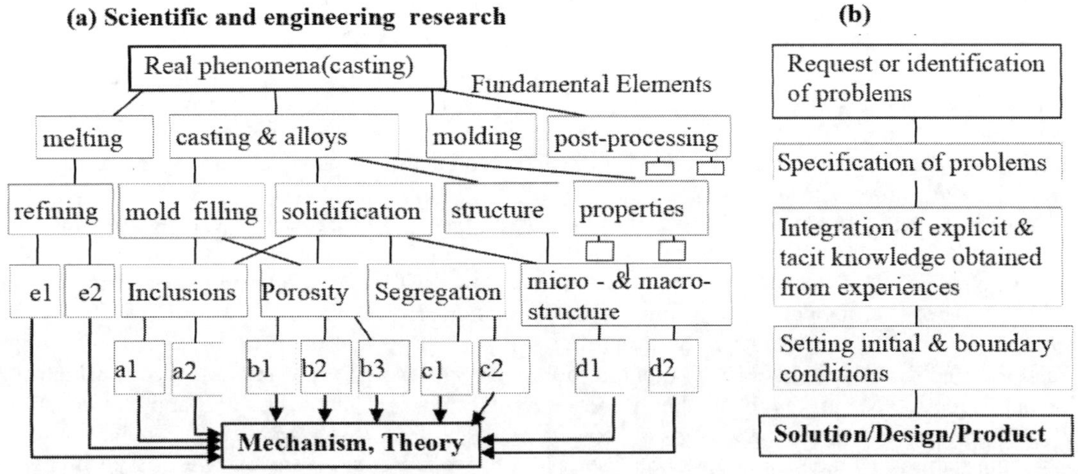

Figure 4. Subdividing of scientific and engineering research(a) and Engineering activity(b)

also find proper initial and boundary conditions such as the thermal resistance between the casting and mold. If we confuse them, we may do a lot of useless work.

Also we should not confuse simulations based on science- and engineering-oriented models and artistic model. The science-oriented model aims to real understanding of phenomena and requires rigorous proving or testing, while the engineering-oriented model aims to solve real problems and not only the accuracy but also cost and time are important. The artistic model aims artistic expression of things and can use all methods including equations which are impossible to happen in the real world.

The simulation based on the science-oriented model is necessary to really understand complex phenomena which cannot be understood by analytical methods and it is ideal to develop the engineering-oriented model from it. Unfortunately, however, there are several barriers and difficulties in conducting the research on the science-oriented model. Some of them are the following:

a) It requires a rigorous proving of the models.

First it requires accurate data for rigorous validation and it is not easy to measure them as described in 3.3. Second, it needs good repeatability of the phenomena concerned. This is also difficult, because it is difficult to perform experiments repeatedly with the same initial and boundary conditions.

Third, it requires for third persons to be able to evaluate the correctness of the work and it is too hard for the third persons in the conventional way. It would be necessary for them to perform the same simulation and compare the results with the reliable data.

b) It is not so attractive works for researchers, in particular for young people, because they should wait long time until the evaluation has been finished.

We should differentiate the engineering-oriented model with the science-oriented one. This is because even the results by the engineering-oriented model agree with limited experimental results, it is not sure that the model is really correct. For example, Niyama criterion[3] is a good example of the engineering-oriented model which was derived from the consideration of unidirectional solidification of steels or Al-alloys. If a proper value is selected depending on casting, defect sites can be estimated with short CPU time. However, it doesn't work always, because this model misses many factors affecting porosity formation such as nucleation pore distribution, gas contents and segregation, solid phase movement, deformation of solidified region, etc. Therefore we cannot say that the model explains the porosity formation mechanism itself.

Also we should pay attention to the level of the engineering-oriented model. For example, it is possible to develop a simulation code without sound engineering-oriented models but using parameters to just agree with limited experimental data. If the simulation results defer from the measured data, suitable parameters to fit the data could be used. Although it could be useful, if it works in some useful range, it is dangerous because it is natural that it fails in many cases. The engineering-oriented model requires more rigorous proving and is more reliable.

One problem on engineering-oriented model is the difficulty of disclosure of its detailed information, because it is usually closely connected to business. However, if its assumption and approximation methods are not published as much as possible, their contribution to the progress of science and engineering is limited. In particular, if a simulation based on an engineering-oriented model is published without proper disclosure of the details, it could be a kind of advertisement even if the results agreed well with measured ones.

We should also distinguish between the simulation based on engineering-oriented models and animation based on the artistic model. Animation is sequential display of phenomena causing illusion of real phenomena. When the results are shown together with measurable numeric data such as temperature, we can evaluate the quality of the animation. However, in the case that it just shows picture it is difficult to evaluate it, because often we don't have real observation results in castings. Therefore the developer should pay attention not to try to develop such animation code which just looks like reality. The professional ethics is very important for simulation providers and researchers.

4.2.2 Integration of existing knowledge and solving complex engineering problems.

Although there is a fundamental question such as " can we really solve the engineering problems by integrating knowledge obtained by such conventional research ? or "can we reproduce the complex phenomena by integrating existing knowledge ?", anyway we should do our best to integrate the knowledge. However, we have the following difficulties:

a) There already exist and are still increasing a huge amount of knowledge related to casting and they are dispersed in various fields and their correctness is not sure. In addition, most of the research results are obtained under limited conditions, namely, simple shape, limited initial and boundary conditions as described before. Actual conditions of the problems are different from those. How can we select proper knowledge?

b) Evaluation of correctness of information is not easy, because it requires wider knowledge beyond the knowledge engineers or researchers usually have.

c) There haven't been enough scientific research on how to integrate and solve complex problems. Today it is mainly just up to each engineers and they often follow the way they got from experienced people with using old and limited knowledge.

4.2.3 Existence of many phenomena which are not so well understood.

There are still many process-related phenomena or subjects which require correct and clear understanding not only qualitatively but also quantitatively. This includes the porosity formation in spheroidal cast iron, nucleation of porosity, gas and inclusion entrapment mechanisms, phenomena at mold-melt boundary, etc.

Although a lot of existing information collected in handbooks[1] is useful, it is not enough from the view point of quantitative estimation of defects. Such knowledge has been obtained mainly from experiences. Namely the causes have been derived mainly by conjectures based on observed phenomena such as morphology of defects, location and casting conditions including casting process, mold coating or mold sand, etc. Therefore, usually the predictions based on the conjectures haven't been tested enough. Further, they often don't show quantitative data of casting conditions such as temperature and pouring velocity. Therefore, it is very difficult to re-produce similar conditions and confirm the correctness of the explanations.

Anyway, most of conventional works are qualitative and have not clear proof of correctness with clear evidence. Therefore even a possible cause of defect is found by using such knowledge, concrete measures have to be found by trial and error, while it is useful to see directions of the trial and error.

However, scientific research on such subjects is not easy because of the following reasons:

a) It is difficult to perform experiments repeatedly with the same initial and boundary conditions.

In the real phenomena the alloys and molds are not constant because their base metals and raw materials vary with time, and processing such as melting and sand preparation and molding vary with time and operators. Pouring temperature and velocity are also fluctuating with time. However if research was carried out under a simple system having clear but limited initial and boundary conditions, then can we guarantee that such works have considered all essential factors in real casting ?

In particular, in this context it is very important to accumulate case studies which identify clear causes of the defect change with time such as mentioned at figure 3. From such works we could be able to find out missing factors to solve the problems. Although some of scientific works might have been done in large foundries, they have not been disclosed. It is more difficult to conduct such works in usual research organizations, because such real situations are very difficult to realize.

b) There are not enough human resources who can conduct such works requiring a wide range of knowledge.

For example, although mold flow and solidification have been carried out mainly by researchers belonging to metallic materials, porosity defects problems in sand casting requires also knowledge of sand mold, namely heat and mass transfer and chemical reaction in molds, and chemical reaction between melt and mold or coating. Solving porosity problems in cast iron requires mechanical behaviour of casting and mold. Therefore it is difficult for researchers specializing only in metallic alloys to handle such subject. It is also not easy for researchers who specialize in numerical simulation to conduct complicated experimental works.

4.2.4 Existence of many properties data necessary for the understanding and simulation.
As mentioned in section 3.1.1 and 3.1.2 there are many necessary data. Although some data can be measured at laboratories, some data should be measured in foundry plants or in research organizations under similar conditions. However, it is not easy.
4.2.5 Black box problem.
With wider use of simulation, more user-friendly code have been developed and tend to become a kind of black-box, because the user-friendliness often means easy operation without deep thinking. Further, as mentioned in section *3.2*, the detail of algorithms of various complex phenomena such as phenomena at melt-mold interface or melt -gas interface are often not disclosed. Such codes are kind of black-box.

It might be no problem, if many problems can be solved with such simulation codes. However, it is not expected in a decade or so to realize such situation. If they are used to rely on black-box-like codes, they do not need to accumulate the comprehensive knowledge as the era when problems were solved by experience, and lose the ability to deal with unexpected situations which often occur in practice.

5. What we should do in the future?
Let us consider how we solve difficulties such as described above and how to develop science and engineering which can solve complex real problems. Some of the suggestions might be the following:

5.1 Promotion of simulation based on the science-oriented model

This requires us to solve problems such as discussed in 4.2.1. Some suggestions are as follows:
a) Special awards and funding

As described in 4.2.1, developing the scientific simulation needs a lot of work and requires very hard evaluation. If its appreciation in the academic world is the same with other conventional research, no one would try such research. Therefore we need to figure out special awards for it.

Further, it requires more funding than conventional research. This is because, for example, we cannot soon reproduce the whole real process of porosity formation during pouring and solidification, considering all phenomena. Therefore we have to proceed step by step in the conventional way shown in figure 4, but more quantitatively, for example, measuring the initial and boundary conditions as much as accurately not only for simple-shaped casting but also more complicated-shaped one. It requires more funding resources than conventional research.
b) Reconsideration of the way of publication of the science-oriented model and engineering-oriented model

As described before, publication is a must for the science-oriented model but not easy. One of the reasons is that the publication by paper is very expensive, because it requires a lot of pages. However, it may be solved if we use a web system.

In order to properly evaluate scientific works for publishing it may be recommended that the evaluation is carried out by a team consisting of specialists from the most relevant fields for the scientific simulation. Although the team can evaluate the scientific quality and correctness of the theory, complete evaluation is very hard for them. Therefore, the final evaluation should be done after gathering results of comparison with reliable data by third persons who used the simulation codes provided.

The evaluation of engineering simulation should be made by the similar team mentioned above but including experienced engineers, who evaluate it from the viewpoints of application, integration and engineering appropriateness. However, rigorous evaluation such as that for the scientific simulation is not required.
c) Encourage to disclose reliable data which can be used for validation of simulation results. It should include not only the data for simple shaped casting but more importantly for rather complex castings. So far, it was difficult to do such works, because they haven't been so well appreciated in academic worlds. We need to appreciate such works and reconsider the evaluation view points for such works.

In particular, direct observation of moving phenomena of small bubbles and oxides during casting and dynamic observation of formation of porosity defects under clear initial and boundary conditions are important. These require not only advanced X-ray techniques but also new method such as neutron beams, because conventional X-ray do not have enough resolution.

d) Encourage works to make clear the effects of fluctuation of materials, initial and boundary conditions on the final results.

Because it is very difficult to measure, for example, physical properties of every molds used in the industry, it is very important to investigate the effects of minor change in chemical composition or density on the properties and create their estimation methods. For this purposes, it is necessary to investigate how they vary with time or plants, requiring to develop methods to measure various values in process.

5.2 Promotion of works on error accumulation and propagation in integrated process simulation or ICME.

It is necessary to estimate the accuracy of the final results and to develop to decease the error. It would be necessary to introduce the concept of provability. Although the accuracy would be decreased with improving the modelling, an error due to input data would remain as shown in figure 3.

5.3 Developing of methods to solve complex real problems

Although conventional knowledge is integrated into various books such as[1, 8] which describe from basic to application or explain classified items, they don't contain the latest knowledge which increases rapidly. Further, they are inconvenient for the problem solving, because how to use such knowledge is up to individuals. Simulation codes are also a kind of knowledge-integrated products but it is not yet complete problem-solving tool.

One solution might be creating new books with using ICT, including simulation. In the new books contents are classified according to problems. For example, it shows the problem solving procedures such as those described in the porosity defect problems in section 2. If a reader cannot well understand the words or requires more detailed information, then the reader clicks the word and immediately can get detailed information, including more fundamental explanations and reliable original papers having scientific evidences. Simulation works well on the understanding, in particular, on the evaluation of simulation results, etc.

Such problem solving document could help to avoid using simulation codes as a black box and the tendency to simply use the default input data.

5.4 Promotion to investigate the real casting phenomena in detail and quantitatively during a long period while problems happen suddenly or occasionally

It is not certain whether fundamental knowledge we have been expecting to know is sufficient to rebuild the complex phenomena or not. There might be something missed and we have to find them. Further, eventually we should evaluate the overall accuracy of the casting simulation.

Although such research might have been performed in large foundries such as automotive foundries where one kind of casting are producing in large quantities with the same line, it is very difficult for small and medium-sized foundries producing various castings in rather small quantities. Further, such works done in industry are not usually disclosed to public, and quality or correctness are not known. Therefore they are not contributing to the progress of engineering problem solving.

Therefore it is hoped to create such a collaborative joint project that some teams consisting of foundry engineers, researchers, simulation specialists, etc. perform the research in foundry plants which are special in the sense that a good measuring system is installed.

6. Concluding remarks

Similar discussion such as described here is possible for many other cases. Although the issues discussed here could be solved if we have enough resources, the biggest problem we face would be the decrease in research resources, particularly in developed countries. This is because there are many more economically attractive and immersing research fields such as robotics, bioengineering, micro-technology, energy-related issues, each country has to distribute their limited resources to various fields.

To solve this problem, we need a better system where people can use their best abilities and collaborate with each other. Further, we shouldn't forget simulation codes are engineering products similar to cars and airplanes, requiring continuous and endless improvement.

Therefore we need to always learn and apply the knowledge in practices. It requires collaborative learning system. Learning requires flour dialogues: 1) Dialogue with problems and phenomena, 2) Dialogues with existing knowledge, experience, mentors such as teachers and experienced engineers, 3) Dialogues with colleagues and 4) Dialogues with one selves(reflection).

This MCWASP conference is a wonderful opportunity for such learning and could be improved from such view points. For example, although it is good to have all fields related to solidification, if the room is divided into individual field and attendees just listen to their own specific talk and discuss them, it is difficult to integrate them. One solution would be to organize for example, session of porosity defects where all phenomena related porosity defects are presented and discussed to find how to solve porosity defects problems. Then we can better see how many factors are related, what are unknown and what we should do as the next step

Acknowledgments

I myself have learned a lot from this conference and from people I met, and sincerely thank to all of them including my colleague, students, industry people, researchers in various organizations, etc.

Reference

[1]Reuter H und Schneider P 1971 *Gussfehler-Atlas* (Dusseldorf: Giesserei-Verlag G.M.B.H.)
Rowley M T 1999 International Atlas of Casting Defects (Schaumburg:AFS)
Anselman G W et al. 1974 Analysis of Casting Defects (Schaumburg:AFS)
Goodrich G M et al. 2008 Casting Defect Handbook: Iron and Steel (Schaumburg:AFS)

[2] Wlodawer R 1967 *Gelenkte Erstarrung von Stahlguss* (Dusseldorf: Giesserei-Verlag G.M.B.H.)
Wlodawer R 1977 *Gelenkte Erstarrung von Gusseisen* (Dusseldorf: Giesserei-Verlag G.M.B.H.)

[3] Niyama E, Uchida T, Morikawa M and Saito S, A Method of Shrinkage Prediction and its Application to Steel Casting Practice, 1982 *AFS Int. Cast Met. J.*, **7**(3), 52-63

[4] I Ohnaka I, Zhu J-D, Sako N, Sugiyama A and Ye C-K 2011 Challenging issues in computer simulation of casting *Int.Cast Metals Re. J.* **24** 133-138

[5] Kron K, Bellet M, Ludwig A, Pustal B, Wendt J and Fresriksson H, Comparison of numerical simulation models for predicting temperature in solidification analysis with reference to air gap formation 2004 *Int. J. Cast Metals Research*, **17** 295-310

[6] Kubo K, Recent Trends of casting simulation systems 2004 *J. Japan Foundry Engineering Society*, **76** 1022-1030

[7] Ohnaka I Sato A, Sugiyama A and Kinoshita F Mechanism and Estimation of Porosity Defects in Ductile Cast Iron 2008 *Int.Cast Metals Re. J.* **21** 11-16

[8] Flemings M C 1974 *Solidification Processing* (New York McGraw-Hill)
Kurz W and Fisher D J 1984 *Fundamentals of solidification* (Vermont Trans Tech Pub.)
Minkoff I 1986 *Solidification and Cast Structure* (Chichester John Wiley & Sons)
Campbell J 1991 *Castings* (Oxford Butterworth Heinemann)
Stefanescu D M 2009 *Science and Engineering of Casting Solidification* (New York Springer)

IOP Conf. Series: Materials Science and Engineering **84** (2015) 012035 doi:10.1088/1757-899X/84/1/012035

Integrated Computational Materials Engineering and Modelling of Shape Casting Processes – Needs, Benefits, Limitations and Hurdles

M C Schneider, J C Sturm, W Schaefer, E Hepp and V Gurevich

MAGMA Giessereitechnologie GmbH, Aachen, Germany

E-mail: M.Schneider@magmasoft.de

Abstract. In this paper, the industrial needs and potential benefits of ICME for shape castings are described from the point of view of a commercial provider of casting process simulation tools. At the same time, the paper addresses the challenges, limitations, and hurdles regarding the extent to which ICME is or can be adopted on an industrial scale for shape cast components. The discussion is backed by concrete examples illustrating the advantages and limitations of integrating models and simulation in the design chain of cast components, in the design and analysis of the processing route of a casting, as well as over the different length scales through which the structures and corresponding behavior of the cast material are determined. The biggest impact of the ICME approach will only be apparent when the increased accuracy or level of detail provided by the methodology is truly necessary, because it leads to changes in design decisions for the product or manufacturing process.

1. Introduction

In the last several years, Integrated Computational Material Engineering, or ICME, has been a buzz word in the world of materials science and engineering. In its 2008 report [1], the Committee on Integrated Computational Materials Engineering of the National Research Council described ICME as "the integration of materials information, captured in computational tools, with engineering product performance analysis and manufacturing-process simulation." The NRC's definition of ICME explicitly addresses the coupling between material – process – performance being captured in computational tools, but what does this mean for modelling of shape casting processes? What are the needs of industry and which benefits can they expect from ICME, what limitation and barriers exist for extending the extent of ICME, and which consequences does or should ICME have for those working in the field of modelling of shape casting processes?

From the authors' point of view and as shown in figure 1, the concept of ICME for shape castings can be broken down roughly into three different scenarios: integration of materials information across varying scales up to the cast component level; integration of materials information across the various processing steps which lead to a finished shape casting; and integration of materials information from the manufacture of a cast component into the analysis of its performance. Each of these scenarios can be important on its own, and depending on the application and the problem to be solved, two or all three may be required. On the other hand, the problem to be solved also determines the required level of detail and depth of integration within any of the three described scenarios.

In this paper, two examples showing the use of ICME for shape cast components are briefly presented. In the discussion of the examples, the authors attempt to illustrate and discuss the limitations and barriers to a more extensive integration in the sense of ICME. To conclude, the authors summarize what they see as the current barriers and limitation for more extensive use of ICME and contrast these with the perceived industrial needs.

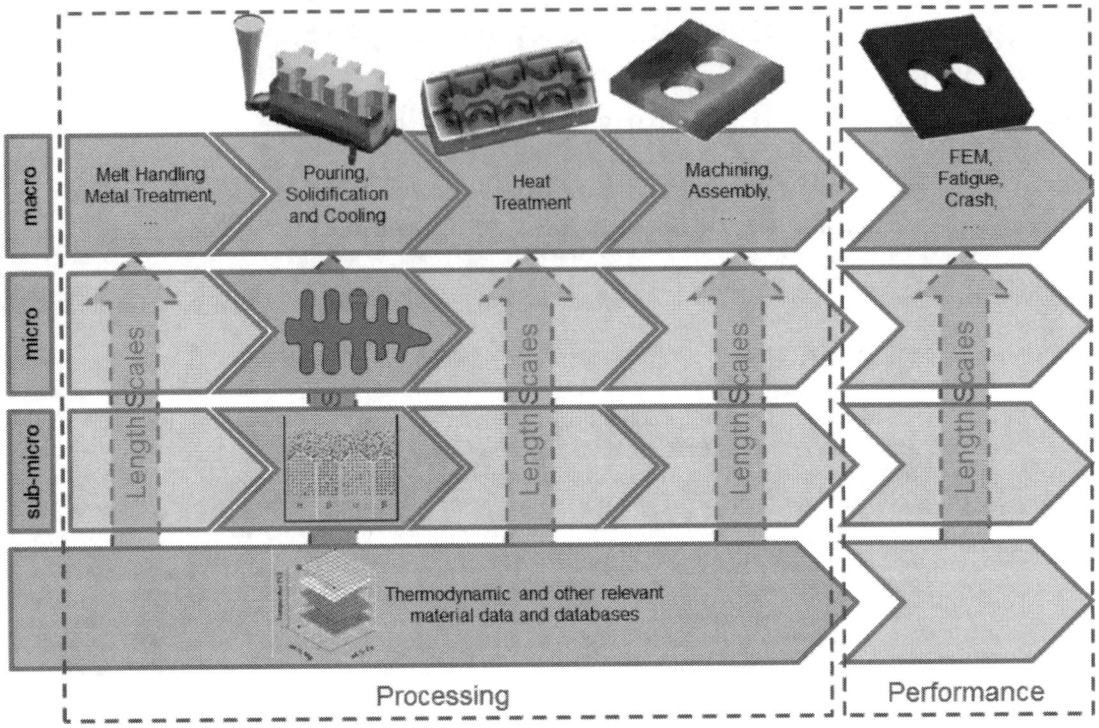

Figure 1. Schematic illustrating the principles of ICME for a shape casting process, including three integration scenarios (dotted lines) across length scales, along the manufacturing processes and from manufacturing to performance analysis.

2. Through process integration over varying length scales.

Integration over the length scales involved in determining the structures and material behavior of the metal in a shape cast component, including thermodynamics and thermochemistry, can be considered part of ICME. However, it is important to consider that a cast component may undergo a number of processing steps to achieve its final shape and properties, i.e. casting and cooling, heat treatment, machining, etc. In the first example presented here, the modelling of the changes in the microstructure of a heat treated cast iron component to create austempered ductile iron (ADI) are discussed [2].

The focus of the modelling activities was on the optimization of the heat treatment process to obtain the desired ADI structure in a robust and economical way. The project objectives meant that the primary focus was on the changes in microstructure during austenitization, quenching and austempering, which needed to be coupled with a macroscopic consideration of the heat flow during processing. Figure 2 illustrates schematically the methodology used for modelling of microstructural changes during the transformation of ferrite to austenite in a eutectic cell during austenitization. This transformation is modelled as being diffusion-controlled, with local thermodynamic equilibrium assumed at the phase boundaries. Rather than a detailed microscopic description geometrically, a representative eutectic cell was modelled for each macroscopic computational cell.

In addition, a so-called para-equilibrium approach was adopted [2], where the diffusion of all alloying elements other than carbon was neglected and consequently a Fe-C section of the multicomponent phase diagram used. In this case, the para-equilibrium assumption could be justified by previous work [3]. Naturally, it would have been possible to extend this model to a more detailed consideration of multicomponent thermodynamics and diffusion, with corresponding costs in model complexity and computational resources. As the authors state, the criteria for selecting the modelling approach were robustness, availability of data and ease of integration. In this case, the costs of a more detailed microscopic consideration were carefully weighed by the authors against the potential benefits of a more accurate description of the microscopic phenomena on results at the macroscopic level.

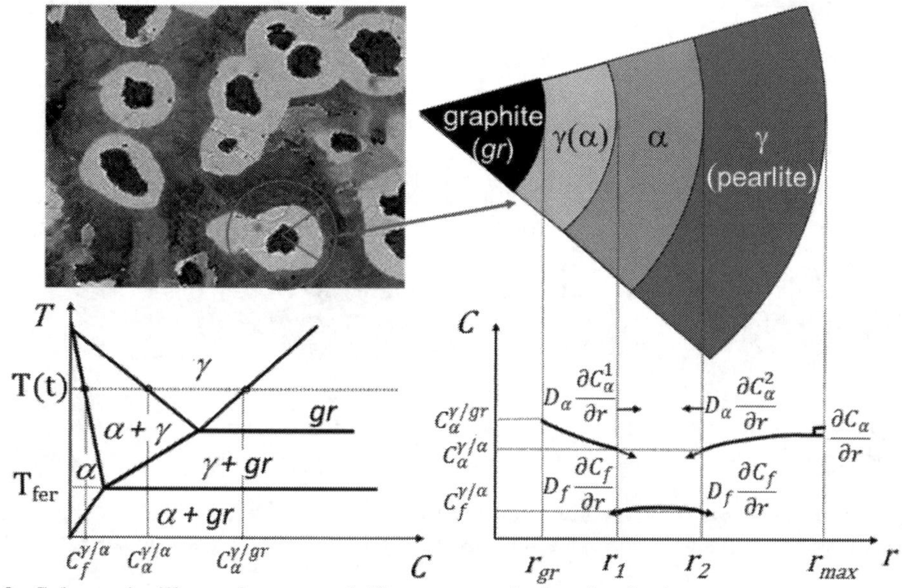

Figure 2. Schematic illustrating a modelling approach for the ferrite to austenite transformation during austenitization heat treatment of ductile iron to form ADI: a) typical ductile iron microstructure with eutectic cell marked in red; b) schematic of the phase distributions in the eutectic cell; c) Fe-C section of the multicomponent phase diagram; d) concentration profiles across the eutectic cell during the transformation (from [2]).

The heat treatment of the component is carried out to modify the microstructure in the cast iron to obtain ADI. This also means that the starting point for a heat treatment simulation of the component needs to consider information about the as-cast microstructure distribution in the casting. In the model of ADI heat treatment, the predicted local graphite nodule count and the ferrite/pearlite fraction distributions were taken from a simulation of the solidification and cooling of the casting and used as the starting point for the heat treatment simulation [2]. As shown in figure 3, different initial nodule count distributions in the casting (exaggerated) lead to greatly varying rates of austenitization and dissolved carbon content in the casting during the austenitization process.

This example clearly illustrates that two of the principles describing ICME are required in the modeling of the ADI manufacturing process: coupling varying length scales (also integrating thermodynamic considerations) and integrating different manufacturing steps. However, the level of detail needed from smaller scales and/or the degree of coupling necessary between length scales is very dependent on the questions to be answered through using modelling and simulation.

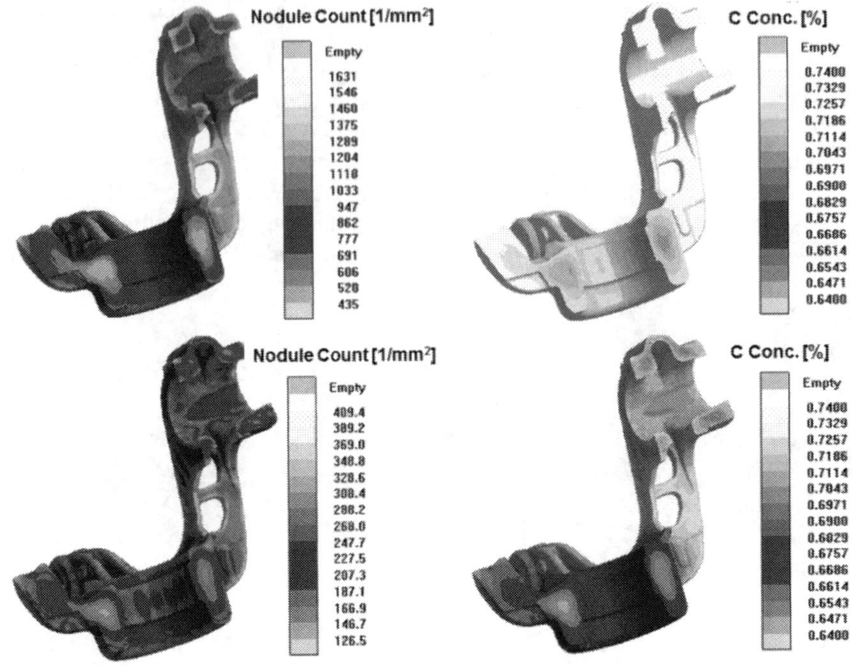

Figure 3. Simulation results illustrating the effects of solidification microstructure (nodule count) on the rate of austenitization and carbon content during production of ADI: a) higher nodule count, b) lower nodule count.

3. Integration of process and performance simulation

As can be illustrated by numerous examples, assuming that a casting has uniform, homogenous properties and is free of residual stresses can easily lead to an under- or over-dimensioning of the component, with corresponding penalties in poor performance or addition of unnecessary weight in the design. This means that a consideration of the relationship between material – process – performance can be of critical importance when designing or analyzing (FEM, crash, fatigue life) the design of a cast component [4]. In the example considered here, the influence of the local microstructure in a casting was taken into account in the prediction of the fatigue life of a ductile iron component.

Models integrated in macroscopic casting process simulation for the prediction of the distribution of local microstructures and the corresponding static mechanical properties of ductile iron castings exist for some time [5]. However, the coupling of dynamic mechanical properties (fatigue life) to microstructure for these materials first required extensive experimental effort to determine microstructure dependent S-N curves for the material, illustrated schematically in figure 4 [6].

As shown in the figure, to characterize the relationship between microstructure and fatigue life, samples were prepared from castings with different wall thicknesses and varying process conditions to obtain varying microstructures. The machined samples were then subjected to fatigue testing to determine the corresponding S/N or ε/N curves. After testing, the samples where characterized (using image analysis) through 22 parameters including the number, shape and size of the graphite nodules and the ferrite/pearlite content. Finally, an extensive variance and regression analysis was used to determine the microstructural parameters which had the largest influence on the fatigue life of the material and to obtain a correlation between these parameters and the fatigue life. In the investigation described here, the local pearlite content was determined to be of primary influence (see figure 4).

IOP Conf. Series: Materials Science and Engineering **84** (2015) 012035 doi:10.1088/1757-899X/84/1/012035

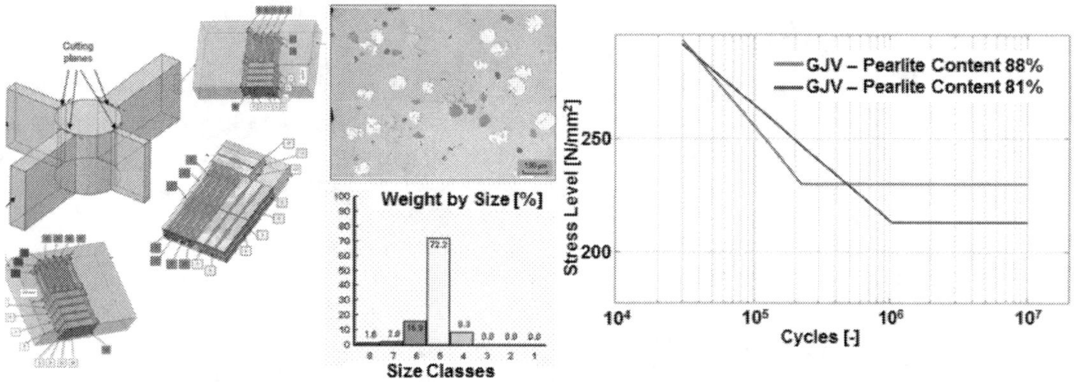

Figure 4. Experimental determination of fatigue life for cast iron with varying microstructures: a) sample positions for an experimental casting with different wall thicknesses; b) example of image analysis used to determine microstructural parameters, here graphite nodule size; c) two experimentally determined S-N curves for a compacted graphite iron grade with differing pearlite contents [6].

The description of the durability as a function of pearlite content was coupled to the local microstructures predicted through a casting process simulation and exported for use in the finite-element and durability analysis of a bracket used as a demonstrator component. At the top of figure 5 the predicted fatigue strength of the component and the location of the actual crack initiation in laboratory tests of the component are visualized. The bottom of figure 5 shows that a classical fatigue analysis of the component would predict the start of cracking at a different location than observed. Due to the variations in the predicted pearlite content of the casting, the most highly stressed regions of the bracket have a predicted higher fatigue strength than the regions where crack initiation was observed. The fatigue analysis considering the local material properties more accurately predicted the real crack initiation site (see figure 5).

This example clearly shows the potential benefits of the ICME approach to integrate material, process and performance analyses to improve the predictability of performance in the design of a cast component. Note, however, that the development of the necessary material data for the correlation of fatigue behavior to corresponding microstructural characteristics required significant experimental effort. Other factors, such as a weakening of the material through defects (e.g. porosity or nonmetallic inclusions) that almost certainly play a role in the fatigue life of many cast iron components have been completely neglected in this analysis and would require further experimental characterization. This means that the complexity of the real situation has, in this case, only been partially accounted for in the model considerations.

4. Needs, Benefits, Limitations, Barriers to ICME

The previous examples have illustrated the potential benefits of the ICME approach for shape cast components. In industry, especially the demand for lighter and lighter products to achieve reduced energy consumption and emissions means that the design of castings which are economical, durable and fulfill all the in-use requirements will result in increased needs for durable high strength alloys but also for robust designs and production processes. Accurate analysis tools can play an important role is meeting these challenges by shortening development times. The ICME approach in materials development, in casting process layout and in performance analysis provide even more powerful tools to fulfill these needs.

However, there are also inherent limitations in the ICME approach and barriers to its further adaptation, some of which are discussed in the following subsections.

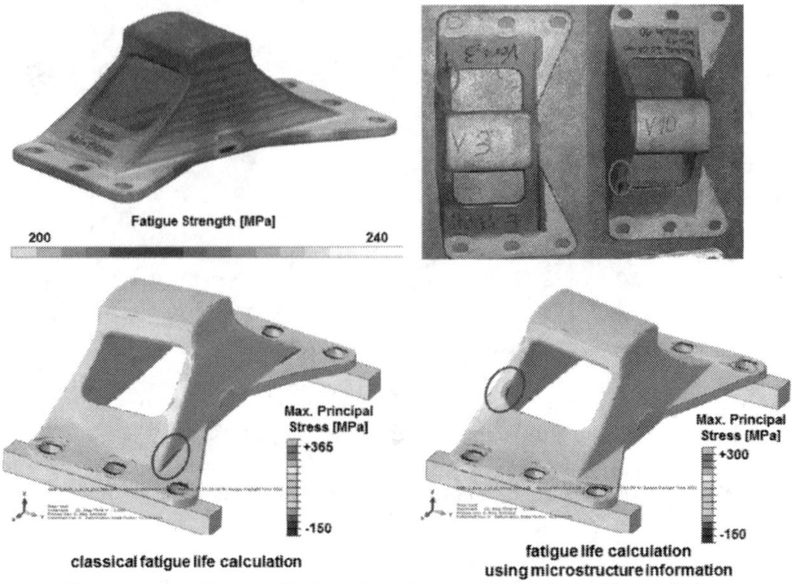

Figure 5. Fatigue life analysis of a ductile iron bracket casting: a) predicted fatigue strength based on the local predicted pearlite content of the microstructure; b) brackets with observed crack initiation sites; c) results of classical fatigue life calculations for the bracket; d) fatigue life prediction considering the local microstructure for the bracket [6].

4.1. Time to Answer – Computational Expense

Not only are the technical demands on cast components growing, there is also a continuous demand to shorten development times. One the one hand, this speaks for the consequent use of the ideas of ICME in design, in order to significantly reduce the number of physical trials required to develop a casting design and the corresponding manufacturing process. On the other hand, this also means that the time available for doing detailed model analyses is limited. Therefore, one barrier to the adaptation of the ICME approach in process or component design are the computational times required e.g. for complex simulations integrating detailed models for different length scales up to the component level. Alone the integration of thermodynamic equilibrium calculations into the simulation of the solidification of an industrial steel grade have been reported to approximately double simulation times, despite simplifications made in the frequency with which the equilibrium calculations were carried out [7].

In many cases, reaching calculation times on the scale of minutes, hours or overnight at the casting level means that the often adopted approaches of e.g. volume-averaging, homogenizing over length scales or making model simplifications (e.g. phase diagram linearization), etc. become a necessity for an effective use of simulation in product or process design. This has been illustrated through the assumption of para-equilibrium e.g. for the ADI heat treatment simulation presented above. It seems obvious that this also means that in many cases it will be difficult to generalize models to be applicable to a wide range of industrially relevant multicomponent, multiphase alloys. Adaptations will always have to be made on a case-by-case basis.

4.2. Lack of Data – Costs of Measurements

The example shown above for the integrated fatigue life prediction of a ductile iron component shows that in many cases extensive (and expensive) experimental investigations are required to develop the necessary material data to allow a complex integration of material models with process and performance simulations. Anyone who has worked in the fields of modelling and simulation will have quickly realized how difficult it typically is to find good, current and reliable data as input to models. These costs can be another barrier for a more extensive adoption of ICME for shape castings.

For complex new materials under development to satisfy future requirements, for example in power generation, aspects of ICME are already being used, see e.g. [8, 9]. Unfortunately, the focus in these modelling and measurement efforts is often on the material's behavior in use, for example creep strength at increased operating temperatures, but the processing of the materials to manufacture the corresponding components is not considered. In this case, the knowledge and tools are lacking to optimize the processing of the corresponding cast components and correctly predict the solidification behavior and the development of defects in the casting (e.g. creep or other mechanical properties measured up to near-solidus temperatures). This leads to considerably increased costs in manufacturing as well as a slower adoption of the new material in the production of critical components.

One hope for ICME could be that expensive physical experimentation can be replaced by numerical experiments which deliver the same information. Clearly, this can only take place if the corresponding models have been verified and validated to represent the behavior of the real material. Also, the requirements on computational costs for carrying out the numerical calculations must be lower than the actual physical experiments.

4.3. Real Complexity – Production Variability

The essence of the statement "Essentially, all models are wrong, but some are useful" by G.E.P. Box [10] can be used to describe a limitation and potential barrier to further integration of models in the framework of ICME. All models make simplifying assumptions and idealizations in order to make the problem tractable. This means it is extremely important to carefully weigh the benefits of a tighter integration of materials information into a process or performance simulation against the uncertainties introduced by other modelling assumptions – the weakest link in a chain of models or simulations will determine the accuracy of the whole approach. Of course, as scientists and engineers, this is a factor that is always in the minds of those working in the modelling community.

A related factor, which is often forgotten both in the modelling community as well as in the practical use of modelling and process simulation, is that industrial casting processes exhibit variability in production. Even for the high pressure die casting process, which is largely mechanized and automated, variations are observed in casting quality from part to part [11]. Scrap rates in metal casting are seldom one-hundred percent, but typically range from a few to sub-percent levels. From this point of view, a perceived or expected improvement in the accuracy of a model through a higher level of integration in the framework of ICME needs to be critically weighed against the expected "natural" variations in the processing and performance of the component produced.

4.4. Ease of Use – Single Source

There are some efforts to try to define and develop a generic methodology for ICME which would allow a "plug and play" standardized, modular modelling platform that can be adapted to a specific material, manufacturing process chain and product [12]. Although on the surface this idea is attractive, alone the complexity of the different materials to be addressed (plain carbon, low alloy and high alloy steels; grey, ductile, compacted graphite and austempered ductile irons; aluminum, magnesium and copper alloys; etc.) and the need for model homogenization and shortest possible simulation times seems to speak against a generic platform that would allow a "sticking together" of different models.

In addition, it is the authors' experience that in industry there is a desire to work with as few sources as possible when it comes to modelling and simulation of e.g. the complete manufacturing process of a component. One reason for this is that, despite dedicated interfaces, the transfer of information from one program to another is always connected with additional effort. Another important aspect is that using different tools also means having to deal with different sources or suppliers when questions or problems arise in use. Here, the advantages of using different, coupled tools from different sources needs to significantly outweigh any disadvantages which arise through extra effort required through the interfacing and support of the programs.

4.5. Commercial Interests – Protection of Investments

A final hurdle to a wider use of ICME is that in many cases there are conflicts of commercial interests in the integration of various tools. Clearly, any commercial provider of simulation tools is interested in protecting his market position and the investment he has made in developing his capabilities. Any advantages of an easier integration within ICME will be weighed against the fact that this ease of integration may make a tool interchangeable.

Similarly, if a company using simulation tools finances e.g. extensive measurements or computations to better characterize and understand a material of interest to them, they do so in order to increase their competitive advantage in the market. Understandably, in most cases this company will have a justified interest in protecting the developed information or modelling capabilities and will have very little interest in seeing this information be diffused through open interfaces or databases.

5. Conclusions

In this paper, the authors have tried to illustrate through two different examples that ICME is already being developed and is being used today in the commercial modeling and simulation of shape castings. However, despite the needs and benefits for further development of the approach, there are a number of limitations and barriers to a further adoption of ICME on the commercial level.

No matter whether along design or process chain or over different length scales, the overriding consideration for effective ICME has to be a careful consideration regarding the level of integration needed to ensure that design decisions (component design, process design, material selection) can be made with the required level of confidence. As proposed by Allen et al. [13], the key is not to eliminate uncertainty in design decisions (through a more detailed coupling of models), but rather to leave it where it does not affect the design decisions made. This means that in each case, the trade-offs in the cost of more extensive modelling, computation and integration of tools needs to be balanced against the perceived value of the information or increased accuracy provided. In the design process, this also means that it is important to develop solutions that are robust and insensitive to uncertainties, whether these uncertainties have their source in the models or in inherent production variations [13].

From this point of view, the way in which information is passed from level to level over length scales or along a process or design chain needs to be considered on a case-by-case basis, to find the right level of integration for providing the right information with the necessary accuracy to obtain all of the real and perceived benefits of the ICME approach.

References

[1] Committee on Integrated Computational Materials Engineering, National Research Council 2008 National Academies Press Washington, D.C.
[2] Hepp E, Hurevich V and Schaefer W 2012 *IOP Conf. Ser.: Mater. Sci. Eng.* **33** 012076
[3] Fridberg J, Torndahl L E, and Hillert M 1969 Jernkonteorets annaler **153**(6) 263
[4] Egner-Walter A and Hartmann G 2007 ATZ Sonderbeilage **109**(12) 2
[5] Wessen M and Svensson I L 1994 Met Mater Trans A **27A** 1996
[6] Thomser C, Bodenburg M and Sturm J C 2014 Giesserei **101**(11) 26
[7] Schneider M C, Gu J P, Beckermann C, Boettinger W J and Kattner U R 1997 Metall Mater Trans **28A** 1517
[8] Rajek H J 2005 PhD Thesis TU Graz, Austria
[9] http://www.cost.eu/COST_Actions/mpns/Actions/536
[10] Box G E P and Draper N R Empirical Model-Building and Response Surfaces Wiley 424
[11] Flender E and Hartmann G Giesserei **92**(03) 38
[12] Schmitz G, 2014, 1st International Workshop on Software Solutions for ICME, Aachen/Rolduc
[13] Allen J K, Panchal J H, Mistree, F, Gautham B P and Singh A K 2014 1st International Workshop on Software Solutions for ICME, Aachen/Rolduc

IOP Conf. Series: Materials Science and Engineering **84** (2015) 012036 doi:10.1088/1757-899X/84/1/012036

Application of Numerical Optimization to Aluminum Alloy Wheel Casting

J Duan[1], C Reilly[1], D M Maijer[1], S L Cockcroft[1] and A B Phillion[2]

[1] The Department of Materials Engineering, The University of British Columbia, Vancouver, BC, Canada V6T 1Z4
[2] School of Engineering, The University of British Columbia, Kelowna, BC, V1V 1V7

Abstract. A method of numerically optimizing the cooling conditions in a low-pressure die casting process from the standpoint of maintaining good directional solidification, high cooling rates and reduced cycle times has been developed for the production of aluminum alloy wheels. The method focuses on the optimization of cooling channel timing and utilizes an open source numerical optimization algorithm coupled with an experimentally validated, ABAQUS-based, heat transfer model of the casting process. Key features of the method include: 1) carefully designed constraint functions to ensure directional solidification along the centerline of the wheel; and 2) carefully formulated objective functions to maximize cooling rate. The method has been implemented on a prototype production die and the results have been tested with plant trial test.

1. Introduction

The aluminum wheel industry is very competitive and manufacturers are constantly under pressure to improve quality and to reduce cost. This is achieved by increasing production rates while attempting to also maintain or reduce scrap rates. One of the defects that is a significant challenge to manufacturers is shrinkage porosity. Shrinkage porosity occurs when liquid or semi-solid metal is cut off from a source of liquid metal, which is needed to compensate for the volume change associated with solidification (the volume change is in the range of 5 to 7% depending on the aluminum alloy). In the case of wheel casting, achieving controlled progress of the solidification front, or "directional solidification", is the key to avoiding shrinkage porosity. As illustrated in **Figure 1**, good directional solidification is achieved when solidification starts high in the inboard-rim flange, progresses down through the rim, and then across the spokes, ending just below the hub in the sprue. To achieve this solidification pattern, both proper die design and careful management of the cooling conditions within a cast cycle are required.

Historically, the design of a die and the associated operational parameters, were determined based on a combination of experience and trial-and-error optimization. The methodology typically involves long design lead times, prototyping, several preproduction trials and can result in high scrap rates and less than optimum production rates. The move to adopt water-cooling in place of air cooling in order to reduce cycle times, improve productivity and improve wheel fatigue performance represents a significant challenge to the foundry engineer as there is a limited experiential base to draw on. This manuscript describes a methodology in which the trial-and-error optimization of the operational parameters (die cooling timing) is eliminated from the preproduction trial process. It combines a casting process model with an advanced optimization algorithm that is applied to achieve the optimum set of operating conditions that produce a combination of minimum cycle time and limited or no shrinkage porosity in a given die design.

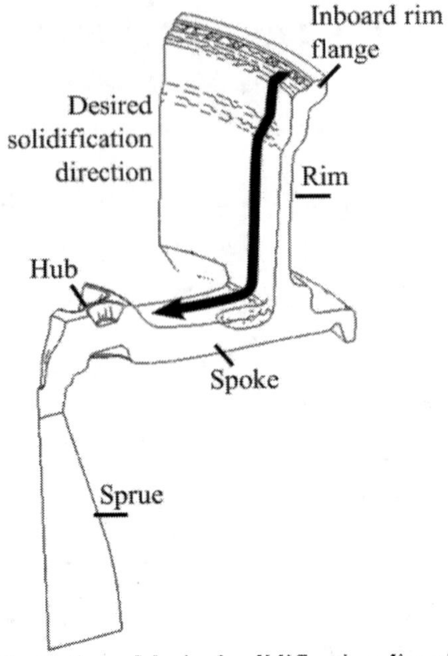

Figure 1. Illustration of desired solidification direction in wheel

2. Background

Numerical design optimization provides a systematized and versatile procedure for arriving at a design solution. It can offer a reduction in design time, removes experiential biases from the design process and virtually always yields some design improvement [1]. Numerical optimization first found application in structural design [2, 3] and is now increasingly applied in other areas such as, cast product and casting process design [4-7].

The numerical optimization process is essentially a procedure for finding the minimum or maximum of an objective function, which mathematically describes some attribute of the process or product being optimized. Additionally, constraints are normally added to account for additional limitations imposed on the design. The general mathematical expressions of the objective function and the constraint functions are presented in Equations (1) and (2).

$$\text{max (or min)} \quad f(x) \quad \boldsymbol{x} = (x_1, x_2, \ldots, x_n) \tag{1}$$

$$\text{subject to} \quad g_i(x) = c_i \quad \text{for } i = 1, \ldots, \mathrm{n} \tag{2}$$

$$h_j(x) \leq d_j \quad \text{for } j = 1, \ldots, \mathrm{m}$$

where $\boldsymbol{x}$ is a vector containing all the design variables that may be modified in order to achieve the optimization goal, $f(x)$ is the objective function, and $g_i(x)$ and $h_j(x)$ are equality and inequality constraints, respectively.

The objective and constraint functions are generally evaluated based on the results of a numerical model. In this work, a comprehensive and experimentally validated model of a low pressure die casting (LPDC) process for the production of A356 automotive wheels has been used. Details related to the casting process model have been discussed previously in the literature [8, 9]. The current work presents the application of the optimization algorithm using two versions of the casting process model (see Sections 0 and 0) in order to highlight both the utility of the optimization algorithm and the importance of the process boundary conditions in the model utilized within the optimization algorithm.

3. Architecture of the optimization tool

Figure 2 shows the overall structure of the optimization algorithm or "tool", which consists of three major modules. The first one is the casting process model, which has been developed in the commercial finite-element package ABAQUS™. The casting process model simulates the casting process and calculates the thermal field within the wheel and die. Due to the large thermal mass of the die, each change in cooling timing requires several casting cycles for the full effect of the change to propagate through the die and for a new cyclic steady-state condition to be achieved. Once the new steady state condition is achieved, the necessary data is obtained from the ABAQUS output file and processed with the analysis module developed in Python. Specifically, the analysis module evaluates the objective and the constraint functions and periodically, based on a set of criteria, evaluates the sensitivity of these functions to the design variables. This information is then passed to the optimizer. The optimizer is based on an open-source Python package (scipy.optimize), which has a FORTRAN-based core wrapped in Python. The optimizer determines if the design criteria, subject to the imposed constraints has been reached. If the criteria are not met, it then determines how to change the design variables, which are then passed back to the casting process model. If met, the optimal design parameters are output. The whole process iterates until an optimum solution is reached or until the user-defined maximum number of iterations is reached.

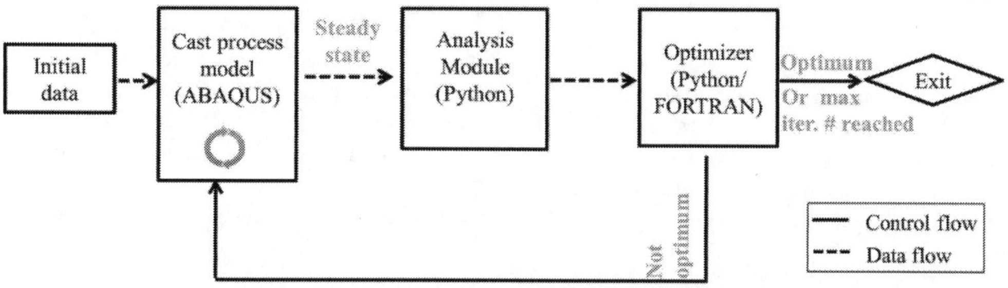

Figure 2. Architecture of the optimization tool

The strategy used to move from one iteration to the next in the design variable space distinguishes one algorithm from another. The package utilized in this study uses the line search method. The line search approach first finds a descent direction along which the objective function is reduced and then computes a step length that determines how far the update should move along that direction. In the line search method, the design variables are updated as follows [1]:

$$x_{k+1} = x_k + \alpha S \tag{3}$$

where x_k is the design at the kth iteration and x_{k+1} is the new design at the $(k+1)^{th}$ iteration, S is the search direction, and α is the step length.

The descent direction is determined using the quasi-Newton method in the current algorithm. The Quasi-Newton method is reported to have better convergence behavior than the Steepest Descent method, and it is more efficient than Newton's method, as it does not require information on second-order derivatives. Detailed discussion of these three methods can be found in reference [1]. The step length can be determined either exactly, by finding the minima of the objective function in the given direction S, or inexactly, by finding the value of α that leads to an acceptable descent amount.

4. Application to a 2D Axisymmetric Prototype Die Model

The die structure used in the application presented in this paper is proprietary and cannot be disclosed. Hence, only the wheel and the approximate locations of the water-cooling channels are shown in

™ ABAQUS is a trademark of Dassault Systèms

IOP Conf. Series: Materials Science and Engineering **84** (2015) 012036 doi:10.1088/1757-899X/84/1/012036

Figure 3. There are seven cooling channels in the die design: three in the top die (TD_CC_1 ~ 3), two in the side die (SD_CC_1 ~2), and two in the bottom die (BC_CC_1 ~ 2).

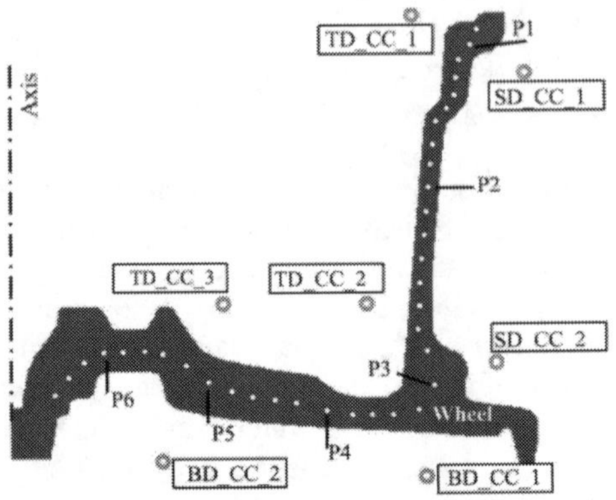

Figure 3. Cross-section of the wheel geometry with approximate locations of the seven cooling channels and showing the locations used for evaluation of the objective (P1-6) and the constraint (white dots) functions.

The design optimization strategy is based on a combination of maximizing the cooling rate in the wheel while achieving directional solidification. The former is desired to both increase productivity (reduce the overall cycle time) and produce a finer structure with better overall fatigue performance. The latter is needed to eliminate/reduce shrinkage porosity, which, if present, would negate the positive benefits of a reduced cycle time on fatigue performance.
The objective and constraint functions applied in the present work are given in Equations (4-5).

$$f(\boldsymbol{x}) = \sum_{i=1}^{6} a_i \times (-\dot{T}_i^{calculated}) \quad (4)$$

$$h_j^1(\boldsymbol{x}) = t_{j+1}^{575°C} - t_j^{575°C} \geq 0\,, \ j = 1,2,3, \dots 35 \quad (5)$$

$$h^2(\boldsymbol{x}) = (t_{dieOpen} - t_{solidification})/t_{dieOpen} \geq 0 \quad (6)$$

where $f(\boldsymbol{x})$ is the objective function, $\boldsymbol{x}$ is a vector containing the design variables (the on and off timings of the 7 water cooling channels), a_i are constants used to adjust the relative weights of the objective function, $\dot{T}_i^{calculated}$ is the calculated average cooling rate within the solidification range at the ith point, $h_j^1(x)$ is the first type of constraint imposed at the jth constraint point, $h^2(x)$ is a second the type of constraint function, $t_{dieOpen}$ is the time the die opens, and $t_{solidification}$ is the time taken for the cast to solidify to the middle of the hub (a solid shell forms at this time to maintain the wheel structure).
Careful selection of the objective and constraint functions is crucial to obtaining good results. In Equation (4), the cooling rate is maximized when the optimization algorithm minimizes $f(\boldsymbol{x})$. The calculated cooling rates are assessed at six points distributed at different regions within the wheel, indicated by points P1 through P6 in Figure 3. The first type of equality/inequality constraints are based on the time to reach 575°C, which is the eutectic temperature of A356 [10]. As shown in Equation (5), the constraints are formulated to require that the difference in time between two adjacent points, $j+1$ and j, to reach 575°C is greater than or equal to zero. The constraint functions are evaluated at 36 points, which are shown as the white dots in Figure 3. j = 1 is located at the top of the inboard rim flange, the first place to solidify, and j = 36 is located at bottom of the hub, the preferred last place to solidify. Referring back to Equation 5, the j+1 point is located further along the path than

the jth point in the desired direction for progress of the solidification front – i.e. upstream in the direction of the compensatory flow necessary to feed solidification shrinkage. This constraint is utilized to ensure directional solidification.
A second type of equality/inequality constraint is applied that requires the wheel to be solidified before the die opens (Equation (6)). The solidification requirement constraint is normalized to bring it into the same magnitude range as the other constraints as numerical difficulties often arise when one constraint function is of a different magnitude, or changes more rapidly, than the others and dominates the optimization process [1].

4.1 First Application – Base-Case Version of the Model

The optimization tool was first applied to a 2-D axisymmetric base-case version of the casting process model. The large computational overhead associated with repetitively running the thermal model necessitated that the initial application be based on a 2-D axisymmetric version of the LPDC process model in order to minimize computational time during development of the optimizer. The model includes the different stages in the casting process – i.e. die filling, wheel cooling and solidification, die open, wheel removal and die closing. A few additional features are listed below:

1. The wheel and die material properties are temperature dependent;
2. Die filling is not simulated however, the interfacial cooling between the die and the wheel is activated based on the expected filling sequence, to approximate the effect of filling on heat transfer;
3. The initial distribution of temperature in the wheel is assumed to vary from 620 °C at the top of the rim to 700 °C at the inlet from the sprue (this is done to approximate the heat loss from the wheel to the die as the liquid metal is filling the die cavity);
4. The die/wheel interface boundary conditions vary with temperature to account for gap formation during solidification and are also different for the different die sections – top, side and bottom dies – to account for differences in the evolution of the gap at the interface;
5. The heat transfer in the water cooling channels is assumed to be constant when on and is described using a single-phase convection heat transfer coefficient; and
6. Each model is capable of being run in a cyclic mode – i.e. the initial condition for the die at the beginning of each cycle uses the results from the previous one – and therefore can obtain the cyclic steady state condition for a given set of process cooling conditions.

The results showing the solidification patterns obtained for two sets of cooling timings are shown in **Figure 4**. Each row of images shows color contours of the temperature on a cross-sectional slice of the wheel at different times in the cycle for a given set of cooling conditions. The 572 and 575 °C isotherms are added on each contour to delineate the lines representing approximately 75 and 50% solid fraction for A356, respectively. The top row shows the results using the initial cooling times, which are based on trial-and-error optimization using the model. As can be seen at 49s there is a region of potential liquid/semi-solid encapsulation in the top of the rim indicated by the 50% fraction solid contour, which could result in solidification shrinkage. This is a well-known problem area in the water-cooled version of the LPDC process (*note: this is occurring in an area that will be accurately described using the 2-D axisymmetric model*). There is however, no liquid encapsulation predicted in the rim-spoke junction (73s), which is also an area prone to solidification shrinkage [9,11]. The results obtained with the numerical optimizer are presented in the lower row of images. As can be seen, these results show the elimination, or reduction, in the tendency to form shrinkage porosity in the top of the rim while maintaining similar solidification time.

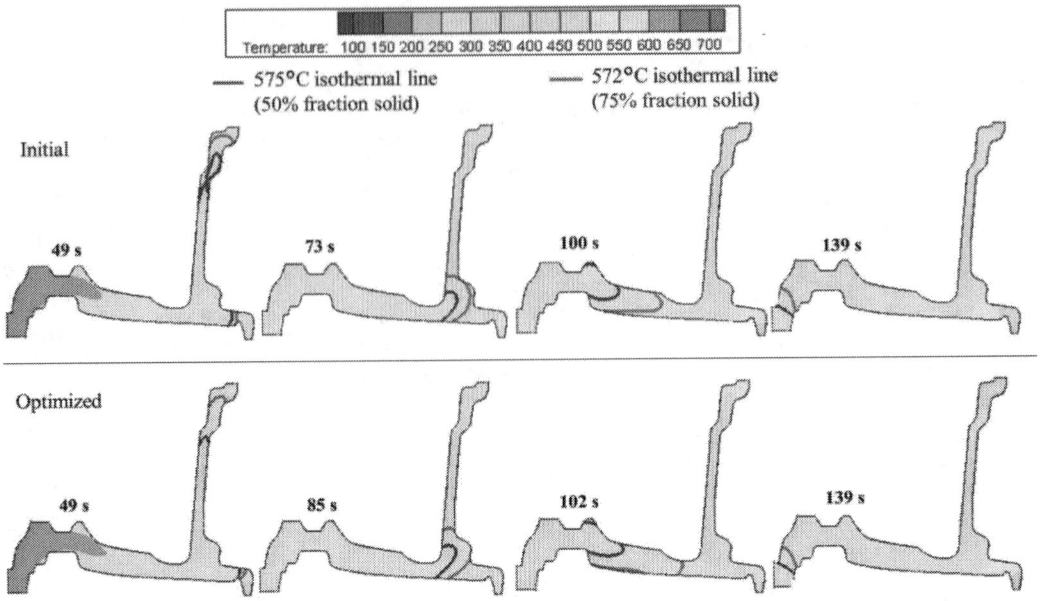

Figure 4. The initial (top) and the numerically optimized (bottom) solidification sequence in a water-cooled die.

Based on the optimized cooling timing, a casting trial was run at a commercial wheel plant (*note: the water-cooling timing actually used in the industrial casting trial was further refined using a 3-D model of the LPDC casting process to address limitations in the 2-D axi-symmetric model*). The results from this trial revealed fine, distributed, shrinkage porosity persistent in the upper rim – see Figure 5.

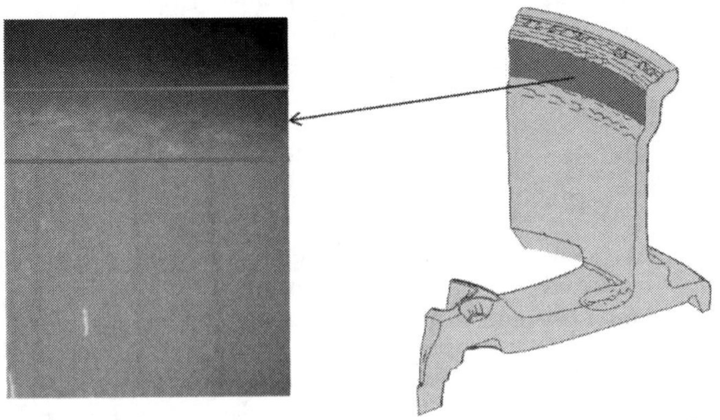

Figure 5. X-ray image showing distributed shrinkage porosity in the top rim at the top of the wheel rim

4.2 Second Application – Updated Wheel/Die Interface and Water Channel Boundary Conditions

To better understand why the industrial trial failed to produce a shrinkage free wheel, the thermal model used in the optimization tool was re-examined to assess the veracity of some of the key boundary conditions. On going work at UBC has revealed that: 1) the behaviour of the wheel die interface during a casting cycle is complex with gap formation in some regions, the development of pressure (no gap) in others and both gap formation followed by gap closure and pressure development in still others[12]; and 2) the heat transfer behaviour in the cooling channels is also complex exhibiting transient periods of two phase boiling channel flow when the water first enters the channel and again when the flow is terminated to the channel [13]. The results of re-running the model with improved boundary conditions are presented in Figure 6. The upper set of images in Figure 6 (labeled "initial") shows the results obtained with the previously optimized timing applied in the model with the updated boundary conditions. The lower image shows the results of re-running the optimizer to obtain a new optimum (labeled "optimized"). Focusing first on the initial results, the model now predicts shrinkage porosity in the upper rim consistent with what was observed in the industrial trial – i.e. the encapsulation of the 75% solid fraction isotherm would lead to distributed, relatively fine, shrinkage porosity. *Note: some shrinkage was also predicted in the spoke approximately mid-way between the rim and the hub, which was not observed in the wheel. This is an artefact of the 2D axi-symmetric analysis, as the shrinkage is not predicted using a 3D version of the model.* Thus, the initial boundary conditions lacked the sophistication needed for the optimizer to reach a valid solution. Turning to the optimized results, re-running the optimizer failed to produce a solution that was free of encapsulation / shrinkage in the upper rim owing to the elongated 75% solid fraction isotherm; however, it did reduce the severity of the encapsulation. The prediction of mid-spoke shrinkage was also eliminated. The conclusion from this result was that the current design of the dies from the standpoint of size and placement of the cooling channels in the vicinity of the top rim would not yield a wheel free of porosity without the addition of a riser to the top of the rim.

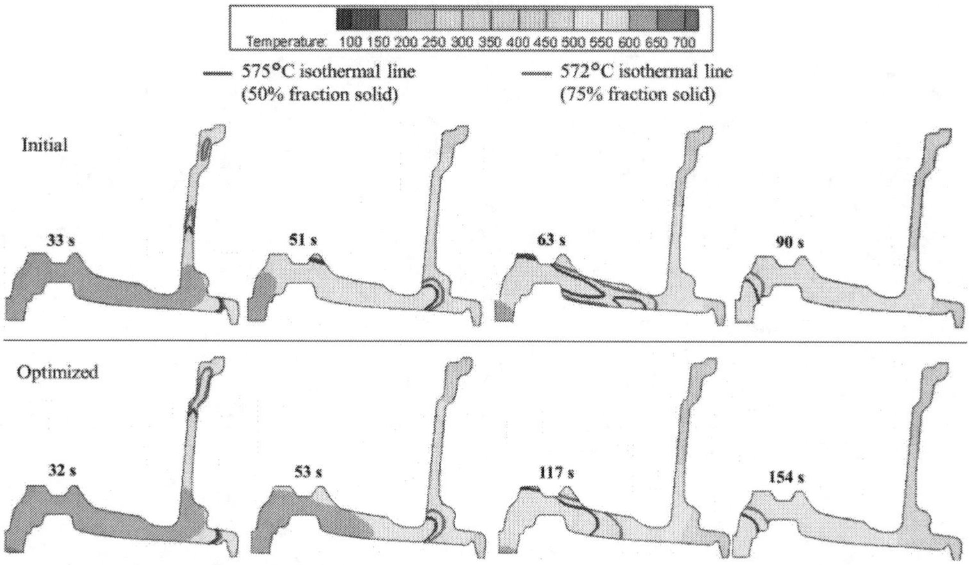

Figure 6. The initial (the upper row) and the optimized solidification conditions (the lower row) for the model run with the updated boundary conditions.

5. Conclusions

This paper discusses the development and the application of an optimization methodology to an LPDC process used for the commercial production of A356 automotive wheels. The optimization methodology has been successfully applied to optimize the on and off cooling timings of up to 7 water-cooling channels both from the standpoint of eliminating shrinkage based porosity associated with encapsulation of liquid and reducing the overall solidification time (process cycle time). The initial application of the results of the optimizer in an industrial setting revealed an inconsistency in the prediction of porosity in the upper rim area with what was observed in the cast wheel. One of the primary reasons for the inconsistency was the overly simplified boundary conditions used in the thermal model that was utilized within the optimizer. However, reapplication of the optimizer on a model with updated boundary conditions revealed limitations in the underlying design of the dies with respect to the placement and size of some of the cooling channels. A key and perhaps obvious finding of the study is that the utility of numerical optimizers applied to casting processes in an industrial setting hinges critically on an accurate description of the process conditions. Looking forward, design of dies for the production of wheels will ultimately require the development and application of topological optimization algorithms in conjunction with process timing optimization, a formidable computational challenge.

References

[1] Vanderplaats G N 1984 *Numerical Optimization Techniques for Engineering Design: With Applications* (McGraw-Hill, New York).
[2] Ellis J A and West P E 1970 A structural optimisation technique, *Comput. -Aided Des.* **2** 24-28.
[3] Petiau C 1991 Structural optimization of aircraft *Thin Wall Struct.* **11** 43-64.
[4] Ebrahimi S A, Tortorelli D A, Dantzig J A 1997 Sensitivity analysis and nonlinear programming applied to investment casting design *Appl. Math. Model.* **21** 113-123.
[5] Morthland T E, Byrne P E, Tortorelli D A, Dantzig J A 1995 Optimal riser design for metal castings *Metall.Mater.Trans.B* **26** 871-885.
[6] Singh D PK, Mallinson G D, Panton S M 2002 Applications of optimization and inverse modeling to alloy wheel casting *Numer. Heat. Tr. A-Appl* **41** 741-756.
[7] Singh D P K, Mallinson G D, Panton S M, Palle N Die design strategy for improved productivity and quality in die casting 1999 *Trans. Am. Foundrymen's Soc.* **107** 127-133.
[8] Reilly C, Duan J, Yao L, Maijer D M, Cockcroft S L 2013 Process Modeling of Low-Pressure Die Casting of Aluminum Alloy Automotive Wheels *JOM.* **65** 1111-1121.
[9] Zhang B, Maijer D M, Cockcroft S L 2007 Development of a 3-D thermal model of the low-pressure die-cast (LPDC) process of A356 aluminum alloy wheels *Mat. Sci. Eng. A-Struct* **464** 295-305.
[10] Thompson S, Cockcroft S L, Wells M A 2004 Advanced light metals casting development: solidification of aluminium alloy A356 *Mater. Sci. Tech.* **20** 194-200.
[11] Zhang B, Cockcroft S, Maijer D, Zhu J, Phillion A 2005 Casting defects in low-pressure die-cast aluminum alloy wheels *JOM.* **57** 36-43.
[12] Wei X 2014 *Thermal mechanical analysis of interfacial behavior in aluminum alloy wheel casting process* (Master thesis, The University of British Columbia).
[13] Moayedinia S 2014 *Quantification of cooling channel heat transfer in the low pressure die casting modeling* (Master thesis, The University of British Columbia).

Integrated modelling of transitions in mechanical conditions during casting and heat treatment

J Thorborg , J Klinkhammer and M Heitzer
MAGMA Gießereitechnologie GmbH, Aachen, Germany

E-mail: J.Thorborg@magmasoft.de

Abstract. The mechanical material behaviour of a cast component changes significantly during casting and heat treatment. The big difference in temperature levels during the different process steps causes different deformation mechanisms to be active. The thermal gradients promote transient stresses that can lead to inelastic deformations, residual stresses and in some cases to defects in the final part. It is a big challenge to make a reasonable transition in the mechanical model, and hence material data, when modelling several different coupled process steps. It is important to use an integrated approach where the transition is included in the full load history of the part. When industrial examples are considered, the sequence of process steps typically also changes the thermal and mechanical boundary conditions significantly e.g. going from being mechanically constrained during casting to being supported point-wise during the heat treatment process. This change includes mapping of results and obtaining equilibrium in a new global system, where the further reaction forces from the supports must be handled with contact conditions to e.g. predict deformations due to gravity during solution heat treatment.
The work presented in this paper is focused on modelling the mechanical fields, taking into account the changes in the mechanical material model at different temperature levels, and the transition in mechanical behaviour when the microstructure is changing during the different steps of the heat treatment process. The approach used is based on a unified model where creep effects are considered at high temperature and rate effects are included in general during cooling. Proposals are made to include cooling rate sensitivity, annealing and precipitation hardening via modification of mechanical properties in the different process steps.

1. Introduction

Casting of aluminum parts for the automotive industry is often followed by a heat treatment process, where the mechanical properties are improved by modifying the as-cast microstructure in a sequence of thermal steps. In this way, it is possible to obtain a material with increased ductility and higher strength compared to the as-cast conditions. The process view of the casting and heat treatment processes can be used to illustrate the thermal history of the part, see figure 1, left. The different steps can have significant influence on the stress levels and the distortion that can build up during e.g. elevated temperatures.

The most widely used aluminum casting alloys are based on different contents of Si, Cu and Mg. For the hypoeutectic alloys, the microstructure consists of a primary α-Al phase and Al-Si eutectic. The additional components Mg and Cu are either contained in the α-Al in solid solution or

IOP Conf. Series: Materials Science and Engineering **84** (2015) 012037 doi:10.1088/1757-899X/84/1/012037

precipitated as intermetallic phases e.g. Mg_2Si or Al_2Cu. The content of Mg and Cu in solid solution is temperature dependent according to the phase diagram. This temperature dependency is exploited during solution treatment to form a super-saturated solution of Mg and Cu, which is maintained during quenching to allow a controlled precipitation of intermetallic phases during artificial aging.

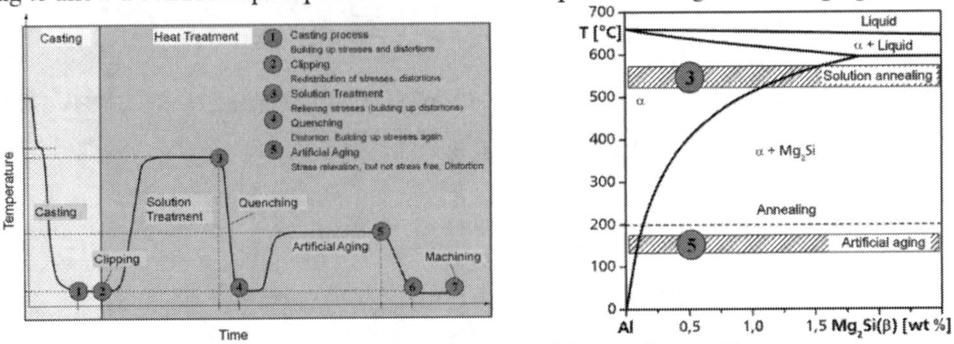

Figure 1. Left: Process view showing the thermal history of the casting and heat treatment processes. Right: The phase diagram indicating the path for generating the super-saturated solid solution.

During the casting and the heat treatment process steps, stresses and deformations develop depending on the thermo-mechanical loading of the material. High thermal gradients can lead to significant stresses that promote the risk of cracks and cause inelastic strains, which can lead to residual stresses in the final part. To get an assessment of the thermo-mechanical loading of the material it is important to perform a numerical simulation where the different process steps are modeled. However, when these process steps are analyzed using numerical simulation, the mechanical properties are typically only assumed to be temperature dependent and at elevated temperature also strain rate dependent, [1]. Hence, the evolution of the mechanical properties with respect to microstructure is in general not taken into account. Nevertheless, the reason for doing heat treatment is to improve the performance of the material, which will also affect the numerical analysis. It is the purpose of this paper to include the evolution of the mechanical properties in the numerical modeling of the casting and heat treatment processes. In this paper it has not been the intention to model the microstructure and relate this to the mechanical properties, but rather to identify what is important to include in order to make the first steps in having local properties that depend on the thermal history, i.e. temperature level and cooling rate.

2. Modelling the evolution of mechanical properties

Thermo-mechanical modeling of the casting and heat treatment process is a challenge even without considering the evolution of the material data. The main concern is to model the response of the material at different temperature levels, on different time scales and sometimes with different strain rates, which is governed by different deformation mechanisms. To have a reasonable compromise between the stability in the results when doing calculations in different process steps and agreement between simulations and performed creep and tensile tests, a unified creep formulation is used as the fundamental constitutive law, [2]. The model is based on Norton's power law and includes by that, strain rate sensitivity and the possibility to describe creep at elevated temperature, see (1) below.

$$\dot{\varepsilon}^{in} = A \exp\left(\frac{-Q}{RT}\right)\left(\frac{\sigma}{\sigma_{ref}}\right)^{m} \tag{1}$$

The Arrhenius expression scales the response according to the temperature dependency and the model is therefore applicable in a wide temperature range. The reference stress σ_{ref} describes the isotropic strain hardening by a classical power law,

$$\sigma_{ref} = \sigma_{0,ref}\left(1+\frac{Eh}{n\sigma_{0,ref}}\right)^{n}, \tag{2}$$

where the hardening parameter h is based on the inelastic strain to capture the effect of hardening when e.g. dislocations are piling up and annealing when the temperature is elevated, allowing diffusion processes to be active. The model is governed by two temperature dependent properties, the initial reference stress $\sigma_{0,ref}$ and the hardening parameter n. The response of the creep equation and the hardening law can be illustrated by the classical creep curve and the stress-strain curve in figure 2 left and right, respectively. It should be noted that the stress response of the material is not directly given by the reference stress curve illustrated by the dashed line in figure 2 right, but a similar response where the level of the response depends on the strain rate, see the full line in the same graph.

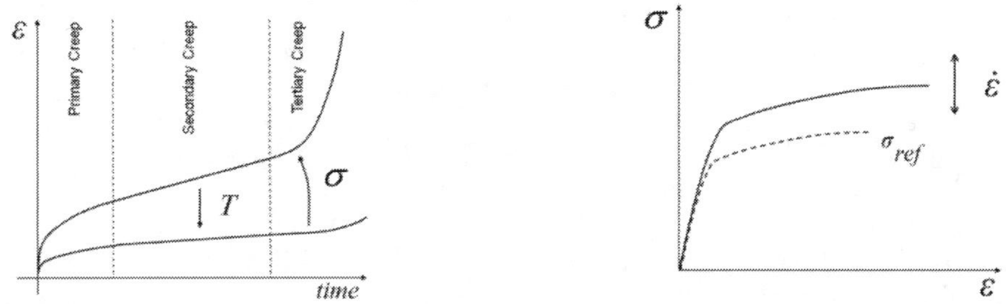

Figure 2. Left: The creep response of the material model. Right: The hardening response of the material model.

The following subsections will give a short summary of how the different process steps influence the mechanical fields and the mechanical properties due to changes in the microstructure. The proposed implementations are used on industrial examples in the simulation sections.

2.1. Cooling rate sensitivity during casting and quenching

The stresses mainly build up during the cooling and quenching steps, where thermal gradients promote different thermal contraction and corresponding internal constraints, which can lead to stresses in the material. For high cooling rates, not only the thermal gradients influence the level of the internal stresses but also the mechanical properties are reported to depend on the cooling rate. Kessler et.al. [3] have performed a series of measurements in a quenching deformation dilatometer to determine the properties at different temperature levels for different cooling rates for a wrought aluminum alloy. It was seen by the measurements that the examined alloy had a clear cooling rate dependency at the lower and intermediate temperature levels. The same dependency has been adapted for the considered alloy in this paper, A356, and two of the resulting curves are shown in figure 3.

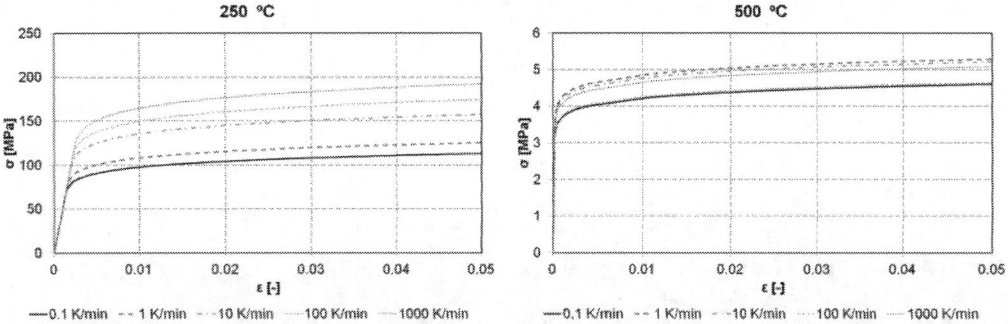

Figure 3. Influence of the cooling rate on the mechanical properties for the considered alloy in this paper, A356, for two different temperatures.

In the simulations, the influence of the cooling rate has been included by using the locally calculated cooling rate to modify the input data for the hardening law according to the curves shown in figure 3. That is, the initial reference stress depends locally on the cooling rate. In this way the hardening response of the material will be cooling rate sensitive and by that follow the curves in figure 3 in the simulations. The effect from this modification is mainly expected to be seen in the quenching process but of course it also has an influence during cooling in the casting process.

2.2. Annealing during casting and solution treatment

During mechanical loading at lower temperatures, plastic deformation leads to strain hardening of the material, which can be measured by a classical tensile test. The response is typically described by a power law, which can be used to fit the tensile measurements for many alloys, see equation (2). At elevated temperatures the mechanical response is no longer making a contribution to additional strain hardening and it is observed that, depending on temperature and time, the already existing strain hardening vanishes. This type of softening is called annealing[1], which is caused by dislocations starting to diffuse to e.g. the grain boundaries and by that removing the dislocation obstacles that caused the hardening effects at lower temperatures. In this work it is proposed to use the inelastic strain rate to both describe the strain hardening and the annealing of the material, by multiplying the contribution from the inelastic strain increment to the strain hardening parameter by a temperature dependent factor. The strain hardening variable used in the strain hardening law, equation (2), is updated according to the equation below

$$\Delta h = f_h \, \Delta \varepsilon^{in}, \qquad (3)$$

where Δh is the increment in the strain hardening parameter h.

In this way it is possible to add inelastic strain to the strain hardening parameter at low temperatures, resembling strain hardening, and subtract inelastic strain from the hardening parameter at higher temperatures, resembling annealing. This functionality plays an important role during solution treatment, where the material is more or less stress free with no strain hardening at the end of solution treatment. Of course, annealing also has an influence on the mechanical behavior during the casting process which is also seen in the simulations.

The temperature dependent factor f_h in the hardening evolution law, equation (3), is assumed to follow the curve given in figure 4. The curve has been adjusted to give a classical update at lower and intermediate temperature levels, where $\Delta h = \Delta \varepsilon^{in}$, and to remove hardening at higher temperature levels to resemble annealing. The adjustments of the curve have been done with simulations, but of course the goal is to couple this adjustment to measurements.

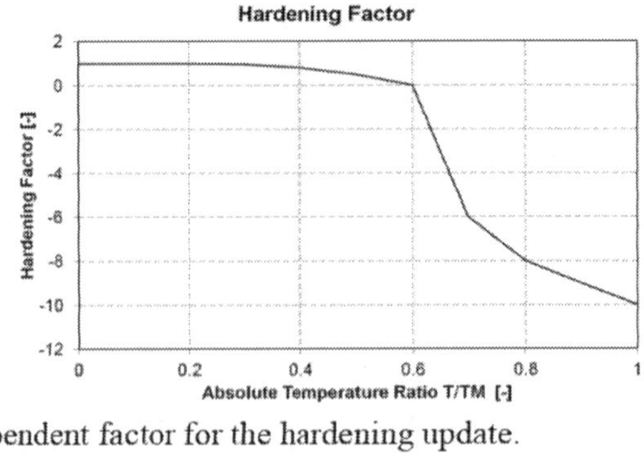

Figure 4. Temperature dependent factor for the hardening update.

[1] In literature annealing is used for cold worked metals and divided into the processes: recovery, recrystallization and grain growth, [4]. For cast alloys we use the term annealing to describe the combined effect of increasing ductility and reducing strength.

2.3. Precipitation hardening during artificial aging

The final step of the heat treatment process is used to strengthen the material due to precipitation hardening. However, the effect of this step is quite sensitive to the process temperature and the time the part is heated. According to measurements presented in [5], the considered A356 alloy has an increased yield strength and tensile strength when precipitation hardened at 150 °C and 205 °C for 3-5 hours. For a higher aging temperature, 260 °C, the material starts to overage already after approx. 1 hour, and after 5 hours the influence of precipitation hardening has vanished.

The influence of precipitation hardening is included in the simulations by modifying the mechanical data that describes the yielding behaviour. The aging temperature and aging time is used to evaluate the influence on the yield curves and hence used to scale the initial reference stress in the hardening function, equation (2). The curves presented in figure 5 illustrate the input to the simulations presented in the next section. It is assumed that the quench process was fast enough to maintain the super-saturated solid solution of alloying elements to make an optimal initial condition for the artificial aging process.

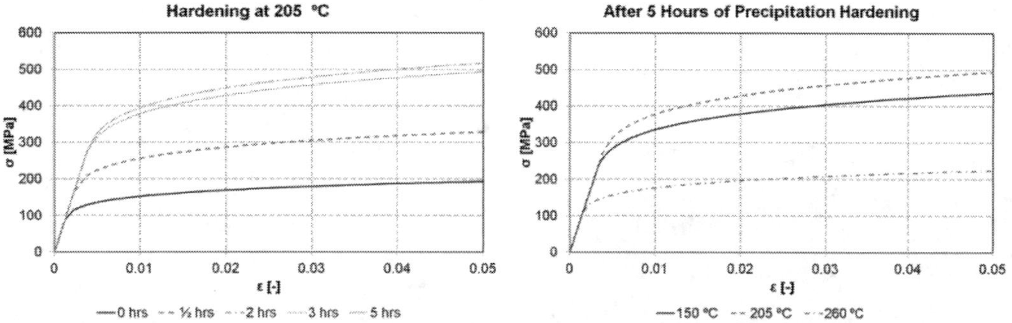

Figure 5. Influence of precipitation hardening on the yield curves, left: different aging times at 205 °C and right: for different temperature levels after 5 hours.

3. Modelling gravitational load during solution treatment

During solution treatment, structural parts can experience a considerable amount of deformation due to their own weight. To achieve a prediction of a components final distortion, it is necessary to consider all relevant steps of the manufacturing process. This imposes some challenges for the simulation of the integrated process. The importance of an appropriate simulation technique stems from a relatively new trend in the automotive industry to apply distortion engineering to structural parts, by designing support frames in a way that exploits the distortion capability during solution heat treatment in order to control the deformation so that subsequent straightening operations are minimized [6].

Typically the parts are placed in a support frame, which means that the deformed parts need to be positioned into the supports and at the same time the coordinate system changes compared to the casting process. The positioning must be done in a way that the system, including contact between the part and the supports under the influence of gravity, becomes statically determinate.

The gravitational load is extremely small compared to the thermal load in heating or cooling steps so that the solution of a linear sub-step in a Newton-Raphson scheme is close to a singular solution of the equation system and the detection of global convergence, i.e. the fulfillment of the force equilibrium after several Newton-Raphson iterations, is very sensitive to these conditions. The first Newton-Raphson iteration requires appropriate initial conditions for the contact algorithm, and if the system is statically indeterminate, subsequent iterations will either reveal a rigid body motion or will simply diverge. Furthermore, if one wants to account for the thermal influence of the supports and/or their deformation, a new enmeshment of the calculation domain and mapping of state and history variables from the end of casting to the beginning of heat treatment is needed. This will require additional equilibrium iterations at the start of the heat treatment simulation. Altogether, it is extremely import to have sufficient and well described boundary conditions to get convergence and 'high' accuracy in the solution.

4. LPDC wheel example

Several simulations have been performed to examine the influence of different process settings and the influence of cooling rate sensitivity, strain hardening scale factor and finally precipitation hardening during artificial aging. The initial conditions for the heat treatment simulations are based on a casting simulation with the implemented modifications activated. The considered example in figure 6 to 8 is an aluminum rim with an approximate diameter of 420mm and width of 205mm. The wall thickness is in the interval of 10-30mm.

4.1. Influence of different solution treatment temperatures

Due to the relatively high temperature level during solution treatment, stresses are relaxed to zero during this step and hardening vanishes due to annealing. This is shown for 3 different solution treatment temperature levels, where the v. Mises stress level is shown in the right graphs and the hardening variable in the left side of figure 6. The 3 different temperatures are: T_{sol} = 525 °C, T_{sol} = 500 °C and T_{sol} = 475 °C.

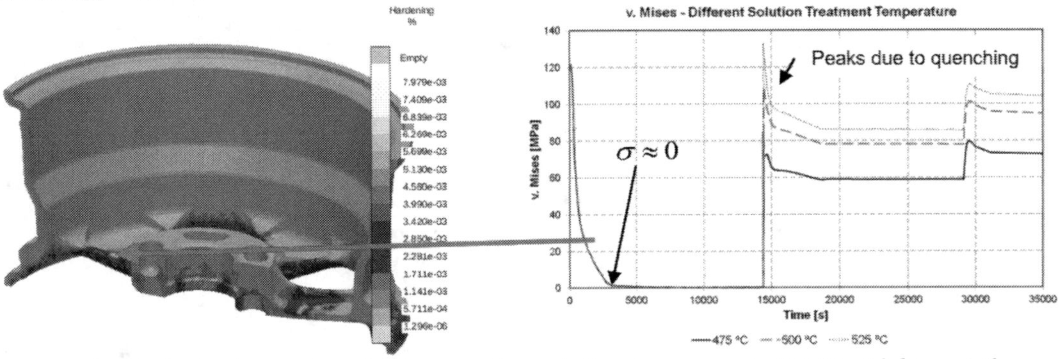

Figure 6. Left: Vanishing hardening distribution after solution treatment at 525 °C. Right: v. Mises stress level as function of time at the selected point for 3 different solution temperatures.

4.2. Influence of different quench media – end of quenching

Quenching after solution treatment is a compromise between sufficiently fast cooling to get an optimal precipitation hardening afterwards and a low cooling rate to reduce the thermal gradients and by that the transient and residual stresses [7]. The residual stress distributions after quenching are shown for 3 different quench media in figure 7. As expected the residual stress level is highest for water quenching with T_{water} = 40 °C compared to water quenching with T_{water} = 95 °C and air quenching.

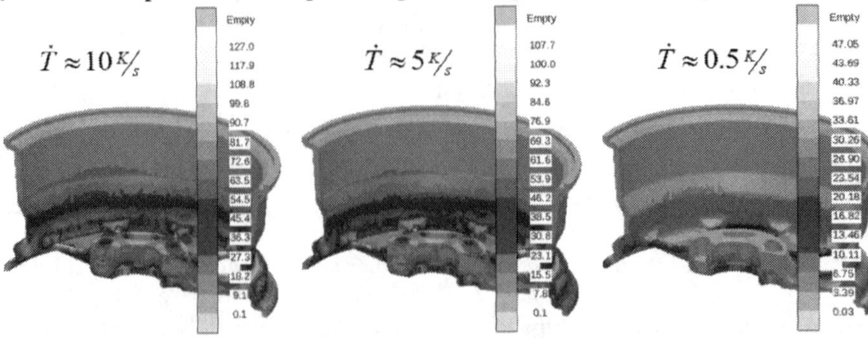

Figure 7. v.Mises stress level at the end of quenching. Left: Water quenching T_{quench} = 40 °C, Middle: Water quenching T_{quench} = 95 °C and Right: Air cooling T_{quench} = 25 °C.

4.3. Influence of aging temperature and time

Finally, the aging process is considered for 3 different aging temperatures, T_{aging} = 150 °C, T_{aging} = 205 °C and T_{aging} = 260 °C. The influence on the yield stress level is shown at the considered point for the different temperature levels in figure 8. The yield stress evolution is governed by the temperature level and aging time according to the graphs in figure 5. At the lowest aging temperature, the influence of precipitation hardening is delayed, while at 205 °C the precipitation hardening is quite early and stable during the rest of the process. It is clearly seen that at T_{aging} = 260 °C the material overages very quickly.

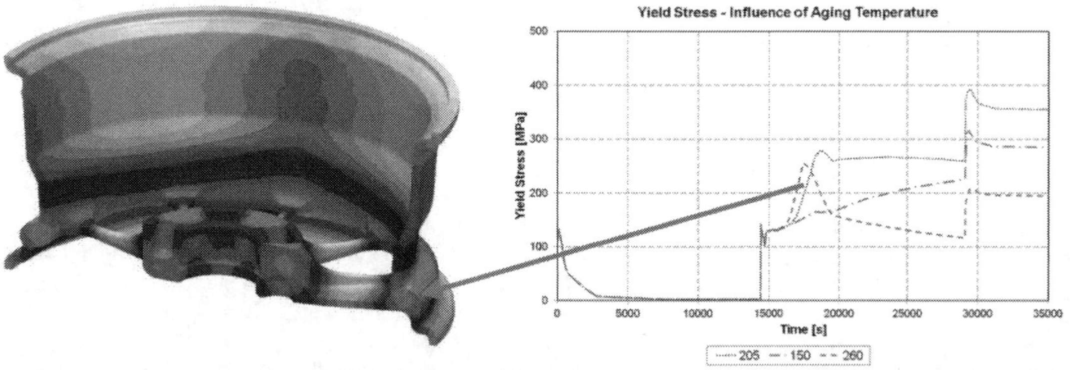

Figure 8. Left: Temperature profile of the considered heat treatment process. Right: Evolution of the yield stress at the considered point for the 3 different aging temperatures.

5. Deformation of structural parts during solution heat treatment

Figure 9 shows a rear spoiler with a fictitious arrangement of support pins from below. The part is approximately 900mm long, 50mm high and has a thickness of 2-5mm. Figure 10 shows the vertical displacement for different support scenarios, where the numbers indicate the active pins. The two selected results show the deformation after 30 minutes of solution annealing at 480 °C. It is clearly seen how the different number of supports affects the results.

Figure 9. Rear spoiler with fictitious arrangement of support pins.

Figure 10. Deformations during solution treatment for different support systems.

In the example, the pins are modeled as rigid and fixed in space using contact boundary conditions [8]. Also note that pins 2, 4 and 6 slightly surround the front edge in order to fix the part against sliding as can be seen in figure 9.

6. Conclusion and future work

The work presented in this paper shows the influence of including the evolution of local mechanical properties in the casting and heat treatment processes. Three modifications have been implemented and tested on industrial cast aluminum parts for the automotive industry, a wheel and a rear spoiler. The motivation has been to describe the influence of cooling rate sensitivity, annealing and precipitation hardening to get a consistent description of the mechanical material behavior. For the wheel example, the focus has been on different process conditions and for the rear spoiler the focus has been on the distortions during solution treatment due to the gravitational forces. The used material data are based on measurements and the applied modification due to cooling rate and precipitation hardening are based on information from literature. No coupling was done directly to the microstructure evolution. Future work will be focused on evaluating measurements of properties at different states. Typically, one has measurements of mechanical properties after fabrication, i.e. as-cast conditions, and after heat treatment. Ideally, the samples for determination of the as-cast properties should be remelted and cooled down to the testing temperature with an appropriate cooling rate corresponding to the casting process, which in the case of A356 can be sand casting, die casting or high pressure die casting. If the samples are just heated to the testing temperature, the measured mechanical properties actually refer to the heating phase of the solution treatment step. Data should be recorded with each sample going through the course of casting and heat treatment until reaching the testing temperature and time. Extending the knowledge of the material behavior with further measurements, hopefully also makes it possible to make a more direct relationship to microstructure calculations. Mapping of results between the process steps and handling of gravitational loads and supports to avoid a singular system are presented and discussed.

References

[1] Thorborg J, Hattel JH and Bellini A 2006 Thermomechanical conditions in heat treated aluminium cast parts in Casting *Proc. from the 11th Int. Conf. on, Modeling of Casting, Welding and Advanced Solidification Processes* vol. 11, eds C A Gandin and M Bellet (Warrendale, PA: TMS) pp. 193-200

[2] Thorborg J, Klinkhammer J and Heitzer M 2012 Transient and residual stresses in large castings, taking time effects into account *Proc. from the 13th Int. Conf. on, Modeling of Casting, Welding and Advanced Solidification Processes* vol. 13, eds Ludwig A, Wu M and Kharicha A (IOP Conf. Ser.: Mater. Sci. Eng. 33) 012050

[3] Reich M and Kessler O 2012 Quenching Simulation of Aluminum Alloys Including Mechanical Properties of the Undercooled States *Materials Performance and Characterization* **1** 1-18

[4] Dieter G E 1988 Mechanical Metallurgy - SI Metric Edition adapted by D Bacon (London, McGraw-Hill) pp 233-236

[5] Charlie R. Brooks, Principles of Heat Treating of Nonferrous Alloys, Heat treating vol 4, 10 Edition, ASM International, 1991

[6] Karsten O, Schimanski K, von Hehl A and Zoch, H-W 2011 Reasons and influencing factors on distortion within a process chain of an aluminium die casting component in the automotive industry *3rd Int. Conf. on Distortion Engineering* ed H-W Zoch and T Lübben (Bremen) pp 179-186

[7] Manente A and Timelli G 2011 Optimizing the Heat Treatment Process of Cast Aluminium Alloys *Recent Trends in Processing and Degradation of Aluminium Alloys* ed Z Ahmad (Rijeka, InTech) 197-220

[8] Klinkhammer J and Thorborg J 2009 Modeling of Mechanical Contact Conditions in Casting *Proc. from the 12th Int. Conf. on Modeling of Casting, Welding and Advanced Solidification Processes* vol 12

Modelling the solidification of ductile cast iron parts with varying wall thicknesses

M Bjerre[1], N S Tiedje[1] J Thorborg[2] & J H Hattel[1]

[1] Department of Mechanical Engineering, Technical University of Denmark
[2] Magma GmbH

E-mail: makabj@mek.dtu.dk

Abstract. In the present paper modelling the solidification of cast iron parts is considered. Common for previous efforts in this field is that they have mainly considered thin walled to medium thickness castings. Hence, a numerical model combining the solidification model presented by Lesoult et al. [1] with a 2D FE solution of the heat conduction equation is developed in an in-house code and model parameters are calibrated using experimental data from representative castings made of ductile cast iron. The main focus is on the influence of casting thickness and resulting local cooling conditions on the solidification pattern and the relation to formation of degenerate graphite.

1. Introduction

Ductile cast iron (DCI) is widely used in the automotive industry but also for parts in wind turbines (bed frame, main shaft and rotor hub) due to the combination of good mechanical properties and low cost. In heavy sections, however, degenerate graphite morphologies might form reducing ductility and fatigue strength of the part. A wide range of factors have been found to influence the formation of degenerate graphite as reviewed by Källbom et al. [2] and the mechanisms describing its formation are still not fully understood.

One of the factors known to be important is the thermal conditions. Slow cooling rates are often associated with the formation of degenerate graphite and the degenerate morphology mainly appears at the thermal centre of heavy sections. The cooling curves measured from heavy castings usually also show distinct features not seen in ordinary DCI castings, [3], such as up to three arrests associated with primary solidification, initial limited eutectic growth and bulk eutectic growth.

In this paper, we validate a model combining a temperature solver and the microstructural model originally proposed by Lesoult et al. [1] and further developed by Pedersen et al. [4]. This is done by comparing temperature curves and nodule counts from the simulation to experimental data from bars with cross sections of 34.7 mm $\times$ 34.7 mm, 52 mm $\times$ 52 mm and 65 mm $\times$ 65 mm. We then use our validated model to simulate a thick-walled part of cross section 300 mm $\times$ 300 mm in order to study the solidification pattern and thermal gradients the part experiences during solidification as the thermal conditions are at present believed to play a major role for the formation and distribution of porosities, [5], and may also be significant for formation of degenerate graphite. The model does not describe these phenomena but the overall solidification pattern including temperature fields is nevertheless of interest in this relation.

2. Experimental data

A series of experiments were carried out where bars of different thermal moduli were cast from a high-silicon melt. The experiment is presented elsewhere, Vedel-Smith et al. [6], and we will only present it briefly: Bars in three sizes were produced using a scalable model where the length of the bar is 3 times the length of the two remaining sides. The side lengths were 34.7 mm, 52 mm and 65 mm producing bars of thermal moduli 8 mm, 10 mm and 15 mm respectively. The dimensions of the bars allow us to treat the bars as two dimensional objects in the sense that it is reasonable to assume that the thermal variations in the length direction of the bar are negligible. The bars were cast from a hypereutectic high-silicon melt, see table 1 and the temperature was measured at the centre of each bar using a k-type thermocouple. Microstructural data in terms of nodule number densities were obtained from castings of side lengths 52 mm and 65 mm, however not from those in which the temperature was measured. The compositional variations between the castings were however small and will have limited effect on the conclusions to be drawn here.

Table 1. Alloy composition [%wt], carbon equivalent (CE) and casting temperature [6].

CE	C	Si	Casting T [°C]
4.57	3.31	3.81	1387

3. Numerical Model

A two dimensional thermal model has been set up for casting and mould and it is coupled to a microstructural model that describes the solidification of the casting. The casting domain is a quarter of a bar cross section, see Fig. 1, and is discretised using 10 by 10 elements for the smallest bar and 20 by 20 for the largest. The total size of the domain is 150 mm by 150 mm. Convective cooling is applied on the two sides which represent the outside of the mould while adiabatic boundary conditions are applied at the remaining two sides representing internal symmetry lines.

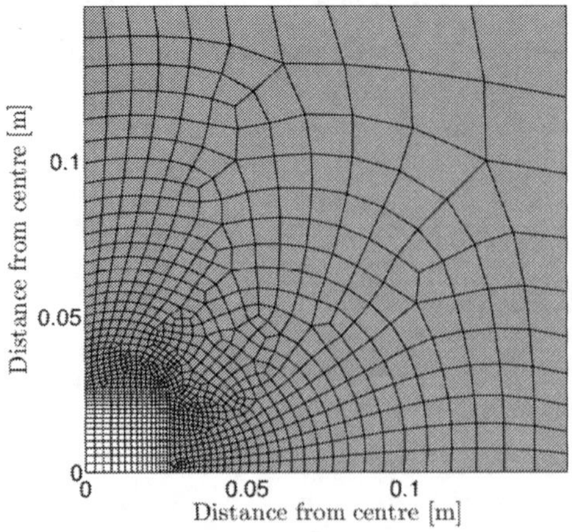

Figure 1. Calculation domain for the casting with side lengths of 52 mm. The shaded area and the white area represent the mould and the casting, respectively. Adiabatic boundary conditions are applied on left and bottom boundaries and convective cooling on the right and top boundaries. The mesh was generated using ABAQUS.

Microstructural model

The microstructural developments during solidification are described using the model developed by Lesoult et al. [1, 7]. In the case of a hypereutectic melt, the model predicts the nucleation of primary graphite forming as the first solid phase in the melt. The nucleation is described by a modified version of Oldfield's model in which the rate of nucleation, ΔN_i, is determined by the change of the undercooling with respect to the graphite liquidus, ΔT_L^g, and the remaining liquid volume, V^{liq}, in the element:

$$\mathrm{d}N_i = \begin{cases} A_n(\Delta T_L^g)^{n-1} \cdot V^{\text{liq}} \frac{\mathrm{d}\Delta T_L^g}{\mathrm{d}t}\mathrm{d}t & \text{if} \quad \frac{\mathrm{d}\Delta T_L^g}{\mathrm{d}t} > 0 \\ 0 & \text{else} \end{cases} \tag{1}$$

where A_n is a constant related to the exponent n, which is set to 1 in this and previous works [7, 4]. The nodules nucleated at time point i then represent a family of nodules which will grow by the same rates. This is a reasonable assumption as two nodules nucleated simultaneously in the same volume element will experience the same growth conditions with this model.

As the volume reaches the austenite liquidus determined by the phase diagram, the nodules are immediately encapsulated in an austenite shell of a predefined thickness δr^γ as is any new nodule nucleated from this point on. As soon as the eutectic solidification has started the nodules grow only by diffusion of carbon through the austenite shells, while the shell is in direct contact with the melt. The dynamic equations for the radius of the graphite nodule, r^g, and the austenite shell, r^γ, during eutectic solidification are described by

$$\frac{\mathrm{d}r^g}{\mathrm{d}t} = \frac{\rho^\gamma}{\rho^g} D_C^\gamma \frac{1}{1 - w_C^{\gamma/g}} \frac{r^\gamma}{(r^\gamma - r^g) r^g} (w_C^{\gamma/l} - w_C^{\gamma/g}) \tag{2}$$

$$\frac{\mathrm{d}r^\gamma}{\mathrm{d}t} = \frac{D_C^\gamma r^g}{(r^\gamma - r^g) r^\gamma} \frac{w_C^{\gamma/l} - w_C^{\gamma/g}}{w_C^{l/\gamma} - w_C^{\gamma/l}} \left(1 + \frac{\rho^\gamma - \rho^g}{\rho^g} \frac{w_C^{l/\gamma} - w_C^{\gamma/l}}{1 - w_C^{\gamma/g}}\right) + \Delta(r^\gamma) \tag{3}$$

with

$$\Delta(r^\gamma) = \begin{cases} \frac{1}{m_C^\gamma} \frac{\partial T}{\partial t} \frac{\Phi \cdot \frac{1}{3}\left(\rho^l ({}^o r)^3 - \rho^g (r^g)^3 - \rho^\gamma ((r^\gamma)^3 - (r^g)^3)\right)}{(r^\gamma)^2 \rho^\gamma (w_C^{l/\gamma} - w_C^{\gamma/l})} & \text{if} \quad V^\gamma \leq 0 \\ 0 & \text{if} \quad V^\gamma > 0 \end{cases}$$

$$\Phi = \frac{\rho^l (1 - f^\gamma) + \rho^\gamma f^\gamma k_C}{\rho^l (1 - f^\gamma) + \rho^\gamma f^\gamma}$$

where ρ^ϕ is the density of phase ϕ, $w_C^{\phi/\theta}$ is the carbon content in phase ϕ in equilibrium with phase θ. D_C^γ is the diffusion coefficient of carbon diffusing in austenite. In $\Delta(r^\gamma)$, m_C^γ is the slope of the austenite liquidus, k_C is the partition coefficient and ${}^o r$ is the average distance between nodules. Further, f^γ is the fraction of the off-eutectic volume which is austenite.

During solidification, eutectic spheres will increasingly impinge reducing the actual growth compared to the rates described by Eq. 2 and 3. This is taken into account by multiplying the growth rates of the total eutectic and graphite volume in the element by the Avrami correction factor $\psi = 1 - f^s$, f^s being the fraction solid, [7],

$$\frac{\mathrm{d}V^{\text{Eut}}}{\mathrm{d}t} = \sum_i \Delta N_i (r^\gamma)^2 \frac{\mathrm{d}r^\gamma}{\mathrm{d}t} \psi \qquad \frac{\mathrm{d}V^g}{\mathrm{d}t} = \sum_i \Delta N_i (r^g)^2 \frac{\mathrm{d}r^g}{\mathrm{d}t} \psi \tag{4}$$

where the sum is over the nucleation events and V^{Eut} and V^g are the total eutectic and graphite volume of an element respectively. Nucleation and growth of nodules reduce the carbon content of the melt which in some cases is compensated for by sufficient growth of the eutectic austenite keeping the carbon content of the liquid at the austenite liquidus. In other cases, especially when the cooling is strong, the growth of eutectic austenite will not be sufficient. The model then predicts the growth of off-eutectic austenite dendrites to keep the melt in equilibrium with austenite giving a growth rate of off-eutectic austenite of

$$\frac{\mathrm{d}f^\gamma}{\mathrm{d}t} = \frac{-X_C}{(1-k_C)w_C^l} + \frac{\Phi}{(1-k_C)w_C^l}\frac{1}{m_C^\gamma}\frac{\mathrm{d}T}{\mathrm{d}t} \tag{5}$$
$$X_C = \frac{-\rho^\gamma w_C^l[1-\Phi]\left(\frac{\mathrm{d}V^{\mathrm{Eut}}}{\mathrm{d}t} - \frac{\mathrm{d}V^g}{\mathrm{d}t}\right) - (\rho^\gamma - \rho^g)[(1-\Phi)w_C^l + w_C^{\gamma/g} - w_C^{\gamma/l}]\frac{\mathrm{d}V^g}{\mathrm{d}t}}{\rho^l V^\circ - \rho^g V^g - \rho^\gamma(V^{\mathrm{Eut}} - V^g)}$$

where w_C^l is the carbon content of the liquid phase.

Thermal model

Heat transfer is modelled using a 2D finite element thermal solver. The description of the heat transfer coefficient (HTC) across the casting-mould interface and the coupling to the microstructural model is done as described by Pedersen et al. [4]: The HTC is made dependent on the fraction solid of the surface elements of the casting, with an initial value of 1500 W/m^{2}K ending at 300 W/m^{2}K when the fraction solid reaches 0.4. The latent heat from solidification is released mainly through a direct source term but also partly through a modified ρc_p in the heat conduction equation: The heat conduction equation with constant material properties is given as

$$\rho c_p \frac{\mathrm{d}T}{\mathrm{d}t} = k\nabla^2 T + \dot{Q}''' \tag{6}$$

where ρ in this case represents an average density in the element and $\dot{Q}'''$ is the latent heat release rate by solidification:

$$\dot{Q}''' = \Delta H \cdot \left(\frac{\mathrm{d}V^{\mathrm{Eut}}}{\mathrm{d}t}\rho + \rho^\gamma \frac{\mathrm{d}f^\gamma}{\mathrm{d}t}V^{\mathrm{off}} + \rho^\gamma f^\gamma \frac{\mathrm{d}V^{\mathrm{off}}}{\mathrm{d}t}\right)\frac{1}{V^t} \tag{7}$$

where ΔH is the specific latent heat of the casting alloy and V^t is the element volume. Inserting the equation for $\frac{\mathrm{d}f^\gamma}{\mathrm{d}t}$ in $\dot{Q}'''$ and rearranging we have

$$\widetilde{\rho c_p}\frac{\mathrm{d}T}{\mathrm{d}t} = k\nabla^2 T + \tilde{\dot{Q}}''' \qquad \text{with} \tag{8}$$
$$\widetilde{\rho c_p} = \left(\rho c_p - \Delta H \rho^\gamma \frac{\Phi V^{\mathrm{off}}}{(1-k_C)w_C^l m_C^\gamma}\frac{1}{V^t}\right)$$
$$\tilde{\dot{Q}}''' = \Delta H \rho^\gamma \cdot \left(\frac{\mathrm{d}V^{\mathrm{Eut}}}{\mathrm{d}t} - \frac{X_C \cdot V^{\mathrm{off}}}{(1-k_C)w_C^l} + f^\gamma \frac{\mathrm{d}V^{\mathrm{off}}}{\mathrm{d}t}\right)\frac{1}{V^t}$$

Thus we include the latent heat released from the off-eutectic austenite when solving for the temperature, even though the amount of solidified material is yet unknown. After having calculated the rate of change of temperature the change in the off-eutectic austenite solid fraction can be obtained. Unless otherwise noted, the model constants are as in Ref. [4].

4. Validation of the model

A series of simulations was performed in order to obtain agreement between experimental data and simulations. The initial temperature of the melt in the simulation was set to 1325°C as this is the highest temperature registered in the experiment [6]. The nucleation parameter A_n was along the work of others, [7, 4], set to values around $A_n = 10^{11}$ m^{-3}K^{-1}. Since the three bars were cast from the same melt, the parameters that are used for the melt must be the same for all three.

Figure 2 and 3 show the temperature at the centre of the casting as a function of time for each of the three bar sizes from experimental and simulated data. The figure for the 52 mm casting shows data for two values of A_n to illustrate the sensitivity of the cooling curve to this parameter as commented on below. The solidification times for the 34.7 mm and 52 mm castings are well reproduced in the simulated data despite too short eutectic plateaus which in the case of the 65 mm casting results in a solidification time that is too small. The figures also show that the initial cooling conditions seem to be well simulated for the 34.7 mm bar as there is a good correspondence between the initial slopes of the temperature curves, but as size increases the experimental data show faster cooling than the simulated. Further, the undercooling just before the eutectic plateau is not reproduced in any of the simulations.

The differences in the shapes of the cooling curves and in initial cooling rate might originate partly from imprecise modelling of the decrease of the mould's ability to extract heat from the casting. In green sand moulds this phenomenon is closely linked to the water evaporation and vapour transport as the mould heats up, [8], which might not be well modelled through a dependency on the fraction solid of the surface element as it is done in this study and also done conventionally. Further, as discussed by Pedersen et al. [4], it is assumed that there is no nucleation barrier for the formation of off-eutectic austenite in the model which might not be realistic. In the model, off-eutectic austenite will start forming as soon as the liquidus line is reached and thus the timing of the release of latent heat from this process might not be correct. This will contribute to differences between experiment and simulation.

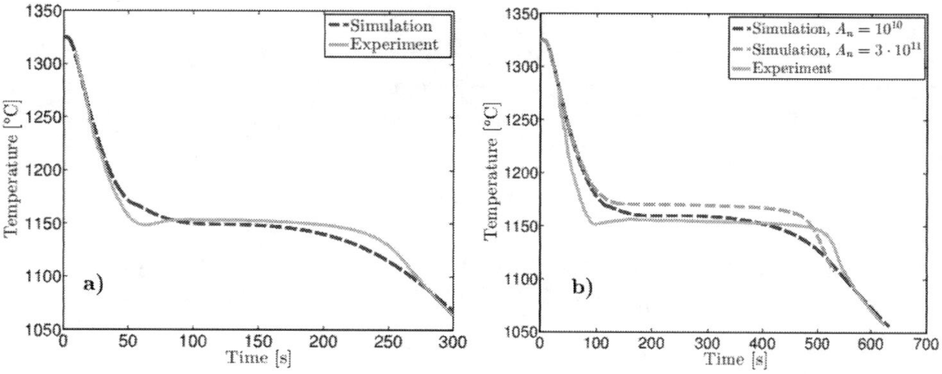

Figure 2. Temperature at the cross section centre of the **a)** 34.7 mm and **b)** 52 mm bars obtained by simulation and experiment. The initial cast temperature was 1325°C and the nucleation constant $A_n = 10^{10}$ m^{-3}K^{-1} for the 34.7 mm casting while results for $A_n = 3 \cdot 10^{11}$ m^{-3}K^{-1} were also included for the 52 mm case.

The nucleation constant A_n greatly influences the temperature of the eutectic plateau in this model and the results obtained using two different values are used to illustrate the following points. High values of A_n leads to a higher nucleation rate and thus to an increased amount of carbon removed from the liquid before reaching the austenite liquidus which will result in a higher temperature for the eutectic plateau on the cooling curve. This is illustrated by the curve for $A_n = 3 \cdot 10^{11}$ m^{-3}K^{-1} in the simulation of the 52 mm bar, Fig. 2 b). Thus, even though the simulations with $A_n = 10^{10}$ m^{-3}K^{-1} predict the level of the eutectic plateau quite accurately it is at the cost of using a quite low nucleation constant. This leads to low nodule number densities.

IOP Conf. Series: Materials Science and Engineering **84** (2015) 012038 doi:10.1088/1757-899X/84/1/012038

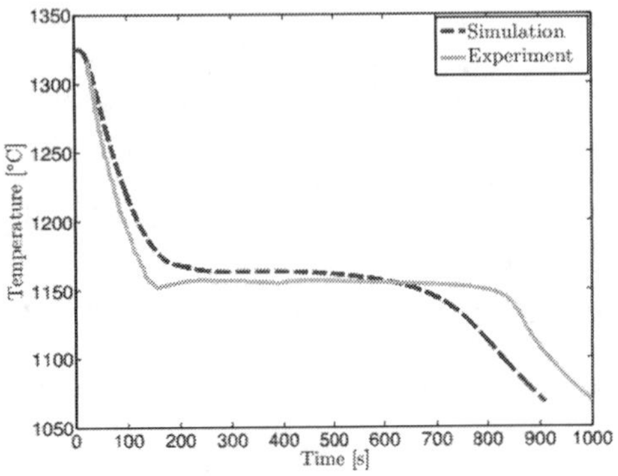

Figure 3. Temperature at the centre of the 65 mm bar cross section obtained by simulation and experiment. The initial cast temperature was 1325°C and the nucleation constant $A_n = 10^{10}$ m^{-3}K^{-1}.

Table 2 shows the nodule number densities from experiment and simulation at the centre and the upper right corner of the casting as seen from Fig. 1. Microscope images from the corner and the centre of the 52 mm casting is presented in Fig. 4. Degenerate graphite was found at the centre of both the 52 mm and the 65 mm casting. This observation will bias the nodule count from the centre as a single degenerate nodule will be segmented by the image analysis software as multiple individual nodules. This limits the value of the this data from the casting centre as basis for validation. The three dimensional size distribution was obtained from a two dimensional distribution using the Schwartz-Saltykov method [9].

The simulations with $A_n = 10^{10}$ m^{-3}K^{-1} predict much lower density than what was found from the experiment and this is a direct consequence of the low A_n as can be seen by comparing to the results with $A_n = 3 \cdot 10^{11}$ m^{-3}K^{-1} which show values comparable to those of the experiment. This at the same time raised the temperature of the eutectic plateau to 15°C above the experimental curve in the case of the 52 mm bar illustrating the trade off between a good fit to the cooling curve and to the nodule number density that has to be made with the present model.

Table 2. Nodule count (nodules pr m^3) at the centre of the castings compared to calculated nodule counts at the centre and near the corner of a casting cross section respectively. Data are from the 52 mm and 65 mm castings and only nodules with a diameter greater than 5 μm have been included.

	52 mm		**65 mm**	
	Corner	Centre	Corner	Centre
Experiment	$4.08 \cdot 10^{13}$	-	$2.22 \cdot 10^{13}$	-
Simulation $A_n = 10^{10}$ m^{-3}K^{-1}	$2.18 \cdot 10^{12}$	$1.34 \cdot 10^{12}$	$2.01 \cdot 10^{12}$	$1.20 \cdot 10^{12}$
Simulation $A_n = 3 \cdot 10^{11}$ m^{-3}K^{-1}	$3.91 \cdot 10^{13}$	$2.16 \cdot 10^{13}$	$3.97 \cdot 10^{13}$	$1.97 \cdot 10^{13}$

5. Solidification of thick-walled part

We have validated the model with reasonable success and now expand the simulation domain to contain a quarter of a bar with cross section of 300 mm by 300 mm to simulate the solidification of a thick-walled part. Even though the model does not include the formation

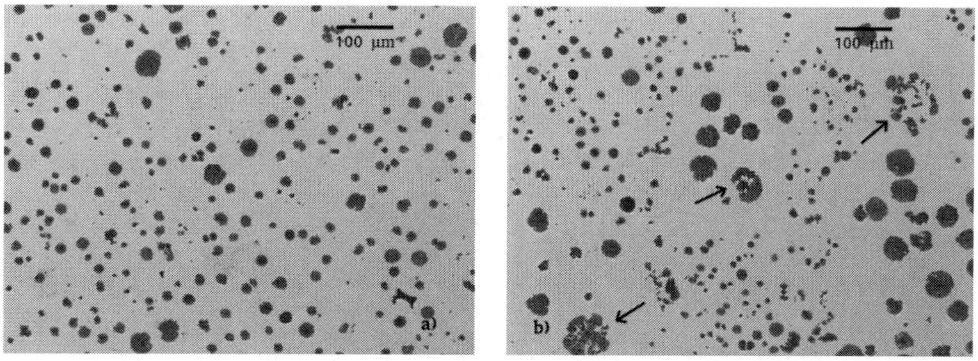

Figure 4. Example of graphite morphology from **a)** the corner and **b)** the centre from the 52 mm casting. Examples of degenerate graphite are marked with arrows.

of degenerate graphite morphologies and the cooling curves from castings involving degenerate graphite are affected by the formation of this type of graphite, [3], the fundamental features of the solidification pattern of thick-walled castings are likely to be the same.

Figure 5 shows the temperature and fraction solid in the elements along the diagonal of the cross section. The solidification starts by the formation of a solid shell at the interface to the mould which over time advances towards the centre of the casting. Initially, large temperature gradients exist throughout the casting but as the solidification proceeds a thermally homogeneous zone is established in the central part of the casting which to a large extend is still liquid. The central zone of the casting is maintained close to the eutectic temperature over long time intervals with only limited amount of solid present in the form of encapsulated nodules. As the formation of degenerate graphite is associated with low cooling rates the central zone in the thick-walled casting is indeed where one would expect degenerate graphite to form.

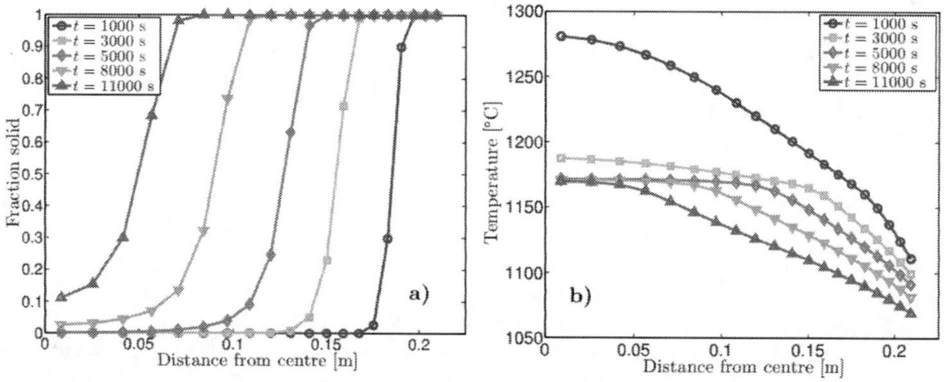

Figure 5. Fraction solid **a)** and temperature **b)** at different time points along the diagonal of the 300 mm bar cross section for $A_n = 10^{10}$ $m^{-3}K^{-1}$.

In Fig. 6 the temperature and fraction solid profiles along the diagonal of the 34.7 mm ba at various points in time are plotted as a comparison to Fig. 5. The 34.7 mm casting shows a distinctively different solidification pattern. While there is no clear solidification front in the 34.7 mm as large parts of the cross section are solidifying more or less simultaneously, the thick-walled 300 mm bar first creates a solid shell and then establishes a zone with only small thermal gradients which solidify homogeneously. The 52 mm and 65 mm bars show elements of both solidification patterns by initially creating a solid shell leaving the centre of the castings to solidify simultaneously.

This corresponds well to the observation of degenerate graphite found at the centre of the 52 mm casting presented in Fig. 4 as low cooling rates promote the formation of this type of morphology. Also, analysis of the porosity distribution in the castings suggests shell formation in the 65 mm casting, [6], in agreement with the present simulations.

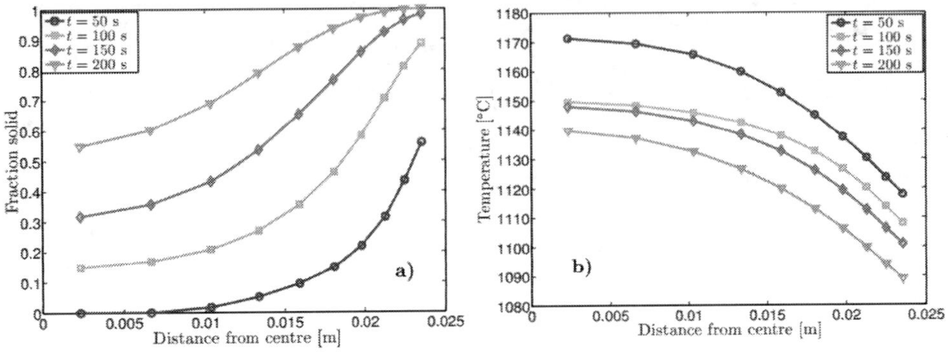

Figure 6. Fraction solid **a)** and temperature **b)** at different time points along the diagonal of the 34.7 mm bar cross section. The cooling curve for the centre of the casting is shown Fig. 2 a).

6. Conclusion

A solidification model for ductile cast iron was presented and the solidification castings of varying wall thicknesses was simulated. The simulation was validated against experimental data in terms of cooling curves and nodule number densities with reasonable agreement however also stressing the parameter sensitivity of the microstructural model applied. The analysis of the solidification pattern showed the formation of a long term thermally homogeneous zone at the centre of the thick-walled 300 mm by 300 mm casting with solidification only proceeding slowly from the edge towards the centre. The small 34.7 mm by 34.7 mm casting displayed a fundamentally different solidification pattern where no clear solidification front exists. The solidification pattern appears closely related to the development of the thermal gradients with thermally homogeneous zones associated with the centre of large castings relevant to the formation of porosities and degenerate graphite.

References

[1] Lesoult G, Castro M and Lacaze J 1998 *Acta Metallurgica* **46** 983–995
[2] Källbom R, Hamberg K and Björkegren L 2005 Chunky graphite formation and influence on mechanical properties in ductile cast iron *Component Design by Castings, VTT Symposium 237* (Espoo) pp 63–86
[3] Sertucha J, Suárez R, Asenjo I, Larrañaga P, Lacaze J, Ferrer I and Armendariz S 2009 *ISIJ International* **49** 220–228
[4] Pedersen K M, Hattel J H and Tiedje N 2006 *Acta Materialia* **54** 5103–5114
[5] Vedel-Smith N, Rasmussen J and Tiedje N 2015 *Journal of Materials Processing Technology* **217** 262–271 ISSN 09240136
[6] Vedel-Smith N K and Tiedje N S 2014 Feeding Against Gravity with Spot Feeders in High Silicon Ductile Iron *10th International Symposium on the Science and Processing of Cast Iron - SPCI10* (Mar Del Plata)
[7] Lacaze J, Castro M and Lesoult G 1998 *Acta Metallurgica* **46** 997–1010
[8] Kubo K and Pehlke R D 1986 *Metallurgical Transactions B* **17** 903–911
[9] Underwood E E 1970 *Quantitative Stereology* (Reading, Mass.: Addison-Wesley)

Numerical Simulation of Cast Distortion in Gas Turbine Engine Components

A.A.Inozemtsev[1], A.S. Dubrovskaya[2], K.A. Dongauser[2], N.A. Trufanov[3]
[1]Managing Director & General Designer Aviadvigatel OJSC, Perm, Russia
[2]Casting simulation department, Aviadvigatel OJSC, Perm, Russia
[3]Head of Department Computation Mathematics and Mechanics, PNRPU, Russia

E-mail: dubrovskaya-as@avid.ru

Abstract. In this paper the process of multiple airfoil vanes manufacturing through investment casting is considered. The mathematical model of the full contact problem is built to determine stress strain state in a cast during the process of solidification. Studies are carried out in viscoelastoplastic statement. Numerical simulation of the explored process is implemented with ProCAST software package. The results of simulation are compared with the real production process. By means of computer analysis the optimization of technical process parameters is done in order to eliminate the defect of cast walls thickness variation.

1. Introduction

One of the most important tasks of mathematical modeling of physical processes is comprehensive study of certain aspects of modern engineering industry. The result is creation of cost-effective innovative production technologies which allow not only to save expensive raw materials, energy and labour resources, but also improve the quality of parts manufactured and minimize the rejects.

Among wide range of aviation parts a turbine vane is unrivaled in terms of load degree and purpose criticality. Hence high requirements arise to its metallurgical and strength quality. Design and technological features of hollow vanes have serious effect on gas-dynamic characteristics of an engine. Operating conditions of thin-wall vanes in modern gas turbine engines require their high geometrical precision. That's why particular attention must be paid to the issues of vanes design and casting.

The process of gas turbine engines vanes manufacturing through investment casting is accompanied by multifactor physical phenomena. The application of mathematical modeling method for processes that accompany parts manufacturing allows to reduce costs since the development of gate system design and temperature-time process parameters is carried out not on real expensive casts, but in virtual space of a mathematical model. Low cost and short terms of carrying out computer experiments, and also big volume and visualization of information received on process progress and quality of a future cast make computer simulation the essential research tool [1].

2. Exploring the technological problem

A vane cast manufactured by investment casting method may differ from a design model in terms of its geometrical characteristics. Cooling of a cast is accompanied by changing of its dimensions due to temperature and structural shrinking deformations which show up in form of linear shrinkage and curvature of individual features axes or shape distortion. High heat-shrinking deformations lead to distortion of a casting mould, a core, to shape distortion of a cast itself. That is why to receive a product with set geometry, one should take into account the possibility of deformation in advance.

Applying predistortions to cast geometry allows to receive a cast matching a design model with set precision. For this purpose it is necessary to know in advance the mechanical behaviour of a cast, a mould, and a ceramic core. A ceramic core forms the internal cavity of a casting; that is why its distortion during tapping and the following solidification may affect the shape of a future cast causing defects in form of walls thickness variation from pressure and suction side along the whole span. Selection of a technological mode for thin-wall vanes casts manufacturing process with minimal distortions of a ceramic core is the essential engineering task.

Let us study the process of hollow vane cluster manufacturing from heat-resistant nickel-base superalloy. The cluster is casted in a special-purpose machine of solidification. Solidification is proceeding in ceramic mould obtained by precise casting. The Ceramic core forms a blade hollow spaces.

Let us consider simulation of stress-strain state (SSS) which appears during manufacturing of hollow vane cluster. In a real production process non-contact optical 3D-scanning revealed casts geometry deviations from design models. Maximum deviations were found on the pressure side of the cast near the internal edge. This deviation and concomitant airfoil walls thickness variation are probably caused by the core distortion during the cluster manufacturing.

It is necessary to carry out the analysis of casting and cast solidification process in order to confirm the hypothesis about core distortion and find a technological mode with minimal distortion. The study was carried out through mathematical modeling. Figure 1 a-c shows the real wax pattern of a cast and pouring gate system, the design model and the discrete model respectively.

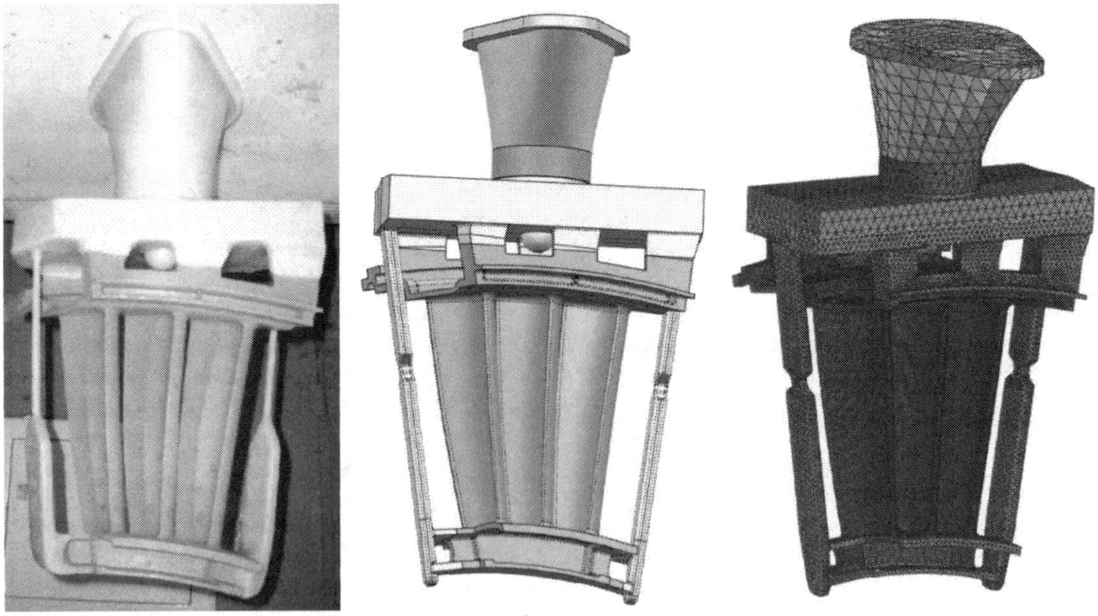

a b c

Figure 1. Hollow vanes cluster: a) wax pattern, b) design model, c) finite-element model

Vertical pouring gate system was used with metal supply to radial platform wedge from two sides. Mould heater temperature was 1000°C, tapping time – 5 sec.

3. Mathematical statement of the problem

In the course of the study the full contact problem of SSS determination is resolved in the statement "deformed mould – deformed cast – deformed cores". Studies are carried out in viscoelastoplastic statement.

IOP Conf. Series: Materials Science and Engineering **84** (2015) 012039 doi:10.1088/1757-899X/84/1/012039

Full deformation looks as follows:

$$\hat{\varepsilon} = \hat{\varepsilon}^{el} + \hat{\varepsilon}^{vp} + \hat{\varepsilon}^{th} + \hat{\varepsilon}^{tr} \tag{1}$$

Hence the following physical relations:

$$\hat{\sigma} = {}^4R \cdot\cdot \hat{\varepsilon} = {}^4R \cdot\cdot \left(\hat{\varepsilon} - \hat{\varepsilon}^{th} - \hat{\varepsilon}^{el} - \hat{\varepsilon}^{vp} - \hat{\varepsilon}^{tr}\right) \tag{2}$$

where ${}^4R$ is quartic elastic constants tensor.
Geometrical relations look as follows:

$$\hat{\varepsilon} = \frac{1}{2}\left[\nabla \bar{u} + (\nabla \bar{u})^T\right] \tag{3}$$

Balance equation is taken as follows

$$\nabla \cdot \hat{\sigma} = 0 \tag{4}$$

Temperature deformation:

$$\hat{\varepsilon}^{th} = \alpha\left(T - T_0\right) \tag{5}$$

α is temperature expansion coefficient, $\hat{I}$ .is identity tensor.
Liquid-solid phase solidification shrinkage is considered as follows:

$$\hat{\varepsilon}^{tr} = \frac{\beta_{tr}}{3} g_{tr}(\bar{x}, t)\hat{I} \tag{6}$$

g_{tr} is volume fraction of liquid-solid phase, transforming into a new phase, and β_{tr} is coefficient of volume shrinkage associated with phase transformation.
Viscoplastic behavior law is taken as follows:

$$\hat{\varepsilon}^{el} = \frac{3}{2}\frac{\dot{\varepsilon}_{eq}}{\sigma_{eq}}(\hat{\sigma} - \sigma\hat{I}) \tag{7}$$

To compute equivalent deformation rate, Perzyna model is used that allows to consider linear hardening:

$$\dot{\varepsilon}_{eq}^{vp} = \frac{1}{\mu}\left\langle \frac{\hat{\sigma}_{eq}}{\sigma^*} - 1 \right\rangle^p \tag{8}$$

where σ^* is yield stress, σ_{eq} are von Mises equivalent stresses, μ is viscous parameter, p is strain rate sensitivity coefficient, σ is average stress.
At liquid and solid interface (ls) normal and tangential stresses look as follows:

$$\sigma_{ls\,\tau}^{l} = \sigma_{ls\,\tau}^{s} = 0, \quad \sigma_{lsn}^{l} = \sigma_{lsn}^{s} \tag{9}$$

For contact interfaces of a casting, mould and a core conditions of a contact are set by the following relations (the indices 1 and 2 denote the contact body):
In case of full cohesion and slippage with static friction:

$$\bar{u}_i = \bar{u}_j \Rightarrow \sigma_{in} = \sigma_{jn}, \; \sigma_{i\tau} = \sigma_{j\tau}, \tag{10}$$

while $\left|\sigma_\tau\right| < q\left|\sigma_n\right|$

IOP Conf. Series: Materials Science and Engineering **84** (2015) 012039 doi:10.1088/1757-899X/84/1/012039

In case of slippage with sliding friction:

$$u_{in}=u_{jn},\ u_{i\tau}\neq u_{j\tau}\Rightarrow \sigma_{in}=\sigma_{jn},\ \sigma_{i\tau}=\sigma_{j\tau},\ \sigma_n<0, \tag{11}$$

while $|\sigma_\tau|=q|\sigma_n|$

In case of full detachment:

$$u_{in}\neq u_{jn}\Rightarrow \sigma_n=\sigma_\tau=0 \tag{12}$$

For surface of a mould installed on the bottom of a casting installation the following fastening conditions are true:

$$u_z=0 \qquad \bar{x}\in\Gamma_0 \tag{13}$$

For free boundaries of a mould, a cast and a ceramic core the following is true:

$$\hat{\sigma}\cdot n=0 \tag{14}$$

4. Numerical simulation results

The application task of a real casting process simulation does not have an analytical solution. Thus, we may need to simulate casting process numerically. For several years Aviadvigatel has been using a universal program software named ProCAST for simulating the process of pouring and solidification of gas turbine engine component castings. ProCAST is a system of casting numerical simulation based on FEM method.

Figure 2 shows shape distortion in the cast and cores after solidification, distortion in the cast is increased 20 times, in cores – 50 times.

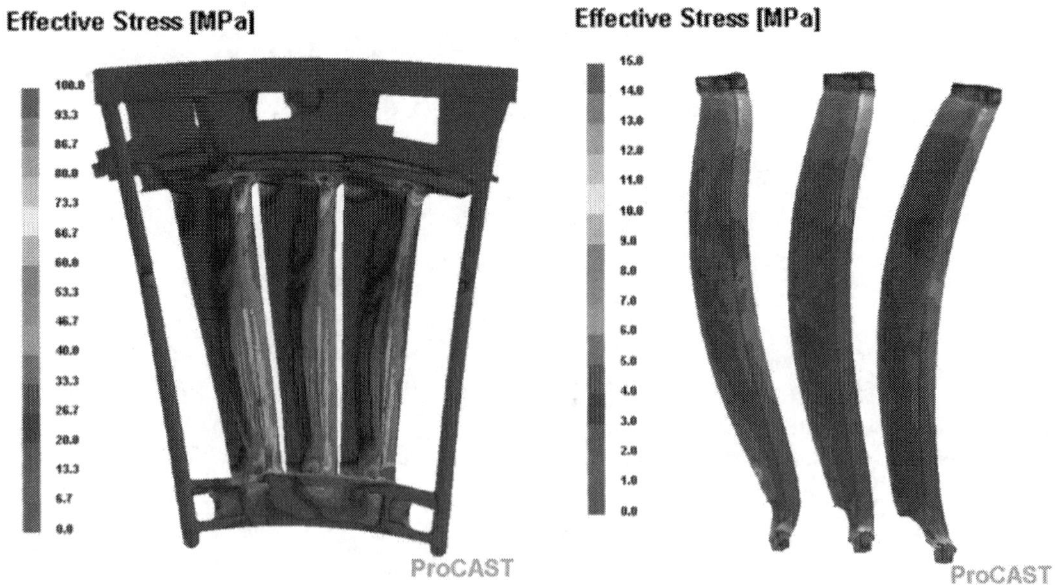

a б

Figure 2. Results of simulation of the cast SSS after solidification: a) shape distortions in the cast (×20); б) shape distortions in the core (×50)

From Figure 2 the conclusion can be made about significant displacement of a core in the medium part and its influence on final geometrical shape of the cast. Figure 3.a shows core displacement in the central section of the cast, 5 times increased. Results of the actual cast measurement in respective section using non-contact 3D-scanning are given in Figure 3.b. Results of computer simulation and the cast measurements showed the displacement of the ceramic core towards the pressure side. It caused the effect of cast walls thickness variation.

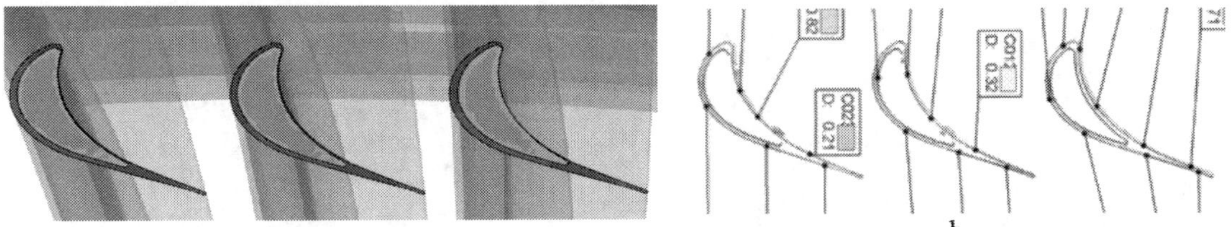

a b

Figure 3. Displacement of the ceramic core towards the pressure side of the airfoil a) – predicted by numerical simulation (mm×5); b) – received at the actual cast measurement, mm

To eliminate the casting defect in form of walls thickness variation it was decided to carry out a number of numerical experiments on simulation of hollow vanes cluster manufacturing process with different technological parameters and to select optimal ones.

5. Optimization of technological parameters

In the numerical study of the multiple airfoil vane manufacturing process maximal core displacement towards the pressure side was 0.35 mm.

It was decided to explore the influence of melt tapping on the core displacement, according to the studies undertaken in [6]. It was suggested to reduce tapping time from 5 to 3 seconds. The rest of production parameters remained unchanged.

Computer simulation results showed that maximum distortion of the core was 0.3 mm, which is 0.05 mm less than maximum distortion found in production process simulation at alloy tapping in 5 seconds. Thus, the increase of melt feed rate is rational for reduction of cast distortion.

Next suggestion was to increase the mould heater temperature during mould preheating. Two technological modes were suggested: mould heater heating up to 1100°C and up to 1200°C. Tapping time was 3 seconds.

For mould temperature 1100°C maximum core distortion was 0.23 mm. At heating up to 1200°C the distortion did not exceed 0.14 mm. Results of the numerical experiments carried out are given in Table 1.

Table 1. Cast wall thickness deviations at different technological parameters during vane casting

Number of numerical experiment	Alloy tapping time	Mould temperature	Maximum displacement
1	5 s	1000 C	0.35 mm
2	3 s	1000 C	0.30 mm
3	3 s	1100 C	0.23 mm
4	3 s	1200 C	0.14 mm

IOP Conf. Series: Materials Science and Engineering **84** (2015) 012039 doi:10.1088/1757-899X/84/1/012039

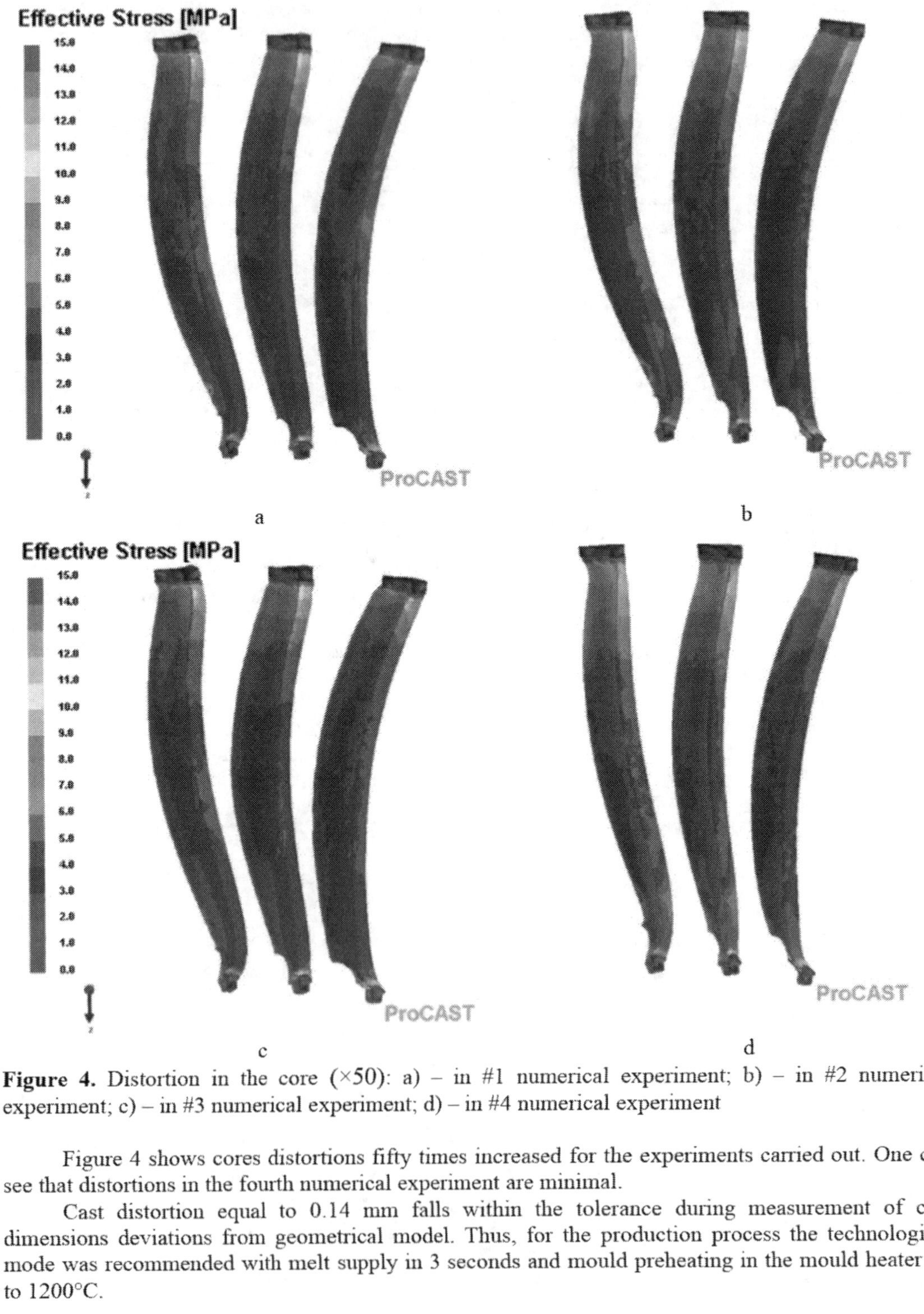

Figure 4. Distortion in the core (×50): a) – in #1 numerical experiment; b) – in #2 numerical experiment; c) – in #3 numerical experiment; d) – in #4 numerical experiment

Figure 4 shows cores distortions fifty times increased for the experiments carried out. One can see that distortions in the fourth numerical experiment are minimal.

Cast distortion equal to 0.14 mm falls within the tolerance during measurement of cast dimensions deviations from geometrical model. Thus, for the production process the technological mode was recommended with melt supply in 3 seconds and mould preheating in the mould heater up to 1200°C.

Comparison of a casting model and resultant blade using three-dimensional scanning indicates how obtained results are essential. Results of the vane clusters measurement in respective section using non-contact 3D-scanning are given in Figure 5. Figure 5.a shows result of cluster measurement produced by regular technology. Average casting distortion is about 0.58 mm. Figure 5.b shows result of cluster measurement produced by optimized production conditions. Average casting distortion is 0.19 mm. Thus application of this technology for the hollow vane cluster manufacturing allowed reducing distortion of cores more than three times.

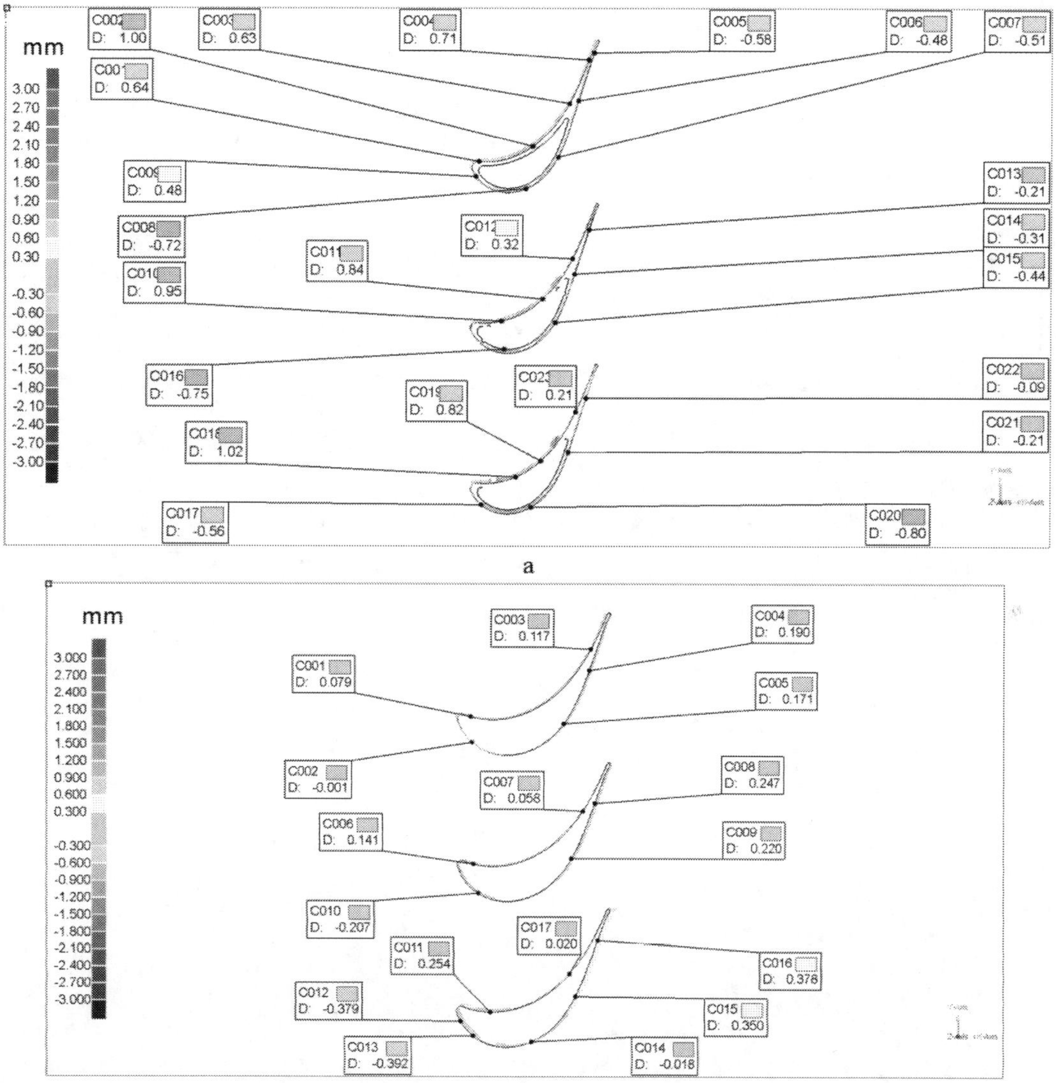

b

Figure 5. Displacement of the ceramic cores received at the actual casts measurement a) – fore regular technology b) – fore optimized production conditions

So, solution of the stress-strain evaluation task enables to determine temperature and time parameters with which deformation will be minimal. This helps to deal with such defects as variations of wall thickness in hollow airfoils.

6. Conclusions

Mathematical modeling of hollow vanes cluster manufacturing process was carried out. The problem included the processes of casting, crystallization and calculation of the cast SSS during solidification.

Deviations of the deformed grid from initial cast geometry were compared with the real cast distortion, found by non-contact optical 3D-scanning. High level of correlation of distortions received by numerical simulation with real cast distortions was noted.

After the cast measurements, wall thickness variation was found, caused by the displacement of the core towards pressure side. This data confirmed the results of mathematic modeling.

A number of numerical experiments was carried out to select technological parameters of multiple airfoil vane manufacturing process in order to reduce the displacement of the core towards pressure side.

As a result of the numerical study, it was suggested to change alloy tapping time and the heated mould temperature, which allowed to reduce distortion of cores more than three times and to bring airfoil thickness variation on pressure and suction side into the tolerance zone.

References

[1] Monastyrsky A V, Monastyrsky V P and Levitan E M 2007 *Cast. Prod.* **9** 29-34
[2] Dantzig J A and Rappaz M 2009 *Solidification* (Lausanne: EPFL Press)
[3] Campbell J 2011 *Complete Casting Handbook: Metal Casting Processes, Techniques and Design* (Oxford: ELSEVIER)
[4] Inozemtsev A A, Dubrovskaya A S, Dongauzer K A and Trufanov N A 2014 *ISIJ Int.* **54.2** 254-58
[5] Dubrovskaya A S and. Dongauzer K A 2011 *PSTU: News. Appl. Math. Mech.* **9** 81-102
[6] Kablov E N 2001 *Cast Blades of Gas Turbine Engines. Alloys, Tehnologies, Coatings* (Moscow: MISIS)

Two- and three- dimensional studies of dendritic morphology in magnesium alloy by means of synchrotron X-ray microtomography and cellular automaton modelling

M Yang[1], Z Guo[1 *] and S Xiong[1,2]

[1] School of Materials Science and Engineering, Tsinghua University, Beijing, 100084, China
[2] State Key Laboratory of Automobile Safety and Energy, Tsinghua University

E-mail:zhipeng_guo@mail.tsinghua.edu.cn

Abstract. Magnesium is the lightest structural material. As one of the dominant microstructure features, dendritic pattern determines the mechanical behaviour and performance of magnesium alloys. Dendritic topological observation was carried out on Mg-based alloy using synchrotron X-ray micro-tomography and the microstructure pattern of α-Mg dendrite was obtained. It was found that the α-Mg dendrite grew with eighteen primary stems, of which six lay in the (0001) basal plane, and the other twelve in the $(10\bar{1}0)$ plane. An according numerical model based on the cellular automata method was developed. By defining a specific capturing functional mechanism, simulation of α-Mg dendrite in 3-D with eighteen branches was successfully achieved. The simulation results show that the model could reasonably describe the evolution of the dendritic microstructure and the subsequent dendrite morphology agrees well with that observed in the synchrotron X-ray tomography experiment.

1. Introduction

With demands for light-weighted, decreasing of fuel consumption and CO_2 emission, conserving finite oil reserves, magnesium alloys have achieved increasing applications in automobile industry in recent years, such as steering wheels, gear boxes, instrument panels and air intake systems, etc. [1-3]. The mechanical properties of magnesium alloy castings are greatly affected by the microstructure of magnesium alloys, especially the dendritic structure. It's of great importance to understand the dendritic morphology of magnesium alloys.

A great number of studies on the dendritic morphology of face-centered cubic (FCC) and body-centered cubic (BCC) alloy system have been carried out in two and three dimensions [4, 5]. It is widely accepted that the preferred growth orientation is typically along $\langle 100 \rangle$ in these alloys. The two-dimensional morphology of aluminum alloys exhibits four-fold symmetry, and the three-dimensional dendrite has six branches. For the hexagonal close-packed (HCP) alloy system, different growth orientations have been reported, such as $\langle 10\bar{1}0 \rangle$ [6, 7] or $\langle 0001 \rangle$ [6] for zinc, and $\langle 10\bar{1}0 \rangle$ for H_2O [4].

For magnesium alloys (HCP alloy system), very limited studies have been performed to investigate the 3-D morphology of the dendrite. Pettersen *et al* [8, 9] studied the AZ91 magnesium alloy under directional solidification using SEM and EBSD and found that dendrites have two different stem directions i.e. $\langle 22\bar{4}5 \rangle$ and $\langle 11\bar{2}0 \rangle$. Wang et al [10] obtained the three-dimensional morphology of directionally solidified Mg-9wt.%Al alloy by synchrotron X-ray tomography, and suggested that α-Mg(Al) grew along $\langle 11\bar{2}0 \rangle$ in the {0001} basal plane. The dendrites have plate-like morphology and exhibit six-fold symmetry in the {0001} basal plane. Wang *et al* [11] studied the three-dimensional equiaxed dendritic morphology of Mg-Zn alloys, which was similar to the $\langle 11\bar{2}0 \rangle$-type stems in reference [8].

Numerical modeling and simulation is widely used to predict the microstructure evolution during solidification of magnesium alloy [12]. Böttger *et al* [13, 14] integrated a hexagonal anisotropy function into the phase-field model (PF), and simulated the three-dimensional dendritic growth of magnesium, which had six prism orientations in the {0001} basal plane and two basal orientations in the ⟨0001⟩ directions. Wu *et al* [15] simulated the dendrites growth of AM50 alloy by cellular automaton model (CA) and the results were consistent with [13, 14]. Eiken [16] investigated the growth texture evolution of magnesium by PF model. They found that grains grew faster in $\langle 11\bar{2}0 \rangle$ orientation and retarded in ⟨0001⟩ orientation. The grain arrangement is plate-like within the basal plane.

As seen, for the magnesium alloys the $\langle 11\bar{2}0 \rangle$-type stems in the basal plane have been accepted as one type of the primary growth directions. However, the preferred growth directions in the non-basal plane have not been clearly determined, i.e. discrepancy exists between the experiment observed $\langle 22\bar{4}5 \rangle$ direction and the simulated ⟨0001⟩ direction.

In this paper, experiment methods including SEM, synchrotron X-ray micro-tomography, combining with a cellular automaton model were applied to study the 2-D and 3-D microstructure of Mg-30wt.%Gd alloy. The purpose was to shed some lights on determining and thus understanding the full structure of the magnesium alloy dendrites in 3-D.

2. Experimental methods

2.1. Materials

An Mg-30wt.%Gd alloy was selected because of the better absorption contrast. The eutectic temperature and liquidus temperatures of the alloy are 542°C and 587.2 °C, respectively. The alloy was prepared as follows. Firstly, magnesium (99.95 wt. %) and gadolinium (99.99 wt. %) were placed in a mild steel crucible and heated by an electric resistance furnace. The mixture in the crucible was melted and mixed at 800 °C under a protective gas of 99.7% N_2 and 0.3% SF_6. The melt was then poured into a permanent mould which was preheated to 300 °C and held for 30 min [17, 18]. A cylindrical specimen of 10 mm in diameter and 30 mm in length was machined from this object. Secondly, the specimen was sealed in a quartz tube with argon as protective atmosphere, then it was remelted and quenched between the liquidus and eutectic temperatures. Cylindrical specimens of 1.0 mm in diameter and 5 mm in length were machined from the specimen for tomography experiments.

2.2. Synchrotron X-ray micro-tomography

Tomography experiments were carried out at the BL13W1 beam line of the Shanghai Synchrotron Radiation Facility (SSRF). A high-speed CCD camera was used to record the transmitted intensity of a nearly monochromatic X-ray beam when the sample was continuously rotated over 180° [19]. An X-ray energy of ~21.5 keV was used to penetrate the specimens, and 900 projections were taken between 0° and 180° using an exposure time of 4s per projection. Dark field images without X-rays, and flat field images with X-rays but without the sample were also recorded for further image-processing [19, 20]. The distance between the specimen and the camera was 20 cm. The reconstructions resulted in volumes of 2048^3 voxel with a voxel size of $(0.74\ \mu m)^3$.

2.3. Image processing

A software package PITRE [21] was used to convert the images from raw data to reconstructed tomographic slices and from 32 bit to 8 bit. The dendritic microstructure was segmented from sub-volumes of about 400×400×400 voxels (296μm^3) after filtering and thresholding (figure 1).

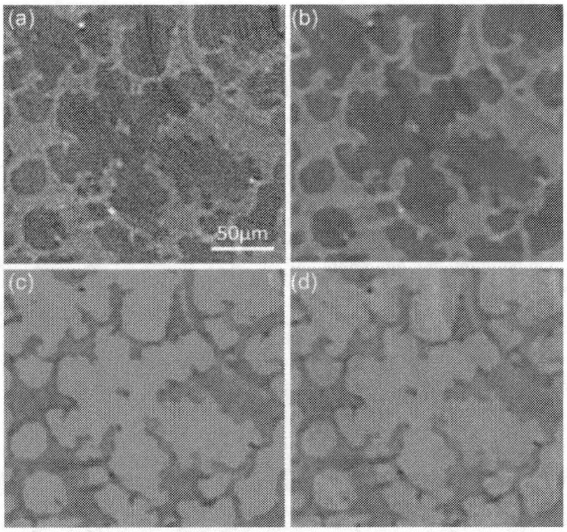

Figure 1. Key steps of data processing including (a) cropping (b) 3D median filter (c) thresholding, and (d) segmentation.

3. Results and discussion

3.1. The 2-D microstructure

Based on the two-dimensional characterization, some interesting dendritic microstructure can be observed. Figure 2 shows the optical micrographs (OM) and simulated dendrites of magnesium alloy. The model was based on our previous work [12]. By using a special neighborhood configuration with the square CA grid, and the capturing rules proposed by BELTRAN-SANCHEZ and STEFANESCU, modeling of dendritic growth was achieved [12]. As observed in figures 2a-c designated by "1", the primary dendrite exhibits a typical six-fold symmetric shape with six branches growing out from the nuclei. Figure 2d indicates the simulated equiaxed dendrites under cooling rate of 80 K/s which is roughly equal to that in the quenched condition. According to figure 2d, the simulated dendrites exhibit six-fold symmetric shape as designated by "1" (figure 2a-c). It is noted that not all the dendrites of magnesium alloy are fully simulated in figure 2d. With the six-fold symmetric morphology, six separated particles with five branches (designated by "2") can be found in figure 2a and 2c. Instead of six-fold symmetric structure, most dendrites develop a low symmetrical or even disordered morphology as four-branch dendrites designated by "4", tortoise shell-like dendrites designated by "5" and five-branch dendrites of separated particles designated by "3".

3.2. Three-dimensional microstructure

3.2.1. Synchrotron tomography. Figure 3a shows the 3D morphology of one of the reconstructed magnesium dendrites from synchrotron X-ray micro-tomography. Some planes and directions are used to cut the dendrite to help understanding the dendritic microstructure. Figure 3b indicates the morphology on S_0, which has six branches i.e. d_1~d_6 and the angles between the d_x-type directions are 60°. For the two-dimensional projections on S_1 (figure 3c), which contain the three planes across g_1

and g_4, g_2 and g_5, or g_3 and g_6, the dendritic microstructure have two branches on S_0 and other four branches next to them (in figure 3d). It is noted that the S_1-type planes are perpendicular to plane S_0 and magnesium dendrites have eighteen branches. To achieve a better understanding of the dendritic morphology, two other planes i.e. S_2 and S_3 (figure 3e and 3g) are added in the dendrite. For the projections on S_2 and S_3, the dendrites have six and four branches respectively, as observed in figure 3f and 3h.

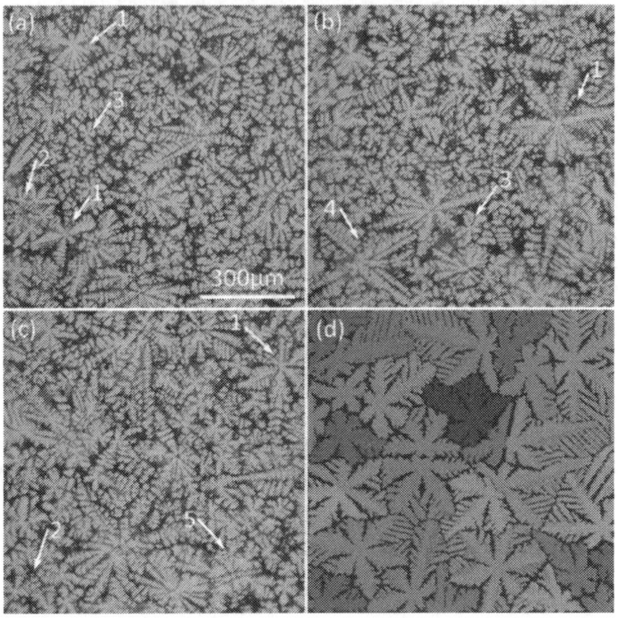

Figure 2. Optical micrographs (a-c) and simulated results (d) of dendritic morphology of magnesium alloy.

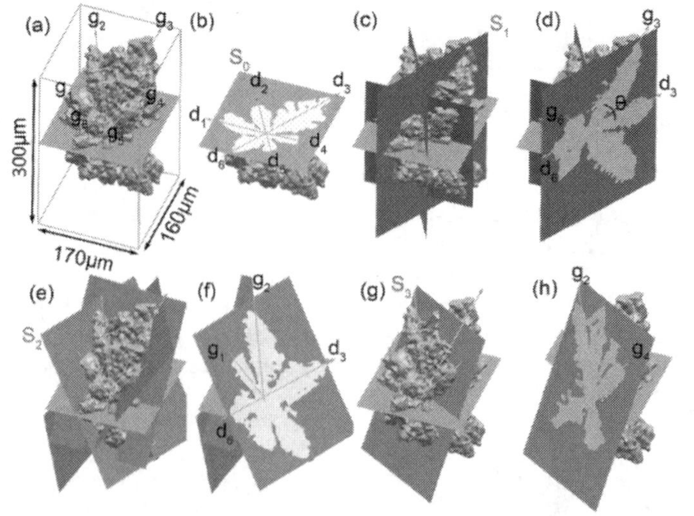

Figure 3. Three-dimensional dendritic microstructure of magnesium alloy (a) and cross profiles with different sections: (a-b) section S_0; (c-d) section S_1; (e-f) section S_2; (g-h) section S_3.

As demonstrated in figure 3, the dendrites exhibit six branches on planes S_0, S_1, and S_2. Only the dendrites on S_0 are symmetric in six-fold, and the plane is designated as the basal plane. It is equivalent to {0001} crystallographic plane. The d_x-type direction is equivalent to $\langle 11\bar{2}0 \rangle$ crystallographic orientation. The dendrites on S_1 or S_2 are not symmetric in six-fold, and the angles between the branched on the basal plane and the non-basal plane are also different from each other. Based on orientation measurements of thirty dendrites, the angle between g_3 and d_3, namely θ, is determined to be 54°±3°. Then the g_x-type direction is equivalent to $\langle 11\bar{2}3 \rangle$ crystallographic orientation, which is close to the reported $\langle 22\bar{4}5 \rangle$ direction in literature [8]. The dendrites on S_3, which have four branches, are consistent with the "4" dendrite in figure 2b. In this way, the 3D dendritic morphology of magnesium alloy grows along $\langle 11\bar{2}0 \rangle$ and $\langle 11\bar{2}3 \rangle$ orientations, and these dendritic morphologies and patterns in figure 2 can be explained and understood.

3.2.2. Three-dimensional simulated results. Based on the reconstructed three-dimensional dendritic morphology of magnesium alloy, a model describing eighteen-branch growth of magnesium alloy is constructed. As indicated in figure 4, it has six prime orientations along $\langle 11\bar{2}0 \rangle$ in the basal plane and twelve orientation along $\langle 11\bar{2}3 \rangle$ in the non-basal plane. A corresponding neighbourhood configuration was defined based on this schematic growth model and our previous work [15].

Figure 4. Schematic diagram of three-dimensional growth model of magnesium alloy.

Table 1. Parameters and physical properties of Mg-30wt.%Gd alloy used in the present simulation [12, 15]

Property	Mg-30wt.%Gd
Initial concentration of Gd (wt. pct)	30.0
Solute partition coefficient	0.6
Liquidus slope (K/wt pct)	-2.1
Gibbs-Thamson coefficient (m·K)	6.2×10^{-7}
Liquidus temperature (K)	5.87×10^{2}
Solidus temperature (K)	5.42×10^{2}
Solute diffusion coefficient in liquid (m^2/s)	1.8×10^{-9}
Solute diffusion coefficient in solid (m^2/s)	1.0×10^{-12}

The dendritic growth of Mg-30wt.%Gd alloy was simulated to verify the established three-dimensional Cellular Automaton model. A single nucleus was planted at the center of a calculation domain, i.e. 200×200×200 grids with a grid size of 2 μm. The temperature field was uniform with a cooling rate of 80 K/s. The parameters and physical properties of Mg-30wt%Gd alloy used in this simulation are listed in Table 1 [12, 15]. Figure 5a and 5b demonstrate the simulated three-dimensional dendritic morphology and solute map of Mg-30wt.%Gd alloy. It can be observed that there is a strong dominance of eighteen orientation in the simulated 3D dendrite. For the *xy*-plane shown in figure 5c, the primary dendrite arms grow along six prism orientation with the angle of 60° between arms, the same as the six-fold symmetric morphology cut by S_0 indicated in figure 3b. For the *xz*-plane (figure 5d), the dendrite arms grow along six arms i.e. two arms in the *xy*-plane and four arms next to them as the six branches morphology cut by S_1 (figure 3d). Compared with the reconstructed three-dimensional dendrite of magnesium alloy in figure 3, the simulated results with eighteen arms dendrite agree well with the experimental results.

The effect of the cooling rates on the three-dimensional dendritic morphology of magnesium alloy were also studied. A single nucleus was planted at the center of a calculation domain i.e. 200×200×200 grids with a grid size of 1 μm, three different cooling rates of 30, 50 and 80 K/s. Figure 6 shows the simulated equiaxed dendritic growth of Mg-30wt.%Gd alloy under different cooling rate. It can be observed that with the increase of the cooling rate, the dendrite grew into a more well-developed morphology and the branches of dendrite arms became enhanced. For the primary arms on the non-basal plane, only a few secondary arms branch under the cooling rate of 30 K/s (figure 6d), while secondary arms develop sufficiently under the cooling rate of 80 K/s (figure 6f). An increasing cooling rate results in a higher growth velocity, and then lead to a higher accumulation of the solute at the solidification front (figure 6d-6f). Hence, the primary dendrite arms grow faster with an increasing cooling rate [15].

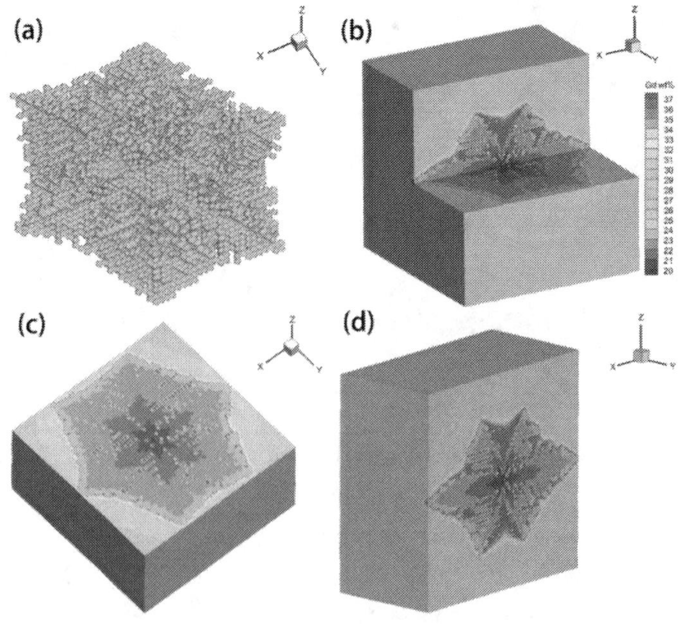

Figure 5. Simulated dendritic growth of Mg-30wt.%Gd with eighteen-branch structure.

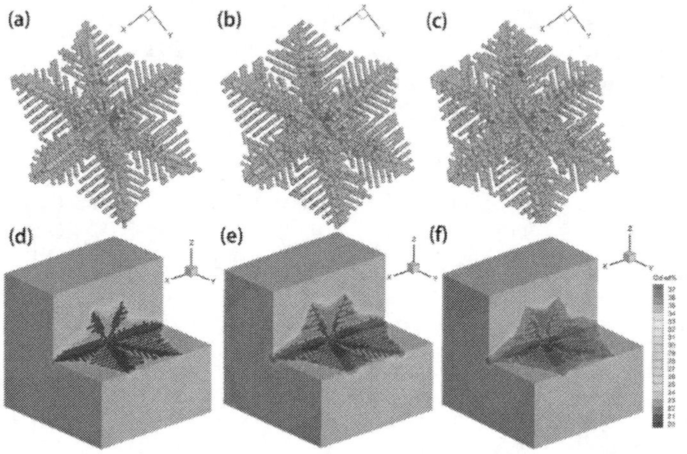

Figure 6. Simulated equaxied dendritic growth of Mg-30wt.%Gd alloy under different cooling rates: .(a, d) 30 K/s; (b, e) 50 K/s; (c, f) 80 K/s

4. Conclusions

(1) The 2-D and 3-D dendritic microstructure of magnesium alloy was studied using SEM and the synchrotron X-ray micro-tomography. It was found that the α-Mg dendrite has eighteen branches including six growing along the $\langle 11\bar{2}0 \rangle$ direction in the basal plane and the other twelve growing along the $\langle 11\bar{2}3 \rangle$ in the $(10\bar{1}0)$ plane.

(2) Based on the observed dendritic morphology, a three-dimensional cellular automaton model was developed to simulate the dendritic growth of magnesium alloy. Simulated dendrites grow into eighteen branches which are agree with the experimental results.

Acknowledgments

This work was financially supported by the National Natural Science Foundation of China, the National Science and Technology Major Project of Ministry of Science and Technology of China, and the Tsinghua University Initiative Scientific Research Program under contract nos. 51275269, 2012ZX04012011 and 20121087918. The authors would also like to thank the Shanghai Synchrotron Radiation Facility for the provision of beam time.

References

[1] Friedrich H and Schumann S 2001 *J. Mat. Proc Tech.* **117** 276
[2] Kulekci M K 2008 *Int J. Adv. Manuf. Tech.* **39** 851
[3] Mordike B L and Ebert T 2001 *Mat. Sci. Eng.* A **302** 37
[4] Asta M, Beckermann C, Karma A, Kurz W, Napolitano R, Plapp M, Purdy G, Rappaz M and Trivedi R 2009 *Acta Mater.* **57** 941
[5] Dantzig J A and Rappaz M 2009 *Solidification* (Boca Raton, FL: Taylor & Francis)
[6] Audero M A and Biloni H 1972 *J. Cryst. Growth* **12** 297
[7] Weinberg F and Chalmers B 1952 *Can. J. Phys.* **30** 488
[8] Pettersen K, Lohne O and Ryum N 1990 *Metall. Trans.* A **21** 221
[9] Pettersen K and Ryum N 1989 *Metall. Trans.* A **20** 847
[10] Wang M Y, Williams J J, Jiang L, Carlo F De, Jing T and Chawlab N 2011 *Scripta Mater.* **65** 855
[11] Wang M Y, Xu Y J, Jing T, Peng G Y, Fu Y N and Chawla N 2012 *Scripta Mater.* **67** 629
[12] Wu M W and Xiong S M 2012 *Acta Metall. Sinica.* (English Letters) **3** 169

[13] Böttger B, Eiken J and Steinbach I 2006 *Acta Mate.* **54** 2697
[14] Böttger B, Eiken J, Ohno M, et al. 2006 *Adv. Eng. Mater.* **8** 241
[15] Wu M W and Xiong S M 2012 *T. Nonferr. Metals Soc.* **9** 2212
[16] Eiken J 2009 *Int. J. Cast Metals Res.* **22** 86
[17] Zou Z W and Xiong S M 2009 *Mater. Sci. Forum* **610** 765
[18] Wang X F and Xiong S M 2013 *Corros. Sci.* **66** 300
[19] Limodin N, Salvo L, Boller E, Suéry M, Felberbaum M, Gailliègue S and Madi K 2009 *Acta Mater.* **57** 2300
[20] Tolnai D, Townsend P, Requena G, Salvo L, Lendvai J and Degischer H P 2012 *Acta Mater.* **60** 2568
[21] Chen R C, Dreossi D, Mancini L, Menk R, Rigon L, Xiao T Q and Longo R 2012 *J. Synchrotron Radiat.* **19** 836

Heat transfer and stress evolution behaviours of an aluminium alloy low pressure shell casting

D Q Zhang, J X Zhou[1] and T Chen
State Key Laboratory of Material Processing and Die & Mould Technology, Huazhong University of Science and Technology, Wuhan, 430074, Hubei, China

E-mail: zhoujianxin@hust.edu.cn

Abstract. Considering the solidification, demoulding and heat treatment processes in low pressure casting, relatively complete processes of an aluminium alloy shell casting are simulated to investigate the heat transfer feature and stress behaviours variation of casting in each multi-processes stage. FDM is used to discrete thermal conduction model when studying the heat transfer process, while FEM is adopted to solve the elastic-plastic model when studying the stress behaviour variation. When matching the two models, we map the finite difference mesh to finite element mesh. Three different temperature conditions, namely 300℃, 400℃ and 500℃, are simulated when we research the influence of demoulding temperature on stress behaviour. The simulation results demonstrate that the higher demoulding temperature is, the greater casting deformation and the smaller stress value are. The final casting stress status and the initial heat treatment temperature have a different relationship as for different parts of casting.

1. Introduction

Aluminium alloys have been utilized extensively in the automotive, aircraft and vessel etc. because of their superior mechanical properties. ZL101A is one of the commonly used aluminium alloy. It has good casting capability which can be strengthened by heat treatment. This kind of material is widely used in Shell parts. Stress of the casting is extremely easy to be produced and even the deformation and crack will appear when the thickness of casting is uneven. And as a means to improve the performance of aluminium alloy shell, the subsequent heat treatment also can lead to the emergence of the thermal stress. Even if the process of treatment is not proper, casting may still be discarded. So there are many scholars separately studying the stress distribution in the casting process [1-4] or in the heat treatment process [5-12], but few of them researching the stress in the whole process by combining the casting process with heat treatment process. Therefore the paper attempts to study the influence on the stress distribution and deformation in different initial heat treatment temperature or demoulding temperature. Firstly, the numerical simulation of solidification will be carried out for aluminium alloy shell casting; Secondly, when the temperature decreases to 500, 400, or 300℃, the demoulding stress will be calculated; Lastly, the numerical simulation of heat treatment will be made at initial temperature of 500, 400, or 300℃.

2. Mathematical model

2.1. Heat transfer model

The heat transfer model solves 3D transient conduction equation, using Finite difference method.

---

[1] J. X. Zhou, Male, Professor, Ph.D. His research interests mainly focus on the computer applications in foundry, especially on casting process simulation and informatization for foundry enterprise.
Email: zhoujianxin@hust.edu.cn

$$\rho c\frac{\partial T}{\partial t}=\lambda\left(\frac{\partial^2 T}{\partial x^2}+\frac{\partial^2 T}{\partial y^2}+\frac{\partial^2 T}{\partial z^2}\right) \tag{1}$$

Where ρ is the density, c is the specific heat capacity, λ is the thermal conductivity, T is the temperature, t is the time, x, y and z are the coordinates.

2.2. Stress evolution model

Thermal elastic-plastic model of the stress model can be divided into elastic stage and plastic stage. Material complies with the generalized Hooke's law at elastic stage and the expression is Eq2. But the state of strain depends on the paths of loading at plastic stage, and then it only builds the relationship between stress increment and strain increment at instantaneous state, namely the increment theory and the expression of Eq3. In practical engineering applications, usually we use the bilinear model instead of the nonlinear model for plastic stage in order to simplify the thermal elastic-plastic model for disposing plastic state of complex material. Then C^{ep} can be represented as Eq4. Yield criterion is used to determine whether the material turn into a plastic state from an elastic state. The expression is Eq5.

$$\sigma=C^e\varepsilon \tag{2}$$

$$d\sigma=C^{ep}d\varepsilon \tag{3}$$

$$C^{ep}=mC^e+(1-m)C^p \tag{4}$$

$$\overline{\sigma}=\frac{1}{\sqrt{2}}[(\sigma_1-\sigma_2)^2+(\sigma_2-\sigma_3)^2+(\sigma_3-\sigma_1)^2] \tag{5}$$

Where σ is the stress, ε is the strain, C^e is the constitutive tensor of elastic, $d\sigma$ is stress increment, $d\varepsilon$ is strain increment, C^{ep} is constitutive forth order tensor of elastic-plastic. Eq3 contains an assumption, a physical parameters and mechanical parameters are very small with the change of temperature and time in an incremental step, and it can be negligible. Where C^p is constitutive tensor of plastic, $\overline{\sigma}$ is equivalent stress, σ_1, σ_2 and σ_3 are the principal stresses.

3. Simulation process

3.1. Casting model

Figure 1 shows the 3D model of the casting. The green part is casting and the blue part is casting system. The grid of the model is shown in b). The process is low pressure casting. The thickness of the shell casting is in uneven distribution. The stress produced in casting and heat treatment can bring about deformation. It has a great influence on product reliability.

Figure 1. 3D model of the casting. a)model; b) finite difference grid; c) finite element mesh.

3.2. Material parameters

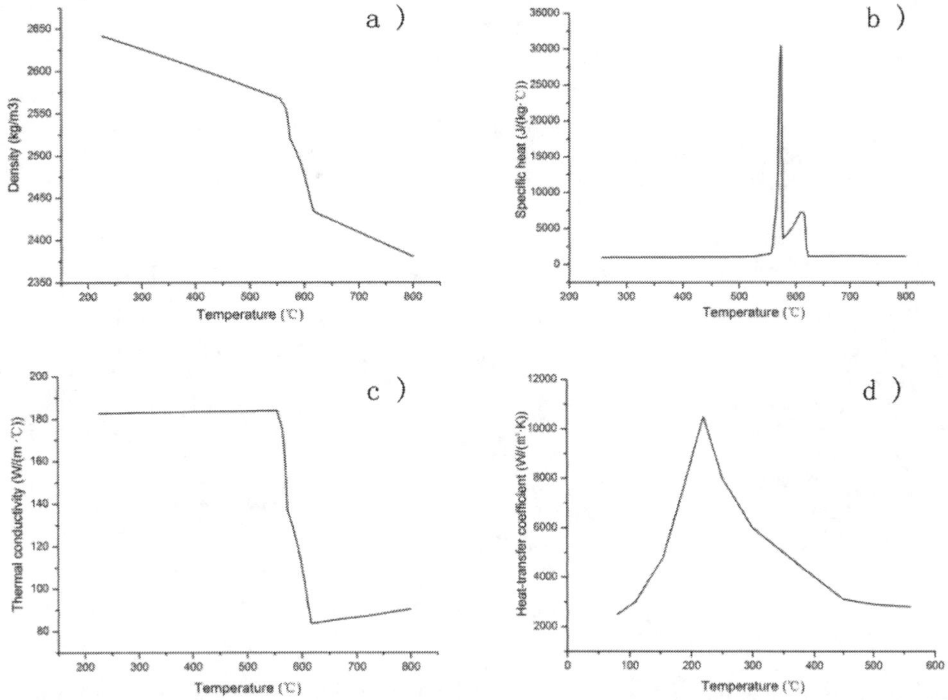

Figure 2. Thermal physical parameters of ZL101A.

Figure 2 shows the thermal physical parameters of ZL101A. Density, specific heat, thermal condition and heat transfer coefficient are shown in a), b), c) and d) respectively. These parameters are calculated by J-Matpro[13] or obtained by reviewing literature [14]. The mould is metal and the density is 7.86g/cm^3, the thermal conduction is 58.6w/(m • ℃), the specific heat is 518.8 J/(Kg • ℃).

3.3. Process parameters

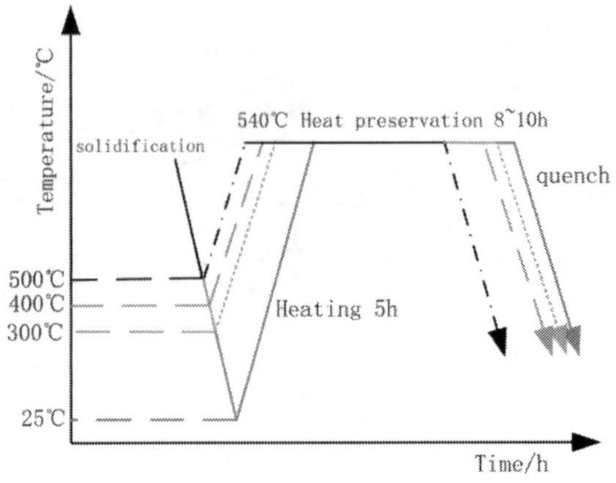

Figure 3. Process curves.

Figure 3 shows the threes process curves of heat treatment. When the temperature of casting decreases to 500℃, 400℃, 300℃ after solidification, demoulding will be done, then the casting will be raised to 540℃ with the heating rate of 101℃/h and quenching in the water of 80℃ after thermal insulation at 540℃ for 8-10h. This article simulates the process of filling solidification, demoulding and heat treatment by FDM, calculates the stress of the whole process by FEM, and the influence on the performance in different demoulding temperature is analyzed.

4. Results and discussion

4.1. Solidification stage

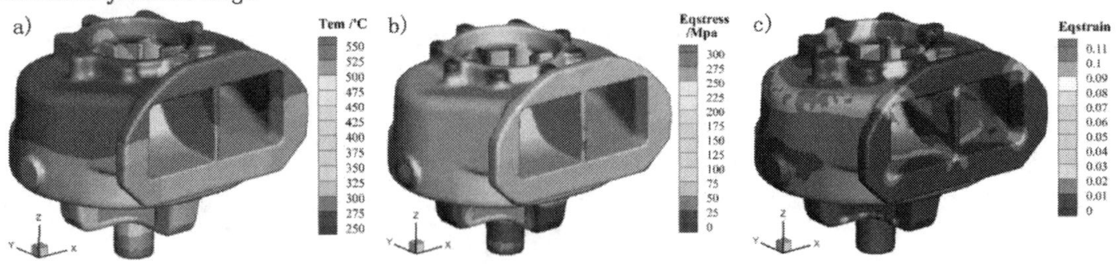

Figure 4. Temperature, stress and strain distribution after solidification. a) temperature, b)stress, c)strain.

Figure 4 shows the temperature, stress and strain after solidification. The temperature at the inlet is the highest and the temperature at the top of the thin-wall casting part is the lowest (Figure 4 a)). On the whole, the temperature of the upper casting is lower than temperature of the runner on the bottom. The reason is that its way of casting is low pressure casting. The equivalent stress distribution of solidification as shown in Figure 4 b) is about from 0Mpa to 280Mpa, and most of the casting stress is 150Mpa. But stress concentration would be appeared at some corners; on the contrary, the stress of thick parts is very small. The equivalent strain distribution is very uneven, this means that the castings will be out of shape, which has a great influence on the casting quality. If we only consider the heat treatment stress and ignore the solidification stress, the final prediction accuracy of the performance would be greatly reduced.

4.2. Stress and strain curves

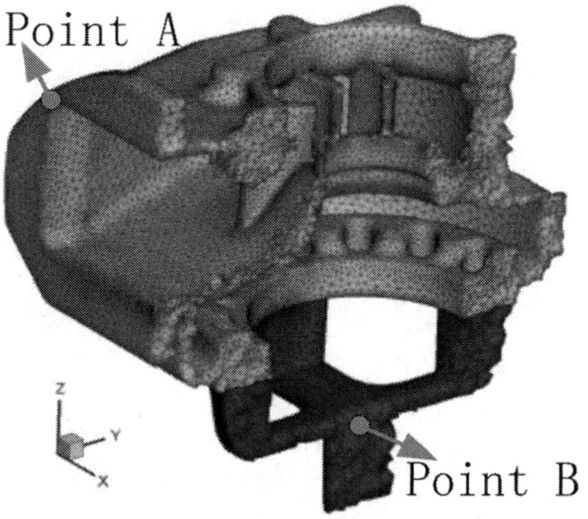

Figure 5. Position of point A and B.

In order to analyze stress and strain in the whole process, two points (Figure 16) on the casting are taken to draw curves. The curves of the whole process are too long, so here the whole curves of stress and strain are divided into three parts: 0-100s (Figure 6 a) and d)), 100s-4000s (Figure 6 b) and e)), 4000s-47000s (Figure 6 c) and f)). Point A is at the corner of casting while point B is at the centre of the runner.

The residual stress and strain distribution of solidification are uneven, such the two points (Figure 6 a) and d)). If the residual stress of solidification process is not considered, then the initial stress state would be assumed to be uniform and zero for heat treatment. It has a great influence on the stress distribution for the casting.

For demoulding stage, the stress and strain will be mutant when demoulding, Such as Figure 6 a) and d). The stress and strain of point A and point B have different trends at different demoulding temperature. For point A, the stress and strain have minimal values at temperature demoulding of 400℃. But the minimal stress and strain of point B appear at demoulding temperature of 300℃. The main reason is that the point A is in the thin wall while the point B is in the sprue center of thick wall.

For heating stage, the stress increase and stress of 500 demoulding temperature is smallest at point A. But the stess of 400 demoulding temperature is smallest at point B (Figure 6 b)). The strain of point A is difference for the three kinds of condition, but that of point B is approximate equality for the three kinds of condition ((Figure 6 e))). The reason is that the point B is at the centre of the runner, the strain and stress is very little, but the point A is far from the centre.

For the cooling stage, the stress of both points all increase first and then reduce, last approximately remain the constant. But the strain of two points is different. The strain of point A increase first and then reduce, however, the strain of point B reduce first and then increase. The reason is that the main factors influencing the value of stress and strain are cooling rate and temperature. The cooling rate is very great and then it reduces. For the thin wall, the influence of cooling rate on the stress and strain is dominant in the beginning. The influence of temperature on them is dominant with the loss of the cooling rate, so the values increase first and then reduce. For the thick wall, the cooling rate is not very great, and it will be postponed relative to that of thin wall. The influence of temperature on strain is dominant in the beginning, so the values reduce first and then increase.

IOP Conf. Series: Materials Science and Engineering **84** (2015) 012041 doi:10.1088/1757-899X/84/1/012041

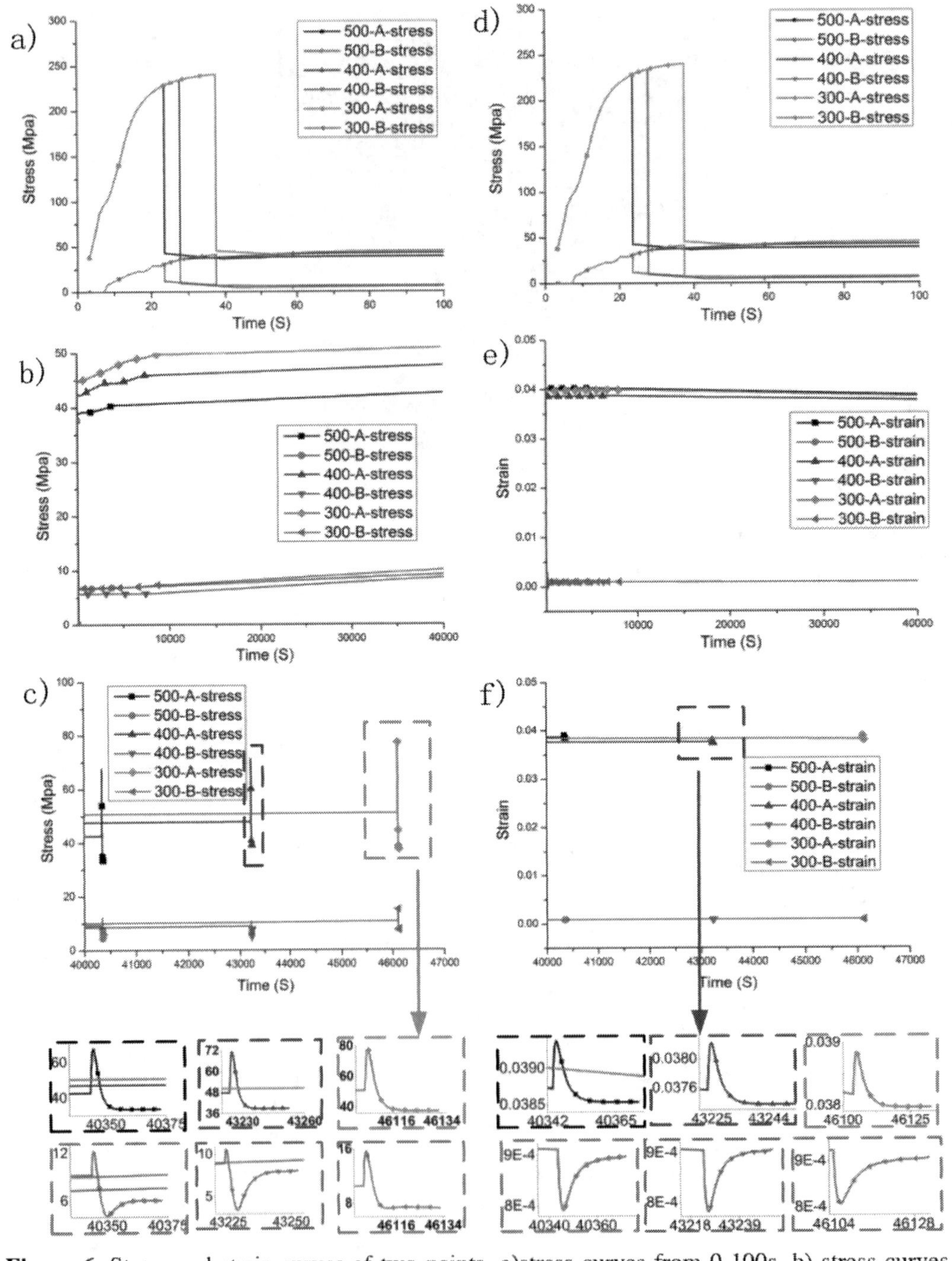

Figure 6. Stress and strain curves of two points. a)stress curves from 0-100s, b) stress curves from 100-40000s, c) stress curves from 40000-47000s, d) strain curves from 0-100s, e) strain curves from 100-40000s, f) strain curves from 100-40000s.

Table 1. The values of stress and strain at different stage.

Stress(Mpa)/strain		Demoulding temperature 500		Demoulding temperature 400		Demoulding temperature 300	
		Point A	Point B	Point A	Point B	Point A	Point B
Before demoulding	Stress	228.4182	30.42318	234.5103	35.5647	240.445	40.9189
	Strain	0.069600	0.001860	0.072170	0.00198	0.07465	0.00209
After demoulding	Stress	42.53665	11.79172	36.22975	9.5876	45.2694	7.34954
	Strain	0.042880	8.81E-4	0.040240	9.62E-4	0.04256	0.00109
Final stage	Stress	33.46335	6.00246	39.09574	8.00276	37.6050	7.25448
	Strain	0.038530	8.98E-4	0.037370	9.06E-4	0.03797	9.08E-4

Table 1 shows the values of stress and strain for the two points at different stage. For the same demoulding temperature, the stress and strain of point A are greater than that of point B before demoulding, and the decrement of stress and strain after demoulding is also greater than that of point B. For the same point, there are different rules for thick wall and thin wall casting. The higher the demoulding temperature is, the greater the stress is and the smaller the strain is for thick wall after demoulding(point B). But the stress and strain is smallest when the demoulding temperature is 400 for thin wall (point A). At last, the stress at the demoulding temperature of 500℃ has the minimal value. Bnd the strain at the demoulding temperature of 400 has the minimal value. For the point B, the final casting stress and strain at the demoulding temperature of 500℃ has the minimal value.

5. Conclusion

The solidification, demoulding, and heat treatment of an aluminium alloy shell have been investigated in this paper. The simulation results at three different demoulding temperatures are compared and the following conclusions can be drawn:

1) The higher the demoulding temperature is, the smaller the whole stress of casting before demoulding is, but the bigger the whole stress of casting after demoulding is.

2) For the same point, The higher the demoulding temperature is, the greater the stress is and the smaller the strain is for thick wall after demoulding

3) the stress at the demoulding temperature of 500℃ has the minimal value. Bnd the strain at the demoulding temperature of 400 has the minimal value. For the point B, the final casting stress and strain at the demoulding temperature of 500℃ has the minimal value

Acknowledgements

This research was financially supported by National Science & Technology Key Projects of Numerical control (2012ZX04012-011) and State Key Laboratory of Materials Processing and Die & Mould Technology Research Project (2014-02).

References

[1] Zhu H, Kang J W, Huang T Y and Liu X H 2006 *SPECIAL CASTING & NONFERROUS ALLOYS* **26** 411-413
[2] Hu S C and Ma W C 2008 *Materials for Mechanical Engineering* **32** 82-86
[3] Zhao Z W 2011 *NUMERICAL SIMULATION OF THERMALSTRESS AND PREDICTION OF HOT TEARING INA356 ALLOY CASTINGS (*Harbin Institute of Technology: Department of Materials Engineering*)*
[4] Sobolev S L 2014 *Physics Letters A* **378** 475-479
[5] Su X, Wang G, Zhang Y T, Li J F and Rong Y M 2013 *Physics Procedia* **50** 360-367
[6] Hasemann G, Schneilbel J H and George E P 2012 *Intermetallics* **21** 56-61
[7] Dolan G P, Robinson J S 2004 *Journal of Materials Processing Technology* **153-154** 346-351
[8] Koç M, Culp J and Altan T 2006 *Journal of Materials Processing Technology* **174** 342-354
[9] Wyatt J E, Berry J T and Williams A R 2007 *Journal of Materials Processing Technology* **191** 170-173
[10] Robinson J S, Tanner D A, Truman C E, Paradowska A M and Wimpory R C 2012 *Materials Characterization* **65** 73-85

[11] Yang X, Zhu J, Nong Z, Lai Z and He D 2013 *Computational Materials Science* **69** 396-413
[12] YANG X, ZHU J, LAI Z, LIU Y, HE D and NONG Z 2013 *Transactions of Nonferrous Metals Society of China* **23** 1751-1760
[13] http://www.jmatpro.cn/.
[14] Hou Z L, Yao S, Wang T L, Zhang X G and Jin J Z 2008 *Journal of materials and heat treatment* **29** 157-161

Gas-liquid Two Phase Flow Modelling of Incompressible Fluid and Experimental Validation Studies in Vertical Centrifugal Casting

J X Zhou[1], X Shen, Y J Yin, Z Guo, H Wang
State Key Laboratory of Material Processing and Die &Mould Technology, Huazhong University of Science and Technology, Wuhan, China

E-mail: zhoujianxin@hust.edu.cn

Abstract. In this paper, Gas-liquid two phase flow mathematic models of incompressible fluid were proposed to explore the feature of fluid under certain centrifugal force in vertical centrifugal casting (VCC). Modified projection-level-set method was introduced to solve the mathematic models. To validate the simulation results, two methods were used in this study. In the first method, the simulation result of basic VCC flow process was compared with its analytic solution. The relationship between the numerical solution and deterministic analytic solution was presented to verify the correctness of numerical algorithms. In the second method, systematic water simulation experiments were developed. In this initial experiment, special experimental vertical centrifugal device and casting shapes were designed to describe typical mold-filling processes in VCC. High speed camera system and data collection devices were used to capture flow shape during the mold-filling process. Moreover, fluid characteristic at different rotation speed (from 40rpm, 60rpmand 80rpm) was discussed to provide comparative resource for simulation results. As compared with the simulation results, the proposed mathematical models could be proven and the experimental design could help us advance the accuracy of simulation and further studies for VCC.

1. Introduction

Vertical centrifugal casting (VCC) is a special casting technique could significantly improve the quality of thin-walled castings. During the process of mold-filling and solidification, the high centrifugal-force and the gravity are applied to the molten metal in the spinning die. The free surface changes become more intense by bi-action of two direction forces, particular under the higher rotation speed condition. So the mold-filling process of VCC is more complex than other casting process, and the numerical simulation is challenged.

In primer research, the numerical simulation of mold-filling in VCC was used to the Solution Algorithm (SOLA)-Volume of Fluid (VOF) method. Jia et al developed analytic method to analyse the relationship between filling velocity and filling distance, and the relationship between the maximum rotational angular velocity and diameter and filling distance[1]. Xu D M et al. proposed a mathematical model based on SOLA-VOF method for mold-filling processes under centrifugal force conditions, and

[1]Corresponding author: Male, Professor, Ph.D. His research interests mainly focus on the computer applications in foundry, especially on casting process simulation and informatization for foundry enterprise.
E-mail: zhoujianxin@hust.edu.cn

simulated the free surface developments in a rotating cylindrical container[2]. A foundry CAE system based on the SOLA-VOF method for centrifugal casting was developed by Zeng X W and was applied to the industry [3]. Humphreys et al[4-5] established numerical models to analyse morphological changes of fluid free surface, and validate it through water simulation test results. Wu S P et al[6] studied the filling process in VCC process by hydraulic experiment. Prasad et al[7] used different viscosity of transparent water and 140EP oil to study the flow shape affected by processes parameters for horizontal centrifugal casting, and also applied the STAR-CD to simulate the fluid behaviour in cold die and verify the result with the experiment[8]. Combined with water simulation and theoretical analysis method, Li et al[9-10] qualitatively analysed the flow pattern of VCC filling process with advanced high-speed cameras, and systematically discussed impact factors of melt flow pattern and casting defects of titanium alloy by processes parameters in the following practical experiment studies. Mukunda et al[11], Watanabe et al[12], Shimizu et al[13], and etc. studied actual production process problems of centrifugal casting for different casting alloy.

However, above simulation and experiment researches have weakness on exploring complex two phase flow and geometry of VCC. On one hand, the free surface of fluid and gas becomes more complex because of the bi-action of centrifugal and gravity, and the gas pressure would be important influential factors on the shape of fluid in VCC. In this paper, mold-filling process is modelled by the combination of the Projection and the Level Set. The Projection algorithm is applied to calculate the governing equation of the flow filed in VCC, and the Level set method is adopted to deal with the interface transformation. On the other hand, because of hard prospecting and data recording under rotation condition, few primer researches give comprehensive fluid information such as fluid pressure, velocity etc. In this paper, we built specific experiment instruments and plan to create new data recording schemes in the future studying.

2. Mathematical models

2.1. The governing equation of the flow filed in vertical centrifugal

As casting process, the mold-filling process of VCC is combined with the transformation of the free surface. The momentum and mass governing equations, i.e. the incompressible Navier-Stokes equations that describe the motion of the Newtonian fluids are available as well.

The based governing equations of casting mold-filling are written as[2]:

$$\vec{F}-\frac{1}{\rho}\nabla P+\mu\Delta\vec{U}=\frac{\partial\vec{U}}{\partial t}+\left(\vec{U}\cdot\nabla\right)\vec{U} \tag{1}$$

$$\nabla\cdot\vec{U}=0 \tag{2}$$

Where $\vec{F}$ is the acceleration velocity component, ϱ is the fluid density, P is the pressure, μ is the constant viscosity, $\vec{U}$ <u, v, w>is the velocity field in the Cartesian coordinate system, t is time.

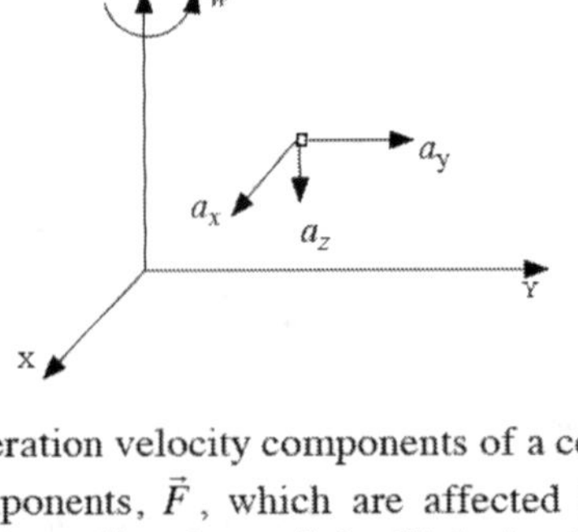

Figure 1. The acceleration velocity components of a cell in each direction.

The acceleration velocity components, $\vec{F}$, which are affected by the centrifugal force and the Coriolis force[14], are a_x, a_y, a_z in three directions of the Right-handed coordinate system (see fig.1). The sum of the vector of acceleration velocity in each direction is written as:

$$a_x = \omega^2 x + 2\omega v_r \tag{3}$$

$$a_y = \omega^2 y - 2\omega u_r \tag{4}$$

$$a_z = g \tag{5}$$

Where ω is the angular velocity, u_r and v_r refer to the relative velocity in the *X*, *Y* direction, *g* is the gravity acceleration velocity.

2.2. Free Surface Treatment

In this paper, the Level Set method is applied to simulate the interface movement between the gas and the liquid in the mold filling process[15].As depicted in fig. 2, consider a grid configuration on a domain Ω separate into two disjoint subsets Ω1 and Ω2, such that $\Omega = \Omega 1 \cup \Gamma \cup \Omega 2$, where Ω1 is filled liquid, Ω2 is gas, and Γ(t) is the interface between Ω1 and Ω2. Assume that the domain Ω is represented by a level function $\phi(x, y, z, t)$, and the interface Γ(t) is the set of function $\phi(x, y, z, t)$ is equal to zero:

$$\phi(x,y,z,0) = \begin{cases} -d(x,y,z,\Gamma) & \text{if } (x,y,z) \in \Omega 1 \\ 0 & \text{if } (x,y,z) \in \Gamma \\ d(x,y,z,\Gamma) & \text{if } (x,y,z) \in \Omega 2 \end{cases} \tag{6}$$

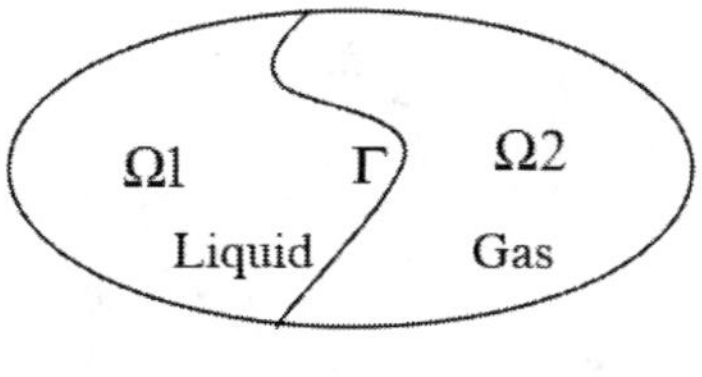

Figure 2. The interface of the computational domain.

Time step t_n, $\Gamma(t)$ is from calculating the function $\phi(x, y, z, t)$. By updating the Γ, we could track the fluid-gas interface. At the time *t*, the microelement on the free surface consists with the function listed below:

$$\phi(x(t), y(t), z(t), t) = 0 \tag{7}$$

Then,

$$\frac{\partial \phi}{\partial t} + \vec{U}(\nabla \phi) = 0 \tag{8}$$

For specific fluid field, the Level Set function is written as this format:

$$\frac{\partial \phi}{\partial t} + u\frac{\partial \phi}{\partial x} + v\frac{\partial \phi}{\partial y} + w\frac{\partial \phi}{\partial z} = 0 \tag{9}$$

Where *u*, *v* and *w* refer to the velocity component in the Cartesian coordinate system.

2.3. Projection method

The Projection method states that any vector field $\vec{U}^*$ could be decomposed into the sum of divergence-free vector $\vec{U}$ and a weighted gradient field $\nabla P/\varrho$ for scalar function *P* and positive function ϱ. Considering the Eq. 1 and Eq. 2. The momentum equation is discretized to explicit scheme:

$$\frac{U^{n+1} - U^n}{\Delta t} + U^n \cdot \nabla U^n = -\frac{1}{\rho}\nabla P^n + \mu \Delta U^n + F \tag{10}$$

Define tentative velocity U^*. Now the Eq.10 could be written as:

$$\frac{U^{*}-U^{n}}{\Delta t}+U^{n}\cdot\nabla U^{n}=\mu\Delta U^{n}+F \tag{11}$$

$$\frac{U^{n+1}-U^{*}}{\Delta t}=-\frac{1}{\rho}\nabla P^{n+1}. \tag{12}$$

Take the divergence on each side, and then apply the mass equation $\nabla\cdot U_{n+1}=0$. The equation becomes a pressure Poisson Equation (PE):

$$\frac{\nabla\cdot U^{*}}{\Delta t}=\nabla\cdot\frac{\nabla P^{n+1}}{\rho} \tag{13}$$

With Neumann boundary conditions on the domain's boundaries and on the solid objects:

$$n\cdot\frac{\nabla P}{\rho}=n\cdot(U_{bc}^{*}-U_{bc}) \tag{14}$$

During the numerical computation, the Level Set function must be re-initiated so as to keep the definition of level function $\phi(x, y, z, t)$[16]. Hamilton-Jacobi equation would be solved to update the function ϕ in this paper[17-18]:

$$\begin{cases}\dfrac{\partial\phi}{\partial t}=\mathrm{sgn}(\phi_0)(1-|\nabla\phi|)\\ \phi_0=\phi(x,y,z,0)\end{cases} \tag{15}$$

Where $sgn(\phi_0)$ is signum function, ϕ_0 is the value of the preceding time step. Eq.15 should be iterated to reach a steady state, and $\phi(x, y, z, t)$ would be updated in this time step.

3. Numerical and analytic solutions examples

To verify the mathematical model and numerical algorithm of VCC, an example is carried out in this paper. In fig. 3, the height of the computation casting model is 80mm. We put the emphasis on inspecting the effect of the centrifugal force and the transformation of the free surface. Assume that the molten metal has finished the pouring process before the die spinning, and the height of the liquid is 60mm at the initial time in the cavity. The liquid is aluminium with a purity of 99.999%, and the density is 2350kg/m^3. The density of the gas is 1.2kg/m^3. The kinematic viscosity of the liquid aluminium and the gas is 4.255x10-7 m^2/s and 1.48x10-6 m^2/s respectively. In this paper, the uniform grid is applied, and the spatial step is 1mm.

(a)Test model (b)Inwall of the cavity(mm)

Figure 3. The Model of VCC.

3.1. The Shape of the Free Surface

Theoretically, the cavity rotates with the angle velocity of ω by the vertical axis, i.e. *Z* axis in the governing equations, so the liquid in the cavity would rotates with the constant angular velocity ω in the steady state. Assume that the pressure in the cell *(r, z)* is *P*, and the weight density is γ. Every cell in a steady-state flow field satisfies the equilibrium equation[19]:

$$dP = \frac{\gamma}{g} w^2 r dr - r dz \tag{16}$$

This equation is just the base pressure equation that we study the state of relative rest. As known, we should consider the pressure gradient $dP = 0$ in the isobaric face. Then,

$$dz = \frac{w^2}{g} r dr \tag{17}$$

Take the integral on each side:

$$z = \frac{w^2 r^2}{2g} + c_1 \tag{18}$$

Where, c_1 is a constant. If we draw a curve by r and z, which r is treated as independent variable, while z as dependent variable, it would be a parabola. It illustrates that the isobaric surface of the free surface in the spinning die is paraboloidal surface of revolution, as shown in fig. 4.

Figure 4. The free surface in 3D model.

(a)0.02s (b)0.25s (c)0.70s

Figure 5. The distribution of free surface at a time .

Fig. 5 shows the distribution of free surface at a time (during the pouring process at 0.02s, 0.25s and 0.7s). The height of the molten metal near the mold increases, while the height of the molten metal in the rotational center of the mold decreases. At last, the free surface is formed, and it appears that the side near the wall is high and the side far from the wall is low.

For the restriction of accuracy of mesh, the free surface may have burr in the computation zone with jag. But the distribution of the liquid phase calculated by the Projection-Level Set method coincides with the physical truth.

3.2. The Distribution of the Pressure

This paper observes and studies the distribution of pressure in the cross section which the normal direction is the z-axis. For a designated section, the item dz is equal to 0 in Eq.16. Then,

$$dP = \frac{\gamma}{g} w^2 r dr \tag{19}$$

Take the integration on each side [20], we obtain:

$$P = \frac{\gamma w^2 r^2}{2g} + c_2 \tag{20}$$

Where, c_2 is a constant. In this equation, p is only related to r, and with the increase of r, it shows a quadratic curve change. In the spinning die, it also indicates that, the pressure distribution of the molten metal is parabola. As illustrated in fig.6, the numerical and analytic solutions are presented, and the change regularity of which are homologous and tallied with Eq.20.

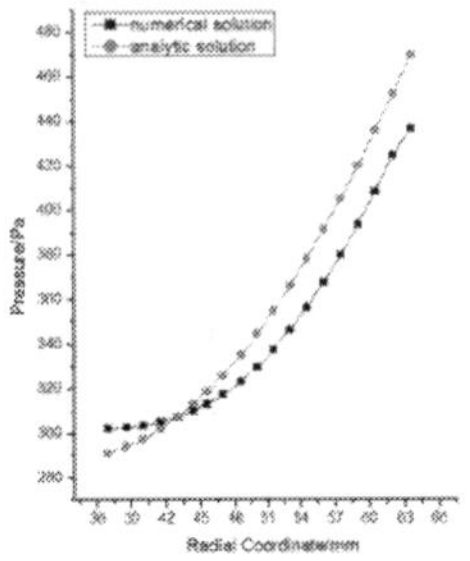

Figure 6. The numerical and analytic solutions in the selected section.

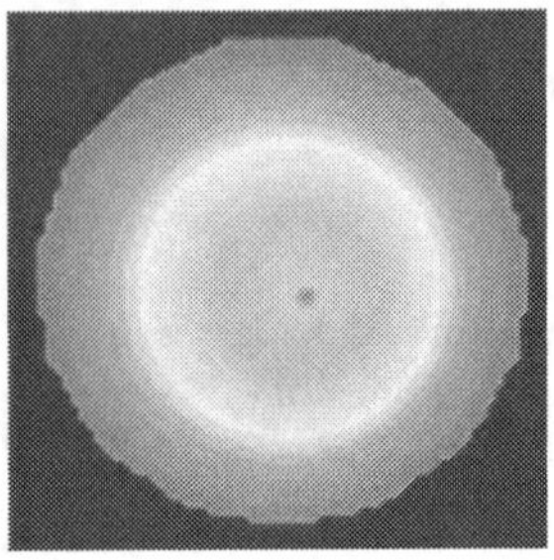

Figure 7. The contour of pressure in the selected section.

At flow steady state time, the radical pressure distribution of same investigate points (19 points along with the radial coordinate) is compared. Since the analytic solution (calculated by equation (20)) ignores viscous forces of molecules of the fluid, the pressure gradient would generally be larger than actual phenomenon. Relative error between numerical and analytic solutions: the maximum error is 7.8%; the minimum error is 0, and the average error is 4.1%. The error value is acceptable, and the numerical solution could describe the equation (20).For observation purposes, this paper selects a horizontal section to draw the contour of pressure, which the height is 25mm. Fig.7 shows that the pressure and the axis center are radially symmetric.

4. Water simulation experiment and discussion

This article puts emphases on shape change of the flow during mold-filling. The schematic of the experimental platform is shown in fig. 8. The actual equipment is shown in fig. 9. The major devices of the experiment are die casting mold, high speed camera, the centrifugal workbench, console, acquisition system, motor, etc. Experimental media are water, carbon ink, plotting paper and so on. The carbon ink is used as a marker of water simulation tracer. . The die casting mold is transparent organic glass, its three-dimensional size is as shown in fig.10. The high speed camera used in the experiment is professional high speed camera of Motion Pro Y4 - S1 type of the IDT Company.

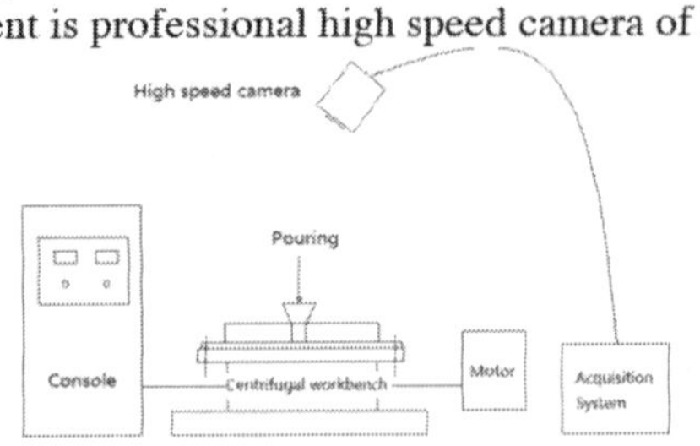

Figure 8. The schematic of experimental platform.

Figure 9. Actual equipment.

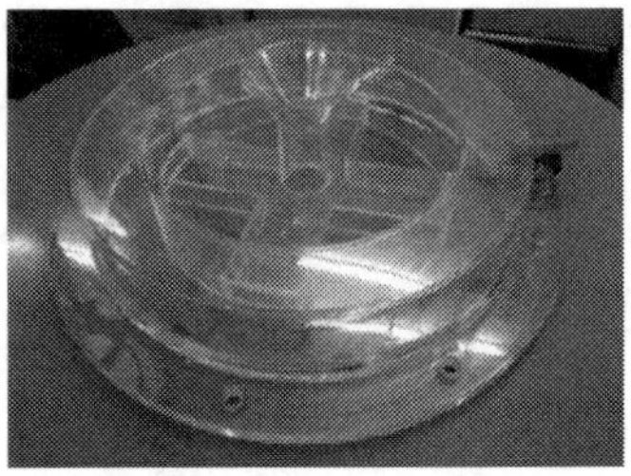

Figure 10. Installation of die casting mold.

The centrifugal workbench is driven by motor through the pulley and the level error is about ±1mm. The motor is three-phase asynchronous motor which is controlled by the frequency converter to make speed stable. The speeds are acquired through laser reflection principle to get signal of centrifugal spinning. This measurement is characterized by high resolution, wide scope of application and high reliability. The console is used to control pouring, motor, centrifugal rotary table and display speed. The centrifugal force is associated with the centrifugal rotation. Therefore, the rotation speed is the main experimental parameter, which is set as 40rpm, 60 rpm, 80 rpm and the according fluid flow during filling is filmed respectively.

Simulation and experiment discussion

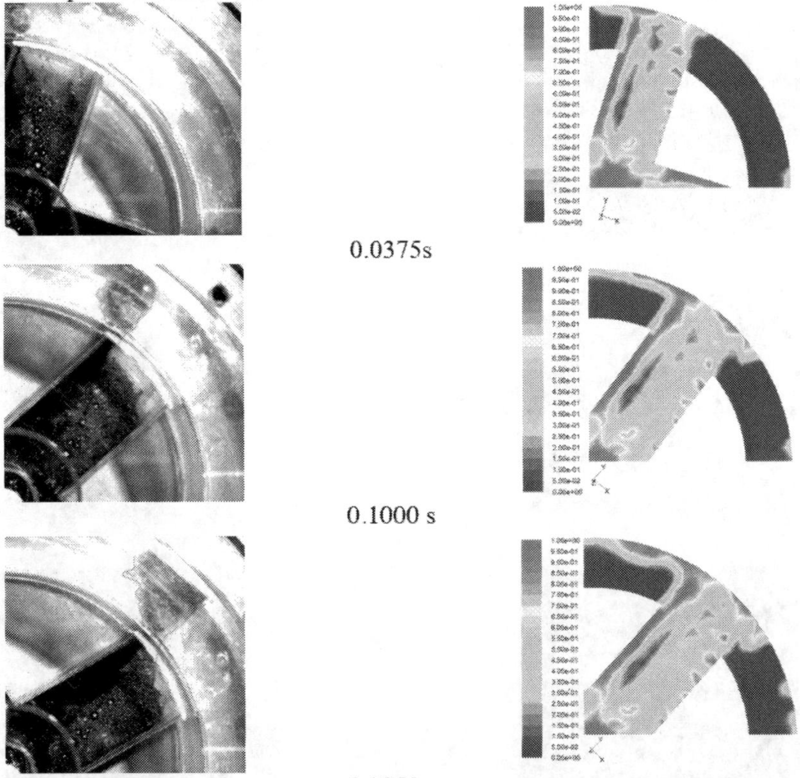

0.2375s

Experiment Simulation

Figure 11. Top view at cross gate at different filling time

0.16s

0.18 s

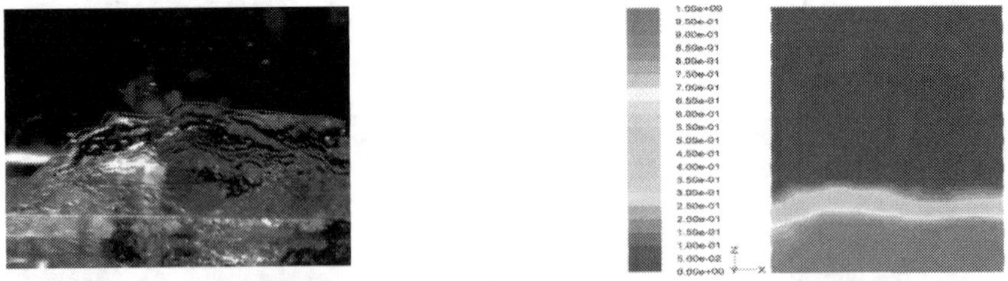

0.19 s

0.20s

Experiment Simulation

Figure 12. Front view at the mold outer wall at different filling time

With same parameter setting, the proposed models and numerical results (based on Fluent) is presented to compare with experiment results. Due to spatial confined, this paper simply enumerates partial data of experiment and simulation as shown in fig. 11 and fig. 12. Before pouring process, the mold is put on the centrifugal system and is in a stable operation at 80 rpm (because of paper length limit, we just discuss it to present 40rpm and 60 rpm). Fig. 11 shows that the fluid flows through the runner gate and present wall-flow on account of centrifugal force during pouring process. The simulation results are basically meeting the tendency in experimental results. And it results in that most of the fluids only flow to one side, which is just the opposite direction of mold rotation. Meanwhile, gas flows into mold centre along with the radius direction. But at the runner gate, partial fluids hinder the gas escape. It comes into being a liquid gas mixture because of severe turbulence. Besides, the fluid front at initial time is random like turbulence and is more likely to exist in the centre of runner gate, this is the result of interaction between liquid and gas. Eventually, the fluids gather in outer mold. On a negative note, the accurate and precise gas-liquid interface is difficult to trace owing to swift-running fluid.

In fig. 12, the fluids from runner gate crash into the mold outer wall and splash as wave crest. At the same time, from the experiment figure, it can be seen that the fluid deflect from a straight course of runner gate. Combined with the above analysis results from fig. 11, it can be known that the inertial effect cause the fluid to tend to flow along original orientation from runner gate. In addition, the gas-liquid interface is also affected by the fluid flow form, especially in collision between liquid and mold. That will bring about casting defects in modern casting process.

Therefore, in order to acquire high-quality castings, the principle of smooth filling in centrifugal system is very important. And it needs to be able to avoid collisions between liquid and mold. At last, the structure optimization of casting mold must be given enough thought to the centrifugal force.

5. Summary

In this paper, The Projection-Level Set method is proposed to solve the flow filed governing equation to simulate the mold-filling in VCC. Contrast to the analytic solution, the reliability of the proposed method is acceptable. On the other hand, experiment instrument and simple water simulation experiment are designed. By intuitively and qualitatively analysing, the simulation results could reflect the experimental phenomena. The filling sequence and shape of fluid in simulation is basic familiar with the experiment. In the future studying, the systematic experimental schemes and more comprehensive data (fluid pressure, velocity, etc.) recorders should be developed to deeply discuss the simulation results with quantitative compassion.

Acknowledgments

This research was financially supported by National Natural Science Foundation of China (No.51305149) and (No.51475181) and State Key Laboratory of Materials Processing and Die & Mould Technology Research Project (2014-02).

References

[1] Jia J, Sheng W B, Guo J J, Su Y Q and Ding H S 1999 *Special casting & nonferrous alloys* **3** 1-4
[2] Xu D M, An Y and Li X 2003 *Chinese Journal of Mechanical Engineering* **39** 146-150
[3] Zeng X W, Chen L L and Liu R X 2004 *Foundry* **53** 310-313
[4] Humphreys N J, McBride D, Shevchenko D M, Croft T N, Withey P, Green N R and Cross M 2013 *Applied Mathematical Modelling* **37** 7633-7643
[5] McBridec D, Humphreys N J, Croft T N, Green N R, Cross M and Withey P 2013 *Computers & Fluids* **16** 63-72
[6] Wu S P, Li C Y, Guo J J, Su Y Q, Lei X Q and Fu H Z 2006 *Trans. Nonferrous Met. China* **16** 1035-1040.

IOP Conf. Series: Materials Science and Engineering **84** (2015) 012042 doi:10.1088/1757-899X/84/1/012042

[7] Prasad K S K, Mysore M and Pudukottah M 2010 *Frontiers of Materials Science in China* **1**, 103-110

[8] Prasad K S K, Mysore M, Pudukottah M and Sekhar M 2011 *Metallurgical and Materials Transactions B* **1** 144-155

[9] Li C Y 2010 *Rare Metal Materials and Engineering* **3** 388-392

[10] Li C Y, Wu S P, Guo J J, Su Y Q, Bi W S and Fu H Z 2006 *Journal of Materials Processing Technology* **176** 268-272

[11] Mukunda P G, Shailesh R A and Rao S S 2010 *Metals and Materials International* **1**137-143

[12] Watanabe K, Miyakawaa O, Takadab Y, Okunob O and Okabec T 2003 *Biomaterials* **10** 1737-1743

[13] Shimizu H, Habua T, Takadab Y, Watanabec K, Okunob O and Okabed T 2002 *Biomaterials* **11** 2275-2281

[14] Information on http://en.wikipedia.org/wiki/Coriolis_effect.

[15] Osher S and Fedkiw R 2002 Springer-Verlag Press, NY

[16] Li F Y, Shu C W, Zhang Y T and Zhao H K 2008 *Journal of Computational Physics* **17** 8191-8208

[17] Enright D, Losasso F and Fedkiw R 2005 *Computers & Structures* **83** 479-490

[18] Osher S and Sethian J A 1988 *Journal of Computational Physics* **79** 12-49

[19] Group of sixteen colleges 1978 *teaching material*, North of China

[20] Zhang J 2011 *Master dissertation*, Huazhong University of Science and Technology

IOP Conf. Series: Materials Science and Engineering **84** (2015) 012043 doi:10.1088/1757-899X/84/1/012043

The Through Process Simulation of Mold filling, Solidification, and Heat Treatment of the Al Alloy Bending Beam Low-pressure Casting

Yajun Yin, Jianxin Zhou[1], Zhao Guo, Huan Wang, Dunming Liao and Tao Chen

State Key Laboratory of Materials Processing and Die & Mould Technology, Huazhong University of Science & Technology, Wuhan, 430074, China

E-mail:zhoujianxin@mail.hust.edu.cn

Abstract: The research on the simulation for the through process of low-pressure casting and heat treatment is conducive to combine information technology and advanced casting technology, which will help to predict the defects and mechanical properties of the castings in the through process. In this paper, we focus on the simulation for through process of low-pressure casting and heat treatment of ZL114A Bending beam. Firstly, we analyze the distribution of the shrinkage and porosities in filling and solidification process, and simulate the distribution of stress and strain in the late solidification of casting. Then, the numerical simulation of heat treatment process for ZL114A Bending beam is realized according to the heat treatment parameters and the corresponding simulation results of temperature field, stress, strain, and aging performance are given. Finally, we verify that simulation platform for the through process of low-pressure casting and heat treatment can serve the production practice perfectly and provide technical guidance and process optimization for the through process of low-pressure casting and heat treatment.

1. Introduction

The through process or whole process of casting generally consists of many continuous complicated process such as mold design, alloy pouring and heating treatment and so on. Each of them has mutual effect and may seriously affect the casting quality. For example, the design defects in mold design have an important impact on the pouring process and usually result in a lot of reject castings. In turn, successful mold filling require the rationality and feasibility of mold design. And much of alloy pouring process may produce the defects of shrinkage porosity and shrinkage cavity in castings. Then, heating treatment will be adopted to eliminate the casting defects.

In order to obtain these high-quality castings, it is essential for complex structures of castings to conduct process design analyses. And there are lots of methods [1-3] in which the numerical calculation applied for this process can quickly and accurately simulate the forming process and predict tissue defects of castings, thereby the optimization result of process design can be achieved. More researches were focused on a particular process above such as YUWEN et al [4] and Choudhari

[1] Corresponding author: Male, Professor, Ph.D. His research interests mainly focus on the computer applications in foundry, especially on casting process simulation and informatization for foundry enterprise.
E-mail: zhoujianxin@mail.hust.edu.cn

et al [5] and Zhang [6]. Moreover, casting filling process is tightly coupled with solidification process. Therefore, the influences of coupled process in whole process of casting were also undertaken, including Paul et al [7], Anglada et al [8], Miller et al [9] and so on. However, at the present stage, the requirements of the Foundry Engineering Technology is to better serve the actual production, since the forming process of a cast that includes more continuous physical and chemical changes, and the ultimate mechanical properties or casting quality is affected by the combined effects of these processes rather than just a single liquid metal filling process or solidification process or heat treatment process, etc. Accordingly, this paper is based on professional and advanced whole process simulation platform for casting forming and heat treatment. The whole process of aluminum casting ZL114A of bending beam by low pressure, covering filling, solidification and heat treatment process, have been numerically calculated. And a comprehensive analysis and forecast results for the tissue forming defects and mechanical properties in the process have been achieved and expected to provide the subsequent analysis of the whole process reference.

2. Process scheme of aluminium alloy casting by low-pressure

2.1. Geometric model

Figure 1(a) shows a sketch of the prototype. The overall size of this casting is 826mm × 169mm × 330mm. According to the technological and structural characteristics of aluminum alloy castings by low-pressure, its requirements must be to fill molten metal smoothly and to ensure that aluminum alloy molten metal is not easily oxidized. Therefore the pouring gating system is set as the expansion system, the flow gate is conformal, the size of cross gate is 60mm × 50mm and the diameter of rising pipe is 70mm. Then a three-dimensional model with pouring gate system is established by three-dimensional software as shown in Figure 1(b).

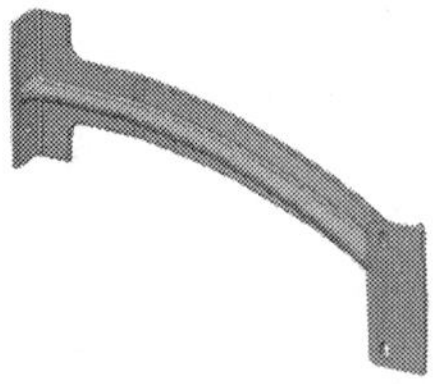

(a) Sketch of the prototype

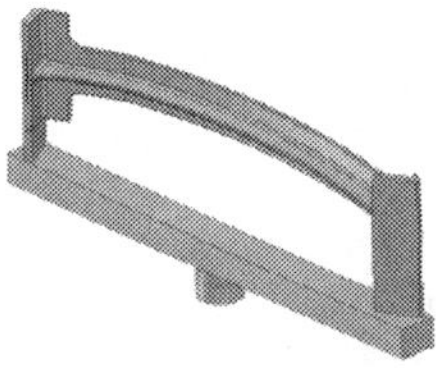

(b) Assembly model with pouring gate system

Figure 1. Prototype and assembly model of aluminum alloy bending beam.

2.2. Flow process diagram of simulation

In this paper, the main casting process, including filling process, solidification process and heat treatment, have been analyzed in the following. First, by requirements of casting, the casting materials and geometric model are carefully designed. Then the data of geometric model is applied into the filling process simulation in which the information of temperature and fluid flow field are obtained. On one hand, the information can be fed back to the data of mold design. On the other hand, it also flows into the next process, i.e. solidification process. Subsequently, based on the previous results, the solidification process is simulated. Likewise, the temperature and stress-strain field are also obtained, and it affects the mold design and guides the heat treatment. Finally, a high-quality casting process is designed. And the flow process diagram of simulation is shown in Figure 2.

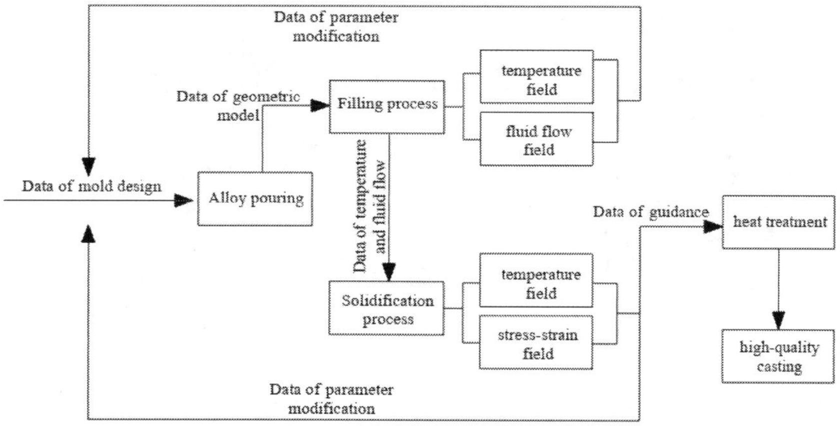

Figure 2. Flowchart diagram of the through process simulation.

3. Mathematical models and boundary conditions

3.1. Mathematical models

Since this paper is based on the numerical simulation of the whole process platform, the whole process from cast form to heat treatment has been numerically simulated, including the flow field, the temperature field and the stress field. The mathematical models mainly consist of equation of continuity, momentum equation, equation of heat conduction and stress-strain equation [9-10].

3.2. Boundary conditions

The casting material of aluminum alloy is ZL114A, the physical parameters of which are as follows (1), (2) and (3) and shown in Table 1.

$$\rho(kg/m^3) = 2681.35699 - 0.20259 \times T \quad (1)$$

$$\lambda(\mathrm{W/(m \times K)}) = 152.0237 + 0.04223 \times \mathrm{T} \quad (2)$$

$$C_p(J/(kg \cdot K)) = 885.66921 + 0.44541 \times \mathrm{T} \quad (3)$$

Table 1. Physical parameters of calculation.

Parameters	value	Parameters	value
Coefficient viscosity/(m^2/s)	2×10^{-6}	Heat transfer coefficient($W/m^2 \bullet K$)	962.7
Latent heat/(kJ/kg)	418.6	Threshold of solid fraction	0.625
Liquidus temperature(°C)	615	Liquid shrinkage rate	0.00005
Solidus temperature(°C)	555	Shrinkage rate of phase transition	0.03
Pouring temperature(°C)	720	Young's modulus (GPa)	67.8
Pouring time(s)	3	Poisson's ratio	3.72

The heat treatment conditions of casting are as follows:

(i) The cast aluminum is heated to 540 °C and kept warm in 10 hours, then cooled down at the temperature of 80 °C in water.
(ii) Aging treatment in 165 °C and heat preservation of 8 h in air.
(iii) It is a free state at entire process of heat treatment and cooling treatment.
(iv) According to the simulation, the creepage is considered during heating.
(v) The elastic-plastic model is applied in the simulation process.
(vi) The relationship between strength and cooling rate in quenching are as follow (4) and (5) (R represents the average cooling rate between 450 ~ 200 °C).

$$\sigma_m(MPa) = 29.55 \times \log(R) + 289.5 \tag{4}$$

$$\sigma_s(MPa) = 30.45 \times \log(R) + 233.12 \tag{5}$$

4. Simulation results and analysis

4.1. Mold filling process

Filling process is one of the key to casting quality, which involves a liquid metal flow and temperature fields coupled. And its various parameters such as filling velocity and temperature directly affect the cast form. So it is necessary to first study the filling process. Meanwhile, the effects of casting pouring system and bending beam structure on filling velocity and temperature should also be researched. Each stage of Correlated Color Temperature for filling process is shown in figure 3(a), (b), (c).

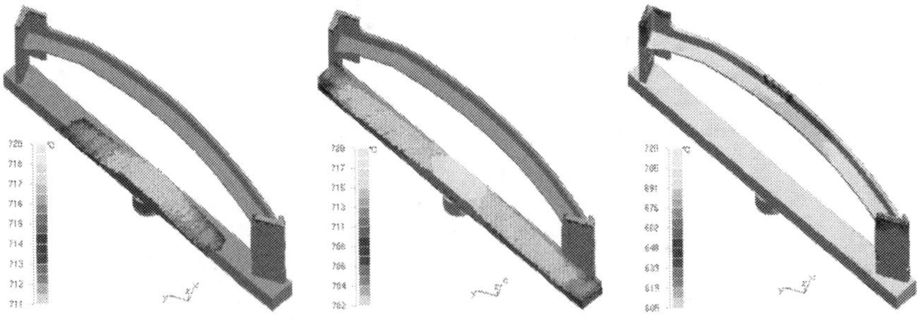

(a) t=0.51s (b) t=1.12s (c) t=2.65s

Figure 3. Correlated Color Temperature of filling process.

Figure 3 shows that, in the beginning of filling process, high-temperature liquid metal at low pressure is slowly rising from the riser tubes and smoothly split within the cross gate, which reduces the gas entrapment and inclusion for high temperature liquid metal in a certain way. Especially for aluminum alloy casting, this method has the advantage of preventing oxidized. Subsequently, the temperature drop of liquid metal in frontier is excessive with respect to other zones. But the low-temperature liquid metal is heated by follow-up high-temperature liquid metal, thus promoting that the molten metal fill forward in small temperature range until the entire cast is fully filled. Meanwhile, in the junctional area between cross gate and flow gate, the temperature drop is obvious as well as in both ends of the bending beam castings. For this situation, there are numerous and diversified reasons, which include the sudden change in the flow direction of the molten metal, the occurrence of turbulent flow and the more heat loss. And these reasons also bring about that the high temperature liquid metal flow forward in transient delay and cannot provide a continuous flow of heat for the leading edge of the molten metal, resulting in a big temperature drop of molten metal at the leading edge. In addition, at the final filling region of the bending beam casting, the temperature is significantly less than the other regions.

4.2. Solidification process

4.2.1. Cooling rate of solidification.

When the filling process is finished, the molten metal becomes solidified simultaneously. And the solidification process is influenced by a number of factors [11-12], such as filling process analyzed above and temperature variation. From the macro perspective, the solidification sequence is determined by the filling process and temperature variation, which contribute to composition segregation, shrinkage, cracks in solidification process; from the micro perspective, the microstructure forming in the solidification process is determined by the nucleation process.

At last, the mechanical properties of the casting are affected by these factors, wherein the cooling rate of solidification process is a key problem. To study the law, the cooling curves are plotted and analyzed in terms of four different points selected from the casting, as shown in Figure 4(a), (b).

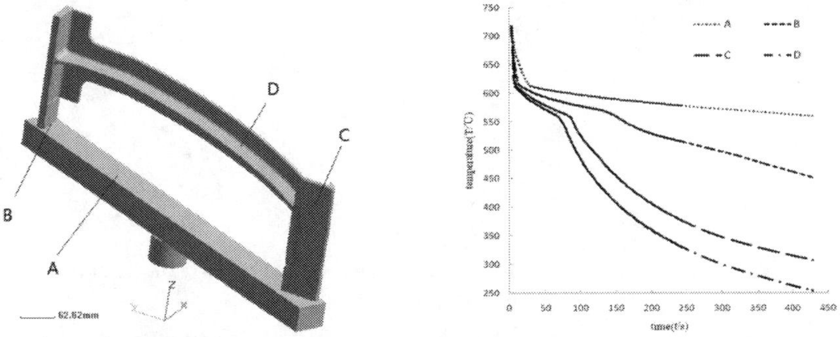

(a) Different points (b) Cooling curves

Figure 4. Cooling curve of solidification process.

In figure 4, all the four cooling curves of aluminum alloy castings show distinct variation law. As soon as the filling process come to an end, the casting temperature decline rapidly. When the temperature go down to about 620 degrees that is close to the liquidus temperature of the casting, all the cooling curves appear to horizontal but different degrees. In the First place, it is the coordinates of A at the cross gate in which the trend of the cooling curve is more smooth and linear with time. in the next place, the coordinates of B is located at the junctional section between the cross gate and flow gate, its cooling curve is relatively flat and easy. But when the temperature decreased to about 560 degrees that is close to solidus temperature, the inflection point appear. Then the trend of cooling curve tends to be declined; finally, the coordinates of C and D are separately located at the top end and central region of the bending beam, in which the trends of two cooling curves coincide with each other and also show a gradual lesser extent. Likewise, when the temperature is then decreased to about 560 degrees, there is a sharp decline until approximately 250 degrees. It is analyzed that, the above variation in the same kind of aluminum alloy castings is mainly determined by the process scheme of aluminum alloy castings in low-pressure casting. on the one hand, in the low-pressure casting, the casting mold is filled with molten metal from the down to top, forming the solidification process in proper sequence of central section, top end , junctional section and cross gate that is from top to down. On the other hand, owing to the sequence of solidification process, the casting produce the macrosegregation, and the later solidification location of the mold cannot be fed with molten metal, thus resulting in a different cooling curve.

4.2.2. Stress and strain distribution of solidification process

In the late stage of solidification, different cooling rate in the portions of the casting result in uneven distribution of the temperature. And the cooling metal is to shrink and be blocked that produce shrinkage stress. This stress can cause the cracks of the casting and a more serious impact on the serve life. Therefore, to provide guidance for the subsequent heat treatment of castings, it is necessary to analyze the stress and strain distribution of solidification process. Because of the correspondence between stress and strain, the stress distribution is analyzed only. It is shown in figure 5(a), (b) for different times of stress distribution.

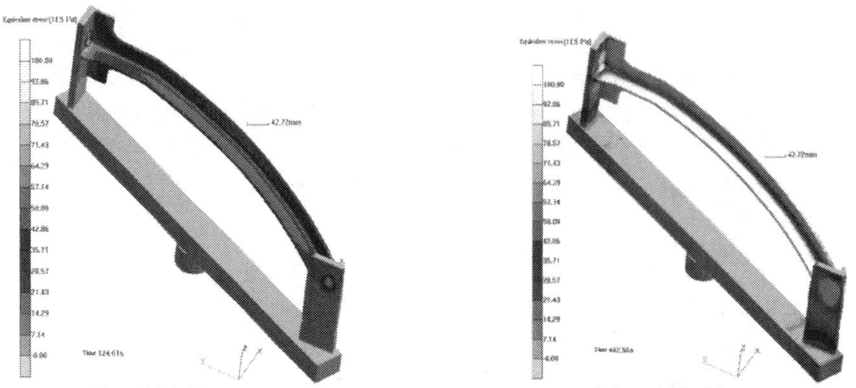

(a) t=124.01s (b) t=442.66s

Figure 5. Stress distributions in the various stages of the solidification.

In figure 5, the stress distribution is ever-changing with time and similar to the distribution of shrinkage porosity and shrinkage cavity above. In the early stages, the stresses gradually appear in the boundary of the bending beam. Subsequently, there also appear stresses from outside to inside and to extend to the bending beam casting ends. Meanwhile, the stress is gradually increasing and the maximum stress emerges in the place where the stresses appear in the beginning stages. It is because that the stresses and shrinkage properties are all affected by the temperature changing with time in the solidification process. So it is assumed that the shrinkage properties or mechanical properties are influenced by the stress in the late stage of solidification. And a follow-up heat treatment is necessary to relieve the casting stress and improve the mechanical properties.

4.3. Heat treatment process

In the actual production there are more or less casting defects in filling process and solidification process, although the optimization results which was obtained by analysis of the previous coupled process are helpful for controlling defects. To further eliminate the casting defects, appropriate heat treatment is indispensable [13].

In general, a cast after filling process and solidification process is directly heated by accumulation of experience or little instruction from previous processes, thus resulting in a higher defective index. Namely, the whole process parameters cannot be seperated from each other for a high quality casting. Besides, the continuation of the flow of process parameters in whole casting process also can promote the increase of efficiency. Here, a professional simulation platform of whole process is developed to continuously simulate the heat treatment. In that way, the temperature and stress-strain field of the bending beam above is applied into the simulation of heat process. Heat treatment in this article consists of heating process, cooling process and aging treatment.

IOP Conf. Series: Materials Science and Engineering **84** (2015) 012043 doi:10.1088/1757-899X/84/1/012043

4.3.1. Heating process

In the heating process, the temperature field of casting is changing with time and its surface temperature is not uniformity, which produce the stress and deformation. So the temperature field and corresponding stress or deformations need to be researched in this part, as shown in fig. 6(a), (b).

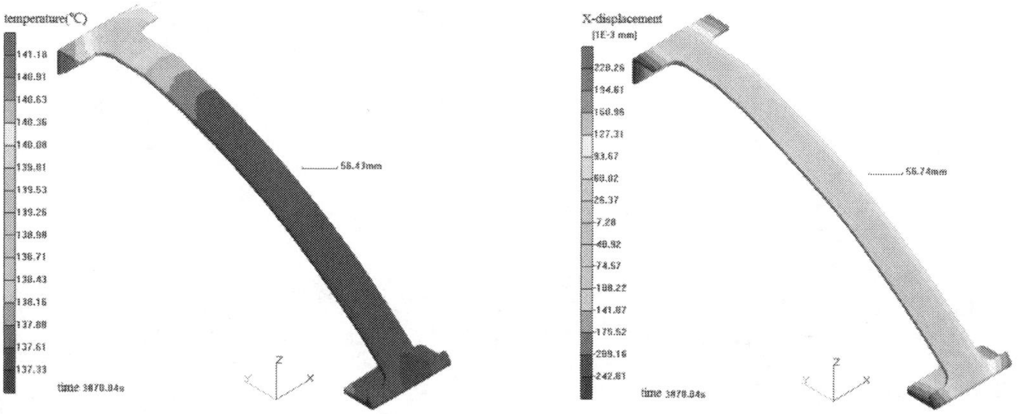

(a) Temperature field (b) X-displacement

Figure 6. Temperature field and its corresponding deformations in heating process.

In fig.6, when the casting is heated to a temperature between about 137 degrees and 141 degrees, the corresponding X-displacement varies from -0.243mm to 0.228mm and the results also show that there exist obvious symmetry variations for the displacement. That is because the difference in temperature is not big, though the temperature field is not uniformity. At this moment, the heat-up time is about 3870s, just in its beginning stage. In general, the maximum and minimum X-displacements decrease to certain values that can be ignored at last. It shows that the casting deformations caused by stress are relieved. From the above, it is concluded that the temperature field distribution is similar to the X-displacement on the whole, that is to say, the absolute values of temperature and X-displacement are big in the end and small in the middle. There are two reasons for this distribution law. First, at the beginning the high temperature is located at the edge of casting, and then the heat energy is transferred from edge to inside. But the total energy is on the increase. Second, as the temperature rises, the casting deformations are transferred in the opposite direction. And the maximum deformations at the edge of casting are accumulated through this transfer process, thus forming the distribution law.

4.3.2 Cooling process

Similar to the heating process, temperature reduction also cause the casting stress and deformations. But there are different reasons for this temperature field and X-displacement distribution in cooling process, as shown in figure 7(a), (b).

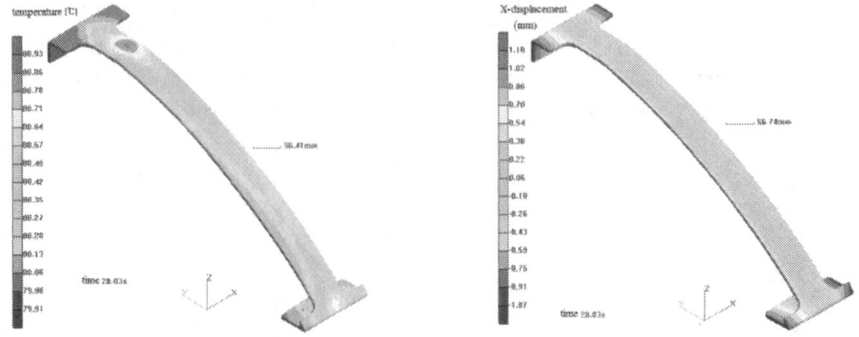

(a) Temperature field (b) X-displacement

Figure 7. Temperature field and its corresponding deformations in cooling process.

In fig.7, different from the temperature field in the heating process, the high temperature in cooling process is located at the inside of casting. And the corresponding maximum X-displacement is still located at the edge of casting. As the temperature decreases, the heat energy is transferred from inside to edge and the casting deformations or X-displacement caused by casting stress is on the increase and transferred in the same opposition. And the maximum deformations at the edge of casting are also accumulated through this transfer process. As shown by the numerical results, when the casting is cooled down to a temperature between about 80 degrees and 80.9 degrees, the corresponding X-displacement varies from -1.07mm to 1.18mm. The absolute values of temperature are big in the middle and small in the end. On the contrary, the absolute values of X-displacement are big in the end and small in the middle.

4.3.3. Aging treatment

Generally speaking, the hardness and strength of cast by quenching process are not immediately to be maximized, especially for the aluminium alloy. To obtain the high-quality casting, the aging treatment is to be needed after heating process and cooling process. The tensile and yield strength of bending beam of aluminum alloy casting which is processed by aging treatment are shown in figure 8(a), (b).

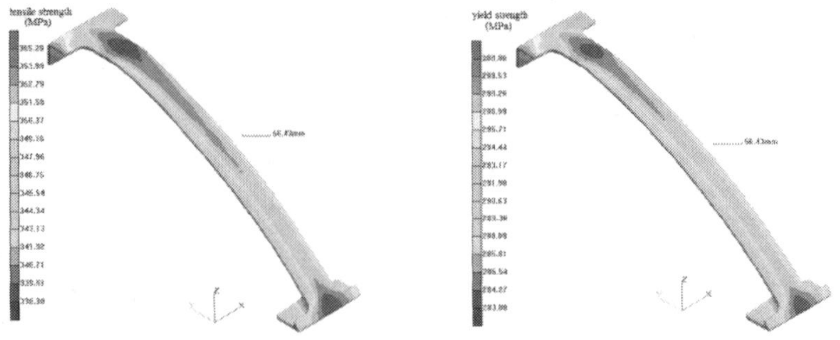

(a) Tensile strength (b) Yield strength

Figure 8. Tensile and yield strength of bending beam of aluminum alloy casting.

In fig. 8, the distribution of tensile strength is similar to the temperature field in the cooling process. But the values of tensile strength are just the opposite. Its maximum tensile strength is approximately 355MPa and located at the edge of casting, and the minimum value is approximately 338MPa and located at the inside of casting. It is assumed that there exist some corresponding relationships between the tensile strength and the temperature field and the tensile strength is improved through aging treatment. As such, the distribution of yield strength is similar to tensile strength and the values of yield strength are approximately limited on range from 283MPa to 300MPa.

5 Conclusions

By the advanced simulation platform of casting and heat treatment, the whole process of aluminum alloy cast by low-pressure and heat treatment are simulated and all forming factors in various stages of the whole process are analyzed. The conclusions are obtained as follows:

(1) In filling process, all the sudden change in the flow direction of the molten metal, the occurrence of turbulent flow and the more heat loss contribute significantly to the obvious temperature drop, and the temperature at the final filling region is significantly less than the other regions;

(2) In solidification process, the distribution of shrinkage porosity and shrinkage cavity caused by filling and solidification process has been predicted; in the late stage of solidification, the stress distribution is similar to the distribution of shrinkage porosity and shrinkage cavity. Meanwhile, the stress and shrinkage properties are all affected by the temperature changing with time in the solidification process;

(3) In heating process, the absolute values of temperature and X-displacement are big in the end and small in the middle; in cooling process, the absolute values of temperature are big in the middle and small in the end. On the contrary, the absolute values of X-displacement are big in the end and small in the middle. There exist some corresponding relationships between the tensile strength and the temperature field and the tensile strength is improved through aging treatment. As such, the distribution of yield strength is similar to tensile strength and the values of yield strength are approximately limited on range from 283MPa to 300MPa.

Acknowledgements

This research was financially supported by National Science & Technology Key Projects of Numerical control(2012ZX04012-011) and State Key Laboratory of Materials Processing and Die & Mould Technology Research Project (2014-02).

References

[1] Zhang L Q and Wang R J 2013 *The International Journal of Advanced Manufacturing Technology* **65** 517
[2] Jiang W M and Fan Z T 2014 *China Foundry* **11** 119
[3] Zhang B, Cockcroft S L and Maijer D M 2005 *JOM* **11** 36
[4] Yuwen X X, Chen L and Han Y J 2012 *Energy Procedia* **17** 1864
[5] Choudhari C M, Narkhede B E and Mahajan S K 2014 *Procedia Materials Science* **6** 786
[6] Zhang P, Zhang Y K, Liu L G, Ren X J, Zhang Y, Fang Y and Yang Q X 2012 *Computational Materials Science* **52** 61
[7] Cleary P W 2010 *Applied Mathematical Modelling* **34** 3189
[8] Anglada E, Meléndez A, Maestro L and Domiguez I 2013 *Procedia Engineering* **63** 75
[9] Miller A E and Maijer D M 2006 *Materials Science and Engineering A* **435-436** 100
[10] Zhang B and Maijer D M 2007 *Materials Science and Engineering A* **464** 295
[11] Choong D L and Suk J Y 2014 *Metals and Materials International* **4** 601
[12] Teng X., Mae H, Bai Y and Wierzbicki T 2009 *Engineering Fracture Mechanics* **8** 983
[13] Choong d L 2013 *Materials Science and Engineering: A* **559** 496

IOP Conf. Series: Materials Science and Engineering **84** (2015) 012044 doi:10.1088/1757-899X/84/1/012044

Multi-scale simulation of ductile iron casting

J Kubo
CAPCAST INC., 1-29-1, Gion, Shimotsukeshi, Tochigi 329-0434, Japan

E-mail: j-kubo@capcast.co.jp

Abstract. It has been well known that addition of rare earth elements was indispensable for production of ductile cast iron. The addition reduces solidification shrinkage. However there are still ambiguities in understanding the mechanism and predicting the casting defects. A possible explanation for the reduction of shrinkage is that the addition of rare earth promotes expansion of castings due to the graphite formation, which is related to cooling rate. In this study, the effect of rare earth addition was considered as a function of cooling rate and solidification shrinkage rate was determined by the macro-scale simulation. In the meso-scale and the micro-scale simulations, distribution of graphite nodules was compared with the experimental results. The multi-scale simulation result well reproduced the experimental results.

1. Introduction

Rare earth element is commonly added into the melt for production of ductile cast iron [1,2,3]. It is well realized that the resources of rare earth element are limited and the cost is relatively high, comparing to that of magnesium. It is of interest to know how the added rare earth elements influence the solidification structure, the defect formation (i.e. shrinkage defect) and consequently mechanical properties. One of the solutions is the prediction of the solidification structure including the casting defects by simulation in foundry. Since the addition of rare earth elements influences graphite nucleation and growth (micro-scale), interaction of graphite particles and dendrites (meso-scale) and shrinkage defect and residual stress (macro-scale).

This paper demonstrates an approach of multi-scale simulation using a commercial software code for predicting shrinkage defects. Figure 1 shows a conception for the multi-scale analysis for the cast iron in this study [4]. FEM [5] is used to know melt flow, solidification and stress in the macro-scale. In meso-scale, CAFE model which is originally developed by [6] is used to know microstructure and porosity. MCAFE model [7,8] is used to know dendritic structure in the micro-scale.

2. Macro-scale simulation predicting shrinkage defect by FEM mesh

2.1. Analysis model and conditions

The model used in the FEM analysis is shown in Figure 2. The mold is designed to produce two test pieces with 15 and 25 mm in thickness. The mesh size is approximately 3mm. A specimen which is equivalent to FCD 450 with the amount of rare earth (0.5% and 1.0%) is considered. It is assumed that molten cast iron (1350C) is assumed to be poured into mold.

In the model, the solidification shrinkage rate is treated as a function of cooling rate to represent the influence of rare earth elements on solidification manner. In general, volume of casting such as aluminium alloys and steel alloys monotonically shrinks during solidification. On the other hand, shrinkage of the ductile cast iron during solidifying is rather complicated. Due to volume expansion of graphite growth, the shrinkage depends on solidification condition. In this model, the influence of rare earth element addition and cooling rate is included by considering graphite nucleation.

In ductile cast iron, the following continuous equation is considered (Fig. 3) [9]. The first term corresponds to solidification shrinkage. It is generally negative, but the magnitude can be reduced by the graphite formation in the cast iron. The second term represents the melt flow in the mushy region (interdendritc liquid metal feeding). The third term gives formation rate of shrinkage porosity. As a result, the shrinkage rate depends on the graphite growth.

$$\left(\frac{\rho_S}{\rho_L}-1\right)\frac{\partial f_L}{\partial t}-\left(\frac{\partial f_L u}{\partial x}+\frac{\partial f_L v}{\partial y}+\frac{\partial f_L w}{\partial z}\right)+\frac{\partial f_v}{\partial t}=0 \qquad (1)$$

In the macro-scale simulation, test pieces with two different rare earth amount were analysed. The same parameters were used except the relationship between solidification shrinkage rate and cooling rate.

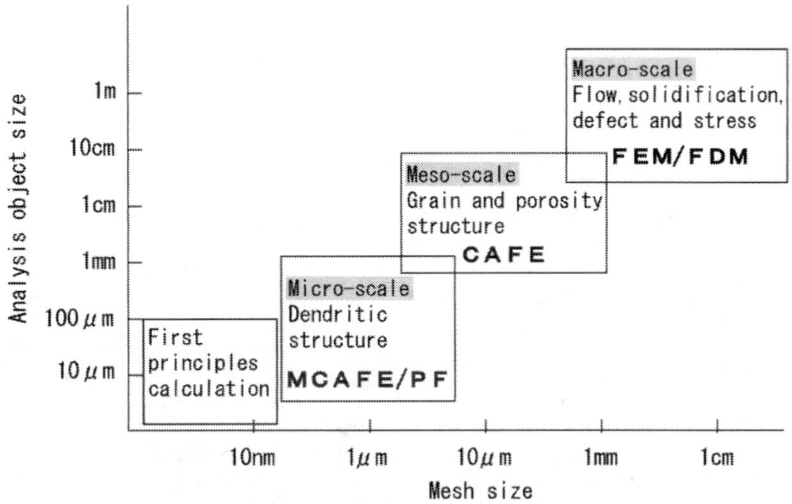

Figure 1. Multi-scale simulation of casting

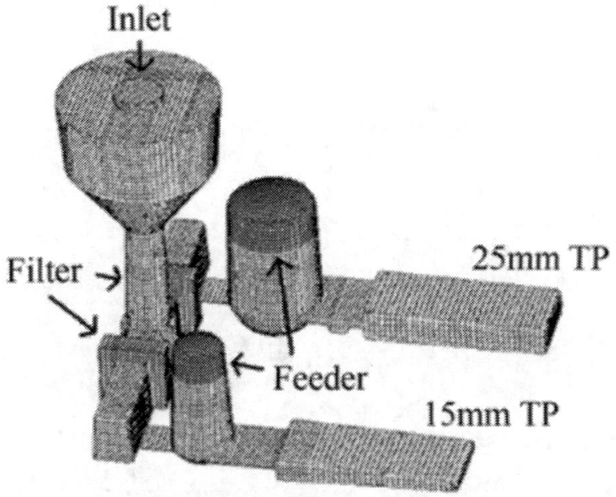

Figure 2. FEM mesh model

$$\underbrace{\boxed{\left(\frac{\rho_S}{\rho_L}-1\right)}\frac{\partial f_L}{\partial t}}_{①} - \underbrace{\left(\frac{\partial f_L u}{\partial x}+\frac{\partial f_L v}{\partial y}+\frac{\partial f_L w}{\partial z}\right)}_{②} + \underbrace{\frac{\partial f_v}{\partial t}}_{③} = 0$$

① The amount of solidification shrinkage

② Interdendritic liquid metal feeding

③ The amount of porosity growth

Figure 3. Mechanism of shrinkage defect occurrence

2.2. Result and discussion

Figure 4 shows the longitudinal center section of test pieces of the experimental results and the calculated results. The predicted positions of the shrinkage defects essentially agree with those observed in the cast specimen. In the test piece of 15mm thickness, the shrinkage occurred near the feeder for 0.5% and 1% of rare earth content. In this case, the amount of rare earth elements up to 1% does not affect the position of the shrinkage defect.

In the test piece of 25mm thickness with 0.5% of rare earth elements, although the shrinkage defect in experimental specimen appears at the opposite side of feeder, the shrinkage defect is observed at the center. In the experimental specimen with 1.0% of rare earth elements, the shrinkage defect was completely avoided. The calculation also shows no shrinkage defect is produced in the test pieces. The experimental results clearly shows that the shrinkage defect is not likely to form with increasing amount of the rare earth elements. This tendency can be predicted by considering the influence of rare earth on the nucleation and growth of the graphite.

Quantitative comparison of the shrinkage defect volume between the experimental and the calculated results is shown in Figure 5. The volume of shrinkage defect in the experiments was measured by the Archimedes method. The calculated results qualitatively agree with the experimental results.

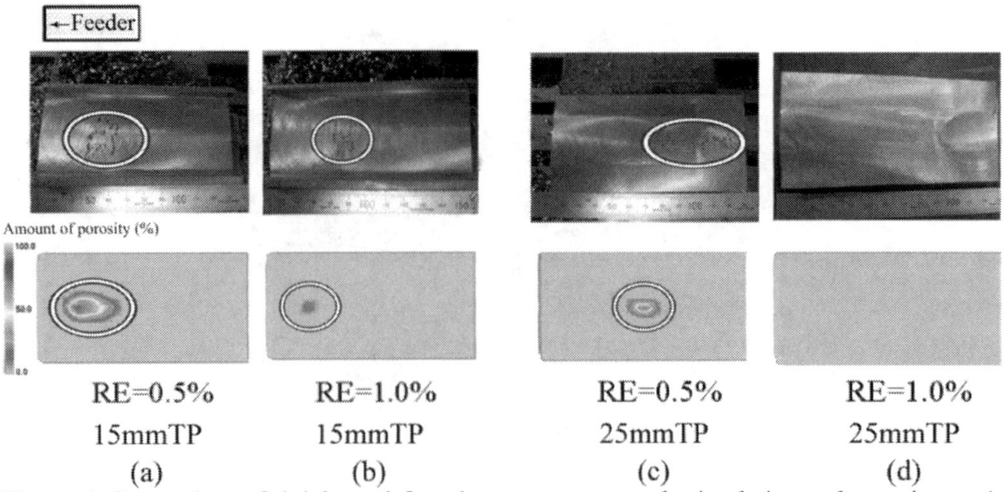

Figure 4. Comparison of shrinkage defects between macro-scale simulation and experiment. (a) 15mmTP RE=0.5%, (b) 15mmTP RE=1.0%, (c) 25mmTP RE=0.5%, (d) 25mmTP RE=1.0%.

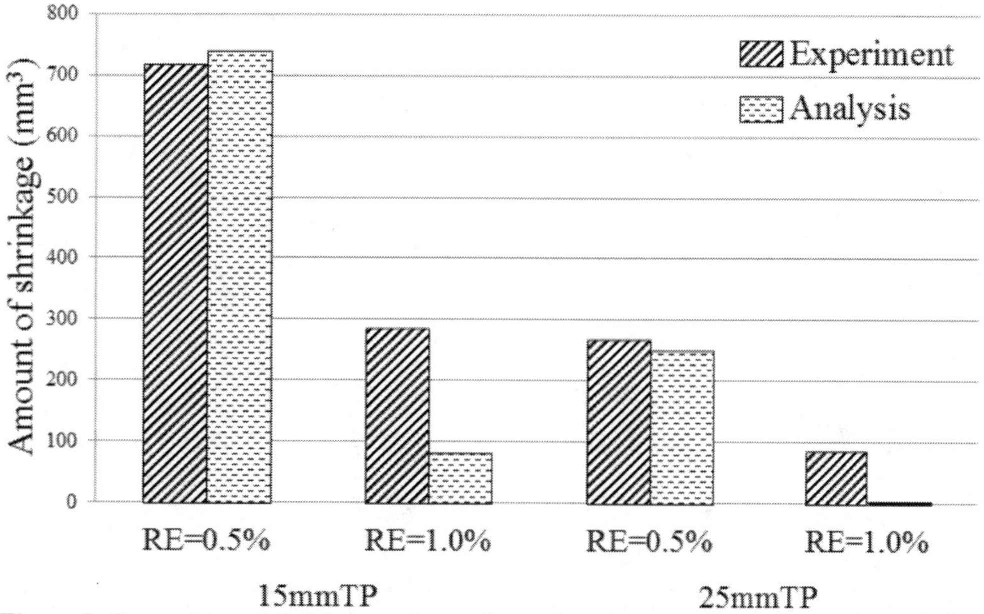

Figure 5. Comparison of shrinkage volume of test plates between simulation and experiment

3. Meso-scale simulation predicting shrinkage defect by CAFE method

3.1. Analysis model and conditions

In the meso-scale simulation, only two-dimensional model at the center section of the test pieces was analysed. The model was constructed by 200 μ m rectangular mesh. The grain structure was calculated by the CAFE method. In the calculation, the temperature profile, the solid fraction and the porosity volume, which was obtained by the macro-scale simulation, were introduced. The same parameters were used in CAFE model for the two different rare earth contents.

3.2. Result and discussion

Figure 6 shows comparison of the shrinkage defects with macro (grain) structure between the meso-scale simulation and the experiment. In the macro-scale simulation, the shrinkage defect was observed as the porosity distribution as shown in Figure 4. On the other hand, in the meso-scale calculation, shape and size of the shrinkage defect were indicated more precisely because 200μm rectangular mesh was small enough to represent shrinkage pore by vacant cells directly. Figure 6 shows the good agreement in the shape and size of the shrinkage defects between the calculated results and experimental one.

The positions of shrinkage are consistent from the macro-scale simulation to the meso-scale simulation. The effect of rare earth elements on the appearance of the shrinkage defects is also similar to that of the macro-scale calculation. Because the porosity volume calculated by the macro-scale calculation was distributed to the meso-scale cells, the vacant cells were produced at the same position.

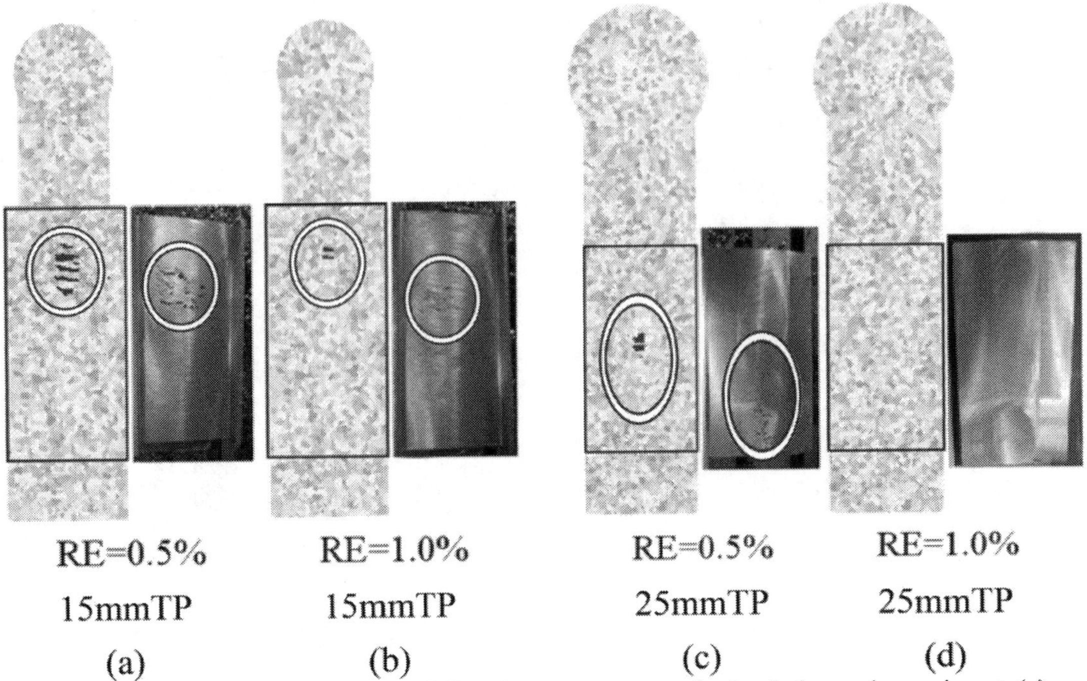

Figure 6. Comparison of shrinkage defects between meso-scale simulation and experiment. (a) 15mmTP RE=0.5%, (b) 15mmTP RE=1.0%, (c) 25mmTP RE=0.5%, (d) 25mmTP RE=1.0%.

4. Micro-scale simulation predicting micro structure by MCAFE

4.1. Analysis model and conditions

Location for the micro-scale simulation was 30mm from the tip of the plate, and the calculation domain was two-dimensional (1.3mm x 1.0mm). This model was constructed by 2μm x 2μm rectangular meshes. This simulation was calculated by MCAFE method [8] using the temperature profile and the solid fraction obtained by the macro-scale simulation. Then the graphite nodule number was given by optimizing the parameters of the nucleation density of the graphite particles. The same parameters were used for the two different rare earth contents.

4.2. Result and discussion

Figure 7 compares the microstructure obtained by the MCAFE calculation with the microstructure obtained by the experiments. In order to analize these results quantitatively, the images shown in Figure 7 were digitize. Then the number, average particle diameter and area ratio of the graphite nodules were measured by image processing. The number of graphite nodule in the test piece of 15mm thickness was larger than that in the test piece of 25mm thickness regardless of amount of rare earth. The reason is that the graphite nucleation frequency increases with increasing cooling rate. The test piece of 15mm thickness solidified earlier than 25mm thickness. Consequently the supercooling for the graphite nucleation in the test piece of 15mm thickness became larger and more nucleation of graphite nodule occurs. The average graphite nodule diameter in the test piece of 25mm thickness was larger than that in 15mm thickness. This tendency could be found in both the calculated result and the experimental one. The graphite area ratio ranged from 9 to 10% in both the calculated result and the experimental one. The graphite area ratio in the test piece of 25mm thickness was slightly larger than that in 15mm thickness.

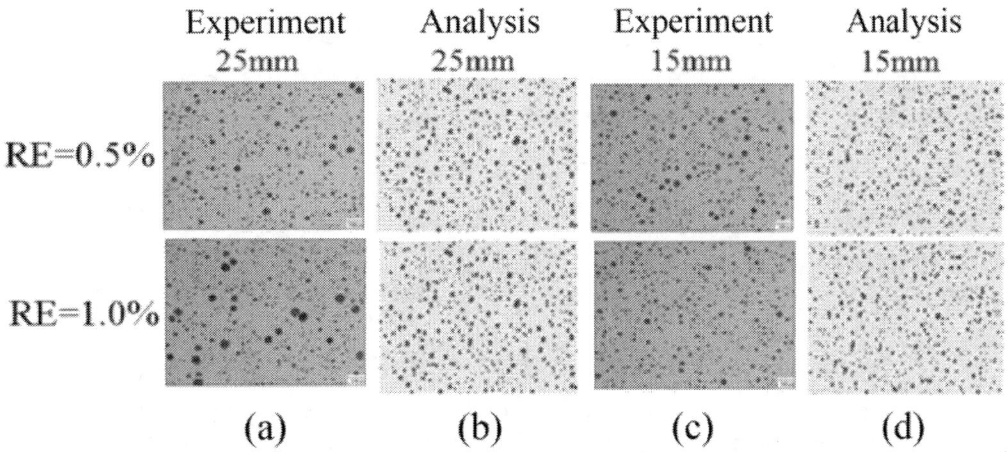

Figure 7. Comparison of micro structure between MCAFE simulation and experiment. (a) 25mm Experiment, (b) 25mm Analysis, (c) 15mm Experiment, (d) 15mm Analysis.

5. Conclusion

Multi-scale simulation for the plate ductile iron castings was performed. The results are summarized as follows:

(1) Taking the effect of rare earth elements on the volume expansion due to the graphite formation into consideration, the shrinkage volume was calculated by the macro-scale simulation using FEM. In the test piece of 25mm thickness with addition of 1% of rare earth elements, no shrinkage defects were observed in both the simulation and the experiment. Amount of the shrinkage volume in the test pieces, obtained by the simulation is coincided well with the experimentally obtained value.

(2) In the meso-scale simulation using CAFE method with 200μm x 200μm rectangular meshes, the calculated results sufficiently reproduced shape and size of the shrinkage defects, comparing to the experimental results.

(3) In the micro-scale simulation using MCAFE method with 2μm x 2μm rectangular meshes, number density, mean diameter and area ratio of the graphite nodules coincided qualitatively with the experiment results.

(4) On the basis of the macro-scale simulation, the meso-scale and the micro-scale simulations are useful for knowing the microstructure and the shrinkage defects in the ductile iron castings.

References

[1] Morrogh H, Grant J W 1947 J. Iron and Steel Inst., **155** 321
[2] Gagnebin P A, Millis K D and Pilling N B 1949 The Iron Age, Feb., 77
[3] Stefanescu D 1988 ASM Handbook Volume 15: Casting , ASM, Metals Park, Ohio
[4] Kubo J, Okada K. 2013 Japan Foundry Engineering Society 162nd Meeting **5** 50
[5] Kubo K, Deki N, Kubo J 2012 13th World Conference on Investment Casting T7
[6] Rappaz M, Gandin Ch.-A 1993 Acta Met. Mater. **41** 345
[7] M F Zhu, C P Hong 2001 ISIJ Int. **41** 436
[8] Kubo K 2011 Journal of Japan Foundry Engineering Society, **83** 485, 533
[9] Kubo K, Pehlke R D 1985 Metallurgical Transactions **16B** 359

Cellular automaton simulation of ferrite-austenite transformation in low-carbon steels

D An[1], S Pan[1,2], Q Yu[1], C Lin[1], T Dai[1], B W Krakauer[3], M Zhu[1]

[1] Jiangsu Key Laboratory for Advanced Metallic Materials, School of Materials Science and Engineering, Southeast University, Nanjing, Jiangsu 211189, China
[2] Engineering Training Center, Nanjing University of Science and Technology, Nanjing, Jiangsu 210094, China
[3] AO Smith Corporate Technology Center, Milwaukee, WI 53224, USA

E-mail: zhumf@seu.edu.cn

Abstract. A two-dimensional (2-D) cellular automaton (CA) model is adopted to simulate ferrite (α)-austenite (γ) transformation in low-carbon steels. In the model, the preferential nucleation sites of austenite, the driving force of phase transformation coupled with thermodynamic parameters, solute partition at the α/γ interface, and carbon diffusion in both the α and γ phases are taken into consideration. The model is able to simulate the ferrite-to-austenite transformation during isothermal heating in the ferrite-austenite two-phase region, and the austenite-to-ferrite transformation during continuous cooling. The influences of cooling rate and α grain size on the final carbon concentration field are investigated. The results show that after isothermal heating at 815°C for 300 s, the carbon concentration in both the α and γ phases reach the respective equilibrium values. The simulated microstructures compare well with the SEM images. After cooling to room temperature, the carbon distribution is more uniform when cooled at 1.2°C/s than at 7°C/s. The carbon distributions for different α grain sizes cooled at 1.2°C/s are similar. The simulation results are used to understand the mechanisms of the experimental phenomena of an enameling steel.

1. Introduction

In addition to the casting process, the subsequent heat treatment also impacts significantly the microstructures of the casted metal materials, and thereby the final mechanical properties. It is therefore necessary to study the phase transformation during different heat treatment processes. Among these transformations, ferrite-austenite transformations occupy a significant position in current research because of their fundamental role in solid-solid phase transformations. Extensive experimental work [1-5] has been carried out to study the mechanisms of ferrite-austenite phase transformation. Due to the limitation of the analytical tools, however, the microstructure evolution and solute distribution during solid-solid phase transformation could hardly be observed in real-time.

With the recent development of numerical models, simulations using different methods, such as the phase field (PF) method [6,7], the Monte Carlo (MC) method [8,9] and the cellular automaton (CA) method [10-13], can provide insight into the microstructure evolution and solute diffusion during the ferrite-austenite transformation. The present authors proposed a two-dimensional (2-D) CA model to simulate the ferrite (α)-austenite (γ) transformation and carbon diffusion during different heat treatment processes [14]. In this paper, the previously proposed CA model is applied to simulate the evolution of microstructure and carbon concentration field during continuous cooling with different cooling rates after isothermal heating in the ferrite-austenite two-phase region. The influence of cooling rate and α grain size on the carbon distribution is investigated. The simulation results are used to understand the mechanisms of the experimental phenomena that a cold-rolled low-carbon enameling steel exhibits different mechanical properties after different heat treatment processes.

2. Experimental phenomena

The material used in the experiment is a hot-rolled enameling steel, containing 0.07 *wt.*% C, 0.43 *wt.*% Si, 1.22 *wt.*% Mn, 0.04 *wt.*% P, and 0.001 *wt.*% S, produced by thin-slab continuous casting and direct hot rolling. The steel was then subject to the heat treatment processes, including isothermal heating at 815°C for 5 min and subsequent air cooling (~7°C/s) or sand cooling (~1.2°C/s). Other samples with the same composition as listed above, but having different grain sizes produced by different rolling processes were sand cooled after heated at 815°C for 5 min to investigate the influence of the α grain size on mechanical properties.

Figure 1 presents the stress-strain curves of samples with a similar average α grain size of about 7 μm after isothermal heating at 815°C for 5 min and then cooled by air cooling and sand cooling. As shown, the sample sand cooled has a higher yield strength but a lower tensile strength compared to the sample air cooled. In addition, the yield platform of the sample sand cooled is longer and more obvious than that of the sample air cooled.

Figure 2 shows the stress-strain curves of samples with different α grain sizes after isothermal heating at 815°C for 5 min and then cooled by sand cooling. The sample with the finer α grain size exhibits both higher yield strength and tensile strength compared to the sample with the coarser α grains. However, the yield platforms of the both samples are similar.

It is known that the yield platform is closely related to the Cottrell atmospheres, and the sufficient carbon diffusion contributes to the formation of the Cottrell atmospheres. Therefore, a previously proposed CA model [14] is applied to simulate the α - γ transformation and carbon diffusion during different heat treatment processes to verify the mechanisms of the experimental phenomena exhibited in figures 1 and 2.

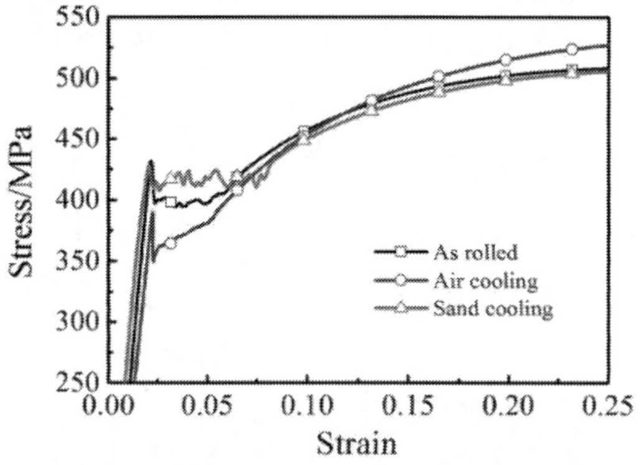

Figure 1. The stress-strain curves of the samples as rolled, and after isothermal heating at 815°C for 5 min and then air cooled and sand cooled.

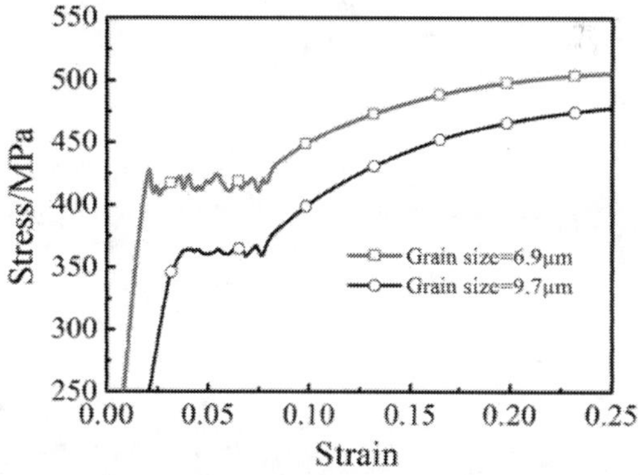

Figure 2. The stress-strain curves of the samples with different α grain sizes after isothermal heating at 815°C for 5 min and sand cooled.

3. Model description and numerical algorithm

The experimental material is a multi-component Fe alloy. For simplicity, a pseudo-binary Fe-C alloy is adopted in the present simulations. The initial carbon concentration used in the simulation is taken as u_0 = 0.375 *mol.*% C. This composition will produce the equilibrium γ phase fraction of 0.327 at 815°C, which is identical with that of the experimental steel. The phase transformations of both α→γ during isothermal heating and γ→α during cooling are simulated, while the coarsening of ferrite grains is neglected. The heat treatment processes used in the simulations are divided into two parts:

(1) Isothermal heating: The heating temperature is 815°C and holding time is 300 s. During isothermal heating, α→γ phase transformation, solute carbon partition at the α/γ interfaces, and carbon diffusion in both the α and γ phases take place;

(2) Continuous cooling from 815°C to room temperature: The temperature decreases from 815°C to room temperature at the cooling rates of 1.2°C/s and 7°C/s, corresponding to sand cooling and air cooling, respectively. During cooling, γ→α phase transformation, carbon partition at the α/γ interfaces, and carbon diffusion take place. In the present model, the phase transformation of austenite to pearlite is not considered. Considering the kinetic effect on phase transformation, the end temperatures of γ→α phase transformation are taken as 650°C and 600°C for the cooling rates of 1.2°C/s and 7°C/s, respectively. When the temperature is cooled down below the end temperature, the retained austenite is assumed to be transformed to supersaturated ferrite, while carbon diffusion continues to room temperature.

According to the interface dynamic theory [15] and the assumption of dilute solution, the velocity of interface migration, v_n, is calculated by

$$v_n = M_{eff} \cdot \left\{ -RT_M \left(k_e - 1\right)\left[u^\alpha(T) - u_e^\alpha(T)\right] / V_m \right\} \tag{1}$$

where V_m is the molar volume, R is the gas constant, M_{eff} is the interface mobility, T_M is the allotropic transformation temperature, $u^\alpha(T)$ is the actual carbon concentration in ferrite at temperature T, and k_e is the equilibrium partition coefficient, defined as $k_e = u_e^\gamma(T)/u_e^\alpha(T)$, where $u_e^\alpha(T)$ and $u_e^\gamma(T)$ are the equilibrium concentrations in ferrite and austenite at temperature T, respectively. The relevant thermodynamic data are obtained from the phase diagram calculation using Pandat [16], a thermodynamic phase diagram calculation software. Equation (1) indicates that the interface velocity v_n is determined by both the interface mobility M_{eff} and the local concentration difference $u^\alpha(T) - u_e^\alpha(T)$, where $u^\alpha(T)$ is determined by carbon diffusion.

The interface mobility, M_{eff}, is calculated by [17]

$$M_{eff} = M_0 \exp(-Q / RT) \tag{2}$$

where M_0 is the pre-exponential factor, Q is the activation energy. In the present model, the value of M_{eff} is taken as $4.2\times10^{-7}\exp(-14700/RT)$ $m^4/J{\cdot}s$ [18].

During ferrite-austenite phase transformation, solute partition at the α/γ interfaces is considered according to $u^\gamma = k_e u^\alpha$. Defining the mean carbon concentration as $u = \phi u^\alpha + (1-\phi)u^\gamma$, where ϕ is the α phase fraction, carbon diffusion in 2D is calculated by

$$\frac{\partial u}{\partial t} = \frac{\partial}{\partial x} \cdot \left[D(\varphi) \frac{\partial u^\alpha}{\partial x} \right] + \frac{\partial}{\partial y} \cdot \left[D(\varphi) \frac{\partial u^\alpha}{\partial y} \right] \tag{3}$$

where $D(\varphi)$ is the diffusion coefficient associated with the fraction of ferrite, calculated by

$$D(\varphi) = \varphi D^\alpha + k_e (1-\varphi) D^\gamma \tag{4}$$

where D^α and D^γ are the carbon diffusivities in the α and γ phases, equaling $2.2\times10^{-4}\exp(-122500/RT)$ m^2/s and $1.5\times10^{-5}\exp(-142100/RT)$ m^2/s [6], respectively.

In the present CA model, the computational domain is divided into a lattice of 236 ×348 square cells with cell size of Δx = 0.3 μm, corresponding to a real domain of 72 μm × 105 μm. Each cell is characterized by the following variables: (1) grain orientation, I; (2) α phase fraction, φ (the cell represents the α phase or γ phase, when $\varphi = 1$ or $\varphi = 0$, respectively); (3) the symbol $\varphi_{int} = 1$ indicates the cells at the α/γ interfaces; and (4) mean carbon concentration, u. The solution sequence for the simulation is as follows: (1) assign the initial state for each cell; (2) track α/γ interface and calculate the increment of ferrite fraction for the interface cells; (3) update the states of all cells in each time step; (4) calculate carbon diffusion in both the α and γ phases; (5) repeat steps (2)-(4) for the different heat treatment processes. The time step is determined by $\Delta t = \Delta x^2/4.5D^{\alpha}$, and the zero-flux boundary condition is adopted. Other details about the implementation of the numerical algorithms used for the simulations of phase transformations and carbon diffusion can be found elsewhere [14].

4. Results and discussion

4.1. Effect of cooing rate

At the beginning of the simulation, the computational domain is initialized with a uniform carbon concentration $u = u_0$ and α grains with various grain orientations. The number of initial α grains is taken as 112 according to the metallography, corresponding to the average α grain size of 6.9 μm. Based on the experimental observations, the nucleation sites of austenite are preferentially distributed at the ferrite grain boundaries.

During the isothermal heating at 815°C, α→γ phase transformation and carbon diffusion takes place. Figure 3 (a) shows the simulated γ phase fraction varying with time. At the early stage of α→γ transformation, the increasing velocity of γ phase fraction is slow. Then, the γ phase fraction increases evidently within 0.1 s~10 s, which implies that the α→γ transformation might take place mainly during this period. After holding about 11 s, the γ phase fraction approaches gradually the equilibrium value of 0.327. Figure 3 (b) shows the simulated mean carbon concentrations in the α phase and γ phase varying with time. As shown, in the early stage around $t < 0.4$ s, the mean carbon concentration in the α phase decreases dramatically. This is because the carbon solubility in the α phase is lower than that in the γ phase. During the α→γ phase transformation, carbon partition at the α/γ interface takes place. The carbon concentration in the α phase is absorbed by the formed γ phase. After around $t > 1$s, the carbon concentration in the α phase gradually approaches the equilibrium value of 0.0082 *mol.*%. On the other hand, the variation of the mean carbon concentration in the γ phase exhibits a different trend. At the beginning, the mean carbon concentration in the γ phase increases with time and reaches the maximum value of 10.48 *mol.*% when the time is around 0.03 s. Then, it decreases remarkably. After around 10 s, the carbon concentration in the γ phase approaches the equilibrium value of 1.13 *mol.*%. The increase of the carbon concentration in the γ phase at the early stage could be explained by the competition between phase transformation rate and carbon diffusion. At the early stage of the α→γ transformation, the phase transformation rate is slow as indicated in figure 3 (a). The rate of carbon diffusion from the α phase to the γ phase is higher than the growth rate of the γ phase, leading to the mean carbon concentration in the γ phase increasing in this period.

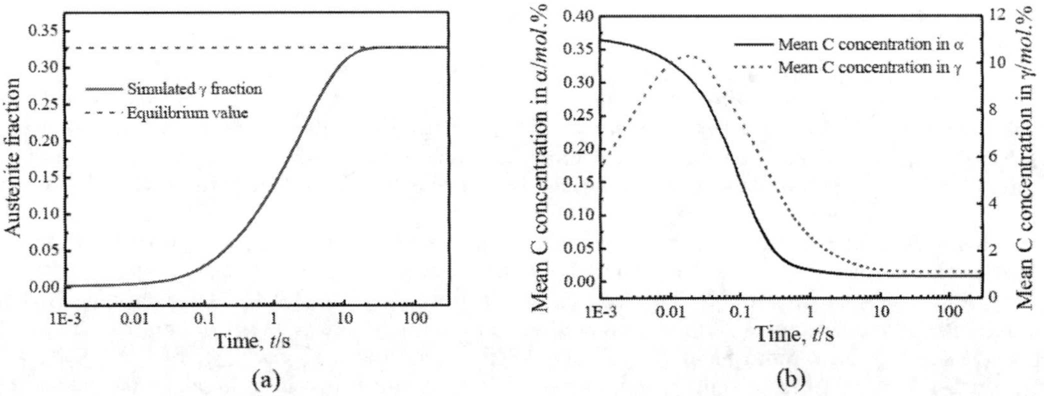

Figure 3. Time history of (a) γ phase fraction, and (b) mean carbon concentration in α and γ phases.

Figure 4 (a) presents the simulated carbon concentration field of an Fe-0.375 *mol.%* C alloy after isothermal heating at 815°C for 300 s. The regions of the high and low carbon concentration are the γ phase and α phase, respectively. It is noted that the carbon concentrations are nearly uniform in the both α and γ phases and reached their respective equilibrium values. Figure 4 (b) shows the SEM image obtained by quenching the experimental sample after heating at 815°C for 300 s to retain the austenite morphology. It is noted that the simulated morphology of austenite grains is somewhat coarser than that observed experimentally, which might be due to the following two main reasons. The current CA model adopts the instantaneous nucleation model for austenite nucleation, while austenite might nucleate continuously during the experimental ferrite-austenite transformation. In addition, the simulation is performed in a 2-D domain, while the dimensionality in the sample is 3-D. Nevertheless, the simulated microstructure agrees reasonably with the experimental observation.

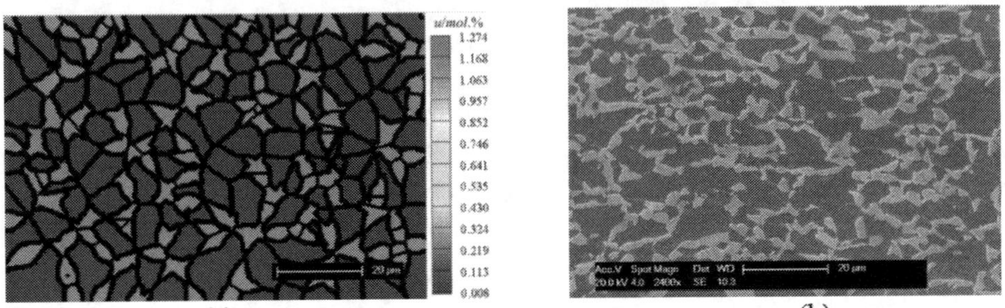

(a) (b)

Figure 4. (a) Simulated carbon concentration field and (b) SEM image of an Fe-0.375 *mol.%* C alloy after isothermal heating at 815°C for 300 s, and then quenching.

Figure 5 presents the evolution of the carbon concentration field of an Fe-0.375 *mol.%* C alloy during cooling at 1.2°C/s after isothermal heating at 815°C for 300 s. The lower 8 levels and the upper 5 levels of the color legends show the carbon concentrations in the α and γ phases, respectively. According to the model assumption, for the slow cooling rate, γ→α transformation and carbon diffusion take place when the temperature is above 650°C, while only carbon diffusion is considered between 650°C to room temperature. As shown in figures 5 (a) and (b), the carbon concentration is higher in the areas adjacent to the α/γ interfaces, due to the fact that the newly formed α phase rejects carbon atoms during the γ→α transformation. Driven by the concentration gradients, carbon diffusion occurs in both α and γ phases. When the temperature decreases to 650.1°C (figure 5 (c)), the carbon distribution in the ferrite matrix appears to be nearly uniform, indicating that the rate of carbon diffusion in ferrite is sufficient due to the slow γ→α transformation with a low cooling rate of 1.2°C/s. Then, the small quantity of the retained austenite is transformed to be the supersaturated ferrite and carbon diffusion continues. When cooled to room temperature, the carbon concentration field is nearly uniform as shown in figure 5 (d). The maximum and minimum concentrations are 0.375 *mol.%* C and 0.374 *mol.%* C, respectively, indicating a sufficient carbon diffusion when cooled at 1.2°C/s.

Figure 6 presents the evolution of the carbon concentration field of an Fe-0.375 *mol.%* C alloy during cooling at 7°C/s after isothermal heating at 815°C for 300 s. In this case the end temperature of γ→α transformation is assumed to be 600°C. Comparing figures 6 (a) and (b) with figures 5 (a) and (b), it is found that at the similar temperatures during cooling, the fraction of retained austenite is higher and the carbon concentration in ferrite is less uniform for the case of 7°C/s than for the case of 1.2°C/s. Figure 6 (c) shows the carbon concentration field when the temperature reaches 602.3°C. The high concentration regions distributed at grain boundaries illustrate the retained austenite that is transformed to the supersaturated ferrite when the temperature is below 600°C. Then, carbon diffusion continues as the temperature is further decreased to room temperature. The maximum and minimum carbon concentrations at room temperature obtained from cooling at 7°C/s are 2.845 *mol.%* C and 0.279 *mol.%* C, respectively. Comparing figure 5 (d) with figure 6 (d), it is obvious that the carbon concentration field cooled at 1.2°C/s is more uniform than that cooled at 7°C/s. Apparently, the cooling process with lower cooling rate provides more sufficient time for carbon diffusion, which is beneficial for carbon diffusion to the locations of dislocations to form high concentration Cottrell atmospheres. It is known that high concentration Cottrell atmospheres contribute to pinning dislocations, leading to an obvious yield platform in the stress-strain curve, and an increased yield strength [3]. However, Cottrell atmospheres have less effect on the tensile strength and the tensile strength could be increased by increasing cooling rate [19, 20]. Thus, it can be understood that the sample by sand cooling has a more obvious yield platform and a higher yield strength, while a lower tensile strength compared to the sample by air cooling, as shown in figure 1.

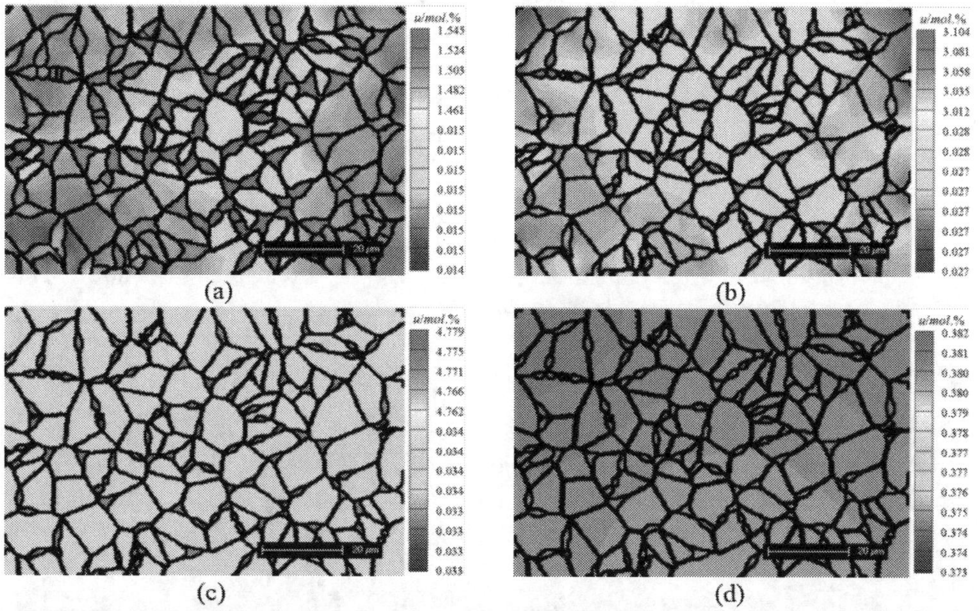

Figure 5. Simulated evolution of the carbon concentration field of an Fe-0.375 *mol.*% C alloy during cooling at 1.2°C/s for: (a) 20.7 s, 790.2°C; (b) 70.4 s, 730.5°C; (c) 137.4 s, 650.1°C; (d) 658.3 s, 25.0°C.

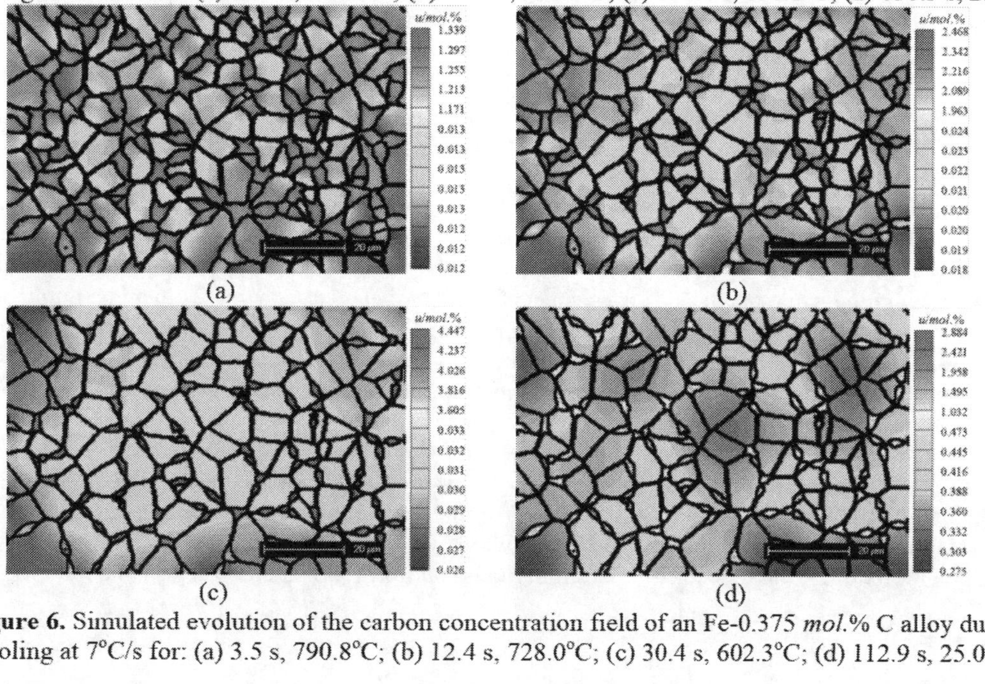

Figure 6. Simulated evolution of the carbon concentration field of an Fe-0.375 *mol.*% C alloy during cooling at 7°C/s for: (a) 3.5 s, 790.8°C; (b) 12.4 s, 728.0°C; (c) 30.4 s, 602.3°C; (d) 112.9 s, 25.0°C.

IOP Conf. Series: Materials Science and Engineering **84** (2015) 012045 doi:10.1088/1757-899X/84/1/012045

4.2. Effect of grain size

To investigate the influence of grain size on the carbon concentration field, simulations are performed for an Fe-0.375 *mol.*% C alloy with different α grain sizes after isothermal heating at 815°C for 300 s and then cooled at 1.2°C/s. The results are presented in figure 7. The simulation conditions of figure 7 (a) are identical with those used in figure 5. For the simulation of figure 7 (b), the number of initial α grains is set as 60, corresponding to the average α grain size of 9.7 μm. Other conditions are also identical with those used in figures 5 and 7 (a). As shown, the carbon concentration fields with different α grain sizes have the similar uniformity, indicating that the cooling process at 1.2°C/s provides sufficient time for carbon diffusion in both cases with different grain sizes. Thus, the α grain size has less influence on the formation of high concentration Cottrell atmospheres when cooled at 1.2°C/s. As discussed above, the yield platform is mainly affected by the carbon distribution and the formation of high concentration Cottrell atmospheres. Accordingly, it is understandable that the two samples with different α grain sizes have similar yield platform widths when cooled by sand cooling as shown in figure 2. On the other hand, it is well known that fine grain size will contribute to increasing both yield strength and tensile strength [3,4,21-26]. Thus, as shown in figure 2, the sample with the average α grain size of 6.9 μm exhibits both higher yield strength and tensile strength than that of the sample with the average α grain size of 9.7 μm.

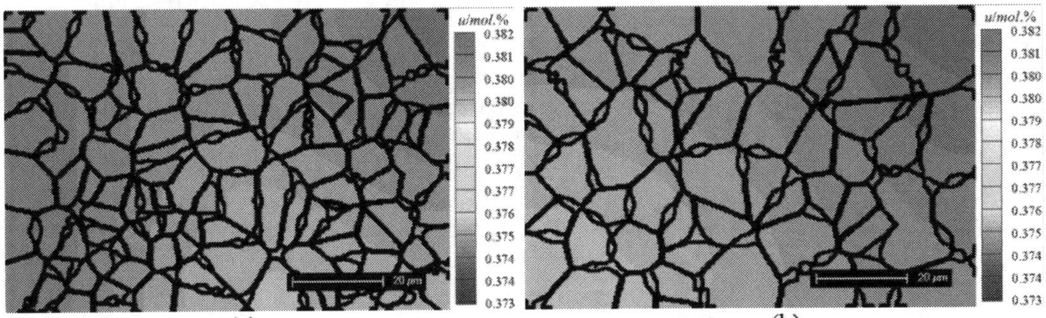

(a) (b)

Figure 7. Simulated carbon distribution of an Fe-0.375 *mol.*% C alloy after cooling to room temperature at 1.2°C/s with different average α grain sizes: (a) 6.9 μm; (b) 9.7 μm.

5. Conclusions

A two-dimensional cellular automaton model is applied to simulate the ferrite (α)-austenite (γ) transformation in a low-carbon steel during different heat treatment processes. The α→γ transformation during isothermal heating at 815°C and the γ→α transformation during continuous cooling at 1.2°C/s and 7°C/s are simulated. During the isothermal heating, the γ phase fraction increases evidently during the period of 0.1 s ~ 10 s. The mean carbon concentration in the α phase decreases with time while the mean carbon concentration in the γ phase increases to a maximum at the early stage of α→γ transformation and then decreases gradually, approaching the equilibrium value. After holding for 300 s, the carbon concentrations in both the α phase and γ phase reach their respective equilibrium values. The simulated microstructures agree well with the SEM images. After subsequent cooling to room temperature, the sample cooling at 1.2°C/s exhibits a uniform carbon distribution, while a non-uniform carbon concentration field is obtained when cooled at 7°C/s. The uniformities of the carbon concentration fields with different α grain sizes are similar in the case of cooling at 1.2°C/s. The simulation results are used to understand the experimental phenomena that the sample sand cooled exhibits a longer and more obvious yield platform than samples air cooled, while the samples with different grain sizes have similar yield platforms when cooled by sand cooling.

Acknowledgments
The authors wish to thank CompuTherm Company, USA, for generously providing the Pandat software license. This work was financially supported by the AO Smith Corporate Technology Center, USA, and NSFC (Grant No. 51371051).

References

[1] Oliveira F L G, Andrade M S and Cota A B 2007 *Mater. Charact.* **58** 256
[2] Liu Y C, Wang D J, Sommer F and Mittemeijer E J 2008 *Acta Mater.* **56** 3833
[3] Jiang M, Yang X F, Pan S Y, Krakauer B W and Zhu M F 2012 *J. Mater. Sci. Technol.* **28** 737
[4] Jiang M, Zheng F J, Huang X F, Dai T, Krakauer B W, Zhu M F 2012 *IOP Conf. Series: Materials Science and Engineering (Schladming)* vol 33 (IOP Publishing) 012044
[5] Jimenez-Melero E, Blonde R, Sherif M Y, Honkimaki V and van Dijk NH 2013 *Acta Mater.* **61** 1154
[6] Militzer M, Mecozzi M G and Sietsma J 2006 *Acta Mater.* **54** 3961
[7] Yamanaka A, Takaki T and Tomita Y 2012 *ISIJ Int.* **52** 659
[8] Tong M M, Li D Z and Li Y Y 2005 *Acta Mater.* **53** 1485
[9] Xiao N M, Tong M M, Lan Y J, Li D Z and Li Y Y 2006 *Acta Mater.* **54** 1265
[10] Kumar M, Sasikumar R and Nair P K 1998 *Acta Mater.* **46** 6291
[11] Zhang L, Zhang C B, Wang Y M, Wang S Q and Ye H Q 2003 *Acta Mater.* **51** 5519
[12] Lan Y J, Li D Z and Li Y Y 2004 *Acta Mater.* **52** 1721
[13] Zheng C W, Raabe D and Li D Z 2012 *Acta Mater.* **60** 4768
[14] An D, Pan S Y, Huang L, Dai T, Krakauer B and Zhu M F 2014 *ISIJ Int.* **54** 422
[15] Bos C and Sietsma J 2007 *Scripta Mater.* **57** 1085
[16] Chen S L, Daniel S, Zhang F, Chang Y A, Yan X Y, Xie F Y, Schmid-Fetzer R and Oates W A 2002 *CALPHAD* **26** 175
[17] Krielaart P and van der Zwaag S 1998 *Mater. Sci. Technol.* **14** 10
[18] Hillert M and Hoglund L 2006 *Scripta Mater.* **54** 1259
[19] Sung H K, Shin S Y, Hwang B, Lee C G and Lee S 2013 *Metall. Mater. Trans. A* **44A** 294
[20] Hu J, Du L X, Wang J J, Gao C R, Yang T Z, Wang A Y and Misra R D K 2013 *Metall. Mater. Trans. A* **44A** 4937
[21] Mao X P, Huo X D, Sun X J and Chai Y Z 2010 *J. Mater. Process. Technol.* **210** 1660
[22] Yu H, Chen Q X, Kang Y L and Sun Y 2005 *Mater. Charact.* **54** 347
[23] Yi H L, Du L X, Wang G D and Liu X H 2008 *J. Iron. Steel. Res. Int.* **15** 76
[24] Bruce L 1997 Structure/Property Relationships in Irons and Steels *Materials Selection and Design, Vol 20, ASM Handbook*, ASM International, p 357
[25] Tomota Y, Narui A and Tsuchida N 2008 *ISIJ Int.* **48** 1107
[26] Patra S, Hasan Sk Md and Chakrabarti D 2012 *Mater. Sci. Eng.* A **538** 145

Modeling of microporosity formation during solidification of aluminum alloys

T Wang, D An, Q Zhang, T Dai, M Zhu
Jiangsu Key Laboratory for Advanced Metallic Materials, School of Materials Science and Engineering, Southeast University, Nanjing, Jiangsu 211189, China

E-mail: zhumf@seu.edu.cn

Abstract. A two-dimensional (2D) multi-phase cellular automaton (MCA) model is adopted to simulate the dendrite and microporosity formation during solidification of aluminium alloys. The model involves three phases of liquid, gas, and solid. The effect of liquid-solid phase transformation on the nucleation and growth of porosity, the redistribution and diffusion of solute and hydrogen, and the effects of surface tension and environmental pressure are taken into account. The growth of both dendrite and porosity is simulated using a CA approach. The diffusion of solute and hydrogen is calculated using the finite difference (FD) method. The simulations can reveal the interactive and competitive growth of dendrites and micropores, and the microsegregation of solute and hydrogen. The porosity nuclei with large size are able to grow preferentially, while the growth of the small porosity nuclei is inhibited. Gas pores grow spherically when it is enveloped by liquid. After touching with dendrites, the shapes of pores become irregular. An increased initial hydrogen concentration reduces the incubation time of porosity nucleation, but increases the final percentage of porosity and the average porosity size at the eutectic temperature. With cooling rate decreasing, the competitive growth between gas pores becomes more evident, leading to non-uniform porosity sizes, and more irregular morphology of the porosities with larger size. The simulation results are compared reasonably well with the experimental data reported in literature.

1. Introduction

Castings of lightweight materials, such as aluminum and magnesium alloys, are widely used in the automotive and aerospace industries since they play a significant role in improving the performance and fuel efficiency [1]. The formation of microporosity in castings is, however, considered to be one of the primary detects affecting their mechanical properties, which is mainly caused by the insufficient liquid feeding and gas evolution during solidification [2].

Extensive work has been performed to simulate the microporosity formation during solidification of aluminum and other alloys over the last decades. Most models, including analytical solutions, criteria functions, Darcy's law coupled with the conservation and continuity equations, and gas diffusion-controlled pore growth models, mainly focus on the prediction of the amount of porosity in castings, but without reproducing graphical morphology [3-10]. See *et al.* [11] compared deterministic (DE) model, continuum stochastic (CS) model and cellular automaton (CA) model for the prediction of microporosity in aluminum-silicon alloys. They considered that CA model provided an excellent tool for further exploration of the complex interaction between porosity formation and the developing microstructures. Lee *et al.* [12-15] developed the CA approach coupled with hydrogen diffusion controlled growth models to simulate the meso-scale morphologies of grain structure and porosity in two and three dimensions. In the models the internal pore pressure was compared with the surface pressure to determine the pore growth. Dong *et al.* [16], Conley *et al.* [17] and Bayani *et al.* [18] proposed CA based models to simulate the evolution of grains or dendrites and micro-shrinkages during solidification. Karagadde *et al.* [19] adopted an explicit enthalpy scheme coupled with a level-set (LS) method to capture the final pore morphology during the solidification of an aluminum alloy. The present authors [20, 21] proposed a 2D multi-phase cellular automaton-finite difference (MCA-FD) model, to simulate dendrite and hydrogen porosity formation during solidification of Al alloys. The pore growth was determined based on the difference of the local hydrogen saturation and hydrogen concentration. The model is able to reasonably describe a complete time-dependent interaction of dendrite solidification, the nucleation and growth of porosities, solute and hydrogen partitions at the interfaces, and the diffusion of both solute and hydrogen.

In this paper, the previously proposed MCA-FD model is adopted to simulate the nucleation and growth of microporosity, the interactions between dendrites and micropores, and the microsegregation evolution of hydrogen during solidification of an Al-7 *wt.*% Si alloy. The effects of the initial hydrogen concentration and cooling rate on the formation of microporosity are investigated.

2. Governing equations and numerical algorithm

2.1. Model description

The focus of the present work is to simulate the microporosity formation caused by the difference of hydrogen solubility between the liquid and solid phases. The effects of buoyancy, Marangoni flow, and external forces are not considered. Therefore, the formation of gas porosities and dendrites are driven by pure diffusion. In addition, the temperature in the computational domain is assumed to be uniform and cooled down with a constant cooling rate. The simulation is stopped when the temperature drops to the eutectic temperature.

2.2. Pore nucleation

A stochastic model is implemented to assign the locations of pore nuclei randomly. The distribution function dn/dS used for evaluating the number of gas pore nuclei is defined as

$$\frac{dn}{dS}=\begin{cases}\dfrac{N_{max}}{(S_{max}-S_{min})}; & (C_l(H)/S_l(H)>S_n)\\ 0; & (C_l(H)/S_l(H)\le S_n)\end{cases} \tag{1}$$

where N_{max} is the maximum pore nucleation density, S_{max} and S_{min} are the maximum and minimum pore nucleation saturation, respectively [10, 15]. S_n is the critical saturation criterion for porosity nucleation. $C_l(H)$and $S_l(H)$ are the local hydrogen concentration (mol/m^3) and the local hydrogen saturation in liquid, respectively. When the local hydrogen concentration in liquid reaches a state of supersaturation, a pore nucleus with a certain radius might occur. According to Sievert's law, the hydrogen saturation in liquid, $S_l(H)$, of Al-Si alloys can be calculated by [10]

$$S_l(\mathrm{H})=\sqrt{\frac{P_g}{P_{ref}}10^{\left(\frac{-2760}{T}+2.96-0.0119W_{Si}\right)}} \tag{2}$$

where P_g is the internal pressure of a gas pore, P_{ref} is the reference pressure defined as the standard atmospheric pressure that is taken as 1 atm in this work, T is the temperature, and W_{Si} is the mass fraction of solute Si. The internal pressure is affected by two factors: the atmospheric pressure and surface energy pressure, and the effects of the metallostatic pressure and shrinkage pressure are not considered. Therefore, P_g in the two dimensions is given by: $P_g=P_{ref}+\lambda_{LG}/r_p$, where λ_{LG} is the surface tension between liquid and gas, and r_p is the radius of the gas pore. Based on the above equations, the radii of gas pore nuclei formed under different hydrogen saturations can be calculated.

2.3. Microporosity and dendrite growth

The CA technique is adopted to simulate the growth of both microporosity and dendrite. The two-dimensional computational domain is divided into a lattice of square cells. Each cell is characterized by different variables, including hydrogen concentration, solute concentration, crystallographic orientation of a dendrite, temperature, phase fractions (solid, liquid, and gas), and cell states including liquid cell ($f_s+f_g=0$), gas cell ($f_g=1$), solid cell ($f_s=1$), the gas/solid (G/S) interface cell ($f_s+f_g=1$), the gas/liquid (G/L) interface cell ($0<f_g<1$, and $f_s=0$), the solid/liquid (S/L) interface cell ($0<f_s<1$, and $f_g=0$), and the gas/liquid/solid (G/L/S) interface cell ($0<f_s+f_g<1$), where f_g and f_s are the gas fraction and solid fraction of each cell, respectively. The cells that belong to a same pore share the characteristics of this pore, such as the hydrogen concentration, the internal pressure and the volume of the pore.

Assuming that a gas pore denoted as A obeys the ideal gas law, the gas volume increment of the G/L interface cell (i, j) that belongs to the pore A, $\Delta V_A^{(i,j)}$, can be calculated by

$$\Delta V_A^{(i,j)} = R_H \cdot \frac{G_g^{(i,j)}}{\sum G_g^{(i,j)}} \cdot \frac{RT}{P_g} \tag{3}$$

where R_H is the quantity of hydrogen absorbed by all G/L interface cells of the pore A, R is the gas constant, $G_g^{(i,j)}$ is the geometrical factor of the G/L interface cell, (i, j), and $\sum G_g^{(i,j)}$ is the sum of all the geometrical factors of all G/L interface cells in pore A. The geometrical factor G_g is related to the state of neighboring cells and is defined by

$$G_g = \min\left[1, \frac{1}{2}\left(\sum_{m=1}^{4} S_m^I + \frac{1}{\sqrt{2}}\sum_{m=1}^{4} S_m^{II}\right)\right]; S^I, S^{II} = \begin{cases} 0 & (f_g < 1) \\ 1 & (f_g = 1) \end{cases} \tag{4}$$

where S^I and S^{II} indicate the state of the nearest neighbor cells and the second-nearest neighbor cells, respectively, and f_g is the gas fraction of neighbor cells. The geometrical factor defined by equation (4) is used to account for the effect of neighbor cell state on pore growth. Equation (4) also reflects the effect of the second nearest neighbor cells is weaker than those of the nearest cells.

Therefore, the gas fraction of the G/L interface cell in pore A at time t_n can be defined as

$$f_g^{(i,j),t_n} = \sum_{n=1}^{N} \frac{\Delta V_A^{(i,j)}}{V_c} \tag{5}$$

where N is the iteration number, and V_c is the volume of cell. After a time step interval, the gas fraction of a G/L interface cell of pore A is updated using equation (5). When the gas fraction of the G/L interface cell becomes 1 through N time iterations, the cell state is changed from the G/L interface to gas pore. Then, the newly formed gas cell captures its liquid neighboring cells to be the new G/L interface cells. The gas fractions of these new interface cells are also calculated using equation (5) in the next time step, and thus the pore growth continues.

The dendrite growth is also simulated by the CA approach. The governing equations and numerical algorithm for the simulation of dendrite growth can be found elsewhere [20].

2.4. Solute and hydrogen diffusion

During the process of dendrite growth and gas pore formation, a solute gradient exists ahead of the S/L interface, while a hydrogen gradient exists ahead of both the S/L and G/L interfaces due to the solute and hydrogen partitions at the interfaces. The solute and hydrogen gradients lead to the solute and hydrogen diffusion in the domain. The partitions and diffusions of solute and hydrogen are calculated by

$$\frac{\partial C_i(X)}{\partial t} = D(X) \cdot \nabla^2 C_i(X) + R(X) \tag{6}$$

where $C_i(X)$ is the concentration, $D_i(X)$ is the diffusion coefficient, X represents solute silicon or hydrogen, the subscript i represents solid or liquid, respectively. $R(Si)$ indicates the amount of solute rejected at the S/L interface, and $R(H)$ includes hydrogen partitions at the S/L interface and the G/L interface. The solute silicon and hydrogen partitions between liquid and solid at the S/L interface are considered according to

$$C_s^*(X) = k_X C_l^*(X) \tag{7}$$

where $C_s^*(X)$ and $C_l^*(X)$ are the interface silicon or hydrogen concentrations in solid and liquid phases, respectively, and k_X is the partition coefficient. Therefore, silicon and hydrogen partitions at the S/L interface in one time step can be obtained by

$$R(X)_{S/L} = \begin{cases} \Delta f_s\left(C_l^*(Si)(1-k_{Si})\right); (X = Si) \\ \Delta f_s\left(C_l^*(H)(1-k_H)\right); (X = H) \end{cases} \tag{8}$$

On the other hand, the amount of hydrogen absorbed by the pore, $R(H)_{G/L}$, is evaluated by

$$R(H)_{G/L} = \sum_j (1 - f_g^{(i,j)} - f_s^{(i,j)})(C_l(H) - S_l(H))V_c \tag{9}$$

where j represents all G/L interface cells of a same pore, and $f_g^{(i,j)}$, $f_s^{(i,j)}$, $C_l(H)$ and $S_l(H)$ are the gas fraction, solid fraction, hydrogen concentration, and hydrogen saturation in liquid, respectively, at the G/L interface cell.

Equation (6) is solved using an explicit finite difference schedule. It is considered that the diffusion coefficient of hydrogen in liquid is about two orders of magnitude larger than that of solute Si, two different time steps are thus used in the simulations to increase the computational efficiency. The time step for the calculations of the solute Si diffusion, the growth of both dendrites and gas pores, is determined by

$$\Delta t = \min\left\{ \frac{\Delta x^2}{4.5 D_l(Si)}, \frac{1}{4.5 \Delta f_{s,\max}} \right\} \tag{10}$$

where $D_l(Si)$ is the diffusion coefficient of solute Si in liquid, and $\Delta f_{s,max}$ is the maximum increment of solid fraction in one time step. On the other hand, the time step for hydrogen diffusion, Δt_H is determined by $\Delta t_H = \Delta x^2 / 4.5 D_l(H)$, where $D_l(H)$ is the diffusion coefficient of hydrogen in liquid. In the simulations, if the ratio of two time steps is $N_t = \Delta t / \Delta t_H$, the calculation of hydrogen diffusion is run for N_t times using the time step Δt_H. Then, the diffusion of solute silicon, and the growth of dendrites and gas pores are calculated using Δt for one time step.

2.5. Coupling growth of microporosity and dendrite

In the present work, the CA approach [20] is used to simulate the dendrites growth. The solute Si and hydrogen rejected from the newly formed solid cells are calculated by equation (8) and then $R(X)_{S/L}$ is added into the remaining liquid in the same S/L interface cell and its surrounding neighbouring cells to keep the conservation of solute Si and hydrogen in the domain. Then, the diffusions of solute and hydrogen are calculated by equation (6). With dendrites growing, the local hydrogen concentration increases. Equation (1) is used to estimate the possibility of gas pore nucleation. If a gas pore is nucleated, the increasing gas fraction is calculated by equations (2)–(5) and (9), and the radius of the gas pore is updated. This series of calculations are repeated until the end of simulation. The physical parameters used in this work can be found elsewhere [20].

3. Results and discussion

3.1. Evolution of microporosities and dendrites

Figure 1 shows the evolution of porosities and dendrites of an Al-7 wt.% Si alloy with a cooling rate of 8°C/s and an initial H concentration of 0.6 mol/m^3. The calculation domain consists of a 300×300 mesh with a uniform mesh size of 3 μm.. The pore nucleation parameters N_{max}, S_{max}, S_{min}, and S_n are taken as 1× 10^{11} m^{-3}, 1.6, 2.3, and 1.2, respectively. These parameters are taken as constant in all simulations in the present work. Locations for each pore nucleation and solid seeds with various crystallographic orientations are assigned randomly in the domain. As shown in figure 1, with the temperature decreasing, dendrites grow with developing the main arms along their crystallographic orientations, and side branching of the secondary arms. Meanwhile, the H concentration increases in liquid and it is higher in the area close to dendrites (figure 1 (a)). According to equation (2), the solubility of hydrogen decreases with decreasing temperature. When the local H concentration in liquid is higher than the supersaturation for porosity nucleation, several gas pore nuclei with different radii appear (figure 1 (b)). Owing to the effect of surface tension, the gas pores grow spherically in liquid until they have impinged upon dendrites (figure 1(c)). After the gas pores encounter the dendrites, the shapes of gas pores become irregular (figure 1(d)). Since the pore nucleation and growth consume H atoms, the H concentration in liquid, shown in figure 1(d), is lower than those in figures 1 (a)-(c), and even lower than the initial concentration of 0.6 mol/m^3, except the bottom right of the domain, due to the fact that no gas pore appears in the area. It also can be seen that some pores with relatively smaller size barely grow, even though the local H concentration around them is actually higher than that around the pores grow obviously. This is because the smaller the pores, the higher the internal pressure. Thus, the smaller pores have higher H solubility than that of the larger pores (equation (2)), leading to a smaller gas volume increment (equations (3) and (9)). Therefore, the pores with larger size are able to grow preferentially, while the growth of the small pores is limited.

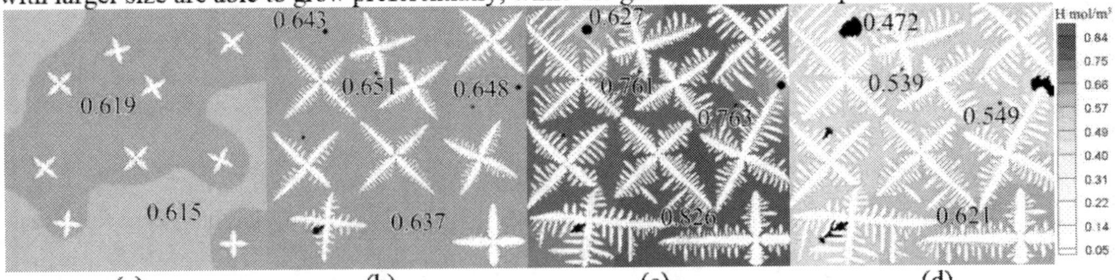

Figure 1. Evolution of microporosity, dendrite and H concentration field of an Al-7*wt.*%Si with a cooling rate of 8°C/s and an initial H concentration of 0.6 mol/m^3 at different temperatures: (a) 608.5°C, F_s=4.0%, F_g=0; (b) 601.1°C, F_s=20.0%,F_g=0.16%; (c) 590.8°C, F_s=43.5%, F_g=0.41%; (d) 572.8°C, F_s=49.3%, F_g=1.44%. (Numbers in the figures show the local H concentration in liquid phase. F_s and F_g are the total solid fraction and percentage of porosity in the domain, respectively).

3.2. Effect of initial hydrogen concentration

Figure 2 shows the simulated H concentration fields and morphologies of dendrites and microporosities at the eutectic temperature with different initial H concentrations of 0.5 mol/m^3 and 0.7 mol/m^3. The other conditions are identical with those of figure 1. It can be seen that porosity size becomes larger with the higher initial H concentration. It is understandable that with a higher initial H concentration, the amount of hydrogen atoms rejected from the growing dendrites increases (equation (7)), leading to the increase of the local H concentration. More hydrogen atoms in liquid are available for pore growth (equations (8) and (9)). Therefore, the higher initial H concentration will produce the pores with larger size. Since the large pores absorb more hydrogen atoms, the H concentration in liquid at the eutectic temperature for the case of a higher initial H concentration is even lower than that of the lower initial H concentration as shown in figure 2.

Figure 3 represents the percentage of porosity varying with solidification time with a cooling rate of 8°C/s and different initial H concentrations. Other simulation conditions are identical with those of figure 1. It is observed that the porosity percentage increases slowly at the early stage, then increases rapidly, and slows down again at the late stage. It is obvious that there is an incubation period for pore nucleation, the higher the initial H concentration, the shorter the incubation time. It is understandable that the higher initial H concentration results in a higher local H concentration in liquid, which shortens the time required to reach the supersaturation for pore nucleation. It also can be seen from figure 3 that the final percentage of porosity of the case with higher initial concentration is larger than that of the lower one. Figure 4 shows the average pore radius at the eutectic temperature as a function of initial H concentration with a cooling rate of 8°C/s. It can be seen the average pore radius at the eutectic temperature increases with increasing initial H concentration when the initial H concentration is higher than 0.4 mol/m^3. However, when the initial H concentration is lower than 0.4 mol/m^3, the average pore radius is relatively small. It seems that there is a critical level of the initial H concentration, below which gas pores could barely grow.

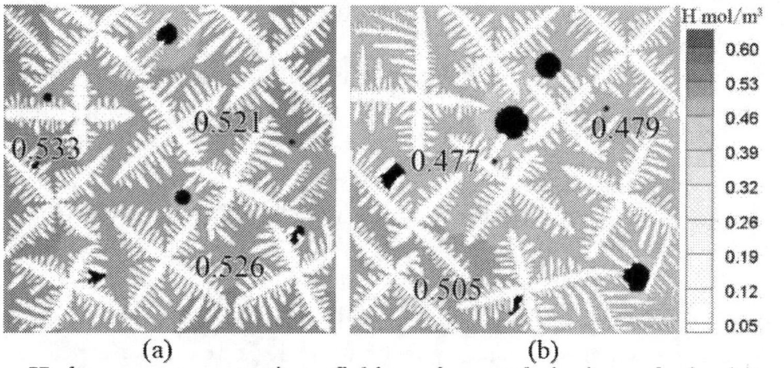

Figure 2. Hydrogen concentration field and morphologies of dendrite and microporosity at the eutectic temperature with a cooling rate of 8°C/s and different initial hydrogen concentrations of (a) 0.5 mol/m^3, and (b) 0.7 mol/m^3.

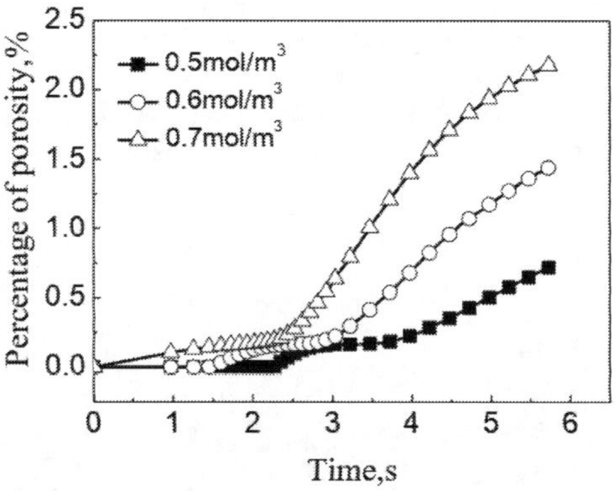

Figure 3. Percentage of porosity as a function of time with a cooling rate of 8°C/s and various initial H concentrations.

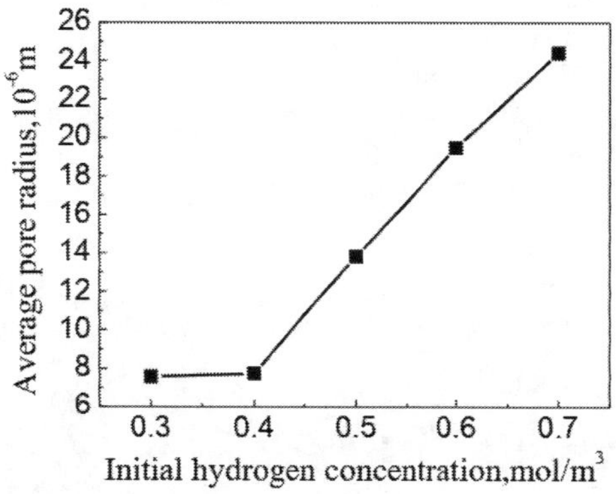

Figure 4. Average pore radius at the eutectic temperature as a function of initial H concentration with a cooling rate of 8°C/s.

3.3. Effect of cooling rate

Figure 5 shows the simulated H concentration field and morphologies of dendrites and microporosities at the eutectic temperature with different cooling rates of 2°C/s, 8°C/s and 15°C/s. Other simulation conditions are identical with those of figure 1. It is observed that with a higher cooling rate, the

IOP Conf. Series: Materials Science and Engineering **84** (2015) 012046 doi:10.1088/1757-899X/84/1/012046

dendrites become finer, and the sizes of finally formed pores are smaller and more uniform. These results are coincident with experimental observations [22, 23]. Dendrites grow rapidly with finer and longer arms at a higher cooling rate. Thus, the complex dendrite network could be formed sooner, which blocks hydrogen diffusion in liquid. Therefore, less hydrogen atoms are available at the G/L interface, leading to the growth of porosities to be inhibited. Moreover, a higher cooling rate also reduces the time for hydrogen diffusion and pore growth. Since less hydrogen atoms are consumed by gas pores at a higher cooling rate, the H concentration in liquid at the eutectic temperature increases with increasing cooling rate as shown in figure 5.

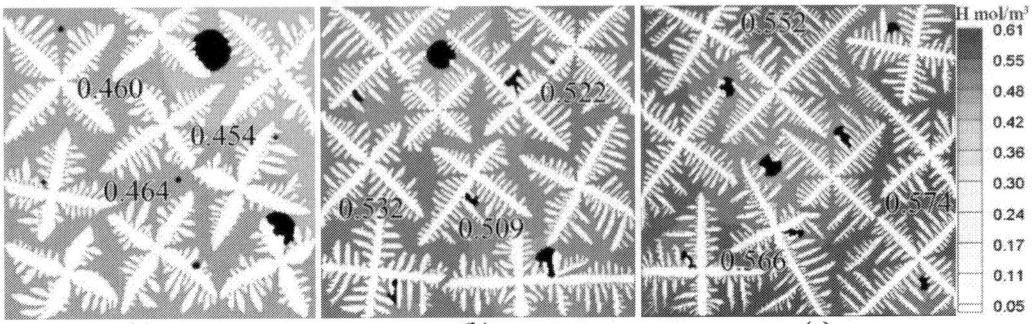

(a) (b) (c)

Figure 5. Hydrogen concentration fields and morphologies of dendrite and microporosity with an initial H concentration of 0.6 mol/m^3 and different cooling rates: (a) 2°C/s, (b) 8°C/s,(c) 15°C/s.

Figure 6 shows the percentage of porosity as a function of temperature with different cooling rates. The initial H concentration is 0.6 mol/m^3. Other simulation conditions are identical with those of figure 1. As shown, with increasing cooling rate, gas pores nucleate and start to grow at lower temperatures. The final percentage of porosity tends to decease as the cooling rate increases, which can explain the reason why the H concentration in liquid at the eutectic temperature shown in figure 5 (c) is higher than that in figure 5 (a) and (b). The tendency of the porosity percentage varying with cooling rate shown in figure 6 is consistent with the experimental data obtained by Fang and Granger [24], and Lee and Hunt [25], as well as the simulation results obtained by Felicelli *et al.* [6], and Carlson *et al.* [7]. Figure 7 represents the maximum pore radius as a function of cooling rate with an initial H concentration of 0.6 mol/m^3. It is found that with decreasing cooling rate, the final maximum porosity radius decreases obviously. This implies that at a slow cooling rate, the pore nuclei with relative larger size tends to grow preferentially, while the growth of small pore nuclei is limited, leading to the formation of non-uniform porosities as shown in figure 5 (a).

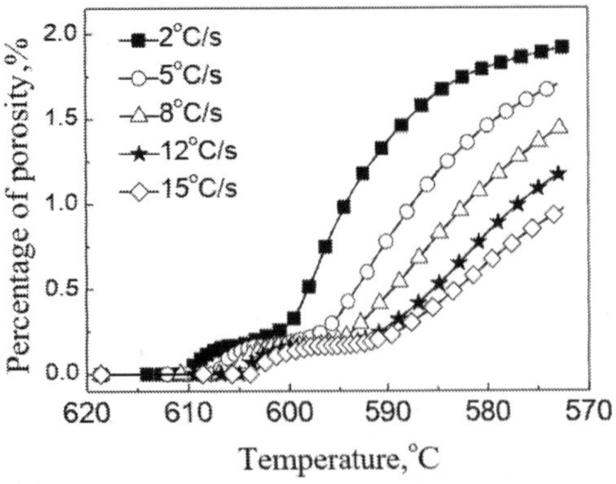

Figure 6. Percentage of porosity as a function of cooling rate with an initial H concentration of 0.6 mol/m^3.

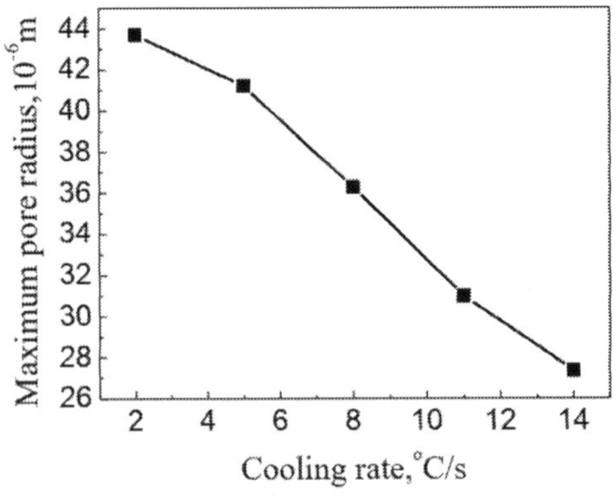

Figure 7. Maximum pore radius as a function of cooling rate with an initial H concentration of 0.6 mol/m^3.

4. Conclusions

A previously proposed 2D multi-phase cellular automaton-finite difference (MCA-FD) model is adopted to simulate the evolution of porosities and dendrites during solidification of an Al-7*wt.*%Si alloy. The nucleation and growth of porosities, dendrite solidification, solute and hydrogen partitions at the interface, and the diffusion of both solute and hydrogen are involved in the model. The interactive and competitive growth of dendrites and porosities, and the microsegregation of solute and hydrogen are observed. It is found that the pores with relatively larger size grow preferentially, while the growth of small porosity nuclei is limited. The gas pores grow spherically until they have impinged upon the solid dendrites.

The effects of the initial H concentration and cooling rate on microporosity formation are investigated. The results show that as the initial H concentration increases, the incubation time of porosity nucleation tends to decrease, while the final percentage of porosity and the average porosity size at the eutectic temperature increases. With cooling rate decreasing, the percentage of porosity increases, and the competitive growth between porosities becomes more obvious. Thus, a low cooling rate leads to non-uniform porosity sizes and increased maximum porosity size. The simulation results agree reasonably well with the experimental data.

Acknowledgment

This work was supported by the National Natural Science Foundation of China (Grant No. 51371051).

References

[1] Ambroziak A and Korzeniowski M 2010 *Arch. Civ. Mech. Eng.* **1** 5
[2] Kim J R and Abbaschian R 2011 *J. Mater. Sci.* **46** 6213
[3] Li K D and Chang E 2004 *Acta Mater.* **52** 219
[4] Ferreira I L, Lins J F C, Moutinho D J, Gomes L G and Garcia A 2010 *J. Alloy Compd.* **503** 31
[5] Voller V R 1998 *Can. Metall. Quart.* **37** 169
[6] Felicelli S D, Wang L, Pita C M and Obaldia E E D 2009 *Metall. Mater. Trans.* B **40B** 169
[7] Carlson K D, Lin Z, and Beckermann C 2007 *Metall. Mater. Trans.* B **38B** 541
[8] Pequet C, Gremaud M and Rappaz M 2002 *Metall. Mater. Trans.* A **33A** 2095
[9] Spittle J A, Almeshhedani M and Brown S G R 1994 *Cast Met.* **7 (1)** 51
[10] Atwood R C, Sridhar S, Zhang W and Lee P D 2000 *Acta Mater.* **48 (2)** 405
[11] See D, Atwood R C and Lee P D 2001 *J. Mater. Sci.* **36** 3423
[12] Lee P D and Hunt J D 2001 *Acta Mater.* **49** 1383
[13] Lee P D, Atwood R C, Dashwood R J and Nagaumi H 2002 *Mater. Sci. Eng.* A **328** 213
[14] Atwood R C and Lee P D 2003 *Acta Mater.* **51** 5447
[15] Lee P D, Chirazi A, Atwood R C and Wang W 2004 *Mater. Sci. Eng.* **365** 57
[16] Dong S Y, Xiong S M and Liu B C 2004 *Mater. Sci. Technol.* **20** 23
[17] Conley J G, Huang J, Asada J and Akiba K 2000 *Mater. Sci. Eng.* A **285** 49
[18] Bayani H, Mirbagheri S M H, Barzegari M and Firoozi S 2014 *J. Mater. Res. Technol.* **3** 55
[19] Karagadde S, Sundarraj S and Dutta P 2012 *Comput. Mater. Sci.* **65** 383
[20] Zhu M F, Li Z Y, An D, Zhang Q Y and Dai T 2014 *ISIJ Int.* **54** 384
[21] Li Z Y, Dai T and Zhu M F 2013 *Acta Metall. Sin.* **49** 1032
[22] Lashkari O, Yao L, Cockcroft S and Maijer D 2009 *Metall. Mater. Trans.* A **40A** 991
[23] Zhao L 2012 *PhD Thesis* (Nanjing: Southeast University)
[24] Fang Q T and Granger D A 1989 *AFS Trans.* **97** 989
[25] Lee P D and Hunt J D 1997 *Acta Mater.* **45** 4155

Effect of heat treatment on the microstructure and yield strength of a cold-rolled enameling steel

D Peng[1], Q Yu[1], C Lin[1], D An[1], T Dai[1], B W Krakauer[2] and M Zhu[1]

[1] Jiangsu Key Laboratory for Advanced Metallic Materials, School of Materials Science and Engineering, Southeast University, Nanjing, Jiangsu 211189, China
[2] AO Smith Corporate Technology Center, Milwaukee, WI 53224, USA

E-mail: zhumf@seu.edu.cn

Abstract. The mechanisms of yield strength reduction of a cold-rolled enameling steel after enamel-fire annealing at 760°C by air cooling, and the effect of the tempering process on the microstructure and yield strength, are studied by combining experiments and thermodynamic calculations. The results show that after heat treatment at 760°C and air cooling, the lump phase, enriched with the element carbon, appears along the ferrite grain boundaries, which leads to yield strength reduction. After tempering at 200°C~400°C, the lump phase disappears gradually and is transformed to lamellar pearlite as the tempering temperature increases, resulting in the yield strength increasing.

1. Introduction

Enameling steels, produced by continuous casting and hot-rolling processes, and then subject to enamel-fire annealing (EFA), are used in industries that require chemical, corrosion, and weather resistance, and where colour or appearance is important [1]. The mechanical properties of enameling steels are not only impacted by casting, but also by the subsequent EFA process. When used for thin-walled pressure vessels, high yield strengths are crucial for safety [2]. However, during the EFA process that consists of heat treating in the range of 750~900°C, the yield strengths of the enameling steels decrease, requiring sufficient steel thicknesses to meet the safety requirements and the associated added expense. In an effort to reduce these added expenses, studies have been carried out to investigate the mechanisms of yield strength reduction during the EFA. Jiang et al. [3, 4] studied the strengthening mechanisms of both the hot-rolled and cold-rolled enameling steels by establishing empirical equations to relate the microstructure and yield strength. The results show that grain refinement strengthening and solid solution strengthening are the main strengthening mechanisms. The yield strength reduction results primarily from an increase in grain size and a decrease in dislocation density. By comparing the effects of different strengthening mechanisms, Mao et al. [5] reported that the precipitation hardening of nanometer TiC particles has the greatest contribution to the yield strength increase of a hot-rolled steel. Isasti et al. [6] observed that the addition of the element Mo to Nb-bearing steels provides a refinement of the precipitates and enhances the contribution of precipitate strengthening. The grain size contribution is found to be the most important strengthening factor, followed by dislocation density and precipitation.

In this work, a cold-rolled enameling steel produced by thin-slab continuous casting are exposed to various heat treatment processes to simulate the EFA process used in industry. The microstructures and yield strengths of the enameling steel after different heat treatment processes were studied by combining experiments and thermodynamic calculations.

2. Experiment

A cold-rolled enameling steel was investigated in the present work. The chemical composition is given in table 1. The EFA was simulated by heat treating the samples with a size of 200×80×1.2 mm in a box type furnace. The samples were isothermally heated at 760°C for 5 min, and then air cooled (~5°C/s) or sand cooled (~2°C/s) to room temperature. The samples air cooled were then subjected to a tempering process in the temperature range of 200°C~400°C for 20 min.

Table 1. Chemical composition of the enameling steel (wt.%).

C	Mn	Si	S	P	N	Al	Cu	Cr	Mo	Nb	Ti	Fe
0.07	1.22	0.43	0.001	0.051	0.006	0.038	0.1	0.04	0.01	0.002	0.021	bal.

The tensile specimens were prepared according to the standard for sheet tensile specimens (ASTM A370-05). Tensile tests were performed at a crosshead speed of 6 mm per min with a CMT5105 tensile machine. The microstructures were observed with a XL30 scanning electron microscope (SEM). Transmission electron microscopy analysis was performed on thin foils with a TECNAI G2 20 transmission electron microscope (TEM). The thin foils were prepared by mechanically thinning to 50 μm, and then perforated by twin-jet electropolishing.

The thermodynamic calculations of the equilibrium phases and element solubilities were performed using Pandat that is a phase equilibrium calculation software of multi-component alloys [7].

3. Results and discussion

3.1. Microstructure and yield strength

Figure 1 shows the SEM images of the samples as rolled and after isothermal heating at 760°C for 5 min and then quenching. As shown, the as rolled microstructure is ferrite and free cementite. In the quenched microstructure, free cementite nearly vanishes, and small amounts of martensite appear at the ferrite grain boundaries. The martensite shown in figure 1 (b) represents the morphology of austenite before quenching.

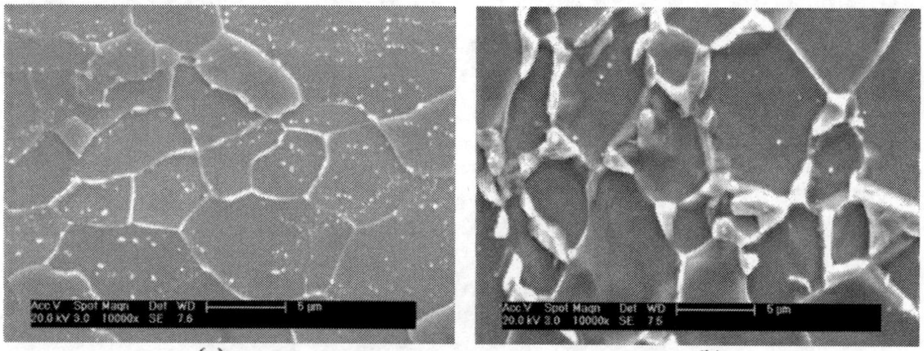

(a) (b)

Figure 1. SEM images of the samples (a) as rolled and (b) after isothermal heating at 760°C for 5 min and then quenching to retain the austenite morphology.

Figure 2 shows the SEM images of the samples after isothermal heating at 760°C for 5 min and then sand cooled or air cooled to room temperature. It is seen that the main differences between the two microstructures are the amount and features of the phases distributed at the ferrite grain boundaries. In the microstructure of the sand cooled sample, there are small amounts of pearlite and lump phase at the ferrite grain boundaries (figure 2 (a)). However, in the case of the air cooled sample, nearly no pearlite appears, and the amount of lump phase is higher than in the sand cooled sample as shown in figure 2 (b). By comparing figure 2 with figure 1 (b), it is apparent that the pearlite and lump phase along the ferrite grain boundaries are transformed from the retained austenite during cooling.

The samples air cooled were then tempered at 200°C~400°C for 20 min. The SEM images of the tempered samples are presented in figure 3. It is noted that the microstructure after tempering at 200°C (figure 3 (a)) remains nearly unchanged compared to the sample air cooled (figure 2 (b)), except for the appearance of extremely small amounts of lamellar pearlite at the ferrite grain boundaries. After tempering at 300°C, the transformation from lump phase to pearlite proceeds further, and the two phases coexist along the grain boundaries (figure 3 (b)). When the tempering temperature is 400°C, the lump phase almost completely transformed into lamellate pearlite as shown in figure 3 (c).

(a) (b)

Figure 2. SEM images of the samples after isothermal heating at 760°C for 5 min and then (a) sand cooled and (b) air cooled.

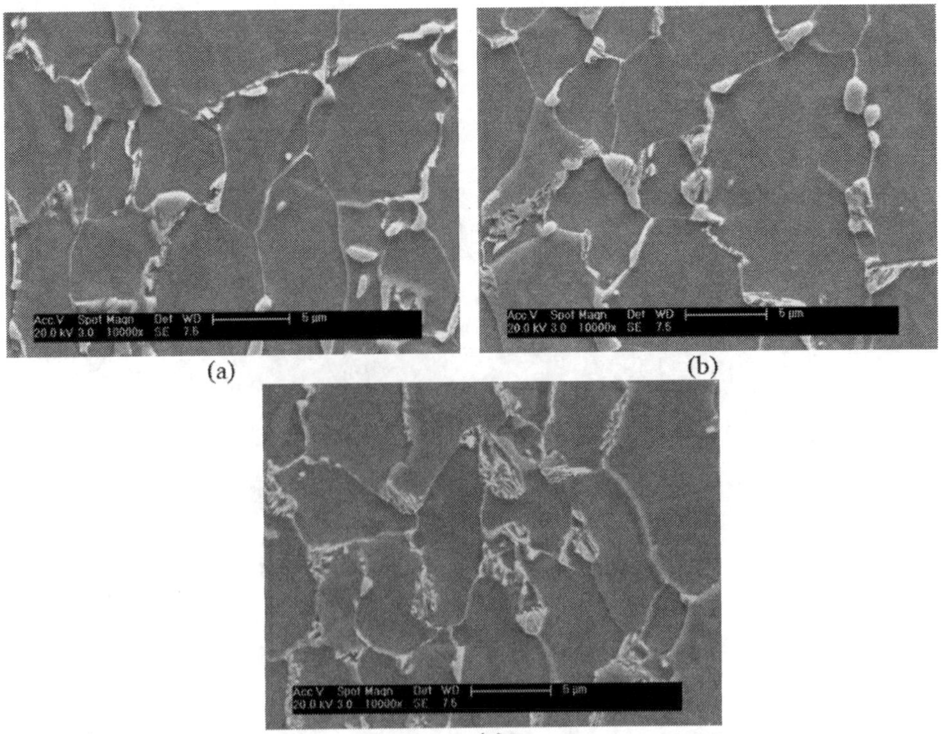

(a) (b) (c)

Figure 3. SEM images of the samples tempered at different temperatures of (a) 200°C; (b) 300°C; and (c) 400°C for 20 min after heat treatment at 760°C and then air cooled.

Figure 4 shows the stress-strain curves of the samples as rolled and after heat treatment at 760°C and then sand cooled or air cooled. It is found that the sample after heat treatment by sand cooling presents the higher yield strength and longer yield platform than the as rolled sample. However, for the sample after heat treatment by air cooling, the yield strength is decreased and the yield platform becomes short compared to the as rolled sample as shown in figure 4.

Figure 5 shows the stress-strain curves of the air cooled samples before and after tempering at different temperatures. It is found that for the air cooled samples, after tempering at various temperatures for 20 min, the yield strength gradually increases and the yield platforms become longer with increasing tempering temperature.

Table 2 presents the yield strengths of the samples as rolled, after heat treatment at 760°C and then sand cooled and air cooled, and tempering at different temperatures. The yield strengths of the samples after heat treatment at 760°C by sand cooling and air cooling are ~ 50 MPa higher and ~ 40 MPa lower than that of the as rolled sample, respectively. After the air cooled samples are tempered at different temperatures, the yield strengths increase with increasing the tempering temperature. When the tempering temperature is 400 °C, the yield strength is higher than that of the samples as rolled, sand cooled, and air cooled by ~ 60 MPa, 10 MPa, and 100 Mpa, respectively.

It is known that yield strength and yield platform features are closely related to the formation of high concentration Cottrell atmosphere by the interstitial elements C and N pinning dislocations [8, 9]. Thus, it is considered that the different yield strengths obtained from the different heat treatments might be caused by the impact of different heat treatment processes on the formation of high concentration Cottrell atmosphere and related to the amount of lump phase distributed at grain boundaries.

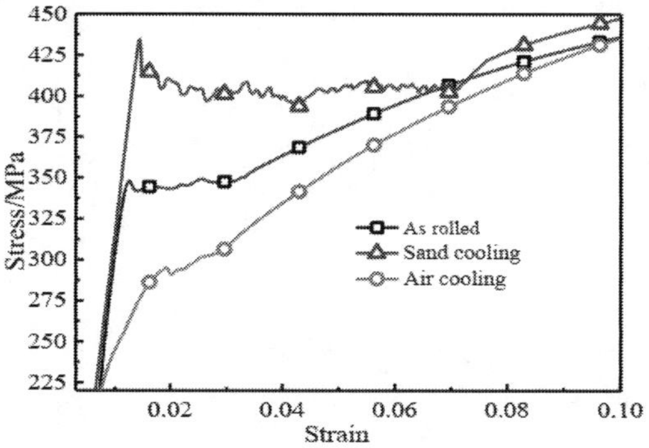

Figure 4. Stress-strain curves of the samples as rolled and after heat treatment at 760°C and then sand cooled and air cooled.

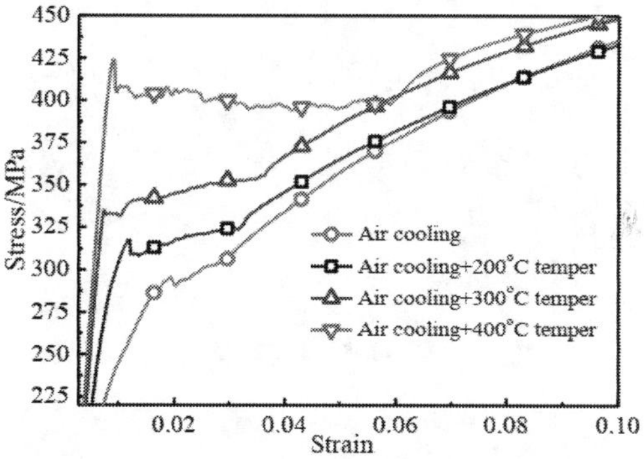

Figure 5. Stress-strain curves of the air cooled samples before and after tempering at different temperatures.

Table 2. Yield strengths of the samples as rolled, after heat treatment at 760°C and then sand cooled and air cooled, and tempering at different temperatures.

	As-rolled	Sand cooled	Air cooled	Tempering temperature		
				200°C	300°C	400°C
Yield strength/MPa	336±4	389±4	291±12	309±10	324±8	397±6

3.2. Analysis of the lump phase

According to the microstructures and yield strengths of the samples exposed to different heat treatments presented in section 3.1, it is noted that the yield strength is closely related to the volume fraction of lump phase distributed at the ferrite grain boundaries. Thus, the detailed analysis of the lump phase was performed by applying SEM and TEM analysis, and by applying thermodynamic calculations using Pandat software.

Figure 6 and table 3 show the energy dispersive spectrometer (EDS) results of the lump phase and the ferrite matrix of the sample after heat treatment at 760°C and then air cooled. It is found that the average carbon concentration in the lump phase is higher than that in the ferrite matrix.

Figure 7 shows the morphology and the selected area electron diffraction (SAED) pattern of the lump phase. It is found that there is a sharp interface between the lump phase and the ferrite matrix. Evident dislocation tangles are observed inside the lump phase. According to the diffraction pattern shown in figure 7 (b), the lump phase has a bcc lattice. Therefore, it considered that the lump phase is a type of bainite-ferrite phase [10], namely, a supersaturated ferrite phase.

Figure 8 shows the fractions of ferrite and austenite, the solubilities of elements C, Mn, and Si in ferrite, austenite, and precipitates, respectively, varying with temperature; the data was calculated using Pandat. It is seen that the phase transformation from ferrite to austenite starts at ~700°C, and finishes at ~890°C. Thus, the heat treatment temperature at 760°C is within the ferrite-austenite two-phase region. The solubility of element C in ferrite is significantly lower than in austenite. When the matrix is the ferrite phase, most carbon atoms are tied up as precipitates as shown in figure 8 (b). The solubility of Mn in ferrite is slightly lower than in austenite when the temperature is around 500~700°C, and gradually decreases with decreasing temperature(figure 8 (c)), while the solubility of element Si is nearly identical in austenite and ferrite as shown in figure 8 (d).

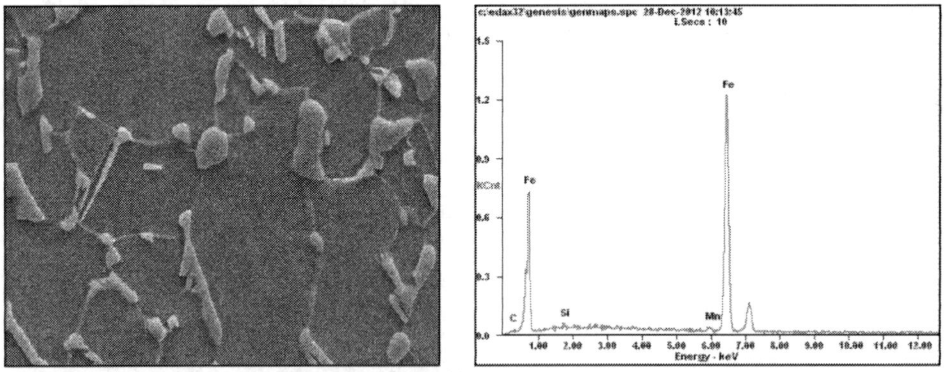

(a)

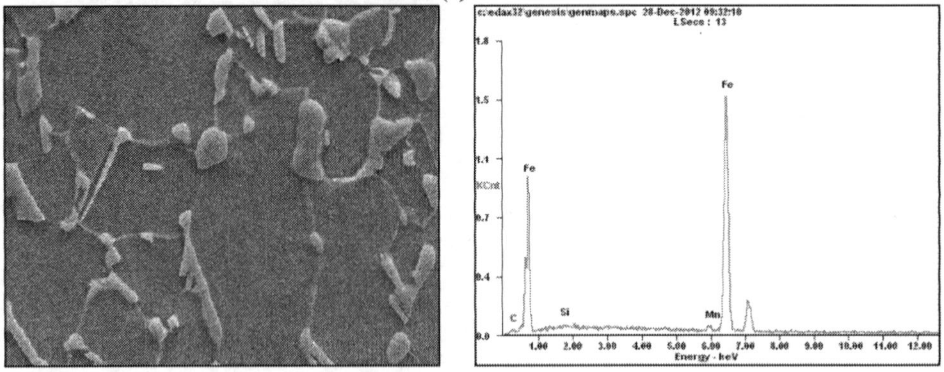

(b)

Figure 6. EDS analysis of the sample after heat treatment at 760°C and then air cooled: (a) lump phase; (b) ferrite matrix.

Table 3. EDS analysis of the sample after heat treatment at 760°C and then air cooled.

Content (wt.%)	C	Si	Mn	Fe
Lump phase	6.00	0.75	1.94	91.31
Ferrite matrix	3.45	0.64	1.83	94.08

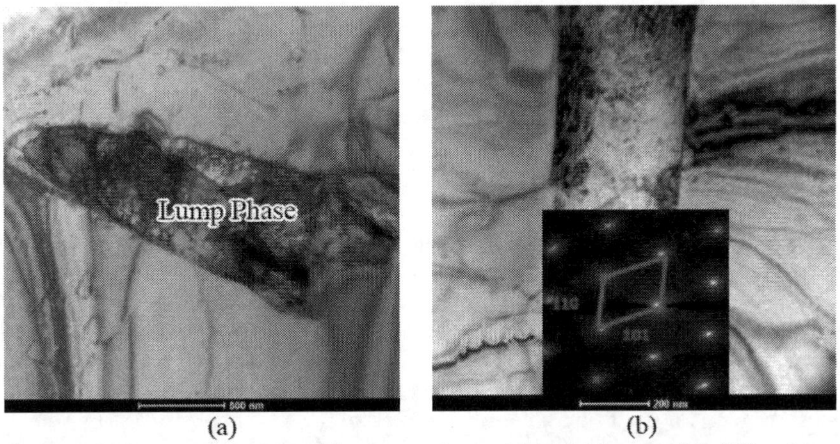

Figure 7. Morphology and SAED of the lump phase: (a) TEM image; (b) diffraction pattern.

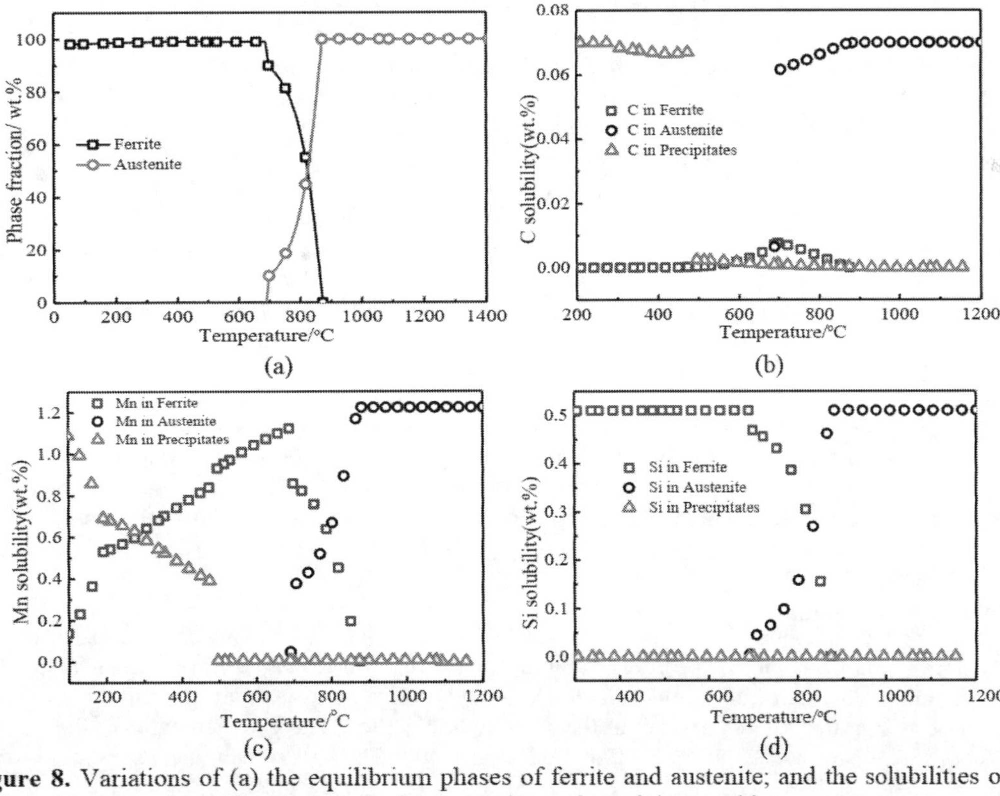

Figure 8. Variations of (a) the equilibrium phases of ferrite and austenite; and the solubilities of (b) carbon; (c) manganese; (d) silicon in ferrite, austenite, and precipitates with temperature.

Based on the above analysis, during isothermal heating at 760°C, within the ferrite-austenite two-phase region, ferrite-to-austenite phase transformation takes place, and the formed austenite is primarily distributed along the ferrite grain boundaries. Since the solid solubilities of interstitial elements, C and N, in austenite are higher than that in ferrite, while the diffusivities of interstitial atoms in austenite are lower than that in ferrite [11], these interstitial elements are enriched in the austenite grains. During the subsequent cooling processes from 760°C to room temperature, austenite-to-ferrite phase transformation occurs. When the temperature is cooled down to a certain temperature, the retained austenite is transformed to pearlite or supersaturated ferrite through allotropic transformation. It is known that austenite to pearlite phase transformation is controlled by carbon diffusion. When the cooling rate is relative slow, more time is available for phase transformation and carbon diffusion. Thus, it is understandable that in the sand cooled sample there are small quantities of pearlite and lump phase at the ferrite grain boundaries, while in the air cooled sample, nearly no pearlite is observed, and the amount of lump phase is higher than that in the sand cooled sample, as shown in figure 2. Since the lump phase at the grain boundaries is enriched with the element carbon, the carbon atoms cannot diffuse sufficiently to pin the dislocations and to form high concentration Cottrell atmospheres, leading to the reduction of dislocation strengthening [12]. In addition, the microsegregation of carbon atoms at grain boundaries reduces the solid solubility in the ferrite matrix, which weakens the contribution of solid solution strengthening [13]. Dong An *et al.* [14] simulated the carbon concentration distribution after the cooling processes from 760°C to room temperature at the cooling rates of 2°C/s and 5°C/s using a cellular automaton (CA) model. It is found that after cooling at 2°C/s, the carbon concentration field is nearly uniform, while a higher cooling rate of 5°C/s results in a non-uniform carbon concentration field. Accordingly, the air cooled sample presents a lower yield strength and a shorter yield platform than that of the sand cooled sample, and even than that of the as rolled sample, due to that fact that the fast cooling rate is unfavourable for the formation of high concentration Cottrell atmospheres and weakens the solid solution strengthening component.

When the air cooled samples are then tempered at 200°C~400°C for 20 min, carbon diffusion takes place, which is beneficial for the formation of high concentration Cottrell atmospheres. As the diffusivities of the interstitial elements are temperature dependent, when the sample is tempered at 200°C, the diffusion of the interstitial atoms from boundaries to the locations of dislocations is still limited. The simulation results by An *et al.* [14] show that tempering at 200°C for 20 min has less effect on the carbon concentration field. Thus, after tempering at 200°C, the yield strength is only slightly increased, and the yield platform becomes slightly longer compared to the air cooled sample. With the tempering temperature increasing, the diffusivities of the interstitial elements increase, leading to more sufficient diffusion of the interstitial atoms to form high concentration Cottrell atmospheres. Therefore, yield strength increases and the yield platform becomes longer with increasing tempering temperature, as shown in figure 5. In addition, more sufficient carbon diffusion at higher tempering temperature also promotes the transformation of the non-equilibrium lump phase (supersaturated ferrite) to the equilibrium pearlite phase. Thus, as tempering temperature increases, the lump phase is gradually transformed to pearlite. After tempered at 400°C for 20 min, the lump phase at the grain boundaries nearly completely transformed to lamellate pearlite as shown in figure 3 (c).

4. Conclusions

A cold-rolled enameling steel was isothermally heated at 760°C for 5 min and then sand cooled or air cooled to room temperature. The microstructure of the sample sand cooled consists of the ferrite matrix and a small quantity of pearlite and lump phase distributed at the ferrite grain boundaries. However, in the microstructure of the sample air cooled, nearly no pearlite forms, and the amount of lump phase is significantly higher than in the sample sand cooled. The yield strengths of the samples sand cooled and air cooled are 50 MPa higher and 40 MPa lower than the as rolled sample, respectively.

After tempering at 200°C~400°C for 20 min, the lump phase in the samples air cooled is gradually transformed to pearlite with increasing tempering temperature. When the tempering temperature is

400°C, the lump phase almost completely decomposed into lamellate pearlite. Meanwhile, the yield strength gradually increases with increasing tempering temperature. After tempering at 400°C, the yield strength is higher than that of the samples as rolled, sand cooled, and air cooled by ~60 MPa, ~ 10 MPa, and ~100 MPa, respectively.

The results of the SEM and TEM analysis and the thermodynamic calculations show that the lump phase is a supersaturated ferrite phase enriched with carbon. During the air-cooling process from the ferrite-austenite two-phase region to room temperature, less time is available for austenite-to-ferrite and austenite-to-pearlite phase transformation and carbon diffusion. The retained austenite is thus transformed to the supersaturated ferrite (lump) phase through allotropic transformation. In this case, the carbon atoms cannot diffuse sufficiently to pin the dislocations to form high concentration Cottrell atmospheres, leading to yield strength reduction and yield platform shortening. When the samples air cooled are tempered at 200°C~400°C, carbon diffusion takes place, promoting the formation of high concentration Cottrell atmospheres. This results in the yield strength increasing. The effect of tempering on yield strength increases with increasing tempering temperature, due to the fact that the diffusivities of the interstitial elements increase with temperature.

5. Acknowledgements
The authors wish to acknowledge CompuTherm Company, USA, for generously providing us with the license of Pandat software. This work was financially support by the AO Smith Corporate Technology Center, USA, and NSFC (Grant No. 51371051).

6. References
[1] Leila S, Hossein S and Alireza M 2007 *Mater. Sci. Eng. A* **458** 88
[2] Son Y K, Lee C J, Lee J M and Kim B M 2012 *Trans. Nonferrous Met. Soc. China.* **22** s838
[3] Jiang M, Zheng F J, Huang X F, Dai T, Krakauer B W and Zhu M F 2012 *IOP Conf. Series: Materials Science and Engineering (Schladming)* vol 33 (IOP Publishing) 012044
[4] Jiang M, Yang X F, Pan S Y, Krakauer B W and Zhu M F 2012 *J. Mater. Sci. Technol.* **28** 737
[5] Mao X P, Huo X D, Sun X J and Chai Y Z 2010 *J. Mater. Process. Technol.* **210** 1660
[6] Isasti N, Badiola D J, Taheri M L and Uranga P 2014 *Met. Mater. Int.* **20** 807
[7] Chen S L, Daniel S, Zhang F, Chang Y A, Yan X Y, Xie F Y, Schmid F R and Oates W A *Calphad* **26** 175
[8] Ha K F 1983 *A microscopic theory of mechanical properties in metal* (Beijing: Science press) p 76
[9] Weng Y Q 2008 *Ultra-fine grained steels* (Beijing: Metallurgical industry press) p 262
[10] Yang Z G, Yang J B, Fang H S and Bai B Z 2003 *Mater. Sci. Forum* **426** 1261
[11] Gamsjager E, Fischer F D and Svoboda J 2002 *J. Eng. Mater. Technol.* **125** 22
[12] Scheu C, Stergar E, Schober M, Cha L, Clemens H, Bartels A, Schimansky F P, and Cerezo A 2009 *Acta Mater.* **57** 1504
[13] Portavoce A and Treglia G 2014 *Acta Mater.* **65** 1
[14] An D, Pan S Y, Huang L, Dai T, Krakauer B W and Zhu M F 2014 *ISIJ Int.* **54** 422

Numerical study of macrosegregation in a large steel ingot with multiple pouring process

Zhenhu DUAN[1], **Xuan DU**[2], **Houfa SHEN**[1], **Baicheng LIU**[1]

[1]Key Laboratory for Advanced Materials Processing Technology, Ministry of Education. School of Materials Science and Engineering, Tsinghua University, Beijing 100084, China
[2]CITIC Heavy Industries Co., Ltd., Luoyang, 471039, China

E-mail: shen@tsinghua.edu.cn

Abstract: Multiple pouring (MP) process with a higher concentration in the beginning and a lower concentration at the last is generally utilized to reduce macrosegregation in the large steel ingot production. However, it is insufficient to demonstrate the practical efficiency of MP process from the point view of theory. In this paper, the whole filling and solidification processes were investigated to analyze the formation of macrosegregation in a large steel ingot. Firstly, a filling transportation model considering the entire MP process was presented. The carbon concentration variation through the outlet of the tundish at different times was predicted and provided as the boundary conditions for the mold filling and the solidification processes simulation. Secondly, the macrosegregation model was integrated into the filling transportation model to investigate the influence of the MP on the macrosegregation in the ingot through the comparison of the MP and the single pouring process (SP). Finally, the top section of a 438-t steel ingot was cut off and the carbon concentration was measured to validate the numerical results. The results indicated that the MP process has a certain favorable effect on reducing the positive segregation in the top region of the ingot.

1. Introduction

Large steel ingots weighing more than 100 tons are widely used in the heavy equipment for power, shipbuilding, petrochemical, metallurgy and other industries, such as steam turbine rotor, hydraulic turbine shaft, pressure shell of atomic reactors, and roller for large mill housing [1]. For a long time, many steel plant workers and researchers have been devoted to improve the quality of these products through the casting practice and laboratory investigation. The defect of macrosegregation encountered in the large steel ingots has been become a great concern in recent years. Some special technologies have been come up to reduce the macrosegregation in castings [2-3]. At the same time, great efforts have already been made in the last few decades in the modelling of macrosegregation [4-8]. However, the industrial application of these models is still limited to the steel ingots less than 100 tons regardless of pouring process. As the large steel ingots of 300~600 tons concerned, few researchers have attempted to take both the filling and solidification processes into account because of the large size of the calculation domain and the limitation of the hardware resource. The numerical simulation in a 360 tons steel ingot was carried out by Liu et al. Whereas the filling in the tundish and mold should be introduced in detail [9]. A marked reduction of macrosegregation was obtained in the ingot body for 570 tons steel ingot produced by MP process [10]. The MP process of a 292 tons steel ingot was simulated by Tu et al. And the favorable initial carbon distribution in the mold was found after MP process, however, the formation of the macrosegregation after filling was not presented [11]. The

IOP Conf. Series: Materials Science and Engineering **84** (2015) 012048 doi:10.1088/1757-899X/84/1/012048

objective of the present research is to analyze the macrosegregation with the carbon concentration variation during the filling process. The influence of the MP process on the final macrosegregation in a large ingot is numerical simulated with validation of an industrial experiment.

2. Process description

The whole MP process with three stages of the filling, holding and draining is shown in figure 1. In the first stage (figure 1a), the refined molten steel is poured into the tundish from the first ladle with higher concentration until the liquid steel reaches at the highest critical height in the tundish, then the tundish is closed and keeps standstill for a while. In the second stage (figure 1b), the outlet and inlet of the tundish are opened at the same time, the molten steel from sequential ladles with lower concentrations comes into the mold, and the level of the melt in the tundish remains certain height by controlling the flow rates through the inlet and outlet of the tundish. In the third stage (figure 1c), when the total molten steel are poured into the tundish completely, the bath level reduces to the lowest critical height, then the inlet and outlet of the tundish is closed, and the whole pouring process is over. After filling, molten steel solidifies in the mold and macrosegregation may form during solidification.

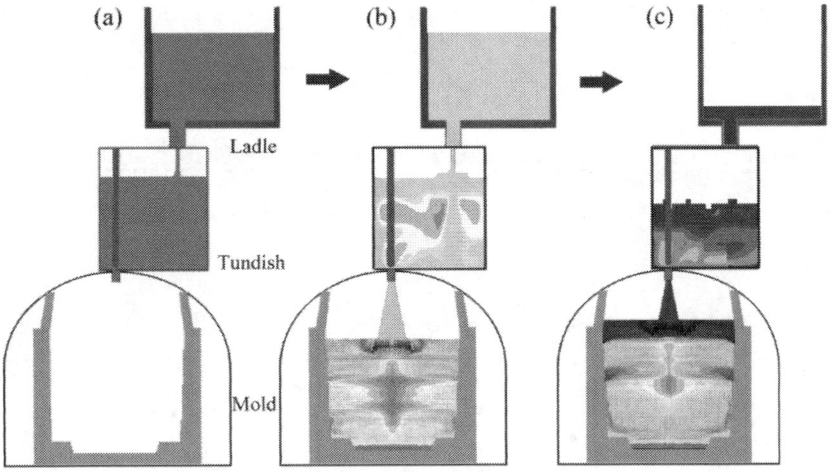

Figure1. Schematic of MP process for a large steel ingot: (a) the filling stage, (b) the holding stage, (c) the draining stage.

3. Mathematical model

A turbulence model and a continuum model were integrated to investigate the macrosegregation in steel ingot with MP process. Table 1 shows the CFD turbulence transportation model based on the multi-phases and VOF methods for unsteady, turbulent and gas-liquid two phase flow in the tundish. The Realizable k-ε model was employed in the far-wall field zones, which has superior performance for flows involving rotation, boundary layers under strong adverse pressure gradients, separation, and recirculation. The commercial software Fluent was applied to simulate the solute mixing behavior in the tundish. The mass continuity, momentum, energy as well as the chemical species conservation equations were solved, and the carbon distribution in the tundish and the carbon variation through the outlet of the tundish during the MP process at different times were simulated.

Furthermore, the previous continuum macrosegregation model in reference [8] was incorporated into the filling transportation model to simulate the filling and solidification processes in the mold. The fluid flow in the mold was much weaker compared to that in the tundish, because the melt was sprayed into the mold during the filling with vacuum. The main assumptions and simplifications for the macrosegregation model were employed as follows: a) two dimension in the calculation; b) laminar and Newtonian fluid flow in the liquid phase; c) rigid and stationary in the solid; d) the Boussinesq assumption in the buoyancy term of the momentum equation. The conservation equations

IOP Conf. Series: Materials Science and Engineering **84** (2015) 012048 doi:10.1088/1757-899X/84/1/012048

are listed in table 2. The moving grid method with expanded calculation domain was applied during the mold filling process. The volume of liquid steel through the output nozzle of the tundish at each time step was calculated and equivalently added to the free surface of the molten steel in the mold with boundary conditions of the velocity and the carbon distribution on the affected surface. Once the new added volume of the molten steel at a calculation time step was larger than one layer of elements in the mold, the free surface was set upwards by one layer until the end of the MP process. However, the detailed dripping process of the liquid steel from the output nozzle of the tundish to the free surface of molten steel in the mold was simplified during MP process.

According to the actual MP process, three ladles with different carbon concentrations were poured in succession from the tundish to the mold for the production of steel ingot with average Fe-0.44%C (C_0 = 0.44) in order to counteract the negative segregation at the bottom and the positive segregation at the riser. The weight of the molten steel in each ladle was 150 tons and the carbon concentrations from the first to the third ladle were 0.51%, 0.44%, 0.37%, respectively. The weight of ingot was about 438 tons with the height of 5.1 m and the average width of 3.765 m. The teeming temperature was 1560 □, and the total pouring time was about 5200 s. The heat transfer coefficients for the interfaces at the mold/ingot and mold/ambient were assumed as constant values of 1000 $W \cdot m^{-2} \cdot K^{-1}$ and 100 $W \cdot m^{-2} \cdot K^{-1}$, respectively. A small heat transfer coefficient of 5 $W \cdot m^{-2} \cdot K^{-1}$ was used at the interface between the casting and powder at the mold top. The time step was 0.01 s for the fluid flow simulation in the tundish. The time step was 1 s in the mold filling process and then 2 s at the later stage of the solidification simulation. The mesh size was 30 mm with a square and tetrahedron shape for the mold and tundish, respectively.

Table 1: turbulence transportation model: the mathematical model for the flow in tundish

Mass	$\frac{\partial a_q}{\partial t} + \mathrm{u} \cdot \nabla \cdot a_q = \frac{1}{\rho_q} \sum_{p=1}^{n} \left(m_{pq} - m_{qp} \right)$	(1)
Momentum	$\rho \frac{\partial \mathrm{u}}{\partial t} + \rho \nabla \cdot (\mathrm{uu}) = -\nabla p + \mu_1 \nabla^2 \mathrm{u} + \rho g$	(2)
Realizable k-ε	$\rho \frac{\partial k}{\partial t} + \rho \nabla \cdot (\mathrm{u}k) = \left(u + \frac{u_{eff}}{\sigma_k} \right) \nabla^2 k + G_k - \rho \varepsilon$ $\rho \frac{\partial \varepsilon}{\partial t} + \rho \nabla \cdot (\mathrm{u}\varepsilon) = \left(u + \frac{u_t}{\sigma_\varepsilon} \right) \nabla^2 \varepsilon + \rho C_1 E \varepsilon - \rho C_2 \frac{\varepsilon^2}{k + \sqrt{v\varepsilon}}$	(3)
Energy	$\rho \frac{\partial T}{\partial t} + \rho \nabla \cdot (\mathrm{u}T) = \frac{K}{c_p} \nabla^2 T$	(4)
Species	$\rho \frac{\partial C}{\partial t} + \rho \nabla \cdot (\mathrm{u}C) = \nabla \cdot (D \nabla C)$	(5)

Table 2: continuum model: the mathematical model for the macrosegregation in mold

Mass	$\nabla \cdot \mathbf{u} = 0$	(6)
Momentum	$\rho \frac{\partial \mathbf{u}}{\partial t} + \rho \nabla \cdot (\mathbf{uu}) = -\nabla p + \mu_1 \nabla^2 \mathbf{u} - \mu_1 K^{-1} \mathbf{u} + \rho_b \mathbf{g}$	(7)
Density in buoyancy term	$\rho_b = \rho[1 - \beta_T (T - T_0) - \beta_C (C_1 - C_0)]$	(8)
Energy	$\rho \frac{\partial T}{\partial t} + \rho \nabla \cdot (\mathrm{u}T) = \frac{K}{c_p} \nabla^2 T + \rho \frac{\mathrm{L}}{c_p} \frac{\partial g_s}{\partial t}$	(9)
Species	$\frac{\partial C}{\partial t} + \nabla \cdot (\mathbf{u}C) = \nabla \cdot (g_1 D_1 \nabla C) + \nabla \cdot [g_1 D_1 \nabla (C_1 - C)] - \nabla \cdot [\mathbf{u}(C_1 - C)]$	(10)
Permeability	$K = \frac{\lambda_2^2}{180} \frac{g_1^3}{(1 - g_1)^2}$	(11)
Temperature and liquid concentration	$T = T_m + m_1 C_1$	(12)
Solid volume fraction and temperature	$g_s = \frac{1}{1 - k_p} \frac{T - T_1}{T - T_m}$	(13)

In above equations, a_q is the volume fraction of the molten steel, m_{pq} the mass fraction of the molten steel, m_{qp} the mass fraction of the air; u denotes the velocity, ρ the density, t the time, p the pressure, μ_l the liquid viscosity, g the gravity, T the temperature, K the thermal conductivity (equation (4) and (9)), c_p the specific heat, L the latent heat, g_s the solid volume fraction, g_l the liquid volume fraction, C the species concentration, k_p the partition coefficient, T_m the melting temperature of the pure iron, m_l the liquidus slope of the phase diagram. The Boussinesq approximation (equation (8)) was used to model the thermo-solutal convection, where ρ_b denotes the density in the buoyancy term, β_T and β_C are the thermal and the solutal expansion coefficients; T_0 and C_0 are the reference values for the temperature and concentration, respectively. The permeability of the mushy region, K, is defined by the Carman-Kozeny relation (equation (11)), which depends on the liquid volume fraction g_l and the secondary dendrite arm spacing λ_2.

4. Results and discussion

4.1. Carbon distribution in the tundish and mold during filling

Figure 2 shows the carbon distribution in the tundish at different times during the whole MP process. Before 1600 s, the carbon concentration in tundish is homogenous, i.e., 0.51% which is same as that in the first ladle melt. When the molten steel from the first ladle has completely been poured into the tundish, the molten steel with less species from the sequential ladles begins to mix with the original molten steel in the tundish. This mixing behavior is carried on mainly by the way of convection owing to the violent flow induced by the impinging of the plunging jet from the ladle. In the holding stage as indicated in figure 2 (a-f), the molten steel with lower carbon concentration is transferred to the corners of the tundish, and several swirls with higher carbon concentration form in the middle of the tundish. When it comes to the draining stage and all ladles of the molten steel are poured into the tundish completely, the bath level in the tundish decreases to the lowest critical height because the molten steel flows out through the tundish outlet without supplement as shown in figure 2 (g-h). The carbon distribution is almost the same and remains higher than that of molten steel in the third ladle. Another aspect should be mentioned that the carbon concentration through the outlet of the tundish comes from the following two parts: the circumfluence flow from the middle of the tundish and the plunging jet flow directly to the outlet. This carbon concentration at different times through the outlet of tundish has been calculated and used as the boundary conditions in the simulation of mold filling and solidification stage.

Figure 3 (a-d) shows both the filling and the solidification processes with different times in the mold. There is no carbon variation (0.51%C) in the mold before 1600 s during the filling stage. In the impact region of the upper mold, the carbon mixing behavior is carried on by the way of the forced convection; the carbon distribution pattern takes on a kind of V-shape with the deepest height of about 1 m. Another phenomenon can be found that the low carbon solute from the top region is transferred to the lower region along the solidified shell growing from the mold wall. At the same time, the low carbon dilutes with the high carbon in the inner region of the mold mainly by the way of the diffusion, and thus the carbon distribution gradually decreases from the center to the side in middle of the mold. During the third ladle filling at times of 3750 s and 4700 s as shown in figure 3 (c-d), the impact area of the fresh steel on the free surface becomes narrow and the impact depth increases with the molten steel level rising in the mold. Moreover, an intensive turbulence forms around the center region on the free surface. Finally, the initial carbon concentration in the mold after MP process presents obviously low in the hot top section and high in the solidified shell near the mold bottom.

Figure 3 (e-h) shows the carbon distribution evolution and the fluid flow during the later solidification stage after pouring process. High carbon distribution is presented at the lowest bottom of the ingot because the high carbon species with 0.51 wt. %C from the first ladle melt has been captured by the solidified metal at the beginning of solidification. With the solidification proceeding, the carbon distribution in the center of the liquid steel increases, and finally the serious positive segregation is presented in the hot top region of the ingot. It should be mentioned that if the carbon mixing in the tundish was ignored, the carbon concentration of ingot is different compared with that considering the tundish filling especially during the pouring process and finally may be little different according to [9].

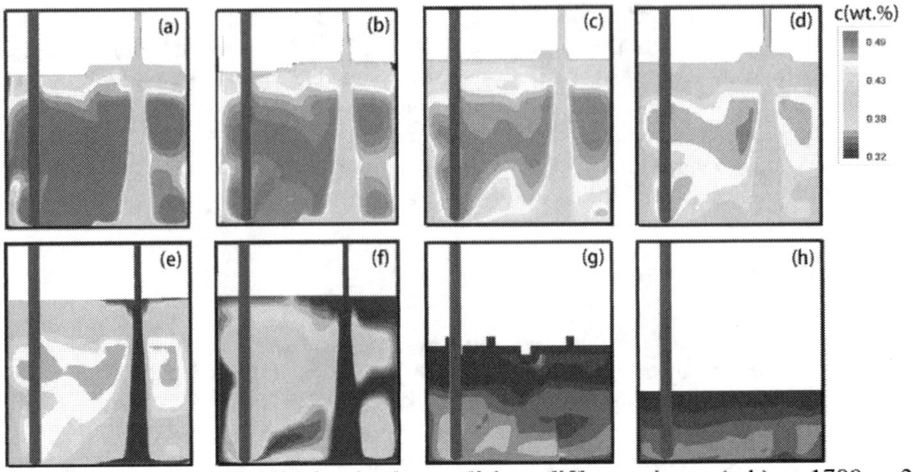

Figure 2. Evolution of carbon distribution in the tundish at different times. (a-h) at 1700 s, 2000 s, 2500 s, 3000 s, 3500 s, 4300 s, 4900 s and 5200 s.

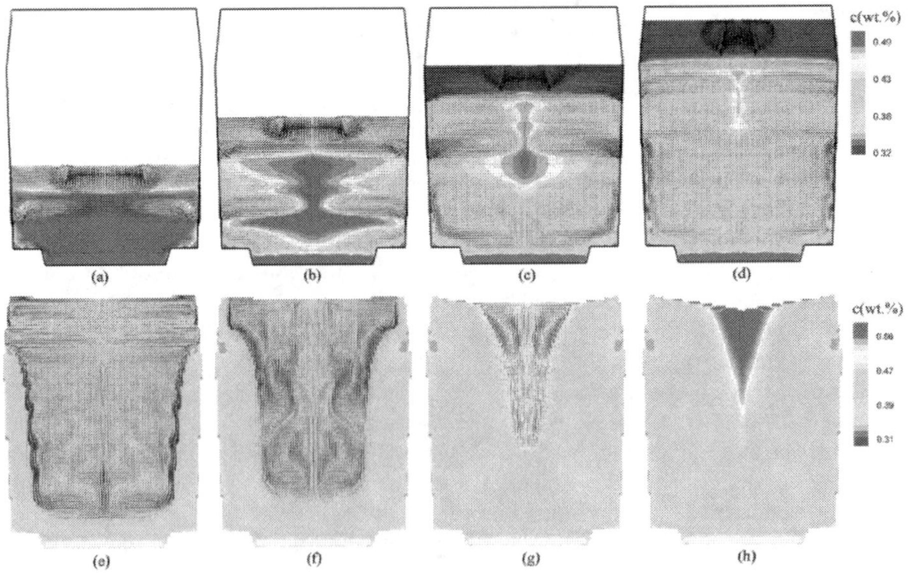

Figure 3. Fluid flow and carbon distribution at different times in the mold. (a) 1750 s, (b) 2250 s, (c) 3750 s, (d) 4700 s, (e) 15000 s, (f) 36000 s, (g) 108000 s, (h) final solidification stage.

In order to validate the present numerical model, a trial steel ingot weighting 438 tons was produced in the steel plant of the CITIC Heavy Industries Co., Ltd. as illustrated in figure 4 (a). After the steel ingot had been cooled and stripped out from the mold, the riser part of the steel ingot was cut off and the specimens were taken along these transverse sections. The concentration of these samples was measured using spectral analysis.

Figure 4 (b-d) shows the comparison between the predicted and measured result of the carbon concentration along the transverse sections (A-C) in the hot top of ingot. The distance of these sections (A-C) from bottom of the riser is about 0.7 m, 0.6 m, 0.5 m, respectively. In general, the agreement between the predicted concentration and the measured tendencies can be considered good. The predicted positive segregation around the riser center region is underestimated as compared to the experimental results. However, the difference between them can be accepted in industrial applications. Besides, negative segregation in the bottom of ingot was not simulated successfully due to the fact that the motion of equiaxed grains is not considered in current continuum model.

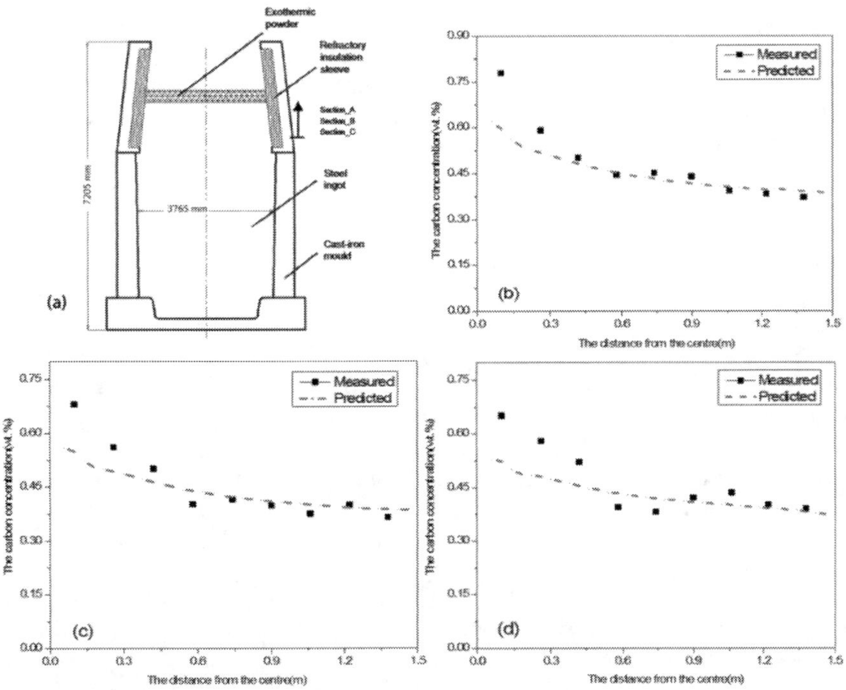

Figure 4. (a) Schematic showing 438 tons ingot mold and carbon measurement positions. Measured and predicted carbon concentration at (b) section A, (c) section B and (d) section C.

4.2. Comparison of the macrosegregation between the MP and SP
In order to investigate the effect of MP technique on the final macrosegregation in ingot, two cases with the MP and the SP (single pouring) process were simulated and compared. The SP process is referred to the MP process with the uniform carbon concentration of the molten steel in the sequential ladles.

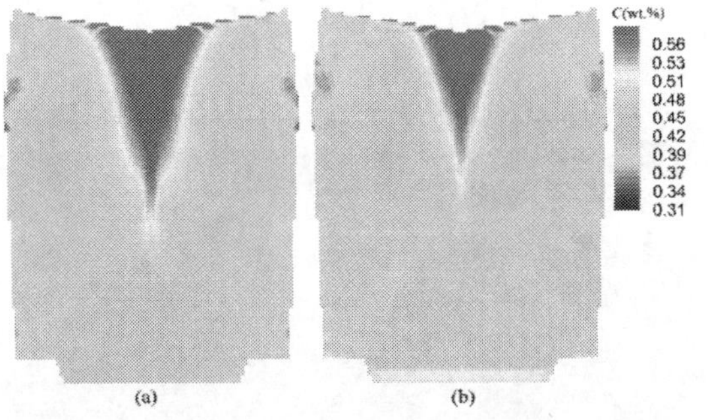

Figure 5. Comparison of the macrosegregation between the SP (a) and the MP (b).

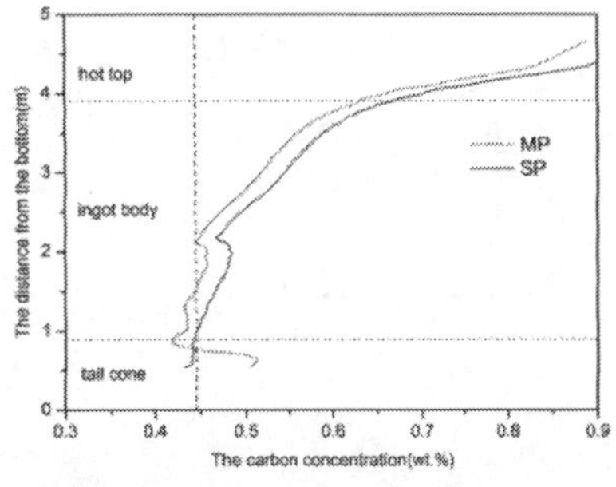

Figure 6. Comparison of the carbon concentration along the ingot center line.

Figure 5 shows the comparison of the final macrosegregation in ingot between the case 1 (SP) and case 2 (MP). The overall patterns of macrosegregation are nearly similar for two cases, namely, the strongly positive segregation in the riser. It can be seen from the width and the depth of the positive segregation zone as illustrated in figure 5 that the positive segregation in the top center region of the ingot for the case 1 is more serious than the case 2. It can be explained simply by the reason that the low carbon species from the last ladle remains for a longer time in the riser with MP. Figure 6 shows the comparison of the carbon distribution profile along the ingot center line between the two cases. It can be illustrated that the carbon concentration at the same center line's position with SP is higher than that with MP except for the tail section of the ingot. The highest carbon is shifted upper into the hot top riser of the ingot with MP, that is to say, the macrosegregation degree in the steel ingot body is released to some extent with MP. Therefore, the comparison of the simulation results between the two cases demonstrates that the MP technique has some certain favorable effects on suppressing the positive segregation in ingot. However, the macrosegregation is still severe in the ingot body and the optimized MP process should be adapted in the future research work.

5. Conclusions

A filling transportation model has been realized to describe the carbon mixing behavior and distribution in the tundish during the multiple pouring (MP) process. The macrosegregation model was integrated into the filling transportation model, and the numerical results of the macrosegregation with MP process have been validated with the carbon measurements in a 438 ton steel ingot. Comparison results show that the multiple pouring process has some certain favorable effect on reducing the final macrosegregation as compared to the conventional constant concentration pouring process.

Acknowledgments

This work was financially supported by the National Basic Research Program of China (No. 2011CB012900) and the National Science and Technology Major Project of the Ministry of Science and Technology of China (2012ZX04012011).

References

[1] Tanaka Y and Ishiguro T 1997 *physics state solid* (a) 305-160
[2] Liu D R, Kang X H, Sang B G, Li D Z 2011 *Acta Metallurgical Sinica (Engl. Lett.)* **24** 54-64
[3] Sang B G, Kang X H, Li D Z 2010 *Journal of Materials Processing Technology* **210** 703-711
[4] Wu M H, Ludwig A 2009 *Acta Materialia* **57** 5621-5631
[5] Ni J, Beckermann C 1991 *Metallurgical and Materials Transaction* B **22** 349-361
[6] Vreeman C J, Krane M J M, Incropera F P 2000 *International Journal Heat Mass Transfer* **43** 677-686
[7] Ahmad N, Combeau H, Desbiolles J L, Jalanti T, Lesoult G, Rappaz J, Rappaz M, Stomp C 1998 *Metallurgical and Materials Transactions* B **22** 617-630
[8] Li W S, Shen H F, Liu B C 2010 *Steel Research International* **81** 994-1000
[9] Liu D R, Sang B G, Kang X H, Li D Z 2010 *International Journal of Cast Metals Research* **23** 354-63
[10] Tateno M 1985 *The Iron and Steel Institute of Japan Meet* **25** 97-108
[11] Tu W T, Zhang X, Shen H F, Liu B C 2014 *China Foundry* **11** 52-58

Impacts of Interface Energies and Transformation Strain from BCC to FCC on Massive-like δ-γ Transformation in Steel

M Yoshiya[1,2], M Sato[1], M Watanabe[1], K Nakajima[1], T Yokoi[1], N Ueshima[1], T Nagira[1] and H Yasuda[1,3]
[1]Department of Adaptive Machine Systems, Osaka University, 2-1 Yamadaoka, Suita, Osaka 565-0871, Japan.
[2]Nanostructures Research Laboratory, Japan Fine Ceramics Center, 2-4-1 Mustuno, Atsuta-ku, Nagoya 456-8587, Japan.
[3]Department of Materials Science and Engineering, Kyoto University, Yoshida-Honmachi, Sakyo-ku, Kyoto, 606-8501, Japan.

E-mail: yoshiya@ams.eng.osaka-u.ac.jp

Abstract. Interface energies of δ/γ, γ/γ, δ/δ, L/δ, and L/γ interfaces, at first, as a function of misorientation were evaluated with an aid of atomistic simulations with embedded atom method. Then, under geometric constraints where grains or interfaces compete each other to minimize overall free energy, effective interface energies for those interfaces were quantified. It is found that neither the minimum nor effective δ/γ interface energies, 0.41 or 0.56 J/m^2, respectively, is significantly higher than those of other interfaces including liquid/solid interfaces, but the δ/γ interface energy is significantly high for the small entropy change upon δ-γ massive-like transformation, resulting in significantly higher undercooling required for γ nucleation in the δ phase matrix than in solidification. Detachment of δ-phase dendrite tips away from γ-phase dendrite trunks can be explained only from a viewpoint of interface energy if small misorientation is introduced at the δ/γ interface from the perfect lattice matching between BCC and FCC crystal structures. Examining the BCC-to-FCC transformation strain on the γ nucleation in the massive-like transformation, the γ nucleation is prohibited 170 K or more undercooling is achieved unless any relaxation mechanism for the transformation strain is taken into account.

1. Introduction

Recent *in situ* observations of solidifications to BCC-structured δ phase of carbon steel and following phase transformation to FCC-structured γ phase reported that, when the sample was furnace-cooled at a cooling rate of 0.33 K/s or more, the solid-state phase transformation was instantly completed, within 1 s in a millimetre-dimension window, instead of the peritectic reaction of δ + L → γ suggested by the Fe-C phase diagram even if carbon content is more than peritectic composition [1-3]. The δ-γ massive-like transformation required much higher undercooling than it is expected for the peritectic reaction. This led to the idea that the difficulty in nucleation of γ phase in the solidified δ matrix is, at least in part, responsible for the high undercooling. The difficulty of the nucleation can be easily expected since lattice matching between BCC and FCC is poor, which can be quantified by δ/γ interface energy. Since the massive-like phase transformation may lead to fracture upon continuous casting or may allow solute elements to diffuse through less dense BCC lattice than in FCC lattice at high temperature, it is crucial to understand the underlying mechanism of the massive-like phase transformation.

Experimental studies have reported δ/γ interface energy and α/γ interface both of which are identical to each other if magnetic energy contribution and temperature effects are put aside since α phase also has a BCC crystal lattice. However, the number of reports are much fewer than those for homo-phase interfaces, i.e., grain boundaries such as δ/δ and γ/γ ones. Furthermore, the reported values of δ/γ [4] and α/γ [5] interfaces are different from each other and experimental conditions methods for measuring grain boundary energies and the hetero-interface energies are also different from those for liquid/solid interface energies. Thus, it is not straightforward to examine interface energy dependence on nucleation in general on the same ground. Besides, it is not clear what kind of representative values of those interface energies should be compared in order to discuss the difficulty of the nucleation.

In order to quantify the difficulty of nucleation of γ phase in the solidified δ matrix in comparison with other sorts of nucleation of iron, at first we attempt in this paper to quantify the difficulty of δ-nucleation on the same ground based on interface energies evaluated in our previous studies [6,7]. Then, fragmentation of dendrite tip by re-melting at δ/γ interface [2] is discussed based on the interface energies. Finally, influence of the BCC-to-FCC transformation strain, an alternative factor, on the nucleation of γ phase [8], is also quantitatively examined using a simple geometric model of a γ nucleus and a δ matrix.

2. Methodology

Atomistic simulation for hetero-phase interfaces in addition to homo-phase interfaces were performed to obtain interface energies as a function of misorientation between neighbouring grains [6]. The static energy minimization technique using GULP code [9] was employed to optimize atomic coordinations near the interfaces using embedded atom method (EAM) [10-12] to describe interatomic bonding. In other words, temperature effects were excluded from the interface energies for simplicity so that they can be quantitatively compared on the same ground upon examining their influence on nucleation. Nishiyama-Wassermann (NW) orientation relationship [13,14] was assumed for the best lattice matching at δ/γ interface and further modifications proposed by Rigsbee and Aaronson [15] to improve the lattice matching were employed.

Although the minimum δ/γ interface energy is 0.41 J/m^2 [6], it is still unclear what misorientation should be given to δ/δ or γ/γ interface to compare either one with δ/γ interface energy since both of their minimum interface energies at the best lattice matching are, needless to say, zero. On the other hand, upon discussing grain growth, non-zero substantial values of δ/δ or γ/γ interface energies are used in many cases. Thus, in order to enable us to compare the hetero-phase interface energies including those for δ/γ, L/δ, and L/γ interfaces with the homo-phase interface energies including δ/δ and γ/γ on the same ground, representative values of the interface energies were evaluated under the same geometric constrains which mimic the one present in multi-phase multi-grain system in reality [7]. Phase-field modelling (PFM) [16-18] was employed to quantify a representative value for each interface and a converged value is determined for each interface as an effective interface energy, since morphologies of grains are determined only by free energy defined after influence of artificial initial morphologies vanished in the course of grain growth-like simulations by PFM [7]. For L/δ and L/γ interface energies, it makes little sense to define misorientation in this context, and thus interface between a crystal slab oriented as in the aforementioned δ/γ interface wetted with liquid region is defined as L/δ or L/γ interface and the effective interface energies of L/δ and L/γ interfaces were determined by statistical averaging over up to results obtained using six initial configurations solely from atomistic simulations.

To examine the influence of BCC-to-FCC lattice strain on the nucleation, a simple model is assumed so that it can be quantitatively compared with the results in preceding sections on the influence of interface energy. A homogeneous nucleation is assumed and it is assumed that a nucleus is strained without any strain in the matrix side. Bulk modulus was calculated using the same EAM potential and the strain energy term is added to the free energy change of nucleation.

3. Results and Discussion

3.1. Interface Energy

In NW orientation relationship, (111) plane of δ phase is parallel to (011) plane of γ plane. Introducing periodic lattice ledges and misfit dislocations introduced at the δ/γ interface following prescription proposed by Rigsbee and Aaronson resulted in inclination of the interface while it significantly reduce bond distortion as shown in Fig. 1(a). Bond distortion near the interface is confined to a several atomic layer at most even for the boundary of those dissimilar lattices. In spite of the good lattice matching at the d/g interface, its interface energies are found to be as high as 0.41 J/m^2 even when misorientation between two crystal slabs is absent. Table 1 summarizes the minimum interface energies for hetero-phase boundaries and homo-phase boundaries [6,7]. Needless to say, the minimum interface energies for homo interfaces are zero since there is actually no lattice discontinuity at the interfaces. It should be emphasized that the δ/γ interface energies is not significantly greater than those of liquid/solid interfaces, L/δ and L/γ. Thus, care must be taken upon discussing the difficulty of γ nucleation compared with those in solidification.

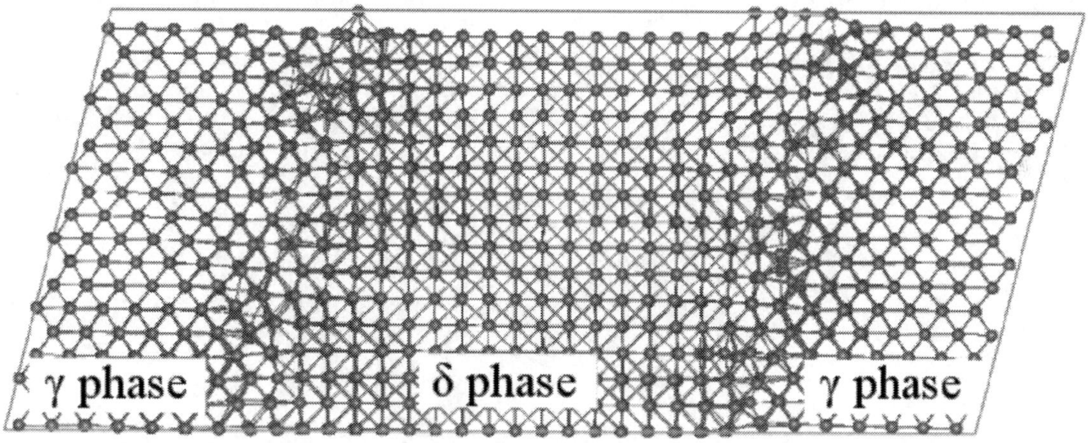

Figure 1. Optimized atomic configuration in a supercell containing two δ/γ interfaces projected to a two-dimensional plane. Ledges are on the projected plane and misfit dislocations are introduced parallel to the projected plane in the figure.

Effective interface energies computed with an aid of phase-field modelling taking into account the geometric constraints that grains or interfaces experience in practice during grain growth are shown together in the table. For the competition among grains or interfaces, volume free energy change is excluded from the total free energy used in the phase-field modelling and the competition is driven only by interface energies. To our surprise, the effective δ/γ interface energy does not increase as much as those of homo-phase interfaces, δ/δ and γ/γ, do. Close monitoring of temporal evolution of the interface energies under competition among grains or interface, it is found that the increase of the δ/γ interface energy is suppressed at the sacrifice of δ/δ and γ/γ interfaces which have much lower interface energies thereby capable of decreasing overall total free energy of a system. Slightly higher effective δ/δ interface energy than those for γ/γ interface can be attributed to less symmetrical crystal structure of BCC than FCC.

Table 1. Minimum and effective interface energies for various interfaces [6,7]. (*) Note that L/δ and L/γ interface energies were obtained solely by atomistic simulations with several initial configurations.

Interface	Hetero-Phase			Homo-Phase	
	δ/γ	L/δ	L/γ	δ/δ	γ/γ
Minimum [J/m^2]	0.41	0.29 ± 0.03	0.34 ± 0.03	0.0	0.0
Effective [J/m^2]	0.56 ± 0.03	0.29 ± 0.03$^{(*)}$	0.34 ± 0.03$^{(*)}$	0.44 ± 0.01	0.37 ± 0.01
Difference [J/m^2]	0.15	-	-	0.44	0.37

3.2. Impact of Interface Energy on Nucleation and Dendrite Tip Detachment

Based on the effective energies, the difficulties of γ nucleation from δ matrix is quantitatively examined using a classical nucleation theory with a simplified geometric model shown in Fig. 2. In the model, nucleation of any phase takes place on an interface between phase-A and phase-B, and when phase-A and phase-B are identical to each other and there is no misorientation between the two phases, the model is reduced to a homogeneous nucleation model with a nucleus of a perfect sphere. It is assumed that contact angles of a nucleus on either side of the interface is determined only by interface energies between two phases in the matrix and a nucleating phase as if they were in equilibrium. To account for driving force of nucleation, it is assumed that entropy change upon phase transformation is independent of temperature, using values at the peritectic temperature obtained directly from a thermodynamic databook or interpolation or extrapolation of values from the databook [19,7].

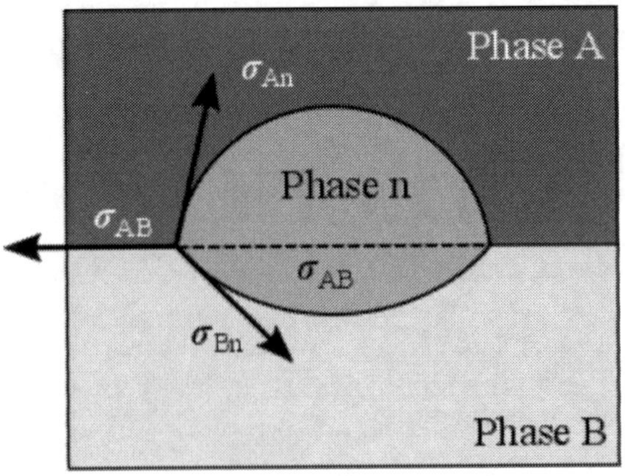

Figure 2. Nucleation model used. Dihedral angle of a nucleaus and contact angle with matrix phases are determined by interface energies between phase i and j, σ_{ij}.

Through numerical analysis [6,7], it is found, in the case of homogeneous nucleation, that critical radius of γ nucleation in δ matrix at a given undercooling is more than one orders of magnitude larger than that for solidification to δ or γ phase. At a given critical radius, more than one orders of magnitude greater undercooling is needed for the γ nucleation than in the δ or γ solidification.

Furthermore, the energy barrier at respective critical radius differs by about three orders of magnitude, suggesting that δ-γ transformation is prohibitive. As shown in Table 1, effective δ/γ interface energy is not significantly higher than L/δ or L/γ interface energy, with about 0.2 J/m^2 difference. This can be attributed to the significant difference in entropy change upon phase transition: From liquid phase to δ phase or γ phase, it is -7.165 or -7.668 J/mol K at 1665 K, respectively, which is by more than one orders of magnitude greater than that for phase transformation from δ phase to γ phase at the same temperature, -0.503 J/mol K. According to a classical nucleation theory [20,21], critical radius r^* and energy barrier for nucleation ΔG^* are given by

$$r^*(\Delta T) = -2\sigma_{ij}/\Delta T \Delta S_{i\to j} \text{ and } \Delta G^*(\Delta T) = -16\pi\sigma_{ij}^3/3\Delta T^2 \Delta S_{i\to j}^2,$$

respectively, where ΔT is undercooling, σ_{ij} is interface energy between phases i and j, $\Delta S_{i\to j}$ is entropy change upon phase transformation from i to j. As evident in these equations, not only interface energy but also entropy change upon phase transformation determines critical radius, energy barrier for nucleation, and undercooling required for the nucleation. Thus, it is concluded that the difficulty of γ nucleation in δ matrix upon the δ-γ massive-like phase transformation is arisen from the relatively higher interface energy for the entropy change for the solid state phase transformation.

When γ nucleation takes place on a δ/δ grain boundary, a part of γ/γ grain boundary with in a nucleus vanishes upon nucleation and thus it would be advantageous for the γ nucleation upon the massive-like phase transformation. Using the effective δ/δ interface energy in Table 1, energy barrier for the γ nucleation is reduced by a factor of 0.44. Furthermore, it is found that the γ nucleation on δ/γ interface, i.e. the γ nucleation on an existing γ nucleus, is more energetically advantageous. This indicates concurrent γ nucleation after high undercooling is achieved to trigger initial γ nucleation [7]. Microstructure of the solidified δ phase can be changed by controlling experimental conditions for solidification such as cooling rate. Besides, taking full advantage of grain boundary segregation or micro-segregation between dendrite trunks or arms, δ/δ grain boundary energy can be modified, although it is almost impossible to modify entropy change upon phase transition unless crystal structure or chemical composition is drastically changed. Therefore, it is suggested that controlling δ/δ interface energy through modifying solidified microstructure, thereby changing nucleation rate, is one of the principal ways to control the δ-γ massive-like phase transformation.

According to careful *in situ* observation [22], when δ/γ interface climbs up dendrite trunks and temperature is held, dendrite tips are simultaneously detached at the δ/γ interface away from the bottom of dendrites. Liquid is present between the detached d-phase dendrite tips and remaining g-phase dendrite trunks. It is as if the reverse peritectic reaction, γ → δ + L, takes place. If the δ/γ interface is at its best lattice matching, this phenomenon cannot be explained in terms of interface energy since δ/γ interface energy is 0.41 J/m^2 at its best lattice matching while the sum of interface energies of L/δ and L/γ interfaces is 0.63 J/m^2. Even when effective δ/γ interface energy, 0.56 J/m^2, is used, it cannot be explained solely from interface energies. However, close look at misorientation dependence of the δ/γ interface energy, if linear dependence of the interface energy on misorientation is assumed, only small misorientation by 4.19 deg. enables the dendrite tip detachment even if other factors are neglected [6]. The misorientation of 4.19 deg. is actually smaller than rotational orientation difference between NW orientation relationship and that of yet another well-known orientation relationship, Kurjumov-Sachs (KS) orientation relationship, 5.26 deg. Therefore, unless δ/γ interface climbs up the dendrite trunks with nearly the best lattice matching, the detachment of dendrite tip can be explained solely in terms of interface energy. The detachment may promote refinement of solidified microstructure which, in turn, drastically changes resultant mechanical properties of the casted steel. Since it is likely that the misorientation between δ-phase dendrites tip and γ-phase dendrite trunks depends again on the solidification to δ phase, controlling solidification process may enable to further tailor the solidified microstructure.

3.3. Impact of Transformation Strain on Nucleation

In addition to interface energies and entropy change upon phase transformation, one more factor may govern the δ-γ massive-like phase transformation. That is the transformation strain from BCC to FCC lattices built up by the same number of the same atoms. Although exact value of the inevitable transformation strain changes with temperature due to thermal expansion, the temperature effect is excluded to enable direct comparison with other results in preceding subsections on the same ground. When the strain effect is taken into account, free energy change upon nucleation in the case of homogeneous nucleation is given by

$$\Delta G\left(\Delta T,C,r\right)=-\frac{4\pi r^3}{3}\Delta T\Delta S_{i\to j}+4\pi r^2\cdot\sigma_{ij}+\frac{4\pi r^3}{3}\cdot\frac{1}{2}C\varepsilon^2$$

where C is an elastic constant and ε is a strain. Third term on the right hand side of the equation is the strain term. Bulk moduli of both BCC δ phase and FCC γ phase were calculated using the same EAM interatomic interaction. It is assumed that the nucleating γ phase is uniformly strained without straining the δ matrix phase for simplicity. No further phenomena such as relaxation of strained γ nucleus by dislocations or creep deformation was taken into account for clarity and simplicity, which is plausible based on the experimental fact that relaxation of the strained γ nuclei follows the δ-γ massive-like phase transformation spending longer elapsed time than the δ-γ massive-like transformation itself according to *in situ* X-ray diffraction measurements [22]

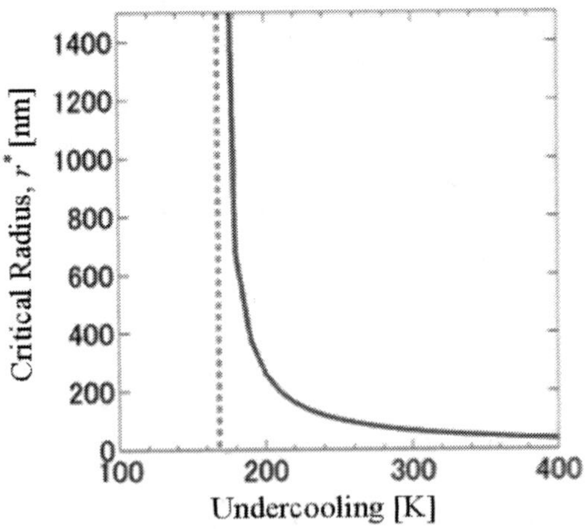

Figure 3. Critical Radius as a function of undercooling when the strain term is taken into account using bulk modulus as an elastic constant.

Figure 3 shows critical radius for homogeneous γ nucleation as in δ matrix a function of undercooling. One of the major differences from previous subsection where only effects of interface energy are taken into account is not that high undercooling is required for the nucleation but that the nucleation is prohibited due to the strain term up to about 170 K of undercooling. This can be easily understood from the equation of the free energy change: the strain term is also proportional to volume of a nucleus in addition to volume free energy change, the first term on the right hand side of the equation. Thus, simply increasing size of a nucleus does not lead to the negative ΔG unlike the interface term, but larger undercooling more than $C\varepsilon^2/2\Delta S_{i\to j}$ is required to realize the nucleation.

Although in reality an elastic constant such as Young's modulus decreases with increasing temperature, overall picture would not change. Thus, any means including aforementioned relaxation to decrease the strain is essential to give energetical advantage for the nucleation. In other words, if any microprocess that lowers the energy barrier for the nucleation due to the strain term is prescribed, then it would realize or promote the γ nucleation of the δ-γ massive-like phase transformation. Moreover, here lattice strain is assumed to be independent of the interface energy, which is not true in reality. For more quantitative assessment, strain energy needs to be determined in correlation with interface energy. Further details on the strain effect is addressed elsewhere [8].

4. Summary

By the combinational use of atomistic simulations and phase-field modelling, δ/γ interface energy was evaluated in comparison with those of other interfaces including homo-phase grain boundaries and hetero-phase interfaces including liquid/solid interfaces. It turned out that the minimum δ/γ interface energy, 0.41 J/m^2 is not significantly higher than those of L/δ and L/γ interfaces when two phases having dissimilar crystal structures are at the best lattice matching on the atomic level. Even under geometric constranits where grains or interfaces compete to minimize total free energy of a system, the δ/γ effective interface energy is, 0.56 J/m^2, not significantly higher than those of other interfaces. However, the δ/γ interface energy is very high for the very small entropy change upon the phase transition compared with those for solidification. This results that very high undercooling is needed for the δ/γ massive-like phase transformation. However, results indicate that once initial γ nucleation is succeeded at high undercooling, then concurrent γ nucleation on existing γ nuclei is promoted by its reduced energy barrier. Neither minimum nor effective δ/γ interface energy justifies detachment of dendrite tips leading to fragmentation and, in turn, promoted heterogeneous nucleation in liquid, but if very small magnitude of misorientation, smaller than orientation difference between KS- and NW-orientation relationships, is allowed, then the detachment by re-melting at the δ/γ interface can be explained from an energetical viewpoint. Finally, γ nucleation in δ-phase matrix without involving liquid phase is prohibited until more than 170 K of undercooling is achieved. Thus, any mechanism to relieve the transformation needs to be introduced to enable nucleation in realistic undercooling range.

Acknowledgments: The study was supported by Grant-in-Aid for Scientific Research (S) and Grant-in-Aid for Scientific Research on Priority Areas "Atomic Scale Modification" (No. 474) from the Ministry of Education, Culture, Sports, Science and Technology (MEXT), Japan.

References

[1] Yasuda H, Nagira T, Yoshiya M, Uesugi M, Nakatsuka N, Kiire M, Sugiyama A, Uesugi K, Umetani K 2011 *IOP Conf. Ser.: Mater. Sci. Eng.* **27** 012084. DOI: 10.1088/1757-899X/27/1/012084

[2] Yasuda H, Nagira T, Yoshiya M, Sugiyama A, Nakatsuka N, Kiire M, Uesugi M, Uesugi K, Umetani K, Kajikawa K 2012 *IOP Conf. Ser.: Mater. Sci. Eng.* **33** 012036

[3] Kiire M, *Master Dissertation*, Osaka University, Japan, 2011

[4] Yin H, Emi T, Shibata H 1998 *ISIJ Int.* **38** 794

[5] Van Vlack L H 1951 *Trans. AIME* **192** 251

[6] Yoshiya M, Nakajima K, Watanabe M, Nagira T, Yasuda H 2015 *Mater. Trans.*, under review

[7] Yoshiya M, Watanabe M, Nakajima K, Nagira T, Yasuda H 2015 *Mater. Trans.*, under review

[8] Sato M, Watanabe M, Ueshima N, Yokoi T, Yoshiya M, Nagira T, Yasuda H, in preparation

[9] Gale J D, Rohl A L 2003 *Mol. Simulat.* **29** 291

[10] Daw M S, Baskes M L 1983 *Phys. Rev. Lett.* **50** 1285

[11] Daw M S, Baskes M L 1984 *Phys. Rev. B* **29** 6443

[12] Dai X D, Kong Y, Li J H, Liu B X 2006 *J. Phys.: Condens. Matter* **18** 4527

[13] Nishiyama Z 1934 *Sci. Rept. Tohoku Univ.* **23** 368

[14] Wassermann G 1933 *Arch. Eisenhüttenwes.* **16** 647
[15] Rigsbee J M, Aaronson H I 1979 Acta Met. **27** 351
[16] Kobayashi R 1993 *Physica D* **63** 410
[17] Kim S G, Kim W T, Suzuki T 1999 *Phys. Rev. E* **50** 7186
[18] Steinbach I, Pezzolla F, Nestler B, Seeβelberg M, Ptieler R, Schmitz G J, Rezende J L L 1996 *Physica D* **94** 135
[19] "*NIST-JANAF Thermochemical Tables*", Fourth Edition, (The American Chemical Society and The Americn Institute of Physics for The National Institute of Standards and Technology, New York, USA, 1998), Part II, pp.1221-1226, and references therein
[20] Kurz W, Fisher D J "*Fundamentals of Solidification*", Forth Revised Edition, (Trans Tech Publications, Uetikon-Zuerich, Switzerland, 1984), Ch. 2.1.
[21] Danzig J A, Rappaz M "*Solidification*", (EPFL Press, Lausanne, Switzerland, 2009), Ch. 7.2.
[22] Yasuda H, Nagira T, Yoshiya M, Nakatsuka N, Ueshima N, Sugiyama A, Uesugi K, Umetani K, unpublised
[23] Kurjumov G V, Sachs G 1939 *Z. Phisik* **64** 325

Cellular Automaton Modeling of Dendritic Growth Using a Multi-grid Method

Y Natsume and K Ohsasa

Department of Materials Science and Engineering, Akita University, 1-1 Tegata Gakuen-chou, Akita 010-8502, Japan

E-mail: natsume@gipc.akita-u.ac.jp

Abstract. A two-dimensional cellular automaton model with a multi-grid method was developed to simulate dendritic growth. In the present model, we used a triple-grid system for temperature, solute concentration and solid fraction fields as a new approach of the multi-grid method. In order to evaluate the validity of the present model, we carried out simulations of single dendritic growth, secondary dendrite arm growth, multi-columnar dendritic growth and multi-equiaxed dendritic growth. From the results of the grid dependency from the simulation of single dendritic growth, we confirmed that the larger grid can be used in the simulation and that the computational time can be reduced dramatically. In the simulation of secondary dendrite arm growth, the results from the present model were in good agreement with the experimental data and the simulated results from a phase-field model. Thus, the present model can quantitatively simulate dendritic growth. From the simulated results of multi-columnar and multi-equiaxed dendrites, we confirmed that the present model can perform simulations under practical solidification conditions.

1. Introduction

Recently, with the advances in numerical simulation techniques and computers, numerical models have been developed for the simulation of dendritic growth during solidification. Models based on a cellular automaton (CA) method can quantitatively reproduce most solidification microstructures observed experimentally within an acceptable computational time. CA models have been developed for models of grain growth [1-3] and of dendritic growth [4-8]. However, in the model of dendritic growth, a smaller grid size needs to be used, which is usually less than 1 μm, because the smaller radius of curvature at the dendrite tip has to be simulated. Therefore, it is difficult to perform large scale simulations of dendritic growth due to high calculation cost. In order to overcome this difficulty, methods using adaptive mesh [9] and multi-grid methods [6] have been reported. The adaptive mesh technique is effective for reducing the computational time, but the programing is not simple. On the other hand, the programing for the multi-grid method is comparatively simple. Thus, in the present work, we adopted the multi-grid method as an approach for reducing the computational time and carried out two-dimensional CA simulations of dendritic growth. In the present model, different grid sizes were used for the calculations of temperature, solute concentration, and solid fraction fields; in other word, this is a triple-layered multi-grid method. In order to evaluate the validity of the present model, we performed simulations for comparison with the Lipton-Glicksman-Kurz (LGK) model [10], and experimental secondary dendrite arm spacing (SDAS) [11], and phase-field (PF) simulation [12].

IOP Conf. Series: Materials Science and Engineering **84** (2015) 012050 doi:10.1088/1757-899X/84/1/012050

Additionally, in order to investigate the capabilities of the model, the present model was used to simulate the formation of both multi-columnar and multi-equiaxed dendrites under practical solidification conditions.

2. Model

2.1. Temperature field

In the present model, heat conduction was not directly calculated because the calculation region is small. Instead of heat conduction, the calculation domain was cooled at a constant cooling rate R_C. The latent heat released during solidification was considered. The governing equation to calculate the temperature field is given as follows:

$$\frac{\partial T}{\partial t} = R_C + \frac{\Delta H}{\rho C_p}\frac{\partial f_S}{\partial t} \tag{1}$$

where T is temperature, t is time, ΔH is latent heat, C_p is specific heat and f_S is the solid fraction in each grid.

2.2. Concentration field

The governing equation for solute diffusion in solid and liquid phases is given as follows:

$$\frac{\partial c_i}{\partial t} = \nabla \cdot (D_i \nabla c_i) + c_i(1-k)\frac{\partial f_S}{\partial t} \tag{2}$$

where c_i is the solute concentration and D_i is the diffusion coefficient, and the subscript i indicates solid or liquid, respectively. For the solid/liquid interface, the solute concentration c_i, and the diffusion coefficient, D_i are defined as $c_i = c_L$ and $D_i = D_L$ between interface and liquid, and $D_i = D_S$ between interface and solid. And the solute distribution between solid and liquid is given by the following relationship:

$$c_S = kc_L \tag{3}$$

where k is the equilibrium partition coefficient, and c_S and c_L are the solute concentrations of solid and liquid phases, respectively.

2.3. Solid fraction

The driving force for dendritic growth is directly related to the difference between the equilibrium solute concentration of liquid, c_L^*, and the actual solute concentration of liquid, c_L, at the solid/liquid interface. In the local equilibrium condition, the equilibrium solute concentration of liquid is given by:

$$c_L^* = c_0 + \frac{T^* - T_L}{m_L} + \frac{\Gamma K[1 - 15\varepsilon \cos 4(\theta - \theta_0)]}{m_L} \tag{4}$$

where c_0 is the initial solute concentration, T^* is the equilibrium temperature at the interface, T_L is the equilibrium liquidus temperature, m_L is the slope of liquidus, Γ is the Gibbs-Thomson coefficient, K is the interface curvature, ε is the anisotropy parameter, θ is the growth angle between the normal to the interface and the x-axis and θ_0 is the angle between the preferential growth direction and the x-axis. The interface curvature is calculated by the following equation:

$$K = \left[2\frac{\partial f_S}{\partial x}\frac{\partial f_S}{\partial y}\frac{\partial^2 f_S}{\partial x \partial y} - \left(\frac{\partial f_S}{\partial x}\right)^2 \frac{\partial^2 f_S}{\partial y^2} - \left(\frac{\partial f_S}{\partial y}\right)^2 \frac{\partial^2 f_S}{\partial x^2}\right] \Bigg/ \left[\left(\frac{\partial f_S}{\partial x}\right)^2 + \left(\frac{\partial f_S}{\partial y}\right)^2\right]^{\frac{3}{2}} \tag{5}$$

The equilibrium solute concentration of liquid is compared with the actual solute concentration of liquid that is calculated by Eq. (2). The increase (or decrease) in solid fraction of an interface cell can be given as follows:

$$\Delta f_S = \frac{c_L^* - c_L}{c_L^*(1-k)} \quad (6)$$

When the sum of the solid fraction in an interface cell becomes one, that is $f_S = 1$, this interface cell changes its state to solid. When $f_S = 0$, this interface cell changes its state to liquid. The capture rule of the interface cell used in the present model was based on the CA approach. When an interface cell has been fully solidified (melted), its neighboring liquid (solid) cells are captured as interface cells.

2.4. Multi-grid method

Generally, the grid size of the used solid fraction field has been the same as that of the solute concentration field. Thus, the multi-grid method of dual-layer for temperature and concentration fields has been adopted. As a new approach, we used a different size grid between the solute concentration and the solid fraction fields in the present model; in other word, this is a triple-grid type CA model. Figure 1 shows a schematic illustration of the multi-grid system used in the present model. The temperature field is divided into the largest grid (Layer-1), the solute concentration field is divided into the second largest grid (Layer-2), and the solid fraction field is divided into the smallest grid (Layer-3). The grid sizes of Layer-1, Layer-2, and Layer-3 are defined as Δx_{L1}, Δx_{L2}, and Δx_{L3}, respectively. The relationships among Δx_{L1}, Δx_{L2}, and Δx_{L3} are $\Delta x_{L1} = n_{12}\Delta x_{L2}$ and $\Delta x_{L2} = n_{23}\Delta x_{L3}$, that is, number of cells in Layer-2 in Layer-1 is $(n_{12})^2$ and number of cells in Layer-3 in Layer-2 is $(n_{23})^2$.

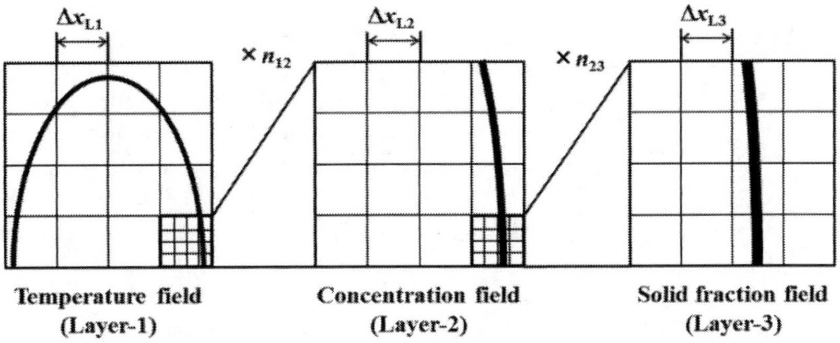

Figure 1. Multi-grid method used in the present model.

The solid fraction field is directly related to the morphology of the solid/liquid interface because it is used in the calculation of interface curvature. Thus, the small grid size is preferred. On the other hand, the grid size of temperature and solute concentration fields is expected to be large because a larger time step is required for the simulations of the dendritic growth under the practical solidification conditions. Using the triple-grid approach, the smaller radius of curvature at the dendrite tip could be accurately simulated by calculating the grid size of Layer-3, and the larger time step obtained from the grid size of Layer-1 and Layer-2 can be used in the simulation. In order to reduce error, the temperature value of each grid of Layer-2 divided into a grid of Layer-1 is given by a linear approximation of the neighboring grids of Layer-1. The relationship between Layer-2 and Layer-3 is also the same.

3. Results and discussion

3.1. Grid dependency and comparison with LGK model

In order to evaluate the grid dependency of the present model, we carried out the simulations for a single dendrite of Fe- 0.3wt.%C alloy growing at constant melt undercooling, ΔT, with various grid sizes. Table 1 shows the material properties of the Fe-C alloy used in the simulations. The calculation domain was 0.5 x 1 mm, and the initial solid was set on the center of the left wall of the calculation domain. In this calculation, the released latent heat was not considered.

Table 1. Materials properties of Fe-C alloy.

Equilibrium partition coefficient, k	**0.17**
Liquidus slope, m_L (K/wt%C)	**-84.9**
Diffusion coefficient in solid, D_S (m^2/s)	**6×10^{-9}**
Diffusion coefficient in liquid, D_L (m^2/s)	**2×10^{-8}**
Gibbs-Thomson coefficient, Γ (mK)	**1.9×10^{-7}**
Density, ρ (kg/m^3)	**7.0×10^3**
Specific heat, C_p (J/kg/K)	**795.0**
Latent heat, ΔH (J/m^3)	**1.93×10^9**
Melting point of pure Fe, T_m (K)	**1811**
Peritectic temperature, T_p (K)	**1766**

Figure 2 shows the change in the tip growth velocity for various grid sizes of Layer-2 with values of n_{23} ranging from one to five at ΔT = 2 K. In the case of n_{23} = 1 ($\Delta x_{L2} = \Delta x_{L3}$), the tip growth velocity rapidly decreases when Δx_{L2} becomes larger than 1.0 μm. On the other hand, for n_{23} = 2 and n_{23} = 3 ($\Delta x_{L2} > \Delta x_{L3}$), the tip growth velocity nearly maintains the converged velocity (52 μm/s) until the value of Δx_{L2} reached approximately 2.0 μm, and gradually decreases after that. For n_{23} = 5, the tip growth velocity obtained for all of Δx_{L2} is significantly smaller than the converged velocity. Figure 3 shows the simulated dendrite morphologies for the various Δx_{L2} and n_{23} at a growth time of 2.0 s. In case of n_{23} = 1, although the necking of the root of the primary arm was observed when Δx_{L2} was small, it was not observed as Δx_{L2} increased. This difference of growth morphology is related to the change in the tip growth velocity. In case of n_{23} = 2, 3, and 4, necking was observed except for the case where Δx_{L2} = 4.0 μm and n_{23} = 4. The dendrite morphologies were in good agreement with the dendrite morphology for the smallest Δx_{L2}. In case of n_{23} = 5, the size of dendrites was smaller than that for the other values of n_{23}.

Next, we compared the present model and the LGK model. In the LGK model, the thermal diffusion due to the latent heat release has been considered. But, in the present model, there is no thermal diffusion term in Eq. (1). In the condition of a lower melt undercooling, the melt undercooling far away from the dendrite tip is kept almost constant [6]; the thermal gradient is very small. Thus, we assumed that the thermal diffusion could be negligible. Figure 4 shows the comparison of tip growth velocity and tip radius of curvature predicted by the LGK model and the present model for various amounts of melt undercooling. The simulated tip growth velocities were lower than those predicted by the LGK model. These changes in tip growth velocity over undercoolings predicted by the two models show the same trend. The agreement of the simulated tip radius of curvature appears to be very good. Therefore, we confirmed that the simulation of dendritic growth can be quantitatively performed by the present model. It is especially effective to choose the values of parameter n_{23} from 2 to 4.

IOP Conf. Series: Materials Science and Engineering **84** (2015) 012050 doi:10.1088/1757-899X/84/1/012050

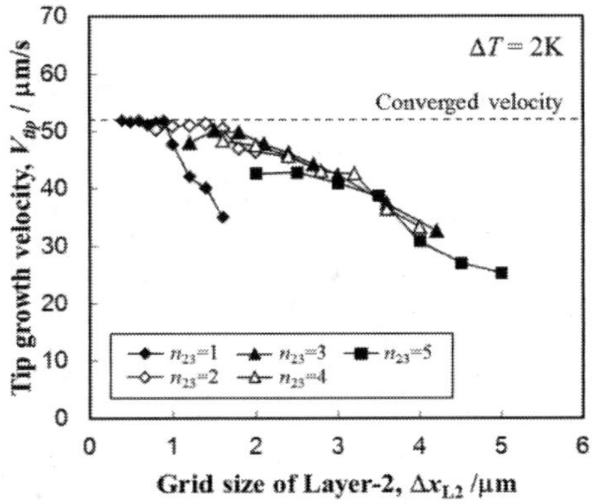

Figure 2. Changes in tip velocity with various values of Δx_{L2}.

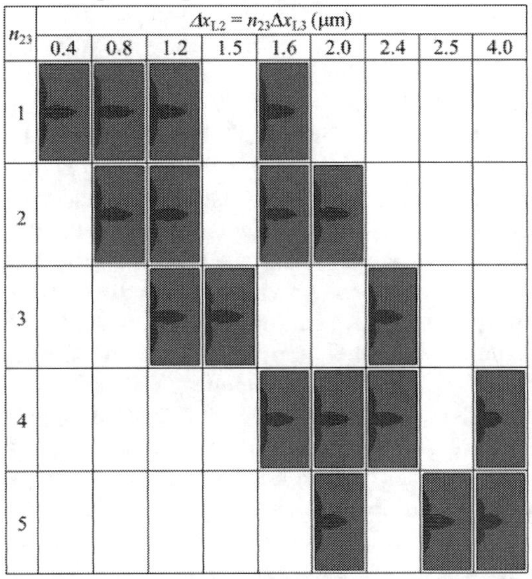

Figure 3. Simulated dendrite morphologies for various values of Δx_{L2} and n_{23}.

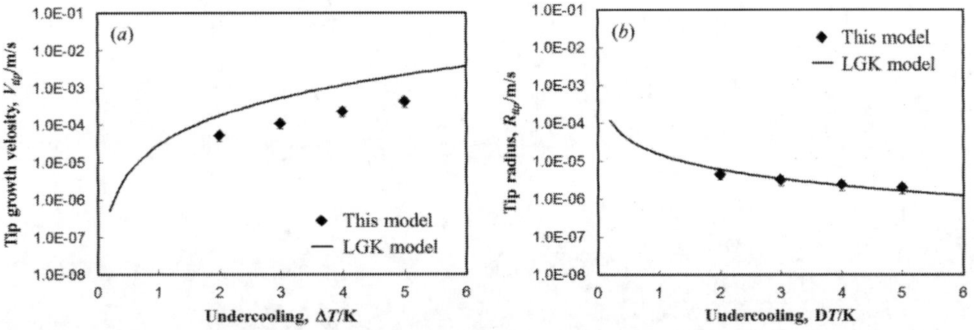

Figure 4. Comparison of (*a*) tip growth velocity and (*b*) tip radius with melt undercoolings predicted by the present model and LGK model.

In order to evaluate the reduction of computational cost, we examined the computational time required to perform the simulations for n_{23} = 1, 2, and 3 under a condition of Δx_{L3} = 0.5 μm. Because the calculation domain of 0.5 x 1.0 mm is used for the whole of conditions, in the case of n_{23} = 1, the number of total grids is the largest and the time step is the smallest. As a result, the computational time of the simulation for n_{23} = 2 and 3 were 1/11 and 1/50, respectively, of that for n_{23} = 1. Therefore, the triple-grid method developed in the present model is very effective in reducing the computational cost.

3.2. Comparison with experimental SDAS and PF simulation

For the quantitative evaluation of the present model, we predicted the SDAS for Fe- 0.1wt.%C, Fe- 0.3wt.%C and Fe- 0.4wt.%C alloys using the present model, and compared our results with the experimental data by Okamoto *et al.* [11] and with the PF simulation by Ode *et al.* [12]. The calculation domain was 2.01 x 0.255 mm, and grid sizes of Δx_{L1} = 15 μm, Δx_{L2} = 1.5 μm, and Δx_{L3} = 0.5 μm were used. Initial planar solid was set on the bottom of the calculation domain, which corresponds to the primary dendrite arm. In this simulation, the released latent heat was also considered. Constant cooling rates from 2 K/s to 50 K/s were used. The values of the SDAS were estimated at the end of solidification before the peritectic temperature (1766 K). Figure 5(a) shows the relationship between the SDAS and the cooling rate in the simulations. For comparison, the experimental data of Fe- 0.41wt.%C- 6.71wt.%Cr and Fe- 0.24wt.%C- 14.2wt.%Cr alloys and the results of PF simulation of Fe- 0.1wt.%C alloy are also shown in Fig. 5(a). We compared our results with the experimental data for Fe-C-Cr ternary alloys because it was difficult to obtain the experimental data for Fe-C binary alloys. First, we compared it with the experimental data of Fe- 0.41wt.%C- 6.71 wt.%Cr alloy; the simulated results for Fe- 0.4wt.%C alloy were in good agreement with the experimental data. The carbon content of these alloys is very similar. Next, we compared it with the experimental data of Fe- 0.24wt.%C- 14.2 wt.%Cr alloy; the experimental values of the SDAS were smaller than those of every simulation. This discrepancy is presumably due to the fact that the Cr content added to an alloy has affected the SDAS of the experimental alloy, and this result may be expected that the present model can quantitatively predict the SDAS for a multi-component alloy. We will investigate the validity of a model for multi-component alloys with the present multi-grid approach in near future. In the comparison with the PF simulation for the same carbon content, that is, Fe- 0.1wt.%C, the simulated results obtained from the present model were in fairly good agreement with those predicted by the PF model. Figure 5(b) shows the relationship between the SDAS and the local solidification time obtained from the simulations of the present model and the PF simulation for Fe-C binary alloys. The simulated results obtained from both models show a good agreement. From the simulated results obtained from the present model, we estimated the exponent of the relationship between the SDAS and the local solidification time. The value of the exponent was 0.45, which corresponds to the slope of line shown in Fig. 5(b). This value of 0.45 is larger than the value of 1/3 predicted analytically [13]. However, it has been reported that the experimental values of the exponent of some Fe-base alloys or steels can be larger than 1/3 [14, 15].

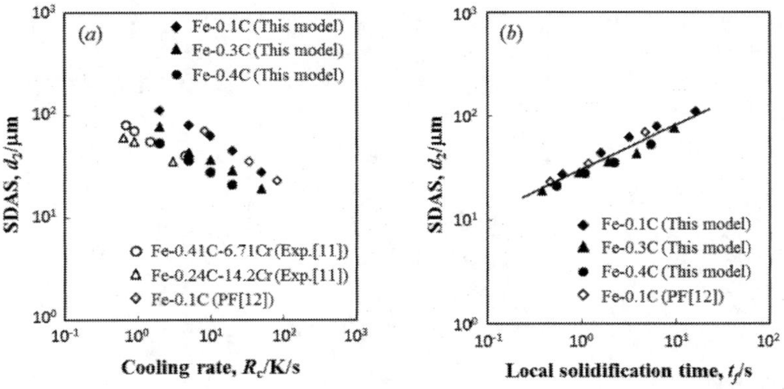

Figure 5. Relationship between SDAS and (*a*) cooling rate, (*b*) local solidification time for Fe-C alloys.

3.3. Large-scale simulations of columnar and equiaxed multi-dendritic growth

The present model was used to simulate the formation of both multi-columnar and multi-equiaxed dendrites in Fe- 0.15wt%C and Fe- 0.3wt%C alloys under practical solidification conditions.

The simulations of multi-columnar dendrites were performed within a calculation domain of 3.0 x 2.01 mm and the grid sizes were $\Delta x_{L1} = 15$ μm, $\Delta x_{L2} = 1.5$ μm, and $\Delta x_{L3} = 0.5$ μm. The five initial solids were placed on the left wall of the calculation domain at equal intervals. The temperature gradient, *G*, was 5 K/mm and the cooling rates were 1 K/s and 2 K/s. The temperature of a grid of Layer-1 reached the peritectic temperature and was maintained at that temperature because the peritectic reaction is not considered in the present model. The simulations were stopped when the temperature of the whole domain reached the peritectic temperature. Figure 6 shows the simulated solute concentration distribution of columnar dendrites for Fe -0.15wt.%C and Fe -0.3wt.%C alloys. Comparing the simulated dendrites growing at cooling rate of 1 K/s (Fig. 6(a1) to (a3)) with those at 2 K/s (Fig. 6(b1) to (b3)) for Fe- 0.15wt.%C alloy, the growth of a large number of secondary arms can be observed in the columnar dendrites growing at a higher cooling rate of 2K/s. Comparing the growth of columnar dendrites of Fe- 0.15wt.%C alloy with those of Fe- 0.3wt.%C alloy at a cooling rate of 1 K/s (ex. Fig. 6(a3) and (c3)), it is obvious that the simulated dendrite morphologies reflect the influence of the initial composition.

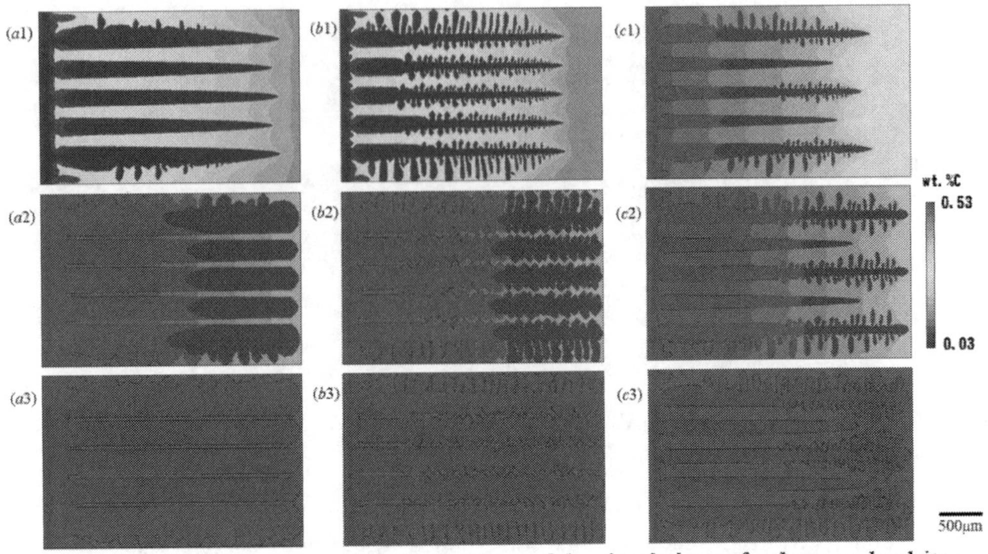

Figure 6. Simulated solute concentration distributions of the simulations of columnar dendrites.
(*a*1)-(*a*3) Fe- 0.15wt.%C, G = 5 K/mm, R_C = 1 K/s: Elapsed time = (*a*1) 16 s, (*a*2) 36 s, and (*a*3) 46 s.
(*b*1)-(*b*3) Fe- 0.15wt.%C, G = 5 K/mm, R_C = 2 K/s: Elapsed time = (*b*1) 8.5 s, (*b*2) 18 s, and (*b*3) 23 s.
(*c*1)-(*c*3) Fe- 0.30wt.%C, G = 5 K/mm, R_C = 1 K/s: Elapsed time = (*c*1) 18 s, (*c*2) 22 s, and (*c*3) 34 s.

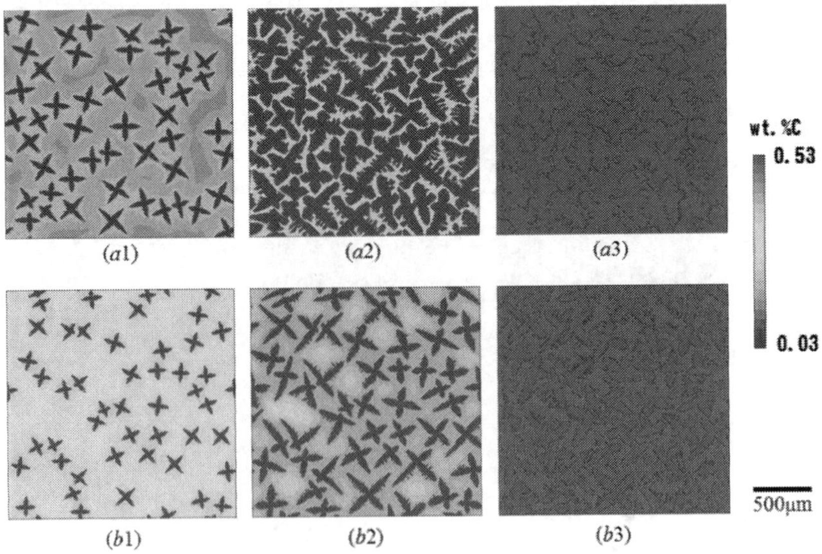

Figure 7. Simulated solute concentration distributions of the simulations of equiaxed dendrites.
(*a*1)-(*a*3) Fe- 0.15wt.%C, R_C = 10 K/s: Elapsed time = (*a*1) 0.37 s, (*a*2) 1.72 s, and (*a*3) 3.13 s.
(*b*1)-(*b*3) Fe- 0.30wt.%C, R_C = 10 K/s: Elapsed time = (*b*1) 0.47 s, (*b*2) 0.92 s, and (*b*3) 1.86 s.

The simulations of multi-equiaxed dendrites were performed in a calculation domain of 2.01 x 2.01 mm and the grid sizes were Δx_{L1} = 15 μm, Δx_{L2} = 1.5 μm, and Δx_{L3} = 0.5 μm. The fifty initial solids having random preferred growth orientations were randomly and simultaneously distributed in the calculation domain. The preferred growth orientations were given in the range from -45 to 45 degrees with respect to the horizontal direction. The domain that was assumed to have uniform temperature was initialized at the liquidus temperature and alloy composition, and was cooled at a constant cooling rate of 10 K/s. The simulations were stopped when the temperature of the entire domain reached the peritectic temperature. Figure 7 shows the simulated solute concentration distribution of equiaxed dendrites for Fe- 0.15wt.%C and Fe- 0.3wt.%C alloys. In the early stage of solidification, the primary dendrite arms developed along their crystallographic orientations. As solidification proceeds, the growth and coarsening of secondary dendrite arms occur. The influence of the initial composition is seen as the simulation of multi-columnar dendrites.

These simulations were carried out to investigate the computational performance of the present model; the simulations of multi-columnar dendrites and multi-equiaxed dendrites required a computational time of about one week and two days on a PC Core i7 with 1-CPU 3.4GHz of 6 cores using Open MP parallelization. However, if an approach like a multi-grid method is not used, huge amounts of computational time would be required to perform these large-scale simulations under practical solidification conditions.

4. Conclusions

A 2-D CA model with a multi-grid method was developed to simulate a dendritic growth. In the present model, we used a triple-grid for temperature, solute concentration and solid fraction fields as a new approach of the multi-grid method. In order to evaluate the present model, we carried out simulations of single dendritic growth, secondary arm growth, multi-columnar dendritic growth and multi-equiaxed dendritic growth. From the simulation results of the present model using the triple-grid, we confirmed that the present model can perform a quantitative simulation of dendritic growth at dramatically reduced computational cost. These simulated results have shown that a practical simulation of dendritic growth in a short computational time can be performed by the introduction of the multi-grid method using triple-grid.

References

[1] Gandin C A and Rappaz M 1994 *Acta Metall. Mater.* **42** 2233
[2] Gandin C A and Rappaz M 1996 *Acta Metall. Mater.* **44** 3339
[3] Gandin C A and Rappaz M 1997 *Acta Metall. Mater.* **45** 2187
[4] Sanchez L B and Stefanescu D M 2003 *Metall. Mater. Trans.* A **34** 367
[5] Sanchez L B and Stefanescu D M 2004 *Metall. Mater. Trans.* A **35** 2471
[6] Zhu M F and Stefanescu D M 2007 *Acta Mater.* **55** 1741
[7] Pan S and Zhu M F 2010 *Acta Mater.* **58** 340
[8] Wang W, Lee P D and McLean M 2003 *Acta Mater.* **51** 2971
[9] Wei L, Lin X, Wang M and Huang W 2012 *Physica B* **407** 2471
[10] Lipton J, Glicksman M E and Kurz W 1984 *Mater. Sci. Eng.* **65** 57
[11] Okamoto T, Matsuo S and Kishitake K 1977 *Tetsu-to-Hagane* **63** 936
[12] Ode M, Kim S G, Kim W T and Suzuki T 2001 *ISIJ Int.* **41** 345
[13] Kurz W and Fisher D J 1998 *Fundamentals of Solidification* chapter 4 pp 85-89
[14] Suzuki A, Suzuki T, Nagaoka Y and Iwata Y 1968 *J. Jpn. Inst. Met.* **32** 1301
[15] Suzuki A and Nagaoka Y 1969 *J. Jpn. Inst. Met.* **33** 658

A benchmark for the validation of solidification modelling algorithms

E Kaschnitz[1], S Heugenhauser[1] and P Schumacher[2]

[1] Österreichisches Gießerei-Institut, Parkstraße 21, 8700 Leoben, Austria
[2] Lehrstuhl für Gießereikunde, Montanuniversität Leoben, Franz-Josef-Straße 18, 8700 Leoben, Austria

E-mail: erhard.kaschnitz@ogi.at

Abstract. This work presents two three-dimensional solidification models, which were solved by several commercial solvers (MAGMASOFT, FLOW-3D, ProCAST, WinCast, ANSYS, and OpenFOAM). Surprisingly, the results show noticeable differences. The results are analyzed similar to a round-robin test procedure to obtain reference values for temperatures and their uncertainties at selected positions in the model. The first model is similar to an adiabatic calorimeter with an aluminum alloy solidifying in a copper block. For this model, an analytical solution for the overall temperature at steady state can be calculated. The second model implements additional heat transfer boundary conditions at outer faces. The geometry of the models, the initial and boundary conditions as well as the material properties are kept as simple as possible but, nevertheless, close to a realistic solidification situation. The gained temperature results can be used to validate self-written solidification solvers and check the accuracy of commercial solidification programs.

1. Introduction

During a recent research project, significant differences in the results of numerical simulations using different commercial simulation programs were noticed by the authors. Especially the release of heat at solidification of an alloy varied considerably with the selected programs. This fact led to a literature search for a benchmark model configuration describing a simple solidification problem. Whereas different benchmark studies are found for relatively complex simulations of fluid flow (e.g. [1-4]), little is published for the simple diffusive transport of heat including the change of state. While analytical solutions are available for one-dimensional semi-infinite configurations [5], no numerical solutions for a heat conduction problem with release of heat of fusion have been found.

Two simple models, close to a realistic solidification situation, were set up to check several commercial solvers similar to a round-robin test with regard to their accuracy to compute the temperature distribution. The first model is a solidifying aluminum cube in an isolated cube-shaped copper mold; this configuration is in principle an adiabatic calorimeter. This has the advantage that the final common temperature of the aluminum and copper after an infinite time can be analytically calculated. The second model is the same geometrical configuration, but with heat transfer at two of the outer walls in order to check the influence of boundary conditions.

The models were computed with the following software programs: MAGMASOFT version 5.2 [6], FLOW-3D version 10.1 [7], ProCAST version 2013.0 [8], WinCast version [9], ANSYS Workbench version 14.5 [10], and OpenFOAM (in a customized version) [11]. For the following, the software

IOP Conf. Series: Materials Science and Engineering **84** (2015) 012051 doi:10.1088/1757-899X/84/1/012051

programs are – according to the rules of a round-robin test – anonymized and given arbitrary numbers from 1 to 6.

Based on the results of the inter-comparison, the support of the software manufacturer of these two programs with noticeably deviating results was contacted. One of the programs was updated immediately (within some days) for a software bug; the second program needed a much tighter convergence for the release of heat of fusion compared to the standard adjustment, resulting in smaller time steps and longer computation time.

2. Models

Figure 1 shows a sketch of the geometry of the two models consisting of a mold with an overall dimension of 50 mm x 50 mm x 50 mm; a second cube of melt measuring 20 mm x 20 mm x 20 mm is placed in the center. At zero time the mold has a temperature of T_{mold}=200°C and the melt T_{melt}=700°C. The heat transfer coefficient at the faces between the two materials is constant at a value of $HTC_{internal}$=100 W m^{-2} K^{-1}. Three measuring points for temperature (P1, P2, and P3), marked as green dots, are defined in order to compare the simulation results among the different software programs. A cooling time of 1200 s is specified to ensure a temperature very close to equilibrium for both metals at the end of the simulation.

The first model has adiabatic boundary conditions at all outer walls. In this case, the equilibrium temperature of the melt and mold after an infinite time can be analytically calculated. The temperature measurement point P1 is in the center of the aluminum melt with the coordinates x=25 mm, y=25 mm, z=25 mm and the point P2 is located in a corner of the copper mold with the coordinates x=48,5 mm, y=48,5 mm, z=1,5 mm.

The second model has heat transfer boundary conditions at two opposite outer walls (x_{min} and x_{max} in fig. 1). A heat transfer coefficient of 2000 W m^{-2} K^{-1} to the surrounding ($HTC_{external}$), which is kept at 20°C, is defined, while the other four walls are kept adiabatic. One temperature measurement point is again P1 in the center, the second point P3 is on the horizontal line of symmetry at the coordinates x=42,5 mm, y=25 mm, z=25 mm.

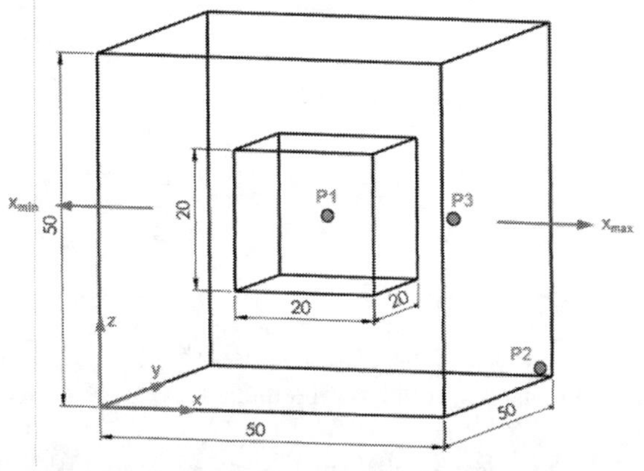

Figure 1. Geometry of the simulation models.

Table 1 and 2 show the thermophysical properties of the materials used. Intentionally, the properties are close to copper and aluminum alloy, but simplified. The values of density ϱ, specific heat capacity cp and thermal conductivity λ are kept constant in the entire temperature range. The fraction solid is a linear function of temperature between liquidus and solidus temperature.

Table 1. Thermophysical properties of the mold material

Property	Value
Density ϱ_{mold}	8900 kg m^{-3}
Specific heat $c_{p,mold}$	380 J kg^{-1} K^{-1}
Thermal conductivity λ_{mold}	320 W m^{-2} K^{-1}

Table 2. Thermophysical properties of the melt material

Property	Value
Density ϱ_{melt}	2700 kg m^{-3}
Specific heat $c_{p,melt}$	1000 J kg^{-1} K^{-1}
Thermal conductivity λ_{melt}	150 W m^{-2} K^{-1}
Liquidus temperature $T_{liquidus}$	638°C
Solidus temperature $T_{solidus}$	520°C
Latent heat of fusion L_{fusion} (in addition to specific heat)	364 826 J kg^{-1}

3. Analytical solution for the adiabatic model

Comparing the enthalpies in the initial state (T_{mold}=200°C and T_{melt}=700°C) and in complete equilibrium state leads to an equilibrium temperature $T_{equilibrium}$:

$$T_{equilibrium} = \frac{V_{mold}\rho_{mold}c_{p,mold}T_{mold}+V_{melt}\rho_{melt}(c_{p,melt}T_{melt}+L_{fusion})}{V_{mold}\rho_{mold}c_{p,mold}+V_{melt}\rho_{melt}c_{p,melt}}, \qquad (1)$$

where V_{mold} is the volume of the mold, V_{melt} is the volume of the melt. The equilibrium temperature of the adiabatic model results in $T_{equilibrium}$ = 244.7652°C.

4. Numerical solutions

The energy conservation equation in combination with Fourier's law of heat transport under consideration of the release of heat leads to the equation for the change of temperature T with time t

$$\frac{\partial}{\partial t}(\rho c_p T) = \nabla(\lambda \nabla T) + \rho L_{fusion}\frac{f_{solid}(T)}{\partial T}\frac{\partial T}{\partial t}, \qquad (2)$$

where ρ is density, c_P is heat capacity, λ is thermal conductivity, L_{fusion} is heat of fusion, and f_{solid} is the fraction of solid. Some of the used programs implement the release of heat of fusion by a modified heat capacity function.

This equation was numerically approximated by all used software programs. To achieve a realistic accuracy of the results similar to a "typical" software user, default software parameters (e. g. convergence criterions) were used as far as possible for the calculations. In relation to the relative simple model, the meshes were extremely fine. The temperatures at the defined positions in the models are recorded as a function of time for the inter-comparison.

Figure 2 shows the temperatures at point P1 and point P2 as a function of time for the adiabatic model; fig. 3 shows temperatures at point P1 and P3 for the model with external heat transfer. These results have been obtained by one of the commercial solvers.

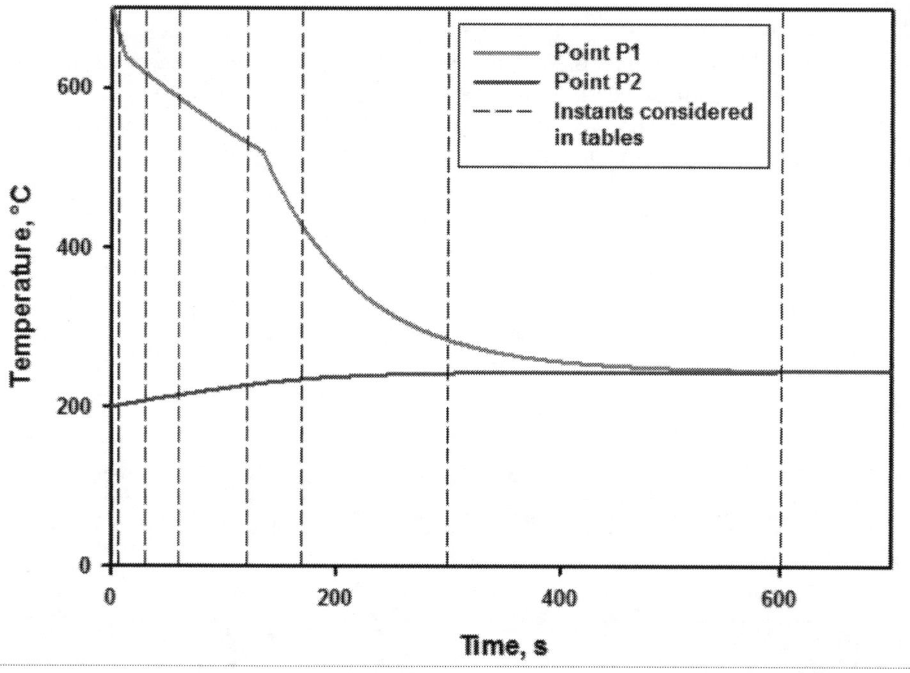

Figure 2. Temperature as a function of time at two measuring points of the adiabatic model.

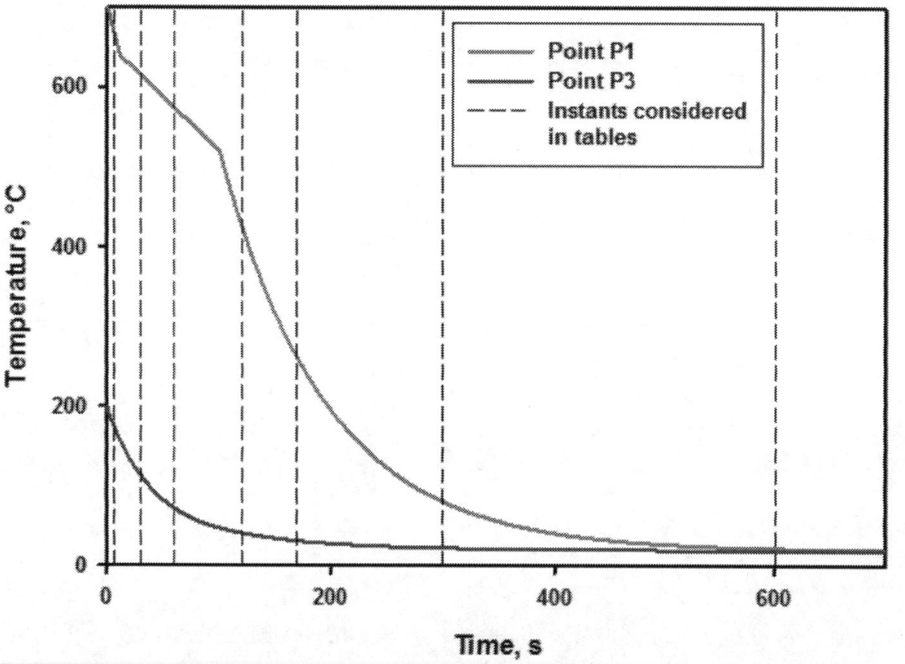

Figure 3. Temperature as a function of time at two measuring points of the model with external heat transfer.

5. Results and comparison

The temperature results as function of time at the "virtual thermocouple" points of the individual simulation results were fitted to be able to compare the results of the individual simulation programs at exactly the same instants of time. For the case of available temperature data close to the selected instant, the two neighboring points were fitted by a linear function. For wider gaps, the data in the liquid range, the solidification interval, and the solid range were fitted according to the equation:

$$T = \mathrm{a} + \mathrm{b}e^{-ct} \quad (3)$$

where a, b, and c are constants. The deviation of the fitted data from the simulated temperatures is very small (in the Milli-Kelvin range) except close to zero time and in the vicinity of the beginning and the ending of solidification, where the temperature traces show characteristic kinks. The regions of noticeable deviations were omitted. Tables 3 and 4 show the results of the simulated temperatures of the adiabatic model at selected times at point P1 and P2, as well as the arithmetic mean and the standard deviation.

Table 3. Adiabatic model: Temperatures in the center of the melt (point P1) at selected times

	Temperature results of simulation program number							
Time (s)	1 (°C)	2 (°C)	3 (°C)	4 (°C)	5 (°C)	6 (°C)	Arithmetic mean (°C)	Standard deviation (°C)
6	669.65	669.46	669.23	669.34	669.45	669.75	669.43	0.16
30	618.41	618.81	619.50	617.72	618.92	619.44	618.67	0.66
60	586.73	587.05	588.18	585.98	586.92	587.70	586.97	0.79
120	532.05	532.27	533.74	531.30	531.78	532.92	532.23	0.92
170	426.40	426.29	431.26	423.88	424.73	428.77	426.51	2.86
300	284.88	284.58	286.55	283.77	284.33	286.70	284.82	1.05
600	245.90	245.97	246.08	245.95	246.19	246.65	246.02	0.12
1200	244.67	244.77	244.73	244.91	245.02	245.04	244.82*	0.14

*Exact result: $T_{equilibrium}$ = 244.77°C

Table 4. Adiabatic model: Temperatures at the corner of the mold (point P2) at selected times

	Temperature results of simulation program number							
Time (s)	1 (°C)	2 (°C)	3 (°C)	4 (°C)	5 (°C)	6 (°C)	Arithmetic mean (°C)	Standard deviation (°C)
6	201.40	201.41	201.39	201.44	201.43	201.41	201.41	0.02
30	207.65	207.67	207.58	207.75	207.76	207.66	207.68	0.08
60	214.72	214.77	214.59	214.95	214.93	214.75	214.79	0.15
120	226.93	227.00	226.71	227.28	227.29	226.99	227.04	0.24
170	234.64	234.75	234.48	234.99	235.05	234.77	234.78	0.24
300	242.45	242.57	242.46	242.74	242.83	242.62	242.61	0.17
600	244.60	244.70	244.67	244.85	244.95	244.87	244.75	0.14
1200	244.67	244.77	244.72	244.90	245.02	245.05	244.81*	0.14

*Exact result: $T_{equilibrium}$ = 244.77°C

Tables 5 and 6 show the results of the simulated temperatures of the second model with external heat transfer at selected times at point P1 and P3, the arithmetic mean of the temperatures and the standard deviation.

Table 5. Model with external heat transfer: Temperatures at the center of the melt (point P1) at selected times

Time (s)	Temperature results of simulation program number 1 (°C)	2 (°C)	3 (°C)	4 (°C)	5 (°C)	6* (°C)	Arithmetic mean (°C)	Standard deviation (°C)
6	668.95	668.86	668.36	668.86	668.73	669.46	668.75	0.23
30	613.44	615.13	615.47	615.16	615.11	617.83	614.86	0.81
60	572.93	574.45	575.78	574.43	574.09	583.61	574.34	1.02
120	424.79	427.11	434.27	426.71	424.48	520.20	427.47	3.97
170	260.10	260.74	266.65	260.35	259.01	346.86	261.37	3.02
300	81.32	81.04	83.67	80.08	80.47	140.85	81.32	1.40
600	22.61	22.56	22.80	22.28	22.52	36.48	22.55	0.19

*Results excluded from mean value and standard deviation.

Table 6. Model with external heat transfer: Temperatures at the center of the melt (point P1) at selected times

Time (s)	Temperature results of simulation program number 1 (°C)	2 (°C)	3 (°C)	4 (°C)	5 (°C)	6* (°C)	Arithmetic mean (°C)	Standard deviation (°C)
6	175.37	175.08	175.25	174.97	175.22	189.45	175.18	0.16
30	113.26	112.60	112.66	111.94	112.99	166.90	112.69	0.49
60	72.25	71.64	71.61	71.26	72.04	143.92	71.76	0.39
120	41.30	41.06	41.06	40.94	41.25	110.40	41.12	0.15
170	31.47	31.39	31.46	31.34	31.45	89.76	31.42	0.06
300	22.69	22.68	22.77	22.62	22.68	53.32	22.69	0.05
600	20.11	20.11	20.13	20.10	20.11	25.67	20.11	0.01

*Results excluded from mean value and standard deviation.

Tables 7 shows the results of the simulated times ($t_{liqidus}$ and $t_{solidus}$) to reach the characteristic kinks at start and end of solidification ($T_{liqidus}$ and $T_{solidus}$) of both models at point P1.

Table 7. Time from start to the characteristic kinks at $T_{liqidus}$ and $T_{solidus}$ (see fig. 2 and 3) in the center of the melt (point P1)

	Time results of simulation program number 1 (s)	2 (s)	3 (s)	4 (s)	5 (s)	6 (s)	Arithmetic mean (s)	Standard deviation (s)
				Adiabatic model				
$t_{liqidus}$	12.35	12.21	12.33	12.68	12.21	12.00	12.30	0.21
$t_{solidus}$	134.12	134.01	135.88	133.02	133.56	135.26	134.31	0.98
				Model with external heat transfer				
$t_{liqidus}$	12.01	11.61	11.79	11.77	11.51	12.22*	11.74	0.17
$t_{solidus}$	99.84	100.11	101.57	100.26	99.72	120.00*	100.30	0.66

*Results excluded from mean value and standard deviation.

6. Conclusions

Two simple heat transfer models describing diffusive heat transport and solidification were set up to check several commercial software solvers similar to a round-robin test to check their accuracy to compute temperature distributions. According to the rules of a round-robin test, no judgement of the quality or accuracy of the included solvers is given. Five of six solvers gave similar results with differences in the range of up to several degrees Celsius. One of them had been bug-fixed, the results of the fixed version are included, the original results have been omitted. The sixth solver deviates quite significantly when applying external heat transfer, the reason for this was not unveiled by the software manufacturer.

Following conclusions can be drawn:

- Commercial solvers can have (rarely, but it is happening) software bugs influencing the results even of simple solidification models. This could be due to frequent version changes.
- Differences between commercial solvers are in the range of up to several degrees Celsius, especially after the end of solidification. This could be due to the numerical treatment of the release of the heat of fusion in correlation to the computing speed optimization.
- For the model with external heat transfer, one of the programs was removed from the calculation of arithmetic mean and standard deviation due to a significant discrepancy.
- The models can be employed to validate self-written solidification solvers and check the accuracy of commercial solidification programs.

Acknowledgements

This work was supported by the „Austrian Research Promotion Agency (FFG)“, Vienna under contract no. 838821. Contributions of A. Vakhrushev, Lehrstuhl für Simulation und Modellierung metallurgischer Prozesse, Montanuniversität Leoben, Austria, A. Harrison, LKR Leichtmetallkompetenzzentrum Ranshofen GmbH, Ranshofen, Austria, and M. Pintore, Lehrstuhl für Umformtechnik und Gießereiwesen, Technische Universität München, Germany are greatly appreciated.

References

[1] Le Bars M and Worster M G 2006 *J. Comput. Phys.* **216** 247-63
[2] Pardeshi R, Voller V R, Singh A K and Dutta P 2008 *Int. J. Heat Mass Tran.* **51** 3999-3409
[3] Wang X and Fautrelle Y 2009 *Int. J. Heat Mass Tran.* **52** 5624-33
[4] Wittig K and Nikrityuk P A 2011 *IOP Conf. Series: Mat. Sci. Eng.* **27** 012024
[5] Carslaw H S and Jaeger J C 1959 *Conduction of Heat in Solids* (Oxford: Clarendon Press) pp 282-96
[6] http://www.magmasoft.de/en/index.html
[7] http://www.flow3d.com
[8] https://www.esi-group.com/software-services/virtual-manufacturing/casting-simulation-suite
[9] http://www.rwp-simtec.de/eng_cn/homeEN.html
[10] http://www.ansys.com
[11] http://www.openfoam.org

FEM stress analysis of the cooling hole of an HPDC die

T Kimura[1], H Yamagata[2] and S Tanikawa[3]
[1] Graduate School of Engineering, Gifu University, 1-1 Yanagido, Gifu City, Japan
[2] Faculty of Engineering, Gifu University, 1-1 Yanagido, Gifu City, Japan
[3] Kotobuki Kinzoku Kogyo Co., Ltd., Kitayama Houkouji, Nishio City, Japan

E-mail: t3122010@edu.gifu-u.ac.jp

Abstract. Cracking at a cooling hole is a typical die failure mode in a high-pressure die-casting (HPDC) die. We simulated the thermal distortion of a die considering the HPDC machine deflection and revealed a stress concentration at the cooling hole. The stress concentration at the cooling hole changes after injection or after spraying and blowing air. The cooling hole top remains in a compression stress state 5, 10, and 20 mm deep from the die surface, but the stress amplitudes are higher when the depths are shallower. It was suggested that cracking takes place due to the high compressive stress and that the shear stress assists the propagation of the initiated crack. On the other hand, the stress condition at the R portion of the cooling hole is always a tensile state, but the mean stress and stress amplitude values were not found to be in the range that causes fatigue fracture. It was demonstrated that the developed analysis is valuable in designing the cooling hole of an HPDC die.

1. Introduction

The life of a die-casting die is shortened by typical cracking such as heat checking and gross cracking. It was reported [1] that half of the cracking takes place at the cooling holes or at the cavity corners. It is considered that the main reasons for the cracking are thermal stresses and the die material properties. However, since heating is inevitable in die-casting dies, it is not easy to suppress the generation of the thermal stresses. Therefore, it is important to predict the thermal stresses and to minimize them during the die design stage. Die temperature control is also important. In casting, too low of a die temperature causes a shortage of molten metal, and too high of a die temperature causes seizure. The die is heated by the molten metal and is cooled by the cooling spray as well as the cooling water provided. The cooling spray is easy to adjust at the production site after the installation of the completed die. On the other hand, since internal cooling through cooling holes is very efficient, a cooling hole is positioned in the high-temperature region of the die with the top of the hole positioned in the vicinity of the cavity surface. However, if the cooling hole is positioned too close to the die surface, the surface temperature becomes too low, and the low strength between the cooling hole top and the cavity surface is likely to cause cracking. In general, we can use two types of cooling circuits. One is spot cooling, which normally has the cooling hole positioned vertically in relation to the cavity surface, where the localized position of the cavity is preferentially cooled. The other is line cooling, which has the cooling hole parallel to the cavity surface, where a wide area of the cavity surface can be cooled. In this analysis, we investigate the spot cooling case.

As mentioned above, it is important to manage the die temperature and to prevent cracking near the cooling hole. The depth of the cooling hole greatly influences the heat extraction. However, there are few reports on the relationship between the depth of the cooling hole and cracking. Tamura et al. [2, 3] tried to predict the cracking from the cooling hole using computer-aided engineering (CAE). They observed the effect of the cooling hole depth on the thermal stresses, but it remained to be an introduction of the prediction method. In normal die design, the designer intentionally challenges and determines the limit depth based on his experience. However, this is not always successful.

Our recent CAE analyses combined with actual experimental observations revealed issues in high-pressure die casting (HPDC), such as the relationship between the HPDC machine distortion and the

IOP Conf. Series: Materials Science and Engineering **84** (2015) 012052 doi:10.1088/1757-899X/84/1/012052

die deflection [4], flashing caused by the thermal die deflection [5], warpage of the products [6], and heat checking at the die surface [7]. These studies observed that the modeling range and the restraint conditions influence the deflection. The modeling analysis including the entire HPDC machine and the injection system also simulated how plunger rod misalignment influences the product quality [8].

In this study, we carry out a similar approach that reproduces the following three elements: the die-locking force, the casting pressure, and the thermal stress. First, the casting and solidification simulation reveals the temperature distribution of the die. A comparison of the measured die surface temperature with the calculation confirms the validity of the simulation. Then, finite element method (FEM) stress analysis evaluates the stresses around the cooling hole.

2. Simulation

The analysis modeling and the simulation were carried out in accordance with the actual casting equipment and casting conditions. The product was a flat-shaped JIS-ADC12 part. The HPDC used an oil-based die release agent. The boundary conditions were similar to those in the previous report [7], some of which are introduced below.

2.1. Die temperature distribution

The steady-state die temperature distribution was obtained using the casting and solidification simulation software ADSTEFAN Ver. 2013. Figure 1 shows the computer-aided design (CAD) model of the die inserts. Figure 2 shows the positions of the cooling holes. Cooling holes 1, 2, 4, and 5 had a diameter of 17 mm, and cooling holes 3 and 6 had a diameter of 13 mm. The distances between the tops of cooling holes 1–6 and the die surface were 14, 12, 13, 12, 10, and 11 mm, respectively. The depths of cooling holes 4 and 5 were 10 mm and 12 mm, respectively, in the actual die. These original designs did not show any cracking of the cooling holes. Therefore, the calculation intentionally added cooling hole 5 with a depth of 5 and 20 mm and cooling hole 4 with a depth of 3 mm and 20 mm to generate high-stress conditions.

Figure 3 shows the assembled model. The FEM mesh had 1 mm element edge lengths and a total of 30,827,082 elements. The injected temperature of the molten metal was 580°C. The material properties [4] were the same as in the previous report [6]. The temperature dependence of the specific heat is listed in Appendix 2. The die insert material was JIS-SKD61, and the die holder was FCD500. Some of the material properties are listed in the table of Appendix 1. A comparison of the actual die surface temperature after the spray and blown air with the simulated temperature determined the heat transfer coefficient between the product and the die to be 41.9 kW/ (m^2 °C).

In the actual operation, the oil-based die release agent and the blown air cooled the die surface. The oil-based die release agent together with the water cooling kept the die surface temperature relatively high, enabling a short cycle time. It was assumed that the heat of the die was homogeneously removed from the surface into an external air environment of 100°C. A heat transfer coefficient of 0.47 kW/(m^2 °C) was used during the spraying and air blowing.

The circulating water cooled the die. The temperature of the cooling water was set to 25°C. The heat transfer coefficient between the die and the cooling water was adjusted for the surface temperature of the cooling hole top to be around 60°C.

Two temperature distributions of the die were prepared to calculate the thermal stresses. One was the temperature distribution "after injection," which is the highest temperature case where the injection pressure transfers the molten metal heat to the die. The other is the state where the product is released from the die, and then the die release agent is sprayed. Hereinafter, it is referred to "after spraying and blowing air." This die temperature also includes the stages of heat removal during die opening and the releasing of the casting.

One shot of the actual casting process includes 5 stages. Repeating 20 shots calculates the steady-state temperature distribution of the die. The curing time is usually the time until the solidification of the biscuit, but the curing time here was set to 1 s in order to consider the heat transfer time after injection.

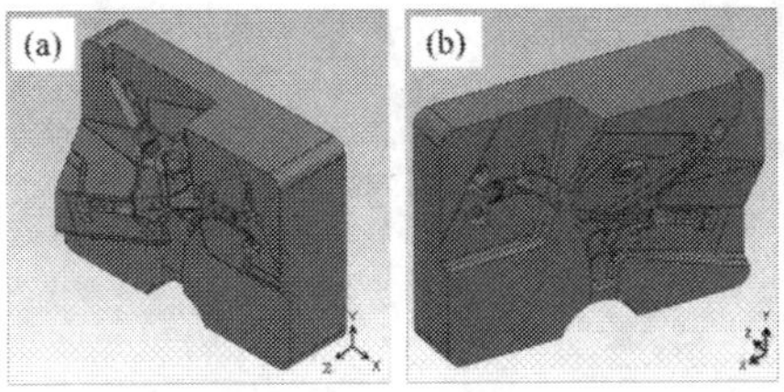

Figure 1. CAD models of (a) movable and (b) stationary die inserts.

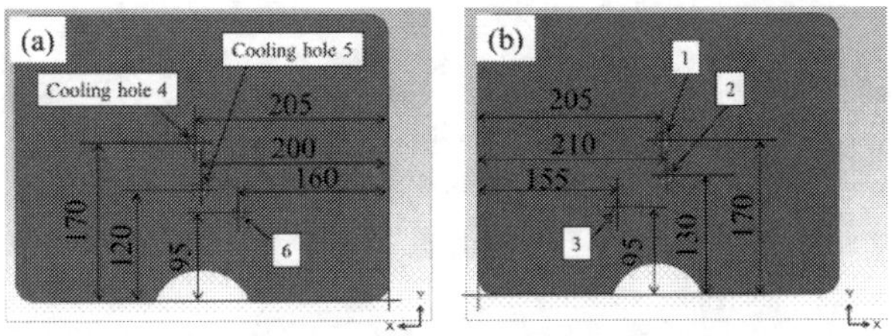

Figure 2. Cooling holes (1–6) observed at the back sides of (a) movable and (b) stationary die inserts. [mm]

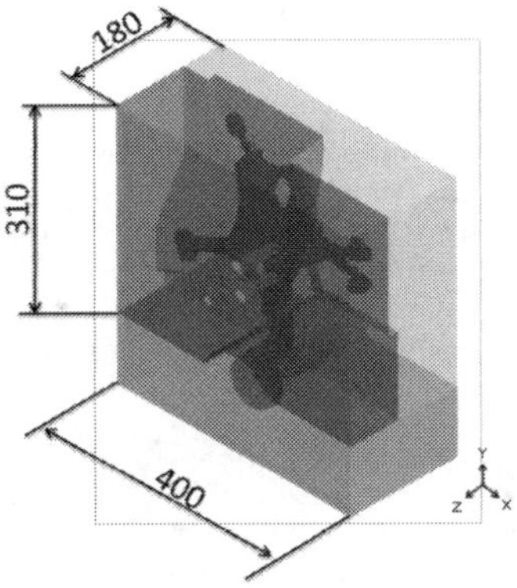

Figure 3. Assembled model for casting solidification simulation. [mm]

2.2. Temperature mapping and deflection analysis

The FEM stress analysis was executed with ANSYS Workbench Ver. 14.0. The temperature distribution obtained was overlaid onto the FEM mesh of the die inserts. Then, the stress was calculated.

The other materials such as structural steel and S55C were represented according to the actual materials used in the machine. Typical material properties are listed in Appendix Table A1. The boundary conditions were set by reproducing the contact conditions among the parts and the constraint conditions, such as the fixation of the parts and the component temperatures [4].

In these calculations, we used two types of constraints for the die. One corresponds to the state after injection, and the other refers to the state after spraying and blowing air. The model after injection includes the die inserts as well as the entire HPDC machine. The total model includes 140,242 tetrahedral second-order elements. The die-locking force and the casting pressure were applied. The die-locking force was applied from the toggle of the movable platen with a load of 2450 kN [4]. A casting pressure of 60 MPa was vertically and homogeneously applied to the cavity surface.

The case after spraying and blowing air did not include the die-locking force or the casting pressure because of the die opening. The model after spraying and blowing air reproduced the die insert, the die holder, and the back plate. The reverse side of the back plate was fixed. The calculation was only performed at the movable die side. The modified Goodman diagram evaluated the fatigue life of the cooling holes

3. Results

3.1. Matching the simulated temperature with the actual measurement

Figure 4 shows the simulated temperature distribution after injection. Figure 5 shows the case after spraying and blowing air. Note that the color contours are different. The analysis revealed that the surface temperature accurately simulates the actual thermograph.

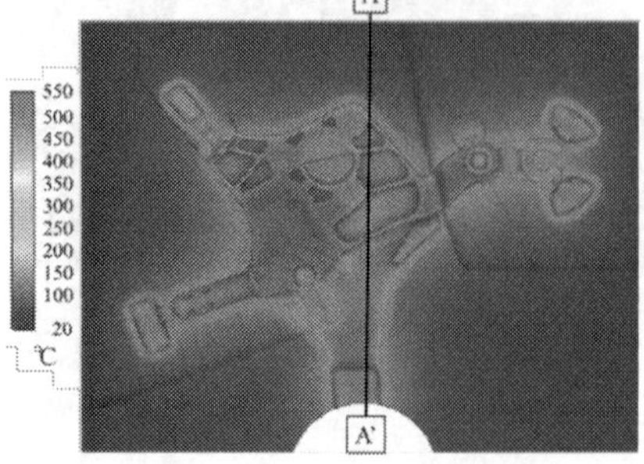

Figure 4. Temperature distribution of cavity surface after injection.

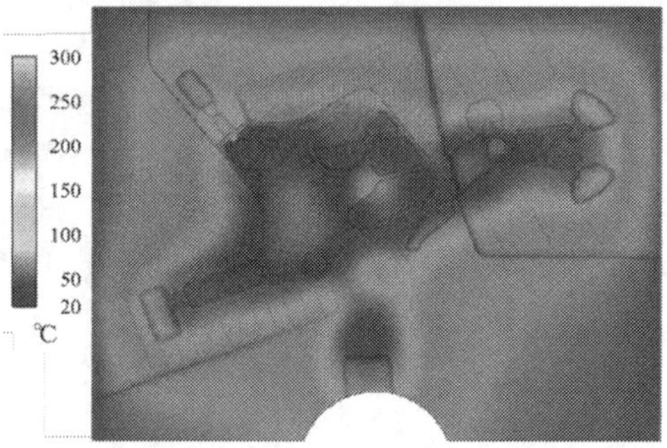

Figure 5. Temperature distribution of surface after spraying and blowing air.

3.2. Stress around the top of the cooling hole

Figure 6 shows the temperature distributions at cross section A-A' in Figure 4 (a) after injection and (b) after spraying and blowing air. This cross section is selected as to include cooling hole 5. Figure 7 shows the temperature change along the solid line in Figure 6. The diagram shows the die temperature (up to a 10 mm depth) and shows the surface temperature of the cooling hole (beyond 10 mm). The heat transfer coefficient between the die and the cooling water was adjusted to make the surface temperature of the cooling hole top around 60°C. The variation in the surface temperature of the cooling hole beyond 10 mm comes from the constant heat transfer coefficient.

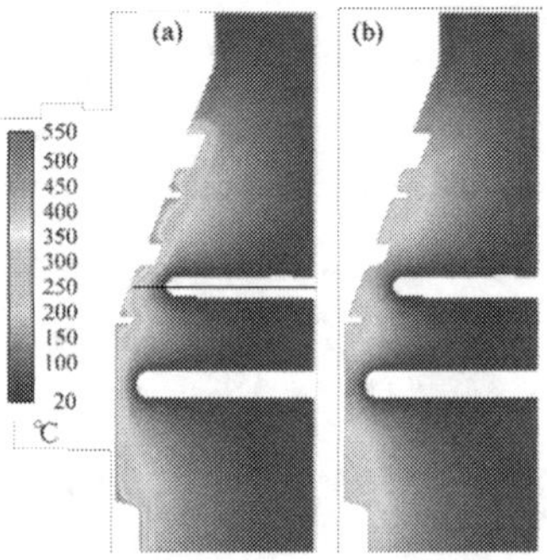

Figure 6. Die temperature distribution at cross section (A-A') (a) after injection and (b) after spraying and blowing air.

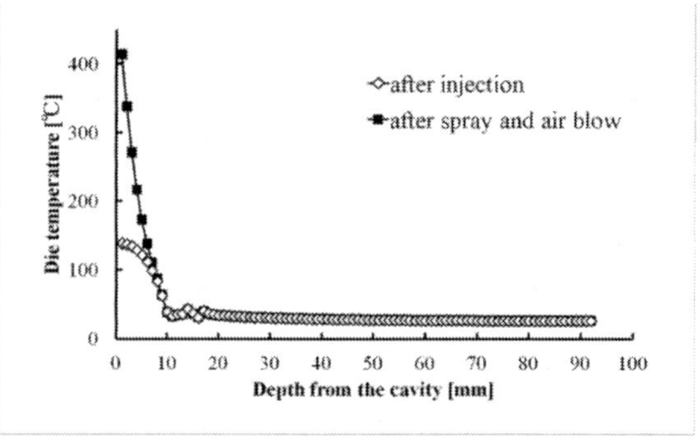

Figure 7. Die temperature vs. depth of the cooling hole top.

Figure 8 shows the stress distribution corresponding to Figure 6. The cooling holes have multi-axial stresses. High stress values were observed around the cooling hole tops. The observation of the maximum principal stress revealed that the stress takes place along the cooling hole surface. Also, the minimum principal stress at the cooling hole top was vertical to the cooling hole surface. It was found to not cause cracking.

In order to estimate the compressive stress at the cooling hole top, Figure 9 shows the minimum principal stress values at the cooling hole surface. It also includes the cases with depths of 5 mm and 20 mm. The stress change at the decreased depth is relatively small after injection. However, the compressive stresses drastically increase with a decrease in the cooling hole depth after spraying and blowing air. The 5 mm deep case has a high compressive stress of 500 MPa.

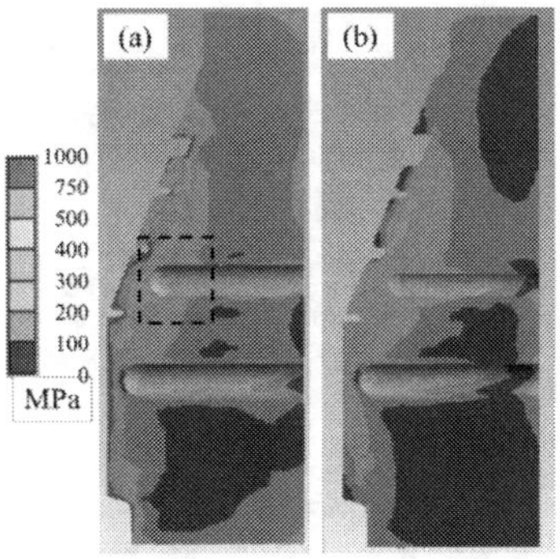

Figure 8. Mises stress distributions at the cross section (A-A') (a) after injection and (b) after spraying and blowing air.

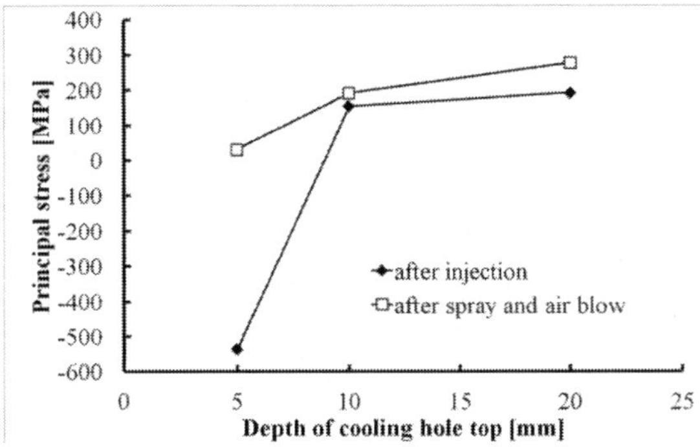

Figure 9. Effect of the depth of cooling hole 5 on the principal stress at the hole top.

Figure 10 shows the maximum principal stress distribution near the broken line in Figure 8(a) after injection. A high tensile stress of 550 MPa was concentrated at corner R (indicated by the arrow). A high tensile stress also existed along the depth direction on the cooling hole surface.

Figure 11 indicates the maximum principal stress values after injection and after spraying and blowing air. It also includes the cases with depths of 5 mm and 20 mm. Tensile stresses were observed after injection and after spraying and blowing air. It was found that the shallower the hole depth, the lower the stress. However, the stress amplitude is too small to cause fatigue failure.

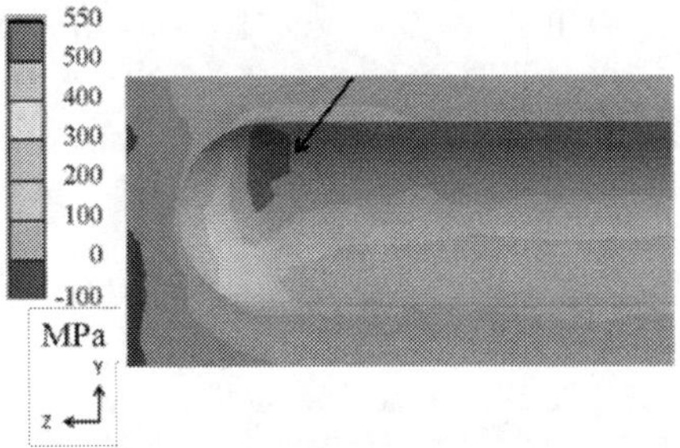

Figure 10. Cross-sectional view of the principal stress at cooling hole 5 after injection.

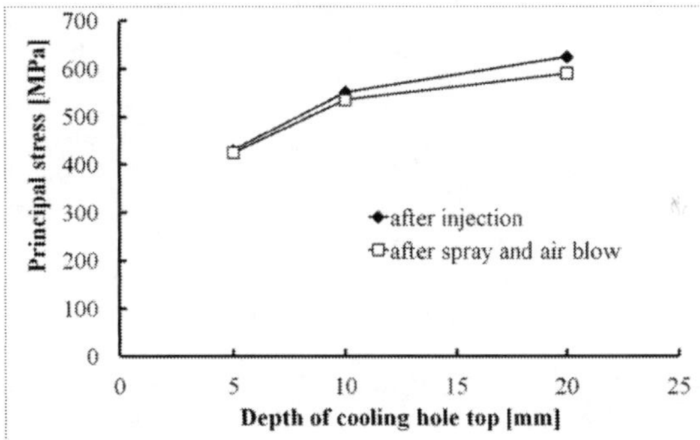

Figure 11. Effect of the depth of cooling hole 5 on the principal stress at corner R.

The stress amplitude and the mean stress at cooling hole 5 were obtained. Figure 12 shows the fatigue limit diagram. The upper and lower limit stresses are necessary for calculating the stress amplitude and the mean stress. The stress along the hole surface will initiate and open a crack. The minimum principal stress at the cooling hole top was along the hole surface. The lower limit stress was chosen for the case after injection, and the upper limit stress was chosen after spraying and blowing air. The yield limit line and the modified Goodman line are illustrated. For the yield limit line, a 0.2% proof stress of 1300 MPa at room temperature was chosen. The modified Goodman line was given using a fatigue stress of 675 MPa at 10^6 cycles obtained by rotational bending testing at room temperature and an ultimate tensile strength of 1750 MPa at room temperature. These material data of the die steel SKD61 (HRC 50) were quoted from the steel manufacturers [9, 10]. In the fatigue limit diagram, fatigue failure was not predicted at all cooling hole depths.

Figure 13 illustrates the fatigue limit diagram for cooling hole 4. It is observable that the 3 mm depth case has an increased tensile stress, and it is positioned near the fatigue limit line.

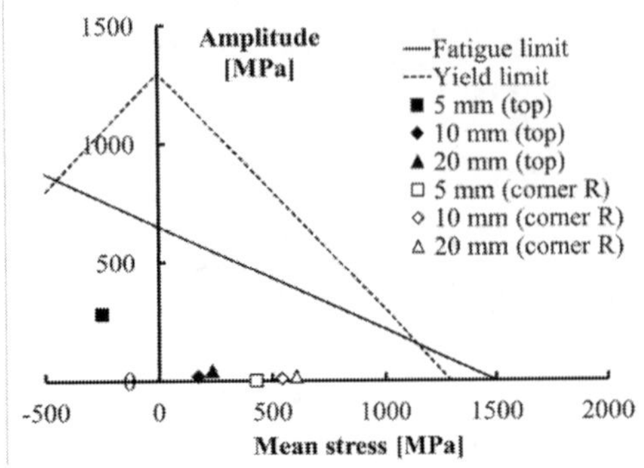

Figure 12. Fatigue limit diagram for cooling hole 5.

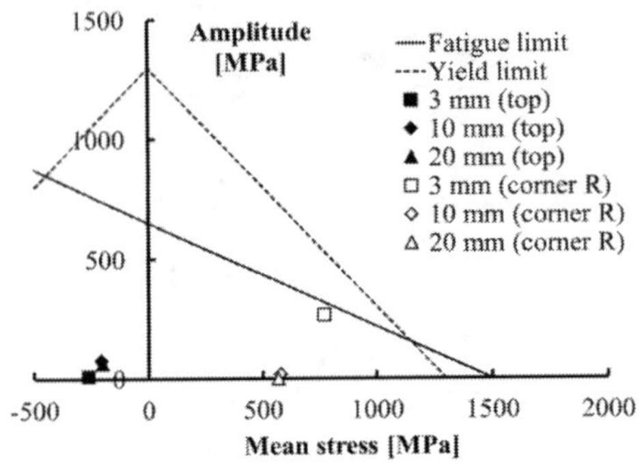

Figure 13. Fatigue limit diagram for cooling hole 4.

4. Discussion

The depth of the cooling hole influences the stress. A shallow cooling hole generates a compressive stress due to the heat input from the die cavity. Particularly, at a depth of 5 mm for cooling hole 5, a high compressive stress appears at the area between the cooling hole top and the cavity surface. The observed value of 500 MPa shown in Figure 9 is below the yield stress. However, if the hole surface has a low roughness value, the stress concentration causes cracking even under compressive stress. In this case, we can predict the probability that the compressive stress triggers mode-II fatigue fracture and that the shear stress helps propagate the initiated crack. The shear stress at the cooling hole top is as small as 40 MPa but can assist the growth of the initiated crack.

Corrosion is liable to take place at the cooling hole, so we also have to consider the corrosion as a factor that can cause fracture. Also, the surface roughness of the cooling hole is not normally controlled. Poor surface roughness will cause stress concentrations, which can decrease the strength.

A tensile stress of 500 MPa was observed at the cooling hole corner R. It was suggested that the stress causes fatigue failure. However, the stress amplitude is small because of the small stress difference between the states after injection and after spraying and blowing air. Hence, it is concluded that fatigue fracture will not take place.

At cooling hole 4 (Figure 13), a depth of 3 mm was simulated as an extreme case. There is a potential that the portion (open square) causes fatigue fracture due to the stress value being near the fatigue limit.

It was demonstrated that the clamping force from the die-casting machine and the injection pressure are small, so these do not influence the calculation results. The depth and location of the cooling hole greatly influence the probability of cracking. The results in Figures 12 and 13 had quite different stress states. Therefore, in order to pursue the best cooling performance without die failure, the design process using CAE is crucial to locating the cooling hole close to the surface.

5. Conclusion

We investigated the probability of cracking at the cooling hole of the die for a flat-shaped HPDC part. The FEM analysis observed the effect of thermal stresses on the stress concentration at the top and at the corner R of the cooling holes. The stress state changes after injection and after spraying and blowing air. At the top of the cooling hole, these stresses had a compressive character. The stress amplitude takes a higher value the shallower the depth. It was suggested that a high compressive stress initiates a crack, and the shear stress assists propagation. On the other hand, the stress at the corner R had a tensas demonstrated that the developed analysis is valuable in designing the cooling hole of an HPDC die.

Appendices

Table A1. Material properties for die and machine.

	Structural steel	SKD61	FCD500	S55C
Density [kg/m^3]	7850	7800	7100	7800
Coefficient of thermal expansion [1/K]	1.20E-05	1.50E-05	1.00E-05	1.20E-05
Young's modulus [N/m^3]	2.00E+11	2.06E+11	1.73E+11	2.05E+11
Poisson's ratio	0.3	0.3	0.3	0.3

References

[1] Japan Die Casting Association and The Materials Process Technology Center 2000 *The materials process technology center research report* No 541 (Tokyo: The Materials Process Technology Center) p 27

[2] Tamura Y, Nakatsu H, Nagasawa M, Kataoka K and Yamashita Y 2008 Development of a new die casting die steel with excellent heat cracking resistance *Japan Die Casting Congress Transactions (Yokohama, Japan, 6-8 November 2008)* p 1

[3] Tamura Y and Nagasawa M 2004 Digital engineering of the material for die casting *Japan Die Casting Congress Transactions (Yokohama, Japan, 4-6 November 2004)* p 1

[4] Yamagata H, Shigenaga Y, Tanikawa S and Nikawa M 2013 Deflection simulation of a die installed in a high-pressure die-casting machine *Transactions of the JSME* C **79-806** pp 3920–9

[5] Yamagata H, Funahashi H, Tanikawa S and Nikawa M 2014 Prediction of flash by computer simulation of die distortion in JIS ADC12 *J. JFS* **86-2** pp 121–6

[6] Yamagata H, Kurokawa K, Tanikawa S and Nikawa M 2014 Computer simulation for prediction of thermal distortion of JIS ADC12 flat-shape die-casting *Transactions of the JSME* **80-813** pp 1-11

[7] Yamagata H, Kimura T and Tanikawa S 2014 Deflection analysis of die and die-casting machine *Japan Die Casting Congress Transactions (Yokohama, Japan, 13-15 November 2014)* pp 93–100

[8] Yamagata H and Kimura T 2014 CAE analysis of the deviation of the plunger rod in the shot system of a high pressure die-casting machine *ALTOPIA* vol 44 No 6 (Tokyo: Kallos) pp 31–3

[9] Hitachi Metals, Ltd. 2013 *YSS hot work tool steel* catalogue p 11

[10] Daido Steel Co., Ltd. 2012 *DHA1-A* catalogue

IOP Conf. Series: Materials Science and Engineering **84** (2015) 012053 doi:10.1088/1757-899X/84/1/012053

Equiaxed and columnar dendrite growth simulation in Al-7Si-Mg ternary alloys using cellular automaton method

Rui Chen, Qingyan Xu[1] and Baicheng Liu

Key Laboratory for Advanced Materials Processing Technology (Ministry of Education), School of Materials Science and Engineering, Tsinghua University, Beijing 100084, China

E-mail: scjxqy@tsinghua.edu.cn

Abstract. In this paper, a modified cellular automaton (MCA) model allowing for the prediction of dendrite growth of Al-Si-Mg ternary alloys in two and three dimensions is presented. The growth kinetic of S/L interface is calculated based on the solute equilibrium approach. In order to describe the dendrite growth with arbitrarily crystallographic orientations, this model introduces a modified decentered octahedron algorithm for neighborhood tracking to eliminate the effect of mesh dependency on dendrite growth. The thermodynamic and kinetic data needed for dendrite growth is obtained through coupling with Pandat software package in combination with thermodynamic/kinetic/equilibrium phase diagram calculation databases. The effect of interactions between various alloying elements on solute diffusion coefficient is considered in the model. This model has first been used to simulate Al-7Si (weight percent) binary dendrite growth followed by a validation using theoretical predictions. For ternary alloy, Al-7Si-0.5Mg dendrite simulation has been carried out and the effects of solute interactions on diffusion matrix as well as the differences of Si and Mg in solute distribution have been analyzed. For actual application, this model has been applied to simulate the equiaxed dendrite growth with various crystallographic orientations of Al-7Si-0.36Mg ternary alloy, and the predicted secondary dendrite arm spacing (SDAS) shows a reasonable agreement with the experimental ones. Furthermore, the columnar dendrite growth in directional solidification has also been simulated and the predicted primary dendrite arm spacing (PDAS) is in good agreement with experiments. The simulated results effectively demonstrate the abilities of the model in prediction of dendritic microstructure of Al-Si-Mg ternary alloy.

1. Introduction

Due to an excellent combination of castability, high corrosion resistance and comprehensive mechanical properties of Al-7Si-Mg ternary casting alloy, typically A356, A357, ZL114, it has been widely used in the automotive and aerospace industries as structural component[1,2]. Since the characteristic of dendrite has a dominant effect on the mechanical properties of castings, its understanding of the dendritic microstructural formation is of great importance to control the desirable microstructure, and thereby to modify the performance of castings.

[1] Correspondent: Qingyan Xu, Tel: 86-10-62795482, Fax: 86-10-62770190, E-mail: scjxqy@tsinghua.edu.cn.

Numerical simulation has provided a powerful tool to study the dendrite structure during solidification, and cellular automaton (CA) model, as an effective method with high calculation efficiency and clear physical meaning, has been widely applied in the investigation of dendrite growth process[3~13]. However, most of the existing CA models are confined to describe the dendrite of binary alloys with well-defined crystallographic axes, while in practice, most commercial alloys are multi-component with three or more elements, and the solidification process involves complex polycrystalline growth with various crystallographic orientations. Besides, the work on three-dimensional dendrite simulation using CA model is also far from satisfaction and further effort is still needed. In addition, carrying out the simulation combining with the practical industrial production and enhancing the practicability of CA model is one of the further developments[14].

In this paper, through coupling with Pandat software package in combination with thermodynamic/kinetic/equilibrium phase diagram calculation databases, a modified cellular automaton (MCA) model allowing for the prediction of dendrite growth of ternary alloys in two and three dimensions is presented. This model has been first used to simulate Al-7Si binary dendrite for model validation, and then applied to Al-7Si-0.5Mg ternary alloy simulation for the investigation of the different solute distribution behavior between Si and Mg as well as the effects of solute interactions on dendrite growth. Finally, to show the model ability in practical application, it is applied to simulate the equiaxed dendrite growth and columnar dendrite growth of Al-7Si-0.36Mg alloy.

2. Mathematical equations and numerical algorithms

2.1. Solute distribution calculation

To describe mass transfer of Al-7Si-Mg ternary system, the solute fields of Si and Mg need to be calculated separately. In this model, the effect of solute interaction between Si and Mg elements on solute diffusion coefficient is considered, and the solute diffusion is calculated based on the following equation

$$\frac{\partial w_i^\phi}{\partial t} + \nabla \cdot J_i = w_i^\phi (1-k_i)\frac{\partial f_S}{\partial t} \tag{1}$$

where w is the composition with a superscript ϕ, denoting the liquid or solid phase, and the subscript i refers to the alloying element Si or Mg. k_i is the composition and temperature dependent equilibrium partition coefficient which is obtained through coupling with equilibrium phase diagram database. f_S is the solid fraction. J_i is the diffusion flux of element i given by Fick-Onsager law as:

$$J_i = -\sum_{j=\mathrm{Si,Mg}} D_{ij}^\phi \nabla w_j^\phi \tag{2}$$

where D_{ij}^ϕ is the solute diffusivity matrix for phase ϕ. Due to the fact that Si and Mg atoms are substitutional, D_{ij}^ϕ can be expressed as[15]

$$D_{ij}^\phi = \sum_{v=\mathrm{Al,Si,Mg}} (\delta_{vi} - x_i^\phi) x_v^\phi M_v^\phi \left(\frac{\partial \mu_v^\phi}{\partial x_j^\phi} - \frac{\partial \mu_v^\phi}{\partial x_{\mathrm{Al}}^\phi} \right) \tag{3}$$

where x_v^ϕ is the mole fraction of component v, and the parameter M_v^ϕ is the atomic diffusion mobility of element v in the given phase ϕ. M_v^ϕ can be determined from the experimental data by means of the Einstein relation $D_v^*=RTM_v^\phi$ (D_v^* is the tracer diffusivity of element v), or can also be obtained by using the kinetics database. δ_{ji} is the Kronecker delta and it equals to 1 when $j=i$, otherwise $\delta_{ji}=0$. The term $\partial\mu_v^\phi/\partial x_j^\phi$ corresponding to the thermodynamic factor can be obtained by thermodynamic database.

2.2. Interface growth kinetics and solid fraction increment

Assuming that local thermodynamic equilibrium exists at the interface due to the influence of constitutional undercooling and curvature undercooling, the interface equilibrium temperature is characterized by the following equation neglecting the kinetic undercooling

$$T(t) = T_{\mathrm{L}}^{\mathrm{liq}}(w_{\mathrm{Si}}^{\mathrm{L}}, w_{\mathrm{Mg}}^{\mathrm{L}}) - \Delta T_{\mathrm{C}} - \Delta T_{\mathrm{R}} \tag{4}$$

where $T_{\mathrm{L}}^{\mathrm{liq}}(w_{\mathrm{Si}}^{\mathrm{L}}, w_{\mathrm{Mg}}^{\mathrm{L}})$ is the liquidus temperature obtained from the multi-component phase diagram at a given set of concentration $(w_{\mathrm{Si}}^{\mathrm{L}}, w_{\mathrm{Mg}}^{\mathrm{L}})$. ΔT_{C} and ΔT_{R} are the constitutional and curvature undercooling, respectively. In order to consider the influence of different alloying elements on the constitutional undercooling, the commonly used method is to superpose the effects of each alloying element. Therefore, the constitutional undercooling can be expressed as:

$$\begin{aligned}\Delta T_{\mathrm{C}} &= T_{\mathrm{L}}^{\mathrm{liq}}(w_{\mathrm{Si}}^{\mathrm{L}}, w_{\mathrm{Mg}}^{\mathrm{L}}) - T_{\mathrm{L}}^{\mathrm{liq}}(w_{\mathrm{Si}}^{\mathrm{L}^*}, w_{\mathrm{Mg}}^{\mathrm{L}^*}) \\ &\approx \frac{\partial T_{\mathrm{L}}^{\mathrm{liq}}(w_{\mathrm{Si}}^{\mathrm{L}}, w_{\mathrm{Mg}}^{\mathrm{L}})}{\partial w_{\mathrm{Si}}^{\mathrm{L}}}(w_{\mathrm{Si}}^{\mathrm{L}} - w_{\mathrm{Si}}^{\mathrm{L}^*}) + \frac{\partial T_{\mathrm{L}}^{\mathrm{liq}}(w_{\mathrm{Si}}^{\mathrm{L}}, w_{\mathrm{Mg}}^{\mathrm{L}})}{\partial w_{\mathrm{Mg}}^{\mathrm{L}}}(w_{\mathrm{Mg}}^{\mathrm{L}} - w_{\mathrm{Mg}}^{\mathrm{L}^*})\end{aligned} \tag{5}$$

in which the term $\partial T_{\mathrm{L}}^{\mathrm{liq}}(w_{\mathrm{Si}}^{\mathrm{L}}, w_{\mathrm{Mg}}^{\mathrm{L}})/\partial w_i^{\mathrm{L}}$ (i denotes Si or Mg) is the slope of liquidus surface with respect to the solute element i (*i.e.*, m_i^{L}) at the composition $(w_{\mathrm{Si}}^{\mathrm{L}}, w_{\mathrm{Mg}}^{\mathrm{L}})$. $w_i^{\mathrm{L}*}$ is the local equilibrium composition at the interface.

The curvature undercooling can be expressed approximately by assuming that the two principal curvatures are equal

$$\Delta T_{\mathrm{R}} = \frac{1}{\Delta S_{\mathrm{F}}} \cdot \kappa \sum_{i=1}^{2} \left(\gamma(\mathbf{n}) + \frac{\partial^2 \gamma(\mathbf{n})}{\partial \theta_i^2} \right) \tag{6}$$

in which ΔS_{F} is the melting entropy, and κ is the local principal curvature of S/L surface. θ_i is the standard spherical angle of the interface normal $\mathbf{n}$, and $\gamma(\mathbf{n})$ is the anisotropic surface energy function. The function $\gamma(\mathbf{n})$ describing the dependence of the surface energy on normal direction $\mathbf{n}$ in the case of cubic symmetry is given by

$$\gamma(\mathbf{n}) = \gamma_0(1-3\varepsilon)\left[1 + \frac{4\varepsilon}{1-3\varepsilon}(\cos^4\theta_1 + \sin^4\theta_1(1 - 2\sin^2\theta_2\cos^2\theta_2)\right] \tag{7}$$

where γ_0 is the isotropic interfacial energy with the average Gibbs-Thomson coefficient given by $\Gamma = \gamma_0/\Delta S_{\mathrm{F}}$, ε is the strength of anisotropy, $\theta_1 = \arccos\left(\frac{\partial f_s}{\partial x}\Big/\sqrt{(\frac{\partial f_s}{\partial x})^2 + (\frac{\partial f_s}{\partial y})^2}\right)$, $\theta_2 = \arccos\left(\frac{\partial f_s}{\partial z}\Big/\sqrt{(\frac{\partial f_s}{\partial x})^2 + (\frac{\partial f_s}{\partial y})^2 + (\frac{\partial f_s}{\partial z})^2}\right)$. For two dimension, the detailed calculation of local curvature undercooling ΔT_{R} can be found in Ref.[11].

For the interface curvature of a cell with solid fraction $f_{\mathrm{S}}(0<f_{\mathrm{S}}<1)$, the counting method is used with the expression given by

$$\kappa = \frac{1}{\Delta x}\left[1 - 2\left(f_{\mathrm{s}} + \sum_{i=1}^{N} f_{\mathrm{s}}^i\right)\Big/(N+1)\right] \tag{8}$$

where Δx is the mesh size and f_{s}^i is the solid fraction of the neighboring cells, and N is total number of the first layer neighboring cells which is equal 26 in three dimension and 8 in two dimension.

Assuming that on the length scale of an interfacial cell, the concentrations are uniform, and the solid and liquid mixed well with a local equilibrium, therefore the lever rule can be adopted to calculate the increment of solid fraction. The solute conservation of each solute element during each time step for interfacial cells yields the following defining equation for solid fraction variation

$$w_{\mathrm{Si}}^{\mathrm{L}^*} - w_{\mathrm{Si}}^{\mathrm{L}} = w_{\mathrm{Si}}^{\mathrm{L}^*}(1 - k_{\mathrm{Si}})\Delta f_{\mathrm{S}} \tag{9}$$

$$w_{\mathrm{Mg}}^{\mathrm{L}^*} - w_{\mathrm{Mg}}^{\mathrm{L}} = w_{\mathrm{Mg}}^{\mathrm{L}^*}(1 - k_{\mathrm{Mg}})\Delta f_{\mathrm{S}} \tag{10}$$

The analytical solution expressions for interfacial equilibrium concentrations (w_{Si}^{L*}, w_{Mg}^{L*}) and Δf_S can be directly obtained by a combination of Eq.(4)~Eq.(10).

2.3. Coupling Pandat software package

In the present model, the approach for microstructural simulation of Al-7Si-Mg ternary alloys is coupled with Pandat software package in combination with thermodynamic/kinetic/equilibrium phase diagram calculation databases to obtain the required data for simulations[16]. The atomic diffusion mobility and thermodynamic factor needed for the calculation diffusion matrix are obtained through thermodynamic and kinetic databases respectively, and equilibrium phase diagram data including liquidus temperature $T_L^{liq}(w_{Si}^L, w_{Mg}^L)$, the partition coefficients (k_{Si},k_{Mg}) and the slope of liquidus surface (m_{Si}^L, m_{Mg}^L) as a function of liquid compositions(w_{Si}^L, w_{Mg}^L) is generated by PanEngine[16].

2.4. Algorithmic description for 3-D polycrystals with various crystallographic orientations

In CA models, due to the influence of mesh induced anisotropy, the traditional capture process will make the dendrite tend to grow along the axis independent of the original crystallographic orientation. In practice, however, the growth orientations of the dendrites in the undercooled melt vary from each other, as shown in Fig.1a. In order to deal with different crystallographic orientations for polycrystalline growth in three dimension, a space coordinate conversion algorithm is adopted. As demonstrated in Fig.1a, any individual dendrite is assigned to a local coordinate system (*i.e.*, x_0, y_0, z_0 axis in Fig.1), which is chosen to be parallel to the [100] directions of the dendrite arm. Therefore, the relative coordinate conversion from (x_0, y_0, z_0) to (x, y, z) can be uniquely determined by a rotation matrix (define as Rotate$_1$, Rotate$_2$, Rotate$_3$) which is expressed in the terms of three Euler angles (α, β, γ) as shown in Eq.11, Eq.12 and Fig.1b.

$$\text{Rotate}_1=\begin{pmatrix}\cos\gamma & -\sin\gamma & 0\\ \sin\gamma & \cos\gamma & 0\\ 0 & 0 & 1\end{pmatrix}\quad \text{Rotate}_2=\begin{pmatrix}\cos\beta & 0 & \sin\beta\\ 0 & 1 & 0\\ -\sin\beta & 0 & \cos\beta\end{pmatrix}\quad \text{Rotate}_3=\begin{pmatrix}1 & 0 & 0\\ 0 & \cos\alpha & -\sin\alpha\\ 0 & \sin\alpha & \cos\alpha\end{pmatrix} \tag{11}$$

$$\begin{pmatrix}x\\ y\\ z\end{pmatrix}=\text{Rotate}_1\cdot\text{Rotate}_2\cdot\text{Rotate}_3\begin{pmatrix}x_0\\ y_0\\ z_0\end{pmatrix} \tag{12}$$

For two dimension condition (*i.e.*, α=0, β=0), the relationship of Eq.12 can be expressed as

$$\begin{pmatrix}x\\ y\end{pmatrix}=\begin{pmatrix}\cos\gamma & -\sin\gamma\\ \sin\gamma & \cos\gamma\end{pmatrix}\begin{pmatrix}x_0\\ y_0\end{pmatrix} \tag{13}$$

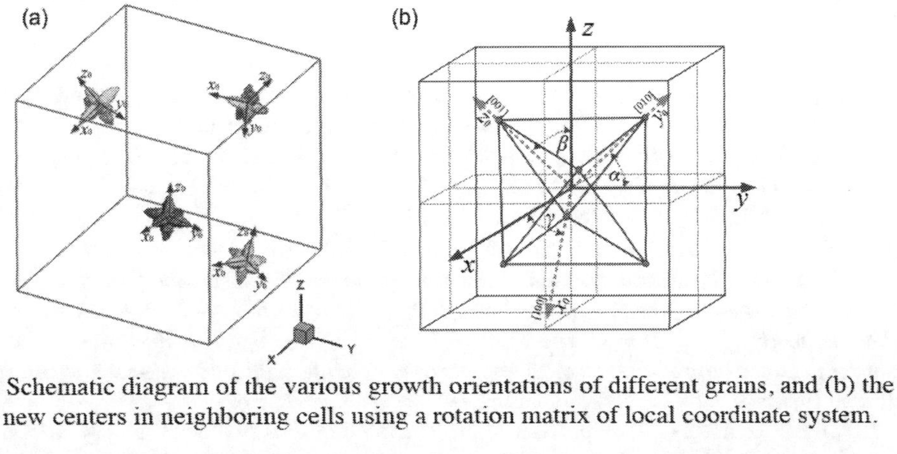

Figure 1. (a) Schematic diagram of the various growth orientations of different grains, and (b) the definition of new centers in neighboring cells using a rotation matrix of local coordinate system.

The underlying capture approach is similar to the 3-D decentered octahedron growth or 2-D decentered square growth technique. In the paper, the 3-D dendrite capture process was analyzed in detail and the 2-D capture rule can be found in Ref.[11]. A misoriented grain with the preferential angle (α, β, γ) nucleates at the center and the cell turns into interface and begins to grow into an octahedron shape. At each time step, the evolution of distance L(t) from the center to the vertex of the octahedron is related to the increment of solid fraction Δf_S , and can be calculated based on the following equation

$$L(t+\Delta t) = L(t) + \sqrt{2}\Delta x \cdot \Delta f_S \tag{14}$$

Once the value L(t) is calculated, the coordinates of the six vertexes of the octahedron in the local coordinate system is known, thus their corresponding coordinates in the (x, y, z) coordinate system can be easily obtained through coordinate conversion. Once the octahedron grows large enough and the vertexes touch any of the neighboring liquid cells, the touched cell will be changed to interface state, meanwhile is assigned with the same preferential orientation as the nuclei's. A new octahedron is generated for the new interface cell and can keep growing along the same direction as the nuclei'.

3. Verification of the CA model

Model validation was carried out in this section by comparing the simulated results with classic analytical model for Al-7Si binary alloy. The physical parameters used in simulations were referred to the literature[17]. This simulation was carried out in a domain of 0.3mm×0.3mm×0.3mm with a cell size Δx=2μm. A single solid seed with initial composition $k_{Si}w_{Si}^0$ and the preferential crystallographic orientation (0°, 0°, 0°) was set at the center of the domain. Various constant undercooling ΔT (defined as $T_L^{liq}(w_{Si}^0)$-T^*in which $T^*$is the imposed temperature) in the range of 2K to 8K was used to simulate the dendrite growth. Fig.2 shows the simulated dendrite morphologies at the undercooling of 2K and 6K. It can be seen that the dendrite under 2K exhibits a branchless needle shape while the dendrite under 6K appears well-developed side branching, being consistent with the fact that higher undercooling tends to make the dendrite appearing side branching[18].

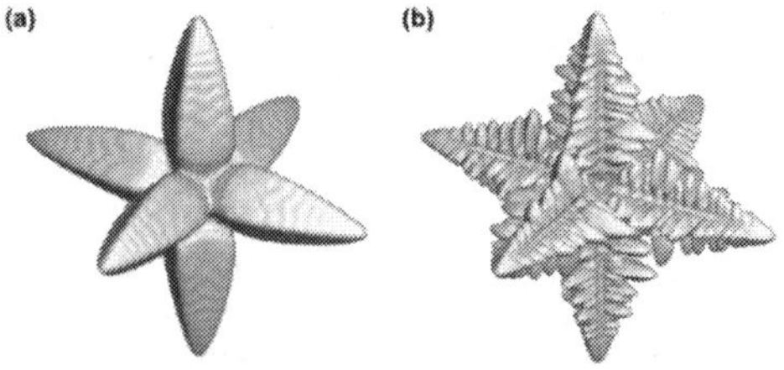

Figure 2. Dendrite morphologies under different undercooling, (a)ΔT=2K, t=4.17s, (b)ΔT=6K, t=0.98s.

The secondary dendrite arms grow perpendicular to the primary arms and the initial secondary dendrite arm spacing λ_2, which develops immediately behind the primary dendrite tip, has a random spacing. For the dendrite tip morphology characteristics, Langer and Müller-Krumbhaarhas proposed the scaling law (Eq.15) based on the detailed numerical analysis of the wavelength of instabilities along the sides of the dendrite tip[19]

$$\lambda_2/R=2.1\pm0.03 \tag{15}$$

where R is the steady-state dendrite tip radius. In order to measure the tip radius of the simulated dendrite, the parabola fitting approaches is used, and the 3-D isosurface plot is depicted for f_s=0.5. For ΔT=6K, the parabolic line is evaluated as $y = 0.1434x^2 + 0.032x + 5.973$ as shown in Fig3, and the tip radius is determined as R=3.49μm. The secondary dendrite arm spacing λ_2, near the dendrite tip was approximately 8.0μm. Therefore, the value of λ_2/R equals to 2.29, which is close to the theoretical value in Eq.15.

In order to make a comparison with the predictions of the Lipton-Glicksman-Kurz (LGK) analytical model, the steady-state tip growth velocity under different undercooling was measured. Since the selection parameter σ^* varies with the anisotropy parameter ε predicted by microsolvability theory, according to the 3-D linearized solvability theory, the value of σ^* for 3-D dendrite is 0.085 corresponding to ε=0.03 as used in the simulations. Fig.4 compares the simulated dendrite tip velocity with the values calculated by LGK model as a function of undercooling. As expected, the tip velocity increases with the undercooling increasing, and both of the model predictions follow the same trend with a reasonable agreement.

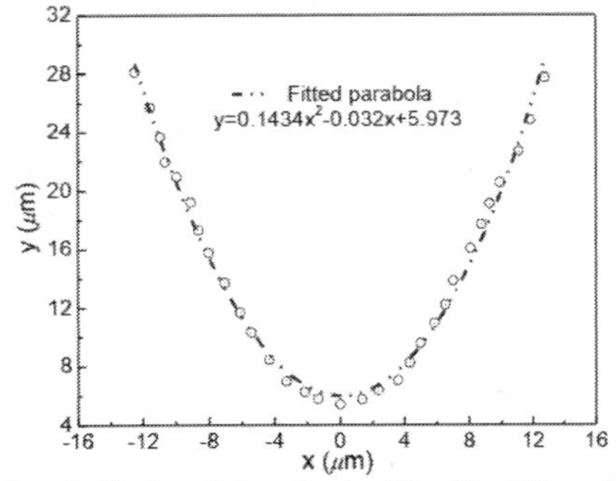

Figure 3. The fitted parabola for 3-D dendrite tip in Fig.2b. The circles indicate the S/L interface position near the dendrite tip, and the dash line is the fitted parabola.

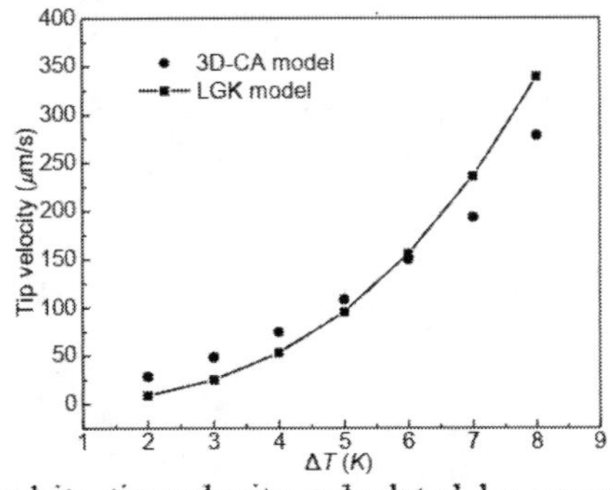

Figure 4. The steady-state dendrite tip velocity calculated by present model and theoretical LGK model as a function of undercooling.

4. Solute distribution behavior and interactions between Si and Mg elements

In this section, simulations were performed for Al-7Si-0.5Mg alloy in a domain containing 200×200×200 cells with a cell size Δx=2μm, and isothermal solidification is adopted with a constant temperature T^*=881K. The effect of interactions between alloying elements on diffusion coefficients was considered. Fig.5 depicts the dendrite morphologies and solute distribution both for Si and Mg elements. Since the diffusivity of solute is much smaller than the dendrite growth velocity, the solute discharged from the newly solidified solid phase can not sufficiently diffuse to the bulk liquid. Therefore, it can be clearly seen from the Fig.5 that both Si and Mg elements are piled up near the dendrite interface as a result of solute redistribution. It is found that the maximum concentration of Si in the region between the dendrite arms is approximately 7.9wt.% while the value for Mg is 0.54wt.%. If defining $(w_i^L - w_i^0)/w_i^0$ as the degree of solute enrichment for i element, the value for Si is 12.8% which is greater than that of Mg (8%). This can be accounted by the reason that the liquid diffusion coefficient of Mg is approximately two times of the diffusion coefficient of Si, *i.e.*, $D_{Mg}^L \approx 2D_{Si}^L$, which means that the accumulated Mg element ahead dendrite interface is easier to diffuse out. Furthermore, the difference in equilibrium partition coefficient (k_{Si}≈0.13, k_{Mg}≈0.24) also contributes to this phenomenon. The larger equilibrium partition coefficient results in the dissolution of more solutes by the solid phase, and less solutes discharged from the newly solidified solid phase into the liquid phase.

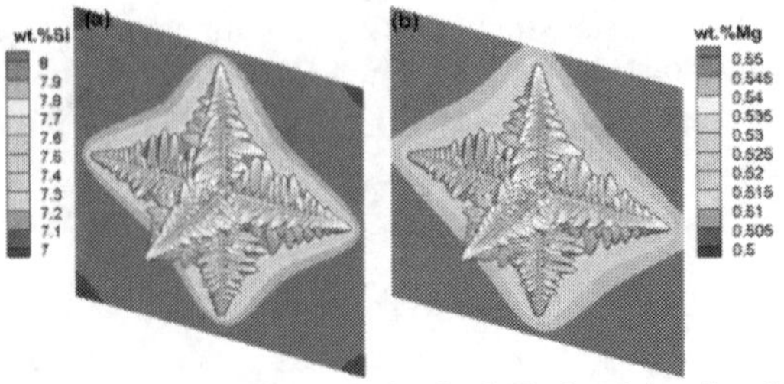

Figure 5. Morphologies and solutes distribution of single 3-D dendrite for Al-7Si-0.5Mg(wt.%) alloy solidified at a constant temperature of 881K.

Fig.6 shows the diffusion matrix in 2-D slice both for diagonal (D_{sisi} and D_{mgmg}) and off-diagonal (D_{simg} and D_{mgsi}). It is found that in the liquid region, the diffusion coefficients are concentration dependent, and the off-diagonal terms are negative which means the attractive interaction between the alloying solutes. It can be seen that the values of off-diagonal terms (D_{simg} and D_{mgsi}) and diagonal term (D_{mgmg}) at the interface is smaller than that far from the interface, while the diagonal term (D_{sisi}) at the interface is larger than that at other positions. Fig.7 represents the diagonal diffusion coefficients of Al-7Si-0.5Mg alloy varying with the concentration of Si and Mg at the isothermal temperature of 881K. As is shown, the diffusion coefficient D_{mgmg} decreases with the increasing of Si and Mg. For diffusion coefficient D_{sisi}, it increases approximately from 2.385×10^{-9} m^2/s to 2.405×10^{-9} m^2/s as the Si varies from 7wt.% to 8wt.%, while it decreases from 2.385×10^{-9} m^2/s to 2.380×10^{-9} m^2/s with the Mg increasing from0.5wt.% to 0.8wt.%. Therefore, it can be seen that the diffusion coefficient D_{sisi} is more sensitive to the change of Si composition and that is why the value of D_{sisi} near the dendrite interface is larger than that far from the interface in Fig.6a.

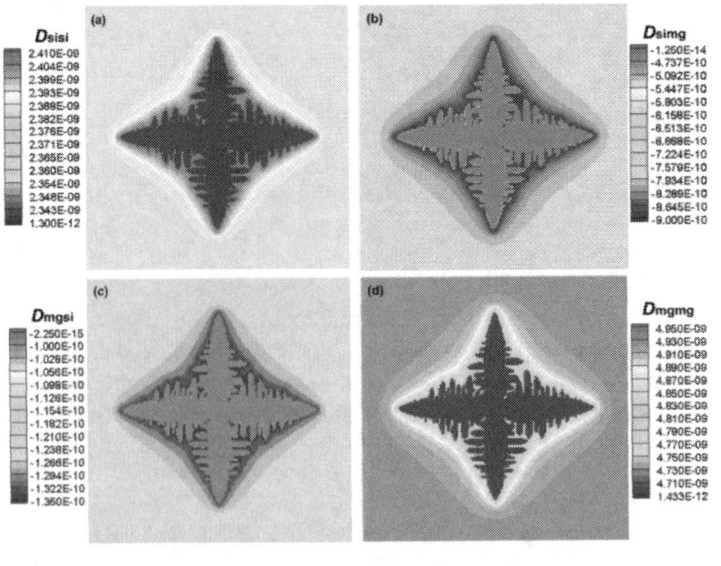

Figure 6. Concentration dependent diffusion matrix profile for Al-7Si-0.5Mg alloy solidified at the constant temperature of 881K. (a)D_{sisi}, (b)D_{simg}, (c)D_{mgsi}, (d)D_{mgmg}.

In Fig.6, we can also clearly see that the off-diagonal diffusion coefficients are smaller than that of diagonal term by an order of magnitude. To illustrate the influence of solute interaction on dendrite growth, Fig.8 shows the comparison of the solute concentrations at the dendrite tip in the liquid as a function of dendrite growth time by taking or without taking the solute interaction into consideration. It is interesting to note that the concentration of Si is a little lower in the case with solute interaction than that without solute interaction, while for Mg element, the concentration at dendrite tip is higher with consideration of solute interaction than the case without solute interaction. A brief analysis of this

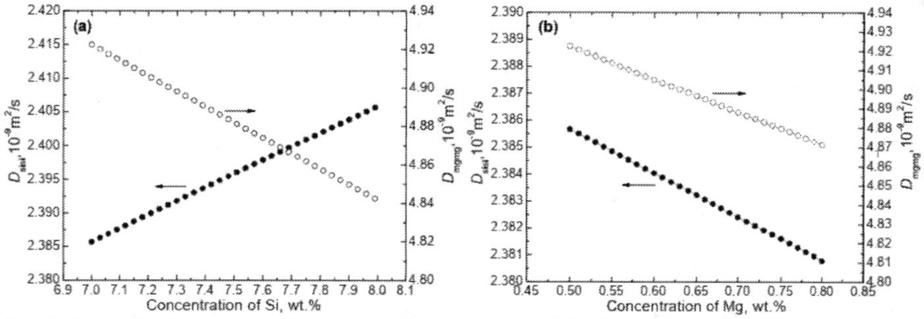

Figure 7. Variation of diagonal diffusion coefficients D_{sisi} (black circles), D_{mgmg} (white circles) in the liquid as a function of concentration of Si and Mg at the temperature of 881K. (a) Si varies from 7wt.% to 8wt.% while Mg identically equal to 0.5wt.%. (b) Mg varies from 0.5wt.% to 0.8wt.% while Si equals to 7wt.%.

phenomenon can be described as: attractive interactions tend to drive both of the Si and Mg elements toward the solid-liquid interface by the concentration gradient of the other solute, slowing down the diffusion behavior from the interface to the bulk liquid, which tends to make solute concentration at the dendrite tip higher when considering solute interaction. The higher solute concentration at the dendrite tip when considering solute interaction leads to a decrease in the undercooling and a further decrease in the tip growth velocity. Through measuring the average growth velocity during the solidification time of 0.2s~0.7s, the value in the condition of solute interaction is 183μm/s and in the condition of no solute interaction, it is 191μm/s. Hnece the lower growth rate when the effects of solute interaction is taken into consideration makes it more time available for the solute ahead of the dendrite to diffuse out. Besides, due to the composition of Si is much higher than Mg in the Al-7Si-0.5Mg alloy, the contribution of cross diffusion term (D_{simg}) to Si solute distribution is much smaller compared with the contribution of D_{mgsi} to the distribution of Mg solute. Both of the reasons result in the difference of the solute distribution behavior of Si and Mg observed in Fig.8.

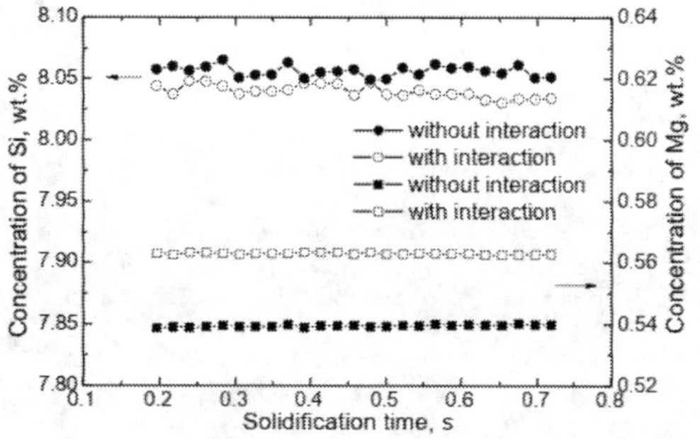

Figure 8. Steady state composition of Si (circles) and Mg (squares) ahead of the dendrite tip varying with the solidification time by with or without considering off-diagonal terms in the diffusion matrix.

5. Application of the model in practical solidification

5.1. Equiaxed dendrite growth for various cooling rates

Since most practical industrial casting processes of Al-7Si-Mg casting alloy mainly involve equiaxed polycrystalline growth in random crystallographic orientations, this case is simulated using this model for Al-7Si-0.36Mg (wt.%) alloy. The temperature profiles for simulation were directly obtained from the measured cooling curves as shown in Fig.9, and the temperature was assumed to be uniform in the whole calculation domain. The corresponding experimental procedure as well as results including cooling curves and metallographical microstructure will report in another paper. Two dimensional simulations were performed in a domain with the size of $7.2\times7.2mm^2$. Fig.10 shows the simulated final dendrite microstructure of sample 2, sample 4, sample 6, and sample 7. It can be seen that with the increasing of cooling rate, the nucleation density increases and the secondary dendrite arm spacing (SDAS) decreases, both of which result in the refinement of the microstructure.

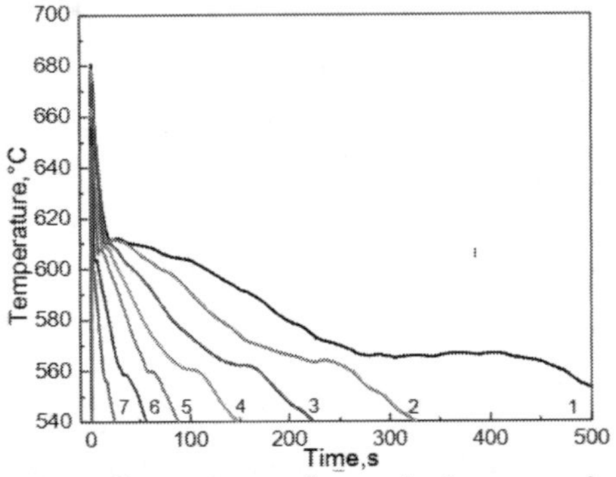

Figure 9. Experimental measured cooling curves of sample 1 to sample 7 for Al-7Si-0.36Mg alloy. The cooling rate increases from sample1 to sample7.

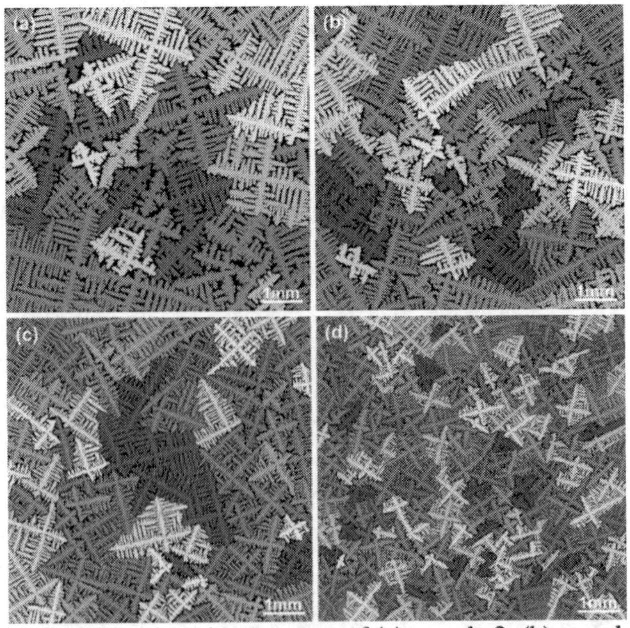

Figure 10. The simulated final dendrite microstructure of (a) sample 2, (b) sample 4, (c) sample 6, (d) sample 7. Different colours represent different dendrites.

The scale of microstructure under different cooling conditions are analyzed by measuring the secondary dendrite arm spacing which exerts a marked influence on the mechanical properties of castings. From Fig.11 which describes the plot of the SDAS versus average cooling rate R_c, it can be seen that the numerical results from the model agree well with the experimental ones, and the fitted formula, *i.e.*, SDAS=42.8$R_c^{-0.302}$, is quite in agreement with the experimental result as reported in Ref.[20] that SDAS in A356/A357 alloys fits well the empirical equation SDAS=39.4$R_c^{-0.317}$, proving the reliability of present model in predicting SDAS.

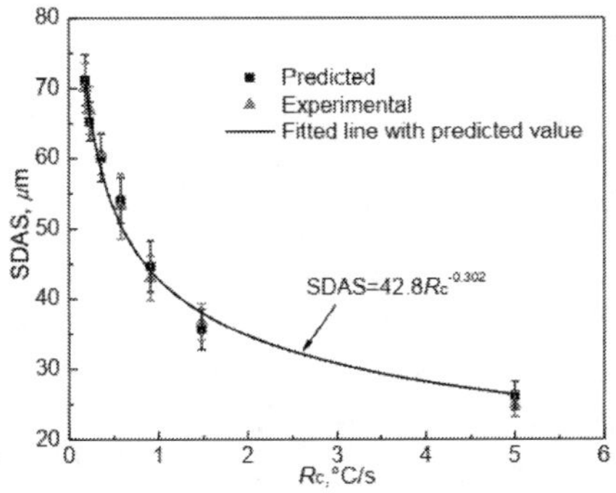

Figure 11. Comparison between the predicted secondary dendrite arm spacing and the experimental ones solidified in different cooling conditions.

5.2. Primary dendrite spacing prediction of columnar dendrite

Columnar dendrite is the most frequently observed microstructure in directional solidified products, and the primary dendrite spacing, as an important characteristic microstructural length scale, determining the segregation of alloying elements as well as the size and distribution of eutectic particles and solidification defects, shows a significant effect on the mechanical properties of solidified products. Therefore, it is of particular importance to investigate the columnar dendrite evolution and to give a precise prediction of primary dendrite spacing. For a given solidification history, the dendrite spacing mainly controlled by the temperature gradient G and dendrite growth velocity v. Hence, both experimental and simulation investigation on dendrite spacing evolution for Al-7Si-0.36Mg alloy during directional solidification is carried out. The experimental results will published in another paper. The simulations were performed in a domain of 6.25mm×15mm with a mesh size of 5μm. At the beginning of the simulation, a certain amount of crystal seeds with <100>directions aligned with the coordinate axes were nucleated at the bottom of the domain. Fig.12 presents the simulated steady state columnar dendrite arrays for v=200μm/s, v=150μm/s, v=100μm/s, v=50μm/s at a given temperature gradient G=15K/mm. It can be seen that during initial growth stage, developed tertiary arms emitted from the secondary arms are observed as a result of too large initial dendrite spacing λ_{ini}, *i.e.*, λ_{ini} exceeds the upper limit λ_{max}. However, most of the initial tertiary arm spacing between adjacent dendrite arms is lower than the lower limit λ_{min}, and competitive growth occurs by blocking some tertiary arms, and some become the stable primary dendrite arms. After a period of competitive growth, the dendrite array reaches a stable state without dendrite spacing varying. Fig.13 shows the comparison between the CA prediction, analytical model predictions[21~23], and the experimental results for the average primary dendrite spacing λ_{ave}. It is observed that CA model prediction show quite good agreement with the experimental results, which is much better than the agreement between analytical models' predictions and experiments. The comparisons reveal that the present CA model has a high accuracy in predicting primary dendrite spacing in directional solidification.

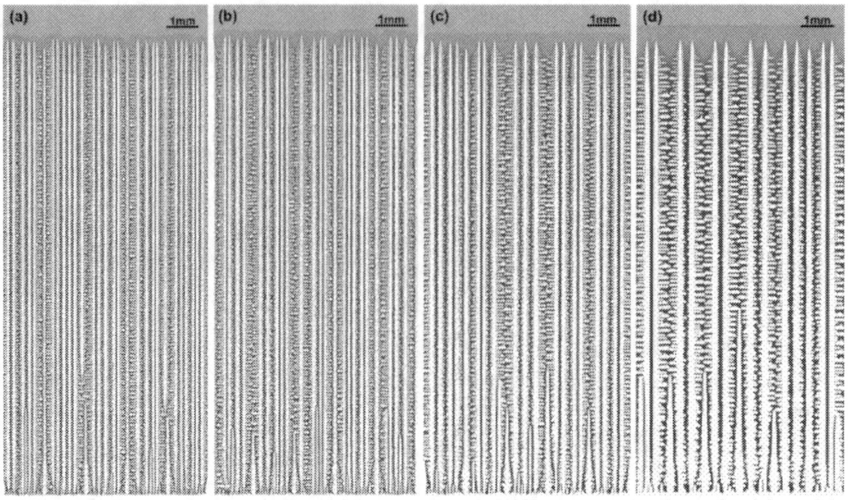

Figure 12. Simulated steady state dendrite arrays at different growth velocities (a)v=200μm/s, (b)v=150μm/s, (c)v=100μm/s, (d)v=50μm/s.

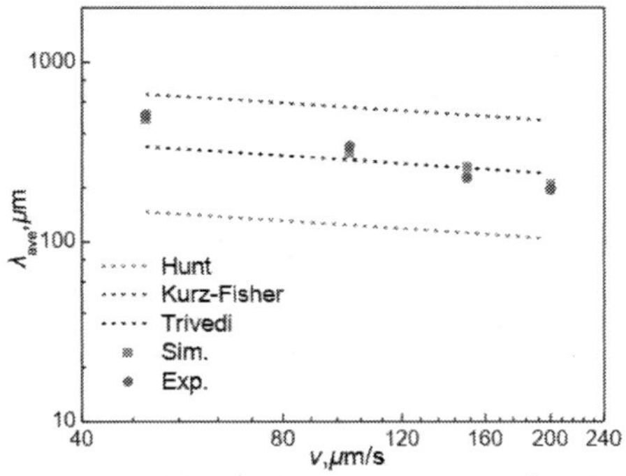

Figure 13. Comparison of the average primary dendrite spacing λ_{ave} between the CA simulations, analytical predictions and the experimental results.

6. Conclusions

A cellular automaton model for the prediction of the ternary dendrite growth in two and three dimensions during solidification processes is developed. The model was coupled with thermodynamic/kinetic/equilibrium phase diagram calculation databases incorporated in the Pandat software package and the solute interactions between alloying elements was considered. To eliminate the influence of mesh-induced anisotropy on dendrite growth, a modified decentered octahedron for neighborhood tracking is introduced. First, detailed model validations have been performed with the simulated steady-state tip parameters being in good agreement with the LGK predictions and the wavelength of instabilities theory. The model is then applied to Al-7Si-0.5Mg dendrite simulation for investigation of Si and Mg solute distribution behavior and the effects of solute interactions on dendrite growth. It shows that Si element is easier to segregated due to smaller diffusion coefficient and smaller equilibrium partition coefficient compared with Mg element. The diffusion coefficient is concentration dependent and interactions between Si and Mg elements can affect the dendrite growth behavior. Finally, this model is applied to simulation of equiaxed dendrite growth and columnar dendrite growth for Al-7Si-0.36Mg alloy in practical solidification process. The predictions of secondary dendrite arm spacing and primary dendrite spacing shows a reasonable agreement with the experimental results. The simulated results effectively demonstrated the abilities of this model in the prediction of dendritic microstructure in ternary alloys.

Acknowledgments
The authors gratefully acknowledge the financial support of the National Basic Research Program of China(Grant NO. 2011CB706801), National Natural Science Foundation of China(Grant Nos. 51374137, 51171089), and National Science and Technology Major Projects (Grant Nos. 2012ZX04012-011, 2011ZX04014-052).

References
[1] Li P T, Liu S D, Zhang L L and Liu X F 2013 *Mater. Design* **47** 522
[2] Zhu M, Jian Z Y, Yang G C and Zhou Y H 2012 *Mater. Design* **36** 243
[3] Wang W, Lee P D and McLean M 2003 *Acta Mater.* **51** 2971
[4] Beltran-Sanches L and Stefanescu D M 2003 *Metall. Mater. Trans. A* **34** 367
[5] Zhu M F and Stefanescu D M 2007 *Acta Mater.* **55** 1741
[6] Zhang X F, Zhao J J, Jiang H X and Zhu M F 2012 *Acta Mater.* **60** 2249
[7] Yu J, Xu Q Y, Cai K and Liu B C 2007 *Acta Metall. Sin.* **43** 731
[8] Shi Y F, Xu Q Y, Gong M and Liu B C 2011 *Acta Metall. Sin.* **47** 620
[9] Huo L, Han Z Q and Liu B C 2009 *Acta Metall. Sin.* **12** 1414
[10] Wu M W and Xiong S M 2010 *Acta Metall.Sin.* **46** 1534
[11] Chen R, Xu Q Y and Liu B C 2014 *J. Mater. Sci. Technol.* **30** 1311
[12] Chen R, Xu Q Y and Liu B C 2014 *Acta Phys.Sin.* **63** 188102
[13] Zhang H, Xu Q Y, Shi Z X and Liu B C 2014 *Acta Metall.Sin.* **50** 345
[14] Zhao J Z, Li L and Zhang X F 2014 *Acta Metall.Sin.* **50** 641
[15] Zhang R J, He Z, Wang X Y and Jie W Q 2005 *Mater. Lett.* **59** 2765
[16] Cao W, Chen S L, Zhang F, Wu K, Yang Y, Chang Y A, Schmid-Fetzer R and Oates W A 2009 Calphad **33** 328
[17] Shi Y F. 2013 *PhD Dissertation*, Tsinghua University, Beijing
[18] Yin H, Felicelli S D and Wang L 2011 *Acta Mater.* **59** 3124
[19] Langer J S and Müller-Krumbhaar H 1978 *Acta Metall.* **26** 1681
[20] Wang Q G and Cáceres C H 1996 *Mater.Sci.Forum* **242** 159
[21] Hunt J D. *Solidification and Casting of Metals*, The Metal Society, London, 1979
[22] Kurz W and Fisher D J 1981 *Acta Metall.* **29** 11
[23] Trivedi R 1984 *Metall.Trans.A* **15** 977

IOP Conf. Series: Materials Science and Engineering **84** (2015) 012054 doi:10.1088/1757-899X/84/1/012054

Mechanical Simulation of the Localized Deformation in the Aluminum Foams: A Three-dimensional (3D) Structure Based Study

Zhu Kai[1], Guo Enyu[2], Zhou Wenqian[1], Shuai Sansan, Jing Tao[1,*], Hou Hongliang[3], Xu Yanjin[3]

[1]School of Materials Science and Engineering, Tsinghua University

[2]Materials Science and Engineering, School for Engineering of Matter, Transport, and Energy, Arizona State University, Tempe, USA

[3]Metal Forming Technology Department, Beijing Aeronautical Manufacturing Technology Research Institute

E-mail: jingtao@mail.tsinghua.edu.cn

Abstract Metal-foam materials have been used increasingly in industry for their low-density, high-toughness and high impact resistance properties. Understanding the macro-scale mechanical properties of these materials is essential to evaluate their actual performance and thus to optimize the structures and properties accordingly. Synchrotron radiation X-ray microtomography technique is a promising method to study 3D structures at small length scales, which provides high spatial resolution and allows the researchers to observe the change of structures/features in situ without destroying the original objects. In this work, the real 3D structure of closed-cell aluminum foam was obtained by using synchrotron radiation X-ray microtomography. The reconstructed 3D model of the foam was further utilized as input for the subsequent mechanical study to investigate the localized deformation behaviors and evolution process of the foam under longitudinal quasi-static uniaxial compressive loading. By analyzing the simulated results, it is demonstrated that the deformation bands always initiate and propagate along the cell walls which are finally folded upon loading. And the large spherical cells are more susceptible to yielding, as well as to the stress concentration than the cells with other shapes. This finding is consistent with the experimental results.

1. Introduction

Cellular foam has been widely used as the key functional material in transportation applications, military tools and thermal insulation components due to the combination of unique performances such as high stiffness, yield strength, thermal and humidity insulation, superior energy absorption, as well as low densities and economical concerns [1-8]. Closed-cell aluminum foam is a type of cellular metal foams which exhibits attractive mechanical properties, excellent impact stress wave attenuation and significant energy absorption behavior. Hence, closed-cell aluminum foam gradually becomes an important filling material for ultra-weight sandwich panels, integral armors and vehicle parts etc. [9-13]. Several manufacturing strategies such as melt route or powder metallurgical route have been explored and applied for producing the closed-cell aluminum foam with desired performance [12].

Due to the remarkable properties of closed-cell aluminum foam, extensive investigations have been carried out both experimentally and numerically to study the effect of characteristics, such as the

IOP Conf. Series: Materials Science and Engineering **84** (2015) 012054 doi:10.1088/1757-899X/84/1/012054

relative density, cell shape and cell wall microstructure on the mechanical properties and then to reveal the deformation mechanisms. For example, it has been reported that factors such as cell morphologies, imperfections and microstructures of the matrix materials in closed-cell aluminum foam, as well as the loading styles such as uniaxial compression, tension and indentation can affect the mechanical properties and deformation behavior [1, 14-18]. Besides that, finite element method has also been applied to simulate and predict the flow behavior of closed-cell aluminum foam [13, 19-24]. In these studies, the effect of cell shape, closed-cell foam structure and the relative density of the closed-cell foam were revealed [13, 19, 24]. The mechanical properties of the cell wall such as elastic modulus, offset yield stress and power-law hardening exponent on the compressive behavior were also studied numerically [20-22]. Up to now, most of those previous numerical studies have either conducted on 2D models which were based on the conventional X-ray microtomography technique [2, 5], or performed on the ideal/simplified 3D models in which the unit cells are taken as perfect sphere or circle which is not the case in the real material. However, the 2D models or ideal/simplified 3D models could not capture the real whole structures/features of the closed-cell foam due to the opaqueness of the material and complex spatial topologies of the cell structure.

To overcome the shortages in the 2D models or ideal/simplified 3D models, we aim to develop a combined experimental and numerical method which is based on the synchrotron radiation X-ray microtomography to understand the deformation behavior and evolution process of the closed-cell aluminum foam under uniaxial compression in this work. The evolution and distribution of stress and strain during the deformation process are of our concerns. More importantly, synchrotron X-ray radiation microtomography is adopted to characterize the initial cell structure of the material in three dimensions. Especially, the obtained 3D data is further employed as the input model in ABAQUS to establish a real 3D-structure based mechanical model with which the mechanical response of the material to the external loading conditions being revealed.

2. Materials and methods

2.1 Materials and specimens preparation
The tested closed-cell aluminum foam material was fabricated via a melt route by blowing compressed air into the molten aluminum alloy through a specially designed pipe which was inserted deep in the furnace. The detail of the fabrication process has been presented elsewhere [25]. The density of the foam (weight divided by the volume of the sample) was calculated to be 0.52 g/cm^3. Before characterization, the cylindrical specimens were prepared by the electro-discharge machining from a large foam plate with a dimension of 275mm×250mm and 35mm thickness. The dimensions of the mechanical tested specimens were 7.0mm in diameter and 7.0mm in height. Then, the specimens were ground and polished from both top and base cross sections and were used for the optical observation of the cell morphologies.

2.2 Mechanical testing and 2D microstructure examination
Mechanical test was performed on a universal servo-hydraulic mechanical testing machine which recorded the displacement and load data simultaneously. Before the quasi-static uniaxial compression test, the parallel platen surfaces were lubricated to reduce the effect of frictions resulted from the contacting surfaces between the specimen and the platen. The displacement rate applied to the top surface of the specimen was 2 mm/min which corresponded to the initial strain rate of $4.8\times10^{-3}s^{-1}$. In order to detect the localized deformation behavior at different strain levels under the compressive loading process, four specimens were pre-strained to 0%, 10%, 30% and 50% respectively. Subsequently, the samples were cut into two parts along the longitudinal section by the electro-discharge machining. The machined sections of the parts were ground and polished for the optical observation. Figure 1 shows the uncompressed specimen, the scheme of compression tests in this research.

Figure 1. Images of the compressive specimens and schematic of the uniaxial compressive experiment before compression (a) the specimen (b) experimental set-up

2.3 Synchrotron radiation X-ray microtomography observation and 3D model reconstruction

The synchrotron radiation X-ray microtomography observations were conducted at beamline BL13W1 of the Shanghai Synchrotron Radiation Facility (SSRF). According to the specimen dimensions (7mm in diameter, 7mm in height), a pixel size of 3.7μm was chosen for the observation. A tunable monochromatic X-ray energy of 28keV was utilized. The distance between the sample and the detector was 6mm which enhanced the edge detection. A YAG:Ge scintillator screen was used to convert the X-rays to visible light. This was coupled with a 2048×2048 pixels with an Optique Peter CCD camera. In total, 1200 projections were collected for a 180° rotation using an acquisition time of 1s per projection. The 2D projections were reconstructed to a 3D volume using a filtered-backprojection algorithm with the software PITRE.

For the 3D reconstructions, the dataset was segmented using Amira™ (now Avizo). Subsequently, the extracted surface were smoothed and optimized. Aiming at studying the deformation behavior of models with different scales, a localized region was chosen to reproduce a simplified 3D model. Figure 2 shows the reconstructed 3D models of the closed-cell aluminum. After the 3D model was obtained, a surface mesh generation process was conducted. These surface meshes were further used to generate tetrahedron meshes with high quality. All the meshing processes were performed by using Amira software. Figure 3 shows the 3D models after meshing. In total, a number of 231256 and 172294 tetrahedron elements for figure 4 (a) and (b), respectively, were generated.

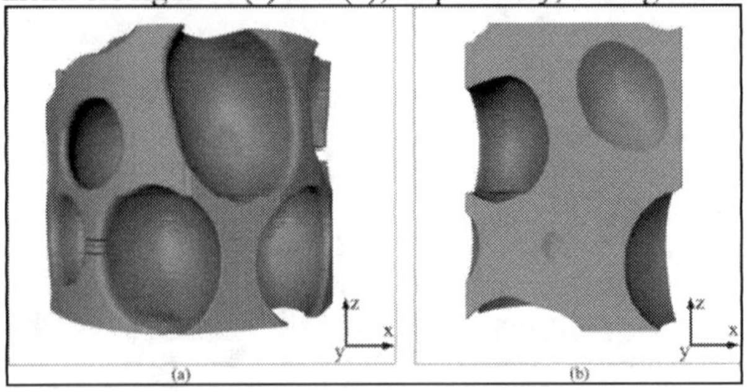

Figure 2. Results of the reconstructed models (a) the whole model (b) the simplified model

IOP Conf. Series: Materials Science and Engineering **84** (2015) 012054 doi:10.1088/1757-899X/84/1/012054

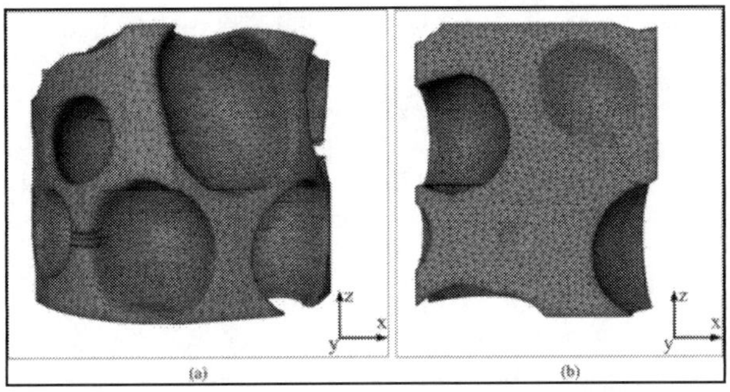

Figure 3. Resulted 3D models after meshing (a) the whole model (b) the simplified model

In this study, the intrinsic mechanical properties of the master aluminum alloy were evaluated by additional mechanical experiments. The Young's modulus, Poisson ratio and yield stress of the master aluminum alloy were measured to be 68 GPa, 0.33, and 143.6 MPa respectively. The platen contacting the sample was treated as rigid part. Frictionless contact-surface between the specimen and the rigid parts was applied to reflect the experimental boundary conditions. An explicit algorithm self-contained in the ABAQUS was utilized to simulate the deformation process. Figure 4 gives the flow chart of this 3D finite element model which indicates the detailed investigation steps.

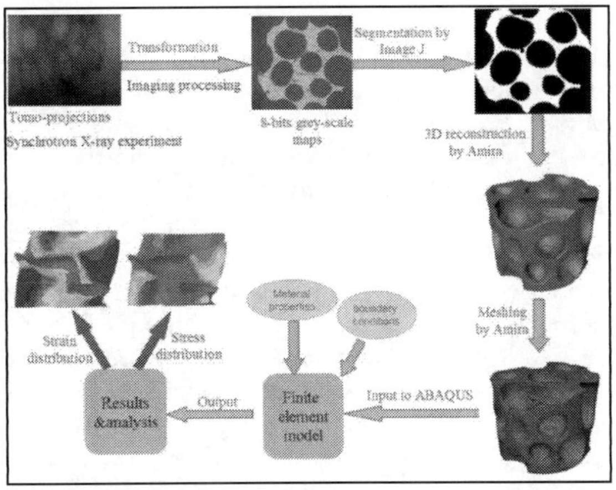

Figure 4. Flow chart of the 3D-structure based finite element model

3. Results and discussion

3.1 Localized deformation behavior revealed by experiments

Figure 5 shows the original and deformed results in the cross sections of samples which were subjected to different strain levels of 0%, 10%, 30% and 50% for (a), (b), (c) and (d) respectively. It is clear that the cell configurations have small curvatures on the walls, as shown in figure 5(a). Furthermore, most of the cells distribute uniformly with a nearly equivalent size of around 2.0 mm in diameter. Subsequently, it can be seen that the cell has a slight bending plastic deformation at the strain of 10% as indicated by the blue rectangle frame in figure 5(b). At this strain level, the morphology of the cell tends to be elliptical due to the applied perpendicular force on the top surface. The plastic bending of the cell wall associate with shearing further takes place and the cell wall is severely distorted (blue rectangle in fame in figure 5(c)), resulting from the initial bending area when the strain level is increased to 30%. Finally, it is observed that the cell wall is torn by shear stress concentrated along the cell wall when the strain is 50% in figure 5(d). The experimental results clearly demonstrate that the failure progress of the specimen is moving forward step by step and that a group of cells work collectively to allow large localized deformation, which is consistent with the aforementioned research results [15].

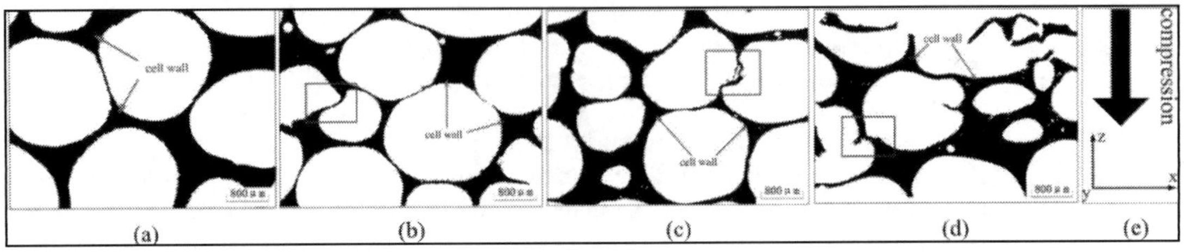

Figure 5. Binarized images of the cell deformation patterns of the closed-cell aluminum foam subjected to different strain levels (a) 0% (b) 10% (c) 30% and (d) 50%, the loading direction is indicated in (e).

3.2 Simulation results and analysis

We choose ABAQUS/explicit module to model and characterize the status of large deformation in closed-cell aluminum foam. Figure 6 and figure 7 exhibit the evolution of equivalent plastic strain and Misses stress in the 3D model at stages with different strains, respectively. For comparison, the initial state of the sample is shown in figure 6(a) and figure 7(a) where no equivalent plastic strain and Misses stress are found. Detailed examination of the deformation process upon loading indicates that the plastic strain localization starts to occur along the cell wall and the large spherical cell (see the regions within the red closed curves in figure 6) is more susceptible to yielding than the others. This deformation behavior was also observed in the other study of compressive deformation mechanisms in closed-cell aluminum foam [2]. These yielding regions can potentially be treated as bulked sites, which have been found in the experimental results, as shown in figure 5. After strain initiation, the strain spreads out along the shear bands as shown in figure 6 (c), which leads to fracture at last. Therefore, it is clearly that the cell shape which is one of the dominant reasons has an important influence on the closed-cell foam deformation behavior. The neighborhood cell is converted into elliptical morphologies due to the compatible deformation with the cell which has been yielding already (see the regions within the white closed curves) and this phenomenon was also observed in some other studies [2]. The occurrence of the plastic deformation band is believed to be caused by stress concentration (see figure 7(b)) within these deformation bands. Accompanying with the increase of stress value in these regions, the appearance of yielding happens earlier than that of other sites which are still at the elastic stage. Interestingly, we also find that the stress starts to concentrate in those regions at the beginning which will be fully folded or torn at the end of deformation and then progressively spreads out to the other regions, as shown in figure 7(b-d). The inhomogeneous distribution of Misses stress in figure 7 matches well with the distribution of the deformation behavior shown in figure 6.

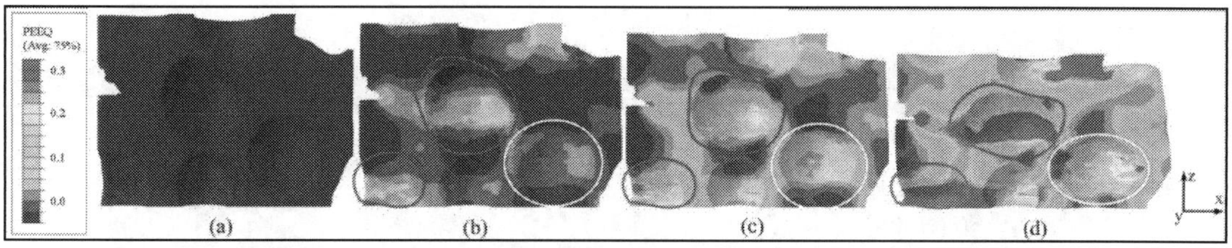

Figure 6. The distribution of the simulated equivalent plastic strain field in the whole model at different strain levels: (a) 0% (b) 5% (c) 10%, and (d) 20%

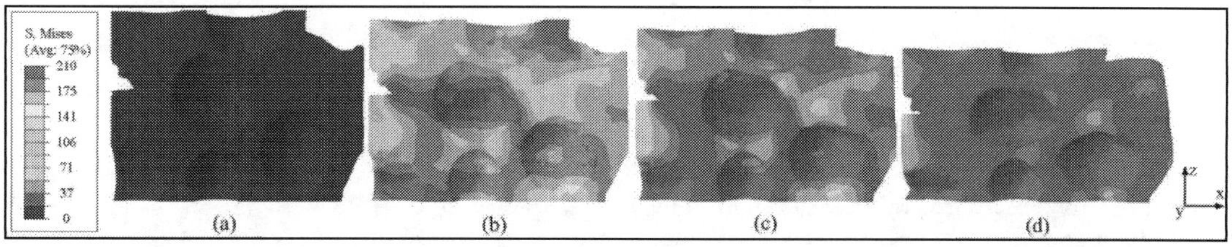

Figure 7. The distribution of the simulated Misses stress field in the whole model at different strain levels: (a) 0% (b) 5% (c) 10%, and (d) 20%

Figure 8 and figure 9 show the distribution of the strain and stress, respectively, in the simplified model using ABAQUS/explicit module. It can be seen that permanent deformation starts to take place in the cell with relatively thinner wall at the beginning (see the regions within the red closed curves), as shown in figure 8 (b). The reason for this initial deformation might be that the cell wall is too thin to sustaining the external loading. Therefore, the thin cell wall is detrimental to the mechanical properties of the closed-cell foam because that it is more susceptible to bending than the other cells with thicker cell walls. After one cell gets yielding, the deformation bands spread out along the junctional cell walls fast due to the occurrence of the compatible deformation in the neighborhood cell (see the regions within the black closed curves), as shown in figure 8 (c). It is should be noted that the cell could experience limited deformation under external loading (see the regions within the white closed curves) when the major radius of this ellipsoid shape cell is parallel to the external compressive direction. The distribution of the simulated equivalent plastic strain field shows that the permanent deformation is first formed in the cell walls with large curvature and then spreads out to the other regions. This progressive collapse process is consistent with the other study [6]. Similar with the results in figure 6 and 7, the stress distribution in figure 9 is consistent with the strain distribution in figure 8.

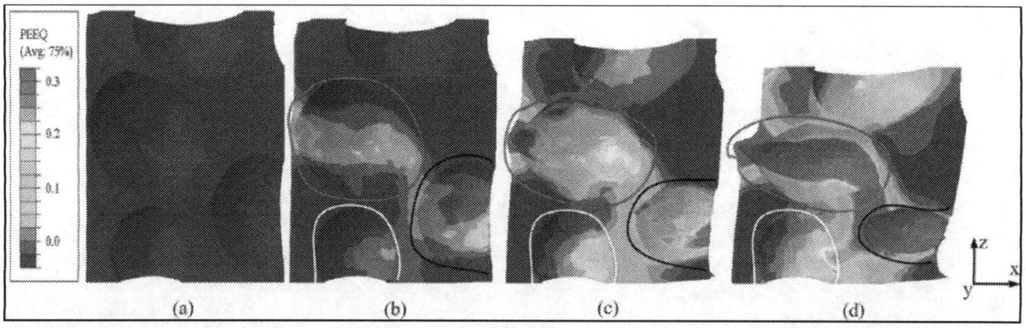

Figure 8. The distribution of the simulated equivalent plastic strain field in the simplified model at different strain levels: (a) 0% (b) 5% (c) 10%, and (d) 20%

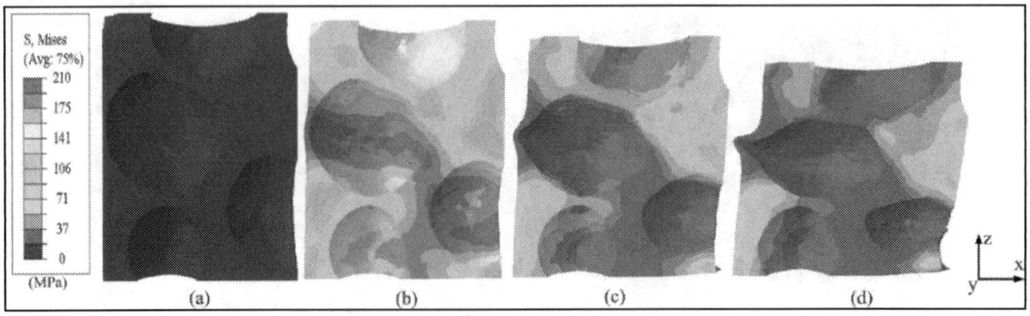

Figure 9. The distribution of the simulated Misses stress field in the simplified model at different strain levels: (a) 0% (b) 5% (c) 10%, and (d) 20%

4. Conclusions

In this research, the localized deformation behavior and its evolution process are investigated both experimentally and numerically. It is found in the experiments that the morphology of the cell tends to be elliptical continuously under the external compressive loading and that the cell walls are tore by the concentrated shearing stress with the progress of the deformation process. A real 3D structure based model has been established by incorporating into the 3D data obtained by synchrotron X-ray radiation microtomography observation. The simulated results reveal that the cell shape is the dominated factor which controls the plastic deformation initiation of the closed-cell aluminum foam under compression. The initial deformation region always occurs in the thin cell wall with large curvature then spreads out to the other regions because of the following compatible deformation. This study also reveals that the deformation evolution process of the closed-cell aluminum foam occurs by the progressive collapse of the layer-by-layer cells.

Acknowledgements

This work was financially support by Innovation Platform for Though Process Modeling and Simulation of Advanced Materials Processing Technologies Project of China (2011ZX04014-52), the National Science Foundation of China (51175292), the Doctoral Scientific Fund Project of the Ministry of Education of China (20130002110012).

References

[1] Sugimura Y, Meyer J, Bart-Smith H, Grenstedt J and Evens A G 1997 *Acta Mater.* **12** 5245
[2] Bart-Smith H, Bastawros A.-F., Mumm D R, Evens A G, Sypeck D J and Wadley H N G 1998 *Acta Mater.* **10** 3583-92
[3] Papka S D and Kyriakides S 1998 *Acta Mater.* **8** 2765
[4] Ivañez I and Sanchez-Saez S 2013 *Composites Structures.* **106** 716
[5] Kádár Cs, Maire E, Borbély A, Peix G, Lendvai J and Rajkovits Zs 2004 *Mater Sci Eng A.* **397-389** 321
[6] Santosa S and Wierzbicki T 1998 *Computers and Structures.* **68** 343-344
[7] Ma Y L, Su X Y, Pyrz R and Rauhe J Ch 2013 *Acta Mechanica Solida Sinica.* **26** 2012
[8] Ruan D, Lu G, Ong L S and Wang B 2007 *Composites Science and Technology.* **67** 1217
[9] Gama B A, Bogetti T A, Fink B K, Yu C J, Claar T D, Eifert H H and Gillespie Jr J W 2001 *Composites Structures.* **52** 381-382
[10] Langdon G S, Karagiozova D, Theobald M D, Nurick G N, Lu G and Merrett R P 2010 *Int J Impact Eng.* **37** 638
[11] Pinnoji P K, Mahajan P, Bourdet N, Deck C and Willinger R 2010 *Int J Impact Eng.* **37** 274
[12] Liu H, Cao Z K, Yao G C, Luo H J and Zu G Y 2013 *Materials and Design.* **47** 483
[13] Kadkhodapour J and Raeisi S 2014 *Computational Materials Science.* **83** 137
[14] Motz C and Pippan R 2001 *Acta Mater.* **49** 2463-2470

[15] Markaki A E and Clyne T W 2001 *Acta Mater.* **49** 1677
[16] Ramamurty U and Kumaran M C 2004 *Acta Mater.* **52** 181-182
[17] Wang D Q, Xue W W, Meng X J and Shi Z Y 2005 *J Mater Sci.* **40** 3475-3480
[18] Li Z B, Zheng Z J, Yu J L and Tang L Q 2012 *Mater Sci Eng A.* **550** 222-226
[19] Konstantindis I Ch, Papadopoulos D P, Lefakis H and Tsipas D N 2005 *Theoretical and Applied Fracture Mechanics.* **43** 157
[20] Jeon I, Katou K, Sonoda T, Asahina T and Kang K J 2009 *Mechanics of Materials.* **41** 60-73
[21] Jeon I, Asahina T, Kang K J, Im S and Lu T J 2010 *Mechanics of Materials.* **42** 227-236
[22] Saadatfar M, Mukherjee M, Madadi M, Schröder-Turk G E, Garcia-Moreno F, Schaller F M, Hutzler S, Sheppard A P, Banhart J and Ramamurty 2012 *Acta Mater.* **60** 3064-3065
[23] Nammi S K, Myler P and Edwards G 2010 *Materials and Design.* **31** 712
[24] Liu Y, Gong W W and Zhang X 2014 *Computational Materials Science.* **91** 223
[25] Zhang H M, Chen X, Fan X L and Li Y X 2012 *China Foundry.* **3** 215-220

IOP Conf. Series: Materials Science and Engineering **84** (2015) 012055 doi:10.1088/1757-899X/84/1/012055

In-situ observation of the dynamic of peritectic coupled growth using the binary organic system TRIS-NPG

J.P. Mogeritsch, A. Ludwig

Department of Metallurgy, Chair for Simulation and Modelling Metallurgical Processes, Montanuniversitaet Leoben

johann.mogeritsch@unileoben.ac.at

Abstract. This paper shows the evolution from banded structure to peritectic coupled growth and describes the mechanism which leads to such microstructure formation. Experimental investigations on binary organic TRIS-NPG (Trishydroxymethyl-aminomethane, Neopentylglycol) alloys at concentrations in the peritectic region were carried out in order to study the formation of peritectic microstructures. The experiments were done vertically with relatively large rectangle glass tubes so that slow natural convection occurred. Each time a sample was hold unmoved in the vertical micro Bridgman-furnace for 2 hour to establish a constant temperature gradient. Afterwards, the samples were moved at several pulling rates through the furnace and a camera recorded the dynamic of the solid/liquid interface. The in-situ observations show layered structures for concentrations within the hyper-peritectic region. Carefully evaluations of the results exhibit, that (i) the solid/liquid interface is curved in observation direction toward the glass walls, (ii) the initial existing peritectic phase grows at micrometer-sized liquid channels inside the primary phase toward to the solid/liquid/glass wall junction, (iii) afterwards the peritectic phase spreads along the liquid-primary phase boundary and forms a so-called band, then (iv) the primary phase gets largely overgrown and unstable peritectic coupled growth (PCG) forms; (v) this PCG is inherent unstable and disappears and occurs in several cycles; until (vi) finally stable growth of the primary phase prevails.

1. Introduction

The interesting characteristic in peritectic systems is the possibility of forming peritectic phase β above the peritectic temperature T_P. In direct solidification experiments with a large temperature gradient/solidification rate (G_T/V) ratio the microstructure is governed by solute transport and nucleation kinetic. This can enable layered microstructure patterns like bands, islands or peritectic coupled grow (PCG). Trivedi [1] published a theoretical model under purely diffusive condition to explain banded microstructures. The model predicts bands in the two phase region $c_{p\alpha} < c_0 < c_{p\beta}$ (hypo-peritectic composition) with an oscillating solid/liquid interface above and below the peritectic temperature T_p. Karma et al. [2] introduce a quantitative model of banding in purely diffusive and strongly convective regimes which explains experimentally observed bands within the two phase region $c_{p\beta} < c_0 < c_L$ (hyper-peritectic composition). Mazumder et al. [3] investigated Sn-Cd and Pb-Bi alloys and observed a new morphology of island bandings consisting of the peritectic phase inside the continuous matrix of the primary phase. They developed a physical mechanism for microstructure formation which is supported by a numerical model. In contrast to the banded structure Lee et al. [4] found eutectic-like two-phase microstructures in Ni-Al alloy. Trivedi and Park [5] did experimental investigations with Sn-Cd alloys to evaluate the role of nucleation on periodic microstructure formation. They found out that under diffusive growth conditions the microstructure depends on the growth competition of the nucleated phase and the existing phase. In extra thin samples only a single nucleus of the new phase forms which grows laterally. In large samples multiple nuclei can occur. Thus, the nucleation distance and the competition between the lateral and parallel growth of both phases determines the microstructure pattern [5]. Lo et al. [6] carried out corresponding experiments in combination with phase field simulations. They found that island banding disappears or acts as a transient microstructure which leads to coupled growth. Whereby, two preconditions must be satisfied: (i) a stable range of coupled growth spacing must exist; and (ii) the mean distance between islands must fall approximately within this stable range. Otherwise the lamellae will be erased.

To deepen our understanding of layered structures we performed investigations of the solid/liquid interface dynamics and the occurring microstructure formations by using transparent organic model substances. Experiments were carried out in extra thin rectangle glass samples (inner dimensions 2.0 x 0.1 mm^2) in order to suppress convection and relatively large rectangle glass samples (inner dimensions 6.0 x 0.6 mm^2) which enables slow natural convection, with pulling velocities close to the stability limit of both solid phases. The experimental results for the extra thin samples show bands, islands and isothermal peritectic coupled growth in the hypo-peritectic region and are published in [7-12]. Large rectangle samples exhibit layered structures in the hyper-peritectic region. The observed layered structures exhibit initial band formations which leads to PCG. Ludwig et al. [13] published the results for c_0 = 0.52 mol fraction NPG, at V = 0.13 µm/s and in the present paper, we report in-situ observations from banded structures to isothermal PCG for c_0 = 0.53 mol fraction NPG, at V = 0.16 µm/s.

2. Experimental Set-up

The solidification experiments were done with an alloy consisting of the organic components TRIS (Trishydroxymethyl-aminomethane) and NPG (Neopentylglycol) [14-16]. Both components [17, 18] show high temperature non-faceted phases which solidify in a planar, cellular or dendritic manner like metals. The peritectic region goes from $c_{p\alpha}$ = 0.47 to c_L = 0.54 mol fraction NPG with the peritectic concentration at $c_{p\beta}$ = 0.51 mol fraction. Detail information of the TRIS-NPG phase diagram are given in [17, 19]. Investigations were carried out in steps of 0.1 mol fraction NPG over the whole peritectic region.

For the experiment, filled large rectangle glass sample (inner dimensions 6.0 x 0.6 mm^2) was placed into the preheated micro Bridgman furnace [19] and held at least 120 minutes to establish a thermal equilibrium and a planar solid/liquid interface. Afterwards, the sample was pulled with a constant pulling rate through the furnace for at least 9 hours. The experimental investigations were done with a constant temperature gradient of G_T = 6.6 K/mm and pulling velocities that ranges from V = 0.09 μm/s to 1.9 μm/s for each selected concentration. Different formations of bands, island bands and peritectic coupled growth were found for all concentrations within the hyper-peritectic region (0.52 - 0.54 mol fraction). In this paper only the experimental result of an alloy with 0.53 mol fraction NPG is presented.

The micro Bridgman-furnace was made up of two brass parts fixed on symmetrical ceramic shelters to create a 7 mm adiabatic gap, see Fig. 1. The temperature within the brass parts was controlled by electrical resistant heaters and measured with Pt 100 temperature sensors placed inside each brass block. A slot (6.5 x 1.2 mm^2) was milled into the brass blocks to fix the glass sample. The sample was illuminated through glass windows in the ceramic plates at the adiabatic zone in order to observe the morphology of the solid/liquid interface with a ZEISS microscope in combination with a camera. In this paper we define the glass wall next to the illumination as the backward side and the glass wall close to the camera as the front side. During solidification images were recorded with Δt = 30 s simultaneously with relevant temperatures. The movement of the glass sample through the furnace was PC-controlled. Details on alloy preparation, filling and sealing of the sample, as well as on the Bridgman-furnace can be found at [6-12, 19].

IOP Conf. Series: Materials Science and Engineering **84** (2015) 012055 doi:10.1088/1757-899X/84/1/012055

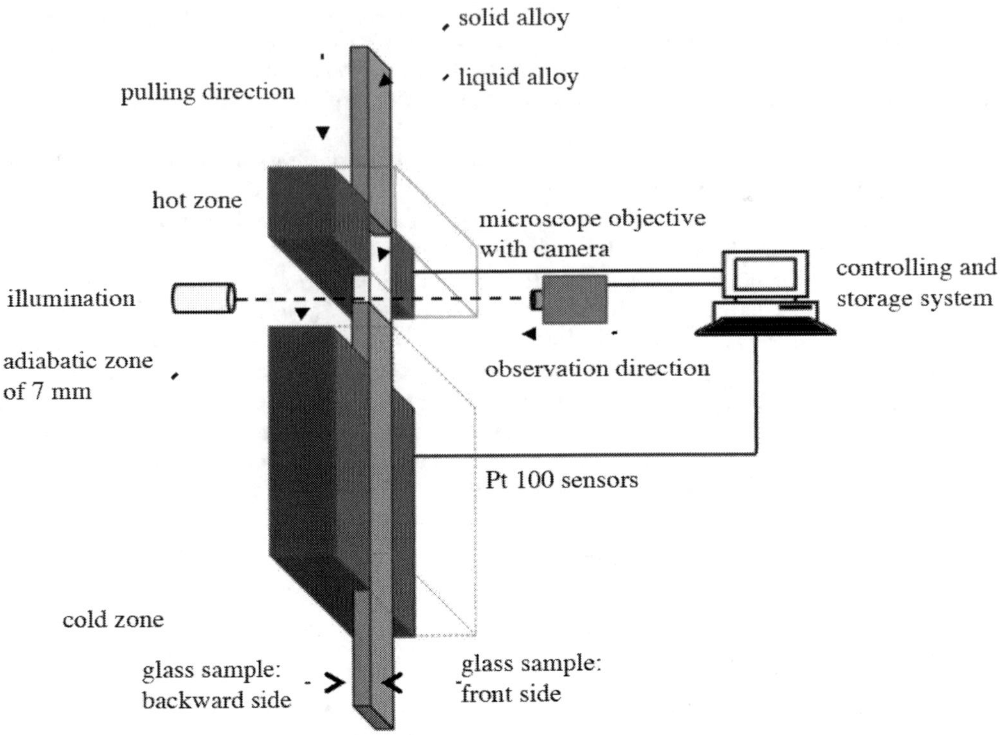

Figure 1: Sketch of the micro Bridgman-furnace. The brass parts of the front ceramic shelter (not shown in the sketch) and the solid alloy within the glass sample are full colored. The brass parts on the front ceramic shelter and the liquid alloy are shown transparent. The glass sample has inner dimensions of 6.0 x 0.6 mm^2 and a wall thickness of 0.4 mm.

3. Results

The initial microstructure exhibits a polycrystalline solid/liquid (s/ℓ) interface. Below this first interface, the polycrystalline grains of α phase are surrounded by liquid until further down a second interface is visible, the transition from the α-ℓ region to the β-ℓ region. The final microstructure formation after approximately 9 h is shown in Fig. 2 and illustrates the following different pattern from bottom to top:

- The initial polycrystalline form of the grains is still visible below the initial second interface.
- Above the initial second interface there are some small NPG enriched liquid inclusions (droplets), which are leftover initially surrounding the α phase.
- Formation of one layered structure in form of a β phase band.
- Unstable heart-like shaped β phase surrounded by α, formed by unstable PCG.
- Change into another tulip-like unstable PCG which reveal a much finer spacing.
- Disappearing of α lamellae and formation of a second β island band.
- Another cycle of unstable PCG starting again with relative large α regions. Quickly the thickness of the α lamellae decreased and those of β increases.
- This time PCG disappears by first an increase and then an abrupt disappearing of β, whereby now no β band forms but multiple β lamellae "survive" and start to form newly PCG. During these two last cycles, the β lamellae have again kind of tulip-like shape and instabilities which resemble 2-λ oscillations.

IOP Conf. Series: Materials Science and Engineering **84** (2015) 012055 doi:10.1088/1757-899X/84/1/012055

- PCG structures finished abrupt, except for few isolated PCG (left side of Fig. 2) which disappear after three additional cycles.
- The s/ℓ interface slightly curved normal to the observation direction, but strongly in observation direction. This becomes obvious by grey diffuse line below the sharp s/ℓ interface. The grey line accompanies the sharp s/ℓ interface since the formation of the β band. As discussed in a former publication of the authors [11] curving of the interface can be understood by segregation originated by the thermo-solutal convection within the sample.

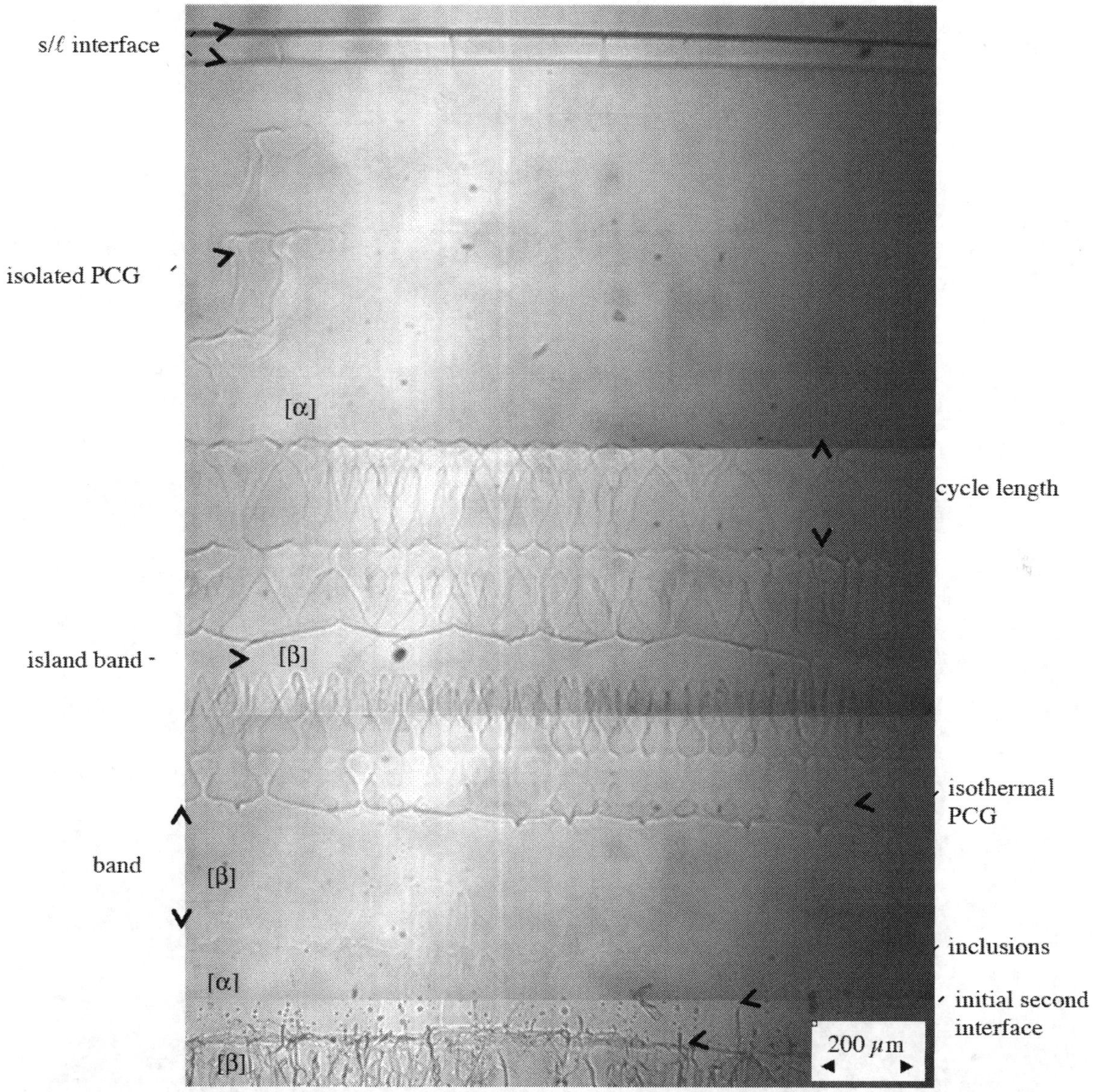

Figure 2: Microstructure formation of the organic compound TRIS-NPG at pulling rates close to the constitutional undercooling. Note that the shadowy black points were objects located outside of the sample. The figure is a compilation of three pictures taken at around 5, 7 and 9 hours after start pulling.

IOP Conf. Series: Materials Science and Engineering **84** (2015) 012055 doi:10.1088/1757-899X/84/1/012055

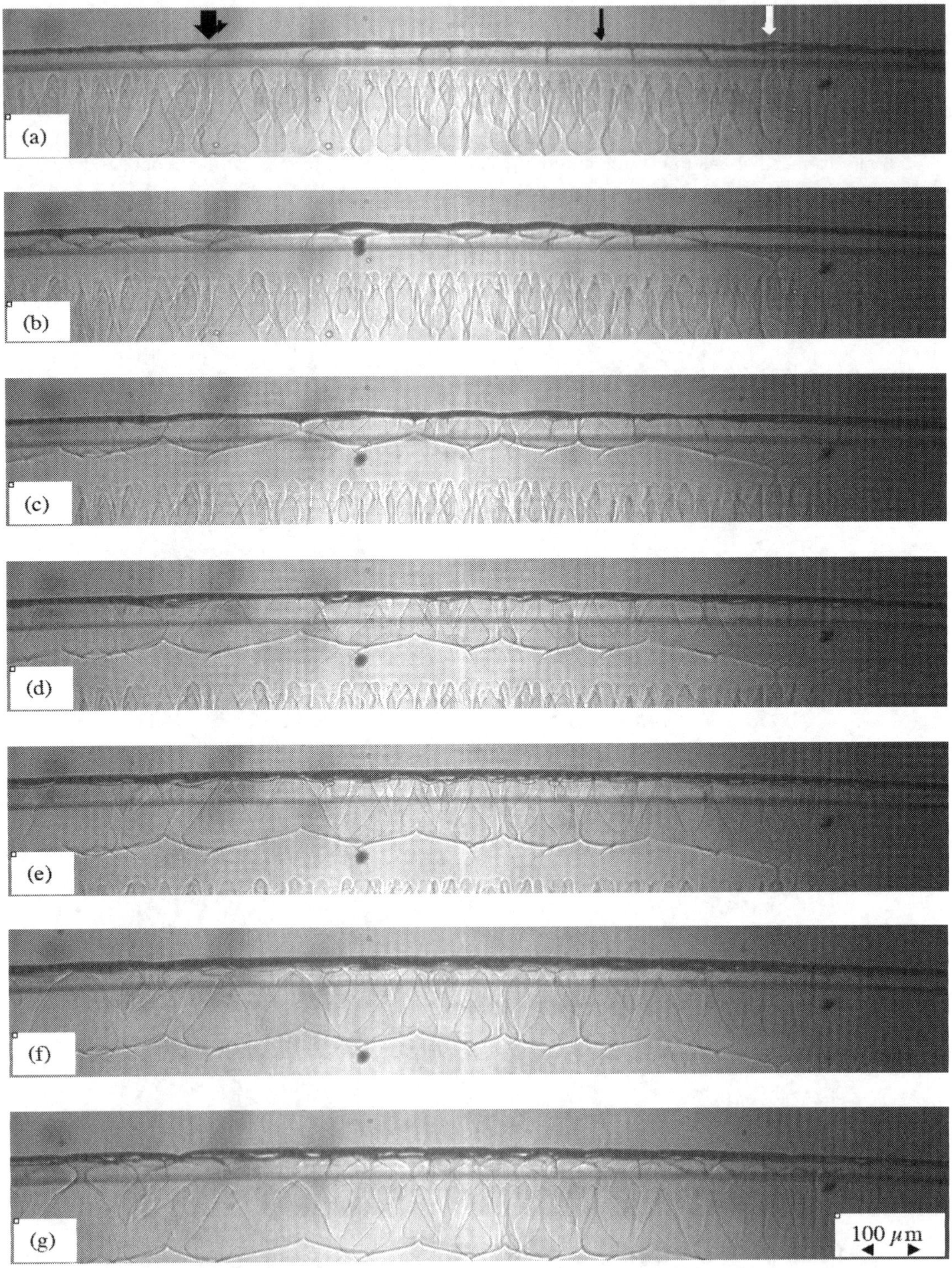

Figure 3: Details of one oscillation cycle. The sequence shows the dynamic of the s/ℓ interface with a time step of 540 s.

4. Discussion

Both solid phases are transparent and the contrast is only given by phase boundaries, impurities and inclusions. After the sample is thermally equilibrated for 120 minutes the alloy is totally molten and transparent above the polycrystalline s/ℓ interface. Below this temperature, the solid phases α and β consist of polycrystalline grains (not shown). Below the s/ℓ interface the crystals of α phase are surrounded by liquid until further down a second interface indicates the transient from α to β phase. The huge amount of grains and inclusions occur mostly blurred and diffuse. In microscopy, the depth of field is very short, here only 55 μm, compared to the illumination through a 600 μm thick layer of organic compounds. As a consequence, sharp grain boundaries can only be obtained by focusing on the front section of the solid alloy, approximately 100 μm after the front glass wall.

With the beginning of pulling the α -phase starts to grow. Since the growth rate is smaller than the pulling rate the interface recoils, namely from liquidus towards solidus. Simultaneously, the second interface moved with the sample almost the whole withdrawal distance. Delayed in time, the β-phase grows then from the moving second interface along liquid channels though the α -ℓ region up to the initial position of the second interface. At that time, β spreads perpendicularly to the sample axis and forms a β-band. At this stage in time, the α and β phases are more or less polycrystalline but the liquid surrounding the grains has either vanished or transferred to liquid channels, or to round or elongated liquid inclusions (droplets).

The growth of the β band goes hand in hand with the appearing of a diffusive gray line. We assume that this blurry line is the glass wall s/ℓ interface junction. Please note, the diffusive gray line occurs below the polycrystalline s/ℓ interface. As we have never achieved to well focus the diffusive line, we believe that the s/ℓ interface reveals some curvature when approaching the glass wall. From studies with extra thin rectangle glass samples, the strong curvature of the s/ℓ interface close to the glass wall is well known. Also it is conceivable that the contact line between the rear glass wall and the s/ℓ interface has a similar appearance. However, we were not able to clearly observe the rear contact line. Indeed, sometimes we have the impression that the blurry line might be representing two overlaying blurry lines rather than one.

Careful observation of the dynamics of the interface exhibits that from the afore mentioned burry line, a well-focused sharp line travels towards the s/ℓ interface in the bulk of the sample. This observation suggests that β growth from the band at the glass wall towards the sample center. Shortly before the β phase envelops completely the remaining α phase, PCG occur. The first cycle exhibits heart-like shaped β lamellae structures, Fig. 2. The next cycle exhibit that the lamellar spacings increase first slightly and later form tulip-like shaped formations. Subsequently, the amount of lamellae increases approximately 2 to 3 times and overlaps in such a way that a β island band occurs.

Although the β phase overgrows nearly the α phase a new cycle of PCG occurs. The lamellae spacing of the new cycle in nature is non-uniform and shows a transition from heart-like shape growth to tulip-like or even 2-λ oscillation growth, see cycle three in Fig. 2. All ß lamellae spacing in common is the increasing of the horizontal growth velocity shortly before the lamellae is nearly erased. Afterwards, a last fourth cycle of PCG with 2-λ oscillation lamellae spacing and heart-like shaped comes into being. The layered solidification structure finished abrupt, except for 3 additionally isolated layered structures cycles. The final solidification pattern shows a planar α phase s/ℓ interface.

The dynamics of the s/ℓ interface during the formation of the third PCG cycle is shown in more details in Fig. 3a-g. The picture sequence covers the time period from 11,160 s to 14,400 s with a time step of 540 s. All pictures reveal a slightly curved dark s/ℓ interface followed by a grey blurry line some 40 μm behind. The grey blurry line is the result of the curvature of the s/ℓ interface in observation direction as mentioned above. The dark dot at the right side is an impurity outside the glass sample.

IOP Conf. Series: Materials Science and Engineering **84** (2015) 012055 doi:10.1088/1757-899X/84/1/012055

The sequence starts with the growing of the β island band. Beginning from the left side toward the center the first picture (Fig. 3a) illustrates the following important features: (i) fine curved lines which follow the s/ℓ interface curvature. One of it is indicated with a thick black arrow. We interpret these fine lines as grain boundaries separating grains of the same solid phase. It is quite important to notice, that during the course of a cycle these fine lines continue to exist. (ii) It is also observable that the grain boundaries are the origin of the appearance of patches of the other solid phase. Such a position is indicated with a small black arrow. (iii) The far right side of the pictures shows an already wavy s/ℓ interface. Beginning with small patches a new solid phase starts to grow, marked with a white arrow. As can be seen in Fig. 3b-c those patches appear from the s/ℓ interface and successively spread over the whole interface. The blurry character of the phase boundary assumes that the patches are growing close to the center of the sample. We recognize this new phase in Fig. 3 a-b as α phase.

Fig. 3c-d shows that the growing α phase spreads perpendicular to the sample axis and toward the observation direction in such a way that they form nearly a new band. Additionally, PCG is visible within the α phase patches which are supposed to be located some distances away from the growing α phase, probably in the center of the sample. Then having spread nearly over the whole interface the α phase goes through the focus which means that the phase is now close to the front wall. At the same time, the β lamellae spacing of the PCG beyond the α phase increase and new patches at the s/ℓ interface becomes visible, Fig. 3d. From the interpretation of the PCG, it can be concluded that the β phase grows perpendicular and in observation direction. PCG exhibits a tulip-like shape but considering the three dimensionality of the sample a cone-like shape is more precise. In Fig. 3d-f the β phase reach down to the front side and the phase boundaries are within the focus. We understand this circumstance that the β phase is between the center and the front side of the sample.

In Fig. 3g the growth of a new phase is observable. The s/ℓ interface gets wavy and new patches are visible. We interpret this as the growing of the α phase similar to the beginning of the cycle, Fig. 3a.

5. Conclusions

Experimental results in binary organic TRIS-NPG alloy have been presented that exhibit complex microstructure formation in the presence of convection. The main finding of the present work is the in-situ observation of cycles of band formation and unsteady peritectic coupled growth. Note that the evaluation of the interface dynamic with the recorded video is much easier than presenting a sequence of pictures. However, careful observation of the dynamics of the s/ℓ interface shows the transformation from banded structure to an unsteady peritectic coupled growth. In the presence of convection the s/ℓ interface becomes curved toward the sample wall. The corresponding concentration profile ahead of the solidification front enables the growth of the peritectic phase close to the sample wall which leads to a α phase matrix covered by a β phase (probably both in the foreground and in the background). This circumstance leads to a competitive growth in parallel and lateral solidification direction. This is visible in form of heart-like shaped and tulips-like shaped β phase lamellae. Beginning with small patches, equal to the tip of the cone, the β phase spreads out in lateral und parallel direction. The increasing volume fraction of the β phase leads to a single island. As soon as the α phase is preferred the tulip-like cones are quickly reduced to small patches and the solidification pattern starts again with an increasing volume fraction of the β phase. This cycle is visible by instabilities between the patch and the maximum size of the cone. The solidification mechanism shows a relatively large volume fraction of β but the phase cannot fully cover the α phase. Finally, the α phase grows stable in form of a planar s/ℓ interface.

Acknowledgement
This work was supported in part by the European Space Agency ESA through means of the ESA map project METCOMP and in part by the Austrian Space Agency ASA within the frame of the MET-TRANS-ISS FP project.

References
[1] Trivedi R., 1995 Metall. Mater. Trans. **26A** 1583-90
[2] Karma A., Rappel W.J., Fuh B.C., and Trivedi R., 1998 Metall. Trans. **29A** 1457–70
[3] Mazumder P., Trivedi R. and Karmer A., 2000 Met. Mat. Trans. **31A** 1234-46
[4] Lee J.H., and Verhoeven J.D., 1994 J. Cryst. Growth **144** 353-366
[5] Trivedi R. and Park J.S, 2002 J. Cryst. Growth **235** 572–88
[6] Lo TS, Dobler S, Plapp M, Karma A and Kurz W, 2003 Acta Mater. **51** 599-11
[7] Mogeritsch J P, and Ludwig A, 2011 *IOP Conf. Series*: Mater Sci. Eng. **27** 012028
[8] Ludwig A, and Mogeritsch J P, 2011 in *Solidification Science and Technology, Proc*. John Hunt Int. Symp. 233-42
[9] Ludwig A, Mogeritsch J P, and Grasser M, 2009 Trans. Indian Institute of Metals **62** 433-36
[10] Mogeritsch J P, and Ludwig A, 2012 in *TMS Ann. Meeting*, Symp. Materials Research in Microgravity 48-55
[11] Ludwig A., Mogeritsch J.P., 2014 Trans Tech Publications, Materials Science Forum Vols. **790-791** 317-22
[12] Mogeritsch J P, Eck S., Grasser M. and Ludwig A, 2010 Material Science Forum **649**. 159-64
[13] Ludwig A, and Mogeritsch J P, Acta Mater. submitted
[14] http://en.wikipedia.org/wiki/Tris
[15] Information on http://www.sigmaaldrich.com
[16] Information on http:// www.merck.de
[17] Barrio M., Lopez D.O., Tamarit J.L., Negrier P., Haget Y., 1995 J. Mater. Chem. **5** 431-39
[18] Barrio M., Lopez D.O., Tamarit J.L., Negrier P., Haget Y., 1996 J. Solid State Chem. **124** 29-38
[19] Mogeritsch J P, 2013 *PhD thesis*. Montanuniversitaet Leoben, Austria

A numerical benchmark test for continuous casting of steel III

B Šarler[1,2] and R Vertnik[1,3]
[1] Laboratory for Simulation of Materials and Processes, Institute of Metals and Technology, Lepi pot 11, SI-1000 Ljubljana, Slovenia
[2] Laboratory for Multiphase Processes, University of Nova Gorica, Vipavska 13, SI-5000 Nova Gorica, Slovenia
[3] Štore Steel d.o.o., Research, Železarska 3, SI-3220 Štore, Slovenia

E-mail: bozidar.sarler@imt.si, robert.vertnik@store-steel.si

Abstract. This paper represents a continuation of numerical results regarding the recently proposed industrial benchmark test [1], obtained by a meshless method. A part of the benchmark test, involving turbulent fluid flow with solidification in two dimensions, was elaborated in [2]. A preliminary macrosegregation upgrade was presented in [3], and in [4], a first three dimensional test was performed. Previous tests were bound to calculations in mold and sub-mold regions only. In the present paper, reference calculations in two dimensions are presented for the entire strand. The physical model is established on a set of macroscopic equations for mass, energy, momentum, species, turbulent kinetic energy, and dissipation rate. The mixture continuum model is used to treat the solidification system. The mushy zone is modeled as a Darcy porous media with Kozeny-Karman permeability relation, where the morphology of the porous media is modeled by a constant value. The incompressible turbulent flow of the molten steel is described by the Low-Reynolds-Number k-ε turbulence model, closed by the Abe-Kondoh-Nagano closure coefficients and damping functions. Lever microsegregation model is used. The numerical method is established on explicit time-stepping, collocation with scaled multiquadrics radial basis functions with adaptive selection of its shape on non-uniform five-nodded influence domains. The velocity–pressure coupling of the incompressible flow is resolved by the explicit Chorin's fractional step method. The advantages of the method are its simplicity and efficiency, since no polygonisation is involved, easy adaptation of the nodal points in areas with high gradients, almost the same formulation in two and three dimensions, high accuracy and low numerical diffusion.

1. Introduction

Continuous casting [5] is the most common process in the production of steel. The process starts by pouring the molten metal into to the water cooled mold, where cooling intensity is high enough to solidify steel around the inner surface of the mold and generates the solid shell with molten metal in the center of strand. After several minutes, the strand is pulled into the secondary cooling system, which contains water spray systems with much smaller cooling intensity as in the water channels in the mold, and rollers, which support and guide the strand up to the end of the casting machine. The quality of the casted product (round or square billets, blooms or slabs) depends mainly on the process parameters in the mold region, where complex physical phenomena occur. The liquid metal, poured with a high velocity from the submerged entry nozzle (SEM) into the mold, produces turbulent flow with several re-circulating zones. Large heat fluxes, extracted from the mold, are a consequence of very high flow rates of the cooling water in the mold channels. It is impossible to measure temperature and velocity field inside the strand due to the very high temperatures of steel and inaccessibility of the region during the process. The numerical models [6] help to better understand the casting behavior and to further improve and optimize the process, particularly in such experimentally sophisticated situations. Various numerical methods have already been used to simulate the described spectra of problems. However, until now, there is a lack of benchmark tests that would enable verification of numerical methods for solving transport phenomena in continuous casting of steel in an ordered way, similar as it was done for binary solidification [7]. We recently tried to fill this gap with proposing a first such benchmark [1], elaborated in detail in a journal paper [2]. First related macrosegregation simulations have been displayed in [3]. The simulations from [2] have been in [4] upgraded to three dimensions. In this paper, the two dimensional turbulent heat transfer, fluid flow, and species transfer solidification discussion in [3] is displayed for the whole length of the strand and not only to the mold and near sub-mold regions. The main aim of the test is to get new reference results and to be able to further compare the results of different numerical methods for the same physical model. The presented physical model considers turbulent fluid flow and macrosegregation of carbon steel. Turbulent flow is modeled by the two-equation eddy-viscosity model with the low-Re corrections. The numerical results are shown for the case with simplified lever microsegregation model of 0.8 wt%C steel. The novel meshless Local Radial Basis Function Collocation Method (LRBFCM) is employed for solving the coupled system of six partial differential equations in two dimensions. The objective of the present paper is to produce a first reference numerical solution of the given problem for the strand of the length 12 m, where solidification is completed. LRBFC was first developed in [8] for diffusion problems, to convection diffusion problems in [9], and to driven cavity and natural convection problems in [10, 11]. The turbulence was tackled in [12, 13]. The melting benchmark [14] was successfully solved in [15] and the macrosegregation benchmark [7] in [16, 17]. The magneto-hydrodynamic effects were tackled in [18] for natural convection in a cavity and in [19] for backward facing step situation.

2. Governing equations

The system of equations, describing the coupled transport phenomena in the continuous casting of steel, is derived based on the mixture continuum formulation by Bennon and Incropera [20] and Reynolds time-averaging approach for modelling incompressible turbulent flow [21]. The time-averaged transport equations for mass, energy, momentum and solute conservation are

$$\nabla \cdot \mathbf{u} = 0 , \tag{1}$$

$$\rho \frac{\partial \mathbf{u}}{\partial t} + \rho \nabla \cdot (\mathbf{u}\mathbf{u}) = -\nabla P + \nabla \cdot \left[2(\mu + \mu_t)\mathbf{S} \right] - \frac{2}{3}\rho \nabla k - \mu_L \frac{K_0 \left(1 - f_L^2\right)}{f_L^3} (\mathbf{u} - \mathbf{u}_S) + \mathbf{F}_b , \tag{2}$$

$$\rho \frac{\partial h}{\partial t} + \rho \nabla \cdot (\mathbf{u}h) = \nabla \cdot (\lambda \nabla T) + \rho \nabla \cdot \left[f_S (h_L - h_S)(\mathbf{u} - \mathbf{u}_S) \right] + \nabla \cdot \left(f_L \frac{\mu_t}{\sigma_T} \nabla h_L \right), \tag{3}$$

$$\rho \frac{\partial C}{\partial t} + \rho \nabla \cdot (\mathbf{u}C) = \nabla \cdot (f_S D_S \nabla C_S + f_L D_L \nabla C_L) + \rho \nabla \cdot \left[(\mathbf{u} - \mathbf{u}_S)(C_L - C) \right] + \nabla \cdot \left(\frac{f_L \cdot \mu_t}{\sigma_C} \nabla C_L \right) \tag{4}$$

with $\mathbf{u}$, P, h, T and C for velocity, pressure, enthalpy, temperature and solute concentration, respectively, and ρ, μ, μ_t, f, λ, D are standing for density, molecular dynamic viscosity, turbulent dynamic viscosity, volume phase fraction, thermal conductivity, and diffusivity of the material. In above equations, subscripts L and S denote liquid and solid phase. $\mathbf{S}$ stands for the strain-rate tensor. The fourth term in equation (2) represents the Darcy term, where K_0 is the morphology constant of the porous medium and $\mathbf{u}_S$ is velocity of the solid phase. The fifth term in equation (2) represents the buoyancy forces, defined as

$$\mathbf{F}_b = \rho\left[\beta_T \mathbf{g}\left(T - T_{ref}\right) + \beta_C \mathbf{g}\left(C_L - C_{ref}\right)\right] \tag{5}$$

where β_T, β_C, T_{ref} and C_{ref} stand for thermal expansion coefficient, solute expansion coefficient, reference temperature and reference concentration, respectively. The turbulent dynamic viscosity is defined as $\mu_t = \rho c_\mu f_\mu k^2 / \varepsilon$, where k and ε are kinetic energy and dissipation rate, respectively. They are calculated by the following transport equations

$$\rho\frac{\partial k}{\partial t} + \rho\nabla\cdot(\mathbf{u}k) = \nabla\cdot\left[\left(\mu_L + \frac{\mu_t}{\sigma_k}\right)\nabla k\right] + P_k - \rho\varepsilon + \rho D + \mu_L\frac{K_0\left(1-f_L^2\right)}{f_L^3}k\,, \tag{6}$$

$$\rho\frac{\partial \varepsilon}{\partial t} + \rho\nabla\cdot(\mathbf{u}\varepsilon) = \nabla\cdot\left[\left(\mu_L + \frac{\mu_t}{\sigma_\varepsilon}\right)\nabla \varepsilon\right] + \rho\left(c_{1\varepsilon}f_1 - c_{2\varepsilon}f_2\varepsilon\right)\frac{\varepsilon}{k} + \rho E + \mu_L\frac{K_0\left(1-f_L^2\right)}{f_L^3}\varepsilon\,. \tag{7}$$

In equations (3,4,6,7) c_μ, f_μ, $c_{1\varepsilon}$, f_1, $c_{2\varepsilon}$, f_2, σ_T, σ_C, σ_k and σ_ε are the closure coefficients, while D and E are additional source terms of the low-Reynolds turbulent model [21]. The constitutive temperature-enthalpy relationship h_S and h_L in equation (3) are defined as

$$h_S = \int_{T_{ref}}^{T} c_{pS}dT\,,\quad h_L = \int_{T_{ref}}^{T_S} c_{pS}dT + \int_{T_S}^{T} c_{pL}dT + h_m = h_S(T) + \int_{T_S}^{T}\left(c_{pL} - c_{pS}\right)dT + h_m\,, \tag{8,9}$$

respectively. The liquid fraction f_L, concentration of solute in liquid phase C_L, partition ratio k_p, and liquidus temperature T_L are determined based on the lever rule microsegregation model, i.e.

$$f_L = 1 - \frac{1}{1-k_p}\frac{T - T_L}{T - T_m}, C_L = \frac{C}{1 + f_S\left(k_p - 1\right)}, k_p = \frac{C_S}{C_L}, T_L = T_m + \left(T_e - T_m\right)\frac{C}{C_e} \tag{10,11,12,13}$$

where T_m, T_e and C_e are melting temperature, eutectic temperature and solute concentration in eutectic alloy, respectively. They are obtained from the phase diagram for the iron-carbon system.

3. Solution procedure

Our solution procedure is based on the novel meshless technique that follows the following steps. We seek the solution of the equations for the velocity, pressure, temperature, concentration, turbulent kinetic energy, and dissipation at time $t + t_0$ by assuming known fields $\mathbf{u}$, p, T, C, k, and ε at time t_0 and known boundary conditions at time $t > t_0$. The coupled set of mass conservation equation (1) and momentum conservation equations (2) are solved by the fractional step method [22], where the continuity of the mass (1) is considered by constructing the pressure Poisson's equation. The governing equations are discretized by using the explicit time discretization. This leads to the following algorithm, summarized through the following 7 steps: (I) The intermediate velocity components are calculated without the pressure gradient, (II) the pressure Poisson equation is treated, (III) the velocity components are corrected by the pressure gradient, (IV), after the solution of the velocity field, the transport equation (3) for energy is solved, (V) the temperature field is calculated from the enthalpy field, calculated in previous step, using the inverse of the constitutive temperature-enthalpy relationships (8,9), i.e. $T = T(h)$, (VI) species concentration (4), turbulent kinetic energy (6) and dissipation rate (7) fields are calculated, and finally (VII) the turbulent viscosity, solute concentrations (11,12) and liquidus temperature (13), are updated for each node and the solution is set ready for the next time step. All derivatives of the transport equations are calculated by the LRBFCM [8], where the collocation is made locally on overlapped sub-domains. On each sub-domain, the scalar function Φ (standing for temperature, species concentration, velocity components, pressure, turbulent kinetic energy, and turbulent dissipation rate) is represented over a set of (in general) non-equally spaced nodes $\mathbf{p}_n$; $n = 1,2,...,{}_lN$ in the following way $\Phi(\mathbf{p}) \approx \sum_{k=1}^{K} \psi_k(\mathbf{p})\alpha_k$, $\psi_k(\mathbf{p}) = \left[r_k^2 + c_k^2\right]^{1/2}$, where ψ_k stands for the multi-quadric radial basis shape functions [23], α_k for the coefficients of the shape functions, K the number of the shape functions, c_k the shape parameter, and r_k the radial distance between two points in the influence domain. Five nodes are used in each of the influence domains. The radial distance is scaled as $r_k^2 = (p_x - p_{xk)})^2 / p_{xk\max}^2 + (p_y - p_{yk)})^2 / p_{yk\max}^2$. The higher the shape parameter is, more accurate are the results. However, higher value of shape parameter induces ill-conditioning of the collocation matrix. The shape parameter of the multiquadrics is respectively adaptively scaled in each influence domain in such a way that the condition number of the local collocation matrix does not exceed 10^{10}. The detailed procedure of calculating the derivatives with the LRBFCM can be found in [2].

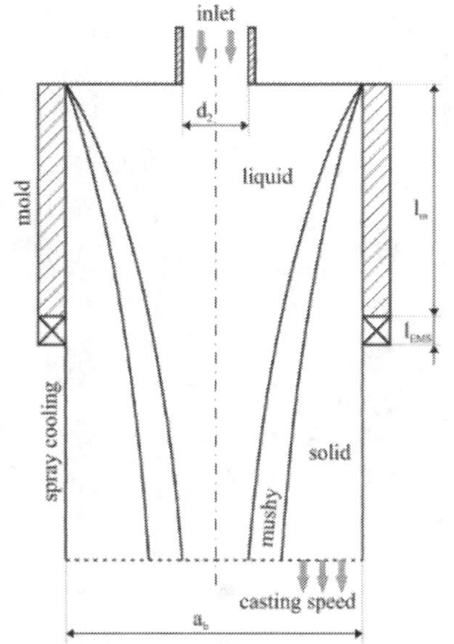

Figure 1. The upper part of geometry with typical dimensions.

Figure 2. Considered boundary condition types.

4. Description of the benchmark test

The proposed benchmark represents a simplified 2D model of the continuous casting process. The geometry of the simplified casting machine is shown in Fig. 1. A narrow side cross-section of a slab is assumed. The computational domain represents half of the longitudinal section of the slab with length 12 m, shown in Fig. 2. This simplification is used mainly for straightforward verification of the developed numerical model, and in order to reduce the CPU time of the simulations. The dimensions of the casting geometry of the slab are narrow side width $a_b = 0.18$ m, diameter of the SEN $d_2 = 0.035$m, and mould height $l_m = 0.8$m.

The material properties of steel are temperature dependent. However, for the benchmark purpose, constant values are used for each phase. They are $\rho = 7200\text{kgm}^{-3}$, $c_p = 700\,\text{Jkg}^{-1}\text{K}^{-1}$, $\lambda = 30\,\text{Wm}^{-1}\text{K}^{-1}$, $\mu = 0.006\,\text{kgm}^{-1}\text{s}^{-1}$, $h_m = 250000\text{Jkg}^{-1}$, $K_0 = 1.6\times10^{8}$, , $D_S = 1.6\times10^{-11}\text{m}^2\text{s}^{-1}$, $D_L = 1.0\times10^{-8}\text{m}^2\text{s}^{-1}$, $\beta_T = 0.0001\text{K}^{-1}$, $\beta_C = 0.004$. The inlet carbon concentration is $C_0 = 0.8\,\text{wt\%}$, inlet temperature is $T_0 = 1800\text{K}$, and the slab casting velocity is $u_{cast} = 1.1$ m/min, respectively.

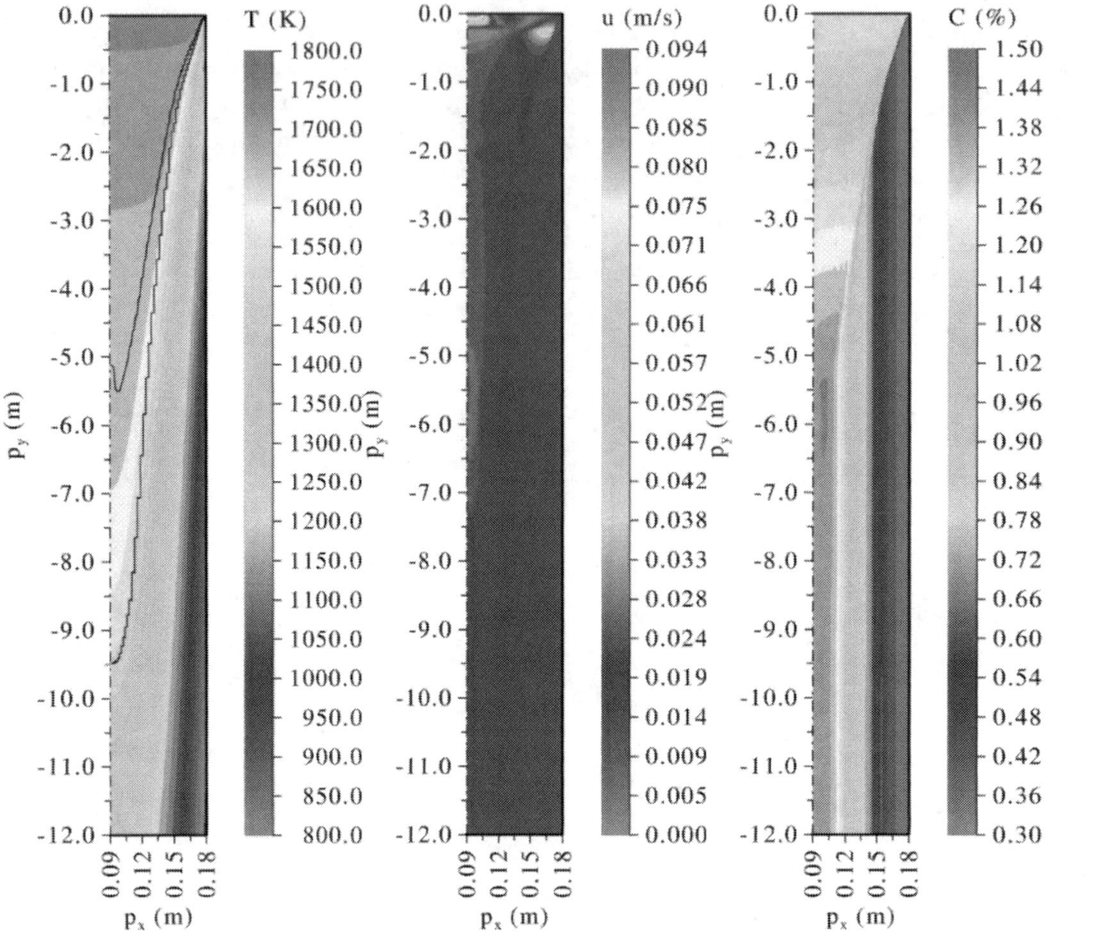

Figure 3: Left - Temperature field. Center - absolute velocity field. Right - carbon segregation field.

The boundary conditions for the velocity, temperature, turbulent kinetic energy, dissipation rate, and species concentration, are set as follows

SEN outlet: Constant values are prescribed for all variables $u_{inlet,x} = 0$, $u_{inlet,y} = a_b u_{c,y} / d_2$, $k_{inlet} = 1.5\left(I_t u_{inlet,y}\right)^2$, $\varepsilon_{inlet} = \dfrac{c_\mu^{0.75} k_{inlet}^{1.5}}{0.07 d_2}$, $T_{inlet} = T_0$, $C_{inlet} = C_0$, where $u_{c,y}$ is the casting velocity in the casting direction, and I_t is the turbulent intensity calculated from the following relation

$$I_t = 0.16\mathrm{Re}^{-1/8}, \mathrm{Re} = \rho u_{inlet,y} d_2 / \mu . \qquad (14,15)$$

Normal gradient of pressure is set to zero, i.e. $\dfrac{\partial P}{\partial y} = 0$.

Slab end: The pressure outlet is prescribed, where the following boundary conditions are used

$$P=0\,,\frac{\partial u_x}{\partial y}=0\,,\frac{\partial u_y}{\partial y}=0\,,\frac{\partial k}{\partial y}=0\,,\frac{\partial \varepsilon}{\partial y}=0\,,\frac{\partial T}{\partial y}=0\,,\frac{\partial C}{\partial y}=0\,. \tag{16}$$

Top surface (meniscus): At the meniscus, the symmetry boundary conditions are used (free surface flow). The normal derivative of all variables is set to zero, except the vertical velocity u_y, which is set to zero.

Moving walls: The walls with the solidified steel move with casting velocity along the casting direction. At the walls, where the liquid phase exists, the no-slip boundary conditions for the velocity are set. In the mould, the Robin boundary condition is used, with the surface heat transfer coefficient equal to $2000\,\mathrm{Wm^{-2}K^{-1}}$. Below the mould, at the secondary cooling system, the heat transfer coefficient is equal to $800\,\mathrm{Wm^{-2}K^{-1}}$. The normal derivative of the solute concentration is set to zero.

5. Numerical results

The low-Re turbulence model by Abe-Kondoh-Nagano (AKN) [23] is selected, where the normal distance in closure coefficients is calculated up to the wall. The simulation was performed on the node arrangement with 84707 nodes. The steady-state solution shown in Figure 3 is approached by a false transient calculation using a fixed time-step of 10^{-3} s. The results of the simulation, which runs about one week on 20 contemporary cores, are shown in Fig. 3.

6. Conclusions

This paper represents a continuation of numerical benchmarking [1,2,3,4] proposed for the continuous casting of steel. The geometry and material properties of the continuous casting process are simplified in order to enable straightforward performance comparisons with various numerical methods. In the paper, our in-house developed meshless numerical method, based on the local radial basis function collocation, is used to get the steady state solution of the turbulent fluid flow with heat transfer, solidification, and macrosegregation of carbon along the whole longitudinal geometry of the two-dimensional strand. A symmetric half of the narrow side of the slab was considered. The results were obtained without electromagnetic forces [25] in the momentum equation, and with a simplified lever microsegregation model. The benchmark will be further upgraded and adapted, based on the feedback from the colleges in the field of the numerical modelling of solidification processes. A journal paper with tabulated results of the present simulations, similar as [2], is underway. A web page of the benchmark results with systematically tabulated results of different numerical methods will be established, similar to [7].

Acknowledgement: The financial support of Slovenian Grant Agency and Štore Steel company under grant L2-6775 is kindly acknowledged.

References

[1] Šarler B, Vertnik R and Mramor K 2012 MCWASP XIII, IOP Conf.Series: Mat.Sci.Eng. 33 012012.
[2] Vertnik R and Šarler B 2014 *Eng. Anal. Bound. Elem.* **45** 45-61
[3] Šarler B and Vertnik R 2014 *Materials Science Forum* **790-91** 279-284
[4] Vertnik R and Šarler B 2014 ICASP IV, IOP Conf. Series: Mat.Sci.Eng (to appear)
[5] Irwing W R 1993 Continuous Casting of Steel, (London: The Institute of Materials)
[6] Thomas B G 2002 Modeling of the Continuous Casting of Steel - Past, Present Future, www.brimacombecourse.org/pdf/2001_Thomas.pdf
[7] Bellet M, Combeau H, Fautrelle Y, Gobin D, Rady M, Arquis E, Budenkova O, Dussoubs B, Duterrail Y, Kumar A, Gandin C A, Goyeau B, Mosbah S and Založnik M 2009 *Int. J. Therm. Sci.* **48** 2013-16
[8] Šarler B and Vertnik R 2006 *Comput. Math. Appl.* **51** 1269-82
[9] Vertnik R and Šarler B 2006 *Int. J. Numer. Method Heat Fluid Flow* **16** 617-640
[10] Mramor K, Vertnik R and Šarler B 2013 *Comput. Mater. Cont.* **36** 1-21
[11] Kosec G and Šarler B 2009 *Int. J. Numer. Method Heat Fluid Flow* **18** 868-82
[12] Vertnik R and Šarler B 2009 *Comput. Model. Engin. Sci.* **44** 65-95
[13] Vertnik R and Šarler B 2011 *Advances in Applied Mathematics and Mechanics* **3** 259-279

[14] Gobin D and Le Quéré P 2000 *Comput. Assist. Mech. Eng. Sci.* **7** 289-306
[15] Kosec G and Šarler B 2009 *Comput. Model. Eng. Sci.* **47** 191-216
[16] Kosec G Založnik M, Šarler B and Combeau H 2011 *Comput. Mater. Cont.* **22** 169-59
[17] Kosec G and Šarler B 2014 *Eng. Anal. Bound. Elem.* **45** 36-44
[18] Mramor K, Vertnik R and Šarler B 2013 *Comput. Model. Eng. Sci.* **92** 327-352.
[19] Mramor K, Vertnik R and Šarler B 2014 *Eng Anal. Bound. Elem.* doi: 10.1016/j.enganabound.2014.04.013
[20] Bennon W D and Incropera F P 1988 *Numer. Heat Transf. A-Appl.* **13** 277-96
[21] Wilcox, D C 1993 *Turbulence modeling for CFD* (California: DCW Industries, Inc.)
[22] Chorin A J 1967 *J. Comput. Phys.* **2** 12-26
[23] Buhmann M D 2003 *Radial Basis Function: Theory and Implementations* (Cambridge: Cambridge University Press)
[24] Abe K, Kondoh T and Nagano Y 1994 *Int. J. Heat Mass Transf.* **37** 139-51
[25] Mramor K, Vertnik R and Šarler B 2014 Mat. Technol. 48 281-288.

Temperature and Pressure Evolution during Al Alloy Solidification at Different Squeeze Pressures

Junwen Li, Haidong Zhao and Zhenming Chen

School of Mechanical and Automotive Engineering, South China University of Technology, Guangzhou 510640, China

E-mail: hdzhao@scut.edu.cn

Abstract. Squeeze casting is an advanced and near net-shape casting process, in which external high pressure is applied to solidifying castings. The castings are characterized with fine grains and good mechanical properties. In this study, a series of experiments were carried out to measure the temperature and pressure histories in cavity of Al-Si-Mg direct squeeze castings with different applied solidification pressures of 0.1, 50, 75, and 100 MPa. The evolution of the measured temperatures and pressures was compared and discussed. The effect of pressure change on formation of shrinkage defects was analyzed. Further the friction between the castings and dies during solidification was calculated. It is shown that the applied squeeze pressure has significant influence on the friction at die and casting interfaces, which affects the pressure evolution and transmission. The results could provide some benchmark data for future thermal-mechanics coupled modeling of squeeze castings.

1. Introduction

Squeeze casting is a special fabrication technique in which solidification is promoted under high pressure. In the casting process, the high pressures applied during solidification can eliminate both gas and shrinkage porosity. The increased cooling rate caused by intimate contact between the solidifying alloy and die results in the formation of fine grains with small dendritic arm spacing. Consequently, casting soundness and mechanical properties are improved. Given the advantages of squeeze casting mentioned above, it is particularly used to fabricate high integrity near-net-shape components with high performance [1, 2].

In the process, the applied pressure improves contact between castings and dies and changes liquidus, hence affecting heat transfer and temperature distribution in the casting system. Meanwhile, solid formation and volume shrinkage have important effect on pressure transmission from plunger or punch [3]. Therefore, thermal-mechanics behavior is the essential problem for squeeze casting process. Several researchers have established mathematical models to simulate the behaviour [4-6]. However, due to the difficulty and complexity of measuring in-cavity pressures of castings with high forced pressures, the pressure in high pressure die casting were measured and reported by only few researchers [3, 7] and that in squeeze casting has not been reported yet. Hence, the models were only verified with the porosity results rather than validated directly with experiment measurement.

In this study, a series of experiments were carried out to measure the temperature and pressure histories in cavity of Al-Si-Mg direct squeeze castings with different applied solidification pressures. The evolution of the measured temperature and pressure was analyzed and discussed. It is shown that

IOP Conf. Series: Materials Science and Engineering **84** (2015) 012057 doi:10.1088/1757-899X/84/1/012057

the applied squeeze pressure has significant influence on the casting solidification and the friction of the die and casting interfaces.

2. Experimental works

Figure 1 shows schematic of the experimental squeeze castings. The experiments were conducted with a 1000kN hydraulic press with a PLC control. The materials used were AlSi7%Mg0.4% alloys and the die was made of H13 steels. The solidification was carried out at gravity condition and the forced pressures of 50, 75, and 100 MPa. The temperature changes in both casting centre and die side at 75 and 150 mm height were recorded with K type thermocouples. A quartz pressure sensor (Kistler 6175A2) designed for application up to temperatures of 700 and pressures of 200 MPa [8] was set to the casting bottom surface to evaluate the pressure transmission from the top punch. The castings were cut along the symmetric cross-sections in vertical direction for assessment of inside macro morphology. Also for each casting, a sample was obtained from the location of thermocouple TC1, and the sample microstructures were observed with a Leica microscope.

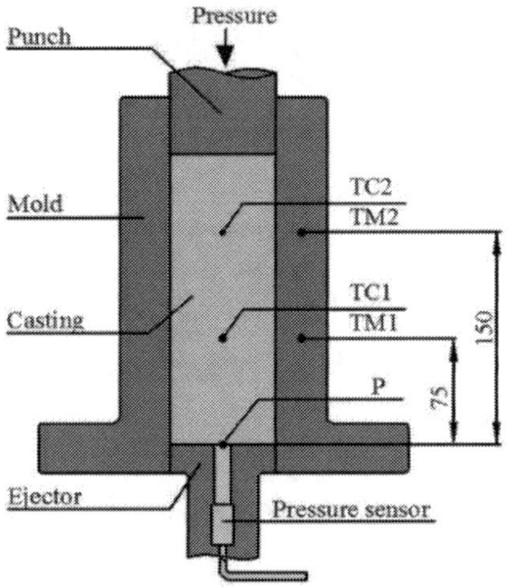

Figure 1. Schamatic of thermo-mechanical measurement in squeeze casting

3. Results and discussion

Figure 2 exhibits the halves of experimental castings at the different solidification pressures. It is obvious that there exists a concave shrinkage cavity at the top of the gravity casting whereas the castings with applied pressures have flat top surfaces. The roughness of castings with squeeze pressures is finer compared to the gravity casting. Figure 3 shows the microstructures of the casting samples, which are consisted of the alpha cells and eutectics. It is indicated that the castings with the forced pressures are characterized with finer alpha grains and less eutectics than the gravity one.

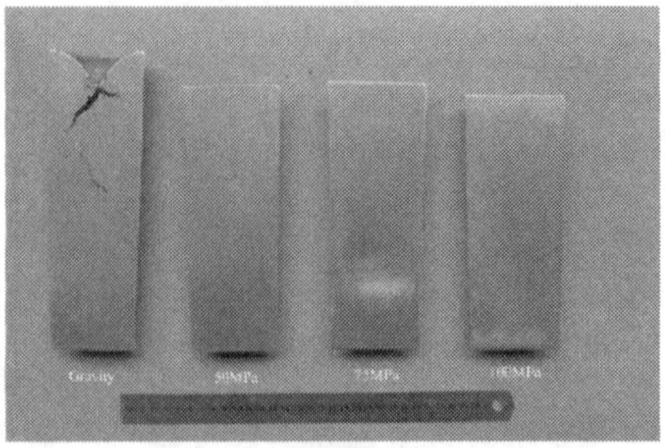

Figure 2. Experimental squeeze castings under different applied pressures

Figure 3. The metallurgical photography of the casting samples

3.1. Temperature and pressure evolution

3.1.1. Gravity solidification

Figure 4 shows the curves of temperatures of castings and dies as well as the pressure variation during solidification at the gravity condition. The pressure measured from the casting bottom surface kept unchanged and was of near zero. According to the temperatures change at the TC1 and TC2 points, the temperature of liquid melt after pouring dropped to the range about the alloy liquidus (635℃). The phase transformation of primary alpha-cell formation happened from 12.6 to 63.8 seconds during which the temperatures dropped slowly. From 63.8 to 116.6 seconds after pouring the casting temperature kept constant at 595℃. It is obvious that within the period the eutectic transformation occurred. After the transformation, the casting temperatures decreased quickly. Due to the effect of heat transfer from the casting to the bottom platen, the temperature drop at location TC1 closer to the platen was larger than that at TC2. From Fig. 4 the die temperatures increased dramatically before 28 seconds, and then went up quite slowly due to the formation of air gap at the casting and die interface caused by the solidification shrinkage.

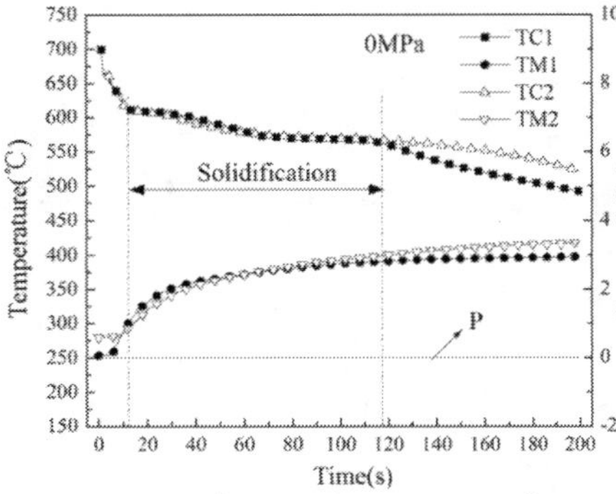

Figure 4. Temperature and pressure curves under gravity condition

3.1.2. Solidification at 50MPa

The evolution of the measured temperatures and pressures of the squeeze castings at 50 MPa was shown in Fig. 5. The casting solidification happened from 5.6 to 65.4 seconds. Compared with the data in Fig. 4, it is clear that with the squeeze pressure the solidification time decreased and the cooling rate increased. This is considered to be contributed to two factors. First, the pressure forced the solidified casting to deform and hence have an intimate contact with the die, increasing the heat flux through the interface. Second, the pressure made the eutectic point in equilibrium Al-Si diagram move to higher Si contents [9] and fraction of eutectic phases in the alloy was decreased, which could also be seen from by the microstructures in Fig. 3a and 3b. The die temperatures in Fig. 5 increased quickly and finally had almost same values with those of the castings, which indicated intensive heat flux from casting to die when using squeeze pressure.

The pressure curve in Fig. 5 shows that after start of the forced pressure from the upper punch at 17.2 seconds, the pressure at the casting bottom increased quickly and reached to the peak value of 35.35 MPa. At the beginning of the casting solidification, the solidified shells encapsulated the casting were quite thin and could not against the forced pressure so that the solidifying casting was pressed to expand in radial direction and had well contact with die. The contact increased the friction between them and decreased the transmission of pressure from the upper punch. As a result, the measured pressure declined gradually after reaching the peak and had a minimum pressure of 18.01 MPa at 54.8 seconds. Along with the solidification, the growing solids increased the casting strength and prevented the further deformation under the forced pressure. Meanwhile, the solidification shrinkage made the casting detach from the side die surfaces, forming air gap at the interface. Then the friction declined and the measured pressure increased after the decrease.

It was pointed out by Stangeland [10] that the Al-7%Si-0.4%Mg alloys had a maximum solidification shrinkage at temperature of 560.8℃ and the solid fraction of 88%, coinciding with the temperature when the minimum pressure was measured in Fig. 5. We considered that the tendency forming shrinkage defects increased at the period with the lower measured pressure due to large solidification shrinkage and insufficient squeeze pressure inside the casting, illustrated by the shade region in Fig. 5.

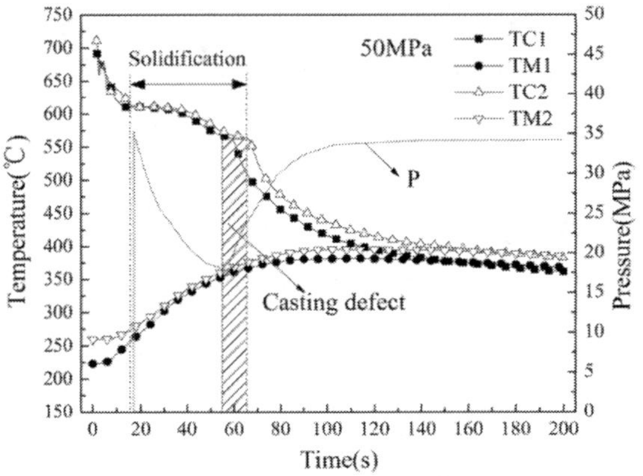

Figure 5. Temperature and pressure curves under 50MPa applied pressure

3.1.3. Solidification at 75MPa

Figure 6 displays the measured temperatures and pressures at the squeeze pressure of 75 MPa. The evolution of the temperature and pressure is similar with those in Fig. 5. The casting solidification lasted 50 seconds (from 19.6 to 69.6 seconds) a little less than that at 50 MPa. Compared to Fig. 5, the eutectic transformation time in Fig. 6 is shorter because of the larger forced pressure.

It is shown in Fig. 6 that the measured pressure at 75 MPa condition has the peak value of 48.38 MPa larger than that at 50 MPa. It should be noted that the experiments were performed with a press, and the velocities of the upper punch (please see Fig. 1) moving downwards with the different internal pressures set by PLC were different causing the different start times of measured pressure in Fig. 5 and 6. Since the larger squeeze pressure of 75 MPa could induce more intimate contact at the casting and die interface by forcing the solidifying casting to expand in radial direction, the measured pressure dropped dramatically to a minimum value of 17.25 MPa less than that in Fig. 5. Similar to Fig. 5, the pressure increased from the later period of solidification and then kept at a stable pressure of 33.97 MPa. Compared to the results in Fig. 5, the duration from 560.8℃ (at which the alloy has maximum shrinkage ratio) to the end of solidification marked by the shade area in Fig. 6 was shorter decreasing shrinkage defects at the larger squeeze pressure.

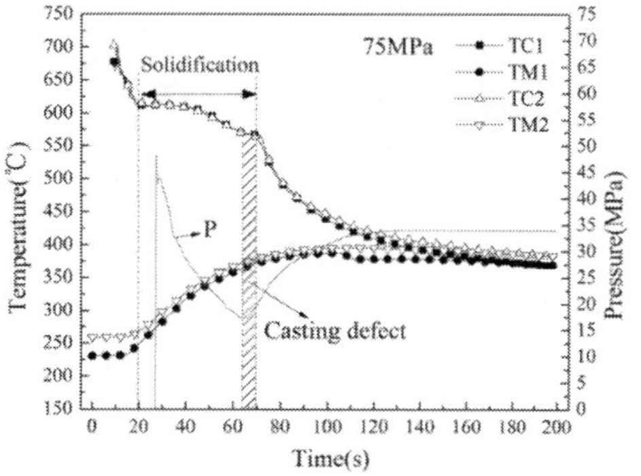

Figure 6. Temperature and pressure curves under 75MPa applied pressure

3.1.4. Solidification at 100 MPa

The measured temperatures and pressures at the forced pressure of 100MPa were plotted in Fig. 7. Compared with the solidification at 50 and 75 MPa cases, the solidification time at 100 MPa from 4.6 to 51.2 seconds is the shortest indicating large cooling rate. Also the eutectic transformation period usually with constant temperature in cooling curves was not very obvious due to less eutectic fractions under 100 MPa pressures. Although the die temperatures in Fig. 7 show the similar evolution tendency with those in Fig. 5 and 6, however, the die temperatures measured reached approximate 387℃ higher than the other two cases.

As shown in Fig. 7, the measured pressure quickly reached its peak of 78.90MPa. Due to the casting deformation under the forced pressure and the friction between the casting and die, the pressure during the solidification period decreased, similar with the results in Fig. 5 and 6. On contrary to Fig. 5 and 6, after the solidification the pressure continued to drop. The reason is that the 100 MPa pressure could make the casting at high temperature deform and hence keep well contact with the die during the whole casting process, maintaining large friction at the interface. The friction limited the squeeze pressure transmission, and hence the measured pressure declined continuously. Fig. 7 shows that the pressure applied in the solidifying casting is higher over 50 MPa, which could prevent formation of shrinkage defects.

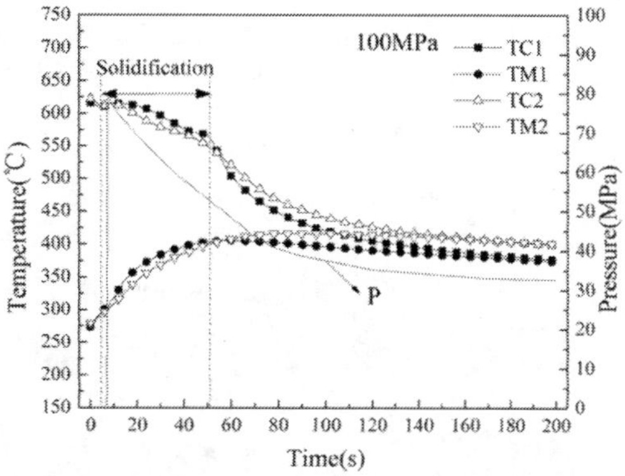

Figure 7. Temperature and pressure curves under 100MPa applied pressure

3.2 Analysis of friction between casting and die

In squeeze casting process, the external pressures pressed castings against die walls inducing their well contact. The contact results in large friction between them, which is detrimental to die service lives. As a result, the lives of squeeze casting dies are usually less than those of low pressure die castings and even high pressure die castings.

Figure 8 shows the force schematics for the experimental castings. The applied forces are given by the set-up of hydraulic press. The support force can be calculated from the measured pressure. Then the friction at the casting and die interfaces can be determined by

$$f = F \times S_u - P \times S_d \tag{1}$$

where f is the friction between casting and die, F is applied pressure from punch, P is pressure measured from the casting bottom surface, S_u and S_d are the areas of casting upper and bottom surfaces, respectively.

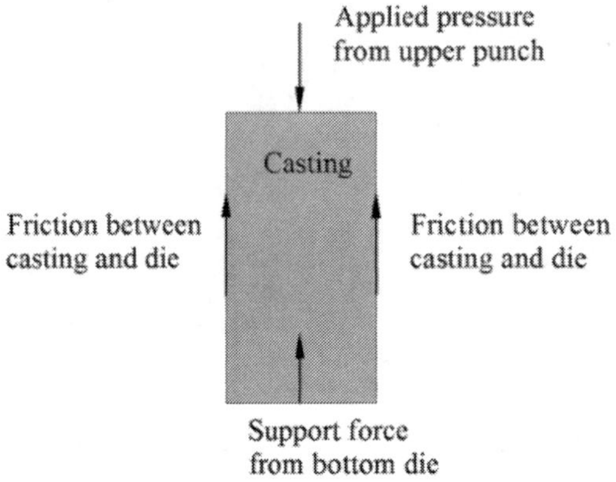

Figure 8. The force analysis during the squeeze casting process

Appling the Eq. 1 to the experimental results, the evolution of friction between the castings and dies is illustrated and compared in Fig. 9. In the figure, the open symbols express the friction during the solidification stages while the filled ones mean that after the solidification. It is obvious that the friction became larger with the increased applied pressures. The frictions at the applied pressures of 50 and 75MPa increased after the start of solidification, and then decreased from the later period of solidification due to the shrinkage. On the contrary, the friction at the 100 MPa pressure increased during the whole solidification process. The results have shown that in thermal-mechanics coupled modeling of squeeze casting process, the friction should be taken into account. As discussed in the literature [1], efficient pressure transmission during squeeze casting process is in favor of preventing or eliminating shrinkage defects, and the applied pressure after casting solidification would not contribute to feeding shrinkage but do harm to die life and waste energy. Therefore, the applied pressure of 100 MPa induced high internal pressure in the solidifying casting preventing shrinkage defects but the pressure dwelling after the solidification was harmful to the die life. It is indicated that applied pressure and its dwelling times should be properly set to prevent shrinkage defect and improve die lives.

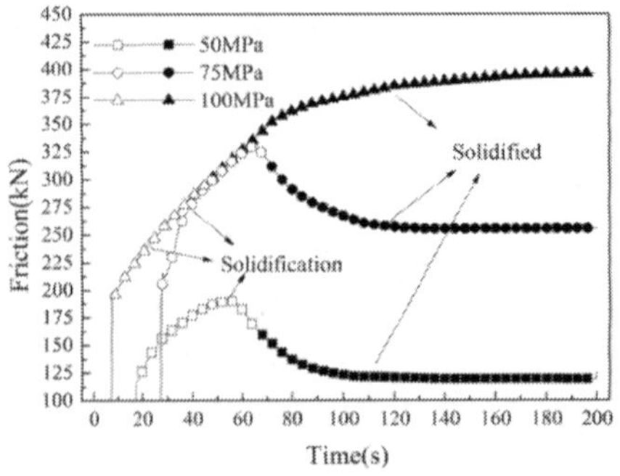

Figure 9. The friction between castings and dies of the squeeze casting processes

4. Conclusion

(1) With increased applied pressures in the squeeze castings, the times of casting solidification and eutectic transformation decreased due to improved contact between the castings and dies as well as decreased eutectic fractions in the alloys.

(2) During the solidification period at 50 and 75 MPa, the measured pressures decreased first, and then increased since the solidification later period because of the declined friction at the casting and die interface. On the contrary, the measured pressure at 100 MPa showed a decrease tendency during the whole fabrication process. The higher squeeze pressure can induce more intimate contact of the casting and die maintaining larger friction between them.

(3) As the applied squeeze pressures increased, the friction was larger limiting the pressure transmission. It is shown that the friction should be considered in thermal-mechanics coupled modeling of squeeze casting process.

Acknowledgement

This research was funded by Guangdong-Ministry Collaborating Project (2012A090300016) and Guangzhou Science and Technology Project (2014Y2-00214).

References

[1] Ghomashchi M R and Vikhrov A 2000 Squeeze casting: an overview *J. Mater. Process. Tech.* **101** 1-9

[2] Clegg A 1993 Squeeze casting-a new process technology for engineer *Foundry Trade Journal* **166** 484-5

[3] Hamasaiid A, Dour G, Dargusch M S, Loulou T, Davidson C and Savage G 2008 Heat-transfer coefficient and in-cavity pressure at the casting-die interface during high-pressure die casting of the magnesium alloy AZ91D *Metall. Mater. Trans.* A **39** 853-64

[4] Postek E W, Lewis R W, and Gethin D T. Finite element modelling of the squeeze casting process *International Journal of Numerical Methods for Heat & Fluid Flow* **18** 325-355

[5] Han Z Q, Zhu W, and Liu B C. Thermo-mechanics model of solidification of squeeze casting 1 Mathematical model and solution methodology *Acta Metallurgica Sinica* **45** 356-362

[6] Zhu W, Han Z Q, and Liu B C. Thermo-mechanics model of solidification of squeeze casting II Numerical simulation and experimental validation *Acta Metallurgica Sinica* **45** 363-368

[7] Liu W H, Liu Y G, Xiong S M, Liu B C, Matsumoto Y, and Murakami M. Comparative study if the cavity pressure of magnesium alloy and aluminum alloys during die casting process *Acta Metallurgica Sinica* **41** 103-107

[8] Quartz Cavity Pressure Sensor (Aluminium), 6175a2, Kistler Instrument Corporation, Armhest, NY, Available at: http://www.intertechnology.com/Kistler/pdfs/Pressure_Model_6175A2.pdf

[9] Epanchistov O G 1972 Structure and properties of metals solidified under high pressure *Russian Casting Prod* **2** 34-7

[10] Stangeland A, Mo A and Eskin D 2006 Thermal strain in the mushy zone for aluminum alloys *Metall. Mater. Trans. A* **37** 2219-29

Modelling of micro- and macrosegregation for industrial multicomponent aluminium alloys

K.Ellingsen[1], D. Mortensen[2] and M. M'Hamdi[1]

[1] SINTEF Materials and Chemistry, Oslo, N-0314, Norway
[2] Institute for Energy Technology, Kjeller, Norway

kjerstin.ellingsen@sintef.no

Abstract. Realistic predictions of macrosegregation formation during casting of aluminium alloys requires an accurate modeling of solute microsegregation accounting for multicomponent phase diagrams and secondary phase formation. In the present work, the stand alone Alstruc model, a microsegregation model for industrial multicomponent aluminium alloys, is coupled with the continuum model ALSIM which calculates the macroscopic transport of mass, enthalpy, momentum, and solutes as well as stresses and deformation during solidification of aluminium. Alstruc deals with multicomponent alloys accounting for temperature dependent partition coefficients, liquidus slopes and the precipitation of secondary phases. The challenge associated with computation of microsegregation for multicomponent alloys is solved in Alstruc by approximating the phase diagram data by simple, analytical expressions which allows for a CPU-time efficient coupling with the macroscopic transport model. In the present work, the coupled model has been applied in a study of macrosegregation including thermal and solutal convection, solidification shrinkage and surface exudation on an industrial DC-cast billet.

1. Introduction

Macrosegregation, the non-homogeneous chemical composition on the length scale of the casting, is the result of several phenomena including thermal convection, solutal convection, shrinkage induced flow and surface exudation. Mathematical models quantifying the formation of macrosegregation are well-established and based on mixing theory or volume-averaging of the conservation laws for mass, solute mass, enthalpy, and momentum [1]. These models need a sub-model, *e.g.* a microsegregation model, relating the enthalpy (or temperature) and total solute concentrations to the liquid fraction and solute concentrations. Realistic predictions of macrosegregation formation formation during casting of aluminium alloys requires an accurate modeling of solute microsegregation accounting for multicomponent phase diagrams and secondary phase formation.

Multi-component microsegregation models have only to a limited extent been coupled to macrosegregation models [2-6] as the models become quite computer time and storage demanding as the total solute concentration in the volume elements change and a unique solidification path cannot be used throughout the whole casting. The solidification path needs to be updated at each time step and for each volume element of the macro model.

IOP Conf. Series: Materials Science and Engineering **84** (2015) 012058 doi:10.1088/1757-899X/84/1/012058

In the present work, the Alstruc model [7], a microsegregation model for industrial multicomponent aluminium alloys, was applied as a micro-module for macrosegregation computations. The module was coupled with the continuum model ALSIM [8-10] which calculates the macroscopic transport of mass, enthalpy, momentum, and solute as well as stresses and deformation during casting. The challenge associated with computation of microsegregation for multicomponent alloys is solved in Alstruc by approximating the phase diagram data by simple, analytical expressions which allows for a CPU-time efficient coupling with the macroscopic transport model. Preliminary assessments demonstrating the usefulness of the new module has been presented elsewhere [11,12]. The module was also used in a simplified DC-casting demonstrating the effect of secondary phase precipitation on macrosegregation formation for a 5xxx alloy [13]. In this study the coupled models are applied in a study on an industrial DC-cast billet including the effects of thermal and solutal convection, solidification shrinkage and surface exudation on the ingot concentration fields. Modelling results are compared with measurements of chemical composition through the billet from the centre to the surface.

2. Model decription

2.1. *The Alsim model*

Alsim [8-10] calculates the development of heat, fluid flow, stresses, and deformation during casting. For aluminium DC casting situations, the model addresses the thermal and fluid flow boundary conditions to a very high level of details regarding contact zones, air gap sizes, and water hit points. This also includes the effect upon the surface heat transfer and associated surface exudation caused by air-gap formation [10, 14]. Alsim also address the formation of macrosegregation.

For a fixed solid (vs=0) the mass and momentum equations are given by:

$$\frac{\partial}{\partial t}\rho + \nabla\cdot\left(\rho^l g^l \mathbf{v}^l\right) = 0 \tag{1}$$

$$\begin{aligned} &\frac{\partial}{\partial t}\left(g^l \mathbf{v}^l\right) + \nabla\cdot\left(g^l \mathbf{v}^l \mathbf{v}^l\right) = -\frac{g^l}{\rho^l}\nabla p^l \\ &+\nabla\left[\nu\left(\nabla\left(g^l\mathbf{v}^l\right) + \left[\nabla\left(g^l\mathbf{v}^l\right)\right]^T\right)\right] + \frac{g^l}{\rho^l}\mathbf{g}(\Delta\rho^l) - \frac{g^l\nu}{K}\left(g^l\mathbf{v}^l\right) \end{aligned} \tag{2}$$

where ρ, ρ^l, ρ^s, g^l, g^s, v^l, p^l, ν, f_i^l , g and K are the mixture density ($\rho = \rho^l g^l + \rho^s g^s$), liquid density (assumed constant except in the buoyancy term using the Boussinesq approximation), solid density, liquid volume fraction, solid volume fraction, liquid velocity, liquid pressure, kinematic viscosity, gravity vector and permeability, respectively. Solidification shrinkage, interfacial friction, and macroscopic viscous stress contribution are included. For solute conservation the following mixture equations are applied:

$$\frac{\partial}{\partial t}(\rho\cdot c_i) + \nabla\cdot\left(\rho^l g^l c_i^l \mathbf{v}^l\right) = \nabla\cdot(\rho D_i \nabla c_i) \tag{3}$$

where c_i is the mixture concentration for element i, and $\rho c^i = \rho^l g^l c_i^l + \rho^s g^s c_i^s$. D_i is the mixture diffusion constant and and $\rho D_i = \rho^l g^l D_i^l + \rho^s g^s D_i^s$. The term in the right-hand side of Eq. (3) is added for numerical stability reasons. Similar for energy conservation the following mixture enthalpy equation is applied:

$$\frac{\partial}{\partial t}(\rho\cdot h) + \nabla\cdot\left(\rho^l g^l h^l \mathbf{v}^l\right) = \nabla\cdot(\lambda\nabla T) \tag{4}$$

where h is the mixture enthalpy and ρ□h = $\rho^l g^l h^l + \rho^s g^s h^s$, and λ is the mixture thermal conductivity ($\lambda = g^l \lambda^l + g^s \lambda^s$).

The mechanical analysis is carried out in the fully solid regions of the ingot as well as in the coherent part of the mushy zone. The upper boundary of the coherent mushy zone corresponds to a solid volume fraction–at coherency (g^s_{coh}) which is an input to the model. Denoting the solid stress $\boldsymbol{\sigma}^s$, the mixture momentum balance in the coherent part of the mushy zone can be expressed by:

$$\nabla \cdot \left[g^s \boldsymbol{\sigma}^s \right] - \nabla\left[(1-g^s) p^l \right] + \rho \mathbf{g} = \mathbf{0} \tag{5}$$

where **g** is the gravity vector. The total solid strain rate tensor is sub-divided into thermal, elastic, and viscoplastic parts and constitutive equations have to be provided for each contribution. Constitutive equations for the viscoplastic deformation of aluminium alloys during solidification are based on the work of Ludwig *et al.*[15] where the mushy zone is considered as a partially cohesive porous solid saturated with liquid

$$\dot{\boldsymbol{\varepsilon}}_{\mathbf{s}}^{\mathbf{p}} = \frac{\dot{\varepsilon}_0 e^{-\frac{Q}{RT}}}{(C\sigma_0)^n} \left[-\frac{1}{9} A_2(g^s) J_1 I + \frac{3}{2} A_3(g^s) g^s \boldsymbol{\tau}^s \right] \left[-\frac{1}{9} A_2(g^s) J_1^2 + 3 A_3(g^s) J_2 \right]^{\frac{n-1}{2}} \tag{6}$$

$$\dot{C} = \alpha(g^s, X) \left(1 - \frac{C}{C^*(g^s, X)} \right) \dot{\bar{\boldsymbol{\varepsilon}}}_s^p \tag{7}$$

J_1 and J_2 are the first and second invariants of the effective stress tensor $\widehat{\boldsymbol{\sigma}}^{\mathbf{s}} = \left(g^s \sigma^s - g^l p^l \mathbf{I} \right) + p^l \mathbf{I}$, $\boldsymbol{\tau}^{\mathbf{s}}$ is the solid deviatoric stress tensor, $\dot{\bar{\boldsymbol{\varepsilon}}}_s^p$ is the effective strain rate, and X is the stress triaxiality. Equation (7) describes the evolution of the partial cohesion of the coherent mushy zone. Expressions for the functions A_2, A_3, a and C^* are all given in Reference 7. Note that for the viscoplastic part of the strain, viscoplastic volumetric deformation is taken into account for temperatures between solidus and coherency, *i.e.*, $\dot{\bar{\boldsymbol{\varepsilon}}}_s^p \neq 0$. In the fully solid regions (g_s=1, A_2=0, A_3=1, C=1), Equation (6) simplifies to the classical flow rule with a power-law creep relation. In the fully solidified regions, strain hardening is also taken into account in the manner described in Reference 5.

2.2. The ALSTRUC model

ALSTRUC [7] calculates the solidification path of AlMgCuFeMnSi alloys in the Al rich corner of the phase diagram. ALSTRUC also handles additions of Cr, V, Sr, Ti, and Zn in commercial alloys. The model uses the well-established assumptions of full miscibility in the liquid and thermodynamic equilibrium at the solid/liquid interface and accounts for solid-state diffusion and particle growth undercooling. A constant total alloy composition is input along with the cooling rate and the final characteristic dendrite arm length. By gradually changing the solid fraction in a step-wise manner, Alstruc computes the solute concentrations in the liquid phase, the temperature, the alloying element concentration profiles and the amount and composition of the various secondary phases in a simplified geometrical representation of the dendrite arm. Phase diagrams are formulated as simple, approximate, analytical expressions. Computer time and storage demanding calls to a thermodynamic data base, or alternatively, use of numerical mapping of such data [4, 6], is thus avoided. The amount and concentration of the secondary phases are calculated at the end of each step by considering solubility products assuming that all secondary phases are precipitated from the liquid phase. The model parameters have been tuned to metallographic investigations of solidification microstructures in samples from numerous experiments [16-21].

2.3. Alsim/Alstruc coupling

The ALSIM/ALSTRUC coupling is carried out through a dynamically linked library (DLL). The solid phase diffusion of the alloying elements in the dendrite arms is neglected (Scheil-Gulliver approximation). This simplification also keeps the computation time and data storage low because the alloy concentration profiles do not need to be traced at the nodal points of Alsim. Input to the micro-model are the Alsim predicted values of the enthalpy and average alloy compositions at the start and end of the time step along with values for the solid fraction, alloying element concentrations in the liquid phase, and temperature at the beginning of the step. To compute the change in solid fraction, the Alstruc micro-model requires the change in liquid composition without considering solidification as input. The latter is computed from the total average concentration at the end of the time step in the following manner:

$$c_i^{l\,Alstruc} = (c_i(t) - f^s(t-1) \cdot c_i^s(t-1)) / (f^s(t-1) - 1) \tag{8}$$

where $c_i(t)$ is the average total concentration at the end of the step, $f^s(t-1)$ *is* the fraction of solid and $c_i^s(t-1)$ the composition of the solid at the beginning of the step. The approach is justified by the different time and length scales of the macroscopic and microscopic concentration changes.

The solid fraction, alloying element concentrations in the liquid phase and temperature at the end of the time step are calculated by integrating differential versions of the Scheil-Gulliver-like equations before precipitation of secondary phases.

3. Model application

The model was applied to study the macrosegregation formation in a DC cast 6xxx 270mm (11 inch) billet produced at Alcoa Lista [22] with alloy composition as shown in Table 1. A fully coupled heat and fluid flow, macrosegregation, stresses and deformation calculation was performed. The simulation included the ingot, the starting block, the mould, the gas, oil and insulation (transition) rings above the mould and the hot-top. Elements are added to the computational domain as the simulation proceeds so the domain increases in size similar to the casting itself. The contact area and the air-gap size against the mould were dynamically calculated by the model and back-coupled to the thermal boundary conditions. Fluid outflow from the semi-solid surface (exudation) is allowed if there is a local air-gap at the current element surface. The minimum air-gap size for exudation was set to 1 micrometer. If the calculated air-gap is larger than 1μm a pressure boundary condition is applied in the solution of the liquid mass and momentum equations. The pressure at the melt surface is equal to zero and the pressure at the local air-gap surface is calculated from the height of the metal column above the surface. This pressure drop is then the driving force for the fluid outflow, while the pressure drop through the semi-solid material is the limiting factor. More details about thermal and mechanical boundary conditions can be found in [22]. Thermal properties for the alloy were calculated by the microstructure model Alstruc [7]. The permeability of the semi-solid material is assumed to follow the Kozeny-Carman relation with $K_0=1{,}39 \cdot 10^{-11}$. Diffusion constants applied in this study was $10^{-8}m^2/s$ and $10^{-12}m^2/s$ for the liquid and the solid, respectively (for all alloy components).The solid mechanical properties were taken from the 6063-alloy in [23]. The parameters for the cohesion mechanical model of the semi-solid material were taken from [7], as measured for a grain-refined Al-Cu-alloy and the upper boundary of the coherent mushy zone was set to a solid volume fraction equal to 0.6.

Table 1. Nominal composition for the alloy.

Conc. [wt%]	Si	Mg	Fe	Mn	Cu
	0.593	0.213	0.170	0.445	0.170

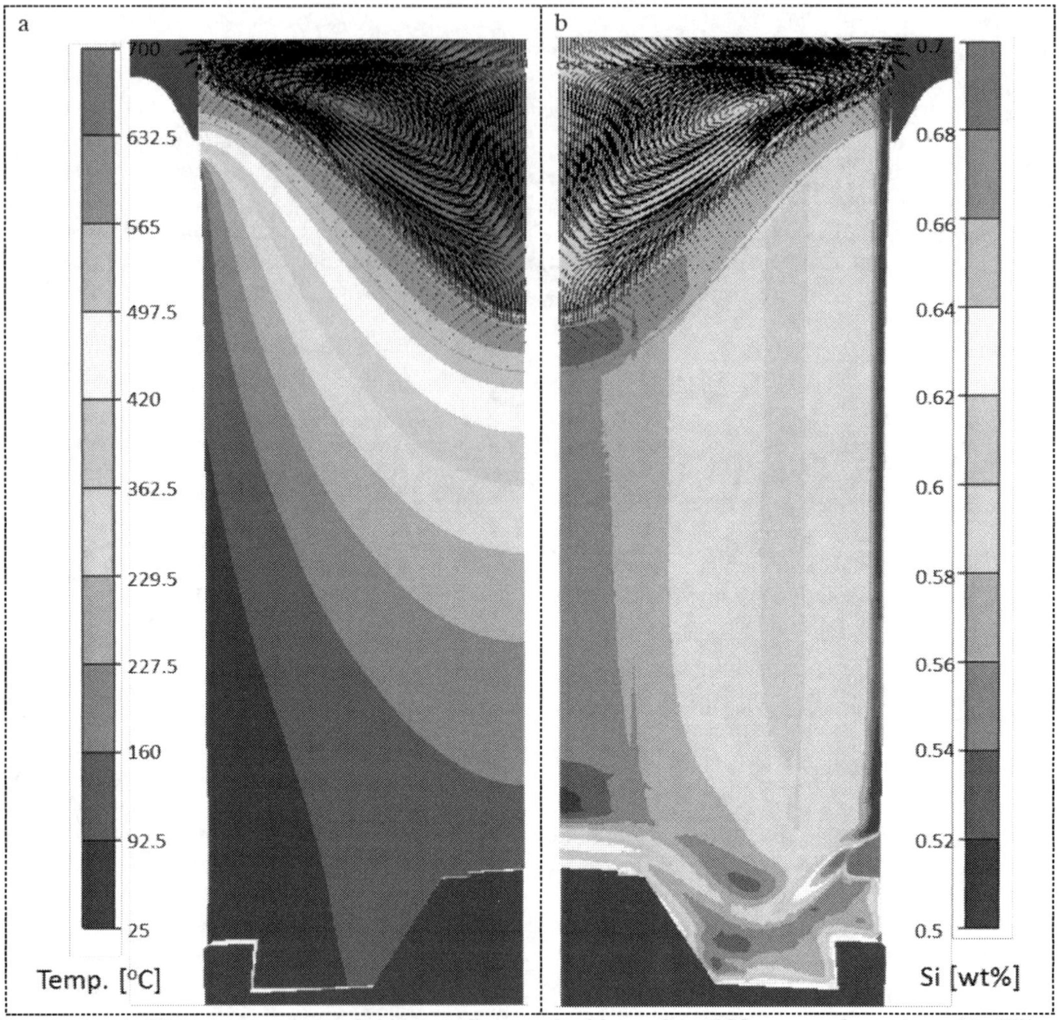

Figure 1: a) The temperature field and flow velocities after 280 sec. of casting. b) Silicon concentration field and flow velocities after 280 sec. of casting. The countour lines show the fraction of solid from 0,1 to 1,0 with a spacing of 0,1.

4. Results

The temperature and flow fields are shown in Figure 1a, and the silicon concentrations field is shown in Figure 1b. Figure 1 also show the air-gap formed against the bottom block and the mould .The metal flows downwards along the solidification front, partly flows into the mushy zone to feed the solidification shrinkage and flows upwards in the ingot center creating two vortices with flow in the clockwise direction. Positive segregation is obtained towards the bottom block is due to solidification shrinkage against a chill surface. Due to initial solidification during the initial holding phase, the impression of the transition ring can also be seen about 20mm up from the bottom block edge. As the bottom block starts to move the continuous casting enhanced air-gap is developing and so is the exudation. In the steady state we find a depleted center due to solidification shrinkage. A positive surface is obtained due to the inverse segregation and exudation. Exudation creates a depleted zone behind the positively segregated surface since melt enriched with Si due to microsegregation is pushed through the mushy zone into the air-gap formed in the mould region. The concentration in the exuded layer was calculated to be 3.46wt% and the thickness of the layer was 0.14 mm. The contribution of different mechanisms to the obtained macrosegregation pattern was investigated in [24], showing that solidification shrinkage is the dominating mechanism.

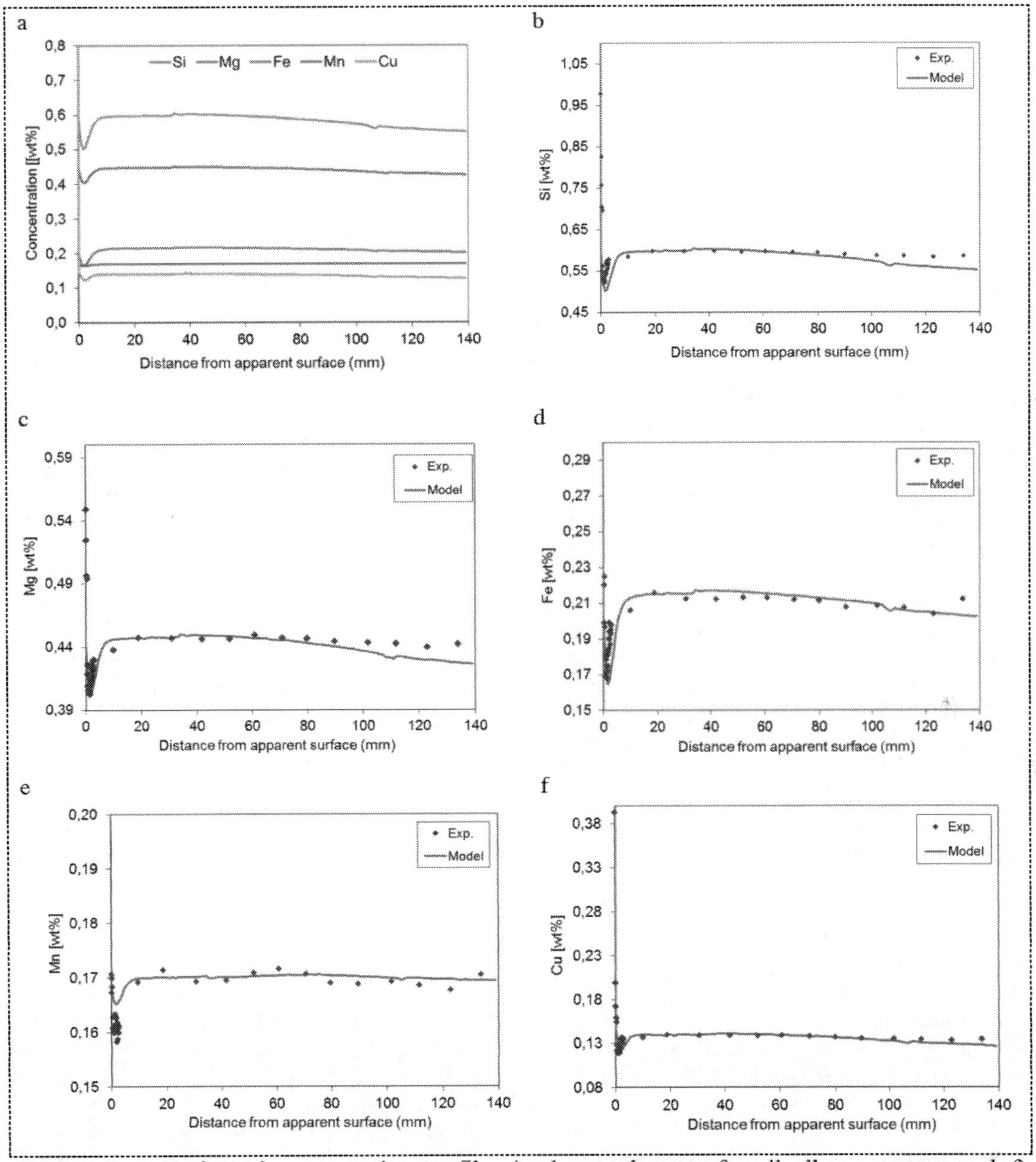

Figure 2: a) Horizontal concentration profiles in the steady statevfor all alloy components. b-f) Comparison between measured concentrations (diamond symbols) [24] and model results (continuous line).

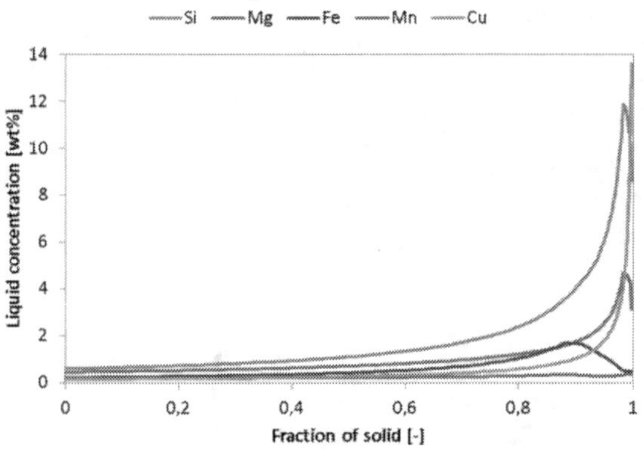

Figure 3: The liquid concentration of alloying elements as a function of solid fraction for the nominal alloy composition.

The calculated horizontal steady state concentration profiles for all alloying elements are shown in figure 2a. All elements have a similar profile as the one discussed for Silicon, with a depleted center, an enriched surface and a depleted zone behind the positively segregated surface. The largest relative segregation is calculated for the elements where the remaining liquid is largely enriched with alloying elements late in the solidification interval (from fraction of solid 0.5). Si has the largest relative segregation and Mn has a relatively flat profile.The concentration development in the liquid during solidification as calculated by the microsegregation model is shown in Figure 3 for the nominal alloy composition. The calculated profiles were compared to available measurements [24] and is shown in Figure 2 b-f. There is a good overall agreement with the measured concentration profiles. The model predicts slightly lower concentrations in the center for Si and Mg (deviation 2-3 %), and the Mn concentration in the depleted region close to the surface is slightly higher than measured concentration. However, the depletion behind the surface is captured by the model due to the coupling with an advanced micromodel. A simplified micromodel with a linearized phase diagram and a constant partition coefficient typically set to ~1 for Mn, would not capture the depletion.

5. Conclusion

In this study macrosegregation in an industrial DC-cast billet including the effects of thermal and solutal convection, solidification shrinkage and surface exudation has been calculated using coupling to an advanced microsegregation model. The proposed approach has been validated by comparing modelled concentration profiles from the centre to the surface of the billet in the steady state with measurements of chemical composition.

Acknowledgements

Funding of this work through the ALSIM/PRIMAL-project is gratefully acknowledged. Partners in ALSIM/PRIMAL are Alcoa Norway ANS, Aleris Rolled Products Germany GmbH, Hydro Aluminium ASA, Stiftelsen SINTEF and Institute for Energy Technology (IFE). The project is supported by the Norwegian research council through the BIA programme.

References

[1] H. Combeau, J.M. Drezet, A. Mo and M. Rappaz 1996: *Metallurgical and Materials Transactions* **27A** 2314–2327
[2] D.K. Banerjee, M.T. Samonds, U.R. Kattner and W.J. Boettinger 1997: *Proc. Int. Conf. Solidification Processing IV* (eds. J. Beech and H. Jones), University of Sheffield 354–357
[3] X. Doré, H. Combeau and M. Rappaz 2000: Acta Materialia, 2000 48, 3951–3962.
[4] S.A. Cefalu and M. J. M. Krane: Material Science and Engineering A, 2003, 359, 91–99.

[5] Q. Du and A. Jacot: Acta Materiala, 2005, 53, 3479–3493.
[6] Q. Du, D.G. Eskin and L. Katgerman: Metallurgical and Materials Transactions A, 2007, 38A, 180–198
[7] A.L. Dons, E.K. Jensen, Y. Langsrud, E. Trømborg and S. Brusethaug 1999: Metallurgical and Materials Transactions A **30A** 2135–2146
[8] D. Mortensen 1999: *Metallurgical and Materials Transactions B* **30B** 119–133
[9] H.G. Fjær and A. Mo 1990: *Metallurgical Transactions B* **21B** 1049–1061
[10] H.J. Thevik, A. Mo and T. Rusten 1999: *Metallurgical and Materials Transactions B* **30B** 135–142
[11] K.Ellingsen, A. L. Dons, M. M'Hamdi, A. Mo, and D. Mortensen 2009: *Int. J. Cast Met.* **22** 220-223
[12] K. Ellingsen, A. L. Dons, M. M'Hamdi, D. Mortensen 2009: *Proc. of the International Conference on Modeling of Casting, Welding, and Advanced Solidification Processes 12* 651-657
[13] K.Ellingsen, Mohammed M'Hamdi, Knut Tveito and Dag Mortensen 2014: *Light Metals, Proceedings of TMS 2014 Annual Meeting & Exhibition* 861-866
[14] D. Mortensen, B.R. Henriksen, M. M'Hamdi and H.G. Fjær 2008: *Light Metals, Proceedings of the TMS 2008 Annual Meeting & Exhibition* 773–779
[15] O. Ludwig 2004: Ph.D. Thesis, Institut National Polytechnique de Grenoble, Grenoble, France.
[16] A.L. Dons 1990: *Zeitschrift für Metallkunde* **81** 484–489
[17] A.L. Dons 1991: *Zeitschrift für Metallkunde* **82** 684–688
[18] A.L. Dons, E.K. Jensen, J. Voje and J. Rødseth 1993: *Proc. Int. Conf. Modelling of Casting, Welding and Advanced Solidification Processes VI* 37–44
[19] A.L. Dons, S. Brusethaug and J. Voje 1998: *Proc. Int. Conf. Modelling of casting, welding and advanced solidification processes – VIII* 195–202
[20] A.L. Dons, L. Pedersen and L. Arnberg 1999*: Materials Science and Engineering A* **271** 91–94
[21] J. Voje and A.L. Dons 2001: *Aluminium* **77** 284–287
[22] D. Mortensen, B.R. Henriksen, J. Hvistendahl, H.G. Fjær 2007: *Light Metals, Proceedings of the TMS Aluminum Committee at the TMS 2007 Annual Meeting & Exhibition* 715-720
[23] W.M. van Haaften 2002: "Constitutive behaviour and hot tearing during aluminium DC casting", PhD-Thesis TUDelft, The Netherlands.
[24] D. Mortensen, M. M'Hamdi, K. Ellingsen, K. Tveito, L. Pedersen and G. Grasmo 2014: Light Metals, Proceedings the TMS Aluminum Committee at the TMS 2014 Annual Meeting & Exhibition 867-872

IOP Conf. Series: Materials Science and Engineering **84** (2015) 012059 doi:10.1088/1757-899X/84/1/012059

Effect of solidified structure on hot tear in Al-Cu alloy

Y Yoshida[1]**, H Esaka**[2] **and K Shinozuka**[2]

[1] Graduate student, National Defense Academy, 1-10-20 Hashirimizu, Yokosuka, 239-8686, Japan

[2] Department of Materials Science and Engineering, National Defense Academy, 1-10-20 Hashirimizu, Yokosuka, 239-8686, Japan

E-mail:boudai.kakashi@gmail.com

Abstract. Hot tear is the one of the biggest problems of cast products of aluminum alloy. The effect of solidified structure on the hot tear has not been clear. Therefore, this study has been carried out to correlate solidified structure and hot tear. Al-2.0 wt% Cu alloy was cast at 750 °C in a mold cavity, which could intentionally form hot tear. To change solidified structure, some amount of refiner was added to the molten alloy. Length of hot tear decreased with increasing the amount of refiner. Further, the area of fine eutectic Al_2Cu increased as amount of refiner increased. These may indicate that probability of healing increased in case of equiaxed structure. Therefore, the length of hot tear decreased with increasing the amount of refiner.
great care should be taken in constructing both.

1. Introduction

Hot tear is one of the biggest problems of cast products of steel, copper and aluminum alloys. Continuous casting of steel is widely popular process in industries, though, there are some problems, such as surface and internal cracks. Once these cracks form, they propagate and usually remain in steel products[1]. Therefore, cracks decrease productivity. It is regarded that some surface cracks and most of internal cracks initiate during solidification. For example, longitudinal surface crack is believed to initiate in the mold and increase its size in secondary cooling zone[2]. Hence, mild and uniform cooling in the mold may be a countermeasure to reduce the longitudinal surface cracks.

In order to reduce crack or hot tear, it is important to know where and when it initiates and how it propagates. There have been many researches in order to understand the formation mechanism of hot tear[3-5]. However, it is still uncertain when a hot tear initiates and how it propagates. Furthermore, the effect of solidified structure on the hot tear has not been clear. Therefore, this study has been carried out to correlate solidified structure and hot tear.

2. Experimental procedure

Figure 1 shows a schematic view of a mold. One of four side plates was a steel plate. The steel plate had two grooves in order to intentionally make hot tear. In addition to this, a ceramic insulator was set in the middle of two grooves to slowly solidify in local. The other three side plates as well as bottom plate were alumina brick. The size of ingot was 30 mm x 100 mm x 40 mm. Al-2.0 wt% Cu alloy was cast at 750 °C in a mold cavity. The liquidus and eutectic temperatures of this alloy are 650 °C and 550 °C, respectively. To change solidified structure, some amount of refiner was added to the molten alloy.

The refiner includes TiB_2 particles and promotes heterogeneous nucleation. The amount of refiner were 0 , 0.03 , 0.1 and 0.3 wt%.

The plane for characterization was 10 mm from the ingot bottom. The solidified structure in the middle of two grooves were observed using a FE-SEM and an optical microscope(OM). In order to characterize the hot tear, the observation with low magnification was used. A length of hot tear in the cross section was measured vertically to a plane set a ceramic insulator. To characterize the solidified structure near the hot tear, backscattered electron(BSE) image was used with high magnification. The grain size was measured using the BSE image near a hot tear. Here, the lineal analysis was adopted.

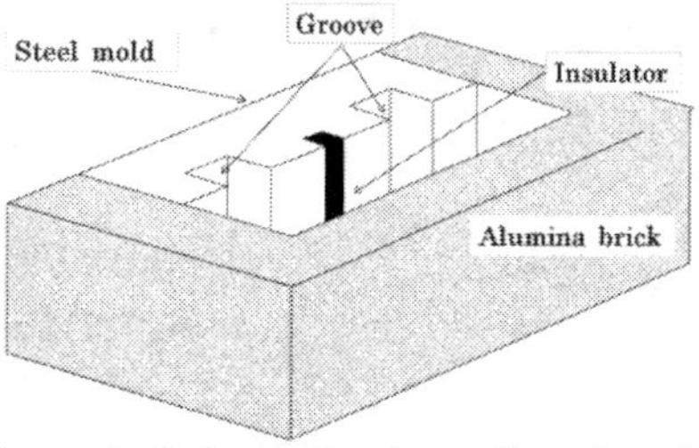

Figure 1. Schematic view of steel mold.

3. Results and Discussions

3.1. Relationship between grain size and length of hot tear

Figure 2 shows OM images of near the hot tear. Top of these photographs is on the steel plate surface and the direction of solidification is from the top to the bottom. Black and wavy lines are hot tears. It became thinner as the distance from the steel plate increased. In case of w = 0 wt% or 0.03 wt%, two major hot tears were recognized on the observation plane. The length and width of hot tears are long and large. On the other hand, in case of w = 0.1 wt% or w=0.3 wt%, there was one small hot tear on the observation plane. The length and width of it is short and small. Especially when the addition of refiner(w) was 0.3 wt%, the hot tear is discontinuous. Here, L is defined as the length between the surface and the hot tear tip.

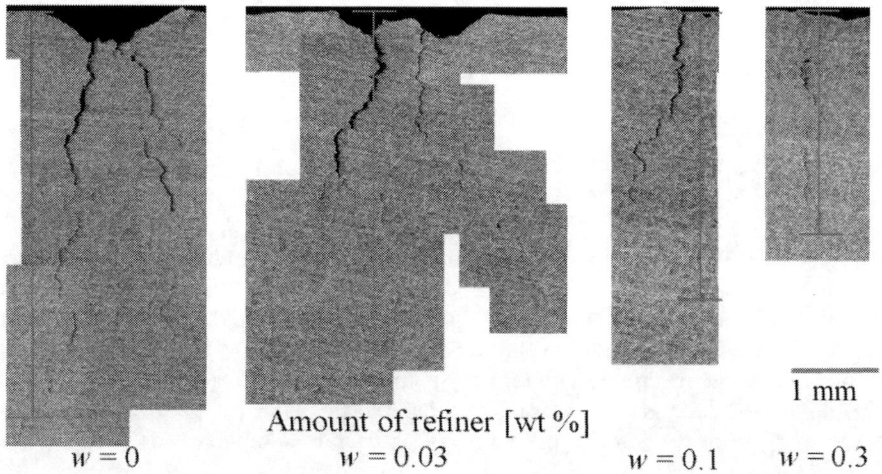

Figure 2. OM images of hot tears and the definition of hot tear length (L).

IOP Conf. Series: Materials Science and Engineering **84** (2015) 012059 doi:10.1088/1757-899X/84/1/012059

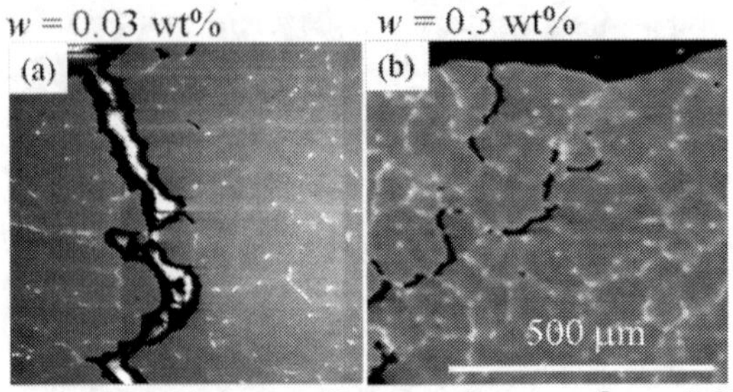

Figure 3. BSE images of enlarged views of hot tear in case of $w = 0.03$ wt% and 0.3 wt%.

An enlarged views of hot tear are shown in figure 3 in case of $w = 0.03$ wt% and 0.3 wt%. In case of $w = 0.03$ wt%(figure 3(a)), the width of hot tear is wide. Since white or bright region is Cu concentrated or eutectic region and they should be interdendritic region, the equiaxed grains can be judged. As w increases, equiaxed grain size decreases, and the hot tear was discontinuous and turned much.

Figure 4 shows the relationship between w and the length of hot tear(L). The hot tear was severe and L is long when no refiner was added to the molten aluminum alloy. L decreased much when the refiner was added and L decreased as w increased. However the slope with respect to w decreased with increasing w. The reasons of these changes may be due to macroscopic solidified structure and grain size. Figure 5 shows the relationship between w and the grain size. In case of w = 0 wt%, since the solidified structure was columnar dendrite, the experimental data was omitted. The grain size gradually decreased as w increased.

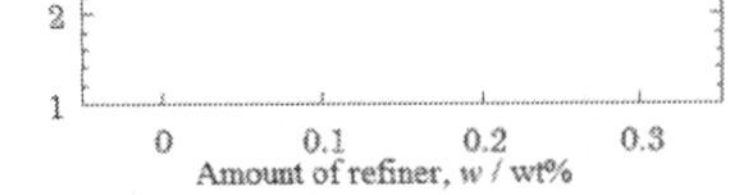

Figure 4. Relation between amount of refiner and length of hot tear.

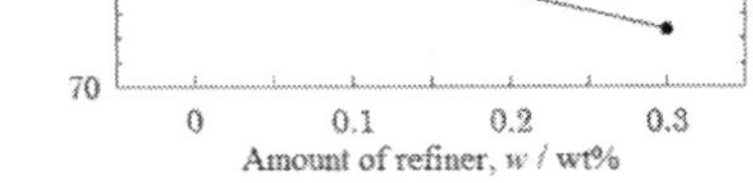

Figure 5. Relation between amount of refiner and size of equiaxed grain.

Figure 6 shows the image of propagation of hot tear along columnar dendrites or equiaxed grains. The solidified structure near a hot tear was columnar dendrites in case of without refiner. Therefore hot tear propagates along interdendritic region easily and rather long and relatively straight hot tear results(figure 6(a)). Thus, the sensitivity to hot tear is high when the solidified structure is columnar dendrite. On the other hand, in case of with refiner, the hot tear propagates along the inter-grain region. Since the packing of equiaxed grains are usually complex, the path of hot tear is wavier than that in case of columnar dendrite. Therefore, the hot tear tends not to initiate nor propagate when the solidified structure is equiaxed(figure 6(b)).

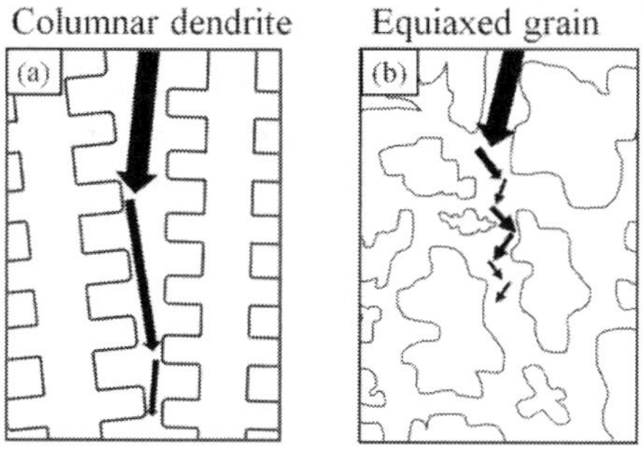

Figure 6. Image of propagation of hot tear along columnar dendrites(a) or equiaxed grains(b).

3.2. Observation of Eutectic structure

Figure 7 shows the BSE image of the solidified structure in case of w = 0.3 wt% near the hot tear. As described before, white regions refer to area where copper concentration was high and it is found that some hot tears are mended. Enlarged views of figure 7 are shown in figure 8. Here, lamellar eutectic structure is observed[6]. The morphologies of lamellar eutectic structure are rod like(i) as well as spherical(ii). The "rod like" lamellar eutectic structure is often observed in the hot tear. This may mend or heal the hot tear. On the contrary, a white and round particle(iii) is recognized. This is a divorced eutectic Al_2Cu[7]. As far as the observation concerned, the morphologies of divorced Al_2Cu were spherical and there is no divorced eutectic Al_2Cu in the hot tear.

Comparing the size of eutectic Al_2Cu, it in divorced eutectic is larger than that in lamellar eutectic. Both lamellar and divorced eutectic are sometimes observed in the same area and this indicates that they may have the same cooling rate. However, the structures are different. This indicates that lamellar eutectic structure solidified rapidly and this suggests that the remaining liquid in the interdendritic region may move to the hot tear due to negative pressure.

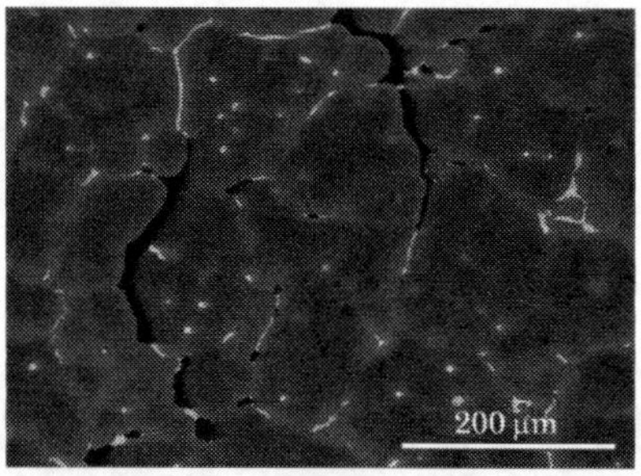

Figure 7. BSE image of the solidified structure in case of w = 0.3 wt% near a hot tear.

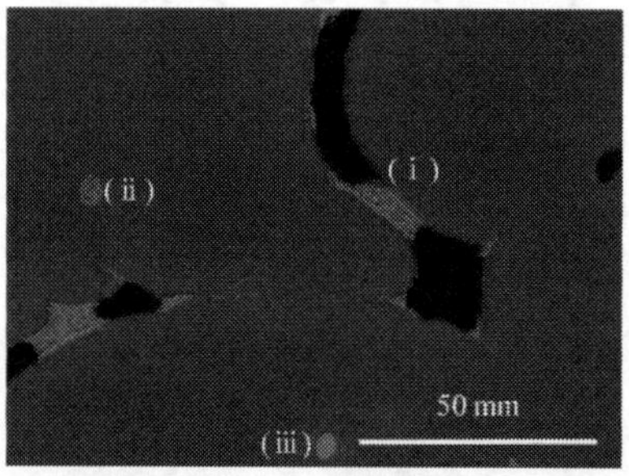

Figure 8. BSE image of lamellar eutectic structure and divorced eutectic Al_2Cu. (i) Rod like lamellar eutectic structure. (ii) Spherical lamellar eutectic structure. (iii) Divorced eutectic structure.

Here, the areal ratio of lamellar eutectic structure(σ) is defined as a following equation:

$$\sigma\ [\%] = \frac{\text{area of lamellar eutectic}}{\text{area of lamellar eutectic} + \text{area of divorced eutectic}} \times 100 \quad (1)$$

σ was measured using an image analysis software. At around middle of hot tear, σ was characterized and indicated in figure 9 as a function of w. As shown, all data are over 70 %. σ increased with increasing w. In other words, σ gradually increased as the equiaxed grain size decreased. A grain boundary may be a route for remaining liquid due to suction. Since number of grain boundaries increases with decreasing grain size, it is reasonable that σ increased as the amount of refiner increased.

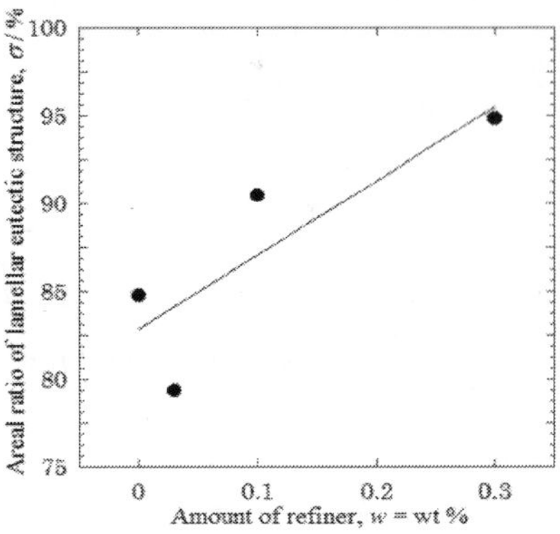

Figure 9. Relation between amount of refiner and areal ratio of lamellar eutectic structure near the middle point of hot tear.

Then, σ was measured along a line indicated in figure 10, where 1 mm apart from the steel plate surface. Figure 11 shows the relation between the distance from hot tear and σ. In case of x □ 0 mm, the values of σ are over 70 %. σ rapidly decreased with increasing x. But, in case of $x > 6$ mm, the values of σ are 0 %. These results indicate that the area where the lamellar eutectic structures are found is limited near the hot tear. In normal solidification condition, a divorced eutectic structure is found in Al-2.0 wt%Cu alloy. Therefore, lamellar eutectic structures are attributed to the suction due to negative pressure in hot tear formation. As shown in figure 11, the range of suction may spread approximately 3 mm from the hot tear.

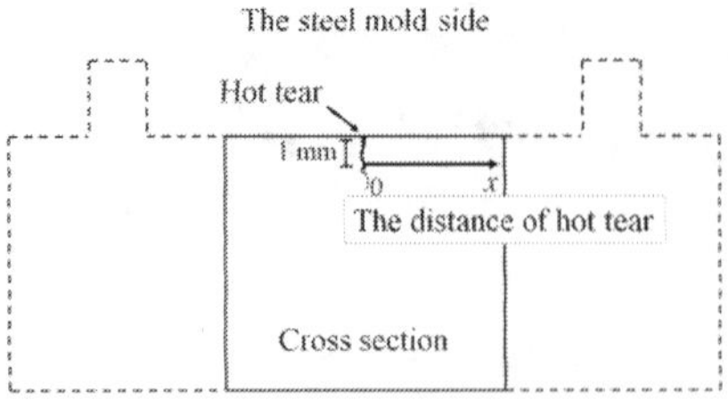

Figure 10. A schematic view of horizontal cross section and the line for measuring σ.

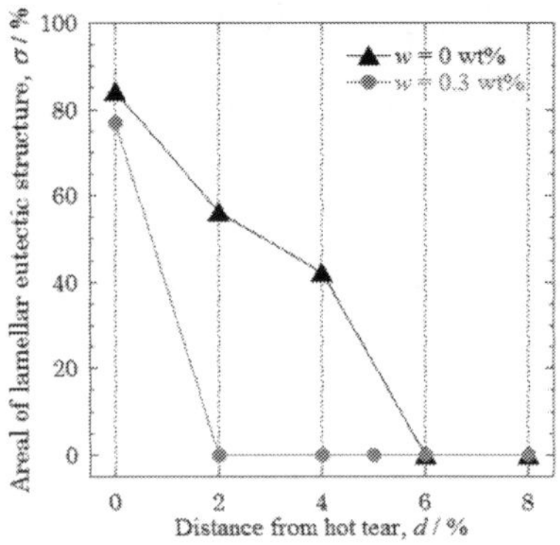

Figure 11. Relation between distance from hot tear and areal ratio of lamellar eutectic structure.

IOP Conf. Series: Materials Science and Engineering **84** (2015) 012059 doi:10.1088/1757-899X/84/1/012059

4.– Conclusions

In order to characterize the hot tear, casting test using Al-2.0 wt%Cu alloy with and without refiner has been performed. Solidified structure has been metallographically observed and following conclusions have been derived.

(1) It was found that the length of hot tear decreased with decreasing the equiaxed grain size.
(2) Hot tear was easily healed as size of equiaxed grain decreased.
(3) The range of suction may spread approximately 3 mm from the hot tear.

References

[1] Morishita M, Abe M, Tokuda K and Yoshida M 2008 *J. Japan Inst Metals* **59** pp417-423
[2] Brimacombe J K and Sorimachi K 1977 *Metall.Trans.B* **8** pp489-505
[3] Watanabe T, Kimura R, Nakazawa T, Chiba H, Tanaka S, Ueki T, Toriyama T and Yoshida M 2008 *J. Japan Inst Metals* **58** pp395-405
[4] Li S, Sadayappan K, and Apelian D 2011 *Int. J Cast Metal Research* **24** pp 88-95
[5] Monroe C and Beckeramann C 2014 *JOM* **66** pp1439-1445
[6] Kurz W and Fisher J D 1992 *Funfamentals of Solidification* (Aedermannsdorf : Trans Tech Publication) chapter 5
[7] Bramfitt B L and Benscoter A O 2002 *Metallographer's Guide* (Ohio : ASM International) chapter 9

IOP Conf. Series: Materials Science and Engineering **84** (2015) 012060 doi:10.1088/1757-899X/84/1/012060

Influence of Supercooling on Formation of Primary Phase

R Yoshimura[1], H Esaka[2] and K Shinozuka[2]

[1] Graduate student, National Defense Academy, 1-10-20 Hashirimizu, Yokosuka, 239-8686, Japan
[2] Department of Materials Science and Engineering, National Defense Academy, 1-10-20 Hashirimizu, Yokosuka, 239-8686, Japan

E-mail:em53024@nda.ac.jp

Abstract. It is reported actual volume fraction of primary phase in alloys is larger than the equilibrium value. Larger volume fraction of the primary phase may cause shrinkage cavities and surface or internal cracks. Although control of solidified structure is important for the quality of cast products, this problem has not been elucidated. Taking these results into account, this study has been carried out in order to comprehend a phenomenon of larger volume fraction of primary phase. Sn-Pb alloy has been used as a test alloy and to examine the relation between supercooling for nucleation and the volume fraction of primary phase has been mainly characterized. Actually, volume fraction of primary phase in Sn-Pb alloy is larger than that of lever rule. It was also observed that the volume fraction of β-Sn decreases with decreasing the supercooling in early stage of solidification. In the final stage of solidification, however, the effect of supercooling on volume fraction of primary phase is small. Furthermore, when the supercooling was low, volume fraction of primary phase was slowly increased.

1. Introduction

Recently, it has been reported that the actual volume fraction of primary β-Sn ($f_{\beta\text{-Sn}}$) is larger than the equilibrium value in Sn-base alloys, which are the candidates for lead-free solders. For example, in Sn-1.27 mass% Ag alloy, the volume fraction of primary β-Sn was 0.784. In these cases, $f_{\beta\text{-Sn}}$ calculated from equilibrium diagram is 0.65. Therefore, the actual $f_{\beta\text{-Sn}}$ was 0.19 larger than prediction. The fact that $f_{\beta\text{-Sn}}$ exceeds the equilibrium value is universal regardless of species of alloys and composition [1]. Larger $f_{\beta\text{-Sn}}$ may cause shrinkage cavities and surface or internal cracks, since the volume of remaining liquid is too small to fill in the volume change due to solidification. Therefore, this problem may critically influence mechanical properties and significantly decrease reliability of solders.

It is considered that $f_{\beta\text{-Sn}}$ depend on the supercooling for nucleation (ΔT) [2]. Therefore, the objective of this study was to clarify the relationship between ΔT and $f_{\beta\text{-Sn}}$. Time-dependent $f_{\beta\text{-Sn}}$ has been characterized, changing ΔT.

2. Experimental procedure

5g of Sn-25mass% Pb alloy was melted in an alumina crucible and kept at 300 °C for 30 minutes. Then, the temperature of the furnace was lowered at a constant cooling rate of 3 °C/min. The temperature of the specimen was directly measured by a K-type thermocouple installed in the crucible. When the temperature was below the liquidus temperature (=197 °C), a stainless steel rod was inserted as a trigger for nucleation at predetermined supercooling (ΔT). The ΔT used in this study ranged 0.1 °C

IOP Conf. Series: Materials Science and Engineering **84** (2015) 012060 doi:10.1088/1757-899X/84/1/012060

to 20 °C. After nucleation, the specimen was fallen into a water bucket at predetermined timing to quench the structure. A typical thermal history is shown in figure 1 and timing of quenching is also indicated: before the peak, peak, 190 °C and 185 °C. Since the cooling rate of quenching was 200 °C /sec, it was possible to realize the solidified structure before the quenching. The specimen was vertically cut at the center and polished. After being etched by a dilute aqua regia, solidified structure was observed by a FE-SEM and $f_{\beta\text{-Sn}}$ before quenching was measured by an image processer.

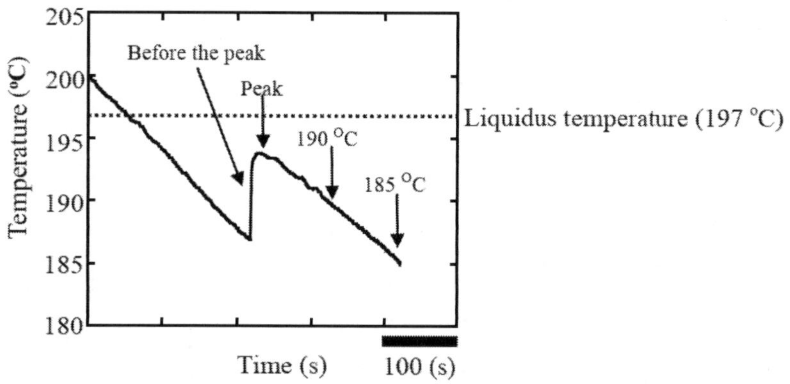

Figure 1. A typical thermal history and Timing of quenching.

3. Experimental results

Two examples of SEM images are shown in figures 2(a) and (b). In case of ΔT=20 °C and quenched at 185 °C, solidified structure is shown in figure 2(a). In case of ΔT=1 °C and quenched at before the peak, solidified structure is shown in figure 2(b). The primary phase was distinguished from size and morphology to measure the volume fraction of primary (β-Sn) that had solidified before quenching, they were realized and marked by hand (figure 2. a2, b2). After this, $f_{\beta\text{-Sn}}$ was measured using an image processing software. For example, $f_{\beta\text{-Sn}}$ in figure 2 a2) and in figure 2 b2) are 0.47 and 0.17, respectively.

The relationship between quenching timing and $f_{\beta\text{-Sn}}$ is shown in figure 3. In early stage of solidification, it was observed that the $f_{\beta\text{-Sn}}$ increased with increasing the supercooling. In case of small ΔT, $f_{\beta\text{-Sn}}$ increased gradually with time. On the other hand, in case of large ΔT, $f_{\beta\text{-Sn}}$ was large just after the nucleation. Finally, $f_{\beta\text{-Sn}}$ was larger than equilibrium value regardless of ΔT.

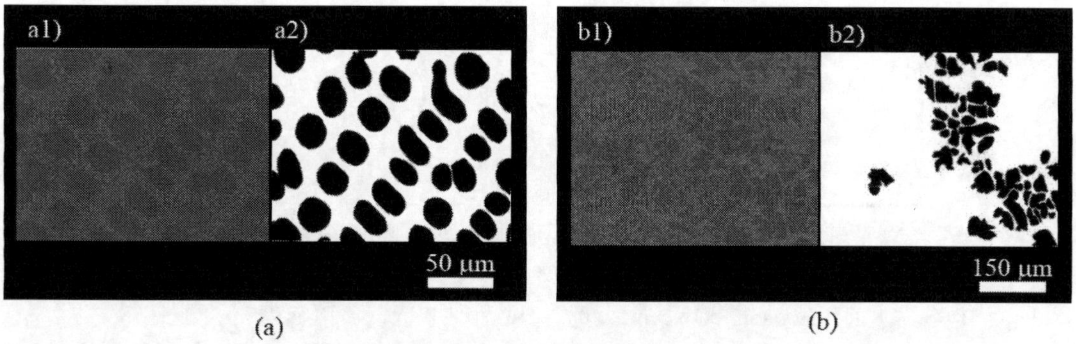

Figure 2. SEM images of solidified structure.
(a) quenched at 185 °C (ΔT=20 °C),
(b) quenched just after nucleation (ΔT=1 °C)

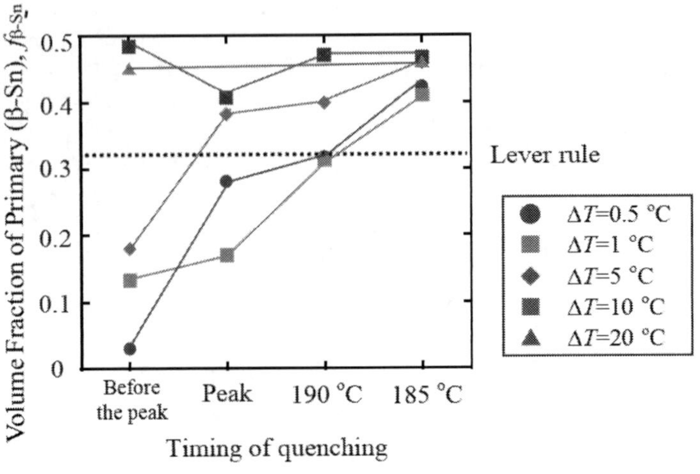

Figure 3. The relationship between $f_{\beta\text{-Sn}}$ and timing of quenching.

Now, the time (t) is defined as the time elapsed from the nucleation for primary β-Sn. The relationship between t and $f_{\beta\text{-Sn}}$ in case of large ΔT is shown in figures 4, (ΔT=5 °C) and (ΔT=10 °C). The relationship between t and $f_{\beta\text{-Sn}}$ in case of small ΔT is shown in figures 5, (ΔT=0.5 °C) and (ΔT=1 °C). In case of ΔT=10 °C, the $f_{\beta\text{-Sn}}$ exceeded equilibrium value within a few seconds. In case of ΔT=5 °C, the $f_{\beta\text{-Sn}}$ exceeded equilibrium value 46 seconds after the nucleation. These results indicate that the $f_{\beta\text{-Sn}}$ exceed quilibrium value in a short time if the molten metal was largely supercooled. Futhermore, the final $f_{\beta\text{-Sn}}$ was 0.09 larger than the lever rule in both cases.

Comparing to the supercooling of 20 or 10 °C, $f_{\beta\text{-Sn}}$ gradually increased in case of 0.5 and 1 °C. At the final stage of solidification, $f_{\beta\text{-Sn}}$ exceeded the equilibrium value, but it was 0.05 larger than the equilibrium value at most.

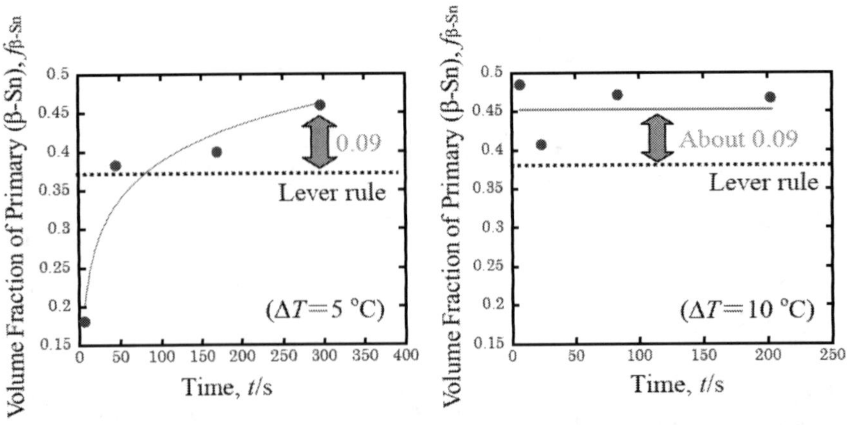

Figure 4. Relationship between t and $f_{\beta\text{-Sn}}$ in case of large ΔT (5 °C, 10 °C).

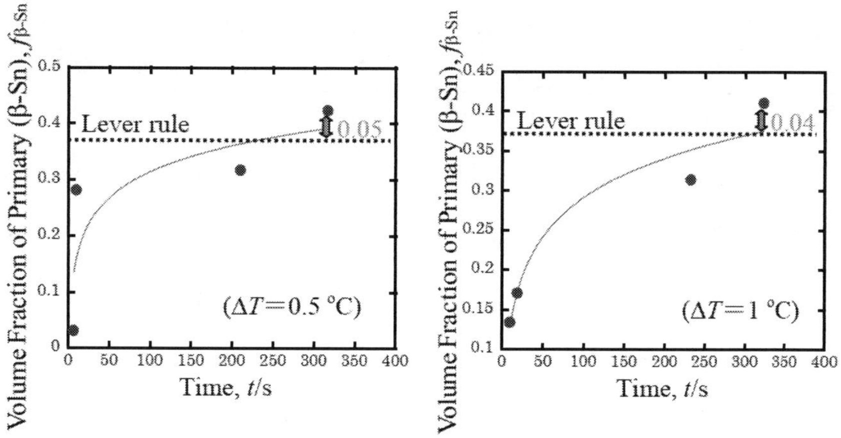

Figure 5. Relationship between t and $f_{\beta\text{-Sn}}$ in case of small ΔT (0.5 °C, 1 °C).

4. Discussions

The present experimental study using a trigger can separate the nucleation and growth of primary β-Sn. As mentioned earlier, the important results are following two points.

(1) $f_{\beta\text{-Sn}}$ increased rapidly, when the ΔT was large. But, $f_{\beta\text{-Sn}}$ gradually increased when the ΔT was low.

(2) Final $f_{\beta\text{-Sn}}$ was larger than equilibrium value regardless of ΔT. But, strictly speaking, $f_{\beta\text{-Sn}}$ was larger with increasing ΔT.

The factors which affect $f_{\beta\text{-Sn}}$ may be nucleation and growth. According to the classical nucleation theory, it is known that the frequency of nucleation for pure substance increases as ΔT increases [3]. The amount of ΔT discussed in the classical nucleation theory may be over 50 °C [4]. In present study, the range of ΔT is 0.5 ~ 20 °C, which is too small for homogeneous nucleation. Actually in this experiment, a stainless steel rod was used for a trigger. The temperature of rod was room temperature and it promoted heterogeneous nucleation on the surface of the rod. Frequency of nucleation was not measured in this experiment, though, it may not depend on the ΔT in the range of this study.

There have been some experimental studies to relate the ΔT and the growth velocity in pure substance. According to these results, it is known that the growth velocity is approximately proportional to ΔT^2 [5]. Therefore, it is reasonable that $f_{\beta\text{-Sn}}$ slowly increased when the ΔT was small because dendrite grew slowly. In a similar way, it is valid $f_{\beta\text{-Sn}}$ is high in the initial stage of solidification in case of large ΔT. The reason that the final $f_{\beta\text{-Sn}}$ increased with increasing ΔT has been uncertain. Further, the reason that the $f_{\beta\text{-Sn}}$ is larger than the equilibrium value has been uncertain.

5. Conclusions

To comprehend the reasons that $f_{\beta\text{-Sn}}$ is larger than equilibrium value, the present experimental study was carried out focusing on relationship between $f_{\beta\text{-Sn}}$ and ΔT. From this study, following results were clarified.

(1) In Sn base alloys, final $f_{\beta\text{-Sn}}$ was larger than equilibrium value regardless of ΔT.

(2) $f_{\beta\text{-Sn}}$ in case of large ΔT tend to be larger than that in case of small ΔT.

(3) Dendrite growth velocity may influence $f_{\beta\text{-Sn}}$ rather than frequency of nucleation.

References

[1] Nagatomo Y, Esaka H and Shinozuka K 2012 *Collected Abstracts of the 2012 Autumn Meeting of Japan Inst. Metals* 28
[2] Takamatsu Y, Esaka H and Shinozuka K 2010 *J. Japan Inst. Metals* **74** 551
[3] Kurz W and Fisher D J 1992 *Funfamentals of Solidification* (Aedermannsdorf : Trans Tech Publication) chapter 2 p 21
[4] Nishizawa T, 2008 *Thermodynamics of microstructure* (Ohio: ASM International) chapter 8 p 199
[5] Powel G L F, Colligan G A, Surprenant V A, and Urquhart A 1977 *Metall. Trans. A* **8A**, p 971

Modelling of macrosegregation in direct chill casting considering columnar-to-equiaxed transition using 3-phase Eulerian approach

J Hao[1], Y J Lin[1], Y Nie[1], M Wu[2] and A Ludwig[2]

[1] Department of Mechanical, Materials and Manufacturing Engineering, University of Nottingham Ningbo China, China

[2] Chair for Modelling and Simulation of Metallurgical Processes, University of Leoben, Austria

E-mail: jane.hao@nottingham.edu.cn

Abstract. A 3-phase Eulerian approach is used to model the macrosegregation during solidification in direct chill (DC) casting of binary bronze (Cu-Sn). The three phases are the melt, the solidifying columnar dendrites and the equiaxed grains. The thermodynamic information of Cu-Sn is included based on published thermodynamic data, which are coupled with the 3-phase solidification model. The occurrence of columnar-to-equiaxed transition (CET), phase interactions, feeding flow, equiaxed sedimentation and their influence on macrosegregation are considered in the model. The model is applied to a laboratory DC casting process of bronze as a benchmark to demonstrate the model potentials. The simulation results of mixed columnar and equiaxed solidification as well as the formation of macrosegregation are presented. The focus of this work is to analyze and discuss the macrosegregation mechanisms by different flow including feeding flow and crystal sedimentation.

1. Introduction

Bronzes (Cu-Sn) are among the oldest engineering materials, and nowadays are still used in many different applications with which high quality materials are required for example electronics. Due to the fact that bronzes have a rather large mushy zone (with the solidification interval of about 200°C), macrosegregation often appears in direct chill (DC) casting process of bronze. Macrosegregation is an inhomogeneous distribution of alloy composition that exists over the entire casting [1]. It decreases the workability of the product and thus has to be removed. Macrosegregation is known to be caused by a combination of microsegregation and relative motion between liquid and solid in the mushy zone, induced by different flow phenomenon of forced convection, thermo-solutal convection, feeding flow, equiaxed sedimentation, etc [2]. Since diffusion in the solid is too slow to remove macrosegregation by heat treatment, it has to be minimized during solidification by choosing appropriate casting parameters.

Several macrosegregation models have been developed by researchers [3-7]. Flemings et al.[3] developed a model to predict the formation of channel-type segregation for multicomponent systems.

The model considered heat, mass and momentum transport in the mushy zone. Beckermann et al.[4] applied a coupled multicomponent solidification model with melt convection to a large industry-scale ingot. Their simulation results qualitatively agreed with the positive segregation observed in the ingot. Ludwig et al. [5] presented a model coupling the ternary phase diagram with the multiphase solidification simulation and apply it to a DC casting of bronze. Goyeau et al.[6] developed a macroscopic model by using a volume averaging technique with local closure, taking into account the spatial variation of the pore scale geometry within the mushy zone. Combeau et al. [7] presented a two-phase model to study the influences of both motion and morphology of equiaxed grains on a steel ingot. However, these models did not distinguish between the columnar phase and the equiaxed phase; the two phases are presented as one solid phase during solidification.

Wu et al. [8] developed a mixed columnar-equiaxed solidification model, which considered the competitive growth of columnar and equiaxed phase, melt convection, equiaxed grain sedimentation, together with their influence on the species transport and macrosegregation. This model successfully predicted the conic negative segregation in the lower region of the ingot and the columnar-to-equiaxed-transition (CET) [9]. The authors have modified and applied the model to study macrosegregation in a laboratory DC casting of bronze [10]. Based on that study this paper presents an insight investigation on the formation mechanisms of macrosegregation by different flow phenomena, especially the effects of equiaxed crystal sedimentation are discussed in details.

2. Model Description

2.1. General Assumptions.

Ternary and binary simulation work on different continuous casting processes have been published and discussed during the last years by the authors [5,8]. The presented simulation results for a binary CuSn DC casting takes into account three phases: the liquid phase, the columnar phase, and the equiaxed phase. The thermodynamic information of Cu-Sn is included based on thermodynamic data [11] and coupled with the multiphase model. Details of the three-phase mixed columnar-equiaxed solidification model can be found in the literature [8,9]. Here, only a brief outline of the model assumptions is given.

- Columnar dendrites, approximated by growing cylinders, start to grow at the mold wall only if the local temperature drops below the liquidus temperature.
- The diffusion field around the columnar cylinder induces a shell-like growth around the cylinder. The growth of columnar is controlled by specific model assumptions as described in the next part.
- Homogenuous nucleation is modelled by initialising a certain number of nucleus in each cell. No further nucleation is allowed after the growth of crystals start.
- Growth of equiaxed grains is included by a diffusion driven growth around a spherical crystal and controlled by specific model assumptions as described in the following.
- Feeding flow is considered by taking into account the density difference between liquid and solid.
- Mechanical interaction between the mush and the flow is calculated via Darcy's low.
- The mushy permeability is modelled by the Blake-Kozeny permeability approach [12].

2.1.1. Blocking Mechanism.

Based on the fact that during solidification the columnar and equiaxed phase grow competing with each other, the solidification model is performed by allowing each phase growing simultaneously with the following blocking mechanisms to ensure physical growth conditions:

(i) The columnar phase is not allowed to grow if the undercooling of the liquid (ΔT) is smaller than 4K [12]. The applied undercooling allows the growth of the equiaxed phase in front of the columnar dendrite tips.

(ii) The columnar phase is not allowed to grow if the equiaxed volume fraction (f_e) exceeds 0.49. Experimental observations [13] indicate that the columnar front is blocked by a certain amount of equiaxed crystals in front of it. Therefore, this condition ensures that the columnar cannot overgrow the equiaxed phase when it has already a certain amount of volume fraction.

(iii) The equiaxed phase cannot grow if its diameter (d_e) is bigger than the space between the primary dendrite arm spacing (λ_1) of the columnar phase and the diameter of the columnar dendrite trunk (d_c) [13]. This condition models the growth of the equiaxed phase within the dendrite trunks; it is thought that an equiaxed crystal is not able to grow further if it's sticking within two dendrite trunks.

(iv) The equiaxed phase is blocked in its movement when the sum of the volume fraction of the equiaxed phase and the columnar phase reaches the packing limit of $f_e + f_c >= 0.637$ [14].

(v) The equiaxed is trapped and forced to move with the columnar when the volume fraction of columnar reach to a certain value, that is, trapping limit of $f_c >= 0.2$ is reached [14].

3. Benchmark Configuration

The presented model is applied to a benchmark simulation of a laboratory DC casting process as displayed in figure 1. For the process simulation a casting velocity of $\vec{u}_{cast} = 1.44$ mm□s^{-1} and a casting temperature of T_{cast} = 1523 K are used. Calculations were done for the binary alloy CuSn6. Since the mold is of a round shape, a 2D axis-symmetric simulation has been performed. Figure 1a displays the configuration of the benchmark, and figure 1b displays the geometry and grid for the simulation. The grid size is 3 mm with quadrilateral mesh. The boundary conditions used in the simulation are marked in the figure: ① indicates the position of the inlet where a pressure inlet is taken, ② assigns the upper part of the mold which is assumed to be insulating, ③ shows the lower part of the graphite mold where h = 3000 W□m^{-2}□K^{-1} and $T_{lower\text{—}mold}$ = 550 K, and ④ displays a water cooling system with h = 3000 W□m^{-2}□K^{-1} and T_{water} = 300 K, and ⑤ shows the position of the outlet with a casting velocity of $\vec{u}_{cast} = 1.44$ mm□s^{-1}. The mold wall is considered to move with casting velocity, therefore a non-slip condition for the columnar phase is applied. As initial conditions, hot melt (T_{init} = 1523 K) at the casting velocity ($\vec{u}_{cast} = 1.44$ mm□s^{-1}) is assumed. The presented results are taken after reaching steady state.

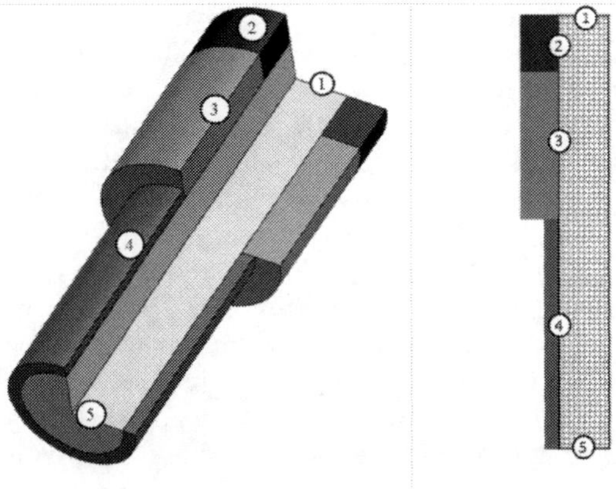

(a) (b)

Figure 1. (a) Configuration of benchmark. (b) Geometry and grid for calculation. The grid size is 3 mm. ① marks the position of the inlet, ② and ③ mark the upper and lower part of the mold, ④ indicates a water cooling system, and ⑤ displays the position of the outlet.

4. Result and Discussion

Two simulation settings have been applied in the study, namely (i) Case A: just feeding flow has been considered to model the shrinkage during solidification induced by the density difference between solid and liquid, and (ii) Case B: in addition to the feeding flow, the flow induced by the settling effect of the equiaxed phase is taken into account by including gravity force. In the following the two cases

IOP Conf. Series: Materials Science and Engineering **84** (2015) 012061 doi:10.1088/1757-899X/84/1/012061

are first presented in each own and afterwards compared. The temperature field, the volume fraction of the equiaxed and columnar phase, the relative velocity between phases and the macrosegregation are shown and discussed.

4.1 Case A: Feeding Flow

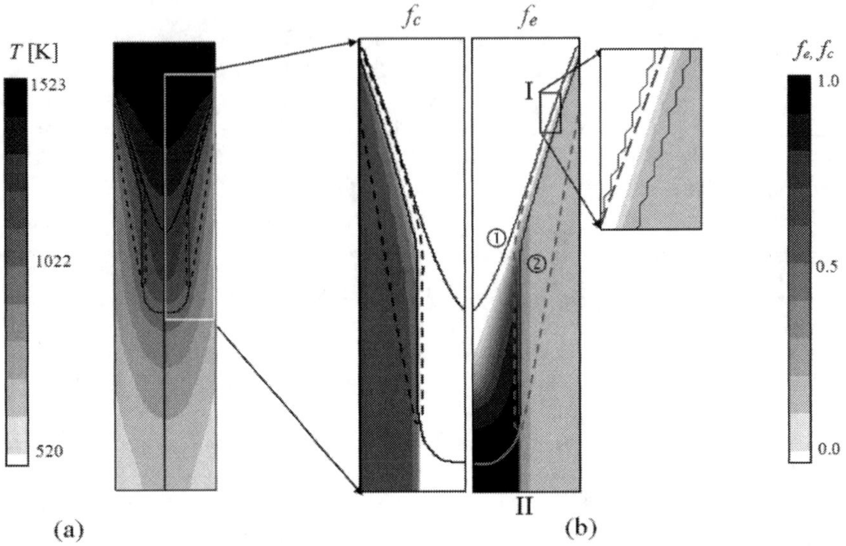

Figure 2. (a) Temperature distribution in the casting. (b) Volume fraction of columnar (left) and equiaxed (right) in the rectangle marked in (a). The solid iso-surface ① shows equiaxed grow region, and dashed iso-surface ② shows columnar grow region.

In this case study only feeding flow due to solidification shrinkage caused by density differences between liquid and solid is considered in the model. figure 2 shows the simulation results of temperature distribution and volume fraction of equiaxed (f_e) and columnar (f_c), including the equiaxed growth region marked by the solid iso-surface ① and the columnar growth region marked by the dashed iso-surface②. In figure 2b it could be seen that at the mold surface where the solidification process starts, the volume fraction of equiaxed can reach about 30%. This is due to the above stated blocking mechanism (i), which delays the growth of columnar phase with 4K undercooling, creating space for equiaxed growth. The competing growth between the two phases can be seen more clearly from the zoomed region (I) that equiaxed starts to grow earlier than columnar. Figure 2b shows that that the volume fraction of equiaxed is kept at 30% from the casting surface to columnar-to-equiaxed transition (CET) marked by (II); this is due to the blocking mechanism (iii) with which the equiaxed growth stops when the its diameter (d_e) is larger than the space between the primary dendrite arm spacing (λ_1) and the diameter of the columnar dendrite trunk (d_c). It should be noted that in reality the equiaxed grains formed near the casting surface would grow along the heat flux direction and change into columnar phase; but in the simulation the equiaxed formed is kept not changed, which is convenient for analyzing the results. In the simulation columnar-to-equiaxed transition (CET) occurs at location (II) marked in figure 2b, where the columnar growth stops because the undercooling is less than 4K again, and thus equiaxed can grow to reach volume fraction 0.49 to block the columnar dentrite tip till the end of solidification; therefore there is only equiaxed growing in the center of the casting.

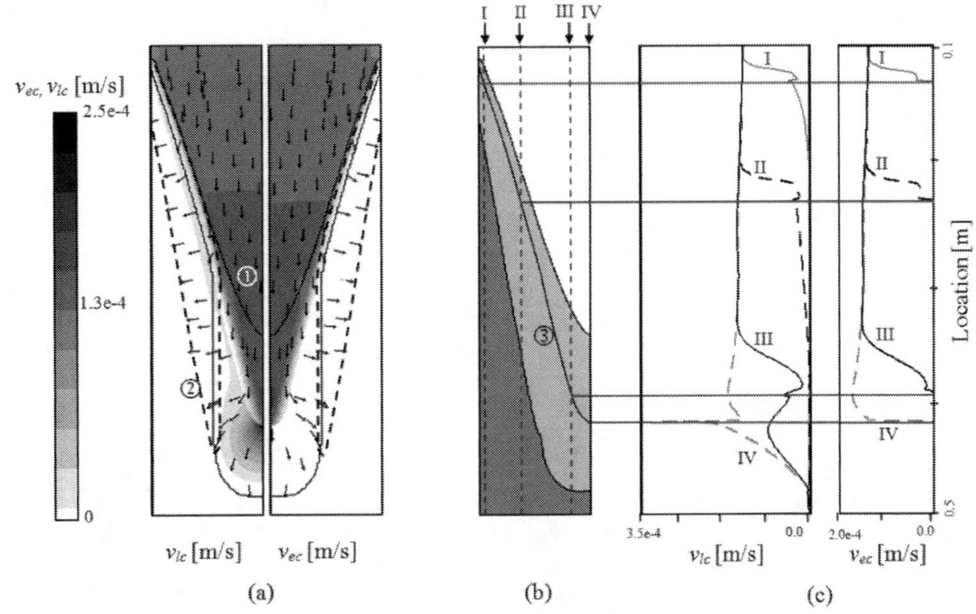

Figure 3. (a) Relative velocity between liquid and columnar v_{lc}, and between equiaxed and columnar v_{ec} in the region marked by the white rectangle in figure 2a. The magnitudes of the relative velocity are shown by the contours and the black vectors display the directions. The solid iso-surface ① shows equiaxed grow region, and dashed iso-surface ② shows columnar grow region. (b) Sketch of the liquid zone (white color), mush zone (light gray), and solid zone (dark gray). ③ marks the iso-surface of packing limit f_e+f_c = 63.7%. (c) Relative velocity plotted from top to bottom in the casting at 4 locations (marked by label I to IV).

Studying the relative velocity fields between the liquid and the solid phases, the different flow behaviour is getting obvious as shown in figure 3. In figure 3a it could be seen that the relative velocity between columnar and liquid (v_{lc}) shows the same flow pattern with the relative velocity between equiaxed and columnar (v_{ec}); this is due to the fact that the equiaxed moves with the melt before they are settled. In the upper part of the casting where no columnar phase exists the relative motion v_{lc} represents the melt flow speed-up compared to the casting speed, which is caused by the solidification-induced shrinkage of the whole casting. In reality there is no equiaxed phase in the upper part of the casting, yet in the simulation it is assumed a certain number of nucleus in each cell so the relative motion v_{ec} is presented. Close to the columnar tip front the relative motion turn directions so that v_{lc} and v_{ec} are now perpendicular to the solid fraction iso-surface and point from the dendrite tip towards their roots; this is induced by that the liquid and equiaxed try to feed the solidifying columnar region.

In figure 3b three different colored regions are displayed: the white area is the region where only liquid presents; the light grey zone shows the mushy zone where all the three phases exist; the dark grey region is full solidified without any liquid left so only columnar and equiaxed phase present. The iso-surface ③ marks the packing limit of $f_e + f_c \geq 0.647$ in blocking mechanism (iv). In figure 3c relative velocities are plotted from top to bottom in the casting at 4 locations marked by (I) to (IV). The curves plotted in figure 3c show that there are some peaks of relative velocity v_{lc} and v_{ec} as marked by the horizontal lines, which shows that the locations of peaks correspond to iso-surface③, the packing limit. These peaks can be explained as following: the mixture of liquid and equiaxed are being sucked into the mushy zone to feed the shrinkage of columnar phase before the packing limit is fulfilled; when the packing limit is reached the equiaxed grains must stop moving, but the solidification still goes on, so the liquid is accelerated to feed the reduced volume. Therefore in figure 3c it could be seen that when the packing limit is reached v_{ec} is reduced to zero, while at the same time v_{lc} shows a positive peak. It also can be observed that v_{ec} has a small peak shortly before reduced to zero; this is because that the sudden increase of liquid velocity gives the equiaxed some momentum.

4.2 Case B: Feeding Flow and equiaxed sedimentation

In this case study both feeding flow and equiaxed grains sedimentation are considered in the model. Figure 4 shows the flow pattern of relative velocity between the liquid and the columnar v_{lc}. Here only v_{lc} is displayed because Case A already shows that the relative velocity v_{lc} and v_{ec} has the same flow pattern. Three zoomed regions (I to III in figure 4a and figure 4c) display the detailed flow pattern with vectors showing velocity directions and contours of magnitudes. It can be seen that the movement of liquid in the equiaxed growth area is mainly influenced by the grain sedimentation. At the beginning of solidification (marked by I), equiaxed starts to grow and settle; thus additional fluid is needed to fulfill the mass conservation, which induces a first small vortex there. Afterwards (marked by II) due to grain sinking the equiaxed crystals are moving along the solidification front at the columnar dendrite tip area and continue to drag the liquid downwards. In the casting centre (marked by III) again a vortex forms transporting smaller grains upwards. In the area of columnar growth (enveloped by iso-surface②) the movement of liquid has the same pattern as in Case A, flowing from the dendrite tips towards the dendrite roots to feed the solidification shrinkage. It can be seen from the magnitude of the velocity field that the effect of sedimentation on liquid motion is remarkable stronger than that of feeding flow.

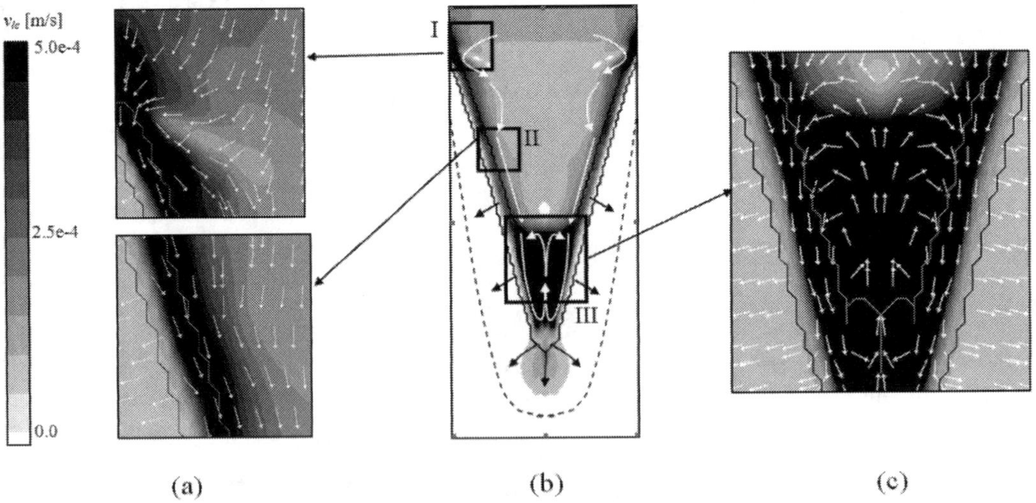

Figure 4. Relative velocity between liquid and columnar v_{lc} in the region marked by the white rectangle in figure 4a. The magnitudes of the relative velocity are shown by the contours with flow patterns indicated by arrows (middle). Three zoomed regions (I to III) display the detailed flow pattern with vectors showing directions (left and right).

Figure 5 compares the relative velocity v_{lc}, temperature and liquid concentration Sn for Case A and Case B at the casting center. This is to explain the simulation results in [10] that less equiaxed formed when sedimentation is included in the model and columnar can grow to the casting center. In figure 5a, it can be seen that the temperature field in Case B is lifted up compared to Case A; this is caused by the upwards movement of liquid which is induced by the vortex of equiaxed grains in casting center. Therefore the temperature is colder in Case B than in Case A in the casting center, which is one of the reasons that the growth of equiaxed is delayed. Another reason is that the vortex of equiaxed grains brings Sn-enriched liquid from the mushy zone into the bulk melt in the casting center, as shown in figure 6b; such increased liquid concentration leads to a lower liquidus temperature, which also delays the growth of equiaxed. Therefore the equiaxed starts to grow at the same time with columnar, and in the center of the casting equiaxed volume fraction is less than 0.49 so columnar is not blocked by the equiaxed.

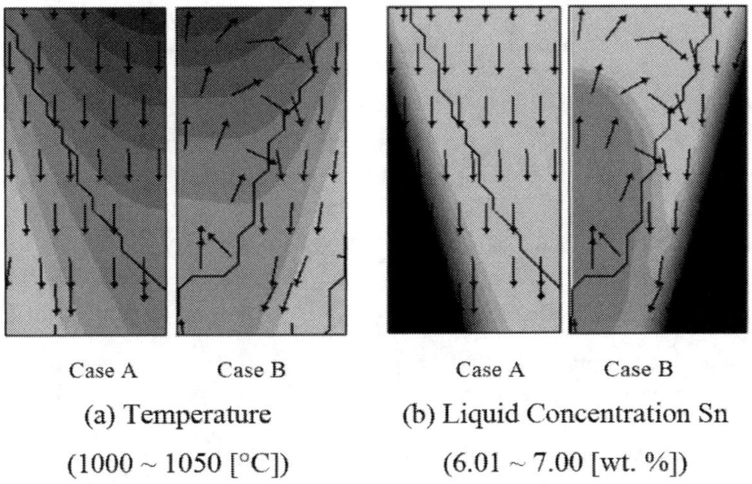

Figure 5. Comparison of (a) temperature and (b) liquid concentration for Case A and Case B at the casting center. Quantities are shown by gray levels with black for maximum and white for minimum with the ranges shown in the brackets. Vectors display directions of relative velocity between liquid and columnar. The black curve shows the start of solidification.

The predicted macrosegregation distribution of Sn in the casting was presented in the previous study [10], which showed that a positive macrosegregation formed in the center of the casting. This is caused by the equiaxed sedimentation induced vortex motion as displayed in figure 5, which brings Sn-enriched liquid from the mushy zone into the bulk melt. The study [10] also showed the macrosegregation along the outlet of the casting; a W-type macrosegregation profile was seen in both the simulation and measurement. In the simulation it is observed that most of the equiaxed grains are settled down not at the casting center, but at the locations slightly off-center where the mixture concentration is strongly reduced by their deposition, leading to the formation of W-type macrosegregation profile. The typically measured Sn profile in the measurements shows the similar macrosegregation pattern; this hints that the reasons behind the W-type macrosegregation might be the vortex movement caused by the equiaxed grains sedimentation. This proposition will be investigated by experiments in future.

5. Conclusion

A three-phase Eulerian solidification model was applied to simulate a binary CuSn6 DC casting process, which has shown the model's potential of calculating the mixed columnar and equiaxed solidification. Two cases were studied: one with feeding flow, and one with both feeding flow and sedimentation of equiaxed grains. The following conclusions are stated:

1. The developed model can be used to simulate the competing growth of columnar and equiaxed during solidification of DC casting process of bronze. Columnar grows from the mold wall towards the center of the casting, while equiaxed grow between the dentrite arms, ahead of the columnar front, and also in the bulk melt. With feeding flow the phase transition CET is predicted.
2. The simulation results show a peak of liquid velocity at the point where the movement of equiaxed grains is stopped. The phenomena behind this are: the mixture of liquid and equiaxed are being sucked into the mushy zone to feed the solidification shrinkage of columnar dentrite; as soon as the movement of equiaxed is stopped, liquid is accelerated to feed the shrinkage.
3. When sedimentation of equiaxed is added to the model, the volume fraction of equiaxed in the center of the casting is not large enough to block the columnar tip growth. The reason of less

equiaxed forming is that the sedimentation of equiaxed brings Sn-enriched liquid from the mushy zone into the bulk melt; such increased liquid concentration in the bulk melts leads to a lower liquidus temperature and hence the growth of equiaxed is delayed.

4. The sedimentation of equiaxed induces a vortex motion of both liquid and equiaxed in the center of the casting. The equiaxed grains are settled down not at the casting center, but at the locations slightly off-center where the mixture concentration would be strongly reduced by their deposition, leading to the formation of W-type macrosegregation profile. Further experiments are expected to support the proposition.

Acknowledgment
The author J Hao acknowledges the financial support by the Christian-Doppler Laboratory "Multiphase Modelling of Metallurgical Processes" during her stay at University of Leoben, Austria. Also authors are grateful to the support of National Natural Science Foundation of China (Grant No. 51305216), and Ningbo Science and Technology Bureau of China (Grant No. 2011B81006).

References

[1] Beckermann C, Gu J P and Boettinger W J 2000 Development of a freckle predictor via Rayleigh number method for single-crystal nickel-base superalloy castings *Metall. Mater. Trans.* A **31** 2545-57

[2] Wu M, Konozsy L, Ludwig A, Schutzenhofer W and Tanzer R 2008 Columnar to equiaxed transition during ingot casting using ternary alloy composition *Steel Res. Int.* **79** 56–63.

[3] Fujii T, Poirier D R and Flemings M C 1979 Macrosegregation in a multicomponent low alloy steel *Metall. Trans.* B **10** 331–9

[4] Gu J P and Beckermann C 1999 Simulation of convection and macrosegregation in a large steel ingot *Metall. Mater. Trans.* A. **30** 1357–66

[5] Ludwig A, Gruber-Pretzler M, Mayer F, Ishmurizin A and Wu M 2005 A way of coupling ternary phase diagram information with multiphase solidification simulations *Mater. Sci. Eng.* A **413–4** 485–9

[6] Rouxa P, Goyeau B, Gobin D, Fichot F and Quintard M 2006 Chemical non-equilibrium modeling of columnar solidification, *Int. J. Heat Mass Transfer* **49** 4496–510

[7] Combeau H, Zalozˇnik M, Hans S and Richy P E 2009 Predication of macrosegregation in steel ingot: influence of the motion and the morphology of equiaxed grain *Metall. Mater. Trans.* B **40** 289–304

[8] Wu M and Ludwig A 2006 A three-phase model for mixed columnar equiaxed solidification *Metall. Mater. Trans.* A **37** 1613–31

[9] Wu M and Ludwig A 2007 Using a three-phase deterministic model for the columnarto-equiaxed transition *Metall. Mater. Trans.* A **38** 1465–75

[10] Hao J, Grasser M, Wu M, Ludwig A, Riedle J and Eberle R 2011 A 3-phase model for mixed columnar-equiaxed solidification in DC casting of bronze *IOP Conference Series: Materials Science and Engineering* **27** 012015

[11] Massalski TB 1990 *Binary Alloy Phase Diagrams* (Ohio: ASM International)

[12] Bird R B, Steward W E and LIGHTFOOT E N 1960 *Transport Phenomena* (New York: John Wiley & Sons)

[13] Hunt J D 1984 Steady state columnar and equiaxed growth of dendrites and eutectic *Mater. Sci. Eng.* **65** 75-83

[14] Wu M and Ludwig A 2009 Modeling equiaxed solidification with melt convection and grain sedimentation-I: Model description *Acta Mater.* **57** 5621-31

IOP Conf. Series: Materials Science and Engineering **84** (2015) 012062 doi:10.1088/1757-899X/84/1/012062

Kinetics of the δ/γ interface in the massive-like transformation in Fe-0.3C-0.6Mn-0.3Si alloys

T Nishimura[1], K Morishita[1], T Nagira[2], M Yoshiya[2] and H Yasuda[1]
[1] Department of Materials Science and Engineering, Kyoto University
[2] Department of Adaptive Machine Systems, Osaka University

E-mail: nishimura.tomohiro.66s@st.kyoto-u.ac.jp

Abstract. In Fe-C alloys, it has been considered that the peritectic reaction occurs during solidification. Recently, the massive-like transformation, in which δ phase undercooled below the peritectic temperature transforms into γ phase, has been observed by X-ray imaging. It is of interest to know how the δ phase transforms into the γ phase, because the transformation is related to deformation of solidifying shell. Time-resolved and in-situ X-ray imaging with frame rate up to 500 fps was performed to know the δ/γ interface motion in the massive-like transformation in Fe-0.3C-0.6Mn-0.3Si alloys (mass%). The moving velocity of the δ/γ interface locally fluctuated. The average velocity ranged from several mm/s to 200 mm/s. However, no relationship between the moving velocity and the undercooling from the peritectic temperature was observed. The interface was not planar and the morphology continuously changed. The results showed that the stress / strain induced though the transformation from the δ phase to the γ phase in the steel solidification was not simply explained by the conventional peritectic reaction.

1. Introduction

The bcc (δ) phase can transform into the fcc (γ) phase below the peritectic temperature (1766 K) in Fe-C alloys with 0.1-0.5 mass%C. Due to the volume change in the transformation has been known to cause deformation of solidifying shell [1]. To avoid defect formation due to the transformation, many efforts have been done to understand the peritectic reaction / solidification [2]. In Fe-C alloys with 0.1-0.5 mass%C, it has been considered that the peritectic solidification, in which the γ phase was produced at the interface between the δ phase and the liquid phase, took place. Thus, the solidification of the Fe-C alloys has been considered on the basis of the peritectic reaction [2].

Radiography using synchrotron radiation X-rays has been developed [3]. For example, dendritic solidification of Sn and Al alloys has been reported [4-5]. Recently, the technique has been extended to observe solidification of Fe-C alloys [6-8] and deformation of semisolid [9]. Since the in situ observation allows us to know how the solidification and the phase transformation proceed, the technique provides direct and valuable information for modelling microstructural evolution and defect formation. In addition, the technique can be used to obtain experimental data for validating model and numerical simulation.

In our previous study [7-8], the massive-like transformation from the undercooled δ phase to the γ phase dominantly occurred, comparing to the peritectic solidification. The γ phase hardly nucleated in the δ+L state and consequently the δ phase continued to grow even below the peritectic temperature. Once the γ phase nucleated in the undercooled δ phase, the massive transformation completed within

1s in the observation area (5 mm x 5 mm). Since the massive-like transformation can induce higher strain rates for the viscoelastic solidifying shell, the influence of the massive transformation on deformation and crack can differ from that of the peritectic solidification [8]. Thus, it is valuable to know the transformation behaviour (*i.e.* moving velocity and morphology of the δ/γ interface) for building a model of the massive-like transformation.

The in-situ observation with a frame of 1 fps in the previous study [7] was too slow to observe the δ/γ interface during the massive transformation. done with a frame rate of 1 fps. This study presnets the moving velocity and the interface morphology for a conventional carbon steel (Fe-0.3C-0.6Mn-0.3Si in mass%) by using in-situ observation with a relatively high rate up to 500 fps. The kinetics of the δ/γ interface in the massive-like transformation is also discussed.

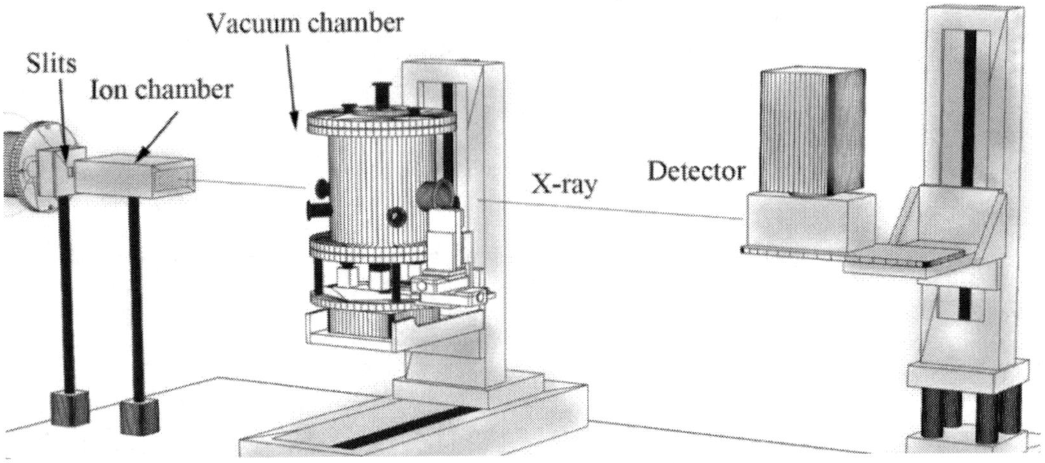

Figure 1. Setup for observing the massive-like transformation of Fe-0.3C-0.6Mn-0.3Si.

Table 1. Experimental conditions for observing.

Composition	Fe-0.3C-0.6Mn-0.3Si (mass%)
Dimension	10 mm x 10 mm x 0.1 mm
X-ray energy	21 KeV
Detector	C-MOS type camera
Pixel size	5 μm x 5 μm
Observation area	5 mm x 5 mm
Frame rate	125 fps, 250 fps, 500 fps
Cooling rate	0.17 K/s, 0.33 K/s, 0.87 K/s

2. Experiments

2.1. Setup for observing

The in situ observation was performed at a beamline of BL20XU in SPring-8, Hyogo, Japan. Figure 1 shows a setup for observing the massive-like transformation. Details of the setup were described in the previous work [6]. An ion chamber for measuring intensity of incident X-ray beam, vacuum chamber in which a specimen with a heating system is set, and an X-ray beam monitor are placed along X-ray beam stream. The monitor consisting of a fluorescent screen, an optical lens and a high speed CMOS-type camera was used. The images with a format of 1024 x 1024 pixels and 12-bit depth were stored. The pixel size was 5 μm x 5 μm and the frame rate was 125 – 500 fps.

2.2. Observation condition

The observation conditions for Fe-0.3C-0.6Mn-0.3Si alloys are listed in Table1. Mother alloy with a composition of 0.3mass%C, 0.6mass%Mn and 0.3mass%Si was made by arc melting of electrolytic Fe, C, Mn and Si. A specimen cell made of Al_2O_3 window plates 0.15mm) and BN retainer plates was used [6]. The specimen with a dimension of 10 mm x 10 mm x 0.1 mm was heated by graphite heaters. X-ray energy used for absorption images was 21 keV.
Procedures for observing the δ/γ interface motion were as follows; (1) the primary δ phase was remained in the bottom part to aviod nucleation undercooling of the δ phase, (2) observation position was set at the center of the specimen, (3) the specimen was cooled at a constant cooling rate ranging from 0.17 K/s to 1.67 K/s, (4) Temperature when the dendrite tips rached at the center of the observation area was assumeed to be the liquidus temperature, (5) Recording the images was quited when the δ/γ transformation was observed.

3. Results

3.1. Average moving velocity of the δ/γ interface

Figure 2 shows typical dendritic growth of the primary δ phase. According to the phase diagram of Fe-C system, the peritectic reaction in which the γ phase is produced from the δ and the liquid phases could start at the peritectic temperature. However, the γ nucleation did not occur and consequently the solidification was nearly competed without the γ phase.

The difficult of the γ nucleation resulted in the massive-like transformation. Figure 3 shows an example of the δ/γ interface motion during the massive-like transformation in the Fe-0.3C-0.6Mn-0.3Si alloy. Cooling rate was 0.83 K/s and frame rate of radiography was 250 fps. Fig.3(a) shows a snap shot of the δ phase before the massive-like transformation. The γ nucleation followed by the massive-like transformation occurred at a undercooling of 60 K from the peritectic temperature. Fig.3(b) shows a snap shot of the δ/γ interface moving from the right to the left. The δ/γ interface positions were indicated by black lines. The interface position was detected by many dark regions, where the Bragg condition was satisfied to the incident X-ray. Namely, the strain / stress were introduced in the γ phase and even the multiple grains of the γ phase were produced from the γ single grain in Fig.3(c). Shape of the δ/γ interface was not planar and fluctuated. For example, the concave shape changed to the convex shape even within the frames (1/250 s). The duration of the δ/γ interface motion in the observation areas was 0.044 s and the moving length was 5 mm. The duration and the length give the average moving velocity of the δ/γ interface in the observation area. The moving velocity in Fig. 3 was 100 mm/s. It should be noted that the local moving velocity of the δ/γ interface remarkably fluctuated and the motion was far from the steady state.

Figure 4 shows relationship between the average moving velocity and the undercooling from the peritectic temperature. The average moving velocity ranged from several mm/s to 200 mm/s in the undercooling ranging from several K to 100 K. There was no clear relationship between the average moving velocity and the undercooling.

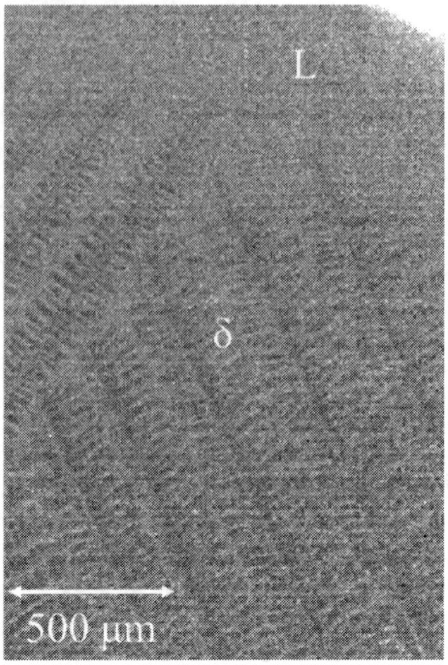

Figure 2. Growth of the δ dendrites.

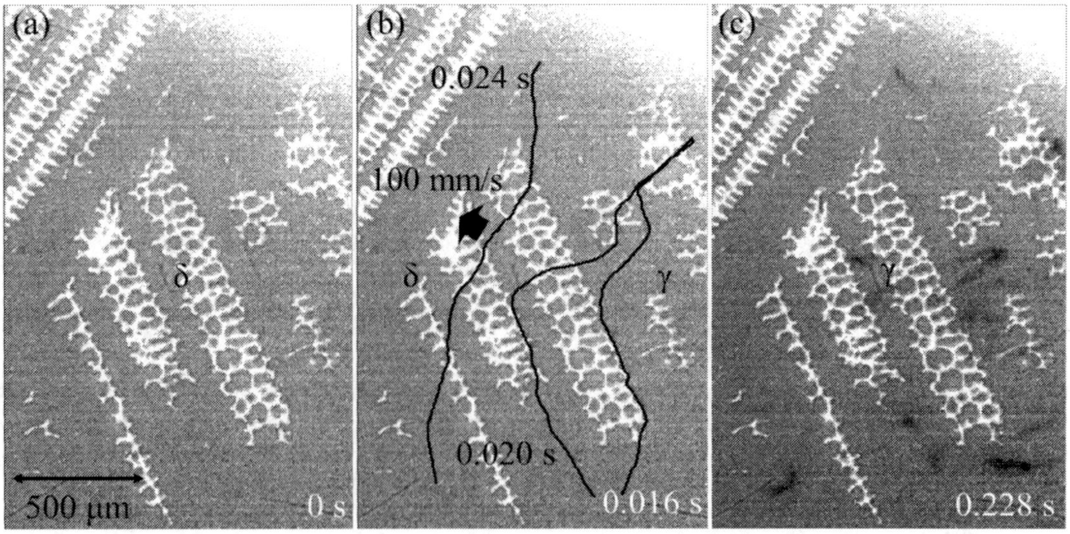

Figure 3. (a)The δ phase before the massive-like transformation. (b)The δ/γ interface. Line: δ/γ interface Frame rate:250 fps (c)The γ phase after the massive-like transformation.

3.2. Local moving velocity of the δ/γ interface

Since the δ/γ interface in the massive-like transformation was not planer and remarkably fluctuated, the average moving velocity does not represent the motion of the δ/γ interface sufficiently. In this study, the local moving velocity was also defied to know the δ/γ interface motion. Fig.5 shows an example to explain the estimation of the local velocity. The line indicated by "line" was the shortest line from A to the interface at the previous frame. It means that the local velocity was estimated to be as low as possible. The local moving velocity was simply calculated by the length of the shortest line and the interval between the frames. Figure 5 shows the local motion (black arrows) of the δ/γ interface (black line). The length of the arrow (black) corresponds to moving distance of the δ/γ interface at the position. The local velocity at the position is calculated from the distance and the interval between the frames. Figure 6 shows histogram of the local velocity. The local moving velocity ranged from several mm/s to 200 mm/s. The results show that the moving velocity of the δ/γ interface locally fluctuated significantly.

4. Discussion

4.1. Contribution of undercooling to the δ/γ interface motion

Since the γ nucleation occurred at an undercooling as high as 60 K. The transformation from the δ phase to the γ phase does not require the solute partition at the δ/γinterface. If the solute partition does not occur and the elastic energy is ignored, the driving force, which is defined by the difference in the Gibbs energies between the δ phase and the γ phase, is a function of the undercooling. Thus, the moving velocity should depend on the undercooling. However, the relationship between the average moving velocity and the undercooling was not observed, as shown in Fig.4. Even the local velocity fluctuated frame by frame. The discrepancy suggested that the solute partition and / or the elastic energy induced by the transformation may modify the driving force and consequently fluctuated the local moving velocity.

4.2. Influence of solute partition at the δ/γ interface

Since diffusivity in the δ phase is much larger than that in the γ phase [10], the solute profile at the δ/γ interface is essentially the same as that at the steady state solidification with planar interface [11]. Thickness of diffusion layer is given by 2D/V (D: diffusivity, V: velocity). The thicknesses for carbon, manganese and silicon are 10^{-8} m, 10^{-10} m and 10^{-10} m, respectively. Here, the diffusivities of carbon, manganese and silicon were 10^{-9}, 10^{-12} and 10^{-11} m^2/s [10]. Since the thicknesses for manganese and silicon is in the order of atomic distance, the two substitutional atoms are hardly partitioned at the δ/γ interface. On the other hand, the thickness for C (interstitial atom) is in the order of 10^{-8} m. It suggested that carbon atoms could be partitioned at the δ/γ interface in the massive-like transformation. The partition of carbon atoms may change the driving force and induce the instability of the planar interface.

4.3. Morphology of the δ/γ interface and solute partition

It is of interest to consider the curved interface from a viewpoint of the solute partition of carbon atoms to know an origin of the fluctuation. As mentioned in the previous section, the diffusivity in the γ phase is much smaller than that in the δ phase. Thus, the instability of the solidifying front due to the solute partition can be used to consider the interface in the massive-like transformation. According to the marginal stability theory [12], a specific wavelength λ given by eq. (1) is introduced when temperature gradient is negligibly small.

$$\lambda = 2\pi\left(\frac{\Gamma}{mG_c}\right)^{\frac{1}{2}}$$

Here, Γ is Gibbs-Thomson coefficient, m is a gradient of liquidus line and G_c is a gradient of solute concentration at the interface. The wavelength roughly gives a wavelength of interface fluctuation.

If the wave length is calculated in Fe-C, Fe-Mn and Fe-Si binary system. The lengths for carbon, manganese and silicon are estimated to be 0.1μm, 2μm and 2μm, using the typical physical properties. The interval of fluctuation observed in the massive-like transformation was in the order of 10μm or 100μm. Thus, the observed interval is not explained by the solute partition at the δ/γ interface. In addition, the transformation from the convex shape to the concave shape is not explained by the instability induced by the solute partition and vice versa.

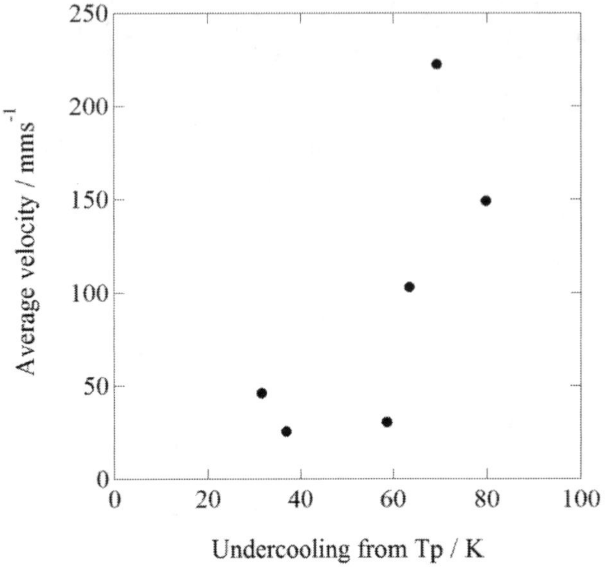

Figure 4. Relationship between the average velocity and the undercooling in Fe-0.3C-0.6Mn-0.3Si.

Figure 5. Sequence image of the local velocity in Fe-0.3C-0.6Mn-0.3Si.

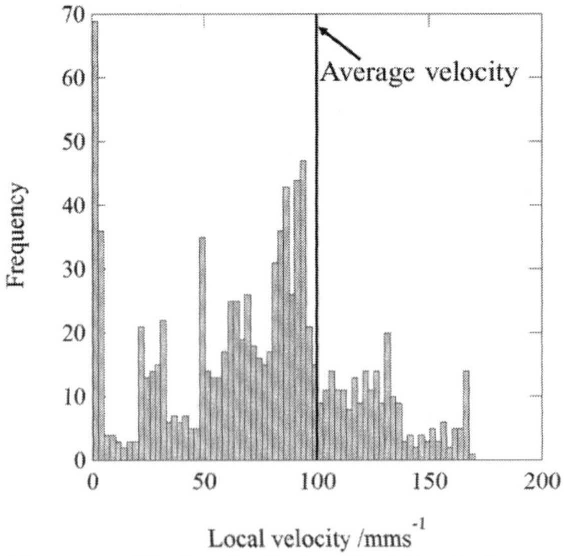

Figure 6. Histogram of the local velocity. Black line is the average velocity.

5. Conclusion

The in situ observation with relatively high time resolution was performed to observe the δ/γ interface motion in the massive-like transformation in Fe-0.3C-0.6Mn-0.3Si.

(1)The average moving velocity ranged from several mm/s to 200 mm/s. Since there were no relationship between the velocity and the undercooling, it was not determined just by the undercooling.

Other factors such as elastic energy or solute partition can also play a role for determining the δ/γ interface motion.
(2)Mn and Si were hardly partitioned at the δ/γ interface while C can be somewhat partitioned. The partition of C may influence the δ/γ interface motion.
(3)The morphology of the δ/γ interface was not planar and fluctuated during motion. The period of the δ/γ interface could not be explained just by the solute partition. Other factors such as elastic energy can also play a role for determining the δ/γ interface morphology.
(4)The local moving velocity ranged from several mm/s to 200 mm/s and fluctuated. The δ/γ interface could not grow in the steady state.

Acknowledgments
This study was in part supported by a Grant-in-Aid for Scientific Research (S) from MEXT. The synchrotron radiation experiments were performed as general projects at beamlines of BL20XU and BL20B2 at SPring-8 (JASRI).

References
[1] Matsumiya T, Saei T, Tanaka J and Ariyoshi J 1982 *Tetsu-to-Hagane* **68** 1782
[2] Takahashi T, Ohsasa K and Tanaka J 1987 *Tetsu-to-Hagane*. **79** 99
[3] Mathiesen R H, Arnberg L, Ramsoskar K, Weitkamp T, Rau C and Snigirev A 2002 *Metall. Mater. Trans.* B **33** 613
[4] Yasuda H, Ohnaka I, Kawasaki K, Sugiyama A, Ohmichi T, Iwane J and Umetani K 2004 *J. Crystal Growth* **262** 645
[5] Yasuda H, Kato S, Shinba T, Nagira T, Yoshiya M, Sugiyama A, Umetani K and Uesugi K 2009 *Int.J.Cast Met.Res.* **21** 125
[6] Yasuda H, Nagira T, Yoshiya M, Nakatsuka N, Sugiyama A, Uesugi K and Umetani K 2011 *ISIJ Int.* **51** 402
[7] Yasuda H, Nagira T, Yoshiya M, Uesugi M, Nakatsuka N, Kiire M, Sugiyama A, Uesugi K and Umetani K 2011 *IOP Conf. Ser.* **27** 012084
[8] Yasuda H, Nagira T, Yoshiya M, Sugiyama A, Nakatsuka N, Kiire M, Uesugi M, Uesugi K, Umetani K and Kajiwara K 2012 *IOP Conf. Ser.* **33** 012036
[9] Nagira T, Gourlay C.M., Sugiyama A, Uesugi M, Kanzawa Y, Yoshiya M, Uesugi K, Umetani K and Yasuda H 2011 *Scr. Mater.* **64** 1129
[10] Ueshima Y, Mizoguchi S, Matsumiya T, Kajioka H 1986 *Metall. Trans.* B **17** 845
[11] Kurz Fisher 2005 *Fundamentals of Solidification 48*
[12] Kurz Fisher 2005 *Fundamentals of Solidification 55*

GPU-accelerated 3D phase-field simulations of dendrite competitive growth during directional solidification of binary alloy

S Sakane[1], T Takaki[1], M Ohno[2], T Shimokawabe[3] and T Aoki[3]

[1] Graduate School of Science and Technology, Kyoto Institute of Technology
[2] Graduate School of Engineering, Hokkaido University
[3] Global Scientific Information and Computing Center, Tokyo Institute of Technology

E-mail:takaki@kit.ac.jp

Abstract. Phase-field method has emerged as the most powerful numerical scheme to simulate dendrite growth. However, most phase-field simulations of dendrite growth performed so far are limited to two-dimension or single dendrite in three-dimension because of the large computational cost involved. To express actual solidification microstructures, multiple dendrites with different preferred growth directions should be computed at the same time. In this study, in order to enable large-scale phase-field dendrite growth simulations, we developed a phase-field code using multiple graphics processing units in which a quantitative phase-field method for binary alloy solidification and moving frame algorithm for directional solidification were employed. First, we performed strong and weak scaling tests for the developed parallel code. Then, dendrite competitive growth simulations in three-dimensional binary alloy bicrystal were performed and the dendrite interactions in three-dimensional space were investigated.

1. Introduction

Control and prediction of columnar solidification structure are very important in casting and solidification. The columnar structure is formed through the competitive growth of multiple dendrites with different preferred growth directions. Walton and Chalmers competitive growth model is well accepted for the selection of growing multiple dendrites [1]. However, recently, an unusual competitive growth phenomenon, which cannot be explained by the conventional Walton and Chalmers model, have been reported [2-5]. The detailed mechanisms behind the new selection are not known except for the two-dimensional cases [6-8].

Phase-field method is the most powerful numerical tool for simulating dendrite growth. However, because the phase-field method is a diffuse interface model and fine numerical meshes are required to accurately express dendrite morphology, most phase-field simulations of dendrite growth performed so far are limited to two-dimension (2D) or single dendrite in three-dimension (3D). To express actual solidification structures, multiple dendrites with different preferred growth directions should be computed at the same time in 3D. Therefore, it is critical to develop a high-performance computational scheme for phase-field simulation [9]. In our previous study, we enabled very-large-scale phase-field simulations of dendrite growth during directional solidification of binary alloy [10]. In the computation, the number of meshes and time steps were approximately $4{,}000^3$ and four million, respectively. Although it was quite a large computation, the results were qualitative and the pulling time for directional solidification was not enough.

The objective of the present study is to develop a 3D phase-field code using multiple graphics processing units (GPUs) with a quantitative phase-field method for binary alloy solidification [11] and moving frame algorithm for directional solidification, and to investigate the dendrite interactions in 3D space.

The organization of this paper is as follows: First, we specify the quantitative phase-field model for the simulation of dendrite growth during the directional solidification of a dilute binary alloy. Next, we describe parallel computation using multi-GPUs and show strong scaling and weak scaling results for a developed code. Then, we describe the simulation of 3D dendrite competitive growth of binary alloy bicrystal and consider the simulation results.

2. Phase-field model

Here, we employed a quantitative phase-field model developed by Ohno and Matsuura [11] for dilute binary alloy solidification. In this model, the phase-field variable ϕ is defined as +1 for solids and −1 for liquids. According to the mixture rule, the local concentration c is defined as $c=[(1+\phi)c_s+(1-\phi)c_l]/2$, where c_s and c_l are the concentrations in the solid and liquid phases, respectively. We assume the relation $c_s^e=kc_l^e$ between the equilibrium concentrations of solid c_s^e and liquid c_l^e at a temperature T_0, where k is the partition coefficient.

The dimensionless supersaturation u is defined as

$$u=\frac{c_l-c_l^e}{c_l^e-c_s^e}. \tag{1}$$

Introducing the frozen temperature approximation during directional solidification [12], the temperature T is given as

$$T=T_0+G\left(z-V_p t\right), \tag{2}$$

where G is temperature gradient along z axis, V_p is pulling speed, T_0 is the temperature at $z=0$ and $t=0$, and t is the time.

The time-evolution equation of the phase-field variable ϕ is expressed as

$$\tau\left[1-(1-k)u'\right]\frac{\partial\phi}{\partial t}=\nabla\left[W^2\nabla\phi\right]+\frac{\partial}{\partial x}\left[W\frac{\partial W}{\partial\phi_x}|\nabla\phi|^2\right]+\frac{\partial}{\partial y}\left[W\frac{\partial W}{\partial\phi_y}|\nabla\phi|^2\right]$$
$$+\frac{\partial}{\partial z}\left[W\frac{\partial W}{\partial\phi_z}|\nabla\phi|^2\right]-\frac{\partial f(\phi)}{\partial\phi}-\lambda^*\frac{\partial g(\phi)}{\partial\phi}(u+u'), \tag{3}$$

where $\phi_x=\partial\phi/\partial x$, $\phi_y=\partial\phi/\partial y$, and $\phi_z=\partial\phi/\partial z$. $\tau=\tau_0 a_s^{\,2}$ is the inverse mobility and $W=W_0 a_s^{\,2}$ is interface thickness. The anisotropy function a_s is expressed as

$$a_s(\nabla\phi)=\bar{a}(1-3\zeta)\left[1+\frac{4\zeta}{1-3\zeta}\frac{\left(\frac{\partial\phi}{\partial x}\right)^4+\left(\frac{\partial\phi}{\partial y}\right)^4+\left(\frac{\partial\phi}{\partial z}\right)^4}{|\nabla\phi|^4}\right], \tag{4}$$

where ζ is the anisotropy parameter. The interpolating functions $f(\phi)$ and $g(\phi)$ are selected as $\partial f(\phi)/\partial\phi=-\phi+\phi^3$ and $\partial g(\phi)/\partial\phi=(1-\phi^2)^2$, respectively. λ^* is associated with the thermodynamic driving force expressed as $\lambda^*=a_1W_0/d_0$, where $a_1=0.8839$ and d_0 is the capillary length defined as $d_0=k\Gamma/\left[|m|(1-k)c_\infty\right]$ with the Gibbs-Thomson coefficient Γ, liquid slope m, and the initial alloy concentration c_∞. In equation (3), u' is added by introducing the frozen temperature approximation and is expressed as

$$u' = \frac{z - V_p t}{l_T}, \tag{5}$$

where l_T is the thermal length defined by $l_T = |m|(1-k)c_\infty/(Gk)$. The dendrite crystal orientations, which express preferred growth direction of each grain, are expressed by the coordinate transformation of $\phi_x = \partial\phi/\partial x$, $\phi_y = \partial\phi/\partial y$, and $\phi_z = \partial\phi/\partial z$ [7].

The diffusion equation is expressed as

$$\frac{\left[1+k-(1-k)\phi\right]}{2}\frac{\partial u}{\partial t} = \nabla\left[D_l q(\phi)\nabla u - j_{AT}\right] + \frac{\left[1+(1-k)u\right]}{2}\frac{\partial \phi}{\partial t} + \nabla\cdot\vec{J}, \tag{6}$$

where the interpolating functions $q(\phi)$ is given as $q(\phi) = [\,kD_s + D_l + (kD_s - D_l)\phi\,]/(2D_l)$, where D_s and D_l are the diffusion coefficients of the solid and liquid phases, respectively. The antitrapping current term j_{AT} is added to remove all the spurious effects at the interface and enable quantitative simulations independent of the interface thickness. j_{AT} is expressed as

$$j_{AT} = -a(\phi)W_0\left[1+(1-k)u\right]\frac{\partial\phi}{\partial t}\frac{\nabla\phi}{|\nabla\phi|}, \tag{7}$$

where we set $a(\phi) = [1 - kD_s/D_l\,]/(2\sqrt{2})$. The side branching is expressed by introducing a fluctuating current $\vec{J}$ of the Gaussian random number with the variance of $2D_l F_u^0 q(\phi)[1+(1-k)u]/(\Delta t \Delta x^3)$, where F_u^0 is the constant noise magnitude, Δt is the time increment, and Δx is the lattice size [13].

Equations (3) and (6) are discretized by using the normal finite difference method. The time is discretized by the first-order forward difference method. For the Laplacians in equations (3) and (6), the isotropic discretization using the nearest and next-to-nearest neighbors and the second-order central difference approximations are used respectively.

3. Parallel computation using multi-GPUs

In the following computations, we employ the TSUBAME2.5 supercomputer at Tokyo Institute of Technology. The TSUBAME2.5 has 1408 compute nodes, each of which are equipped with three NVIDIA Tesla K20X GPU accelerators and two Intel Xeon X5670 central processing units (CPUs). Therefore, the TSUBAME2.5 can be called a GPU rich supercomputer.

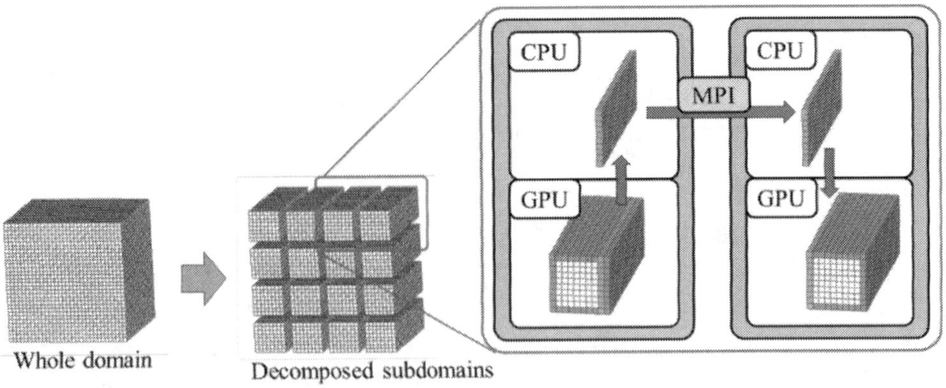

Figure 1. Domain decomposition and boundary data exchange between distributed GPUs.

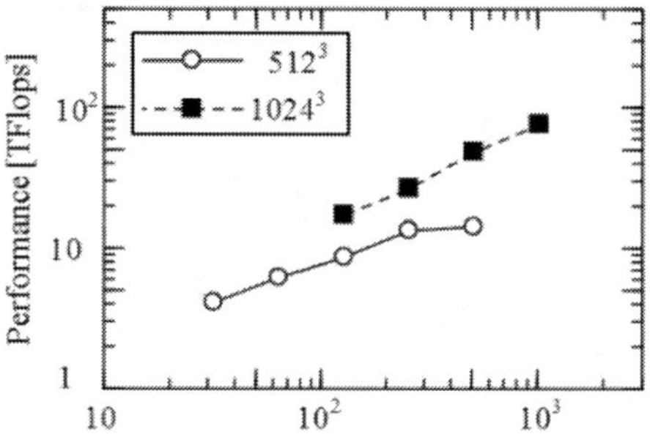

Figure 2. Strong scaling result.

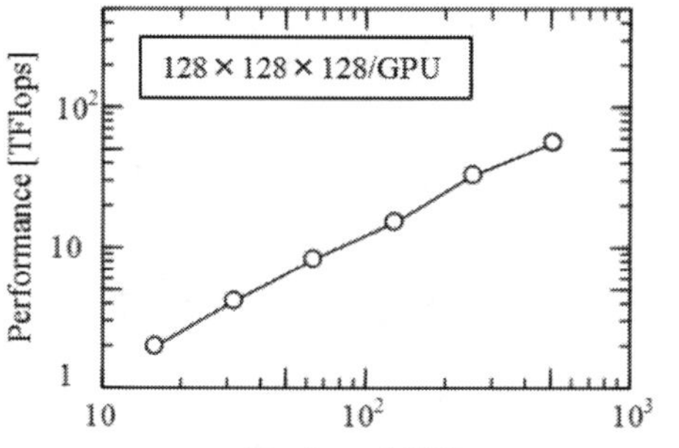

Figure 3. Weak scaling result.

The GPU code is written in compute unified device architecture (CUDA), and message passing interface (MPI) is used for the inter-node communication. Figure 1 shows the image of domain decomposition and boundary data exchange between distributed GPUs. The whole domain is decomposed along the y and z directions, and the decomposed subdomains are assigned to each GPU. Because the domain used in the following simulations is not wide in the x-direction, the 2D domain decomposition shows higher computational efficiency than the 3D domain decomposition.

The connection process of the boundary data is as follows: First, one GPU sends the boundary data to a CPU. Next, the boundary data is exchanged between CPUs in the distributed neighbor subdomain by MPI. One CPU then sends the received boundary data to the GPU. This process is performed in every step.

Figure 2 and 3 shows the results of strong scaling and weak scaling for developed code running on the multiple GPUs in single precision. The strong scaling tests were executed for 512^3 and 1024^3 meshes. In figure 2, the performance increases linearly with increase in the number of GPUs for both of 1024^3 and 512^3 domains. However, the good linearity breaks down at 512 GPUs for 512^3 meshes. Therefore, from these results, we determine the number of GPUs that show the best performance. In addition, it can be observed that the performance is higher for 1024^3 than 512^3 meshes. This is the same as previous study [14]. The weak scaling tests were performed by fixing 128^3 meshes for each GPU. From figure 3, we can observe good linearity with increasing GPU number.

4. Computational conditions

Figure 4 shows the computational and initial conditions for the directional solidification of Al–Cu binary alloy bicrystal. Initially, all domains were filled with liquid Al-3wt%Cu (0.013 at. frac.), and seeds were placed regularly on the bottom as shown in top view. To simulate bicrystal competitive growth, seeds with two different preferred growth directions θ are prepared. As shown in figure 4, the favorably and unfavorably oriented (FO and UO) seeds are placed on the left and right, respectively. The FO seed has $\theta = 0^\circ$ or grows toward heat flow direction, which is parallel to the z-direction, and the UO seeds have $\theta = 10^\circ$ and 30°. Here, secondary arms of the dendrites grow to the x and y directions. The domain is meshed into $n_x \times n_y \times n_z = 512 \times 2048 \times 1536$, and a periodic boundary condition in the x-direction and zero flux boundary condition in the y-direction are imposed. The temperature gradient G along the z-direction was set to 5.0 K/mm. The pulling velocity was also fixed to $V_p = 100\,\mu\text{m/s}$. The other parameters are as follows: $D_l = 3\times10^{-9}$ m²/s, $D_s = 3\times10^{-13}$ m²/s, $k = 0.14$, $\zeta = 0.02$, $\Gamma = 0.24\times10^{-6}\,\text{Km}$, $m = -620$ K/at. frac., $W_0 = 0.625\times10^{-6}\,\text{m}$, $\Delta t = 1.18\times10^{-5}\,\text{s}$, and $\Delta x = \Delta y = \Delta z = 0.5\times10^{-6}\,\text{m}$. The directional solidification is modeled by the moving frame algorithm to reduce the computational domain, where the computational domain shown in figure 4 tracks the dendrite tips to keep the enough diffusion layers in front of the dendrite tip.

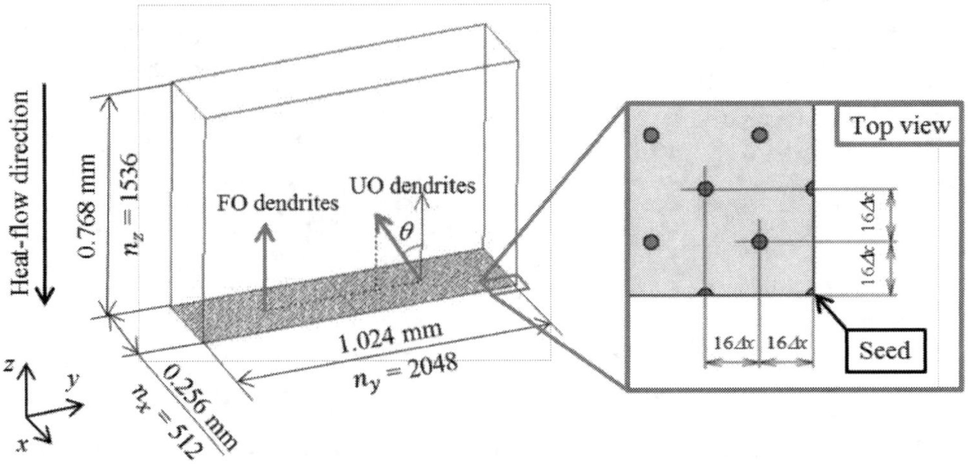

Figure 4. Computational and initial conditions.

5. Computational results and discussion

Figures 5 and 6 show the time slices during the simulations with the crystal orientation $\theta = 10°$ and $\theta = 30°$, respectively. Here, the isosurfaces with $\phi = 0$ are expressed with different colors for the two crystal orientations. After starting the simulations, the bottom of the domain is covered with the solid phase, and many dendrites initiate and grow upwards with selection. At that time, as shown in figures 5(a) and 6(a) the FO dendrites grow faster than the UO dendrites. Especially, the FO dendrites located at the grain boundaries grow faster than the other dendrites. Figure 6(a) shows that the side arms of the FO dendrites at the grain boundary overhang the UO dendrites side because the growth speed of UO dendrites with $\theta = 30°$ is much slower than the FO dendrites. Even in figures 5(e) and 6(e) at 1,000,000 steps, the dendrite selections in FO grain occur and the equilibrium growth condition is not achieved. The UO dendrites migrate to left with dendrite growth, and are blocked by the FO dendrites at the grain boundary. Figure 6(d) shows that the UO dendrite goes into the space between the FO dendrites at that time. This phenomenon cannot be expressed by 2D simulations. The new dendrites of the UO dendrites with $\theta = 30°$ are initiated by the branching as shown by the red circle in figure 6. On the other hand, for the UO dendrites with $\theta = 10°$, the dendrite initiations cannot be observed until one million steps. Therefore, we need much longer computation time to investigate the dendrite competitive phenomena at the grain boundary as performed in 2D simulations [7]. The 384 GPUs are used for these computations and the computation time was about 12 hours. Therefore, it can be concluded that the present 3D simulations can be continued with reasonable computational time.

6. Conclusions

In this study, we developed a 3D phase-field code using multi-GPUs with a quantitative phase-field method for binary alloy solidification and moving frame algorithm for directional solidification. Then, we simulated the dendrite competitive growth during the directional solidification of a binary alloy bicrystal, and demonstrated that the 3D phase-field simulations can be performed within reasonable computational time. It will be a powerful simulation tool for predicting solidification microstructures.

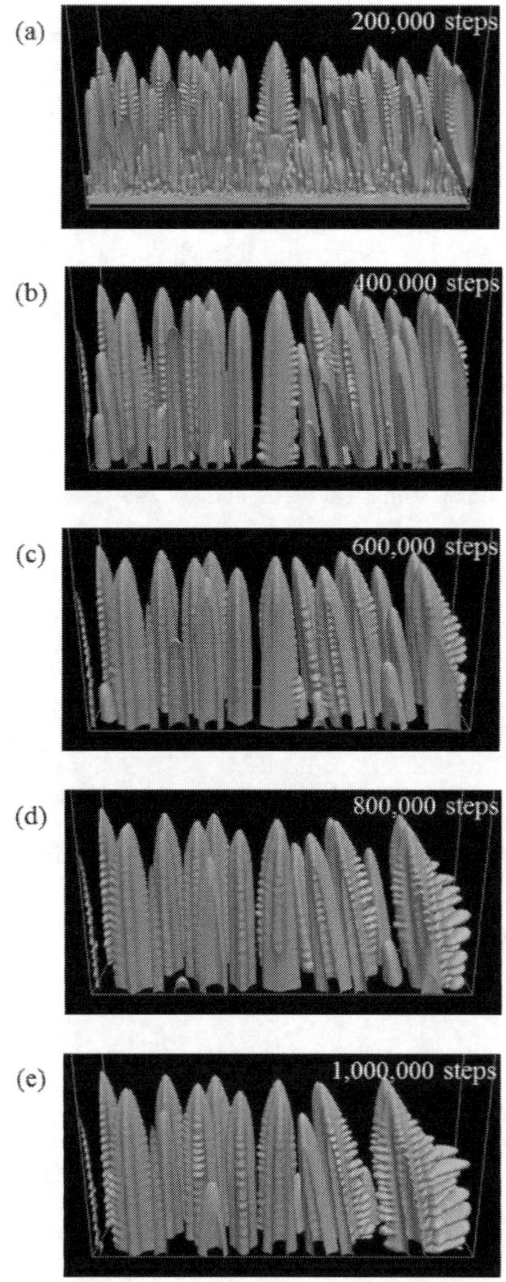

Figure 5. Dendrite morphologies at (a) 200,000, (b) 400,000, (c) 600,000, (d) 800,000, and (e) 1,000,000 steps for the case of the UO dendrite with $\theta = 10°$.

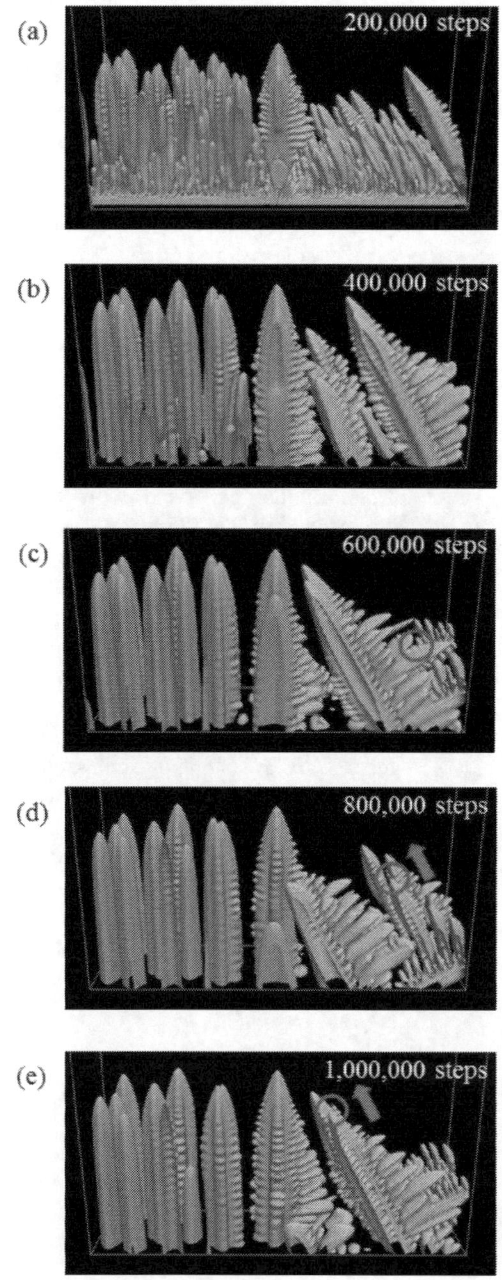

Figure 6. Dendrite morphologies at (a) 200,000, (b) 400,000, (c) 600,000, (d) 800,000, and (e) 1,000,000 steps for the case of the UO dendrite with $\theta = 30°$.

IOP Conf. Series: Materials Science and Engineering **84** (2015) 012063 doi:10.1088/1757-899X/84/1/012063

References

[1] Walton D and Chalmers B 1959 The Origin of the Preferred Orientation in the Columnar Zone of Ingots *Transactions of the Metallurgical Society of Aime* **215** 447–457
[2] D'Souza N, Ardakani M G, Wagner A, Shollock B A and McLean M 2002 Morphological aspects of competitive grain growth during directional solidification of a nickel-base superalloy, CMSX4 *Journal of Materials Science* **37** 481–487
[3] Wagner A, Shollock B A and McLean M 2004 Grain structure development in directional solidification of nickel-base superalloys *Mater. Sci. Eng.* A **374** 270–279
[4] Zhou Y Z, Volek A and Green N R 2008 Mechanism of competitive grain growth in directional solidification of a nickel-base superalloy *Acta Mater.* **56** 2631–2637
[5] Liu Z, Lin M, Yu D, Zhou X, Gu Y and Fu H 2013 Dependence of competitive grain growth on secondary dendrite orientation during directional solidification of a Ni-based superalloy *Metall. Mater. Trans.* A **44** 5113–5121
[6] Li J, Wang Z, Wang Y and Wang J 2012 Phase-field study of competitive dendritic growth of converging grains during directional solidification *Acta Mater.* **60** 1478–1493
[7] Takaki T, Ohno M, Shimokawabe T and Aoki T 2014 Two-dimensional phase-field simulations of dendrite competitive growth during the directional solidification of a binary alloy bicrystal *Acta Mater.* **81** 272–283
[8] Tourret D and Karma A 2015 Growth competition of columnar dendritic grains: A phase-field study *Acta Mater.* **82** 64–83
[9] Takaki T 2014 Phase-field Modeling and Simulations of Dendrite Growth *ISIJ Int.* **54** 437–444
[10] Takaki T, Shimokawabe T, Ohno M, Yamanaka A and Aoki T 2013 Unexpected selection of growing dendrites by very-large-scale phase-field simulation *J. Cryst. Growth* **382** 21–25
[11] Ohno M and Matsuura K 2009 Quantitative phase-field modeling for dilute alloy solidification involving diffusion in the solid *Phys. Rev. E* **79** 1–15
[12] Diepers H -J, Ma D and Steinbach I 2002 History effects duringthe selection of primary dendrite spacing. Comparison of phase-field simulations with experimental observations *J Cryst Growth* **237–239** 149
[13] Echebarria B, Karma A and Gurevich S 2010 Onset of sidebranching in directional solidification *Phys Rev E Stat Nonlinear Soft Matter Phys* **81** 1–17
[14] Shimokawabe T, Aoki T, Takaki T, Yamanaka A, Nukada A, Endo T, Maruyama N and Matsuoka S 2011 *Proceedings of 2011 SC* 1–11

SPH based modelling of oxide and oxide film formation in gravity die castings

K. Ellingsen[1], T. Coudert[2] and M M'Hamdi[1]

[1] SINTEF Materials and Chemistry, Oslo, N-0314, Norway
[2] SINTEF Materials and Chemistry, Trondheim, N-7465, Norway

kjerstin.ellingsen@sintef.no

Abstract. Gravity die casting is an important casting process which has the capability of making complicated, high-integrity components for *e.g.* the automotive industry. Oxides and oxide films formed during filling affect the cast product quality. The Smoothed particle hydrodynamics (SPH) method is particularly suited to follow complex flows. The SPH method has been used to study filling of a gravity die including the formation and transport of oxides and oxide films for two different filling velocities. A low inlet velocity leads to a higher amount of oxides and oxide films in the casting. The study demonstrates the usefulness of the SPH method for an increased understanding of the effect of different filling procedures on the cast quality.

1. Introduction

Gravity die casting is an important casting process which has the capability of making complicated, high-integrity components for *e.g.* the automotive industry. In gravity die castings, the major defects are porosity and entrapped oxide/oxide films and inclusions, which affect the mechanical strength of the cast product [1]. In addition to the design of the mould and the gating system, the ingate velocities significantly affect the number and distribution of the oxides generated from the filling stage. Improvements to product quality and productivity can be brought about through a better control of the die filling. Numerical simulation offers a powerful and cost effective way to study the effectiveness of different die filling procedures and die designs [2 ,3 ,4].

During the casting, liquid metal oxidizes only for the time period the metal is exposed to air, i.e., when there is a free surface. The oxides formed are subsequently transported by the melt flow into the casting. When fronts of an oxidized surface meet and merge, oxide films are formed. The Smoothed particle hydrodynamics (SPH) method has an inherent advantage when tracking free surfaces and following complex flows, and thereby for studies of oxide and oxide film formation and transport. The SPH method is a grid-free, Lagrangian particle method where the continuum equations of fluid dynamics are replaced by particle equations [5]. The particles are the computational framework on which the fluid equations are solved and therefore, SPH automatically follow complex flows.

This study is a preliminary attempt to use the SPH method to study filling of a gravity die. Particles on the free surface are identified at all times of the filling process and the amount of oxides formed at the air exposed surface is estimated. A simple oxide growth model is employed for oxide formation. The subsequent transport of the oxides is analyzed. Furthermore, oxide films formed are identified and tracked as they moved with the melt.

2. Model description

Smoothed particle hydrodynamics (SPH) is a grid-free, Lagrangian method where the fluid is represented by a set of particles. Each particle carries fundamental physical properties, *e.g.* mass, position, velocity, density and other related properties. The value of any function f at a particle is approximated by summing over the properties of its neighbouring particles. The SPH interpolation of the function f of a particle i at position $\boldsymbol{r}_i$ is given by:

$$f(\boldsymbol{r}_i) \approx \sum_j \frac{m_j}{\rho_j} f_j W(\boldsymbol{r}_i - \boldsymbol{r}_j, h) \tag{1}$$

where index j corresponds to any neighbouring particle of particle i, m_j and ρ_j are the mass and the density for particle j, f_j is the value of f for particle j, the function W is an interpolation smoothing kernel, and h is a smoothing length that defines the radius of influence around the particle i. We use the Gaussian kernel function:

$$W_{ij} = \frac{C}{h^d} e^{(-q^2)} \tag{2}$$

where d is the dimension (2 or 3), $q = |\boldsymbol{r}_i - \boldsymbol{r}_j|/h$ and $C = 1/\pi$ for a 2D simulation and $C = \frac{1}{\sqrt{\pi^3}}$ for a 3D simulation. The radius of influence of this function is $3h$. It means that the particle i will not be affected by any neighboring particles farther than $3h$. The gradient of the function f is given by differentiating the interpolation equation (1) to give:

$$\nabla f(\boldsymbol{r}_i) \sim \sum_j \frac{m_j}{\rho_j} f_j \nabla_i W_{ij} \tag{3}$$

where

$$\nabla_i W_{ij} = \nabla W(\boldsymbol{r}_{ij}, h) = \frac{1}{h^2} \frac{\boldsymbol{r}_{ij}}{|\boldsymbol{r}_{ij}|} F(|\boldsymbol{r}_{ij}|, h) \tag{4}$$

and F is the derivative of the kernel function: $F_{(q)} = -2Ce^{-q^2}$.

2.1. *SPH flow solver*

The SPH fluid method is largely described in the literature [6, 7]. The governing equations of fluids in SPH method are based on the Navier-Stokes equations in the Lagrangian form. By using the SPH interpolation, the continuity equation describing the evolution of the fluid density over time can be written as:

$$\frac{d\rho_i}{dt} = \rho_i \sum_j \frac{m_j}{\rho_j} (\boldsymbol{v}_i - \boldsymbol{v}_j) \cdot \nabla W_{ij} + \chi_{ij} \tag{5}$$

where m and ρ_j are the mass and the density for particle j is the density for particle j, t is time and v is velocity. χ_{ij} is a density diffusive term introduced in Refs. [8, 9] which helps eliminate numerical noise.

The momentum equation describes the acceleration of the fluid and can be written as:

$$\rho \frac{d\boldsymbol{v}}{dt} = -\sum_j m_j \left(\frac{p_i}{\rho_i^2} + \frac{p_j}{\rho_j^2} - \frac{\xi}{\rho_i \rho_j} \frac{4\mu_i \mu_j}{\mu_i + \mu_j} \frac{\boldsymbol{v}_{ij} \cdot \boldsymbol{r}_{ij}}{r_{ij}^2 + \eta^2} + \pi_{ij} \right) \nabla W_{ij} + \boldsymbol{F}_i \tag{6}$$

where p is pressure, $\boldsymbol{F}$ is external force, μ_i is the dynamic viscosity of particle i and $\boldsymbol{v}_{ij} = \boldsymbol{v}_i - \boldsymbol{v}_j$. $\eta = 0.1h$ is used to prevent numerical divergence when two particles are approaching each other. ξ is the viscous scaling factor [10].

An equation of state is required to calculate the pressure in Eq.6. The equation of state used here is quasi-compressible form which is calculated by using the density information from Eq.5. The equation of state is given by:

$$p_i = \beta\left[\left(\frac{\rho_i}{\rho_o}\right)^{\gamma} - 1\right] \tag{7}$$

where $\beta = c^2{}_0\rho_0/\gamma$ is the magnitude of the pressure, ρ_0 is the reference density, $\gamma = 7$ for liquid metal, and c is the speed of sound. The value of speed of sound should be large enough to ensure that the density fluctuation is less than 1% or close to the incompressible flow. The solid boundaries are represented by ghost particles as proposed in [8].

2.2. *Oxide and oxide film formation*

Prediction of oxide content is dependent on the ability to predict the amount of liquid metal exposed to air (*i.e.* metal located at the free surface) and the duration of that exposure. In SPH, each particle represents the same specific volume of fluid throughout the simulation and it can carry with it information on composition and the history of that particle. The oxide formation is calculated using the following key elements:

- Liquid metal particles oxidize for any period of time they are on the free surface. They do not oxidize when they are not currently on the free surface. This requires identification of which particles are on the surface at any time and an estimate of their exposed surface area.
- The rate of oxidation is given by a simple ordinary differential rate equation in this Lagrangian framework.
- Oxide already formed is transported by its host SPH particle to wherever it moves in the casting process.

The subsequent distribution of oxide produced by the fluid flow is therefore automatically provided by the method. The oxide content of any particle is given by a rate equation where k_l is the rate constant [11]:

$$\frac{dO_x}{dt} = k_l \tag{8}$$

The free surface is detected using the algorithm proposed in the work of Marrone et al. [12]. An algorithm is proposed to detect the oxide film during the simulation. When the normals of an oxidised surface are pointing against each other, oxide films is considered as entrained with the flow (as shown in Figure 1). Using the surface normal and the particle velocity, each surface can be detected and the algorithm can decide if a particle belongs to an oxide film.

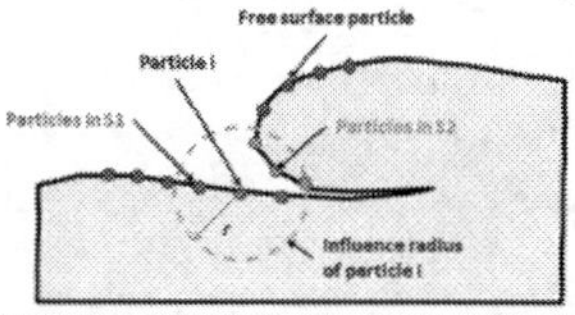

Figure 1: Oxidised surfaces pointing against each other forming an oxide film.

3. Case study

A 2D-model was set up to study filling of a gravity die (Fig. 2). The mould has a built-in pouring cup and ingate. The complete assembly is 150 mm high and 445 mm long. The die cavity has a length of 250 mm and has five steps with length of 50 mm and thickness of 30, 20, 15, 10 and 5 mm respectively. The cavity is slightly inclined to up-hill flow and equipped with a feeder tube. The mould is filled by manually pouring metal into the cup.

The model includes the pouring cup, the ingate, the feeder and the die cavity (the mould is not included). The ghost particle technique was used to describe the solid boundary of the mould and a free slip boundary condition was used. The length of the pouring inlet was set to 16 mm and the center placed 5mm above the pouring cup. Two different pouring velocities were employed: 0,1 m/s and 0,4 m/s. Aluminium was modelled as a perfect liquid (Euler liquid) with a density of 2500 g/cm^3. The particle size was set to 1mm giving 12500 particles in the filled die.

The constant k in the oxidation rate was set to 1, which means that the oxide content is equivalent to the air contact time. Oxide films formed when an oxidized surface folded and merged were identified and tracked as they moved along with the melt flow. The mould filling, the formation and transport of oxides and oxide films was analyzed for the two different inlet velocities.

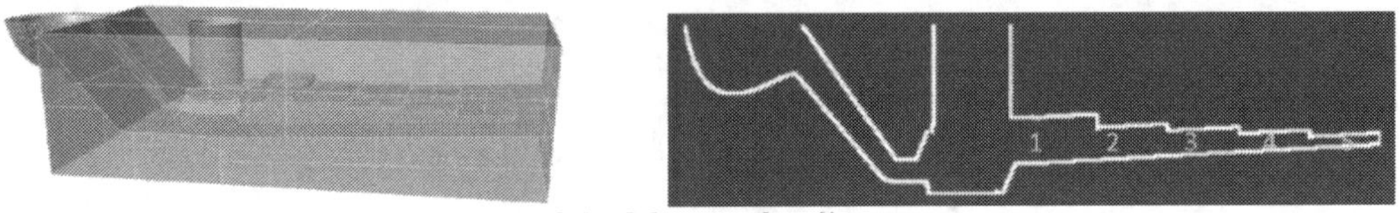

Figure 2. The gravity die and the 2D-model of the gravity die.

4. Results

4.1. High pouring velocity

The total filling time for the case where the pouring velocity is set to 0,4 m/s is 2 seconds. The left panel of Figure 3 shows the flow field at different times during filling. The metal first hits the bottom of the pouring cup, partly flows down the ingate and partly fills up the pouring cup behind the metal jet. The metal flows along the mould bottom towards the steps, but circulates back towards the ingate until the bottom is completely filled up. The metal then enters the step part of the mould that gradually and smoothly fills up. When the steps are completely filled, excess metal is pushed upwards through the feeder. The flow velocity is highest through the ingate and is decreased when the metal hits the bottom of the mould.

Oxides will be formed at the free surface as the metal is exposed to air. The oxide content at different times during filling is shown in the left panel of Figure 4. An increased oxide level is found in the bottom of the mould before entering the steps. These oxides that are formed in the ingate and mould bottom, do not enter the steps, but are pushed up the feeder by the strong upward flow after the steps are filled. However, the free surface of the metal that enters the steps is also oxidized. The oxidized metal ends up in the upper part of the four thickest steps. The metal in step 5 (the thinnest step) has the highest oxidation level due to the long air exposure time of the metal that ends up in this step.

Oxide films are formed where two fronts (free surfaces) of the aluminum melt meet and merge. Oxide films are formed in the pouring cup, in the ingate and in the bottom of the mould where the flow velocities are high and creating melt splashing and circulating flow patterns. The oxide film content is shown in the right panel of Figure 3 at different times during filling. Most of the oxide films formed do not enter the steps, but are effectively transported toward the ingate with the circulating flow and finally pushed up the feeder after the steps are filled. As the steps fills up smoothly, oxide films are not formed within the steps. Some of the oxide films formed in the mold bottom are transported with the melt to the steps, and mostly end up in the upper part of the first and second step.

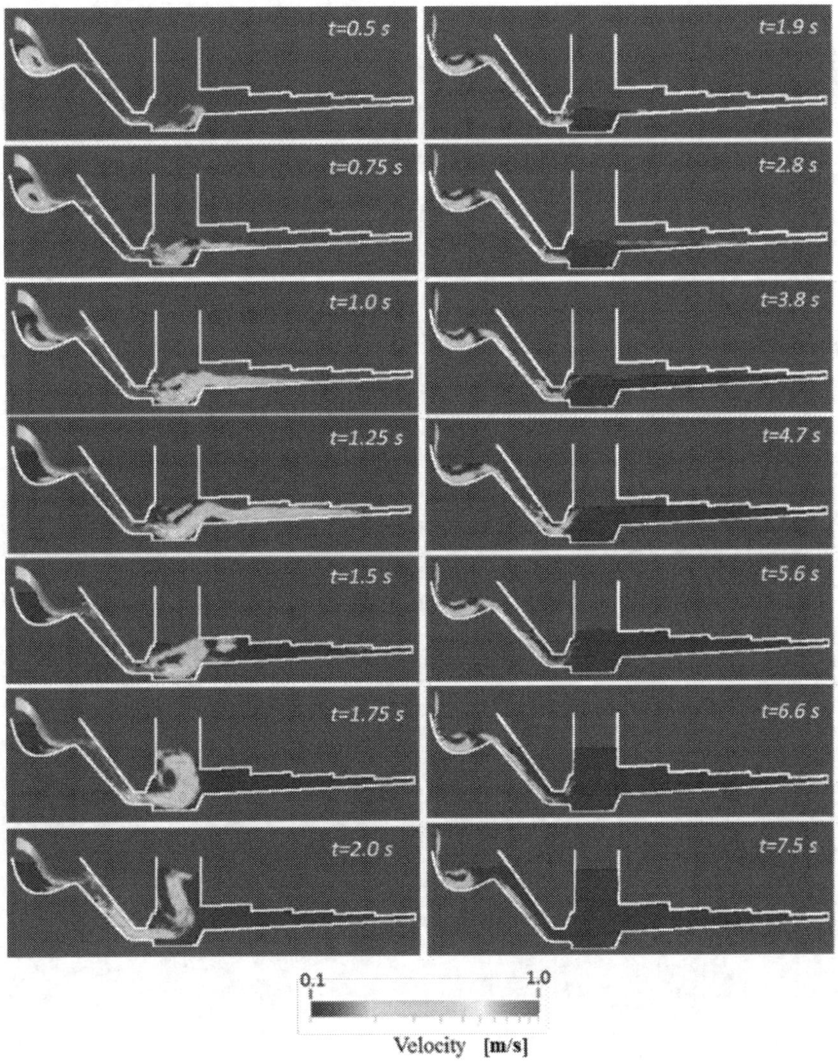

Figure 3. The velocity field (m/s) during filling when the velocity at the inlet is 0,4 m/s and 0,1 m/s in the left and right panel, respectively.

4.2. Low pouring velocity

The total filling time for the case where the pouring velocity is set to 0,1 m/s is 7.5 seconds. The right panel of Figure 3 shows the flow field at different times during filling. The metal first flows along the wall of the pouring cup, it circulates counter clockwise in the cup and when the cup is filled the melt flows down the ingate without filling it up. The melt reaches the bottom of the mould that fills before the melt can enter the step part. The melt velocities are highest in the ingate and at the entrance to the mould. The velocities in the mould bottom are low, but fronts are merging as the melt circulates counter clockwise towards the ingate. When the steps start to fill, the melt also flows from the mould bottom and up the ingate that fills up simultaneously. As the ingate fill up, the melt that enters the mould is pushed upwards along the mould walls to the melt surface. The metal smoothly fills up the step part of the mould. When the steps are completely filled, excess metal fills up the feeder.

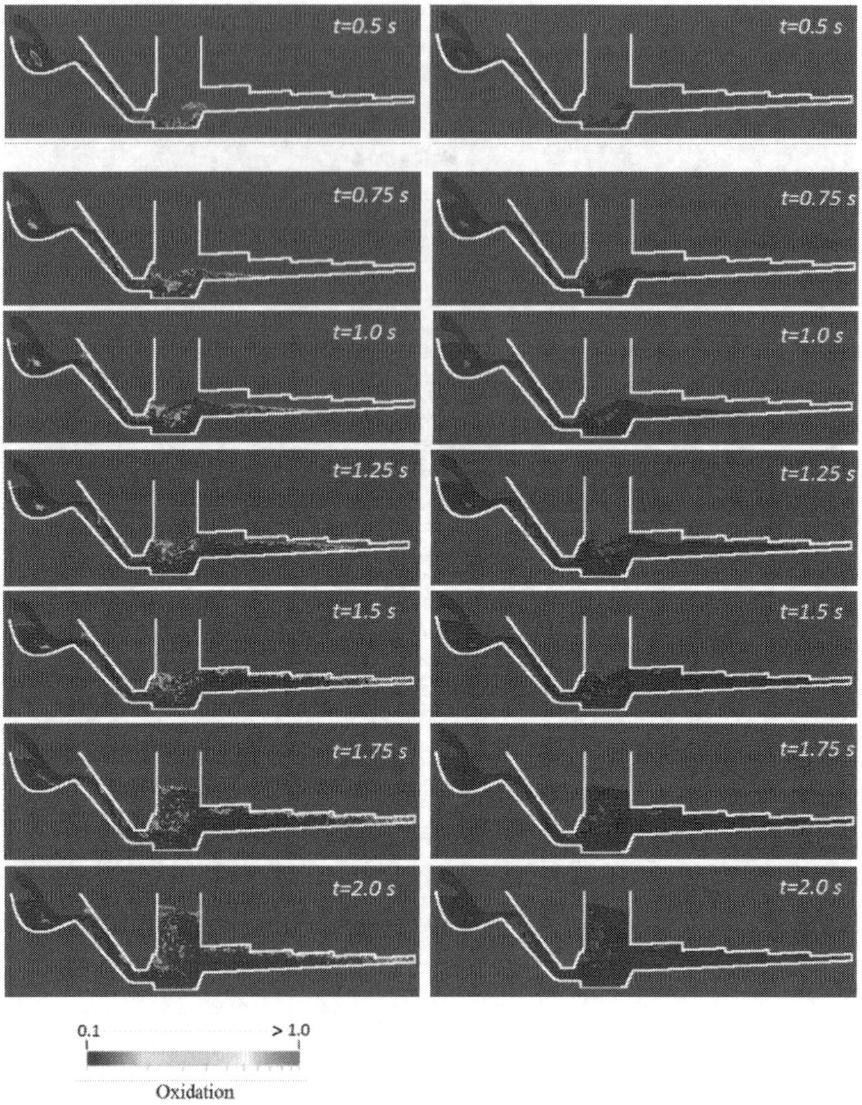

Figure 4. The left panel shows oxidation level during filling when the velocity at the inlet is 0,4 m/s. The right panel shows the corresponding oxide films (marked in red).

The oxide content at different times during filling is shown in the left panel of Figure 5. Elevated oxide content is found in the melt in the bottom of the mould before the melt enters the steps. These oxides are transported towards the melt surface and then flow with the melt into the steps. As the melt enters the steps, the surface is further exposed to air and the oxidation level increases. As the filling time is longer than for the case with high pouring velocity, the exposure time is longer. The metal that fills up the thinnest steps (4 and 5) is highly oxidized. High oxide content is also found in the thicker steps.The oxide film content is shown in the right panel of Figure 4 at different times during filling. Oxide films are formed in the pouring cup and the bottom of the mould due to the circulating flow patterns in these parts. The oxide films are transported to the melt surface in the mould bottom before entering the step part of the mould. As the steps fills up smoothly, oxide films are not formed within

IOP Conf. Series: Materials Science and Engineering **84** (2015) 012064 doi:10.1088/1757-899X/84/1/012064

the steps. Very few oxide films ends up in the thinnest step, but the four remaining steps have a high oxide film content. The oxide films ending up in the feeder are formed after the step part of the mould is filled.

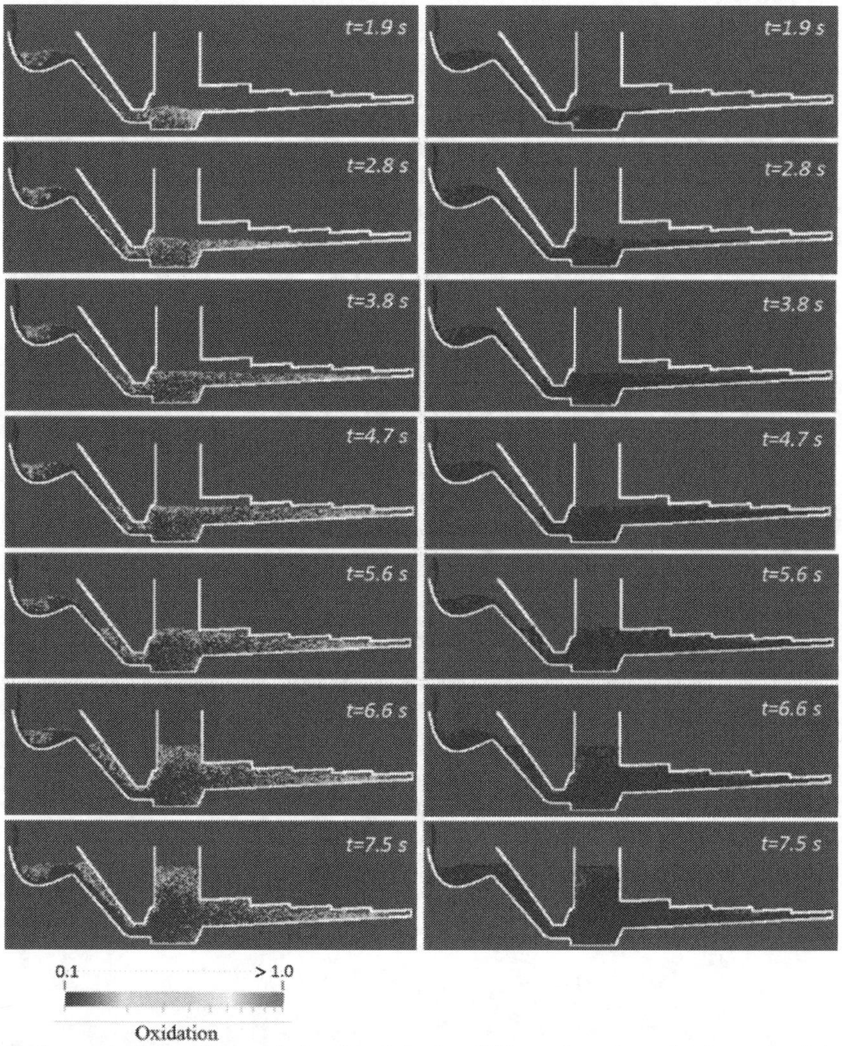

Figure 5. The left panel shows the oxidation level during filling when the velocity at the inlet is 0,1 m/s. The right panel shows the corresponding oxide films (marked in red).

5. Discussion and summary

The oxidization level is higher in the case with low pouring velocity. The free surface has a longer air exposure time when pouring the melt slowly and the filling time is longer. Furthermore, it has been demonstrated that for the higher pouring velocity, the high melt velocity obtained in the mould bottom creates a melt flow pattern that transport oxides effectively towards the feeder. The oxidation level is highest in the thinnest step for both cases as the melt transported to this step has the longest air exposure time. For the case with high pouring velocity the four other steps have a low oxidation level, for the low pouring velocity the oxidation level increases with decreasing thickness and increasing distance from the ingate due to the increasing air exposure time for the free surface of the melt entering the step. The steps in the case with low pouring velocity fill up smoothly and the flow velocity is low. A lower formation rate of oxide films is therefore expected compared to the case with higher pouring velocity. However, the oxide film content is higher in all steps for the low pouring velocity case. This is due to the formation of oxide films in the pouring cup and mould bottom and the subsequent transport towards the step part of the mould. In the case with high pouring velocity, the melt flow patterns leads to transport of oxide films towards the ingate in a circular pattern away from the steps and finally to the feeder. The SPH method is tracking free surface flows with fragmentation and breakup more accurately than is possible with traditional grid based methods [13] and the Lagrangian framework of the method allows tracking of history-dependent properties such as accumulated oxide mass [14]. The preliminary study demonstrates the usefulness and the potential of using the SPH method for studies of melt filling, oxide and oxide film formation and transport. The method can be used to increase the understanding of the effect of different filling procedures on the cast quality.

References

[1] N. R. Green and J. Campbell 1994, *Transactions of the American foundry society* **114** 341 -347
[2] Reilly, M.R. Jolly, N.R. Green, JC Gebelin 2010, *Light Metals, Proceedings of TMS 2010 Annual Meeting & Exhibition*
[3] J. Campbell 2006, *Proceedings of TMS 2006 Annual Meeting & Exhibition* 123-132
[4] X. Yang, X. Huang, X. Dia, J. Campbell and J. Tatler 2004, *Int. Journ. of Cast Met. Res.* **17** 321-331
[5] J.J. Monaghan 1992, *Ann. Rev. Astron. Astrophys.* **30** 543–574
[6] J.J. Monaghan 2005, *Reports on Progress in Physics* **68** 1703
[7] Raden Ahnaf Faqih and Christian Fredy. Naa. 2013, *ArXiv e-prints*..
[8] S. Marrone, M. Antuono, A. Colagrossi, G. Colicchio, D. Le Touzé, and G. Graziani 2011, *Computer Methods in App. Mech. and Eng.* **200** 1526–1542
[9] Jinlian Ren, Jie Ouyang, Tao Jiang, and Qiang Li 2012, *Computational Mechanics* 49 643–665
[10] Paul W. Cleary 2010, *Applied Mathematical Modelling* **34**3189–3201
[11] M. Prakash, H. Joseph, P. Cleary and J. Grandfield 2006, *Proceedings of the Fifth international conference on CFD in the minerals and process industries* (Melbourne, Australia).
[12] S. Marrone, A. Colagrossi, D. Le Touzé, and G. Graziani 2010, *J. of Comp. Phys.* **229** 3652–3663
[13] M. Prakash, P. Cleary and J. Grandfield 2009, *Journal of Mat. Proc. Tech.* **209** 3396–3407
[14] M. Prakash, G. G. Pereira, P. Cleary, P. Roshan and J. A Taylor 2009, *Proceedings of the Seventh international conference on CFD in the minerals and process industries* (Melbourne, Australia, 2009)

Phase field simulation on microstructure evolution in solidification and aging process of squeeze cast magnesium alloy

H W Pan [1], G M Han [1], Z Q Han [1,3] and B C Liu [1,2]

[1]Key Laboratory for Advanced Materials Processing Technology (Ministry of Education), School of Materials Science and Engineering, Tsinghua University, Beijing 100084, China
[2]State Key Laboratory of Automotive Safety and Energy, Department of Automotive Engineering, Tsinghua University, Beijing 100084, China

E-mail: zqhan@tsinghua.edu.cn

Abstract. Phase-field models have been developed to simulate the dendritic growth in pressurized solidification of Mg-Al alloy during squeeze casting and the precipitation of multi-variant β-$Mg_{17}Al_{12}$ phases during the subsequent aging process. For the pressurized solidification, the effects of pressure on the Gibbs free energy and chemical potential of solid and liquid phases, and the solute diffusion coefficient were considered. For the precipitation during aging process, the effects of elastic strain energy, anisotropy of interfacial energy, and anisotropy of interface mobility coefficient were considered. The results showed that the dendritic growth rate tends to increase and the secondary dendrite arms are more developed as the pressure is increased from 0.1 to 100MPa, which showed a good agreement with the experimental results of direct squeeze casting of Mg-Al alloy. The 2D and 3D simulated precipitates had lath shapes with lozenge ends, and the precipitate variants were parallel to the basal plane and oriented in directions with an angular interval of 60 degrees, which is in good agreement with experimental observations.

1. Introduction

Magnesium alloys are important lightweight structural materials and have received increasing attention from electronic, automobile and aerospace industries [1]. Mg-Al series alloys are commonly used because of their good castability and high specific strength. Squeeze casting is an advanced near net shape materials processing technology, in which the casting solidifies under pressure, which causes plastic deformation compensating the internal volume shrinkage [2]. Moreover, the squeeze cast products can be heat treated to further improve the mechanical properties. According to the Mg-Al binary phase diagram, eutectic β phase dissolves during solid solution, and precipitate forms during aging process, which is an important strengthening in Mg-Al based alloys [3]. Due to the advantages of high casting integrity and excellent mechanical properties of products, squeeze casting of Mg-Al alloy gains importance in the lightweighting of automobiles. To understand the solidification mechanism

[3] Correspondent: Zhiqiang Han, Ph.D, Associate professor, School of Materials Science and Engineering, Tsinghua University, Haidian District, Beijing 100084, China.

under pressure and the precipitation process during aging of Mg-Al alloy is of great significance to the design and optimization of squeeze casting and heat treatment process of Mg-Al alloy.

Microstructure evolution in pressurized solidification has been studied experimentally for years. Under superhigh pressure of several GPa, the phase diagrams of alloys are changed significantly [4]. The pressure increases the nucleation density and decreases the solute diffusion coefficient, resulting in fine microstructure [5] and high solid solubility [6]. In squeeze casting, the effect of pressure on heat transfer between the casting and the mold has been studied based on the solidification cooling curves [7, 8]. The enhanced heat transfer due to the applied pressure increases the cooling rate, which reduces the grain size according to Yong et al. [9] and El-khair [10]. Several experimental studies have been performed to gain comprehensive understanding of the morphology and structure of the β phase, and the orientation relationship with the hexagonal close packed (hcp) matrix [11, 12]. The β phase is a stoichiometric intermetallic compound with a composition of $Mg_{17}Al_{12}$ and an α-Mn-type cubic unit cell structure. Results indicate that the precipitate has a lath-shaped morphology, with the majority of precipitates forming with the habit plane parallel to the basal planes (0001) of the matrix [13].

Phase field simulation based on solute diffusion, thermodynamic driving force and ordering potential is considered to be a powerful tool for modeling the dendritic growth during solidification and the morphology evolution of the precipitates during aging. Böttger et al. [14, 15] conducted simulation of equiaxed solidification of Mg-Al alloy based on phase field method. Börzsönyi et al. [16, 17] investigated the effects of time-periodic external forcing on dendritic solidification of CCH3, a kind of liquid crystal, by experiments and phase field modeling, and found that the oscillating pressure and heating can tune the frequency of dendritic side-branching, indicating that thermal and mechanical effects are two dominant factors in interface stability. Li et al. [18] made an early attempt to simulate the precipitates in AZ91 alloy. Later Gao et al. [19] simulated the precipitation of β1 phase in Mg–Y–Nd alloy using a phase field model and discussed the effect of elastic strain energy on the precipitation. Han et al. [20] simulated the precipitation of β-$Mg_{17}Al_{12}$ using a phase field model that considered interfacial anisotropy, and discussed the effects of the two types of solution approximation on the morphology evolution of the precipitate.

The aim of the present work is to use the phase field method to investigate the dendritic growth in pressurized solidification and the growth kinetics of β-$Mg_{17}Al_{12}$ precipitates during aging Mg-Al alloys. Phase field models coupled with thermodynamic calculation for describing the dendritic growth of Mg-Al alloy during squeeze casting and precipitation process are developed, and the simulation results are compared with experimental observations.

2. Model description

2.1. Phase filed model for pressurized solidification

A phase field model based on KKS model [21] is adopted in this work to investigate the pressurized solidification. The total free energy of the system $F(c, \eta, T)$ including chemical free energy and interfacial energy are defined as:

$$F(c,\eta,T) = \int_v \left[f(c,\eta,T) + \frac{\varepsilon^2(\varphi)}{2}(\nabla\eta)^2 \right] dv \tag{1}$$

where F is the total free energy, f is the free energy density of the system, T is absolute temperature, $\varepsilon(\varphi)$ is the gradient energy coefficient related to anisotropic interfacial energy, and φ is an angle used to describe the interfacial anisotropy. The order parameter η is defined as 1 in the bulk solid and 0 in the bulk liquid.

The chemical free energy density of the system is defined as:

$$f(c,\eta,T) = h(\eta) f^{\beta}(c_\beta, T) + \left(1 - h(\eta)\right) f^{\alpha}(c_\alpha, T) + w g(\eta) \tag{2}$$

where c_α and c_β are the molar fractions of Al atoms in the liquid and solid phases, respectively. $f^{\alpha}(c_\alpha, T)$ and $f^{\beta}(c_\beta, T)$ are the chemical free energy density of the liquid and solid phases, respectively. $g(\eta)$ is a double-well potential, and w is the height of the double-well potential. $h(\eta)$ is a monotonous function from 0 to 1. In the present work, $g(\eta)$ and $h(\eta)$ are chosen as:

$$h(\eta) = \eta^3(6\eta^2 - 15\eta + 10) \tag{3}$$
$$g(\eta) = \eta^2(1-\eta)^2 \tag{4}$$

The governing equations of phase field and concentration field are given as follows,

$$\frac{\partial \eta}{\partial t} = M(\varphi)\left(\varepsilon^2(\varphi)\nabla^2\eta - f_\eta\right) \tag{5}$$
$$\frac{\partial c}{\partial t} = \nabla\cdot(D(\eta,T)\nabla c) + \nabla\cdot\left(D(\eta,T)h'(\eta)(c_\alpha - c_\beta)\nabla\eta\right) \tag{6}$$

In the ideal solution approximation, the free energy of a phase depends on the composition and temperature under atmospheric pressure [22]. For the two phases in the Mg-Al system, the general formulas are given as:

$$G_{\mathrm{m}}^{\tau} = c_{\mathrm{Al}}G_{\mathrm{Al}}^{\tau} + (1-c_{\mathrm{Al}})\,G_{\mathrm{Mg}}^{\tau} + RT(c_{\mathrm{Al}}\ln c_{\mathrm{Al}} + (1-c_{\mathrm{Al}})\,\ln(1-c_{\mathrm{Al}})\,) \tag{7}$$

where G_{Al}^{τ} and G_{Mg}^{τ} are the Gibbs free energy of pure Al and Mg in the phase of τ, respectively. c_{Al} is the molar fractions of Al and R is the gas constant.

The Gibbs free energy under pressure is calculated based on the quasi-harmonic approximation and equation of state (EOS) [23] shown as follows, in which P_0 is the reference pressure.

$$G(T,P) = G(T,P_0) + \int_{P_0}^{P} V(T,P')dP' \tag{8}$$

The solute diffusion coefficient is also impacted by pressure, as the viscosity of the melt increases with the increase of pressure. The solute diffusion coefficient under pressure can be calculated as follows [24], which means that the solute diffusion coefficient decreases exponentially with the increase in pressure.

$$D^P = D_0\exp(-PV_0/RT) \tag{9}$$

2.2. *Phase field model for precipitation*

The TEM observation [13] showed that the precipitate has 12 variants. Due to the symmetric distribution of the variants in the basal plane, it is adequate to investigate the evolution of three typical precipitate variants (variant-1, variant-2 and variant-3) with an interval of 60 degrees in the simulation, thus three non-conserved quantities (the structural order parameter η_q, q=1-3) and one conserved quantity, molar fraction of the solute Al, are introduced and a phase field model for precipitation is constructed,

$$\frac{\partial \eta_q}{\partial t} = -M(\varphi_q)\frac{\delta F}{\delta \eta_q}, q = 1-3 \tag{10}$$
$$\frac{\partial c}{\partial t} = \nabla\cdot\left(\frac{D(T)}{f_{cc}}\nabla(f_c)\right) \tag{11}$$

The total energy F can be expressed as follows, where the first integral term denotes the free energy, including the contribution of the chemical free energy and the gradient energy. The second term E^{ela} denotes the elastic strain energy.

$$F(c,\{\eta_q\},T) = \int_v\left[f(c,\{\eta_q\},T) + \sum_{q=1}^{3}\frac{\varepsilon^2(\varphi_q)}{2}(\nabla\eta_q)^2\right]dv + E^{ela} \tag{12}$$

In the simulation, the precipitate variants in the direction of 0, 60 and 120 degrees are referred to as variant-1, variant-2 and variant-3, respectively. The eigenstrain tensor of precipitation transformation of variant-q, and its matrix form for variant-1 is denoted by $\varepsilon_{ij}^{0}(1)$.The eigenstrain tensor of the other two variants can be obtained via coordinate rotation transformation as follows [25],

$$\varepsilon_{ij}^{0}(2) = (R\ \ 60°)\varepsilon_{ij}^{0}(1)(R\ \ 60°) \tag{13}$$
$$\varepsilon_{ij}^{0}(3) = (R\ \ 120°)\varepsilon_{ij}^{0}(1)(R\ \ 120°) \tag{14}$$

where $(R\ \ 60°)$ and $(R\ \ 120°)$ are rotation transformation matrices, and $(R\ \ 60°)$ can be obtained as follows,

$$(R\ \ 60°) = \begin{pmatrix} cos60° & sin60° & 0 \\ -sin60° & cos60° & 0 \\ 0 & 0 & 1 \end{pmatrix} \tag{15}$$

Mesh induced anisotropy is a common issue in phase field simulation and it is more serious when interfacial anisotropy has to be dealt with in a simulation [26]. In order to reduce the mesh induced anisotropy, a special numerical technique based on two sets of mesh, a hexagonal and an orthogonal, was employed to simulate the multi-variant precipitation. The hexagonal mesh was used to solve the phase field governing equations, and the orthogonal mesh was adopted to compute the elastic strain energy. The numerical technique based on the two sets of mesh combines the advantages of the hexagonal mesh in dealing with the precipitate orientation and those of the orthogonal mesh in computing the elastic strain energy. Fig. 1(a) shows the layout of the two meshes in two dimensions, where the hexagonal mesh is in red and the orthogonal in blue, and the two sets of mesh were coupled via data interpolation. Fig. 1(b) shows the layout of the hexagonal mesh and orthogonal mesh in three dimensions.

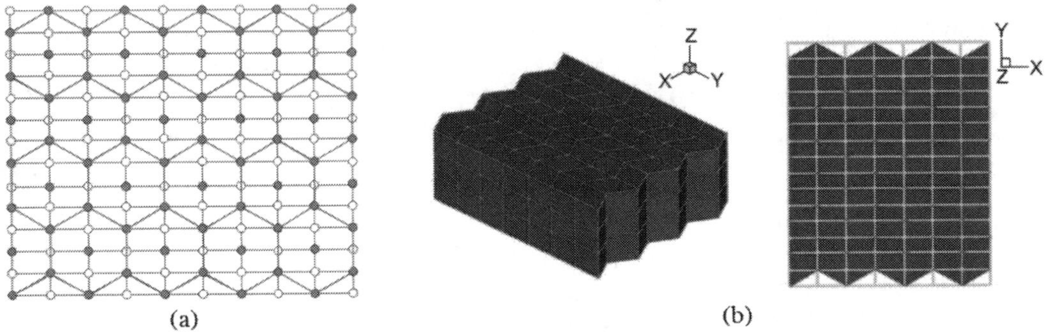

Figure 1. A schematic of the hexagonal and orthogonal mesh employed in simulation, (a) 2D, and (b) 3D

3. Results and discussion

3.1. Pressurized solidification

According to the above calculation method of Gibbs free energy under pressure, the Gibbs free energy of Mg-Al alloy increases with the increase of pressure, which also increases the thermodynamic driving force of the system and expedites the dendritic growth. However the pressure inhibits the solute diffusion as shown in equation (9), which impedes the dendritic growth. The overall effect of pressure on dendritic growth of Mg-9 wt.%Al alloy is shown by the phase field simulation results in Fig. 2, where Δt is the simulation time step, and Fig. 3 shows the solute concentration profile. The simulation is carried out under isothermal condition with the temperature of 851K.

It can be seen from Fig. 2 that the dendritic growth rate under 100MPa tends to increase compared with that under atmospheric pressure and the second dendrite arms are more developed. The solute concentration profile of Fig. 3 shows that due to the fast dendritic growth rate and the decreased solute diffusion coefficient, the solute rejected from the transformation from liquid to solid phase accumulates at the roots of the dendrite arms, which enhances the branch of microstructure.

From the comparison of the morphology of the typical dendrites under atmospheric and 100MPa pressure shown in Fig. 4, the primary dendrite arms under atmospheric pressure are bolder than that under elevated pressure. While with the pressure applied, the primary dendrite arms become slimmer and the secondary dendrite arms are more developed. The simulation results agree well with the experiment results comparing Fig. 2 with Fig. 4.

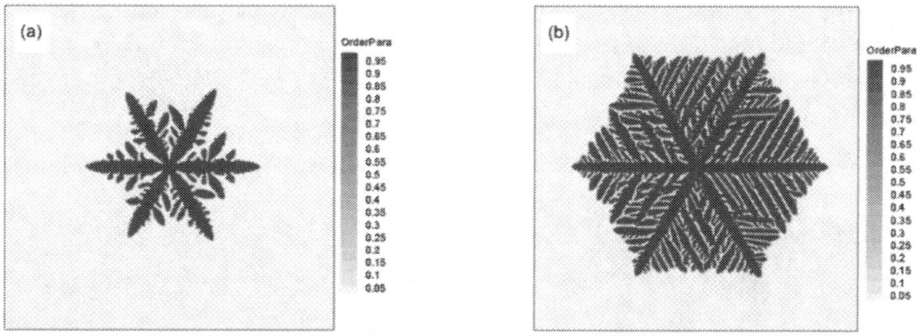

Figure 2. Phase field profile for isothermal solidification simulation at temperature T=851K of dendritic growth of Mg-9 wt.%Al alloy at t=10000Δt, (a) under atmospheric pressure, and (b) under 100MPa

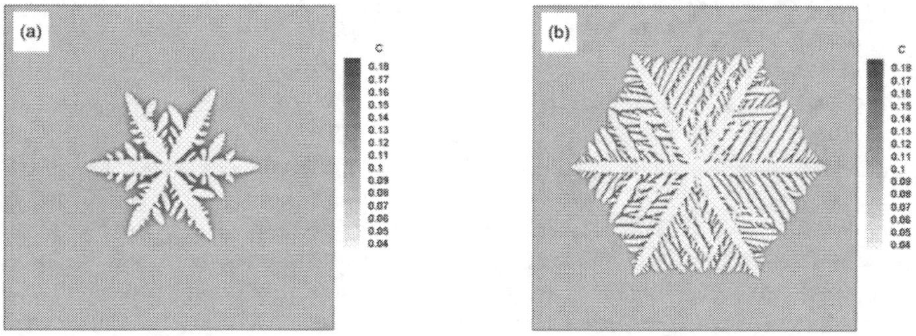

Figure 3. Solute concentration profile for isothermal solidification simulation at temperature T=851K of the dendritic growth of Mg-9 wt.%Al alloy at t=10000Δt, (a) under atmospheric pressure, and (b) under 100MPa

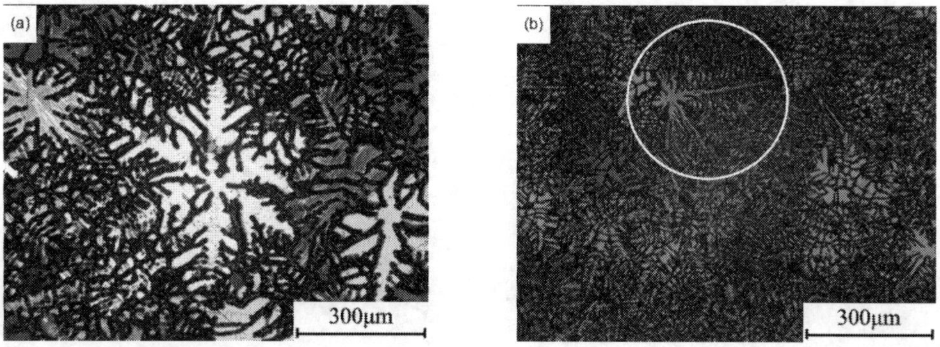

Figure 4. Typical dendrites of AZ91D Mg-Al alloy, (a) under atmospheric pressure, and (b) under 100MPa

3.2. Precipitation process

Fig. 5(a) shows 2D simulation results of the multi-variant precipitates at 1600 seconds, where both the interface anisotropy and the elastic strain energy were taken into account. The blue areas in Fig. 5(a) are the matrix, and the red areas represent the precipitate phase. It can be seen that the precipitate phase has a lath shape with lozenge ends, and the three variants distribute in the directions with an interval of 60 degrees. Fig. 5(b) is the TEM bright-field image obtained from AZ91 aged at 441K for 16 hours, where three typical precipitate variants with lath shape and lozenge ends can be seen in the basal plane. The angle between the variant-1 and variant-2 is about 80 degrees, and the angle between variant-1 and variant-3 is about 120 degrees. The simulation results of the morphology and distribution of the precipitate variants show a good agreement with the experimental observations.

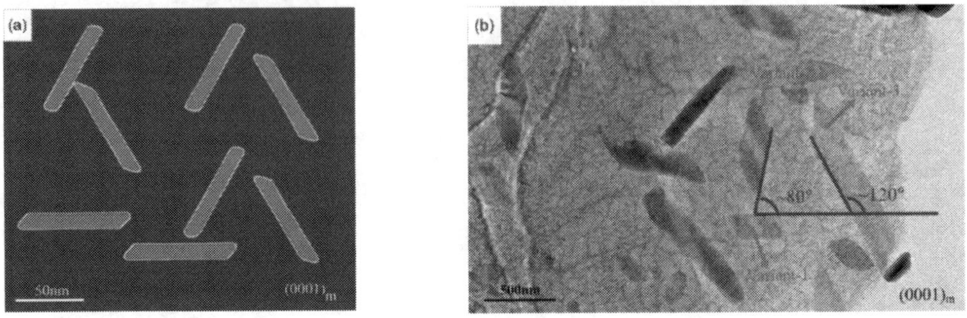

Figure 5. The simulation and TEM observation of multi-variant precipitation, (a)simulation results, and (b) TEM bright-field image of the precipitates for AZ91 aged at 441k for 16 hours

3D simulation on the morphology and distribution of the precipitate variants was carried out as well. In the simulation, a 256 × 256 × 256 hexagonal mesh was employed to solve the phase-field equations, and a 256 × 512 × 256 orthogonal mesh was adopted to calculate the elastic strain energy, which represents a real domain of about 2.22μm × 2.56μm × 2.08μm. The simulation domain is large enough that it is feasible to explore the peak-hardening stage of precipitation during the aging process of this alloy and to predict its properties. In the simulation, the initial compositions of the precipitates and the matrix are 0.4 (molar fraction) and 0.076, respectively. The temperature of the domain was set to 573.15K. To reduce the computation time, parallel computing was employed. The entire parallel numerical calculation was executed on the high-performance cluster computing platform of the Tsinghua National Laboratory for Information Science and Technology.

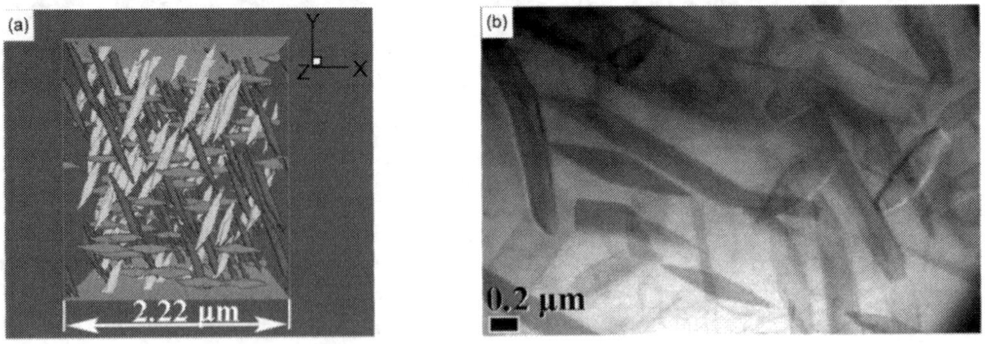

Figure 6. Simulations and TEM observations of multi-variant precipitation in AZ91 aged at 573.15K for 2 hours, (a) simulation results in the xy plane, i.e., (0001) plane of the matrix, t = 2 hours, and (b) TEM image of the precipitates in the basal plane

Fig. 6 shows the 3D simulation results and experimental results for the multi-variant precipitates. In the simulation, the xy plane denotes the (0001) plane of the matrix, and the red, blue, and green laths in Fig. 6(a) are the variant-1, variant-2, and variant-3 precipitates, respectively. It can be seen that many long lath-shaped precipitates with lozenge ends appear in the simulation domain, and the three variants are oriented in different directions with intervals of about 60 degrees. Fig. 6(b) shows TEM images of multi-variant precipitates viewed in the basal plane, for AZ91 alloy aged at 573.15K for 2 hours. The simulation results are in good agreement with the experimental observations in terms of the variant morphology and distribution.

4. Conclusions

A phase field model coupled with thermodynamic calculation for describing the dendritic growth in pressurized solidification of Mg-Al alloy during squeeze casting has been developed, in which the effects of the pressure on the Gibbs free energy and chemical potential of solid and liquid phases, the solute diffusion coefficient were considered.

2D and 3D phase-field models have been developed to simulate the multi-variant β-$Mg_{17}Al_{12}$ phase precipitation in the aging process of Mg-Al alloy, considering the effects of elastic strain energy, anisotropy of interfacial energy, and anisotropy of interface mobility coefficient.

For pressurized solidification, the dendritic growth rate tends to increase and the secondary dendrite arms are more developed as the pressure is increased from 0.1 to 100MPa, which showed a good agreement with the experimental results of direct squeeze casting of Mg-Al alloy.

For precipitation process, multi-variant precipitates were simulated in 2D and 3D. The simulated precipitates had lath shapes with lozenge ends, and the precipitate variants were parallel to the basal plane and oriented in directions with an angular interval of 60 degrees, which is in good agreement with experimental observations.

Acknowledgements

This work is funded by the National Natural Science Foundation of China (Grant No. 51175291), Tsinghua University Initiative Scientific Research Program (Grant No. 2011Z02160), and the State Key Laboratory of Materials Processing and Die & Mold Technology, Huazhong University of Science and Technology. The support of the State Key Laboratory of Automotive Safety and Energy, Tsinghua University under the contract 2013XC-A-01 is also gratefully acknowledged.

References

[1] Mordike B L and Ebert T 2001 *Mater. Sci. Eng.* A **302** 37
[2] Ghomashchi M R and Vikhrov A 2000 *J. Mater. Process. Technol.* **101** 1
[3] Celotto S 2000 *Acta Mater.* **48** 1775
[4] Yu X F, Zhang G Z, Wang X Y, Gao Y Y, Jia G L and Hao Z Y 1999 *J. Mater. Sci.* **34** 4149
[5] Zhao S S, Peng Q M, Li H and Liu B Z 2014 *J. Alloys Compd.* **584** 56
[6] Kagaya H M, Suzuki T, Takaya M S and Soma T 2001 *Physica* B **293** 343
[7] Sun Z Z, Hu H and Niu X P 2011 *J. Mater. Process. Technol.* **211** 1432
[8] Amin K M and Mufti N A 2012 *J. Mater. Process. Technol.* **212** 1631
[9] Yong M S and Clegg A J 2004 *J. Mater. Process. Technol.* **145** 134
[10] Abou El-khair M T 2005 *Mater. Lett.* **59** 894
[11] Nie J F 2004 *Acta Mater.* **52** 795
[12] Nie J F 2006 *Metall. Mater. Trans.* A **37** 841
[13] Hutchinson C R, Nie J F and Gorsse S 2005 *Metall. Mater. Trans.* A **36** 2093
[14] Böttger B, Eiken J, Ohno M, Klaus G, Fehlbier M, Schmid-Fetzer R, Steinbach I and Bührig-Polaczek A 2006 *Adv. Eng. Mater.* **8** 241
[15] Böttger B, Eiken J and Steinbach I 2006 *Acta Mater.* **54** 2697
[16] Börzsönyi T, Tóth-Katona T, Buka A and Gránásy L 1999 *Phys. Rev. Lett.* **83** 2853
[17] Börzsönyi T, Tóth-Katona T, Buka A and Gránásy L 2000 *Phys. Rev.* E **62** 7817
[18] Li M, Zhang R J and Allison J 2010 *Magnesium Technology* 623
[19] Gao Y P, Liu H, Shi R, Zhou N, Xu Z, Zhu Y M, Nie J F and Wang Y Z 2012 *Acta Mater.* **60** 4819

[20] Han G M, Han Z Q, Luo A A, Sachdev A K and Liu B C 2012 *IOP Conf. Series: Mater. Sci. Eng.* **33** 012110
[21] Kim S G, Kim W T and Suzuki T 1999 *Phys. Rev.* E **60** 7186
[22] Lukas H, Fries S G and Sundman B 2007 *Computational Thermodynamics: The Calphad Method* (London: Cambridge University Press)
[23] Brosh E, Makov G and Shneck R Z 2007 *Calphad* **31** 173
[24] Xu R 2005 *Mater. Lett.* **59** 2818
[25] Han G M, Han Z Q, Luo A A, Sachdev A K and Liu B C 2013 *Scripta Mater.* **68** 691
[26] Mullis A M 2006 *Comput. Mater. Sci.* **36** 345

IOP Conf. Series: Materials Science and Engineering **84** (2015) 012066 doi:10.1088/1757-899X/84/1/012066

GPU phase-field lattice Boltzmann simulations of growth and motion of a binary alloy dendrite

T Takaki[1], R Rojas[1], M Ohno[2], T Shimokawabe[3] and T Aoki[3]

[1] Mechanical and System Engineering, Kyoto Institute of Technology, Matsugasaki, Sakyo-ku, Kyoto 606-8585, Japan
[2] Faculty of Engineering, Hokkaido University, Kita 13 Nishi 8, Kita-ku, Sapporo, Hokkaido 060-8628, Japan
[3] Global Scientific Information and Computing Center, Tokyo Institute of Technology, 2-12-1-i7-3, Ohokayama, Meguro-ku, Tokyo 152-8550, Japan

takaki@kit.ac.jp

Abstract. A GPU code has been developed for a phase-field lattice Boltzmann (PFLB) method, which can simulate the dendritic growth with motion of solids in a dilute binary alloy melt. The GPU accelerated PFLB method has been implemented using CUDA C. The equiaxed dendritic growth in a shear flow and settling condition have been simulated by the developed GPU code. It has been confirmed that the PFLB simulations were efficiently accelerated by introducing the GPU computation. The characteristic dendrite morphologies which depend on the melt flow and the motion of the dendrite could also be confirmed by the simulations.

1. Introduction

Melt flow plays a crucial role in the formation process of solidification microstructures. The melt flow transports heat and solute rejected from solid during solidification, and greatly changes the dendritic growth speed and morphology compared to the free dendritic growth, i.e. growth without flow. In addition, the homogeneously nucleated solid seeds and the detached dendrite arms are also transferred by the melt flow, and those form the equiaxed grains. Recently, in situ and real time observations of solidifying metals using synchrotron radiation facility have revealed the mechanical bending and fragmentation of dendrite arms by the melt flow [1-6]. In this way, the melt flow occurring during solidification drastically changes the solidification morphologies.

In order to accurately control and predict the solidification microstructures, it is of key importance to express the dendrite or cellular structures with high accuracy. Phase-field (PF) method has emerged as the most powerful numerical tool to simulate dendrite structures [7-10]. The PF method is also applied to the solidification problem including the melt flow [11-18]. However, almost all the studies are focused on the stationary dendritic growth, i.e. without motion. Very few papers treat the simultaneous growth and motion of dendrites using the PF method. Do-Quang and Amberg [19] simulated a settling dendrite due to gravity. They expressed the rigid body motion of a dendrite using two different meshes to compute the dendritic growth and the fluid flow separately. Although the rigid body motion of a growing dendrite could be well represented, it would be difficult to apply this method to the multiple dendrite growth problem and large scale computation. Medvedev et al. [20] developed a dendritic solidification model with translation and rotation by employing multi-phase-field method and lattice Boltzmann method. Although they simulated dendritic growth in shear and duct flows, the accuracy examinations of the model were not performed. Recently, we also developed a phase-field lattice Boltzmann (PFLB) model to simultaneously simulate the dendritic growth and motion of solid due to melt flow [21]. Because of the simplicity in the algorithm of the PFLB, it is relatively straightforward to apply it to large scale parallel computation.

In this study, we develop a graphics processing unit (GPU) code for the PFLB model to make faster PFLB computations. Using the developed code, we simulate the dendritic growth in a shear flow and the settling dendritic growth due to gravity. The PF method is very well suited for the GPU computation [22-24]. Although the present simulations are carried out in two-dimensions and in a single GPU, it can be easily applied to the three-dimensional problem and multi-GPUs computation.

2. Model

In the PFLB model, we employed a quantitative phase-field model for isothermal solidification of a dilute binary alloy [9] and lattice Boltzmann method for the melt flow [25].

2.1. Phase-field method

The phase-field variable ϕ is defined as $\phi = +1$ in solid and $\phi = -1$ in liquid. The solute concentration c is considered by introducing a nondimensional supersaturation u defined by $u = \left(c_L - c_L^e\right)/c_L^e - c_S^e$ where c_L is the concentration in liquid and c_L^e and c_S^e are the equilibrium concentrations of the liquid and solid at a constant temperature T_0. The time evolution equation of ϕ is expressed as

$$\tau(\theta)\left(\frac{\partial\phi}{\partial t} + \mathbf{U}_S\cdot\nabla\phi\right) = \nabla\cdot\left(W(\theta)^2\nabla\phi\right) - \frac{\partial}{\partial x}\left[W(\theta)\frac{\partial W(\theta)}{\partial\theta}\frac{\partial\phi}{\partial y}\right] + \frac{\partial}{\partial y}\left[W(\theta)\frac{\partial W(\theta)}{\partial\theta}\frac{\partial\phi}{\partial x}\right] - \frac{df(\phi)}{d\phi} - \lambda^*\frac{dg(\phi)}{d\phi}u \tag{1}$$

Here, $W(\theta)$ is the interface thickness expressed by $W(\theta) = W_0 a_s(\theta)$ where W_0 is the standard interface thickness, and $a_s(\theta)$ is the anisotropy function expressed by $a_s(\theta) = 1+\varepsilon_4\cos(4\theta)$ which depends on the crystal orientation, θ. $\tau(\theta)$ is the relaxation time considering the anisotropy by $\tau(\theta) = \tau_0 a_s(\theta)^2$. The interpolating functions $f(\phi)$ and $g(\phi)$ are chosen as $df(\phi)/d\phi = -\phi+\phi^3$ and $dg(\phi)/d\phi = (1-\phi^2)^2$, respectively. λ^* is associated with the thermodynamic driving force and is expressed as $\lambda^* = a_1 W_0/d_0$, where $a_1 = 0.88388$ and d_0 is the chemical capillary length. $\mathbf{U}_S$ is the solid velocity. The time evolution equation of u is expressed as

$$\frac{1}{2}\left[1+k-(1-k)\phi\right]\left(\frac{\partial u}{\partial t} + \mathbf{U}\cdot\nabla u\right) = \nabla\cdot\left[D_L q(\phi)\nabla u - \mathbf{J}_{AT}\right] + \frac{1}{2}\left[1+(1-k)u\right]\frac{\partial\phi}{\partial t} - \nabla\cdot\mathbf{J}, \tag{2}$$

where, $k = c_S^e/c_L^e = c_S/c_L$ is the partition coefficient and $q(\phi)$ is the interpolating function selected as $q(\phi) = (kD_S + D_L + (kD_S - D_L)\phi)/2D_L$. D_L and D_S are the diffusion coefficients in liquid and solid, respectively. $\mathbf{J}_{AT}$ and $\mathbf{J}$ are the antitrapping and the fluctuation currents, respectively, and $\mathbf{U}$ is the fluid velocity.

2.2. Lattice Boltzmann method

The lattice Boltzmann equation is given by

$$f_i(\mathbf{x}+\mathbf{c}_i\Delta t, t+\Delta t) = f_i(\mathbf{x},t) - \frac{1}{\tau}\left[f_i(\mathbf{x},t) - f_i^{eq}(\mathbf{x},t)\right] + G_i(\mathbf{x},t)\Delta t, \tag{3}$$

where, τ is the single relaxation time, f_i is the particle distribution function in the ith direction, $\mathbf{x}$ is the position vector, $\mathbf{c}_i$ is the discrete particle velocity, Δt is the time step, and G_i is the discrete force. The equilibrium distribution function f_i^{eq} is expressed as

$$f_i^{eq} = w_i\rho\left[1 + \frac{3(\mathbf{c}_i\cdot\mathbf{U})}{c^2} + \frac{9(\mathbf{c}_i\cdot\mathbf{U})^2}{2c^4} - \frac{3\mathbf{U}\cdot\mathbf{U}}{2c^2}\right], \tag{4}$$

where, w_i is the weighting function depending on the employed model and c is the lattice velocity defined by $c = \Delta x/\Delta t$ using lattice size Δx and time increment Δt. The fluid density ρ and velocity **U** are computed by

$$\rho = \sum_{i=0}^{Q-1} f_i , \tag{5}$$

$$\rho \mathbf{U} = \sum_{i=0}^{Q-1} \mathbf{c}_i f_i , \tag{6}$$

where, Q is the number of discrete velocities. The discrete force G_i in equation (1) is given with second-order accuracy as [26]

$$G_i = \rho w_i \left[3\frac{\mathbf{c}_i - \mathbf{U}}{c^2} + 9\frac{(\mathbf{c}_i \cdot \mathbf{U})\mathbf{c}_i}{c^4} \right] \cdot \mathbf{G} , \tag{7}$$

where, **G** is the external force on the solid-liquid interface, where the phase-field ϕ changes from -1 to +1, to satisfy the non-slip boundary condition. **G** is modeled as a dissipative drag [11]:

$$\mathbf{G}(\mathbf{x},t) = \frac{2\rho v h}{W_0}\left(\frac{1+\phi}{2}\right)^2 (\mathbf{U}_S - \mathbf{U}) , \tag{8}$$

where, v is the kinematic viscosity, h is a constant of h = 2.757 and $\mathbf{U}_s$ is the solid velocity. In the following simulations, the two-dimensional nine-velocity (D2Q9) model is used.

2.3. *Solid motion*

The motion of solids is determined by solving the following set of equations [27; 28]:

$$M_S \frac{d\mathbf{U}_T}{dt} = \mathbf{G}_S , \tag{9}$$

$$\mathbf{I}_S \frac{d\omega_S}{dt} = \mathbf{T}_S , \tag{10}$$

where, M_S is the mass of the solid, $\mathbf{U}_T$ is the translational velocity of the solid, $\mathbf{I}_S$ the tensor for the moment of inertia, ω_S is the angular velocity, and $\mathbf{G}_S$ and $\mathbf{T}_S$ are the total force and torque acting on the solid, respectively. $\mathbf{G}_S$ and $\mathbf{T}_S$ are calculated by

$$\mathbf{G}_S = -\sum_{\mathbf{x}\in\Omega} \mathbf{G}(\mathbf{x},t)\Delta V + \left(1 - \frac{\rho_L}{\rho_S}\right) M_S \mathbf{g} , \tag{11}$$

$$\mathbf{T}_S = -\sum_{\mathbf{x}\in\Omega} (\mathbf{x} - \mathbf{X}_S) \times \mathbf{G}(\mathbf{x},t)\Delta V , \tag{12}$$

where, ΔV is the volume of the computational domain, Ω is the domain in the vicinity of the interface, ρ_L and ρ_S are the density of the liquid and solid phases, **g** is the gravity, and $\mathbf{X}_S$ is the center of mass of the solid. The solid velocity $\mathbf{U}_S$ is given by

$$\mathbf{U}_S = \mathbf{U}_T + \omega_S \times (\mathbf{x} - \mathbf{X}_S) . \tag{13}$$

3. Computations

The dendritic growth in a shear flow and the sedimentation of a growing dendrite due to gravity are simulated using the developed code. In both simulations, the dendrites grow with translation and rotation. An Al-3wt%Cu binary alloy is used in the following simulations and the material parameters are the same as those used in the reference [22].

3.1. Dendritic growth in a shear flow

Figure 1(a) shows the computational domain and the initial condition. The size of the computational domain is set to $L_x \times L_y = 8{,}192\Delta x \times 1{,}024\Delta y$ with a lattice size $\Delta x = \Delta y = 0.5$ μm. A solid seed with radius of $3\Delta x$ is placed at the point of ($200\Delta x$, $512\Delta y$). Bottom and top flow velocities are fixed to U_0 = 6 mm/s and $-U_0/2$, respectively. The initial distribution of flow velocity in x-direction is set to $U(y) = U_0(1\text{-}3y/2L_y)$ and the initial velocity in y-direction is set to $V = 0$ in all domain. Periodic boundary conditions are set to the melt flows at the left and right sides. For phase-field ϕ and the supersaturation u, zero Neumann conditions are set to all boundaries.

Figure 1 shows the time changes of the dendrite morphology and the concentration distribution. The dendrite shape is expressed by the solid black line which is the counter of $\phi = 0$. The color indicates the concentration distribution. It can be observed that the dendrite grows with translation and rotation. The four dendrite arms are designated by a1, a2, a3 and a4. Initially, the arm a1 grows toward x-axis direction and its location changes with time. We can see that the dendrite rotates about 180° every 25,000 steps.

Figure 2(a) shows the variation of area of the dendrite, A_S, during the computation shown in figure 1. The dashed line indicates the variation in A_S of a free growth dendrite computed without flow. It can be observed that the melt flow largely enhances the dendritic growth. Figure 2(b) shows the variations of the dendrite migration velocity in the x-direction, U_S, and the y-coordinate of the center of mass of the dendrite, Y_S. The dendrite migration velocity U_S increases almost linearly until about 75,000 steps and then accelerates. These velocity changes can be explained by the change of Y_S.

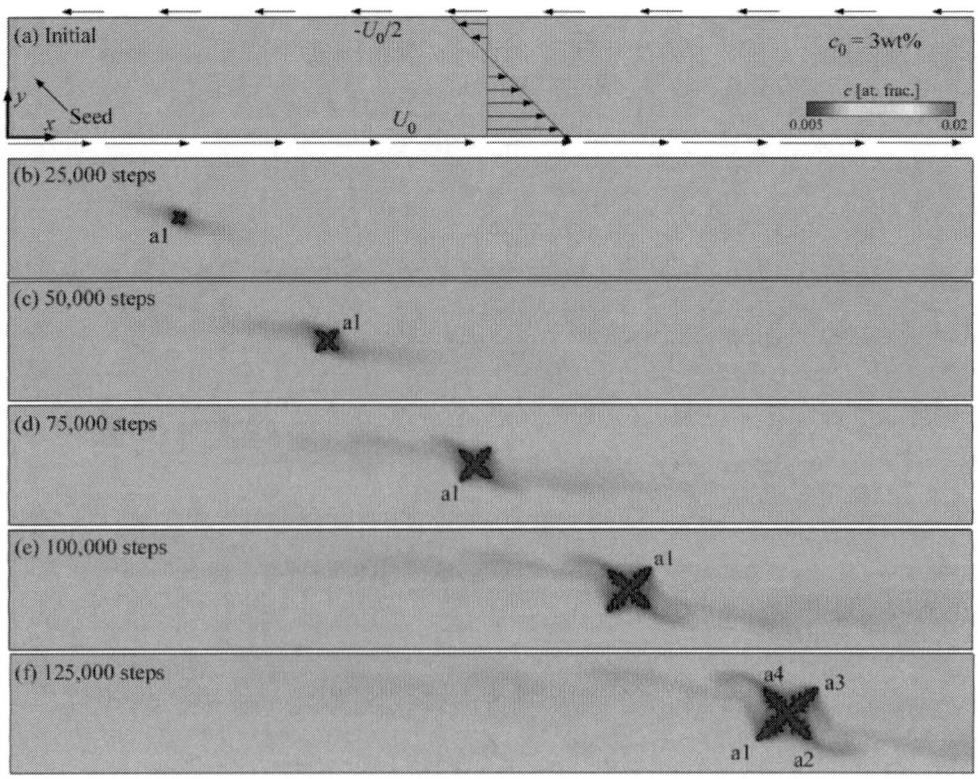

Figure 1. Time slices during dendrite growth in shear flow.

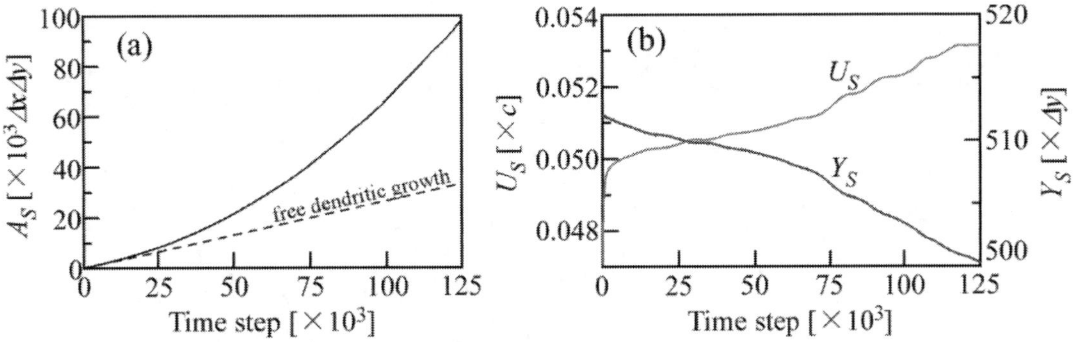

Figure 2. Variations of (a) area of the dendrite, A_S, and (b) velocity in x-direction, U_S, and position in the y-direction, Y_S, of center of mass of the dendrite.

3.2. Dendritic growth during settling

The settling of growing dendrites, which are caused by the density difference between solid and liquid, is a very important phenomenon in solidification and casting [29-34]. Figures 3-5 show the dendrite morphological changes and the changes of concentration field obtained in the dendrite settling simulations under three different initial supersaturations, i.e. u_0 = -0.2, -0.3 and -0.4. The computational domain is set to $L_x \times L_y = 2{,}048\Delta x \times 4{,}096\Delta y$ with a lattice size $\Delta x = \Delta y = 0.5$ μm. A solid seed with radius of $3\Delta x$ is placed at the point of ($1{,}024\Delta x$, $3{,}896\Delta y$). The no-slip condition and the zero Neumann condition for ϕ and u are set to all boundaries. The density ratio, ρ_S/ρ_L, is assumed to be 1.05, and the gravity acts along the y-direction.

From figures 3-5, it can be observed that the dendrites settle straightly until about $y = 3{,}000\Delta y$. The dendrite arm growing downward becomes longer compared to the other three arms, and the growth of the one oriented upward is suppressed by the high concentration field. The dendrite arms with the preferred growth direction along the x-axis slightly grow downward. This nonsymmetrical growth of the four dendrite arms is caused by the melt flow initiated in the settling of the dendrite. This behavior is very similar to the experimental observations of a transparent material [32]. Then, as shown in figure 3-5(b), the rotation of the dendrite starts. The rotation is caused due to the non-symmetry in the left and right sides of the dendrite morphology.

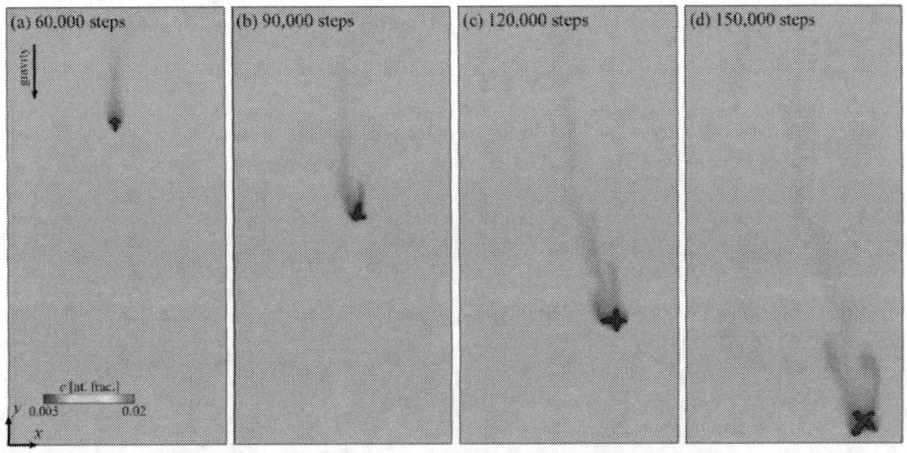

Figure 3. Dendrite settling simulation for u_0 = -0.2.

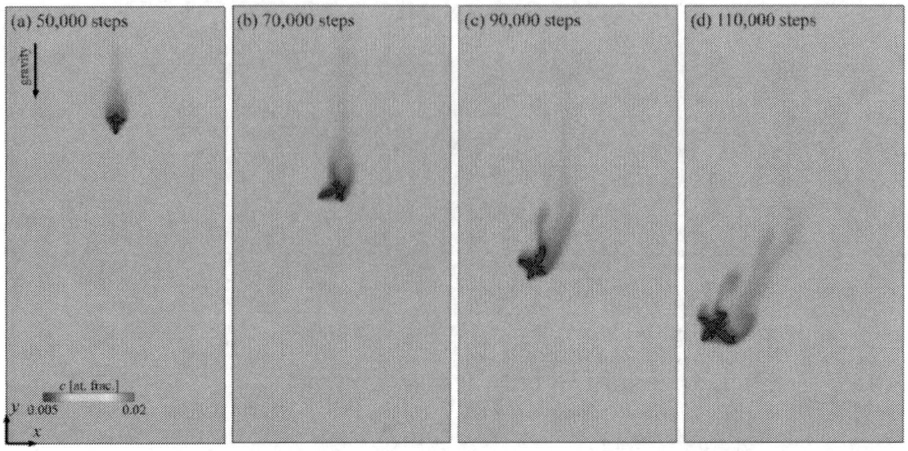

Figure 4. Dendrite settling simulation for $u_0 = -0.3$.

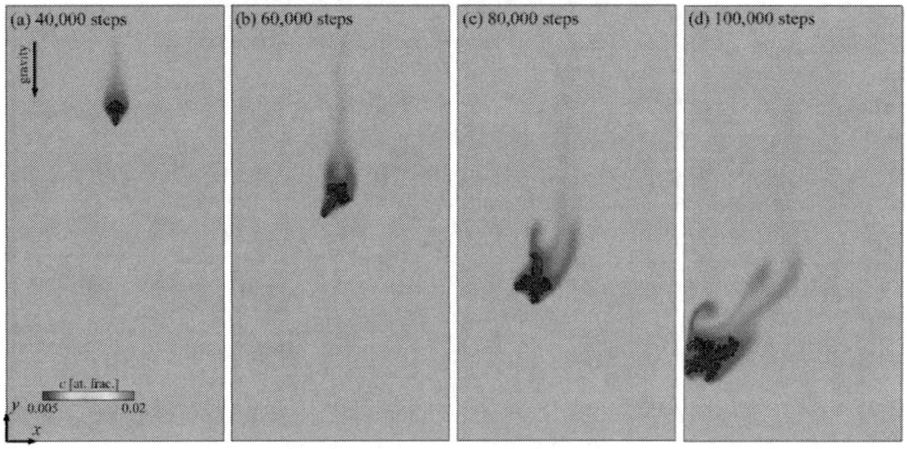

Figure 5. Dendrite settling simulation for $u_0 = -0.4$.

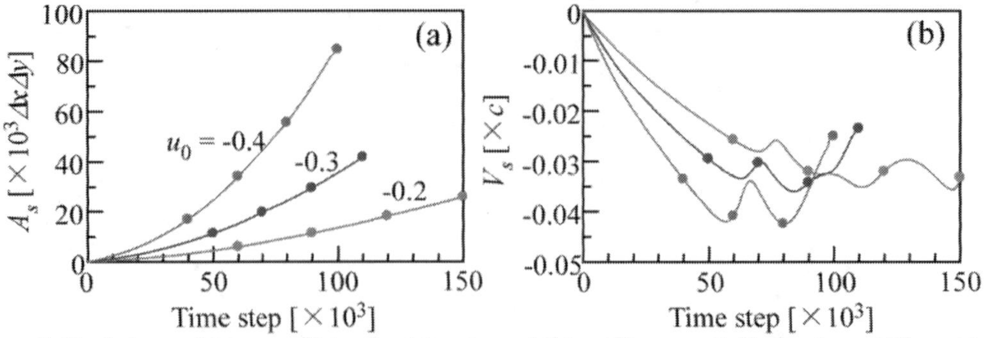

Figure 6. Variations of (a) area of the dendrite, A_s, and (b) settling speed, V_s, for three different initial supersaturations, u_0.

Figure 6 shows the variations of the dendrite area, A_s, and the settling velocity, V_s. The four solid circles on the each curve correspond to the time step in figures 3-5. As can be seen, the growth speed increases as the supersaturation reduces, and that is why the settling speed, V_s, is faster for a smaller supersaturation. Although A_s increases monotonically, V_s fluctuates around the velocity before the solid starts rotating. Therefore, the fluctuations in figure 6(b) are due to the rotations of the dendrite. For u_0 = -0.3 and -0.4, large speed reductions are observed at the last stage of the computations. This is due to the interaction with the boundary as shown in figures 4(d) and 5(d).

The computational time of the shear flow and the settling simulations using 8,388,608 meshes was about 140 minutes for 150,000 steps which includes the data output time every 1,000 steps. Here, single GPU of the NVIDIA Tesla K20X was used. The computational efficiency can be improved even further by modifying the developed code. However, the computational time in the present study is thought to be reasonable for larger computations and systematical investigations.

4. Conclusions

In this study, we have developed a GPU code for the PFLB method, which can simulate the dendritic growth with translation and rotation. The GPU accelerated PFLB method has been implemented using CUDA C. The dendritic growth in a shear flow and settling condition has been simulated by the developed GPU code using a single NVIDIA Tesla K20X. As a result, the characteristic dendrite morphologies and the growth speed differences depending on the melt flow could be observed. It has been also confirmed that the GPU accelerated PFLB computations were enabled in reasonable computational times. By parallelizing the developed GPU code using multiple-GPUs, larger computations and three-dimensional computations of dendritic growth with motion will be possible.

Acknowledgements

This work was supported by a Grant-in-Aid for Scientific Research (KAKENHI) (Grant Number 25289006 and 26220002) from the Japan Society for the Promotion of Science (JSPS) and a Research Promotion Grant from the Iron and Steel Institute of Japan (ISIJ).

References

[1] Ruvalcaba D, Mathiesen R H, Eskin D G, Arnberg L and Katgerman L 2007 In situ observations of dendritic fragmentation due to local solute-enrichment during directional solidification of an aluminum alloy *Acta Mater.* **55** 4287

[2] Yasuda H, Yamamoto Y, Nakatsuka N, Yoshiya M, Nagira T, Sugiyama A, Ohnaka I, Uesugi K and Umetani K 2009 In situ observation of solidification phenomena in Al-Cu and Fe-Si-Al alloys *Int. J. Cast Met. Res.* **22** 15

[3] Billia B, Nguyen-Thi H, Mangelinck-Noel N, Bergeon N, Jung H, Reinhart G, Bogno A, Buffet A, Hartwig J, Baruchel J and Schenk T 2010 In Situ synchrotron x-ray characterization of microstructure formation in solidification processing of Al-based metallic alloys *ISIJ Int.* **50** 1929

[4] Mirihanage W U, Arnberg L and Mathiesen R H *IOP Conference Series: Materials Science and Engineering,2012),* vol. Series 33*)*

[5] Liotti E, Lui A, Vincent R, Kumar S, Guo Z, Connolley T, Dolbnya I P, Hart M, Arnberg L, Mathiesen R H and Grant P S 2014 A synchrotron X-ray radiography study of dendrite fragmentation induced by a pulsed electromagnetic field in an Al-15Cu alloy *Acta Mater.* **70** 228

[6] Reinhart G, Nguyen-Thi H, Mangelinck-Noël N, Baruchel J and Billia B 2014 In situ investigation of dendrite deformation during upward solidification of Al-7wt.%Si *JOM* **66** 1408

[7] Kobayashi R 1993 Modeling and numerical simulations of dendritic crystal growth *Physica D: Nonlinear Phenomena* **63** 410

[8] Karma A and Rappel W J 1996 Phase-field method for computationally efficient modeling of solidification with arbitrary interface kinetics *Physical Review E - Statistical Physics, Plasmas, Fluids, and Related Interdisciplinary Topics* **53** R3017

[9] Ohno M and Matsuura K 2009 Quantitative phase-field modeling for dilute alloy solidification involving diffusion in the solid *Physical Review E - Statistical, Nonlinear, and Soft Matter Physics* **79**

[10] Takaki T 2014 Phase-field modeling and simulations of dendrite growth *ISIJ Int.* **54** 437
[11] Beckermann C, Diepers H J, Steinbach I, Karma A and Tong X 1999 Modeling Melt Convection in Phase-Field Simulations of Solidification *J. Comput. Phys.* **154** 468
[12] Tong X, Beckermann C and Karma A 2000 Velocity and shape selection of dendritic crystals in a forced flow *Physical Review E - Statistical Physics, Plasmas, Fluids, and Related Interdisciplinary Topics* **61** R49
[13] Beckermann C, Li Q and Tong X 2001 Microstructure evolution in equiaxed dendritic growth *Science and Technology of Advanced Materials* **2** 117
[14] Jeong J H, Goldenfeld N and Dantzig J A 2001 Phase field model for three-dimensional dendritic growth with fluid flow *Physical Review E - Statistical, Nonlinear, and Soft Matter Physics* **64** 416021
[15] Tong X, Beckermann C, Karma A and Li Q 2001 Phase-field simulations of dendritic crystal growth in a forced flow *Physical Review E - Statistical, Nonlinear, and Soft Matter Physics* **63** 061601/1
[16] Al-Rawahi N and Tryggvason G 2002 Numerical simulation of dendritic solidification with convection: Two-dimensional geometry *J. Comput. Phys.* **180** 471
[17] Al-Rawahi N and Tryggvason G 2004 Numerical simulation of dendritic solidification with convection: Three-dimensional flow *J. Comput. Phys.* **194** 677
[18] Guo Z, Mi J, Xiong S and Grant P S 2013 Phase Field Simulation of Binary Alloy Dendrite Growth Under Thermal- and Forced-Flow Fields: An Implementation of the Parallel–Multigrid Approach *Metallurgical and Materials Transactions B* **44** 924
[19] Do-Quang M and Amberg G 2008 Simulation of free dendritic crystal growth in a gravity environment *J. Comput. Phys.* **227** 1772
[20] Medvedev D, Varnik F and Steinbach I *Procedia Computer Science,2013),* vol. Series 18*)* pp 2512
[21] Rojas R, Takaki T and Ohno M 2015 Submitting
[22] Takaki T, Ohno M, Shimokawabe T and Aoki T 2014 Two-dimensional phase-field simulations of dendrite competitive growth during the directional solidification of a binary alloy bicrystal *Acta Mater.* **81** 272
[23] Takaki T, Shimokawabe T, Ohno M, Yamanaka A and Aoki T 2013 Unexpected selection of growing dendrites by very-large-scale phase-field simulation *J. Cryst. Growth* **382** 21
[24] Yamanaka A, Aoki T, Ogawa S and Takaki T 2011 GPU-accelerated phase-field simulation of dendritic solidification in a binary alloy *J. Cryst. Growth* **318** 40
[25] Chen S and Doolen G D 1998 Lattice boltzmann method for fluid flows. In: *Annual Review of Fluid Mechanics,* pp 329
[26] Medvedev D, Fischaleck T and Kassner K 2006 Influence of external flows on crystal growth: Numerical investigation *Physical Review E* **74** 031606
[27] Glowinski R, Pan T W, Hesla T I, Joseph D D and Périaux J 2001 A Fictitious Domain Approach to the Direct Numerical Simulation of Incompressible Viscous Flow past Moving Rigid Bodies: Application to Particulate Flow *J. Comput. Phys.* **169** 363
[28] Feng Z G and Michaelides E E 2004 The immersed boundary-lattice Boltzmann method for solving fluid-particles interaction problems *J. Comput. Phys.* **195** 602
[29] Badillo A, Ceynar D and Beckermann C 2007 Growth of equiaxed dendritic crystals settling in an undercooled melt, Part 2: Internal solid fraction *J. Cryst. Growth* **309** 216
[30] Badillo A, Ceynar D and Beckermann C 2007 Growth of equiaxed dendritic crystals settling in an undercooled melt, Part 1: Tip kinetics *J. Cryst. Growth* **309** 197
[31] Lesoult G 2005 Macrosegregation in steel strands and ingots: Characterisation, formation and consequences *Materials Science and Engineering: A* **413–414** 19
[32] Appolaire B, Albert V, Combeau H and Lesoult G 1999 Experimental study of free growth of equiaxed NH4Cl crystals settling in undercooled NH4Cl-H2O melts *ISIJ Int.* **39** 263
[33] Appolaire B, Albert V, Combeau H and Lesoult G 1998 Free growth of equiaxed crystals settling in undercooled NH4Cl-H2O melts *Acta Mater.* **46** 5851
[34] Ramani A and Beckermann C 1997 Dendrite tip growth velocities of settling NH4CI equiaxed crystals *Scripta Mater.* **36** 633

IOP Conf. Series: Materials Science and Engineering **84** (2015) 012067 doi:10.1088/1757-899X/84/1/012067

Study of dendritic growth and coarsening using a 3-D phase field model: Implementation of the Para-AMR algorithm

Z Guo[1*], S M Xiong[1,2]

[1] School of Materials Science and Engineering, Tsinghua University, Beijing, China
[2] Key Laboratory for Advanced Materials Processing Technology, Ministry of Education, Beijing, China

E-mail: zhipeng_guo@mail.tsinghua.edu.cn

Abstract. To efficiently solve the coupled phase field equations in 3-D, an algorithm comprising of adaptive mesh refinement (AMR) and parallel (Para-) computing capabilities was developed. Dendritic growth and subsequent coarsening were studied by employing the model to simulate multi-dendrite growth under isothermal conditions. Quantitative comparison including decrease of interface area (*S*) and nonlinear growing of the characteristic length (ratio between solid volume *V* and surface area S i.e. *V/S*) as time was performed between the simulation results and these predicted by the existing theories. In particular, various mechanisms including growth of lower curvature area in expense of higher curvature one, coalescence of neighbouring dendrite arms and groove advancement at the root of higher order arms for dendritic coarsening were identified and successfully revealed *via* the 3-D phase field simulation. In addition, results showed that the proposed algorithm could greatly shorten the computing time for 3-D phase field simulation and enable one to gain much more insight in understanding the underlying physics during dendrite growth in solidification.

1. Introduction

In solidification of metallic alloys, growth of solid dendrites is always accompanied by coarsening. To reduce the free energy of the system, the coarsening always occurs spontaneously by decreasing the overall curvature or solid/liquid interface area [1,2].

The phase field method is becoming a standard approach to simulate phase transitions, especially for those with complex geometries, e.g. dendrites during solidification [3]. In the phase field approach, an extra equation (the phase field equation) is introduced to characterize the transition between the solid phase and the liquid phase through the variable ϕ, termed the phase field, which is constant in bulk phases (-1 in liquid and 1 in solid) but changes steeply across the solid-liquid interface. The most beneficial feature for phase field methodology is that explicit tracking of the interface can be avoided, which greatly simplifies the computational procedure.

However, to accurately characterize the underlying physics during phase transition, the grid size must be very small in order to have enough resolution to describe the interface. The consequence is then enormous computing overhead must be dealt with during the phase field simulation especially in the study of coarsening where simulation has to be performed in much longer time before the underlying physics could be revealed. Because of such limitation, current literature in studying dendritic coarsening is rather limited and most existing studies were performed in 2-D [4]. Even

IOP Conf. Series: Materials Science and Engineering **84** (2015) 012067 doi:10.1088/1757-899X/84/1/012067

though certain qualitative aspects of physics in dendritic coarsening can be revealed in 2-D, quantitative characterization has to be found through 3-D simulations.

In this paper, based on the block-structured grid configuration as introduced by Berger and co-workers [5] we will first develop an algorithm with both adaptive mesh refinement and parallel computing ability i.e. a Para-AMR approach and then employ this approach to solve the coupled phase field equations in 3-D. Different phase field simulations will then be performed to study the dendritic coarsening.

2. The phase field model and numerical approach

2.1. The phase field model

The 3-D phase field model was derived based on our previous work [6] and the governing equations read:

$$\tau\frac{\partial\phi}{\partial t}=\nabla\cdot\left(W(\vec{n})^2\nabla\phi\right)+\phi\left(1-\phi^2\right)-\lambda\left(1-\phi^2\right)^2(\theta+kU)$$
$$+\frac{\partial}{\partial x}\left(|\nabla\phi|^2W(\vec{n})\frac{\partial W(\vec{n})}{\partial\phi_x}\right)+\frac{\partial}{\partial y}\left(|\nabla\phi|^2W(\vec{n})\frac{\partial W(\vec{n})}{\partial\phi_y}\right)+\frac{\partial}{\partial z}\left(|\nabla\phi|^2W(\vec{n})\frac{\partial W(\vec{n})}{\partial\phi_z}\right) \quad (1)$$

$$\left(\frac{1+k}{2}-\frac{1-k}{2}\phi\right)\frac{\partial U}{\partial t}=\nabla\cdot\left(D\frac{1-\phi}{2}\nabla U-\vec{j}_{at}\right)+\frac{1}{2}\left[1+(1-k)U\right]\frac{\partial\phi}{\partial t} \quad (2)$$

$$\frac{\partial\theta}{\partial t}=\alpha\nabla^2\theta+\frac{1}{2}\frac{L/c_p}{\Delta T_0}\frac{\partial\phi}{\partial t} \quad (3)$$

where τ is the relaxation time, ϕ is phase field, $W(\vec{n})$ is the anisotropic width of the diffuse interface with $\vec{n}$ the unit normal vector pointing out into the liquid, k is the partition coefficient for solute, D and α are solute and thermal diffusivities respectively, L is the heat of fusion per unit volume, c_p is the specific heat of alloy. λ is the scaling parameter defined as

$$\lambda=\frac{15}{16}\frac{RT_M(1-k)}{v_0H|m|}\Delta T_0 \quad (4)$$

where R is the gas constant, T_M is the melting temperature of the solvent, v_0 is the molar volume, H is the energy barrier of the double well potential, and m is the liquidus slope in the phase diagram.

$$\vec{j}_a=-\frac{W}{\sqrt{2}}\frac{c/c_\infty}{[1+k-(1-k)\phi]}\frac{\partial\phi}{\partial t}\frac{\nabla\phi}{|\nabla\phi|} \quad (5)$$

is the so-called "anti-trapping" current developed by Karma [7] to counterbalance spurious effects during the phase field simulation.

$$U=\frac{\dfrac{2c/c_\infty}{1+k-(1-k)\phi}-1}{1-k} \quad (6)$$

and

$$\theta=\frac{T-T_M-mc_\infty}{\Delta T_0} \quad (7)$$

are dimensionless solute concentration and temperature respectively, where c is solute concentration, c_∞ is the initial solute concentration and

$$\Delta T_0=\frac{|m|c_\infty(1-k)}{k} \quad (8)$$

is the equ*ilibrium freezing tem*perature range according to c_∞.

The crystal anisotropy was introduced by taking $\tau = \tau_0 A(\vec{n})^2$ and $W = W_0 A(\vec{n})$ and

$$A(\vec{n}) = 1 + \varepsilon_1 (Q - 3/5) + \varepsilon_2 (3Q + 66S - 17/7) \quad (9)$$

$$Q = \left(\phi_x^4 + \phi_y^4 + \phi_z^4\right) / |\phi|^4 \quad (10)$$

$$S = \phi_x^2 \phi_y^2 \phi_z^2 \quad (11)$$

where $\phi_x = \partial\phi/\partial x$, $\phi_y = \partial\phi/\partial y$, $\phi_z = \partial\phi/\partial z$. ε_1 and ε_2 weight the magnitude of the anisotropy strength favoring the crystalline orientation of [100] and [110], respectively [3].

According to the thin interface analysis in [5], the kinetic effect can be made to vanish by taking

$$\tau = \tau_0 \left[\frac{1}{Le} + k\left(1 + (1-k)U\right) \right] A(\vec{n})^2 \quad (12)$$

where $Le = \alpha/D$ is the Lewis number. The length and time in equations (1) – (3) were further scaled by $W_0 = \lambda d_0 / a_1$ and $\tau_0 = d_0^2 a_2 \lambda^3 / \left(D a_1^2\right)$ to be fully non-dimensional, where $d_0 = \Gamma/\Delta T_0$ is the chemical capillary length, and Γ is the Gibbs-Thomson coefficient. According to the thin interface analysis [7], a1 = 0.8839 and a2 = 0.6267 respectively.

In this work a simple method was applied to characterize the multi-orientation of the dendrites, which has been employed successfully in our previous work [6] for the 2-D case. The orientation of each of the dendrites is specified with a pre-existing solid seed introduced in the computational domain. As the dendrite grows, very small volumes of liquid are added/transformed to solid; the crystal orientation for this new increment in solid is assumed the same as that of the local crystal orientation of the dendrite. In this way, each dendrite acts like an orientation attracting sink which transforms its closest external layer of liquid into solid by imposing its orientation to it.

2.2. *The Para-AMR algorithm*

To solve equations (1) – (3), we have developed a Para-AMR algorithm which is in fact an implementation of the adaptive mesh refinement on a parallel computing architecture.

2.2.1. *The adaptive mesh refinement (AMR)*

In this study, the AMR was realized based on a block-structured grids, which was first introduced by Berger and co-workers [5]. Accordingly, the AMR was performed in three sequential steps including (1) tagging the target cells that need refined, (2) clustering the tagged cells into patch-boxes, and (3) ensuring the coarse/refined grids are properly nested.

To refine the target grids the first step is to tag the cells that need refined. For dendritic microstructure the solid-liquid interface is the location where the phase changes the fastest and thus the position that needs refined. In this work, we employed a gradient criterion to perform the tagging i.e.

$$|\nabla\phi| + \beta_U |\nabla U| + \beta_\theta |\nabla\theta| \geq \xi \quad (13)$$

where β_U and β_θ are weighted parameters for solute and temperature respectively. ξ is a threshold value (always small in magnitude) which has to be retrieved *via* numerical tests according to specific simulation cases.

After certain grids/points have been tagged, the grid refinement would start from the coarsest grid level. A cluster algorithm developed by Berger and co-workers [8] was adopted to separate the tagged points into clusters. The clusters are also called patch-boxes, which in 2-D are rectangles but in 3-D they are boxes with interior boundaries. After the patch-box was generated, grid points/cells would be filled inside and used for data computation. The cluster algorithm would be performed repeatedly until reaching the top grid level, after which the multi-level grid with a hierarchical structure was created.

An extra layer of cells or two will always be added to the coarse grid level to ensure the multi-level grids are properly nested.

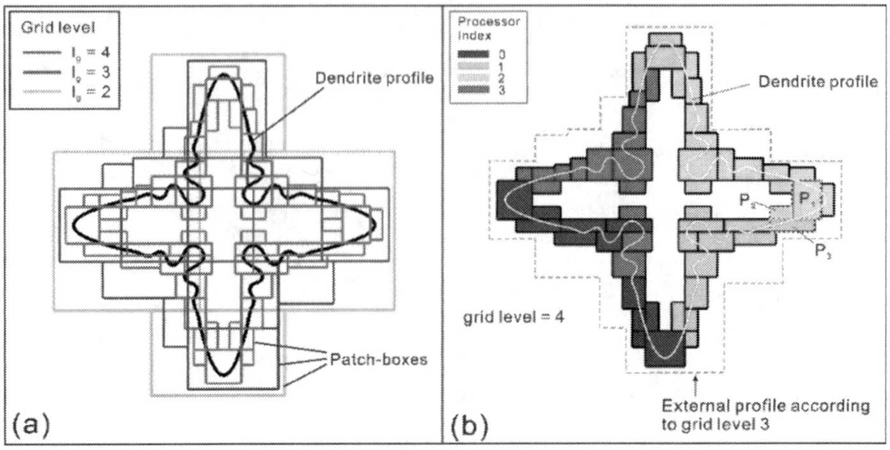

Figure 1. (a) The configuration of the patch-boxes on different grid levels generated during the adaptive mesh refinement and (b) the dispatching of computing work to the parallel processors. The dendrite profile i.e. the solid/liquid interface was indicated in the middle of each figure.

A typical result of the grid structure according to a dendrite profile was shown in figure 1a, in which the location of the patch-boxes on different grid levels (l_g) was shown. The simulation was according to a single dendrite growth in an isothermal domain i.e. θ = -0.2. Other computation parameters were k = 0.15, λ = 10, β_U=0.3 and β_θ= 0 (isothermal). A solid seed was initially planted in the middle of the domain and then growing for 6000 time steps. The phase field profile according to $\phi = 0$ was then extracted and shown in the middle of figure 1a.

As shown, the multi-level grid was properly nested i.e. the patch-boxes belonging to a finer grid always lay inside that of a coarser grid. Besides, the layout of the patch-boxes could perfectly describe the outline profile of the dendrite, especially on the top grid level.

2.2.2. The parallel computing

After the grid was generated, the layout of the patch-boxes including all information regarding to the configuration of geometry was then broadcasted to all processors, based on which the computing work on each grid level was constructed, partitioned and dispatched. Two schemes were applied in this approach to avoid load imbalance. If the number of the patch-boxes was smaller than that of the parallel processors, the algorithm will dispatch the computing work uniformly starting from the root process. On the contrary, if the number of the patch-boxes was larger the computing work would be dispatched by employing a method based on space filling curve [9].

A typical result about the parallel dispatching of the computing work is shown in figure 1b, in which four processors were used to demonstrate the parallelization on grid level 4. Because the number of the patch-boxes was larger than that of the parallel processors the dispatching of the computing work was realized using the space filling curve principle and as seen from figure 1b the patch-boxes were distributed rather uniformly into the four processors (as indexed by 0 - 3).

At each time step after updating the variables, related data has to be communicated, which primarily comprised two steps. The first step was to communicate and update the data at the external boundaries of each patch-box. This could happen either within one processor (e.g. between patch-boxes P_1 and P_2 in figure 1b) or between different processors (e.g. between P_1 and P_3). To simplify this, a layer of cells (also known as ghost cells) were added to the external boundaries of each patch-box to receive the data from its nearest neighbours. Secondly, certain amount of data between different grid levels must be communicated to perform the restriction (i.e. updating values on the coarse grid using those from the fine grid) and interpolation (i.e. updating values on the fine grid using those from the coarse grid) at the interior boundaries.

The code for solving the phase field equations was built on the BoxLib [10] software framework for block-structured adaptive mesh methods, which currently supports a hierarchical programming approach for multi core architectures based on both MPI [11] and OpenMP [12].

3. Numerical tests

To test the accuracy and efficiency of the algorithm, simulation was performed in 3-D on a single dendrite growth scenario with key parameters including $k = 0.15$, $\lambda = 30$, $\theta = -0.12$, $\beta_U = 0.3$ and $\beta_\theta = 0$. The computing domain was 204.8×204.8×204.8, i.e. equivalent to 256×256×256 grid cells if a uniform grid of $dx = 0.8$ was employed. During the test, the number of the grid levels i.e. N_{amr} and the number of the parallel processes i.e. N_p were varied and attention was focused on their effects on the simulation elapsed time. For each simulation, a total of 6000 time steps were calculated.

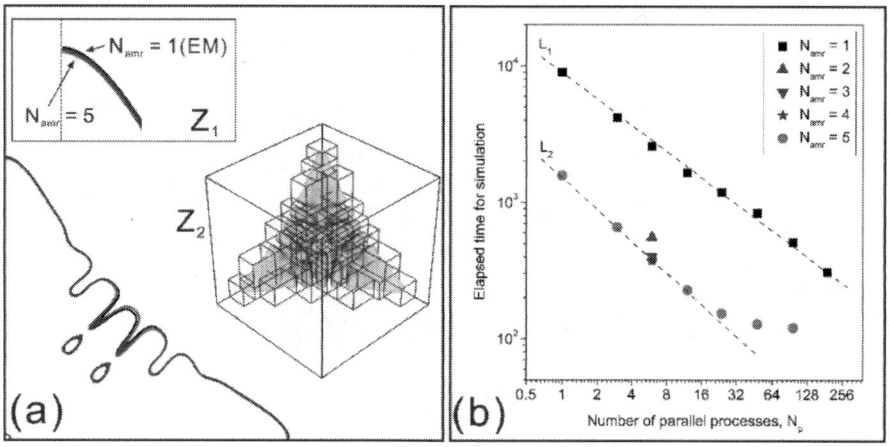

Figure 2. Numerical test results on (a) accuracy and (b) parallel efficiency. The dendrite profiles extracted using the explicit algorithm (EM) and the adaptive mesh refinement (AMR) were compared in detail in the insert "Z_1", together with "Z_2" showing the dendrite morphology in 3-D and the configuration of the patch-boxes on the top grid level.

Figure 2 shows the related result for the numerical test. In figure 2a, the phase field profiles according to $\phi = 0$ were extracted for the case when the explicit method (EM) was applied i.e. $N_{amr} = 1$ and $N_p = 1$ and the case when the AMR was applied i.e. $N_{amr} = 5$ and $N_p = 1$. Because the two profiles were almost overlapping a zoom-in insert "Z_1" was used to show the difference. According to the simulation result, the difference on the tip position of the two profiles was about 0.05%, clearly indicating the accuracy of the algorithm. The morphology of the dendrite in 3-D and the related configuration of the patch-boxes on the top grid level were shown in the insert designated by "Z_2". Figure 2b shows the elapsed time as a function of N_p and as shown, without the adaptive mesh refinement, the total elapsed time was about 9000 seconds when $N_p = 1$ i.e. when the explicit algorithm was applied. Increasing the process number to 48 decreased the elapsed time to 834 seconds, i.e. the simulation was speeded up for about 10.8 times. It is clear that the gained speed-up here is lower than the theoretical value i.e. 48 and data communication between patch-boxes and grid levels should be responsible for this.

The application of the AMR also significantly reduced the elapsed time. As shown by the solid spheres, the elapsed time for the case when $N_{amr} = 5$ and $N_p = 1$ was about 1570 seconds i.e. less than 18% of that when the explicit algorithm was applied. Further increasing the process number to $N_p = 12$ led to another speed-up of about 86%, i.e. 40 times speed-up overall.

From figure 2b, it can be seen that the elapsed time (t_{simu}) changed as a function of the process number in the following manner:

$$\lg t_{simu} = k_c \lg N_p + \lg t_{simu}^{N_p=1} \tag{14}$$

where $t_{simu}^{N_p=1}$ is the elapsed time for the simulation when $N_p = 1$. This relationship i.e. equation (14) was best described by the linear fits as designated by "L_1" and "L_2" for $N_{amr} = 1$ and $N_{amr} \neq 1$ respectively. And according to figure 2b, the slope for these two cases was about -0.64 (L_1) and -0.78 (L_2) respectively, i.e. the algorithm worked better if the AMR was applied. The slope values for both cases are lower than the theoretical magnitude i.e. -1 but comparing with the explicit algorithm, it is clear that with the application of the Para-AMR algorithm the computing efficiency for the 3-D phase field simulation could be increased by two orders of magnitude i.e. simulation time could be reduced from an order of magnitude of 10^4 seconds to 10^2 seconds.

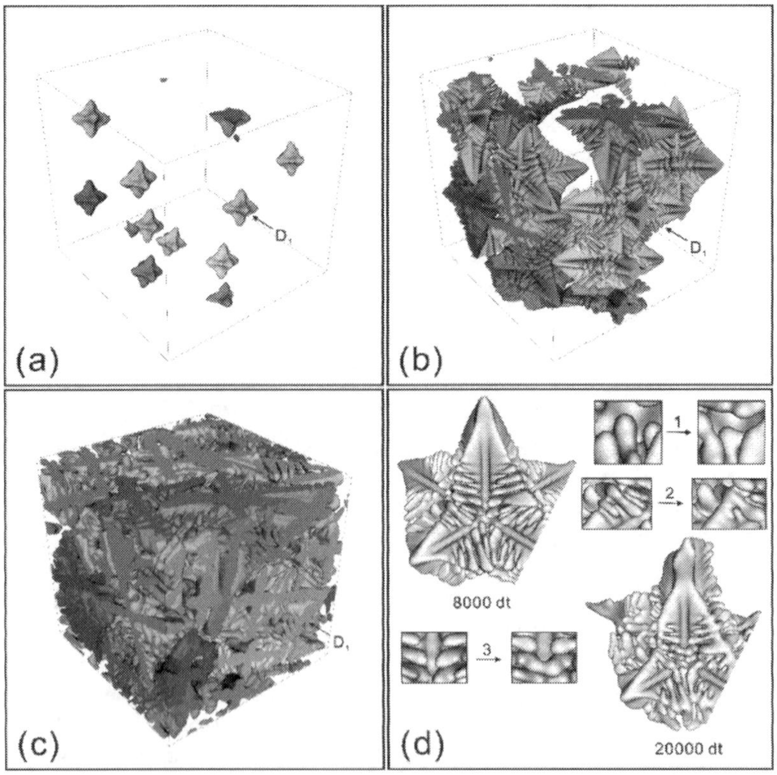

Figure 3. Phase field simulations of the dendritic morphology at time (a) 2000 *dt*, (b) 8000 *dt* and (c) 20,000 *dt*. (d) shows the coarsening of dendrite D_1 from t = 8000 *dt* to 20,000 *dt*. The three coarsening mechanisms including small arm melting, groove advancement and coalescence were highlighted using examples designated by "1", "2" and "3" respectively.

4. Simulation of dendrite growth and coarsening

Simulations were performed to study the multi-dendrite growth and coarsening in 3-D. Similar as the numerical test, five levels of grid were applied during the simulation and 192 cores were implemented for the parallel computation. The system domain was 409.6×409.6×409.6, i.e. equivalent to 512×512×512 cells if an uniform grid of dx = 0.8 was applied. The simulation was initialized by planting 10 solid seeds with random orientations in an isothermal domain with temperature of θ = -0.12. Other calculation parameters stayed the same as those in the numerical test until state otherwise. Periodic boundaries were applied at each wall of the domain. The simulation was kept on going for a total of 140,000 time steps to enable the dendrites to reach a rather mature growing stage i.e. coarsening.

4.1. Multi-dendrite growth and coarsening

Figure 3 shows the phase field simulation of a multi-dendrite growth case. The dendritic morphologies at t = 2000 dt, 8000 dt and 20000 dt were shown in figures 3a, 3b and 3c respectively. As time evolved, dendrites started to grow and then impinge together. Detailed features including side-branching of the high order arms could also be observed. Taking dendrite D_1 for example, the three coarsening mechanisms including small arm melting ("1"), groove advancement at the roots ("2") and coalescence between neighboring arms ("3") could be clearly revealed through the simulation, which strikes great similarity to that reported in literature [13].

4.2. The characteristic length of coarsening

To quantify the coarsening behavior of the dendrite growth, the characteristic length R i.e. the ratio between solid volume ($0 < \phi < 1$) and the surface area was calculated. Figure 4 shows the change of the characteristic length as solidification time evolved. To achieve a better illustration and comparison with existing theories, R^n was plotted as a function of time (t) instead of R.

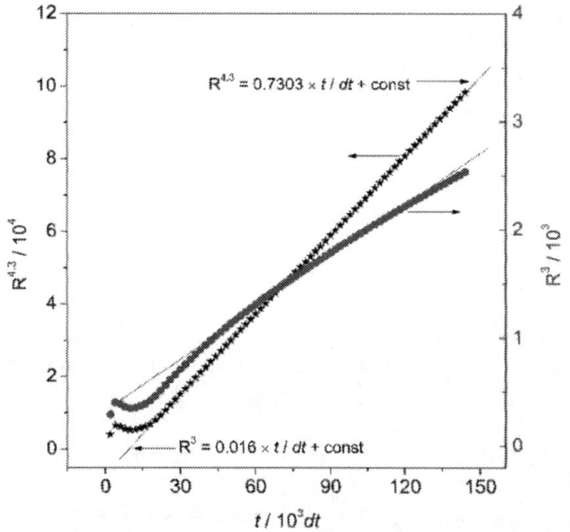

Figure 4. Evolution of the characteristic length R as time evolved during dendrite growth and coarsening. To achieve a better illustration R^n was plotted instead, together with the linear fits according to n = 3 and 4.3.

According to the LSW theory [14], the characteristic length changes in the following manner:

$$R_t^n - R_0^n = k_c t \tag{15}$$

where R_t^n is the characteristic length at time t and k_c is a rate constant related to alloy thermodynamic properties. Through numerical tests, it was found that the best fit for the current simulation occurred when $n = 4.3$ rather than 3, i.e. the commonly accepted value for the LSW theory. However, $n = 4.3$ was quite close to that obtained by Terzi and co-workers [15] through the recent high brilliance X-tomography experiment (in which $n = 4.4$). More evidence is thus required from both modeling and experiment to clarify this difference.

5. Conclusion

(1) A computing scheme with the ability of adaptive mesh refinement and parallel computing (i.e. Para-AMR) has been developed to solve the coupled phase field equations in 3-D. The whole algorithm was realized based on (a) a gradient criterion for grid/point tagging, (b) the automatic clustering algorithm for grid generation, and (c) parallelization i.e. dispatching of computing work based on the principle of space filling curve. Numerical tests revealed that the Para-AMR algorithm, without compromising any accuracy of the phase field simulation, can significantly reduce the computing time in a degree of about two orders of magnitude.

(2) 3-D phase field simulations of multi-dendrite growth were performed and dendrite growth and coarsening were investigated. The three coarsening mechanism including small arm melting, interdendritic groove advancement and coalescence between neighboring arms can be successfully revealed through the simulation. By fitting the growing law during dendritic coarsening in particular the formula

$$R_t^n - R_0^n = k_c t$$

we found that the best fit was achieved when $n = 4.3$ which was higher than that predicted by the classic LSW theory but very similar to that determined through the X-tomography experiments.

Acknowledgments

The authors would like to thank the Natural Science Foundation of China (Project No. 51205229), the Tsinghua University Initiative Scientific Research Program (20121087918) and the Newton International Fellowship Scheme for financial support, and the National Laboratory for Information Science and Technology in Tsinghua University for access to supercomputing facilities. The authors would also like to thank the open source software Visit, based on which the visualization of the simulation results were performed.

References

[1] Rowenhorst D J and Voorhees P W 2012 *Annu. Rev. Mater. Sci.* **42** 105
[2] Bray A J 2002 *Adv. Phys.* **51** 481
[3] Haxhimali T, Karma A, Gonzales F and Rappaz M 2006 *Nat. Mater.* **5** 660
[4] Wang J and Yang G 2008 *Acta Mater.* **56** 4585
[5] Berger M and Colella P 1989 *J. Comput. Phys.* **82** 64
[6] Guo Z, Mi J, and Grant P S 2012 *J. Comput. Phys.* **231** 1781
[7] Karma A 2001 *Phys. Rev. Lett.* **87** 115701
[8] Berger M and Rigoutsos I 1991 *IEEE Trans. Syst. Man Cybern.* **21** 1278
[9] Bader M 2013 Space-Filling Curves: An introduction with applications in scientific computing (Berlin: Springer)
[10] BoxLib https://ccse.lbl.gov/BoxLib/index.html
[11] Gropp W, Lusk E and Skjellum A 1999 Using MPI: Portable Parallel Programming with the Message-Passing Interface (Cambridge: MIT Press)
[12] Chapman B, Jost G and van der Pas R 2007 *Using OpenMP Portable Shared Memory Parallel Programming* (Cambridge: MIT Press)
[13] Limodin N, Salvo L, Boller E, Suery M, Felberbaum M, Gailliegue S and Madi K 2009 *Acta Mater.* **57** 2300
[14] Ardell A J and Ozolins V 2005 *Nature Mater.* **4** 309
[15] Terzi L, Salvo L, Suery M, Dahle A K and Boller E 2010 *Acta Mater.* **58** 20

IOP Conf. Series: Materials Science and Engineering **84** (2015) 012068 doi:10.1088/1757-899X/84/1/012068

An adaptive mesh method for phase-field simulation of alloy solidification in three dimensions

P C Bollada [1], P K Jimack[1] and A M Mullis[2]

University of Leeds: [1] School of Computing, [2] Institute of Materials Research

E-mail: p.c.bollada@leeds.ac.uk

Abstract. We present our computational method for binary alloy solidification which takes advantage of high performance computing where up to 1024 cores are employed. Much of the simulation at a sufficiently fine resolution is possible on a modern 12 core PC and the 1024 core simulation is only necessary for very mature dendrite and convergence testing where high resolution puts extreme demands on memory. In outline, the method uses implicit time stepping in conjunction with an iterative solver, adaptive meshing and a scheme for dividing the work load across processors. We include three dimensional results for a Lewis number of 100 and a snapshot for a mature dendrite for a Lewis number of 40 .

1. Introduction

This paper summarises our mathematical and numerical approach to solving the set of multi-scale, coupled, non-linear PDEs presented by alloy solidification using a phase field. In order to capitalise on routinely available parallel computing both on desktop PCs and high performance computing we adopt strategies that differ from that in which serial code is the end product. The major challenge is to take advantage of both adaptive grid and a multigrid solver in a parallel environment. A key strategy here is to use 'blocks'. Blocks are subsets of the computational domain containing a uniform grid and the adaptive grid strategy operates on the block level. Efficient communication between blocks is enhanced the larger the block, but efficient adaptive meshing is more effective with smaller blocks. We describe our approach to parallel implementation which rests on an ordering scheme akin to Morton ordering. In this approach communication at any level of grid refinement is minimised, but at a cost of communication across levels. In order to reduce communication between blocks we use a single 'ghost' layer of nodes in tandem with compact $3 \times 3 \times 3$ finite difference stencils for computing the PDE operators. Once again there is a balance to be struck since a double ghost layer is more expensive on communication but allows higher order accuracy in stencil operators. The multigrid solver we employ is the Full Approximation Scheme (FAS) which utilises Jacobi smoothing on each layer which, in turn, employs a pointwise Newton solver to approximate the non-linear algebraic terms. The Newton solver, being an extension of the Newton-Raphson method to systems of equations, becomes a generalisation of the Jacobi iterative method to non-linear systems. To find the pointwise Newton correction simply requires knowledge of the contribution of all the variables at any particular computational node. This is relatively easy to implement once the PDEs have been discretised. We employ (and adapt) the open source code PARAMESH to control the parallel communication between blocks so that once the FAS

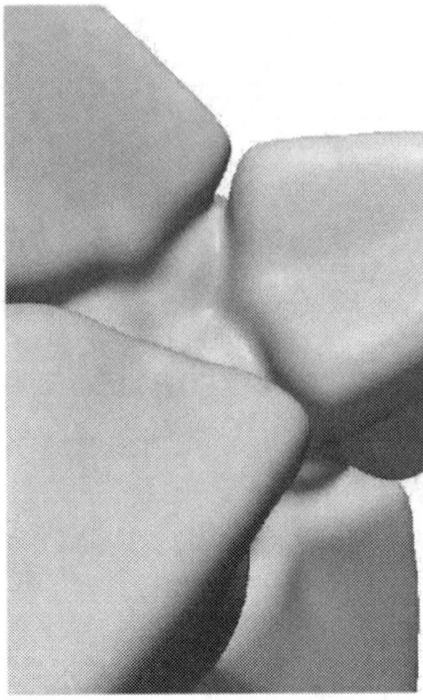

Figure 1. A close up of one of the arms of mature dendrite. An under cooling of $\Delta = 0.525$ at $Le = 40$ was used for this simulation. This took about three days on a 12 core desktop PC and represents a real growth time of about $0.6\mu s$ and size $0.25\mu m$ from an inital seed of about 6nm.

has been implemented the code is readily adapted to a variety of applications - not necessarily phase field. More details of this approach plus validation may be found in [1].

The particular phase field model we employ is an extension of [2], and is based on the three dimensional thermal- phase field model of [3] and two dimensional thermal-solutal phase field model of [4]. Fig. 1 shows a close up of one arm of a mature dendrite from a simulation in which the undercooling, $\Delta = 0.525$ and Lewis of 40. Such a dendrite does not differ in appearance to a simulation without coupling to a evolving temperature field, but a fully coupled model (with temperature field) becomes increasingly demanding on computing resources because of the stiffness of the resulting equations and extent of the temperature field. It is this fact that has generated the computing scheme detailed here.

Finally, we would like to draw the attention of researchers using an anisotropic phase field model in three dimensions to the equation given in Eq 19 which is significantly more efficient to compute in this form than its expanded equivalent of partial differential operators, resulting from, say, a **Maple** evaluation. It also aids the non-linear solver implementation detailed in Sec. 3.4.

2. Governing equations

The non-dimensional equations for the phase field, ϕ, the solute concentration, c and the dimensionless temperature, θ, are given via a specification of the free energy

$$F \equiv \int_V A(\mathbf{n})^2 \nabla\phi \cdot \nabla\phi + f(\theta, \phi)V \tag{1}$$

and the relations

$$\tau(c, \phi)A^2(\mathbf{n})\dot{\phi} = -\frac{\delta F}{\delta\phi} \tag{2}$$

$$\dot{c} = \nabla \cdot \left(K(\phi)\nabla\frac{\delta F}{\delta c} - \mathbf{j}\right), \tag{3}$$

IOP Conf. Series: Materials Science and Engineering **84** (2015) 012068 doi:10.1088/1757-899X/84/1/012068

$$\dot{\theta} \quad = \quad D_\theta \nabla^2 \theta + \dot{\phi}. \tag{4}$$

The solute diffusion parameter is given by

$$K = D_c \frac{(1-\phi)}{2}. \tag{5}$$

The parameter D_c is a diffusion constant, thus $K = 0$ in the solid ($\phi = 1$) and $K = D_c$ in the liquid ($\phi = -1$). D_θ is the temperature diffusion coefficient (assumed constant). The normal to the interface is given by

$$\mathbf{n} = \frac{\nabla \phi}{|\nabla \phi|}, \tag{6}$$

which is well defined around $\phi = 0$, and the anisotropy function for cubic symmetry (growth is preferred along the normals to the faces) is given for three dimensions by [3],

$$A(\mathbf{n}) \quad \equiv \quad A_0 \left[1 + \tilde{\epsilon}\left(n_x^4 + n_y^4 + n_z^4\right)\right] \tag{7}$$

where $\mathbf{n} = [n_x, n_y, n_z]$, $A_0 = 1 - 3\epsilon$, $\tilde{\epsilon} = 4\epsilon/(1 - 3\epsilon)$ and $\epsilon \approx 0.02$ governs the amount of anisotropy. The dimensionless relaxation time function is defined by

$$\tau(c, \phi) \equiv \frac{1}{Le} + Mc_\infty[1 + (1 - k_E)U], \tag{8}$$

where the Lewis number $Le = D_\theta / D_c$ and

$$U \equiv \frac{1}{1 - k_E}\left(\frac{2c/c_\infty}{1 + k_E - (1 - k_E)\phi} - 1\right). \tag{9}$$

Here k_E is the equilibrium partition coefficient, c_∞ is the far boundary condition for c. The anti-trapping current $\mathbf{j}$, appearing in the solute equation, Eq. 3, is prescribed by

$$\mathbf{j} = -\frac{1}{2\sqrt{2}}[1 + (1 - k_E)]U\dot{\phi}\mathbf{n}, \tag{10}$$

The profile of c exhibits a spike at the interface, which can present computational difficulties. Following [4], this is largely overcome by rewriting the solute equation using the variable U:

$$\left(\frac{1 + k_E}{2} - \frac{1 - k_E}{2}\phi\right)\frac{\partial U}{\partial t} = \nabla \cdot \left\{D_c \frac{1 - \phi}{2}\nabla U + \mathbf{j}\right\} + \frac{1}{2}[1 + (1 - k_E)U]\frac{\partial \phi}{\partial t}. \tag{11}$$

The physical temperature field, T, can be recovered by the relationship

$$\theta = \frac{T - T_M - mc_\infty}{L/C_p}, \tag{12}$$

where L and C_p are the latent heat of the phase transition and heat capacity respectively. The slope of the liquidus line is given by $m = ML/[C_p(1 - \kappa_E)]$ and T_M is the melting temperature of the alloy.

Finally the bulk free energy density is given by

$$f(\theta, \phi) \equiv \frac{\phi^2}{2}\left(\frac{\phi^2}{2} - 1\right) + \lambda(\theta + c_\infty U)\left(\phi - \frac{2\phi^3}{3} + \frac{\phi^5}{5}\right). \tag{13}$$

We solve the system of equations 2,4 and 11 plus initial (typically small) solid seed and far boundary conditions

$$\begin{aligned} \phi|_{far} &= -1 \\ U|_{far} &= 0 \quad (\equiv c|_{far} = c_\infty) \\ \theta|_{far} &= -\Delta \end{aligned} \tag{14}$$

where Δ is the given under-cooling. The equation for temperature is a standard diffusion equation with a heating term, $\dot{\phi}$, proportional to the solidification rate. The phase equations, though compact in variational form are complicated in PDE form. However, it is possible to write the resulting PDEs in a reasonably compact form convenient for computation.

$$\tau(c,\phi)A(\mathbf{n})^2\dot{\phi} = \frac{\partial^2\phi}{\partial x^i \partial x^j} g_{ij} - \frac{\partial f}{\partial \phi} \tag{15}$$

where $\tau(c,\phi)$ is given by Eq. 8, $\dfrac{\partial f}{\partial \phi}$ is given by

$$\frac{\partial f}{\partial \phi} = \phi^3 - \phi + \lambda(\theta + c_\infty U)(1 - 2\phi^2 + \phi^4), \tag{16}$$

and g_{ij} defined by

$$\begin{aligned} g_{ii} &\equiv (24X_i - 3)A^2 + (-48X_i^2\tilde{\epsilon} + 12X_i\tilde{\epsilon} - 40X_i + 4)A_0A + 16X_i(X_i\tilde{\epsilon}+1)^2A_0^2 \\ g_{ij} &\equiv \frac{\phi_{,i}\phi_{,j}}{g}\left[24A^2 + (-24X_i\tilde{\epsilon} - 24X_j\tilde{\epsilon} - 40)A_0A + (16(X_j\tilde{\epsilon}+1))(X_i\tilde{\epsilon}+1)A_0^2\right], \quad i \neq j. \end{aligned} \tag{17}$$

with $X_i \equiv \phi_{,i}^2/|\nabla\phi|^2$. However, for the purposes of discretisation we adopt the following form by defining

$$M_{ij} \equiv \frac{\partial\phi}{\partial x^i \partial x^j} - \frac{1}{3}\nabla^2\phi\delta_{ij} \tag{18}$$

so that the phase field equations Eq. 15 becomes

$$\tau(c,\phi)A(\mathbf{n})^2\dot{\phi} = \frac{1}{3}\delta_{ij}g_{ij}\nabla^2\phi + M_{ij}g_{ij} - \frac{\partial f}{\partial \phi} \tag{19}$$

The fully coupled system is then given by Eq 19, Eq. 11 and Eq. 4.

2.1. Parameter values

For the purposes of this paper we choose a selection of parameters to use as default values for the simulations below in Table 1.

3. Discretisation

The approach taken to discretisation is based upon a cell centred finite difference scheme, in that the nodes of the domain are located at the centre of cubic cells. and thus, we use the term 'node' and 'cell centre' inter changeably. One consequence of this is that there are no nodes on the domain boundary, thus making the use of Dirichlet boundary conditions non-trivial. The scheme makes use of the PARAMESH library to support mesh adaptivity in parallel [6, 7]. The meshes obtained by this approach take the form of an oct tree of regular blocks, within which the mesh is uniform, and it is the spatial discretisation on any one of these blocks that we discuss here.

Table 1. Table of parameter values used for the simulations in this paper.

Physical property	Symbol	value
Anisotropy	ϵ	0.02
Boundary concentration	Mc_∞	0.05
Equilibrium partition coefficient	κ_E	0.3
Dimensionless interface width	λ	2
Ratio of solute diffusivity to characteristic diffusivity	D_c	1.2534
Lewis number - D_θ/D_c	Le	40 and 100
Dimensionless Undercooling at the far boundary	Δ	0.25 to 0.80
Initial nuclear radius	R_0	5.0
Computational property	symbol	value
Finest grid size	Δx	0.195 to 0.78
Computation domain size	L^3	800×800×800

3.1. Temporal discretisation

For temporal discretisation we employ backward differential formula using two previous values (BDF2).

3.2. Spatial discretisation

Compact finite difference stencils (3×3×3), are used to discretise the first and second derivatives. Denoting these 27 points by **Q** and defining a generic 27 point Laplacian stencil, W_{abc}, around a point $\mathbf{p} = [i, j, k]$ by

$$\nabla^2 u|_{\mathbf{Q}} = \frac{1}{(\Delta x)^2} \sum_{a=-1}^{1} \sum_{b=-1}^{1} \sum_{c=-1}^{1} W_{abc} u_{\mathbf{p}+[a,b,c]} \tag{20}$$

where Δx is the physical distance between nearest neighbours, we can recover the 7 point Laplacian stencil, built from only the centre node, **p** and the 6 nearest neighbours ($a^2+b^2+c^2 = 1$) by setting the weights

$$W_{abc} = \begin{cases} -6 & a^2+b^2+c^2 = 0 \\ 1 & a^2+b^2+c^2 = 1 \\ 0 & otherwise \end{cases} . \tag{21}$$

However, this stencil is more prone to grid anisotropy than the following 27 point Laplacian stencil (see [5]), with weights

$$W_{abc} = \begin{cases} -128/30 & a^2+b^2+c^2 = 0 \\ 14/30 & a^2+b^2+c^2 = 1 \\ 3/30 & a^2+b^2+c^2 = 2 \\ 1/30 & a^2+b^2+c^2 = 3 \end{cases} \tag{22}$$

The discretisation of Eq 19 is achieved using the Laplacian stencil above and stencils for g_{ij} and M_{ij}. The latter have discretisations that do not have any contribution from the central node. This is, firstly, because g_{ij} is a rational polynomial function of first derivatives of ϕ and, secondly, we deliberately adopt the 7 point stencil for the Laplacian in the definition of M_{ij} for this purpose. For example, M_{11} and M_{12} at grid point, $\mathbf{p} = [i, j, k]$ are discretised

$$M_{11}|_{\mathbf{p}} \equiv \frac{2}{3\Delta x^2}(\phi_{i+1,j,k} + \phi_{i-1,j,k}) + \frac{1}{\Delta x^2}(\phi_{i,j+1,k} + \phi_{i,j-1,k} + \phi_{i,j,k+1} + \phi_{i,j,k-1}) \tag{23}$$

$$M_{12}|_{\mathbf{p}} \equiv \frac{\partial^2 \phi}{\partial x^1 \partial x^2}|_{\mathbf{p}} = \frac{1}{4\Delta x^2}\left(\phi_{i+1,j+1,k} + \phi_{i-1,j-1,k} - \phi_{i+1,j-1,k} - \phi_{i-1,j+1,k}\right) \tag{24}$$

IOP Conf. Series: Materials Science and Engineering **84** (2015) 012068 doi:10.1088/1757-899X/84/1/012068

3.3. Adaptive mesh and block tree structure

The domain is first divided into a number of mesh blocks each of which contains $N \times N \times N$ hexahedral cells, where we typically choose $N = 8$. We employ a domain of 800^3, which is large enough for Lewis numbers of the order of 100. We divide this domain into 4^3 blocks, so that when N=8, each cell is of size 25^3 and refer to this as level 1. The adaptive mesh strategy then imposes a hierarchical sub-division of some or all of these blocks, and their descendants, based upon an oct tree structure. This subdivision aims to concentrate cells where gradients of the oct tree variables are highest and to ensure neighbour blocks differ by at most one level. The finest grid we work with is at level 7, with a corresponding $\Delta x = 25/2^7 = 0.1953125$. We find that the minimum finest level necessary to obtain qualitatively reasonable results is level 5, corresponding to $\Delta x = 0.78125$.

In order to discuss further the tree structure of the blocks we denote any block by its label, i and its contents/properties, $B(i)$:

$$B(i) = [l, s, p, \mathbf{c}, \mathbf{x}] \tag{25}$$

where $l \in [1, n]$ is the level, $s \in [1, 8]$ is the sibling number (i.e. an index for which child of p the block is), p is the parent index, $c_i, i \in [1, 8]$ are the 8 child indices, and $\mathbf{x} = [x, y, z]$ is the Cartesian position coordinates of the block origin. Any one of these properties can be accessed by the block number, i. Some examples: $p(i)$ is the block number of the parent of block i; $\mathbf{x}(p(i))$ is the position of the parent's origin; $c_{s(i)}(p(i)) = i$ is an identity.

A complete specification of all the blocks in the oct tree is then specified by the list:

$$\mathbf{B} = \{B(i), i \in [1, B_N]\} \tag{26}$$

where B_N is the total block number. Moreover, a childless block, i, can be indicated by specifying, $\mathbf{c}(i) = 0$ and so the oct tree also can be specified by a listing of just the leaf blocks

$$\mathbf{B} = \{B(i), i \in [1, B_N] : \mathbf{c}(i) = \mathbf{0}\}. \tag{27}$$

Each block has uniform meshing and the adaptive strategy is further restricted by only allowing blocks at level n adjacent to blocks of $n-1, n$ and $n+1$. Thus, even though a block may be flagged for coarsening, this (latter) restriction often prevents this happening. Blocks are flagged for refinement if, for any point, $\mathbf{p}$, in the block, the following criterion is satisfied:

$$e \equiv \max\{e_\phi|\phi_\mathbf{p} - \phi_{\mathbf{p}-\mathbf{q}}|, e_U|U_\mathbf{p} - U_{\mathbf{p}-\mathbf{q}}|, e_T|T_\mathbf{p} - T_{\mathbf{p}-\mathbf{q}}|\} > \eta, \tag{28}$$

where we use, for tolerance, $\eta \sim 1$ and

$$|\phi_\mathbf{p} - \phi_{\mathbf{p}-\mathbf{q}}| \equiv \sqrt{(\phi_\mathbf{p} - \phi_{\mathbf{p}-[1,0,0]})^2 + (\phi_\mathbf{p} - \phi_{\mathbf{p}-[0,1,0]})^2 + (\phi_\mathbf{p} - \phi_{\mathbf{p}-[0,0,1]})^2}, \tag{29}$$

etc. Typically the weights, e_ϕ, e_U and e_T are chosen to sum to unity. If $e < 0.1\eta$, the block is flagged for derefinement.

For communication between blocks the PARAMESH implementation allocates a block of dimension $(N+2) \times (N+2) \times (N+2)$. The first and last cells in each dimension are ghost cells - an update function may be called at any time in order to populate these guard cells with the corresponding values from the interior of each of the neigbouring blocks. The application of a discrete stencil on any block requires access to neighborouring blocks via the guard cell nodes. When the neighbouring block is coarser the guard cell of the coarse block is found by a weighted average of the 8 surrounding coarse nodes. Using a tri-linear function of x, y and z, gives the weightings, in order of nearest neighbours first

$$\mathbf{w} = \left[\frac{27}{64}, \frac{9}{64}, \frac{9}{64}, \frac{9}{64}, \frac{3}{64}, \frac{3}{64}, \frac{3}{64}, \frac{1}{64}\right]. \tag{30}$$

The process is known as prolongation. The reverse process, of finding a guard cell value for a coarse block when one or more neighbours is refined is known as restriction and is the simple average of the eight nearest, one level finer, cell centres. Both operations, restriction and prolongation using Eq. 30, are also required for multigrid as detailed in the next section.

3.4. Nonlinear smoother

The non-linear system of algebraic equations can be written using the generic vector notation $\mathbf{v}_{\mathbf{p}}^{n+1} \equiv [\phi_{\mathbf{p}}^{n+1}, \theta_{\mathbf{p}}^{n+1}, U_{\mathbf{p}}^{n+1}]$ by

$$\mathbf{A}_{\mathbf{P}}(\mathbf{v}_{\mathbf{Q}}^{n+1}) = \mathbf{0} \tag{31}$$

for each node, $\mathbf{p}$, in the grid and $\mathbf{Q}$ indicates coupling between points $\mathbf{p}$ and neighbouring nodes. Using an iteration method, with $\mathbf{v}_{\mathbf{p}}^{n+1}$ approximated by $\mathbf{v}_{\mathbf{p}}^{n+1,m}$, we define the defect

$$\mathbf{d}_{\mathbf{p}}^{n+1,m} = -\mathbf{A}_{\mathbf{p}}(\mathbf{v}_{\mathbf{Q}}^{n+1,m}). \tag{32}$$

The pointwise Newton update for this iteration is given by

$$\mathbf{v}_{\mathbf{p}}^{n+1,m+1} = \mathbf{v}_{\mathbf{p}}^{n+1,m} - \omega \tilde{\mathbf{d}}_{\mathbf{p}}^{n+1,m}, \tag{33}$$

where $\tilde{\mathbf{d}}_{\mathbf{p}}^{n+1,m}$ is found by solving the 3×3 system

$$\mathbf{J}_{\mathbf{p}}^{n+1,m} \tilde{\mathbf{d}}_{\mathbf{p}}^{n+1,m} \equiv \mathbf{d}_{\mathbf{p}}^{n+1,m} \tag{34}$$

with the 3×3 Jacobian matrix defined by

$$\mathbf{J}_{\mathbf{p}}^{n+1,m} \equiv \frac{\partial \mathbf{d}_{\mathbf{p}}^{n+1,m}}{\partial \mathbf{v}_{\mathbf{p}}^{n+1,m}}. \tag{35}$$

In practice we typically select, $\omega \approx 0.9$ and find that off diagonal terms of $\mathbf{J}_{\mathbf{p}}^{n+1,m}$ are not essential to obtain a convergent iteration.

3.5. Multigrid and parallel implementation

We supplement the Jacobi smoother with the Full Approximation Scheme (FAS) as developed in [8]. Our parallel implementation requires communication between blocks both on the same level and between levels. We identify communication across any one level as being as being more costly and so we implement a depth first traversal of the blocks and then, using this numbering, divide the work load at each level in turn between all processors. For a uniform mesh this strategy results in a near optimum allocation to cores. For adaptive meshes the communication on any level is also optimum, but communication between levels is compromised. See [1] for details.

4. Results

We demonstrate the code with sets of parameters in both 2D and 3D and with two Lewis numbers. The higher Lewis number, more undercooling and higher dimension combine to give a severe challenge to a desktop computer. Ultimately, for higher Lewis numbers, HPC resources are necessary with this approach. This is because the problem becomes stiff and the time step size required for stability becomes smaller and the domain required by the temperature field also becomes larger with larger Lewis number. Fig. 2 shows the evolution of tip radius for 2D $Le = 100$ and, for both $Le = 40$ and $Le = 100$, in 3D. Here, the capillary length,

IOP Conf. Series: Materials Science and Engineering **84** (2015) 012068 doi:10.1088/1757-899X/84/1/012068

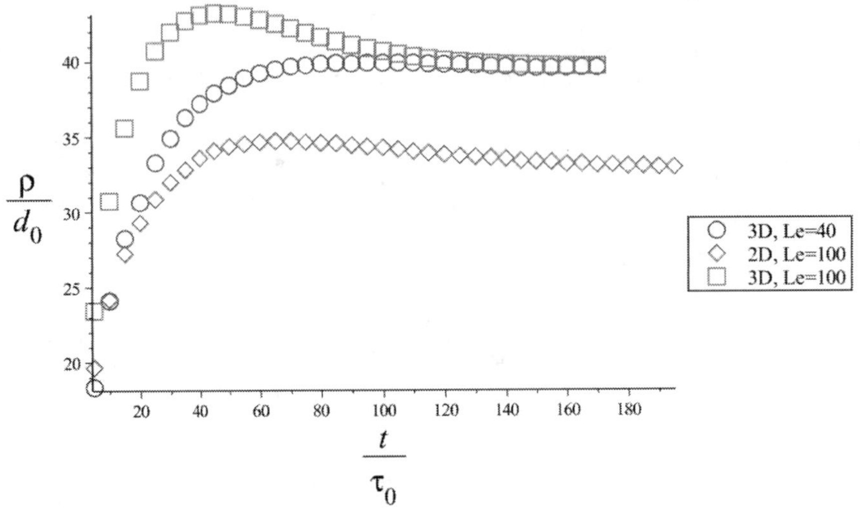

Figure 2. The tip radius for an undercooled dendrite, $\Delta = 0.325$ and $Le = 100$ and $Le = 40$

$d_0 \equiv \Gamma T_M C_p/L^2$, where Γ is the Gibbs Thompson coefficient; T_M the melting temperature; C_p heat capacity; and L the latent heat of fusion. The characteristic time, $\tau_0 = D_c W_0^2/\tilde{D}$ where $\tilde{D} = 10^{-9}\text{m}^2/\text{s}$, $D_c = 1.25$ and interface width $W_0 = \lambda d_0/a_1 = 2.26 d_0$. In our simulation we set both W_0 and τ_0 to unity. A larger domain is needed for the the 2D simulations due to the extent of the temperature profile in 2D for a given Lewis number. So despite the large increase in degrees of freedom in 3D there is some compensation due to this feature. These simulations were run on 12 core desktop PCs and typically take 3 or 4 weeks to achieve steady state. Note that the $Le = 40$ achieves steady tip radius after about $t/\tau_0 = 100$ but at $Le = 100$ steady state is achieved after $t/\tau_0 = 180$. This is generally a feature of the higher lewis number and is an additional significant reason for the difficulty of computing higher Lewis numbers.

5. Conclusion

We have presented a numerical method that permits computation in three dimensions of fully coupled alloy solidification of dendritic growth at the meso scale. The model is restricted to dilute alloys in order to take advantage of a model formulation that compensates for a larger than realistic liquid-solid interface – anti solute trapping. Despite this we opt to use a near realistic value for the interface in order to get good resolution of the tip radius. This is reflected in our parameter, $\lambda = 2$, which corresponds to approximately twice a realistic interface width. The computational methods we employ are summarised as follows:

- finite difference compact stencils
- adaptive mesh
- implicit time stepping
- non-linear Newton-multigrid solver
- parallel code and division of labour across cores

The main difficulty of the above method, partly overcome, is that combining adaptive meshing and multigrid makes it very difficult to allocate an even loading to the parallel cores.

A realistic Lewis number for typical alloy solidification is $Le \sim 10,000$, which corresponds to the ratio in length scales between heat and solute diffusion. We here simulate the model at comparatively modest Lewis numbers, $Le = 40$ and 100, to demonstrate the method. These being accessible on parallel desktop machines within a few weeks of computation time. Early results suggest that not only run-times are longer for higher Lewis numbers but physical (model) time is longer for steady state behaviour to be established.

References

[1] P.C. Bollada, C.E. Goodyer, P.K. Jimack, A.M. Mullis and F.W. Yang 2015 Three dimensional thermal-solute phase field simulation of binary alloy solidification *Journal of Computational Physics*, vol 287, pp 130 – 150

[2] A. M. Mullis, C. E. Goodyer, and P. K. Jimack 2012 Towards a Three-Dimensional Phase-Field Model of Dendritic Solidification with Physically Realistic Interface Width. *Transactions of the Indian Institute of Metals*, vol 65(6): pp 617–621

[3] Alain Karma and Wouter-Jan Rappel 1996/1997. Phase-field simulation of three-dimensional dendrites: is microscopic solvability theory correct? American Crystal Growth 1996 and Vapor Growth and Epitaxy. *Journal of Crystal Growth*, vol 174(14): pp 54 – 64

[4] J. C. Ramirez, C. Beckermann, A. Karma, and H-J J. Diepers May 2004 . Phase-field modeling of binary alloy solidification with coupled heat and solute diffusion. *Physical review. E, Statistical, nonlinear, and soft matter physics*, vol 69(5 Pt 1)

[5] W. F. Spot and G. F. Carey 1996 A high-order compact formulation for the 3d poisson equation. *Numer. Methods Partial Differential Eq.*, vol 12: pp 235 243

[6] Peter MacNeice, Kevin M. Olson, Clark Mobarry, Rosalinda de Fainchtein, and Charles Packer 2000. Paramesh: A parallel adaptive mesh refinement community toolkit. *Computer Physics Communications*, vol 126(3): pp 330 – 354

[7] K. Olson 2006. Paramesh: A parallel adaptive grid tool. in parallel computational fluid dynamics 2005: Theory and applications. *ed A. Deane et al. (Elsevier)*

[8] A. Brandt 1977. Multi-level adaptive solutions to boundary-value problems. *Math. Comp.*, vol 31: pp 333 – 390

IOP Conf. Series: Materials Science and Engineering **84** (2015) 012069 doi:10.1088/1757-899X/84/1/012069

Dual-scale phase-field simulation of Mg-Al alloy solidification

A Monas[1], O Shchyglo[1], D Höche[2], M Tegeler[1] and I Steinbach[1]

[1] Interdisciplinary Centre for Advanced Materials Simulations (ICAMS), Ruhr University Bochum, Universitätsstraße 150, Bochum, 44801, Germany

[2] Institute of Materials Research, Helmholtz-Zentrum Geesthacht, Max-Planck-Straße 1, Geesthacht, 21502, Germany

E-mail: `alexander.monas@rub.de`

Abstract. Phase-field simulations of the nucleation and growth of primary α-Mg phase as well as secondary β-phase of a Mg-Al alloy are presented. The nucleation model for α- and β-Mg phases is based on the "free growth model" by Greer et al.. After the α-Mg phase solidification we study a divorced eutectic growth of α- and β-Mg phases in a zoomed in melt channel between α-phase dendrites. The simulated cooling curves and final microstructures of α-grains are compared with experiments. In order to further enhance the resolution of the interdendritic region a high-performance computing approach has been used allowing significant simulation speed gain when using supercomputing facilities.

1. Introduction

Magnesium alloys are among the most promising structural materials for lightweight applications due to their low density compared to aluminum and steel. In Mg technology the most applied alloying element for technical alloys like AZ91, AM50 (cast) or AZ31 (wrought) is aluminum. The mechanical properties of such Mg-Al alloys are primarily determined by their micro-structure which typically consists of two main phases: the Mg-rich hexagonal close packed (HCP) α-phase and a near stoichiometric $Mg_{17}Al_{12}$ β-phase, and further minor secondary phases due to third alloying elements. From a scientific point of view it remains a challenge to simulate the dendritic solidification structure of the primary HCP-α-phase. In particular the anisotropy function of the interface energy between liquid and the α-phase which determines the growth structure of the dendrite, remains under debate [1, 2]. A second question is the morphology of interdendritic eutectic, which is decisive for the corrosion properties of the alloy [3]. In the present work we apply the phase-field method [4, 5, 6, 7] to simulate the microstructure evolution of Mg-Al alloys during solidification. We apply the anisotropy function as given in [8] which produces a dendrite trunk normal to the basal plane of the HCP crystal. While several authors reported 2-dimensional and 3-dimensional phase-field simulations eg.[8, 9, 2] of Mg-alloy solidification, a new challenge is to reveal the eutectic structure, which is assumed to happen in a divorced mode, but is at least one order of magnitude finer than the primary dendritic structure.

2. Experimental details

The binary Mg-Al test alloys were prepared by permanent mold indirect chill casting as described in [10]. The alloys were melted in an electric resistance furnace under a protective atmosphere of 2% SF_6 and Ar. The composition was chosen to be Mg - 9 wt% Al. After mixing, the melt was held at 1020 K for 10 min and then poured into a standard steel mold preheated to 933 K. In addition, the mold containing the melt and a temperature sensor were held at 933 K for 15 min and lowered into water at a rate of 10 mm/s until the top of the melt was in line with the cooling water level. The size of the cast ingot is 250 mm in length and 105 mm in diameter.

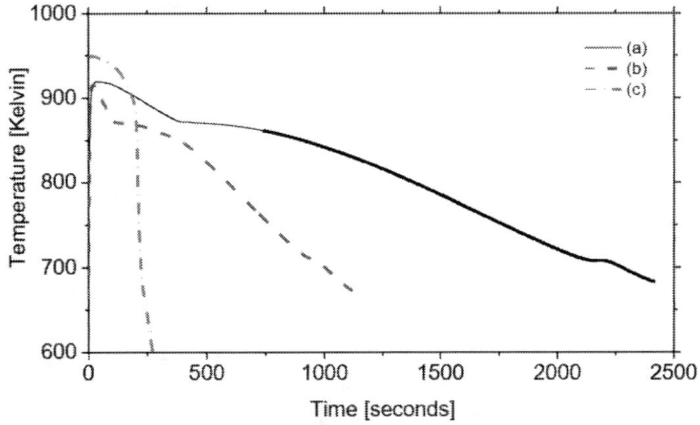

Figure 1. Experimentally measured cooling curves of Mg-9wt.%Al: (a) - furnace controlled cooling, (b) - air cooling, and (c) - submersion in water.

3. Phase-field model

In this study we apply the multi-phase-field method on a meso-scale with locally linearized phase diagram information [8]. In this method each grain of a distinct crystallographic phase is attributed by its own phase-field variable $\phi_p(\vec{x}, t) \in [0, 1]$. Three distinct crystallographic phases are considered: liquid melt, Mg-rich α-phase and a near stoichiometric $Mg_{17}Al_{12}$ β-phase, where each solid grain is attributed by an identifier of the crystallographic phase and its orientation in space (for details see [5, 8]). For the interface between melt and the α-phase an anisotropic interface energy $\sigma_{pq} = \sigma_{pq}(\vec{n})$ as a function of the interface normal $\vec{n}$ with respect to the crystallographic orientation of each grain is used [8]. The interface energy between the melt and β-phase as well as between the solid α- and β-phases assumed isotropic. In order to consider the effect of the external cooling rate on nucleation and growth of the solid phases we solve the heat balance in average over the whole domain with n_{xyz} discrete elements:

$$\rho c_p \dot{T} = \sum_{n_{xyz}} \sum_{p \neq q} \dot{\psi}_{pq} L_{pq} + \dot{Q}_{ext}, \tag{1}$$

where T is the temperature of the system, $\dot{Q}_{ext}$ is the heat extraction rate, ρ and c_p are the mass density and the specific heat which are taken constant throughout the simulation. $\dot{\psi}_{pq}$ are the local transformation rates between pairs of phases which sum up to the total transformation rate of individual phases $\dot{\phi}_p = \sum_{q \neq p}^{N} \dot{\psi}_{pq}$ in the interface or a junction point with N phases (for details see [11]). L_{pq} is the latent heat associated with the individual transformations between phases p and q. In the present study we treat all possible transformations between liquid $\leftrightarrow \alpha$, liquid $\leftrightarrow \beta$ and $\alpha \leftrightarrow \beta$ phases.

Table 1. Parameters of the linearized Mg-Al phase diagram as shown in Fig. 2

intersection concentration	$C_{\alpha\leftrightarrow liq.}$	0.00 %	slope (liquidus)	$m_{liq.\leftrightarrow\alpha}$	-6.90 K/%
intersection temperature	$T_{\alpha\leftrightarrow liq.}$	923.15 K	slope (solidus)	$m_{\alpha\leftrightarrow liq.}$	-27.58 K/%
intersection concentration	$C_{\beta\leftrightarrow liq.}$	45.64 %	slope (liquidus)	$m_{liq.\leftrightarrow\beta}$	3.50 K/%
intersection temperature	$T_{\beta\leftrightarrow liq.}$	760.60 K	slope (solidus)	$m_{\beta\leftrightarrow liq.}$	9.00 K/%
intersection concentration	$C_{\alpha\leftrightarrow\beta}$	34.20 %	slope (solvus)	$m_{\alpha\leftrightarrow\beta}$	22.40 K/%
intersection temperature	$T_{\alpha\leftrightarrow\beta}$	1301.84 K	slope (solvus)	$m_{\beta\leftrightarrow\alpha}$	-103.00 K/%

4. Phase diagram

In order to obtain the phase diagram along with the exact transformation temperatures and slopes for our calculations the CalPhaD (Calculation of Phase Diagrams) method was used via the Thermocalc TCS 4.0 database TCBIN. For convenience the obtained phase diagram was linearized for the simulations as it is shown in Fig. 2

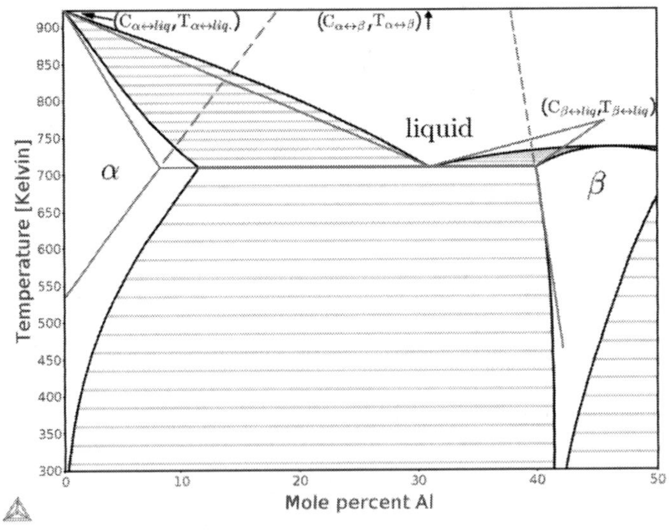

Figure 2. Mg-Al binary phase diagram with linearization (red).

All slopes but $m_{liq.\leftrightarrow\alpha}$ and $m_{\alpha\leftrightarrow liq.}$ are taken at the eutectic temperature of 709.4K. Solidus- and liquidus-slopes are the slopes between the points (923.15K; 0.0%Al) and (709.4K; 31.0%Al) [eutectic point] or (709.4K; 11.6%Al) respectively.

This particular version alters the point of maximum solubility in the α from 11.6% to 7.75% to match the area of the original *liquid* $\leftrightarrow$ α two-phase area in the phase diagram. Thus the solidus slope ($m_{\alpha\leftrightarrow liq.}$) was adjusted from the original -18.43 K/% to -27.58 K/%. Also $C_{\alpha\leftrightarrow\beta}$ and $T_{\alpha\leftrightarrow\beta}$ were adjusted to retain $m_{\alpha\leftrightarrow\beta}$. Consequently the original $\alpha \leftrightarrow \beta$ intersection-point (1229.84K ; 34.83%Al) was adjusted to (1301.84K ; 34.20%Al). All of the used values for the characterization of the phase diagram are summarized in table 1.

Table 2. Nucleation model parameters

particle density for α-phase nucleation K_α	$0.2 \cdot 10^{14} m^{-3}$
particle density for β-phase nucleation K_β	$1.5 \cdot 10^{16} m^{-3}$
a of the normal size-distribution (mean)	$-1.5 \cdot 10^{-7} m$
b of the normal size-distribution (standard deviation)	$1.0 \cdot 10^{-7} m$

5. Nucleation model

In order to capture the dependence of the nucleation density on the cooling rate we employ the so-called "free growth model" by Greer and coworkers [13]. This model is based on a given distribution of seed particles of different sizes. Assuming perfect wetting conditions between the primary α-phase and the seed particles, the hemispheric cap of the α-phase on the seed particles defines the critical undercooling for free growth of the nucleating phase which is inversely proportional to the size of the seed particles [12]. In our simulations we start from a random distribution of the particles with a given density and size distribution over the entire simulation domain initially filled with liquid melt. The utilized normal distribution is given by

$$N(d) = K \cdot \frac{1}{b\sqrt{2\pi}} exp\left[-(d-a)^2/2b^2\right]. \quad (2)$$

Each particle is assigned an effective size 'd' which determines the critical undercooling for its activation. The resulting number of particles is typically higher than the actual number of the activated particles (active nucleation sites). Activation happens if a grid cell, which contains a particle, reaches it's critical undercooling, i.e. if the difference between actual temperature and melting temperature dependent on the local alloy concentration exceeds the local critical undercooling. A similar model has also been applied by Eiken et al. [8]. The parameters of the utilized nucleation model are summarized in table 2. Note, that only the tail of a normal distribution is used, thus a is negative, but only seeds with a positive size are created. We also assume a significantly higher nucleation density of the β-phase in the melt channels between the primary α-phase dendrites.

6. Simulation results

The simulations are performed on two different scales. Although, the primary solidification in three-dimensional domains can be investigated in order to reveal the effect of the cooling rate on the nucleation density of α-phase dendrites (see Fig. 3), it is computationally demanding to study the entire solidification process in 3D over the temperature range of more than $200K$ even with high cooling rates. In contrast, in the 2D simulations it is possible to investigate the entire solidification process starting from the primary α-phase dendritic solidification and continuing with the eutectic solidification of the secondary β-phase upon cooling below the eutectic point.

It shall be noted, that on a scale of primary α-phase dendrites the eutectic structure in the interdendritic region cannot be resolved and as the result the secondary β-phase builds a bulk structure between the primary dendrites as can be seen in Fig. 4. In order to investigate eutectic growth in detail we zoom into the channel between two primary dendrites upon reaching the eutectic temperature. Here we select the scale of the simulations by a factor of 5 finer than the scale of primary α-phase dendritic solidification simulations. The heat extraction rate is assumed constant, which relates to an applied cooling rate (neglecting release of latent heat) $CR = \frac{\dot{Q}}{\rho c_p}$. We note that this cooling rate is different from the cooling rate evaluated from the experiments, due to the difference in definition. We will compare both entities by comparing

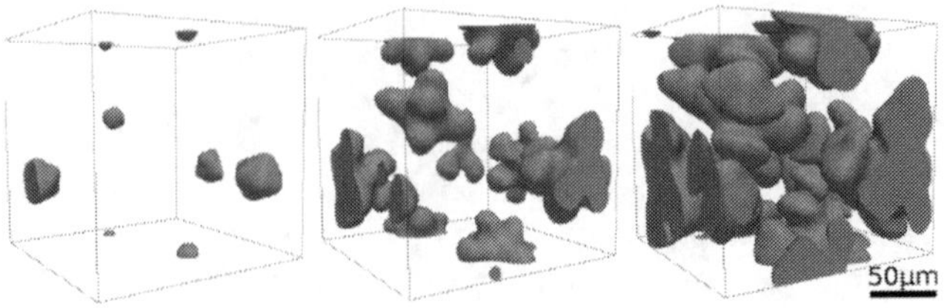

Figure 3. Results of a 3D simulation of primary α-phase nucleation and growth.

the solidification time (see below). All simulations of primary solidification are performed in a domain with periodic boundary conditions initially filled with homogeneous melt. The simulations of divorced eutectic growth start from a concentration profile taken around the melt channel of a 2D simulation of primary α-phase growth. Then a second particle distribution for β-phase nucleation is generated using the parameters given in Table 2.

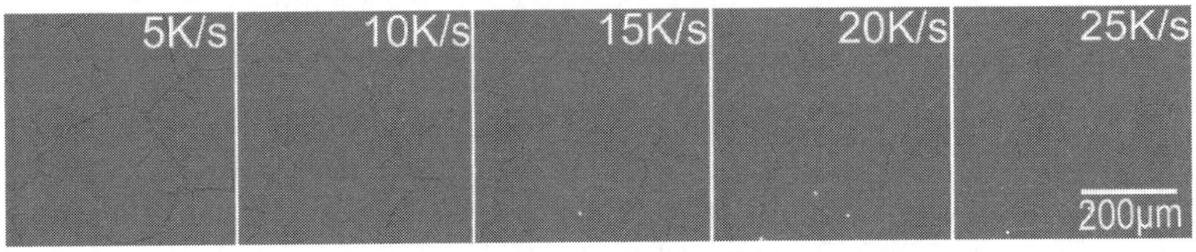

Figure 4. 2D simulated microstructures of Mg-Al alloy solidified at different cooling rates.

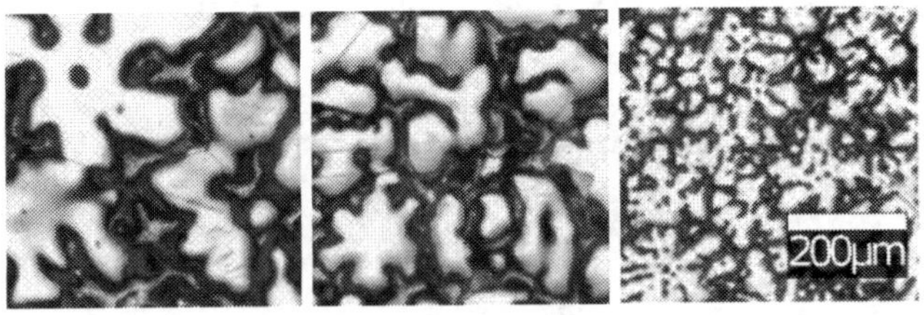

Figure 5. Metallographic images of etched as cast Mg-Al microstructures for different cooling rates. The Mg content is 9 wt.%. The applied cooling rates are (a) furnace controlled cooling, (b) air cooling and (c) submersion in water. The temperature-curves for these experiments are shown in Fig. 1

Simulations of consecutive α- and β-phase nucleation and their divorced eutectic growth were carried out in 2D. The applied cooling rates were varied from $5K/s$ to $25K/s$. The resulting micro-structures are shown in Fig. 4. The solidified microstructure consists of a primary α-phase dendrites (blue) surrounded by the thin channels of β-phase (orange). The volume fraction of the β-phase is about 5% in all simulations. A clear trend of increasing α-phase grain density with increasing cooling rate can be observed which is in agreement with experimental observations (see Fig. 5). This behavior is achieved naturally by the implementation of a free growth model proposed by Greer et al. [13] which allows to reproduce the recoalescence effect responsible for the increased nucleation density at higher cooling rates. The corresponding simulated cooling curves are shown in Fig. 6.

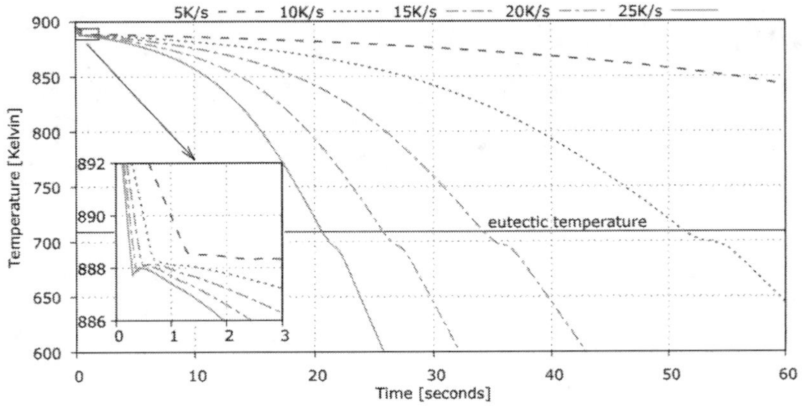

Figure 6. Simulated cooling curves.

The solidification experiment with water cooling takes approximately 20 seconds from fully liquid to fully solid which matches our simulations with an applied cooling rate of 25 K/s. Thus the simulated microstructure shown in Fig. 4(e) corresponds to the experimentally obtained microstructure shown in Fig. 5(c). Comparison between simulation and experiment shows good agreement in grain density and connectivity. The characteristic grain-size in the simulation is about 150 μm which is in correspondence with the experiment for the same estimated cooling rate. Due to the prohibitive computational expense of slower solidification conditions, the lowest simulated cooling rate was 5 K/s and takes about 110 seconds of simulated time to fully solidify. The experimental results for air or furnace controlled cooling were much slower taking 600 and 1700 seconds respectively. However a clear trend of coarser grains with decreasing cooling rate can be observed.

Another important factor is the closed shell β-phase formation which is responsible for increasing the corrosion-resistance of Mg-Al-alloys. Thus good correspondence of our simulation results with experimental observations allows for computationally assisted design of corrosion resistant Mg-alloys. Of course there is a trade off between fine grain structure (better mechanical properties) and low connectivity of primary dendrites (better corrosion properties). The balance between mechanical and corrosion properties can be further tuned and investigated using the proposed method.

In order to investigate in detail the β-phase nucleation and growth kinetics and the formation of a divorced eutectic microstructure with the tertiary α-phase nucleation in the residual melt these simulations were performed in 3D. The results of a divorced eutectic solidification simulation are shown in Fig. 7.

The simulations starts with a residual melt channel with the eutectic composition of 31% and nearly eutectic α-phase at the bottom and the top of the simulation volume. This system is cooled down starting at 712.0 K, slightly above the eutectic temperature. The α-phase is still slowly growing on this scale. As soon as the undercooling is sufficient, nucleation automatically takes place and generates a large number of β-grains in the residual melt. The high nucleation density of $3 \cdot 10^{13}$ particles per m^3 was chosen because it is expected for the particles to be aggregated in the residual melt. With an abundant supply of nucleation sites and our chosen cooling rate of 5 K/s we get around 50 nuclei within the channel. A strong tendency of β-phase growing towards and at the α-liquid-interface can be observed. This is because Al is segregated strongly by the growing α-phase into the melt close to the interface, since the solubility for Al in α is only 11.6% at the eutectic temperature of 709.4 K. This leads to a higher concentration

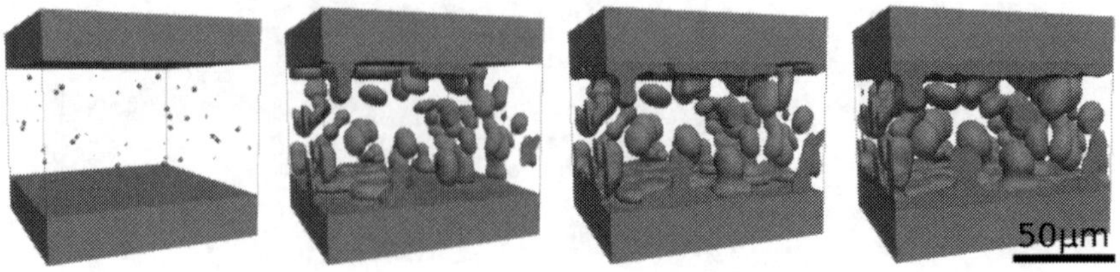

Figure 7. Zoomed in region of the residual eutectic melt channel in between two α-phase (blue) dendrites at different time-steps. Undercooling causes the β-phase (orange) to nucleate at randomly chosen positions in the melt (a) and grow (b-c) until the α-phase boundary is fully covered (d). Then residual melt is depleted and the α-phase can nucleate and grow again.

of Al in the melt close to the slowly growing α-phase. The β-phase, consisting of about 40% Al grows preferably in the area of the increased concentration. Due to this circumstances a rapid, full enclosure of the α-phase by the β-phase can be observed. Thus, further growth of the α-phase is prohibited. Continuing β-phase growth depletes the concentration in the residual melt. It is impossible for the melt at that state to fully solidify as β-phase due to the lack of Al. This leaves the system in a state where a tertiary α-phase nucleation is necessary to fully solidify. Investigations of this behavior will be presented elsewhere.

7. Outlook

We have presented 2D simulations of consequent α- and β-phase evolution, shown in Fig. 4. 3D simulations of the same accuracy will require about two orders of magnitude more memory and computational power, which is currently not feasible on a single compute node.

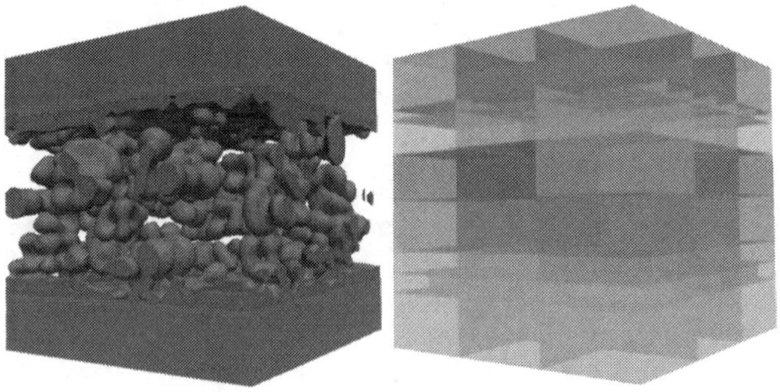

Figure 8. 3D benchmark simulation of the eutectic solidification in α-channel shown in Fig. 7. The domain is subdivided into hundreds of blocks shown to the right with the rank-dependent color-coding.

Therefore, distributed-memory parallelism based on block-structured domain decomposition has been implemented in OpenPhase, which will be used for further 3D simulations on large grids. Fig. 8 shows a benchmark-simulation of eutectic solidification in α-channels. The computational domain of $300 \times 300 \times 300$ grid points is subdivided into hundreds of blocks that are assigned to 128 processors. As the phase-field values can only change in the interface regions, OpenPhase only calculates the updates of the phase-fields in these regions. While this approach drastically

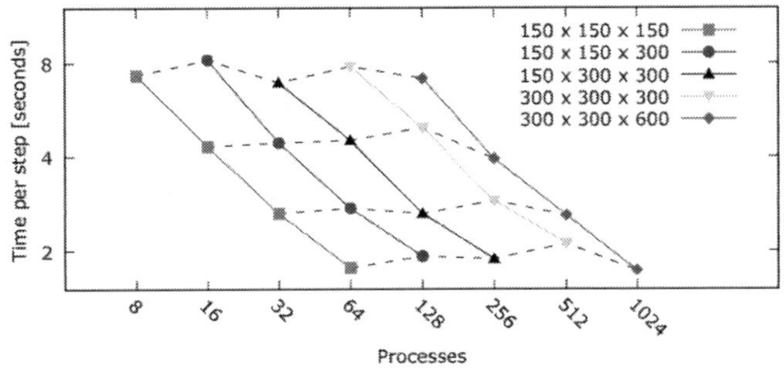

Figure 9. Benchmark for eutectic solidification in α-channels showing combined strong scaling (solid lines) and weak scaling (dashed lines) results. The colors and symbols represent the different simulated domain sizes.

improves the performance, it leads to a very inhomogeneous distribution of the computational load and requires sophisticated load balancing techniques. An over-decomposition of the computational domain allows a graph-partitioner to assign blocks to processors, so that in the worst case the load imbalance is bounded by the computation time on the most expensive block. Therefore, expensive blocks are divided into multiple parts, when the implemented work load model predicts a benefit for the wall clock time for the next time steps. The system is rebalanced at runtime, whenever it is needed. Figure 9 shows performance results on the supercomputer JuRoPA at Forschungszentrum Jülich with up to 1024 processors. Excellent weak scaling is observed: it can be seen that the computation time does not change significantly when the total work load per process is kept constant. This allows the realization of large scale simulations utilizing a parallel cluster without increasing the wall time. For primary solidification perfect weak scaling results were achieved with up to 2048 processes on grids with up to $800 \times 800 \times 1600$ grid points. With this new possibilities of large-scale simulations, we hope to set up a full 3D solidification simulation analogous to the 2D calculations shown in Fig. 4. Also, due to an increased resolution due to higher count of grid points, a simulation can be carried out resolving both: the primary as well as the eutectic solidification on their different length-scales. For a realistic setup a grid of about $600 \times 600 \times 600$ grid points is needed. Thus the time needed for one timestep is estimated to be about 2 seconds with 4096 processes. A calculation of 25 seconds real time would take approximately 11.5 days.[14]

8. Summary

In this paper we presented an approach to simulate the solidification of Mg-Al alloys. A phase diagram based on Thermocalc TCS 4.0 database TCBIN was linearized and adjusted for consistency. The "free-growth" model was implemented into OpenPhase for realistic nucleation behavior of both α- and β-phases. 3D simulations of the first stage of primary solidification were carried out. A full solidification cycle was simulated in 2D including a primary α- as well as secondary β-phase nucleation and growth in one simulation, with close to 200 K of cooling in between the two events. Resulting cooling curves as well as simulated microstructures were compared to experiments. Remarkable agreement has been obtained for characteristic grain size and β-phase closed shell formation. Due to the insuficient resolution a dual-scale approach was employed to model an interdendritic region of eutectic melt prior to final solidification. Divorced eutectic behavior was reproduced showing a need for tertiary nucleation of α-phase due to full shielding of primary dendrites after a short period of time. The possibility for future highly parallelized computations of large scale 3D domains with high resolution at different scales was implemented and tested. Benchmarks show excellent weak scaling properties.

9. Acknowledgements

The authors (AM, OS and IS) would like to thank the Korea Institute of Materials Science (KIMS) for support under the collaboration project "Phase-field simulation of Microstructure Evolution in Ni-based Superalloys and Light-weight alloys". IS and MT like to thank the German Research Foundation (DFG) for support under STE 1116/13-1. DH would like to thank the Helmholtz-Association for financial support.

References

[1] Sun D Y, Mendelev M I, Becker C A, Kudin K, Haxhimali T, Asta M, Hoyt J J, Karma A and Srolovitz D J 2006 *Phys. Rev. B* **73** 024116
[2] Wang M, Xu Y, Zheng Q, Jung T, Wu S and Chawla N 2014 *Met. Mat. Trans. A* **45** 2562
[3] Zhao M C, Liu M, Song G and Atrens A 2008 *Corros. Sci.* **50** 1939
[4] Chen L-Q 2002 *Annu. Rev. Mater. Res.* **32** 113-140
[5] Steinbach I 2009 *Mod. Sim. Mater. Sci. Eng.* **17** 073001-31
[6] Steinbach I 2013 *Annu. Rev. Mater.* **43** 89-107
[7] Steinbach I 2013 *JOM* **65** 1096
[8] Böttger B, Eicken J and Steinbach I 2006 *Acta Mater.* **54** 2697-2704
[9] Böttger B, Eicken J, Ohno M, Klaus G, Fehlbier M., Schmid-Fetzer R, Steinbach I and Bührig-Polaczek A 2006 *Adv. Eng. Mater.* **8** 24147
[10] Elsayed F R, Hort N, Salgado-Ordorica M A and Kainer K 2011 *Mater. Sci. Forum* **690** 65-68
[11] Steinbach I and Pezzolla A 1999 *Physica D* **134** 385-393
[12] Maxwell I and Hellawell A 1975 *Acta Metalurgica* **23** 229-237
[13] Greer A L, Bumm A M, Tronche A,Evans P V and Bristow D J 2000 *Acta Mater.* **48** 2823-2835
[14] Tegeler M, Monas A, and Sutmann G 2015 *Proceedings of the Fourth International Conference on Parallel, Distributed, Grid and Cloud Computing for Engineering* **107** 5 (doi:10.4203/ccp.107.5)

Atomic investigation of steady-state dendrite tips by using phase-field crystal method

Sai Tang, Junjie Li, Zhijun Wang and Jincheng Wang

State Key Laboratory of Solidification Processing, Northwestern Polytechnical University, Youyi Western Road 127, 710072, Xi'an, China

jchwang@nwpu.edu.cn

Abstract. Steady-state dendrite growth is controlled by the states of dendrite tips, thus the morphology and growth kinetics of dendrite tip have always been a highly-studied issue in the field of dendrite growth. A crucial problem is to find out the factors influencing the morphology of dendrite tip, and to clarify their working mechanism. Since dendrite tip is at micrometer or even atomic scale, interface energy may probably plays an important role during the formation and stabilization of dendrite tips. Investigations of this issue at atomic scale is of significant because it can help us to clearly observe the morphology evolution of dendrite tip fundamentally, and specifically reveal the relationship between interface energy anisotropy and dendrite tip morphology. We investigate growth kinetics and morphology of dendrite tips at atomic scale by using phase-field crystal simulation. Atomic scale steady-state dendrite tips are obtained at different interface energy anisotropy and growth driving force. It is shown that the morphology of the forefront of dendrite tip varies periodically with time, which is related to the behavior of atom attachment during the steady-state growth of dendrite tip. Moreover, we demonstrate that growth driving force determines the overall shape of dendrite tip, but exerts little influence on the morphology of the forefront of dendrite tip which actually is determined by interface energy anisotropy. The dendrite tip of high interface energy anisotropy often presents sharp shape which is in favour of the formation of dendrite tip. Nevertheless, the shape of dendrite tip approaches to circle with decreasing magnitude of interface energy anisotropy, indicating that it is more difficult for the formation of dendrite tip under the circumstance of lower interface energy anisotropy.

1. Introduction

Steady-state dendrite growth is controlled by the state of dendrite tips, thus the morphology and growth kinetics of dendrite tip have always been a highly-studied issue in the field of dendrite growth. For the study of dendritic growth, a crucial problem is to find out the factors influencing the morphology of dendrite tip, and to clarify their working mechanism. Macroscopically, Ivantsov solution, experimental results, and phase field simulation illustrate that the product between dendrite tip radius R and growth velocity V or V^2 is a constant. Therefore, R is decreasing nonlinearly with undercooling or supersaturation if assuming V increases linearly with undercooling or supersaturation. That is to say that the overall shape of dendrite tip is affected by macroscopic driving force such as undercooling or supersaturation [1,2].

Moreover, since dendrite tip is at micrometer or even atomic scale, interface energy may probably plays an important role during the formation and stabilization of dendrite tips. As we know, the Wulff theory elucidates the close relationship between interface energy anisotropy and equilibrium crystal shape. The round or sharp corners of equilibrium crystal shape are attributed to the magnitude of interface energy anisotropy. So it is interesting to explore that whether the dendrite tip morphology is also closely related to interface energy anisotropy or not. Microscopic solvability theory suggests that interface energy anisotropy is indispensible during the formation of dendrite tip, and illustrates that the stability criterion constant being related to the steady-state of dendrite tip is a function of interface energy anisotropy. Moreover, phase field simulations have obtained similar results. Phase field simulations have not only reproduced the steady-state growth of dendrite tips at micrometer or larger scale, but also verified the dependence of stability criterion constant on interface energy anisotropy as indicated by the microscopic solvability theory. Nevertheless, these works only have explained the importance of interface energy anisotropy on dendrite tip, failing to tell us clearly how interface energy anisotropy affects dendrite tip.

Solving this problem requires us to conduct research at atomic scale which can help us to clearly observe the morphology evolution of dendrite tip fundamentally, and specifically reveal the relationship between interface energy anisotropy and dendrite tip morphology. A more microscopic research can contribute a better understanding about dendrite tip formation, and atomic scale investigations may provide us with the most direct picture for structure and shape of dendrite tip. However, to observe dendrite growth at atomic scale still faces formidable difficulties and challenges in experimental and numerical studies at present.

To present atomistic resolution analysis of morphology of the dendritic growth front for different strength of anisotropy of the interfacial energy, in this study, we employ a new methodology—phase-field crystal (PFC) model, which has the unique advantage of simulating crystal growth at atomic length scale and at diffusive time scale [3~9]. For the purpose of acquiring a better understanding about dendrite tip growth kinetics and morphology microscopically, we firstly study growth kinetics of steady-state dendrite tip, and then investigate the dependence of dendrite tip morphology on growth driving force and interface energy anisotropy. It is found that interface energy anisotropy determines the morphology of the forefront of dendrite tip, while growth driving force tends to exert its effects on the overall morphology of dendrite tip.

2. Method and simulation details

Based on the classical density functional theory (CDFT) of freezing, Elder et al.[4] invented the phase-field crystal (PFC) model. Like CDFT, the PFC model is formulated as the free-energy function with respect to scaled density related to atom number density,

$$F(\psi) = \int d\vec{x}\{\frac{\psi}{2}[-\varepsilon + (1+\nabla^2)^2]\psi + \frac{\psi^4}{4}\} \quad (1),$$

where ψ is the scaled coarse-grained (ensemble- or time- averaged) atomic number density field, ε is related to the crystalline anisotropy. ψ is periodic in crystal phases, while uniform in liquid phase. Compared with conventional phase-field simulation in which the crystallographic information of solid phase is smeared, the PFC model is capable of visualizing the atom configuration of solid phases with various crystal structures.

The simulation parameter $(\varepsilon, \bar{\psi}_0)$ is chosen based on 2D PFC phase diagram. The crystalline anisotropy increases with increasing ε. The initial density of liquid phase is $\bar{\psi}_0$ which determines the growth driving force. The crystal growth begins with a circular nucleus with radius's length about five times of the lattice constant. The circular nucleus locates at the center of the simulation box and is surrounded by homogenous liquid. The initial density of the nucleus is given by one-mode approximation which reads,

$$\psi = A[\cos(qx)\cos(qy/\sqrt{3}) - \cos(2qy/\sqrt{3})/2] + \bar{\psi}_0, \quad (2)$$

where A is the amplitude of the density waves, and q is the magnitude of the principal reciprocal lattice vectors with the value of $\sqrt{3}/2$. Thus, the lattice constant of hexagonal structure is given by $a = 2\pi / q$. The dimensionless grid size is $dx = \pi / 4$, and the time step is dt 0.75.

3. Results and discussion

Fig.1 describes the growth kinetic and morphology of steady state dendrite growth when $(\boldsymbol{\varepsilon}, \bar{\boldsymbol{\psi}}_0) = (\mathbf{0.3}, -\mathbf{0.34})$. As shown in Fig. 1(a) the shape of smoothed density profiles on the front of solid-liquid interface remains unchanged after dendrite growth reaches steady-state, and the length of the diffusion field in liquid phase converges with time as shown in the inset figure in Fig. 1(a). These aspects suggest that the density diffusion field around dendrite tip can reach a steady-state, allowing the growth velocity and the minimum density on the front of solid-liquid interface to converge after a transient stage, i.e., the dendrite tips grow in a constant velocity after a short transient stage.

Fig. 1(c) describes the change of density minimum in the liquid phase on the front of dendrite tip. The periodic characteristics reflect the growth mechanism. As shown in Fig. 1(d), the peak corresponds to the generation of a new layer in the vicinity of the forefront of dendrite tip, and the valley indicates that the new layer has already been fully occupied by atoms. Our simulation shows

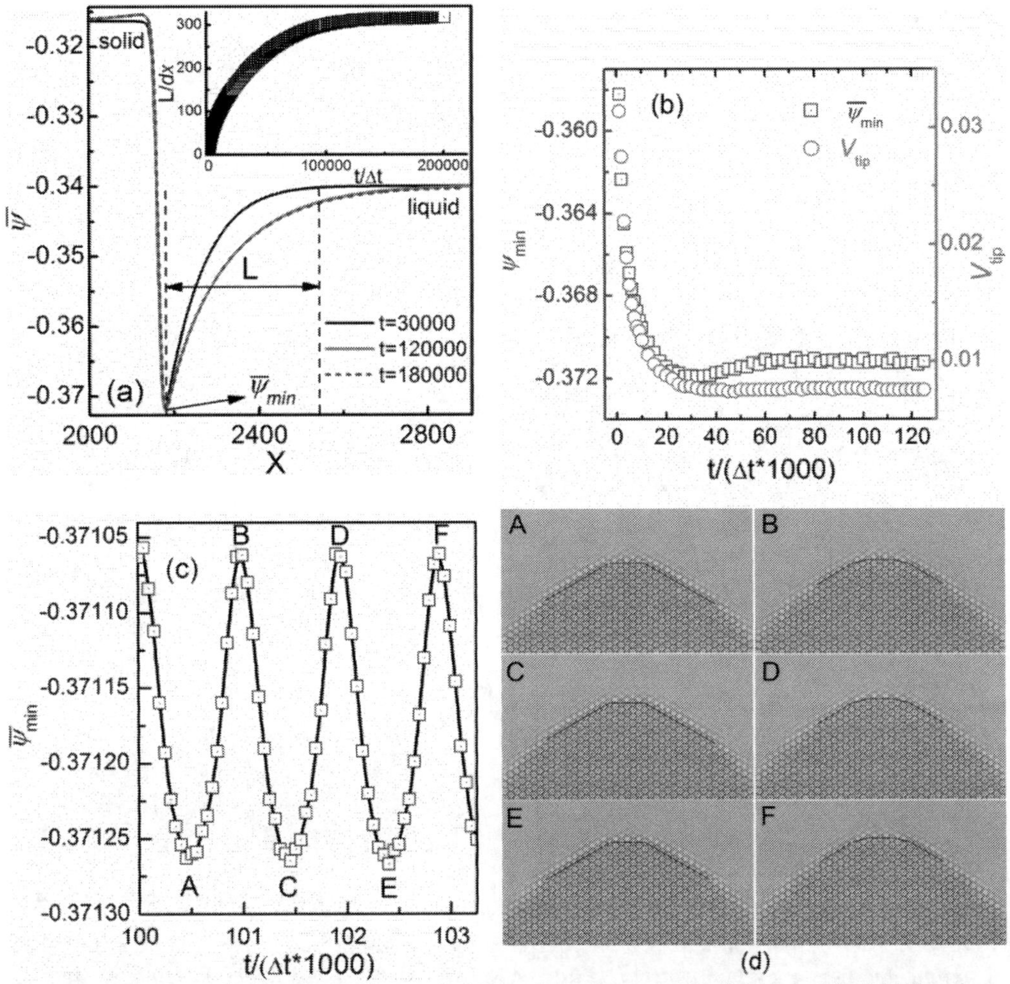

Figure 1. The growth kinetics and morphology of steady-state dendrite growth. (a) The shape of smoothed density profiles along the direction of dendrite arm at various times, where L is the length of the diffusion field, and $\bar{\psi}_{min}$ the minimum density of liquid phase on front of solid-liquid interface. The inset figure in (a) show that L will converge when dendrite growth reaches to steady-state. (b) The change of growth velocity V and $\bar{\psi}_{min}$ with time. (c) The variation of

$\bar{\psi}_{min}$ during a very short time. (d) the morphology of the forefront of dendrite tip corresponding to time A-F in (c).

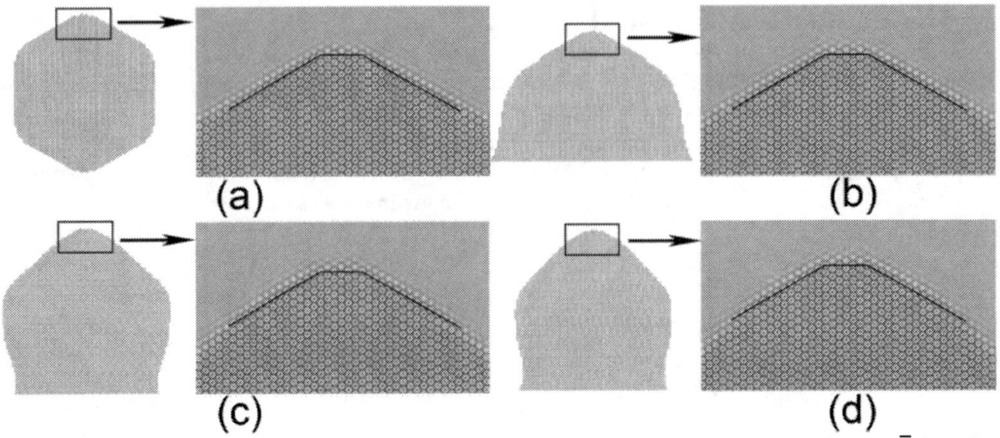

Figure 2. The dendrite morphology of dendrite tip at different growth driving force ($\bar{\psi}_0$) under the same interface energy anisotropy with ε=0.3. (a) $\bar{\psi}_0$=-0.3575, (b) $\bar{\psi}_0$=-0.345, (c) $\bar{\psi}_0$=-0.3425, and (d) $\bar{\psi}_0$=-0.34. The left figure in (a)-(d) is the overall shape of dendrite, while the right one corresponds to the forefront part of dendrite tip.

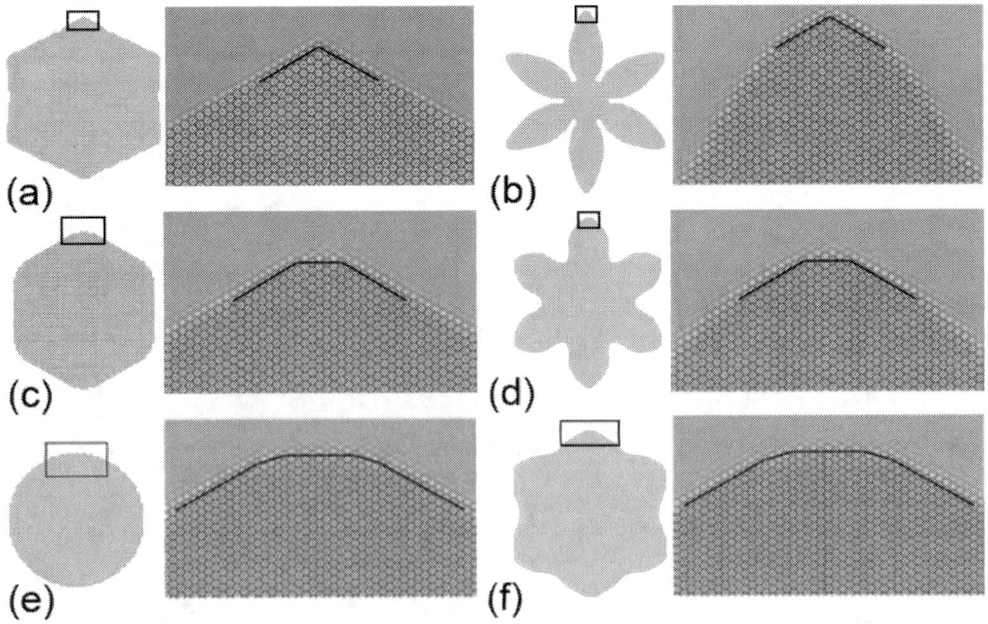

Figure 3. The comparison between the morphology of equilibrium crystal and dendrite tip at different interface energy anisotropy (ε): (a) and (b) ε=0.4; (c) and (d) ε=0.3; and (e) and (f) ε=0.25.

that the morphology of dendrite tips at atomic scale varies between the two circumstances periodically. Experimental study at microscopic scale also find that the morphology of dendrite tip present similar oscillation. For the convenience, we regard the morphology corresponding to valleys as the dendrite tip morphology in this study.

As shown in Fig.2, the morphology of the forefront of dendrite tip doesn't change with growth driving force under the same interface energy anisotropy when ε=0.3, though overall shape of dendrite tip varies substantially with growth driving force. The part of contour lines corresponding to the forefront of dendrite tip overlap, but they begin to separate at the position away from the forefront of dendrite tip. Therefore, though overall radius of dendrite tips shows obvious dependence on growth driving force, the morphology of the dendrite tip is independent of growth driving force. Thus, it is interesting to explore the factors determining dendrite tip morphology at atomic scale.

As shown in Fig. 3, we compare the morphology of the corner of equilibrium crystal with the forefront of dendrite tip. For the case of high interface energy anisotropy, both the equilibrium crystal and dendrite tip present sharp shape. For the case of intermediate interface energy anisotropy, the dendrite tip and equilibrium crystal tip present a small round shape limited by two small facets; For the case of weaker interface energy anisotropy, the round shape at dendrite tip becomes wider compared with Fig. 3(c) and (d). However, no dendrite can be obtained if decreasing interface energy anisotropy further, because the interface energy is too weak to sustain dendrite tip. Though the overall morphology of dendrite tip shows great differences, the local atom configuration at the corners of equilibrium crystal and dendrite tips is almost the same. As we know, the local part of the corner is greatly determined by interface energy anisotropy. Therefore, it is inferred that the morphology of the forefront of dendrite tip is determined by interface energy anisotropy.

4. Conclusion

In this work, through phase-field crystal simulation, we study the steady-state dendrite tip growth kinetics and morphology at atomic scale. It is shown that atomic scale morphology of the forefront of dendrite tip varies periodically with time, which is resulted from the behaviours of atom attachment at dendrite tip. Moreover, being consistent with previous studies, the radius of the overall dendrite tip decreases with increasing driving force, but the radius of the forefront of dendrite tip generally shows little dependence on driving force. We find that the morphology of the forefront of dendrite tip is mainly determined by interface energy anisotropy. Dendrite tip obtained under high interface energy anisotropy often has sharp dendrite with high curvature, which is in favour of dendrite tip formation. For the case of low interface energy anisotropy, the radius of the forefront part of dendrite tip increases steeply with decreasing anisotropy, which is disadvantage for the formation of dendrite.

Acknowledgements

This work has been supported by Nature Science Foundation of China (Grant Nos. 51371151, 51071128), and National Basic Research Program of China (Grants No.2011CB610401).

References

[1] Kurz W and Fisher DJ 1986 *Fundamentals of Solidification* (Rockport, MA: Trans Tech Pub.)
[2] Asta M, Beckermann C, Karma A, Kurz W, Napolitano R, Plapp M, Purdy G, Rappaz M and Trivedi R 2009 *Acta Mater*. **57** 941
[3] Emmerich H, Lowen H, Wittkowski R, Gruhn T, Toth GI, Tegze G and Granasy L 2012 *Advances in Physics*, **61** 665
[4] Humadi H, Ofori-Opoku N, Provatas N, Hoyt JJ 2013 *JOM* **65** 1103
[5] Elder KR, Provatas N, Berry J, Stefanovic P and Grant M 2007 *Phys. Rev. B* **75** 064107
[6] Tang S, Yu YM, Wang JC, Li JJ, Wang ZJ, Guo YL and Zhou YH 2014 *Phys. Rev. E* **89 012405**
[7] Tang S, Wang ZJ, Guo YL,Wang JC,Yu YM and Zhou YH 2012 *Acta Mater*. **60** 5501
[8] Tang S, Backofen R,Wang JC, Zhou YH, Voigt A and Yu YM 2011 *J. Cryst. Growth* **334** 146
[9] Toth GI, Tegze G, Pusztai T, Toth G and Granasy L 2010 *J. Phys.: Condens. Matter* **22** 364101.

IOP Conf. Series: Materials Science and Engineering **84** (2015) 012071 doi:10.1088/1757-899X/84/1/012071

Deterministic side-branching during thermal dendritic growth

Andrew M Mullis[1a]
[1]Institute for Materials Research, University of Leeds, Leeds LS2-9JT, UK.

E-mail: [a] a.m.mullis@leeds.ac.uk

Abstract. The accepted view on dendritic side-branching is that side-branches grow as the result of selective amplification of thermal noise and that in the absence of such noise dendrites would grow without the development of side-arms. However, recently there has been renewed speculation about dendrites displaying deterministic side-branching [see e.g. ME Glicksman, Metall. Mater. Trans A 43 (2012) 391]. Generally, numerical models of dendritic growth, such as phase-field simulation, have tended to display behaviour which is commensurate with the former view, in that simulated dendrites do not develop side-branches unless noise is introduced into the simulation. However, here we present simulations at high undercooling that show that under certain conditions deterministic side-branching may occur. We use a model formulated in the thin interface limit and a range of advanced numerical techniques to minimise the numerical noise introduced into the solution, including a multigrid solver. Not only are multigrid solvers one of the most efficient means of inverting the large, but sparse, system of equations that results from implicit time-stepping, they are also very effective at smoothing noise at all wavelengths. This is in contrast to most Jacobi or Gauss-Seidel iterative schemes which are effective at removing noise with wavelengths comparable to the mesh size but tend to leave noise at longer wavelengths largely undamped. From an analysis of the tangential thermal gradients on the solid-liquid interface the mechanism for side-branching appears to be consistent with the deterministic model proposed by Glicksman.

1. Introduction

One of the most fundamental and all pervasive microstructures produced during the solidification of metals is the dendrite. The dendrite is a prime example of a pattern forming system where complex morphologies arise from initially homogeneous conditions due to the highly non-linear response of the controlling system. The formation of the secondary (and higher order) side-branches has been an area of particular interest as it is these secondary arms that give dendrites their characteristic morphology.

Imaging of dendritic growth in transparent analogue casting systems, such as succinonitrile [1] and xenon [2], show that close to the tip region the dendrite is apparently free from side-branching. Moving away from the tip small oscillations are initiated which grow as they move down the dendrite trunk. Initially the spacing of these perturbations is uniform. However, as amplification of the initial perturbations proceeds the process becomes competitive with faster growing side-arms becoming dominant, leading to a less uniform spacing. Eventually coarsening becomes dominant, with side-arms of higher curvature being preferentially re-melted [3, 4].

A number of measures have been proposed to quantify the side-branch morphology of dendritic structures. The simplest, and often the only measure that can be applied to as-solidified materials, is the secondary arm spacing, λ_2. However, when the growing dendrite can be imaged *in situ* a far greater range of measures may be applied. The most direct is $\bar{z}_{sb}$, which is the mean distance along the trunk (in units of ρ) at which the amplitude of the side-branching becomes equal to ρ, ρ being the radius of curvature at the dendrite tip. Other measures which have been used in the literature [2, 5] to characterise dendritic structures include the fractal dimension, shape parameterization of the inner and outer side-branch envelopes, projection area and contour length.

Probably the most carefully controlled study of dendritic growth to be undertaken is the Isothermal Dendrite Growth Experiment [6] (IDGE), in which the onset of side-branching, as measured by the distance to the first detectable side-branch, was determined for high purity succinonitrile under microgravity conditions. This was determined as being at $(11.8 \pm 1.7)\rho$. In a control experiment, which was identical except that it was performed under terrestrial gravity, the same measurement yielded $(12.7 \pm 2.3)\rho$. Bisang & Bilgram [2] have measured $\bar{z}_{sb}$ for xenon dendrites under terrestrial conditions, finding $\bar{z}_{sb} = (17.5 \pm 3)\rho$. This figure is somewhat higher than that found during IDGE due to the different methodologies applied, the first detectable side-branch having an amplitude significantly less than ρ, so that the two data sets are probably mutually consistent.

The generally accepted theory of side-branch formation during dendritic growth is that side-branches grow due to the selective amplification of thermally induced noise at the solid-liquid interface with models for this process, in the framework of microscopic solvability theory, having been put forward by Langer [7, 8] and by Brener & Tempkin [9]. Such models appear to be supported by a number of phase-field simulations, in which it has been observed that dendrites grow without side-branches unless noise is introduced at the solid-liquid interface [10], with the observed level of side-branching increasing both with the strength of the noise and the undercooling (or equivalently, velocity) at which the dendrite grows [11]. Recently however, based upon a re-evaluation of the IDGE data, Glicksman [12, 13] has proposed that dendrites may undergo deterministic side-branching. The proposition is that capillary mediated differences in the temperature of the solid-liquid interface give rise to heat currents (termed bias-fields) which lead to the onset side-branching even in the absence on thermally induced noise. It is argued in [12, 13] that although the temperature difference along the solid-liquid interface in minute, due to the small length scales involved the thermal gradient may not be. It has also been pointed out [14] that if an anisotropic capillary mediated temperature distribution is calculated for a *parabolic* dendrite, a temperature inversion will arise near the tip for anisotropies exceeding 0.011, that is the lowest temperature is attained not at the tip but at some point downstream of the tip. Moreover, even if the interface temperature decreases monotonically moving back from the dendrite tip, due to the curvature of the interface the tangential thermal gradient is not unidirectional. It is argued that the point on the interface at which the gradient reverses is the point of initiation of deterministic side-branching as here there is an inward heat flow from both the dendrite tip and the downstream region which therefore perturbs the dendrite shape.

Contrary to most phase-field studies, Wang et al. [15] did find evidence of side-branching in the absence of artificially introduced noise for dendrites grown at the high undercooling of $\theta_{sys} = 0.8$, apparently in support of the ideas proposed by [12, 13]. Here θ_{sys} is the dimensionless bulk undercooling, $\theta_{sys} = H\Delta T/C_p$, H being the latent heat on fusion and C_p the specific heat. However, such investigations are potentially fraught with pit-falls. In particular, unless considerable care is taken it is difficult to ensure that the numerical noise is not introduced into the solution via the computational scheme employed. This is particularly the case when explicit solution schemes of the forward Euler type are used with a fixed (stability limited) time-step at high undercooling, wherein the dendrite grows further per time-step, potentially introducing noise which may initiate side-branching.

In this paper we re-evaluate the findings of [15], using a thermal phase-field model formulated using a range of advanced numerical techniques designed to minimise the noise introduced in to the solution by the numerical scheme. In particular, we use an implicit time-stepping scheme with a non-linear multigrid solver which is a particularly efficient way of obtaining low noise solutions to a set of coupled partial differential equations. Typically, a Jacobi or Gauss-Seidel iteration is efficient at smoothing noise at wavelengths comparable to the mesh spacing, h, but very inefficient at smoothing noise at longer wavelengths, which therefore tends to remain in the solution. In multigrid, the restriction of the solution to progressively coarser nested meshes means that noise at all wavelengths is efficiently smoothed. Moreover, the time-step in the implicit solution is limited by the accuracy of the solution, not its stability. Therefore, by employing a dynamically adjusting variable time-step determined by a series of error estimators based upon the norm of the residuals in the solution we can ensure that the same level of accuracy is obtained in all solutions irrespective of the growth rate.

2. Computational Model

The model adopted here is based upon that of [16] in which, following non-dimensionalization against characteristic length and time scales, W_0 and τ_0, the evolution of the phase-field, ϕ, and the dimensionless temperature field, θ, are given by

$$A^2(\psi)\frac{\partial\phi}{\partial t} = \nabla\cdot\left(A^2(\psi)\nabla\phi\right)+\phi(1-\phi^2)-\lambda\theta(1-\phi^2)^2 \\ -\frac{\partial}{\partial x}\left(A(\psi)A'(\psi)\frac{\partial\phi}{\partial y}\right)+\frac{\partial}{\partial y}\left(A(\psi)A'(\psi)\frac{\partial\phi}{\partial x}\right) \tag{1}$$

$$\frac{\partial\theta}{\partial t} = \alpha\nabla^2\theta + \frac{1}{2}\frac{\partial\phi}{\partial t} \tag{2}$$

where, for 4-fold growth, $A(\psi) = 1 + \varepsilon.\cos(4\psi)$, d_0 is the chemical capillary length, k_E is the equilibrium partition coefficient, α is the dimensionless thermal diffusivity and λ is a coupling parameter given by $\lambda = \alpha/a_2 = a_1 W_0/d_0$ with a_1 and a_2 taking the values $5\sqrt{2}/8$ and 0.6267 respectively [17].

The governing equations are discretized using a finite difference approximation based upon a quadrilateral, non-uniform, locally-refined mesh with equal grid spacing in both directions. Second differentials are calculated using a (spatially) forth order accurate 9-point stencil, while a compact 17-point forth order stencil has been used for Laplacian terms, in order to reduce the mesh induced [18] anisotropy. The stencils, and the weights associated with each point therein, are illustrated graphically in Fig. 1. To ensure sufficient mesh resolution around the interface region local mesh refinement (coarsening) is employed when the weighted sum of the gradients of ϕ and θ exceeds (falls below) some predefined value.

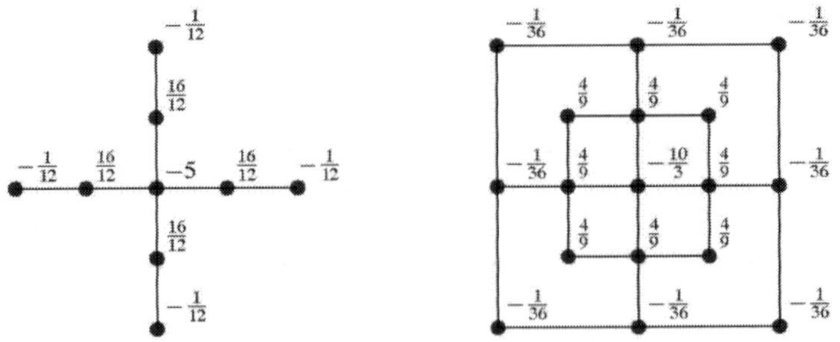

Figure 1. 9 and 17-point finite difference stencils used to evaluate the second differentials and Laplacian terms in Equations (1-2).

An implicit temporal discretization is employed based on the second order Backward Difference Formula with variable time-step, wherein it is necessary to solve a very large, but sparse, system of non-linear algebraic equations at each time-step. Multigrid methods are among the fastest available solvers for such systems and in this work we apply the non-linear generalization known as FAS (full approximation scheme [19]). The local adaptivity is accommodated via the multilevel algorithm originally proposed by Brandt [20]. The interpolation operator is bilinear while injection is used for the restriction operator. For smoothing the error we use a fully-coupled nonlinear weighted Gauss-Seidel iteration where the number of pre- and post-smoothing operations required for optimal convergence is determined empirically. Full details of the numerical scheme are given in [21].

3. Numerical Results

Figure 2 compares the morphology of the solid liquid interface for dendrites grown at θ_{sys} = 0.6875 and θ_{sys} = 0.8000, the other computational parameters for these simulations being α = 1, ε = 0.03 and h = 0.78. The domain size, Ω, is $[-1600:1600]^2$ and the initial seed with radius 5 is located at the centre of the domain. It is apparent that from the plots that spontaneous side-branching is occurring at high undercooling but not at lower undercooling. While superficially these results look like those of [15] there are two important differences. Firstly, although [15] present images (e.g. their Figs. 2a and 3a) showing equiaxed dendrites with 4-fold symmetry, their simulations are in fact conducted on a quarter domain with the equiaxed dendrite being created by rotating and replicating a single primary branch. In contrast, the equiaxed dendrite shown in this work (Fig. 2b) is nucleated from a seed in the middle of the domain. Each of the primary arms shown has therefore grown independently and the fact that each shows an identical branching pattern in itself represents strong evidence of a deterministic branching mechanism as numerical noise, by its random nature, would give rise to different patterns on each primary branch. Secondly, in [15] very low anisotropies are used in order to obtain branching. Strong branching is evident at ε = 0.005, with branching becoming almost indiscernible at ε = 0.02. It is well known that noise induced side-branching is suppressed by strong capillary anisotropy. In contrast the branched dendrite shown in Fig. 2b is grown at the relatively high anisotropy of ε = 0.03. In deed, if we very the anisotropy and determine the dimensionless undercooling, θ_{sys}, required for the onset of spontaneous side-branching we find that θ_{sys} decreases as ε increases, as shown in Fig. 3. This suggests that the mechanism responsible is not the same as operating in the simulations reported by [15] and is not related to the amplification of numerical noise.

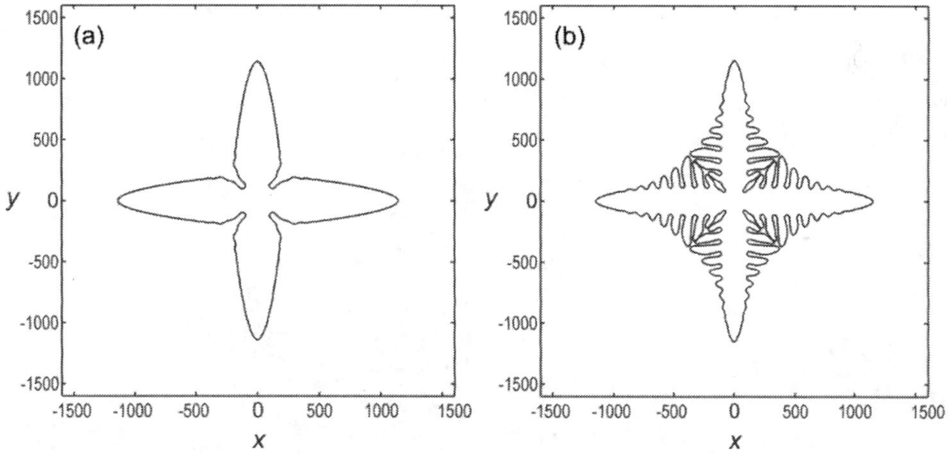

Figure 2. Example phase-field dendrite morphologies at a system undercooling of (a) θ_{sys} = 0.6875, without side-branching and (b) θ_{sys} = 0.8000, displaying spontaneous side branching.

To investigate the mechanism for side-branch formation in these simulated dendrites and to explore the similarities with the deterministic side-branching mechanism proposed in [12, 13] it is instructive to look in some detail at the temperature distribution at the dendrite tip. Figure 4 compares (a) the temperature distribution obtained when an anisotropic surface energy is imposed upon an exactly parabolic needle crystal with (b) the temperature distribution obtained from a phase-field dendrite (b).

In both cases $\varepsilon = 0.03$ and in the phase-field simulation $\theta_{sys} = 0.6875$. The temperature distribution for the parabolic needle was obtained by (i) generating an exact parabola; (ii) calculating its curvature and hence the temperature at its interface; (iii) bringing the shape into thermal equilibrium, treating the temperature calculated in (ii) above as a Dirichlet boundary condition on the solid-liquid interface. It can be observed that in the case of the exactly parabolic needle there is indeed a temperature inversion, with the lowest temperature occurring either side of the tip. In contrast, for the phase-field dendrite no such temperature inversion is observed, with the tip being the region with the lowest temperature. In deed, in no phase-field simulation, irrespective of the undercooling at which the dendrite is grown or the anisotropy imposed upon the system, have we observed a temperature inversion on the solid-liquid interface, with the tip region always being the region lowest temperature.

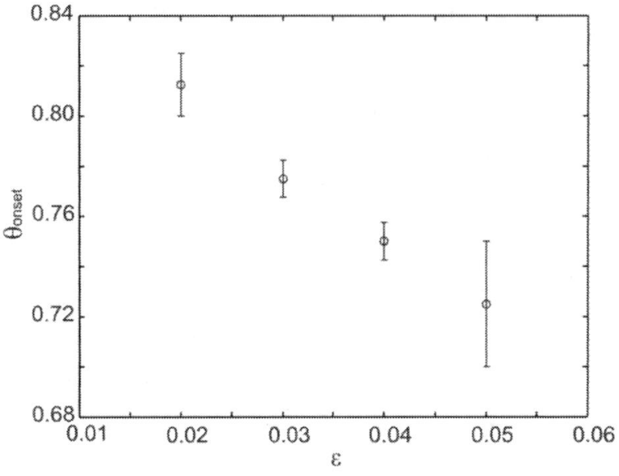

Figure 3. The undercooling, θ_{onset}, at which spontaneous side-branching is first observed as a function of the imposed anisotropy, ε.

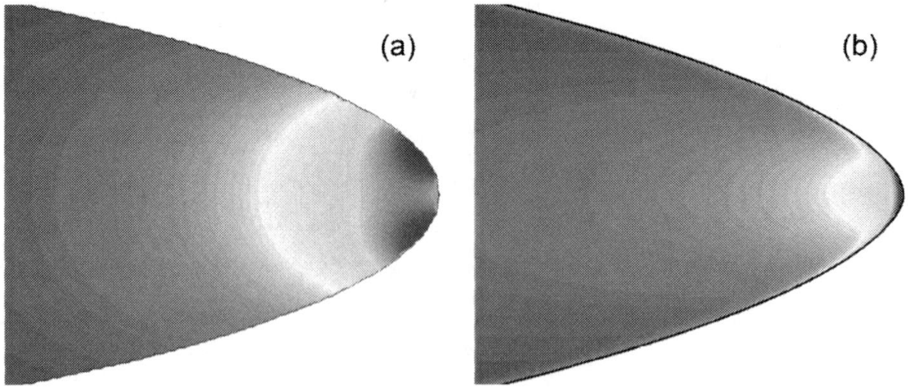

Figure 4. Thermal profile in the tip region of (a) an exactly parabolic dendrite with an anisotropic surface energy ($\varepsilon = 0.03$) and (b) a phase-field dendrite growing with a surface energy anisotropy of $\varepsilon = 0.03$.

This appears to be because the shape of the dendrite tip alters in response to the imposed anisotropy in such a way that a temperature inversion is avoided. Specifically, as the anisotropy is increased the curvature increases more rapidly as one approaches the tip than would be the case for the equivalent parabolic crystal. This can be seen most readily by calculating the actual and parabolic radii of curvature, ρ_a and ρ_{para} respectively, for the phase-field dendrites. Here, if (X, Y) is the locus of the solid-liquid interface (the $\phi = 0$ isoline), ρ_a is given, for a dendrite growing along the x-direction, by

$$K = \frac{\partial^2 X/\partial Y^2}{\left(1+(\partial X/\partial Y)^2\right)^{3/2}}, \quad \rho_a = \frac{1}{|K|} \tag{3}$$

The parabolic tip radius is given by fitting a curve of the form

$$Y^2 = -2\rho_{para}(X - X_0) \tag{4}$$

to the down-stream region of the dendrite, typically the region $2\rho_a$ behind the tip. The procedure for calculating ρ_{para} is given in more detail in [16]. In Figure 5 we plot the ratio ρ_{para}/ρ_a as a function of system undercooling and anisotropy. It can be seen that the ratio ρ_{para}/ρ_a increases rapidly with increasing ε while there is a much smaller variation as a function of undercooling, i.e. that as the anisotropy increases the tip region sharpens more rapidly than would be expected from the near parabolic down-stream shape, thus avoiding a temperature inversion at the tip.

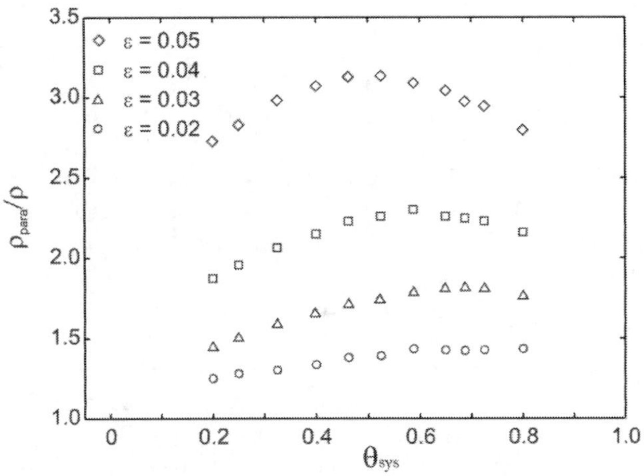

Figure 5. Ratio of the parabolic to the actual tip radius as a function of undercooling and anisotropy. As the anisotropy increases the tip becomes progressively sharper, relative to the down-stream parabolic profile, with the effect that a temperature inversion at the tip is avoided.

However, as pointed out in [12, 13], due to the curvature of the solid liquid interface the tangential thermal gradient evaluated on the interface may not be monotonic, even if the temperature does decrease monotonically moving down-stream from the tip. For the phase-field model this tangential surface thermal gradient may be evaluated from

$$\hat{\Phi} = \frac{\phi_x\theta_y - \phi_y\theta_x}{|\nabla\phi|^2} \tag{5}$$

where the subscripts denote differentiation with respect to x or y respectively. A plot of $\hat{\Phi}$ as a function of the distance downstream from the dendrite tip is shown in Fig. 6, for the dendrite displaying side-branching in Fig. 2b. The similarity in the form of the curve shown in Fig. 6, calculated for a phase-field dendrite, with the profiles calculated analytically by Glicksman [see e.g. Fig. 5 in Ref. 12] is remarkable. In particular, in both cases the curve has a turning point somewhat downstream from the tip and it is this turning point in the tangential thermal gradient which [12, 13] has associated with the deterministic side-branching mechanism. Note that the co-ordinate system used here is different from that in [12, 13] in which an elliptic geometry was considered. Here, the phase-field simulation results in the growth of a near parabolic dendritic crystal. The abscissa on Figure 5 is therefore the distance back from the dendrite tip expressed in units of the actual tip radius, ρ_a. For reference this tip radius is $\rho_a = 6.55$.

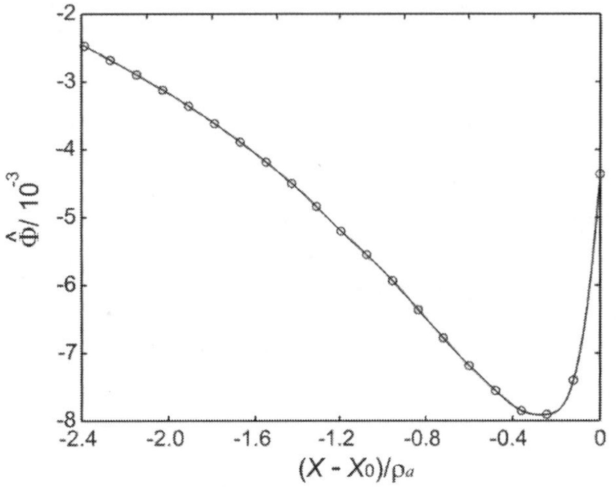

Figure 6. The tangential thermal gradient calculated on the solid-liquid interface for a dendrite with $\theta_{sys} = 0.8$ and which is displaying spontaneous side-branching.

One of the distinguishing features in Fig. 6 is that the turning point in $\hat{\Phi}$ is very close to the tip at $X - X_0 = 0.3\rho_a$. In contrast the first detectable side branch is at $X - X_0 = -28\rho_a$. In order to determine whether the tangential thermal gradients shown in Fig. 6 might potentially be perturbing the dendrite shape very close to the tip and thereby having an effect leading to observable side branching further downstream the following procedure has been adopted. The tip region of the dendrite (from $X - X_0 = 0$ to $X - X_0 = -15\rho_a$) has been fitted with an order n polynomial, where n is 4-5. This is of sufficiently high order to capture the non-parabolic tip shape but not of sufficiently high order to capture perturbations to the tip-shape from side-branching. We have then calculated the residuals between the fit and the original data to identify small perturbations to the dendrite shape. The results of this analysis, $\theta_{sys} = 0.8000$, are shown in Fig. 7, wherein it can be seen that there is indeed a cyclic perturbation to the dendrite tip shape. Moreover, the initial perturbation is located at $X - X_0 = -0.3\rho_a$, co-incident with the minimum in the tangential thermal gradient shown in Fig. 6. We take this to be at least suggestive that the heat flow associated with the tangential surface gradients and the observed side-branching are causally related. We also note that the period of the perturbations in Fig. 7 ($\sim 7.4\rho_a$), is almost exactly half the period of the finally observed side-branching, the mean secondary dendrite arm spacing being $14.7\rho_a$. As an aside we note (not shown due to reasons of brevity) that at lower undercoolings the calculated tangential thermal gradient has a similar form to that shown in Fig. 6, but that the magnitude of the minimum decreases as the undercooling is decreased. This still appears to lead to the occurrence of a shape perturbation within $-3\rho_a \leq X - X_0 \leq -0.1\rho_a$ of the tip. However, unlike the perturbations shown in Fig. 7, as we move downstream from the tip the amplitude of the perturbation is decreasing, rather than increasing. This suggests that there is an amplification condition that needs to be satisfied before deterministic side-branching will be manifest.

5. Conclusions

By using a phase-field model incorporating a range of advanced numerical techniques designed to achieve high solution accuracy with minimal induced numerical noise we have been able to demonstrate that under certain circumstances dendrites may undergo spontaneous, deterministic side-branching. Calculation of the tangential thermal gradients on the phase-field solid-liquid interface and comparison of these with location of the first perturbation on the solid-liquid interface appear to be

IOP Conf. Series: Materials Science and Engineering **84** (2015) 012071 doi:10.1088/1757-899X/84/1/012071

consistent with the capillary mediated 'bias-field' branching mechanism proposed by Glicksman [12, 13]. However, the occurrence of such side-branching appears to be confined to high undercooling/growth velocity. At low velocity the amplitude of the perturbations decreases downstream, eventually decaying away. The mechanism for this is not yet fully understood.

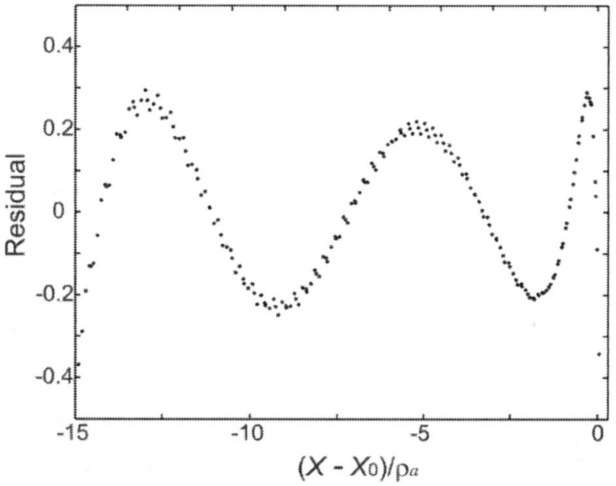

Figure 7. Residual between the actual shape of a phase-field dendrite grown at $\theta_{sys} = 0.8$ and an order 4 polynomial fit to the tip region. This procedure is very efficient at showing up small cyclic perturbations to the tip region. Note that the perturbations begin near $X\text{-}X_0 = -4d_0$, which is the location of the minimum in $\hat{\Phi}$

6. References

[1] Corrigan D P, Koss M B, La Combe J C, DeJager K D, Tennenhouse L A and Glicksman M E 1999 *Phys. Rev. E* **60** 7217.
[2] Bisang U and Bilgram J H 1996 *Phys. Rev. E* **54** 5309.
[3] Kattamis T Z, Coughin J C and Flemings M C 1967 *Trans. Met. Soc. AIME* **239** 1504.
[4] Young K P and Kirkwood D H 1975 *Met. Trans.* **6A** 197.
[5] Li Q and Beckermann C 1998 *Phys. Rev. E* **57** 3176.
[6] Tennerhouse L A, Koss M B, LaCombe J C and Glicksman M E 1997 *J. Cryst. Growth* **174** 82.
[7] Barber M N, Barbieri A and Langer J S 1987 *Phys Rev. A* **36** 3340.
[8] Langer J S 1987 *Phys. Rev. A* **36** 3350.
[9] Brener E and Temkin D 1995 *Phys. Rev. E* **51** 351.
[10] Warren J A and Boettinger W J 1995 *Acta metall. mater.* **43** 689.
[11] Mullis A M 2001 *Acta mater.* **49** 2205.
[12] Glicksman ME 2012 Coherent Capillary-mediated dendritic branching *Proc. MCWASP XIII (Schladming, Austria, 17–22 June 2012) IOP Conf. Series: Materials Science and Engineering* **33** 012097.
[13] Glicksman ME 2012 *Metall. Mater. Trans.* **43A** 391.
[14] Glicksman ME 2012 Oral presentation at *MCWASP XIII (Schladming, Austria, 17–22 June 2012)*.
[15] Wang Z, Wang J and Yang G 2008 *Phys. Rev. E* **78** 042601.
[16] Rosam J, Jimack P K and Mullis A M 2008 *Acta Mater.* **56** 4559.
[17] Karma A 2001 *Phys. Rev. Lett.* **87** 115701.
[18] Mullis A M 2006 *Comp. Mater. Sci.* **36** 345.
[19] Trottenberg U, Oosterlee C and Schuller A 2001 *Multigrid* (Academic Press, London).
[20] Brandt A 1977 *Math. Comp.* **31** 333.
[21] Rosam J, Jimack P K and Mullis A M 2007 *J. Comp. Phys.* **225** 1271.

IOP Conf. Series: Materials Science and Engineering **84** (2015) 012072 doi:10.1088/1757-899X/84/1/012072

Evolution of specific interface area in dendritic alloy solidification

H Neumann-Heyme[1] K Eckert[1] and C Beckermann[2]

[1] Institute for Fluid Dynamics, Technische Universität Dresden, 01062 Dresden, Germany
[2] Department of Mechanical and Industrial Engineering, The University of Iowa, Iowa City, IA 52242, USA

E-mail: hieram.neumann-heyme@tu-dresden.de

Abstract. The specific area of the solid-liquid interface is an important integral measure for the morphological evolution during solidification. It represents not only the inverse of a characteristic length scale of the microstructure, but it is also a key ingredient in volume-averaged models of alloy solidification. Analytical descriptions exist for either pure coarsening or pure growth processes. However, all alloy solidification processes involve concurrent growth and coarsening. In the present study, the kinetics of the solid-liquid interface of a columnar dendrite are studied using a 3D phase-field model. The simulation results are combined with data from recent experiments to study the influence of the cooling rate on the evolution of the interfacial area.

1. Introduction

A key aspect in predicting the microstructure in castings is the detailed knowledge of how geometrical features evolve over time during solidification. Often, local features, such as the secondary dendrite arm spacing, are used for the geometrical characterization of the microstructure. However, they represent incomplete descriptions of the solid structure and their measurement can become difficult during the late stages of solidification, when the structure undergoes fundamental transformations. Alternatively, integral measures, such as the specific area of the solid-liquid interface, can be introduced that more generally characterize the overall morphology [1, 2]. One definition of the specific interface area is the amount of interface area A per volume of the enclosed solid phase V_s,

$$S_s = A/V_s, \tag{1}$$

which may also be considered a characteristic inverse length scale of the microstructure. Another definition is the ratio of the interface area A to the sample volume V containing both solid and liquid phases

$$S_v = A/V = f_s S_s, \tag{2}$$

where $f_s = V_s/V$ is the solid volume fraction. S_v is also referred to as the interfacial area density. S_s and S_v can be measured directly from metallographic sections. Both quantities are key ingredients in volume-averaged (macroscopic) models of alloy solidification and are needed, for example, in modeling of microsegregation (back-diffusion) or melt flow through the mush [3].

In the latter example, the permeability of the mush K_p is directly related to the interfacial area density via the Kozeny-Carman relation $K_p \propto (1-f_s)^3/S_v^2$.

Under isothermal conditions, the evolution of the inverse specific interface area S_s^{-1} is usually described by the following relation for surface energy driven coarsening [1]

$$S_s^{-n} - S_{s0}^{-n} = Kt \quad (3)$$

or

$$S_s^{-1} = \left(S_{s0}^{-n} + Kt\right)^{\frac{1}{n}}, \quad (4)$$

where t, n, S_{s0}, and K are time, coarsening exponent, specific interface area at $t = 0$, and coarsening rate constant, respectively. For volume diffusion-limited coarsening an exponent of $n = 3$ has been firmly established by both experiments and theory. This exponent was first obtained in the context of Ostwald ripening by the LSW theory [4, 5, 6], describing the long-time evolution of a system of dispersed spherical particles. While the LSW theory assumes an idealized geometry and vanishing solid fractions, it has been possible to extend the validity of $n = 3$ to more general geometries [1, 7] and higher solid fractions [8], including morphologies that are initially dendritic. In the latter case, the coarsening rate constant K is known to be a strong function of the solid fraction [8]. While in pure coarsening theories the solid fraction is assumed to remain constant, a model has been developed for the case of concurrent growth and coarsening [9]. While this model is limited to low solid fractions, an exponent of $n = 3$ was obtained even in the presence of solidification.

In contrast to pure coarsening, solidification implies that the solid fraction f_s increases over time. Eventually, the specific interface area becomes strongly affected by coalescence and the theory of Ref. [9] is no longer valid. For processes that involve only growth, but no surface energy driven coarsening, the interfacial area density S_v is often correlated to f_s by

$$S_v = C f_s^p (1-f_s)^q, \quad (5)$$

where C, p, and q are constants. According to Eq. (5), S_v experiences a steep increase during growth, goes through a maximum, and then decreases due to impingement and coalescence of interfaces. Different values for the exponents p and q have been suggested in the literature. Speich and Fisher [10] found that data from a broad range of recrystallization experiments could be described by $p = q = 1$. These exponents were later confirmed by a computational model for the growth and impingement of grains [11]. Other suggestions have been $p = q = 2/3$ [12] and $p = q = 1/2$ [13]. A geometrical model of growing and impinging spheres has demonstrated that the parameters C, p, and q are influenced by the nucleation kinetics and the spatial distribution of the spheres [14]. Hence, generally valid values for C, p, and q are unavailable.

In summary, Eqs. (4) and (5) are useful relations for the specific interface area, but are limited to seemingly opposing cases. While Eq. (3) was developed for the isothermal case ($\dot{\theta} = 0$, f_s =const), where the interface area evolves over time due to coarsening, Eq. (5) is meant to describe situations where f_s varies with time due to growth ($\dot{\theta} \neq 0$, $f_s \neq$const) but the interface area does not change when the solid fraction is held constant. Hence, the question remains how these two models can be combined for situations that involve both growth and coarsening, such as dendritic solidification of alloys.

The direct measurement of the specific interface area during alloy solidification has not been possible until about a decade ago. Now, high-speed X-ray tomography is able to provide time-resolved geometric data during metallic alloy solidification [15, 16]. In addition, recent advancements in computational methods allow for detailed studies of solidification using phase-field simulations. The present work uses a 3D phase-field model to analyze concurrent growth and coarsening during directional solidification of a binary alloy. Experimental data for the specific interface area are extended to cooling rates that are beyond the limit of present X-ray tomography. As a result, we are able to show how the specific interface area kinetics change with cooling rate.

2. Model

To analyze the morphological evolution during growth and coarsening we use a three-dimensional phase-field model of a columnar dendrite (Al-6wt.%Cu). The model setup corresponds to a Bridgman experiment, where dendrites grow in a fixed temperature gradient G that moves at constant velocity V_p. We employ a phase-field model for directional solidification of a binary alloy, based on the frozen gradient approximation of the temperature field, that is discussed in detail in [17]. The model is extended to include finite-rate solute diffusion in the solid [18].

The numerical implementation of the problem is based on the FEM library AMDiS [19, 20], which enables the use of adaptive mesh refinement and efficient parallelization on a HPC infrastructure. A semi-implicit time integration scheme is employed to allow for adaption of the time steps to the different time scales of the interface dynamics during growth and coarsening. The present simulation is for a pulling speed of $V_p = 300\,\mu m/s$ and temperature gradient of $G = 200\,K/cm$. The material data are representative of an Al-Cu alloy and are given by an alloy solute concentration $c_0 = 6\,wt.\%$, liquidus slope $m = -2.6\,K/wt.\%$, partition coefficient $k = 0.14$, and mass diffusivities in the liquid $D_l = 3000\,\mu m^2/s$ and solid $D_s = 0.3\,\mu m^2/s$, respectively. The capillary length is taken as $d_0 = 0.005\,\mu m$ and the surface energy anisotropy coefficient as $\varepsilon_4 = 0.02$. The computational domain covers a 1/8 sector of a full dendrite by using available symmetries. The width of the simulation domain is 70 µm, i.e. one half of the primary dendrite spacing, while the length is 350 µm. No-flux conditions are applied on all boundaries and the initial geometry of the seed at the bottom of the domain is a parabola of revolution. The domain is limited at the top, such that the dendrite tip impinges on the upper wall, and the simulation proceeds by further solidification and coarsening of the previously grown structure (see Fig. 1). Numerical and phase-field parameters were chosen in order to obtain converged results for the steady state dendrite tip undercooling. This value was then used as the initial liquid undercooling in the present simulation. The computations were performed on a HPC cluster using 512 CPU's and took about one week of time. The smallest element size was 0.153 µm and the average problem size was 2.5×10^7 degrees of freedom.

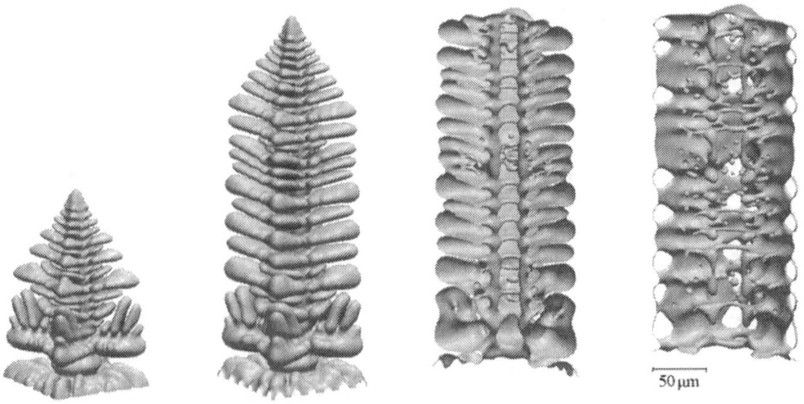

(a) $t = 0.5\,s$ (b) $t = 1\,s$ (c) $t = 2.5\,s$ (d) $t = 7\,s$

Figure 1: Evolution of the dendrite geometry: (a)-(b) full view of the growing dendrite, (c)-(d) cutaway view of half of the dendrite during the coarsening stage.

3. Results and discussion

Fig. 1 shows snapshots of the computed dendrite at different times. The first stage is characterized by a rapid increase of the interface area, while at later times coarsening and coalescence of sidebranches can be observed. At high solid fractions, liquid channels and inclusions are formed inside the solid structure (Figs. 1c, 1d).

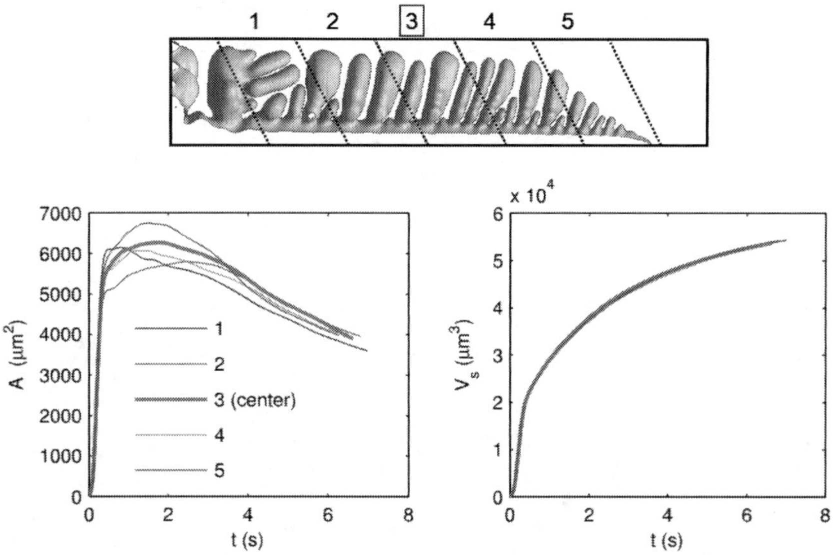

Figure 2: Averaging volumes at different positions along the growth direction (top), and evolution of the averaged interface area and solid volume (bottom).

For the evaluation of the interface area A and the solid volume V_s of the dendrite shown in Fig. 1, five sample volumes are placed along the direction of growth inside the computational domain; see Fig. 2 (top). The size of the sample volumes is chosen small enough to neglect temperature variations within them, but large enough to avoid excessive scatter in the integral measures. The tilted shape of the sample volumes further aids in suppressing scatter by covering an approximate constant number of sidebranches between adjacent volumes. The interface area A and solid volume V_s for each sample volume are plotted in Fig. 2 (bottom) as a function of time, where $t = 0$ refers to the instant when a portion of the interface enters the sample volume. It can be seen that A differs more strongly between the five sample volumes than V_s. The center sample volume is most representative of the average variation in A and is used exclusively in the following analysis.

A scaled undercooling and cooling rate can be defined, respectively, as

$$\theta = \frac{T_l(c_0) - T}{\Delta T_0}, \qquad \dot{\theta} = \frac{-\dot{T}}{\Delta T_0}, \tag{6}$$

where T is the temperature, $\dot{T}$ is the cooling rate, $T_l(c_0) = T_m - |m|c_0$ is the equilibrium liquidus temperature and $\Delta T_0 = |m|c_0(1/k - 1)$ is the equilibrium freezing range. Figure 3 shows the computed solid fraction as a function of the scaled undercooling. As expected, the solid fraction is equal to zero until the scaled undercooling reaches the dendrite tip undercooling ($\theta \approx 0.04$); afterwards, the solid fraction increases sharply with increasing scaled undercooling. This solid fraction variation can be compared to the classical lever rule and Scheil equation predictions, which assume that the dendrite tips are located at the equilibrium liquidus isotherm ($\theta = 0$). In terms of the present nomenclature, the lever rule and the Scheil equation are given, respectively, by

$$f_s = \left[1 + k\left(\frac{1}{\theta} - 1\right)\right]^{-1} \qquad \textit{(lever rule)} \tag{7}$$

and

$$f_s = 1 - \left[1 + \theta\left(\frac{1}{k} - 1\right)\right]^{\frac{1}{k-1}} \qquad \textit{(Scheil equation).} \tag{8}$$

The above equations can also be written in time-dependent form $f_s(t)$ by using the relation $\theta = \dot{\theta}t$, where it is assumed that $T(t = 0) = T_l(c_0)$. Figure 3 shows that, other than for the dendrite tip undercooling effect, the lever rule and the Scheil equation closely bound the $f_s(\theta)$ variation from the phase-field simulation.

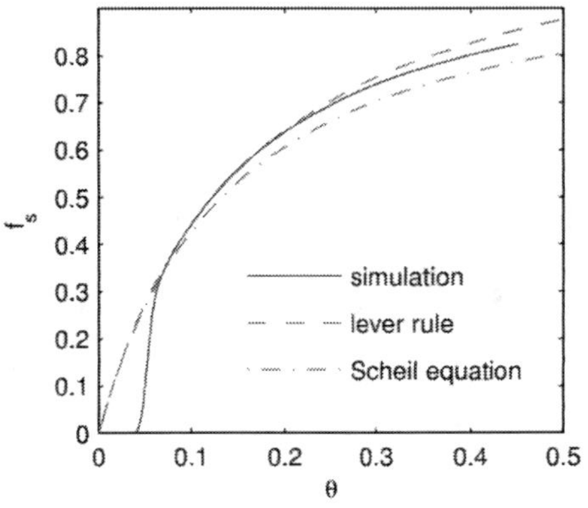

Figure 3: Solid fraction as a function of the scaled undercooling.

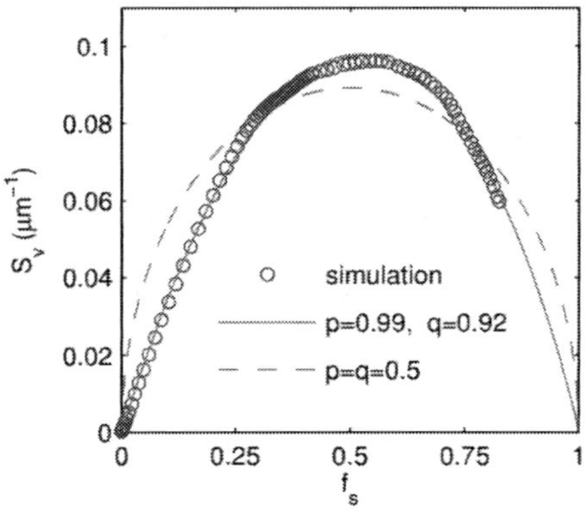

Figure 4: Variation of the interfacial area density with solid fraction.

The computed interfacial area density S_v is plotted in Fig. 4 against the solid fraction. The figure shows that S_v varies in accordance with Eq. (5). By fitting the present data to Eq. (5), it is found that the exponents are equal to $p = 0.99$ and $q = 0.92$, which is close to $p = q = 1$ suggested in Ref. [10]. Clearly, exponents of $p = q = 1/2$ [15] do not fit the simulation results.

The various temporal evolutions of the inverse specific interface area S_s^{-1} are shown and compared in Fig. 5. Figures 5a, 5b, and 5c represent experimental data from three different studies [21, 16, 15], while Fig. 5d provides the results for the present simulation. The plots are ordered by increasing cooling rate. The experimental data for S_s are fit to Eq. (4) in order to determine the exponent n. For a vanishing cooling rate ($\dot{\theta} = 0$), Fig. 5a indicates that the value of $n = 3$ that is expected for pure coarsening is approximately attained. The exponent decreases with increasing cooling rate. For the present simulation with the highest cooling rate (Fig. 5d), an exponent of $n = 3$ is obtained for short times ($t < 2\,\text{s}$), while an exponent of $n = 0.86$ fits the simulation data at longer times ($t > 2\,\text{s}$). The exponent of $n = 3$ during the initial growth stage is in agreement with the finding in Ref. [9] for concurrent growth and coarsening of spheres in the limit of low solid fractions. The solid fraction at $t = 2\,\text{s}$ is equal to 0.5, indicating that the neglect of coalesence of solid is only appropriate up to this fraction. The exponent of $n = 0.86$ observed at higher solid fractions ($t > 2\,\text{s}$) may be explained as follows. By inserting the Scheil equation, Eq. (8), into Eq. (5), assuming that $p = q = 1$ and using $S_s^{-1} = f_s/S_v$, an analytical

IOP Conf. Series: Materials Science and Engineering **84** (2015) 012072 doi:10.1088/1757-899X/84/1/012072

relation for the inverse specific interface area as a function of time can be derived as

$$S_s^{-1} = \frac{1}{C}\left[1 + \dot{\theta}t\left(\frac{1}{k} - 1\right)\right]^{\frac{1}{1-k}}. \tag{9}$$

Comparing this relation to the general coarsening law given by Eq. (4) gives

$$n = 1 - k, \tag{10}$$

where $k = 0.14$ in the present simulation. Figure 5d shows that an exponent of $n = 0.86$ does indeed provide a good fit of the predicted S_s^{-1} at long times. Note that this derivation is only valid for $p = q = 1$ and cannot be applied to the data in Figs. 5b and 5c.

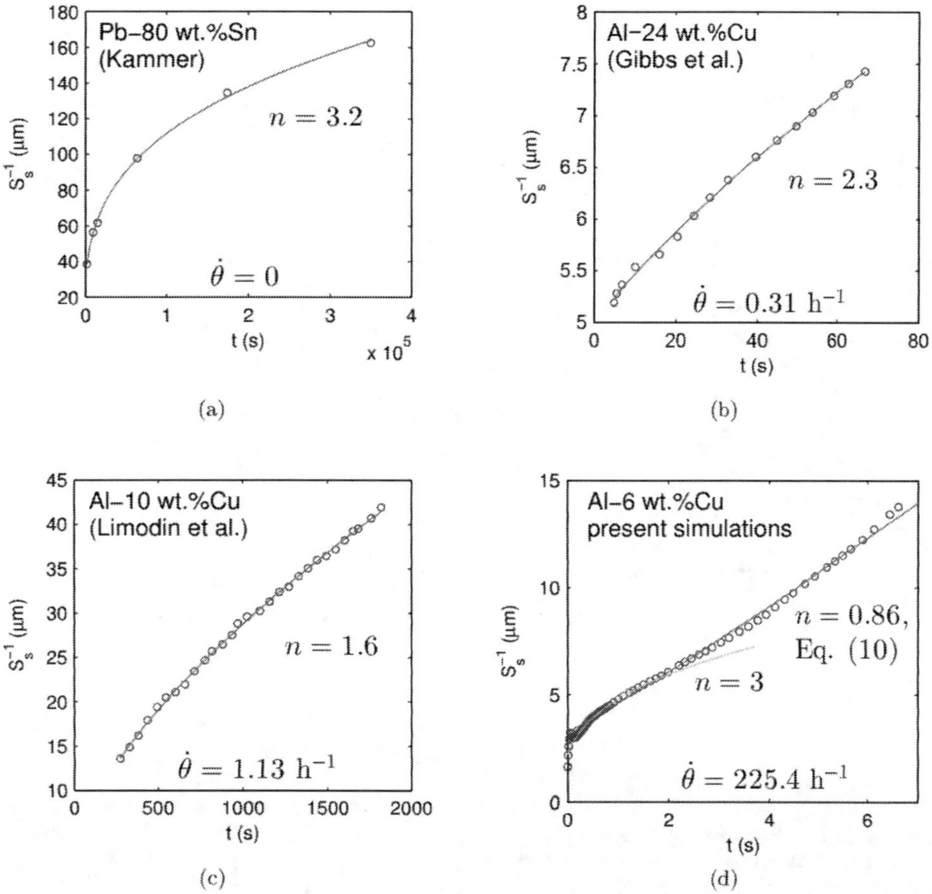

Figure 5: Evolution of the characteristic length scale for increasing scaled cooling rates: experimental data from the dissertation of D. Kammer [21] (a), personal communication with P. Voorhees [16] (b), and Limodin et al. [15] (c); (d) shows the present simulation results.

4. Conclusions

In this work we have studied the kinetics of the solid-liquid interface of a columnar dendrite by performing a 3D phase-field simulation. The computed interface area and volume are integrated over a representative volume element and presented in terms of the inverse specific interface area as a function of time and the interfacial area density as a function of solid fraction. These results are compared to existing models for pure coarsening and pure growth. For the latter case, Eq. (5), exponents close to $p = q = 1$ are obtained, which compares favorably with the exponents suggested by Speich and Fisher [10]. Comparing the present data to a pure coarsening law, Eq. (4), gives an exponent of $n = 3$ at short times and $n = 0.83$ at longer times. The former is in agreement with the concurrent growth and coarsening theory of Ref. [9], while the latter is explained in the special case of $p = q = 1$ and the solid fraction following the Scheil equation. An examination of previous experimental data, together with the present simulation results, reveals that the coarsening exponent decreases with increasing cooling rate. Nonetheless, considerable additional research is necessary to obtain a generally valid relation for the evolution of the specific interface area in alloy solidification. Simulations are underway that investigate the effect of different cooling rates and other alloy characteristics on the interface kinetics.

Acknowledgments
This work was financially supported by the Helmholtz alliance Limtech and NASA (NNX10AV35G and NNX14AD69G). We thank the supercomputing center in Jülich (HDR08) for providing computing time.

References
[1] Marsh S and Glicksman M 1996 *Metall Mater Trans A* **27** 557–567 ISSN 1073-5623
[2] Mendoza R, Alkemper J and Voorhees P 2003 *Metall Mater Trans A* **34** 481–489 ISSN 1073-5623
[3] Ni J and Beckermann C 1991 *Metall Trans B* **22** 349–361 ISSN 0360-2141
[4] Lifshitz I and Slyozov V 1961 *J Phys Chem Solids* **19** 35–50 ISSN 0022-3697
[5] Wagner C 1961 *Z Elektrochem* **65** 581–591
[6] Voorhees P 1992 *Annu Rev Mater Sci* **22** 197–215 ISSN 0084-6600
[7] Mullins W 1986 *J Appl Phys* **59** 1341–1349 ISSN 0021-8979
[8] Marsh S and Glicksman M 1996 *Acta Mater* **44** 3761–3771 ISSN 1359-6454
[9] Ratke L and Beckermann C 2001 *Acta Mater* **49** 4041–4054 ISSN 1359-6454
[10] Speich G; Fisher R 1966 *Recrystallization, grain growth and textures* (ASM, Materials Park, OH)
[11] Price C 1987 *Acta Metall Mater* **35** 1377–1390 ISSN 0001-6160
[12] Cahn J 1967 *T Metall Soc Aime* **239** 610–& ISSN 0543-5722
[13] Ratke L and Genau A 2010 *Acta Mater* **58** 4207–4211 ISSN 1359-6454
[14] Almansour A, Matsugi K, Hatayama T and Yanagisawa O 1996 *Mater T Jim* **37** 1595–1601 ISSN 0916-1821
[15] Limodin N, Salvo L, Boller E, Suery M, Felberbaum M, Gailliegue S and Madi K 2009 *Acta Mater* **57** 2300–2310 ISSN 1359-6454
[16] Gibbs J, Mohan K, Gulsoy E, Shahani A, Xiao X, Bouman C, De Graef M and Voorhees P 2015 (to be published)
[17] Echebarria B, Folch R, Karma A and Plapp M 2004 *Phys. Rev. E* **70** 061604
[18] Ohno M and Matsuura K 2009 *Phys Rev E* **79** 031603
[19] Voigt A and Witkowski T 2012 *Journal Of Computational Science* **3** 420–428 ISSN 1877-7503
[20] Witkowski T 2013 *Software concepts and algorithms for an efficient and scalable parallel finite element method* Ph.D. thesis Technische Universität Dresden
[21] Kammer D 2006 *Three-dimensional analysis and morphological characterization of coarsened dendritic microstructures* Ph.D. thesis Northwestern University

Phase-field modeling of liquid droplet migration in a temperature gradient

S Pan[1, 2], M Zhu[2], M Rettenmayr[3]

[1] Engineering Training Center, Nanjing University of Science and Technology, Nanjing, 210094, China
[2] Jiangsu Key Laboratory for Advanced Metallic Materials, School of Materials Science and Engineering, Southeast University, Nanjing, 211189, China
[3] Chair of Metallic Materials, Otto Schott Institute of Materials Research, Friedrich Schiller University, Löbdergraben 32, 07743 Jena, Germany

E-mail: zhumf@seu.edu.cn

Abstract. The migration of liquid droplets in a solid phase caused by temperature gradient zone melting (TGZM) is simulated by employing a quantitative phase-field (PF) model proposed by Echebarria et al. The PF simulation results are compared with the predictions of an analytical model that describes the droplet migration for both static and dynamic conditions, allowing the direct solution of the time dependent migration velocity of a liquid droplet that is initially located at an arbitrary position in the mushy zone. For dynamic conditions as e.g. during directional solidification, criteria for the critical pulling velocity and critical droplet position are suggested and validated by the PF simulations. When the pulling velocity is lower than the critical pulling velocity, the droplet will migrate through the moving liquidus into the bulk liquid. The droplet velocity gradually increases as it is approaching liquidus. On the other hand, when a pulling velocity higher than the critical pulling velocity is imposed, the droplet will travel through the moving solidus into the fully solid region while the droplet velocity decreases with time. The droplets initially located above the critical position migrate toward liquidus, while the others sink into the bulk solid. The effect of the temperature gradient on the droplet migration kinetics is investigated by both PF simulations and analytical predictions. The results confirm that the upward droplet migration velocity increases, while the time needed for a liquid droplet to move through the entire mushy zone decreases with increasing temperature gradient. The PF simulation results compare well with the analytical predictions.

1. Introduction

During directional solidification in steep temperature gradients, coupled local solidification and remelting in the mushy zone can occur, which is known as temperature gradient zone melting (TGZM). Typical phenomena of TGZM involve liquid droplets, films, or interdendritic pools migrating in the solid phase, or secondary dendrite arms moving in the liquid phase, in both cases toward higher temperatures [1-4]. TGZM influences the solidifying microstructure and microsegregation drastically. Motivated by the scientific and practical importance, numerous experimental and theoretical investigations have been performed to study the kinetics of TGZM and its inference on the solidified microstructures and microsegregation [4-12].

Over the recent decades, in parallel to the advances of rigorous analytical models and more refined experimental techniques, numerical simulation has developed rapidly and has become a powerful approach for studying microstructure evolution [13]. The phase-field (PF) method is an elegant and integrated simulation technique for quantitatively describing arbitrary complex topological changes in phase transformations without the need of explicitly tracking the moving interface. Various investigations on dendritic growth during directional solidification have been performed using the PF method [14-20]. Nevertheless, most PF studies focus on microstructure evolution during solidification, but neglect the effect of melting. Numerical modeling of TGZM phenomena involving both solidification and melting has so far remained scarce.

Recently, an analytical model describing liquid droplet migration caused by TGZM for both static and dynamic conditions was proposed by the present authors [21]. In this paper, the migration of liquid inclusions in a solid phase due to TGZM is studied by PF simulations. The effects of pulling velocity, initial droplet position, and temperature gradient on droplet migration kinetics are investigated. The simulation results are compared with the analytical predictions.

2. Governing equations of the phase-field model

In the present work, a two-dimensional (2D) quantitative PF model with an anti-trapping current for directional solidification of dilute binary alloys, as first proposed by Echebarria *et al.* [14], is adopted to simulate the migration of droplets in a temperature gradient. The model is based on the assumptions of zero diffusivity in the solid (one-sided diffusion) and a low phase transformation velocity. The temperature field is expressed as $T(y)=T_0+G(y-V_p t)$, where T_0 is the temperature at the cold end, G is the temperature gradient, and V_p is the pulling velocity during directional solidification. Based on the assumption of a dilute binary alloy, T_0 is determined as a function of the initial solidus composition, c_s^0, $T_0 = T_m - |m| c_l^0 = T_m - |m| c_s^0 / k$, where T_m is the melting temperature of the pure solvent, m is the liquidus slope, and $c_l^0 = (T_m - T_0)/|m|$ is the equilibrium liquid concentration at T_0. In the 2D Cartesian coordinate system (x, y), the governing equations of the PF variable and dimensionless solute supersaturation are expressed as

$$\tau_0[1-(1-k)\cdot\frac{y-V_p\cdot t}{l_T}]\cdot\partial_t\phi = W_0^2\nabla^2\phi+\phi-\phi^3+\lambda g'(\phi)(U+\frac{y-V_p\cdot t}{l_T}), \tag{1}$$

$$(\frac{1+k}{2}-\frac{1-k}{2}\cdot h(\phi))\cdot\partial_t U = \nabla\cdot(Dq(\phi)\nabla U + a(\phi)W_0[1+(1-k)U]\partial_t\phi\frac{\nabla\phi}{|\nabla\phi|})+[1+(1-k)U]\frac{1}{2}\partial_t h(\phi), \tag{2}$$

where ϕ is the PF variable with ϕ=1 and ϕ=-1 for the solid and liquid phase, respectively, W_0 is the interface thickness, $U=(c_l-c_l^0)/(c_l^0(1-k))$ is the dimensionless solute supersaturation, k is the partition coefficient, c_l is the local composition of the liquid droplet, D is the diffusion coefficient in the liquid. $\lambda = a_1 W_0 / d_0$ is the coupling constant, where $a_1 = 0.8839$, $d_0 = \Gamma / \Delta T_0$ is the solute capillary length, Γ is the Gibbs-Thomson coefficient, and $\Delta T_0 = |m|(1-k)c_l^0$ is the freezing range. $\tau_0 = a_2 \lambda W_0^2 / D$ is the relaxation time, where $a_2 = 0.6267$. $l_T = \Delta T_0 / G$ is the mushy zone length. The interpolation functions, $g(\phi)$, $q(\phi)$, and $h(\phi)$, are expressed as $g(\phi)=\phi-(2/3)\phi^3-(1/5)\phi^5$, $q(\phi)=(1/2)(1+k-(1-k)\cdot h(\phi))$, and $h(\phi)=\phi$, respectively. $\nabla=\{\partial_x,\partial_y\}$ is the 2D gradient operator, and ∂_t is the time derivative.

The simulation system is initialized by setting the domain length, grid size, and the selected parameters. Equations (1) and (2) are solved using an explicit finite difference scheme with the time step determined by $\Delta t=\Delta x^2/(4.5D)$, where Δx is the grid size. The zero-flux boundary condition is applied to numerically calculate both the phase field and concentration field. The physical parameters of succinonitrile-acetone (SCN-ACE) alloys used in the present work are taken from reference [22].

3. Solutions of the analytical model

Allen and Hunt [4] proposed an analytical model that describes liquid droplets migrating in a stationary temperature gradient. The migration velocity of a droplet, V_d, is given by

$$V_d = \frac{dy_d(t)}{dt} = -\frac{D}{c_l^*(1-k)} \cdot \frac{\partial c_l^*(y_d)}{\partial y_d}, \text{ with } \frac{\partial c_l^*(y_d)}{\partial y_d} = -\frac{G}{|m|}, \tag{3}$$

where $y_d(t)$ is the position of the droplet as a function of time t, $c_l^*(y_d)$ is the droplet composition. Since $c_l^*(y_d)$ varies with time, the migration velocity of a droplet, V_d, given by equation (3) cannot be explicitly solved. Thus, an explicit solution for the time dependent V_d is proposed through solving a first-order ordinary differential equation of the variable $y_d(t)$ derived from equation (3). Then, the time dependent position of a liquid droplet, initially located at an arbitrary position in the mushy zone, can be obtained by solving the nonlinear ordinary differential equation [21]. When a non-zero pulling velocity is imposed, the dimensionless position of the droplet, $\tilde{y}_d(\tilde{t})$, is calculated as

$$\tilde{y}_d(\tilde{t}) = \varepsilon\,\tilde{t} - \frac{\varepsilon - 1 - f^{-1}[(\varepsilon - 1 + \tilde{y}_0 \cdot (k-1)\varepsilon)\exp(\varepsilon - 1 + \tilde{y}_0 \cdot (k-1)\varepsilon - (k-1)\varepsilon^2 \cdot \tilde{t})]}{(k-1)\varepsilon}, \tag{4}$$

where $\tilde{y}_0 = y_0/l_T$, $\tilde{t} = tD/l_T^2$ and $\varepsilon = V_p \cdot l_T / D$ are the dimensionless initial droplet position, time, and pulling velocity, respectively. The droplet is located at an arbitrary position in the mushy zone, and satisfies $0 \le \tilde{y}_0 = \tilde{y}_d(0) \le 1$, where $\tilde{y}_0 = 0$ and $\tilde{y}_0 = 1$ represent the dimensionless initial droplet location at the initial positions of the moving solidus and the moving liquidus, respectively. f^{-1} is the inverse function of $f(w) = w \cdot \exp(w)$. The dimensional position of the droplet $y_d(t)$ can be obtained by the dimensional transformation of equation (4) by $\tilde{y}_d = y_d/l_T$, $\tilde{y}_0 = y_0/l_T$, $\tilde{t} = tD/l_T^2$ and $\varepsilon = V_p \cdot l_T/D$. Evaluating the derivative of $y_d(t)$, the solution of the time dependent migration velocity of a liquid droplet, V_d, is obtained by

$$V_d = \frac{dy_d}{dt} = \frac{DG}{(1-k)[|m|c_l^0 - G \cdot (y_d - V_p \cdot t)]}. \tag{5}$$

Equation (5) indicates that the droplet migration kinetics due to TGZM is directly influenced by the pulling velocity. There is a critical pulling velocity $V_{p,cr}$ corresponding to the initial droplet location, y_0. On the one hand, when the pulling velocity, V_p, is lower than $V_{p,cr}$ ($V_p < V_{p,cr}$), the droplet migration velocity, V_d, will be higher than V_p ($V_d > V_p$), and the droplet will migrate through the moving liquidus into the bulk liquid. On the other hand, when a pulling velocity higher than $V_{p,cr}$ is imposed ($V_p > V_{p,cr}$), the droplet migration velocity is lower than the pulling velocity ($V_d < V_p$), and the droplet will travel through the moving solidus into the fully solid region. The critical pulling velocity is calculated by

$$V_{p,cr} = \frac{DG}{(1-k)[|m|c_l^0 - G \cdot y_0]}. \tag{6}$$

According to equation (6), the critical liquid droplet position is defined as

$$y_{0,cr} = \frac{(1-k)|m|c_l^0 V_p - DG}{(1-k)GV_p}, \text{ and } \tilde{y}_{0,cr} = \frac{y_{0,cr}}{l_T}. \tag{7}$$

For a given constant V_p, the migrating droplet with $y_0 > y_{0,cr}$ will migrate toward liquidus and get integrated in the bulk liquid. Inversely, the migrating droplet with $y_0 < y_{0,cr}$ will gradually approach solidus and finally sink into the bulk solid. Accordingly, equations (6) and (7) are the criteria for the critical pulling velocity and critical droplet position, both yielding constant relative positions of the migrating droplet with respect to the moving liquidus and solidus.

4. Results and discussion

4.1. Effect of pulling velocity

PF simulations and analytical predictions with different pulling velocities are performed for SCN-ACE alloys. Figure 1 presents the time evolution of droplet migration velocity, V_d, and droplet location, y_d, obtained by the PF simulation, and the analytical prediction of a SCN-0.6 *wt.*% ACE alloy with V_p=2$\mu m/s$ and G=12000°C/m. At the beginning of the simulation, the domain with a 150×2400 grid and Δx=0.2μm is initialized with a solid matrix. The solidus is initially located at 16μm above the bottom wall. The solid composition below solidus is uniform with a value of 0.6 *wt.*% ACE, while a solid composition gradient, corresponding to the local temperature between solidus and liquidus, is initialized according to the equilibrium phase diagram. This initialization of the composition field in solid corresponds to the experimental situation after a few minutes of equilibration time [7,8]. A liquid droplet with a radius of 8μm is allocated at 179.5 μm above the bottom of the domain, corresponding to $\tilde{y}_0$=0.5, and the initial liquid composition determined from the phase diagram at the respective local temperature. The critical pulling velocity, $V_{p,cr}$, calculated by equation (6) is 4.71 $\mu m/s$. Thus, the pulling velocity used in the simulation of figure 1 is lower than $V_{p,cr}$. It leads to a droplet velocity increasing with time as shown in figure 1 (a). As discussed in section 3, in this case the droplet will approach (and pass through) liquidus. The locations of the droplet, the moving liquidus, T_l, and solidus, T_s, as a function of time, determined for the pre-set V_p=2$\mu m/s$, are plotted in figure 1 (b). As shown, within a simultaneously migrating mushy zone, the droplet migrates from the initial position toward the moving liquidus, due to the fact that V_p is lower than $V_{p,cr}$=4.71$\mu m/s$.

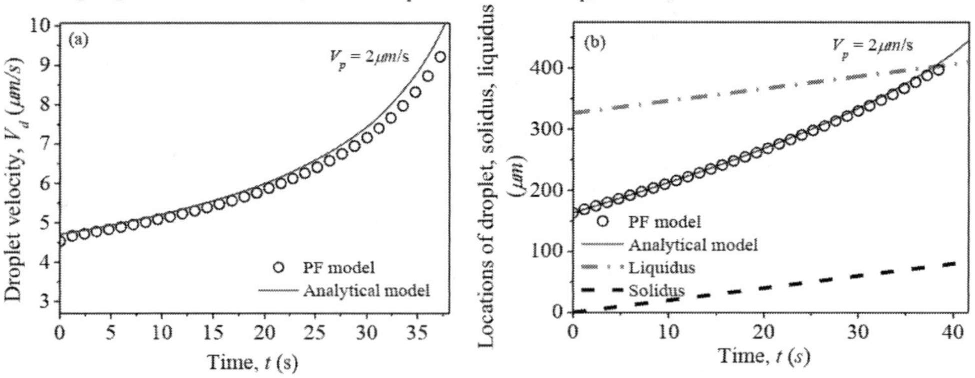

Figure 1. Comparison of PF and analytical model results regarding the time evolution of (a) droplet velocity and (b) droplet location for a SCN-0.6 *wt.*% ACE alloy at V_p=2$\mu m/s$ and G= 12000°C/m, corresponding to $\tilde{y}_0$=0.5 and $V_{p,cr}$= 4.71$\mu m/s$.

PF simulation and analytical calculation have also been performed to study the case of V_p>$V_{p,cr}$. The pulling velocity is set to V_p = 6 $\mu m/s$. Other conditions are identical with those of figure 1. The time evolution of the droplet velocity and location with V_p=6$\mu m/s$, respectively, as obtained by the PF simulation and the analytical model are compared in figure 2 (a) and (b). The profiles of the moving liquidus and solidus *vs.* time, determined by the pre-set V_p=6$\mu m/s$, are also plotted in figure 2 (b). As shown in figure 2 (a), the droplet velocity gradually decreases with time. Figure 2 (b) indicates that the droplet location goes upwards with time, confirming that a droplet always migrates toward the high temperature in a temperature gradient. Yet, the migrating velocity of the droplet is lower than the pulling velocity under the condition of V_p>$V_{p,cr}$, implying that the droplet migrates slower than the mushy zone. Therefore, the droplet gradually approaches the solidus position as shown in figure 2 (b).

It is noted from figures (1) and (2) that the simulated profiles are slightly lower than the analytical predictions. This might be caused by the following reasons. First, the analytical model neglects, while the PF simulation includes, the effect of curvature that exerts a small but perceptible influence on the droplet velocity. Second, the diffuse interface in the PF model inevitably leads to some deviations from the precise thermodynamic equilibrium concentrations, which leads to a minor error that however can accumulate for long migration distances. Despite the mentioned differences between the two models, the agreement is still satisfying.

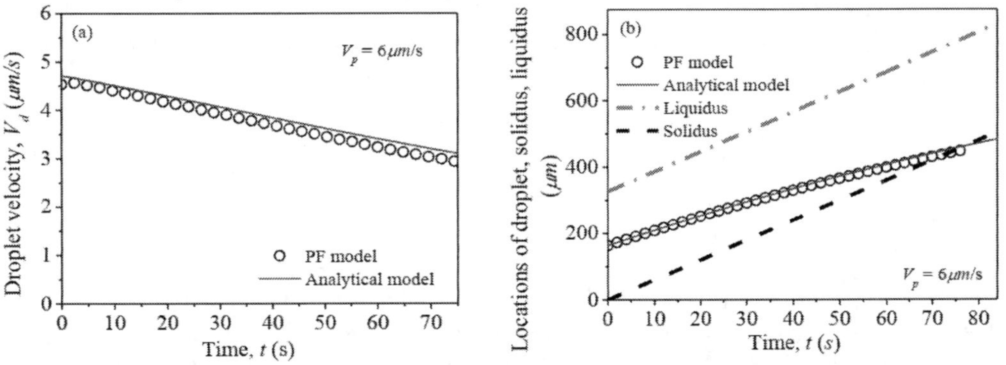

Figure 2. Comparison of PF and analytical model results regarding the time evolution of (a) droplet velocity and (b) droplet location for a SCN-0.6 *wt.*% ACE alloy at V_p=6$\mu m/s$ and G=12000°C/m, corresponding to $\tilde{y}_0$=0.5 and $V_{p,cr}$= 4.71$\mu m/s$.

4.2. *Effect of initial droplet position*

In order to examine the effect of initial droplet position on the migration behaviour of droplets, the PF simulation is performed for five droplets migrating in a solid matrix of a SCN-0.9 *wt.*% ACE alloy at G=10000°C/m. The calculation domain consists of a 500×2300 grid with Δx=0.4μm. A pulling velocity of V_p=2.62 $\mu m/s$ is imposed, yielding $\tilde{y}_{0,cr}$=0.5. Five droplets with an initial radius of 16μm are allocated at $\tilde{y}_{0,cr}$=0.14, 0.25, 0.5, 0.63, and 0.74, respectively. Figure 3 presents the droplet migration illustrated by the concentration field in the entire domain. As shown, the concentrations of both the solid matrix and liquid droplets in the mushy zone decrease along the direction of increasing

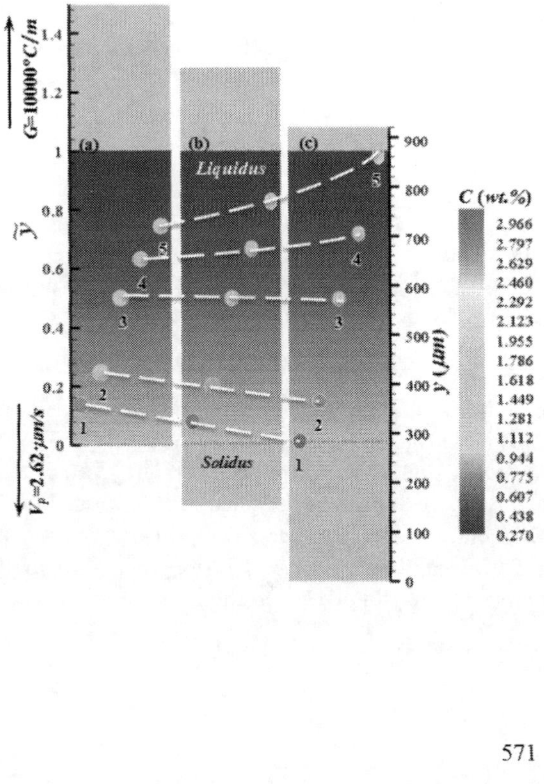

Figure 3. Simulated sequence of liquid droplet positions for a SCN-0.9 *wt.*% ACE alloy at G=10000°C/m and V_p=2.62 $\mu m/s$, corresponding to $\tilde{y}_{0,cr}$ =0.5: (a) 3.2 s, (b) 48 s, (c) 96 s. The initial positions of the five droplets are $\tilde{y}_0$ = 0.14, 0.25, 0.5, 0.63, and 0.74, respectively.

temperature. It is clear that the lower the local temperature, the higher are the equilibrium compositions of both the solid and liquid phases. All droplets with various $\tilde{y}_0$ migrate upwards with time in a fixed coordinate system. The droplet marked as "3" is initially located at the critical droplet position $\tilde{y}_0$=0.5. During the process, the relative position of droplet "3" with respect to liquidus and solidus remains essentially unchanged, while the relative position of other droplets evidently changes. The droplets marked as "4" and "5" initially located at $\tilde{y}_0$ = 0.63, and 0.74, respectively, gradually approach liquidus. When the migration time is around 96 s, droplet "5" is migrating through the moving liquidus into the bulk liquid. On the other hand, the droplets marked as "1" and "2" with $\tilde{y}_0$ =0.14 and 0.25, respectively, gradually approach solidus. When the time is around 96 s, the location of droplet "1" is partly below the solidus.

4.3. *Effect of temperature gradient*

Figure 4 presents the droplet migration velocity varying with time, obtained from the PF simulations and the analytical predictions for a SCN-0.3 *wt*.% ACE alloy with different temperature gradients. The PF simulations are performed in a domain of a 150×2000 grid with Δx=0.2μm at V_p=2$\mu m/s$. At the beginning of the simulations, the domain is initialized with a solid matrix. One liquid droplet with a radius of 8μm is assigned at the initial solidus position, i.e., y_0=0. Corresponding to the temperature gradients of 8000~12000°C/m, the critical pulling velocities calculated by equation (6) are $V_{p,cr}$=4.08~6.12$\mu m/s$, which are higher than V_p. Thus, the droplets will migrate to approach liquidus. Simulations are stopped when the liquid droplet reaches the liquidus position. As shown in figure 4, the droplet migration velocity increases with time for the three temperature gradients. As expected a higher temperature gradient leads to a higher droplet migration velocity, and a shorter time needed for the droplet to reach the liquidus position and a certain velocity. For example, when the temperature gradients are 8000, 10000, and 12000°C/m, the times to reach the velocity of 10$\mu m/s$ are around 60 s, 30 s, and 17 s, respectively.

It is noted that the profiles obtained from PF simulations and analytical solutions for G=12000°C/m show the best possible agreement. With decreasing temperature gradient, the deviation between the profiles obtained from the two models gradually becomes evident. It is understandable that the accumulated deviation between the two models increases with the time needed for a droplet to reach liquidus, which becomes longer for a lower temperature gradient.

Another phenomenon to be noted from figure 4 is that the initial droplet velocity, $V_d|_{t=0}$, is non-zero and it varies with temperature gradient. By substituting t=0 into equation (5), $V_d|_{t=0}$ can be calculated by

$$V_d\,|_{t=0} = \frac{dy_d}{dt}\,|_{t=0} = \frac{DG}{(1-k)[|m|c_l^0 - G\cdot y_0]}. \tag{8}$$

It is found that $V_d|_{t=0}$ calculated by equation (8) is identical with $V_{p,cr}$ by equation (6). Equation (8) indicates that $V_d|_{t=0}$ is proportional to the temperature gradient. As shown in figure 4, the values of $V_d|_{t=0}$ obtained from PF simulations and the analytical model are close to 4.08$\mu m/s$, 5.10$\mu m/s$ and 6.12$\mu m/s$, for G=8000, 10000 and 12000°C/m, respectively.

The case of V_p>$V_{p,cr}$ for various temperature gradients is also investigated by both PF and analytical models. The PF simulations are performed in a domain of 200×2500 grid with Δx=0.2μm, and G= 5000, 6000, 7000°C/m, respectively. As discussed in section 4.1, in the case of V_p>$V_{p,cr}$, the droplet migrates slower than the moving mushy zone, and it will approach solidus. Thus, at the beginning of simulation, one liquid droplet with a radius of 16μm is assigned at the initial liquidus position, i.e., y_0=l_T. According to equation (6), the conditions of y_0=l_T and G=5000~7000°C/m yield critical pulling velocities of $V_{p,cr}$= 8.5~11.9 $\mu m/s$. Thus, a pulling velocity of V_p=15$\mu m/s$ that is higher than the critical pulling velocity is imposed. Simulations are stopped when the liquid droplet reaches the solidus position. Figure 5 shows the time evolution of the droplet migration velocity obtained from the two models for a SCN-0.3 *wt.*% ACE alloy at the conditions of V_p=15$\mu m/s$ and G=5000~7000°C/m. As expected, similar to the case of $V_p < V_{p,cr}$ in figure 4, the droplet migration velocity increases with increasing temperature gradient. For example, when the migration time is 5 s, the droplet migration velocities are around 6.8, 8.3, and 10.1$\mu m/s$, for G=5000, 6000, and 7000°C/m, respectively. As opposed to figure 4, however, during the droplet migration toward solidus, the droplet migration velocity decreases with time for each temperature gradient. In addition, it is noted that for a higher temperature gradient, a shorter time is needed for the droplet to migrate from liquidus to the solidus position. This is because a higher temperature gradient generates a shorter mushy zone length. The mushy zone lengths determined by $l_T = |m|(1-k)c_l^0 / G$ are 392μm, 327μm, and 280μm, for G=5000, 6000, and 7000°C/m, respectively. Accordingly, the time required for a liquid droplet to go through the entire mushy zone decreases with increasing temperature gradient, although the upward velocity of the droplet migration increases with increasing temperature gradient as shown in figure 5.

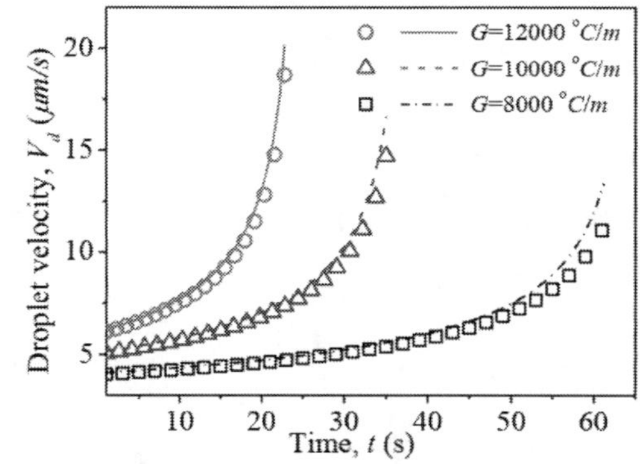

Figure 4. Comparison of PF and analytical model results regarding the time evolution of droplet velocity for a SCN-0.3 *wt.*% ACE alloy at V_p=2μm/s (y_0=0) and various temperature gradients. (Symbols: PF model, lines: analytical model)

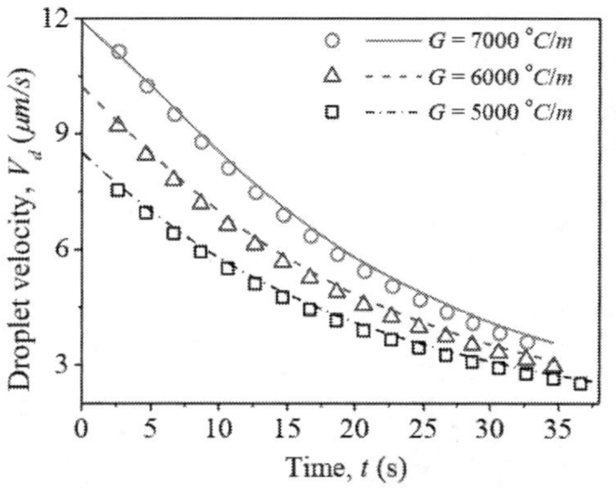

Figure 5. Comparison of PF and analytical model results regarding the time evolution of droplet velocity for a SCN-0.3 *wt.*% ACE alloy at V_p=15μm/s (y_0=l_T) and various temperature gradients. (Symbols: PF model, lines: analytical model)

5. Conclusions

Liquid droplet migration in a solid phase of SCN-ACE alloys caused by TGZM is simulated by employing a quantitative phase-field model for directional solidification. The simulation results are compared with an analytical model that describes the liquid droplet migration during TGZM for both static and dynamic conditions. In the analytical model, criteria are proposed for critical pulling velocity and critical droplet position, both yielding constant relative positions of the migrating droplet with respect to the moving liquidus and solidus.

The effect of the pulling velocity on the droplet migration kinetics is investigated. It is found that when the pulling velocity is lower than the critical pulling velocity, the droplet migrates from the initial location toward the moving liquidus, exhibiting an increasing velocity with time. However, when a pulling velocity higher than the critical pulling velocity is imposed, the droplet velocity gradually decreases with time. While the droplet always moves toward the high temperature direction, for a migration velocity of the droplet that is lower than the pulling velocity, the mushy zone migrates faster than the droplet. As a result, the droplet gradually approaches the moving solidus.

The migration behaviour of droplets with different initial locations has been studied. All droplets with various initial positions migrate upwards with time in a fixed coordinate system. The higher the position where the droplet is initially located, the faster it migrates. For a droplet that is initially located at the critical droplet position, its relative position with respect to the moving liquidus and solidus remains essentially unchanged. The droplets initially located above and below the critical position gradually approach liquidus and solidus, respectively.

The PF simulated evolution of droplet velocity as a function of time is also compared with the results of the analytical model for different temperature gradients. The results show that for the pulling velocity either higher or lower than the critical pulling velocity, the upward droplet migration velocity increases, while the time needed for a liquid droplet to go through the entire mushy zone decreases with increasing temperature gradient. It is found that the initial droplet velocity is non-zero and equals the critical pulling velocity, and it is also proportional to the temperature gradient.

It is found that the simulation data are slightly lower than the analytical predictions. This is considered to be caused by the two main reasons. First, the analytical model neglects, while the PF simulation includes the effect of curvature. Second, the diffuse interface in the PF model leads to some deviations from the precise thermodynamic equilibrium concentrations. Nevertheless, the PF simulation results are found to compare well with the analytical predictions, demonstrating the validity of the analytical model.

Acknowledgement
This work was supported by the National Natural Science Foundation of China (Grant No. 51371051).

References

[1] Rettenmayr M 2009 *Int. Mater. Rev.* **54** 1
[2] Pfann W G 1955 *Trans. AIME* **203** 961
[3] Tiller W A 1963 *J. Appl. Phys.* **34** 2757
[4] Allen D J, and Hunt J D 1976 *Metal. Trans. A* **7** 767
[5] Nguyen-Thi H, Drevet B, Debierre J M, Camel D, Dabo Y and Billia B 2003 *J. Cryst. Growth* **253** 539
[6] Nguyen-Thi H, Reinhart G, Buffet A, Schenk T, Mangelinck-Noël N, Jung H, Bergeon N, Billia B, Härtwig J and Baruchel J 2008 *J. Cryst. Growth* **310** 2906
[7] Buchmann M and Rettenmayr M 2005 *J. Cryst. Growth* **284** (3-4) 544
[8] Bösenberg U, Buchmann M and Rettenmayr M 2007 *J. Cryst. Growth* **304** (1) 281
[9] Combeau H, Appolaire B and Seiler JM 2010 *Nucl. Eng. Des.* **240** (8) 1975
[10] Fischer S, Založnik M, Seiler J-M, Rettenmayr M and Combeau H 2012 *J. Alloys. Compd.* **540** 85
[11] Liu DM, Li XZ, Su YQ, Peng P, Luo LS, Guo JJ and Fu HZ 2012 *Acta Mater.* **60** (6-7) 2679
[12] Liu DM, Li XZ, Su YQ, Rettenmayr M, Guo JJ and Fu HZ 2014 *Appl. Phys. A* **116** 1821.
[13] Asta M, Beckermann C, Karma A, Kurz W, Napolitano R, Plapp M, Purdy G, Rappaz M and Trivedi R 2009 *Acta. Mater.* **57** 941
[14] Echebarria B, Karma A and Plapp M 2004 *Phys. Rev. E* **70** 061604
[15] Chen Y, Bogno A-A, Xiao NM, Billia B, Kang XH, Nguyen-Thi H, Luo XH and Li DZ 2012 *Acta Mater.* **60** (1) 199
[16] Wang Z, Li J, Wang J and Zhou Y 2012 *Acta Mater.* **60** (5) 1957
[17] Li J, Wang Z, Wang Y and Wang J 2012 *Acta Mater.* **60** (4) 1478
[18] Eiken J, Böttger B and Steinbach I 2006 *Phys. Rev. E* **73** (6) 066122
[19] Ghmadh J, Debierre J-M, Deschamps J, Georgelin M, Guérin R and Pocheau A 2014 *Acta Mater.* **74** 255
[20] Amoorezaei M, Gurevich S and Provatas N 2012 *Acta Mater.* **60** (2) 657
[21] Pan S, Zhang Q, Zhu M, Rettenmayr M 2015 *Acta Mater.* **86** 229
[22] Ceynar D L and Beckermann C 2001 *J. Cryst. Growth* **222** 380

Mesoscopic modeling of columnar solidification and comparisons with phase-field simulations

M Založnik[1], A Viardin[2], Y Souhar[1], H Combeau[1], M Apel[2]
[1] Institut Jean Lamour, CNRS – Université de Lorraine, F-54011 Nancy CEDEX, France
[2] Access e.V., Intzestr. 5, D-52072 Aachen, Germany

E-mail: `miha.zaloznik@univ-lorraine.fr`

Abstract. We use two complementary modeling approaches for the simulation of columnar growth in directional solidification of organic alloys: a phase field model and a mesoscopic envelope model of dendritic growth. While the phase-field method captures the details of the dendritic structure and of the growth dynamics, the mesoscopic model approximates the complex dendritic morphology by its envelope. The envelope growth is deduced from the velocities of the dendrite tips, calculated by an analytical LGK-type tip model that is matched to the temperature and concentration fields in the stagnant film around the envelope. The computational cost of the mesoscopic model is several orders of magnitude lower and can bridge the gap between phase-field and macroscopic models. We demonstrate the applicability of the mesoscopic model to columnar growth and we discuss in particular its capabilities to predict the primary dendrite arm spacing.

1. Introduction

The patterns and the dynamics of the nonlinear growth of solidification microstructures are studied with much success using phase field methods. Phase-field models have been developed into efficient tools giving quantitative results. However, as in any model, its application is subject to certain limitations, mainly due to the necessary computing power. The mesoscopic solidification model of Steinbach, Beckermann et al. [1, 2] offers a complementary tool for the study of dendritic solidification. It can be applied at larger scales, in three dimensions and can include fluid flow at acceptable computing cost. Computational efficiency is achieved at the price of several simplifications. The complex dendritic structure is approximated by its envelope and by a field of phase fractions in the interior of the envelope. The growth of the envelope is deduced from the velocities of the dendrite tips, calculated by an analytical LGK-type tip model that is matched to the concentration fields in the vicinity of the envelope. This means that the selection of the dendrite-tip operating state is not predicted by the model. Information from the microscopic scale is required to determine the tip selection parameter σ^* and the matching of the tip model. This information can be obtained indirectly from comparisons with experiments or, in a more rigorous manner, directly form phase-field simulations for simplified configurations. Validation of different mesoscale models by phase-field simulations has been made for example: on dendrite interaction for the envelope model [2], on confined equiaxed growth for the CAFE model [3], or for free equiaxed and for columnar growth for the DNN model [4].

It was already demonstrated that the mesoscopic model can accurately predict envelope shapes of equiaxed dendrites [1, 2, 5] as well as the dynamics of their interactions [2]. The first application of the mesoscopic model to columnar growth was developed by Delaleau et al. [6]. Using an arrangement of columnar grains identical to the one observed in the experiment, they successfully reproduced in-situ X-ray experiments of solidification in thin samples by their 2D model. The question whether the mesoscopic model can address the mechanism of spacing selection in columnar growth was first addressed recently by Založnik et al. [7], who were able to obtain correct predictions of the steady-state primary arm spacing in a transparent organic alloy. Spacing selection is determined by both, the operating state of a single dendrite tip, i.e. its tip radius, and the solutal interaction between neigboring dendrites [8]. While the mesoscopic model should give an accurate representation of solutal interactions, the tip selection, a consequence of the anisotropy of the interface energy, is strongly simplified.

In this paper we go beyond the steady-state spacing [7] and we investigate the dynamics of primary-spacing adjustments in columnar growth. We compare 2D phase-field and mesoscopic simulations of directional solidification of a transparent organic alloy. We show that although the stable steady-state primary arm spacing can be correctly reproduced by the mesoscopic model, the dynamics of the spacing adjustments is different than in the phase-field simulations.

2. The mesoscopic envelope-field model

The core idea of the mesoscopic envelope model [1, 2] is the description of a dendritic grain by its envelope – a virtual smooth surface that links the tips of the actively growing dendrite branches. The velocity of the envelope growth can thus be calculated from the velocities of the dendrite tips. The growth of the dendrite tips is controlled by the solute flux that they reject into their surroundings and is therefore determined by the local conditions in the proximity of the envelope. The branched dendritic structure inside the envelope is only implied and its details are not resolved; the interior of the envelope is rather described in a volume-averaged sense by a phase-fraction field and other volume-averaged quantities. The phase change that determines the evolution of the structure inside the growing envelope is controlled by the solute exchange with the surroundings of the grain.

The model used to describe the kinetics of the dendrite tips is based on the analytical solution of the growth of a single isothermal parabolic tip in an infinite undercooled melt. The classical Ivantsov solution of this problem, recast by Cantor & Vogel [9], gives the supersaturation Ω_δ of the melt at a finite distance δ from the tip with a curvature radius R_{tip} as a function of the tip Péclet number, $\mathrm{Pe}_C = R_{\mathrm{tip}} V_{\mathrm{tip}}/(2D_{\mathrm{l}})$. To determine unique values for the tip velocity and radius, an additional tip selection criterion is required. In the case of a solutally controlled tip (limit of high Lewis numbers) the selection criterion is of the type $R_{\mathrm{tip}}^2 V_{\mathrm{tip}} = \mathrm{const.}$ The tip velocity and radius are then given by the system of the Cantor-Vogel function (here for a 2-dimensional tip) and the tip selection criterion, as a function of the local supersaturation Ω_δ:

$$\Omega_\delta = \sqrt{\pi \mathrm{Pe}_C}\, \exp\left(\mathrm{Pe}_C\right) \left[\mathrm{erfc}\left(\sqrt{\mathrm{Pe}_C}\right) - \mathrm{erfc}\left(\sqrt{\mathrm{Pe}_C \left[1 + \frac{2\delta}{R_{\mathrm{tip}}}\right]}\right)\right] \tag{1}$$

$$R_{\mathrm{tip}}^2 V_{\mathrm{tip}} = \frac{d_0 D_{\mathrm{l}}}{\sigma^*}. \tag{2}$$

$d_0 = \Gamma/(m_{\mathrm{L}} C_{\mathrm{l}}^*(k_{\mathrm{p}} - 1))$ is the solutal capillary length, σ^* is the selection parameter, Γ is the Gibbs-Thomson coefficient, m_{L} is the liquidus slope, k_{p} is the binary equilibrium partition coefficient, and C_{l}^* is the concentration in the liquid at the tip interface (i.e. the equilibrium liquid concentration at the given local temperature T: $C_{\mathrm{l}}^* = (T - T_{\mathrm{f}})/m_{\mathrm{L}}$, where T_{f} is the melting temperature of the pure solvent). Equations (1) and (2) together give the local tip velocity at any point on the grain envelope as a function of the local supersaturation Ω_δ. The latter is defined as $\Omega_\delta = (C_{\mathrm{l},\delta} - C_{\mathrm{l}}^*)/(C_{\mathrm{l}}^*(k_{\mathrm{p}} - 1))$ and is determined by the equilibrium liquid concentration at the envelope C_{l}^* and the concentration of the liquid adjacent to the grain at a distance δ in the direction normal to the envelope, $C_{\mathrm{l},\delta}$. The matching distance δ is a model parameter that is best chosen at the order of the diffusion length $\delta \sim D_{\mathrm{l}}/V_{\mathrm{tip}}$ [5]. In the mesoscopic model the tip growth directions are prescribed, since the key physics for selection of the preferential growth directions is not included. Knowing the local tip velocity and direction, the envelope velocity is calculated: $\vec{v}_n = V_{\mathrm{tip}}\vec{n}\cos\theta$, where θ is the angle between the normal to the envelope and the closest of the tip growth directions (i.e. the one forming the smallest angle with the normal).

To track the front that represents the envelope we use the sharp interface tracking method by Sun & Beckermann [10]. This method represents the front by a contour of a continuous indicator function. The function is propagated by a phase-field like propagation equation. The transport at the mesoscopic scale is described by volume-averaged equations, valid in the whole domain. They are equivalent to the classical volume-averaged macroscopic models [11]. The phase change is described at a microscopic scale, where Scheil assumptions are applied. The diffusion of solute in the solid phase is neglected at the mesoscopic scale. This leads to the following conservation equations for solute mass in the liquid and the solid phase:

$$g_{\mathrm{l}}\frac{\partial C_{\mathrm{l}}}{\partial t} = D_{\mathrm{l}}\nabla\cdot(g_{\mathrm{l}}\nabla C_{\mathrm{l}}) + C_{\mathrm{l}}(k_{\mathrm{p}}-1)\frac{\partial g_{\mathrm{l}}}{\partial t}\,; \qquad \frac{\partial(g_{\mathrm{s}}C_{\mathrm{s}})}{\partial t} = -k_{\mathrm{p}}C_{\mathrm{l}}\frac{\partial g_{\mathrm{l}}}{\partial t}\,. \tag{3}$$

g_{l} and g_{s} are the liquid and solid phase fraction respectively and D_{l} is the diffusion coefficient in the liquid. This equation is valid inside and outside the envelope. Inside and on the envelope the liquid is assumed to be in thermodynamic equilibrium, such that $C_{\mathrm{l}} = C_{\mathrm{l}}^* = (T-T_{\mathrm{f}})/m_{\mathrm{L}}$, where T is the imposed temperature field. Eq. (3) thus gives the liquid fraction inside the envelope. Outside the envelope $g_{\mathrm{l}} = 1$ and eq. (3) is reduced to the diffusion equation.

3. The phase-field model

In the present work, the multi-phase field concept based on Steinbach et al. [12, 13] is employed to describe the morphological evolution during the solid-liquid phase transformation. The phase field method relies on a phase field parameter which varies continuously from liquid "l" (ϕ=0) to solid "s" (ϕ=1) along the interface of width η. The evolution of the phase field parameter in time is given by:

$$\dot{\phi}_s = M^a_{sl}(\vec{n})\left[\sigma^a_{sl}(\vec{n})K_{sl} + w\Delta G_{sl}\right] \tag{4}$$

$$K_{sl} = \phi_s\nabla^2\phi_l - \phi_l\nabla^2\phi_l + \frac{\pi^2}{\eta^2}(\phi_s-\phi_l)\,; \qquad w = \frac{\pi}{\eta}\sqrt{\phi_s\phi_l}\,; \qquad \Delta G_{sl} = -f_s(\vec{c}_s)+f_l(\vec{c}_l)+\vec{\mu}(\vec{c}_s-\vec{c}_l)$$

where $M^a_{sl}(\vec{n})$ is the anisotropic interfacial mobility between liquid and solid as a function of the interface orientation (described by the normal $\vec{n}$) and $\sigma^a_{sl}(\vec{n})$ is the effective anisotropic surface stiffness. K_{sl} is related to the local curvature of the interface. The thermodynamic driving force ΔG_{sl} depends on the respective bulk contributions $f_s(\vec{c}_s)$ and $f_l(\vec{c}_l)$, $\vec{\mu}$ is the generalized chemical potential introduced to conserve local concentration due to the constraint of quasi-equilibrium. This constraint postulates that all the reduced chemical potentials are the same for the coexisting phase i.e $\vec{\mu} = \vec{\mu}_s(\vec{c}_s(\vec{x})) = \vec{\mu}_l(\vec{c}_l(\vec{x}))$. The evolution of the concentration field is given by:

$$\dot{\vec{c}} = \nabla(\phi_s D_s\nabla\vec{c}_s) + \nabla(\phi_l D_l\nabla\vec{c}_l), \tag{5}$$

where D_{s} and D_{l} are respectively the diffusion coefficients in the solid and liquid phase. Phase field simulations have been done using the MICRESS software package [14].

4. Results and discussion

We compared the results of the phase-field and the mesoscopic model for several different configurations. The system consisted of a succinonitrile-2 at%acetone alloy solidifying in a fixed gradient of $G_T = 2 \cdot 10^4\,\mathrm{K/m}$ with a cooling rate of $\dot{T} = -1.16\,\mathrm{K/s}$, which corresponds to a pulling speed of $V_{\mathrm{pull}} = 58\,\mu\mathrm{m/s}$. The growth was initiated with regularly spaced nuclei set at the bottom of the domain at an initial supersaturation of 0.20. Several different spacings of the initial nuclei λ_0 and several lateral confinements, i.e. domain sizes were used. The computational domain is a moving frame, travelling with the dendrite tips, leaving a minimum distance ahead of the most advanced tip. The frame height was $500\,\mu\mathrm{m}$ and the tip-to-top side distance was $100\,\mu\mathrm{m}$, i.e. around 4.5 times the diffusion length $D_{\mathrm{l}}/V\mathrm{pull}$. At the vertical sides a zero flux boundary condition was used in the mesoscopic model and a periodic boundary condition in the phase-field model. A fixed concentration C_0 was imposed at the top boundary and a zero derivative, i.e. no flux at the bottom boundary. The thermophysical properties of the alloys are identical as used in [15]. The tip selection parameter σ^* for the mesoscopic model was determined from the steady-state tip radius obtained by the phase field model. For that, we used the tip selection criterion $R_{\mathrm{tip}}^2 V_{\mathrm{tip}} = (d_0 D_{\mathrm{l}})/\sigma^*$, where the tip velocity is known and the liquid concentration at the tip used in the capillary length was determined from the supersaturation $\Omega_0 = (C_{\mathrm{l}}^* - C_0)/((1 - k_{\mathrm{p}})C_{\mathrm{l}}^*)$ given by the Ivantsov model (Eqs. (1) and (2) for $\delta \to \infty$): $\Omega_0 = \mathrm{Iv}(Pe_{\mathrm{tip}}) = \sqrt{\pi \mathrm{Pe}_C}\exp\left(\mathrm{Pe}_C\right)\mathrm{erfc}\left(\sqrt{\mathrm{Pe}_C}\right)$. We obtained $\sigma^*_{\mathrm{SCN-ace}} = 0.0328$.

In order to understand the effect of the spacing selection mechanism, several simulations with different initial spacings between nuclei and different domain widths were performed. Two types of configurations for the initial nucleus spacing were used. In the first case the initial nucleus spacing is larger than the final stable primary arm spacing (i.e. $\lambda_0 > \lambda_1$) and thus new primary branches have to form. In the second case the initial spacing is smaller than the final spacing (i.e. $\lambda_0 < \lambda_1$) and primary branches have to be eliminated.

The evolution of a configuration where $\lambda_0 > \lambda_1$ towards the stable steady state is shown in Fig. 1. As we can see in Figs. 1(a)–1(b), a fast growth of lateral arms in the horizontal direction is followed by branching and by vertical growth of the secondary arms along the temperature gradient (which will become primary arms). In the phase-field simulation a transient is observed in the spacing evolution. At early stage (i.e $t < 4\,\mathrm{s}$), the primary spacing is already small, many small vertical branches are growing. Between 4 and 10 seconds, some of the branches become predominant, which results in an increase of the primary spacing. At $t = 10\,\mathrm{s}$, shown in Fig. 1(d), the spacing is stabilized and tends towards a steady arrangement of the branches (Fig. 1(e)).

With the mesoscopic model, the first stage of growth after the lateral branches meet (which occurs a little later than in the phase-field simulation) can be considered as a growth of an almost planar envelope (Fig. 1(b)). The shape of the envelope is slightly perturbed only by the vertical primary arms growing directly from the nuclei along the left and right walls, as well as by a groove in the middle, which remains due to the excess solute that had slowed down the lateral arms as they were approaching. These initial perturbations then grow. The accumulation of solute in the grooves slows down the front, while the higher supersaturation ahead of the cusps accelerates it. This mechanism also promotes the development of instabilities of the envelope that appear little later, as we can see in Fig. 1(c). The instabilities develop into a wavy envelope front, with a mean wavelength of $\overline{\lambda}_1 = 77\,\mu\mathrm{m}$, which is very close to the mean primary arm spacing of $\overline{\lambda}_1 = 83\,\mu\mathrm{m}$ predicted by the phase-field simulation. A constant average spacing given by a constant number of dendrites is reached at around $t = 8\,\mathrm{s}$ in the phase-field simulation and at around $t = 14\,\mathrm{s}$ in the mesoscopic simulation. The development of a steady-state arrangement of the branches requires more time in both models until the individual tip positions have been adjusted accordingly.

For the second case, where $\lambda_0 < \lambda_1$, shown in Fig. 2, termination of branches can be observed. The spacing between the initial nuclei is too small to allow for the growth of lateral branches (Fig. 2(a)). As we can see in Figs. 2(b)–2(d), the spacing selection proceeds by a competition between the vertical branches. The phase-field model predicts a fast adjustment of the spacing, which resembles the period-doubling instability observed experimentally by Losert

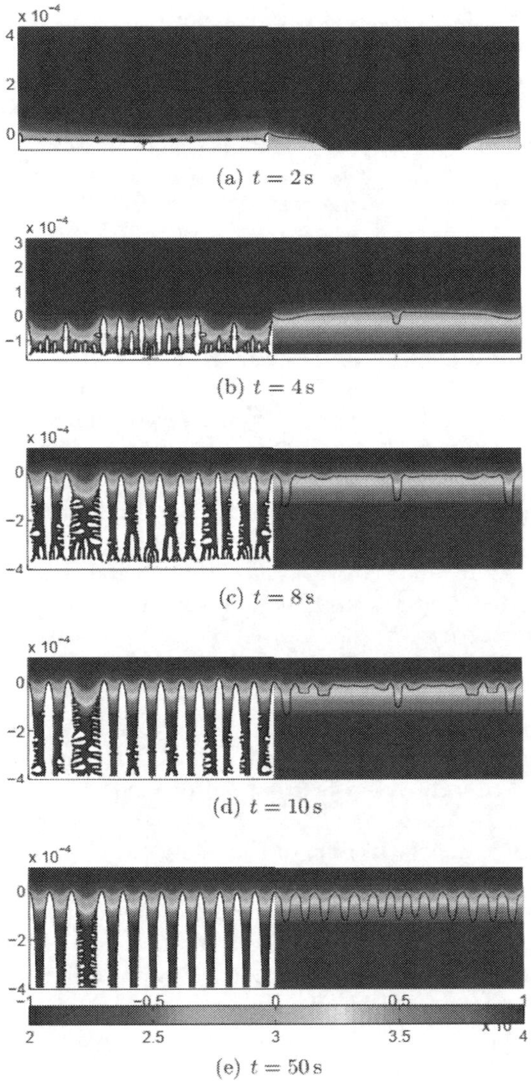

(a) $t = 2\,\mathrm{s}$ (b) $t = 4\,\mathrm{s}$ (c) $t = 8\,\mathrm{s}$ (d) $t = 10\,\mathrm{s}$ (e) $t = 50\,\mathrm{s}$

Figure 1. The evolution of the columnar grains from $\lambda_0 = 1000\,\mu\mathrm{m}$ (two nuclei placed in the bottom corners). Solid phase and liquid concentration calculated by phase field (left), and envelope and liquid concentration calculated by the mesoscopic model (right).

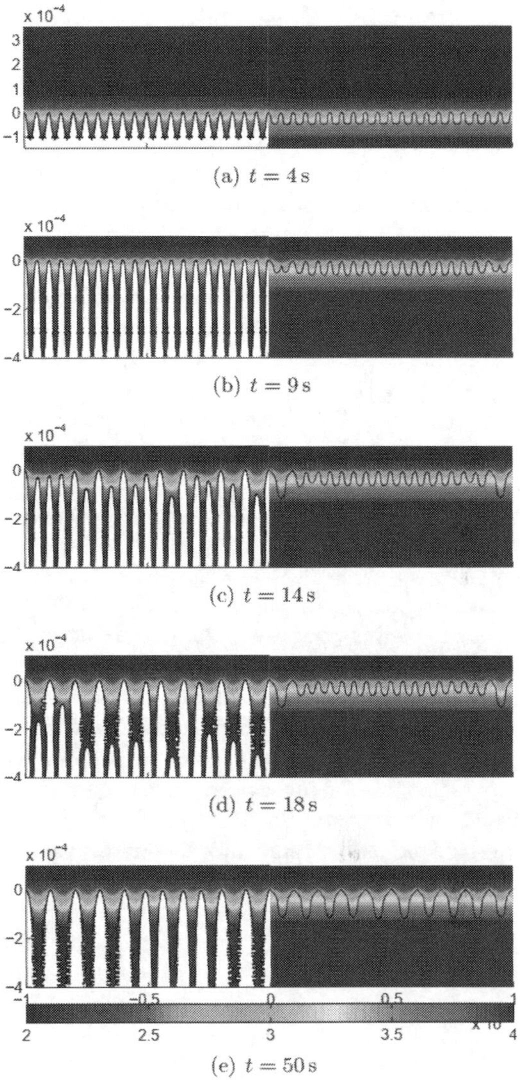

(a) $t = 4\,\mathrm{s}$ (b) $t = 9\,\mathrm{s}$ (c) $t = 14\,\mathrm{s}$ (d) $t = 18\,\mathrm{s}$ (e) $t = 50\,\mathrm{s}$

Figure 2. The evolution of the columnar grains from $\lambda_0 = 50\,\mu\mathrm{m}$ (20 nuclei placed along the bottom). Solid phase and liquid concentration calculated by phase field (left), and envelope and liquid concentration calculated by the mesoscopic model (right).

et al. [16]. In this way the final stable spacing is reached in the time interval between around $t = 12\,\mathrm{s}$ and $t = 18\,\mathrm{s}$ which is comparable to the selection process starting with $\lambda_0 > \lambda_1$. The mesoscopic model predicts an entirely different transition. The adjustment of the spacing proceeds symmetrically from the left and right edges of the domain and branches are eliminated in a cascade. The process takes much longer; starting at around $t = 12\,\mathrm{s}$ and reaching the final spacing at around $t = 37\,\mathrm{s}$. Again, the predicted final mean spacings are very close (Fig. 2(e)):

IOP Conf. Series: Materials Science and Engineering **84** (2015) 012074 doi:10.1088/1757-899X/84/1/012074

$\overline{\lambda}_1 = 91\,\mu\text{m}$ for the phase-field model and $\overline{\lambda}_1 = 100\,\mu\text{m}$ for the mesoscopic model.

In both of these cases we can observe that the initial stage of columnar growth starts with a solute diffusion field close to a perturbed planar front, which is later destabilized, leading to a spacing adjustment. Mesoscopic simulations that were initialized with a planar envelope support this analogy. After an initial destabilization of the planar envelope, protuberances with a small spacing develop. A spacing adjustment in the form of a cascade of branch eliminations, emanating from both borders of the domain sides (similar as in the case shown in Fig. 2) then leads to a stable primary-arm spacing.

We can see that the final stable spacing is not a unique state for a given set of operating conditions (G_T, V_{pull}), but that it also depends on the initial spacing as it was discussed in more detail in [7]. The final spacing is a result of the whole adjustment process, which is sensitive to the initial conditions. This is in line with theoretical analyses [17] and experiments [16] that have shown that a selection mechanism that would uniquely determine the primary spacing does not exist. At a given pulling velocity and temperature gradient a certain range of spacings can be stable. It is nevertheless relevant to compare the spacings predicted by the two models to classical empirical or geometrical models, which give a unique average primary spacing for steady growth as a function of the temperature gradient, the pulling velocity and the solute concentration. These models can provide at least a rough estimation of the spacing. They can be resumed by $\overline{\lambda}_1 = a\sqrt{\ell_{\text{sl}} R_{\text{tip}}}$, where ℓ_{sl} is the total depth of the mushy zone and where the constant a can be alloy-specific [18]. For our configuration different models of this type predict spacings between $32\,\mu\text{m}$ and $150\,\mu\text{m}$; a range across almost an order of magnitude. The predictions of the phase-field and of the mesoscopic model fall easily within these limits, however their range of variation is much smaller, e.g. [7].

For the simulation of grain interactions, for example interactions of a columnar front with equiaxed grains and the columnar-to-equiaxed transition, an important aspect is a faithful representation of the undercooling, and thus of the concentration field, ahead of the columnar front. In the context of modeling two questions can be raised. (i) With what accuracy does the mesoscopic model reproduce the undercooling of the primary tips and of the liquid ahead of the columnar front? (ii) Is it necessary to describe the primary-branch pattern and the mesoscopic interactions in order to accurately predict the undercooling ahead of the columnar zone or could it be sufficient to represent the columnar front by a plane envelope? This question is related to the fact that the columnar front could be just as well described by a smooth envelope, very much like the envelope of secondary branches in equiaxed growth [1, 5]. We try to answer these two questions by comparing the undercooling in the liquid ahead of the columnar grains predicted by the mesoscopic model to the phase-field simulations.

We address question (i) by comparing solute concentration fields in the liquid ahead of the columnar grains in the steady state from phase-field and mesoscopic simulations. In the left part of Fig. 3(a) the concentration field from phase-field and in the right part from the mesoscopic model is shown. The simulations were performed using identical process and model parameters as in the previous section, but a domain width and initial nucleus spacing of $250\,\mu\text{m}$ were used, which gave the same primary-arm spacing. The differences between the concentration fields in the liquid are relatively small. A more detailed representation is given by a plot of the concentration profiles in the liquid ahead of a primary arm in Fig. 3(c). We can see that the differences are barely perceptible. A theoretical Ivantsov solution for an isolated tip is shown for comparison. It shows that the solutal undercooling of the tips of the columnar front is slightly higher than that of an isolated tip. This indicates a slight solutal interaction between the tips. This interaction is captured with good accuracy by the mesoscopic model.

Question (ii) is addressed by a 1D mesoscopic simulation, where the tip kinetics was described by the Ivantsov function using the undercooling at an infinite distance ahead of the columnar tips $\Omega_0 = (C_0 - C_l^*)/(C_l^*(k_p - 1)) = \text{Iv}(Pe_{\text{tip}})$, instead of matching the Cantor-Vogel function

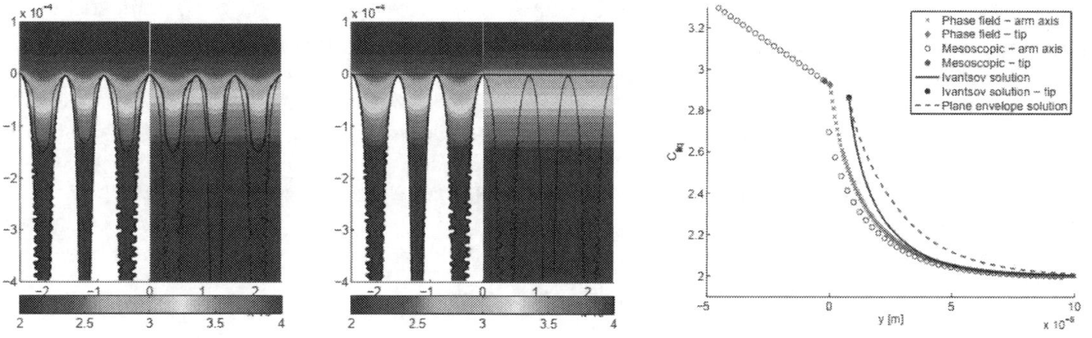

(a) Dendritic (b) Plane envelope (c) Comparison of concentration fields

Figure 3. Comparison between a phase-field simulation and a mesoscopic simulation for the steady state. (a–b) Concentration fields from the phase-field (left) and mesoscopic (right) simulations. (c) Concentration profiles along the axis of a dendrite arm at $x = 0$.

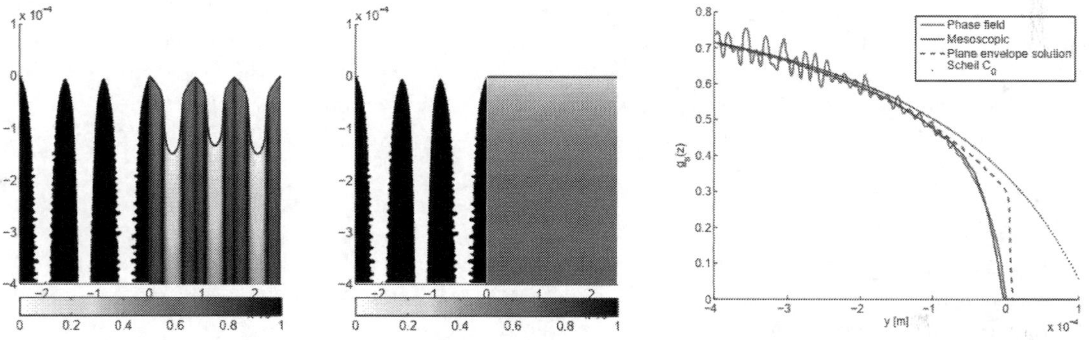

(a) Dendritic (b) Plane envelope (c) Comparison of mean solid-fraction

Figure 4. Comparison between a phase-field simulation and a mesoscopic simulation for the steady state. (a–b) Solid-fraction fields from the phase-field (left) and mesoscopic (right) simulations. (c) Average solid-fraction profiles along the height of the mushy zone.

(Eq. (1)) at a distance $\delta \approx D_\mathrm{l}/V_\mathrm{tip}$, as in the primary-spacing simulations. In practice, a very large matching distance δ, corresponding to around 10 diffusion lengths ($D_\mathrm{l}/V_\mathrm{tip}$), was used in Eq. (1) in order to achieve the Ivantsov limit. The solute concentration fields of the phase-field and of the 1D mesoscopic simulation are compared in Figs. 3(b,c). We can see that the concentration ahead of the plane envelope is slightly shifted and flatter than the concentration profile ahead of the primary tips. The tip undercooling is lower and is equal as in the theoretical Ivantsov solution. The extent of the undercooled region in front of the tips and the undercooling of the liquid are also somewhat smaller than in the phase-field and the mesoscopic solutions.

We also compared the solid fraction in the columnar mushy zone in the three computations. Fig. 4 shows (a–b) solid-fraction fields and (c) the vertical profiles of the solid fraction averaged across the horizontal direction, i.e. $\bar{g}_\mathrm{s}(z) = \int_{-W/2}^{W/2} g_\mathrm{s}(x,z)dx$. All are compared to a so-called closed-system Scheil law, i.e. a solidification path calculated for a constant average concentration C_0 without accounting for mesoscopic segregation effects. We can see that the mesoscopic simulation follows the phase-field solution very closely. The fluctuations due to the presence of secondary arms are of course not reproduced by the mesoscopic model, where the solid-fraction field inside the envelopes is smooth. The plane-envelope solution turns out to be an excellent approximation deep in the mushy zone ($z < -10^{-4}$ m), but it deviates in the tip region ($z > -10^{-4}$ m). In both the phase-field and the mesoscopic simulation with primary branches, the extra-envelope liquid is not in equilibrium and therefore the solidification path in the tip region does not follow a Scheil law as in the plane-envelope solution. The latter also shows a jump of solid fraction at the tips (at the plane envelope), which is equivalent to a Scheil solution truncated at the tip undercooling. All three numerical solutions however differ from the closed-system Scheil solidification path up to deep into the mushy zone. The reason is the diffusion-driven segregation at the mesoscopic scale, which drives solute into the liquid ahead of the tips, increases its solute concentration and thus delays the solidification.

5. Conclusions and perspectives

We demonstrated the applicability of the mesoscopic envelope model to columnar growth in directional solidification. By comparisons to phase-field computations, used as a reference, we showed that the mesoscopic model: (i) correctly reproduces the stable primary spacing; (ii) accurately captures the solute diffusion at the scale of the primary spacing and the solutal undercooling ahead of the columnar tips and (iii) accurately predicts the solidification path in the mushy zone behind the columnar front. Spacing adjustments both by branching and by elimination of branches appear in the mesoscopic model predictions. However, the dynamics of these adjustments is different from phase field predictions and further analyses are needed to understand the transient pathway towards the steady state configuration. Other extensions of this work are 3D computations of columnar growth and the inclusion of fluid flow. Both can be achieved with reasonable computational load.

Acknowledgments

We gratefully acknowledge the support to this research by the German Federal Ministry of Research through the German Space Agency DLR under Contract FKZ 50WM1143. This work was supported by the French State through the program "Investment in the future" operated by the National Research Agency (ANR) and referenced by ANR-11 LABX-0008-01 (LabEx DAMAS).

References

[1] Steinbach I, Beckermann C, Kauerauf B, Li Q and Guo J 1999 *Acta Materialia* **47** 971–982
[2] Steinbach I, Diepers H J and Beckermann C 2005 *Journal of Crystal Growth* **275** 624–638
[3] Carozzani T 2012 Ph.D. thesis Mines ParisTech
[4] Tourret D and Karma A 2013 *Acta Materialia* **61** 6474–6491
[5] Souhar Y, De Felice V F, Založnik M, Combeau H and Beckermann C 2014 *International Conference on Advances in Solidification Processes 4, Old Windsor, 8-11 June 2014* (submitted)
[6] Delaleau P, Beckermann C, Mathiesen R H and Arnberg L 2010 *ISIJ International* **50** 1886–1894
[7] Založnik M, Viardin A, Souhar Y, Combeau H and Apel M 2014 *International Conference on Advances in Solidification Processes 4, Old Windsor, 8-11 June 2014* (submitted)
[8] Steinbach I 2008 *Acta Materialia* **56** 4965–4971
[9] Cantor B and Vogel A 1977 *Journal of Crystal Growth* **41** 109–123
[10] Sun Y and Beckermann C 2007 *Journal of Computational Physics* **220** 626–653
[11] Beckermann C and Viskanta R 1993 *Applied Mechanical Review* **46** 1–27
[12] Steinbach I, Pezzolla F, Nestler B, Seeßelberg M, Prieler R, Schmitz G and Rezende J 1996 *Physica D: Nonlinear Phenomena* **94** 135–147
[13] Eiken J, Böttger B and Steinbach I 2006 *Physical Review E* **73** 066122
[14] MICRESS version 6.1, 2013-2014, www.micress.de
[15] Melendez A J and Beckermann C 2012 *Journal of Crystal Growth* **340** 175–189
[16] Losert W, Shi B Q, Cummins H Z and Warren J A 1996 *Physical Review Letters* **77** 889–891
[17] Warren J A and Langer J S 1993 *Physical Review E* **47** 2702–2713
[18] Steinbach S 2005 Ph.D. thesis RWTH Aachen

Microsegregation in multicomponent alloy analysed by quantitative phase-field model

M Ohno[1], **T Takaki**[2] **and Y Shibuta**[3]

[1] Faculty of Engineering, Hokkaido University, Kita 13 Nishi 8, Kita-ku, Sapporo, Hokkaido 060-8628, Japan
[2] Mechanical and System Engineering, Kyoto Institute of Technology, Matsugasaki, Sakyo-ku, Kyoto 606-8585, Japan
[3] Department of Materials Engineering, The University of Tokyo, 7-3-1 Hongo, Bunkyo-ku, Tokyo 113-8656, Japan

E-mail: mohno@eng.hokudai.ac.jp

Abstract. Microsegregation behaviour in a ternary alloy system has been analysed by means of quantitative phase-field (Q-PF) simulations with a particular attention directed at an influence of *tie-line shift* stemming from different liquid diffusivities of the solute elements. The Q-PF model developed for non-isothermal solidification in multicomponent alloys with non-zero solid diffusivities was applied to analysis of microsegregation in a ternary alloy consisting of fast and slow diffusing solute elements. The accuracy of the Q-PF simulation was first verified by performing the convergence test of segregation ratio with respect to the interface thickness. From one-dimensional analysis, it was found that the microsegregation of slow diffusing element is reduced due to the tie-line shift. In two-dimensional simulations, the refinement of microstructure, viz., the decrease of secondary arms spacing occurs at low cooling rates due to the formation of diffusion layer of slow diffusing element. It yields the reductions of degrees of microsegregation for both the fast and slow diffusing elements. Importantly, in a wide range of cooling rates, the degree of microsegregation of the slow diffusing element is always lower than that of the fast diffusing element, which is entirely ascribable to the influence of tie-line shift.

1. Introduction

Microsegregation is characterized by non-uniform distribution of alloying elements on a scale of dendritic structures [1]. Description and prediction of microsegregation in practical alloys require precise description of multicomponent diffusion during formation process of the solidification microstructure. A number of theoretical and numerical studies on microsegregation have been carried out and they have contributed to understanding of the microsegregation behaviour [2-9]. However, many of the previous works have focused on binary alloy systems and most of the works involve several assumptions and simplifications especially regarding multicomponent diffusion as well as morphology of the growing solid.

Diffusion in the liquid play an important role in solidification when the multicomponent alloy consists of fast and slow diffusing solute elements that are defined in this paper as elements with high and low diffusivities in the liquid, respectively. The diffusivities of fast and slow diffusing elements differ by one order of magnitude in some practical alloys such as carbon steels. In such a case, the concentrations of both the elements at the solid-liquid interface deviate from the tie-line passing through the average concentrations [10]. For example, when the partition coefficients of both the elements are less than 1, the liquid concentration of the slow (fast) diffusing element at the interface increases (decreases) due to coupled solutal diffusion in the liquid phase. In other words, the tie-line during the solidification shifts from the tie-line associated with the average concentration. This is called the *tie-line shift* in this study. The amount of tie-line shift depends on the difference in the liquid diffusivity between the fast and slow diffusing elements. Although effects of the tie-line shift on dendritic growth process has been numerically analysed [11], its influence on the microsegregation has been neither elucidated in a systematic manner nor exploited with a view to controlling microsegregation. Importantly, the amount of tie-line shift is dependent on morphology of the growing phase [10]. Therefore, the detailed description of microstructural process is indispensable for analyses of the influence of tie-line shift on microsegregation.

The phase-field model is a powerful tool to describe microstructural evolution processes in solidification [12-15]. The quantitative phase-field (Q-PF) model, which is constructed based on the thin-interface limit, enables us to carry out quantitative simulations of microstructural evolution processes [16-20]. However, the Q-PF models developed in early works can be applied only to alloy systems without diffusion in the solid phase(s) and therefore these models cannot be utilized for analyses of microsegregation. In order to describe microsegregation behaviour, the Q-PF model was recently extended to deal with alloy systems with an arbitrary value of the solid diffusivity in binary alloys [21, 22]. Moreover, one of the present authors recently developed the Q-PF model which is applicable to non-isothermal solidification in multicomponent alloys with diffusion in the solid [23]. This model can be utilized for detailed analyses of microsegregation behaviour in multicomponent alloys. The aim of this study is to shed light on the influence of tie-line shift on microsegregation behaviour in multicomponent alloys by means of the quantitative phase-field simulation.

2. Quantitative phase-field model

We recently developed the Q-PF model which is applicable to non-isothermal solidification in multicomponent alloys with arbitrary values of the solid diffusivities [23]. In this study, the Q-PF model in Ref. [23] was used to describe the formation of microsegregation during continuous cooling process. We focus on a ternary alloy system consisting of solutes 1 and 2. The temperature is assumed spatially uniform and it is cooled from the liquidus temperature at a constant rate $|dT/dt|$. Then, the microstructural evolution process is described by the following time evolution equations,

$$\tau_0 \xi(u_1,u_2) a_s(\vec{n})^2 \partial_t \phi = W^2 \nabla \cdot \left[a_s(\vec{n})^2 \nabla \phi \right] + W^2 \nabla \left[|\nabla \phi|^2 a_s(\vec{n}) \frac{\partial a_s(\vec{n})}{\partial (\nabla \phi)} \right] + \phi - \phi^3 - \lambda (1-\phi^2)^2 \left(\theta + u_1 + \left(\frac{d_1}{d_2} \right) u_2 \right), \quad (1)$$

$$\frac{1}{2}(1+k_i-(1-k_i)\phi)\partial_t u_i = \nabla \left[\frac{1}{2}(k_i D_{s,i} + D_{l,i} + (k_i D_{s,i} - D_{l,i})\phi)\nabla u_i - \frac{1}{2\sqrt{2}} \left(1 - \frac{k_i D_{s,i}}{D_{l,i}} \right) (1+(1-k_i)u_i) W \partial_t \phi \vec{n} \right] + \frac{1}{2}(1+(1-k_i)u_i)\partial_t \phi - \nabla \cdot \vec{J}_{i,noise}, \quad (2)$$

where ϕ is the phase-field which takes +1 in the solid and −1 in the liquid and continuously changes from +1 to −1 inside the solid-liquid interface. u_i is the dimensionless diffusion field of solute i and it is defined by $u_i = (c_{l,i} - c_{0,i})/[(1-k_i)c_{0,i}]$ with the liquid concentration, $c_{l,i}$, average concentration, $c_{0,i}$ and the partition coefficient, k_i, of solute i. W is the interface thickness, τ_0 is the relaxation time and $\xi(u_1,u_2)$ is given by $\xi(u_1,u_2) = 1+(1-k_1)u_1+(1+(1-k_2)u_2)D_{l,1}d_1/(d_2 D_{l,2})$ where d_i is the chemical capillary length of solute i and $D_{l,i}$ is the liquid diffusivity of solute i. $a_s(\vec{n})$, where $\vec{n} = -\nabla\phi/|\nabla\phi|$, is a function describing the anisotropy of the solid-liquid interfacial energy. We focus on the fourfold crystalline anisotropy and its strength is determined by a parameter ε_4. The specific form of $a_s(\vec{n})$ is found in Ref.

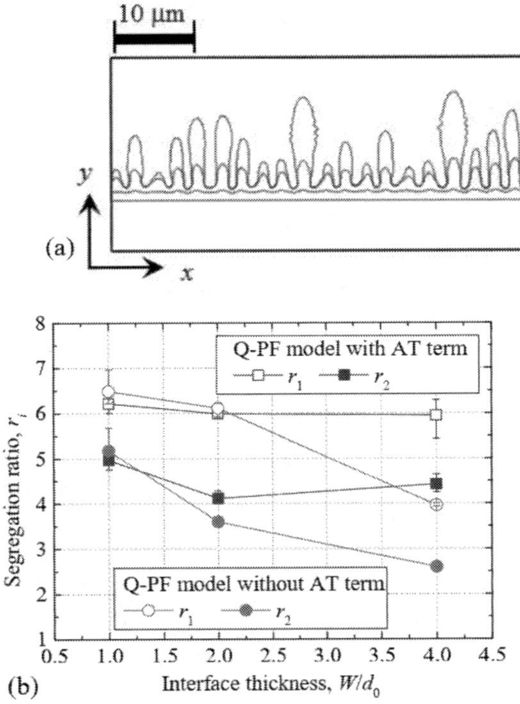

Figure 1. (a) Computational system used in this study and typical growth pattern of the solid calculated for $|dT/dt|$ = 30 K/s, indicating that the instability of planer takes place and the secondary arms develop. (b) Results of convergence test obtained by the Q-PF simulations with and without the antitrapping current. The cooling rate $|dT/dt|$ was set to 30 K/s. Each plot represents the average value of three runs and the error bar indicates the maximum and minimum values in three runs.

[16]. θ is given by $\theta = (T-T_l)/(m_1(1-k_1)c_{0,1}/k_1)$ where T_l is the liquidus temperature of the alloy and m_1 is the liquidus slope of solute 1. λ is the coupling constant and the relation $\tau_0 = a_2\lambda W^2/D_{l,1}$ holds with a constant a_2 (= 0.6267…) when the kinetic coefficient is negligibly small. $D_{s,i}$ is the solid diffusivity of solute i. The last term in equation (2) is the noise term expressing fluctuation of solute diffusion fields which is necessary to simulate the side branching [24,25]. The details of equations (1) and (2) can be found in Ref. [23].

The main concern of this paper is the influence of tie-line shift on the microsegregation behaviour in a ternary alloy. In order to highlight this phenomenon, we focus on a model alloy system rather than practical alloy systems. We employed $k_1 = k_2 = 0.2$ and $m_1 = m_2 = 500$ K/mol. Then, if $c_{0,1}$ is equal to $c_{0,2}$, there is no distinction between the solutes 1 and 2 in terms of the phase equilibria. We chose $c_{0,1} = c_{0,2} = 0.0008$ which yields the solidification range of $\Delta T = 3.2$ K. We focused on this small ΔT to save the computational time but it is sufficient for the present purpose. The capillary lengths were set to $d_0 = d_1 = d_2 = 6.25 \times 10^{-8}$ m and $\varepsilon_4 = 0.02$ was employed as a typical value in metallic alloy systems. Also, for the sake of simplicity, we chose $D_{s,1} = D_{s,2} = 1\times10^{-11}$ m^2/s. The difference between the parameters for the solutes 1 and 2 appears only in the liquid diffusivity. We employed $D_{l,2} = 0.1\ D_{l,1}$ with $D_{l,1} = 1 \times 10^{-9}$ m^2/s. Therefore, the solute 1 corresponds to the fast diffusing element, while the solute 2 is the slow diffusing element. We also carried out the simulation for $D_{l,2} = D_{l,1}$. For convenience, the simulations for $D_{l,2}$=0.1 $D_{l,1}$ and $D_{l,2} = D_{l,1}$ are called cases A and B, respectively.

We employed a two-dimensional system shown in Figure 1(a) where the lengths of x- and y-axes were set to 100 and 25 μm, respectively. The periodic boundary condition was used at x = 0 and 100 μm, while the mirror boundary condition was applied at y = 0 and 25 μm. The plate-shape solid phase was initially put on the edge of y-axis and the initial thickness of this plate was set to 1 μm. As the cooling proceeds, the thickness gradually increases, i.e., the planar interface moves in y direction. Then, depending on the cooling rate, the instability of planar interface occurs and the dendrite arms develop. The dendrite arms developing in y direction are called the secondary arm in this paper. This system was chosen because it is convenient to investigate a response of the microstructural process to different cooling condition and the resulting microsegregation behaviour. We changed the cooling rate $|dT/dt|$ from 1 to 40 K/s and carried out simulations for each cooling condition three times using different seeds for the noise term in equation (2). In addition to the two-dimensional simulation, one-dimensional simulations were conducted.

3. Results and discussion

3.1. Convergence test

The accuracy of the phase-field simulation can generally be improved by decreasing W. However, it leads to increment of the computational cost. Therefore, a value of W should be determined in the light of balance between the accuracy and the computational cost by checking the convergence behaviour of results with decreasing W. It was demonstrated in our previous study [23] that the Q-PF model described in the previous section shows excellent convergence regarding steady-state dendritic growth in ternary alloys. However, our focus in this paper is not steady-state growth and it is thus necessary to check convergence behaviour of simulation results of the microsegregation.

Figure 1(b) shows the segregation ratio, r_i, obtained from the phase-field simulations for different values of W. The segregation ratio, r_i, is defined as $r_i = c_{i,max}/c_{i,min}$ where $c_{i,max}$ ($c_{i,min}$) is the maximum (minimum) value of concentration of solute i in the microstructure just after the completion of solidification. The simulations were performed for $|dT/dt| = 30$ K/s and $D_{l,2} = 0.1\ D_{l,1}$ (case A). It is noted that the secondary arms develop in this cooling condition as shown in Figure 1(a) and hence the microstructure is not exactly the same in each run due to the difference in the noise term. The square plots indicate the results of Q-PF model explained in the previous section, while the circle plots represent those of Q-PF model without the antitrapping (AT) current which corresponds to the term multiplied by $\partial_t\phi$ on the right-hand side of equation (2). It is noted that the result of Q-PF model with AT term does not significantly depend on W, indicating the excellent performance of the present model. However, both r_1 and r_2 gradually decrease with increase in W in the case of Q-PF model without AT term. This comparison clearly demonstrates an importance of the AT term. In this study, we chose $W/d_0 = 2.0$ in the light of balance between the accuracy and the computational cost.

In Figure 1(b), r_2 is lower than r_1. In other words, the microsegregation of slow diffusing element is not significant compared to that of fast diffusing element. This is an important consequence of the tie-line shift as explained in detail below.

3.2. One-dimensional analysis

One-dimensional analysis, in which solidification proceeds merely by motion of the planar interface, should facilitate understanding of important aspects of the tie-line shift. Hence, we first discuss results of one-dimensional simulations in which the system length was set to 25 μm.

Figure 2(a) shows the liquid concentrations at the solid-liquid interface, c_i^*, at each solid fraction calculated for $|dT/dt| = 6$ K/s. For comparison, results of the Brody-Flemings [3], Clyne-Kurz [4] and Ohnaka's models [5] as well as the lever rule and the Scheil model [2] are shown in Figure 2 (a). The solidification time t_f obtained from the Q-PF simulation for the case B was employed to calculate the analytical models. The result of case B coincides with those of three analytical models. In the case A, on the other hand, such agreement was not obtained although not shown here. c_1^* in the case A (open square) is very close to the one in the case B (open circle), while c_2^* (filled square) exhibits quite different behaviour. The high values of c_2^* for $f_s < 0.7$ are associated with the formation of diffusion layer of this slow diffusing element. The concentration of slow diffusing element in the solid is accordingly high compared to the one of fast diffusing element up to $f_s = 0.7$. Then, c_2^* takes slightly lower value than c_1^* in the late stage of solidification. This behaviour is ascribable to the influence of tie-line shift. Figure 2(b) demonstrates the solidification paths at different cooling rates. The average concentrations in the liquid are plotted. The dotted line indicates the relation $c_1 = c_2$ and it is called the

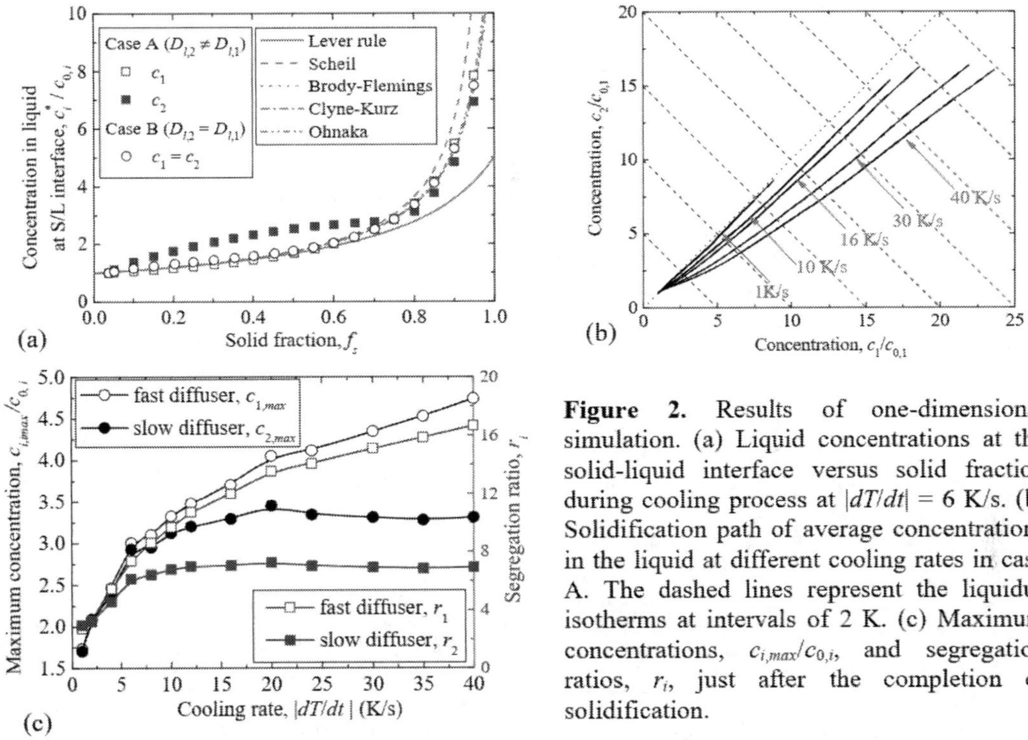

Figure 2. Results of one-dimensional simulation. (a) Liquid concentrations at the solid-liquid interface versus solid fraction during cooling process at $|dT/dt|$ = 6 K/s. (b) Solidification path of average concentrations in the liquid at different cooling rates in case A. The dashed lines represent the liquidus isotherms at intervals of 2 K. (c) Maximum concentrations, $c_{i,max}/c_{0,i}$, and segregation ratios, r_i, just after the completion of solidification.

normal path in that the lever rule, Scheil model, Brody-Flemings, Clyne-Kurz and Ohnaka's models follow this path. The solidification path in the case A gradually deviates from the normal path with increasing cooling rate, which indicates the occurrence of tie-line shift.

Figure 2(c) represents the dependences of $c_{i,max}$ and r_i on the cooling rate in the case A. The results of case B are not shown here because those are almost identical to the results of fast diffusing element in the case A. It is seen that when the cooling rate is higher than about 6 K/s, $c_{2,max}$ is always lower than $c_{1,max}$ and also r_2 is always lower than r_1. In other words, the degree of microsegregation of slow diffusing element is lower than that of fast diffusing element. Furthermore, although $c_{1,max}$ and r_1 gradually increases with increase in the cooling rate, $c_{2,max}$ and r_2 take almost constant values in the region of high cooling rates.

3.3. Two-dimensional analysis

As mentioned in the introduction, the amount of tie-line shift is dependent on morphology of the growing phase [10]. To be more specific, the amount of tie-line shift increases when the curvature radius of solid-liquid interface decreases. Hence, one may expect that refinement of the solidification microstructure should enhance the influence of the tie-line shift on the microsegregation.

Figure 3 (a) shows results of two-dimensional simulations for the case B. The planar interface is stable in the whole solidification process up to about 12 K/s and, thereby, the dependences of $c_{i,max}$ and r_i are almost identical to those of the one-dimensional simulations. However, the instability of planar interface takes place and secondary arms well develop during the cooling process at $|dT/dt|$ > 15 K/s. Then, $c_{i,max}$ slightly decreases with increase in the cooling rate, which is due to the reduction of the arm spacing. r_i takes almost the constant value when the cooling rate is higher than 15 K/s. These results represent effects of microstructural change on the microsegregation in the present system without the tie-line shift.

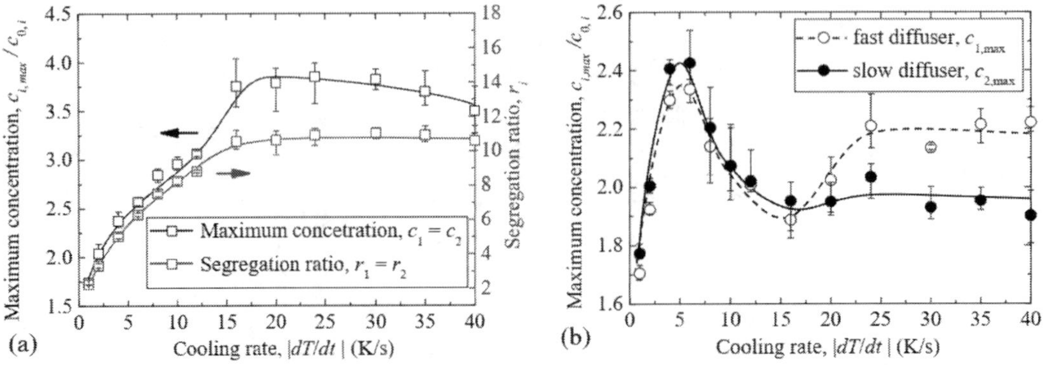

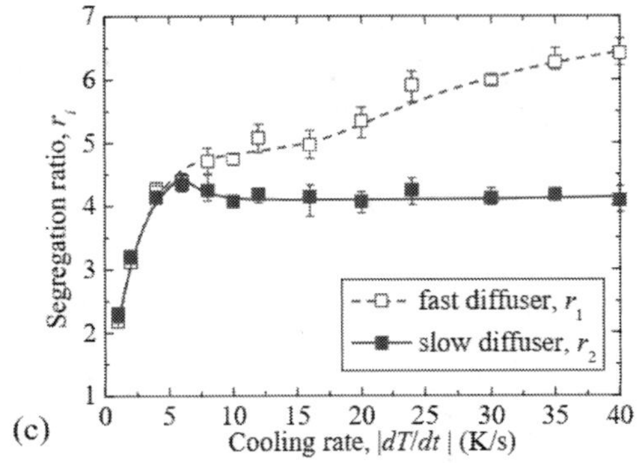

Figure 3. Results of two-dimensional simulations. (a) Maximum concentration and segregation ratio calculated for different cooling rates in case B. (b) Maximum concentrations and (c) segregation ratios calculated for different cooling rates in case A.

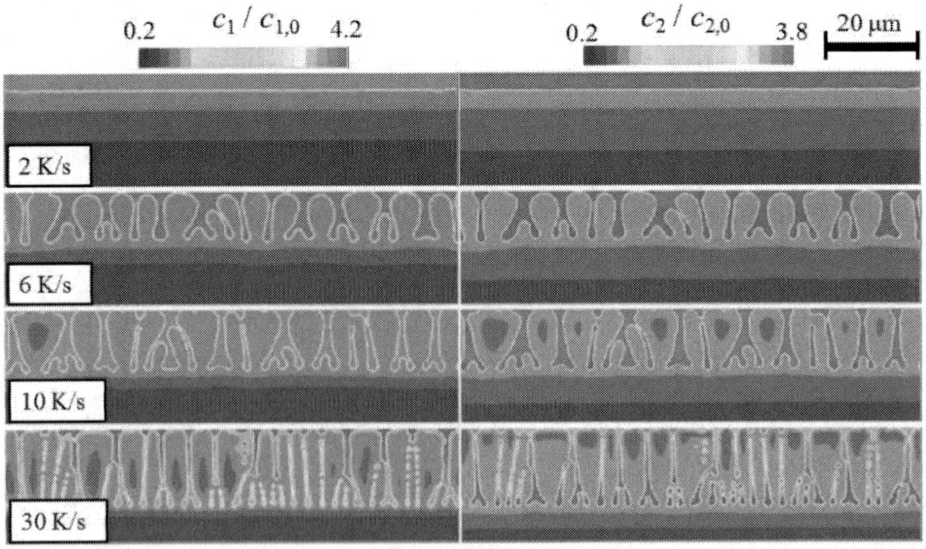

Figure 4. Snapshots of microstructures at f_s = 0.85 calculated for different cooling rates in case A. The snapshots on the left and right-hand sides represent concentration maps of fast and slow diffusing elements, respectively.

Figure 3(b) shows $c_{1,max}$ and $c_{2,max}$ in the case A. They drastically decrease when the cooling rate becomes higher than about 6 K/s. This is because the secondary arms start to develop even at 6 K/s

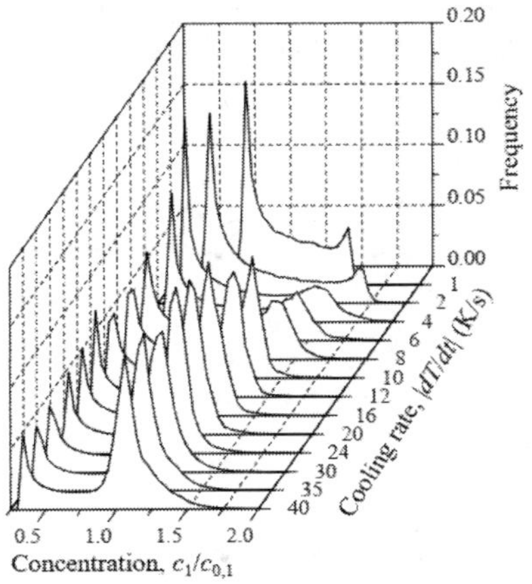

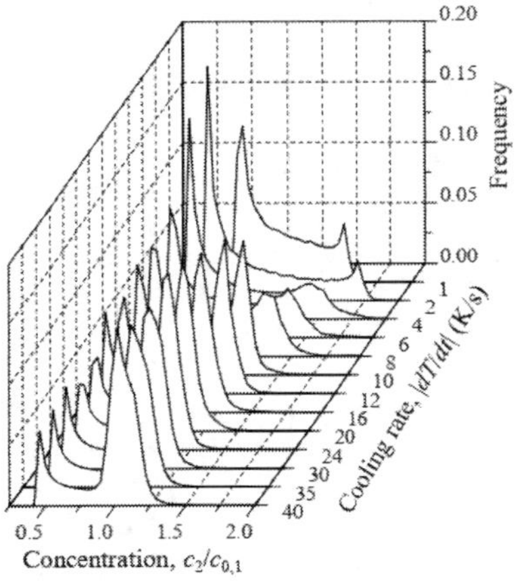

Figure 5. Frequency distributions of (a) fast and (b) slow diffusing elements for different cooling rates.

(see Figure 4) due to the formation of diffusion layer of the slow diffusing element and the microstructure accordingly becomes fine, which facilitates back diffusion in the solid. At $|dT/dt| > 15$ K/s, $c_{1,max}$ slightly increases, while $c_{2,max}$ is almost constant. In Figure 2 (c), also, r_2 takes a constant value when $|dT/dt|$ is higher than 6 K/s. The difference between r_1 and r_2 gradually increases with increase in the cooling rate. Hence, the microsegregation of slow diffusing element is less significant. It is also noted that the segregation ratio of the fast diffusing element in the case A (r_1 in Figure 3(c)) is lower than that that of the case B above $|dT/dt| = 6$ K/s. This is because the refinement of the microstructure (viz., growth of the secondary arm) starts to occur even at low cooling rates in the case A.

Distribution of solute concentrations immediately after the completion of the solidification can be understood in Figure 5 which shows relative frequencies of the solute concentrations in the microstructure in the case A. There are two peaks at each cooling rate in both the figures. The first peak in the low concentration region is related to the redistribution of solutes during motion of the planar interface. The second peak in the high concentration region is associated with the growth of secondary arms. In Figures 5(a) and (b), the height of the first peak gradually decreases and the one of the second peak increases as the cooling rate increases. Importantly, the positions of the first and second peaks for c_2 are closer to the average value $c_{0,i}$ (i.e., $c_i/c_{0,i} = 1$) than those for c_1 at each cooling rate. It indicates that the tie-line shift influences the redistributions of the slow diffusing element during both the planar interface motion and the growth of the secondary arms. As compared to c_1, c_2 is more uniformly distributed near the average value at each cooing rate.

4. Conclusions

Quantitative phase-field (Q-PF) simulations were carried out to investigate the influence of tie-line shift on the microsegregation behaviour in a ternary alloy consisting of fast and slow diffusing elements. We first verified the accuracy of the Q-PF simulation by performing the convergence test of the segregation ratios with respect to the interface thickness. The Q-PF simulations showed the excellent convergence behaviour. Then we carried out one-dimensional analysis. It was found that the microsegregation of slow diffusing element is reduced due to the tie-line shift. The difference between the segregation ratios of the fast and slow diffusing elements gradually increases with increase in the cooling rate. In two-dimensional simulations, the refinement of microstructure, viz., the decrease of the secondary arms spacing occurs at relatively slow cooling rates due to the formation of diffusion layer of the slow diffusing element. It yields reduction of degrees of microsegregation for both the fast and slow diffusing elements. Importantly, in a wide range of cooling rates, the degree of microsegregation for the slow diffusing element is always lower than that of the fast diffusing element, which is entirely ascribable to the influence of tie-line shift.

Acknowledgements
This work is supported by a Grant-in-Aid for Scientific Research (B) (No. 25289266) from MEXT, Japan and ISIJ research project. The authors also acknowledge the Strategic Programs for Innovative Research (SPIRE), MEXT, and the Computational Materials Science Initiative (CMSI), Japan.

References

[1] Ganesan M, Dye D and Lee P D2005 *Metall. Mater. Trans. A* **36A** 2191.
[2] Scheil E 1942 *Z. Metallkd.* **34** 70.
[3] Brody H D and Flemings M C 1966 *Trans. TMS-AIME* **236** 615.
[4] Clyne T W and Kurz W 1981 *Metall. Trans. A* **12A** 965.
[5] Ohnaka I 1986 *Trans. ISIJ* **26** 1045.
[6] Kobayashi S 1988 *Trans. ISIJ* **28** 728 & 535.
[7] Matsumiya T, Kajioka H, Mizoguchi S, Ueshima Y and Esaka H 1984 *Trans. ISIJ* **24** 873.
[8] Nastac L and Stefanescu D M 1993 *Metall. Trans. A* **24A** 2107.
[9] Won Y M and Thomas B G 2001 *Metall. Mater. Trans. A* **32A** 1755.
[10] Coates D E 1973 *Metall. Trans.* **4** 1077.
[11] Rappaz M and Boettinger W J 1999 *Acta Mater* **47** 3025.
[12] Boettinger W J, Warren J A, Beckermann C and Karma A 2002 *Annu. Rev. Mater. Res.* **32** 163.
[13] Steinbach I 2009 *Modelling Simul. Mater. Sci. Eng.* **17** 073001.
[14] Takaki T 2014 *ISIJ Int.* **54** 437.
[15] Takaki T, Shimokawabe M, Ohno M, Yamanaka A and Aoki T 2013 *J. Cryst. Growth* **382** 21.
[16] Karma A and Rappel W J 1998 *Phys. Rev. E* **57** 4323.
[17] Karma A 2001 *Phys. Rev. Lett.* **87** 115701.
[18] Echebarria B, Folch R, Karma A and Plapp M 2004 *Phys. Rev. E* **70** 061604.
[19] Folch R and Plapp M 2005 *Phys. Rev. E* **72** 011602.
[20] Kim S G 2007 *Acta Mater* **55** 4391.
[21] Ohno M and Matsuura K 2009 *Phys. Rev. E* **79** 031603.
[22] Ohno M and Matsuura K 2010 *Acta Mater.* **58** 5749.
[23] Ohno M 2012 *Phys. Rev. E* **86** 051603.
[24] Karma A and Rappel W J 1999 *Phys. Rev. E* **60** 3614.
[25] Echebarria B, Karma A and Gurevich S 2010 *Phys. Rev. E* **81** 021608.

A 4-D dataset for validation of crystal growth in a complex three-phase material, ice cream

P. Rockett[1], S. Karagadde[1,2], E. Guo[1], J. Bent[3], J. Hazekamp[3], M. Kingsley[1], J. Vila-Comamala[4], and P.D. Lee[1,*]

[1] School of Materials, The University of Manchester, Manchester, M13 9PL, U.K
[2] Department of Mechanical Engineering, Indian Institute of Technology Bombay, Mumbai 400076, India
[3] Unilever R&D, Colworth, U. K
[4] Diamond Light Source Ltd., Harwell Science and Innovation Campus, Didcot, OX11 0DE, U.K

* Corresponding Author: Peter.Lee@manchester.ac.uk

Abstract. Four dimensional (4D, or 3D plus time) X-ray tomographic imaging of phase changes in materials is quickly becoming an accepted tool for quantifying the development of microstructures to both inform and validate models. However, most of the systems studied have been relatively simple binary compositions with only two phases. In this study we present a quantitative dataset of the phase evolution in a complex three-phase material, ice cream. The microstructure of ice cream is an important parameter in terms of sensorial perception, and therefore quantification and modelling of the evolution of the microstructure with time and temperature is key to understanding its fabrication and storage. The microstructure consists of three phases, air cells, ice crystals, and unfrozen matrix. We perform *in situ* synchrotron X-ray imaging of ice cream samples using in-line phase contrast tomography, housed within a purpose built cold-stage (-40 to +20°C) with finely controlled variation in specimen temperature. The size and distribution of ice crystals and air cells during programmed temperature cycling are determined using 3D quantification. The microstructural evolution of three-phase materials has many other important applications ranging from biological to structural and functional material, hence this dataset can act as a validation case for numerical investigations on faceted and non-faceted crystal growth in a range of materials.

1. Introduction

Ice cream is a complex multi-phase material that roughly consists of 30% ice, 50% air, 5% fat and 15% sugar solution matrix, by volume, and thus contains the three states of matter – solid, liquid and gas. This investigation seeks to gain an informed understanding of the manner in which phase content alters with progressive temperature variation [1,2]. However, as far as structure is concerned the quality of the product is generally assumed to depend on the size of the constituent air cells and ice crystals, in both cases the smaller the better. Under abusing temperature variations that commonly occur during storage and distribution, the size of these two components are known to increase. In particular, ice crystallization has an important bearing on ice cream shelf life [3]. An understanding of the mechanisms behind ice formation would greatly aid manufacturers in predicting the effects of processing and in the development of formulations. The observation of ice crystallization is beset with technical difficulties given the opaque and delicate labile nature of ice cream. For optical, or electron

IOP Conf. Series: Materials Science and Engineering **84** (2015) 012076 doi:10.1088/1757-899X/84/1/012076

microscopic, relatively extensive alteration to the original composition is involved and in consequence diminishes the value of the resultant measurement. Fortunately, the application of X-ray microtomographic analysis can provide a non-invasive means of observation of three-dimensional structure and in most cases is able to sustain a natural environment. The results discussed in this report were obtained at the Diamond Light Source, using pink beam X-ray tomographic imaging on the Diamond Manchester Branchline (DMB). With filtered undulator harmonics pink beam combines the high intensity required for real time imaging dynamics coupled with a sufficient narrow energy window for high-resolution in-line phase contrast [4].

The motivation for this work is to encourage the application of analytical techniques that would be new to soft solid colloidal materials research but that are commonly employed in other disciplines such as metallic [5–9] and non-metallic [10–12] materials research. Some recent studies have also been performed using laboratory X-ray imaging [13,14]. In this respect the intention is to provide a series of datasets as a gold standard for analysis and model validation. The datasets comprise continuously recorded tomographic reconstructions of ice cream that have undergone phase change structural reorganization during a sequence of heating and cooling cycles. In particular this data will contain information of the dynamics of recrystallization of ice crystals and the coarsening of air cells that result from thermal abuse.

In order to subject a specimen of ice cream to the desired thermal conditions and at the same time provide the appropriate containment and rotational motion required for tomographic projection, a specialized cold-stage was developed for installation on the DMB. The method of temperature reduction was based upon heat extraction by thermoelectric cooling devices and temperature variation on controlled heat input from miniature resistance heaters. The translational and rotational motion instrumentation available on the DMB was incorporated in the cold-stage design to accurately manipulate the ice cream specimen. To match the intrinsic field of view of the beamline instrumentation the ice cream specimens were contained within polyetherimide (PEI) tubes of 3 mm inner diameter with 0.2 mm wall thickness and 10.5 mm length. With this arrangement the specimens could be subjected to temperature variation in the range from ambient values to -40°C with an accuracy of 0.1°C.

2. Methods:

2.1. Ice Cream preparation details

500 ml blocks of 5% fat ice cream were prepared using a scraped surface heat exchanger. The sample tubes were inserted into the ice cream before blast freezing at -35°C and subsequent storage at -25°C. The ice cream filled PEI tubes were cut from the block of ice cream on a bed of dry ice shortly before X-ray analysis.

2.2. Specimen Cold-stage

The specimen cold-stage provides temperature variation from ambient conditions to -40°C with a controlled accuracy approaching 0.1°C. The sample is mounted within a thermally radiating enclosure located within a copper cold block (see Fig. 1). The enclosure is sealed with thermally reflecting aluminium foil X-ray transmission windows at the entrance and exit of the X-ray path. The geometry of the coupling between the cold-block and the rotation spindle is chosen to incorporate existing beam-line and lab systems motion control stages. This approach preserves the proven features of the specimen-mounting arrangements and maintains the accuracy of the specimen transverse location and specimen rotation. In this respect the specimen rotation spindle is mounted on existing rotation stages and vertically loaded into the cold-block assembly. A contact angle of 5° between the spindle cone and a conical recess in the cold block provides a suitably conductive thermal path with accurate location. This geometry also allows spindle rotation with acceptable levels of motion friction.

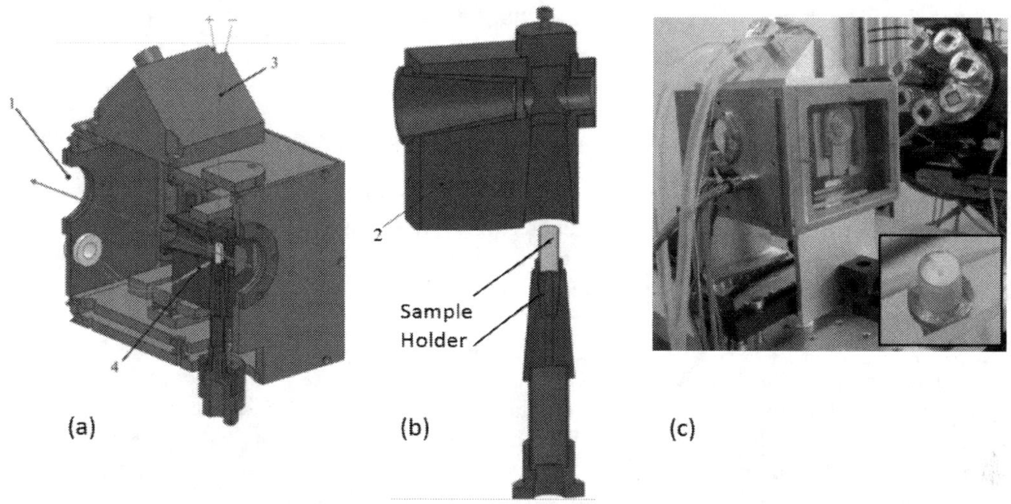

Fig 1: (a) Sectional view of the cold stage assembly (1-X-ray window; 2-Cold block; 3-Outer aluminium casing; 4-Platinum resistance thermometers) (b) Sample mounting and positioning (c) Picture of the cold stage installed in the Diamond Manchester Branchline, with the inset showing a typical sample.

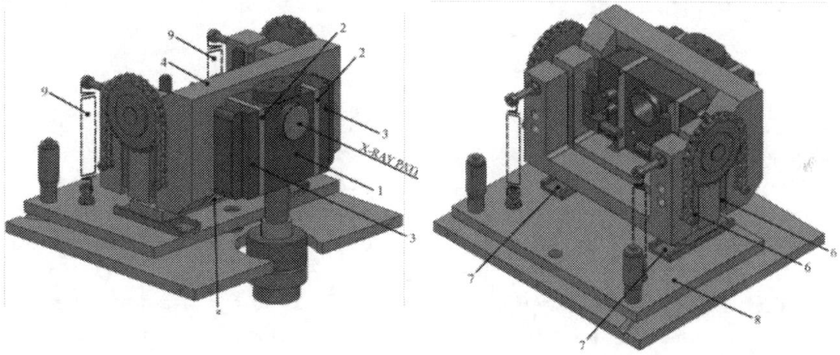

Fig 2: Cold-stage assembly mounted on three-axis slide system. 1-cold block, 2-thermoelectric coolers, 3-water cooled heat exchangers, 4-cooler compression frame, 5-lateral slides, 6-vertical slides, 7-longitudinal slides, 8-pitch and roll plate, 9-counter load springs

The cold block is sandwiched between two 70 W, 40 × 40 mm, planar thermoelectric cooling elements (Marlow Industries, Texas, USA) and held into position under soft compression by two water-cooled heat exchangers that remove heat from the cooling elements (Fig 2). To ensure that the alignment of the rotation stage determines the accurate location of the specimen spindle, the cold-block assembly is freely floating on a three-axis arrangement of miniature linear motion slides. The cold block assembly is mounted on a pitch and roll plate for micrometer alignment of the rotation spindle axis and the cold block recess axis. Two springs are attached between a vertical sliding frame and the pitch/ roll plate to provide adjustable load relief to the rotation spindle via circulating miniature chains.

To provide compatible combinations of image resolution and area the sample containment tubes vary in diameter from 3 – 5 mm. The tubes are fabricated from polyetherimide (PEI, Goodfellow,

Cambridge, UK) with 0.2 mm wall thickness and 10.5 mm length. These fit into a recess in a copper support capsule that conically locates in the top of the rotation spindle. The geometry of the cold-stage system has been structured to allow sample loading either from the top or from the bottom of the system, Fig 1(b).

The cold-stage assembly is housed within an environment chamber. A controlled gas bleed (~0.15 bar) prevents water droplet condensation and maintains the pressure at slightly above ambient so that seals are not required at the rotation spindle entry aperture and any other smaller access ports. Very soft flexible tubing is used for the heat exchanger water coolant and feed into the chamber in an unconstrained manner in order not to restrict the free movement of the cold block assembly. Two P100-type platinum resistance thermometers are inserted in the cold-block, as close as possible to the specimen support capsule, and a K-type thermocouple is inserted through the top of the cold-block assembly to monitor the specimen temperature.

Four 16 W cartridge heaters (Watlow, UK) are inserted at the rear of the cold-block to provide temperature control. The principle of operation is to set the thermoelectric cooler heat out-flux to a fixed value and to control the temperature of the cold-stage to the programmed set point with heat in-flux determined under computer control. The temperature controller (ITC503, Oxford Instruments, UK) is provided with 64 W of heater power and three input channels for sensor monitoring. Two channels were calibrated for P100-type platinum resistance thermometer inputs and one channel was calibrated for K-type thermocouple input. The system could be operated in either a local or remote control. The remote temperature control was implemented with a National Instruments 4-Port Ethernet Interface for RS232 (ENET-232/4). In this manner variable temperature sweep profiles could be programmed.

2.3. Synchrotron X-ray computed microtomography (SXMT)
Synchrotron X-ray tomography of the ice cream microstructure was performed on the Diamond Manchester Branchline at the Diamond Light Source, operating with a pink beam. The pink beam provides high flux undulator radiation with relatively small energy dispersion so that both high temporal and spatial resolution are available to perform 4D in-line phase contrast imaging [4]. A 2560 × 2160 pixel PCO Edge 5.5 CMOS camera optically coupled to a single crystal $CdWO_4$ scintillator resulted in a resolution of 1.6 μm / pixel. For the tomography, a total of 3600 projections were obtained over a 180° rotation with an exposure time of 0.1 s. For optimum image resolution the distance between the sample and the scintillator was ~13 mm. Back-projection reconstruction of the recorded tomographic projection datasets was performed using a combined Fourier-wavelet algorithm to suppress ring artefacts [15]. The reconstructed volume was 2560 x 2568 x 2160 voxels. The 2D images were processed and analysed with ImageJ. 3D visualization was made using Avizo®, FEI-VSG, France.

2.4. Experimental procedure
For long-term storage the ice cream samples, contained within 3 mm diameter PEI tubes, were stored in a -80°C freezer. Typically, a sample was mounted in the cold stage and stabilized at -20°C for ten minutes, after which a tomographic scan was recorded to determine the initial state. Following this the temperature was increased, at a ramp rate of 0.3°C/min, to -8°C and held there for 10 minutes. The sample was then cooled, at the same ramp rate, back to the base temperature of -20°C, after which a tomographic scan was recorded. The thermal cycle was repeated for the same sample but with the highest temperature now set at -7°C. This procedure of repetitive thermal cycling, as shown in Fig 3, contains features that are common to the temperature abuse of ice cream associated with incorrect storage handling.

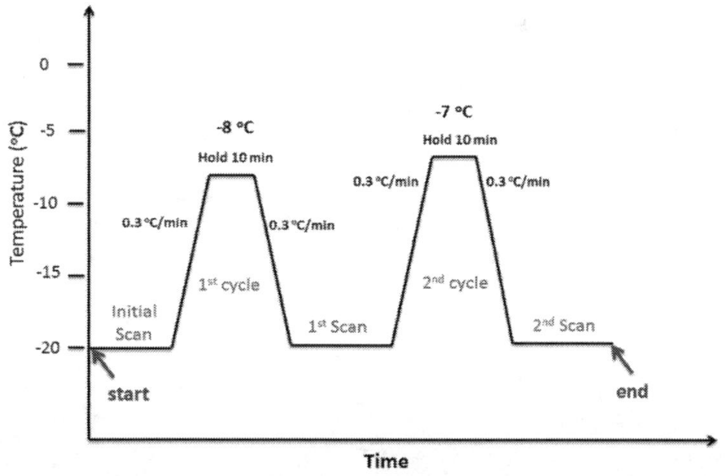

Fig 3: Schematic of the thermal cycling of the ice cream sample and the actions taken at each stage

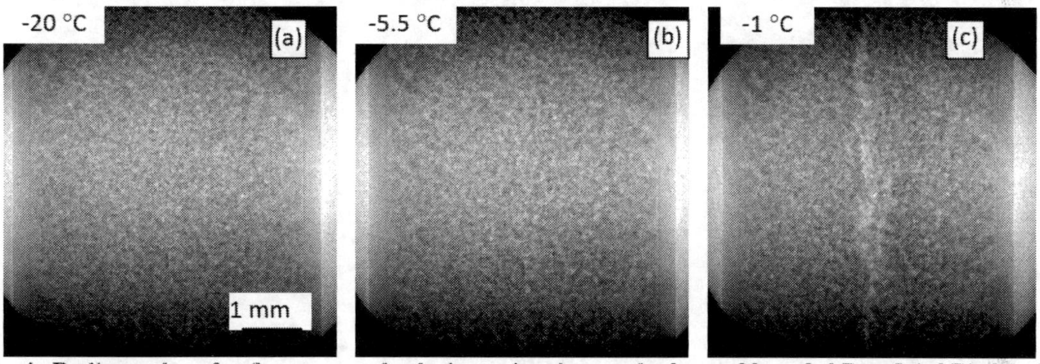

Fig 4: Radiographs of a 3 mm sample during a heating cycle from -20 to 0 °C at 0.1 °C/min: the images were taken at temperatures (a) -20, (b) -5.5 and (c) -1 °C

3. Results

3.1. Radiographic observations

In order to qualitatively capture the influence of thermal cycling, X-ray radiographic observations were performed by mounting the cold-stage inside the Phoenix vltomelx X-ray equipment. Continuous radiographs were acquired during a thermal cycle but with a slower cooling rate of 0.1 K/min. Figure 4 shows the radiographic images of a 3 mm ice cream sample at three temperatures when heated to 0°C from -20°C at 0.1°C/min. During heating, the overall structure remained unchanged until about -6°C. Upon further heating gradual expansion and an increase in air cell volume was noticed. From the literature [2] the melting point of ice cream is around -3°C, and heating beyond -3°C resulted in partial melting and formation of a separation/void within the ice cream sample (Fig 4c). Further cycling has shown that the formation of a void is permanent, and thus showing that -3°C is a critical temperature to be avoided for thermal abuse. Note that in the radiographs, dark grey indicates a denser phase such as sugar/protein matrix and ice-crystals whereas the white indicates air-cells. The lighter shade of grey corresponds to fat/liquid phase. These observations were performed only for a qualitative understanding, whereas the tomographic reconstruction is known to enhance the contrast and enable better segmentation of the image, allowing identification of the distinct phases.

3.2. Tomographic observations and quantification

Based on the observations from the radiography study, thermal cycles were performed and tomographic scans were acquired at -20 °C as outlined in section 2.3. The images were reconstructed [15] and further processed with Avizo® (FEI-VSG, France). Figure 5 shows the sequence of images and a magnified region. The dark grey (black) colour corresponds to a phase with lowest density. Hence in these images the black colour represents air-cells, which are separately shown in the magnified views 5(d-f). The reconstructed volumes are then used to separate the air-cells from the rest of the matrix. The reconstructed images are filtered using Anisotropic diffusion filter, and thresholded to distinguish air cells from the rest of the sample. The connected air cells are separated using watershed separation algorithm [16]. Figures 5(g-i) show 3D rendering of separated air cells. It is evident that the thermal cycling has introduced significant coalescence and growth of air cells.

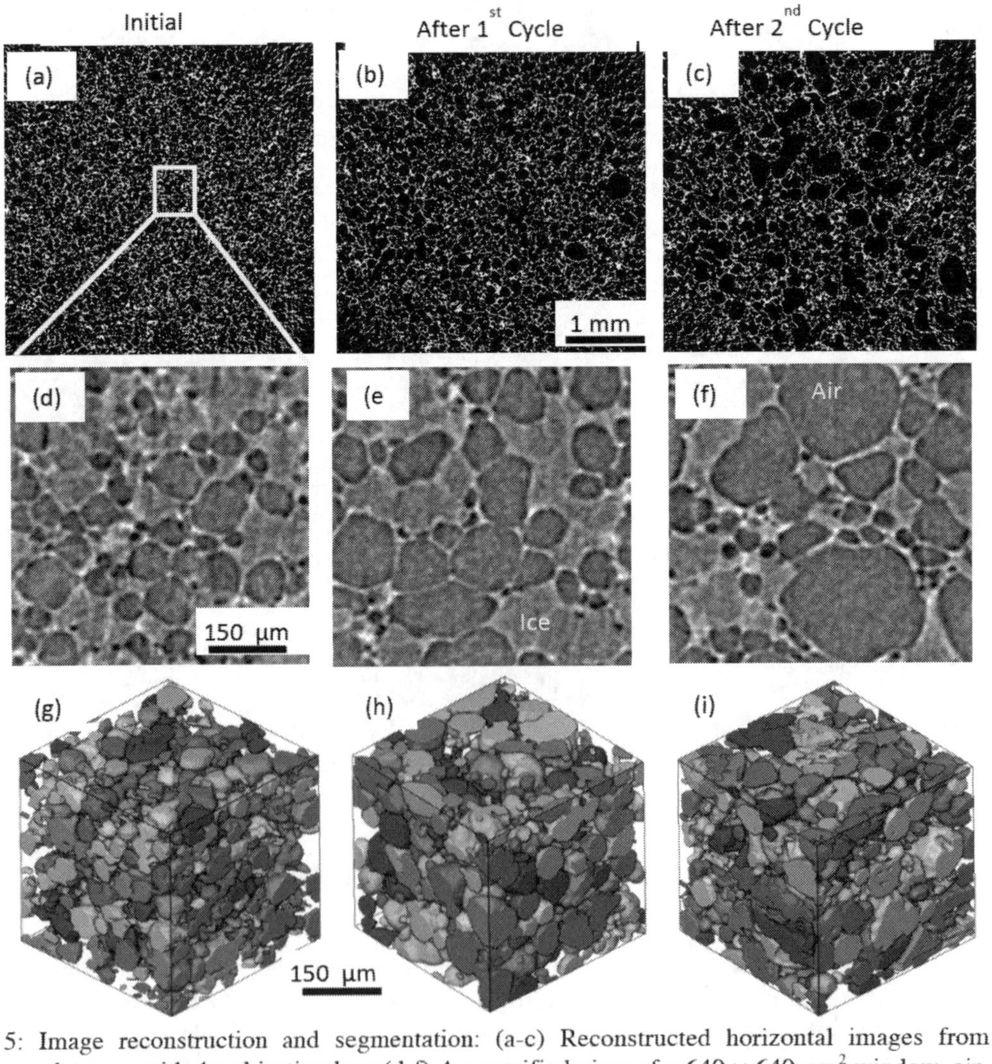

Fig 5: Image reconstruction and segmentation: (a-c) Reconstructed horizontal images from the tomography scan with 4× objective lens (d-f) A magnified view of a 640 × 640 μm^2 window, air-dark grey & ice crystals-light grey (g-i) Separated air-cells rendered in 3D during each stage

The volume fraction of air cells and their size distribution can be measured upon further quantification on a 1 mm^3 volume. The size distribution is shown in Fig. 6, with the mode in pore size increasing after the first cycle from 75 to 135 μm equivalent pore diameter, with the number density in air bubbles continuous decreasing from initially 1976 to 1396 and 1263/ mm^3 after the first and second cycles respectively. This quantifies the effect of thermal cycling on the ice cream sample.

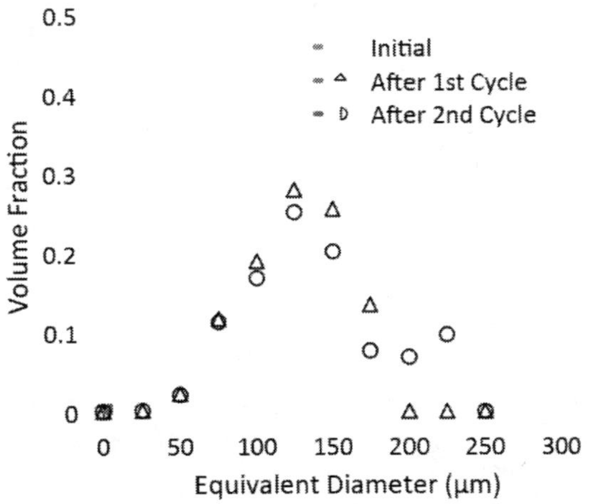

Fig 6: Size distribution of air cells in a 1 mm^3 volume as a function of cycling. The volume fraction of air cells during each stage is indicated in the legend.

Further analysis is required to investigate the effect of thermal cycling on the ice-crystals performed over a large number of cycles to represent industrial production and usage. We believe that the technique and capabilities presented in this work will enable these investigations not only on the ice cream but on a range of thermally controlled multi-phase soft materials.

4. Conclusions

This paper highlights a capability of capturing 4D observations on a multi-phase soft solids such as ice cream using synchrotron X-ray microtomography. The purpose built cold-stage is able to accurately control temperatures in the range of -40 to 20 °C. and allow simultaneous rotation of the specimen for tomographic imaging whilst maintaining a sealed environment. The results have shown successful tomographic imaging of ice cream samples subjected to thermal cycling from a base temperature of -20°C. Two thermal cycles are performed on the sample from -20 up to -7 °C at a constant rate of 0.3 °C/min. This has resulted in an increase in the equivalent diameter of air cells from 75 μm to 135 μm, with the number density decreasing from 1976 to 1263 /mm^3. The technique established here can be further utilized to not only understand the behaviour individual crystals with high resolution imaging but also to extend the methodology to analogous systems that are critical for food and polymer processing.

Acknowledgements

This work was made possible by the facilities and support provided by Unilever R&D, Colworth U.K, the Diamond Manchester Collaboration (beamtime MT11078-1) and the Research Complex at Harwell (EPSRC-EP/I02249X/1).

IOP Conf. Series: Materials Science and Engineering **84** (2015) 012076 doi:10.1088/1757-899X/84/1/012076

References

[1] Goff H D and Hartel R W 2013 *Ice Cream* (Springer)

[2] Clarke C 2012 *Science of Ice Cream* (Royal Society of Chemistry)

[3] Cook K L K and Hartel R W 2010 Mechanisms of Ice Crystallization in Ice Cream Production *Compr. Rev. Food Sci. Food Saf.* **9** 213–22

[4] Rau C, Weitkamp T, Snigirev A A, Schroer C G, Benner B, Tuemmler J, Guenzler T F, Kuhlmann M, Lengeler B, Krill III C E, Doebrich K, Michels D and Michels A 2002 Tomography with high resolution vol 4503 pp 14–22

[5] Lee P D and Hunt J D 1997 Hydrogen porosity in directional solidified aluminium-copper alloys: In situ observation *Acta Mater.* **45** 4155–69

[6] Kareh K M, Lee P D, Atwood R C, Connolley T and Gourlay C M 2014 Revealing the micromechanisms behind semi-solid metal deformation with time-resolved X-ray tomography *Nat Commun* **5** 4464

[7] Guillemot G, Gandin C-A and Bellet M 2007 Interaction between single grain solidification and macrosegregation: Application of a cellular automaton—Finite element model *J. Cryst. Growth* **303** 58–68

[8] Asta M, Beckermann C, Karma A, Kurz W, Napolitano R, Plapp M, Purdy G, Rappaz M and Trivedi R 2009 Solidification microstructures and solid-state parallels: Recent developments, future directions *Acta Mater.* **57** 941–71

[9] Yasuda H, Yamamoto Y, Nakatsuka N, Nagira T, Yoshiya M, Sugiyama A, Ohnaka I, Umetani K and Uesugi K 2008 In situ observation of nucleation, fragmentation and microstructure evolution in Sn-Bi and Al-Cu alloys *Int. J. Cast Met. Res.* **21** 125–8

[10] Bhreasail Á N, Lee P D, O'Sullivan C, Fenton C H, Hamilton R, Rockett P and Connolley T 2012 In-Situ Observation of Cracks in Frozen Soil using Synchrotron Tomography *Permafr. Periglac. Process.* **23** 170–6

[11] Appolaire B, Albert V, Combeau H and Lesoult G 1998 Free growth of equiaxed crystals settling in undercooled NH4Cl-H2O melts *Acta Mater.* **46** 5851–62

[12] Farup I, Drezet J and Rappaz M 2001 In situ observation of hot tearing formation in succinonitrile-acetone *Acta Mater.* **49** 1261–9

[13] Pinzer B R, Medebach a., Limbach H J, Dubois C, Stampanoni M and Schneebeli M 2012 3D-characterization of three-phase systems using X-ray tomography: tracking the microstructural evolution in ice cream *Soft Matter* **8** 4584

[14] Van Dalen G 2012 A Study of Bubbles in Foods by X-Ray Microtomography and Image Analysis *Microsc. Anal.* 26(2):S8–12 (AM)

[15] Titarenko S, Titarenko V, Kyrieleis A and Withers P J 2009 A ring artifact suppression algorithm based on a priori information *Appl. Phys. Lett.* **95** 071113

[16] Kareh K M, Lee P D and Gourlay C M 2012 In situ , time-resolved tomography for validating models of deformation in semi-solid alloys *IOP Conf. Ser. Mater. Sci. Eng.* **33** 012037

Dynamical microstructure formation in 3D directional solidification of transparent model alloys: in situ characterization in DECLIC Directional Solidification Insert under diffusion transport in microgravity

N Bergeon[1], F L Mota[1], L Chen[1], D Tourret[2], J M Debierre[1], R Guérin[1], A Karma[2], B Billia[1], and R Trivedi[3].

[1]IM2NP, Aix-Marseille Université, CNRS UMR 7334, Marseille, France
[2]Physics Department, Northeastern University, Boston, USA,
[3]Department of Materials Science & Engineering, Iowa State University, USA

Email : nathalie.bergeon@im2np.fr

Abstract : To clarify and characterize the fundamental physical mechanisms active in the dynamical formation of three-dimensional (3D) arrays of cells and dendrites under diffusive growth conditions, in situ monitoring of series of experiments on transparent model alloy *succinonitrile* – 0.24 wt% *camphor* was carried out under low gravity in the DECLIC Directional Solidification Insert on-board the International Space Station. These experiments offered the very unique opportunity to in situ observe and characterize the whole development of the microstructure in extended 3D patterns. The experimental methods will be first briefly described, including in particular the observation modes and the image analysis procedures developed to quantitatively characterize the patterns. Microgravity environment provided the conditions to get quantitative benchmark data: homogeneous patterns corresponding to homogeneous values of control parameters along the whole interface were obtained. The sequence of microstructure formation will be presented as well as the evolution of the primary spacing which is one of the most important pattern characteristic. Time evolution of this primary spacing during the microstructure development will be analysed to identify the mechanisms of spacing selection and adjustment; the importance of the macroscopic interfacial curvature will be pointed out.

1. Introduction

The study of solidification microstructure formation and selection is very important in the engineering and processing of advanced new materials [1]. The interface patterns formed by solidification largely govern mechanical and physical properties of materials. The two most typical microstructures are cells and dendrites, and their characteristics profoundly influence the properties of a wide range of commercial metallic alloys. The correlation between microstructure and processing conditions can be best examined by directional solidification experiments in which all parameters can be accurately and independently controlled.

The evolution of the spatial pattern of cells or dendrites occurs under dynamic conditions of growth in which the forming pattern reorganizes into a rather periodic array so that *in situ* and real time observation of the solid-liquid interface is a precious tool to get a detailed knowledge of the entire time-evolution of the interface pattern. This explains the extensive use of transparent organic analogs that behave like metallic alloys regarding solidification but are transparent to visible light so that classical optical techniques are sufficient for their observation [2].

A major impediment in the study of three-dimensional pattern evolution in ground-based experiments is the strong influence of convection on cellular and dendritic patterns. Extensive ground-based studies carried out in metallic and organic bulk samples have clearly established that fluid flow modifies the structure of the solute boundary layer that influences the pattern development [3,4]. Directional solidification experiments under low gravity conditions provide a unique setting to investigate microstructure development in spatially extended sample geometries with negligible convection and under well-controlled conditions of growth rate, temperature gradient and alloy composition. Consequently, those experiments provide unique benchmark data to validate the predictions of theoretical and computational models in a purely diffusive growth regime. The study presented here was thus conducted using the Directional Solidification Insert (DSI) dedicated to *in situ* and real time characterization of the dynamical selection of the solid-liquid interface morphology in *bulk* samples of transparent materials, which was developed by the French Space Agency (CNES) in the frame of the DECLIC project (Dispositif pour l'Etude de la Croissance et des LIquides Critiques). The DECLIC facility of CNES was launched with 17-A shuttle flight (August 2009) as part of a joint NASA/CNES research program and is installed on the International Space Station in microgravity environment. The main instrument monitoring is performed from the CADMOS center, the French User Support and Operation Centre (USOC) in Toulouse. Taking advantage of provided tele-science capabilities, scientists have the possibility to follow in near real-time conditions and to remotely control experiments. The commissioning of the DSI was realized in December 2009 and six runs of two to three weeks each were performed from April 2010 to April 2011. The DSI was then brought back on ground and two runs of two weeks have been performed at the CNES center with similar control parameters than experiments in μg so that convection influence could be addressed.

After a brief description of the experimental procedure and methods of analysis of images, results will be described and discussed in terms of pattern formation and evolution.

2. Experimental procedure

The DECLIC-DSI insert mainly contains 2 elements: the Bridgman furnace and the experimental cartridge. Complete descriptions of DECLIC and its inserts can be found in [5-6].

The experimental cartridge includes the quartz crucible and a system of volume compensation made of stainless steel that is useful to accommodate the specimen volume variations associated to phase changes. The cylindrical crucible has an inner diameter of ten millimeters and a length that enables about ten centimeters solidification, thus allowing the study of the whole development of extended 3D patterns from the initial stage to the steady state. The crucible is equipped with a flat glass window at the bottom and a lens immersed in the melt at the top. The main observation mode (Axial observation) takes advantage of the complete axial transparency of the cartridge provided by these last two elements: the light coming from LEDs passes through the cartridge from the bottom to the top, therefore crossing the interface. The optical imaging system formed by the immersed lens and a following relay lens forms the image of the interface on a CCD camera. On the same cartridge axis, an interferometer is also set but it will not be detailed here (more information on its use can be found in [7]). The interface can also be observed from the side (Transverse observation).

In this study, we used a Succinonitrile (SCN) – 0.24 wt% Camphor alloy. In order to fill the crucible, SCN purified by NASA by both distillation and zone melting was used. The alloy was then prepared by adding the solute. All procedures for sample preparation were carefully realized under vacuum to avoid humidity contamination. Once sealed, the cartridge was inserted inside the Bridgman furnace.

A thermal gradient ranging from 10 to 30 K.cm^{-1} is imposed by regulated hot and cold zones, respectively, located above and below the adiabatic area where the interface is positioned. Upward solidification is achieved by pulling the crucible down into the cold zone of the furnace at a rate ranging from 0.1 to 30 μm/s. For a defined alloy concentration, varying the thermal gradient or more conveniently, the pulling rate, leads to the development of various interface morphologies, ranging from planar front (high gradient – low pulling rate) to cellular and finally dendritic pattern (low gradient – high pulling rate). In the experiments detailed here, the thermal gradient is set to 19 K/cm and pulling velocities range from 0.5 to 12 μm/s.

3.Analysis methods

In situ and continuous observation leads to a huge amount of images so that adequate and as systematic as possible procedures had to be developed and validated to extract relevant quantitative data. Quantitative characterization of the patterns consists in measuring the evolution, as a function of time and growth parameters, of the parameters that describe the interface morphology, such as the primary spacing, order/disorder level of pattern, 3D-shape and the tip radius. Concerning shape of structures and tip radius, measurements are based on interferometric analysis that is not described here. For more information on shape reconstruction by interferometric analysis in DECLIC-DSI, one may refer to Bergeon *et al.* [7].

Image treatment and analysis procedures have been developed on Visilog® Image Analysis Software (Noesis, France) to facilitate exploitation of results. The first step is to obtain a binary image that is then used as an input for specific procedures dedicated to the determination of the primary spacing distribution and the characterization of the pattern order/disorder and type (Minimal Spanning Tree analysis [8,9] and number of first neighbors [9]). A more detailed description of these procedures is given in [9]. In this article, analyses are limited to characterization of primary spacing evolution. Primary spacing can be considered as the centre to centre distance between two neighboring cells. Its determination requires beforehand cell detection and identification. Each cell center is then determined together with the identification of the first neighbors so that center to center distances can be calculated. The spacing histogram is then drawn to determine the average spacing and the standard deviation, the minimal and maximal limits.

4. Results

Figure 1 illustrates the microstructures obtained for pulling rates ranging from 0.5 μm/s to 12 μm/s. In all cases, the pictures correspond to the interface microstructure after about 40 mm of growth. The microstructure is cellular with large spacing at low pulling rate, for example at 0.5 μm/s (figure 1(a)) or 2 μm/s (figure1(b)) and the deep cells observed at 4 μm/s (figure 1(c)) become dendritic (sharpening of the tips, anisotropy of the tip shape) at 8 μm/s as observed on the zooms of figure 1(d) and 1(e). The spacing decreases when the pulling rate increases and this decrease is faster at low pulling rates.

The initial stages of microstructure formation are similar regardless the pulling rate. They are here described on the basis of the experiment at $V_P = 4$ μm/s. The starting interface corresponds to $V_P = 0$ (figure 2(a)). After triggering of the pulling, morphological instability initiates by forming linear ridges along sub-boundaries that finally underline a rather complex array (figure2(b)). Between these defects, in interface areas that are still smooth, some poxes may be present: they can be described as circular undulations of the interface. Ridges underlining sub-grain boundaries progressively emit new ridges in their transverse direction and simultaneously, a uniform corrugation that corresponds to the initial visible wavelength of morphological instability invades the interface (figure 2(c)). The amplitude of all these interface modulations starts to increase, channels are forming but it is still difficult to identify cells (figure 2(d)). At this stage, the interface dynamics is extremely fast and pattern disorder high. There is then a progressive decrease of disorder and a clear pattern of cells eventually emerges (fig. 2(e)). The dynamics then slows down and is limited to progressive size adjustment and array ordering (figure 2(f)).

The evolution of spacing was measured as a function of time for each pulling rate. Two examples of such data are given in figure 3 with the spacing variation as a function of solidified length (namely the "pulling length" = $V_P$xt), for V_P = 0.5 μm/s and V_P = 4 μm/s. At the very beginning of microstructure formation, it is not possible to determine the spacing with center-to-center methods as microstructure does not form distinct cells (images 2(c) and 2(d) for example). Therefore the three first measurements are performed using a FFT (Fast Fourier Transform) based method. At low pulling velocity (for example 0.5 μm/s), the spacing progressively increases to reach its steady-state value. At higher velocity (for example 4 μm/s), the spacing displays an overgrowth of cells at the beginning of solidification before a slow decrease to reach a steady-state value; this dynamics is consistent with previous observations on thin samples of succinonitrile-based alloys, for example by Seetharaman et al. [10].

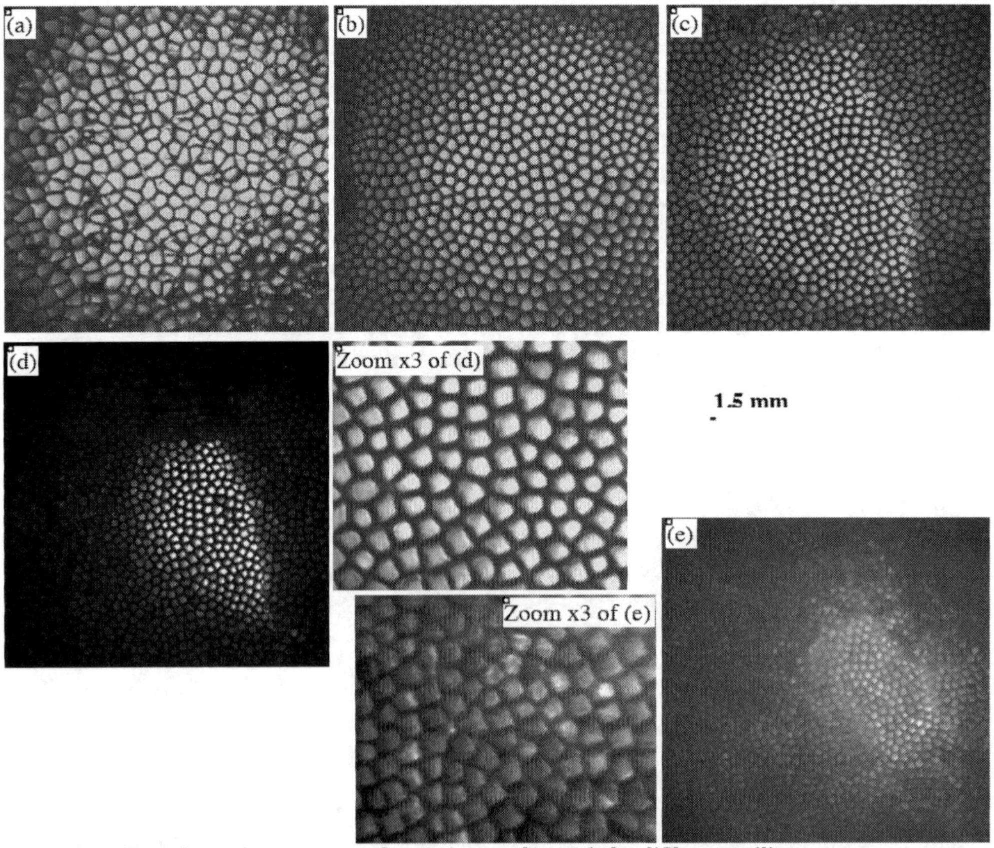

Fig. 1: Examples of interface microstructure after ~ 40 mm of growth for different pulling rates
(a) V_P = 0.5 µm/s (b) V_P = 2 µm/s (c) V_P = 4 µm/s (d) V_P = 8 µm/s (e) V_P = 12 µm/s
(SCN – 0.24 wt% camphor ; G = 19 K/cm)

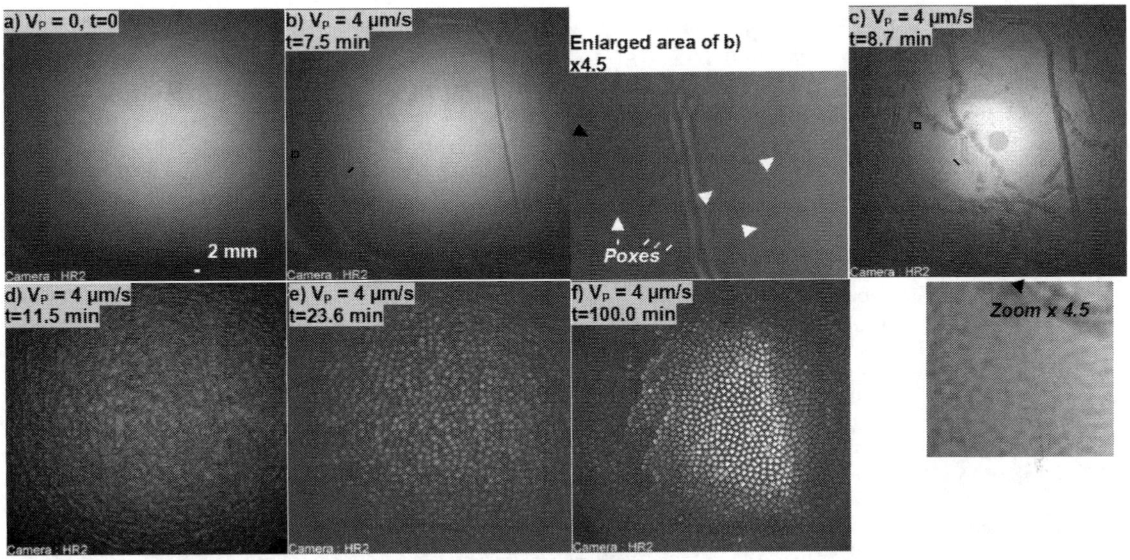

Fig. 2: Microstructure formation and evolution for V_P = 4 µm/s (SCN - 0.24 wt% Camphor, G = 19 K/cm)

Classically, two mechanisms are considered that adjust the spacing. To increase the spacing, smallest cells slightly slow down and are finally eliminated by overgrowth by their neighbours so that the number of cells decreases: "elimination" in fig. 4a. To decrease the spacing, the largest cells split into two daughter cells so that the number of cells increases: "tip-splitting" in fig. 4b. Even if these two mechanisms have been observed in our experiments, detailed analyses revealed that their occurrence is not spatially homogeneous and additional phenomena have to be taken into account. Let us focus for example on the pulling velocities 0.5 and 4 μm/s.

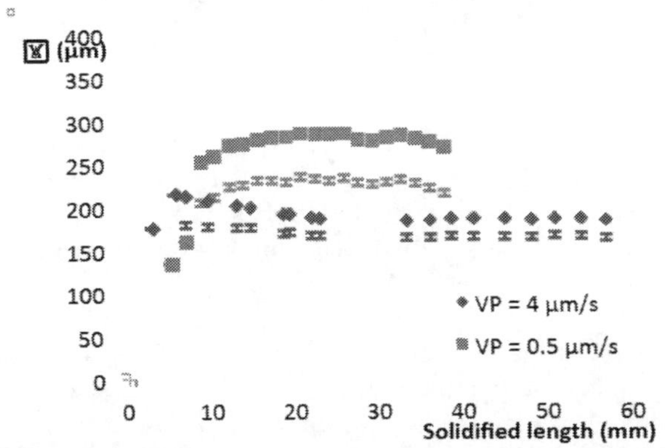

Fig. 3: Primary spacing evolution as a function of the solidified length for V_P = 4 µm/s and V_P = 0.5 µm/s (SCN - 0.24 wt% Camphor, G = 19 K/cm)

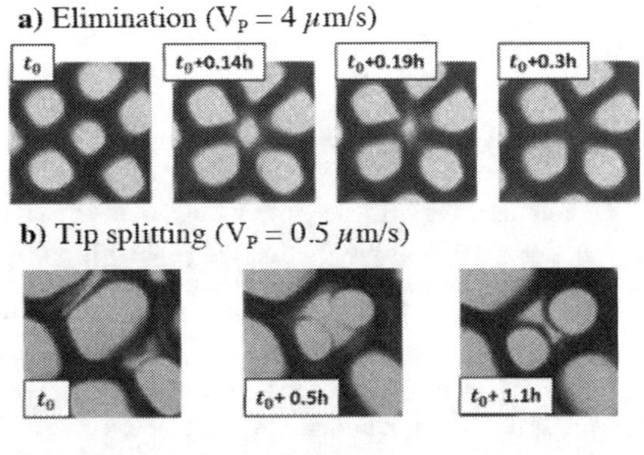

Fig. 4: Mechanisms of spacing adjustment (SCN - 0.24 wt% Camphor, G = 19 K/cm)

For $V_P = 0.5\ \mu m/s$, the interface is macroscopically convex (fig. 5a). This curvature induces a glide of cells along the interface slope from the center towards the border, as for example observed in thin samples by Bottin-Rousseau et al. [11] for experiments with curvature purposely induced. During this gliding, cell size increases so that cells eventually split and the number of cells increases; inversely, the gliding drives the cells towards the border where they disappear and the number of cells decreases. The average steady-state spacing results from the permanent balance between these two phenomena. Homogeneous elimination is not the process for spacing adjustment whereas the spacing increases to reach its steady-state value.

For $V_P = 4\ \mu m/s$, the interface is macroscopically concave (fig. 5b). This curvature induces a glide of cells along the interface slope from the border towards the centre. As it was previously described by Weiss et al. [12] for ground experiments in the same cylindrical geometry, cells are then generated at the border ("source"). During their gliding, their size tends to decrease so that elimination process occurs. In our case, it appeared that even if the gliding is mainly radial, elimination mainly occurs along sub-boundaries that are present in this experiment. The average steady-state spacing results from the permanent balance between the source of cells at the border and the sink along the sub-boundaries. Homogeneous tip-splitting is not the process for spacing adjustment whereas the spacing decreases to reach its steady-state value.

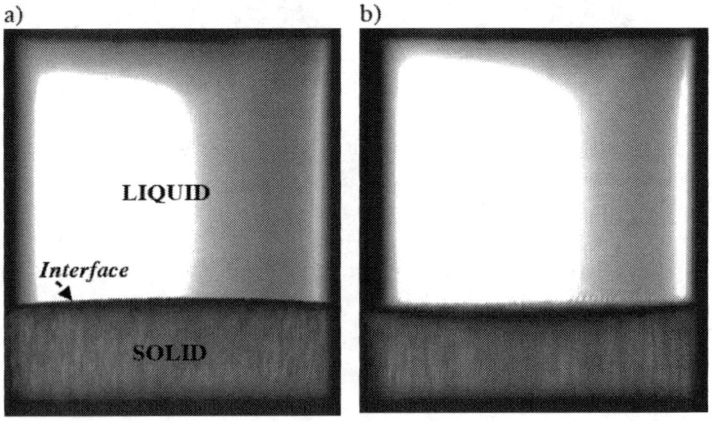

Fig. 5: Side view of the stationary interface for a) $V_P = 0.5\ \mu m/s$ and b) $V_P = 4\ \mu m/s$

(SCN - 0.24 wt% Camphor, G = 19 K/cm)

5. Conclusion

In this article, we presented some results obtained in microgravity using the Directional Solidification Insert of the DECLIC facility on board ISS. The quality of the experimental system that provides very clear images of the interfacial microstructure during its whole evolution has been demonstrated. Microgravity environment provided the conditions to get quantitative benchmark data: homogeneous patterns corresponding to homogeneous values of control parameters along the whole interface were obtained. Long solidifications have been performed to get the whole dynamics and mechanisms of microstructure formation or change, spacing adjustment, pattern ordering. The analyses of mechanisms of spacing adjustment reveal a critical influence of the macroscopic interfacial curvature that induces pattern gliding. The dynamical coupling between the evolution of the macroscopic curvature of the front and the spacing adjustment has been shown theoretically to have an important effect on the stability limit of both cellular [13] and eutectic [14] patterns formed during directional solidification. However, how this coupling affects spacing selection in a situation where the front curvature is maintained by a radial component of the temperature gradient remains poorly understood theoretically. Phase-field simulations of the present experiments that address this question are currently underway.

Acknowledgments

The authors express their gratitude to CNES (Centre National d'Etudes Spatiales) and NASA (National Aeronautics and Space Administration) for the support received in the scientific projects MISOL3D (Microstructures de SOLidification 3D) and DSIP (Dynamical Selection of Interface Patterns). A.K. and D.T. were supported by NASA awards NNX07AK69G and NNX11AC09G.

References

[1] Billia B and Fecht HJ 2001 *A World without Gravity – Research in Space for Health and Industrial Processes* Ed. Fitton B, Battrick B (ESA SP 1251, European Space Agency, Noordwijk)

[2] Billia B and Trivedi R (1993) *Handbook of crystal growth vol 1B* Ed Hurle DTJ (North-Holland)

[3] Jamgotchian H, Bergeon N, Benielli D, Voge P, Billia B and Guérin R 2001 *Phys. Rev. Lett.* **87** 166105

[4] Schenk T, Nguyen Thi H, Gastaldi J, Reinhart G, Cristiglio V, Mangelinck-Noël N, Klein H, Härtwig J, Grushko B, Billia B and Baruchel J 2005 *J. Crystal Growth* **275** 201

[5] Marcout R, Raymond G, Martin B, Cambon G, Zappoli B, Duclos F, Barde S, Beysens D, Garrabos Y, Lecoutre C, Billia B, Bergeon N and Mangelinck N "DECLIC: a facility to investigate fluids and transparent materials in microgravity conditions in ISS" *Proc. of the 57th International Astronautical Congress (2–6 Oct, 2006 ,Valencia, Spain).*

[6] Pont G, Barde S, Zappoli B, Duclos F, Garrabos Y, Lecoutre C, Beysens D, Billia B, Bergeon N, Mangelinck-Noël N, Marcout R and Blonde D "DECLIC: a facility to study crystallization and critical fluids" *Proc. Of the 60th International Astronautical Congress (12–16 Oct, 2009 , Daejeon, Republic of Korea).*

[7] Bergeon N, Weiss C, Mangelinck-Noël N and Billia B 2009 *Trans. Indian Inst. of Met.* **62** 455

[8] Dussert C, Rasigni G, Rasigni M, Palmari J, Llebaria A. 1986 *Phys. Rev. B* **34** 3528

[9] Bergeon N, Ramirez A, Chen L, Billia B, Gu J and Trivedi R 2011 *J. Mater. Sci.* **46** 6191

[10] Seetharaman V, Eshelman MA and Trivedi R 1988 *Acta Metall.* **36** 1175

[11] Bottin-Rousseau S and Pocheau A 2001 *Phys. Rev. Lett.* **87** 076101

[12] Weiss C, Bergeon N, Mangelinck-Noël N and Billia B 2009 *Phys. Rev. E* **79** 011605

[13] Karma A and Pelce P 1990, *Phys. Rev. A* **41** 6741

[14] Akamatsu S, Plapp M, Faivre G and Karma A, *Phys. Rev. E* **66** 030501

IOP Conf. Series: Materials Science and Engineering **84** (2015) 012078 doi:10.1088/1757-899X/84/1/012078

Localization of shear strain and shear band formation induced by deformation in semi-solid Al-Cu alloys

T Nagira[1], S Morita[2], H Yasuda[2], C M Gourlay[3], M Yoshiya[1], A Sugiyama[4] and K Uesugi[5]

[1]Department of Adaptive Machine Systems, Osaka University, 2-1 Yamadaoka, Suita, Osaka, 565-0871, Japan

[2] Department of Materials Science and Engineering, Graduate School of Engineering, Kyoto University, Yoshidahonmach, Sayo-ku, 606-8501, Japan

[3] Imperial College London, Department of Materials, London, SW7 2AZ, UK

[4] Department of Engineering, Osaka Sangyo University, 3-1-1 Nakagaito, Daitou, Osaka, 574-8530, Japan

[5] Japan Synchrotron Radiation Research Institute, 1-1-1, Kouto, Sayo-cho, Sayo-gun, Hyogo, 679-5198, Japan

E-mail: nagira@ams.eng.osaka-u.ac.jp

Abstract. In situ observation of deformation in globular Al-Cu samples at a solid fraction of ~50% and a global shear strain rate of 10^{-1} s^{-1} was performed using time-resolved X-ray imaging. The solid particle motion during shear was quantitatively analysed. The force was transmitted though the contacts between solid particles over a long distance parallel to the shear plane (18 mean grain size, *d*) after only a 1d increment of the Al_2O_3 push-plate motion. On the other hand, the distance of transmitted force in the perpendicular direction to the shear plane was restricted to approximately 11d even for a high displacement of the Al_2O_3 push-plate. A relatively high shear strain rate became localized at the shear domain after a small amount of deformation (a 1d increment). The solid fraction decreased in the region of localized shear strain rate. The shear band width, where the shear strain was localized and the solid fraction decreased, remained mostly unchanged over a 4 d increment of Al_2O_3 push-plate motion.

1. Introduction

Deformation of semi-solid alloys induced by solidification shrinkage and external forces applied by several casting processes often causes casting defects including macrosegregation, porosity, and cracking [1–4]. For example, segregation bands, containing positive segregation and porosity result from the significant stress applied on semi-solid microstructure in high pressure die casting [1–2] and centrifugal casting [3]. The formation mechanism of casting defects is closely related with various mechanical phenomena in the mushy zone. It has been reported that the rheology of semi-solid metallic alloys shows agglomeration/disagglomeration [5–7], deformation of solid particles [6–8], and granular phenomena [1–2]. Moreover, the rheology is dependent on the liquid viscosity, the shear rate, and microstructural features (solid fraction, solid morphology, and grain size) [9–10]. The most direct approach for understanding the complexity of semi-solid deformation is to directly observe the deformation behavior at the grain scale. Recently, techniques for in situ observation of solidification phenomena and casting defect formation using synchrotron X-ray radiography have been developed to understand microstructural evolution and to build physical models [9–16]. In situ observation of deformation in semi-solid metallic alloys has been performed using a similar technique [17–26]. In our previous studies, the microstructure, deformed at 10^{-3} s^{-1} global shear strain rate in semi-solid Al-15wt.%Cu alloys at 48% solid was characterized [19–20]. The shear-induced dilation occurred by rearrangement including rotation and translation of solid particles in response to a force chain network, resulting in the shear band with decreased solid fraction [19–20] Examination of the solid motion showed that localization of the high shear strain rate occurred at the shear domain [20]. This paper demonstrates the in-situ observation of deformation at the faster global shear strain rate of 10^{-1} s^{-1} in semi-solid Al-15wt.%Cu alloys and a solid fraction of 48% using a time-resolved X-ray imaging. The developing processes of the localized shear strain rate and the shear band are studied by quantitative image analysis of solid motion during shear. The influence of the high global shear strain rate on the solid motion during shear is also examined.

2. Experimental Procedures

The ingot with a composition of Al-15Cu-0.5Ti-0.25B (mass %) was produced by arc-melting. Samples were cut to dimensions of 10 × 10 mm with 200 μm thickness. Figure 1 shows a schematic of the sample cell for deformation of semi-solid Al-Cu alloys. The sample was placed in the mold which consists of an Al_2O_3 plate with a thickness of 200 μm. The mold and Al_2O_3 window plates were retained by BN plates. The sample cell was placed in the furnace with the graphite heater [16]. The semi-solid specimen with the globular morphology was produced by partially remelting the fine equiaxed microstructure and isothermally holding for ~5 min. Deformation was applied by pushing the Al_2O_3 push-plate upward. Shear deformation is induced around the corner of the Al_2O_3 push-plate. The temperature was kept at a given temperature during the experiment. The Al_2O_3 push-plate displacement rate was 2900 μm/s, corresponding to a global shear strain rate of 10^{-1} s^{-1}.

Experiments were performed at beamline BL20XU of the SPring-8 synchrotron facility, Hyogo, Japan. The X-ray energy was set at 15 keV to obtain sufficient solid-liquid contrast. The observation area was 5.2 × 5.2 mm, and the pixel size was 5.2 × 5.2 μm. An X-ray imaging detector with a CMOS-type high-speed camera, an optical lens, and a phosphor screen of Ce-doped YAG, was used to record high-speed images at 250 fps [26].

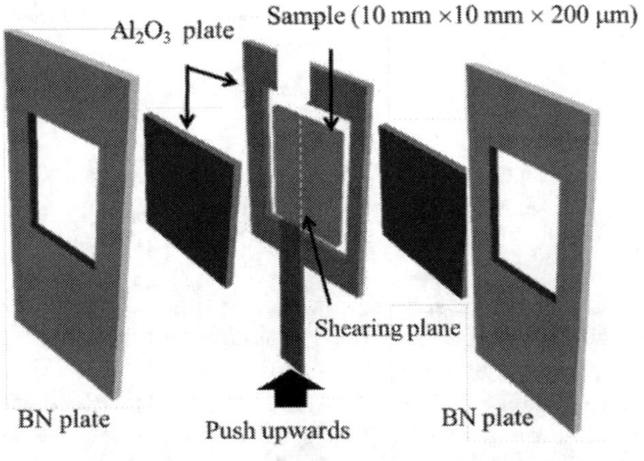

Figure 1. Schematic of sample cell for deformation of semi-solid Al-Cu alloys.

The solid motion during shear was quantitatively examined using a two dimensional image analysis. The transmission image was divided into domains, whose size is approximately two times larger than the mean solid particle size (~2d). The solid velocities in the x_1 direction, u_1, and in the x_2

direction, u_2, were estimated using a correlation pattern between the microstructures before and after a 1d increment of the Al_2O_3 push-plate motion. Further details are shown in reference [20]. The strain rate in the x direction, ε_{11}, the strain rate in the y direction, ε_{22}, and the shear strain rate, ε_{12}, where 1 and 2 indicate positive x and y directions, are given by

$$\varepsilon_{11} = \frac{\partial u_x}{\partial x} \tag{1}$$

$$\varepsilon_{22} = \frac{\partial u_y}{\partial y} \tag{2}$$

$$\varepsilon_{12} = \frac{1}{2}\left(\frac{\partial u_x}{\partial y} + \frac{\partial u_y}{\partial x}\right) \tag{3}$$

The divergence of solid velocity, $div(\vec{u})$, which corresponds to a change in solid fraction, is expressed by

$$div(\vec{u}) = \frac{\partial u_x}{\partial x} + \frac{\partial u_y}{\partial y} \tag{4}$$

Mass conservation gives the following relationship.

$$\frac{\partial f_s}{\partial t} + f_s div(\vec{u}) = 0 \tag{5}$$

where, f_s is the solid fraction.

Here, the initial solid fraction is assumed to be uniform. Since the solid fraction is almost uniform at the initial stage, the assumption is roughly valid. A positive value of the divergence shows a decrease in the solid fraction.

The 3D initial solid fraction, g_S, prior to shear was calculated from the intensity of the transmitted X-ray beam though regions of 100% liquid, I_L, 100% solid, I_S, and semi-solid, I_{SL}. Further details are explained in the reference [19]. In a region, D, the solid fraction is expressed by the following equation.

$$[g_S]_D = \frac{ln\, I_{SL} - ln\, I_L}{ln\, I_S - ln\, I_L} \tag{6}$$

The calculated 3D solid fraction of Al-Cu alloys prior to shear was approximately 48 %.

3. Results and Discussion

3.1. Solid motion during shear

Figure 2(a) shows radiograph of the semi-solid microstructure prior to shear. The average projected grain size is d = 120 μm. The solid particles were partially overlapped in the thickness direction. The increment of the Al_2O_3 push-plate was normalized by the mean solid particle size (d). The deformed microstructures after 1.3d, 3.3d, and 6.9d increments of the Al_2O_3 push-plate motion are shown in Figures 2(b–d), respectively. It can be seen that the liquid-filled spaces have dilated between many solid particles in the shear domain due to the impingement between solid particles and the rearrangement of solid particles including translation and rotation. The solid velocity vector fields for the same increment of deformation are shown in Figures 2(e–g). The distribution of the solid velocity vector was similar regardless of the Al_2O_3 push-plate displacement. The solid particles translated upward ahead of the Al_2O_3 push-plate. The magnitude of solid velocity vectors gradually decreased with distance from the Al_2O_3 push-plate front. The solid velocity vectors gradually changed from the upward to the right toward the shear domain. The overall flow pattern was similar to that in the in situ experiments of semi-solid Al-Cu alloys at the low global shear strain rate of 10^{-3} s^{-1}[18–20]. The x- and y-positions in solid velocity vectors, which are located farthest from the origin, are plotted against the Al_2O_3 push-plate displacement in Figure 3. Note that the origin is defined as the right top corner of the Al_2O_3 push-plate prior to shear, as shown in Figure 3. The solid particles with the velocity vector existed at 18d y-position after only a one grain size (1d) increment of the Al_2O_3 push-plate motion. As most of solid particles translated together in the direction of the Al_2O_3 push-plate motion ahead of the Al_2O_3 push-plate, the force was easily transmitted over a long distance parallel to the shear plane owing to the impingement between solid particles. The y-position increased with increasing the Al_2O_3 push-plate displacement and reached 31d, which is the maximum y-position in the sample, at 3.7d Al_2O_3 push-plate displacement. On the other hand, the x-position increased as the Al_2O_3 push-plate displacement rose to approximately 4d, above which it remained almost unchanged. The average value of the x-position was 11.3d for Al_2O_3 push-plate displacement higher than 4d.

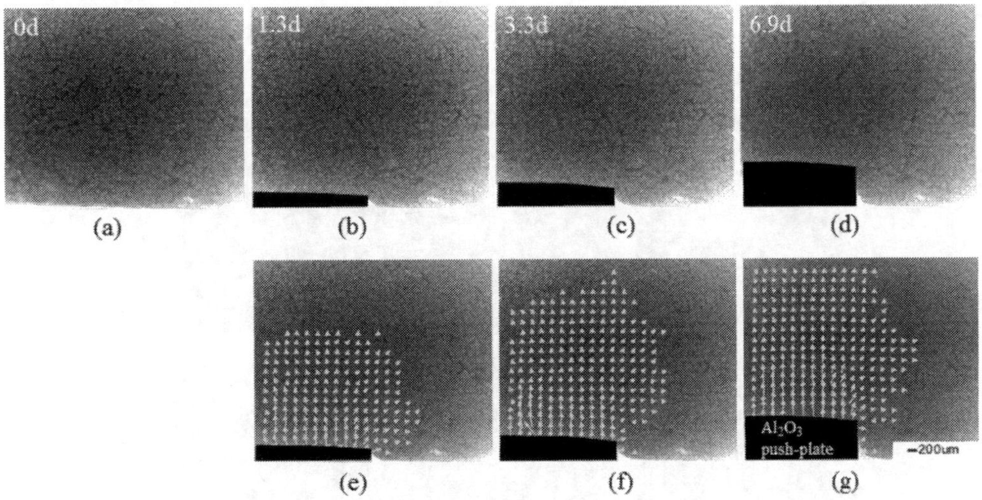

Figure 2. (a–d)Deformation of semi-solid Al-Cu alloys. (e–g)Solid velocity vector distributions after 1.3d, 3.3d, and 6.9d increments of Al_2O_3 push-plate motion.

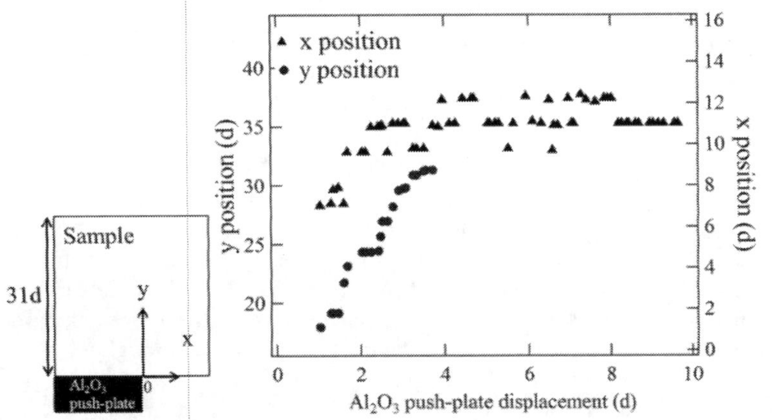

Figure 3. x- and y- positions in the solid velocity vectors which are located farthest from the origin plotted against Al_2O_3 push-plate displacement (d).

The diverging vectors existed especially at the shear domain, which indicates the local dilation, as shown in Figures 2(e–g). An example of local dilation at the shear domain during a 1.2d increment of the Al_2O_3 push-plate motion is shown in Figures 4(a–c). Figures 4(d–f) show the same images after image processing, where boundaries of solid particles are traced black and liquid region between solid particles are defined as gray. The sequence of images clearly showed that the liquid-filled space is enlarged between solid particles. Five solid particles, labeled A-E, are highlighted that play an important role in dilation. The translation of solid particle A induced by the Al_2O_3 push-plate motion pushed B and E. The rotation and translation of B and E led to push C and D to the right, which opened up the liquid-filled space. These findings indicate that the rearrangement of solid particles including translation and rotation caused the displacement of solid particles in the direction perpendicular to the shear plane, as shown in Figure 3. Past research [24] on the in-situ observation of deformation at higher solid fraction and high global shear strain rate of 10^{-1} s^{-1} found that the rearrangement including rotation and translation of solid particles was highly restricted, resulting in strong localization of the shear strain in the narrow region. In this study, no significant influence of shear strain rate on the solid motion was observed because of the lower solid fraction.

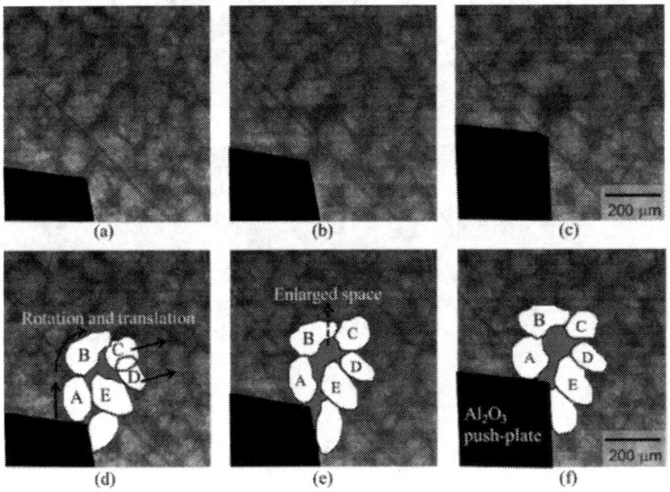

Figure 4. A local dilation event every 0.6d increment of the Al_2O_3 push plate motion.

3.2. Localization of shear strain and formation of a shear band

Figures 5(a–c) show the distribution of the shear strain rate, calculated with equation (3), after 1.3d, 3.3d, and 6.9d increments of the Al_2O_3 push-plate motion. Shear deformation occurs in the blue-shaded regions, which corresponds to negative values of shear strain rate. Even in the initial stage of deformation (1.3d Al_2O_3 push-plate displacement), the relatively high shear strain was localized at the upper right region of the Al_2O_3 push-plate, as shown in Figure 5(a). The localized shear strain region gradually developed in the direction parallel to the shear plane as the Al_2O_3 push-plate displacement increased (Figures 5(b–c)). The localization of shear strain occurred throughout of the shear plane at approximately 4d Al_2O_3 push-plate displacement, which is consistent with that where the x- and y-positions reached the maximum value, as shown in Figure 3. The average shear strain rates in the localized regions for 1.3d, 3.3d, and 6.9d increments of the Al_2O_3 push-plate was calculated to be -1.5, -2.0, and -2.2 s^{-1}, respectively. Those values were approximately 10 times as high as the overall global shear rate. The distribution of the divergence of solid velocity, calculated with equation (4), which corresponds to a change in solid fraction, f_s, is shown in Figures 5(d–f). The red-shaded region (a positive value of the divergence) shows the decrease in solid fraction. At the Al_2O_3 push-plate front, the solid fraction increased. On the other hand, the red-shaded regions were distributed at the upper right region of the Al_2O_3 push-plate. Namely, the solid fraction decreased in the localized shear strain region.

Figure 6 shows the width of shear band plotted against the Al_2O_3 push-plate displacement. The shear band width was defined as the region of coupled localized shear strain and decreased solid fraction. The shear band width increased as the Al_2O_3 push plate displacement increased up to approximately 4d and then it remained almost constant. The average shear band width was approximately 12d for Al_2O_3 push-plate displacements above 4d. Past research [20] on the in-situ observation of deformation showed that the shear band width was approximately 10d at the lower global shear strain rate of 10^{-3} s^{-1}. It is likely that the global shear strain rate has no significant influence on the shear band width in the globular sample at relatively lower solid fraction.

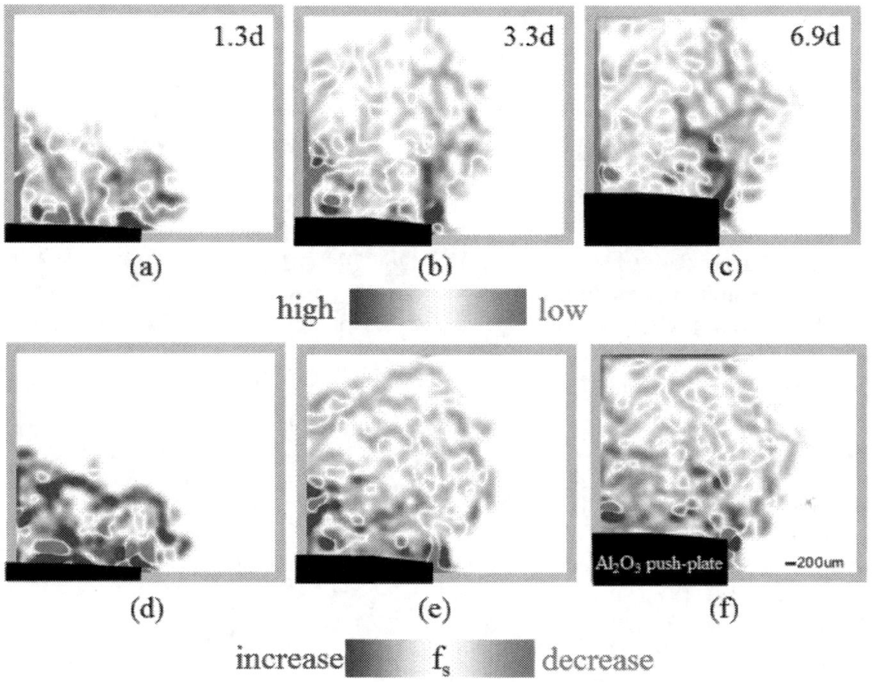

Figure 5. (a–c)Shear strain rate and (d–f) divergence of solid velocity, which correspond to change in solid fraction after 1.3d, 3.3d, and 6.9d increments of the Al_2O_3 push-plate motion.

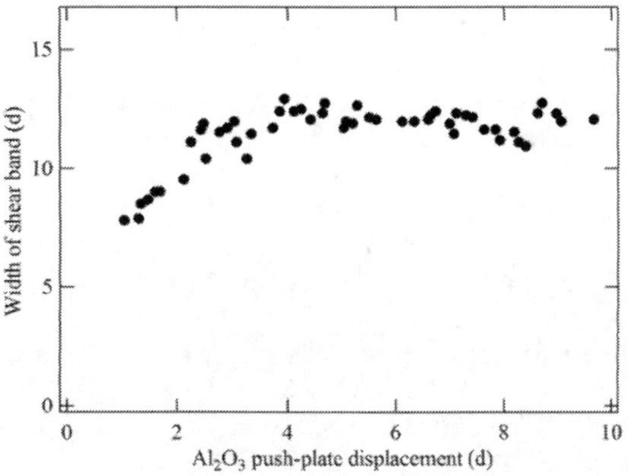

Figure 6. Shear band width plotted against Al_2O_3 push-plate displacement(d).

4. Summary

Time-resolved X-ray imaging has been used to directly observe the dynamics of deformation in semi-solid Al-Cu alloys at a solid fraction of 48% and a global shear strain rate of 10^{-1} s^{-1}. The quantitative analysis of solid motion during shear was performed.

(1)The distribution of the solid velocity vector showed that the force was transmitted a long distance in the direction parallel to the shear plane (18d y-position) after only a 1d increment of Al_2O_3 push plate motion. On the other hand, the distance in the direction perpendicular to the shear plane is limited to approximately 11d even after a high increment of the Al_2O_3 push-plate.

(2)Only a 1d increment of the Al_2O_3 push plate motion was sufficient to cause the localization of the shear strain rate at the shear domain. The localized shear strain region fully developed throughout the shear plane at 4d Al_2O_3 push-plate displacement. The shear strain rates at the localized area were an order of magnitude higher than the global shear strain rate. The decrease in the solid fraction occurred at the localized shear strain.

(3)At more than 4d of Al_2O_3 push plate displacement, the shear band width remained constant (at approximately 11d), which was similar to that obtained in the lower global shear strain rate of 10^{-3} s^{-1}.

Acknowledgements

The synchrotron experiments were performed at beamline 20XU of SPring-8 (Proposal Nos. 2009A(B)0014, 2010A1420, 2011A1209. 2011B1096, 2012A1110, 2012B1173, 2013B1242, and 2014B1090) with the approval of the Japan Synchrotron Radiation Research Institute (JASRI). This work was partially supported by Grants-in-Aid for Scientific Research (KAKENHI) (Grant Nos. 24560910, 24246124, and 24226018) from JSPS.

References

[1] Gourlay C M and Dahle A K 2007 *Nature* **445** 70
[2] Gourlay C M, Meylan B and Dahle A K 2008 *Acta Mater.* **56** 3403
[3] Kang I and Ohnaka I 1997 *J. JFS* **69** 119
[4] Dantzig J A and Rappaz M 2009 *Solidification* (Italy: EPFL Press) p 567
[5] Tzimas E and Zavaliangos A 2000 *Mater. Sci. Eng. A* **289** 228
[6] Metz S A and Flemings M C 1969 *AFS Trans.* **77** 329
[7] Flemings M C 1991 *Metall. Trans.* **22A** 957
[8] Tzimas E and Zavaliangos A 1991 *Acta Mater.* **47** 517
[9] Sumitomo T, StJohn D H and Steinberg T 2000 *Mater. Sci. Eng. A* **289** 18
[10] Dahle A K and StJohn D H 1999 *Acta Mater.* **47** 31
[11] Mathiesen R H, Arnberg L, Mo F, Weitkamp T and Snigirev A 1999 *Phys. Rev. Lett.* **83** 5062
[12] Yasuda H, Ohnaka I, Kawasaki K, Sugiyama A, Ohmichi T, Iwane J and Umetani K 2004 *J. Cryst. Growth* **262** 645
[13] Mangelinck-Noel N, Nguyen-Thi H, Reinhart G, Schenk T, Cristiglio V, Dupouy M D, Gastaldi J, Billia B and Hartwig J 2005 *J. Physics D* **38** A28
[14] B Li, H D Brody, D R Black, H E Burdette and C Rau 2006 *J. Physics D* **39** 4450
[15] Yasuda H, Yamamoto Y, Nakatsuka N, Yoshiya M, Nagira T, Sugiyama A, Ohnaka I, Uesugi K and Umetani K 2009 *Int. J. Cast Met. Res.* **22** 15
[16] Yasuda H, Nagira T, Yoshiya M, Nakatsuka N, Sugiyama A, Uesugi K and Umetani K 2011 *ISIJ Int.* **51** 402
[17] Terzi S, Salvo L, Suery M, Limodin N, Adrien J, Maire E, Pannier Y, Bornert M, Bernard D, Felberbaum M, Rappz M and Boller E 2009 *Scripta Mater.* **61** 449
[18] Nagira T, Gourlay C M, Sugiyama A, Uesugi M, Kanazawa Y, Yoshiya M, Uesugi K, Umetani K and Yasuda H 2011 *Scripta Mater.* **64** 1129

[19] Gourlay C M, Dahle A K, Nagira T, Nakatsuka N, Nogita K, Uesugi K and Yasuda H 2011 *Acta Mater.* **59** 4933
[20] Nagira T, Yokota H, Morita S, Yasuda H, Yoshiya M, Gourlay C M, Sugiyama A, Uesugi K and Umetani K 2013 *ISIJ Int.* **53** 1195
[21] Fonseca J, O'Sullivan C, Nagira T, Yasuda H and Gourlay C M 2013 *Acta Mater.* **61** 4169.
[22] Nagira T and Yasuda H 2014 *In-situ Studies with Photons, Neutrons and Electrons Scattering II*, (Switzerland: Springer) p 231
[23] Gourlay C M, O'Sullivan C, Fonseca J, Yuan L, Kareh K M, Nagira T and Yasuda H 2014 *JOM*, **66** 1415
[24] Nagira T, Morita S, Yokota H, Yasuda H, Gourlay C M, Yoshiya M, Sugiyama A, Uesugi K, Takeuchi A and Suzuki Y 2014 *Metall. Mater. Trans. A* **45** 5613
[25] Zabler S, Ershov A, Rack A, Garcia-Moreno F, Baumbach T and Banhart J 2013 *Acta Mater.* **61** 1244
[26] Goto S, Takeshita K, Suzuki Y, Ohashi H, Asano Y, Kimura H, Matsushita T, Yagi N, Isshiki N and Yamazaki H 2001 *Nucl. Instr. and Meth. A* **467-468** 682

Synchrotron X-ray Tomographic Quantification of Deformation Induced Strain Localisation in Semi-solid Al-15wt.%Cu

B. Cai[1], S. Karagadde[1], T.J. Marrow[2], T. Connolley[3], P.D. Lee[1]

1 School of Materials, University of Manchester, Manchester, UK
2 Department of Materials, University of Oxford, Oxford, UK
3 Diamond Light Source Ltd., Harwell Science and Innovation Campus, Didcot, UK

Email: peter.lee@manchester.ac.uk

Abstract Uniaxial compression and indentation of a semi-solid Al-15wt.%Cu alloy was investigated by high speed synchrotron X-ray microtomography, quantifying the microstructural response of a solidifying alloy to applied strain. Tomograms were continuously acquired whilst performing deformation using a precision thermal-mechanical rig on a synchrotron beamline. The results illustrate how defects and shear bands can form in response to different loading conditions. Using digital volume correlation, the global and localised strains were measured, providing quantitative datasets for granular flow models of semi-solid deformation.

1. Introduction

Semi-solid processing (SSP) is being considered over conventional casting processes for some aerospace and automotive components, as it is thought to give improved mechanical properties. However, defects are still known to occur [1]. Many of the early studies, such as that by Spencer et al [2], looked at the shear-thinning behaviour in semi-solid Sn-Pb. Since that time, numerous studies have been conducted to determine the constitutive properties of partially solidifying alloys using deformation methods such as rotation viscometer, compression, and indentation. The authors of these studies have proposed a range of behaviours, ranging from shear-thinning (thixotropic) [2], shear-thickening [3], and treatment as a continuum 'spongy-like' solid network saturated with liquid [4]. However, these prior studies did not link material behaviour to the detailed kinetics occurring at the microstructural scale during deformation, in which there is a complex interaction of evolving solid and liquid in a solidifying alloy.

One method of overcoming this limitation of linking defect formation to microstructure was demonstrated by Lee and Hunt [5] who developed an X-ray Temperature Gradient Stage for performing in situ radiographic observations of defect formation. Recent years have seen the development of X-ray in situ observation by various groups to study semi-solid deformation, both via radiography [6–8] and tomography [9–12]. Gourlay and co-workers [7,13] designed a direct shear cell to use in situ radiography to study granular deformation in a semi-solid. Cai et al [12] and Kareh et al [14] used parallel plate compression together with high speed synchrotron X-ray microtomography (XMT) to investigate the occurrence of dilatancy in semi-solid dendritic and globular microstructures, respectively. An extrusion cell was also coupled with high speed XMT to investigate deformation-induced flow in a semi-solid alloy [11]. In addition to performing qualitative in situ observations of deformation-induced microstructural changes, there is a need to quantify the strain localization induced by deformation in semi-solids. This has been made possible by advanced image processing methods, for example, strain fields have been measured in 2D via a particle tracking technique [13] and in 3D using digital volume correlation (DVC) [12].

In this study, we quantify material flow at a microstructural level in 3D in semi-solid equiaxed dendritic structures for uniaxial compression and indentation tests. This was achieved using high speed synchrotron X-ray tomography combined with digital volume correlation, allowing a quantitative investigation of deformation induced strain localization and failure mechanisms.

2. Materials and Methods

An equiaxed dendritic Al-15wt%Cu alloy with a grain size of ~600 μm was used. Semi-solid deformation tests (both compression and indentation) were carried out using the bespoke P2R thermo-mechanical rig, with air-bearing continuous rotation built into the load chain [12]. The experiments were performed at the I12 beamline, Diamond Light Source, with a monochromatic, 53 keV X-ray beam. X-ray imaging was performed using the beamline's custom-built imaging modules, consisting of single crystal scintillators lens-coupled to a CMOS high speed camera.

For the compression test, a Phantom V7.3 CMOS camera (Vision Research, USA) was used, providing a field of view (FOV) of 800×600 pixels (9.8 ×7.3 mm) and a pixel size of 12.25 μm. The cylindrical sample (3 mm in diameter and 4 mm height) was heated in the furnace to 555°C (liquid fraction of 30±2%), and held isothermally for 10 min during which slight coarsening occurred. The sample was then compressed at a displacement rate of 5 μm/s. After deforming for 20 s, high speed X-ray tomographic imaging was started, and 24 datasets were captured over 96 s (i.e. one tomogram every 4 s, each 720 projections collected over 180° rotation).

For the indentation test, a Miro 310M CMOS camera (Vision Research, USA) was used, offering 1280×800 pixels (5.12×3.2 mm) FOV and 4 μm pixel size. A cylindrical (3 mm in diameter and 3 mm height) sample was used. The 30° conical indenter was made of alumina. The sample was heated to 555 °C with liquid fraction of 28% and held for 10 min. The deformation was then started at a speed of 5 μm/s. Continuous scans were taken at an interval of 18 s, acquiring 900 projections in 9 s over 180°. Nine datasets were captured in total.

All captured datasets were reconstructed with a filtered back projection algorithm, including ring artefact removal [15]. 3D image visualization was performed using Avizo 8 (Visualization Science Group, France). DaVis Strain Master (Version 8.0.7, LaVision, GmBH, Germany) was used to perform digital volume correlation (DVC) on the time-resolved 3D datasets, providing a full-field displacement field and 3D strain map [12,16,17]. The samples used in this study have a complex dendritic structure providing the texture required for DVC tracking via contrast between the α-Al primary phase and Cu-rich interdendritic liquid. The compression datasets were analysed by a final 24 × 24 × 24 pixel subset with 50% overlap. Hence a displacement matrix grid of 12 pixels or 147 μm was obtained. The indentation datasets were analysed using 64 × 64 × 64 interrogation subsets with 50% overlap, offering a 32 pixel displacement matrix (128 μm). Displacement vectors with a correlation coefficient less than 0.5 were deleted. Figure 1 schematically illustrates the application of digital volume correlation for the determination of the strain field on the time-revolved X-ray tomographic datasets. The displacement fields (V_m, m = 1, 2, 3···) were calculated from the nearby tomographic dataset (step 1→step 2, step 2→step 3···). Then the measured displacements were accumulated (i, i+ii, i+ii+iii ...) to calculate the integrated displacement fields (V_n (n = I, II, III...)) in the data series. The integrated displacement fields were used to calculate the strain tensor ($\square_{ij}$ ($i, j = x, y, z$) by the finite difference method.

$$\varepsilon_{ij} = \frac{\partial V_i}{\partial j}$$

The strain tensor was then used to calculate three principal strain components. The first component of the principal strain (ε_1) is used to demonstrate the strain field.

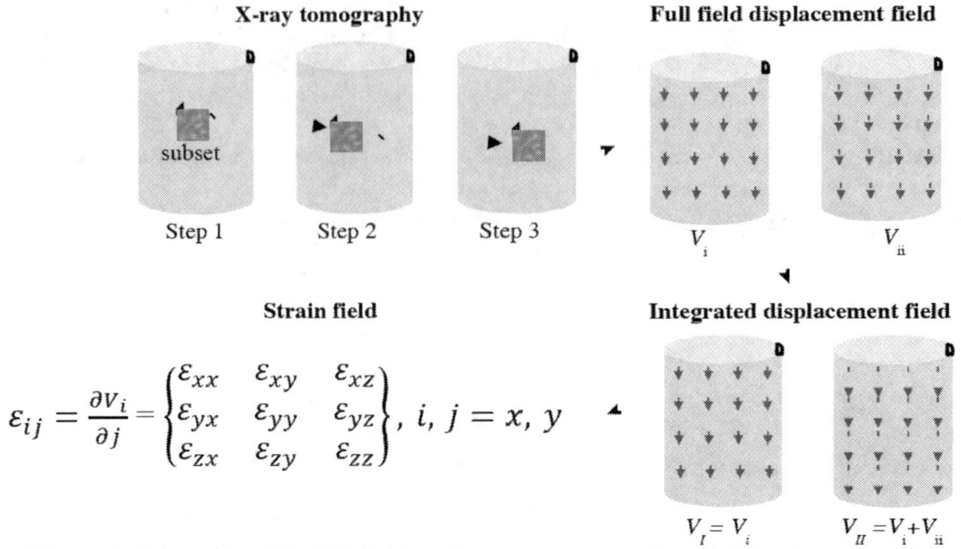

Figure 1. Schematic of the full-field strain measurement using digital volume correlation on time-revolved X-ray tomographic datasets

3. Results and Discussion

3.1. Semi-solid compression

Parallel plate compression is one of the standard methods of measuring the properties of solid materials, and it has also been used to determine the rheology properties of semi-solid alloys. In this study, a semi-solid Al15wt%Cu specimen (liquid fraction of 30%) was compressed at 5 μm/s. The results are shown in Fig. 2 for four sequential steps, with the solid coloured red and the liquid yellow. The initial microstructure consists of equiaxed-dendritic grains with homogeneously distributed Cu-enriched interdendritic liquid surrounding them (Fig. 2a).

As compression proceeds, the sample becomes slightly barrelled (Fig. 2b, $\varepsilon = 4.8\%$ as measured tomographically by the change of sample height). The barrelling is due to the friction between the platens (made of pyrophyllite) and the semi-solid sample. It can be seen that the liquid channel opened slightly at the middle region of the sample with liquid flowing into the opening. As compression continues, the liquid channels enlarge (Fig. 2c, $\varepsilon = 7.3\%$ and 2d, $\varepsilon = 12.6\%$), due to dilatancy. Voids nucleate at the periphery of the sample (Fig. 2b & 2c), growing along the liquid channels and penetrating all the way into the centre of the sample (Fig. 2d). These results support the three stage growth of voids hypothesised in [12]: (1) initial void closure, (2) incubation and (3) final rapid growth.

This increase in liquid-filled interstices and the initiation of voids during semi-solid compression is due to dilatancy, as discovered by Reynolds in 1885 for granular solids [18]. Several studies have confirmed the role of dilatancy in semi-solid alloys. Tzimas and Zavaliangos [19] reported the occurrence of dilatancy in semi-solid equiaxed alloys during compression. Gourlay et al [20] used a rotation viscometer and revealed that partially solidified alloys can exhibit Reynolds' dilatancy. Recently, dilatancy has also been imaged by X-ray radiographic experiments [21]. Our study demonstrates that dilatancy in the semi-solid alloys was caused by the rearrangement of grains responding to the applied deformation. As the grains translate and rotate, the gaps between them increase, and interdendritic liquid first fills the dilated interstices. When liquid is no longer able to feed the increased dilation, gas is drawn in from the edges to form voids along dilated liquid channels.

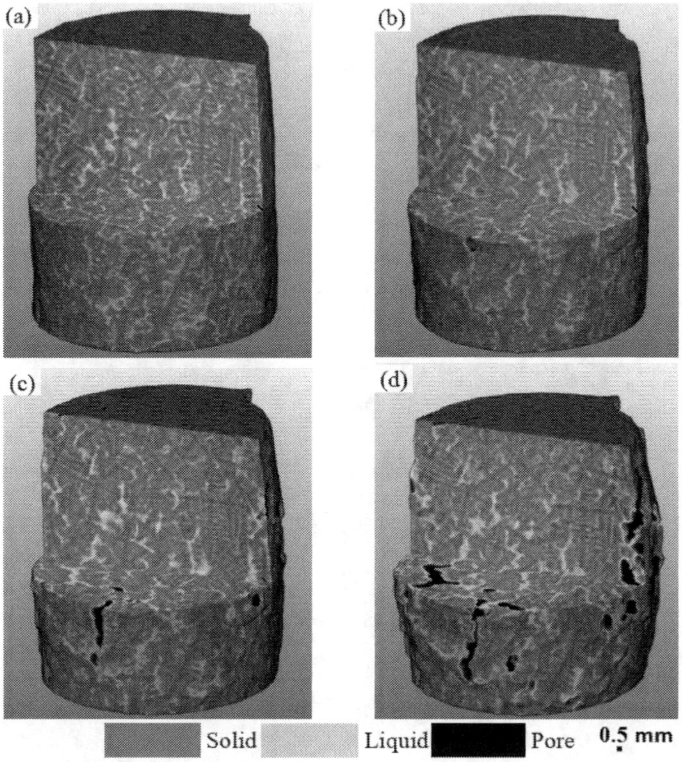

Figure 2. X-ray tomograms of semi-solid Al-15 wt.%Cu alloy (fraction liquid of 30 ± 2%) at four stages of compression: (a) $t = 24$ s, $\varepsilon = 1.2\%$; (b) $t = 52$ s, $\varepsilon = 4.8$ %, (c) $t = 76$ s, $\varepsilon = 7.3\%$; and (d) $t = 116$ s, $\varepsilon = 12.6\%$

The distribution of the principal strain (ε_1) as measured by DVC is shown on the outer surface and cross-section of the specimen in Fig. 3, at four stages of compression. The iso-surface of ε_1 equal to 20% is rendered in red in the other half of the specimen, showing the localisation of the strain into shear bands. The highest strain levels are first observed in the regions where the interdendritic channels widen between grains, followed by subsequent void / damage formation, often connected to the surface of the sample (compare Figs. 2d & 3d). Once voids form, there is no texture for DVC tracking. Hence, a cylindrical mask of 2.7 mm diameter was used to cut out those regions. Nevertheless, Fig. 3 reveals the extent of strain localisation during semi-solid compression. Going through the successive compression steps, initially the strain is homogeneous (Fig. 3a, axial strain 3.6%). However, when compressed to $\varepsilon = 7.3\%$, localisation starts to occur as the grains contact, rotating and shearing (Fig. 3b), although ε_1 larger than 20% is still negligible. When compressed to $\varepsilon = 10.2\%$ (Fig. 3c), dilatant effects become even stronger, and there is a considerable increase of the fraction of ε_1 larger than 20% with the on-set of what appears to be shear banding. At the final stage of deformation ($\varepsilon = 12.6\%$), ε_1 is connected to the edge of the cylindrical mask, with strong bands of localised strain (Fig. 3d). The inhomogeneity of strain distribution during semi-solid compression is directly related to the deformation response at the grain scale: (1) the movement and interplay of solid grains causing contact and rotation of the primary alpha phase grains, as well as (2) bending and fragmentation of the dendritic structures [12].

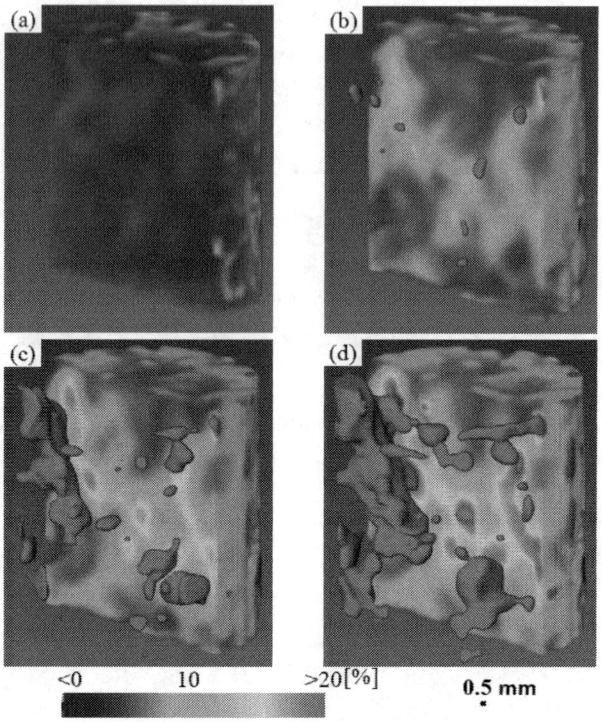

Figure 3. 3D rendering of the first principal strain during semi-solid compression (red iso-surface represents ε_I=20%) at axial strains of: (a) 3.6%; (b) 7.3%; (c) 10.2%; and (d) 12.6%

3.2. Semi-solid indentation

In the second example we investigate the behaviour of semi-solid (fraction liquid 30%) Al-15wt.%Cu during indentation. Indentation is a standard method of measuring the hardness and other properties of solids, but little has been published on the behaviour of semi-solids during this type of deformation [19]. Indentation has been demonstrated to produce highly localized deformation in the vicinity of the indenter in highly ductile alloys even at room temperature [11], providing an interesting comparison to the aforementioned uniaxial compression.

Figure 4 shows results of indenting a semi-solid at an indenter displacement rate of 5 μm/s at four indentation penetration depths. The equiaxed dendritic microstructure of the initial sample when held isothermally at 555±2 °C is clearly visible in Fig. 4a. After only 90 μm of indentation, the highly localised strain around the indenter is visible by the localised opening of liquid channels. The localised contact of the indenter on the grains pushed the grains downwards and sideways, whereas the liquid channels and grain locations far below the indenter appear unaffected. Applying the force with an indenter causes the dilatant behaviour to be highly localised, much like walking on wet sand causes dilatant expansion of the sand only where you stepped, leaving dry footprints behind.

As the indenter travels deeper into the sample, the force chains expand downwards and sideways to the periphery of the sample (Fig. 4c), with liquid channels opening up all the way to the edges. In Fig. 4d the liquid channel on the right hand side of indenter increases (black arrow), whilst the just-opened liquid channel on the left hand side (blue arrow in Fig. 4c) closes. In addition, the secondary dendrites within the equiaxed grain directly below the indenter compact. This shows the interplay of dilatant effects and deformation of the solid phase, with both plastic strain and granular flow occurring.

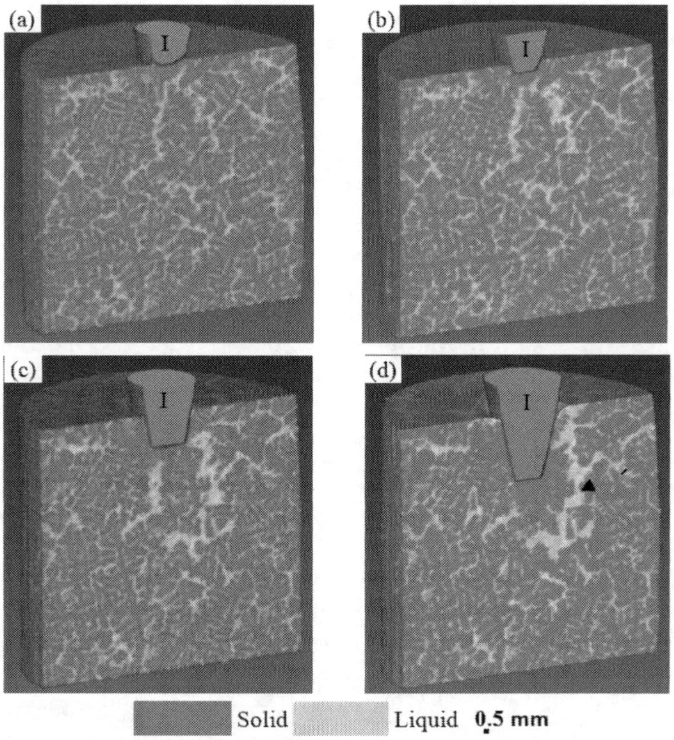

Figure 4. Semi-solid Al-15wt%Cu at subsequent stages of indentation: (a) $t = 0$ s, I = 0 μm; (b) $t = 18$ s, I = 90 μm; (c) $t = 54$ s, I = 270 μm; (d) $t = 126$ s, I = 630 μm

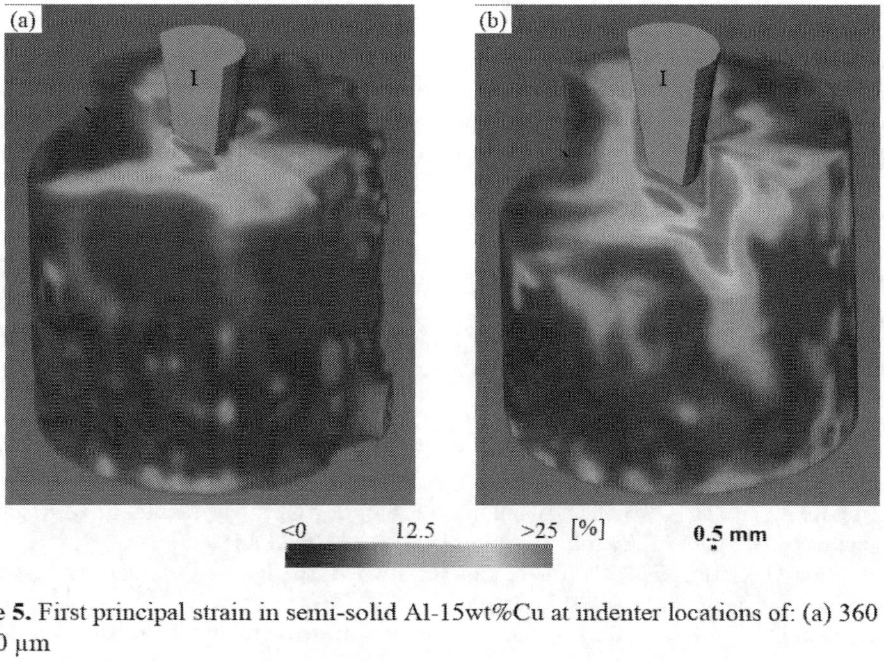

Figure 5. First principal strain in semi-solid Al-15wt%Cu at indenter locations of: (a) 360 μm, (b) 630 μm

The principal strain (ε_1) field is shown at two stages of indentation in Fig. 5, as calculated using digital volume correlation. The distribution of ε_1 at 360 μm indentation is shown in Fig. 5a, with the majority of the strain localised directly below the indenter. As the indenter presses further into the sample (630 μm, Fig. 5b) the principal strain increases further. However, ε_1 now rises not only beneath the indenter, as the grains contact one onto the next, the force chains and strain extend all the way to the sides and bottom of the sample, but in a very heterogeneous manner.

4. Conclusions

The use of in situ X-ray tomography combined with digital volume correlation has allowed the formation of strain localisation and defect formation to be quantitatively studied during two typed of deformation of an Al15wt.%Cu alloy (semi-solid compression and indentation). The results demonstrate that equiaxed dendritic semi-solids behave in both a granular flow regime (exhibiting strong dilatancy), and also undergo plastic deformation. These experiments show the role of microstructure upon the relative importance of these different types of deformation, depending on both the way the deformation is applied, and the overall strain. This data can serve as both the input geometries and validation for models of microstructural deformation during solidification and semi-solid processing.

Acknowledgements

This work was financially supported by the EPSRC (EP/I02249X/1) and the European Union (RFSR-PR-10005 DDT). We thank Diamond Light Source for providing beamtimes (experimental visits EE7604-1 and EE9018-1), and Dr. Robert Atwood at beamline I12 for his help. We are grateful to Vision Research UK for the loan of the Miro 310M camera. B.C. acknowledges financial support from General Electric and the China Scholarship Council.

References

[1] Atkinson H. Prog Mater Sci 2005;50:341.
[2] Spencer DB, Mehrabian R, Flemings MC. Metall Trans A 1972;3:1972.
[3] Kumar P, Martin CL, Brown S. Metall Trans A 1993;24:1107.
[4] Nguyen TG, Favier D, Suéry M. Int J Plast 1994;10:663.
[5] Lee PD, Hunt JD. Acta Mater 1997;45:4155.
[6] Phillion AB, Hamilton RW, Fuloria D, Leung ACL, Rockett P, Connolley T, Lee PD. Acta Mater 2011;59:1436.
[7] Gourlay CM, Dahle AK, Nagira T, Nakatsuka N, Nogita K, Uesugi K, Yasuda H. Acta Mater 2011;59:4933.
[8] Zabler S, Ershov A, Rack A, Garcia-Moreno F, Baumbach T, Banhart J. Acta Mater 2013;61:1244.
[9] Terzi S, Salvo L, Suéry M, Limodin N, Adrien J, Maire E, Pannier Y, Bornert M, Bernard D, Felberbaum M. Scr Mater 2009;61:449.
[10] Puncreobutr· C, Lee PD, Hamilton RW, Cai B, Connolley T. Metall Mater Trans A 2012;44:5389.
[11] Cai B, Karagadde S, Rowley D, Marrow TJ, Connolley T, Lee PD. Scr Mater 2015.
[12] Cai B, Karagadde S, Yuan L, Marrow TJ, Connolley T, Lee PD. Acta Mater 2014;76:371.
[13] Nagira T, Morita S, Yokota H, Yasuda H, Gourlay CM, Yoshiya M, Sugiyama A, Uesugi K, Takeuchi A, Suzuki Y. Metall Mater Trans A 2014;45:5613.
[14] Kareh KM, Lee PD, Atwood RC, Connolley T, Gourlay CM. Nat Commun 2014;5:4464.
[15] Titarenko S, Withers PJ, Yagola A. Appl Math Lett 2010;23:1489.
[16] Mostafavi M, Collins DM, Cai B, Bradley R, Atwood RC, Reinhard C, Jiang X, Galano M, Lee PD, Marrow TJ. Acta Mater 2015;82:468.
[17] Eastwood DS, Yufit V, Gelb J, Gu A, Bradley RS, Harris SJ, Brett DJL, Brandon NP, Lee PD, Withers PJ, Shearing PR. Adv Energy Mater 2014;4:1.

[18] Reynolds O. Philos Mag J Sci 1885;20:470.
[19] Tzimas E, Zavaliangos A. Acta Mater 1999;47:517.
[20] Gourlay CM, Dahle A. Nature 2007;445:70.
[21] Gourlay CM, Nagira T, Dahle AK, Nakatsuka N, Uesugi K, Yasuda H. IOP Conf Ser Mater Sci Eng 2012;27:012086.

IOP Conf. Series: Materials Science and Engineering **84** (2015) 012080 doi:10.1088/1757-899X/84/1/012080

Observation of dendritic growth under the influence of forced convection

O Roshchupkina, N Shevchenko, S Eckert
Helmholtz-Zentrum Dresden-Rossendorf,
Bautzner Lanstrasse 400, 01328 Dresden, Germany

E-mail: o.roshchupkina@hzdr.de

Abstract. The directional solidification of Ga–25wt%In alloys within a Hele-Shaw cell was visualized by X-ray radioscopy. The investigations are focused on the impact of melt convection on the dendritic growth. Natural convection occurs during a bottom up solidification because lighter solute is rejected during crystallization. Forced convection was produced by a specific electromagnetic pump. The direction of forced melt flow is almost horizontal at the solidification front. Melt flow induces various effects on grain morphology primarily caused by convective transport of solute, such as a facilitation of the growth of primary trunks or lateral branches, dendrite remelting, fragmentation or freckle formation depending on the dendrite orientation, the flow direction and intensity. Forced flow eliminates solutal plumes and damps local fluctuations of solute. A preferential growth of the secondary arms occurs at the upstream side of the dendrites, whereas high solute concentration at the downstream side inhibits the formation of secondary branches.

1. Introduction

Electromagnetic stirring by rotating or travelling magnetic fields can be used as an effective tool for flow control in solidifying metal alloys [1, 2]. Recent publications show that the application of such methods during solidification can lead to homogeneous, fine-grained, globular structures without macrosegregation [2, 3]. Nevertheless, the interaction between solidification and melt flow is rather complex and still needs to be fully understood. A better founded knowledge requires new diagnostic tools which allow in situ observation of the process and provide high resolution data of the flow field, the concentration field and the solidified structure. Moreover, access to different length scales, as for example dendrite morphology and flow structure, is of particular interest. There exist plenty of ex-situ investigations considering the melt flow effects on the dendritic growth [3,4]. Moreover, many in situ experiments with transparent materials as model liquids have been carried out using diverse optical diagnostic techniques [5-8]. However, these model liquids can be extrapolated to metal alloys only to a certain extent due to significant differences in thermo-physical properties.

During the last decade in situ X-ray radioscopy became an important method for visualization of solidification processes in metallic alloys [9,10]. Koster et al. used this method for the investigation of melt convection effects in liquid Gallium - Indium alloys and they were able to reproduce qualitative flow patterns [10]. Shevchenko et al. [11] have shown that the forced melt flow damps local fluctuations of solute concentration. Also a preferential growth of the secondary arms at the upstream side of the primary dendrite trunks was observed. The present investigation is focused on the visualization of the dendritic growth during the bottom-up solidification of a Ga-25wt%In alloy under the influence of forced convection of different strength. We study the effect of fluid flow on the concentration profiles, the dendrite morphology and growth velocities.

2. Experimental setup

The solidification experiments were carried out at Helmholtz-Zentrum Dresden-Rossendorf. The detailed description of the experimental setup can be found elsewhere [11-15]. Ga–25wt%In alloy ($T_{Liquidus}$=25 °C, $T_{Solidus}$=15.3 °C) used in our experiments was prepared from high-purity Ga (99.99%) and In (99.99%). Than it was melted in a furnace and filled into a 30x30 mm^2 Hele-Shaw quartz cell with a 150 µm gap parallel to the direction of the X-ray beam (see Figure 1b). A linear array of Peltier elements on the bottom part of the solidification cell cools the alloy and an electric heater mounted at the upper part of the solidification cell heats up the alloy. The temperature gradient across the cell and the cooling rate are controlled by a simultaneous regulation of Peltier cooler and electric heater. A set of miniature K-type thermocouples were contacted to the surface of the cell to monitor the temperature. The cooling rate was 0.01 K/s and the temperature gradient was kept ~ 0.9±0.1 K/mm along the whole experiment. The electromagnetically driven flow was produced by a specific pump, which represents a rotating wheel consisting of two parallel discs with a set of permanent magnets (NdFeB) with alternating polarization at their inner sides. The top part of the solidification cell was placed between the disks. The rotation speeds of the magnetic wheel of 30 and 80 revolutions per minute (rpm) were chosen.

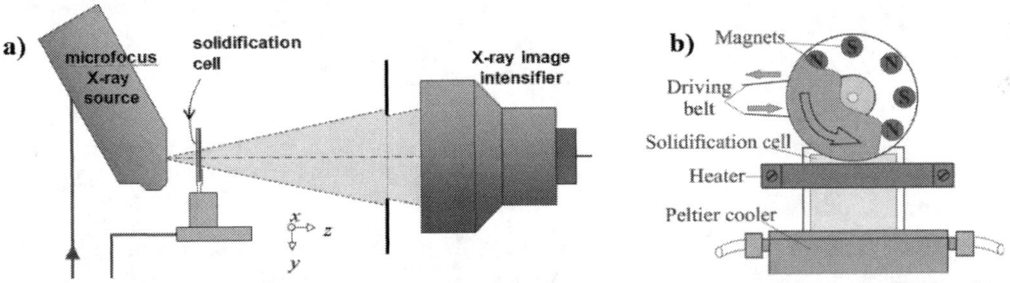

Figure 1. Experimental setup: (a) X-ray diagnostic system; (b) square Hele-Shaw solidification cell equipped with electric heater, Peltier elements and magnetic pump for generation of the melt flow.

The solidification experiments were carried out using a micro-focus X-ray tube (XS225D from Phoenix|X-ray). The observation window was approximately 9x12 mm^2. The images were captured with a scan rate of 50 half frames per second. The images were integrated over a period of 1 s to reduce the noise level of single images. The tube voltage was 63 kV and the electrical current was 140 µA. The solidification cell was positioned between the X-ray tube and detector as demonstrated in Figure 1a.

The analysis of the solidification front velocity and primary and secondary arm spacing follows the algorithm proposed by Boden et al. [12]. The concentration profiles were achieved from the measurements of the local brightness in the images.

3. Results and discussions

The Ga-25wt%In alloy was solidified in vertical direction starting from the bottom of the solidification cell. Prior to the solidification experiment the alloy was additionally heated up to 20 K above the liquidus temperature to avoid segregations in the melt. Figure 2 shows a sequence of images demonstrating a growth process of In dendrites captured at different time steps. Here it should be noticed that no special preparations were made in our experiments to control the nucleation at the bottom of the solidification cell. This in turn results in the development of ensembles of dendrites with different orientations [11].

Figure 2a is captured before the magnetic wheel was switched on. This image illustrates the natural convection. On the corresponding image one can see well developed solutal plumes which are ejected at the solid-liquid interface since their density is lighter than the initial melt composition. The rising plumes are accompanied by a downward flow of In-rich melt in the intermediate regions. This leads to an inhomogeneous horizontal concentration profile along the solidification front (see the black curve

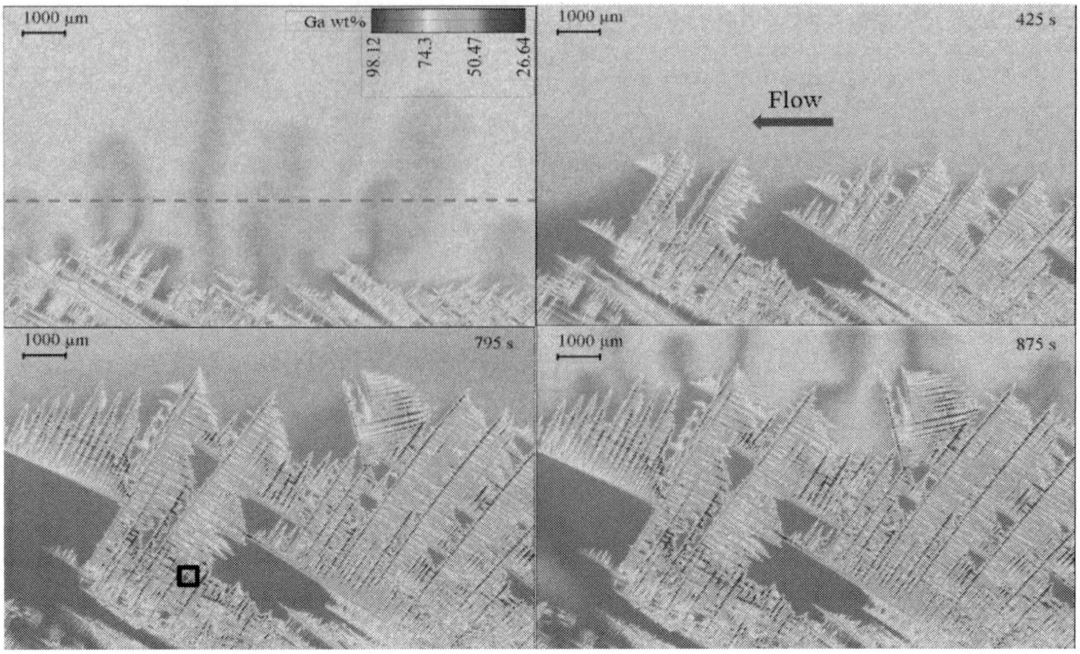

Figure 2. Image sequence showing the solidification of the Ga - 25wt%In alloy under the forced convection captured at different time steps: (a) 155 s, (b) 425 s, (c) 795 s and (d) 875 s. The magnetic wheel was rotated with 30 rpm.

in Figure 3a). Figures 2b and 2c are captured at two different time steps after the magnetic wheel was switched on. In this experiment the magnetic wheel was rotated contra-clockwise with 30 rpm. The direction of the induced flow is almost horizontal with respect to the solidification front and pointing in clockwise direction. The horizontal velocity component reaches values of ~ 0.3 mm/s. For the chosen rotation speed the solutal plumes are strongly damped and the final concentration profile along the solidification front becomes smooth (see the red curve in Figure 3a). Slight increase of the Ga concentration on the left-hand side of the red curve is caused by smeared plumes running along the solidification front from right to the left. The solidification front velocity decreases gradually from ~ 12.9 μm/s down to ~ 4.5 μm/s during exposure to magnetic pump as it can be seen in Figure 3b. This can be explained by an increasing Ga concentration near the solidification front with exposure time (high Ga concentration near the solidification front is especially pronounced in Figure 5c).

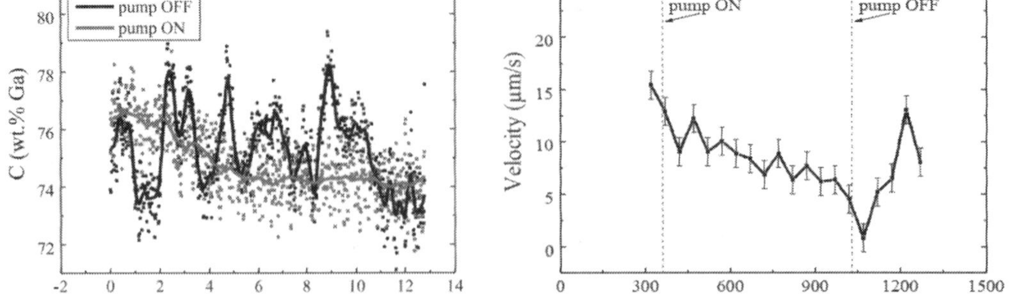

Figure 3. a) Concentration profiles along the solidification front 10 s before (black) and 30 s after (red) switching on the magnetic pump across the line highlighted in Figure 2a. b) Solidification front velocity. The magnetic wheel was rotated with 30 rpm.

Another interesting effect can be observed during solidification: uneven growth of primary dendrites at the beginning of the experiment leads to the formation of Ga-rich zones near the solidification front which develop into distinct segregation freckles as it can be seen on the left-hand side of Figures 2b and 2c. Also a competitive growth between secondary branches and primary dendrites leads to an increase of the primary arm spacing. A similar competition effect is observed for higher order arms: for example the secondary arm spacing is increasing. Moreover, preferential growth of the secondary arms is observed at the upstream side of the dendrites (i.e. on the right hand side of the dendrite trunks). In contrast, high solute concentration at the downstream side strongly influences the formation of secondary arms: in case when dendrites are inclined against the upcoming flow the secondary arms at the downstream side are fully eliminated; in case when the dendrites are inclined towards the upcoming flow only few secondary arms develop and create tertiary arms. Furthermore, the inclination angles of the growing dendrites change by several degrees under the influenced of the forced flow. After the magnetic pump is switched off the solute plumes re-appear (see Figure 2d) and the solidification velocity increases (see Figure 3b).

To study the influence of the intensity of the forced flow on the solidification pattern in the next experiment the magnetic wheel was rotated clockwise with 80 rpm. Figure 4 shows a sequence of images captured at different time steps.

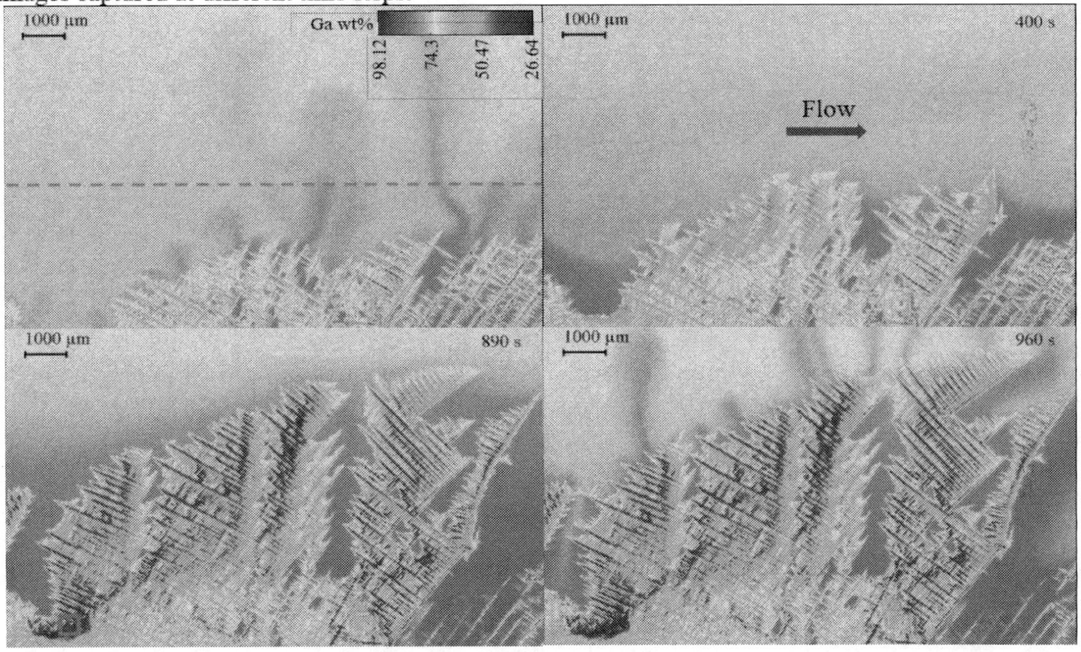

Figure 4. Image sequence showing the solidification of the Ga - 25wt%In alloy under the forced convection captured at different time steps: (a) 260 s, (b) 400 s, (c) 890 s and (d) 960 s. The magnetic wheel was rotated with 80 rpm.

Figure 4a demonstrates the time moment before the magnetic wheel was switched on. The dendritic structure and the flow field are similar to those already shown Figure 2a. As in the previous experiment the plumes are completely suppressed by the induced forced flow after switching on the magnetic wheel (see Figures 4b and 4c). Figure 5a shows corresponding Ga concentration profiles before (black) and after (red) the magnetic pump was switched on. It can be seen that the forced flow induced by higher rotation speed of the magnetic pump flattens the Ga concentration along the solidification front. The solidification front velocity decreases exponentially from ~ 14.2 μm/s down to ~ 2.5 μm/s (see Figure 5b) when the magnetic pump is switched on. Such a fast and strong decrease can be explained by the accumulation of Ga-rich solute with time near the solidification front. In comparison to previous case a pronounced Ga-rich layer develops near the solidification front (see Figure 4c).

Also Ga-rich zones are formed near the solidification front due to uneven growth of dendrites. They evolve with time into distinct segregation freckles (see the right-hand side part of Figure 4b and

4c). The induced strong flow acts like a carrier of In material for growing dendrites. As a result secondary arms grow preferably at the upstream side of the dendrites (i.e. on the left hand side of the

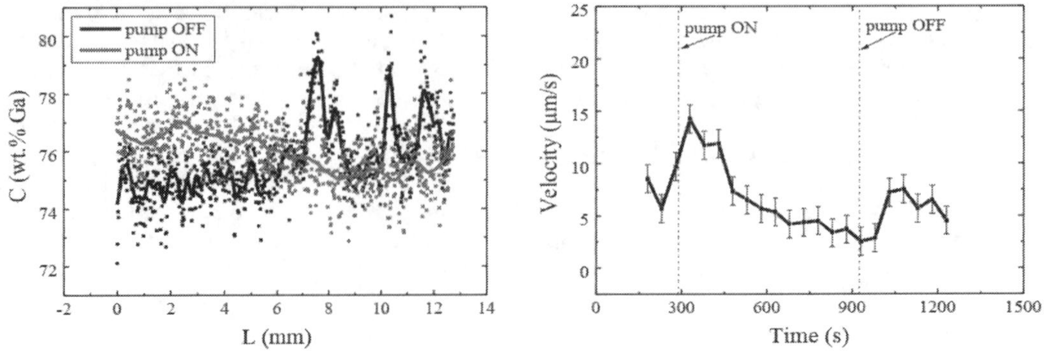

Figure 5. a) Concentration profiles along the solidification front 30 s before (black) and 10 s after (red) switching on the magnetic pump across the line highlighted in Figure 4a. b) Solidification front velocity. The magnetic wheel was rotated with 80 rpm.

dendrite trunks); while at the downstream side due to high solute concentration only some of the secondary arms develop or are completely of the primary trunk towards the upcoming flow (see the discussion in the first experiment). Increasing arm spacing due to competitive growth between secondary and primary arms is also observed. Moreover, the solute plumes re-appear again after the magnetic pump is switched off (see Figure 4d).

Despite multiple similarities certain differences between the two experiments can be clearly identified: first, at higher rotation speed (at 80 rpm) regions with high In concentration are built up (see Figures 2c and 4c for comparison). The average In concentration was measured in both experiments in the most In-rich areas of 0.52x0.52 mm^2 size (the corresponding areas are marked as squares in Figures 2c and 4c). A value of ~ 47.5wt% was found for 30 rpm, while a value of ~ 63.6wt% occurs in the second experiment at 80 rpm (the integration regions are marked in Figures 2c and 4c). Moreover, the forced convection reduces the solidification front velocity much stronger in the second experiment. Also the concentration profiles along the solidification front are getting more flat.

4. Conclusions

The bottom-up solidification of Ga-25wt%In alloys within a Hele-Shaw cell was investigated by the X-ray radioscopy. This method allows in situ observation of the flow pattern, concentration field and dendrite growth. We study the effect of forced convection on the growth behavior of the Indium dendrites. The application of the electromagnetically driven flow displaces the typical pattern of natural convection: depending on the intensity of the induced forced flow the local fluctuations of solute concentration are partially or completely damped. The forced flow induces different effects on grain morphology, such as the uneven growth of primary trunks or lateral branches, remelting of single dendrites and also of lager dendrite ensembles, freckle formation, changes the inclination angle of the dendrites and leads to an increasing arm spacing. These effects are primarily governed by the convective redistribution of solute. A more detailed analysis of the melt flow effects is the subject of ongoing work.

Acknowledgments

The research is supported by the German Helmholtz Association in form of the Helmholtz-Alliance "LIMTECH".

References

[1] Steinbach S and Ratke L 2005 *Mater. Sci. Eng.* A **413-414** 200
[2] Eckert S *et al* 2013 *Eur. Phys. J. Spec. Top.* **220** 123
[3] Räbiger D, Krößig C, Eckert S and Gerbeth G 2012 *J. Iron Steel Res. Int.* **19 (I-1)** 341
[4] Bergman M I, Fearn D R, Bloxham J and Shannon M C 1997 *Metall. Mater. Trans.* A **28** 859

[5] Ramani A and Beckermann C 1997 *Scripta Mater.* **36** 633
[6] Peppin S S L, Huppert E and Worster M G 2008 *J. Fluid Mech.* **599** 465
[7] Murray B T, Wheeler A A and Glicksman ME 1995 *J. Cryst. Growth* **154** 386
[8] Zhang M and Maxworthy T 2002 *J. Fluid Mech.* **470** 247
[9] Mathiesen R H *et al* 2008 *J. Cryst. Growth* **310** 2906
[10] Koster J N, Seidel T and Derebail R 1997 *J. Fluid Mech.* **343** 29
[11] Shevchenko N *et al* 2015 *J. Cryst. Growth* in press
[12] Boden S, Eckert S, Willers B and Gerbeth G 2008 *Metall. Mater. Trans.* A **39** 613
[13] Shevchenko N, Boden S, Gerbeth G and Eckert S 2013 *Metall. Mater. Trans.* A **44** 3797
[14] Boden S, Eckert S and Gerbeth G 2010 *Mater. Letters* **64** 1340
[15] Shevchenko N, Boden S, Eckert S, Borin D, Heinze M and Odenbach S 2013 *Eur. Phys. J. Spec. Top.* **220** 63

IOP Conf. Series: Materials Science and Engineering **84** (2015) 012081 doi:10.1088/1757-899X/84/1/012081

Modeling of percolation of globular-equiaxed grains in Al-Cu alloys

Christophe Mondoux[1], **Hervé Combeau**[2], **Jean-Luc Desbiolles**[1], **Michel Rappaz**[1]

[1] Department of Materials Science and Engineering, Computational Materials Laboratory (LSMX), Ecole Polytechnique Fédérale de Lausanne, station 12, 1015 Lausanne, CH
[2] Institut Jean Lamour, Departement SI2M, UMR CNRS 7198 Nancy-Université UPVMetz, Ecole des Mines de Nancy, Parc de Saurupt CS 14234, F-54042 Nancy cedex, France

E-mail: `christophe.mondoux@epfl.ch`

Abstract. A new mesoscopic model able to describe percolation of globular-equiaxed grains, highly relevant for hot tearing formation in castings, has been developed. This model is inspired from the granular model of Sistaninia *et al.* [1–5] that considers polyhedral grains based on a Voronoi tessellation of space. In the new mesoscopic model, the set of tetrahedra that define the grains are further subdivided into smaller tetrahedra called columns and solute diffusion is considered in both the solid and liquid phases. This allows to obtain smoother shapes of the grains and to better describe a progressive coalescence of the grains. In addition, it gives the possibility of having liquid pockets in the last-stage solidification (which is not possible in the case of polyhedral grains) and thus correctly describe the solid fraction at which the grain structure is percolated. The shape of the grains has first been validated with a multiphase-field method. The relevant microstructural features, such as the percolation state of an Al-Cu alloy, were then deduced from the model. This information will then be used to refine the 3D granular models of Sistaninia *et al.* [1–5] which account for stress development and liquid feeding, in order to create more advanced predictive tools for hot tearing formation.

1. Introduction

Hot tearing is a major defect of solidification and welding processes that occurs in metallic alloys while they are in the semi-solid state [6, 7]. The two main phenomena leading to hot tearing are: a lack of liquid feeding at high solid fractions and tensile/shear strains in the mushy zone that tend to pull apart the solid network. The strains, transmitted through the partially coherent solid in the mushy zone, are induced by contraction of the solid in a thermal gradient, and mechanical constraints imposed by the geometry of components. At low solid fraction ($g_s < 0.9$), the permeability of the mushy zone is large enough so that liquid feeding can heal possible grain boundary openings, forming a so-called *healed hot tear*. At high solid fraction ($g_s > 0.95$) the grains form a continuous solid network possessing sufficient mechanical resistance and the mushy material behaves as a ductile solid. However, there is a critical zone for hot tearing (for solid fractions in the range of $0.9 < g_s < 0.95$) in which grain boundaries openings, due to the deformation localized at those which have still a liquid film, cannot be fed by liquid flow because of the low permeability of the mushy zone [8]. Therefore, it is important to know when the solid grains coalesce and percolate to form a fully coherent solid, i.e., when

the material can transmit stresses and strains but also withstand thermal contraction without rupture [8,9]. Coalescence is defined as the disappearance of a liquid film in between two distinct grains, forming a solid grain boundary, while percolation is defined as the gradual transition of isolated grains or clusters surrounded by a continuous liquid film to a continuous solid network across a domain, even if some liquid regions remain [9, 10].
Coalescence is strongly influenced by the interfacial energies. In the case of a thin liquid film that separates two coalescing grains, as the thickness of the liquid film approaches the solid/liquid interface thickness, $\delta_{s\ell}$, the excess free energy deviates from the initial value of $2\gamma_{s\ell}$, where $\gamma_{s\ell}$ is the energy of the solid/liquid interface. When the liquid has completely disappeared, the excess free energy is equal to the grain boundary energy γ_{gb} [9]. In the case $\gamma_{gb} > 2\gamma_{s\ell}$, the substitution of two solid/liquid interfaces by a grain boundary is associated with an increase of the excess interface energy: a bridging undercooling, ΔT_b, is thus necessary to coalesce the two interfaces (repulsive case). On the other hand, in the case $\gamma_{gb} < 2\gamma_{s\ell}$, no undercooling is needed to coalesce the two interfaces (attractive case). The bridging undercooling needed for the coalescence of a repulsive grain is given by [9]:

$$\Delta T_b = \frac{\gamma_{gb} - 2\gamma_{s\ell}}{\Delta S_f} \frac{1}{\delta_{s\ell}} \tag{1}$$

where ΔS_f is the volumetric entropy of fusion. For a binary alloy, the situation is more complex, because the solute element can alter the grain boundary energy but also mainly because coalescence is influenced by the solute concentration in the liquid film. The rejection and diffusion of solute in the liquid parallel to the film as well as back-diffusion perpendicular to the impinging interfaces can thus play an important role [9]. Coalescence will occur when the temperature and the concentration of the liquid film reaches a so-called *coalescence line*, which is parallel to liquidus line in the phase diagram, but shifted down by ΔT_b.

2. Model

We propose a new mesoscopic model inspired from the 3D polyhedral granular model developed in [1–5], in order to obtain smoother grain morphologies compared to those reported in [1–5] and to predict more accurately the percolation state for a large population of grains. As for the granular model, a Voronoi tessellation is generated from randomly distributed nucleation centers. However, instead of considering the pyramids in 3D (sectors in 2D) having a nucleation center as a summit and the Voronoi facets as bases (see Fig. 1(a) and 1(b)), each of them is subdivided into several columns (see Fig. 1(c) and 1(d)). The liquid composition at the interface $c^*_{\ell,i}$, in column i, possessing a mean curvature $\overline{\kappa}_i$ at a given temperature T (supposed uniform over the calculation domain) is given by:

$$c^*_{\ell,i}(T,\kappa) = -\frac{T_m - T - \Gamma_{s\ell}\overline{\kappa}_i}{m_\ell} \tag{2}$$

where T_m is the melting point of pure aluminum, $\Gamma_{s\ell}$ the Gibbs-Thompson coefficient, and m_ℓ the liquidus slope. Altough small for globular grains (a few mK), the curvature contribution is still necessary because it stabilizes the s-ℓ interface, as expected, by inducing a flow of solute from low to high curved regions.
Within each column i, backdiffusion in the solid is calculated radially by using a Landau transformation of the domain $r \in [0, r^*_i(t)]$ (where r is the distance and r^*_i the distance of the interface from the nucleation center of the grain in the column i) into a reference domain $\eta \in [0, 1]$, as introduced by Voller and Sundarraj [11]. The concentration of the solid $c_{s,i}$ becomes:

$$c_{s,i}(r,t) \rightarrow c_{s,i}(\eta,t) \quad \eta = \frac{r}{r^*_i(t)} \tag{3}$$

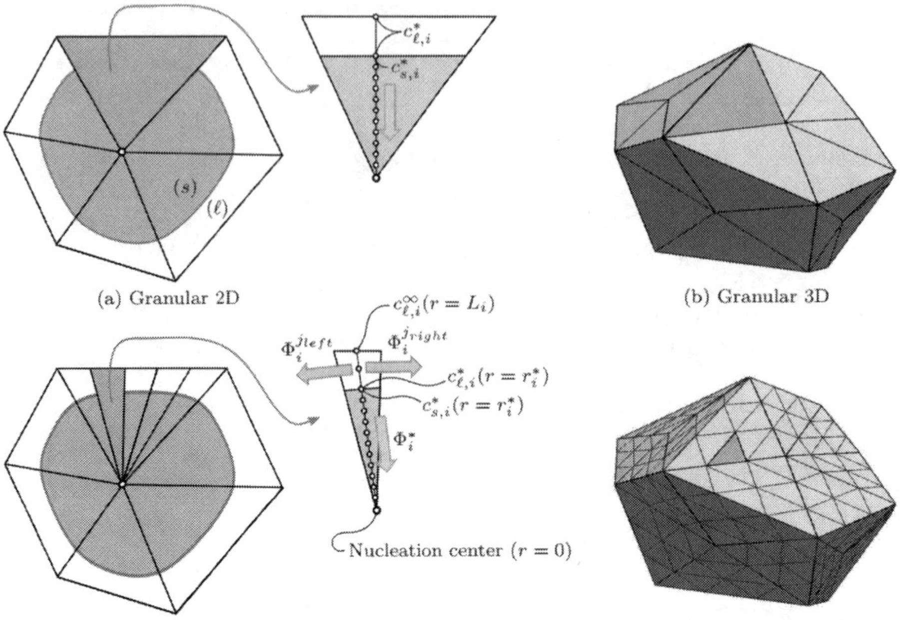

Figure 1: Schematic representation of the polyhedral granular model of Sistaninia *et al.* [1–5] compared to the present smooth granular model.

The diffusion equation in the solid thus becomes:

$$\left(\frac{\partial c_{s,i}}{\partial t}\right)_\eta = \frac{D_s}{r_i^{*2}}\frac{\partial^2 c_{s,i}}{\partial \eta^2} + \left(\frac{\eta v_i^*}{r_i^*} + \frac{(P-1)D_s}{\eta r_i^{*2}}\right)\frac{\partial c_{s,i}}{\partial \eta} \tag{4}$$

where v_i^* is the velocity of the interface, D_s the diffusion coefficient of the solid phase and P the dimension of the problem (3 for 3D and 2 for 2D). Eq. (4) is solved using an implicit finite difference scheme. By integrating the concentration of the solid, one can thus find the flow of solute pumped by the solid at the interface, Φ_i^*:

$$\Phi_i^* = \frac{d}{dt}\int_0^{r_i^*(t)} c_{s,i}dV = \int_0^{r_i^*(t)} \frac{\partial c_{s,i}}{\partial t}dV + c^*_{s,i}\frac{dr_i^*}{dt}S_i^* \tag{5}$$

where S_i^* is the area of the surface separating the liquid and the solid parts of the column i.
In the liquid part of the column, the liquid concentration, $c_{\ell,i}$, is assumed to be a piecewise parabolic-linear function:

$$c_{\ell,i}(r) = \begin{cases} c^{\infty}_{\ell,i} + \left(c^*_{\ell,i} - c^{\infty}_{\ell,i}\right)\left(\frac{\lambda_i-(r-r_i^*)}{\lambda_i}\right)^2 & \text{if} \quad r < r_i^* + \lambda_i \\ c^{\infty}_{\ell,i} & \text{if} \quad r \geqslant r_i^* + \lambda_i \end{cases} \tag{6}$$

where $c^{\infty}_{l,i} = c_{l,i}(r = L_i)$ is the concentration at the Voronoi facet and λ_i the diffusion layer thickness. Before the diffusion layer reaches the Voronoi facet, $c^{\infty}_{\ell,i}$ is set to the nominal concentration and λ_i can be found by knowing the average concentration in the liquid part of the column $\langle c_{\ell,i}\rangle$. Once the diffusion layer has reached the Voronoi facet, λ_i is known and equal to $L - r_i^*$ while $c_{\ell,i}^{\infty}$ is unknown and can be found from $\langle c_{\ell,i} \rangle$. Note that the approximation of the profile in the liquid by a piecewise parabolic-linear function does not correspond to the solution of the diffusion equation in cylindrical or spherical coordinates. However, when the diffusion layer has reached the Voronoi facet, the derivative at $r = L_i$ is zero and it is thus consistent with the fact that there is a symmetric column on the other side of the Voronoi facet imposing a no-flux condition.

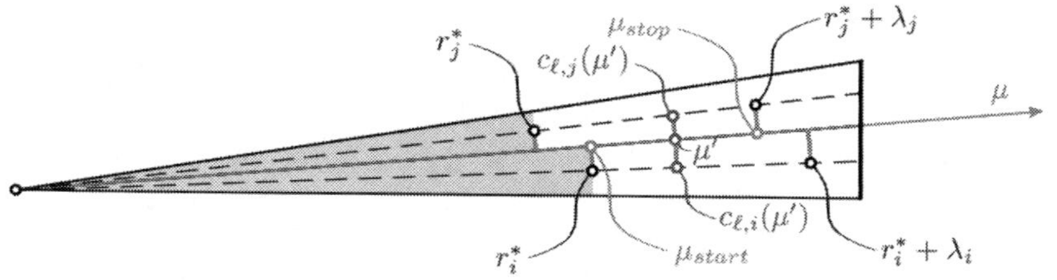

Figure 2: Schematic representation of the calculation of diffusion in the liquid between two adjacent columns.

Then, the flow of solute exchanged between two columns is calculated. Let us define μ as the distance with respect to the nucleation center along the surface separating the two columns (see Fig. 2). The flow of solute is evaluated by calculating the difference in liquid concentration between the columns perpendicularly to the μ-axis. In 2D, the flow of solute leaving the liquid part of column i and going into the adjacent liquid part of column j is given by:

$$\Phi_i^j = -\int_{\mu_{start}}^{\mu_{stop}} D_\ell \frac{c_{\ell,j}(\mu') - c_{\ell,i}(\mu')}{d_{\mu'}} dS_{lat} \tag{7}$$

where D_ℓ is the diffusion coefficient in the liquid, S_{lat} corresponds to the lateral surface through which solute is exchanged (in 2D, $dS_{lat} = d\mu'$) and $d_{\mu'}$ the distance between the positions at which the concentrations are evaluated in the adjacent columns. μ_{start} and μ_{stop} are defined as follows: the two positions corresponding to r_i^* and r_j^* are projected onto the μ axis. The projected position that is the furthest away from the nucleation center is set as μ_{start}. On the other hand, μ_{stop} is defined as the position of the projection of $r_i^* + \lambda_i$ and $r_j^* + \lambda_j$ that is the closest to the nucleation center. In 3D, the method is similar but the surface of the facet separating the two columns as well as the angle with the normal of the facet are taken into account in Eq. (7).
The change of the average concentration in the liquid is thus given by substracting the flow exchanged with the solid part of the column and with the liquid parts of the adjacent columns:

$$\frac{d}{dt}\int_{r_i^*}^{L_i} c_{\ell,i}\, dV = -\left(\Phi_i^* + \sum_{j=\text{neighbour}} \Phi_i^j\right) \tag{8}$$

Please note that the back-diffusion in the solid is purely 1-dimensional in this model, i.e., no direct exchange of solute between the solid parts of the columns and between the solid and liquid parts of different columns. In fact, the solute exchanged is assumed to take place only between the liquid parts of adjacent columns.
The velocity of the interface can then be found by a solute balance at the interface:

$$v_i^* = \frac{1}{(1-k_0)c_\ell^*}\left(D_s\left[\frac{\partial c_s}{\partial r}\right]^* + D_\ell \frac{c_{\ell,i}^* - c_{\ell,i}^{\infty}}{h_i}\right) \tag{9}$$

IOP Conf. Series: Materials Science and Engineering **84** (2015) 012081 doi:10.1088/1757-899X/84/1/012081

k_0 is partition coefficient and $\left[\frac{\partial c_s}{\partial r}\right]^*$ is estimated numerically. h_i corresponds to $\lambda_i/2$ until the solute layer has reach the Voronoi facet. When this occurs, the position (and thus the velocity) of the interface can be deduced from an overall solute balance at the level of one column. Once the velocity is known, the whole calculations can be made again starting from Eq. (4) for the next timestep.
Near the end of solidification, once two solid-liquid interfaces get in close contact, when the liquid layer thickness becomes lower than $\delta_{s\ell} = 1$nm, the liquid layer thickness is fixed to $\delta_{s\ell}$ and the concentration of the liquid is supposed to be uniform within the liquid part of each column and is not anymore linked to the phase diagram with Eq. (2). Essentially, the liquid concentration will decrease (by backdiffusion of solute in the solid perpendicular to the interface and diffusion in the liquid from adjacent columns parallel to the interface). Once the coalescence line is reached, the two interfaces have coalesced.

3. Results

3.1. Validation

First, the model has to be validated with a more refined model, such as the multiphase-field method [12]. A multiphase-field formulation, based on the work of Meidani *et al.* [13], has been developed in order to describe the evolution of the shape of the grains and also, unlike the standard phase-field method, to consider grain boundary energies, by assigning a different phase field variable to each grain. The validation has been performed in simple cases where a single grain grows in a square (2D) or cubic (3D) domain (see Fig. 3).

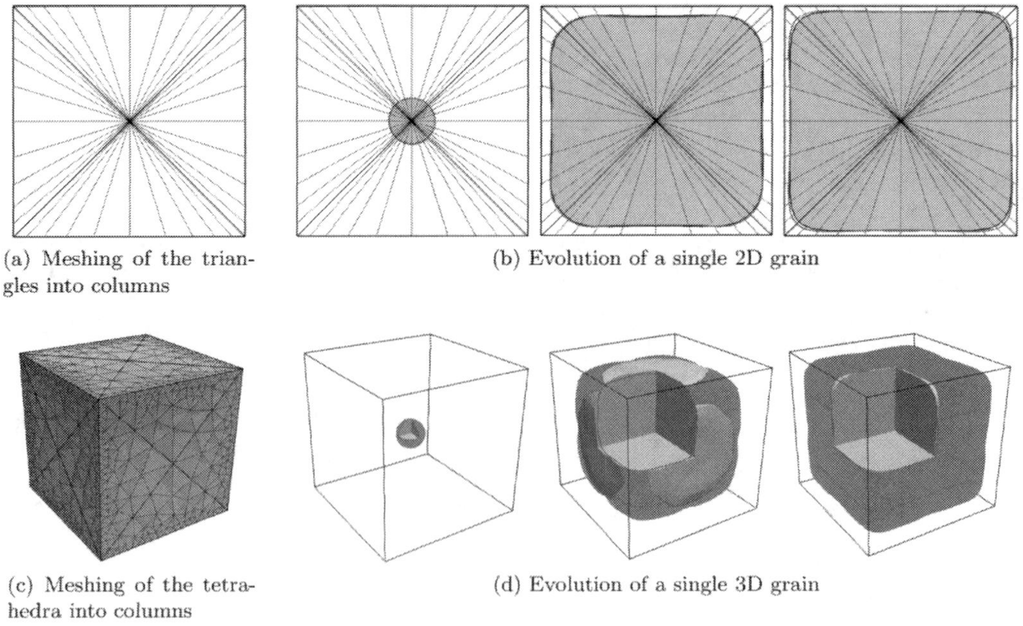

(a) Meshing of the triangles into columns

(b) Evolution of a single 2D grain

(c) Meshing of the tetrahedra into columns

(d) Evolution of a single 3D grain

Figure 3: (a) and (c) Meshing of the triangles/tetrahedra into smaller columns. (b) and (d) Comparison of grain evolution computed by the mesoscopic model (in gray) and the multiphase-field model (in red).

Note that the meshing in both 2D and 3D is more refined closer to the edges, i.e. the places where a higher curvature is expected near the end of solidification. The mesoscopic model is in a good agreement with the multiphase-field model in both 2D and 3D. In fact, the mesoscopic model captures correctly the physics of solute diffusion, since the shapes of the grains are really close, as long as there is no strong intability of the solid-liquid interface (i.e., small grain size) [14]. In addition, as the multiphase-field method, the mesoscopic model predicts the fact that, at the center of the flat parts of the grains, the interface is slightly negatively curved towards the center of nucleation of the grain.

3.2. 2D results

It is then interesting to follow the evolution of a large number of grains. In Fig. 4, one can observe the solidification of a set of 1024 randomly nucleated grains (possessing an average size of 100μm) in 2D for an Al-1wt%Cu alloy cooled at -1°C/s. The progressive formation of grain clusters can be tracked following two different criteria. With the contact criterion, two grains belong to the same cluster when the thickness of the liquid film separating them attains $\delta_{s\ell}$ (*contact clusters*). With the coalescence criterion, two grains belong to the same cluster when they are locally coalesced (*coalescence clusters*). The *grain contact* onset time is defined as the time at which the first two grains get into contact.

As shown in figure 5, the number of contact clusters decreases faster than the coalescence clusters, since the coalescence criterion is more restrictive. This can be seen in particular at 534°C, where the solid structure becomes fully percolated for both criteria but there is a single cluster only with the contact criterion. In reality at 548°C, the eutectic would have filled the remaining liquid space (if the second phase can nucleate), but the model allows to find the solidification of eutectic-free alloys. Another way of tracking the progressive percolation of grains is to observe the evolution of the solid liquid interfacial area per unit volume, $S_{s\ell}/V$. This value has been

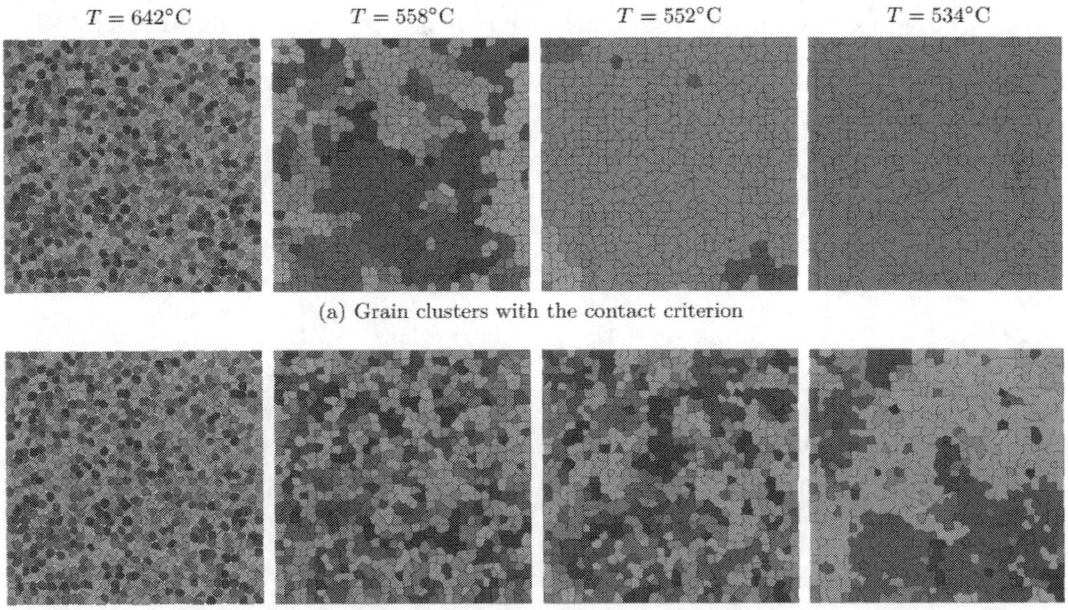

(a) Grain clusters with the contact criterion

(b) Grain clusters with the coalescence criterion

Figure 4: 2D percolation of an Al-1wt%Cu specimen containing 1024 grains. The average grain size was set to 100μm. Each grain cluster is represented by a different color and the grains clusters are either defined by (a) the contact criterion and (b) the coalescence criterion. The solid/liquid interface is represented by black segments.

IOP Conf. Series: Materials Science and Engineering **84** (2015) 012081 doi:10.1088/1757-899X/84/1/012081

normalized by the average grain size, d_{grain}, in order to obtain $\mathcal{S}_{s\ell} = \frac{S_{s\ell} d_{\text{grain}}}{V}$. The evolution of $\mathcal{S}_{s\ell}$ of the model as a function of the solid fraction initially increases because of grain growth, but then decreases abruptly as a solid fraction $g_s = 0.975$ when grain coalescence becomes significant. It is interesting to note that, until significant coalescence takes place, the evolution of $\mathcal{S}_{s\ell}$ closely follows the predictions of a regular arrangement of hexagonal grains.

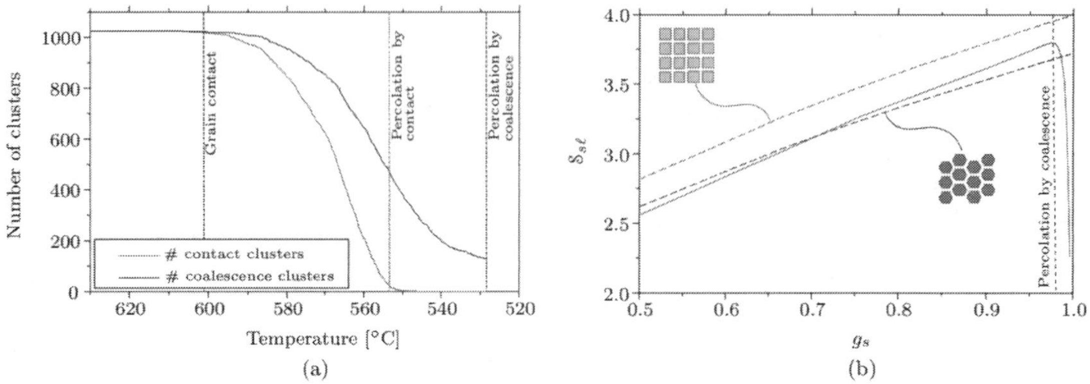

Figure 5: (a) Evolution of the number of clusters as a function of temperature (the grain contact temperature and the temperatures at which the structure is percolated following the two criteria are also indicated). (b) Evolution of $\mathcal{S}_{s\ell} = \frac{S_{s\ell} d_{\text{grain}}}{V}$ as a function of the solid fraction, g_s. The mesoscopic model result is shown in red, while the analytical predictions for a regular arrangement of square and hexagonal grains are in green and blue respectively.

3.3. 3D results

The model was extended in 3D and one preliminary result is shown in Fig. 6, where the progressive percolation of grains can be observed as solidification proceeds. It is interesting to note how the liquid pockets (possessing a negative mean curvature) are formed at the end of

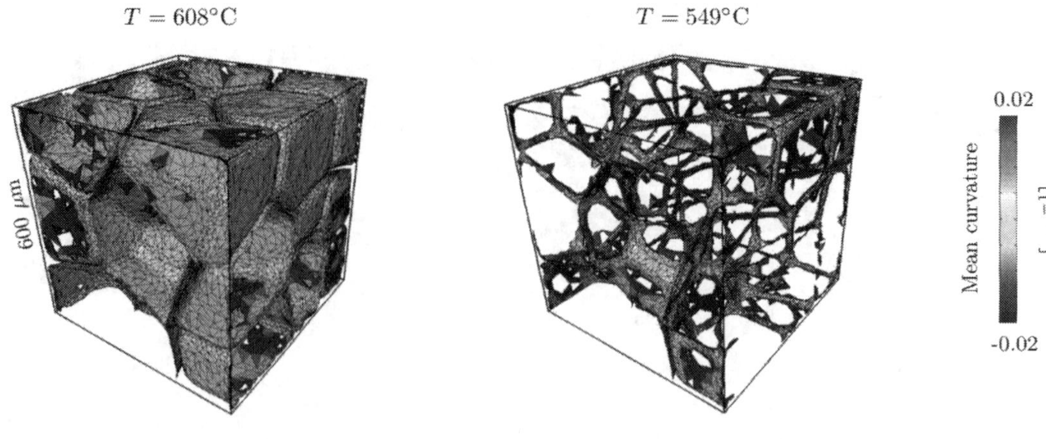

Figure 6: Simulation of the solidification of 3x3x3 grains of an Al-3wt%Cu cooled at -0.05°C/s. The grain size was fixed to 200μm.

solidification. The formation of the liquid pockets in fact is an indication of coalescence, since, when two grains coalesce, part of their solid/liquid interface disappears and only some liquid eventually remains at the corner of the grains.

4. Conclusion

The new mesoscopic model has been successfully validated with a multiphase-field model, in both 2D and 3D for simple domain geometries. This indicates that the assumptions of the mesoscopic model, in particular neglecting direct exchange of solute between the solid parts of the columns and the assumption of a piecewise parabolic-linear profile in the liquid, are reasonable.
In 2D, the mesoscopic model was able to correctly describe the percolation transitions of a large number of grains and currently a benchmark of the percolation transition values for different alloys and cooling conditions is being performed in order to assess the influence of the nominal composition and cooling rates. In 3D, the model predicts the formation of liquid pockets but need to be compared with 3D experimental results.

5. Acknowledgments

The authors gratefully acknowledge the support of the Swiss National Fund (Grant Nos. 200020-132848 and 200020-146879).

References

[1] Phillion A, Vernède S, Rappaz M, Cockcroft S and Lee P D 2009 *Intern. Journ. of Cast Met. Res.* **22** 240-3
[2] Sistaninia M, Phillion A, Drezet J M and Rappaz M 2011 *Met. and Mat. Trans. A* **42** 239-48, 2011
[3] Sistaninia M, Phillion A, Drezet J M and Rappaz M 2012 *Acta Mater.* **60** 3902-11
[4] Sistaninia M, Phillion A, Drezet J M and Rappaz M 2012 *IOP Conf. Ser.: Mat. Sci. and Eng.* **33** 012070
[5] Sistaninia M, Phillion A, Drezet J M and Rappaz M 2012 *Acta Mater.* **60** 6793-803
[6] Campbell J 2003 *Castings (second edition)* Elsevier
[7] Dantzig J A and Rappaz M 2009 *Solidification* EPFL Press
[8] Rappaz M, Drezet J M and Gremaud M 1999 *Met. and Mat. Trans. A* **30** 449-55
[9] Rappaz M, Jacot A and Boettinger W 2003 *Met. and Mat. Trans. A* **34** 467-79
[10] Vernède S and Rappaz M 2006 *Phil. Mag.* **86** 3779-94
[11] Voller V R and Sundarraj S 1993 *Mat. Sci. and Tech.* **9** 474-82
[12] Steinbach I, Pezzolla F, Nestler B, Seeßelberg M, Prieler R, Schmitz G and Rezende J 1996 *Physica D: Nonlin. Phen.* **94** 135-47
[13] Meidani H and Jacot A 2011 *Acta Mater.* **59** 3032-40
[14] Diepers H J and Karma A 2004 *Solidif. Proc. and Microstr.: A Symp. in Honor of Prof. W. Kurz* **369**

Three-dimensional Dendritic Needle Network model with application to Al-Cu directional solidification experiments

D. Tourret[1,2], A. Karma[1], A. J. Clarke[2], P. J. Gibbs[2], S. D. Imhoff[2]

[1] Northeastern University, Department of Physics and Center for Interdisciplinary Research on Complex Systems, Boston, MA, USA

[2] Los Alamos National Laboratory, Materials Science and Technology Division, Los Alamos, NM, USA

E-mail: `dtourret@lanl.gov`

Abstract. We present a three-dimensional (3D) extension of a previously proposed multi-scale Dendritic Needle Network (DNN) approach for the growth of complex dendritic microstructures. Using a new formulation of the DNN dynamics equations for dendritic paraboloid-branches of a given thickness, one can directly extend the DNN approach to 3D modeling. We validate this new formulation against known scaling laws and analytical solutions that describe the early transient and steady-state growth regimes, respectively. Finally, we compare the predictions of the model to *in situ* X-ray imaging of Al-Cu alloy solidification experiments. The comparison shows a very good quantitative agreement between 3D simulations and thin sample experiments. It also highlights the importance of full 3D modeling to accurately predict the primary dendrite arm spacing that is significantly over-estimated by 2D simulations.

1. Introduction

Since the early descriptions of steel dendrites by Grignon [1] in 1775 and Chernoff [2] in 1880, much has evolved in our understanding of dendritic microstructures, including the physical origin of their morphological complexity, and the laws underlying their growth dynamics [3]. However, the schematic drawings of dendritic microstructures contained in these early works (some of which are reproduced in Fig. 1) already capture salient features of the dendritic network organization, such as dendritic arm spacings, crucial to final mechanical properties.

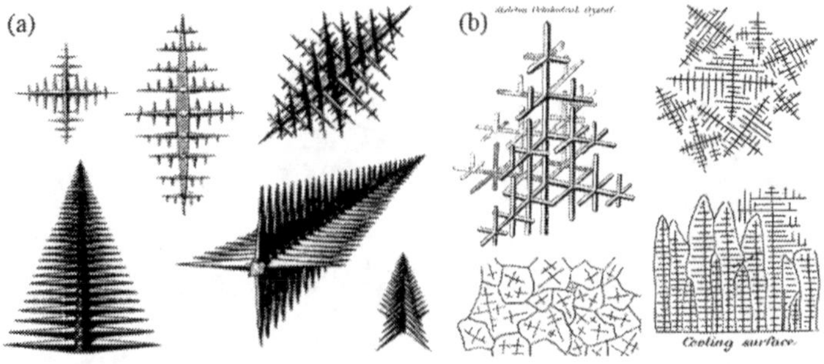

Figure 1. Schematic drawings of steel dendrites after (a) Grignon [1] (1775) and (b) Chernoff [2] (1880).

Recently, a new multi-scale approach to simulate dendritic solidification was proposed [4,5] that conceptualizes the dendritic network as a hierarchical structure of primary, secondary, and higher order branches, with each arm represented as a thin needle much like in the drawings of Fig. 1. Additionally, this approach provides a rigorous theoretical footing for describing the needle dynamics on the dendritic grain scale by integrating out the interface dynamics on the scale of the dendrite tip radius. In concentrated alloys that solidify at low undercooling, this *Dendritic Needle Network* (DNN) approach reproduces analytical and phase field predictions of crystal growth dynamics in both transient and steady-state regimes [5], with much faster simulations than the well established phase field approach [6], since it does not explicitly track the evolution of the complex solid-liquid interface pattern on the dendrite tip scale. This new approach aims at bridging the modeling scale gap that exists between phase field [6] at the scale of a dendrite (or a small dendritic array), and continuum averaged approaches, such as cellular automaton based models [7] at the scale of a technological casting process. In this article, we present the first implementation of the multi-scale DNN model in three dimensions (3D) and its application to the directional solidification of an Al-12at.% Cu alloy, which we compare quantitatively to experimental data from *in situ* X-ray radiography.

2. Modeling

2.1. DNN model in two dimensions

As detailed in Ref. [5], the evolution of a binary alloy solid-liquid interface at a given temperature, T_0, in a diffusive regime with negligible solid state diffusion follows

$$\partial_t u = D\nabla^2 u, \quad (1)$$

$$\left[1-(1-k)u_i\right]v_n = D\partial_n u|_i, \quad (2)$$

$$u_i = d_0 f(\theta)\kappa, \quad (3)$$

where $u \equiv (c_l^0 - c)/[(1-k)c_l^0]$ is a reduced solute field, with c the solute concentration field, c_l^0 the equilibrium concentration at the temperature T_0, and k is the solute partition coefficient at the solid-liquid interface. Equation (1) represents the diffusion of solute in the liquid with a diffusion coefficient D. Equation (2) is a statement of solute mass conservation at the interface, giving the normal velocity of the solid-liquid interface, v_n, as a function of the value of u on the liquid side of the interface, u_i, and its normal gradient, $\partial_n u|_i$. Equation (3) is the Gibbs-Thomson condition for equilibrium at the interface, with $f(\theta)$ the orientation dependence of the capillary length, $d_0 = \Gamma/\left[|m|(1-k)c_l^0\right]$, of the alloy of Gibbs-Thomson coefficient Γ and liquidus slope m, and κ is the local curvature of the interface [8]. For isothermal solidification at T_0, the far-field condition $c \to c_\infty$ reads $u \to \Omega$ with a solute supersaturation $\Omega \equiv (c_l^0 - c_\infty)/[(1-k)c_l^0]$.

In our specific regime of interest, *i.e.* at low supersaturation in concentrated alloys, dendritic microstructures exhibit very sharp needle-like branches similar to those schematized in Fig. 1, and one can write a solute mass balance at an intermediate length scale (Fig. 2a) much larger than the dendritic tip radius, ρ, and much smaller than the diffusion length, $l_D = D/V$, with V the tip growth velocity. At this scale, the needle crystal is infinitely sharp, so that $u_i \approx 0$ from Eq. (3) and the solute locally follows Laplace equation $\nabla^2 u = 0$ in place of Eq. (1). This

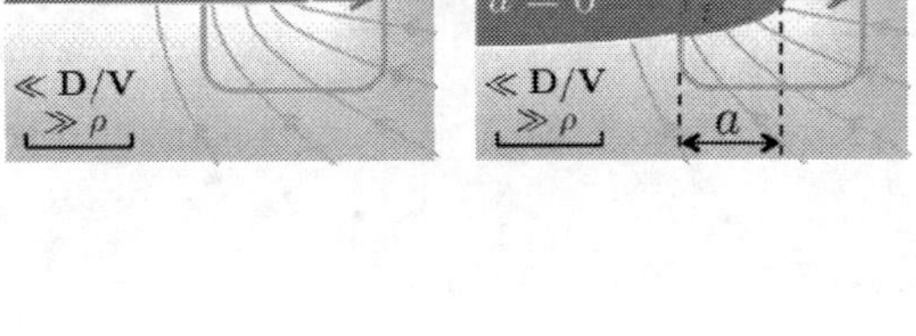

Figure 2. Schematics of a dendritic tip at an intermediate scale between the tip radius ρ and the much larger diffusion length D/V for a sharp (a) and a parabolic (b) needle crystal.

IOP Conf. Series: Materials Science and Engineering **84** (2015) 012082 doi:10.1088/1757-899X/84/1/012082

problem has an analytical solution that exhibits a square-root singularity of amplitude $\mathcal{F}$ of the u-field close to the tip as $\partial u/\partial y|_{y=0} \sim \mathcal{F}/\sqrt{d_0(x_t - x)}$, where $+x$ is the needle growth direction, and $(x, y) = (x_t, 0)$ are the tip coordinates [9]. Since that singularity is similar to that of the stress field in linear elasticity at the tip of a crack in mode III fracture, one can calculate its amplitude $\mathcal{F}$ using an integral along a coutour Γ_e around the tip (see Fig. 2a) similar to the classical J-integral of fracture mechanics [10]. Combining this expression of the normal gradient of u around the tip with the fact that further behind the tip, the normal velocity of the solid-liquid interface is essentially 1D and follows $v_n \approx D\,\partial u/\partial y|_{y=0}$, one can integrate the equation of the solid-liquid interface at this scale, yielding a parabolic shape of radius ρ following [5]

$$\rho V^2 = 2D^2\mathcal{F}^2/d_0. \tag{4}$$

Together with the microscopic solvability condition at the scale of the tip radius ρ [11, 12]

$$\rho^2 V = 2Dd_0/\sigma, \tag{5}$$

with σ the tip selection constant of the alloy, those two equations prescribe uniquely the instantaneous tip velocity, $V(t)$, and radius, $\rho(t)$, of a dendritic tip. While Eq. (5) was introduced as a steady-state solution, it was shown later that $\rho^2 V$ reaches a constant value very early during the transient development of a dendritic branch [13]. Hence, the product $\rho^2 V$ is kept constant and the model captures the evolution of the surrounding solute field and the transient interactions between individual dendrite arms through the evolution of $\mathcal{F}(t)$.

For directional solidification within a frozen temperature field with $T(x) = T_0 + G(x - V_p t)$ with a pulling velocity V_p, if we set T_0 as the solidus temperature, Eq. (3) changes to $u_i = d_0 f(\theta)\kappa + (x - V_p t)/l_T$ with a thermal length $l_T = |m|(1-k)c_l^0/G$ corresponding to the applied thermal gradient G, and the equilibrium along a needle reads

$$u_i = (x - V_p t)/l_T \tag{6}$$

instead of $u_i = 0$ for isothermal solidification at T_0. Thus the prefactor of v_n in Eq. (2) is not ≈ 1 anymore, and the dynamics of a needle tip depends on the value of u at its tip, u_t, as [5]

$$\rho V^2 = 2D^2\mathcal{F}^2 \Big/ \left\{ [1 - (1-k)u_t]^2 \, d_0 \right\} \tag{7}$$

$$\rho^2 V = 2Dd_0 \,/\, \{ [1 - (1-k)u_t]\, \sigma \} \tag{8}$$

with the far-field condition on u now corresponding to its liquidus value $u \to 1$ as $x \to \infty$.

2.2. Extension to a thick needle crystal

The model summarized above, and described in details in Ref. [5], cannot be readily extended to 3D, because no analytical solution exists in 3D for a Laplacian field around an infinitely thin needle abstracted as a line at a fixed concentration, such as the one yielding the square-root singularity of the solute gradient normal to the needle in the tip region in 2D. Therefore, first we developed a new formulation of the 2D DNN model for a needle of finite thickness with u imposed within a parabolic shape of radius ρ (Fig. 2b).

The integration of the solute conservation at the interface $D\partial u/\partial n \approx v_n$ in the tip region over a length $a \ll l_D$, *i.e.* over a contour Γ_i between $x = x_t - a$ and $x = x_t$ as in Fig. 2b, yields

$$D \int_{\Gamma_i} \frac{\partial u}{\partial n} \, \mathrm{d}\Gamma_i = \int_{-y_a}^{y_a} V \, \mathrm{d}y = 2V\sqrt{2\rho a}, \tag{9}$$

with $y_a \equiv y_i(x_t - a) = \sqrt{2\rho a}$. Thus, Eq. (9) can be rewritten in a form

$$\rho V^2 = \frac{1}{8a}\left(D\int_{\Gamma_i}\frac{\partial u}{\partial n}\,\mathrm{d}\Gamma_i\right)^2, \tag{10}$$

that relates the product ρV^2 to the square of the flux intensity factor defined by

$$\mathcal{F} \equiv \frac{1}{4\sqrt{a/d_0}}\int_{\Gamma_i}\frac{\partial u}{\partial n}\,\mathrm{d}\Gamma_i \tag{11}$$

with the identical Eq. (4) defined in the previous section, together with the product $\rho^2 V$ still following Eq. (5) with a value of σ given by solvability theory [14]. In order to integrate the value of $\mathcal{F}(t)$, now defined by Eq. (11), we write the Laplace equation in a moving frame of velocity V, $D\nabla^2 u = -V\partial u/\partial x$, at a scale $\ll D/V$. Integrating $\nabla^2 u$ over a surface Σ bounded by the contours Γ_e and Γ_i shown in Fig. 2b, we can use the divergence theorem to write

$$\int_{\Gamma_i}\frac{\partial u}{\partial n}\,\mathrm{d}\Gamma_i = \int_{\Gamma_e}\frac{\partial u}{\partial n}\,\mathrm{d}\Gamma_e + \frac{V}{D}\iint_{\Sigma}\frac{\partial u}{\partial x}\,\mathrm{d}\Sigma\ , \tag{12}$$

where, for consistency with the sharp needle model, we have the contour normal $\mathbf{n}$ pointing outward of the closed contour $(\Gamma_e + \Gamma_i)$ along Γ_e (see Fig. 2a), while pointing outward of the dendrite, *i.e.* inward the contour, along Γ_i (see Fig. 2b). Finally, the "thick" DNN model follows the exact same growth laws as the sharp needle model, with a constant value of $\rho^2 V$ and a product $\rho V^2 \sim \mathcal{F}(t)^2$. The only difference resides in the new definition of the flux intensity factor $\mathcal{F}(t)$ as Eq. (11), where the contour integral term can be integrated over a chosen outer contour Γ_e using Eq. (12).

2.3. Extension to three dimensions

The model thus defined can be directly extended for a 3D dendrite growing as a paraboloid of revolution with a circular cross section of radius $r_i(x) = \sqrt{2\rho(x_t - x)}$. In 3D, Γ_e and Γ_i are surfaces joining at $x = x_t - a$, and Σ is the volume enclosed within the closed surface $(\Gamma_e + \Gamma_i)$. The mass conservation $D\partial u/\partial n \approx v_n$ in the tip region over a length a now yields

$$D\iint_{\Gamma_i}\frac{\partial u}{\partial n}\,\mathrm{d}\Gamma_i = \int_0^{2\pi}\int_0^{r_a} V\,r\,\mathrm{d}r\,\mathrm{d}\theta = 2\pi a\rho V, \tag{13}$$

with $r_a = \sqrt{2\rho a}$ the radius of the paraboloid at a distance a behind the tip, *i.e.* at $x = x_t - a$. Thus, we can define the flux intensity factor

$$\mathcal{F} \equiv \frac{1}{2\pi a}\iint_{\Gamma_i}\frac{\partial u}{\partial n}\,\mathrm{d}\Gamma_i, \tag{14}$$

and write the flux condition at the tip that replaces Eq. (4) as

$$\rho V = D\mathcal{F}, \tag{15}$$

which follows from Eqs. (13) and (14) . Similarly as in two dimensions, the flux term over the surface Γ_i can be calculated over a surface Γ_e that intersects the paraboloid at $x = x_t - a$ with

$$\iint_{\Gamma_i}\frac{\partial u}{\partial n}\,\mathrm{d}\Gamma_i = \iint_{\Gamma_e}\frac{\partial u}{\partial n}\,\mathrm{d}\Gamma_e + \frac{V}{D}\iiint_{\Sigma}\frac{\partial u}{\partial x}\,\mathrm{d}\Sigma\ , \tag{16}$$

The growth dynamics of the needle tips can thus be calculated by combining the new flux balance Eq. (15) with the solvability condition Eq. (5), which is similar as in 2D with a tip selection parameter σ that still only depends on the strength of surface tension anisotropy, but with a different value as compared to 2D [14].

In the case of directional solidification, similarly as in 2D, we replace the equilibrium condition along the needle $u_i = 0$ by $u_i = (x - V_p t)/l_T$ and the non-vanishing prefactor of v_n in Eq. (2) appears in the needle growth laws, now combining Eq. (8) and

$$\rho V = D\mathcal{F}/\left[1 - (1-k)u_t\right]. \qquad (17)$$

3. Results and Validation

First, we test the steady-state growth regime predicted by the different versions of the DNN model — namely 2D "sharp" (Sec. 2.1), 2D "thick" (Sec. 2.2), and 3D (Sec. 2.3) — against the exact analytical solution for a parabola/paraboloid established by Ivantsov [15, 8], relating uniquely the tip solute supersaturation Ω to the Péclet number $P \equiv \rho V/(2D)$. In order to do so, we let one needle grow within an imposed Ω and measure the value of P once ρ and V reach steady values. Additionally, we define ρ_s and V_s the steady-state values of respectively ρ and V that satisfy the solvability condition Eq. (5) or Eq. (8) together with the Ivantsov solution.

In 2D (Fig. 3), as previously explained in Ref. [5], the sharp-needle formulation tends to a first order approximation $\Omega \approx \sqrt{\pi P}$ of the exact Ivantsov solution $\Omega = \mathrm{Iv}_{2\mathrm{D}}(P) = \sqrt{\pi P}\, e^P \operatorname{erfc}(\sqrt{P})$, hence limiting the scope of the approach to $\Omega \ll 1$. In contrast, the DNN formulation with a parabolic needle of finite thickness tends to the exact 2D Ivantsov solution. Therefore, in addition to making the extension to 3D possible, the thick needle formulation has the advantage of increasing the range of Peclet number that can be quantitatively modeled. This is because this formulation provides a more accurate description of the solute field, and because the calculation of $\mathcal{F}(t)$ by Eq. (12) introduces a correction in V from the moving frame expression of Laplace equation around the tip.

In 3D (Fig. 4), the predictions of the DNN model for a paraboloid follows the steady-state Ivantsov solution $\Omega = \mathrm{Iv}_{3\mathrm{D}}(P) = P\, e^P \int_P^\infty (e^{-s}/s)\, ds$ — significantly different from $\mathrm{Iv}_{2\mathrm{D}}(P)$ plotted for comparison in Fig. 4. This difference between 2D and 3D also has a consequence in terms of appropriate computational parameters. The DNN model is built on the assumption

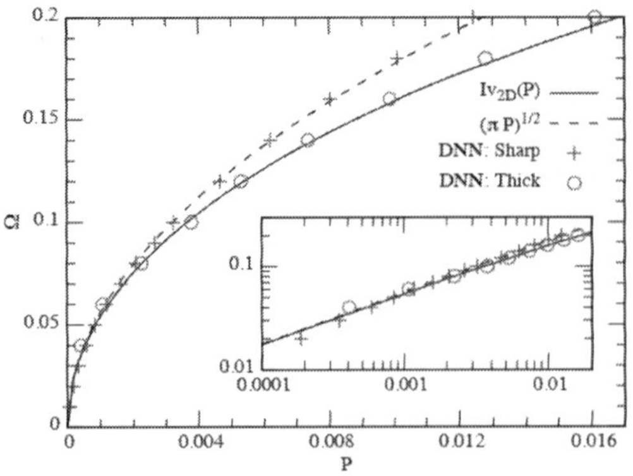

Figure 3. Steady-state predictions of 2D isothermal DNN simulations using a sharp needle (Fig. 2a) with $\Delta x = 10\rho_s$, and using a thick parabolic needle (Fig. 2b) with $\Delta x = \rho_s$.

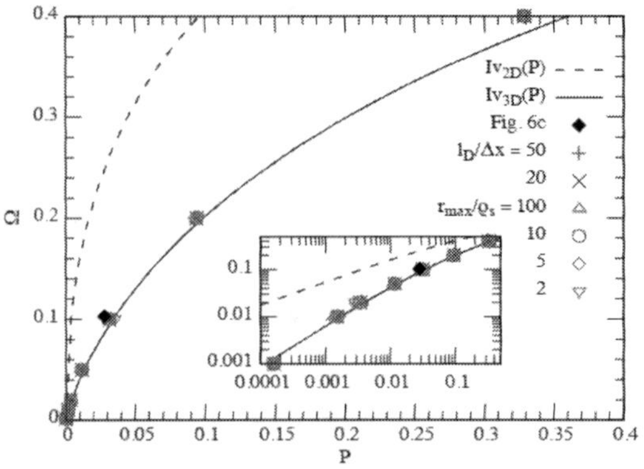

Figure 4. Steady-state predictions of 3D isothermal DNN simulations for $r_{max}/\rho_s = 10$ with $l_D/\Delta x = 50, 20$, and for $l_D/\Delta x = 20$ with $r_{max}/\rho_s = 100$, 10, 5, 2, 1.

IOP Conf. Series: Materials Science and Engineering **84** (2015) 012082 doi:10.1088/1757-899X/84/1/012082

of a scale separation between the tip radius ρ and the diffusion length $l_D = D/V$. As seen in Fig. 4, for a given Ω the ratio $\rho/(D/V) \sim P$ is lower in 2D than in 3D. Thus in 3D, the choice of grid spacing $\Delta x \sim \rho$ is more severely bounded by $\Delta x \ll l_D$ than in 2D. However, this is not such a stringent restriction, as the model reaches accurate results up to $l_D/\Delta x \approx 20$ (Fig. 4).

In the broader context of interactions among multiple branches, when a needle is stopped by diffusive interactions, the constancy of $\rho^2 V$ would imply $\rho \to \infty$ as $V \to 0$. While this is not a problem with a sharp needle as u is only imposed on a sharp line, this may lead to numerical instabilities when u is imposed on the whole thickness of a needle. A way to avoid these instabilities is to bound the thickness of a needle to a given radius r_{max}. Fig. 4 shows that using a r_{max} as low as the steady-state tip radius ρ_s does not significantly affect the results.

We have also tested the DNN predictions of the early transient growth of an equiaxed crystal. In 2D, the growth of an idealized crystal with four sharp branches has an analytical solution with the branches length following $L \sim t^{3/5}$ [16]. This solution was reproduced by phase-field [13] and DNN [5] calculations. In 3D, dimensional analysis validated by phase field calculations yields $L \sim t^{2/3}$ [13]. The results of the 3D DNN simulations in Fig. 5 for $\Omega \approx 0.006$ (*i.e.* $P \approx 0.02$) show that the model also captures accurately the transient equiaxed growth regime.

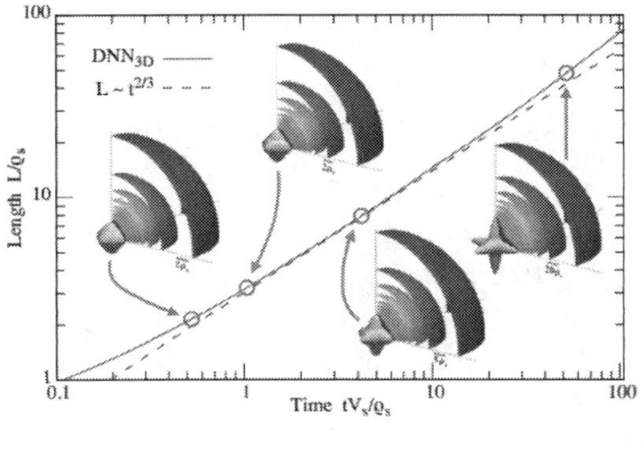

Figure 5. Early growth of a six-branch equiaxed crystal predicted by 3D DNN simulations, compared to a line of slope 2/3, according to the analytical law $L \sim t^{2/3}$ [13]. The plot is built by continuity from simulations with $\Delta x = 0.05\rho_s$ (for $tV_s/\rho_s < 2$), $\Delta x = 0.1\rho_s$ ($2 < tV_s/\rho_s < 10$), and $\Delta x = 0.5\rho_s$ ($tV_s/\rho_s > 10$). The insets show snapshots of the isovalues of u with steps of 0.0005.

4. Application to Al-Cu directional solidification experiments

4.1. Experiments

Aluminum-copper alloys are good candidate for *in situ* observation of dendritic solidification in metals and they have been extensively used for X-ray radiography [17, 18]. We prepared an alloy of nominal composition Al-12at.%Cu, *i.e.* c_∞ = 24.3wt%Cu, by arc-melting in an inert argon atmosphere. Thin samples ($\approx$ 200 μm) were obtained by Electrical Discharge Machining and mechanical grinding. The resulting foil was put into a boron nitride crucible with a $\approx$ 0.1 mm thinned window and 2 mm diameter observation hole. The (sample + crucible) assembly was inserted in a steel rod instrumented with thermocouples and independently heated by two induction coils located above and below the sample position. Directional solidification was achieved by stabilizing the temperature gradient with a melted alloy, then reducing the imposed temperatures at the location of the coils at a controlled cooling rate. Synchrotron X-ray imaging was performed at the Sector 32 Insertion Device beamline at Argonne National Laboratory's Advanced Photon Source with a monochromatic beam at 18 keV. X-rays pass through the metal sample and impinge upon a scintillator that converts them to visible light. A mirror positioned behind the scintillator reflects the images to a CCD camera recording the 1024×1280 pixel images with a pixel size of $\approx$ 1.34 μm. The experiment presented here and compared to DNN simulations in Fig. 6 corresponds to a measured temperature gradient $\approx$ 72 K/cm and a measured isotherm velocity $\approx$ 100 μm/s, both expressed along the gravity axis (vertical in Fig. 6).

4.2. Simulations

We simulate the Al-Cu directional solidification experiments with DNN simulations, both in 2D (sharp-needle formulation [5]) and in 3D. The alloy parameters are $m = -3.4$ K/wt%Cu, $k = 0.17$, $D = 3.5\ 10^{-9}$ m^2/s, $\Gamma = 2.36\ 10^{-7}$ Km [8], and $\sigma = 0.083$ in 2D and $\sigma = 0.06$ in 3D, corresponding to an interfacial anisotropy strength $\approx 1.2\%$ [14]. With a theoretical steady-state radius $\rho_s \approx 1.4$ μm in 2D and 1.8 μm in 3D, we solve the diffusion and needle growth equations on a finite difference grid of spacing $\Delta x = 2\rho_s$, using an explicit Euler time scheme with time step $\Delta t = \Delta x^2/(4D)$ in 2D and $\Delta x^2/(6D)$ in 3D. The integration contour Γ_e in 2D is a square of side $6\Delta x$ centered on each needle tip. In 3D, the integration surface is a cuboid centered on the tip with dimensions $2a = 6\Delta x$ in the growth direction x and $(r_a + 4\Delta x)$ in normal directions. The width along the needles is bound to $r_{max} = 3\rho_s$. Sidebranching events are triggered at a distance $d = (15 \pm 5)\rho$ behind the tip each time a needle grows by a distance d, as described in Ref. [5], with an initial length $l_0^{SB} = \sqrt{2\rho d} + \rho$. An orientation angle α of the crystal with respect to the vertical axis in the (x, y) plane is introduced by adding a y-dependent term on the imposed $u_i = (x - V_p)/l_T - y\tan(\alpha)$ along the needles, in order to match the corresponding experimental crystal orientation $\alpha \approx 30°$ (Fig. 6a). Hence, the equivalent pulling velocity and temperature gradient along the growth axis x are input as $G = 72 \times \cos(\alpha) \approx 62$ K/cm and $V_p = 100/\cos(\alpha) \approx 115$ μm/s. The simulation domain measures $\approx 3000 \times 2500$ μm^2 in 2D and $\approx 3000 \times 2500 \times 100$ μm^3 in 3D (half sample thickness with symmetry boundary conditions in z), over a total time of 180 (2D) and 120 (3D) seconds, both starting with 100 evenly spaced needles — in the $(z = 0)$ plane in 3D — with their tips located at the liquidus temperature.

4.3. Results

Fig. 6 shows the experimental observation (a) compared to DNN results in 2D (b) and 3D (c); all panels are represented at the same scale. The 2D calculation predicts an average primary spacing $\Lambda_{2D} \approx 1020$ μm (b), much larger than the experimentally measured $\Lambda_{exp.} \approx 258$ μm (a), while 3D simulations yields an excellent agreement with experiments as $\Lambda_{3D} \approx 237$ μm (c), with Λ measured normal to the growth direction. The steady-state tip supersaturation of Fig. 6c is also reported in Fig. 4. The corresponding simulations run in $\approx$ 1 hour (2D) and $\approx$ 4.5 hours (3D) on one Nvidia Tesla C2050 Graphic Processing Unit. The results of these simulations clearly show that the 3D DNN model provides a powerful tool to make quantitative predictions of complex dendritic microstructures with time efficient calculations.

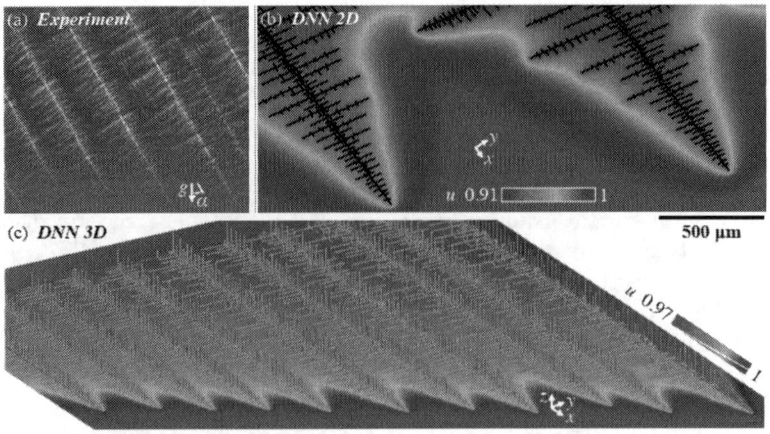

Figure 6. Directional solidification of an Al-12at.%Cu (Al-24.3wt%Cu) alloy at $G = 62$ K/cm and $V_p = 115$ μm/s: (a) X-ray image; (b) 2D simulation; (c) 3D simulation (all represented at the same scale). [*Corresponding movies attached.*]

5. Summary and outlook

We presented the first three-dimensional extension to the previously proposed Dendritic Needle Network (DNN) multi-scale approach to dendritic solidification [5]. The model generally describes both transient and steady-state growth regimes. We compared model predictions with synchrotron X-ray imaging of an Al-Cu alloy during directional solidification. The 3D DNN predictions show excellent agreement with measured primary dendrite arm spacing, whereas the 2D modeling predicts significantly larger spacing than the experiments.

The present modeling approach bridges the length-scale gap between detailed solid-liquid interface pattern formation models, *e.g.* phase-field, and macroscopic average models of grain growth, *e.g.* cellular automata. The possibility to model and explore both intragrain and intergrain dynamical phenomena in complex dendritic microstructures should pave the way for a deeper understanding of microstructure selection mechanisms, such as for dendrite arm spacings, grain texture, and casting defects, crucial to the properties of industrial materials.

The works in progress and outlook for DNN modeling include studies on: scaling laws for dynamics of dendritic secondary sidebranches [19]; the Columnar-to-Equiaxed transition [20]; columnar grain growth competition [21]; and coupling the model with solute-driven fluid flow.

Acknowledgments

The model development, implementation and validation by DT and AK were supported by NASA awards NNX11AC09G and NNX14AB34G within the CETSOL project. The experiments and numerical application by DT, AJC, PJG, SDI were supported by AJC's Early Career award from the U.S. DOE, Office of Basic Energy Sciences, Division of Materials Sciences and Engineering and Los Alamos National Laboratory, operated by Los Alamos National Security, LLC under Contract DE-AC52-06NA25396 for the U.S. Department of Energy. The authors gratefully acknowledge K. Fezzaa (ANL) for assisting with the synchrotron X-ray imaging experiments and J.W. Gibbs (LANL) for comments on the article. Use of the Advanced Photon Source, an Office of Science User Facility operated for the U.S. DOE Office of Science by Argonne National Laboratory, was supported by the U.S. DOE under Contract DE-AC02-06CH11357.

References

[1] Grignon M 1775 *Mémoires de physique sur l'art de fabriquer le fer, d'en fondre et forger des canons d'artillerie; sur l'histoire naturelle, et sur divers sujets particuliers de physique et d'économie* (Paris: Delalin) pp 656.
[2] Chernoff D 1880 *Proc. Inst. Mech. Eng.* 286.
[3] Kurz W 2006 *Mater. Sci. Forum* **508** 313.
[4] Tourret D, Karma A 2012 *IOP Conf. Ser.: Mater. Sci. Eng.* **33** 012095.
[5] Tourret D, Karma A 2013 *Acta Mater.* **61** 6474.
[6] Boettinger WJ, Warren JA, Beckermann C, Karma A 2002 *Annu. Rev. Mater. Res.* **32** 163.
[7] Gandin Ch-A 2010 *C. R. Phys.* **11** 216.
[8] Kurz W, Fisher DJ 1992 *Fundamentals of solidification* 3rd ed (Aedermannsdorf: Trans Tech Pub).
[9] Derrida B, Hakim V 1992 *Phys. Rev.* A **45** 8759.
[10] Rice JR 1968 *J. Appl. Mech.* **35** 379.
[11] Langer JS 1987 *Chance and Matter*, ed J Souletie, J Vannimenus, R Stora (Amsterdam: North-Holland) 629.
[12] Ben Amar M, Brener E 1993 *Phys. Rev. Lett.* **71** 589.
[13] Plapp M, Karma A 2000 *Phys Rev Lett 2000* **84** 1740.
[14] Barbieri A, Langer JS. Phys Rev A 1989;39:5314.
[15] Ivantsov GP 1947 *Dokl. Akad. Nauk. SSSR* **58** 567.
[16] Almgren R, Dai WS, Hakim V 1993 *Phys. Rev. Lett.* **71** 3461.
[17] Nguyen-Thi H, Salvo L, Mathiesen RH, *et al.* 2012 *C. R. Phys.* **13** 237.
[18] Baruchel J, Di Michiel M, Lafford T, *et al.* 2013 *C. R. Phys.* **14** 208.
[19] Couder Y, Maurer J, González-Cinca R, Hernández-Machado A 2005 *Phys. Rev.* E **71** 031602.
[20] Zimmermann G, Sturz L, Billia B, *et al.* 2014 *Mater. Sci. Forum* **12** 790.
[21] Tourret D, Karma A 2015 *Acta Mater.* **82** 64.

IOP Conf. Series: Materials Science and Engineering **84** (2015) 012083 doi:10.1088/1757-899X/84/1/012083

Lamellar eutectic growth with anisotropic interphase boundaries

S Akamatsu[1,2], S Bottin-Rousseau[1,2], G Faivre[1,2], S Ghosh[3], M Plapp[3]

[1] CNRS, UMR 7588, INSP, 75005 Paris, France

[2] Sorbonne Universités, UPMC Univ Paris 06, UMR 7588, Institut des Nanosciences de Paris, 75005 Paris, France

[3] Condensed Matter Physics, Ecole Polytechnique, CNRS, 91128 Palaiseau, France

E-mail: akamatsu@insp.jussieu.fr

Abstract. We present a numerical study of the effect of a free-energy anisotropy of the solid-solid interphase boundaries on the formation of tilted lamellar microstructures during directional solidification of nonfaceted binary eutectic alloys. We used two different methods – phase-field (PF) and dynamic boundary-integral (BI) – to simulate the growth of periodic eutectic patterns in two dimensions. For a given Wulff plot of the interphase boundary, which characterizes a eutectic grain with a given relative orientation of the two solid phases, the lamellar tilt angle depends on the angle between the thermal axis $\mathbf{z}$ and a reference crystallographic axis. Both PF and BI results confirm the general validity of a recent approximate theory which assumes that, at the trijunctions, the surface tension vector of the interphase boundary is parallel to $\mathbf{z}$. In particular, a crystallographic locking of the lamellae onto a direction close to a deep minimum in the Wulff plot is well reproduced in the simulations.

1. Introduction

In many nonfaceted eutectic alloys, two-phase microstructures that are delivered by coupled-growth front patterns during directional solidification (DS) present a marked morphological dependence on the crystal orientation of the two solid phases [1, 2, 3]. In spite of their pratical importance, such crystallographic effects have been largely left aside in previous theories and models. Let us consider the case of lamellar eutectics. Experimentally, it has been known for a long time that eutectic lamellae can grow tilted with respect to the thermal-gradient axis $\mathbf{z}$, and that the lamellar tilt is eutectic-grain dependent (a eutectic grain is a region of uniform crystal orientation of the solid phases α and β [4]). On the basis of in situ experimental observations [5, 6, 7], two categories of eutectic grains have been previously distinguished (Fig. 1a), namely, *floating* (eutectic) grains (Fig. 1b1), within which the growth dynamics is well described by the standard theory of regular eutectics, and *locked* grains, within which the growth direction of the eutectic lamellae is essentially aligned to a certain crystallographic plane, and is generally inclined (or tilted) with respect to the main growth axis $\mathbf{z}$. Lamellar patterns with a strong tilt angle are usually observed when the two solid phases exhibit an epitaxial orientation relationship and interphase-boundary planes with a low free energy [3]. Recently, an approximate theory has been formulated that relates the value of the lamellar tilt to the surface free-energy anisotropy of the α-β interphase boundaries (interfacial anisotropy) in the solid [6]. It has been proposed in coherence with in situ observations indicating that, even for lamellae that grow at a large tilt angle, the solid-liquid interface shape exhibits a mirror symmetry with respect to the mid-plane of the lamellae, similar to the one observed for well-aligned lamellae. For isotropic solid-liquid interfaces, a symmetric shape is obtained under the condition that the contact angles of the solid-liquid interfaces at the trijunctions are the same on both sides of a lamella (Fig. 1b2) [6, 7]. For anisotropic interphase boundaries, the symmetric-pattern (SP) approximation thus entails that the Cahn-Hoffman surface tension vector $\vec{\sigma}$ [8] is aligned with **z** (it is assumed that, in a nonfaceted alloy, the anisotropy of the solid-liquid interfaces has a negligible effect on the lamellar growth dynamics).

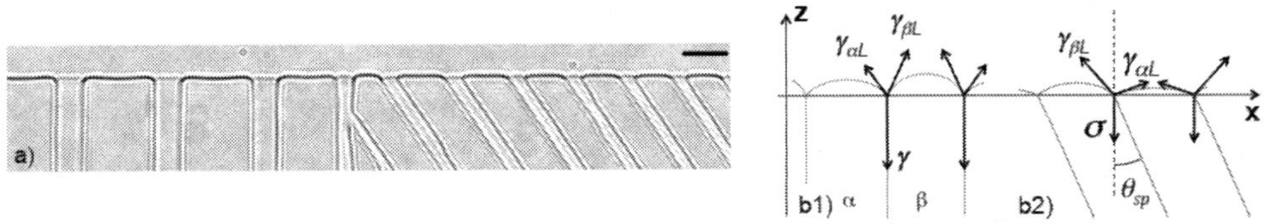

Figure 1. a) Two eutectic grains, one of the floating (lef-thand side) and the other of the locked (right-hand side) type, during real-time DS of a slightly hypoeutectic CBr_4-C_2Cl_6 alloy in a 12-μm thin sample ($V = 0.5\mu ms^{-1}$; bar: 20 μm). Sketches: b1) steady lamellar pattern with isotropic interfaces ; b2) tilted lamellar pattern in a locked eutectic grain under the SP approximation ($\vec{\sigma} \parallel \mathbf{z}$).

Here, we present a test of the SP approximation by numerical simulations. We carried out two-dimensional calculations using two different simulation methods, namely, (i) a sharp-interface code, which uses the dynamic boundary-integral (BI) formalism previously developed by Karma and Sarkissian [9], and (ii) a multi-phase-field (PF) model, based on recent grand-canonical formulations of alloy solidification [10, 11]. Both methods demonstrate that the tilt angle follows quite closely the SP approximation. In addition, a strong locking of the lamellae onto the direction of the peaked minimum of the anisotropy function is well reproduced in both BI and PF simulations.

2. The symmetric-pattern approximation

The relative orientation of the α and β crystals (constant within a grain) determines the interphase boundary free energy and its anisotropy. In two dimensions (ideally, thin-sample DS), the grain orientation and the inclination of the interphase boundary are specified each by a single angle with respect to the thermal axis. Let $\hat{n}$ be the unit normal vector of the interphase boundary, and θ the angle between $\hat{n}$ and the **x** axis ($\hat{t} = -d\hat{n}/d\theta$ is the tangent vector). For a reference grain orientation, the anisotropic interphase boundary energy is given by

$$\gamma_{\alpha\beta}(\theta) = \bar{\gamma}_{\alpha\beta} a_c(\theta), \tag{1}$$

where $\bar{\gamma}_{\alpha\beta}$ is a constant and $a_c(\theta)$ is a dimensionless function. The Cahn-Hoffman $\vec{\xi}$ and $\vec{\sigma}$ vectors [8] are defined by

$$\vec{\xi} = \gamma_{\alpha\beta}\hat{n} - \gamma'_{\alpha\beta}\hat{t} \ ; \ \vec{\sigma} = \gamma_{\alpha\beta}\hat{t} + \gamma'_{\alpha\beta}\hat{n} \tag{2}$$

where $\gamma'_{\alpha\beta} = d\gamma_{\alpha\beta}/d\theta$. The Wulff plot is given by $\vec{r}(\theta) = \gamma_{\alpha\beta}(\theta)\hat{n}$ and the minimum-energy, or Wulff shape, by the vector $\vec{\xi}(\theta)$. For a given eutectic grain, we search the steady-state growth pattern as a function of the orientation angle θ_R of the bicrystal with respect to **z**. The interphase energy becomes

$$\gamma_{\alpha\beta}(\theta) = \bar{\gamma}_{\alpha\beta} a_c(\theta - \theta_R), \tag{3}$$

where $\theta_R = 0$ is chosen such that an interphase orientation of minimal energy is aligned with **z**. In a sharp-interface model (with isotropic solid-liquid interfaces, and no diffusion in the solid), the local equilibrium at the trijunction is given by a Young-Herring equation, that is,

$$\gamma_{\alpha L}\hat{t}_{\alpha L} + \gamma_{\beta L}\hat{t}_{\beta L} + \vec{\sigma} = 0. \tag{4}$$

As stated above, a strictly symmetric shape is thus possible only if $\vec{\sigma}$ (which is not parallel to the interphase boundary) is aligned with **z** (Fig. 1b2), which writes

$$\gamma(\theta - \theta_R)\sin\theta + \gamma'(\theta - \theta_R)\cos\theta = 0. \tag{5}$$

For a fixed eutectic-grain orientation θ_R, this is a nonlinear equation for the interface orientation, which can easily be solved numerically for any $\gamma(\theta)$ function. We will note θ_{sp} the SP-approximation value of the tilt angle, solution of Eq. (5), and θ_t the steady-state tilt angle obtained in the numercial simulations.

3. Results and discussion

We simulated a model binary alloy with a symmetric eutectic plateau (and parallel liquidus and solidus lines), equal α- and β-liquid capillary lengths, at eutectic concentration. Temperature gradient, growth velocity, and lamellar spacing (close to the minimum undercooling spacing) are held constant. In the BI model, the interfacial anisotropy is incorporated in the calculation of the position of the trijunctions, using the Young-Herring condition (Eq. 4). In the PF model, the anisotropy of the interphase boundary is directly incorporated in the free-energy functional [11].

We implemented first a standard anisotropy function with a 4-fold symmetry [$a_c(\theta) = 1 - \epsilon_4 \cos 4\theta$]. The results for $\epsilon_4 = 0.04$ are shown in figure 2a. Both sets of simulation data follow quite closely the SP approximation, up to differences that do not exceed a degree. The value of θ_t passes through a maximum for an orientation that depends on the anisotropy function (as expected, symmetric, non-tilted patterns are found when either a minimum or a maximum of γ is aligned with **z**). The tilt angle always remains much smaller than θ_R, which is characteristic of a weak crystallographic effect. Note that the values of the capillary length, the thermal length, and the lamellar spacing are actually slightly different for BI and PF, which demonstrates that the influence of all of these parameters is weak in realistic conditions.

We then reproduced a strong lamellar-locking effect by using an anisotropy function with a deep minimum, of the form $a_c(\theta) = 1 - \epsilon_g \exp\left[-(\theta/w_g)^2\right]$ (the smoothed gaussian cusp allows us to circumvent some obvious numerical difficulties [12]). Setting $\epsilon_g = 0.2$ and $w_g = 0.1$ entails the appearance of two (quasi) facets in the Wulff shape of the interphase boundary (see inset in Fig. 2b). In this case, the $\vec{\xi}(\theta)$ plot has self-intersections. The Wulff shape is given by the inner convex part of that plot. The other parts, often called "ears", consist of three segments delimited by turning points. The interface stiffness $\tau(\theta) = \gamma_{\alpha\beta}(\theta) + \gamma''_{\alpha\beta}(\theta)$ is negative only on the middle segment (a flat interface within this range is unstable with respect to the formation of a hill-and-valley structure [13]), but all orientations located on the "ears" are missing from the physically observable convex equilibrium shape. The SP approximation then predicts three distinct parts in the θ_{sp} vs θ_R curve: (i) an essentially linear, strongly locked branch with a slope close to 1, which runs from $\theta_R = 0$ to a first $\tau = 0$ turning point (noted θ_l); (ii) a symmetric-pattern branch along which there is no anisotropy effect for θ_R ranging from a second turning point (noted θ_u) to $\pi/2$; (iii) an intermediate, unstable branch. In the $[\theta_u, \theta_l]$ interval, two kinds of lamellar patterns (one is locked, the other is isotropic) are possible for a given eutectic-grain orientation.

Both BI and PF simulations describe the two separate branches (figure 2b) and demonstrate the existence of a strong locking effect over a large orientation range, as predicted by the SP

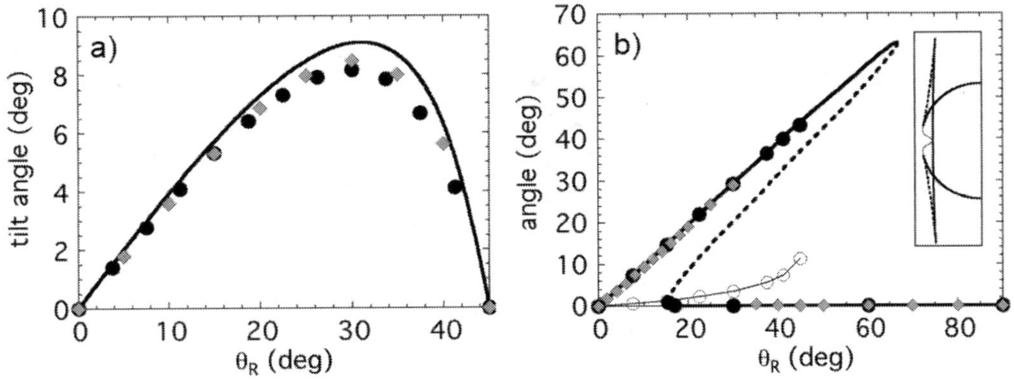

Figure 2. Lamellar tilt angle θ_{sp} predicted by the SP approximation (thick line), and θ_t calculated with the BI (filled circles) and the PF codes (diamonds) as a function of the eutectic-grain orientation angle θ_R. a) weak 4-fold anisotropy function; b) anisotropy function with a peaked gaussian minimum (see text). Open circles: angle of the surface tension vector $\vec{\sigma}$ with **z** (BI data). Inset: half view of the Wulff plot (thin line) and shape (thick line). Dotted lines: Herring-unstable orientations.

approximation. The BI simulations clearly reproduce a hysteretic behavior, which is expected for a bistable system. It should be noted, however, that the approach of the limit tilt angle (about 50 degrees) along the locked branch also corresponds to a steep, apparently diverging increase of the angle of the $\vec{\sigma}$ vector with respect to the thermal axis (Fig. 2b). This indicates that, at large tilt angles, the SP approximation, while it still correctly predicts the locked-tilt angle, becomes inaccurate with regards to other aspects of the dynamics. In contrast with the BI simulations, no hysteresis is observed in the PF simulations.

We can draw an interesting conclusion, of practical relevance, from these results. It is related to the recent development of an experimental method called *rotating directional solidification* (RDS), during which solidification is performed in a fixed thermal gradient by slowly rotating a thin alloy sample about an axis perpendicular to the **xz** plane [6, 7]. The crystal orientation of a given eutectic grain (fixed in the sample reference frame) with respect to the main solidification axis is then varied continuously while solidification proceeds. In the case of eutectic growth, a lamellar pattern then adapts its shape in a quasistationary way. It was demonstrated in Ref. [6] that, under the SP approximation, and with suitably chosen RDS settings, the solidification microstructure is then homothetic to the Wulff shape of the interphase boundaries in the sample plane, and that, accordingly, the morphological analysis of a RDS pattern gives direct access to a two-dimensional section of the corresponding Wulff plot. The present study shows that this method is accurate, within a few percents, which is quite satisfactory as compared to other ordinary sources of experimental uncertainties.

4. Conclusion

We performed time-resolved numerical simulations of tilted lamellar patterns during directional solidification of a model eutectic alloy in the presence of an anisotropy of the interphase boundaries. The simulations confirm the semi-quantitative validity of the SP approximation that assumes that the surface tension vector of the interphase boundary at the trijunctions is parallel to the main growth axis. The most salient result is that a strong locking phenomenon associated with the existence of a sharp minimum in the Wulff plot (this corresponds to a facet in the equilibrium shape of the interphase boundary) is well reproduced by the simulations.

Another interesting conclusion, of practical relevance, is the establishment of the RDS method as a good tool for extracting quantitative information on the Wulff plot of interphase boundaries during a solidification experiment. Moreover, anisotropy functions with more detailed features than the test ones used in the present work can be implemented in the simulation codes, and could serve to reproduce "real" Wulff plots measured experimentally, or calculated by molecular-dynamics methods [14]. Further simulations are planned to study the dependence of the lamellar tilt on the lamellar spacing, and the influence of a non-symmetric phase diagram. Simulations in large systems will be needed to study the dynamics of "sawtooth" patterns [5], which have been observed experimentally, and may be associated with a bistable behavior in strongly anisotropic cases. Phase-field simulations of eutectic-growth patterns with anisotropic interphase boundaries will be extended to three-dimensional, bulk-solidification geometries.

Acknowledgments

We thank Alain Karma for sharing with us his original BI code. This work was financially supported by Centre National d'Etudes Spatiales (CNES), France.

References

[1] Hogan L M, Kraft R W and Lemkey F D 1971 *Adv. Mater. Res.* **5** 8
[2] Llorca J and Orera V M 2006 *Progress in Materials Science* **51** 711
[3] Hecht U, Witusiewicz V T, Drevermann A and Rex S 2005 *Trans. Indian Inst. Met.* **58** 545
[4] Akamatsu S, Moulinet S and Faivre G 2001 *Metall Mater Trans* A **32** 2039
[5] Caroli B, Caroli C, Faivre G and Mergy J 1992 *J Cryst Growth* **118** 135
[6] Akamatsu S, Bottin-Rousseau S, Şerefoğlu M and Faivre G 2012 *Acta Materialia* **60** 3199
[7] Akamatsu S, Bottin-Rousseau S, Şerefoğlu M and Faivre G 2012 *Acta Materialia* **60** 3206
[8] Hoffmann D W and Cahn J W 1972 *Surface. Science* **31** 368
[9] Karma A and Sarkissian A 1996 *Met. Trans.* A **27** 635
[10] Plapp M 2011 *Physical Review E* **84** 031601
[11] Choudhury A and Nestler B *Physics Review E* **85** 021602
[12] Debierre J-M, Karma A, Célestini F and Guérin R 2003 *Phys Rev E* **68** 041604
[13] Herring C 1951 *Physical Review* **82** 87
[14] Kokotin V and Hecht U 2014 *Computational Materials Science* **86** 30

Eutectic morphology evolution and Sr-modification in Al-Si based alloys studied by 3D phase-field simulation coupled to Calphad data

J Eiken and M Apel

Access e.V., Intzestraße 5, 52072 Aachen, Germany

E-mail: j.eiken@access.rwth-aachen.de

Abstract. The mechanical properties of Al-Si cast alloys are mainly controlled by the morphology of the eutectic silicon. Phase-field simulations were carried out to study the evolution of the multidimensional branched eutectic structures in 3D. Coupling to a Calphad database provided thermodynamic data for the multicomponent multiphase Al-Si-Sr-P system. A major challenge was to model the effect of the trace element Sr. Minor amounts of Sr are known to modify the silicon morphology from coarse flakes to fine coral-like fibers. However, the underlying mechanisms are still not fully understood. Two different in literature most discussed mechanisms were modelled: a) an effect of Sr on the growth kinetics of eutectic silicon and b) the formation of Al_2Si_2Sr on AlP particles, which consumes most potent nucleation sites and forces eutectic silicon to form with lower frequency and higher undercooling. The phase-field simulations only revealed a successful modification of the eutectic morphology when both effects acted in combination. Only in this case a clear depression of the eutectic temperature was observed. The required phase formation sequence L→ fcc-(Al)→ AlP→ Al_2Si_2Sr→ (Si) determines critical values for the Sr and P content.

1. Introduction

Hypoeutectic Al-Si alloys offer excellent casting characteristics, good ductility and corrosion resistance. A drawback is the coarse flaky morphology of the unmodified eutectic (Si) phase which deteriorates the mechanical properties and favors cracking [1]. Minor Sr addition of the order of 100-400 ppm is known to significantly improve the properties by modifying the (Si) morphology to refined fibers. Despite the technical experience with Sr modification, the underlying mechanisms by which Sr affects the eutectic morphology are still controversially discussed as reviewed in [2-3]. The most-widely accepted theory is that adsorbed Sr atoms hinder the growth of (Si) in the preferred growth direction and trigger multidirectional twinning. [4, 5]. Another well-approved hypothesis is that Al_2Si_2Sr precipitates neutralize the fine dispersed AlP particles as nucleation sites for (Si), thus reducing its nucleation frequency [6, 7]. We considered both hypotheses in our phase-field model and performed 3D microstructure simulations to test which of both aspects dominates the modification of the eutectic structure. As basis for our phase-field study we chose the prototypic cast alloy AlSi7 whose solidified structure consists of roughly 50% fcc-(Al)-(Si) eutectic. Simulations were run for varied P and Sr levels based on thermodynamic data evaluated from a recently new assessed Calphad database for the Al-Si-P-Sr system [8, 9].

2. The phase field model

2.1 The basic phase-field model and choice of parameters

Microstructure simulation were run using the multicomponent multiphase-field model by Eiken et al. [10] as implemented in the software MICRESS® [11]. Mobility correction and anti-trapping are formulated in the thin interface limit [12] and profile specific FD-correction terms minimize the numerical discretization error [13]. The size of the 3D calculation domain was set to 140 µm x 40 µm x 100 µm with a grid resolution of Δx=0.4 µm, revealing 8.75 million numerical cells. The multiple phase-field equations described the evolution of the phases: liquid, fcc-(Al), (Si), AlP and Al_2Si_2Sr based on the phase-specific interfacial energies: $\sigma^0_{L/fcc} = 0.165$ Jm^{-2} [14], $\sigma^0_{L/(Si)} = 0.352$ Jm^{-2} [14], $\sigma^0_{L/AlP} = 0.1$ Jm^{-2}, $\sigma^0_{L/Al2Si2Sr} = 0.2$ Jm^{-2}, $\sigma^0_{fcc/fcc} = 0.15$ Jm^{-2}, $\sigma^0_{fcc/(Si)} = 0.38$ Jm^{-2}. Solute redistribution and diffusion was solved for the components Si, P and Sr applying Arrhenius functions for the temperature dependent diffusion coefficients, preprocessed from the mobility database TC-MobAl1[15]: $D_0^{Si,L} = 1.34\cdot10^{-7}$ m^{2}s^{-1}, $Q^{Si,L} = 3\cdot10^4$ Jmol^{-1}, $D_0^{Si,fcc} = 0.89\cdot10^{-4}$ m^{2}s^{-1}, $Q^{Si,fcc} = 1.36\cdot10^5$ Jmol^{-1}, $D_0^{Sr,L} = 1.3\cdot10^{-7}$ m^{2}s^{-1}, $Q^{Sr,L} = 3\cdot10^4$Jmol^{-1}. Diffusion coefficients of P were estimated to be equal to those of Sr, as no specific data was available. Since the heat diffusion length was large compared to the domain size, the temperature could be assumed to be homogeneous. The temperature evolution with time was calculated from the balance of heat extraction and growth-related latent heat release averaged over the calculation domain [16]. The simulation started at a temperature of T = 615 °C from 100% liquid phase with a constant heat extraction rate of $\dot{q}$= 1.8 10^{-5}Js^{-1}m^{-3}. All thermodynamic quantities such as driving forces, local undercooling, interface concentrations, heat capacity and enthalpy were evaluated as function of local composition and actual temperature from a Calphad dataset for the Al-Si-Sr-P system [9, 15], as illustrated in Fig. 1.

2.1 Modelling of heterogeneous nucleation

An embedded model for heterogeneous nucleation generates new seeds at particular nucleation sites when the local undercooling, as function of composition and temperature, reaches a critical value [16]. In the present simulations, fcc-(Al) nucleates as primary phase with a critical undercooling of ΔT_{fcc} = 2°C in one of the domain corners. Later, it nucleates again as eutectic phase on the (Si) surface. Finely dispersed AlP particles nucleate with a critical undercooling of ΔT_{AlP} = 2°C and a minimal distance of Δx = 20µm on randomly distributed sites in the melt. The growing AlP particles in turn represent potential sites for nucleation of the phases (Si) and Al_2Si_2Sr. The required undercooling is evaluated according to the free growth concept for heterogeneous nucleation as function of the particle size by

$$\Delta T_\alpha = 2\sigma^0_{\alpha L}/(\Delta s_{\alpha L}\, r_{AlP}) \quad \text{with} \quad r_{AlP} = \sqrt[3]{4\phi_{AlP}/(3\pi)}\,\Delta x, \tag{1}$$

where $\sigma^0_{\alpha L}$ denotes the mean surface energy and $\Delta s_{\alpha L}$ the enthalpy of fusion of the nucleating phase α, which may be either (Si) or Al_2Si_2Sr. The particle radius r_{AlP} is evaluated under assumption of spherical geometry from the local phase fraction ϕ_{AlP} as long as it is too small to be accurately resolved on the numerical grid. Whether (Si) or Al_2Si_2Sr nucleates on a particular AlP particle depends on which of the phases first reaches the required undercooling. In the absence of free AlP particles (Si) can alternatively nucleate on another single nucleation site when it reaches a higher undercooling of $\Delta T_{(Si)}$ = 10°C.

2.2 Modelling of anisotropy and effects of Sr on (Si)-growth kinetics

The interfacial properties of the solid phases against the liquid were modelled anisotropic under consideration of their specific lattice symmetry, e.g. the interfacial energy of fcc-(Al) was modelled by a standard weak cubic anisotropy function with a coefficients of δ_σ=0.02 and the interfacial mobility with a coefficient of δ_μ=0.05. Of greater significance for the evolution of the eutectic structure is the facetted growth of (Si), which is governed by the slowest-growing and lowest-energy {111}-facets. We adopted the modelling of the facetted (Si)-mobility μ_{Si} from [17] and used the same, but reciprocal, anisotropy function $a_{Si}(\theta)$ to model the interfacial stiffness σ^*.

$$\mu(\theta) = \mu^0 \cdot a(\theta) \text{ and } \sigma^*(\theta) = \sigma^0 \cdot a(\theta)^{-1} \quad (2)$$

$$\text{with } a(\theta) = \delta_n + (1\text{-}\delta_n)\cdot|\tan\theta|\cdot\tanh(|\tan\theta|^{-1}) \text{ and } \theta = \operatorname{acos}(\vec{n}\cdot\vec{f}_n). \quad (3)$$

θ denotes the angle between the interfacial normal vector and the nearest {111}-facet vector $\vec{f}_n$. The direct modelling of the interfacial stiffness σ^* instead of the interface energy σ avoids negative values in forbidden growth directions.

A special issue of eutectic (Si) in Al-Si is that internal twinning plays an important role and can be affected by Sr-addition. (Si)-flakes commonly consist of multiple parallel twins, i.e. the crystal orientation is repeatedly mirrored back and forth with respect to the same (111)-twin plane. This arrangement is particularly stable in unmodified Al-Si-alloys due to the twin-plane re-entrant edge mechanism (TPRE), which favors growth in ‹112›-directions parallel to the (111)-twin plane [3, 5]. To consider the internal twinning in our simulations, we developed an effective anisotropy function which features not only the {111}-facets of the original crystal, but additionally the mirrored {111}-facets of the potential twin crystals. The potency for twinning in one or multiple directions can be adjusted by the choice of the facet coefficients. Table 1 gives the facet vectors and the sets of facet coefficients used in this study. Set 1 reflects the standard growth mode according to the TPRE mechanism. It favors twinning with respect to a single (111)-plane and promotes growth in the six ‹112›-directions within this plane. Set 2 models the Sr-modified growth kinetics. It is assumed that adsorbed Sr atoms at the (Si)-surface hinder growth in the preferred ‹112›- direction and promote frequent multidirectional twinning with respect to all four potential twin {111}-planes [3-5]. In this study, Set 2 is applied to model the interface mobility in all Sr-containing alloys, independent of the precise Sr concentration. The reference mobility value was defined as $\cdot 10^{-13} m^4 J^{-1} s^{-1}$ in both cases. The interfacial stiffness remained unaffected by the Sr addition. Fig. 1 illustrates the two different facetted anisotropies for (Si).

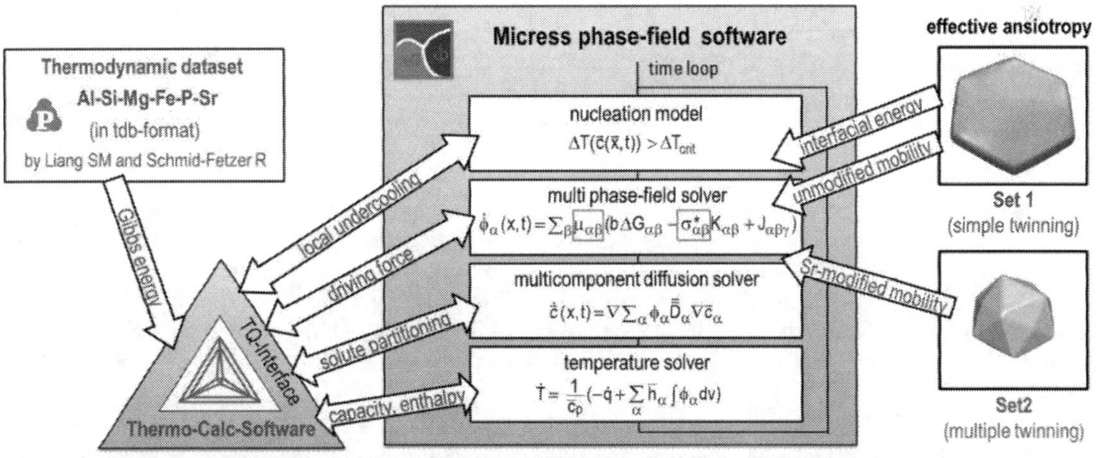

Figure 1 The MICRESS® software simulates nucleation and growth of the multiple solid phases as well as growth-related solute redistribution and latent heat release. Thermodynamic data is evaluated from a CALPHAD database as function of local composition. A dedicated effective anisotropy function models the impact of strontium on the (Si)/liquid interface kinetics due to impurity induced twinning.

IOP Conf. Series: Materials Science and Engineering **84** (2015) 012084 doi:10.1088/1757-899X/84/1/012084

Table 1: Facet vectors and coefficients defined in the effective anisotropy function for eutectic (Si)

$\vec{f}_i$	basic {111}-facets				additional {111}-facets of potential twin crystals											
x	1	-1	1	1	-5	1	1	5	1	1	5	1	-1	5	-1	1
y	1	1	-1	1	1	-5	1	1	5	-1	1	5	1	-1	5	1
z	1	1	1	-1	1	1	-5	1	-1	5	-1	1	5	1	1	5
δ_i (Set 1)	0.001	0.6			0.6			0.8			0.8			0.8		
δ_i (Set 2)	0.4	0.4			0.4			0.4			0.4			0.4		

3. Simulated morphology evolution

The following three simulations with varying strontium levels are representative for different modes of eutectic solidification in hypoeutectic Al-Si cast alloys:

3.1 Simulation of AlSi7 + 5ppm P without Sr addition

The first simulation example illustrates unmodified solidification of the hypoeutectic AlSi7 without Sr addition. Phosphorus has been considered as unavoidable trace element. The P level of 5 ppm was selected as typical value within the purity range of a commercial alloy. Figure 2a shows the micro-structure shortly before onset of the eutectic growth. Small AlP particles have already formed in the melt, especially in the vicinity of primary fcc-(Al) where the P content is highest due to segregation. These AlP particles act as inoculants and trigger frequent nucleation of eutectic (Si). During the initial decoupled growth (Si) evolves as blocky polyhedrons, forming the centers of the numerous eutectic cells (Figure 2b). When the blocky (Si) is almost completely encapsulated by either primary or newly nucleated fcc-(Al), it starts to branch. Both phases then grow in a coupled eutectic mode until end of solidification. Figures 2c and 2d illustrate the evolution of the coarse flaky (Si) morphology around the blocky centers, which is characteristic for unmodified eutectic growth.

3.2 Simulation of AlSi7 + 5ppm P + 50ppm Sr

The second simulation represents an example of unsuccessful modification due to unsufficient Sr addi-tion. The effect of strontium on the interface kinetics of the eutectic (Si) has been modeled by using the modified anisotropy coefficients (Set 2) given in table 1.The simulation results in figures 3a-d reveal that the mere modification of the interface kinetics is not sufficient to achieve a modification of the resulting eutectic morphology. Due to the low Sr content, Al_2Si_2Sr does not become stable early enough to compete with (Si) on the potent nucleation sites. Despite the altered interface kinetics no significant refinement or structural modification can be observed. The temperature curves in section 4 will show that the kinetic growth undercooling is too low to activate the Sr-induced twinning.

3.3 Simulation of AlSi7 + 5ppm P + 100ppm Sr

The last simulation example eventually describes successful eutectic modification. It points out that a refined fibrous (Si) structure is only obtained when Sr affects both, growth and nucleation of the (Si) phase. All process conditions are identically adopted from the previous simulation and also the effect of strontium on the (Si) growth kinetics is modelled in exactly the same way. The only difference to the previous simulation is that the Sr level is increased to 100 ppm. This causes Al_2Si_2Sr to become stable prior to (Si) and thus to first nucleate on the AlP particles as shown by figure 4a. It is empha-sized that the nucleation sequence has not been given as input, but results from evaluation of the local undercooling evaluated from the thermodynamic database as function of composition. Due to the fact that the Al_2Si_2Sr precipitates have already occupied all potent nucleation before onset of the eutectic growth, (Si) nucleates retarded on the single alternative nucleation site. Because of the high undercool-ing, eutectic fcc-(Al) directly nucleates on its surface (figure 4b) and both phases than grow in coupled eutectic mode. Figures 4c and 4d illustrate the refined coral-like growth of the (Si) fibers, which is characteristic for Sr-modified eutectic growth.

Figures 5a-c give the (Si) morphologies after 70s in comparison for the three different Sr levels.

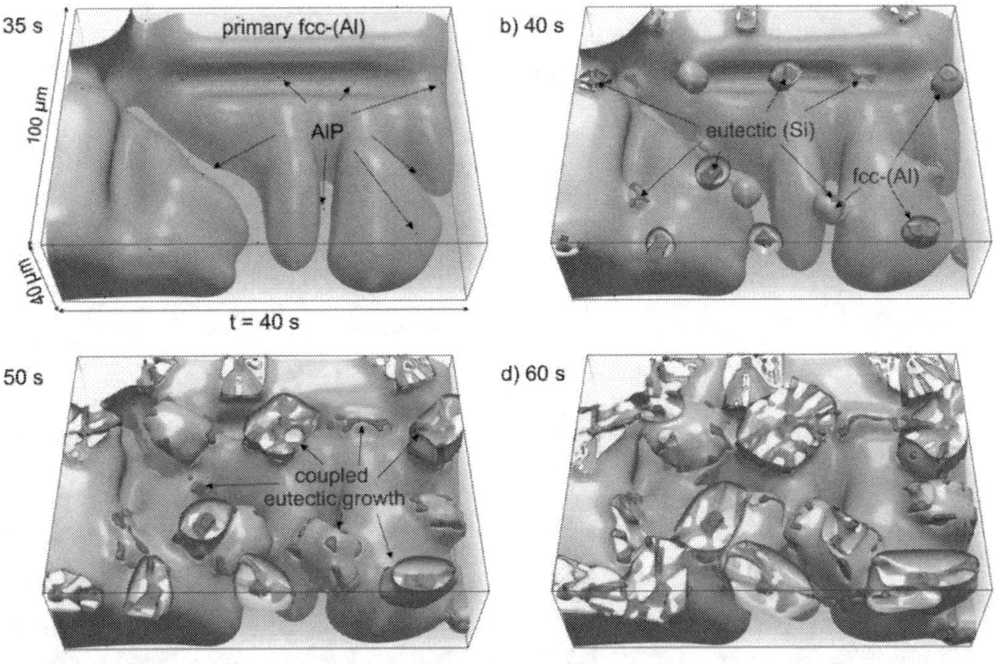

Figure 2 Simulation of unmodified eutectic solidification for AlSi7 + 5 ppmP. (Si) nucleates on numerous AlP particles. It becomes encased by fcc-(Al) and branches during coupled eutectic growth.

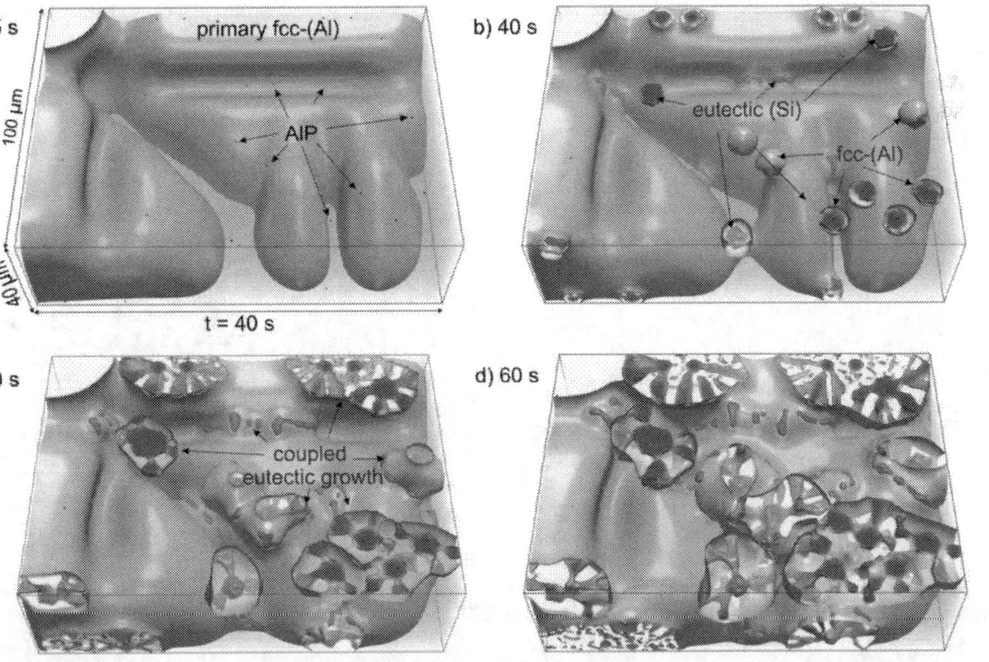

Figure 3 Simulation of unsuccessful eutectic modification for AlSi7 + 5 ppm P + 50 ppm Sr.
The Al_2Si_2Sr phase does not become stable early enough to prevent nucleation of (Si) on AlP particles.

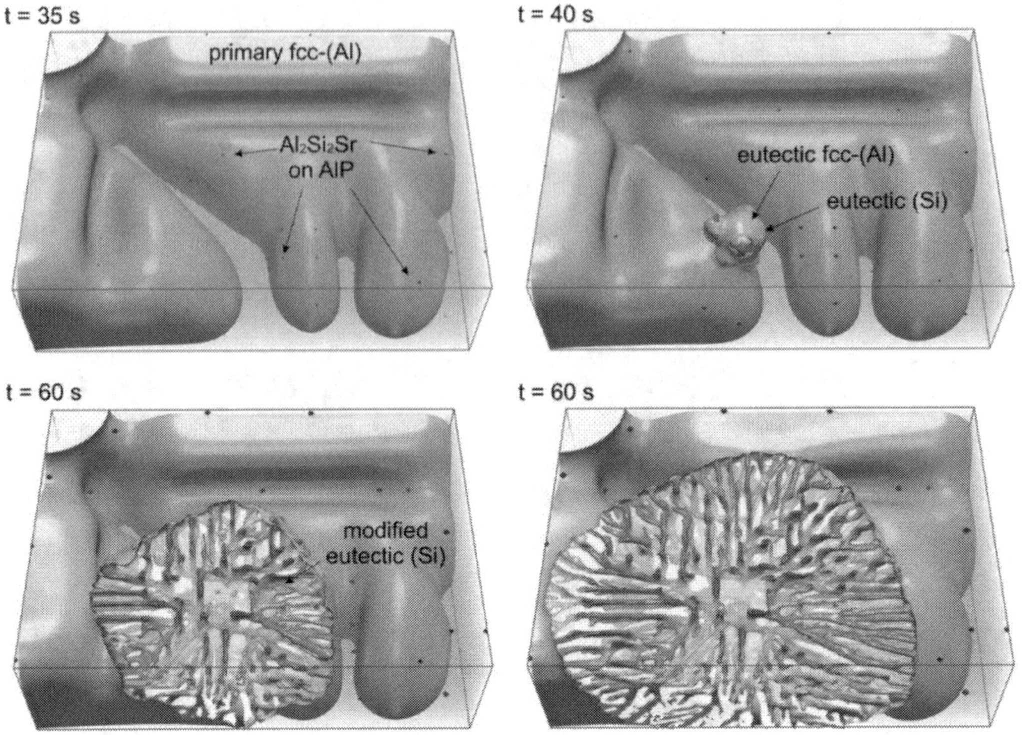

Figure 4 Simulation of successful eutectic modification for AlSi7 + 5 ppmP + 100 ppm Sr.
Fibrous coral-like growth is obtained when Sr affects both nucleation and growth of the (Si) phase.

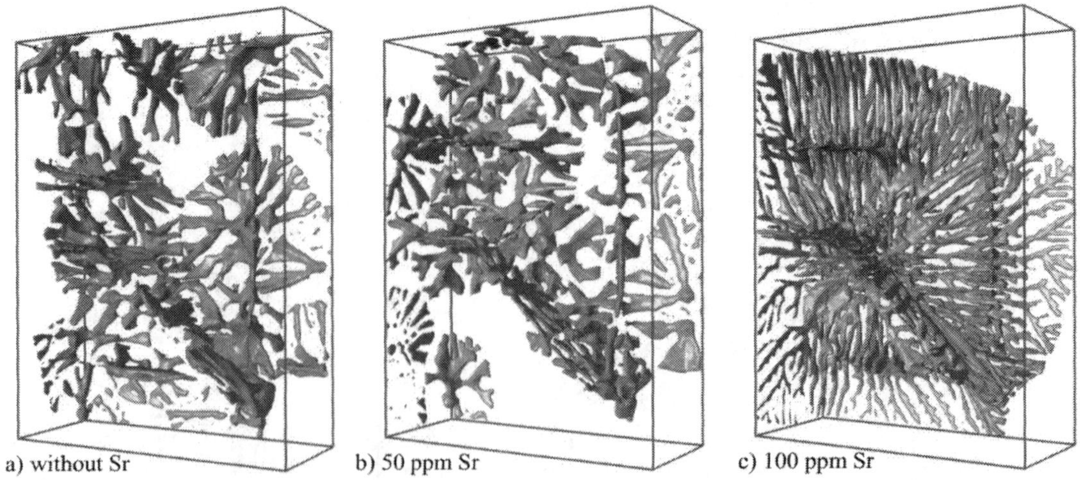

Figure 5 Comparison of the (Si) morphologies at t= 70s simulated for three different Sr levels.
The unmodified structure is characterized by multiple eutectic cells with blocky (Si) centers and coarse flaky (Si) branches. The structure obtained for 50 ppm is almost identical, only the shape of the blocky cell centers exhibits a slightly different shape. The increased Sr level of 100 ppm eventually revealed the targeted modification towards coral-like (Si)-growth with refined fibrous morphology.

3. Discussion of the simulated temperature curves

The modification of the eutectic structure was found to be correlated to a change in the temperature evolution. Figure 6 shows the simulated temperature curves for the previously described examples in comparison. Identically in all three simulations, the temperature continuously decreases during the primary growth of fcc-(Al). Differences start with the onset of the eutectic transition: The first example of unmodified eutectic solidification without Sr addition yields a relative low eutectic nucleation undercooling due to the fact that (Si) forms on potent AlP inoculants. Latent heat release then leads to recalescence. During coupled eutectic growth, the temperature remains almost constant, as the heat extraction is completely compensated by latent heat release. Eventually, on completion of the eutectic solidification, the latent heat release stops and the temperature rapidly drops according to the heat extraction rate.

It is apparent that the temperature curve does not change, when the Sr-addition is insufficient, while the successful modification by 100 ppm Sr yields clearly decreased temperatures, both during eutectic nucleation and growth. This is in agreement with the empirical observation that the depression of the eutectic temperature is a major indicator for successful modification. In fact, the simulations give evidence that the increased growth undercooling is an indispensable prerequisite for eutectic modification. While the thermodynamic impact of Sr on the eutectic temperature is completely negligible, the altered nucleation was found to be the dominating factor. An inverse correlation between growth undercooling and nucleation frequency was observed: When (Si) nucleates with high frequency on AlP, the individual eutectic cells grow relatively slow, i.e. with low undercooling close to local thermodynamic equilibrium. The growth process is in this case controlled by diffusion, and the interface kinetics – no matter if modified or not – are of minor significance.

This is different in the example with an increased Sr content of 100 ppm, where Al_2Si_2Sr successfully blocks the AlP particles and (Si) nucleates with higher undercooling on the alternative site. Since the single eutectic cell has to grow much faster than multiple cells in order to release the same amount of latent heat, the kinetic undercooling is increased and the process is shifted to interface-control. This brings the Sr modified interface kinetics into effect: The reduced mobility further amplifies the initial depression of the eutectic temperature which leads to a general refinement, while the Sr-induced twinning modifies the (Si) morphology to multidirectionally-branched fibers.

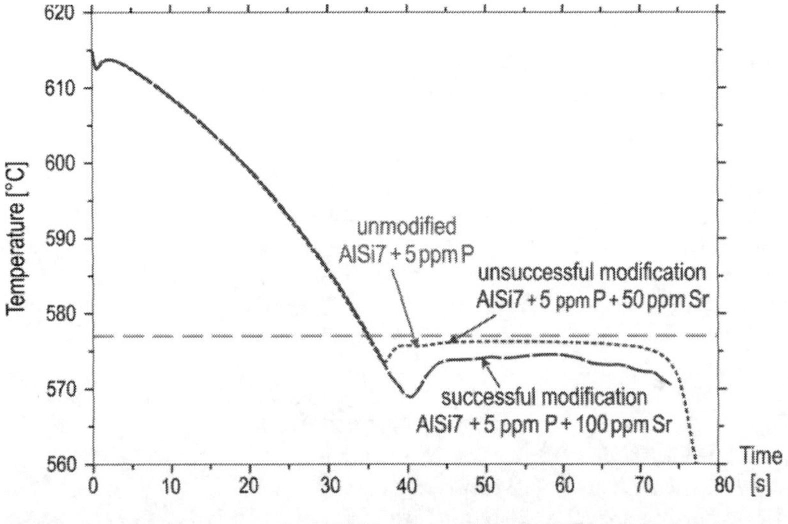

Figure 6 Simulated temperature curves for AlSi7 + 5ppm P with different Sr levels.
In agreement with experimental observations, successful modification of the eutectic morphology reveals an increased nucleation undercooling and a depression of the eutectic temperature.

3. Conclusions

The phase-field method was successfully applied to reproduce both unmodified and Sr-modified eutectic solidification of AlSi7. A major issue had been to bridge the gap from the internal Sr-affected twinning to the outer facetted growth and branching behavior of the eutectic (Si). This problem was solved by formulation of effective anisotropy functions which implicitly account for the mirrored twin facets. The twinning potential in different directions was adjusted by the choice of the facet parameters. Alloys without Sr additions were modelled with standard one-directional twinning and preferred growth in six coplanar ‹112›-directions according to the TPRE mechanism. Applied in simulation of AlSi7 + 5 ppm P this reproduced a coarse blocky and flaky eutectic structure, characteristic for unmodified eutectic growth.

A second set of coefficients was defined to model Sr-induced multi-directional twinning with uniform, but reduced amplitudes in all possible ‹112›-directions. The altered interface mobility was applied to all Sr containing alloys, but only showed an effect above a critical Sr-level of about 80 ppm. In simulations on AlSi7 + 5 ppm P with subcritical Sr levels, the eutectic (Si) nucleates with high frequency on AlP particles and then evolves diffusion-controlled close to local equilibrium. In the absence of kinetic growth undercooling, the altered interface kinetics does not become effective and thus no modification of the eutectic structure is obtained. This is different for AlSi7 + 5 ppm P with supercritical Sr addition whereAl_2Si_2Sr forms prior to (Si) and blocks the AlP particles. (Si) nucleates under this condition with increased undercooling and reduced frequency, and then evolves interface-controlled with high velocity. The Sr-altered interface kinetics, once brought into effect, further amplifies the depression of the eutectic temperature and eventually results in the refined coral-like fibrous growth, characteristic for Sr-modification. From these results, it can be concluded that the effects of Sr on both nucleation and growth of eutectic (Si) have to be considered to efficiently control Sr-modification in hypoeutectic Al-Si alloys.

Acknowledgement
The present study was funded by the German Research Foundation DFG in the framework of the AIF/DFG project PAK499, grant number AP196-6/1 and Schm 588/39.

References
[1] Lasko G, Apel M, Carré A, Weber U, Schmauder S 2012 *Adv Eng Mater* **14 (4)** 236-247 (20)
[2] Makhlouf MM, Guthy HV 2001 *J. Light Metals* **2 (1)** 199-218
[3] Mazahery A, Shabani MO 2014 JOM **66 (5)** 726-738
[4] Lu SZ and Hellawell A 1987 *Metall Trans A* **18 (10)** 1721-1733
[5] Shamsuzzoha M and Hogan LM 1989 *J Mater Sci* **24** 2849-2859
[6] Cho Y and Dahle AK 2009 *Metall Mater Trans* A **40 (5)** 1011-1012
[7] Dahle AK, Nogita K, McDonald SD, Dinnis C, Lu L 2005 *Mater Sci Eng A* **413–414** 243–248
[8] Liang SM and Schmid-Fetzer 2014 Acta Mater **72** 41-56.
[9] Eiken J, Apel M, Liang SM, Schmid-Fetzer R 2014 Impact of P and Sr on Solidification Sequence and Morphology of Hypoeutectic Al-Si Alloys (submitted)
[10] Eiken J, Böttger B, Steinbach I 2006 *Phys. Rev. E* **73** 066122
[11] MICRESS®, the microstructure evolution simulation software, *www.micress.de*
[12] Carré A, Böttger B, Apel M 2013 *J of Crystal Growth* **380** 5–13
[13] Eiken J 2012 *Proc of MCWASP XIII IOP Conference Series-Materials Science and Engineering* **33 (1),** 012105 DOI: 10.1088/1757-899X/33/1/012105
[14] Jones H 2007 *Metall Mater Trans A* **38 (7)** 1563-1569
[15] Thermo-Calc software, *www.thermocalc.com*
[16] Böttger B, Eiken J, Steinbach I 2006 *Acta Mat* **54 (10)** 2697-2704
[17] Steinbach I, Pezzolla F, Prieler R 1995 *Proc of MWASPC VII, Ed by M Cross and J Campbell Warrendale (TMS)*, 695-703

A model for eutectic growth in multicomponent alloys

A V Catalina[1], P W Voorhees[2], R K Huff[1] and A L Genau[3]
[1] Caterpillar Inc., 14009 Old Galena Road, Mossville, IL 61552, USA
[2] Northwestern University, 2220 Campus Dr, Evanston, IL 60208, USA
[3] The University of Alabama at Birmingham, 1075 13th Street South, Birmingham, AL 35294-4440, USA

Email: Catalina_Adrian_V@cat.com
Phone: +1 309 578 6314

Abstract. A model for eutectic growth of multicomponent alloys is proposed. Among several other modifications brought to the classic theory of eutectic growth, the model introduces the phase fraction change as the main mechanism for maintaining an isothermal solid/liquid interface during solidification of eutectic alloys. A comparison between the theoretical results and experimental measurements for the Fe-C based alloys showed a good agreement for alloys of composition not too far removed from the eutectic point. The analysis also showed that an earlier proposed criterion for spacing selection of binary irregular eutectics, based on volume fraction of the faceted phase, can be applied to multicomponent alloys as well. The investigation performed on the effect of Si on the growth kinetics of Fe-C-Si eutectic alloys showed that the growth coefficient μ decreases with the increase of Si content.

1. Introduction/Background

The Jackson and Hunt (JH) [1] classical treatment of the binary eutectic growth starts with solving the solute diffusion equation in front of a planar solid/liquid (s/l) interface moving at a constant velocity, V, in the z-direction:

$$\frac{\partial^2 C}{\partial x^2}+\frac{\partial^2 C}{\partial z^2}+\frac{V}{D_L}\frac{\partial C}{\partial z}=0 \tag{1}$$

where C is concentration and D_L is the diffusion coefficient in the liquid. For a eutectic microstructure (Fig. **1**a) consisting of alternating lamellae of α and β phases the appropriate boundary conditions are:

$$\begin{aligned}&-\text{periodicity}: && C(x+\lambda)=C(x) && \\ &-\text{symmetry}: && \partial C/\partial x=0 && \text{for } x=0 \text{ and } x=\lambda/2 \\ &-\text{far-field}: && C=C_{eut}+\Delta C_\infty && \text{for } z\to\infty\end{aligned} \tag{2}$$

and therefore the general solution of Eq.(1) is given by:

$$C(x,z)=C_{eut}+\Delta C_\infty+B_o\cdot\exp\left(-\frac{V\cdot z}{D_L}\right)+\sum_{n=1}^{\infty}B_n\cdot\cos\left(\frac{2n\pi}{\lambda}\cdot x\right)\cdot\exp(-\omega_n\cdot z) \tag{3}$$

where C_{eut} is the eutectic composition, ΔC_∞ is the difference between the eutectic composition and the composition far from the interface, λ is the lamellar spacing ($\lambda/2=S_\alpha+S_\beta$), and the exponent ω_n can be approximated as $\omega_n \cong 2n\pi/\lambda$ at small growth rates when $\dfrac{\lambda V}{2D_L} << 2\pi$ [1,2]. The Fourier coefficients B_o and B_n have to be determined from the mass balance of solute at the s/l interface which, by ignoring the density differences between phases as in [1], can be expressed as:

$$-\frac{D_L}{V}\cdot\frac{\partial C}{\partial z}\bigg|_{z=0} = \begin{cases}(1-k_\alpha)\cdot C_I(x) & \text{for } \alpha-\text{phase } (0 \le x \le S_\alpha)\\ (1-k_\beta)\cdot C_I(x) & \text{for } \beta-\text{phase } (S_\alpha \le x \le S_\alpha + S_\beta)\end{cases} \tag{4}$$

where $C_I(x)$ is the liquid composition at the s/l interface (*i.e.*, at $z = 0$), and k_α and k_β are the partition coefficients between the liquid and the α- and β-phase, respectively.

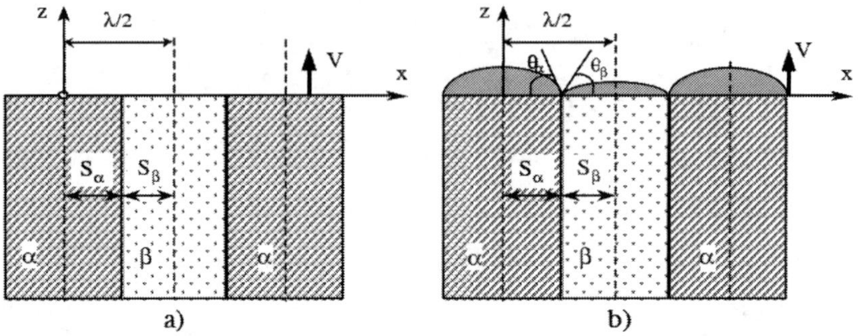

Fig. 1 The eutectic interface geometry: a) planar s/l interface assumed for the diffusion field calculation; b) the solid shaded areas account for the interface curvature.

As formulated in Eq.(4) the mass balance equation cannot be easily applied to calculating the Fourier coefficients because of different values of k_α and k_β. Consequently, it was suggested in JH treatment that $C_I(x)$ can be replaced by the eutectic composition (C_{eut}) thus allowing B_o and B_n be calculated as:

$$B_o = \frac{C_o^\alpha \cdot S_\alpha - C_o^\beta \cdot S_\beta}{S_\alpha + S_\beta} \tag{5}$$

$$B_n = \frac{2}{(n\pi)^2}(S_\alpha + S_\beta)\frac{V}{D_L}C_o \sin\left(\frac{n\cdot\pi\cdot S_\alpha}{S_\alpha + S_\beta}\right) \tag{6}$$

where C_o^α and C_o^β are the differences between C_{eut} and the maximum solute solubility in the solid α- and β-phase, respectively, and $C_o = C_o^\alpha + C_o^\beta$. Also S_α and S_β are half the width of the α- and β-lamellae, respectively. With the composition thus defined and by accounting for the effect of s/l interface curvature, the average interface undercooling for the α- and β-phase can be calculated as [1]:

$$\Delta T_\alpha = m_\alpha\left[\Delta C_\infty + B_o + \frac{2V}{D_L}C_o\frac{(S_\alpha + S_\beta)^2}{S_\alpha}P\right] + \frac{\Gamma_\alpha \sin(\theta_\alpha)}{S_\alpha} \tag{7a}$$

$$\Delta T_\beta = m_\beta\left[-\Delta C_\infty - B_o + \frac{2V}{D_L}C_o\frac{(S_\alpha + S_\beta)^2}{S_\beta}P\right] + \frac{\Gamma_\beta \sin(\theta_\beta)}{S_\beta} \tag{7b}$$

where m_α and m_β are the liquidus slopes, Γ_α and Γ_β are the Gibbs-Thomson coefficients of the α- and β-phase, respectively, θ_α and θ_β are the contact angles at the triple point junction (see Fig.1b), and P has the expression $P = \sum_{n=1}^{\infty}\frac{1}{(n\pi)^3}\sin^2\left(\frac{n\pi S_\alpha}{S_\alpha + S_\beta}\right)$.

IOP Conf. Series: Materials Science and Engineering **84** (2015) 012085 doi:10.1088/1757-899X/84/1/012085

An isothermal s/l interface would require $\Delta T_\alpha = \Delta T_\beta$ but this equality is not necessarily obeyed by Eq.(7a) and Eq.(7b) because no adjustment mechanism has been considered in the JH model. It was in fact pointed out by JH that treating the ratio S_β/S_α as a constant (*i.e.*, constant phase fractions) was a drawback of the model and that a minor change of the phase fractions would have a considerable effect on the magnitude of B_o while negligible for the other terms involving S_α and S_β. Therefore, it was considered that the isothermal condition can still be imposed by eliminating B_o between Eq.(7a) and Eq.(7b). The resulting undercooling was of the form:

$$\Delta T = K_1 \cdot \lambda \cdot V + K_2 / \lambda \tag{8}$$

where K_1 and K_2 are material constants of which expressions are defined in ref.[1]. Eq.(8) contains three variables (*i.e.*, ΔT, V, and λ) and therefore requires and additional condition for the direct correlation of pairs of variables. Consequently, JH made use of the *extremum condition* according to which the system selects the particular spacing, λ_{ex}, which minimizes the undercooling and maximizes the growth velocity. The result of this assumption is the following group of equations:

$$\lambda_{ex}^2 V = K_2 / K_1 \qquad \lambda_{ex} \Delta T = 2K_2 \qquad \Delta T^2 / V = 4K_1 K_2 \tag{9}$$

While the drawback of constant phase fraction treatment could be easily circumvented in the case of binary eutectic alloys through the elimination of the sensitive term B_o, the complexity increases dramatically when a third element is added to the system [3]. Donaghey and Tiller (DG) [4] relaxed the simplifying assumption of JH model related to constant phase fraction and mass balance at the s/l interface but the inconvenience of their treatment is that the Fourier coefficients must be determined numerically from an infinite linear system of equations. The actual phase fraction can then be calculated from the extremum condition. The possibility of adding a third element to the system is also illustrated in [4]. A more recent investigation on DG solution was published by Ludwig and Leibbrandt [5] showing that the extremum condition leads to relationships similar to those of JH (Eq.(9)).

The last relationship in Eq.(9) is mostly used in solidification modeling and by commercial casting simulation software packages in the form:

$$V = \mu \cdot \Delta T^2 \tag{10}$$

with $\mu = (4 \cdot K_1 \cdot K_2)^{-1}$ being defined as the eutectic growth constant (or coefficient) of the alloy. While this relationship along with the others in Eq.(9) closely match the experimental measurements for regular eutectics, such as Al-Cu or Pb-Sn for instance, they do not readily apply to alloys of commercial importance such as Fe-C or Al-Si (*i.e., irregular eutectics*) which exhibit both lamellar spacing and undercooling much higher than the theoretical predictions [6,7,8,9,10]. For this latter type of eutectics the average spacing, $\bar{\lambda}$, follows a relationship of the type:

$$\bar{\lambda} = \phi \cdot \lambda_{ex} \tag{11}$$

and the equivalent of Eq.(9) is:

$$\bar{\lambda}^2 V = \phi^2 K_2 / K_1 \qquad \bar{\lambda} \Delta T = \left(\phi^2 + 1\right) K_2 \qquad \Delta T^2 / V = (\phi + 1/\phi)^2 K_1 K_2 \tag{12}$$

According to the treatment of Magnin and Trivedi (MT) [2] the coefficient ϕ takes the values of 5.4, 1.8, and 3.2 for binary Fe-$C_{graphite}$, Fe-Fe_3C, and Al-Si system, respectively. A more recent treatment of Catalina and Stefanescu (CS) [11] showed that for all the above binary systems the value of ϕ can be related to the volume fraction of the faceted phase (f_β) in the microstructure:

$$\phi = f_\beta^{-\frac{1}{2}} = (1 - f_\alpha)^{-\frac{1}{2}} \tag{13}$$

In both MT and CS treatments the value of ϕ was determined by comparing the theoretical predictions against the experimental measurements. The advantage of Eq.(13), however, is that it relates the eutectic spacing to a microstructure feature and the validity of this relationship in a multicomponent system will be investigated in this paper.

In this paper we propose a theoretical model to describe the eutectic growth in multicomponent alloys. While the mathematical derivation follows the general line of JH treatment, several modifications are made in order to account for the density difference between the phases, precipitation of faceted phases with fixed chemical composition, as well as the change of the volume fractions of the eutectic phases with respect to chemistry and solidification conditions. It will be shown that, according to the proposed model, the phase fraction adjustment affects the composition of the boundary layer ahead the solid/liquid interface and is also the main factor responsible for maintaining an isothermal interface. The presentation of theoretical results will also focus on the influence of alloying elements on the spacing selection and growth velocity of eutectic microstructures. Available experimental measurements will also be presented for comparison. Particular attention will be given to the Fe-$C_{graphite}$ based alloys, *i.e.*, commercial gray cast iron.

2. Model description

2.1. The concentration field

We consider directional solidification of a multicomponent alloy that solidifies with a eutectic microstructure of which one of the phases (α) is non-faceted while the other (β) is a faceted phase of fixed chemical composition. The problem set-up is identical to that of JH (Fig.1) except the number of alloying elements in the system. The diffusion equation therefore is:

$$\frac{\partial^2 C_j}{\partial x^2}+\frac{\partial^2 C_j}{\partial z^2}+\frac{V}{D_{L,j}}\frac{\partial C_j}{\partial z}=0 \qquad (14)$$

where C_j is the concentration and $D_{L,j}$ is the liquid diffusion coefficient of the alloying element j. The same boundary conditions as in Eq.(2) are considered for each element and therefore, for solidification at low Peclet number, the general solution is:

$$C_j(x,z)=C_j^{\infty}+B_{o,j}\cdot\exp\left(-\frac{V\cdot z}{D_{L,j}}\right)+\sum_{n=1}^{\infty}B_{n,j}\cdot\cos\left(\frac{2n\pi}{\lambda}\cdot x\right)\cdot\exp\left(-\frac{2n\pi}{\lambda}\cdot z\right) \qquad (15)$$

where C_j^{∞} is the initial concentration of element j (*i.e.*, far field) and λ is the lamellar spacing. Because the solid β-phase is considered a faceted phase of fixed chemical composition, the mass balance equation at the s/l interface that also accounts for the density differences can be formulated as:

$$-\frac{D_{L,j}}{V}\cdot\frac{\partial C_j}{\partial z}\bigg|_{z=0}=\begin{cases}C_{I,j}(x)-\eta_\alpha\cdot k_{\alpha,j}\cdot C_{I,j}(x) & \text{for } \alpha-\text{phase } (0\le x\le f_\alpha\lambda/2)\\ C_{I,j}(x)-\eta_\beta\cdot C_{\beta,j}^{S} & \text{for } \beta-\text{phase } (f_\alpha\lambda/2\le x\le\lambda/2)\end{cases} \qquad (16)$$

where $C_{I,j}(x)$ is the liquid concentration of element j at the s/l interface (*i.e.*, at z = 0), $k_{\alpha,j}$ is the partition coefficient of j between the liquid and the α-phase, $C_{\beta,j}^{S}$ is the concentration of j in the β-phase, f_α is the volume fraction of the α-phase in the solid microstructure, η_α and η_β are quantities accounting for the density differences between the phases [11]:

$$\eta_i=\rho_i/\rho_{liq}\,,\; i=\alpha,\beta \qquad (17)$$

where ρ_{liq} is the density of the liquid phase and ρ_i is the density of the solid phase i (i = α,β).

As already pointed out earlier, Fourier coefficients cannot be immediately calculated based on Eq.(17) because of the discontinuity of the $C_{I,j}(x)$ multipliers at $x=f_\alpha\lambda/2$. Therefore we consider a simplified formulation of Eq.(16) in the form:

$$-\frac{D_{L,j}}{V}\cdot\frac{\partial C_j}{\partial z}\bigg|_{z=0}=\begin{cases}C_{I,j}(x)-\eta_\alpha\cdot k_{\alpha,j}\cdot\overline{C}_{\alpha,j}^{l} & \text{for } \alpha-\text{phase } (0\le x\le f_\alpha\lambda/2)\\ C_{I,j}(x)-\eta_\beta\cdot C_{\beta,j}^{S} & \text{for } \beta-\text{phase } (f_\alpha\lambda/2\le x\le\lambda/2)\end{cases} \qquad (18)$$

where $\overline{C}_{\alpha,j}^{l}$ is the average liquid concentration of element j at the s/l interface. Note that, as opposed to JH treatment, the approximation made in Eq.(18) is only for the concentration in the solid α-phase (*i.e.*, $k_{\alpha,j}\cdot\overline{C}_{\alpha,j}^{l}$). At this time $\overline{C}_{\alpha,j}^{l}$ is not known but it can be calculated by making use of Eq.(15) and Eq.(16). From Eq.(15) one obtains:

$$\overline{C}^{l}_{\alpha,j} = C^{\infty}_{j} + B_{o,j} + S_n / f_\alpha \tag{19}$$

$$\overline{C}^{l}_{\beta,j} = C^{\infty}_{j} + B_{o,j} - S_n / (1 - f_\alpha) \tag{20}$$

where $\overline{C}^{l}_{\beta,j}$ is the interface concentration of j in front of the β-phase and the quantity S_n is given by:

$$S_n = \sum_{n=1}^{\infty} B_n \frac{\sin(n\pi f_\alpha)}{n\pi} \tag{21}$$

Also, the integration of Eq.(16) on the range $x = [0, \lambda/2]$ gives:

$$\eta_\alpha \cdot k_{\alpha,j} \cdot (f_\alpha \cdot B_{o,j} + S_n) = (1 - f_\alpha \eta_\alpha k_{\alpha,j}) \cdot C^{\infty}_{j} - (1 - f_\alpha) \cdot \eta_\beta \cdot C^{s}_{\beta,j} \tag{22}$$

and therefore the concentrations $\overline{C}^{l}_{\alpha,j}$ and $\overline{C}^{l}_{\beta,j}$ can be calculated by making use of Eq.(19), Eq.(20), and Eq.(22). The result is:

$$\overline{C}^{l}_{\alpha,j} = \frac{C^{\infty}_{j} - (1 - f_\alpha) \cdot \eta_\beta \cdot C^{s}_{\beta,j}}{\eta_\alpha \cdot k_{\alpha,j} \cdot f_\alpha} \tag{23}$$

$$\overline{C}^{l}_{\beta,j} = \frac{(\eta_\alpha \cdot k_{\alpha,j} - 1) \cdot C^{\infty}_{j} + (1 - f_\alpha) \cdot \eta_\beta \cdot C^{s}_{\beta,j} + \eta_\alpha \cdot k_{\alpha,j} \cdot B_{o,j}}{(1 - f_\alpha) \cdot \eta_\alpha \cdot k_{\alpha,j}} \tag{241}$$

With $\overline{C}^{l}_{\alpha,j}$ thus given by Eq.(23) the Fourier coefficients $B_{n,j}$ can now be calculated from Eq.(18) as:

$$B_{n,j} = \frac{\lambda \cdot V}{D_{L,j}} \left(\eta_\beta \cdot C^{s}_{\beta,j} - C^{\infty}_{j}\right) \cdot \frac{1}{f_\alpha} \cdot \frac{\sin(n\pi f_\alpha)}{(n\pi)^2} \tag{25}$$

Also, by replacing Eq.(25) in Eq.(21) and then in Eq.(19) the coefficient $B_{o,j}$ can be obtained as:

$$B_{o,j} = \frac{(1 - \eta_\alpha \cdot k_{\alpha,j} \cdot f_\alpha) \cdot C^{\infty}_{j} - (1 - f_\alpha) \cdot \eta_\beta \cdot C^{s}_{\beta,j}}{\eta_\alpha \cdot k_{\alpha,j} \cdot f_\alpha} - Pe_j \cdot A_j \cdot \frac{P_\alpha}{f_\alpha^2} \tag{26}$$

where Pe_j, A_j and P_α are defined as:

$$Pe_j = \frac{\lambda \cdot V}{D_{L,j}} \tag{27}$$

$$A_j = \eta_\beta C^{s}_{\beta,j} - C^{\infty}_{j} \tag{28}$$

$$P_\alpha = \sum_{n=1}^{\infty} \frac{\sin^2(n\pi \cdot f_\alpha)}{(n\pi)^3} \tag{29}$$

With the Fourier coefficients $B_{n,j}$ and $B_{o,j}$ given in Eq.(25) and Eq.(26) the concentration field given by Eq.(15) is now completely defined. It must be pointed out that the coefficients $B_{o,j}$ and $B_{n,j}$ determined in the present treatment, especially $B_{o,j}$, are different from those obtained by JH. Eq.(26) shows that $B_{o,j}$ also depends on Pe_j and the initial concentration of the element j. The dependency on the phase fraction change is discussed in the next section.

2.2. Calculation of the growth undercooling and phase fraction change during solidification

The average undercooling $\Delta\overline{T}_\alpha$ and $\Delta\overline{T}_\beta$ in front of the α-phase and β-phase, respectively, can be expressed as:

$$\Delta\overline{T}_\alpha = m_{\alpha,1} \cdot \left(C_{eut} - \overline{C}^{l}_{\alpha,1}\right) + \sum_{j=2}^{N_{elems}} m_{\alpha,j} \cdot \left(C^{\infty}_{j} - \overline{C}^{l}_{\alpha,j}\right) + \frac{K^{\alpha}_{r}}{\lambda} \tag{30}$$

$$\Delta\overline{T}_\beta = m_{\beta,1} \cdot \left(C_{eut} - \overline{C}^{l}_{\beta,1}\right) + \sum_{j=2}^{N_{elems}} m_{\beta,j} \cdot \left(C^{\infty}_{j} - \overline{C}^{l}_{\beta,j}\right) + \frac{K^{\beta}_{r}}{\lambda} \tag{31}$$

where N_{elems} is the number of the alloying elements in the composition of the alloy. It was considered in Eq.(30) and Eq.(31) that $j = 1$ represents the main element in the composition, such as *Carbon* in *Fe-C* based alloys for instance. C_{eut} in the above equations is therefore the concentration of the main element in the multicomponent alloy at the eutectic composition. Also, the quantities K_r^{α} and K_r^{β} represent, for a certain spacing λ, the contribution of the interface curvature to the total undercooling and are defined as [1]:

$$K_r^i = \frac{2 \cdot \Gamma^{i/l} \cos(\theta_i)}{f_i}; \quad (i = \alpha, \beta; \ f_\beta = 1 - f_\alpha) \tag{32}$$

where $\Gamma^{i/l}$ is the Gibbs-Thomson coefficient of the phase i, and θ_i the contact angle of the phase i at the three-phase junction (see **Fig. 1**b).

As defined in Eq.(30) and Eq.(31) the undercooling $\Delta\overline{T}_\alpha$ and $\Delta\overline{T}_\beta$ are not necessarily equal (*i.e.*, isothermal s/l interface) for any growth velocity and eutectic spacing. As pointed out by JH [1], it is the phase fraction change (PFC) that insures an isothermal interface and therefore an explicit dependency on PFC must be considered in the model. In the present treatment this dependency is developed based on the first order expansion in the vicinity of the eutectic point. Therefore, if f_α^o is the volume fraction of the α-phase at the equilibrium eutectic point and $\Delta f_\alpha = -\Delta f_\beta$ is the PFC at the imposed solidification conditions, the corresponding changes of any quantity $q = q(f_\alpha)$ can be expressed as:

$$q(f_\alpha) \cong q^o + q' \cdot \Delta f_\alpha \tag{33}$$

where $q^o = q(f_\alpha^o)$ and $q' = q'(f_\alpha^o)$ is the first derivative of q with respect to f_α at f_α^o. If the expansion defined in Eq.(33) is applied to all the appropriate quantities in Eq.(30) and Eq.(31) then Δf_α can be determined by imposing the condition of isothermal s/l interface (*i.e.*, $\Delta\overline{T}_\alpha = \Delta\overline{T}_\beta$). The result is:

$$\Delta f_\alpha = \frac{m_{\alpha,1}\left(C_{eut} - \overline{C}_{\alpha,1}^{l,o}\right) - m_{\beta,1}\left(C_{eut} - \overline{C}_{\beta,1}^{l,o}\right) + \sum_{j=2}^{N_{elems}}\left[m_{\alpha,j}\left(C_j^\infty - \overline{C}_{\alpha,j}^{l,o}\right) - m_{\beta,j}\left(C_j^\infty - \overline{C}_{\beta,j}^{l,o}\right)\right] + \frac{K_r^{\alpha,o} - K_r^{\beta,o}}{\lambda}}{\sum_{j=1}^{N_{elems}}\left(m_{\alpha,j} \cdot \overline{C}_{\alpha,j}^{l'} - m_{\beta,j} \cdot \overline{C}_{\beta,j}^{l'}\right) - \frac{K_r^{\alpha'} - K_r^{\beta'}}{\lambda}} \tag{34}$$

Eq.(34), along with the previous equations defining the composition field, shows that the Δf_α that ensures an isothermal s/l interface depends on the material parameters (*e.g.*, liquidus slopes, partition, diffusion and Gibbs-Thomson coefficients), initial composition of the liquid alloy, interphase spacing λ, as well as the imposed growth velocity V. Eq.(34) also shows that Δf_α can be calculated for any pair of λ and V and is not restricted to the extremum condition as in the DG model [4].

2.3. Application to multicomponent Fe-$C_{graphite}$ based alloys

For this alloy system *carbon* is the main alloying element (*i.e.*, $j=1$). Also, the solid solution α is the austenite while the faceted β-phase is graphite. The partition coefficient of carbon between the liquid and austenite is defined as $k_{\alpha,1} = C_E / C_{eut}$, where C_E is the maximum solubility of carbon at the eutectic temperature (T_{eut}) and C_{eut} is the concentration of carbon in the multicomponent alloys at the eutectic composition. Both C_E and C_{eut} as well as T_{eut} depend on the concentration of the other elements in the composition of the alloy and can be calculated by using the relationships given in ref.[12]. Because at the eutectic temperature the quantity $k_{\alpha,1} \cdot \overline{C}_{\alpha,1}^l$ appearing in Eq.(18) is in fact C_E, it follows that f_α^o can be calculated from Eq.(23) for $\overline{C}_{\alpha,1}^l = C_{eut}$. Furthermore, it is considered that the β-phase is made entirely out of carbon (*i.e.*, $C_{\beta,1}^s = 100wt\%$) and therefore $C_{\beta,j(j>1)}^s = 0$. Consequently, it can be shown that the change Δf_α (see Eq.(34)) for this system can be calculated as:

$$\Delta f_\alpha = \frac{\frac{P_\alpha^o}{f_\alpha^{o2}\left(1-f_\alpha^o\right)}\sum_{j=1}^{N_{elems}} m_{\beta,j}A_j Pe_j - \sum_{j=2}^{N_{elems}}\left(m_{\alpha,j}-m_{\beta,j}\right)\cdot\left(1-\frac{1}{\eta_\alpha\cdot k_{\alpha,j}\cdot f_\alpha^o}\right)C_j^\infty - \frac{K_r^{\alpha,o}-K_r^{\beta,o}}{\lambda}}{-\sum_{j=1}^{N_{elems}}\left(m_{\alpha,j}-m_{\beta,j}\right)H_{o,j} - U_o\cdot\sum_{j=1}^{N_{elems}} m_{\beta,j}A_j Pe_j + \frac{K_r^{\alpha'}-K_r^{\beta'}}{\lambda}} \quad (35)$$

with $H_{o,j} = \dfrac{\eta_\beta C_{\beta,j}^S - C_j^\infty}{\eta_\alpha\cdot k_{\alpha,j}\cdot f_\alpha^{o2}}$ and $U_o = \dfrac{f_\alpha^o\left(1-f_\alpha^o\right)\sum_{n=1}^{\infty}\dfrac{\sin\left(2n\pi f_\alpha^o\right)}{(n\pi)^2} + \left(3f_\alpha^o - 2\right)P_\alpha^o}{f_\alpha^{o3}\left(1-f_\alpha^o\right)^2}$.

Also, the undercooling at the s/l interface is given by:

$$\Delta\overline{T} = \sum_{j=2}^{N_{elems}} m_{\alpha,j}\left(1-\frac{1}{\eta_\alpha\cdot k_{\alpha,j}\cdot f_\alpha^o}\right)\cdot C_j^\infty + \frac{K_r^{\alpha,o}}{\lambda} + \left(\frac{K_r^{\alpha'}}{\lambda} - \sum_{j=1}^{N_{elems}} m_{\alpha,j}\cdot H_{o,j}\right)\cdot\Delta f_\alpha \quad (36)$$

Eq.(36) shows that the undercooling depends on Δf_α and the initial composition of the multicomponent alloy (C_j^∞) in addition to all the other material parameters considered in the previous models.

A graphical representation of $\Delta\overline{T} = f(\lambda)$ and $\Delta f_\alpha = f(\lambda)$ shows that both $\Delta\overline{T}$ and Δf_α go through their minimum values for about the same value of $\lambda = \lambda_{ex}$ (see Fig. 2). Therefore, in order to calculate λ_{ex} one may find convenient to apply the extremum condition to Δf_α instead to $\Delta\overline{T}$. By using the pertinent data for the Fe-C based systems and considering only the terms with relevant order of magnitude in Eq.(35) it can be shown that λ_{ex} can be approximated as:

$$\lambda_{ex} \cong \frac{f_\alpha^o\cdot(1-f_\alpha^o)^{\frac{1}{2}}\cdot(K_r^{\beta,o}-K_r^{\alpha,o})^{\frac{1}{2}}}{\left(P_\alpha^o\cdot\sum_{j=1}^{Nelems}\frac{m_{\beta,j}\cdot A_j}{D_{L,j}}\right)^{\frac{1}{2}}\cdot V^{\frac{1}{2}}} \quad (37)$$

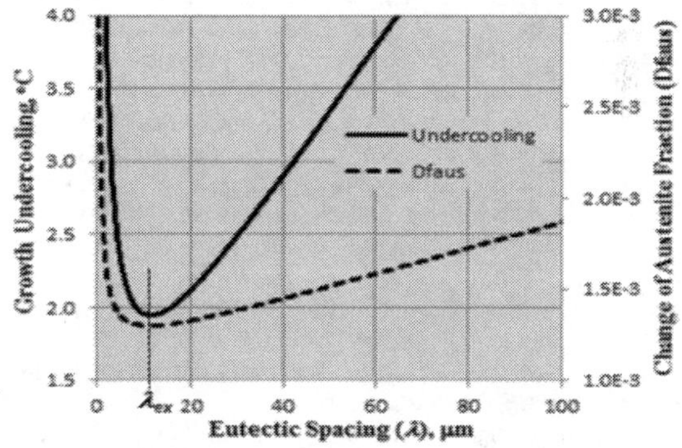

Fig. 2 Calculated variation with spacing of undercooling and change of austenite fraction for Fe-C-2.0wt%Si eutectic alloy solidifying at a velocity of 0.5 µm/s.

Also, by accounting for Eq.(37), Eq.(11) and Eq.(13) the approximate average spacing would be:

$$\overline{\lambda} \cong f_\alpha^o\cdot(K_r^{\beta,o}-K_r^{\alpha,o})^{\frac{1}{2}}\cdot\left(P_\alpha^o\cdot\sum_{j=1}^{Nelems}\frac{m_{\beta,j}\cdot A_j}{D_{L,j}}\right)^{-\frac{1}{2}}\cdot V^{-\frac{1}{2}} \quad (2)$$

thus showing that the well-known relationship $\lambda^2\cdot V = const.$ of the eutectic growth (see Eq.(9) and Eq.(12)) is also recovered by the proposed model. shows that the relationship $V = \mu\cdot\Delta T^2$ is also recovered by the model.

3. Theoretical predictions *vs.* experimental measurements for Fe-C based systems

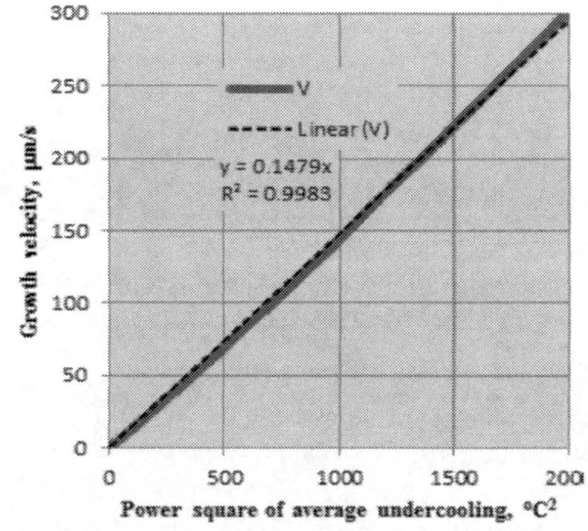

Fig. 3 Growth velocity as function of ΔT^2 for Fe-C-2.0wt%Si eutectic alloy (theoretical

A comparison between the calculated and measured lamellar spacing for binary Fe-$C_{graphite}$ eutectic alloy is presented in Fig. 4. The figure shows that a very good agreement between model predictions and measurements is achieved up to a velocity of about 6.5-7.0 µm/s which corresponds to the limit of eutectic coupled growth [13] after which the microstructure becomes degenerate [8] and exhibits higher spacing [14]. A good agreement between theoretical calculations and measured undercooling was also reported in ref. [11] for a similar but simpler model which did not account for the phase fraction change. That is not shown here due to the economy of space. Experimental measurements for small additions of other elements to pure Fe-C eutectic are also reported in ref.[14]. The comparison with theoretical results is shown in Fig. 5 for eutectic alloys containing 0.1wt%Si and 0.5 wt%Si. Similar to the binary Fe-C system, the agreement is good for solidification velocity smaller than 10 µm/s.

Experimental measurements on directionally solidified specimens containing levels of Si and Mn comparable to those in commercial cast iron were very recently reported by Rivera *et al.* [15]. The composition of the specimens is presented in Table 1 while the measured and theoretical spacing is shown in Table 2.

Of the specimens presented in Table 1, #2, #3, and #4 are hyo-eutectic, #1 is slightly hyper-eutectic, and Sorel Pig is highly hyper-eutectic. Table 2 shows that for the hypo-eutectic specimens the theoretical calculations of the pacing are in good agreement with the measurements. The agreement is also good for specimen #1 for the lower experimental spacing. However, for the Sorel Pig sample which is quite removed from the eutectic composition the theoretical spacing is only half the value of the measured one. This may be an indication that the proposed model is not valid for alloys of composition too far removed from the eutectic point.

An important application of the model would be the calculation of the eutectic growth coefficient, μ, shown in Eq.(10). The calculated influence of Si content on the value of μ is shown in Fig.6. It can be observed from Fig.6 that, according to the model, μ decreases linearly with the increase of Si content in the eutectic alloy.

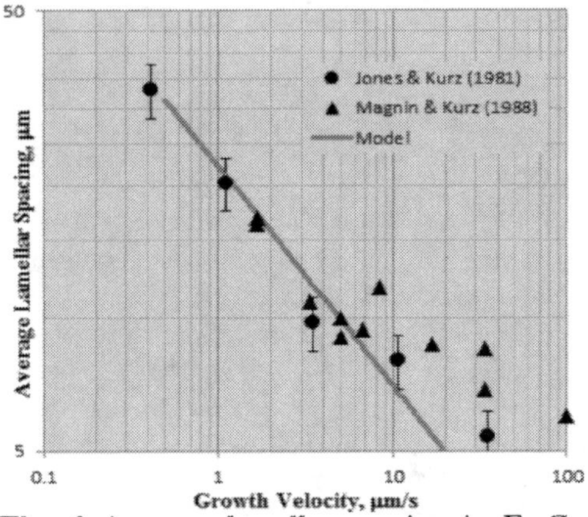

Fig. 4 Average lamellar spacing in Fe-$C_{graphite}$ system (experimental data from refs.[8], [14]).

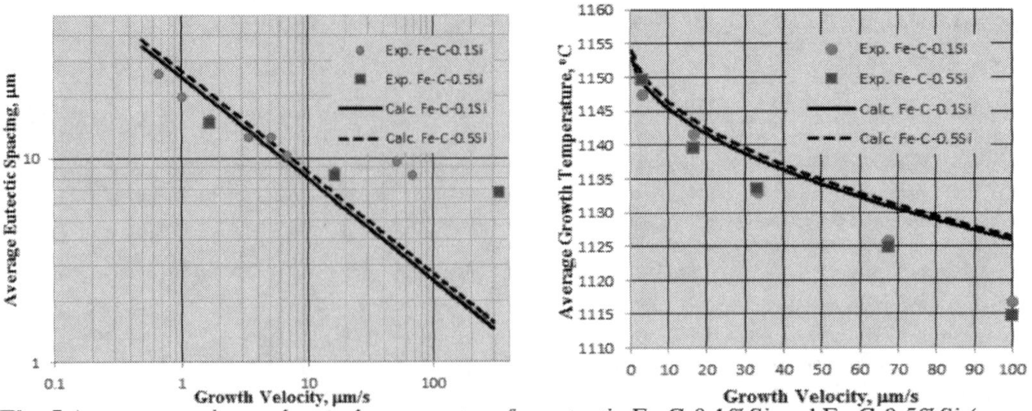

Fig. 5 Average spacing and growth temperature for eutectic Fe-C-0.1%Si and Fe-C-0.5%Si (exp. data from ref.[14]).

Table 1. Chemistry of the specimens used in ref.[15].

Specimen ID	Sorel Pig	#1	#2	#3	#4
C	4.52	3.83	3.42	3.57	3.45
Si	0.21	1.67	2.07	2.01	2.08
Mn	0.025	0.075	0.095	0.593	0.669
P	0.008	0.017	0.015	0.012	0.016
S	0.009	0.007	0.01	0.009	0.011
Cr	0.045	0.049	0.05	0.054	0.054
Mo	0.057	0.06	0.06	0.06	0.06
Ni	0.149	0.141	0.135	0.136	0.138
Cu	0.065	0.08	0.086	0.095	0.09
$C_{equivalent}$	4.59	4.39	4.11	4.23	4.13

Table 2. Experimental and theoretical spacing for the specimens presented in Table 1.

Specimen ID	V (μms^{-1})	Exp. average spacing (μm)	Exp. std. dev. (μm)	Calc. average spacing (μm)
Sorel Pig	0.5	70.6	15.4	35.6
#1	1	32.8 (48.2)*	6.5	29.4
#2	0.5	47.2	6.7	51.8
#3	0.5	45	7.3	46.2
#4	0.5	43.8	7.7	44.4

*) a bimodal spacing was reported in [15] due to the presence of austenite dendrites

4. Conclusions

A model for eutectic growth of multicomponent alloys has been proposed. As presented, the model applies to alloy systems of which one of the eutectic phases exhibits a fixed chemical composition, such as Fe-C or Al-Si based type of alloys, but a more general formulation is also possible. The model introduces the phase fraction change as the main mechanism responsible for maintaining an isothermal s/l interface during solidification. A comparison between the theoretical results and available experimental data for the Fe-C based system showed good agreement for alloys of chemistry not too far from that of the eutectic point. An earlier proposed criterion for spacing selection of binary irregular eutectics was also successfully tested for the case of multicomponent alloys. Calculations of the eutectic growth coefficient μ for the ternary Fe-C-Si alloy showed that Si content decreases the value of μ. The influence on μ of all the other elements can be analysed based on the proposed model. The model is simple and can accommodate any number of alloying elements as long as they do not form additional precipitates.

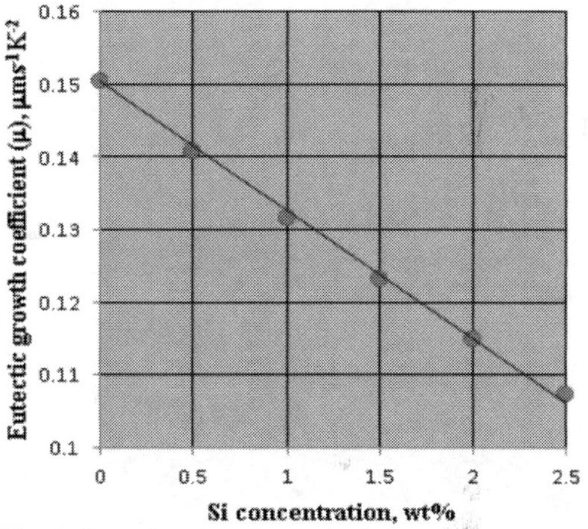

Fig.6 Calculated influence of Si content on the magnitude of the eutectic growth coefficient, μ

Acknowledgements

This work was supported by USA Department of Energy and Caterpillar Inc. through project DE-EE0005980.

References

[1] Jackson K A and Hunt J D 1966 *Trans. Metall. Soc. AIME* **236** 1129-42
[2] Magnin P and Trivedi R 1991 *Acta metal. mater* **39**(4) 453-467
[3] McCartney D G, Hunt J D, and Jordan R M 1980 *Metall. Trans. A* **11A** 1243-49
[4] Donaghey L F and Tiller W A 1968/1969 *Mat. Sci. Eng.* **3** 231-239
[5] Ludwig A and Leibbrandt S 2004 *Mat. Sci. Eng.* A **375-377** 540-546
[6] Toloui B and Hellawell A 1976 *Acta Metallurgica* **24** 565-573
[7] Elliot R and Glenister S M D 1980 *Acta Metallurgica* **28** 1489-94
[8] Jones H and Kurz W 1981 Z. *Metallkunde* **72** 792-797
[9] Hogan L M and Song H 1987 *Metall. Trans. A* **18A** 707-713
[10] Magnin P, Mason J T, and Trivedi R 1991 *Acta metall. mater.* **39** (4) 469-80

[11] Catalina A V and Stefanescu D M 2010 *Proc. The 8th Pacific Rim int. conf.on modelling of casting & solidification processes* (MCSP8, Incheon, Korea), ed. Choi J K, Hwang H Y, and Kim J T (Korea Inst. of Ind. Tech.) pp.125-132

[12] Stefanescu D M and Katz S 2008 *ASM Handbook* **15** (ASM International), 41-55

[13] Jones H and Kurz W 1980 *Metall. Trans. A* **11A** 1265-1273

[14] Magnin P and Kurz W 1988 *Metall. Trans. A* **19A** 1955-1963

[15] Rivera E A, Christiansen T, Genau A, and Catalina A V 2015 *Proc. Advances in the Science and Engineering of Casting Solidification*, TMS 2015 (submitted)

IOP Conf. Series: Materials Science and Engineering **84** (2015) 012086 doi:10.1088/1757-899X/84/1/012086

Benchmark experiments and numerical modelling of the columnar-equiaxed dendritic growth in the transparent alloy Neopentylglycol-(d)Camphor

L Sturz[1,a], **M Wu**[2,3,b], **G Zimmermann**[1,c], **A Ludwig**[2,d] **and M Ahmadein**[4,e]

[1] ACCESS e.V., Intzestr. 5, 52072 Aachen, Germany
[2] Chair for Modeling and Simulation of Metallurgical Processes, University of Leoben, Austria
[3] Christian-Doppler Laboratory for Advanced Process Simulation of Solidification & Melting, University of Leoben, Austria
[4] Department of Production Engineering and Mech. Design, Tanta University, Egypt

[a] L.Sturz@access.rwth-aachen.de, [b] Menghuai.Wu@unileoben.ac.at,
[c] G.Zimmermann@access.rwth-aachen.de, [d] A.Ludwig@unileoben.ac.at,
[e] M.Ahmadein@f-eng.tanta.edu.eg

Abstract. Solidification benchmark experiments on columnar and equiaxed dendritic growth, as well as the columnar-equiaxed transition have been carried out under diffusion-dominated conditions for heat and mass transfer in a low-gravity environment. The system under investigation is the transparent organic alloy system Neopentylglycol-37.5wt.-%(d)Camphor, processed aboard a TEXUS sounding rocket flight. Solidifications was observed by standard optical methods in addition to measurements of the thermal fields within the sheet like experimental cells of 1 mm thickness. The dendrite tip kinetic, primary dendrite arm spacing, temporal and spatial temperature evolution, columnar tip velocity and the critical parameters at the CET have been analysed. Here we focus on a detailed comparison of the experiment "TRACE" with a 5-phase volume averaging model to validate the numerical model and to give insight into the corresponding physical mechanisms and parameters leading to CET. The results are discussed in terms of sensitivity versus numerical parameters.

1. Introduction

Support of advanced castings by process simulations is of interest to understand structure formation and to enhance product quality, as well as to minimize material and energy consumption. The as-cast structure (grain-structure, dendrite morphology, element segregation and other casting defects) of the product is linked to mechanical properties, although some features of the structure can be further improved by subsequent processes like heat-treatment or hot isostatic pressing. For a given alloy, the as-cast structure is a result of the competition between nucleation and growth of the different phases emerging subsequently during cooling towards ambient conditions. In technical multi-component alloys the primary solid phase formed from the melt often exhibits a dendritic columnar or equiaxed grain structure. The type of grain structure (columnar, equiaxed or mixed) is dependent on the alloy properties, experimental conditions like thermal gradient and solidification advancement, melt flow driven by uneven density field of the melt or by volume shrinkage during solidification, sedimentation and buoyancy of solid or mushy particles in the melt and finally the probability to form and activate nuclei or other sources for subsequent equiaxed dendrite growth in the melt [1].

The complex interaction between time-dependent nucleation and solidification conditions on one hand and time-dependent convective melt flow can be reduced by low-gravity experiments, in which convective flow, sedimentation and buoyancy are minimized or completely suppressed. Accordingly, the investigation of dendrite growth, grain formation and the Columnar-to Equiaxed Transition (CET) under diffusive conditions for heat and mass transport can be simplified significantly. Comparative reference experiments on ground may then shed light on the gravity effects and help to properly validate casting simulation tools.

Columnar growth is mandatory in single crystal growth and also used for casting of some turbine blades, where anisotropic properties of the product improve for example creep resistance at elevated service temperature [2]. Polycrystalline equiaxed growth is intended for more isotropic properties or smaller grain sizes. The CET in grain structure is generally not desired in the final product, since it may induce a change of material properties, but its investigation helps identifying suitable processing conditions for columnar and equiaxed growth.

Here we recall the main experimental results from the low-gravity experiment "TRACE" (TRAnsparent Alloys in Columnar and Equiaxed Solidification) [3-5], executed on the sounding rocket mission TEXUS-47 in 2009 with about 7 minutes of low-gravity time. The experiments were focusing on dendrite growth and the CET in the binary hypoeutectic transparent organic model system Neopentylglycol (NPG)-37.5 wt.-% (d)Camphor (DC). Allowing for in-situ and real-time observation of the solidification phenomena by standard optical methods, a detailed comparison of this benchmark experiment to numerical models was intended.

A 5-phase volume averaging model [6,7] was used to calculate the columnar tip front propagation, front undercooling, appearance of equiaxed dendrite crystals, CET, etc. The numerical results will be evaluated by comparison with the experiments. For the uncertainty of some physical properties or process parameters, e.g. the dendrite selection constant σ^* and the maximum nucleation density N_0, a numerical parameter study is performed to analyze the influence of them on the solidification result. Actually an optimal agreement can be achieved by adjusting those parameters.

2. Experimental section

The experiment has been described in detail in [3-5]. Figure 1 shows a schematic drawing of the main parts of the experimental set-up. The evacuated experimental volume of 20 mm width, 16 mm height and 1 mm thickness was filled with the molten alloy NPG-37.5 wt.-% DC. The phase-diagram of the alloy system is given in [8]. In contrast to earlier publications [3, 4], the liquidus-temperature T_L of the alloy is directly taken from the CALPHAD-description of the calorimetric data [8] and given as 67.1°C. This seemingly minor increase by 3.1°C (when compared to the former value of 64.0°C) has impact on the calculation of growth and nucleation undercooling related to T_L. The experiment is controlled by a peltier cooler and a resistance heater, both were close to but not directly in contact with the bottom and the top of the experimental cell, respectively. Optical observation is carried out perpendicular to the height and width-axes of the cell. Macro-images of cell-size are acquired, as well as detail images with roughly ten times smaller field-of-view with two different CCD-cameras and corresponding optical resolutions. Polycrystalline columnar dendritic growth was started on ground well before lift-off of the sounding rocket with a cooling-rate of -1/300 Ks^{-1} applied to cooler and heater. After steady-state columnar growth, the CET was achieved in transient regime by setting the cooling-rates simultaneously to -1/30 Ks^{-1} at the beginning of the low-gravity period; 60 s after lift-off of the sounding rocket. A gravity level lower than ±4 mg in all directions was obtained within the low-gravity period of 410 s from t=60 s to t=470 s. The experimental sequence for the cooler and heater temperatures is indicated in figure 2 with the corresponding gravity levels. The sequence is simplified in the graph and in the simulations for the first 200 s, where in contrast in the experiment an inverted temperature field is active for thermal mixing (figure 6). Times are normalized to the numerical simulation configuration (sec. 3), where the end of the microgravity period is located at time t=6000 s.

The effect of change of gravity levels in figure 2 is ignored in the simulations here, zero-g is used throughout the complete simulation time. This is justified by the relatively strong viscous damping of fluid flow in the thin sheet of 1mm thickness.

Temperatures were measured within the bulk material by means of five thermocouples type K of diameter 0.25 mm, being inserted 2.5 mm through one side of the cell. From image-processing of the macro-images and the thermocouple data the position, velocity and temperature of the columnar and later equiaxed dendritic interface were measured. After heterogeneous nucleation of equiaxed dendrites on unknown substrates their density and nucleation temperatures were determined as a function of time. Volume fractions of liquid, equiaxed and columnar phases were determined at the interface. Main results are published elsewhere [2, 3], here we focus on the determination of nucleation parameters and the dendrite selection constant σ^*, as input or comparison to the numerical modelling. Detail images could not resolve the dendrite tip radii with sufficient accuracy, but were used to estimate the morphological factors for columnar and equiaxed dendrites in numerical simulations (Tab. 2).

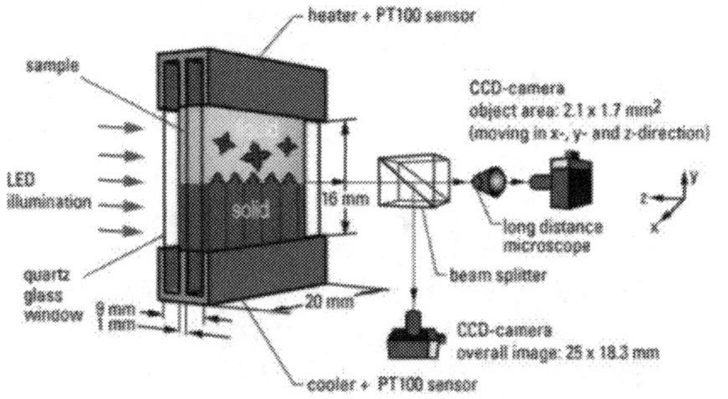

Figure 1: Schematic drawing of the main parts of the experimental setup.

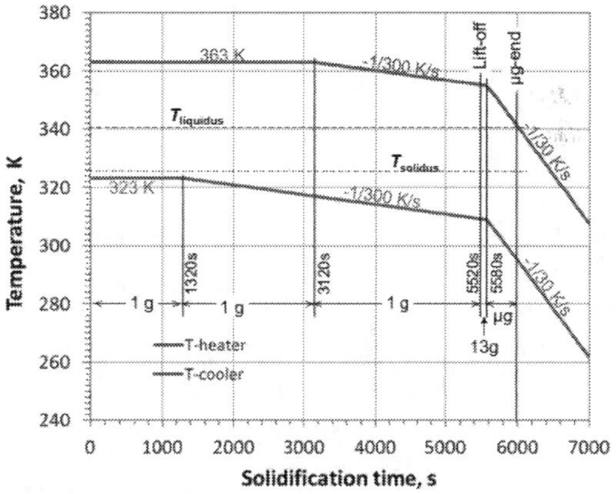

Figure 2: Imposed temperatures at the heater and cooler in simulations, similar to exp.

As shown for example by Mangelinck-Noël et al. [9] by x-ray in-situ observations of equiaxed dendrite growth in Al-Ni alloys, a series of experiments at different cooling rates can be used to estimate their nucleation parameters, assuming specific probability functions for the nucleation temperatures. Here, we obtain the nucleation temperature distribution experimentally from a single transient experiment by increasing the cooling rates and measuring nucleation positions and interpolating thermocouple temperatures. The resulting distribution is given in [3], relative to the previously used value of T_L=64.0°C. The nucleation densities of active particles are calculated from the number of nucleated equiaxed dendrites within the given temperature interval and the average volume (width x depth x height from columnar front to equiaxed dendrite at highest position) with nucleation events. For comparison and input to numerical simulations, we assume a Gaussian probability function f for the nucleation undercooling ΔT:

$$f(\Delta T) = \frac{N_0}{\Delta T_\sigma \sqrt{2\pi}} e^{-\frac{1}{2}\left(\frac{\Delta T - \Delta T_N}{\Delta T_\sigma}\right)^2}, \quad (1)$$

with the maximum nucleation density N_0, the mean nucleation temperature ΔT_N and the standard deviation of the distribution $\Delta T\sigma$. By referencing the nucleation temperatures to T_L=67.1°C, the Gaussian fit (1) of the nucleation density can be plotted, see figure 3.

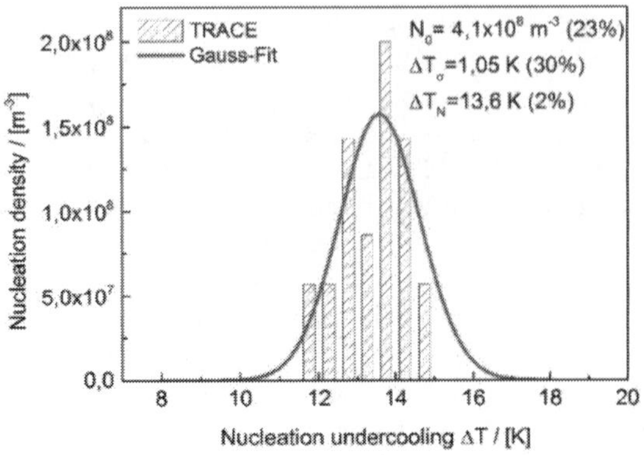

Figure 3: Nucleation density distribution and fitted Gaussian probability function. Number in brackets give the relative fit-errors.

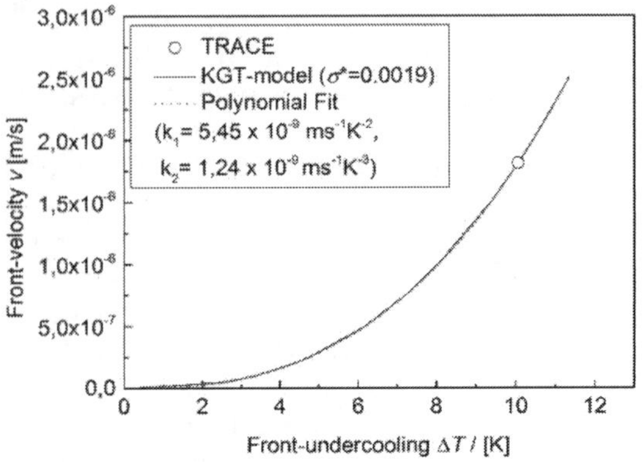

Figure 4: Kinetic law for columnar growth from experiment (single data point), the KGT-model [10] fit and a polynomial fit according to (2).

Furthermore, in numerical simulations a polynomial fit for the kinetic law in columnar growth

$$v_{tip,columnar}(\Delta T) = k_1 \Delta T^2 + k_2 \Delta T^3 \quad (2)$$

to the KGT [10] model is used. Figure 4 shows the steady-state columnar growth velocity and undercooling, experimentally determined before lift-off at the end of the low cooling-rate phase with -1/300 Ks^{-1}, as well as the fit to the KGT-model (with the selection constant σ^* as fit-parameter and other parameters given in Tab. 1) and the polynomial fit, according to (2). Due to these estimates from experimental results given above all simulations are carried out using $\Delta T_N = 13.6$K, $\Delta T_\sigma = 1.05$K, $k_1 = 5.45\text{x}10^{-9}$ ms^{-1}K^{-2} and $k_2 = 1.24\text{x}10^{-9}$ms^{-1}K^{-3}.

Three comparative simulations have been carried out; as reference case-1 the maximum nucleation density $N_0 = 4.1\text{x}10^8$ m^{-3} obtained from the fit and a significantly lowered value of σ^*=0.0076 (when compared to the standard value used in the LGK-model [11] $\sigma^*=1/4/\pi^2$=0.0253) are used. The value of σ^*=0.0019 fitted to experimental data (figure 4) could not be adapted numerically yet for the kinetic law of equiaxed dendrites (LGK), nevertheless the effect of increased dendrite selection constant (to σ^*=0.0253) is investigated in case-2. Finally, the effect of an increased N_0 = 10.0 x 10^8 m^{-3} is investigated in case-3. This is due to the fact, that the fitted value for N_0 is related to an increase in the cooling rate to -1/30 Ks^{-1} and a probably higher value for N_0 could be expected, when higher cooling rates would be applied. Moreover the average volume was used in the calculation of the nucleation density, leading to some errors and a value of N_0 = 5.0 (±1.0) x 10^8 m^{-3} was found experimentally as the maximum nucleation density in time in the available low-gravity period [3]. A summary of the three simulation cases with their nucleation and kinetic, as well as morphological parameters and thermo-physical properties of the alloy are given in Tab.1 and Tab.2.

Model features

A 5-phase mixed columnar-equiaxed solidification model with dendritic morphology was described previously [7]. The growth of the crystals is presented by the expansion of the grain envelope, whose profile connects the primary, secondary or tertiary dendrite tips to form a 'natural' enclosure of the equiaxed grains or columnar trunks. The growth of the grain envelopes and the solidification of the interdendritic melt, i.e. the liquid-to-solid phase transition, are treated differently. The growth of the envelopes is determined by dendrite growth kinetics. The semi-empirical Kurz-Giovanola-Trivedi (KGT, [10]) model is fitted by the polynomial given in equation (2) and is used for the growth of columnar primary dendrite tips. The Lipton-Glicksman-Kurz (LGK, [11]) model is applied for the growth of columnar secondary dendrite tips and equiaxed primary dendrite tips, based on different

IOP Conf. Series: Materials Science and Engineering **84** (2015) 012086 doi:10.1088/1757-899X/84/1/012086

values of σ^*, as given above. The solidification occurs inside the grain envelope, which is driven by the supersaturation of the interdendritic melt and governed by the diffusion in the interdendritic melt region. The model comprises actually only 3 'hydrodynamic' phases: the liquid melt, the equiaxed crystals, and the columnar grains, denoted as l-, e- and c-phases and have corresponding volume fractions; f_l, f_e, and f_c. However, two additional phase regions exist within each of the equiaxed and the columnar crystal envelopes: the solid dendrites and interdendritic melt inside crystal envelopes. Consequently, the system encompasses 5 'thermodynamic' phases: the solid equiaxed dendrite and interdendritic melt within the equiaxed grain envelope, the solid columnar dendrite and interdendritic melt within the columnar crystal envelope, and the extradendritic melt. Interaction, i.e. competition, between columnar and equiaxed growths, and both hard blocking [12] and soft blocking [13] mechanisms for the columnar-to-equiaxed transition (CET) are considered. The origin of the equiaxed crystals is modelled by a heterogeneous nucleation law, equation (1), with three fit parameters taken from experiment. The maximum equiaxed grain nucleation density is either taken from the fitted experimental data (case-1&case-2) or from an assumed data (case-3) to investigate its effect.

The solidification cell filled with NPG-37.5wt.-% DC is simplified to a 2D grid with volume element size of 0.25×0.25 mm^2 as shown in figure 5. The grid is provided by 5 measuring points (thermocouples) to record the cooling curves analogous to experiment. Similar to experiment, a temperature gradient is applied to the simulation grid by superimposing the experimental temperature control at the cooler and heater (figure 2) on the corresponding grid boundaries, with a heat transfer coefficient of 50 Wm^{-1}K^{-1}. The conservation equations of mass, enthalpy, species, and number density are solved sequentially at each time using CFD software package, ANSYS-Fluent version 14.5.0 based on the control volume method. A time step size of 0.2 s was used to achieve solution convergence over a time of ~7000 s.

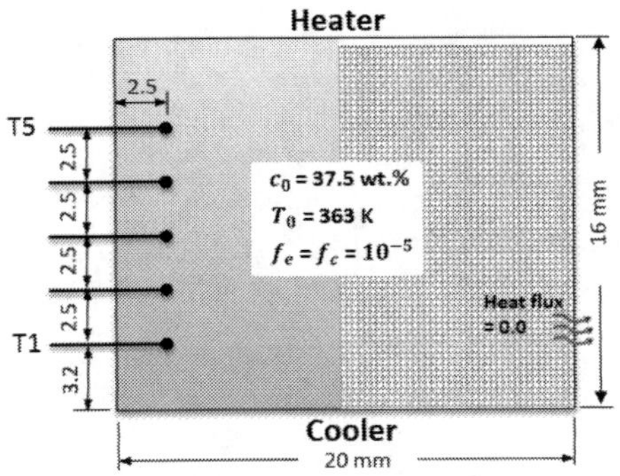

Heat transfer to cooler/heater: 50 Wm^{-1}K^{-1}

Figure 5: 2D grid of the solidification cell with boundary and initial conditions.

Table 1. Thermo-physical alloy properties.

Alloy conc. [wt.-% DC]:	c_0 = 37.5
Eutectic conc. [wt.-% DC]:	c_E = 45.3 [8]
Liquidus-temp. [°C]:	$T_{liq.}$ = 67.1°C (340.25K) [8]
Solidus-temp. [°C]:	$T_{sol.}$ = 52.8°C (325.95K) [8]
Liquidus slope [K/wt.-%]:	$m_{liq.}$ = -1.84 [8]
Partition coef. [-]:	k = 0.085 [8]
Diffusion coef. [m^2/s]:	D_L = 9.7×10^{-11} [14]
	D_S = 8.0×10^{-20} [*]
Thermal capacity [J/kg·K]:	$c_{p,L}$ = 2400, $c_{p,S}$ = 2650 [8]
Thermal cond. [W/m·K]:	λ_L = 0.125, λ_S = 0.265 [16]
Density [kg/m^3]:	ρ = 1035 [5]
Latent heat fusion [kJ/kg]:	L = 33.4 [8]
Gibbs-Thomson coef. [Km]:	Γ = 7.8×10^{-8}, adapted from [15]
Selection constant [-]:	σ^*=0.0019 (figure 4)

(*) assumed

Table 2. Nucleation, kinetic and morphological parameters for modeling.

Nucleation parameters:	ΔT_N = 13.6K, ΔT_σ=1.05 K (figure 3), N_0 (see below)
Kinetic parameters columnar:	k_1 = 5.45x10^{-9} ms^{-1}K^{-2}, k_2 = 1.24x10^{-9}ms^{-1}K^{-3}(figure 4)
Shape factor / circularity col.:	0.798 / 0.50 (estimated from experiment)
PDAS / SDAS columnar:	400 μm / 50 μm (estimated from experiment)
Shape factor / sphericity equiaxed.:	0.58 / 0.84 (estimated from experiment)
Case-1:	σ^*=0.0076, N_0 = 4.1 x 10^8 m^{-3}
Case-2:	σ^*=0.0253, N_0 = 4.1 x 10^8 m^{-3}
Case-3:	σ^*=0.0076, N_0 = 10.0 x 10^8 m^{-3}

3. Results and discussion

Figure 6 shows the control temperatures T_heater and T_cooler, the temperature evolution at the thermocouples T1-T5 as measured experimentally and the corresponding temperature data from simulation in case-1. As discussed before, the first 200 s in simulations are simplified and the initial isothermal conditions of 363K are applied to the bulk material. The cooling-rates in columnar growth ($t<\mu g$-start) are comparable to the experimental data, although a shift in temperatures from T1 to T5 is observable. Furthermore the application of the higher cooling-rate ($t>\mu g$-start) shows a slower reaction in simulations. Both effects can be explained by non-optimized heat-transfer coefficients at cooler and heater. Figure 7 shows the front position (columnar for $t<\mu g$-start, transient to equiaxed for $t>\mu g$-start). As all simulations use identical kinetic parameters k_1 and k_2 for columnar growth, the results are identical in that regime. The corresponding constant shift and delayed response is due to the effects explained in the cooling-curves (figure 6).

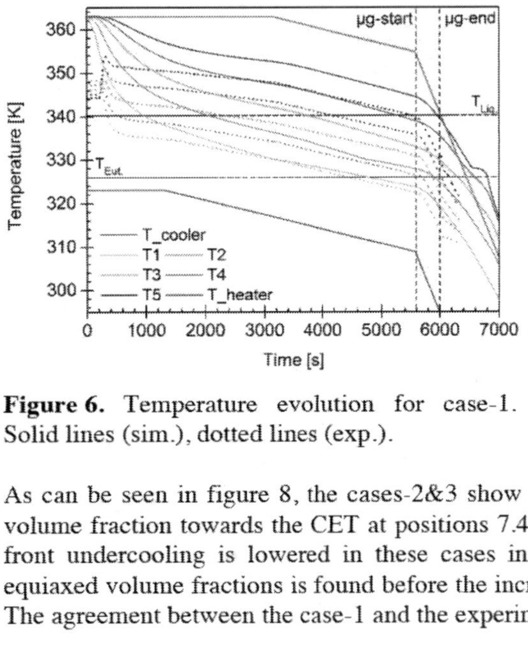

Figure 6. Temperature evolution for case-1. Solid lines (sim.), dotted lines (exp.).

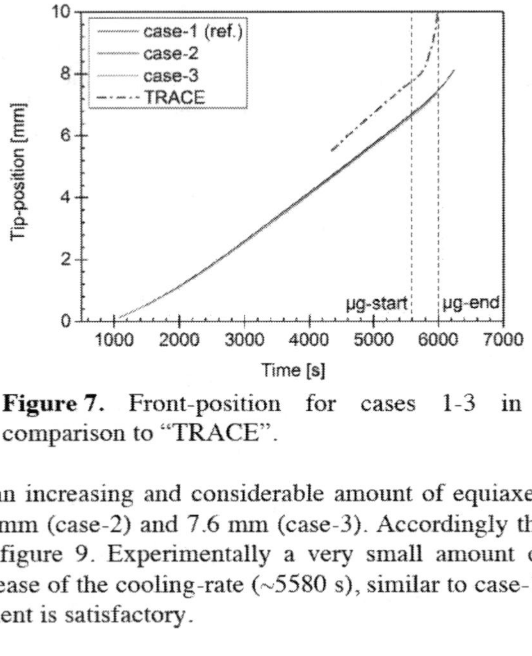

Figure 7. Front-position for cases 1-3 in comparison to "TRACE".

As can be seen in figure 8, the cases-2&3 show an increasing and considerable amount of equiaxed volume fraction towards the CET at positions 7.4 mm (case-2) and 7.6 mm (case-3). Accordingly the front undercooling is lowered in these cases in figure 9. Experimentally a very small amount of equiaxed volume fractions is found before the increase of the cooling-rate (~5580 s), similar to case-1. The agreement between the case-1 and the experiment is satisfactory.

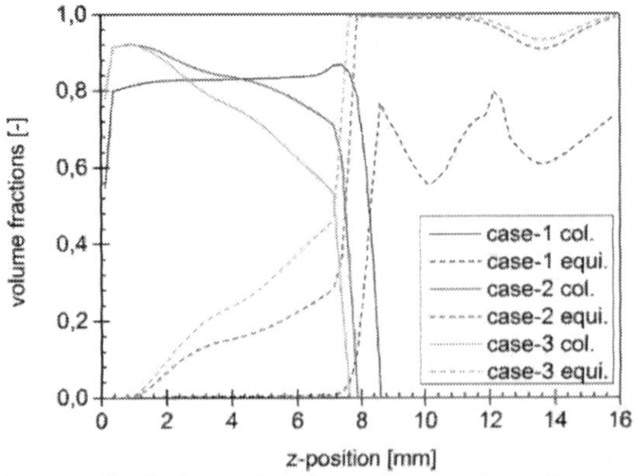

Figure 8. Volume fractions (here only columnar and equiaxed) for cases 1-3.

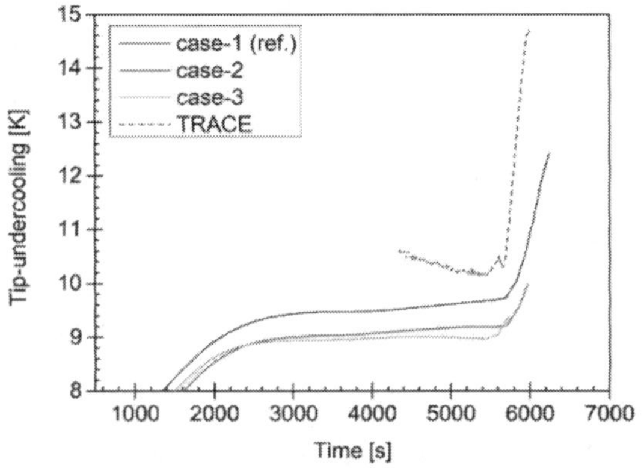

Figure 9. Front undercooling for cases 1-3 in comparison to experimental data.

Furthermore an oscillation in equiaxed volume fraction is achieved in case-1. Interestingly, such behaviour is also found in the experiment "TRACE+" with NPG-20.0wt.-%DC, flown on TEXUS-49 with the CET not fully achieved [17].

The evolution of the columnar and equiaxed volume fractions in time for case-1 are shown in figures 10&11 in log-scale. CET is achieved at t=6270 s and position z=8.4 mm. The significant increase in the equiaxed volume fraction and thus finally the CET is observable between t=5520 s and t=6270 s in the simulation. Experimentally we find a similar situation, showing columnar growth at t=5480 s in figure 12 and the CET happened at t=5990 s in figure 13. The experimental position of CET is at z=9.5 mm.

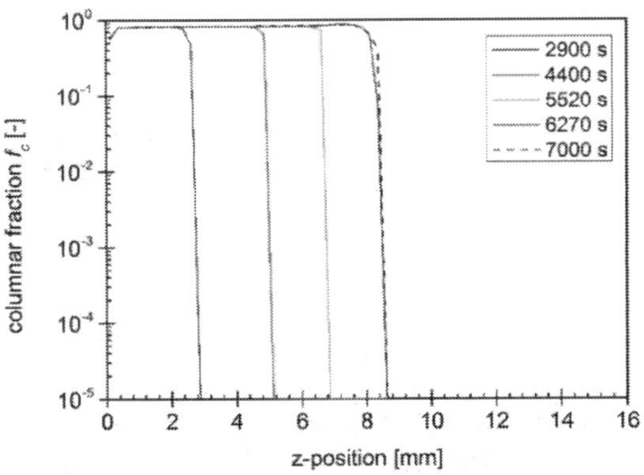

Figure 10. Evolution of columnar volume fraction in time for simulation case-1.

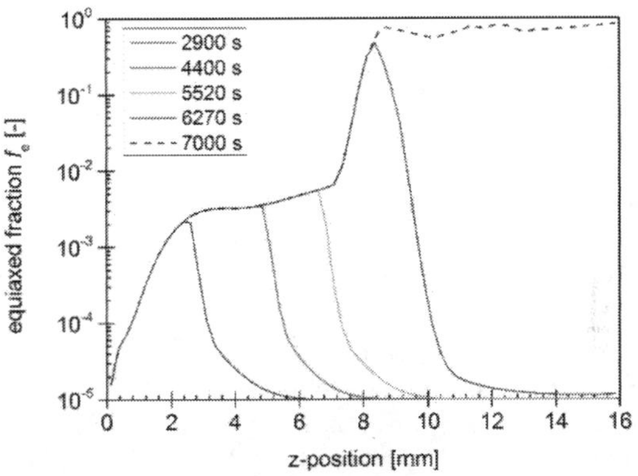

Figure 11. Evolution of equiaxed volume fraction in time for simulation case-1.

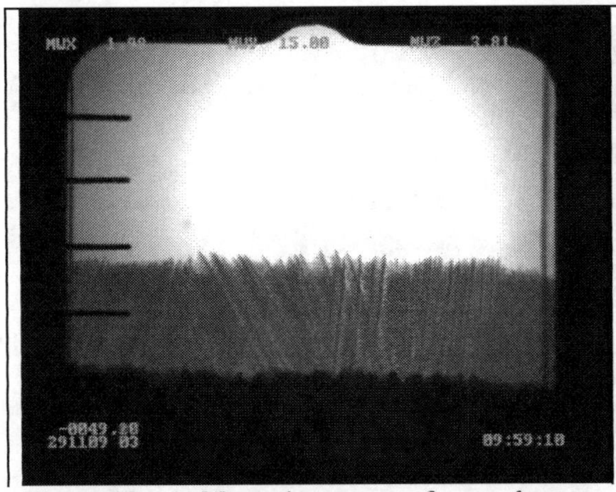

Figure 12. Macro-image of columnar solidification in "TRACE" at t=5480 s (50 s before rocket lift-off, 110 s before increase of cooling-rate=start of μg-time). The lower dark front corresponds to the eutectic temperature.

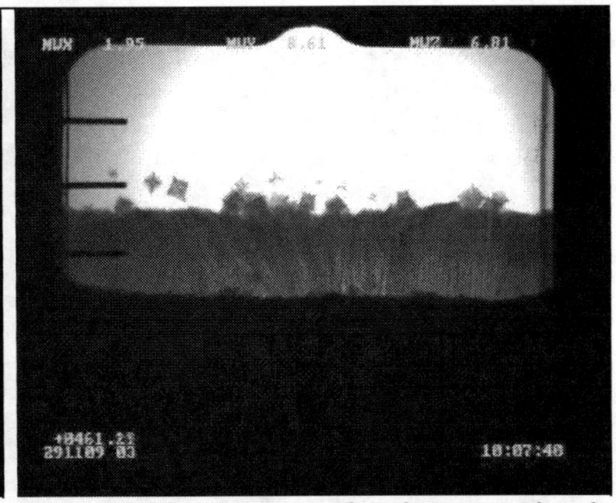

Figure 13. Macro-image of columnar-equiaxed solidification in "TRACE" at t=5990 s (10 s before end of μg-time). The lower dark front corresponds to the eutectic temperature.

4. Summary

Results from the low-gravity experiment "TRACE" investigating columnar and equiaxed dendritic solidification in the transparent alloy NPG-37.5wt.-%DC were compared to predictions from a 5-phase-volume averaging model. By adapting the columnar growth kinetics, morphological parameters, thermo-physical properties and nucleation parameters from experimental analysis a fairly good agreement was achieved for front undercooling and the position of CET in time and space. Further improvement is expected by adjusting heat transfer coefficients and the dendrite selection constant σ^*. For the later a four times higher value than determined experimentally was used and the CET-position was found to be at 8.4 mm (9.5 mm in experiment). The results from simulation case-2 show a later CET in space and time and a higher undercooling, when decreasing σ^* from the standard KGT/LGK-value of 0.0253 to 0.0076. A similar effect is observed when decreasing the maximum nucleation density (case-3).

Acknowledgement
This work is financially supported by the FWF Austrian Science Fund (P23155-N24) and the German BMBF (50WM0843).

References

[1] Spittle J A 2006 *Int. Mater. Rev.* **51** 247
[2] Reed R C 2006 *Superalloys: Fundamentals and Applications*. Cambridge University Press
[3] Sturz L and Zimmermann G 2011 *Proc. 20th ESA Symposium on European Rocket and Balloon Programme and Related Research* **SP-700** (Hyeres: France) 459
[4] Sturz L and Zimmermann G 2011 *J. Physics Conf. Series* **327** Article no. 012002
[5] Ahmadein M Wu M Sturz L Zimmermann G and Ludwig A , subm. to *IOP: Materials Science and Engineering*.
[6] Wu M and Ludwig A 2009 *Acta Mater.* **57** 5621.
[7] Wu M Fjeld A and Ludwig A 2010 *Comput. Mater. Sci.* **50** 32.
[8] Witusiewicz V T Sturz L Hecht U and Rex S 2004 *Acta Mat.* **52 (19)** 5519.
[9] Mangelinck-Noël N, Spinelli J E , Gandin Ch-A, Reinhart G, Nguyen-Thi H and B. Billia 2011 *Materials Science and Engineering* **27** 012090.
[10] Kurz W Giovanola B and R Trivedi 1986 *Acta Metall.* **34 (5)** 823.
[11] Lipton L, Glicksman M and Kurz W 1984 *Mater. Sci. Eng.* **65** 57.
[12] Hunt J D 1984 *Mater. Sci. Eng.* **65 (1)** 75.
[13] Martorano M A Beckermann C and Gandin Ch-A 2003 *Metall. Mater. Trans. A Phys. Metall. Mater. Sci.* **34A (8)** 1657.
[14] Witusiewicz V T Sturz L Hecht U and Rex S 2014 *Journal of Crystal Growth* **386** 69.
[15] Bayram U Aksöz S and Marasli N 2012 *Journal of Crystal Growth* **338** 181.
[16] Bayram U Aksöz S and Marasli N 2012 *Thermochimica Acta* **531** 12.
[17] Sturz L, Hamacher M and Zimmermann G 2013 Proc. 21th ESA Symposium on European Rocket and Balloon Programme and Related Research **SP-721** (Thun: Ch) 285.

IOP Conf. Series: Materials Science and Engineering **84** (2015) 012087 doi:10.1088/1757-899X/84/1/012087

Modelling of Columnar-to-Equiaxed and Equiaxed-to-Columnar Transitions in Ingots Using a Multiphase Model

N Leriche[1], H Combeau[1], Ch-A Gandin[2] and M Založnik[1]

[1] Université de Lorraine – Institut Jean Lamour, UMR CNRS 7198, 54011 Nancy, France
[2] Mines ParisTech – CEMEF, UMR CNRS 7635, 06904 Sophia Antipolis, France

E-mail: herve.combeau@univ-lorraine.fr

Abstract. We present a new method to handle a representative elementary volume (REV) with a mixture of columnar and equiaxed grains in ingot castings in the framework of an Eulerian volume averaged model. The multiscale model is based on a previously established fully equiaxed model. It consists of a three-phase (extra-granular liquid, intra-granular liquid and solid) grain-growth stage coupled with a two-phase (solid and liquid) macroscopic transport stage accounting for grain and nuclei movement.
In this context, we take into account the formation of a columnar structure and its development using a simplified front-tracking method. Columnar solidification is coupled with the growth of equiaxed grains ahead of the columnar front. The particularity of the model is the treatment of concurrent growth of mixed columnar and equiaxed structures only in the volumes that contain the columnar front. Everywhere else, the structure is considered either fully columnar or fully equiaxed. This feature allows for reasonable computational times even in industrial size castings, while describing the solutal and mechanical blocking phenomena responsible for the Columnar-to-Equiaxed Transition.
After a validation of the model, we discuss the numerical results for a 6.2-ton industrial steel ingot by comparison with experimental measurements. Final maps for macrosegregation and grain structures size and morphology are analysed. Furthermore, we quantify the impact of nuclei formation through fragmentation along the columnar front on the result.
An attempt at predicting the occurrence of the Equiaxed-to-Columnar Transition in the later phases of the process is also made.

1. Introduction

The following work is based on a multiphase volume-averaged model developed by Combeau and Založnik, which makes use of a numerical splitting technique [1,2]. Several studies of this research group about industrial ingots have already been presented concerning macro-segregation and grain morphology [3-5]. However, the columnar and equiaxed structures were only distinguished by defining a zone close to the mold where the grains were set to be fixed.

Efforts have recently been made to predict, thanks to multiphase models, the competition in industrial ingots between a columnar structure growing from the mold and transported equiaxed grains [6,7]. This multiphase model has been able to reproduce qualitatively the centerline macro-segregation in a 2.45-ton steel ingot. However, no comparison with CET observation was undertaken and the description of the equiaxed grain morphology remained simplified.

In the work presented thereafter, a complete model has been developed, including the growth of the columnar structures from the mold, their interaction with the equiaxed grains and especially the Columnar-to-Equiaxed-Transition (CET). The columnar grains can also grow from a packed equiaxed zone which leads to an Equiaxed-to-Columnar Transition (ECT).

The origin of the equiaxed grains in a steel ingot can be attributed to various causes including the detachment of grains formed in contact with the mold during filling or fragmentation of existing dendrites. This is the reason why different mechanisms for the formation of equiaxed grains have been investigated in the present model.

2. Description of the model

The present model is based on the fully coupled equiaxed model developed by Combeau and coworkers [1,2]. This model has been recently extended to take into account the presence of an inter-granular liquid [8] within the grain envelopes following the ideas of Rappaz and Thevoz [9]. Thus, the grain structure can be schematized as in Figure 1 where it was chosen to use a square envelope shape (ocathedral in 3D).

In this framework, the goal of this model is to take into account the apparition of columnar structures and their growth concurrently with the development of moving equiaxed grains. Based on previous models in the literature [10,11,12] we chose to use the same envelope shape and morphological parameters to model the columnar and equiaxed structures. In Figure 2, each columnar trunk is schematized by a regular assembly of sub-structures with a square envelope, arranged on a cubic lattice with characteristic length λ_1. The length $\ell(t)$ gives access to the propagation of the columnar front in the volume.

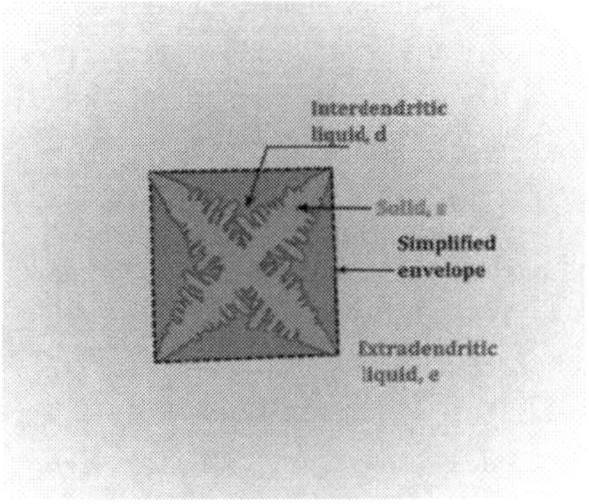

Figure 1. Schematic representation of a grain and its envelope together with the different phases.

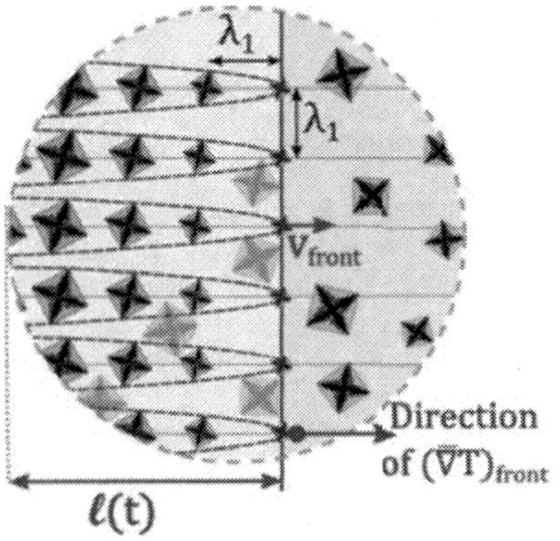

Figure 2. Schematic of a mixed volume containing both (blue, left) columnar and (red, right) equiaxed grains.

The main difference between the two types of structures lies in their origins. The columnar sub-structures are propagated by a front which initially originates at the mold and is connected to it, while the equiaxed grains take their origin in the dendrite fragments or nuclei. The position of the columnar front needs thus to be tracked during solidification. To this end, we make use of a method similar to the one developed by Ludwig and Wu [13]. Therefore, each representative volume can be located (i) behind the columnar front, (ii) contain the growing front or (iii) only contain liquid plus possibly equiaxed grains. In order to propagate the front further to neighboring volumes, a maximal length ℓ_{ref} of the front is defined in each averaging volume, similarly to the procedure in ref[13]. We considered that the direction of propagation of the columnar trunks is aligned with the local thermal gradient in the liquid ahead of the front $(\vec{\nabla}T)_{front}$. Consequently, ℓ_{ref} is simply the length of the volume along the direction of $(\vec{\nabla}T)_{front}$. The velocity v_{tip} of the front at the columnar tips as well as the growth velocities of both the equiaxed envelopes and the columnar "sub-envelopes" are all calculated according to the LGK model [14].

In our model, a mixed volume containing both columnar and equiaxed structures as the one in Figure 2 is made of six phases. Three phases are associated with the equiaxed grains: the volume fraction of the solid phase g_s^{eqx} as well as the inter- and extra-granular liquid fractions g_d^{eqx} and g_e^{eqx} respectively. Similarly, we can define the variables g_s^{col}, g_d^{col} and g_e^{col} for the columnar structures such that: $g_s^{eqx} + g_d^{eqx} + g_e^{eqx} + g_s^{col} + g_d^{col} + g_e^{col} = 1$. In order to simplify the calculations, we suppose that the equiaxed grains present in a mixed volume containing the columnar front are fixed and only the liquid phase is allowed to move in these volumes. This approximation is only made for a layer of mixed volumes. The movement of equiaxed grains is accounted for ahead of the columnar front where columnar grains are not present. In the volumes containing only equiaxed structures, two flow regimes are considered. Where the local grains fraction g_{grain}^{eqx} is larger than the packing limit g_{env}^{pack}, the solid phase is fixed. The flow of the intragranular liquid is then described by a momentum equation for porous media including a Darcy term for the drag interactions. The permeability of the porous solid is modeled by the Kozeny-Carman law. For local grains fractions smaller than the packing limit $g_{gain}^{eqx} < g_{env}^{pack}$, the solid phase is considered to be in the form of free-floating equiaxed grains.

In accordance to most volume-averaged models of the literature, two phenomena responsible for the Columnar-to-Equiaxed Transition (CET) are considered:

1) Solutal blocking of the front. Its description is intrinsic to our model. Indeed, when the equiaxed grains grow, they enrich the equiaxed extra-granular liquid and thus reduce the chemical undercooling used to calculate v_{tip}. This liquid medium becomes solutally well-mixed, in which case v_{tip} is negligible and the columnar tips are effectively blocked.

2) "Mechanical" blocking. It occurs when the equiaxed grain fraction g_{grain}^{eqx} in a mixed volume containing the front reaches a critical value g_{lim}. This parameter has been used in previous studies with values ranging from 0.2 [15] to 1 [16]. In the present work, we used $g_{lim} = 0.5$ according to the geometric criterion proposed by Hunt [17].

At the top of industrial steel ingots, it is common to observe an oriented dendritic structure [3]. It is believed that this structure originates from a packed equiaxed zone when the sources of nuclei and fragments have been depleted [4]. As a consequence, this phenomenon can be assimilated to the restart of a columnar structure during what we call here the Equiaxed-to-Columnar Transition (ECT). Previously, it was necessary to artificially "renucleate" new equiaxed grains to be able to solidify the top of the ingots even with a non-uniform nucleation law [1]. We propose here a more physical way to deal with these zones according to the ECT idea: when the remaining liquid in front of an equiaxed packed zone is emptied of grains and nuclei, a new columnar structure is initiated in the volumes directly adjacent to the packed zone. These columnar trunks will interact with further remaining equiaxed grains and this will possibly lead to a subsequent CET.

The sources of these equiaxed grains have already be mentioned. However, most of the models applied to industrial ingots use either a three-parameter model [13] or a simple uniform model [1] to simulate the volumic heterogeneous nucleation of $n_0[grains.m^{-3}]$ equiaxed grains at undercooling $\Delta T_n[K]$. Yet, it remains unclear if there are a significant number of nucleation sites in non-inoculated alloys [18]. Another potentially important source of grains is through fragmentation of the columnar dendrites, especially in industrial ingots [19]. That is why the goal of this work is also to study the effect of a surface injection model at the columnar front. During the growth of the front, we consider a constant surface flux of equiaxed grains φ_0 $[grains.m^{-2}s^{-1}]$ which will be added to the adjacent liquid volumes that do not contain columnar structures. Because the initial sizes of the fragments are unknown, they will all be fixed to 1 μm, i.e. the same value as for the grains formed by heterogeneous nucleation. The only condition for the surface injection is that the quantity of solid in the mushy columnar zone is greater than the solid formed by the grains being injected.

When modeling the volumic heterogeneous nucleation of the grains, we account for the movement of the nuclei and suppose that they move at the same velocity as the liquid.

3. Case definition and experimental parameters

The present study concerns a 6.2-ton steel ingot cast by Ascometal Industries [3]. The experimental ingot has a square cross-section and a slightly conical shape. In the framework of our model, we consider a simplified 2D axisymmetric geometry of equal mass which is shown in Figure 3. Details concerning the thermal boundary conditions, the thermo-physical properties as well as experimental measurement of the morphologies and macrosegregation can be found elsewhere [3-5].

The ingot exhibits along the mold a roughly uniform columnar zone, between 7 and 11 cm thick. The ingot was cast with a hot-top consisting of a 3 cm thick insulating material. Shortly after the beginning of solidification, an exothermic powder is released besides the hot-top and produces heat by a chemical reaction which lasts for 5 minutes. This exothermic reaction causes significant remelting in the hot-top region and is believed to promote the breakdown of the columnar structure by fragmentation. Therefore, we could expect a thinner columnar zone in the hot-top and probably very few columnar structures originating from an ECT scenario.

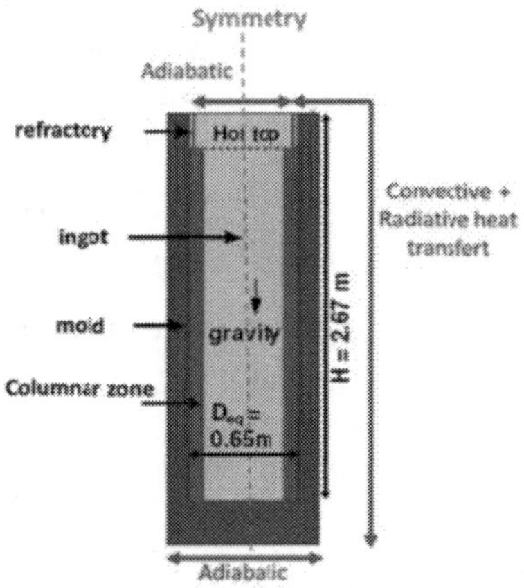

Figure 3. Geometry and boundary conditions for the 6.2-ton steel ingot.

Concerning the new parameter of the model, λ_1, a value of 1mm was chosen which corresponds to the average measured primary dendritic spacing in the columnar zone. It is noteworthy that the value of g_{lim} had only a very limited impact on the results in our case. The other parameters of the model are reported in a previous study on the same ingot [5]. Most notably the packing fraction for equiaxed grains was $g_{env}^{pack} = 0.4$, the alloy was simplified as a binary Fe-1.01 wt.%C and no pouring superheat was considered; i.e. the initial temperature of the liquid steel, T_0, is equal to the liquidus temperature $T_{liq}^0 = 1475\ °C$. The initial mold temperature is set to $25\ °C$.

The numerical domain was discretized into approximately 8600 cells for an average cell size of 1 cm. The time step is variable from $10^{-3}\ s$ to $0.1\ s$. The code is not parallelized and was run on a single Intel Xeon X5550 2.66 Ghz processor. The clock time for a typical simulation is about 60 hours.

4. Results and discussion

Results for volumic injections of equiaxed grains are showed in Figure 4. No fragmentation (surface injection at the columnar front) is considered and only a single class of nuclei n_0 in the range $[10^5; 10^9]$ $[nucl.m^{-3}]$ is activated at the liquidus temperature, i.e. with $\Delta T_N \approx 0\ K$. For comparison and to highlight the role of grains motion, simulation are carried out (wo) without and (w) with movement of the equiaxed grains. The black thick lines indicate the CET whereas the dashed zones bordered by thick white lines indicate columnar zones which have formed following an ECT event. The colour map indicates the equiaxed grain fraction g_{grain}^{eqx} in the whole ingot, characterizing notably the extent of the mixed columnar/equiaxed zones.

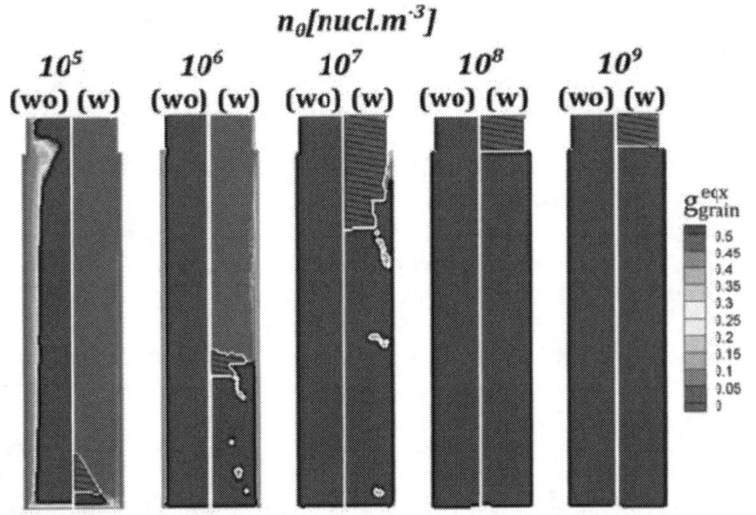

Figure 4. Predicted maps of the final equiaxed grain fraction g^{eqx}_{grain} as well as (thick black lines) the CET for increasing nuclei densities from 10^5 to 10^9 $[nucl.m^{-3}]$. We have assumed no nucleation undercooling. The columnar zones predicted following the ECT are shown as hatched zones bordered by white lines. For each case, the (wo) left map corresponds to fixed solid phases and the (w) right map to moving equiaxed grains.

Without accounting for grain motion and for values of n_0 greater than 10^7 $grains.m^{-3}$, the columnar zone is negligible and the ingot is predicted to have a fully equiaxed structure. This is shown in Figure 4 with $g^{eqx}_{grain} > 0{,}5$. For lower grain densities, we notice a roughly uniform columnar zone, except at the top of the ingot. For $n_0 = 10^5$ $grains.m^{-3}$, sweeping of colder yet not significantly enriched liquid at the bottom of the ingot near the beginning of solidification favours the growth of equiaxed grains. This explains why the front is blocked at the bottom first, leading to a thinner columnar zone. The columnar front is then blocked upwards in the ingot. In the hot-top region, complex thermal conditions and the interaction with the advancing front set up an enriched, cold liquid which is trapped in a clockwise convection cell. At the junction between the hot-top and the mold, this first liquid meets the relatively poorer and hotter liquid from the core of the ingot. Consequently, higher thermal gradients arise at this junction and favour the columnar structure which explains the wedge-like shape of the front there. Considering that without solid movement a computational volume cannot be emptied of equiaxed grains or nuclei, no ECT event is predicted in these cases. The macrosegregation profiles assuming a fixed solid are similar to previous results [5]. An example is shown in Figure 5. Here the average composition, $\bar{C}$, is normalized with the nominal composition C_0. The dashed curve represents the axial central segregation profile for $n_0 = 10^8$ $grains.m^{-3}$ without solid movement. Except for a negative segregation

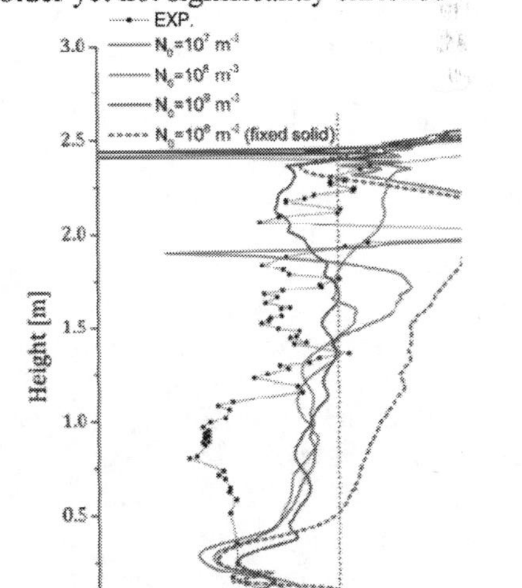

Figure 5. Comparison of the axial central segregation with measurements, where $C_{rel} = 100 \times \frac{(\bar{C}-C_0)}{C_0}$ [%].

at the bottom of the ingot and an inversion at the top, a strong positive segregation is predicted which is not in agreement with the experimental results. For other values of n_0, the results without solid movement are qualitatively similar. Especially, the position and height of the main positive segregation are the same, even though the intensity of this segregation can differ.

When taking into account the movement of the equiaxed grains, we notice a very dissimilar repartition of the structures in Figure 4. Starting from $n_0 = 10^6 grains.m^{-3}$, we see a clear distinction between a lower zone where the columnar thickness is very thin and an upper part which is entirely columnar. As n_0 increases, the equiaxed zone becomes larger and its boundary with the columnar zone moves upward. The top of the ingot always remains columnar. This can be explained by the fact that the motion of the equiaxed grains is controlled primarily by sedimentation. The grains first appear in the vicinity of the columnar front where the undercooling is maximal. Then, the grains grow rapidly and slide down along this region, thus participating to the formation a packed layer at the bottom of the ingot. Because of liquid convection, the whole ingot quickly becomes undercooled (except the top part) and the majority of the nuclei are activated. The newly formed grains directly fall at the bottom of the ingot and favour the growth of the packed layer. The local packing times (the times it takes for the grain in the equiaxed zone to become locally packed) are short and the sedimentation ends at $t \approx 300\ s$ in all cases. This value is to be compared with the total solidification time for the ingot, $t \approx 9000\ s$.After that, virtually no equiaxed grains remain in the liquid as evidenced by the negligible values of g_{grain}^{eqx} in the upper part of the ingot. Once the equiaxed grains have packed and nearly all the nuclei are activated, the columnar structure is free to develop in the remaining liquid. Remarkably, an ECT takes place at the top of the packed layer. The columnar structures originating from this ECT subsequently meet those growing from the mold. This is seen in Figure 4 with $n_0 = 10^6\ [nucl.m^{-3}]$ where a dashed blue region and a plain blue region are connected. The segregation profiles along the axis are shown in Figure 5 for three different densities of nuclei. The results are similar to previous calculations [5]. The best fit of the experimental results is found for $n_0 = 10^9\ nucl.m^{-3}$ but fails predicting the CET and is still not satisfying when considering the measured segregation profile.

The results presented in Figure 5 remain qualitatively valid when considering different nucleation undercoolings ΔT_N. A three parameter heterogeneous nucleation model [20], based on a Gaussian distribution, has also been tried. Comparable results were obtained, albeit with a more cone-shaped equiaxed zone.

It can be concluded that a volumic injection of nuclei fails to predict the observations. The volumic source of nuclei can be connected with heterogeneous nucleation or grains formed at the mold surface during the filling stage and then detached. It is already known that the fragmentation of dendritic columnar structures is another important source of equiaxed grains in industrial castings [19]. Thus, we investigate the influence of fragmentation on the results. The surface injection model mentioned earlier is now used for different flux values $\varphi_0\ [grains.m^{-2}s^{-1}]$ at the columnar front. The results are summarized in Figure 6 where no volume densities of nuclei were considered for these calculations. All calculations include grain movement as simulations with fixed solid do not produce realistic results. This stems from the fact that the fragments appear only in the liquid besides the front and need to be carried by the liquid to fill the centre of the domain. It is first noticeable that for $\varphi_0 = 10^2 grains.m^{-2}.s^{-1}$, the equiaxed zone has a much more pronounced cone shape than in previous volumic injection scenario.

The fraction of equiaxed grains entrapped in the columnar structures is also not negligible and the predicted structure can become considerably mixed especially at the top of the ingot, where remelting takes place. When the fragmentation flux is increased, a thin columnar zone is predicted in the lower part of the ingot whereas there is still a thick columnar zone in the upper part of the ingot. Nevertheless, the columnar zone at the bottom is not negligible with a CET at about 5 cm from the mold providing that $\varphi_0 \geq 4.10^2 grains.m^{-2}.s^{-1}$. The upper part is not entirely columnar in contrast with the case with the volumic density of nuclei scenario and the front is blocked at mid-radius. The columnar zone directly besides the hot-top is thinner because of the thermal conditions and the fact that the columnar structures can re-melt due to the re-heating induced by the exothermic material present in the hot-top [3].

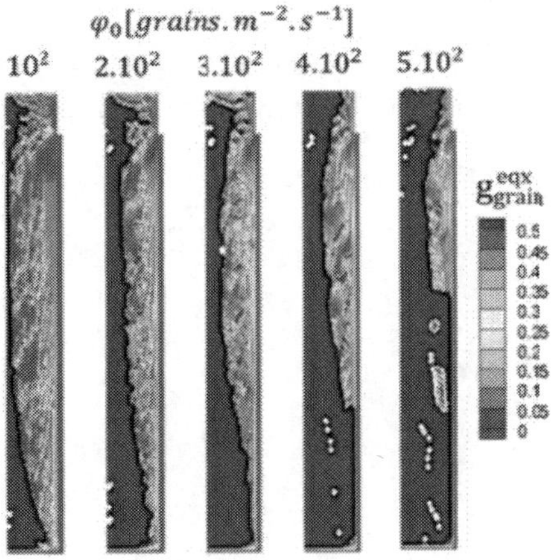

Figure 6. Predicted final equiaxed grain fraction g^{eqx}_{grain} as well as the CET and ECT for increasing fragmentation fluxes at the columnar front.

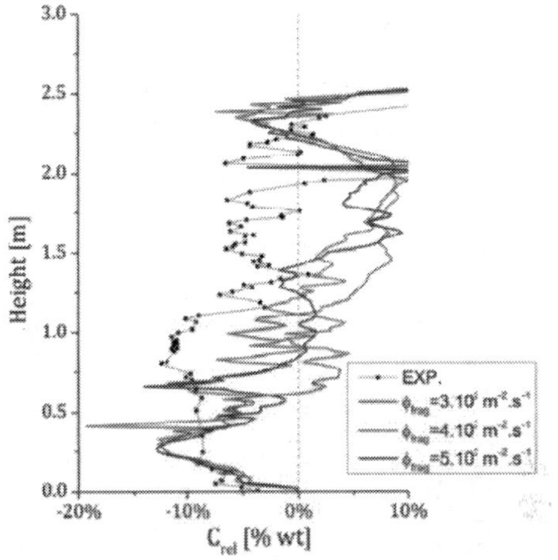

Figure 7. Comparison of the axial central segregation with the measurements where $C_{rel} = 100 \times \frac{(\bar{C}-C_0)}{C_0}$ [%]and for different fluxes.

The CET predicted which fits better the experiment can be explained in terms of packing behaviour of the equiaxed grains. As soon as the columnar structures appear at the mold, fragmentation takes place, producing new equiaxed grains which will at first instantly sediment along the columnar front. These first grains will not only contribute to the growth of a flat packed zone at the bottom, but also to the quick build-up of a vertical packed zone next to the front whose height is exactly that of the thin columnar region. This build-up takes place in about 200 s and causes the CET. The remaining fragments as well as the new ones coming from the front in the upper regions will then fill the rest of the ingot with a normal succession of horizontal layers. The whole ingot is filled with equiaxed grains in about 1300 s explaining why the columnar zone is thicker in the upper parts (except near the hot-top). As evidenced in Figure 6, there are few columnar structures originating from an ECT event contrary to the previous nuclei density model. Only dispersed spots of columnar zones can be seen in the core of the ingot.

The macrosegregation profiles at the centreline are reported in Figure 7 for the three highest fluxes. As we can see, the results at the bottom of the ingot are comparable to the calculations with the volumic nucleation model. In the top half of the ingot the results deviate from the experiment and a positive segregation is found. This corresponds roughly to the position where the columnar zone is too thick.

5. Summary and conclusions

Modelling of the columnar structure in ingot casting, its interaction with moving grains and the resulting CET and ECT have been implemented in a formerly purely equiaxed model [2].

Simulations were carried out for a 6.2-ton industrial steel ingot cast by Ascometal Industries. Two different origins of the equiaxed grains were studied: either a classical heterogeneous nucleation model or the injection of fragments in the liquid ahead of the columnar front. A parametric study for the nucleation densities suggests that this mechanism cannot accurately predict the experimental CET in steel ingots because the source of grain in the liquid are eventually exhausted and the top of the ingot exhibit a purely columnar structure. The use of a fragmentation mechanism seems to give better results for the structures repartition most notably at the top of the ingot.

Both type of injection scenario give comparable results for the axial segregation. For example, using $\varphi_0 = 5.10^2 grains.m^{-2}.s^{-1}$ yields final equiaxed densities in the central part between $n_{eqx} = 10^6$ and $n_{eqx} = 10^7 grains.m^{-3}$. We can see on Figures 5 and 7 that calculations for $\varphi_0 = 5.10^2 grains.m^{-2}.s^{-1}$ and $n_0 = 10^7 grains.m^{-3}$ indeed show similar profiles for the axial segregation. The main difference is noticeable in the top third of the ingot, where the predicted structures are very dissimilar.

It has been reported in the literature [21] that during the filling stage of the ingot, numerous grains grow at the contact of the mold, then become detached and survive in the liquid as fragments or inoculants for the heterogeneous nucleation law. This points to the fact that both mechanisms could operate at the same time. However, our calculations with both nucleation and fragmentation do not exhibit significantly better results at the moment. We can therefore conclude that more work needs to be done to better understand the relative importance of these two mechanisms.

References

[1] Combeau H, Založnik M, Hans S and Richy P E 2009 *Metall. Mater. Trans.* B **40** 289
[2] Založnik M and Combeau H 2010 *Comp. Mater. Sci.* **48** 1-10
[3] Kumar A, Demurger J, Wendenbaum J, Založnik M and Combeau H 2012 *International Conference on Ingot Casting,Rolling and Forging, Aachen, Germany*
[4] Combeau H, Kumar A, Založnik M, Poitrault I, Lacagne G, Gingell A, Mazet T, Lesoult G 2012 *International Conference on Ingot Casting,Rolling and Forging, Aachen, Germany*
[5] Kumar A, Založnik M, Combeau H 2010 *Int. J. Adv. Eng. Sci. Appl. Math.* **2(4)** 140-148
[6] Li J, Wu M, Ludwig A and Kharicha A 2014 *Int. J. Heat Mass Transfer* **72** 668
[7] Li J, Wu M, Ludwig A, Kharicha A and Schumacher P 2014 *Materials Science Forum* **790** 121
[8] Bedel M 2014 Doctoral Thesis, Université de Lorraine, Institut Jean Lamour, Nancy, France
[9] Rappaz M and Thévoz Ph. 1987 *Acta Metall.* **35** 1487-97
[10] Martorano M A, Beckermann C and Gandin Ch-A 2003 *Metall. Mater. Trans.* A **34** 1657
[11] Dagner J, Friedrich J and Müller G 2008 *Comp. Mater. Sci.* **43** 872
[12] Ciobanas A I and Fautrelle Y 2007 *Journal of physics D: Applied Physics* **40** 3733
[13] Ludwig A and Wu M 2005 *Mater. Sci. Eng.* A **413-414** 109-114
[14] Lipton J, Glicksman M E and Kurz W 1984 *Mater. Sci. Eng.* **65** 57-63
[15] Biscuola V B and Martorano M A 2008 *Metall. Mater. Trans.* A **39** 2885
[16] Mirihanage W U and Browne D J 2011 *Shape Casting: The 4th Int. Symp.held at the TMS 2011 Annual Meeting & Exhibition, San Diego, California, USA*
[17] Hunt J D 1984 *Mater. Sci. Eng.* **65** 75-83
[18] Morando R, Biloni H, Cole G S and Bolling G F 1970 *Metallurgical Transactions* **1** 1407
[19] Paradies C J, Smith R N and Glicksman M E 1997 *Metall. Mater. Trans.* A **28** 875
[20] Rappaz M 1989 *Inter. Mater. Rev.* **34** 93-123
[21] Chalmers B 1963 *Journal of the Australian Institute of Metal* **8** 255

IOP Conf. Series: Materials Science and Engineering **84** (2015) 012088 doi:10.1088/1757-899X/84/1/012088

Conditions for CET in a gamma TiAl alloy

R P Mooney[1], J Lapin[2], A Klimová[2] and S McFadden[1]

[1] Trinity College Dublin, Dublin 2, Ireland.
[2] Institute of Materials and Machine Mechanics, Slovak Academy of Sciences, Racianska 75, 831 02 Bratislava, Slovak Republic.

Email: Shaun.McFadden@tcd.ie

Abstract. The solidification of gamma TiAl alloys is of interest to the aerospace and automotive industries. A gamma TiAl multicomponent alloy: Ti–45.5Al–4.7Nb–0.2C–0.2B (at. %) has been the focus of a study to investigate the solidification conditions that led to a Columnar to Equiaxed Transition (CET) in a directional solidification experiment where traditional Bridgman solidification was combined in series with the power down method. In this paper, a numerical modelling result (a locus plot of columnar growth rate and temperature gradient) from this experiment is superimposed onto CET maps generated using an established analytical model for CET from the literature. A parametric study is carried out over suitable ranges of nucleation undercooling and nuclei density values. The predicted CET positions are compared with the experimentally measured CET position. Reasonable agreement is found at low levels of nuclei density. The paper concludes with estimates for the solidification conditions (nuclei density and nucleation undercooling) that led to the CET.

1. Introduction

Gamma titanium aluminide (γ–TiAl or TiAl) alloys are of interest to the aerospace and automotive industries on account of their low density and good mechanical properties at high temperature [1][2]. Applications for TiAl alloys include (in aircraft engines): low pressure turbine (LPT) blades, stator vanes, and radial diffusers; and (in motorcar engines): turbocharger rotors and valve trains [3]. Currently, a 2nd generation TiAl alloy is in service in the LPT of General Electric's 'GEnx' aero-engine which is used to power Boeing's 787 Dreamliner aircraft [4]. Rising fuel costs have motivated further development of 3rd generation TiAl alloys so that the maximum application temperature is increased from 700 °C (for 2nd generation TiAl) to 850 °C by using higher quantities of molybdenum and niobium [5].

The alloy: Ti–45.5Al–4.7Nb–0.2C–0.2B (at. %)—referred to hereafter as alloy 455—is part of this study to investigate Columnar to Equiaxed Transition (CET) in transient directional solidification experiments carried out in a Bridgman furnace where traditional Bridgman solidification was combined with the power down method [6]. CET refers to the growth of columnar dendrites in a casting that are subsequently blocked by the nucleation and growth of equiaxed dendrites. CET is an unwanted phenomenon in cast components since two distinct microstructures may lead to discontinuous mechanical properties.

Hunt [7] provides an analytical model for predicting CET in steady-state directional solidification. CET is predicted to occur at a critical temperature gradient, the calculation of which is based on dendrite growth kinetics and, ultimately, nucleation undercooling and the number of heterogeneous nuclei per unit volume in the alloy melt. The Hunt model can be used to create a 'CET map' of growth rate versus temperature gradient. Previous studies [8][9][10] have applied transient modelling data to Hunt CET maps to examine the location and solidification conditions that can lead to CET.

In this article, Hunt's analytical model is applied in a parametric study that examines the effect of various levels of nucleation undercooling and number of nuclei per unit volume on the critical temperature gradient for CET. The applied model incorporates the same growth kinetics as the numerical model used to simulate columnar growth in the power down experiment that yielded a CET in alloy 455. The numerical model result, specifically, the locus of growth rate and temperature gradient at the columnar front, is superimposed on CET maps determined using the Hunt model. The intersection of the locus plot with the CET map provides an estimate for the position of CET. Suitable combinations of nucleation undercooling and nuclei density that satisfy the experimentally measured CET position are then determined.

2. Methodology

2.1. Directional solidification experiments using a gamma TiAl alloy

Lapin et al. [6] provide a detailed description of directional solidification experiments carried out in a Bridgman furnace using alloy 455. A brief description of these experiments is provided here.

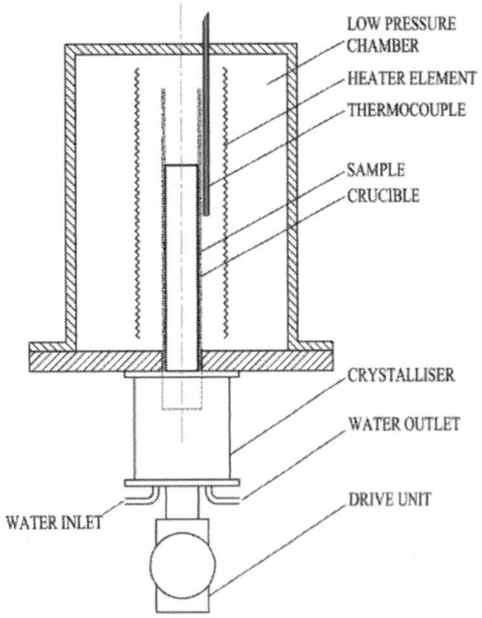

Figure 1. Schematic of the Bridgman furnace apparatus used in the power down experiment.

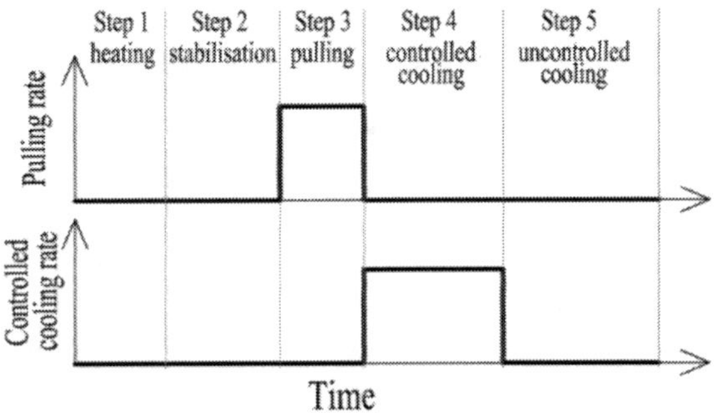

Figure 2. Typical pulling rate and controlled cooling rate signals plotted against time for each step in the experiment procedure.

Figure 1 shows a schematic drawing of the furnace arrangement. Figure 2 shows a plot of pulling rate and controlled cooling rate versus time for each step of the experimental procedure. In these experiments, traditional Bridgman solidification was combined in series with the power down method. The cylindrical sample (150 mm in length and 10 mm in diameter) was initially located in the furnace as shown in figure 1, i.e., its base was aligned with the top of the water-cooled crystalliser. The sample was heated (in Step 1) until the thermocouple read 1720 °C, and allowed to stabilize for 300 s (in Step 2). To induce directional columnar growth, the sample was pulled a distance of 20 mm over at time of 72 s into the crystalliser (in Step 3, Bridgman solidification). Next, controlled cooling began (in Step 4, power down method) at a predefined cooling rate, as measured by the thermocouple, until the entire sample was solidified. The sample was then allowed to cool to room temperature without control (in Step 5). Four experiments were carried out at controlled cooling rates of 15, 20, 30 and 50 °C/min. The results for the sample cooled at 30 °C/min are of interest in this paper.

2.2. Numerical model (for columnar growth)

A hybrid one-dimensional thermal model of solidification, known as the Bridgman Furnace Front Tracking Model (BFFTM) [11], is applied. This model uses an explicit finite difference control volume (CV) method to solve the following heat equation (adapted from [12] and [13]) for temperature in a cylindrical rod of cross-sectional area A, and perimeter p, moving in the axial (x) direction at a pulling rate u,

$$\frac{\partial}{\partial t}(\rho c T) = \frac{\partial}{\partial x}\left(k\frac{\partial T}{\partial x}\right) - u\rho c\frac{\partial T}{\partial x} - \frac{hp}{A}(T - T_\infty) + E \quad , \tag{1}$$

where ρ, c and k are the density, specific heat capacity and thermal conductivity of the rod material, respectively; T_∞ is the temperature of the surrounding heat source (or sink), h is the heat transfer coefficient at the circumference of the rod, and E is the latent heat generated per unit volume. The CVs are disc-shaped so that the numerical domain fully encompasses the sample. In this way, radial heat flow from the heat source (in the hot zone) or to the heat sink (in the cold zone) is accounted for. The heat transfer coefficients were estimated as a function of axial position through a dedicated characterization study [14] of the same furnace used in the experiments described above.

The growth of columnar grains in alloy 455 is estimated using the dendrite growth model of Kurz et al. [15] (KGT model) using the liquidus slope and partition coefficient of a constitutionally similar binary alloy, namely: Ti–46at.%Al, (after Rebow et al. [16]). Using this approach, the following power law curve fit relates columnar growth rate (v_{tip}) to columnar dendrite tip undercooling (ΔT_{tip}): $v_{tip}=C(\Delta T_{tip})^b$, where $C = 2.63\times10^{-6}$ m/s·°C^b and $b = 2.79$. Latent heat release is calculated based on a Thermocalc [17] prediction for solid fraction versus temperature in the full multicomponent alloy, details of which can be found in reference [14]. The solidification conditions at the columnar dendrite tip can be estimated using the model, for example, columnar tip growth rate, temperature and temperature gradient. Since the front position is tracked and the thermal profile is known, it is also possible to estimate the extent of constitutionally undercooled liquid ahead of the front, i.e., the difference between the front position and the position of the equilibrium liquidus temperature isotherm.

2.3. Analytical model (for CET)

Hunt [7] presents a 1-dimensional analytical model that predicts columnar, equiaxed or mixed microstructure for steady-state directional solidification using the Bridgman method. An analysis is presented that uses an intuitive approach to calculate the extended volume fraction of equiaxed grains. The model predicts 'fully equiaxed' growth when,

$$G_{tip} < 0.617 N_0^{1/3}\left\{1 - \frac{(\Delta T_n)^3}{(\Delta T_{tip})^3}\right\}\Delta T_{tip} \quad , \tag{2}$$

where G_{tip} is the temperature gradient at the columnar dendrite tip, N_o is the nuclei density, i.e., the total number of available heterogeneous substrate particles per unit volume, ΔT_n is the nucleation undercooling, and ΔT_{tip} is the columnar dendrite tip undercooling. In Hunt's analysis, ΔT_{tip} is calculated via an empirical function of growth rate, v_{tip}, after Burden and Hunt [18]. The above expression is derived from an assumption based on a probabilistic analysis of columnar and equiaxed grain geometry, where 'fully equiaxed' growth corresponds to an extended volume fraction of equiaxed grains equal to 0.66, and 'fully columnar' growth corresponding to volume fraction of equiaxed grains equal to 1% of this value. The results of the analysis are presented in plots of columnar growth rate versus temperature gradient (known as CET maps), showing the predicted equiaxed, mixed (equiaxed and columnar), and fully columnar regions for a given alloy composition, nuclei density and nucleation undercooling. Hunt concludes the analysis stating that, at low growth rates equiaxed growth depends on the efficiency of grain refiners, and that, at high temperature gradients the number of nucleation sites is more important.

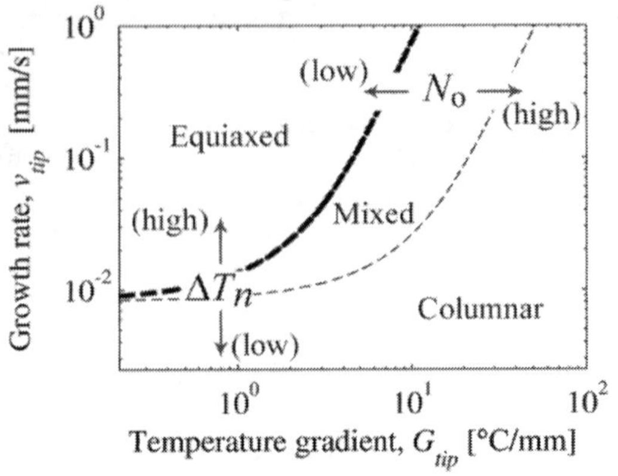

Figure 3. A CET map produced using the Hunt model [7]; the effect of using high and low values for nucleation undercooling (ΔT_n) and nuclei density (N_o) is shown.

This result is illustrated in the CET map shown in figure 3. The thick dashed line is a plot of G_{tip}, calculated from equation (2), over a range of growth rates, while the thin dashed line is a plot of G_{tip} for the 'fully columnar' case. Any combination of growth rate and temperature gradient in the area above the thick line is expected to produce fully equiaxed microstructure and, similarly, any combination of these values in the area below the thin line should produce a fully columnar microstructure. In between these lines a mixed equiaxed and columnar microstructure is expected. High values of N_o tend to shift the plots to the right and low values shift the plots to the left; whereas high values of ΔT_n shift the y-axis intercept of each plot upwards and low values shift the y-axis intercept downwards. The above described criterion for CET has become known as *mechanical* blocking [19] and is currently the most widely used in the literature [20].

3. Results

In a previous study by Mooney et al. [21], optical microscopy was used to reveal the primary β-phase (dendritic) microstructure in the four samples from the directional solidification experiments described in Section 2.1. A CET was discovered in the sample cooled at a rate of 30 °C/min. A microstructural analysis (for all samples), along with numerical modelling results using the model described in Section 2.2, is provided.

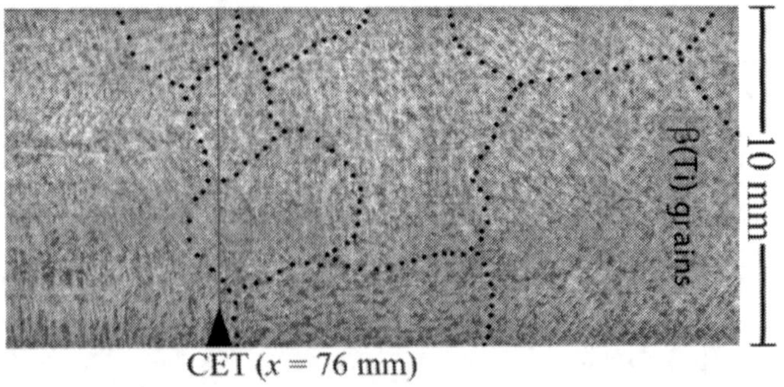

Figure 4. Section image of the CET (at x = 76 mm) given by the primary β-phase microstructure from the sample cooled at 30 °C/min in directional solidification experiments [6] using alloy 455, reproduced from a study by Mooney et al. [21].

For the purposes of this article, we are concerned with the microstructural analysis, specifically, the CET position, and numerical modelling results from reference [21] for the sample that contained the CET only. Figure 4 shows a section image of the sample that contained the CET. The CET was measured at an axial position of 76 mm.

3.1. Numerical modelling results

Figure 5 shows a plot—obtained using the numerical BFFTM—of the locus of growth rate and temperature gradient at the columnar front, with annotations showing the position of the columnar front in the sample. In this plot, the local peak (at a front position of approximately 25 mm) corresponds to the end of the Bridgman stage of the procedure (step 3) and the beginning of power down solidification stage (step 4). The simulation is valid up to the position of CET (at 76 mm) since only columnar growth is simulated. At this location the simulated growth rate was 0.128 mm/s, the temperature gradient was 1.089 °C/mm, and the tip undercooling was 4 °C.

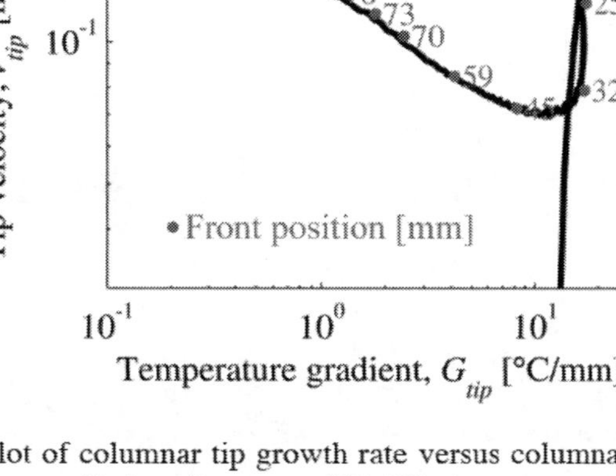

Figure 5. Simulated locus plot of columnar tip growth rate versus columnar tip temperature gradient for the sample solidified at a controlled cooling rate of 30 °C/min.

3.2. Analytical modelling results

CET maps were produced using the Hunt model for a range of combinations of nucleation undercooling values (from 0 °C to 4 °C) and nuclei densities (within a range from 0.01 /mm^3 to 100 /mm^3); 45 CET datasets were produced in total. The KGT dendrite growth kinetics model for alloy 455 was used to relate the columnar dendrite tip growth rate to ΔT_{tip} in the Hunt model. In effect a

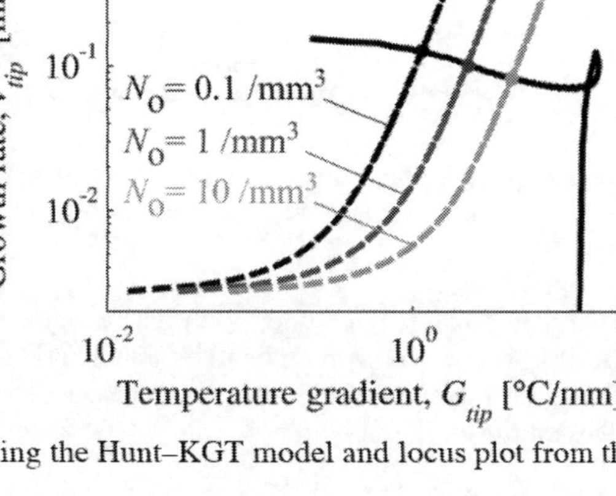

Figure 6 CET maps using the Hunt–KGT model and locus plot from the numerical model.

IOP Conf. Series: Materials Science and Engineering **84** (2015) 012088 doi:10.1088/1757-899X/84/1/012088

'Hunt–KGT' model is implemented.

Figure 6 shows three CET curves {given by equation (2) with ΔT_n = 1 °C} where the nucleation density varies between three levels: N_o=0.1, 1 and 10 /mm^3. The numerically estimated locus of temperature gradient, G_{tip}, and growth rate, v_{tip}, at the columnar front, from figure 5, is superimposed onto figure 6. The intersection of the G_{tip}–v_{tip} locus with a CET curve provides a viable CET prediction based on the specific values for the parameters ΔT_n and N_o. Table 1 shows the results from this exercise for all combinations ΔT_n and N_o in the range of values given above.

Table 1. Predicted front positions for CET (in mm), given by the intersection of the locus of temperature gradient and growth rate at the columnar dendrite tip with the Hunt–KGT CET map, for various nucleation undercooling and nuclei density values.

N_o [/mm^3]	Nucleation undercooling, ΔT_n [°C]								
	0	0.5	1.0	1.5	2.0	2.5	3.0	3.5	4.0
0.01	78.7	78.7	78.8	78.8	79.0	79.3	79.7	80.2	–
0.1	75.7	75.7	75.8	76.0	76.4	77.0	77.8	78.9	80.3
1	69.2	69.2	69.5	70.6	71.0	72.6	74.6	76.9	79.4
10	55.0	55.1	55.6	56.8	59.5	64.0	69.1	74.0	78.3
100	41.7	41.8	42.2	43.3	45.7	50.3	59.8	70.5	77.4

(We note that in the simulations detailed in reference [21], the columnar dendrite tip undercooling did not exceed 4 °C before the CET occurred; therefore we expect ΔT_n to be limited by this value.) Figure 7 shows the predicted CET positions from Table 1 plotted against nucleation undercooling for each nuclei density value; after Martorano et al. [22].

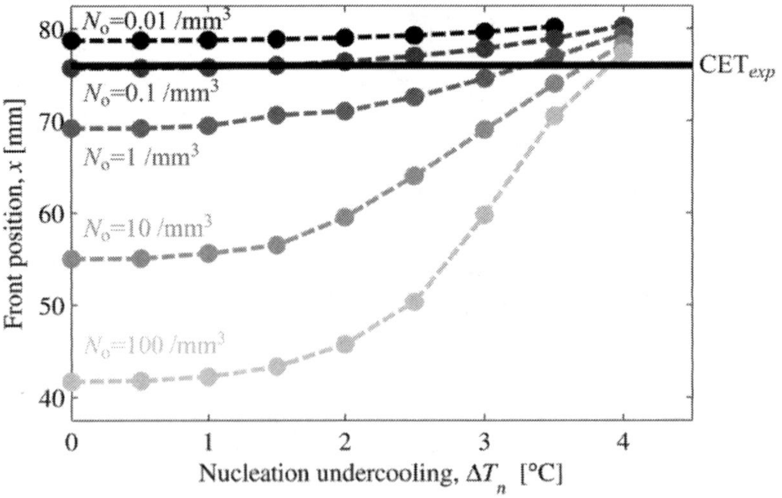

Figure 7. Predicted position of CET (dashed lines) using the Hunt–KGT model at various nucleation undercoolings and nuclei densities, and the measured CET position from the power down experiment.

4. Discussion

Figure 7 summarises the results in table 1. Each data point was calculated at the point of intersection of the BFFTM simulated G_{tip}–v_{tip} locus and the analytically calculated Hunt–KGT curve. In addition to these estimated CET locations, the measured CET location is shown at x = 76mm. At a value of N_o = 0.1 /mm^3 there is agreement with the experimentally measured CET position in the range of nucleation undercoolings between 0 °C to 2 °C. At the lower value of N_o = 0.01 /mm^3, the predicted CET is slightly overestimated for all nucleation undercoolings up to 4 °C. At higher values of N_o (N_o = 1 /mm^3 and greater) there is agreement with the measured CET position, however, agreement occurs at specific values of nucleation undercooling, for example, with N_o=10 /mm^3 agreement occurs ΔT_n=3.7 °C. In any case, we can practically rule out high values for N_o since the alloy used in the experiment was not deliberately inoculated.

Following these results, a nuclei density of 0.1 /mm^3 is a reasonable estimate for alloy 455, as it suggests that nucleation undercooling was of the order of 2 °C. The Hunt CET maps for a nucleation density of 0.1 /mm^3, over a range of nucleation undercoolings, are shown in figure 8 (a) and (b). Exact agreement is found when ΔT_n has a value of 1.5 °C.

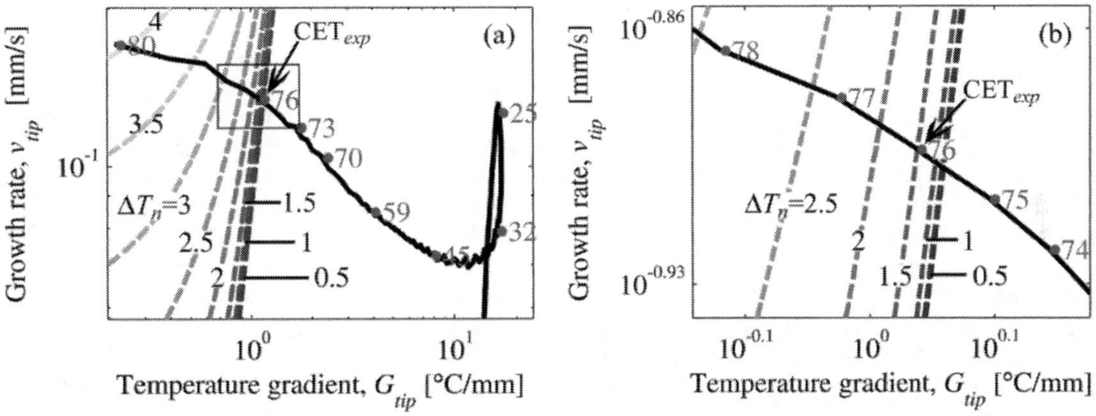

Figure 8. (a) CET maps (dashed lines) using the Hunt–KGT model where the nuclei density is set to 0.1 /mm^3 at various nucleation undercooling values, and superimposed locus plot (solid line) from the numerical model with the front position values (grey text) shown in mm. A portion (black rectangle) of this plot is enlarged in part (b) of this figure.

The BFFTM is appropriate for low Biot number (Bi) scenarios, specifically, where Bi<0.1. The Biot number was 0.03 or less in the simulations presented here. This means that the temperature at the center of the rod was very nearly equal to the temperature at its circumference. The numerical model does not treat convection in the melt. Since we are using a vertical Bridgman furnace, the density gradient is parallel with the gravity vector and, therefore, has a stabilising effect on the melt. Any solutal gradient in the melt that may have a destabilising effect is not treated in the current model. Further discussion on the assumptions of the numerical model can be found in reference [14].

5. Conclusions

The performance of an experimental apparatus was simulated using a numerical model and the simulated results were compared to those from an analytical model for CET. The CET position was measured from the final microstructure and a reasonable estimate of the nucleation parameters was obtained. The main outcomes from this study are outlined below:

- A combined Bridgman, power-down experiment was conducted on a gamma TiAl alloy [6] and a CET was observed when the cooling rate in the power-down phase was 30 °C/min.
- A Bridgman Furnace Front Tracking Model was applied to the experiment and this allowed the transient thermal history to be predicted by simulation.
- Hunt's analytical model for CET [7] with KGT dendritic growth kinetics was applied.

- Using the Hunt–KGT model, a parameters variation analysis focused on a range of nucleation undercooling (ΔT_n=0 °C to 4 °C) and nuclei density (N_o=0.01 /mm^3 to 100 /mm^3) values.
- The locus of growth rate and temperature gradient, given by the numerical simulation, was superimposed onto the CET curves from the Hunt–KGT model. The intersection of the locus with individual CET curves gave the predicted position of CET for each combination of ΔT_n and N_o.
- The predicted CET positions were compared with the experimentally measured CET location.
- A nuclei density of the order of 0.1 /mm^3 gave reasonable agreement between the experiment and model results over a range of nucleation undercooling from 0 °C to 2 °C.

Acknowledgments
This work was carried out as part of the GRADECET (GRAvity DEpendence of Columnar to Equiaxed Transition in TiAl Alloys) research project. The authors (R. P. Mooney and S. McFadden) thank the European Space Agency PRODEX programme for funding the research (agreement number: 4000107132) under the management of the Irish Space Delegation at Enterprise Ireland. The authors (J. Lapin and A. Klimová) acknowledge the financial support of the Slovak Grant Agency for Science under the contract VEGA 2/0149/13 and the Slovak Academy of Sciences under the contract of MVTS funding of ESA project GRADECET. Finally, the authors would like to acknowledge U. Hecht and L. Sturz of ACCESS e.V, Aachen, for their interesting and helpful comments on this paper.

References
[1] Loria E A 2000 *Intermetallics* **8** 9–11 1339–1345
[2] Dimiduk D M 1999 *Mater. Sci. Eng. A* **263** 2 281–288
[3] Appel F, Paul J D H and Oehring M 2011 *Gamma Titanium Aluminide Alloys: Science and Technology*
[4] Bewlay B P, Weimer M, Kelly T, Suzuki A and Subramanian P R 2013 *MRS Proc.* **1516** 49–57
[5] Clemens H and Smarsly W 2011 *Adv. Mater. Res.* **278** 551–556
[6] Lapin J, Gabalcová Z, Hecht U, Mooney R P and McFadden S 2014 *Mater. Sci. Forum* **790–791** 193–198
[7] Hunt J D 1984 *Mater. Sci. Eng.* **65** 1 75–83
[8] Kartavykh A V, Ganina S, Grothe D, Lemoisson F and Herfs W 2010 *Mater. Sci. Forum* **649** 223–228
[9] McFadden S, Browne D J, Sturz L and Zimmermann G 2010 *Mater. Sci. Forum* **649** 361–366
[10] Murphy A G, Reinhart G, Nguyen-Thi H, Abou Jaoude G S and Browne D J 2013 *J. Alloys Compd.* **573** 170–176
[11] Mooney R P, McFadden S, Rebow M and Browne D J 2012 *Trans. Indian Inst. Met.* **65** 6 527–530
[12] Carslaw H S and Jaeger J C 1959 *Conduction of Heat in Solids* 2nd ed.
[13] Bejan A 1993 *Heat Transfer* 1st ed.
[14] Mooney R P, McFadden S, Gabalcová Z and Lapin J 2014 *Appl. Therm. Eng.* **67** 1–2 61–71
[15] Kurz W, Giovanola B and Trivedi R 1986 *Acta Metall.* **34** 5 823–830
[16] Rebow M, Browne D J and Fautrelle Y 2010 *Mater. Sci. Forum* **649** 243–248
[17] Thermo-Calc Software AB 20132013. [Online]. Available: http://www.thermocalc.com/Thermo-Calc.htm
[18] Burden M H and Hunt J D 1974 *J. Cryst. Growth* **22** 2 99–108
[19] Dong H B and Lee P D 2005 *Acta Mater.* **53** 3 659–668
[20] Spittle J A 2006 *Int. Mater. Rev.* **51** 4 247–269
[21] R P Mooney U Hecht Z Gabalcová J Lapin and S McFadden 2015 *Kov. Mater.* **53** 3 187
[22] Martorano M A, Beckermann C and Gandin C- A 2003 *Metall. Mater. Trans. A Phys. Metall. Mater. Sci.* **34 A** 8 1657–1674

IOP Conf. Series: Materials Science and Engineering **84** (2015) 012089 doi:10.1088/1757-899X/84/1/012089

An Efficient 3D Stochastic Model for Predicting the Columnar-to-Equiaxed Transition in Alloy 718

L. Nastac

The University of Alabama, Tuscaloosa, AL, 35487, USA

email: lnastac@eng.ua.edu

Abstract. A three-dimensional (3D) stochastic model for simulating the evolution of dendritic crystals during the solidification of alloys was developed. The model includes time-dependent computations for temperature distribution, solute redistribution in the liquid and solid phases, curvature, and growth anisotropy. The 3D model can run on PCs with reasonable amount of RAM and CPU time. 3D stochastic mesoscopic simulations at the dendrite tip length scale were performed to simulate the evolution of the columnar-to-equiaxed transition in alloy 718. Comparisons between simulated microstructures and segregation patterns obtained with 2D and 3D stochastic models are also presented.

1. Introduction

A two-dimensional (2D) stochastic model has been proposed for the modelling of the evolution of dendritic morphologies during solidification [1]. The significance of the stochastic modeling approach is that the evolution of simulated micro- structures can be directly visualized and compared with the actual microstructures from experiments at two different scales: dendrite grain characteristics such as grain size and location and size of the columnar-to-equiaxed transition can be visualized at the meso-scale, while dendrite morphology (including dendrite tip, various dendrite arm spacings, microsegregation patterns) can be viewed at the microscale.

One of the challenges associated with stochastic modeling is reduction in mesh anisotropy. Basically, because of mesh anisotropy, dendrites will grow aligned with the axis of the mesh or at 45 degrees independently of the initial crystallographic orientation, which can be called anisotropy in growth direction. Furthermore, dendrites aligned with the axis would have narrower tips than predicted by classic theories, which can be termed anisotropy in growth kinetics. To reduce the mesh anisotropy, a formulation to correct the curvature in 2D was proposed and successfully validated [2]. A detailed discussion about correcting the grid anisotropy using the phase field approach is presented in [3].

Detailed comparison of predictions with measured microstructures requires the 2D stochastic model to be extended to 3D. This is why several 3D models for simulating dendritic growth have been proposed in the literature including phase field approach [3, 4] and Lattice Boltzmann and Cellular Automaton approach [5]. To be efficient, these models require parallel computations and can be used only for relatively small domains. A 3D Cellular Automaton approach for modeling the evolution of dendritic grains has also been developed [6]. This 3D dendritic grain model is relatively fast but it cannot simulate the evolution of dendrites and of microsegregation patterns at the dendrite tip level.

The main goal of the current contribution is to utilize an efficient 3D stochastic mesoscale dendritic nucleation and growth model, which is based on the recently developed 3D stochastic mesoscale model presented in [12], to predict the columnar-to-equiaxed transition in alloy 718. The 3D model includes computations for temperature, solute, curvature and anisotropy at the dendrite tip level. Another goal of the paper is to compare the 2D and the 3D predictions in terms of microstructure evolution (including the columnar-to-equiaxed transition) and segregation patterns during the solidification of alloy 718.

2. Description of the Stochastic Dendritic Solidification Model

The mathematical representation of the dendritic solidification process of a binary alloy is considered in a restricted 3-D domain (***Ω***). A cross section of this 3D domain is shown in Fig. 1. Here, $\vec{n}$ is the interface normal vector, $\overline{K}$ is the mean curvature of the interface, and the curve ***Γ*** represents the solid/liquid (S/L) interface which evolves in time and has to be found as part of the solution.

The solidification of binary alloys is governed by the evolution of the temperature ($T(x, y, z, t)$) and concentration ($C(x, y, z, t)$) fields that have to satisfy several boundary conditions at the moving S/L interface as well as the imposed initial and boundary conditions on the computational domain. The equations that describe the physics of the solidification process in 3-D Cartesian coordinates are as follows (see the 2D equations in [1, 6-8]).

- Temperature (*T*) in ***Ω*** (heat transfer equation):

$$\rho c_p \frac{\partial T}{\partial t} = \frac{\partial}{\partial x}\left(K \frac{\partial T}{\partial x} \right) + \frac{\partial}{\partial y}\left(K \frac{\partial T}{\partial y} \right) + \frac{\partial}{\partial z}\left(K \frac{\partial T}{\partial z} \right) - \rho L \frac{\partial f_L}{\partial t} \quad \text{with } f_L = 1 - f_S \qquad (1)$$

where t is time, ρ is the density, K is the thermal conductivity, c_p is the specific heat, L is the latent heat of solidification, f_L is the liquid fraction, f_S is the solid fraction, and x, y and z are the domain coordinates.

- Concentration (*C*) in ***Ω*** (solute diffusion equation):

In the liquid phase (C_L): $$\frac{\partial C_L}{\partial t} = \frac{\partial}{\partial x}\left(D_L \frac{\partial C_L}{\partial x} \right) + \frac{\partial}{\partial y}\left(D_L \frac{\partial C_L}{\partial y} \right) + \frac{\partial}{\partial z}\left(D_L \frac{\partial C_L}{\partial z} \right) \qquad (2)$$

In the solid phase (C_S): $$\frac{\partial C_S}{\partial t} = \frac{\partial}{\partial x}\left(D_S \frac{\partial C_S}{\partial x} \right) + \frac{\partial}{\partial y}\left(D_S \frac{\partial C_S}{\partial y} \right) + \frac{\partial}{\partial z}\left(D_S \frac{\partial C_S}{\partial z} \right) \qquad (3)$$

where D_L and D_S are the interdiffusion coefficients in the liquid and solid, respectively.

Local equilibrium at the S/L interface on $\Gamma(t)$ (here, "*" means at interface):

$$C_S^* = k\, C_L^* \qquad (4)$$

IOP Conf. Series: Materials Science and Engineering **84** (2015) 012089 doi:10.1088/1757-899X/84/1/012089

- Solute conservation at the S/L interface:

$$V_n^* \, C_L^* \,(k-1) = \left(-D_L \left(\frac{\partial C_L}{\partial x} + \frac{\partial C_L}{\partial y} + \frac{\partial C_L}{\partial z} \right) + D_S \left(\frac{\partial C_S}{\partial x} + \frac{\partial C_S}{\partial y} + \frac{\partial C_S}{\partial z} \right) \right) \cdot \vec{n} \tag{5}$$

where V_n^* is the normal velocity of the interface and $\vec{n}$ denotes the normal to the S/L interface that is pointing into the liquid (see Fig. 1).

- The interface temperature (T^*) is defined as (assuming local equilibrium with both phases):

$$T^* = T_L^{EQ} + \left(C_L^* - C_0\right) m_L - \Gamma \, \overline{K} \, f(\varphi, \theta) \tag{6}$$

where T_L^{EQ} is the equilibrium liquidus temperature of the alloy, m_L is the liquidus slope, $\overline{\boldsymbol{K}}$ is the mean curvature of the S/L interface, Γ is the Gibbs-Thomson coefficient, and $f(\varphi,\theta)$ is a coefficient that accounts for growth anisotropy, where φ is the growth angle (*i.e.*, the angle between the normal and the growth axis) and θ is the crystallographic orientation angle.

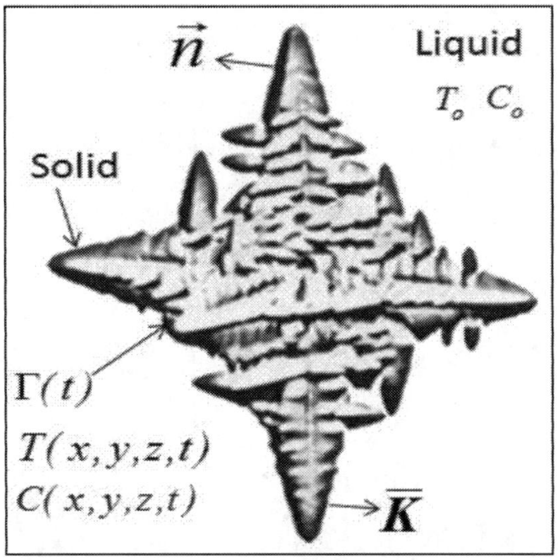

Figure 1. A 3D domain for dendritic solidification

In Eq. (6), second term in the right side is the constitutional undercooling and the last term in the right side is the curvature undercooling. The interface temperature is also affected by the kinetic undercooling. The kinetic undercooling is not accounted for in this model since its effect becomes significant only at very high solidification velocities (*i.e.*, in the rapid solidification regime). Also, the coefficient $f(\varphi,\theta)$ in Equation (6) assumes an axisymmetric approximation where the anisotropy and the interface shape are independent of the polar angle ϕ in the x-y plane perpendicular to the growth axis [2].

The solidification process is governed by Eqs. (1) to (6) and a stochastic model for nucleation and growth. The numerical procedures for calculating the nucleation and growth, temperature and concentration fields as well as the growth velocity of the S/L interface are described in details in [9]. It consists of a regular network of cells that resembles the geometry of interest. The model is characterized by (a) geometry of the cell; (b) state of the cell; (c) neighborhood configuration; and (d) several transition rules that determine the state of the cell. In this work, the geometry of the cell is a cube. Each cell has three possible states: "liquid", "interface", or "solid". The selected neighborhood configuration is based on the cubic von Neumann's definition of neighborhood, that is the first order configuration and it contains the first six nearest neighbors. Solidification behavior depends to a great extent on the transition rules. In this model, the change of state of the cells from "liquid" to "interface" to "solid" is initiated either by nucleation or by growth of the dendrites.

An explicit finite difference scheme is used for calculating the concentration fields in the liquid and solid phases. Zero-flux boundary conditions were used for cells located at the surface of the geometry. The solution algorithm includes the "interface" cells by multiplying the concentration in the liquid by the liquid fraction and the concentration in the solid by the solid fraction of the particular interface cell. Also, during each time-step calculation and for each "interface" cell, the previous values of the liquid and solid concentrations are updated to the current values of the interface liquid and solid concentrations calculated with Eqs. (4), (5) and (7). The calculation of the interface liquid concentration, C_L^* can be obtained from Eq. (6) as

$$C_L^* = C_O + \frac{\left(T^* - T_L^{EQ} + \Gamma \overline{K} f(\varphi,\theta)\right)}{m_L} \tag{7}$$

The procedures for calculating $\overline{K}$ and $f(\varphi,\theta)$ are described below.

The average interface curvature for a cell with the solid fraction f_S is calculated with the following expression:

$$\overline{K} = \frac{1}{a}\left(1 - 2\frac{f_S + \sum_{k=1}^{N} f_S(k)}{N+1}\right) \tag{8}$$

where N is the number of neighboring cells and a is the mesh size. In the present 3D calculations, N = 26, that contains all the first order neighboring cells (including the diagonal cells). Equation (8) is a simple counting-cell technique that approximates the mean geometrical curvature (and not the local geometrical curvature). An improved curvature model is presented in [8]. A comparison of 2D simulations obtained with these curvature models was also provided in [8], based on the work presented in [10, 11].
The anisotropy of the surface tension (see Eq. (6)) is calculated as [2]:

$$f(\varphi,\theta) = 1 + \frac{4\delta}{1-3\delta}\left(\cos^4(\varphi-\theta) + \frac{3}{4}\sin^4(\varphi-\theta)\right) \quad \text{with } \varphi = a\cos\left(\frac{V_z}{\left[(V_x)^2 + (V_y)^2 + (V_z)^2\right]^{1/2}}\right) \tag{9}$$

where φ is the angle between the normal direction to the solid-liquid interface and the [100] direction (growth axis), V is computed with Eq. (12) in [12], θ is calculated with Eq. (7) in [12] and δ accounts for the degree of anisotropy. For cubic symmetry, δ = 0.047 [2].

The algorithm is presented in [12]. The simulation software was written in Visual Fortran 90. The model output consists of screen plotting at any chosen time of *C*, *T*, or color indexes of all cells. The values of *C*, *T*, and color indexes of all cells are also saved at selected times on the computer disk.

3. Simulation Results and Discussion

Thermo-physical properties of the alloys used in simulations are presented in Table 1. The domain size is 10mmx30mm in 2D and 10mmx30mmx2mm in 3D. Newton cooling boundary condition was applied only to the bottom of the computational domain. All other boundaries were perfectly insulated.

The surface heat transfer coefficient used in the present simulations is $h = 3.0\text{x}10^3$ W m^{-2} K^{-1}. Zero-flux solute boundary conditions were applied at the boundaries of the computational domain (*i.e.*, a closed system was assumed). The initial melt temperature was 1400 °C. Also, an initial concentration equal to C_o was assumed everywhere on the computational domain.

Table 1. Thermophysical properties of IN718-5Nb used in simulations [9, 13].

Property	μ_N^C [m^{-2} K^{-2}]	μ_N^E [m^{-3} K^{-2}]	L [J/kg]	ρ [kg/m^3]	K [W/m/K]	c_p [J/kg/K]	T_L [°C]
Value	1x10^6	5x10^8	2.9x10^5	7620	30.1	720	1336
Property	C_o [wt.%]	C_{eut} [wt.%]	k_o -	D_L [m^2 s^{-1}]	D_S [m^2 s^{-1}]	m_L [°C %$^{-1}$]	Γ [K m]
Value	5.0	19.1	0.48	3x10^{-9}	1x10^{-12}	-10.5	3.65x10^{-7}

In Fig. 2, a comparison of 2D and middle cross sections of 3D simulated microstructures during the unidirectional solidification of alloy 718-5 wt. % Nb is presented. The legend in Fig. 2 shows 256 color indexes (*CI*) (shown here in 16 classes, where each class contains 16 different colors), which are used for displaying the preferential crystallographic orientation angle.

The 2D computations were done by using the 2D stochastic dendritic model presented in [1, 7-9]. The competition between nucleation and growth of multiple columnar dendrites is evident in both 2D and 3D cases. The strong growth competition from the sample bottom (>30 dendrites) to 1/4 of the sample height (16 dendrites in 2D and 24 dendrites in 3D) to 1/2 of the sample height (7 dendrites in 2D and 10 dendrites in 3D) to 3/4 of the sample height (5 dendrites in 2D and 7 dendrites in 3D) can be observed. Note also that the columnar morphologies are slightly different in 3D than in 2D.

Figure 2 also shows a comparison between 2D and 3D computations in terms of columnar-to-equiaxed transition (CET). The competition between nucleation and growth of both equiaxed and columnar morphologies can be clearly shown in Fig. 2. It can also be seen from Fig. 2 that the CET starts at an earlier location in 3D than in 2D. The full transition also occurs earlier in 3D than in 2D (about 5/6 of the sample height in 3D and about 11/12 of the sample height in 2D). This is because (i) for a 2D geometry the grain growth is confined in the x-y plane while for the 3D case it is free in the third (z) direction; (ii) more nucleation sites are available in 3D than in 2D and (iii) some dendrites that are visualized in the middle plane of the 3D simulation are cut from a different plane.

Figure 3 shows a comparison of segregation patterns of 2D and cross sections of 3D simulated columnar dendritic morphologies. It can be observed that the segregation values are higher in 2D than in 3D in both liquid and solid phases.

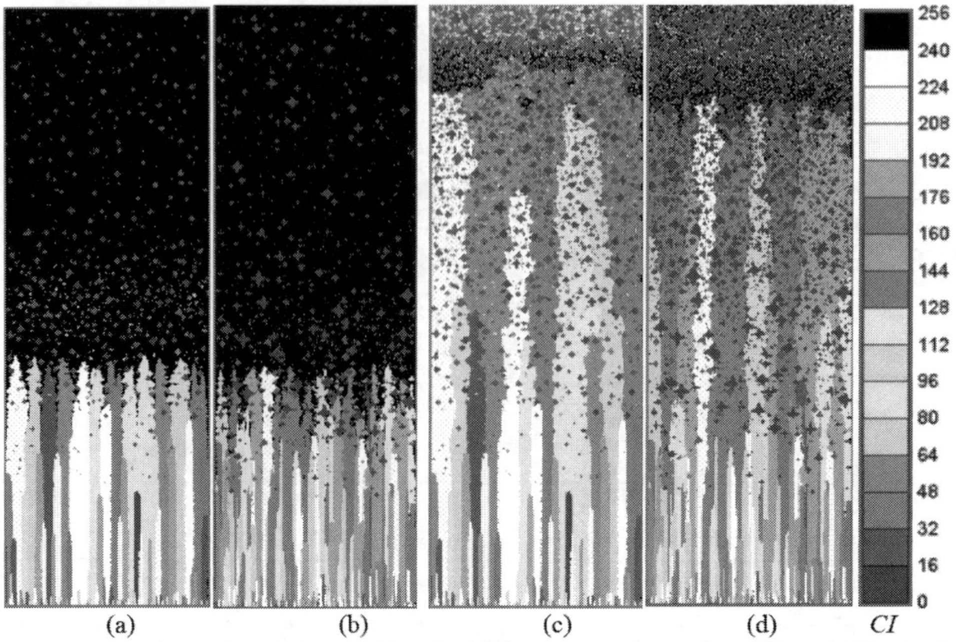

Figure 2. Comparison of 2D ((a) and (c)) and middle cross sections of 3D ((b) and (d)) simulated microstructures during the unidirectional solidification of alloy 718-5 wt. % Nb

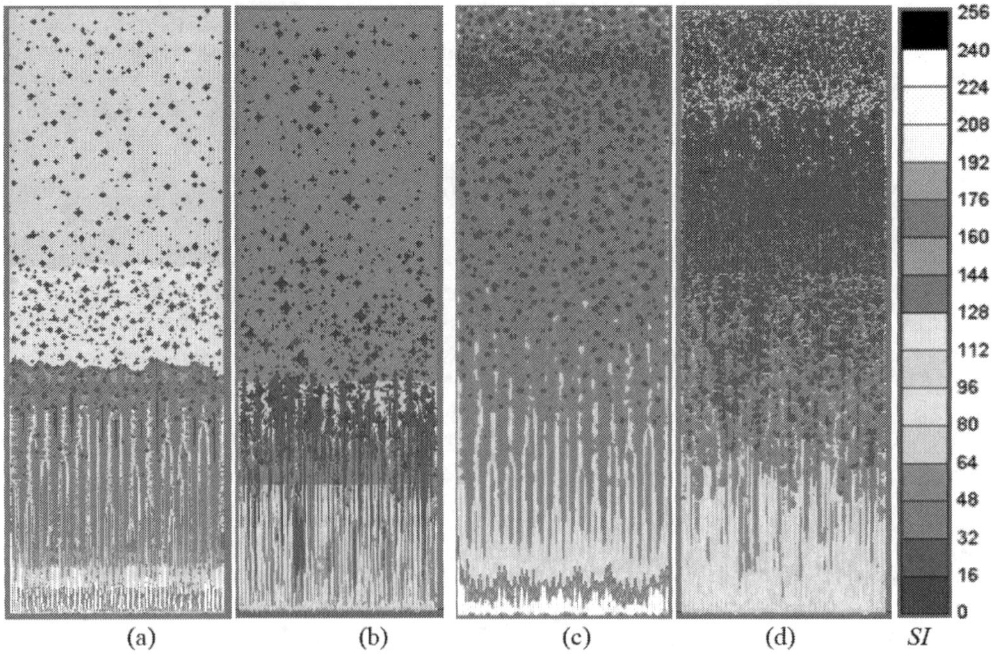

Figure 3. Comparison of 2D ((a) and (c)) and middle cross sections of 3D ((b) and (d)) simulated segregation patterns during the unidirectional solidification of alloy 718-5 wt. % Nb

The Nb segregation index (*SI*) shown in the legend in Fig. 3 is computed as follows:

$$SI = \frac{C_{av} - k * C_o}{C_{eut} - k * C_o} * 255 \quad \text{with} \qquad C_{av} = C_S f_S + C_L f_L \tag{10}$$

Slightly higher segregation values are observed in 2D than in 3D in both liquid and solid phases. This is because the microstructure is finer in 3D than in 2D.

In both Figures 2 and 3, the equiaxed grain density was $fx10^8$ nuclei/m^2 in 2D and 10^{12} nuclei/m^2 in 3D. The conversion factor between 2D and 3D is $f = \pi/4$ [14].

A RAM memory size of 100 bytes/cell is needed in the present computations. CPU-time is related to the domain size and the time-step. The 3D simulation problems shown in Figs. 2 and 3 took less than 1 day on a Dell Precision T7500 desktop.

4. Conclusions

An efficient 3D mesoscale stochastic model was applied to study the evolution of the microstructures including CET and the segregation patterns in alloy 718 - 5wt. % Nb.

Based on the current simulations, it was determined the following:

(i) The 3D columnar dendrites are finer than the 2D ones;
(ii) Segregation values are higher in 2D than in 3D in both liquid and solid phases;
(iii) The CET in 3D starts at an earlier location than in 2D and the full transition occurs earlier in 3D than in 2D;
(iv) The CET length is larger in 2D than in 3D.

References

[1] Nastac L 1999 Numerical Modeling of Solidification Morphologies and Segregation Patterns in Cast Dendritic Alloys, *Acta Met* **47**(17) 4253-62.
[2] Karma A, and Rappel WJ 1998 *Phys. Rev. E* **57**(4) 4323–4.
[3] Jeong JH, Goldenfeld N, Dantzig JA 2001 *Phys. Rev.* E **64** (041602) 1-14.
[4] Eshraghi M, Felicelli SD, Jelinek B 2012, *Journal of Crystal Growth* **354**(1) 129-34.
[5] Gandin CA, Rappaz M 1997 *Acta Met* **45**(5) 2187-95.
[6] Nastac L 1999 A New Stochastic Approach for Simulation of Solidification Morphologies and of Segregation Patterns in Cast Dendritic Alloys, In 4th Pacific Rim International Conference on Modeling of Casting and Solidification Processes (MCSP-4), Ed. Hong CP, Yonsei University, Seoul, Korea.
[7] Nastac L 2000 A Stochastic Approach for Simulation of Solidification Morphologies and Segregation Patterns in Cast Alloys. In Proceedings of the Modelling of Casting, Welding, and Advanced Solidification Processes IX. Engineering Foundation. Aachen, Germany.
[8] Nastac L 2004 Modeling and Simulation of Microstructure Evolution in Solidifying Alloys, Springer New York (ISBN 978-1-4020-7831-6).
[9] Nastac L, Stefanescu DM 1997 *Modelling and Simulation in Materials Science and Engineering* Institute of Physics Publishing **5**(4) 391-420.
[10] Kothe DB, Mjolsness RC, Torrey MD 1991 RIPPLE, A Computer Program for Incompressible Flows with free surfaces, Los Alamos National Lab, LA-10612-MS, Los Alamos, NM, USA.
[11] Chen S, Merriman B, Osher S, Smereka P 1997 *J. Comp. Physics* **135** 8-29.
[12] Nastac L 2014 *Journal of Metallurgical Research & Technology*, EDP Sciences. **111**, 311-319 (DOI: 10.1051/metal/2014016).
[13] Nastac L, Stefanescu DM 1997 *Metallurgical Transactions* **28A** 1582-7.
[14] Kong M, Bhattacharyal RN, James C, Basu A 2005 *Geological Society of America Bulletin* **117**(1-2) 244-9.

IOP Conf. Series: Materials Science and Engineering **84** (2015) 012090 doi:10.1088/1757-899X/84/1/012090

Numerical modelling and in-situ radiographic study of the grain nucleation and growth of inoculated aluminum alloys

Yanjun Li[1], Yijiang Xu [1], Qiang Du[2], Lars Arnberg[1], Ragnvald Mathiesen[3]

[1] Department of Materials Science and Engineering, NTNU, 7491 Trondheim, Norway
[2] SINTEF Materials and Chemistry, Trondheim, Norway
[3] Department of Physics, NTNU, 7491 Trondheim Norway

E-mail: Yanjun.li@ntnu.no

Abstract. To precisely predict the grain size of inoculated aluminium alloy castings has been a big challenge for the researchers in the field of solidification and casting. Up to date, most of grain size prediction models are based on the Free Growth Model, in which the nucleation process is stopped by recalescence. In a previous work [1], we have proposed a new grain size prediction model applicable for solidification of castings without recalescence. In the present work, an in-situ X-ray radiographic study on the grain nucleation and grain growth of inoculated Al-Cu alloys during isothermal melt solidification has been carried out, where the effect of melt convection is minimized. The influences of inoculant particles and cooling rate on the nucleation rate, grain growth rate and final grain size have been quantitatively studied. A comparison between the prediction results and the experimental results is presented.

1. Introduction

Grain refinement by inoculation is an important technique to improve the soundness of aluminium castings. Much research effort has been made in the last decades to develop analytical and numerical models to simulate the heterogeneous nucleation and growth of equiaxed grains in aluminium alloys during solidification [2-11]. The first numerical grain size prediction model for isothermal melt solidification of aluminium alloys has been attributed to Hellawell and Maxwell [2]. In 2000, Greer et al. proposed the Free Growth model [5]. The main progress of this model , in comparison to the Hellawell & Maxwell (H&M) model and other earlier solidification models, is that the heterogeneous nucleation is treated as a deterministic, instead of stochastic, process, where the size of potent nucleant particles was directly related to the undercooling needed for the initiation of free-growing grains. This makes the evaluation of the grain refinement efficiency of grain refiners with different size distributions possible. Similar to the H&M model, the heterogeneous nucleation of grain stops due to the recalescence. The model has been used to simulate the isothermal melt solidification and the simulated grain size is in a good agreement with the measured grain size of standard grain refining test (TP-1) samples, which have an approximately isothermal melt condition during solidification. However, the solidification of most real castings is far away from isothermal melt solidification and recalescence can not always be observed, which makes the application of the Free Growth Model has been limited to the TP-1 type castings. To overcome this limit, in a previous work[1], we proposed a new grain size prediction model based on a new nucleation stopping mechanism, solute stifling mechanism. Such a solute stifling mechanism has been discussed by Shu et al.[10], and Quested et al.[7] but in a less critical way. Although based on isothermal melt solidification, our model has been validated with a DC-casting 5182 alloy, showing a reasonable agreement with experimental results. With the aim to make a more precise evaluation of the model, an isothermal melt solidification experiment has been designed and carried out by using a novel in-situ X-ray radiographic approach.

2. Experimental

The experimental alloy used in this work is Al–10 wt.% Cu alloy prepared by using 5N purity aluminium and 4N purity Cu. After melting, the alloys grain refined with different amounts of Al5Ti1B grain refiners were cast in copper molds of 50mm in diameter and 150mm in height. The cast ingot were then machined and mechanically polished into the final sample dimensions of 50 x 5 x 0.2 ± 0.05 mm. No further heat treatments or coatings were applied to the samples prior to installation into the crucible assembly.

The solidification furnace used for this experiment was a bespoke Bridgman-type gradient furnace (GF), manufactured by the Swedish Space Corporation (SSC) in cooperation with the European Space Agency (ESA) Applications Microgravity Promotion (MAP). By a careful temperature control, desirable temperature gradient can be realized in the sample melt during solidification. The polished thin alloy samples capsuled by glassy carbon are positioned in a sample holder device, which is designed specifically to maximize heat flow through the sample. In this work, the thin plate samples are melted and solidified in a horizontal position, namely, the wide surface of the sample is perpendicular to the gravity direction, which could to a large extent reduce the influence of convection. Constant cooling rates in the range of 0.025~1 K/s were applied for the solidification process. For all the samples, the cooling process started after the samples were completely melted and heated up to 670 °C. An in- situ micro focus X-ray monitoring system was used to record the microstructure evolution during solidification of the castings. The frame capture rate throughout these experiments was 1.5 Hz. A more detailed description of the furnace and experimental procedure can be found elsewhere [12].

3. Model description

The grain size prediction model used in this work is based on the as-cast grain size prediction model reported in [1]. The same free grain growth criterion and the same growth velocity equation of grains (assumed as spherical growth) solely controlled by solute diffusion are applied. The diffusion field surrounding each growing solid grain is at quasi-steady state, namely the concentration profile of the solute keeps the same shape during grain growth.

$$C(r) = C_0 + \frac{R_g}{r}(C^* - C_0) \qquad \text{(Eq. 1)}$$

where C_0 is the initial alloy composition, C(r) is the solute concentration in the liquid with a distance of r to the center of the growing grain, C* is the liquid concentration at the surface of growing grain and R_g is the radius of the grain and R_0 the radius of inoculant particle. The same size distribution function of inoculation particles adopted by Greer et al., [5] is used in the present model.

$$PSD(d) = N_V \frac{\exp\left(-\frac{d}{d_0}\right)}{d_0} \qquad \text{(Eq.2)}$$

where d is the diameter of disc-shaped inoculant particle , d_0 is the characteristic width of the distribution ($d_0 = 0.72\ \mu m$) and N_v is the total number of TiB_2 particles per unit volume of aluminum melt when Al-5Ti-1B grain refiners are added. N_v was calculated according to the empirical rule [5]: for 1 Part Per Thousand (PPT) by weight Al-5Ti-1B addition, N_v is equal to $1.15\times10^{12} m^{-3}$. No tuning was carried out in the present work.

The nucleation process stops due to the solute stifling due to the of the solute segregation layer surrounding growing grains in the melt.

$$\dot{T} - \dot{T}^l(\overline{X}_i^l) = 0 \qquad \text{(Eq.3)}$$

Where $\dot{T}$ is the temperature changing rate, $\overline{X}_i^l$ is the average concentration of element i in the residual liquid of the metal and $\dot{T}^l(\overline{X}_i^l)$ is the change rate of the liquidus of the bulk melt.

4. Results and discussion

Fig. 1 shows the grain structure evolution of Al-10%Cu alloy grain refined by 0.5PPT (0.05wt %) Al5-Ti-1B grain refiners during the in-situ radiographic solidification at two different cooling rates.

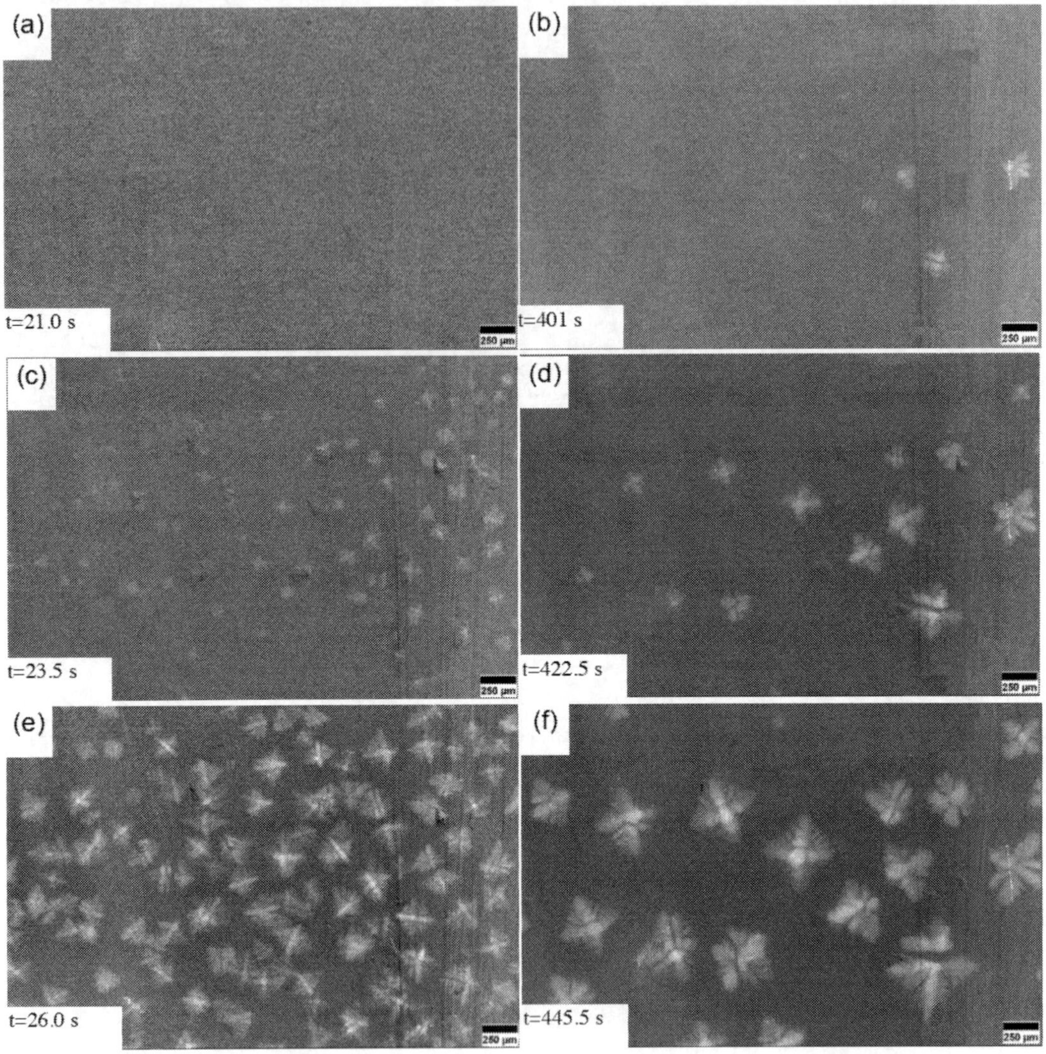

Fig. 1. In-situ X-ray radiographic images showing the grain structure development of Al-10Cu alloy inoculated by 0.5 PPT (part per thousand) Al-5Ti-1B grain refiners, with two different cooling rates, 0.5K/s for images at the left side and 0.025 K/s for the images at the right side. The images were recorded when the temperature of the alloy melt reached 645 °C.

As can be seen, in both cases, equiaxed grains form and grow in the samples and the number of grains of increase with increasing cooling time (temperature decreasing). It has to be noted that the very early stage of the grain growth could not be observed due to the resolution limit. By a careful observation, it can be seen that equiaxed grains appear slightly earlier in the right side than in the left side of the samples, indicating that a small temperature gradient exists in the sample melt. The temperature gradient is measured as about 0.2 K/mm. It is interesting to see that the cooling rate has a strong influence on the heterogeneous nucleation of the grains in the alloy. The final number density of grains in the sample solidified with a cooling rate of 0.5 K/s (Fig. 1e) is much higher than that cooled with a cooling rate of 0.025 K/s (Fig. 1f).

The final grain size of Al-10wt%Cu alloy inoculated by different amount of grain refiners under different cooling rate has been measured. The measured grain size as a function of cooling rate is plotted in Fig. 2. As can be seen, the grain size decreases with increasing cooling rate and increasing addition level of grain refiners. As a comparison, the calculated grain size by the grain size prediction model is also presented in the figure. It is interesting to see that the predicted grain sizes have a good agreement with the experimental results in trend, although the absolute quantity of the predicted grain size is slightly lower. If considering that no tuning was done with the size distribution of TiB_2 particles in the master alloy and it is assumed 100% yield of the added inoculant particles, the prediction results of the model is satisfying. It confirms that our grain size prediction model based on the nucleation stopping by solute stifling mechanism is feasible.

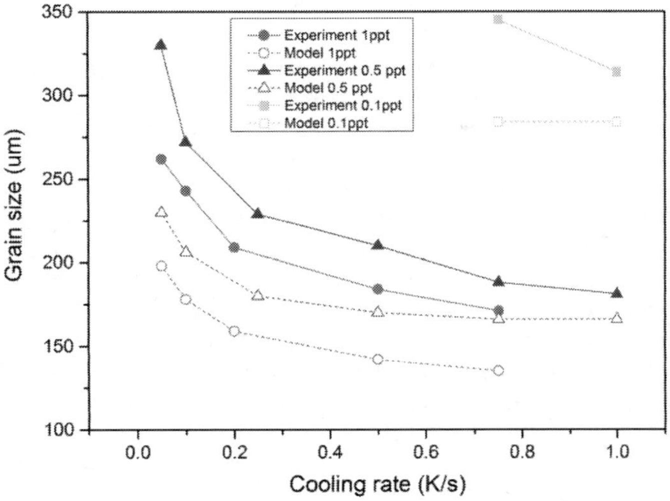

Fig. 2 The measured grain size (based on the final number density of grains) and the predicted grain size of Al-10Cu alloy inoculated by two different level of Al-5T-1B master alloy, 0.1 and 0.5 ppt (part per thousand) under different cooling rates.

By using the in-situ radiographic images, the grain size evolution of the grains during solidification has also been investigated. The equivalent grain radius of the equiaxed grains were obtained by measuring the area of the 2-D projection image of the grains. It has to be noted that the grain size of the grains obtained by such a treatment is larger than the real 3-D equivalent grain size. Fig. 3 shows the evolution of the measured grain size of an equiaxed grain shown in Fig. 1(b) in comparison with the predicted grain size evolution of a grain (nucleated on a inoculant particle of 4μm in radius) during solidification with a cooling rate of 0.025 K/s. As can be seen, the predicted grain size is close to the measured size in the early stage of grain growth, but smaller than the measured grain size in the late stage, which is partly attributed to the over estimation of the grain radius by measuring the 2-D projection image. It has been determined that the grain nucleation stopped within 30 seconds since the first equiaxed grain was observed, namely the maximum nucleation undercooling is less than 0.7K. Within such a short time interval, the predicted grain radius is very close to the real grain size evolution. It suggests that the spherical grain growth is a reasonable assumption for the grain size prediction purpose.

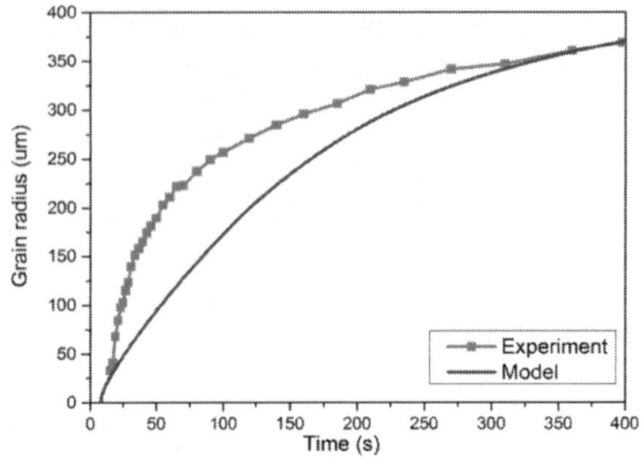

Fig. 3 The evolution of the measured equivalent radius of a single growing grain shown in Fig. 1(b), in comparison to the predicted grain radius evolution of a grain nucleated on an inoculant particle with radius of 4 μm. The measured grain radius is calculated from the measured projection area of the grain.

The thickness of the solute diffusion layer around the growing grains has a strong influence, in terms of solute stifling[1,7], SSN[10] or nucleation free zone (NFZ)[11], on the nucleation of new grains in the melt surrounding the growing grains. Fig. 4 shows the evolution of the predicted ratio between the thickness of solute diffusion layer and the radius of growing grains with different initial grain radius during solidification with cooling rates of 0.025 K/s (Fig.4a) and 0.5 K/s (Fig. 4b). It shows that, until the maximum nucleation undercooling is reached, the ratio δ_c/R_g is less than 10 for the slow cooling rate case, while less than 7 in the fast cooling case. These calculated δ_c/R_g values are much less than the value obtained by the approach by Shu et al[10], but is consistent with our previous analysis in [1,13] on δ_c/R_g, which gives a range between 2.5 and 25.

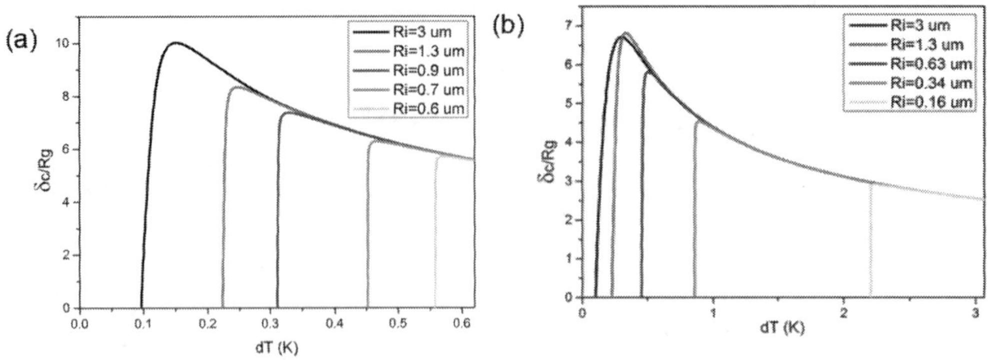

Fig. 4 Evolution of the predicted ratio between the thickness of solute diffusion layer δ_c and the radius of growing grains R_g with different initial grain radius (R_i) during solidification with cooling rates of (a) 0.025 and (b) 0.5 K/s.

IOP Conf. Series: Materials Science and Engineering **84** (2015) 012090 doi:10.1088/1757-899X/84/1/012090

5. Conclusions

An in-situ X-ray radiographic study on the solidification of inoculated Al-10wt%Cu alloy under nearly isothermal melt solidification with constant cooling rate is realized. The evolution of the size, morphology and number density of equiaxed grains during solidification has been quantitatively studied. The influence of cooling rate and addition level of grain refiners has been revealed.

A new grain size prediction model based on a new nucleation stopping mechanism, solute stifling mechanism, has been applied to model the heterogeneous nucleation of the of the experimental alloys during isothermal melt solidification. It shows that the grain size prediction model could well address the influence of cooling rate and addition level of grain refiners and, most importantly the predicted grain size is in a good agreement with the measured grain size.

References

[1] Du Q, Li YJ 2014 *Acta Materialia* **71** 380-389.
[2] Maxwell I, Hellawell A **1975** *Acta Metall* **23** 229.
[3] Easton MA, St John D H 2001*Acta Mater* **49** 1867.
[4] Easton M, St John D 2005 *Metall Mater Trans* A **36** 1911.
[5] Greer AL, Bunn A M, Tronche A, Evans P V, Bristow D J 2000 *Acta Mater* **48** 2823.
[6] Greer A L 2003 *Phil. Trans. R. Soc. Lond.* A **361** 479
[7] Quested T E, Greer A L 2005 *Acta Materialia* **53** 4643.
[8] Rappaz M, Gandin C A 1993 *Acta Metall Mater* **41** 345
[9] Men H, Fan Z 2011 *Acta Materialia* **59** 2704
[10] Shu D, Sun B, Mi J, Grant P 2011 Acta Mater **59** 2135
[11] StJohn D H, Qian M, Easton M A, Cao P 2011 Acta Mater **59** 4907
[12] Murphy A G, Browne D J, Mirihanage W U, Mathiesen R H 2013*Acta Materialia* **61** 4559–4571
[13] Li YJ, Du Q 2014 *Materials Science Forum* **790-791** 179-184

Numerical simulation of mold-filling capability for a thin-walled aluminum die casting

L. Sun[1], E. Subasic[1],J. Jakumeit[1]
[1] Access e.V., Aachen, Germany

E-mail: jakumeit@access.rwth-aachen.de

Abstract. Mold-filling capability is an important property of casting materials. Especially in thin-walled die casting, fast cooling of the melt by contact to the die makes complete filling difficult to ensure. Simulation is an important tool enabling investigation of filling problems, even before the die is manufactured. However, the prediction of misruns is challenging. Flow and solidification have to be computed as closely coupled. The effects of surface tension, the wetting angle and reduced melt flow due to solidification must be modeled with high precision. To meet these requirements, a finite-volume method using arbitrary polyhedral control volumes is used to solve flow and solidification as closely coupled. The Volume-of-Fluid approach is used to capture the phase separation between gas, melt and solid in connection with a High-Resolution Interface-Capturing scheme to obtain sharp interfaces between phases. To model the resistance of the dendrite network to the melt flow, an additional source term in the momentum equation was implemented. The Bolt test was performed for A356 alloys at a range of different casting temperatures. Numerical prediction of incomplete filling in the bottleneck regions agreed well with experimental findings using 3D camera scanning. The simulation enables derivation of the dependence of critical wall-thickness, i.e. the thickness which is fillable, on casting temperature and metallostatic pressure. This could prove useful in predicting filling problems ahead of casting.

1. Introduction

The filling capability of the melt is an important property for the production of solid, thin-walled castings. The mold-filling capability describes the ability of a metal to fill all parts of the geometry, including thin-walled regions. Filling the edges and corners of the mold is difficult for the melt and the penetration of the casting metal is only possible to a certain degree. An established experiment for investigating mold-filling capability is the Bolt test, designed by Engler and Ellerbrok [1] in 1974 at the Foundry Institute Aachen. Two cylinders are in contact lengthwise, forming a narrowing gap in which metal penetrates from both sides. Penetration depth decreases with cylinder height because the metallostatic pressure created by the metal column decreases as cylinder height decreases.
In this work, STAR-Cast [2,3] is used to address this challenging simulation task, which was already successfully applied for predicting misruns in thin-walled aluminum sand and TiAl centrifugal investment castings [4,5]. A brief overview of the underlying methodology is given in the next section. Special attention is paid to those features which are crucial in the prediction of misruns. The third chapter presents the Bolt test experiments, while the fourth chapter contains the validation of the simulation approach achieved by comparing simulation results from the Bolt test with experimental findings. The fifth chapter derives a useful relationship between metallic pressure and critical filling width for an A356 die casting. Finally, there is a short summary.

2. Numerical approach

Simulation of the complete casting process is based on a finite-volume method using the control volumes (CVs) of an arbitrary polyhedral shape. The transport equations for mass, momentum, energy, and phases in integral form are applied to each CV, whereby the surface and volume integrals are approximated using the midpoint rule. Linear equation systems for each variable are solved using either conjugate-gradient or algebraic-multi-grid iterative solvers. The equations describing mass (1), momentum (2), energy (3) and phase conservation (4) in a volume V bounded by the surface S are given below (for details see [6]):

$$\frac{\partial}{\partial t}\int_V \rho dV + \oint_S \rho\vec{v}\, ds = 0, \tag{1}$$

$$\frac{\partial}{\partial t}\int_V \rho\vec{v} dV + \oint_S \rho\vec{v}\vec{v}\, ds = \oint_S T\, ds + \int_V (\rho \vec{f}_b + S_\sigma + S_{rot}) dV, \tag{2}$$

$$\frac{\partial}{\partial t}\int_V \rho h dV + \oint_S \rho h\vec{v}\, ds = \oint_S \vec{q}\, ds + \int_V S_Q\, dV, \tag{3}$$

$$\frac{\partial}{\partial t}\int_V \rho C_i dV + \oint_S C_i\vec{v}\, ds = \int_V S_{Ci} dV, \tag{4}$$

where t is time, ρ is density, $\boldsymbol{v}$ is the velocity vector, $\boldsymbol{T}$ is the Cauchy stress tensor, $\boldsymbol{f}_b$ is the body force, h is the thermal enthalpy, $\boldsymbol{q}$ is the heat flux vector, C_i is the volume concentration of the phase i and $S_{(\sigma/Q/rot/Ci)}$ are source terms described below.

Mass conservation (1), pressure and velocity conservation (2) are coupled via the SIMPLE-algorithm [7, 8]. The transient term is discretized based on an implicitly Euler-segregated concept. Details of discretization and the solution method are available in [6] and are not elaborated here. Temperature distribution is computed using the enthalpy approach (3). For solidification modeling, the volume fraction of solidified liquid is determined using a tabulated fraction solid (f_s) *vs*. temperature curve ($f_s(T)$). Latent heat L is released in proportion to the change in fraction solid ($S_Q = L\, \delta f_s/\delta T$). Details of the method are presented in [8] and are not repeated here.

To calculate surface force correctly, the method used to track the motion of the free surface must provide a sharp interface. The Volume of Fluid (VOF) approach in combination with a High Resolution Interface Capturing (HRIC) scheme is used to solve the problem: the entire fluid domain is considered to be filled with an effective fluid, whose properties vary according to the distribution of volume fractions of melt C_m, solid C_s and gas C_g ($C_m + C_s + C_g = 1$). The transport of melt, solid and gas is computed by solving transport equations for their volume fractions (4) with a source term S_{Ci} for the phase change from melt to solid. To achieve sharpness of interface an HRIC scheme [6, 9] is used, which typically resolves the interface within one cell.

The normal force due to surface tension is treated using the continuum surface force (CSF) model proposed by [10], which defines a volumetric source in the momentum equation that is expressed as:

$$S_\sigma = -\sigma\nabla\left(\frac{\nabla C_{nl}}{|\nabla C_{nl}|}\right)\nabla C_{nl}. \tag{6}$$

To enable a prediction of misruns, flow resistance in the mushy zone must be handled correctly. Assuming that the mushy zone acts like a porous media, the fluid velocity resistance in the mushy zone can be approximated as a pressure drop [11]:

$$-\frac{\partial p}{\partial x_i}=\frac{\mu}{K}\vec{v}+\frac{C_E}{\sqrt{K}}\rho|v|\vec{v}, \tag{8}$$

where μ is viscosity, K is permeability and C_E the Ergun's coefficient. Permeability can be deduced from the Kozeny-Carman equation [12] as $K = ((1-f_s)^3\lambda_2^2)/(180f_s^2)$, where λ_2 is the secondary dendrite-arm spacing (SDAS).

3. Mold-filling capability experiment: Bolt test

Fig. 1 shows the principle behind the Bolt test [1]: two cylinders (radius R), made either as sand or metal cores and are in contact lengthwise, form a narrowing gap (length a) in which metal penetrates from both sides. The penetration depth depends on the height of the metallostatic pressure: the higher the metal column above it, the greater the penetration depth.

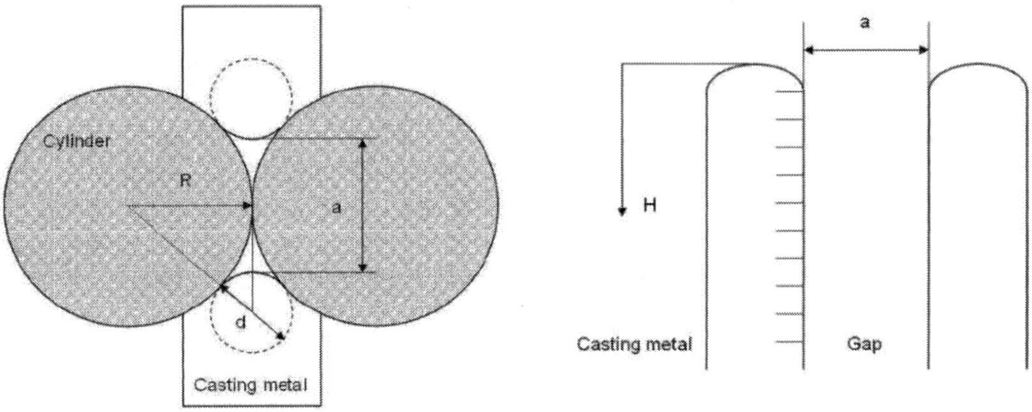

Fig. 1: Bolt test for mold-filling capability measurement

Mold-filling capability is estimated as follows [13]: when metallostatic pressure is equal to capillary pressure $p = \rho * g * H = 2\sigma/d$, the melt stops flowing. Here ρ is the density of metal, g the gravity, H the metallostatic pressure head, σ the surface tension and d the diameter of the round edge of the casting part. The mold filling capability F is the reciprocal of the diameter of the round edge d, $F = 1/d$. Ideally, the diameter of the round edge d can be calculated from the geometrical dimensions of the Bolt tests, $(R + d/2)^2 = R^2 + ¼(d+a)^2$. If so, the following equation can be derived: $F = 1/d = (4R-2a)/a^2$.

Bolt tests are carried out on an MBS KMC-4 tilting permanent mold. The mold consists of two parts, a fixed part and a movable part, both coated with ceramic materials. The mold is preheated to 230°C before melt pouring. Casting trials with AlSi7Mg0.3 (A356) were conducted at various casting temperatures: 720°C, 710°C, 700°C and 690°C in succession. The experimental procedure is as follows: 1.4 kg of overheated melt is taken from the deep part of the melting crucible (to avoid the entrainment of oxides) and transferred into the sprue cup of the Bolt test mold. When the temperature decreases to the desired casting temperature, the tilting of the mold is triggered and the mold tilts 90° at a speed of 7.4°/s. After the melt solidifies, the casting sample is removed from the mold. The mold cools to 230°C, which is measured by two thermocouples attached to the surface of the mold. The casting continues to cool down to room temperature after removal from the mold. This process is repeated several times at casting temperatures of 720°C, 710°C, 700°C and 690°C respectively.

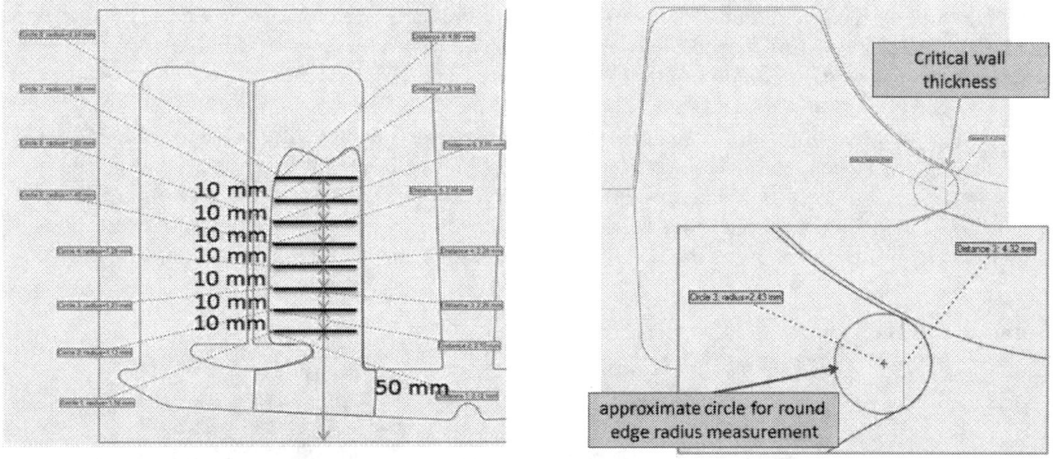

Fig. 2: left: assembly and cutting planes layout with VGStudio Max2.2; right: an example of round edge radius and critical wall thickness measurement

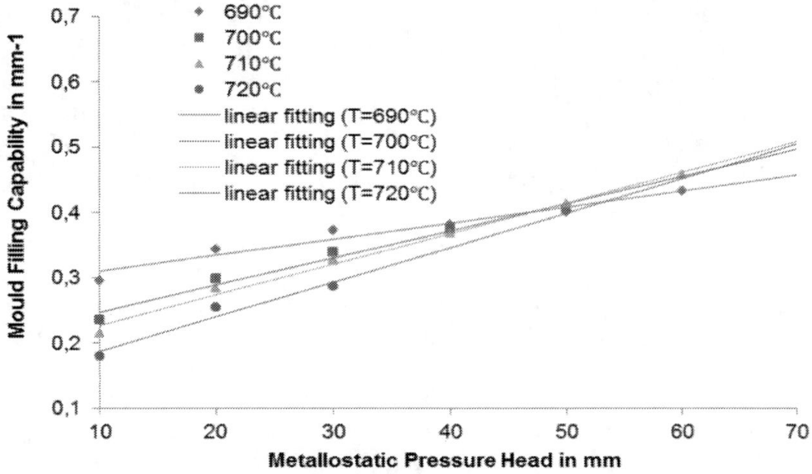

Fig. 3: Relationship between metallostatic pressure head and the mold-filling capability at various casting temperatures

To obtain the data about mold-filling capability, further analysis of the casting results from the Bolt test is necessary. Each casting is cut into four specimens and coated with spray for 3D camera scanning. The software VGStudio Max2.2 is applied to the data measurements from the 3D camera scanning. Fig. 2 shows the measurement methods for the feature values. As well as measuring the casting, it is also necessary to import the original mold shape into the software. Each piece needs to be placed into its original position in the mold before measurement. The bottom surface of the mold is taken as the reference surface. The first cutting position is then 50 mm above the reference surface, and the sample is cut at every 10 mm height increase. Cutting terminates at the end of sample and on average there are 5-8 cuts made on each piece. In this way, it is possible to obtain the penetration contour of the melt at each cross-section. A circle can be generated based on the curvature of this round front contour and its radius can be obtained. As predicted, the majority of these front penetration contours are not perfectly circular and this method could thus provide more precise data. Furthermore, critical wall thickness is also measured through the tangential or intersection point between the front curvature of the penetration contour and the approximate circle.

From the measured position data, it was possible to estimate the metallostatic pressure head. The relationship between the metallostatic pressure head and mold-filling capability is plotted in Fig. 3 (comparison at various casting temperatures). As predicted by the theoretical consideration a linear relation can be identified between mold-filling capability and the metallostatic pressure head. The mold-filling capability increases as the metallostatic pressure head increases.

4. Simulation validation

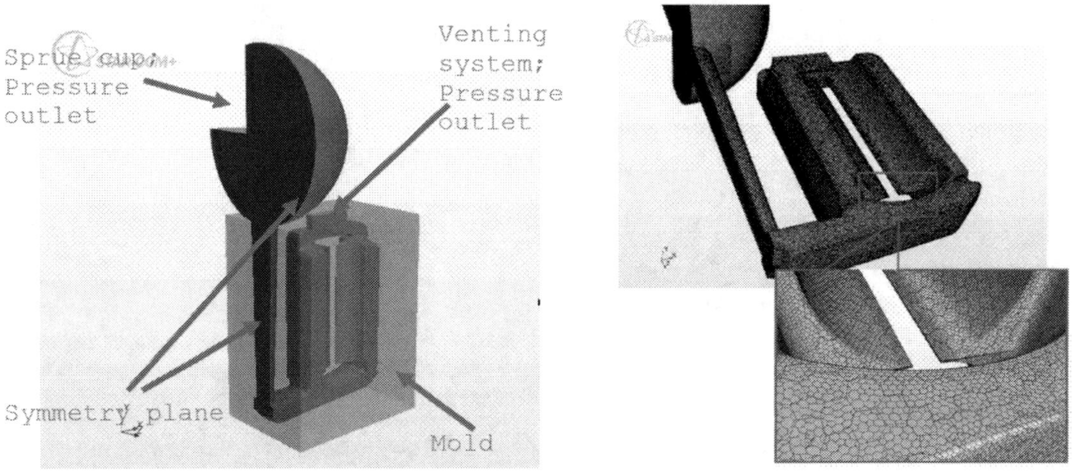

Fig. 4: Bolt test model (left) and applied mesh (right)

The simulation approach described in chapter 2 was applied to the Bolt test geometry shown in Fig. 4. To improve convergence and the stability of the model, the venting system was modified to a larger size. Two regions are created: one solid region for permanent molds, and one fluid region for castings. A coarse mesh is applied to the former, while the latter one, to increase the calculation efficiency, has a finer mesh. For the same reason, only a half part is simulated by the application of a symmetry plane boundary condition. Fig. 4 shows the Bolt test model.

Table 1: Parameters of AlSi7Mg0.3 used in simulation

Properties	Value	Properties	Value
Density	2500kg/m^3	Solidus temperature	542°C
Dynamic viscosity	1.1mPas	Liquidus temperature	610°C
Specific heat capacity	1200J/kgK	Surface tension	0.975N/m
Latent heat	431J/g	Wetting angle	135°
Heat transfer coefficient	2400W/m^{2}K	Thermal conductivity	200W/mK

The pressure outlet boundary condition is applied at the two free surfaces of the venting system and sprue cup shown in Fig. 4. This tilting process is modeled as the rotational motion defined within a user-defined field function. The same tilting velocity of 7.4°/s is applied as in the Bolt test experiments. Since the tilting angle is 90°, the tilting time is 12.2s. The material property parameters of AlSi7Mg0.3 (A356) are taken from the STAR-Cast material database. Table 1 shows the required parameters. By combining DSC analysis and micrographic investigation, the relationship between local solidification time t_{ls} and secondary dendrite arm spacing (SDAS) λ_2 could be determined and can be generalized by the equation $\lambda_2=5.6182 * t_{ls}^{0.4273}$. X37CrMoV5 steel is taken as the mold material. The thermal conductivity of the mold material is 30W/mK.

At the initial conditions, the quantity and position of the melt as well as the casting temperature and mold preheated temperature can be defined as being the same as in the experiments. Specifically, 1.4 kg melt is placed in a sprue cup before tilting at various casting temperature (720°C, 710°C, 700°C and 690°C). The initial temperature of the mold is set to 230°C.

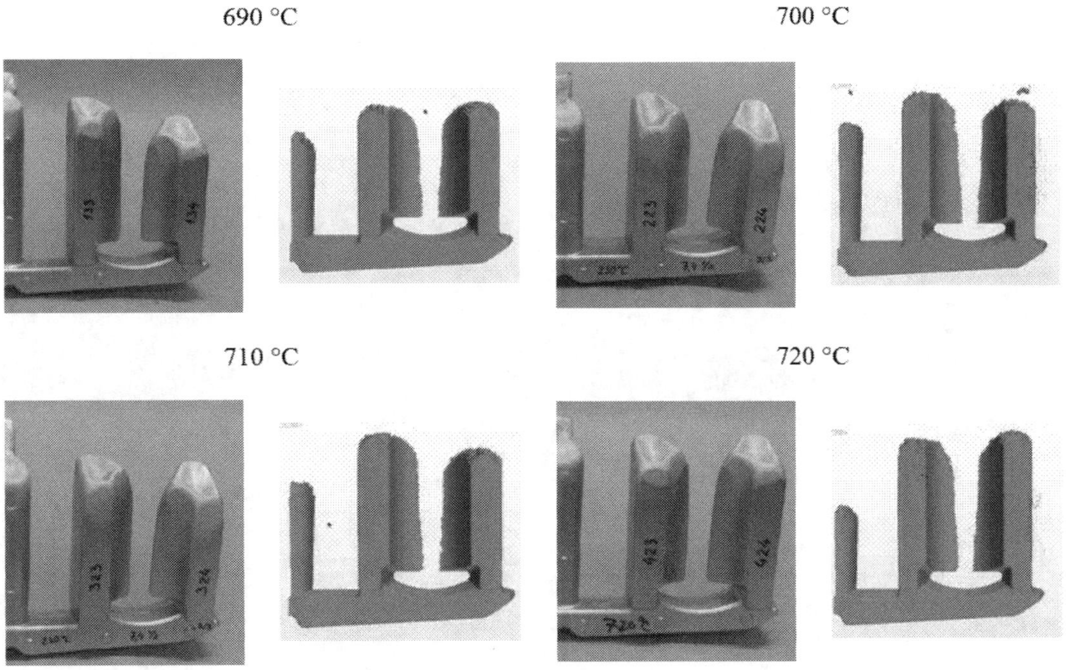

Fig. 5: Comparison of mold filling between experiment (left) and simulation (right) for the two pouring temperatures 690°C, 700°C, 710 °C and 720 °C.

Fig. 5 compares the final castings with the final shape of the melt as predicted by simulation. The general shape of the simulation result agrees well with the experimental findings, although differences can be found in details. The simulation results predict a steeper inside contour and the difference in penetration depth between the upper and lower parts is larger. A more quantitative comparison is provided by Fig. 6, which compares casting temperature dependence on critical wall thickness for two sections. Again, the general behavior is well predicted, while differences can be found in details. Especially at a casting temperature of 720 °C, the experiment shows an increase in critical wall thickness, while thickness continues to decrease in the simulation.

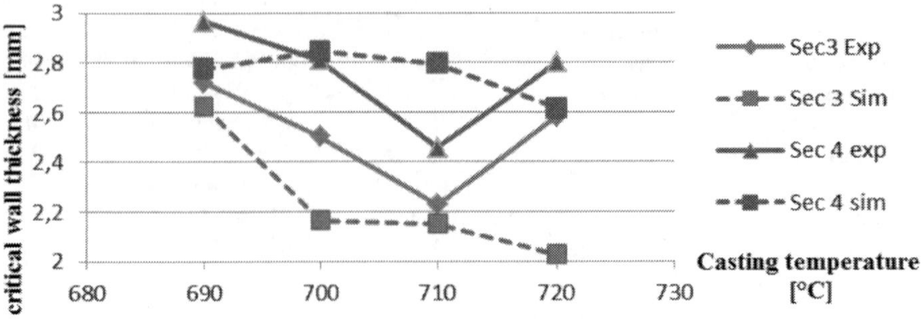

Fig. 6: Comparison of casting temperature dependence on the critical wall thickness between experiment (straight line) and simulation (dotted line)

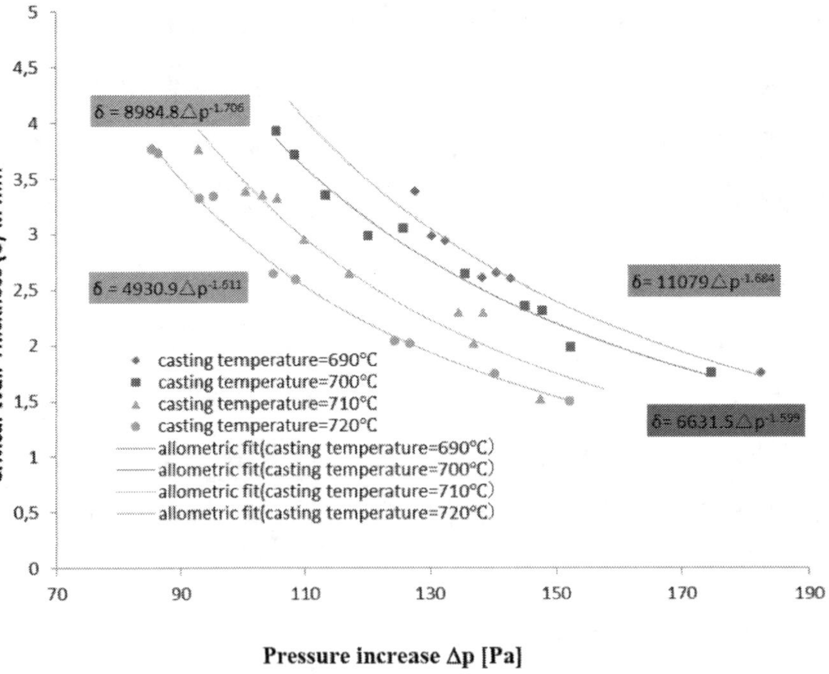

Fig. 7: Relationship between pressure increase and critical wall thickness

5. Impact of local pressure and casting temperature on filling capability

Metallostatic pressure and casting temperature in the range tested have pronounced and similar impact on critical wall thickness. Fig. 7 shows the dependence of critical wall thickness on the pressure increase from the top of the melt to the analyzed region at the four casting temperatures. Greater temperature or higher pressure results in a lower critical wall thickness, i.e. the possibility of filling thinner regions. Regression enabled an analysis of the dependence of critical wall thickness on pressure increase, resulting in the exponential dependence shown in Fig. 7. The equations enable an estimation of critical wall thickness ahead of casting.

6. Summary

The Bolt test was used to investigate the mold-filling capability of the aluminum alloy A356 and to validate misrun prediction capability when using the numerical approach in STAR-Cast for thin-walled aluminum die castings. In general, good agreement between experimental findings and simulation prediction could be found. The shape of the casting and the critical wall thickness agreed well, although differences in details still require further clarification. A dependence of critical wall thickness on metallostatic pressure could be deduced from the simulation result, which may be of use in designing castings.

References

[1] Engler S, Ellerbrok R, 1974 Über das Formfüllungsvermögen der Gusswerkstoffe, Gießereiforschung **26** 49
[2] www.star-cast.com
[3] Jakumeit J, Laqua R, Peric M and Schreck E 2006 Proceedings of Modeling of Casting, Welding and Advanced Solidification Processes XI 63
[4] Jana S., Kättlitz O., Hediger F., Jakumeit J. and Aguilar J., 2012, MCWASP XIII, IOP Publishing, IOP Conf. Series: Materials Science and Engineering, p. 33
[5] Jakumeit J, Subasic E, Bünck M, 2014, Shape Casting: 5th International Symposium, ISBN: 978-1-118-88818-6
[6] Demirdzic I. and Muzaferija S., 1995, Comp. Met. in App. Mech. and Eng. **125**, 235
[7] Ferziger J.H. and Peric M., (ed.) 2002 Computational Methods for Fluid Dynamics, McGraw-Hill, ISBN 2-540-42074-6.
[8] Teskeredzic A., Demirdzic I., Muzaferija S., 2002, Num. Heat Trans., Part B **42**, 437
[9] Muzaferija S. and Peric M., 1999, eds. Mahrenholtz and M. Markiewicz, WIP Press, Southampton, 59
[10] Brackbill J.U., Kothe D.B. and Zemach C., 1992, J. of Computational Physics **100** 335
[11] Sabau A.S. and Viswanath S., 2002, Met. Mat. Trans. B **33B** 131
[12] Poirier D.R., Yeum K. and Mapples A.L., 1987, Metall. Mater. Trans. **A 18**, 1979
[13] Santhi S, 2012 Journal of Metallurgy **2012**, Article ID 624650

IOP Conf. Series: Materials Science and Engineering **84** (2015) 012092 doi:10.1088/1757-899X/84/1/012092

Computational prediction of the refinement of oxide agglomerates in a physical conditioning process for molten aluminium alloy

M Tong[1], S C Jagarlapudi[1], J B Patel[2], I C Stone[2], Z Fan[2] and D J Browne[1]
[1]School of Mechanical and Materials Engineering, University College Dublin, Belfield, Dublin 4, Ireland
[2]Brunel Centre for Advanced Solidification Technology, Brunel University, Uxbridge,Middlesex, UB8 3PH, UK

E-mail: Mingming.tong@ucd.ie

Abstract Physically conditioning molten scrap aluminium alloys using high shear processing (HSP) was recently found to be a promising technology for purification of contaminated alloys. HSP refines the solid oxide agglomerates in molten alloys, so that they can act as sites for the nucleation of Fe-rich intermetallic phases which can subsequently be removed by the downstream de-drossing process. In this paper, a computational modelling for predicting the evolution of size of oxide clusters during HSP is presented. We used CFD to predict the macroscopic flow features of the melt, and the resultant field predictions of temperature and melt shear rate were transferred to a population balance model (PBM) as its key inputs. The PBM is a macroscopic model that formulates the microscopic agglomeration and breakage of a population of a dispersed phase. Although it has been widely used to study conventional de-oxidation of liquid metal, this is the first time that PBM has been used to simulate the melt conditioning process within a rotor/stator HSP device. We employed a method which discretizes the continuous profile of size of the dispersed phase into a collection of discrete bins of size, to solve the governing population balance equation for the size of agglomerates. A finite volume method was used to solve the continuity equation, the energy equation and the momentum equation. The overall computation was implemented mainly using the FLUENT module of ANSYS. The simulations showed that there is a relatively high melt shear rate between the stator and sweeping tips of the rotor blades. This high shear rate leads directly to significant fragmentation of the initially large oxide aggregates. Because the process of agglomeration is significantly slower than the breakage processes at the beginning of HSP, the mean size of oxide clusters decreases very rapidly. As the process of agglomeration gradually balances the process of breakage, the mean size of oxide clusters converges to a steady value. The model enables formulation of the quantitative relationship between the macroscopic flow features of liquid metal and the change of size of dispersed oxide clusters, during HSP. It predicted the variation in size of the dispersed phased with operational parameters (including the geometry and, particularly, the speed of the rotor), which is of direct use to experimentalists optimising the design of the HSP device and its implementation.

IOP Conf. Series: Materials Science and Engineering **84** (2015) 012092 doi:10.1088/1757-899X/84/1/012092

1. Introduction

Significant reductions in energy, production costs and carbon footprint can be achieved by manufacture of high performance automotive components with recycled aluminium alloys instead of primary aluminium alloys. Because scrap aluminium alloys are normally contaminated by harmful elements such as iron, at relatively high levels, a key challenge to the recycling process is the purification of materials. Physically conditioning molten scrap aluminium alloys using high shear processing (HSP) was recently found to be a promising technology for this type of purification. Research at Brunel University [1-6] showed that HSP can refine the solid oxide clusters in molten alloys significantly. These act as effective sites for the nucleation of Fe-rich intermetallic phases. The solidified intermetallic particles can subsequently be removed by the downstream de-drossing process, and hence purify the secondary aluminium alloys by removing the harmful elements (e.g. Fe). Due to the high temperature, opacity of the melt, and the close-coupling feature of the rotor-stator mixing head, it is very difficult to measure the temperature and particularly shear rate of melt inside, and in the close vicinity of the mixing head, without disturbance. Moreover, the shear rate characteristic of laminar melt flow and that of turbulent flow are mathematically formulated in totally different ways. Because the status of fluid flow (e.g. laminar or turbulent) directly depends on operational parameters, it is not easy to experimentally determine the characteristic melt shear rate in the process of HSP. Computational simulation is therefore needed to study the key parameters of the HSP process. By using appropriate mathematical equations that describe the related governing physical mechanisms, and corresponding numerical solution schemes, it is possible to predict computationally a variety of key parameters that are normally difficult to measure experimentally. Moreover, since the input data for such a computer model come from operational parameters and materials' properties that are typical of the HSP process in practice, computational simulation can be used to directly determine the influence of operational parameters on the performance of HSP. The Population balance model (PBM) is a macroscopic model that can formulate the microscopic breakage and agglomeration of a population of a phase that is dispersed in a continuous phase [7-13]. In this paper, we present our PBM modeling of the behaviour of the oxide clusters during HSP.

2. Model details

The key governing equation of PBM is the population balance equation (PBE):

$$\frac{\partial f(v,t)}{\partial t} + \nabla \cdot [\vec{V} f(v,t)] = S(v,t) \tag{1}$$

where f is the number density of dispersed phase (i.e. solid oxide clusters in the case of conditioning molten scrap aluminium alloys with HSP) with internal coordinate v (the volume of dispersed phase). By using the number density, we can directly plot the size distribution of the dispersed phase and mathematically calculate the mean size. $\vec{V}$ is the advection velocity of the dispersed phase. S is the net rate of generation of dispersed phase. We are neglecting, for now, the mechanism of growth and only taking the mechanisms of breakage and agglomeration into account.

Determining the advection velocity of the dispersed phase depends on CFD modelling of melt-solid particle two-phase flow. In the CFD modelling, we use an Eulerian method to formulate the governing continuity, momentum and energy equations for every phase, taking the volume fraction of each phase and the interaction between one phase and another into account.

Once the advection velocity is calculated by the CFD model, the evolution of number density of dispersed phase becomes dependent on the net generation rate in the PBM. In the HSP process, dilute oxide clusters are dispersed in the continuum melt. This mixture is strongly stirred by the mixing head with a rotor-stator design, at a relatively high temperature (e.g. 645 $^{\circ}C$ in [2]). The mechanisms of fragmentation and agglomeration dominate the change of number density of oxide clusters. Therefore, the net generation rate S becomes:

$$S(v) = S^{+}(v) - S^{-}(v) \quad , \tag{2}$$

where S^{+} is the birth rate and S^{-} is the death rate of oxide clusters due to breakage and agglomeration, respectively, and they follow:

$$S^{-}(v) = b(v)f(v) + \int_{0}^{\infty} a(v,v')f(v,t)f(v',t)dv' \quad , \tag{3}$$

$$S^{+}(v) = \int_{v}^{\infty} v'b(v')P(v',v)f(v',t)dv' + \frac{1}{2}\int_{0}^{v} a(v-v',v')f(v-v',t)f(v',t)dv' \quad , \tag{4}$$

where a is the agglomeration rate, b is the breakage rate, and P is the daughter size distribution function for the breakage. It can be seen that the whole PBM modelling is reduced to determining the value of these 3 terms. The first terms of Eq.3 and Eq.4 represent the contribution of mechanism of fragmentation to the change of number density of oxide clusters, while the second terms of Eq.3 and Eq.4 denote the contribution of mechanism of agglomeration to the change of number density of oxide clusters.

In the sub-model for breakage, the shear rate is the dominant factor. The type of melt flow (i.e. laminar or turbulent) and size of oxide clusters (compared with the Kolmogorov length scale) commonly determine the mathematical formulation of the breakage rate [7-10]. For example, the characteristic shear rate depends on the fact that the melt flow falls in the laminar regime or turbulent regime. In the turbulent regime, the mechanism of fragmentation of oxide clusters depends on if the size of oxide clusters is larger than the size of smallest eddies. In the sub-model for the rate of agglomeration, we consider the mechanism of Brownian motion, differential sedimentation, and shear-induced collision of the dispersed phase [7-10].

We use a discrete method to solve Eq.1. The originally continuous profile of size of oxide clusters is firstly discretized into a collection of discrete size bins and Eq.1 is integrated over each bin and then the discrete form of Eq.1 is solved numerically. The discrete oxide cluster size bins follow a geometrical series and the advection velocity of oxides is input from CFD modelling.

3. Configuration of the case study and results of simulation

The case study used in our research is a 2D one. The overall setup of this 2D case study is a very close mimic of the realistic rotor-stator mixer that we are using in experiments. Figure 1 illustrates the geometry of the domain of our 2D case study. This case study corresponds to an infinitely long cylindrical crucible full of molten recycled aluminium alloy, which includes alloy melt and solid oxide clusters. An infinitely long mixing head (with rotor-stator design) is placed at the centre of the crucible to shear the mixture. The impeller of the rotor has 4 flat blades and the cylindrical stator has holes along its circumference. Among the geometrical parameters, the key parameters include the size of the gap between the tip of impeller blades and the internal wall of the stator and the size (and number) of holes in the stator. The diameter of the crucible is 300 mm.

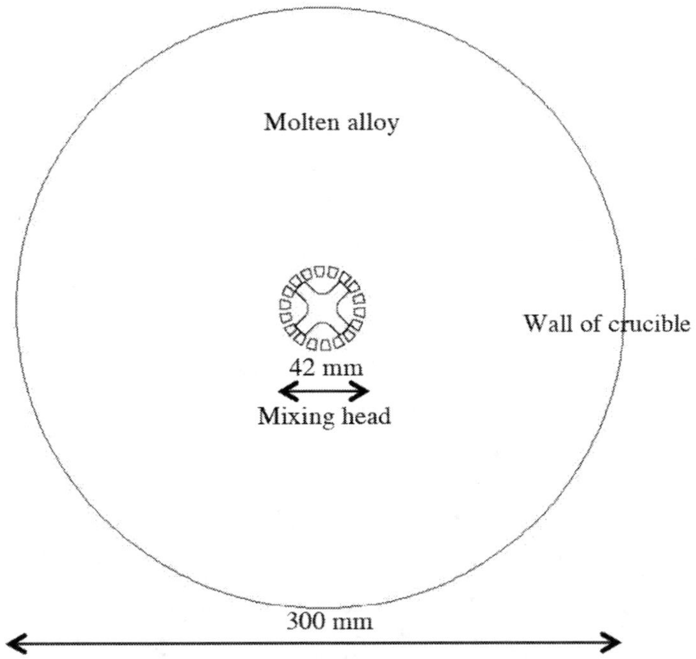

Figure 1. Geometrical setup of a 2D case as base line.

In the design of Fig.1, the width of gap between the internal wall of the stator and the tips of impeller blades is 0.25 mm and there are 18 holes of 2.5 mm in diameter each drilled through the stator. The typical operational rotor speed is 5000 rpm. These 3 values, as well as other geometrical features, exactly follow the real situation of the HSP applied in the laboratory, and we are using this setup as the base line case of our simulation.

At the start of simulation, the melt and oxide clusters are set at $800^{\circ}C$, and the size of oxide clusters are assumed to be mono-dispersed with Sauter mean diameter (SMD) of$10.2\mu m$. At the beginning of HSP, the melt/oxide mixture is stationary and the impeller spins at the designated constant rpm. The overall computation is implemented mainly using the FLUENT module of ANSYS. The CFD solver of FLUENT is used to predict the macroscopic flow features of the melt and oxide clusters, including the heat, mass and momentum transfer, using an Eulerian method. The resultant field predictions are transferred to PBM as its key inputs. Although there are some sub-models for the breakage and agglomeration of dispersed phase available in FLUENT, they are mostly suitable for liquid droplets and gas bubbles rather than solid particles. We program our own user subroutines in C for the breakage/agglomeration rates and load them into the PBM solver of FLUENT, in order to run PBM computation.

4. Results of the base line

We set 5 points in the simulation domain, in order to monitor the evolution of parameters of interest. Points 1-3 are located along the axis of a hole in the stator and Points 4 and 5 are in the middle of the far field and near the wall of the crucible, respectively, as shown in Fig.2.

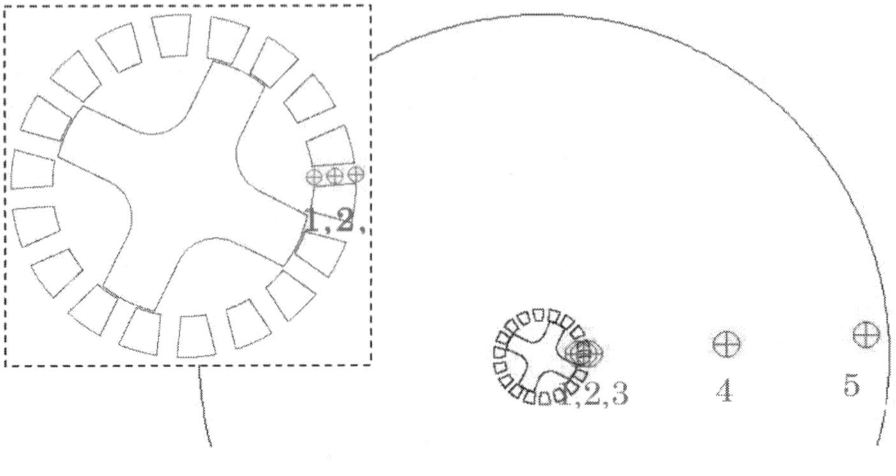

Figure 2. Position of sampling points.

If we plot the predicted temporal evolution of SMD of oxide clusters, as in Fig.3, we can find some interesting phenomena. Firstly as shown by the red curve in Fig.3, at the internal orifice of the hole through the stator, the size of oxide clusters drops very dramatically at the start of HSP and then periodically oscillates towards a plateau value. The period of this oscillation is roughly 0.003 s, which is 1/4 the period of impeller spinning. Considering the 4 fold symmetry of the impeller, this is a good agreement.

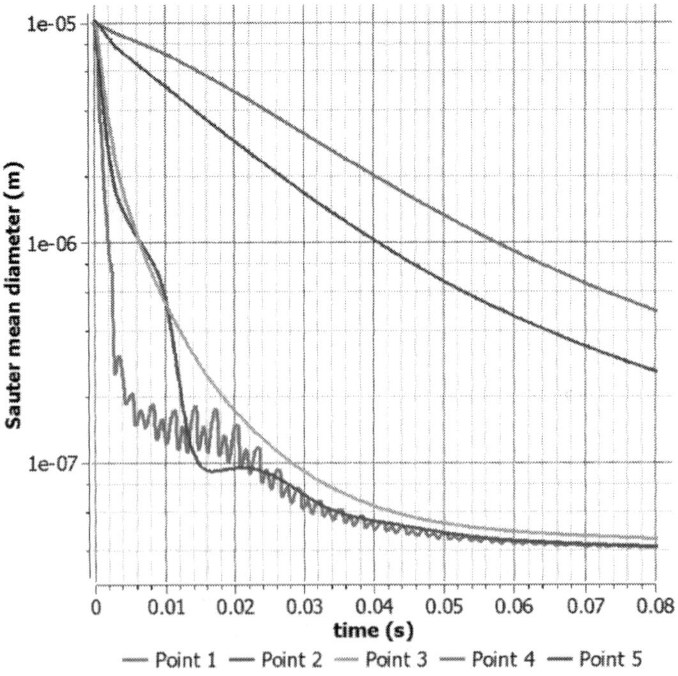

Figure 3. Temporal evolution of SMD of oxide clusters at different sampling positions during HSP in the baseline case.

While checking the size of oxide clusters away from the centre of the impeller, over sampling points 2-5, we can find that the size decreases more and more slowly, particularly beyond the rotor.

In order to explain the reason for the significantly different rate of decrease of cluster size at different positions, we plot the contours of some relevant parameters. In this paper, when plotting parameters at a specific moment of time, we use the data when the diameter of oxide clusters sampled at Point 1 has already reached its plateau value (i.e. 0.08 s).

The spatial distribution of oxide cluster size is illustrated in Fig.4. It shows that the spatial distribution is very non-uniform. In the close vicinity of the mixing head, the oxide clusters are as small as $0.0406\mu m$. In the middle of the far field, the oxide clusters are as large as $0.491\mu m$. They are both much lower than the initial oxide cluster size of $10.2\mu m$.

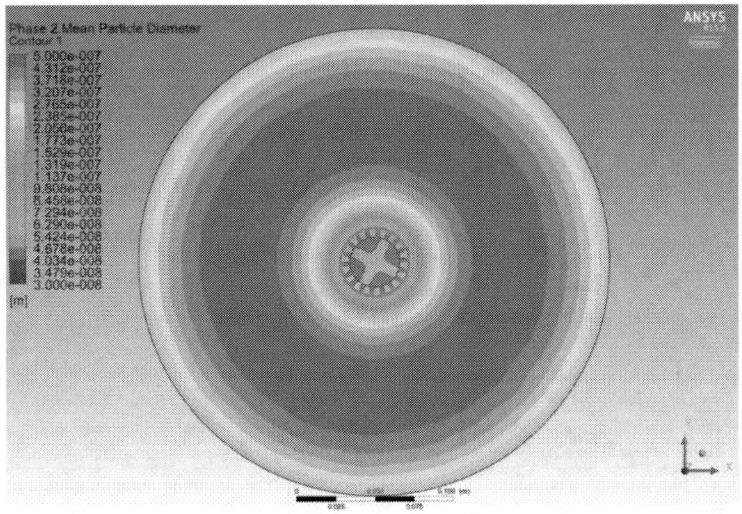

Figure 4. Contour of SMD of oxide clusters in the base line case study at 0.08 s.

Figure 5a shows that, in the far field, the fluid flow is highly turbulent and the turbulence becomes relatively less significant in the close vicinity of the mixing head (Fig.5b). However, the value of turbulent viscosity ratio is generally above 30, except inside the very thin boundary layer at the surface of the solid walls of rotor and stator. This means that the shear induced breakage and agglomeration falls into the turbulent regime. The characteristic shear rate of turbulent flow is dominated by the turbulence dissipation rate as:

$$G = (\frac{\varepsilon_c}{\upsilon_c})^{1/2}, \quad (5)$$

where G is the characteristic shear rate, ε and υ are the turbulent energy dissipation rate and kinematic viscosity of the melt phase, respectively.

Figure 5 (c,d) demonstrate that the turbulence dissipation rate in the close vicinity of the mixing head is significantly higher than that in the far field, by more than a few orders of magnitude. This makes the breakage and agglomeration of oxide clusters in the close vicinity of the mixing head dramatically stronger than that in the far field. The mechanisms of breakage and agglomeration compete with one another, and the net result is a significant refinement of oxide clusters in the overall crucible by HSP in this 2D case. The particle size at sampling point 4 is larger than that at point 5, as shown in Fig.3. This is because there is a narrow circular band of fluid of slightly lower turbulence dissipation rate, which sweeps right through point 4 as shown in Fig.5c. Such relatively low turbulence dissipation rate directly leads to relatively low characteristic shear rate (Eq.5) and hence relatively large size of oxide clusters sampled at point 4. While checking the turbulence dissipation rate of fluid from the mixing head towards the wall of crucible along the radial direction of impeller, although the overall trend of the change of its value is decreasing, the decrease does not have to be monotonic. There can be a few radial positions where the value of turbulence dissipation rate is slightly lower than those at their radially neighboring positions, as separately found by Soos et.al. [14] for example. The number and position of such "cold spots" of turbulence dissipation rate of fluid (along the radial direction in the far field) depends on the agitation of the fluid by the impeller and the interaction between the impeller, stator and wall of the crucible.

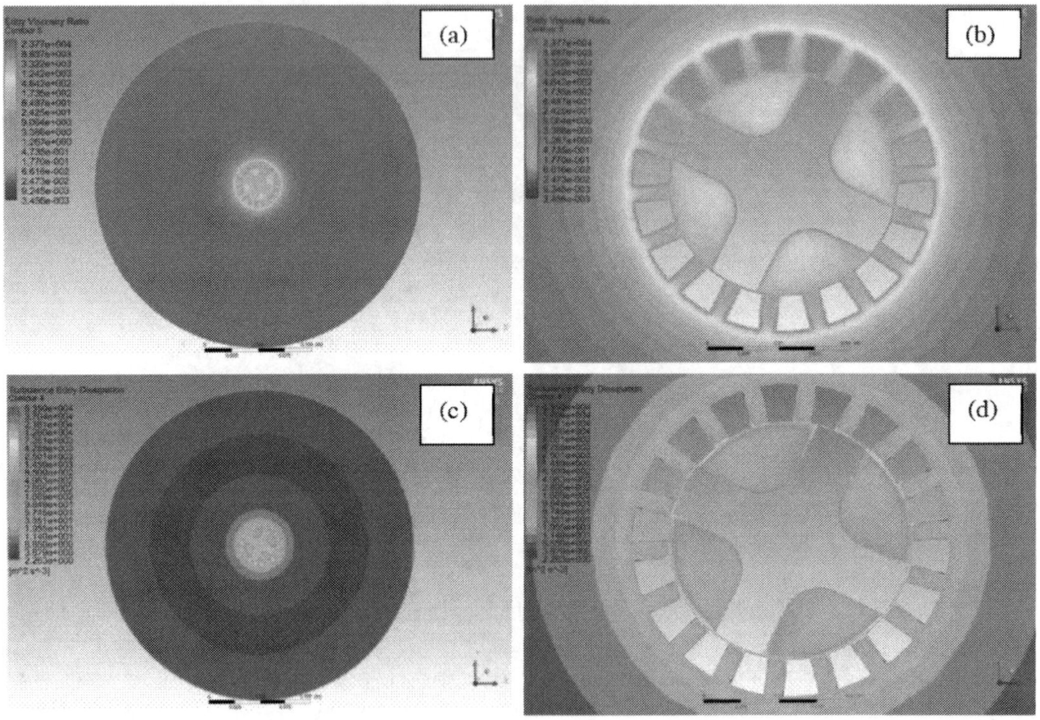

Figure 5. Contour of turbulent viscosity ratio of the melt (a), its local zooming in around the mixing head (b), contour of turbulence dissipation rate of melt (c) and its local zooming in around the mixing head (d) in the base line case study at 0.08 s.

Because the fluid in the close vicinity of the mixing head has been found to have the highest level of characteristic shear rate (in both the 2D simulation of this paper and a separate set of 3D case studies of us), we are paying most of our attention to the breakage of oxide clusters in the close vicinity of the mixing head rather than in the far field. The fluid flow in the far field for sure has significant influence on the macroscopic redistribution of fluid and oxide clusters, but it is our major interest in some other separate publication.

Changing the impeller speed and the geometrical design of the mixing head

In order to investigate the influence of operational rotor speed and geometrical design of the mixing head on the performance of HSP, three different simulations were run, each using a different setup to the reference case study. The rotor speed was reduced from 5000 rpm to 1000 rpm. The size of the gap between the rotor and the stator was increased from 0.25 mm to 2.5 mm. Finally, the number of holes along the circumference of the stator was reduced from 18 to 3, but their diameter was increased from

2.5 mm to 18 mm. Other aspects of the setup of these 3 different cases are as close to those of the base line case as possible. We summarize the simulation results of these 3 different cases and compare them with the result of the base line case, and plot the temporal evolution of oxide cluster size at sampling Point 1, in Fig.6.

Although the size of oxide clusters in the 2.5 mm gap case and the 3 holes case decreases more quickly than the corresponding process of the base line case, the size converges quickly in all of these 3 cases to almost the same plateau value. In the case of running HSP at 1000 rpm, the decrease in size of oxide clusters is significantly slower than the process of base line case and it looks very difficult for the size to converge to the same plateau value as that of the base line case.

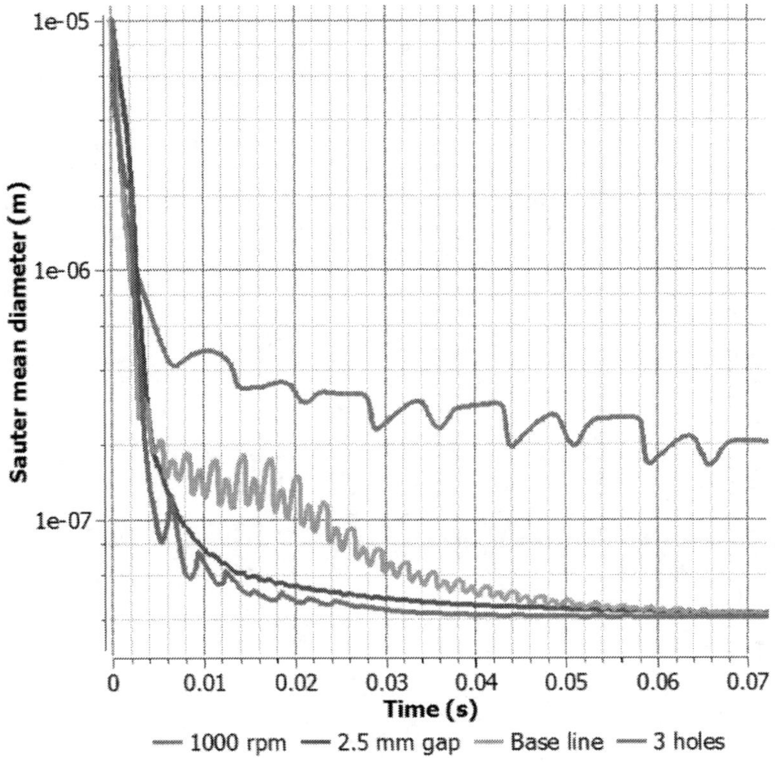

Figure 6. Temporal evolution of oxide cluster size sampled at point 1 with different setups.

5. Analysis and conclusion

Reviewing the simulation results, in conjunction with the mathematical governing equations, a few interesting conclusions can be drawn. Firstly, the turbulence dissipation rate (and hence characteristic shear rate of turbulent flow) in the 2D HSP case is highly non-uniform in space. Therefore, the resultant refinement of oxide clusters is very non-uniform in space accordingly. Most of the significant refinement of oxide clusters is taking place between the rotor and the stator. This means that a reasonably strong macroscopic redistribution of the molten alloy is necessary if we want to refine all the oxide clusters throughout the whole crucible.

Secondly, after changing related operational and geometrical parameters of the case study and comparing the results against the base line case, we have found a very strong influence of impeller speed on the decrease of oxide cluster size. Although in one case the size of the gap between rotor and stator was increased to 2.5 mm, which is 10 times that in the base line case, and in a separate case the size of holes was increased to 15 mm in diameter which is 6 times that in the base line case, the temporal evolution of size of oxide clusters sampled at the internal orifice of the holes of the stator quickly converged to an almost common plateau value in all of these 3 cases. However, when we decreased the rotor speed from 5000 to 1000 rpm, it was found that the decrease in size of oxide clusters becomes significantly slower that the corresponding process in the base line case.

Turbulence dissipation rate is the rate at which turbulent kinetic energy is converted into thermal energy due to the viscosity of a fluid. In Fig.5, we can find that the most significant dissipation of turbulent kinetic energy occurs inside the gap between the tips of the impeller blades and the internal wall of the stator. The value of turbulence dissipation here is higher than anywhere else in the computational domain by a few orders of magnitude. This is true even in the case when there are only 3 relative large holes in the stator (very large openings connecting the inside and outside of the stator), and in the case when the size of the gap is 10 times that of the base line case (very large space between the tips of impeller blades and internal surface of stator). This means that the gap between the tips of impeller blades and internal surface of the stator acts as a sink of turbulent kinetic energy, strongly consuming the kinetic energy fed to the fluid by the agitation of the impeller. The more energy fed into the turbulent fluid flow, roughly the more turbulent kinetic energy is consumed inside this gap. Because the rotational speed of the of impeller is directly related to the energy input rate, it is reasonable to predict that low speed can lead to low turbulence dissipation rate and hence low characteristic shear rate.

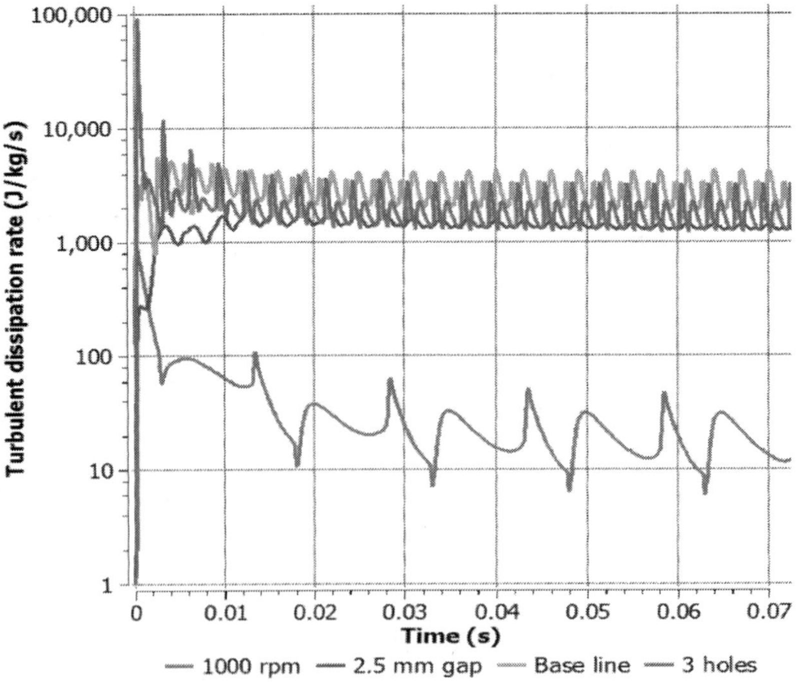

Figure 7. Temporal evolution of turbulence dissipation rate sampled at Point 1 with different setups.

In Fig.7, it is clear that the turbulence dissipation rate sampled at Point 1 in the 1000 rpm case is lower than those of the other cases by nearly 2 orders of magnitude. This directly supports our aforementioned theoretical analysis and we conclude that the energy input rate (related to impeller speed) dominates the performance of the mixing head in terms of refining the oxide clusters.

Acknowledgement
This research is financially supported by the EC FP7 project "High Shear Processing of Recycled Aluminium Scrap for Manufacturing High Performance Aluminium Alloys" (Grant No. 603577).

References

[1] Li H, Wang Y and Fan Z 2012 Enhanced heterogeneous nucleation on oxides in Al alloys by intensive shearing *IOP Conf. Ser.: Mater. Sci. Eng.* **27** p 012047

[2] Li H, Ji S, Wang Y, Xia M and Fan Z 2011 Effect of intensive melt shearing on the formation of Fe containing intermetallics in LM24 Al-alloy *IOP Conf. Series: Materials Science and Engineering* **27** p 012075

[3] Peng G, Wang Y and Fan Z 2013 Effect of Intensive Melt Shearing and Zr Content on Grain Refinement of Mg-0.5Ca-xZr Alloys *Materials Science Forum* **765** pp 336-340

[4] Patel J B, Prasada Rao A K, Jiang B, Zuo Y B, and Fan Z 2012 Melt Conditioned Direct chill casting (MC-DC) process for production of high quality Mg-alloy billets *Mg2012:9th International Conference on Magnesium Alloys and their Applications* pp 731-736

[5] Xia M, Prasada Rao A K and Fan Z 2013 Solidification Mechanisms in Melt Conditioned Direct Chill (MC-DC) Cast AZ31 Billets *Materials Science Forum* **765** pp 291-295

[6] Zuo Y B, Fan Z, Zhu Q F, Lei L and Cui J Z 2013 Modification of a Hypereutectic Aluminium Silicon Alloy under the Influence of Intensive Melt Shearing, *Materials Science Forum* **765** pp 140-144

[7] Wang L, Marchisio D, Virgil R, Fox R 2005 CFD simulation of aggregation and breakage processes in laminar Taylor-Couette flow *Journal of Colloid and Interface Science* **282** pp 380-396

[8] Yeoh G H, Cheung C, Tu J 2014 *Multiphase flow analysis using population balance modelling (bubbles, drops and particles)* Butterworth-Heinemann ISBN:978-0-08-0098229-8

[9] Vanni M 2000 Creeping flow over spherical permeable aggregates *Chemical Engineering Science* **55** pp 685-698

[10] Marchisio D, Vigil R, Fox R 2003 Implementation of the quadrature method of moments in CFD codes for aggregation-breakage problems *Chemical Engineering Science* **58** pp 3337-3351

[11] Liao Y, Lucas D 2009 A literature review of theoretical models for drop and bubble breakup in turbulent dispersions *Chemical Engineering Science* **64** pp 3389 - 3406

[12] Cheng J, Yang C, Mao Z, and Zhao C 2009 CFD Modeling of Nucleation, Growth, Aggregation, and Breakage in Continuous Precipitation of Barium Sulfate in a Stirred Tank *Ind. Eng. Chem. Res.* **48** pp 6992–7003

[13] Maniruzzaman M D, Makhlouf M M 2002 Mathematical Modeling and Computer Simulation of the Rotating Impeller Particle Flotation Process: Part II. Particle Agglomeration and Flotation *Mechanical Engineering Faculty Publications Department of Mechanical Engineering* 4-1-2002

[14] Miroslav Soos, Rene Kaufmann, Raphael Winteler, Martin Kroupa, Beat Luthi 2013 Determination of Maximum Turbulent Energy Dissipation Rate generated by a Rushton Impeller through Large Eddy Simulation, *American Institute of Chemical Engineers* **59** pp 3642-3658

IOP Conf. Series: Materials Science and Engineering **84** (2015) 012093 doi:10.1088/1757-899X/84/1/012093

Intensive riser cooling of a turbine blade steel casting after solidification

J W Kang[1], X K Hao[1], G Nie[1], H M Long[1], B C Liu[1] and C C Zhang[2]

[1] School of Materials Science and Engineering, Key Laboratory for Advanced Materials Processing Technology, Tsinghua University, Beijing 100084, China

[2] Harbin Electric Machinery Co., LTD， Harbin 150040

Emails: kangjw@tsinghua.edu.cn, xiaokun_hao@163.com, peking_126@126.com, niegang2008@163.com, liubc@tsinghua.edu.cn, zhangchengchun@hec-china.com

Abstract. To improve the production efficiency and reduce residual stress and deformation, a new method-Post Solidification Intensive Riser Cooling (PSIRC) was proposed for heavy steel castings. Risers are cooled by forced air from its top surface as the solidification of a casting finishes, so, the risers are turned into cooling passages during the cooling process of a casting with contrast to their function of feeding passages during the solidification process. This method was applied to a hydro turbine blade with 0.9 ton melt. Its cooling speed increased apparently meanwhile with improved temperature distribution uniformity and the hot spot was pushed almost to the casting bottom. The cooling efficiency was improved 40% before shakeout and its residual stress was reduced meanwhile.

1. Introduction

The cooling process of heavy steel castings in sand mold after solidification is usually very slow, maybe over a month or longer for some extra large castings weighing hundreds of tons, which undermines the production efficiency[1]. Furthermore, great thermal gradient appears across their thick sections and between the thick and thin areas, that may cause significant residual stress or deformation or even cracks[2]. In recent years there are more and more heavy castings and they get bigger and bigger, the heaviest one in China consumed 863.5 tons of liquid steel. Therefore, it is necessary to improve the cooling after the solidification of heavy castings. However, the research focus has been the solidification process for long time. For example, to realize sequential solidification of casting, the bottom or isolated hot spots are usually specially cooled by chills, water cooled copper plate or chills, cooling pipes buried in sand mold or core with compressed air [3] or water [4] or even liquid nitrogen [5] though the pipes. These methods only result into more faster cooling of special areas, but more uneven cooling of the whole casting. In production castings are shook by crane to break the wrapping of sand mold so as to facilitate the convection of castings with the atmosphere.

Sometimes water spray or forced wind is blown on the mold surface to extract more heat. Shakeout at high temperature is also a measure. These methods can fasten the cooling of the whole casting, but result into nonuniform temperature field. Rapid but uneven cooling of castings makes them susceptible to cracking, deformation and high residual stress [6], especially for some kind of materials such as martensitic stainless steel undergoing martensitic phase transformation at relatively low temperature. All of the above mentioned methods manage to promote the cooling of a casting from surface to center or from one end to the other end, which falls into a dilemma -high cooling speed, high thermal gradient and high residual stress or deformation. Inner chills can realize the cooling from inside, however; they are too small, which limits their cooling capability. Risers, the biggest hot spots, are conventionally considered as feeders during solidification of castings, no special attention has been paid to their slowest cooling feature after solidification which mainly leads to low production efficiency and uneven cooling. So as to solve this dilemma, the authors proposed a method post-solidification intensive cooling of risers (PSIRC) to realize rapid and uniform cooling of heavy steel castings. The application of this method into a hydro turbine blade casting was investigated in this paper.

2. Application of PSIRC into a hydro turbine blade casting

Hydro turbine blades are used for hydraulic turbines. A turbine blade with size 1460mmx1210mm×48mm, as shown in figure 1, was selected for the investigation of the effect of PSIRC. One riser was designed for its feeding. Two castings were cast in Harbin Electric Machinery Co., LTD, one was under normal cooling after solidification, the other was under PSIRC. The material is martensitic stainless steel ZG0Cr13Ni4Mo (Chinese standard, its composition is listed in Table 1). Resin bonded sand mold was adopted. Each casting required 0.9 ton melt. Pouring temperature was 1560°C.

Table 1. The composition of ZG0Cr13Ni4Mo martensitic stainless steel (%)

C	Si	Mn	P	S	Ni	Cr	Mo	Cu
0.03	0.46	0.52-0.55	0.011	0.012-0.025	4.25	13.46-13.61	0.60-0.65	0.08-0.09

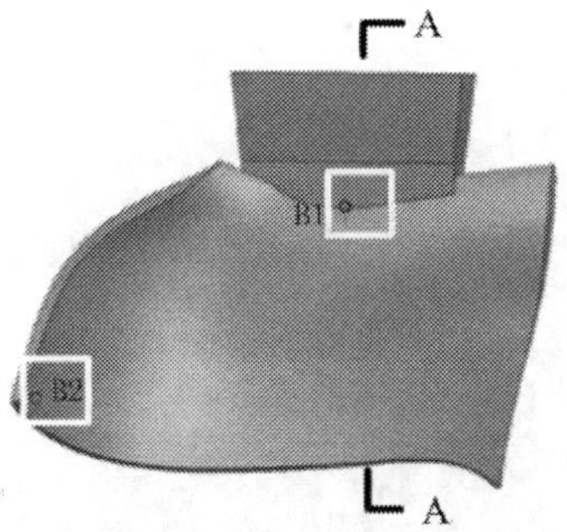

Figure 1. Hydro turbine blade

The functions of riser during solidification and PSIRC are compared in figure 2. During the solidification process, the riser serves as a feeding source with a feeding channel, the heat flows from the riser to the bottom of the blade. As its solidification finished, PSIRC is taken into application. Firstly, clean the riser top and then blow forced air onto the riser top surface to speed its cooling. As PSIRC is adopted, the riser serves as a cooling source, it draws heat out from the riser first, and then from the center of the casting. So the blade can realize fast and uniform cooling. In PSIRC, the cooling of riser reverses the feeding channel during solidification into a cooling channel.

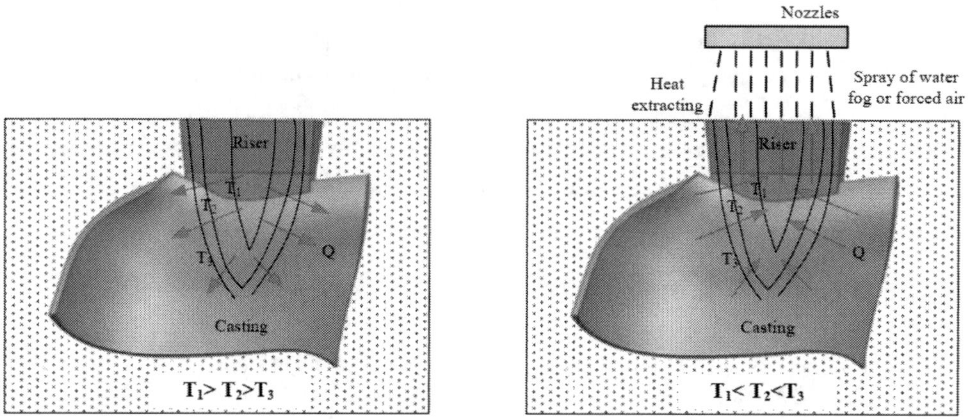

Figure 2. Schematic diagrams of functions of riser during solidification (a) and afterwards with PSIRC (b) for a turbine blade casting

A close-loop PSIRC system was developed, as shown in figure 3. The measured temperature difference between the thinnest B2 and the thickest section B1 (the root of the riser) was sent back to control the operation of an electromagnetic valve to regulate the flow of cooling air on the riser top of the casting. Algorithm was programmed into the control board for the control of forced cooling.

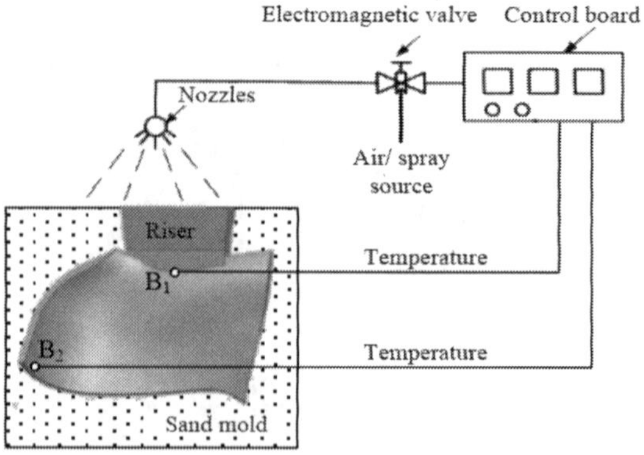

Figure 3. Close-loop PSIRC control system

The onsite experiment is shown in figure 4. As the temperature of B1 dropped to 550°C, 18.3 hours after pouring, the insulation layer of the riser top was cleaned and PSIRC systems began to work with forced air speed 27.9m/s. The cooling speed of B1 was controlled 20°C/h by switching on and off air flow from the nozzle. The nozzle was tilted 100mm atop the riser, as shown in figure 5a. The air speed was measured by a wind gauge and the temperature of the riser top was measured by an infrared camera Flir T250. It can be seen from figure 5b that the forced air flow covered half of the riser top.

(a) thermocouples embedded during molding (b) pouring

Figure 4. Onsite experiment of the turbine blade

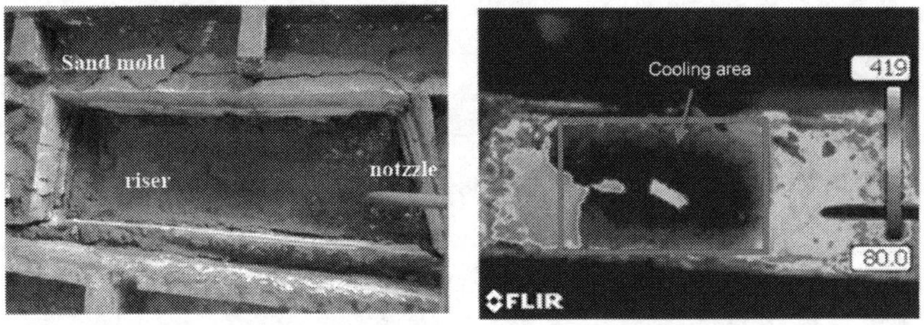

(a) Riser top (b) Infrared image of the riser top

Figure 5. Forced cooling of riser of the hydro turbine blade

3. Results and analysis

3.1 Temperature results

The cooling curves of B1 with the PSIRC method were compared with the tradition one. With the new method, it just took 30 hours to drop to 300℃, while it takes 54 hours for the traditional method. So, 24 hours were saved. The cooling efficiency increased 40%, as shown in figure 6.

The temperature difference of the thickest and thinnest locations of the blade is shown in figure 7. As the PSIRC applied, their temperature difference dropped faster, that means more uniform distribution of temperature and finally less stress and deformation. And after 30hours, the temperature of B2 was higher than that of B1, i.e., heat transferred from B2 to B1.

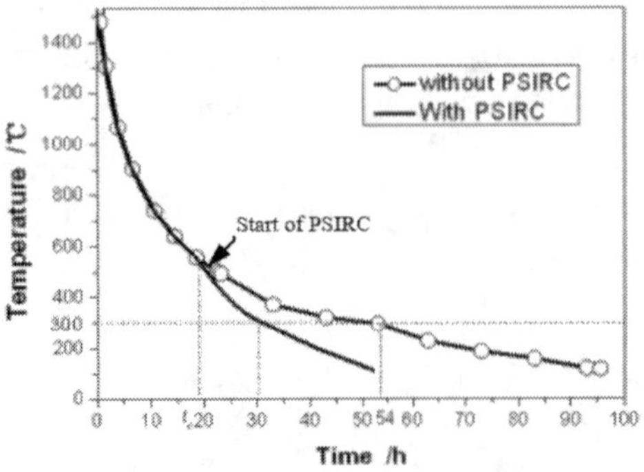

Figure 6. Comparison of the cooling curves of B1 with and without PSIRC

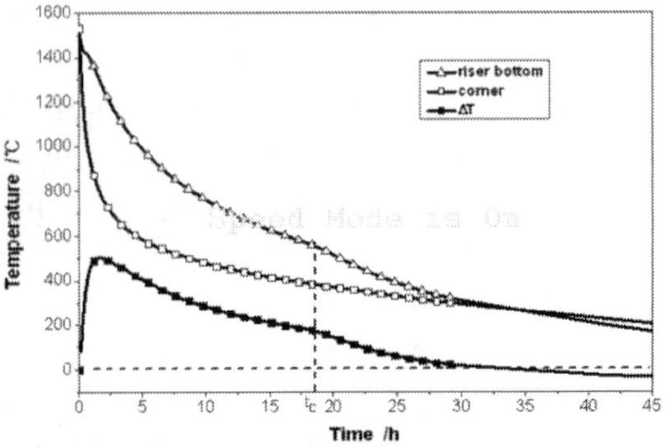

Figure 7. Temperature difference between B1 and B2 of the blade

To study the heat transfer during the casting process, thermocouples were embedded in the sand mold with different distances to the casting, and their cooling curves are shown in figure 8. At the beginning, the temperature of the sand mold dropped with the distance from the casting surface. However, after cooling of 23 hours, the temperatures of P2, P3 and P4 came to be higher than that of P1, P4 higher than P3, P3 higher than P2, P2 higher than P1. That means the thermal gradient at the inner side of the sand mold reversed its direction, pointing to the casting. It indirectly proved that the temperature of casting surface was higher than the core, heat transferred from the casting surface to its center.

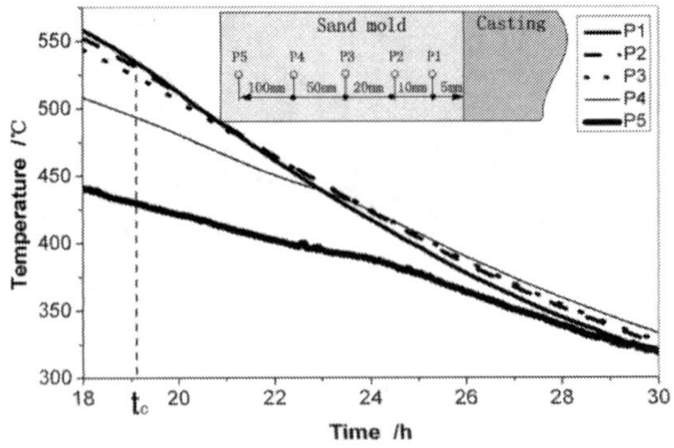

Figure 8. Cooling curves of the points in the sand mold

The casting process of the blade was simulated by using ProCAST. The heat transfer coefficient at the riser top during PSIRC was taken as 200 W/m^{2}K. The heat transfer results are shown in figure 9. The simulated results also proved the effect of the intensive riser cooling. At the beginning of casting process, the casting was suddenly cooled by the sand mold, so the heat flux was pointing to the mold across the casting/mold surface, as shown in figure 9a. As the procession of the cooling, heat transfer from casting to mold came to be very slow, the heat inside of the casting mainly flew from the center of riser to the bottom of the casting, only little heat flew from the riser center to its top sidewall, as shown in figure 9b. As PSIRC was adopted, the hot spot shifted from the riser center to the middle of the casting, so the heat flew from the hot spot toward the top surface of the riser, and some heat still flew to the bottom of the casting, as shown in figure 9c. The total heat released by the casting, 6.18×10^5KJ, was absorbed by the mold just before the adoption of PSIRC. Thirty hours later after pouring, more heat, 1.57×10^5KJ, was released by the casting, it totally escaped to the atmosphere from the riser surface at an average rate of 1.29×10^4KJ/h. Even some heat in mold, 5.6×10^4KJ, was absorbed by the casting and then released from the riser top surface.

The temperature difference between the central line and surface under PSIRC is compared to that without PSIRC, as shown in figure 10. Without PSIRC the temperature at the center was always higher than the surface. However, as PSIRC was adopted, the center's temperature got lower than that at the surface, so, the blade was cooled inside-out. With PSIRC, the hot spot was gradually pushed from the center of the riser to almost the bottom of the casting, as shown in figure 11. Thus, the significant effect of PSIRC is proved by simulated results.

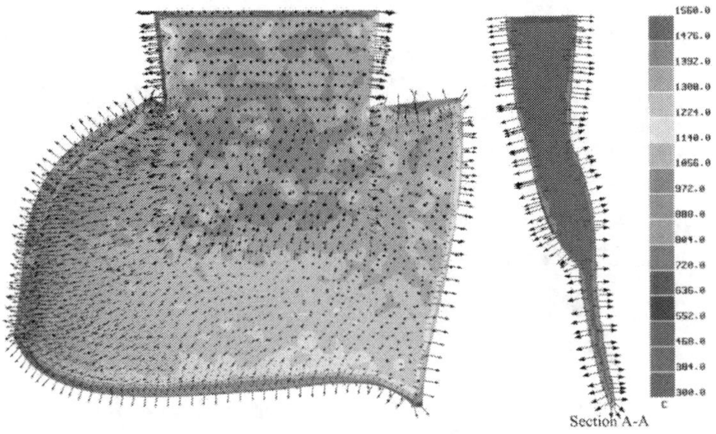

(a) 0.25h

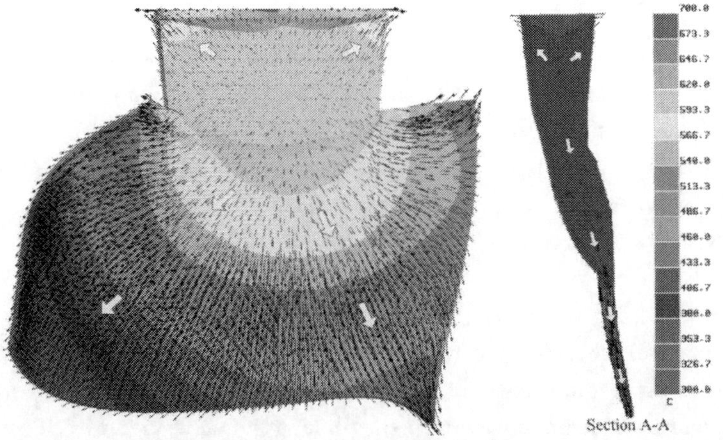

(b) 17.8h before the adding of PSIRC

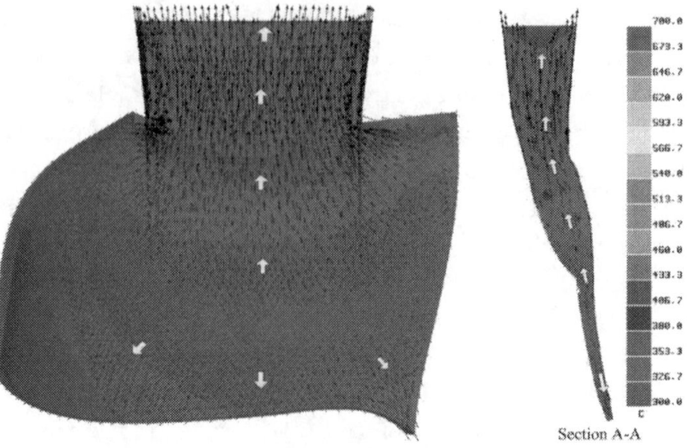

(c) 30h with PSIRC

Figure 9. Heat flow during the casting process

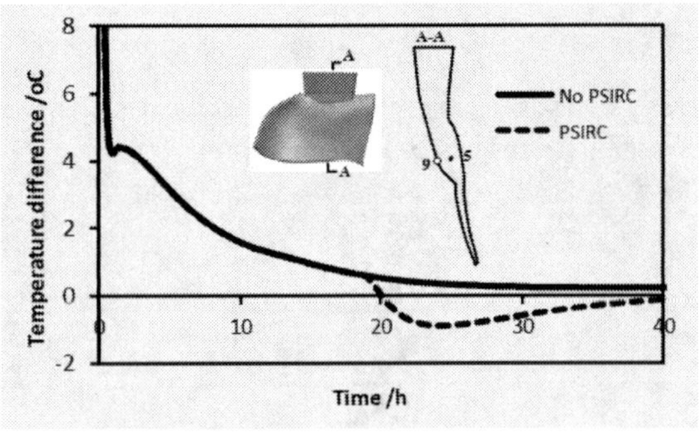

Figure 10. Temperature difference between the center and surface

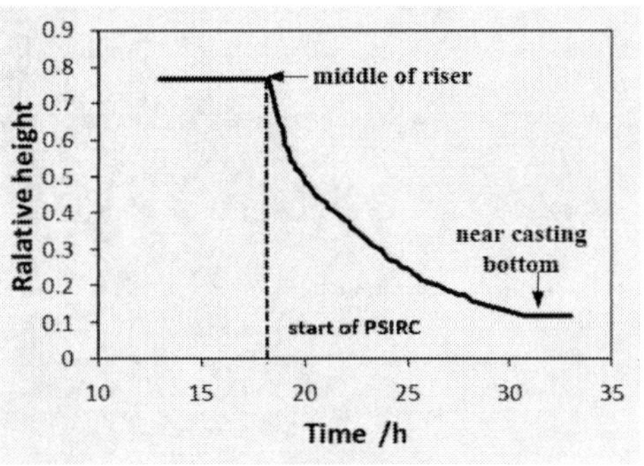

Figure 11. Location of the hot spot during cooling process

3.2 Effect of PSIRC on residual stress

The residual stress at B1 at as-cast status was measured by X-ray method. The residual stress was greatly reduced with PSIRC from 151.4 MPa to 26.3 MPa, which is of significance for the control of residual stress. The reason is the even cooling as PSIRC was applied.

The casting was finally passed though ultrasonic test. Meanwhile, the surface quality, mechanical property and microstructure all met the requirement. Therefore, the PSIRC method is possible and valuable for heavy steel casting production.

3.3 Intermittent intensive cooling

During the intensive cooling of the riser of the blade, there still existed significant thermal gradient along the cooling channel. Therefore, intermittent intensive cooling was designed for the hydro turbine blade casting, 300 minute cooling and 300 minute suspension alternatively, as shown in figure 12. The simulated cooling curves of the points on the center line is shown in Figure13. It can be seen that the temperature difference gets smaller during the suspension of PSIRC, such as from 237℃ to 85℃ for the first cycle. Thus, intermittent intensive cooling of the riser can achieve much more uniform temperature distribution which is helpful for the reduction of residual stress and deformation. The step of the intermittent intensive cooling can be reduced to achieve smaller thermal gradient.

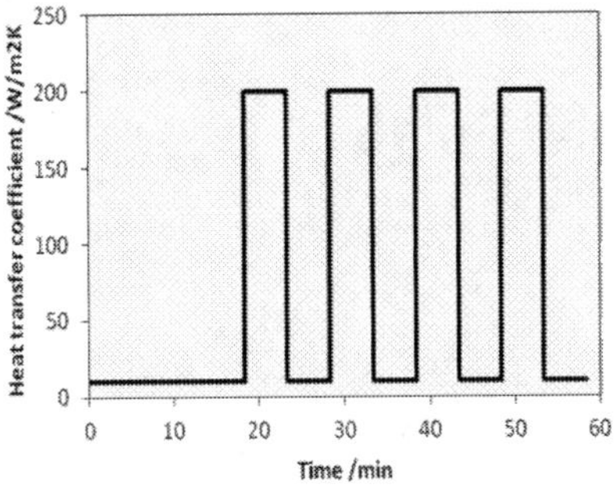

Figure 12. Intermittent cooling schedule of PSIRC

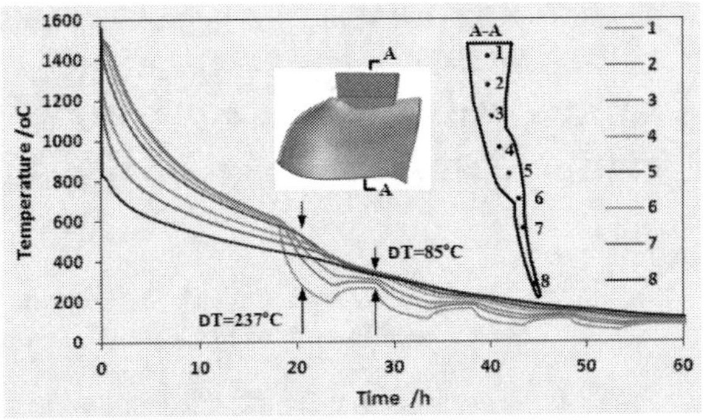

Figure 13. Cooling curves of the points on section A-A during intermittent PSIRC

4. Conclusions

1) Post solidification intensive riser cooling method (PSIRC) was proposed for heavy steel castings to realize fast and uniform cooling, in which risers were cooled by forced air after the solidification of a casting, turning risers into cooling passages during the cooling process of castings in contrast to feeding passages during the solidification process. It may improve production efficiency and reduce residual stress and deformation.

2) One day was saved for a hydro turbine blade casting of 0.9 ton melt before shakeout under PSIRC, its residual stress was greatly reduced. The hot spot was pushed almost to the casting bottom and at the middle height, inside out cooling was achieved. To further improve the effect of PSIRC, intermittent cooling can make more uniform cooling.

Acknowledgement

The project is funded by National Science and Technology Major Project of the Ministry of Science and Technology of China under Project No. 2012ZX04012011.

References

[1] Kang J W, Huang T Y and Liu B C 2008 *China Foundry* **5** 1-6
[2] Yu H L, Kang J Wand Huang T Y 2010 *Frontiers of Materials Science in China* **4(4)** 332-338
[3] Zhao C Z, Chen H M and Yin S B 1997 *Foundry Technology* **18(4)** 3-5
[4] Zhao C Z, Gong J Y, Yang X H and Dong J H 2000 *China Foundry Machinery & Technology* **21(4)** 31-32
[5] Ning Z L, Cao F Yand Sun J F 2007 *China Patent* 200710144925, 2008-06-11
[6] Lee S M and Lee W J 2005 *Journal of materials engineering and performance* **14** 388-394

A numerical study of zone-melting process for the thermoelectric material of Bi_2Te_3

W C Chen[1], Y C Wu[1], W S Hwang[1], H L Hsieh[2], J Y Huang[2] and T K Huang[2]
[1] Department of Materials Science and Engineering, National Cheng Kung University, Tainan, Taiwan
[2] China Steel Corp, Kaohsiung 81233, Taiwan

E-mail: wshwang@mail.ncku.edu.tw

Abstract. In this study, a numerical model has been established by employing a commercial software; ProCAST, to simulate the variation/distribution of temperature and the subsequent microstructure of Bi2Te3 fabricated by zone-melting technique. Then an experiment is conducted to measure the temperature variation/distribution during the zone-melting process to validate the numerical system. Also, the effects of processing parameters on crystallization microstructure such as moving speed and temperature of heater are numerically evaluated. In the experiment, the Bi2Te3 powder are filled into a 30mm diameter quartz cylinder and the heater is set to 800°C with a moving speed 12.5 mm/hr. A thermocouple is inserted in the Bi2Te3 powder to measure the temperature variation/distribution of the zone-melting process. The temperature variation/distribution measured by experiment is compared to the results of numerical simulation. The results show that our model and the experiment are well matched. Then the model is used to evaluate the crystal formation for Bi2Te3 with a 30mm diameter process. It's found that when the moving speed is slower than 17.5 mm/hr, columnar crystal is obtained. In the end, we use this model to predict the crystal formation of zone-melting process for Bi2Te3 with a 45 mm diameter. The results show that it is difficult to grow columnar crystal when the diameter comes to 45mm.

1. Introduction

The green energy technology is always the hot issue in the field of science and engineering. The thermoelectric material plays an important role in green energy technology [1]. The thermoelectric conversion efficiency depends on the figure of merits (ZT value) of itself [2]. In this study, Bi2Te3 is chosen because of its high ZT value at ambient temperature and are frequently applied to thermoelectric module for power generation [3,4]. Bi_2Te_3 showed high anisotropy and said to have higher ZT value perpendicular to the c axis [1,5]. In order to keep its anisotropy to acquire high ZT value, big columnar grain structure should be produced. With big columnar grain, electrical conductivity will be enhanced and thermal conductivity will be reduced. Thus ZT value will be enhanced. In this study, zone-melting which is a unidirectional solidification process is chosen to manufacture the Bi_2Te_3.In the industry, cylindrical ingot with 17mm and 25mm diameter can be mass produced. But the manufacturing time of zone-melting is long which is around a day. So increase ingot's diameter is the way to raise the output of Bi_2Te_3. But temperature field will change with increasing diameter of Bi_2Te_3. With an aim to maintain the quality of bigger ingot, the processing parameters must be adjusted. It is expensive to use experimental method to find the processing parameters leading to big columnar grain because of the high price of Bi_2Te_3. So this study use numerical method to acquire the processing parameter which lead to big columnar grain which make Bi_2Te_3 have high thermoelectric conversion efficiency.

2. Mathematical and Experimental method

The most important part of this study is to validate the simulation system. The simulation system will be validated by temperature field distribution and microstructure. In temperature part, an experiment is conducted to measure temperature profile during the zone-melting process which will be compared with the temperature profile of simulation. A k-type thermocouple is inserted into the quartz tube which is filled with Bi_2Te_3 powder to measure the temperature field during the melting and solidification process as showed by Fig.1. Then, the experimental microstructure will be compared with the results of simulation, too. Following are mathematical methods used in this simulation system.

2.1 Heat transfer module

In this study, convection and mass transfer are not considered. The heat transfer mechanism of zone-melting process is dominated by conduction and radiation. The equations used in heat transfer in this system are as follows

$$\rho C_p \frac{\partial T}{\partial t} = \lambda \nabla^2 T + \rho \Delta H \frac{\partial f_s}{\partial t} \tag{1}$$

$$-k_a(n \cdot \nabla T) = h(T - T_a) + \varepsilon_a \sigma (T^4 - {T_a}^4) \tag{2}$$

This (2-1) is revised from Fourier's fundamental laws of heat transfer to calculate the three-dimension temperature field during zone-melting process, where ρ is the density, C_p is the heat capacity under constant pressure, λ is the thermal conductivity, H is the latent heat and, f_s is the solid fraction of the material used. The (2-2) [6] is the radiation part, derived by energy balance where h is the heat transfer coefficient, ε_a the thermal emissivity of the quartz, σ is the Stefan-Boltzmann constant, and T_a is the effective ambient temperature. The whole heat transfer mechanisms in the system are as Fig 2. The heat is transfer from heaters into the quartz tube and Bi_2Te_3 by radiation. Then the heat is transfer from the Bi_2Te_3 to the cooler and air by radiation or through conduction of Bi_2Te_3 in the cooling process. The parameters used in simulation are listed in Fig 3.

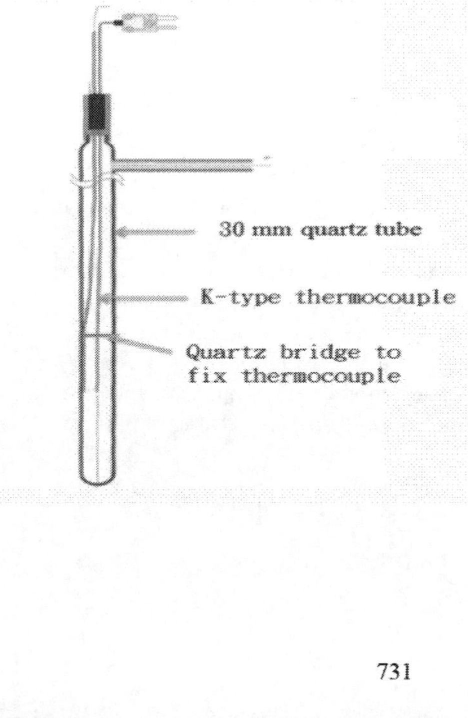

Figure 1. Experimental setup

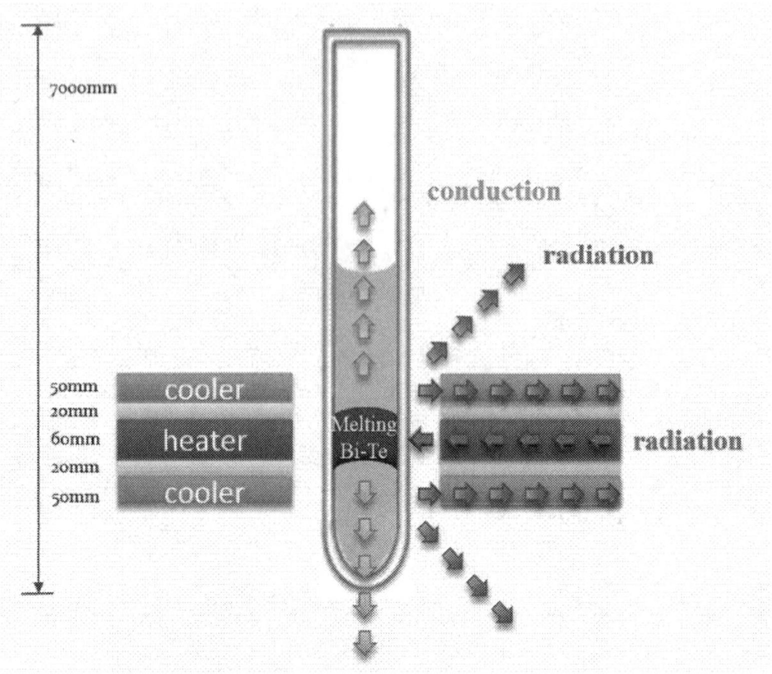

Figure 2. Heat transfer mechanism and dimensions of setup

	Bi-Te	Quartz	Heater	Cooler
Thermal conductivity (W*K/m)	1.48~2.67	1.5~2.5	28.7	330~400
Density(kg/m³)	783	2320	770	8160
Specific heat (kJ*K/kg)	129.3~ 194.4	0.73~ 1.2	0.46	0.47
Latent heat (kJ/kg)	232.9	---------	---------	----------
Solidus(°C)	530	---------	---------	----------
Liquidus(°C)	586	---------	---------	----------
Emissivity	0.8	0.93	0.63	0.7

Figure 3. Parameters used in simulation

2.2 Microstructure (CAFÉ module)

In this study, simulation of microstructure is based on the macroscopic temperature result. Numerical simulation of microstructure is according to CA (cellular automaton) method [7, 8, 9], which is developed by Rappaz and Gandin. CA method use deterministic model to calculate the growth rate of dendrite and use probabilistic model to deal with the distribution of nuclei and direction of growth. This method is based on heterogeneous nucleation model and continuous nucleation model. Then use Gaussian distribution as a relation between nucleation density and undercooling which is governed by (3-3) and (3-4) and showed by Fig. 4.

$$n(\Delta T) = \int_0^{\Delta T} \frac{dn}{d(\Delta T')}\left[1 - f_s(\Delta T')\right]d(\Delta T') \quad (3)$$

$$\frac{dn}{d(\Delta T')} = \frac{n_{max}}{\sqrt{2\pi}(\Delta T_\sigma)} exp\left[-\frac{1}{2}\left(\frac{\Delta T'-\Delta T_n}{\Delta T_\sigma}\right)^2\right] \quad (4)$$

where $n(\Delta T)$ is density of grain, $f_s(\Delta T')$ is solid fraction, n_{max} is the maximum density of nuclei, ΔT_n means nucleation undercooling , and $\frac{dn}{d(\Delta T')}$ is found according to the Gaussian distribution. The calculation of this microstructure module is based on the result of heat transfer module because the undercooling has a large impact on nucleation. When the heater go through the Bi_2Te_3, the Bi_2Te_3 melt firstly. Then melted Bi_2Te_3 will be cooled by radiation and conduction. In this time, undercooling occurs, resulting in the nucleation and grain growth.

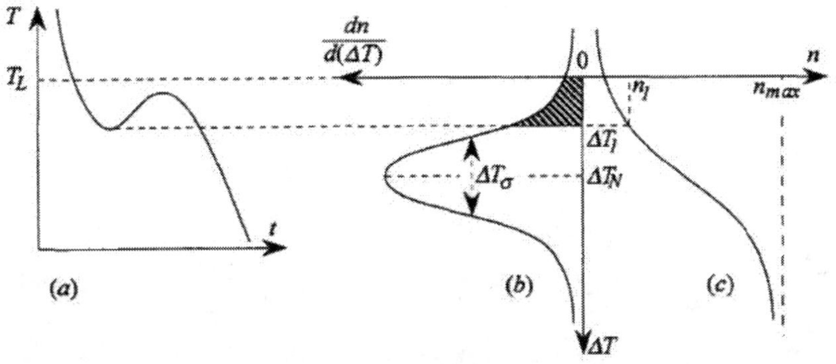

Figure 4. The relation between nucleation density and undercooling.

When the undercooling reaches nucleation conditions, specific CA cells start to nucleate. In a time step δt, temperature of a component reduces δT, then the undercooling increases $\delta(\Delta T)$. At this time, new grain density of nucleation on the liquid metal surface showed as (2-5):

$$\delta n_s = n_s[\Delta T - \delta(\Delta T)] - n_s(\Delta T) = \int_{\Delta T}^{\Delta T+\delta(\Delta T)} \frac{dn_s}{d(\Delta T')} d(\Delta T') \quad (5)$$

In other words, it will give a site and direction of nucleation by random as the basis of following growth of grain. In the heterogeneous nucleation part, the relation between nucleation density and undercooling is explained by Gaussian distribution theory. In the grain growth part, growth rate is calculated using Kurz-Giovanola-Trivedi (KGT) model which is governed by (2-6)

$$\upsilon = a_2 \Delta T^2 + a_3 \Delta T^3 \quad (6)$$

Where υ is the growth rate, a_2 and a_3 are growth kinetic parameters dependent upon the composition of the alloy.

CA cell size(micro-meters)	Nb. Crystallographic orientation	Volume nucleation		
100	5000	ΔT_N	ΔT_σ	N_{max}
		10	0.05	$7.5x10^7$
Nb. Cell in a block	Seed random number	Surface nucleation		
10x10x10	100	ΔT_N	ΔT_σ	N_{max}
		1	0.05	$2.5x10^5$

Figure 5. Parameters used in simulation (ΔT_N : averaged undercooling, ΔT_σ : standard deviation of undercooling, N_{max} : maximum grain density)

IOP Conf. Series: Materials Science and Engineering **84** (2015) 012094 doi:10.1088/1757-899X/84/1/012094

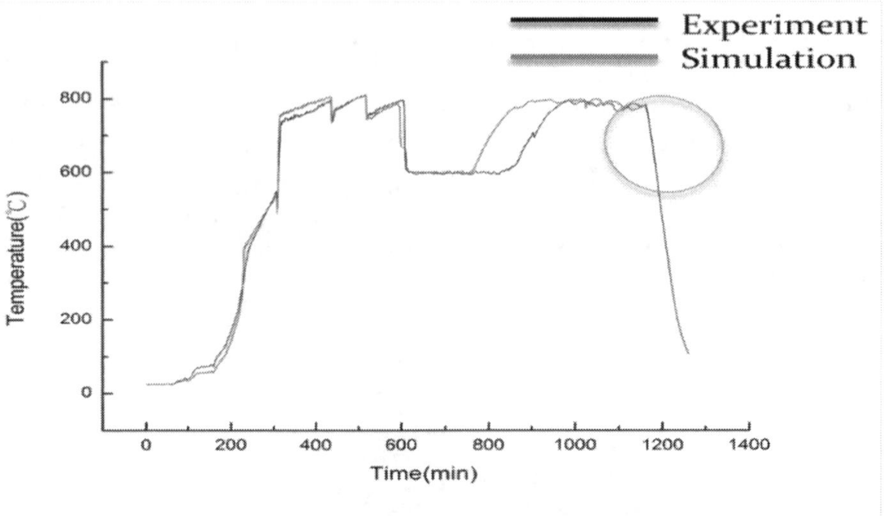

Figure 6. The temperature field of simulation and experiment

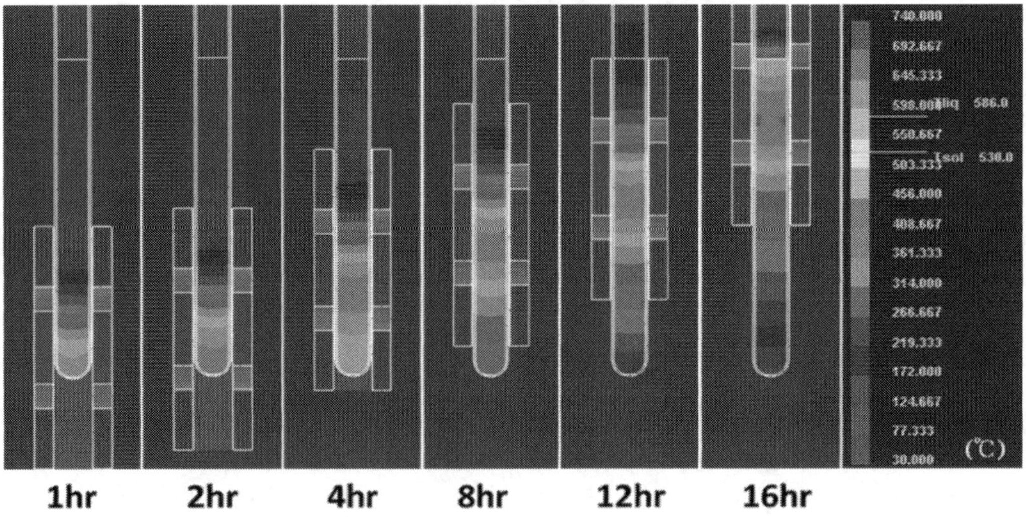

Figure 7. The temperature profile of the whole specimen

3. Results and Discussions

It is important to verify the accuracy of model when using numerical method to simulate manufacturing process. In this study, the experimental parameter are as follows: Heater is heated up to 770℃ in first 30 minutes, then heated to 800℃ from 30 to 60 minutes. After 60 minutes, the heater is fixed at 800℃ to the end of the process. The heater starts to move up along the quartz tube with a moving speed of 1.25 cm/hr from 120 minutes to the end of the process. The parameters of simulation are all the same with the experimental parameters. The results of validation showed on Fig. 6.

The results show that the trend of temperature is nearly the same. There is some area with big difference from 600℃ to 800℃ of heating up part. This part is acceptable because the most concerned part of the curve is the solidification area which is related to the following microstructure. That is, the part pointed by blue circle on the Fig. 6 which is well matched. The solidus temperature and liquidus temperature are 530℃ and 586℃ respectively. In other words, the range between solidus temperature and liquidus temperature is completely located in the well matched part. In other words, the heat transfer module is validated and said to have enough confidence in predicting the temperature field of the zone-melting process. Fig. 7 is the temperature profile of whole specimen with the change of time. Then the validated heat transfer module is applied to the following microstructure module. The morphology of experimental grain is compared to the simulated grain. The results of Fig. 7 and Fig. 8 showed that they both have tendency of columnar grain and the grain size also have the same tendency from the bottom to top of the Bi_2Te_3 ingot. And this microstructure module is been used to predict the morphology at different processing parameters. It's found that when the moving speed is slower than 17.5 mm/hr, columnar crystal is obtained. Then a 45 mm model is build which is using the same grain growth parameters to predict the morphology of grain. The results showed that it is hard to produce big columnar grain when the ingot comes to 45 mm because the cooling is not as homogeneous as the 30 mm manufacturing process.

Figure 8. The longitudinal section morphology of the experimental and simulation with same processing parameters, heater is set to 800℃ with a moving speed of 12.5 mm/hr.

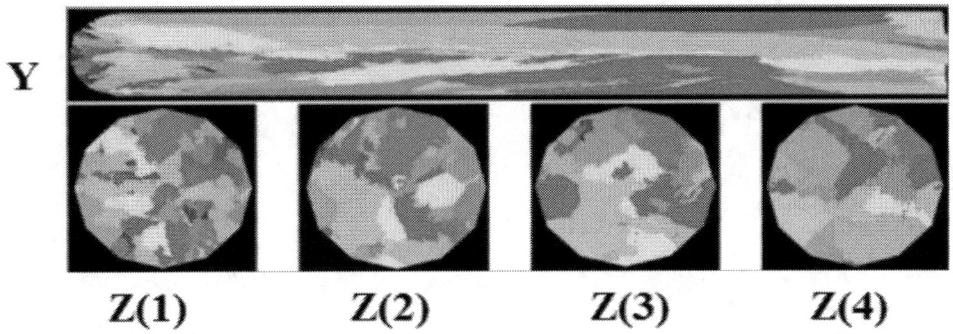

Figure 9. The transverse section morphology of simulation with processing parameters, heater is set to 800℃ with a moving speed of 12.5 mm/hr.

4. Conclusion

A validated numerical model is been built to predict the temperature field and following microstructure of 30 mm Bi_2Te_3 manufacturing by zone-melting process. It's found that when the moving speed is slower than 17.5 mm/hr, columnar crystal is obtained. In the end, we use this model to predict the crystal formation of zone-melting process for Bi_2Te_3 with a 45mm diameter. The results show that it is difficult to grow columnar crystal when the diameter comes to 45mm.

References

[1] Tritt T. M. 2002 Encyclopedia of Materials：Science and Technology 19,ed Buschow K H J(Amsterdam: Elsevier) 1-11
[2] Rowe D M 1995 CRC Handbook of Thermoelectrics (CRC Press)
[3] Rowe D M 1994 MTS **27** 43-50
[4] Elgenk M S, Saber H H and Caillat T 2003 Energy Convers Manage **44** 1755-1772
[5] Snyder G J and Ursell T S 2003 Phys Rev Lett **91** 148301-148303

[6] Siegel R and Howell J R 1992 Thermal Radiation Heat Transfer ,ed Siegel R (Hemisphere Publishing Corporation) 263-311.
[7] Rappaz R and Gandin C A 1993 Acta Metall **41** 345-360
[8] Gandin C A and Rappaz R 1994 Acta Metall **42** 2233-2246
[9] Gandin C A, Desbiolles J L, Rappaz M and Thevoz P Metall. Mater. Trans. A **30A** 3153-3165

Acknowledgments
This study is support by New Materials Research and Development Department, China Steel Corporation in Taiwan, which are gratefully acknowledged.

Simulation and validation of two-phase turbulent flow and particle transport in continuous casting of steel slabs

K Jin[1], B G Thomas[1,4], R Liu[2], S P Vanka[1] and X M Ruan[3]

[1] Department of Mechanical Science and Engineering, University of Illinois at Urbana-Champaign, Urbana, IL 61801, USA
[2] ArcelorMittal Global R&D in East Chicago, 3001 East Columbus Drive, East Chicago, IN 46312, USA
[3] Steelmaking Research Department, Research Institute, Baoshan Iron &Steel Co., Ltd., 889 Fujin Rd., Shanghai 201900, P. R. China

E-mail: bgthomas@illinois.edu

Abstract. In continuous steel casting, argon gas is usually injected at the slide gate or stopper rod to prevent clogging, but entrapped bubbles may cause defects in the final product. To better understand this, the flow of molten steel and the transport and capture of argon gas bubbles have been simulated and compared with plant measurements. First, the flow field was solved with an Eulerian k-ε model of the steel, which was two-way coupled with a Lagrangian model of the large bubbles using a Discrete Random Walk method to include dispersion of bubbles due to turbulence. The asymmetrical flow pattern predicted on the top surface agreed well with nailboard measurements. Then, the motion and capture of over two million bubbles were simulated using two different capture criteria. Results with the advanced capture criterion agreed well with measurements of the number, locations, and sizes of captured bubbles, especially for larger bubbles. The relative capture fraction of 0.3% was close to the measured 0.2% for 1mm bubbles, and occurred very near the top surface. The model presented here is an efficient tool to study the capture of bubbles and inclusion particles in solidification processes.

1. Introduction

Argon bubbles captured during the continuous casting of steel are a major cause of defects, such as blisters and slivers, in rolled steel products. During the casting process, Ar gas is usually injected at the slide gate or stopper rod. The jet of molten steel then carries those Ar bubbles pass through the Submergence Entry Nozzle (SEN) and into the mold cavity. During this transport, the bubbles may pick up inclusions, thereby forming inclusion clusters which lower the steel quality if captured [1, 2]. Bubbles entering the mold region have 3 possible fates: (1) some reach the top surface, pass through the slag layer and escape harmlessly to the atmosphere; (2) some are captured near the meniscus and lead to surface defects; (3) some are captured deep in the caster and cause internal defects.

Argon bubbles also affect the flow pattern, including surface level fluctuations, slag entrainment, particle trajectories, and other phenomena important to steel quality. Previous models of two-phase Ar and steel flow in continuous casting [2–6] have used different methods to achieve the necessary two-way coupling to predict the flow pattern, and reveal the importance of the larger bubbles.

Relatively few studies [2–4] have investigated quantitatively the capture rate and distribution of inclusion particles in continuous casting. Yuan, Thomas and Vanka [3] performed particle capture simulations for small bubbles (less than 40μm) in a thin-slab steel caster using Large Eddy Simulation (LES) where the effect of transient local turbulent eddies on particle transport was automatically included. A removal rate of 8% of small inclusions was predicted, independent of both particle size and density, which suggests that their particles were too small to deviate significantly from the surrounding fluid flow. Zhang and Wang [5] performed 3D one-way coupled $k-\varepsilon$ simulations, adding sink terms at the solidification front to model the momentum loss due to solidification, and a random-walk method [7, 8] to include turbulent dispersion of small (5μm) particles [3] in a simulation of a full-length billet caster. Capture was assumed when particles touched the solidification front, defined by a solid fraction of 30% or 60%. This capture criterion may be reasonable for particles smaller than the PDAS [2–4].

To properly predict the capture rate and location of different Ar bubbles/particles, a suitable capture criterion is needed. Yuan and Thomas [2, 3] developed a particle capture criterion based on a local force balance on particles reaching the solidification front. This model includes the effects of particle size, Primary Dendrite Spacing (PDAS), concentration gradient forces and other effects. This capture criterion was successfully validated with previous measurements, so is used in the current study. Further work with this criterion [2–4] has investigated the entrapment of slag inclusions, alumina clusters, and bubbles during continuous slab casting. They found that to more accurately predict the removal rate of particles (error within ±3%), more than 2500 particles should be injected into the domain. The work presented here investigates different capture criteria ("simple" and the previous criterion [2, 3]) for predicting Ar bubbles with different diameters in a real commercial continuous caster.

2. Plant measurements

Plant measurements of fluid flow and bubble entrapment were conducted on the No. 4 caster, which is a conventional (230×1300 mm) continuous steel slab caster, at Baosteel Shanghai in 2012. Top surface velocities were measured with two sets of nailboard dipping tests, which are widely used to measure mold surface flow [4]. Casting conditions and process parameters are given in table 1. Flow rate of the molten steel through the SEN into the mold is controlled by a slide-gate that moved between the geometric center and the Inside Radius (IR) side of the caster. The slide gate was 70% open at 1.5 m/min, as shown in figure 1.

Table 1. Process parameters

Process Parameters	Value
Mold thickness (L_t)	230 mm
Mold width (L_w)	1300 mm
Casting speed (V_c)	1.5 m/min
Argon volume fraction (α)	8.2 vol%
Steel density (ρ)	7000 kg/m^3
Argon density (ρ_p)	0.5 kg/m^3
Steel viscosity (μ)	6.30×10^{-3} kg/(m·s)
Ar viscosity (μ_p)	2.12×10^{-5} kg/(m·s)
Sample length in casting direction (Δz)	150 mm

Argon gas was injected in the slide gate region to prevent reoxidization and clogging. To evaluate the capture of Ar bubbles during the solidification process, 150×150 mm^2 samples were cut from the center and quarter of the Wide Face (WF), and Narrow Face (NF) of the final steel slab, as shown in figure 2. The outer 3 mm of the surface of each sample was milled away (NF or WF), and a 35X optical microscope was used to count the number of bubbles on that surface and their diameters.

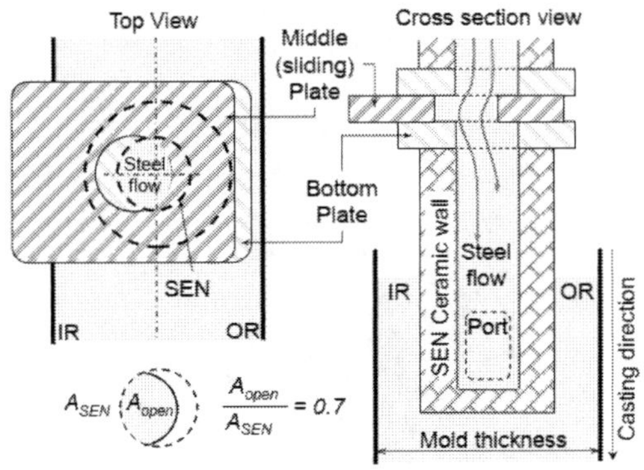

Figure 1. Slide gate configuration

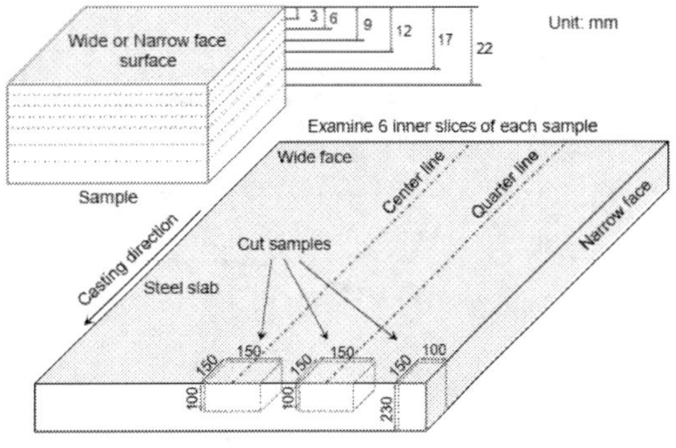

Figure 2. Samples and six examined surfaces

After recording the results, another 3 mm of steel was milled away and the new exposed surface was measured in the same way. This procedure was repeated to examine six layers, located at 3, 6, 9, 12, 17 and 22 mm beneath the outer surface. The symbol s_j is used to represent the distance beneath the outer surface (NF/WF), $s_1 = 3$ mm, $s_2 = 6$ mm, and etc.

3. Computational models and solution procedure

A three-dimensional finite-volume computational model together with Lagrangian particle tracking was applied to study the flow behavior and the transport phenomena of Ar bubbles in a commercial continuous steel caster. First, a steady-state solution of single-phase flow of molten steel was obtained using standard k-ε model. Then, based on that solution, a RANS and Lagrangian Discrete Phase Model (DPM) coupled simulation was used to predict pseudo steady flow, including the effect of Ar gas. The Discrete Random Walk (DRW) model was used to include the effect of turbulence on bubble dispersion. The gas buoyancy affects only the liquid momentum equation, which is a valid approach when gas volume fraction is low (<~12%). Based on this flow field, many bubbles were injected at the SEN inlet to study the transport and capture of Ar bubbles. Two different capture models were implemented into ANSYS FLUENT [7] using User Defined Functions (UDFs) and the results were compared.

3.1. *Bubble size distribution model*

In this work, the total volume of the hot Ar gas V_g was divided into eleven different bubble size groups according to the Rosin-Rammler [9] distribution originally used to describe solid particle distributions, as given in figure 3 and equation (1). The black dashed line is the ideal Rosin-Rammler cumulative distribution F for a mean diameter $\bar{d} = 3$ mm and spread parameter $\eta = 4$.

$$F(d_i) = \frac{V_{g(d<d_i)}}{V_g} = 1 - \exp\left(-\frac{d_i}{\bar{d}}\right)^{\eta} \qquad (1)$$

The blue squares represent the diameter and volume fraction of each group of bubbles of the total Ar volume. The red staircase line is the summation over the volume fraction of bubbles with diameter less than the specified diameter. Note that the Rosin-Rammler distribution passes through the cumulative line of the discrete points. Note also that less than 1% of the bubbles have diameter less than 1 mm.

3.2. *Two-way coupled steel-Ar flow model*

The computation domain contains half of the slide-gate, SEN and mold region (from meniscus surface to 2.5 m below meniscus). The mesh contained ~1 million hexahedral finite volume cells as shown in figure 4. A single-phase steel flow simulation was first carried out with the k-ε model to provide an initial guess for the later two-way coupled steel-Ar flow simulation.

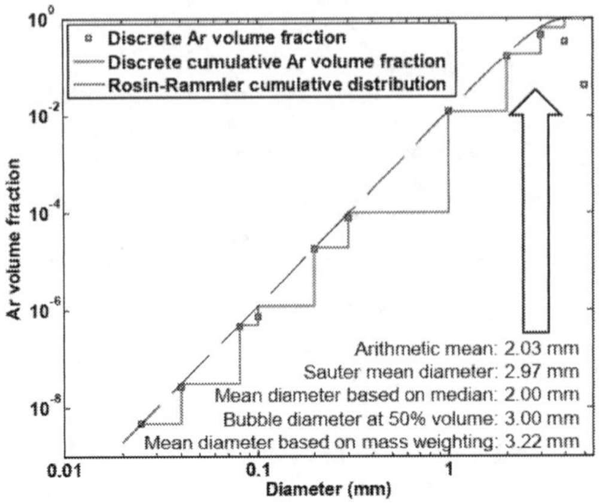

Figure 3. Ar bubbles mean diameters and volume fraction distribution.

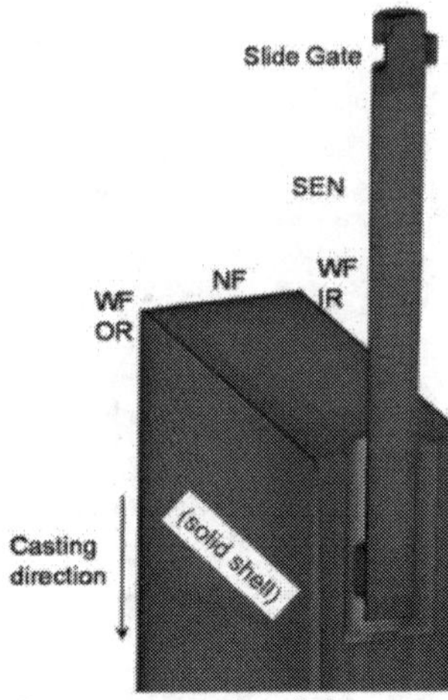

Figure 4. Mesh of the computational domain (Slide gate, SEN and mold)

The solid shell is included in this domain and the shell thickness was determined by $s(mm) = 3\sqrt{t(s)}$. The solid shell (necessary for later study of magnetic field effects) is modeled as a solid zone where the fluid flow equations are not solved. The continuity and momentum equations for liquid steel are:

$$\nabla \cdot (\rho \vec{U}) = S_{\text{mass-sink}} \tag{2}$$

$$\rho \frac{\partial \vec{U}}{\partial t} + \rho (\vec{U} \cdot \nabla) \vec{U} = -\nabla \left(\overline{p} + \frac{2}{3} \rho k \right) + \nabla \cdot \left[(\mu + \mu_t) \nabla \vec{U} \right] + \vec{S}_{\text{momentum-sink}} + \vec{S}_{\text{DPM}} \tag{3}$$

The source term from the DPM model $\vec{S}_{\text{DPM}}$ was included in equation (3) in the two-way coupled simulation step to include the drag of each rising bubble acting on the local fluid [7]. Mass and momentum sinks were added near the shell through UDFs to include the effect of the solidifying shell, as explained elsewhere [4]. The *k-ε* model was used to model turbulence [10].

At the slide gate inlet, fixed uniform velocity inlet boundary conditions $V_{\text{inlet}} = 1.69$ m/s were applied based on the steel flow rate (0.007475 m^3/s) divided by inlet area (0.0044 m^2). The turbulent kinetic energy and its dissipation rate were assumed to be small as 10^{-4} m^2/s^2 and 10^{-5} m^2/s^3. A pressure outlet boundary condition was applied at the domain bottom to include the ferrostatic pressure of the steel (171.5 kPa) based on the distance (2.5 m) below the top surface multiplied by the density and gravity constant. The turbulent kinetic energy and dissipation rate were specified as 10^{-5} m^2/s^2 and 10^{-5} m^2/s^3 for reverse flow from the bottom boundary, respectively. A free slip boundary condition was applied at the top surface. The WF and NF solidification fronts and SEN walls were no slip and no penetration walls.

After the single-phase fluid flow solution was obtained, the Lagrangian DPM tracking and DRW models were used to include the effect of the Ar bubbles on steel flow to correct the flow pattern. The following force balance equation was solved for each individual bubble, of volume V_p.

$$\rho_p V_p \frac{d\vec{u}_p}{dt} = \underbrace{\frac{18\mu}{d_p^2} \frac{C_D}{24} V_p (\vec{u} - \vec{u}_p) \overbrace{\frac{\rho d_p |\vec{u}_p - \vec{u}|}{\mu}}^{\text{Re}_p}}_{\vec{F}_D} + \underbrace{0.5 \rho V_p \left(\frac{D\vec{u}}{Dt} - \frac{d\vec{u}_p}{dt} \right)}_{\vec{F}_v} + \underbrace{\rho V_p \frac{D\vec{u}}{Dt}}_{\vec{F}_p} + \underbrace{\vec{g} V_p (\rho_p - \rho)}_{\vec{F}_b} \tag{4}$$

where the four forces are: drag $\vec{F}_D$, virtual mass $\vec{F}_v$, pressure gradient effect $\vec{F}_p$ and buoyancy/gravity $\vec{F}_b$. The drag force depends on particle Reynolds number Re_p and drag coefficient C_D from Morsi [11].

The first 3 of these forces comprise $\vec{S}_{\rm DPM}$. To save computation time and considering that Ar bubbles < 1 mm comprise <1% of the gas, only large bubbles (1 – 5 mm) were injected and tracked in this two-way coupled calculation. The effect of turbulence on particle dispersion was modeled using the DRW method, where the local fluid velocity $\vec{u}$ is written as $u_i = U_i + u_i'$ where U_i is mean velocity obtained from the RANS model, and u_i' is a Gaussian distributed random velocity fluctuations generated using $u_i' = \zeta\sqrt{\overline{u_i'^2}} = \zeta\sqrt{2k/3}$, where ζ is a random number from a standard normal distribution. The value of ζ was changed to produce new u_i' when the smaller of the eddy lifetime t_e (time scale of an eddy) and eddy crossing time t_{cross} (time for particle to cross an eddy) was reached.

3.3. *Argon bubble tracking and capture models*

After calculating the pseudo steady state two-phase flow solution, ~2.5 million particles were injected into the domain and their trajectories were tracked. The number of bubbles injected with diameter d_i, denoted $N(d_i)$, is determined according to match the plant experiment conditions as follows:

$$N(d_i) = \frac{3\alpha_i}{4\pi(0.5d_i)^3} \cdot \underbrace{\alpha\frac{\Delta z L_t(0.5L_w)}{1-\alpha}}_{V_g} \quad \text{where} \quad \alpha_i = \begin{cases} F(d_i) & i=1 \\ F(d_i)-F(d_{i-1}) & i>1 \end{cases} \tag{5}$$

where V_g is the hot Ar gas volume injected into half of the caster during the time to cast the 150 mm length sample. L_t and L_w are the thickness and width of the mold, respectively. Δz is the sample length which equals 150 mm in this study. α is the total Ar volume fraction which is 8.2%. The volume fractions of bubbles with different sizes are denoted as α_i , and can be determined based on $F(d_i)$ from equation (1). Following these equations, 244,239 bubbles were injected. To make the results statistically more reliable, this bubble tracking calculation was repeated 10 times, meaning ~2.5 million bubbles were tracked for each capture criterion case.

Two different capture criteria were investigated in this work: (1) a "simple" capture criterion which assumes immediate capture when a bubble/particle touches the solidification front; (2) an "advanced" capture criterion, based on the force balance proposed by Yuan and Thomas [2, 4]. For small bubbles less than the PDAS, the particle can enter between the dendrite arms and be captured by entrapment. For bubbles/particles greater than the PDAS, the advanced criterion considers 8 forces acting on a spherical bubble/particle touching three dendrite arms, shown in figure (5). The bubble is suspended and then captured if the tangential drag forces from the flowing fluid that try to rotate it are insufficient to overcome a force balance with the other forces acting on the particle. These forces include those in equation (4), and also the lift force, lubrication force, Van der Waals force, and surface tension gradient force. Details on these forces and this advanced criterion can be found in previous work [2, 4].

3.4. *Plant-sample evaluation model*

To compare with plant measurements, the bubbles expected to be observed on each sample surface must be extracted from the simulation data. A captured bubble k with radius r_k and captured a distance beneath outer surface h_k, can be observed on sample surface j located distance s_j beneath the outer slab surface (WF/NF) only if it satisfies one of the following conditions:

$$\{h_k < s_j \;\; and \;\; h_k + r_k > s_j\} \;\; or \;\; \{h_k > s_j \;\; and \;\; h_k - r_k < s_j\} \tag{6}$$

A post-processing calculation was conducted to check these conditions for each bubble, and to extract the number and average diameter of bubbles observable on each examined sample layer.

The observed diameters then were adjusted to reflect the true bubble size, considering that the examined surface rarely passes through the center of the bubble, so the visible circle is smaller than the true 3D diameter. Specifically, the method proposed by Lekakh et al. [12] was used to convert each visible diameter to its true diameter via $0.785d_{\rm true} = d_{\rm visible}$.

4. Model validation

4.1. Top surface velocity

Figure 6 compares the speed across just beneath the top surface centerline from the pseudo steady-state two-way coupled flow results with the plant nailboard measurements. The predicted cross flow from IR toward OR agrees with the plant measurements. This cross flow was not observed in the single-phase flow results. The plant measurements found even stronger cross flow than predicted. However, only two nailboard measurements were available which comprise only 2 snap-shots of the chaotic flow, so perfect quantitative matching is not expected. More accurate transient computations and further measurements would be needed to improve understanding of the steel-Ar flow pattern.

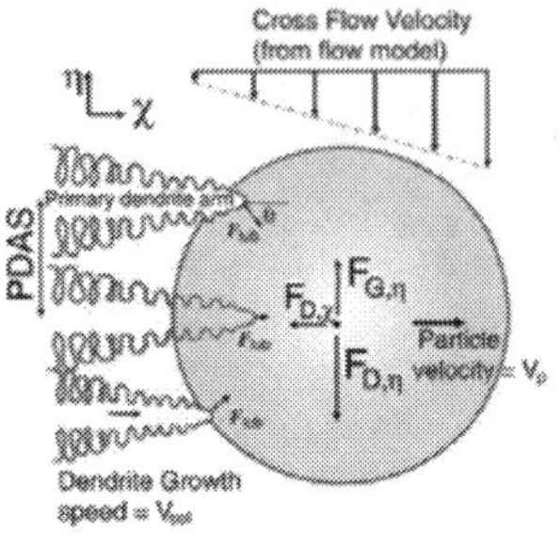

Figure 5. A bubble/particle touching 3 dendrite tips [2]

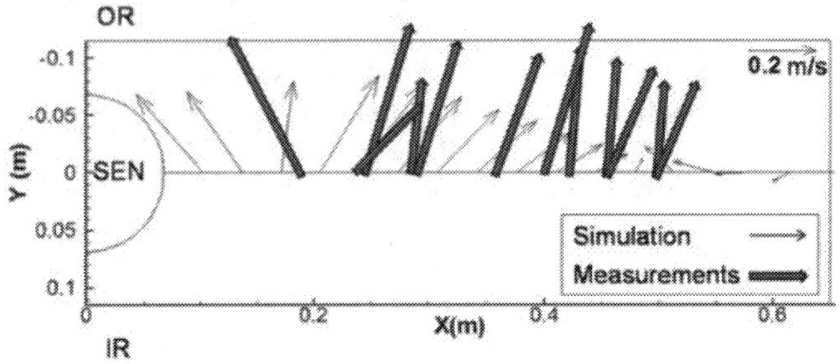

Figure 6. Compare centerline velocity on top surface with plant nailboard measurements

4.2. Capture of small bubbles.

The number and average diameter of bubbles predicted to be captured on the examined sample surfaces are compared with actual measurements in each layer of the plant samples in figure 7. Figures 7(a) and 7(b) show the number of bubbles captured in the central and quarter sample layers, respectively. In the central sample, both criteria under-predicted the number of captured bubbles close to the strand surface. At the quarter region, the advanced capture criterion predicted 20 to 40 bubbles on the first two layers of the examined surfaces, which matches the measurements 35 bubbles. However, the simple criterion over-predicted the number of captured bubbles by 4X, especially in the second layer of the quarter sample. The simple criterion also significantly over predicted the average bubble size, as shown in figures 7(c) and 7(d). The advanced criterion only slightly over-predicted the measured bubble diameters, predicting 0.1 to 0.3 mm, (vs. measured 0.1 to 0.2 mm) in the central sample, and 0.2 to 0.4 mm (vs. measured 0.1 to 0.2 mm) at the quarter sample.

4.3. Capture of large bubbles.

In all of the measured sample layers, ~500 bubbles were observed and only one large bubble (1.4 mm diameter) was observed > 0.5 mm. So, the fraction of all observed captured bubbles that were >1 mm is ψ(1 mm) = 0.2% (1/500). To predict this capture fraction for >1 mm bubbles, table 2 shows: the total number of bubbles injected into the domain during the particle tracking step (column 2); the number of 1 mm bubbles injected (column 3); the total number of bubbles captured in the entire caster (column 4); the number of 1 mm bubbles captured in the entire caster (column 5); the fraction of 1 mm bubbles that were captured (column 6), and the fraction of captured bubbles that were larger than 1mm (column 7). These results show that the advanced capture criterion prediction of 0.3% matches very closely to the plant measurement (0.2%). The simple criterion greatly overpredicts the measurements (by 33X).

Table 2. Capture fractions of 1 mm bubbles

Criteria	$\Sigma N(d_i)$	N(1 mm)	$\Sigma n(d_i)$	n(1 mm)	φ(1 mm)	ψ(1 mm)
Simple	2442390	475640	208944	27799	5.84%	13.3%
Adv.	2442390	475640	137372	432	0.09%	0.3%
Exp.	Unknown	Unknown	~500	1	Unknown	0.2%

5. Results and discussion

The flow pattern calculated in the center plane is shown in figure 8. The standard double-roll flow pattern is modified by the Ar injection. The nozzle jets impinge on the narrow face, and split, sending some recirculating flow upwards and across the top surface towards the SEN. This is met by flow rising up beside the SEN driven by the buoyancy of the Ar gas. Asymmetric swirl caused by the slide gate sends more gas up the inner radius (IR) of the WF, giving complex surface flow with cross flow (figure 6).

Figure 9 shows the capture locations of small bubbles on the WF-IR predicted with the advanced capture criterion, based on the flow field in figure 8. Many 0.3 mm bubbles were captured near the top and SEN side of the caster, with captured bubbles decreasing with distance below the top surface. The horizontal black lines show the location of the examined surfaces in the measurements.

The fraction of each size of bubbles captured is plotted in figure 10. The capture fractions of small bubbles (d_i < 0.1 mm) were ~0.85 (removal fraction ~0.15) for both capture criteria, and were independent of bubble size. This uniformity shows that the capture rate of bubbles smaller the PDAS is governed only by the flow pattern. The capture fraction of large bubbles, on the other hand, decreases dramatically with increasing bubble size. Only 0.1% of 1 mm bubbles are captured with the advanced capture criterion, and no bubble larger than 3mm was captured. The capture fraction drops almost one order of magnitude with increasing bubble size from 1 to 2 mm and from 2 to 3 mm. The removal rate was 99.98% of large bubbles ($1 \leq d_i \leq 5$ mm) , 48.5% of medium bubbles ($0.1 \leq d_i \leq 0.3$ mm) and 16.1% of small bubbles (d_i < 0.1 mm). Previous RANS and LES studies [2] have shown removal rates of 13% and 47% for 0.1 and 0.4 mm slag particles, which agrees closely with the present results.

6. Conclusions

The flow of molten steel, and the transport and capture of argon gas bubbles have been simulated and compared with plant measurements in a continuous slab caster. A two-way coupled Eulerian-Lagrangian model combined with a bubble tracking DRW model has been applied to predict asymmetrical flow on the top surface, which agrees well with nailboard measurements. Two capture criteria were implemented, and the advanced capture criterion showed good agreement with measurements of the

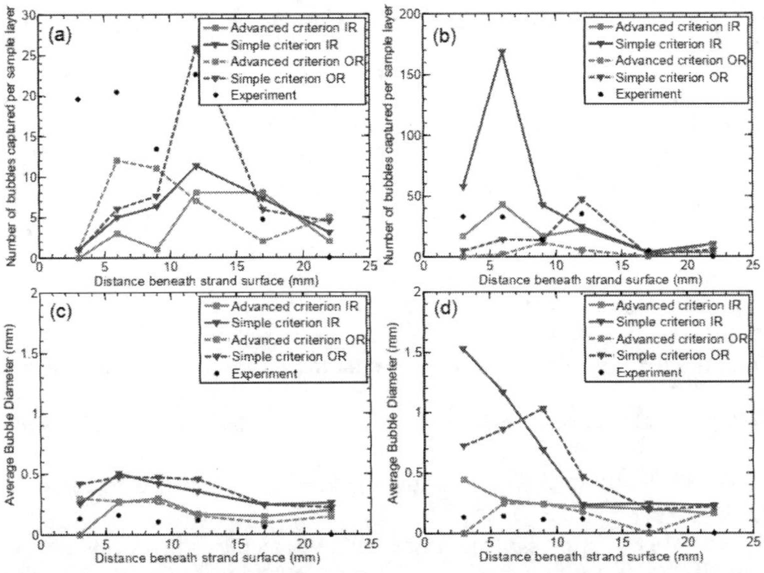

Figure 7. Number of bubbles captured on each sample layer of (a) center and (b) quarter sample; and corresponding average diameter, (c) and (d)

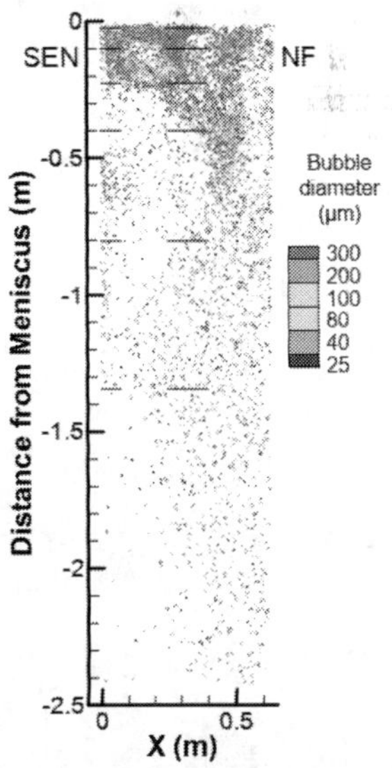

Figure 9. Small bubbles captured on WF-IR with advanced criterion

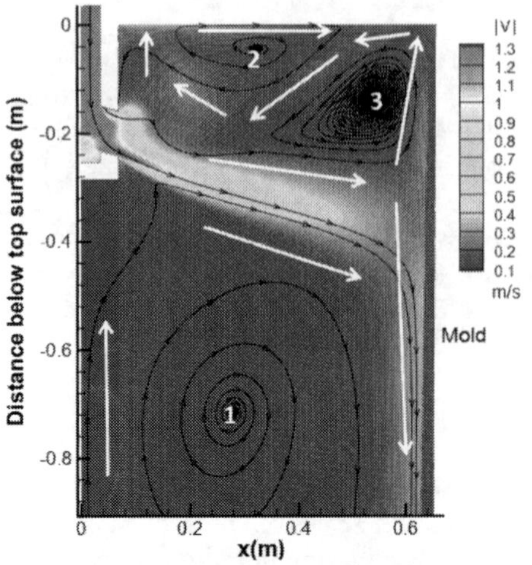

Figure 8. Speed and streamlines in center plane

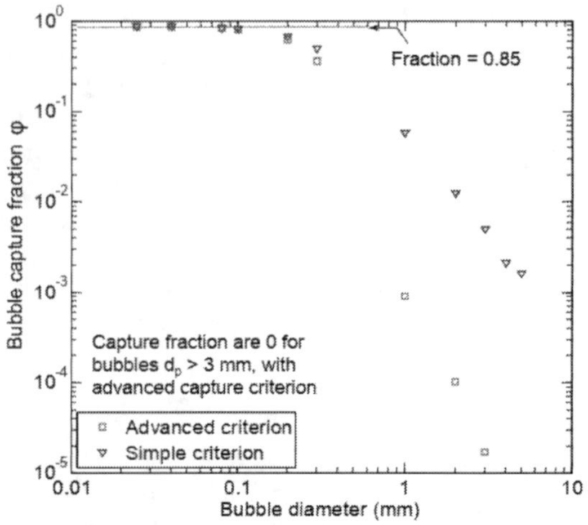

Figure 10. Bubble capture fractions

number, locations, and sizes of captured bubbles, especially for larger bubbles. The capture fraction of 0.3% was close to the measured 0.2% for 1 mm bubbles, and occurred very near the top surface. The model presented here is an efficient tool to study the capture of bubbles and inclusion particles in solidification processes.

Acknowledgments
The authors thank Baosteel, Shanghai, PRC, the National Science Foundation (Grant No. CMMI 11-30882), and the Continuous Casting Consortium, University of Illinois, for support of this project.

References

[1] Zhang L, Aoki J and Thomas B G 2006 Inclusion removal by bubble flotation in a continuous casting mold *Metall. Mater. Trans. B* **37**(3) 361-79
[2] Thomas B G, Yuan Q, Mahmood S, Liu R and Chaudhary R 2014 Transport and Entrapment of Particles in Steel Continuous Casting *Metall. Mater. Trans. B* **45**(1) 22-35
[3] Yuan Q, Thomas B G and Vanka, S P 2004 Study of transient flow and particle transport in continuous steel caster molds: part II particle transport *Metall. Mater. Trans. B* **35**(4) 703-14
[4] Yuan Q 2004 Transient study of turbulent flow and particle transport during continuous casting of steel slabs PhD thesis, University of Illinois at Urbana-Champaign
[5] Zhang L and Wang Y 2012 Modeling the entrapment of nonmetallic inclusions in steel continuous-casting billets *JOM* **64**(9) 1063-74
[6] Yuan Q and Thomas B G 2005 Transport and entrapment of particles in continuous casting of steel *3rd Int. Congress on Sci. and Tech. of Steelmaking* pp 745-62
[7] Ansys FLUENT 12.0 2009 Theory Guide Ansys Inc.
[8] Gosman A D and Ioannides E 1983 Aspects of computer simulation of liquid-fuelled combustors *J. Energy*. **7**(6) 482–90
[9] Rossin P and Rammler E 1933 The laws governing the fineness of powdered coal *J. Inst. Fuel* **7** 29–36
[10] Launder B E and Spalding D B 1974 The numerical computation of turbulent flows *Comput. Method Appl. M.* **3**(2) 269-89
[11] Morsi S A and Alexander A J 1972 An investigation of particle trajectories in two-phase flow systems *J. Fluid Mech.* **55**(02) 193-208
[12] Lekakh S N, Thapliyal V and Peaslee K 2013 Use of an Inverse Algorithm for 2D to 3D Particle Size Conversion and Its Application to Non-Metallic Inclusion Analysis In steel *AISTech Conf. Proc.*

Study of hot tearing and macrosegregation through ingot bending test and its numerical simulation

T Koshikawa[1, 2], M Bellet[1], C A Gandin[1], H Yamamura[3] and M Bobadilla[4]

[1] MINES ParisTech & CNRS, CEMEF UMR 7635, 06904 Sophia Antipolis, France
[2] Nippon Steel & Sumitomo Metal Corporation, Oita Works Equipment Division, 1 Oaza-Nishinosu, Oita City 870-0992, Japan
[3] The Japan Institute of Metals and Materials, 1-14-32, Ichibancho, Aoba-ku, Sendai 980-8544, Japan
[4] ArcelorMittal Maizières, Research and Development, BP 30320, 57283 Maizières-lès-Metz, France

E-mail: koshikawa.4eb.takao@jp.nssmc.com

Abstract. Ingot bending tests are performed on the already formed solid shell of a 450 kg solidifying ingot. It is designed in order to be representative of two defects occurring during secondary cooling in steel continuous casting: hot tearing and macrosegregation. Ingots show the defects which result from the application of bending. In order to understand the different physical phenomena, finite element numerical modelling of the test has been developed, using two different approaches. A first approach consists of a 3D finite element thermomechanical simulation. Hot tearing criteria, based on strain, strain rate and two values for solid fraction are then evaluated. It is demonstrated that such a strain based criterion has an excellent capability to predict the formation of hot tears. A second approach consists of a 2D planar finite element simulation in the median section of the ingot. A two-phase formulation is used, in which the velocity of the liquid and solid phases are distinguished. The simulation shows how solutes are redistributed through the ingot under the effect of bending. Solute-depleted and enriched regions are well reproduced, but peak values of macrosegregation are underestimated.

1. Introduction

The paper focuses on two major defects encountered during secondary cooling in steel Continuous Casting (CC): hot tearing and macrosegregation.

Hot tearing or hot cracking takes place deep in the mushy zone, close to the end of solidification, typically for solid fractions over 0.9 when the material is subject to tensile strains and stresses. In CC this happens during the bending or unbending of the strand, or as a consequence of bulging of the shell between support rolls under the effect of metallostatic pressure. In this study, the macroscopic hot tearing criterion proposed by Won et al. [1] is evaluated. Being interpreted like in [2], this criterion is based on cumulated strain perpendicular to dendrite growth direction. The corresponding strain rate is denoted $\hat{\dot{\varepsilon}}$,. The strain based criterion, HCC_{WYSO}, is then expressed as follows:

$$HCC_{WYSO} = \left(\int_{BTR} \hat{\dot{\varepsilon}} \mathrm{d}t - \hat{\varepsilon}_C \right) \text{with } \hat{\varepsilon}_C = \frac{\varphi}{\left(\hat{\dot{\varepsilon}}\right)^{m^*} BTR^{n^*}} \quad (1)$$

where $\hat{\varepsilon}_C$ is the strain limit which depends itself on the perpendicular strain rate and on the Brittle Temperature Range (BTR). As this BTR is defined by two characteristic solid fractions (typically 0.9 and 1.0), it depends on the solidification path and thus both micro and macrosegregation are expected to affect hot tearing sensitivity. The three other parameters deduced in [1] from non-linear data fitting with several hot tearing experiments are: $\phi = 0.02821$, $m^* = 0.3131$, $n^* = 0.8638$.

Another critical defect in CC products is macrosegregation. It is observed at the centre of the slab. The principal cause is the relative movement of the liquid phase with respect to the solid phase. In CC, as mentioned above, bulging of the solid shell between support rolls and bending and unbending of the strand have an impact on the generation of such relative movement and hence on macrosegregation.

In order to identify those defects, an experiment, so-called ingot bending test, has been developed by several researchers [3-8]. The present tests have been designed and operated by Nippon Steel and Sumitomo Metal Corporation [9, 10]. For analyzing the tests, finite element numerical simulation of the tests has been developed using the commercial package THERCAST®. The location and occurrence of hot tears and macrosegregation are explained and discussed based on the simulations.

2. Ingot bending test

The schematic of the ingot bending test is shown in Figure 1. It has been designed to mimic the thermomechanical solicitations taking place during CC process; for instance, with respect to the evolution of the cooling rate and deformation. The procedure of the experiment is as follows: the molten metal is prepared in a ladle and the temperature is maintained at 1640 °C before pouring. The molten metal is filled into the mold from the top through a tundish. The filling duration is about 70 s. After filling, powder is added at the top of the ingot to limit heat exchanges with the air. The mass of metal is approximately 450 kg. The size of the ingot is 0.16 m thick, 0.5 m width and 0.75 m height. At a precise time after filling, the upper part of one of the two large faces of the mold is removed. The ingot is then deformed during solidification by pushing a horizontal cylindrical tool on the surface of the ingot, its longitudinal axis being located 0.45 m from the bottom of the ingot. The tool velocity and displacement are controlled by means of a hydraulic system and measured during the test. The reaction force is also measured using the time evolution of the hydraulic system pressure. The test conditions referred to as N-1 to N-6 are summarized in Table 1, together with measured nominal compositions in the ladle before filling. Note that a unique alloy composition was targeted for the experimental results reported hereafter. To know temperature evolution during the test, thermocouples are positioned in the ingot and the fixed large face of the mold. Furthermore, the ingot surface temperature has been measured with a pyrometer in test N-4 before and after deformation after partial mold removal. After the experiment, the ingot was cut and micrographs were investigated after appropriate etching to investigate hot tearing occurrence, intensity and location.

3. Thermomechanical modeling

3.1. Standard thermo-mechanical modeling

In the standard thermomechanical modelling approach [11, 12], the mushy zone (composed of liquid and solid phases) is considered as a homogenized continuum so that a unique velocity field is solved for. The numerical simulation is based on thermomechanical stress/strain analysis. It is conducted using the 3D finite element code THERCAST®. The thermal problem and the mechanical problem are solved sequentially at each time increment. The solidification path is defined a priori [13]. The alloy is modeled with a hybrid constitutive equation: over the solidus temperature, the alloy is seen as a non-Newtonian fluid obeying a temperature-dependent viscoplastic law. Below solidus, an elastic-viscoplastic behavior is used with strain hardening considered through a multiplicative law [11, 12].

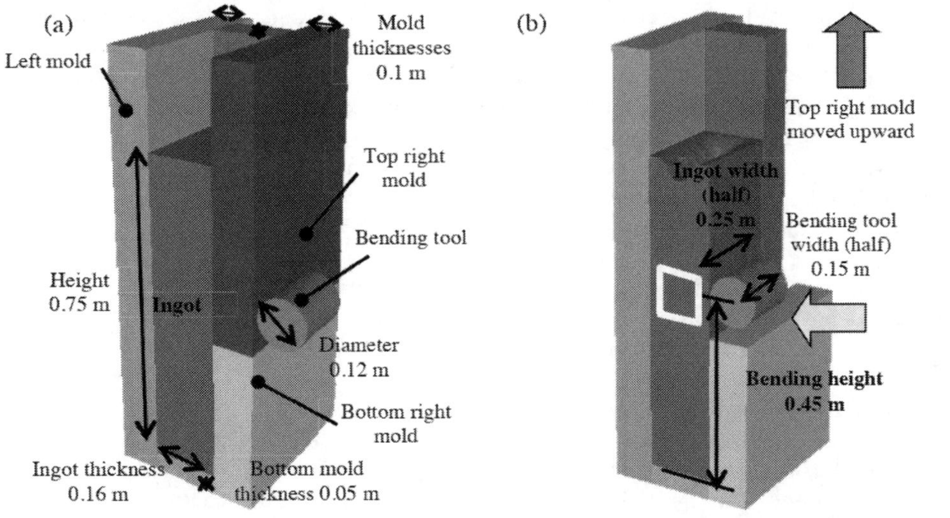

Figure 1. Schematic of the hot tearing experiment developed at Nippon Steel & Sumitomo Metal Corporation showing (a) initial state and (b) bending stage after partial mold removal. The configuration shown is half of the setup, considering symmetry with respect to the central transverse cut through the ingot, mold and bending tool domains.

Table 1. Measured nominal compositions and experimental conditions.

	Nominal composition (mass%)						Mold removal	Bending start	Displace-ment	Velocity
	C	Si	Mn	P	S	Al	(s)	(s)	(mm)	(mm s^{-1})
N-1	0.20	0.22	1.50	0.021	0.007	0.039	590	792	13	2.0
N-2	0.20	0.20	1.47	0.020	0.010	0.038	450	574	6	2.0
N-3	0.19	0.24	1.50	0.022	0.010	0.064	457	517	6	1.0
N-4	0.19	0.21	1.49	0.021	0.010	0.064	300	499	6	0.5
N-5	0.19	0.21	1.50	0.020	0.010	0.041	321	510	6	0.25
N-6	0.20	0.21	1.48	0.021	0.010	0.055	340	510	3	0.25

Concerning solidification, three microsegregation paths are used. The first one is based on the lever rule (LR) approximation, meaning that full thermodynamic equilibrium is assumed in all existing phases at any temperature. The second one is a combination of the partial-equilibrium approximation together with the para-equilibrium approximation (PE+PA) in which an interstitial element (e.g., carbon) has infinite diffusivity in all phases whereas substitutional elements are frozen in solid phases. In addition, the PE+PA approximation accounts for the peritectic reaction taking place in the steel grade investigated in this study. The last solidification path (EX) is extracted from measurements conducted on a simple unidirectional solidification experiment. The details of these approximations and the solidification for alloy similar alloy compositions are given in reference [13, 14].

As illustrated in Figure 1, the 3D numerical domains are described considering the symmetry of the setup. Filling stage is not taken into account so that the thermo mechanical simulation starts 70 s after filling. The initial temperature of the ingot, 1550 °C, is estimated using measurements. It is set to 20 °C for all other domains. The thermal boundary condition in between ingot and mold is defined as an air gap dependent heat resistance. When the ingot is into contact with the mold, the heat transfer coefficient is calibrated with measurement and set to 850 W m^{-2} K^{-1}. When air gap generates, its width is considered and a heat exchange coefficient is computed between the ingot and the mold surfaces. It takes into account radiation and heat conduction in the air. Regarding the ingot top, adiabatic condition is used. After mold removal, the free surface heat exchange as accounts for radiation and convection, with the values for emissivity and convective heat transfer coefficient taken as 0.8 and as 15 W m^{-2} K^{-1}, respectively. This standard thermo-mechanical simulation is carried out for test N-2 to N-6, which is focused on hot tearing study.

3.2. "Two-phase" modeling and macrosegregation analysis

The second approach is an effective "two-phase" modeling in which liquid and solid velocity fields are considered separately and solved concurrently in the mushy zone. It allows simulating solute transport phenomena leading to macrosegregation induced by deformation, shrinkage and advection. It is performed with the 2D finite element code R2SOL. Details of the formulation can be found in references [14, 15].

Concerning solidification path, it is not defined a priori since solute composition varies due to solute transport. To model the microsegregation phenomena, the lever rule approximation is considered for a multi component system assuming constant partition coefficients and liquidus slopes for each solute element. Those physical properties are obtained using Thermocalc with TCFE6 database, shown in Table 2. Fusion temperature is defined as 1538 °C. Regarding with permeability, Carman-Kozeny model is employed. The secondary dendrite arm spacing is predefined as 300 μm measured in the micrograph in case of N-1.

This effective two-phase modeling is applied to test N-1, in which the ingot centre is fully mushy when bending starts (Table 1). The calculation is carried out in two steps. At first, a standard thermomechanical calculation is conducted up to the bending stage. The simulation is then started again using the "two-phase" model. As a consequence, it is assumed that there is no macrosegregation at the beginning of this second two-phase simulation.

Table 2. Partition coefficient and liquidus slope for each solutes.

	C	Si	Mn	P	S
Partition coefficient	0.162	0.639	0.703	0.282	0.027
Liquidus slope (K/wt%)	-81	-12	-5	-29	-34

4. Results and discussion

4.1. Standard thermo-mechanical simulation result

4.1.1. Temperature evolution for test N-4

Figure 2 shows the comparison of temperature evolution in the ingot. There are two thermocouples in the ingot. TC-1 is close to the ingot center (blue line in Figure 2) while TC-2 is located close to the narrow face of the ingot (pink line in Figure 2). In addition, measured surface temperature evolution at the height of the bending tool is plotted (yellow diamond-shape in Figure 2). At 300 s, the mold located in front of the bending tool is removed prior to bending (Table 1). Consequently, the surface temperature increases. The calculated temperature evolution is in good agreement with measured data. Thermal modeling regarding the boundary condition and thermo physical parameters are thus well calibrated from about 300 s. Before this time, cooling curves in Figure 2 reveal overestimation of the predicted temperatures. This is due to the choice not to model the filling stage.

4.1.2. Reaction force comparison for test N-4

Figure 3 shows the comparison of the reaction force evolution during the bending stage for N-4. Good agreement is found between measurements and simulations. This indicates that the thermomechanical modeling works well, leading to good stress and strain field predictions. The microsegregation effect does not clearly appear although the LR approximation shows greater force than the two others. This is linked to the difference of the solid shell thickness at the time when bending is applied.

4.1.3. Hot tearing prediction using strain based criterion

Figure 4a shows a typical micrograph for experiment N-6 close to the bending region identified with a white rectangle in Figure 1b. In case of N-6, the surface deformation due to bending is not so clear. One can see a dendritic microstructure which grew from the ingot surface (right hand side) toward the centre of the ingot (left hand side). Around the center of the ingot, the morphology is changed to equiaxed, which is due to the decrease of the temperature gradient close to the center of the ingot. In the micrograph, hot tears are found in interdendritic regions.

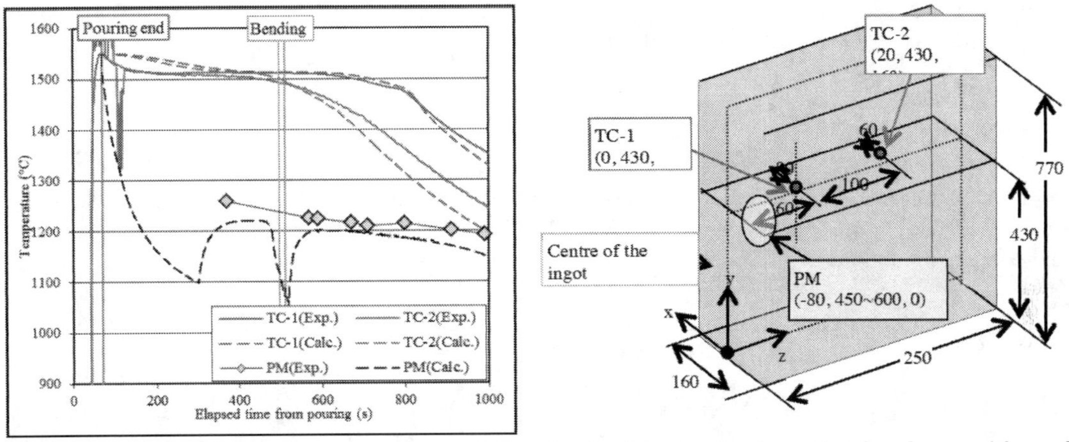

Figure 2. Comparison of temperature evolution in the ingot and schematic showing position of thermocouple TC-1 and TC-2 in the ingot, as well as the position at the ingot surface where pyrometer PM records the temperature after partial mold removal.

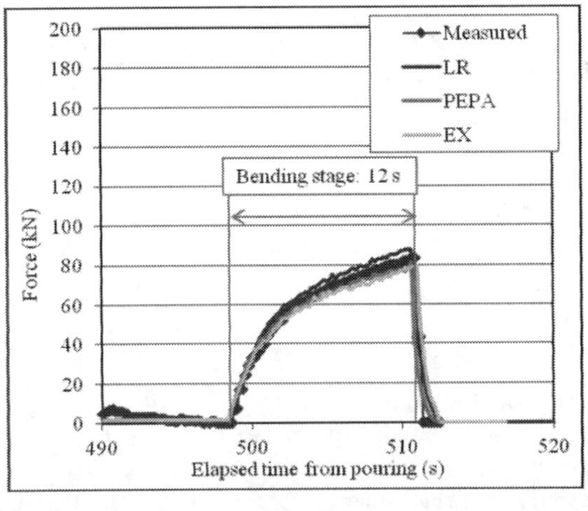

Figure 3. Force evolution during bending stage for trial N-4 deduced from measurements and from simulations using the three microsegregation models.

Figure 4b shows a comparison of the hot tearing sensitivity profile defined with Eq. (1) at the bending tool height for the three microsegregation models. In addition, the crack location is superimposed on the profile. The position of the crack location is in good agreement with two microsegregation cases; PE+PA and EX, with HCC reaching positive values between 0.04 m and 0.05 m from the ingot surface of the punched side. In contrast, the profile using LR shows negative values at the crack position. These results clearly reveal the critical role played by the microsegregation model on hot tearing prediction.

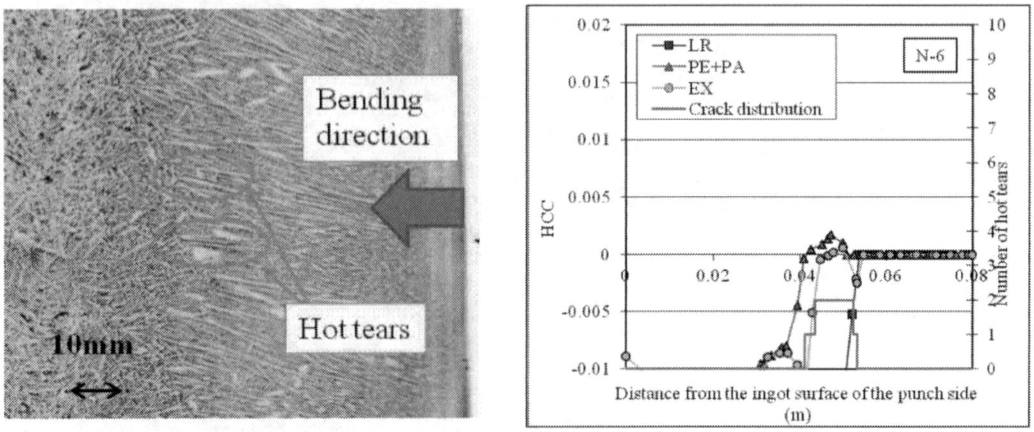

Figure 4. A micrograph of test N-6 and the comparison of the hot tearing sensitivity profile with the location of the hot tears in the bending region. Hot tearing sensitivity is calculated with the three microsegregation models.

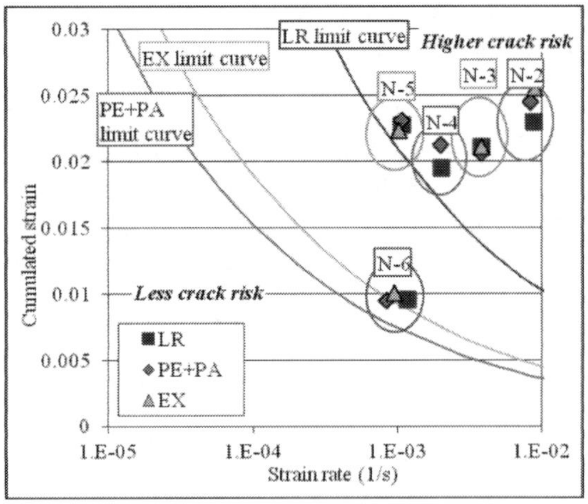

Figure 5. Computed cumulated strain, $\int_{BTR} \hat{\dot{\varepsilon}} dt$, as a function of the maximum perpendicular strain rate calculated when passing through the BTR. Values taken from the HCC_{WYSO} profile at the bending height, an example being shown in Figure 4b. The three continuous curves indicate the strain limit $\hat{\varepsilon}_C$ (Eq. 1).

Figure 5 presents limit curves based on the three microsegregation models. The continuous curves correspond to the strain limit that constitutes the second term of the WYSO criterion in Eq. (1): $\hat{\varepsilon}_C = \varphi \hat{\dot{\varepsilon}}^{-m^*} BTR^{-n^*}$. The different values of the BTR for LR, EX and PE+PA explain the 3 limits. For each curve, the top right domain corresponds to $HCC_{WYSO}>0$, thus presenting a higher crack risk. According to its smaller BTR, the LR limit curve is placed on the top right region of the graph. Calculated values are superimposed on the figure for trials N-2 to N-6. To this end, the profile of the cumulated perpendicular strain and strain rate are drawn along the thickness of the ingot at the bending tool height. The maximum cumulated strain is taken from the profile as well as the corresponding value for the perpendicular strain rate. These are the two variables used to express local values of HCC_{WYSO} from Eq. (1) and then plot the simulation results in Figure 5. The distribution of the results in the graph is linked to the applied displacement and velocity of the bending tool shown in Table 1. It is found that the microsegregation model does not have a major impact on the cumulated strain and maximum strain rate in the BTR since the three points for each trial are very close. Note that, while metallographic investigations of all these trials have revealed the presence of hot tears, only the PE+PA and EX curves give a good response. The simulation using LR does not work as N-6 is not located in the high crack risk region.

4.2. "Two-phase" simulation result

4.2.1. Liquid phase velocity field during bending

Figure 6a shows a micrograph of trial N-1 at the ingot top. It is seen that the different microstructure coming from different cooling effect is observed at the top center. Interpretation is that, due to bending, remaining melt is expelled from the mushy core in the bending region towards the top of the ingot. This is explained in the simulation result presented in Figure 6b. It is found that bending generates an upward flow toward the ingot top region. The boundary of the top is not fully solidified so that the melt goes out and covers the already formed solid shell. Figure 6a clearly shows this effect. It can be seen that the two-phase numerical modeling is able to retrieve this phenomenon.

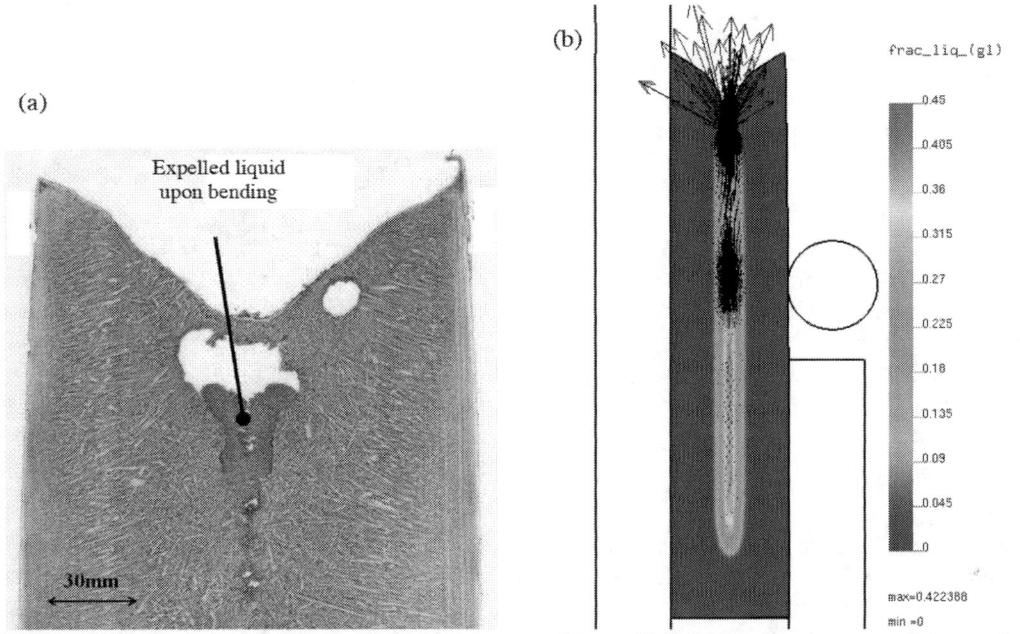

Figure 6. On the left, a micrograph of the ingot top for test N-1 shows the structure of the top ingot resulting from an expulsion of liquid upon bending: some liquid entered the primary shrinkage macrocavity. On the right, simulated relative velocity vectors in the liquid phase are superimposed on the liquid fraction field for trial N-1, showing the same phenomenon during bending.

4.2.2. Comparison for macrosegregation

Figure 7 shows a micrograph of trial N-1 close to the bending region, the print of the bending tool being visible on the right-hand-side of the ingot surface. A white vertical band is seen at the center of the ingot. It reveals clear negative segregation of Mn, which would result from the deformation of the mushy region. A similar profile was obtained for P. Figure 7 also shows the comparison of the relative concentration profile of Mn along the dashed line superimposed to the micrograph at the bending height. The calculated negative segregation is not enough marked compared to the measurement result. There are several possible causes for this difference, among which approximate microsegregation modeling, inaccurate knowledge of rheological parameters, mush permeability, thermodynamic and thermophysical data.

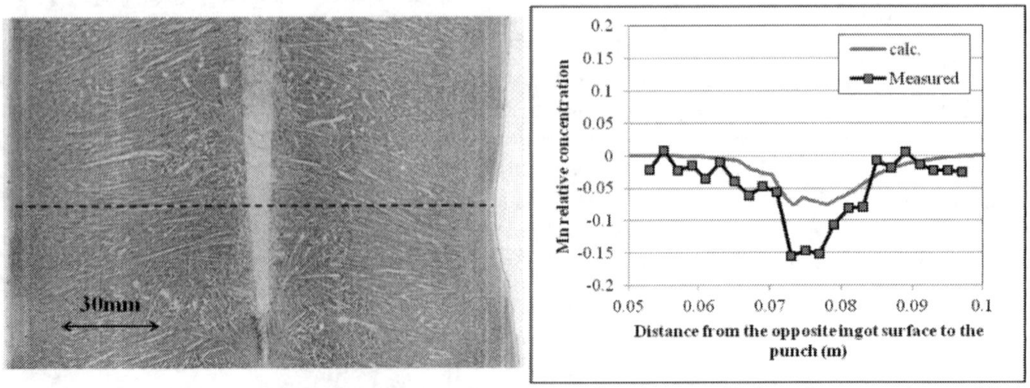

Figure 7. A micrograph of the ingot bending region for the test N-1 and comparison of Mn relative concentration profile at bending height with measurement result by means of EPMA. Simulation results show qualitative agreement with measurements.

5. Conclusion

An ingot bending test has been developed and modeled using two numerical codes. Using a standard thermomechanical model, it is found that strain based criteria have capability to predict crack occurrence and location. In addition, the microsegregation model or solidification path has an impact on the predicted hot tearing sensitivity. Through effective "two-phase" simulation, qualitative agreement with respect to liquid flow and its consequent negative segregation at ingot centre is demonstrated. However, the amplitude of negative segregation is found smaller than the measurement result. Further sensitivity analyses would be useful to understand this discrepancy and reach a better quantitative agreement.

Acknowledgments

This work is supported by Nippon Steel & Sumitomo Metal Corporation in a collaborative project with ArcelorMittal. The authors are deeply grateful to Drs Olivier Jaouen and Frederic Costes from Transvalor SA (Mougins, France), for their kind help and discussion about thermomechanical modeling using THERCAST® software.

References

[1] Won Y M, Yeo T –J, Seol D J and Oh K H 2000 *Mater. Trans. B* **31** 779
[2] Bellet M, Cerri O, Bobadilla M and Chastel Y 2009 *Mater. Trans. A* **40** 2705
[3] Sato H, Kitagawa T, Murakami K and Kawawa T 1975 *Tetsu to Hagane* **61** S471
[4] Miyamura K, Ochi A, Kanamaru K and Kaneko N 1976 *Tetsu to Hagane* **62** S482
[5] Marukawa K, Kawasaki M, Kimura T and Ishimura S 1978 *Tetsu to Hagane* **64** S661
[6] Narita K, Mori T, Ayata K, Miyazaki J and Fujimaki M 1978 *Tetsu to Hagane* **64** S152
[7] Sugitani Y, Nakamura M, Kawashima H and Kawasaki M 1980 *Tetsu to Hagane* **66** S193
[8] Wintz M, Bobadilla M and Jolivet J M 1994 *La Rev. Métall.* **4** 105
[9] Murao T, Kajitani T, Yamamura H, Anzai K, Oikawa K and Sawada T 2013 *Tetsu to Hagane* **99** 94
[10] Murao T, Kajitani T, Yamamura H, Anzai K, Oikawa K and Sawada T 2014 *ISIJ* **54** 359
[11] Bellet M, Jaouen O and Poitrault I 2005 *Int. J. Num. Meth. Heat Fluid Flow* **15** 120
[12] Bellet M and Fachinotti V D 2004 *Comput. Meth. Appl. Mech. Eng.* **193** 4
[13] Koshikawa T, Gandin Ch- A, Bellet M, Yamamura H and Bobadilla M 2014 *ISIJ* **54** 1274
[14] Koshikawa T, Bellet M, Yamamura H and Bobadilla M 2013 *STEELSIM2013 (Ostrava, Czeck Republic, 10-12 September 2013)* topic 5
[15] Fachinotti V D, Le Corre S, Triolet N, Bobadilla M and Bellet M 2006 *Int. J. Num. Meth. Eng.* **67** 1341

One-Way Coupling of an Advanced CFD Multi-Physics Model to FEA for Predicting Stress-Strain in the Solidifying Shell during Continuous Casting of Steel

Johan Svensson[1], Pavel E. Ramírez López[1], Pooria N. Jalali[1] and Michel Cervantes[2]

[1] Casting and Flow Simulation Group, Swerea MEFOS AB
Aronstorpsvägen 1, SE 974 37 Luleå, SWEDEN
[2] Department of Engineering Sciences and Mathematics,
Luleå University of Technology, SE 971 87 Luleå, SWEDEN

E-mail: pavel.ramirez.lopez@swerea.se

Abstract. One of the main targets for Continuous Casting (CC) modelling is the actual prediction of defects during transient events. However, the majority of CC models are based on a statistical approach towards flow and powder performance, which is unable to capture the subtleties of small variations in casting conditions during real industrial operation or the combined effects of such changes leading eventually to defects. An advanced Computational Fluid Dynamics (CFD) model; which accounts for transient changes on lubrication during casting due to turbulent flow dynamics and mould oscillation has been presented on MCWASP XIV (Austria) to address these issues. The model has been successfully applied to the industrial environment to tackle typical problems such as lack of lubrication or unstable flows. However, a direct application to cracking had proven elusive. The present paper describes how results from this advanced CFD-CC model have been successfully coupled to structural Finite Element Analysis (FEA) for prediction of stress-strains as a function of irregular lubrication conditions in the mould. The main challenge for coupling was the extraction of the solidified shell from CFD calculations (carried out with a hybrid structured mesh) and creating a geometry by using iso-surfaces, re-meshing and mapping loads (e.g. temperature, pressure and external body forces), which served as input to mechanical stress-strain calculations. Preliminary results for CC of slabs show that the temperature distribution within the shell causes shrinkage and thermal deformation; which are in turn, the main source of stress. Results also show reasonable stress levels of 10-20 MPa in regions, where the shell is thin and exposed to large temperature gradients. Finally, predictions are in good agreement with prior works where stresses indicate compression at the slab surface, while tension is observed at the interior; generating a characteristic stress-strain state during solidification in CC.

1. Background

Continuous Casting (CC) is the process of solidifying liquid steel into a usable shape (i.e. billets, blooms or slabs) by heat extraction from the melt through two cooling zones: the primary cooling zone consisting of a water cooled copper mould where initial solidification occurs and the secondary cooling zone consisting of a series of water sprays. The process is schematically shown in Figure 1. In the primary zone, the main heat transfer mechanisms are heat conduction through the shell, slag and mould layers as well as convection to the cooling water in the mould; however, radiation also plays often an important role. A complex metal flow is found inside the mould with regions of high turbulence and large temperature gradients due to the feeding of hot steel through a Submerged Entry Nozzle (SEN). Additionally, the entire mould oscillates in order to prevent the liquid metal to freeze against the mould walls; while casting powder is continuously added from the top to insulate and lubricate the process. This powder melts when it comes in contact with the hot metal to form a protective slag cover (i.e. slag bed). The liquid slag also forms a thin layer which acts as a lubricant to reduce friction between the mould and solidified shell. This liquid layer solidifies against the cold mould to form an insulating layer (typically 1-2 millimetres), which controls the heat transfer in the mould. Finally, the secondary cooling zone takes place outside the mould, where the strand is further cooled through a series of water sprinklers at high pressure. The remaining liquid steel at the strand core solidifies within the secondary cooling zone before final cutting.

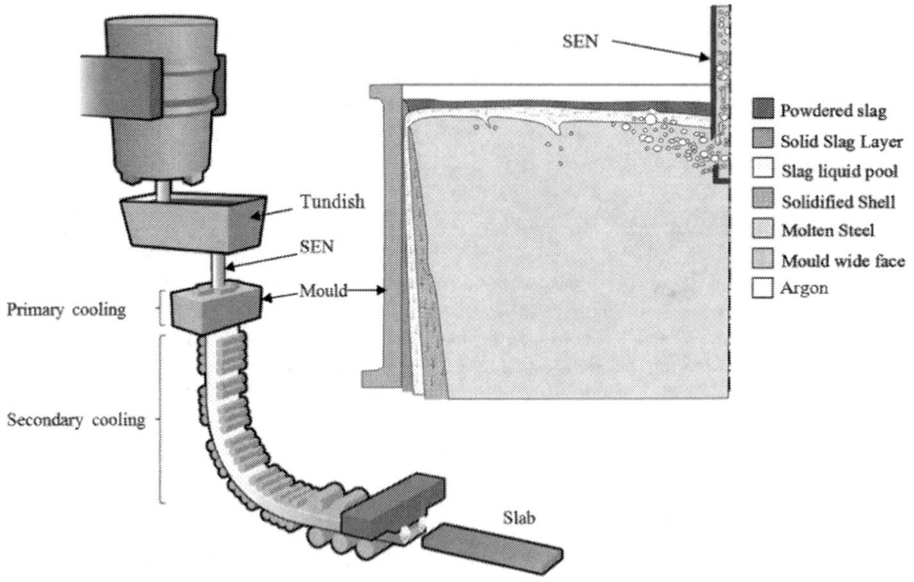

Figure 1. Schematic view of the continuous casting process and a magnified view of the mould interior together with the different phases in the process (e.g. slag, steel, argon and air) based after [1].

A variety of physical phenomena take place during CC such as heat transfer [2], multiphase flow [3], [4], solidification [5] along with mechanical stresses and strains [6-8]. Stresses are induced in the shell as it solidifies and cools causing shrinkage. Part of this shrinkage is due to phase transformations such as $\delta \rightarrow \gamma$ phase [9]. Shrinkage also causes stresses, plastic deformation and surface defects on the finished product, which in the long term can cause processing problems. Moreover, surface defects caused due to high stress concentrations are able to nucleate cracks and in the worst case scenario, breakouts. Breakouts may occur for a variety of reasons, but are often associated to a thin shell not able to withstand the ferrostatic pressure of liquid steel within the strand. This frequently combines with a lack of lubrication inducing high friction between mould and strand; and thereby, hot tearing. Breakouts are still a recurrent problem for steel producers, since the number of defects is higher with increased production [10] due to faster casting speeds (i.e. thin slab casting), higher throughput (i.e. larger slab formats) or casting of "difficult grades" (i.e. peritectics, duplex, trip steels, etc.).

Swerea MEFOS has developed a CFD-CC model able to predict a variety of phenomena including steel/slag interactions, heat transfer and solidification within the oscillating mould in order to simulate problem scenarios and optimize the CC process. Coupling the results from this CFD model to FE analysis allows more accurate predictions of stresses and strains during solidification and opens the door for more realistic breakout predictions.

2. MEFOS CFD - CC numerical model

Swerea MEFOS CFD-CC model predicts the multiphase steel-slag flow in the mould by solving the Navier-Stokes equations combined with the Volume of Fluid approach (VOF) based on the commercial code ANSYS Fluent v.14[1]. The VOF model tracks the volume fraction of steel and slag by solving a unique set of momentum equations for both phases [11]. The volume fraction of the steel phase, ϕ_{steel} is derived by the continuity equation.

$$\frac{\partial}{\partial t}(\phi_{steel}\rho_{steel}) + \nabla(\phi_{steel}\rho_{steel}\vec{v}) = \sum_{p=1}^{n}(\dot{m}_{p,steel} - \dot{m}_{steel,p}) \tag{1}$$

Where $\dot{m}$ is the mass transfer between phases; while the mixture density and viscosity are:

$$\rho_{mix} = \phi_{steel}\rho_{steel} + (1 - \phi_{steel})\rho_{slag} \tag{2}$$

$$\mu_{mix} = \phi_{steel}\mu_{steel} + (1 - \phi_{steel})\mu_{slag} \tag{3}$$

Here, ϕ may take any value between 0 and 1 to represent the steel phase fraction in a cell shared by the two fluids (i.e. steel and slag). A very fine boundary layer mesh (down to 25 μm) is used in order to resolve the slag-metal interface and ultimately, slag film infiltration and shell initiation. The k-ε RNG model with standard wall functions is used to include turbulent effects. The model solves heat-transfer through the copper mould, SEN and one meter of strand below the mould exit. The heat transfer is given by the energy equation:

$$\frac{\partial}{\partial t}(\rho_{mix}H) + \nabla(\vec{v}(\rho_{mix}H + P)) = \nabla\left(K_{eff}\nabla T + (\tau_{eff}\vec{v})\right) \tag{4}$$

where H is the enthalpy, K_{eff} the effective thermal conductivity and τ_{eff} the effective viscosity. Data for specific heat and thermal conductivity is provided for all materials and phases as a function of temperature. Solidification is solved as a function of the heat transfer in the model. The solidification model is based on enthalpy where the liquid fraction in the mushy zone is defined by the lever rule:

$$f_l = \frac{T - T_{solidus}}{T_{liquidus} - T_{solidus}} \tag{5}$$

The pulling velocity of the solidified shell is defined by the temperature at which the steel is assumed to have enough strength to be pulled out in the casting direction or Zero Strength Temperature (ZST). [12]. The model represents one quarter of the mould (due to symmetry along the width and thickness of the slab) and contains 3 million cells including mould, strand (solid and liquid) and slag film/bed (powder, liquid, and solid). Heat is extracted through the water cooled copper mould by a convective heat transfer coefficient determined from the cooling-water flow-rate in the real plants. Figure 1 and 2 illustrate typical calculations with the model. Figure 2 shows a discharging jet of superheated metal leaving the SEN at high speed to dissipate its turbulent energy in the liquid pool; while Figure 3 shows the shell formation in 3D.

[1] *ANSYS Fluent v.14 is a registered trademark of ANSYS INC. 2600 ANSYS Drive Canonsburg, PA 15317. USA.*

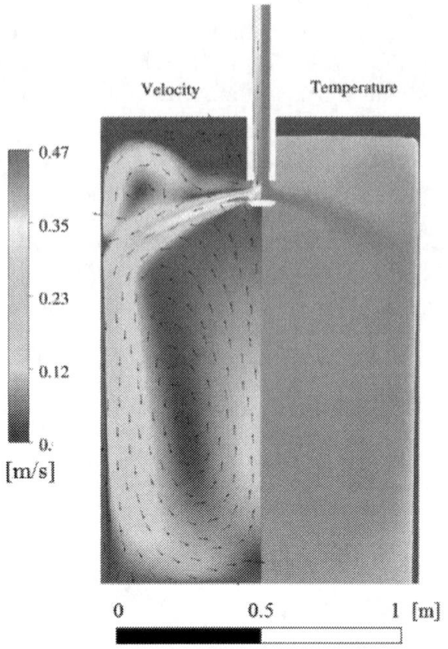

Figure 2. Typical flow pattern within CC mould (left) and temperature profile showing shell growth on the right hand side (right).

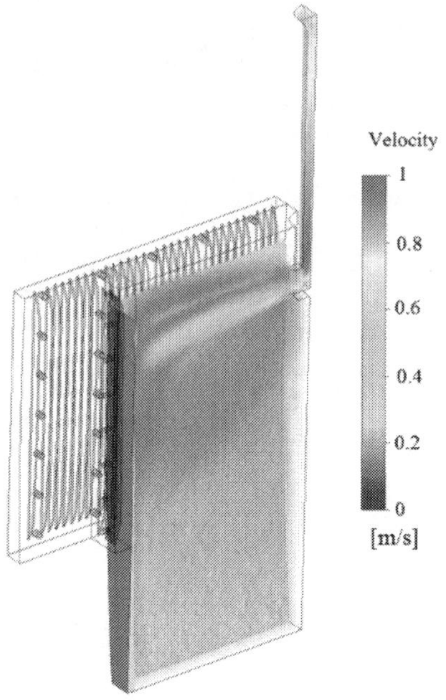

Figure 3. Shell formation in 3D and computational domain with velocity colour map.

3. CFD + FEM coupling

Fluid Structure Interaction (FSI) is an approach to predict the influence of flow on deformation in a body and vice-versa (e.g. where changes in the body geometry affect the flow surrounding it). Several standard solutions are readily available for FSI calculations in commercial software. However, these require a well-defined solid domain that does not change during the calculation process. Unfortunately, there is no such predefined domain during solidification since changes on the fluid flow affect the actual solid formation. This is a major hindrance for applying FSI to modelling of Continuous Casting. Another issue is that the solid-liquid interface often rests within one cell, whilst exporting of grids is often based on full cells to avoid connectivity problems. This results in a poor element representation of the smooth shell from CFD calculations into the FE environment. This occurs since mesh formulations required for CFD and FEM are often dissimilar (e.g. CFD often uses structured grids, while FEM typically uses tetrahedral mesh types). This issue is schematically shown in Figure 4.

FSI is also often referred as *two-way coupling*, where flow results are sent to the FEM model (e.g. *one-way coupling*) and back from the FEM model to the flow solver (e.g. *two-way coupling*). The present work is focused on one-way coupling as a first step towards a feasible FSI approach in the future. In order to translate the smooth shell from MEFOS CC-CFD calculations into FEM, a shell shape has been extracted for a time step where the solution had reached steady state (i.e. equivalent to stable casting conditions). The shell is processed in several steps: 1) CFD results are used to create surface geometries by exporting iso-surfaces in .STL-format, 2) the STL surfaces are combined to form a solid CAD body such that 3) the shell could be meshed with unstructured tetrahedral finite elements, to subsequently 4) map temperatures and loads from the original solution into the new mesh. The procedure is summarized in Figure 5.

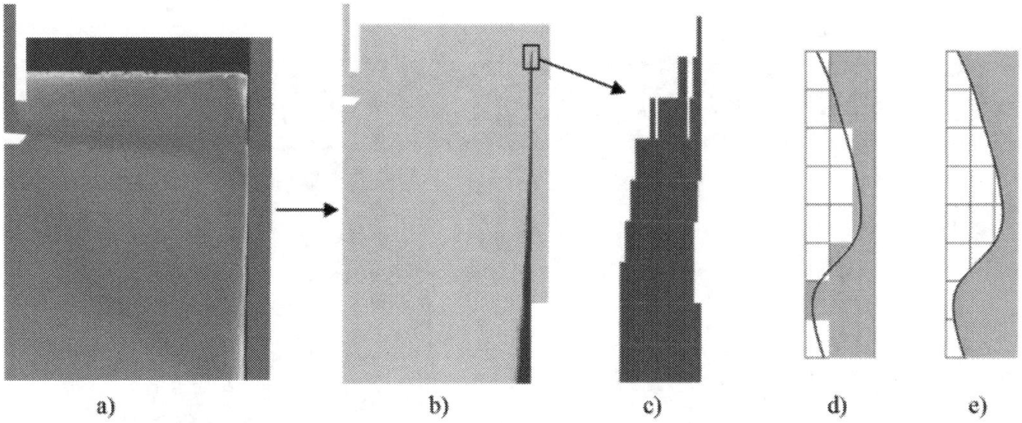

Figure 4. Shell exporting procedure: a) Solidification results in FLUENT, b) Isolated shell (i.e. solid phase), c) Shell tip based on exporting full-cells resulting in poor representation, d) Schematic representation of full-cell exporting issues and e) Original shell profile on CFD results.

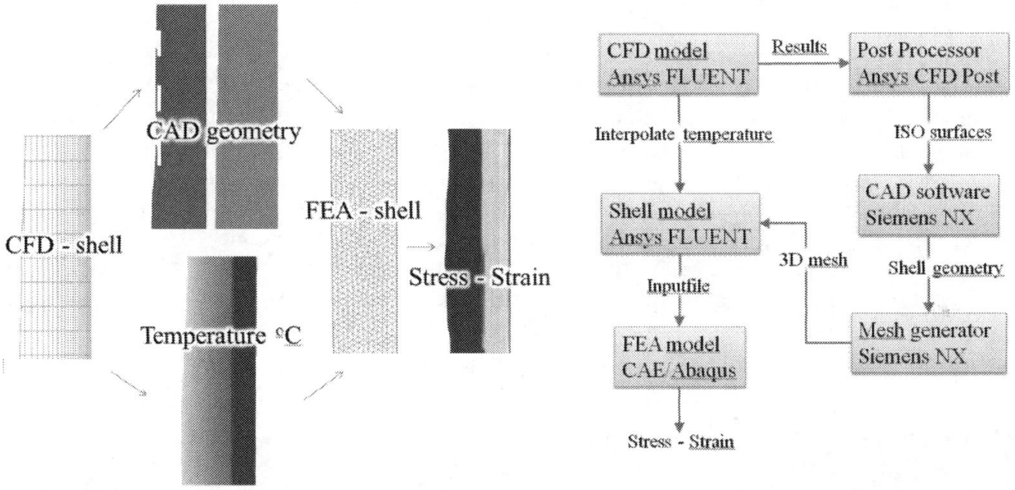

Figure 5. Schematic view of exporting procedure (left) and flow-chart summary (right).

The interior side of the shell is defined by the ZST iso-surface, while the exterior surface is defined by the steel VOF fraction equal to 0.5. The later iso-surface is not fixed in temperature and becomes colder close to the corners as the shell is cooled down from two mould sides. Both iso-surfaces are shown on Figure 6 with a colour map of temperatures. Further processing is required to read the shell exterior and interior surfaces into a FEA software as well as creating a solid body out of them to represent the final solidified shell. The procedure requires reading the STL-files which represent the surfaces with faceted triangular elements that can be read directly into the CAD software Siemens NX8.5[2]. These geometries can; due to some tolerance issues, contain internal defects as seen in Figure **7**a. The holes in the geometry can be eliminated and the faces smoothed with help of the *decimate* command in NX. Although this command modifies slightly the shell thickness; the final geometry becomes a very good approximation of the iso-values defining the original shape.

[2] *NX is a registered trademark of Siemens Product Lifecycle Management Software Inc. Wittelsbacherplatz 2, 80333 Munich, GERMANY.*

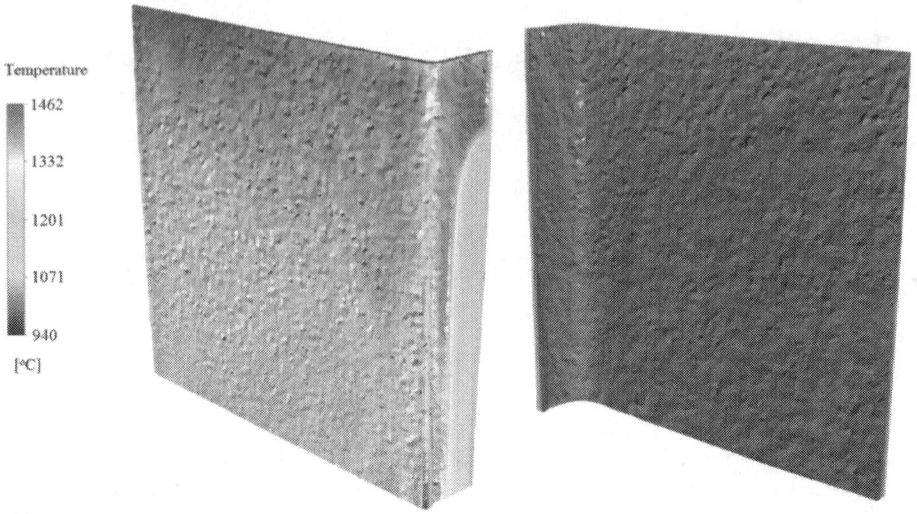

Figure 6. Solid shell in 3D with the temperature map: Slab exterior (left) and interior (right).

The surfaces are stitched together by adding surfaces in the symmetry planes. Finally, a solid body is created as shown in Figure **7**b. This solid body is finally exported with the STEP-format for meshing with unstructured tetrahedral elements in NX. Once the mechanical model is complete, the temperatures provided by the CFD simulation are mapped to the shell. An interpolation file is written containing the nodal temperatures for the entire fluid domain at a given time step. The shell model is then loaded into Fluent together with the interpolating file, so temperatures are mapped to the nodes in the new shell model. Thus, CFD results are used twice; firstly, to produce the shell model and secondly, to map the temperatures in the model. The solid body was meshed with around 36,000 tetrahedral elements. Each node was given the pouring temperature as an initial load for the thermo-mechanical calculation. The final temperature provided by the CFD simulation was set as one single linear-load step for the coupled thermal-stress analysis.

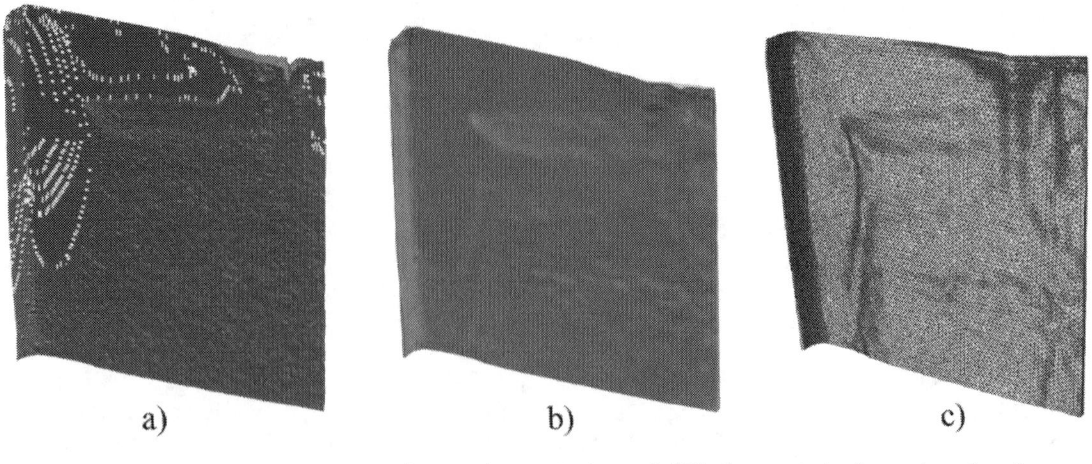

Figure 7. Shell at different stages of exporting procedure: a) STL faces stitched together showing exporting defects, b) Smoothed solid body and c) Final mesh with tetrahedral elements.

4. FE model theory and boundary conditions

The temperature distribution within the solid shell causes shrinkage and displacements, which ultimately lead to stresses and deformations. The total strain due to solidification and the temperature distribution is the sum of the elastic, inelastic (plastic) and thermal strain:

$$\varepsilon = \varepsilon_e + \varepsilon_p + \varepsilon_T \tag{6}$$

Thermal stresses are introduced on the solidified metal when the liquid feeding is unable to compensate for the shrinkage and thermal contractions [4]. The shrinkage over the phase transition can be estimated by the difference in density over the solid phase:

$$\beta = \frac{\rho_S - \rho_L}{\rho_S} \tag{7}$$

where $\rho_{S,L}$ is the density of the solid and liquid phase respectively. Shrinkage is caused by phase transformations since the liquid phase occupies more volume than the solid phase. For a material with a temperature gradient over a length (ΔT), the total strain (ε) can be computed by:

$$\varepsilon = \Delta T \alpha L \tag{8}$$

where α is the thermal expansion coefficient and (L) is the original lenght. The strains for each node in three dimensions are determined by:

$$\varepsilon_{ij} = \frac{1}{2G}\left(\sigma_{ij} - \frac{v}{1+v}\sigma_{kk}\delta_{ij}\right)\alpha T \delta_{ij} \tag{9}$$

and the stresses due to the thermal strain are:

$$\sigma_{ij} = 2G\left[\varepsilon_{ij} + \frac{v}{1-2v}\delta_{ij}\left(\varepsilon_{kk} - \frac{1+v}{v}\alpha T\right)\right] \tag{10}$$

where G is the shear modulus, v is the Poisson's ratio and δ is the delta-Kronecker. Typical low carbon steel was used in the calculations with material data partly from literature and partly from Gleeble measurements. Tensile tests at different strain rates and temperatures were performed at Swerea KIMAB for low carbon steel and Kozlowski's equation [2] was used to complete the material data for FE analysis.

$$\dot{\varepsilon}_p = C \exp\left(\frac{-Q}{T}\right)\left(\sigma - a_\varepsilon \varepsilon_p^{n_e}\right)^n \tag{11}$$

The n and n_e coefficients were obtained by fitting the equation to the experimental stress-strain data available. Stress-strain curves for different temperatures and strain rates are presented in Figure 8 for a temperature range from 1100 to 1700 K under strain rates relevant to continuous casting (0.01s – 1s). As expected, the stress hold by the material decreases with increased temperature as well as decreased strain rate. The shell is constrained in three directions by defining symmetry conditions at the middle wide and narrow sides of the mould, along with fixed translation in the casting direction at the bottom. The shell is free to bend, while the mould wall and ferro-static pressure were excluded. Boundary conditions for the stress analysis are presented on Figure 9. The commercial software CAE/ABAQUS3 was used for the FEA analysis and the problem was solved with a general static approach with a single linear load step.

---

[3] *CAE/ABAQUS is a registered trademark of Dassault Systèmes Inc. Klara Södra kyrkogata 1, SE 11 152 Stockholm, SWEDEN*

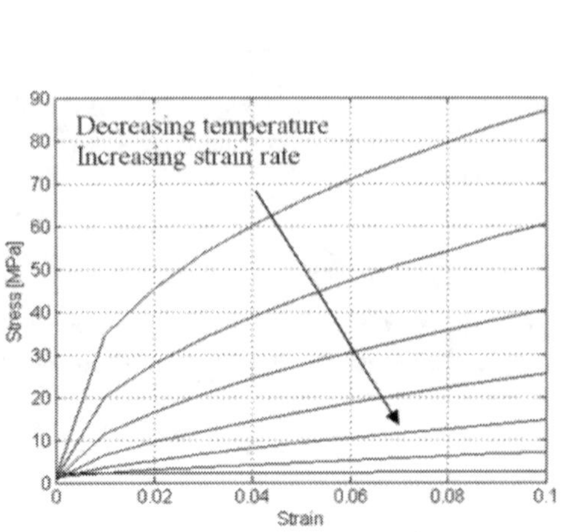

Figure 8. Material properties used in the FEA.

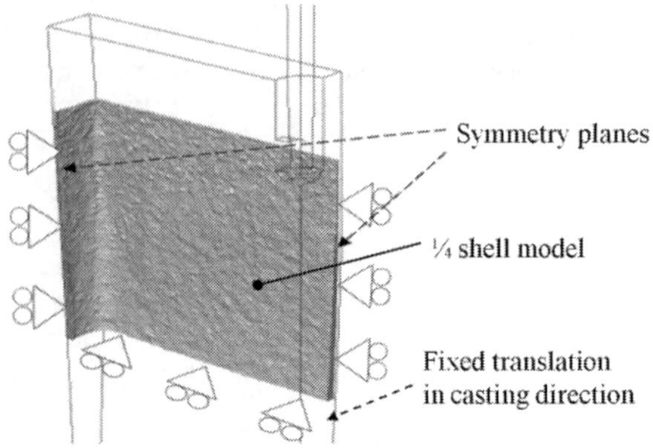

Figure 9. Schematic shell model describing boundary conditions for FEA.

5. Results

CFD+FEM coupling makes possible to account for the effects of slag infiltration, mould oscillation and heat transfer variability on shell formation; as well as predicting subsequent strains-stress accumulation for specific casting conditions (e.g. steel grade, casting speed, slag properties, pouring temperature, etc.) including caster details such as SEN and mould design (e.g. taper, cooling channels, etc.). Results from the three-dimensional ¼ shell section are shown in Figure 10.

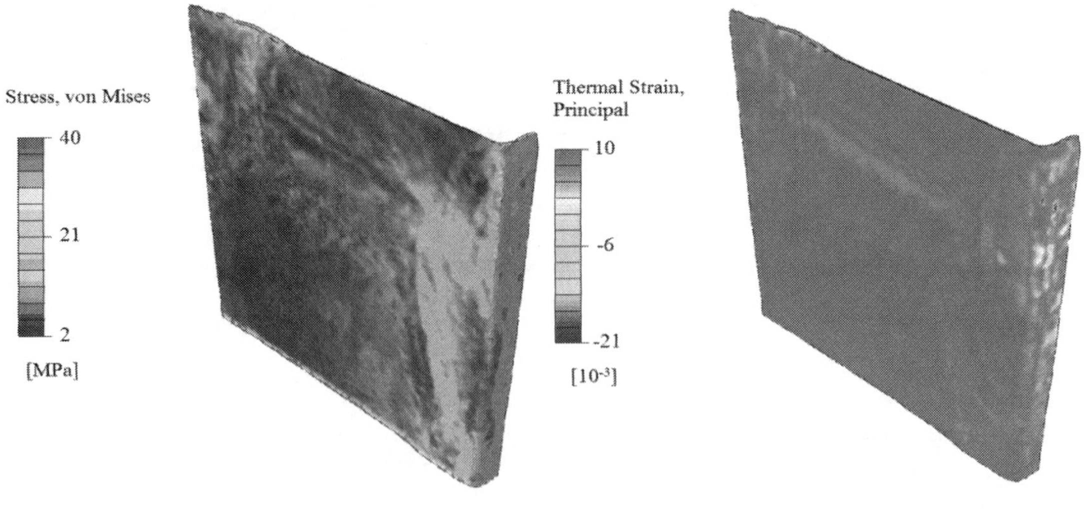

Figure 10. FEM model results showing von Mises stresses (left) and principal thermal strains (right).

A contact condition; defined by the mould wall, would indeed prevent the shell to bend as the results suggest. However, the shell is not in direct contact with the mould in the real caster since the liquid/solid slag layer sits between them. The maximum stresses are limited to a few nodes where the shell is thin. High values are generally found at the bottom, where fixed translation was defined. This should be ignored as it is a response to the boundary condition. The maximum von Mises stress is 40.31 MPa and is found at the narrow face close to the tip, where the shell is thin.

The majority of the shell though, is below a stress magnitude of 20 MPa which is line with expected magnitudes of in-situ experiments by Manchester University in an ongoing RFCS project

[13]. The maximum displacement due to bending is found at the lower corner/symmetry plane of the wide face, where a maximum deflection of 40.4 mm is found. Results show that the shell is not shrinking towards the centre, but bending over the shell/mould interface. When comparing the deformed and the original shape of the bottom face of the shell, it is seen that the corner is more rounded after deformation. The result of thermal strains, Figure 10, shows both compression and tension within the shell. Preliminary results indicate compressive stresses at the exterior and tensile at the interior.

6. Discussion and limitations

As computer performance increases, also does the demand on better computer simulations. This includes the influence of fluid flow in thermo-mechanical analyses as well as shrinkage effects on heat transfer and flow simulations [2]. Simulating the multi-physics phenomena, (i.e. fluid flow, solidification, heat transfer and structure mechanics) during casting has been attempted in previous works [8, 14-16]. However, it lacks a detailed treatment of the slag phase in the calculations which is critical for the CC process. A two-way coupled model would be needed to fully predict the thermo-mechanical behaviour of the solidifying shell.

The simulated effect of thermal shrinkage and stresses is highly dependent on temperature via simulated heat fluxes and material properties. For instance, the heat conductivity and thickness of the slag film affect the heat conducted between the shell and mould. Hence, the heat conduction that defines shell growth is highly dependent on the distance between the solid shell and mould, which is calculated in the FEA. In reality, the shell is shrinking at the same time, which would change the amount of slag into the lubrication film; and consequently the shell thickness; so it can be argued that the change in geometry is affecting the flow field as well. Thus, minor changes in slag film thickness would immediately cause changes in heat transfer between the melt and the copper mould and thereby, affect shell growth. A more physically correct model would be able to couple the mechanical solution back to the fluid simulation and solve the flow with the updated, deformed shell. Ultimately, this would affect the overall thickness of the shell and the temperature distribution in the mushy zone. A two-way coupling may be able to clarify some of these issues; but so far, the model presented is only considering the influence of flow pattern and thermal loads on the stress state. Further investigations can be done to find out how this mechanical behaviour really affects the solidification behaviour into the mould. However, there is no existing FSI approach able to cope with a non-prescribed solid body in the analysis. Moreover, even though the shell is defined as a solid within FLUENT, it is still a fluid with very high viscosity (which can be related to a solid body).

A simple solid shell may be created and used for the purpose of using a standard FSI approach. However, a prefixed shell beats the purpose of the CFD model since the intricate shape of the shell is derived from transient calculations including slag infiltration. An adapting mesh could also be used for the solidification front in FLUENT to increase the model accuracy. However, it increases the computational time considerably. In addition, symmetry conditions have been used in order to decrease the computational time. This may be a source of error as all gradients in the symmetry planes are set to zero. However it is a well-known fact that metal flow in CC is highly asymmetric and unsteady[17]. Thus, the shell is also expected to be irregular and non-symmetric. The material properties used in the mechanical analysis are partly interpolated from Gleeble measurements since it is extremity difficult to obtain material stress-strain data at semi-solid temperatures. This conveys a possible source of error on the input data. Finally, the approach of performing a static solution of the final thermal loads from a pre-defined temperature does not include any temperature history of the nodes during transient solidification. These issues are under current investigation by the authors.

7. Conclusions

A procedure for coupling CFD results to FEA by exporting iso-surfaces, creating a CAD geometry and mapping loads has been successfully developed in the present work. The following conclusions can be drawn from the presented work:

- Transferring of data including features such as temperature, pressure, shape and surface irregularities from CFD to FEM was proven possible.
- The coupling also makes possible to evaluate the stresses and strains within the solid shell based on CC-CFD results for a particular time snapshot during transient calculations.
- Thermo-mechanical results show that stresses are concentrated on areas close to the corner and where the shell is thin, with stress magnitudes in good agreement with existing data.
- Results show compressive stresses at the shell exterior and tensile at the shell interior.
- The magnitude of the strain results indicates plastic deformation of the shell, which provides a characteristic stress map for continuously cast products.

Acknowledgements:
JS would like to thank Mr. Chunhui Luo at Swerea KIMAB for providing data for the FEM analysis and Jonas Dyberg at Dassault systèms Simula Corporation for a CAE/ABAQUS trial license. PRL would also like to thank Hugo Carlsson Stiftelse for supporting this work. The research leading to these results has received funding from the European Union's Research Programme of the Research Fund for Coal and Steel (RFCS) under grant agreement n° [RFSR-CT-2011-00005].

References

[1] SSAB-Communications 2012 *The Steel Book* (Henningsons Tryckeri AB)
[2] Koric S, Hibbeler L C, Liu R and Thomas G B 2010 *Numerical Heat Transfer* B **58** 371-92
[3] Thomas B G 2002 *Metallurgical and Materials Transactions* B **33** 795-812
[4] Thomas B G 2001 *Continuous Casting of Steel* 499-540
[5] Ray K S and Mazumdar 2001 *Solidification Control in Continuous Casting of Steel* **26** 179-98
[6] Bounds S, Moran G, Pericleous K, Cross M and Croft T N *Metallurgical and Materials Transactions* A **31** 515-27
[7] Huespe A E, Cardona A and Fachinotti V 2000 *Computer Methods in Applied Mechanics and Engineering* **182** 439-55
[8] Kristiansson J 1982 *Journal of Thermal Stresses* **5** 315-30
[9] Fredriksson H and Åkerlind U 2006 *Materials Proc. During Casting* (John Wiley & Sons Inc)
[10] Ramirez Lopez P E 2010 PhD Thesis Imperial College London p 170
[11] Ramirez Lopez P E, Björkvall J, Sjöström U, Lee P D, Mills K C, Jonsson B, Janis J, Petäjäjärvi M and Pirinen J 2012 *Ninth Int. Conf. on CFD in the Minerals and Process Industries* (Melbourne, Australia) 1-6
[12] Jalali Nazeem P, Ramirez Lopez P E, Jönsson P, Petäjäjärvi M and Pirinen J 2012 *Scanmet IV* (Luleå, Sweden)
[13] Jonsson T and Ramirez Lopez P E 2011 *DDT "Direct Defect Toolbox"* RFSR-CT-2011-00005
[14] Risso J M, Huespe A E and Cardona A 2006 *Int. Journal for Numerical Methods in Engineering* **65** 1355-1377
[15] Janik M, Dyja H, Berski S and Banaszek G 2004 *Journal of Materials Proc. Technology* **153-154** 578-582.
[16] Wu L, Wang T, Fu Y, Li, J, Cao Z and Li T 2012 *Materials Science and Engineering* **33**
[17] Gupta D, Chakraborty S and Lahiri A K 1997 *ISIJ Int.* **37** 654

Simulations of Macrosegregation with consideration of inclusion effect in solidifying carbon steels

Y F Cao, Y Chen, X H Kang, P X Fu, H W Liu and D Z Li

Shenyang National Laboratory for Materials Science, Institute of Metal Research, Chinese Academy of Sciences, Shenyang, Liaoning, 110016 P. R. China

E-mail: xhkang@imr.ac.cn

Abstract. During casting of steel ingots, the inclusions such as oxide, sulfide will inevitably exist in the melt. These inclusions will flow upward together with light solutes during solidification due to their lower density relative to the steel melt, which therefore causes impacts on the thermo-solutal convection in the melt and final solute distribution. Hence, a macrosegregation model that considers the effects of inclusions on melt flow in the mushy zone is established. Of the new model two important parameters, the inclusion capturing probability by solid, k_p, and the original volume fraction, n_0, are systematically studied in terms of simulations, which shows that decreasing k_p or increasing n_0 leads to stronger ascending flow in the melt. And then as a validation example, the model was used to predict the macrosegregation in a 3.3-ton steel ingot. The prediction demonstrates that with consideration of inclusions, the melt convection strength is enhanced and thus the zones of macrosegregation are expanded comparing to simulations without taking account of inclusions. Further comparison with experiment results indicates that a better agreement of the carbon segregation along the centerline of the ingot can be achieved when considering the inclusion buoyancy.

1. Introduction

Macrosegregation presented as the inhomogeneous solute distributions in very large regions in the final solidified products, is mainly induced by the thermosolutal natural convection, the grain settling or flotation and the shrinkage force of the volume contraction during solidification according to the current theoretical framework. Based on these macrosegregation theories, a large number of numerical simulations have been conducted to predict the chemical variations and inhomogeneities in ingots, such as the multi-component simulation, multiphase simulation, 3D simulation, CAFE simulation and even microstructural level simulation [1-6]. To some extent, these advancing models have a capacity to accurately predicate the formation of macrosegregation in common model alloys, such as Sn-Pb, Sn-Bi, Al-Cu, Ga-In and Ni-base superalloy [4, 7-10], but significant discrepancies between simulations and experiments still persist in solidifying steels [11]. For example, the positive segregation at the hot-top is usually underestimated in simulations, and the shape and segregated carbon content at the bottom negative cone-segregation zone in simulations are much different from experimental measurements. Most significantly, nearly most of the numerical computer simulations cannot reproduce "A-type" channel segregates at the upper-middle part of the steel ingot body. It is definite that the density difference in the melt that leads to the macrosegregation, however, in steel ingots, the difference not only can be resulted by the solute and temperature, but also induced by the solidified equiaxed crystals and the existing inclusions. For the sedimentation of grains, the effects have been well investigated through numerical simulations [2, 3, 12-14], whereas for the effect of impurities and inclusions, it has been neglected for a long term.

Recently, the systematic investigations of experiment and multi-scale simulations of channel segregation in a series of sectioned steel ingots by Li and coworkers [15] have demonstrated strikingly that the A-segregation is actually a zone with enrichments of small Al_2O_3 and MnS inclusions except for the accumulated massive amount of solutes. Their contrast experiments also have shown that decreasing the content of alumina-based inclusions will be beneficial to eliminating the A-segregation in steels. These experiments seem to be right to interpret the channel segregates in steel ingots, because the amount of main solute carbon that drives the melt flow usually is orders of magnitude smaller than that in the often used model alloys (i.e. Pb-Sn, Pb-Bi, Ga-In, Al-Cu and Ni-base superalloy) in experiments and simulations, and hence the density difference in the melt of extensively applied steels cannot drive sufficiently strong convection in the interdendritic region to destabilize the mushy zone and promote the formation of segregation channels. As the inclusion content reaches a high level, the influence of light non-metallic inclusions on the liquid flow, and eventually on the formation of macrosegregation, should be concerned adequately.

Accordingly, a continuum macrosegregation model rather than the multiphase flow model as illustrated in Ref. [15] for describing the transport phenomena in solidification systems, including the effects of inclusions, is presented in this paper. A systematic examination on inclusion parameters of the built model is demonstrated in detail. And then, as an application the developed model is employed to simulate the flow field and macrosegregation formation in a typical 3.3-ton steel ingot.

2. Macrosegregation model

In this paper the two-dimensional calculations are carried out and some main assumptions in current developed model are listed: (1) solidified grains are supposed to be columnar and the solid movement and shrinkage are not considered; (2) the liquid is Newtonian and incompressible while the convection is laminar; (3) the effect of diffusion on the solute transport is also neglected; (4) Boussinesq approximation is used to account for buoyancy driven flow, including the effect of inclusions. On the basis of these hypotheses, the mass, momentum, mixture energy, mixture solute and inclusion conservations are described as follows.

2.1. The mass, energy and solute governing equations

Conservation of mass:

$$\nabla\cdot(\vec{U})=0 \tag{1}$$

$$\vec{U}=(1-f_s)\vec{U_l} \tag{2}$$

Conservation of energy:

$$\frac{\partial[\rho H]}{\partial t}+\nabla\cdot\left(\rho c_p\vec{U}T+\rho\vec{U}\Delta H\right)=\nabla\cdot\left(\lambda\nabla T\right) \tag{3}$$

$$[\rho H]=f_s\rho_s h_s+(1-f_s)\rho_l h_l \tag{4}$$

where ρ and f_s are the mixture density (equal to solid and liquid density of steel) and solid fraction respectively, h_s, h_l, H and ΔH are the solid, liquid, mixture enthalpy and latent heat respectively, c_p and λ are the specific heat and heat conductivity respectively.

Conservation of solute:

$$\frac{\partial[C]}{\partial t}+\nabla\cdot\left(\vec{U}C_l\right)=0 \tag{5}$$

$$[C]=\int_0^{f_s}C_s d\alpha+(1-f_s)C_l \tag{6}$$

where $[C]$, C_s and C_l are the mixture, solid and liquid concentrations respectively, and f_s and $\vec{U}_l$ are the solid fraction and liquid velocity vector during solidification.

2.2. The conservation equations of inclusions and momentums

In this study, alumina particles are assumed with a very small diameter (*i.e.*, < 5~10 μm) so that they can't float up quickly according to Stokes' law, and most of them may move with the liquid. Such fine alumina particles in the melt mostly stem from the adoption of aluminum deoxidation technique during the refining process. This assumption is reasonable because the dissolved oxygen content in steel melt is very low (usually < 5~10 ppm), therefore, the formation of alumina at the solidification stage is negligible. And all the pre-existing alumina particles are distributed evenly in the ingot before solidification. During real solidification, because of the blocking of dendritic networks and trapping of solidified crystals, a certain proportion of alumina cannot be rejected into melt and has neglected contribution to the liquid flow. Therefore, an artificial partition coefficient k_p, similar to the solute partition between solid and liquid, is introduced to depict the probability of the trapped inclusions by solid in a given control element.

Conservation of inclusions:

$$\frac{\partial n}{\partial t} = \nabla(\vec{U} n_l) \tag{7}$$

$$n = \int_0^{f_s} n_s d\alpha + (1-f_s) n_l \tag{8}$$

where n, n_s and n_l are the mixture, solid and liquid volume fractions of inclusions in a specific control element, respectively.

Conservation of u momentum (X direction):

$$\frac{\partial(\rho u)}{\partial t} + \nabla\cdot\left(\rho\vec{U}u\right) = -\frac{\partial P}{\partial x} - \frac{\mu_l}{K}u + \nabla\cdot\left(\mu_l \nabla u\right) \tag{9}$$

Conservation of v momentum (Y direction):

$$\frac{\partial(\rho v)}{\partial t} + \nabla\cdot\left(\rho\vec{U}v\right) = -\frac{\partial P}{\partial y} - \frac{\mu_l}{K}v + \nabla\cdot\left(\mu_l \nabla v\right) + \rho g\left[\beta_T\left(T-T_0\right) + \beta_C\left(C_l - C_0\right)\right] + \rho g \beta_p (n_l - n_0) \tag{10}$$

$$\rho = \rho_p n + \rho_{Fe\text{-}C}(1-n) \Rightarrow \frac{\rho}{\rho_{Fe\text{-}C}} = \frac{\rho_p}{\rho_{Fe\text{-}C}} n + 1 - n = 1 - \left(1 - \frac{\rho_p}{\rho_{Fe\text{-}C}}\right) n \Rightarrow \beta_p = \frac{\rho_{Fe\text{-}C} - \rho_p}{\rho_{Fe\text{-}C}} \tag{11}$$

where β_T, β_C and β_p are the thermal, solute and inclusion expansion coefficient respectively. T_0, C_0 and n_0 are the initial liquidus, solute concentration and inclusion volume fraction, respectively. ρ_p and $\rho_{Fe\text{-}C}$ are the inclusion and steel density respectively. K is the permeability in the mushy zone according to Carman-Kozeny formula:

$$K = \frac{d_s^2 \left(1 - f_s\right)^3}{180 f_s^2} \tag{12}$$

where d_s is the secondary dendritic arm spacing.

2.3. The calculation of solid fraction

The local thermodynamic equilibrium at the solid-liquid interface is guaranteed by assuming that the temperature at the liquid side is equal to the liquidus temperature of the melt:

$$T = T_m + m_l C_l \tag{13}$$

And therefore, in combination with Eqs. (3), (4), (6) and(13) the solid fraction can be eventually obtained, which yields to a quadratic equation of solid fraction f_s:

$$A\cdot\left(f_s^{t+\Delta t}\right)^2 + B\cdot f_s^{t+\Delta t} + E = 0 \tag{14}$$

Here the coefficients are written as,

$$\begin{aligned} A &= (1-k)\rho\Delta H \\ B &= \rho\Delta H(k + kf_s^t - 2) + (k-1)(\rho c_p T_m - [\rho H]^t) \\ E &= (\rho c_p T_m - [\rho H]^t + \rho\Delta H)(1 - kf_s^t) - \rho c_p m_l C_l^t (1 - f_s^t) \end{aligned} \tag{15}$$

where k, T_m and m_l are solute partition coefficient, melting point of iron and liquidus slope respectively, superscripts t and Δt are time and time step, respectively.

2.4. Numerical implementations

In calculations, an explicit scheme of finite volume method is used to solve the coupled equations for momentum, solute, inclusion and heat transports. The solution algorithm for transient fluid flow technique is employed to solve the Navier-Stokes equations. A carbon steel ingot which has a weight of 3.3 ton is used in simulations. The specification of the thermal boundary conditions at outer and inner surfaces and initial conditions can be referred to the paper by Combeau *et* al. [16]. The input parameters for calculations are compiled in table 1. The simulations of the solidification process are performed first with different partition coefficients, k_p and original volume fractions of inclusions, n_0 to investigate the response of these inclusion parameters on the liquid flow intensity. And then the macrosegregation of a steel ingot that takes account of inclusion buoyancy is investigated.

Table 1. The main parameters used in simulations.

Solutal expansion coefficient, β_C	1.4164×10^{-2} wt.%$^{-1}$
Thermal expansion coefficient, β_T	1.07×10^{-4} K^{-1}
Liquidus slope, m_l	−80.45 K wt.%$^{-1}$
Inclusion expansion coefficient, β_p	0.48
The spacing of secondary arm, d_s	500 μm
Dynamic viscosity, μ_l	0.0042 Pa s
Specific heat, C_p	500 kg^{-1}K^{-1}
The latent heat, ΔH	271000 J kg^{-1}
The thermal conductivity, λ	39.3 W m^{-1}K^{-1}
The steel density, ϱ	6990 kg m^{-3}
The melting temperature, T_m	1532 °C
Solute partition coefficient, k	0.314
Inclusion partition coefficient, k_p	0.2, 0.4, 0.6, 0.8
Initial inclusion volume fraction, n_0	0.002, 0.004, 0.006, 0.008, 0.01
Grid size	10 mm × 10 mm
Inclusion density, ϱ_p	3640 kg m^{-3} [17]

3. Numerical Results and discussion

3.1. The effect of inclusion partition coefficient k_p

During solidification of steel ingot, some of small pre-existent inclusions will be captured by the solid dendritic network. This phenomenon as described above in the present model is characterized by a phenomenological partition coefficient, k_p. However, it is difficult to measure the capturing probability in experiments, and thereby a region of the partition coefficient (k_p = 0.2~0.8) are chosen in calculations to investigate the trapping effect induced by solid. In these calculations the original volume fraction of inclusions is supposed to be as high as 0.01 [18]. The contribution of solute carbon on the buoyancy-driven flow is excluded in the simulations for clearly understanding the liquid flow intensity induced by inclusions. Figure 1 shows the simulated liquid flow field and the solid fraction after freezing of 2505 s. At this time almost half of the ingot body has been solidified and the thermal-inclusion buoyancy force induced liquid convection has reached a stable evolution tendency. Initially, due to the chilling of walls, the melt is cooled fast near the wall and a sharp thermal gradient is built up. Hence, the cooled melt resulted from the thermal contraction has a higher density and moves downwards along the chilling wall, while the hotter melt that is relatively lighter in the center moved upwards. With the solidification proceeding, the heat extraction from the walls becomes inefficient due to the thicker solid shell of the ingot. And the thermal boundary layer ahead of the solidification front is therefore thick, which induced a weaker thermal flow in the liquid comparing with that at the onset of solidification. Furthermore, owing to the rejection of inclusions from the solid to liquid, more and more inclusions are accumulated before the solidification front. This lighter substance would thus lead to the liquid flow upwards in and nearby the mushy zone. In contrast to the thermal convective flow, this ascending flow becomes more intense accompanying with the solid growth. Consequently, after solidification for several minutes (about 300 s in current simulation), when the volume fraction of enriched inclusions reaches a certain value and satisfies

$$|\beta_T\left(T-T_{ref}\right)|<|\beta_p(n-n_0)| \tag{16}$$

the flow leaded by inclusion buoyancy force will overwhelm that leaded by thermal buoyancy, and reverses the flow direction of melt in the ingot, that is a clockwise flow on the left half of the entire ingot as shown in figure 1. Afterwards, the overall liquid flow is dominated by the inclusion almost until the end of solidification, playing a similar role of solute carbon on the liquid convection.

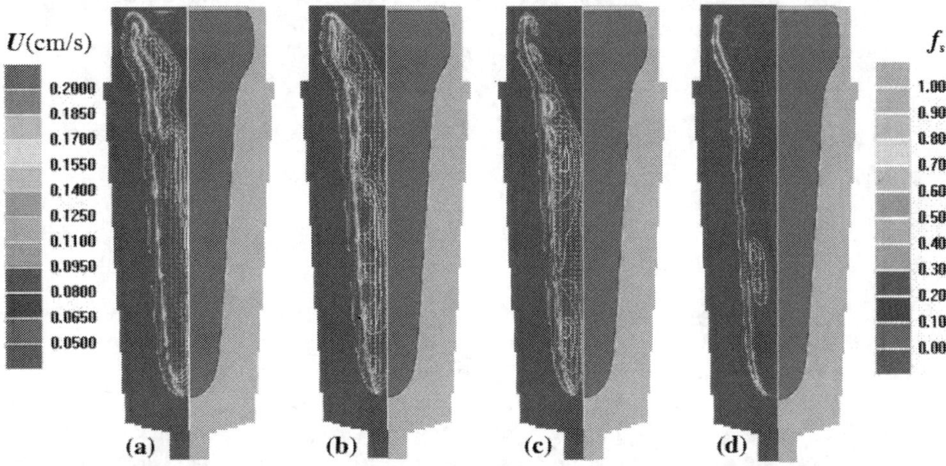

Figure 1. The simulated liquid flow and solidification profiles at 2505 s with different k_p, and n_0 is 0.01. Left: liquid flow velocity; Right: solid fraction. (a) $k_p = 0.2$, (b) $k_p = 0.4$, (c) $k_p = 0.6$, (d) $k_p = 0.8$.

When comparing the simulations between different partition coefficients, k_p, the flow patterns rarely can be observed significant distinctions, but the flow intensity increases with the decrease of k_p, Even though small amounts of inclusions are discharged into the melt, e.g. for the simulation $k_p = 0.8$ shown in figure 1(d), the liquid convection is still dominated by the displacement of inclusions. It should be mentioned that the mushy ranges are rather narrow in all cases, this is caused by neglecting the solute effect in the simulations and heat conduction can proceed very easily compared with solute transport.

In order to quantitatively evaluate the influence of the parameter k_p on convection strength, the interdendritic average velocity |U| in the mush of ingot body, defined as a zone with a solid fraction between 0 and 0.7, are calculated as listed in table 2. Apparently, the flow velocity decreases monotonously with increasing coefficient k_p, but the change is not pronounced. This result further confirms that with the decrease of k_p, more inclusions will not be captured by solid phase and enter the residual melt adjacent to the solidification front. As a result, the larger amounts of light inclusions lead to higher buoyancy force and thus stronger convection.

Table 2. The interdendritic average flow velocity in the ingot body with different k_p at 2505 s.

k_p	0.2	0.4	0.6	0.8
|U| (cm/s)	0.0422	0.0326	0.0238	0.0136

3.2.The effect of initial inclusion volume fraction n_0

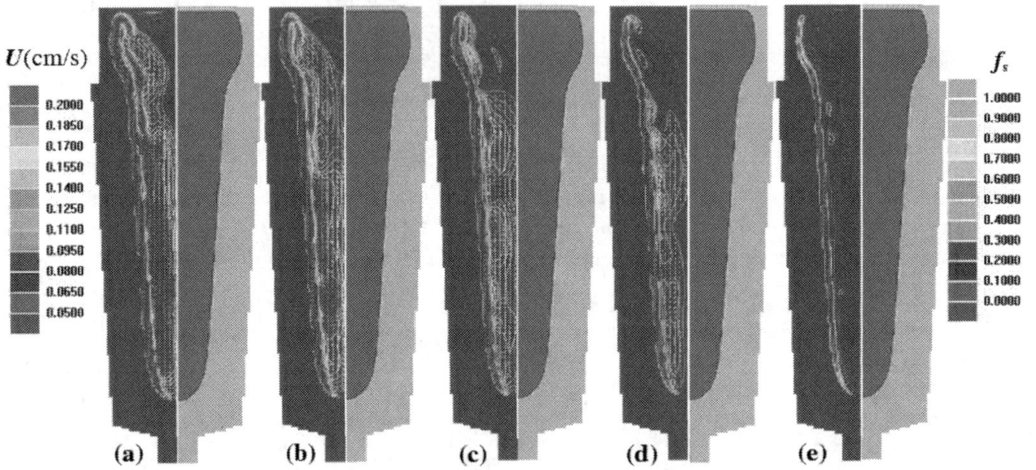

Figure 2. The simulated liquid flow and solidification profiles at 2505 s with different n_0, and k_p is 0.2. Left: liquid flow velocity; Right: solid fraction. (a) $n_0 = 0.01$, (b) $n_0 = 0.008$, (c) $n_0 = 0.006$, (d) $n_0 = 0.004$, (e) $n_0 = 0.002$.

In steelmaking industry, minimizing the inclusion content is always an essential target to prevent cracks or fatigue fractures and even alleviate macrosegregation, specially channel segregation [15]. For the sake of understanding the role that inclusions play, the effect of initial inclusion volume fraction on macrosegregation is investigated. Figure 2 denotes the velocity field and solid fraction at 2505 s in case of inclusion partition coefficient being assumed as 0.2, which indicates obviously that stronger convection is resulted in the bulk melt and the mushy zone as initial inclusion volume fraction increases. Quantitatively, table 3 gives the average melt flow velocity |U| with different n_0.

Table 3. Average velocity of solidification front in ingot body with varied n_0 at 2505 s.

n_0	0.002	0.004	0.006	0.008	0.01
|U| (cm/s)	0.012	0.0205	0.0286	0.035	0.0422

The average velocity variation in table 3 shows that, when n_0 varies from 0.002 to 0.01, the velocity magnitude increases to 0.0422 from 0.012 cm/s, which can be naturally comprehended as the following situations are taken into consideration: Given the same thermal conditions, when solid zone occupies half of the whole ingot, there will be more inclusions advected into the un-solidified zone and more serious accumulation of inclusions forms with increasing n_0, which causes a greater density difference between the interdendritic melt and bulk liquid. Therefore, a stronger driving force of upward flow and higher velocity in the mushy zone are induced.

It is worthy of noticing that except for these two most important parameters of inclusions that affect the liquid flow, the size and shape of an inclusion particle which are usually important characteristics of a solid particle moving in liquid, are not considered in the present simulations. Indeed, the motion of a solid inclusion particle is governed by a balance of buoyancy force, drag force, virtual mass force, Saffman's lift force and other forces and therefore the particle has its own velocity that is locally different from the liquid flow velocity. The moving of particle separately, in turn, will impact the flow of fluid phase, especially the localized flow around the particle. In regard to the shape of a particle, the more complicated shape of the surface, the larger frictional force is acted upon by the liquid; hence the lower moving velocity of melt is dragged. Physically, the particle moving kinetics in liquid dependent on the shape and size actually falls into the category of liquid-solid two phases flow problem, however, in the current model to make the problem to be feasible for simulation, the inclusion and the melt are treated as the same phase but have different densities. Therefore, the divergence stemming from the simplification of the model exists to some extent for inclusion that has a large size and smooth surface. With including this fluid-structure and fluid-inclusion interactions, an improved model combining with an accurate description of the motion of inclusions in steel melt using a Lagrangian approach will be developed in the future work.

3.3. Prediction of macrosegregation in a carbon steel ingot with consideration of inclusions

Based on above examination of the established model, it is then extended to simulate and predict carbon macrosegregation in a Fe-0.36 wt.% C steel ingot. Figure 3 shows the final carbon and inclusion distribution maps with a range of the partition coefficient k_p (0.2~0.8) assuming the original inclusion volume fraction of 0.01. On the whole, the typical positive segregation at the upper ingot, the characteristic conical negative segregation at the bottom and the stripe-like channel segregation in the riser induced by the destabilized mushy zones are all predicted in the four different simulation cases, which agrees well with the simulation results in Ref. [16]. These results further demonstrate that both the positive and negative segregation zones are expanded and become more serious as the inclusion partition coefficient decreases. Similarly, figure 4 provides the predicted final carbon and inclusion segregation patterns with different n_0, where k_p is assumed as 0.2. The predicted segregation pattern is analogous to that shown in figure 3. Moreover, it indicates that the positive segregation at the hot-top and the cone-shaped negative segregation at the bottom are severer with the increase of original inclusion volume fraction. In addition, the inclusion distribution patterns are identical to that of carbon, and inclusion inhomogeneity becomes severer with increasing n_0 and decreasing k_p. The distribution consistency between carbon and inclusion is decided by the assumptions in current applied model, where the similar inclusion behavior with solute is considered and described (see section 2.2).

In order to further quantitatively examine the effect of inclusion partition coefficient and original volume fraction on the carbon macrosegregation, the global extent of macrosegregation *GM* [1] is introduced (Eq. 17) and figure 5 plots the calculation results of *GM* in full ingot and ingot body respectively with various k_p and n_0.

$$GM = \frac{1}{C_0}\left[\frac{1}{V_{casting}}\iiint_{V_{casting}}(C-C_0)^2\,dV\right]^{1/2} \tag{17}$$

Figure 5(a) shows that the severity of *GM* exhibits a dependence on the inclusion partition coefficient closing to a linear relationship, and the extent of *GM* becomes severer as k_p decreases. The decreases of k_p indicates that less inclusions will be trapped by the solid sketch, and more inclusions will segregate into the mushy zone and liquid during solidification. The segregated inclusions, together with carbon, cause a more intensive upward buoyancy force because of their less density and larger driving force. Consequently, the solute can move not only in a larger distance and regions but also with higher quantity. Eventually, larger amounts of solutes are taken away by melt flow from the side and bottom of the ingot into the center and top part; meanwhile solute-lean liquid will enter the bottom from the centerline and a solute depleted region is initiated. Similarly, the large carbon segregation caused by increasing n_0(figures 4 and 5(b)) can be explained due to the larger density difference between interdendritic melt and bulk liquid. In addition, of all the simulation cases the global macrosegregation in full ingot is no except severe than that in ingot body and inclusion buoyancy indeed increases the segregation of carbon compared with case neglecting the inclusion effect as shown in figure 5(b).

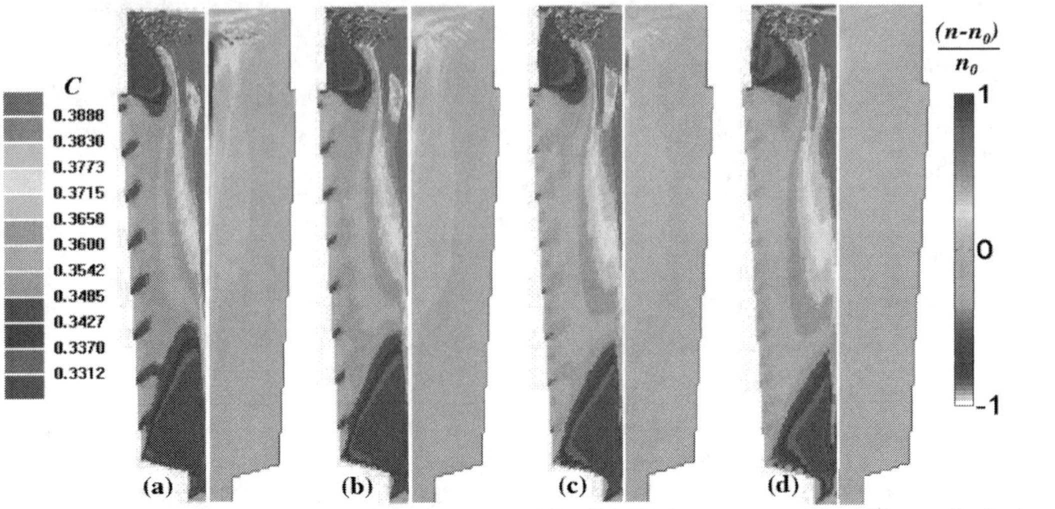

Figure 3. Final carbon (left) and inclusion (right) distribution maps with different inclusion partition coefficients k_p. Here, $n_0 = 0.01$ and (a) $k_p = 0.2$, (b) $k_p = 0.4$, (c) $k_p = 0.6$, (d) $k_p = 0.8$.

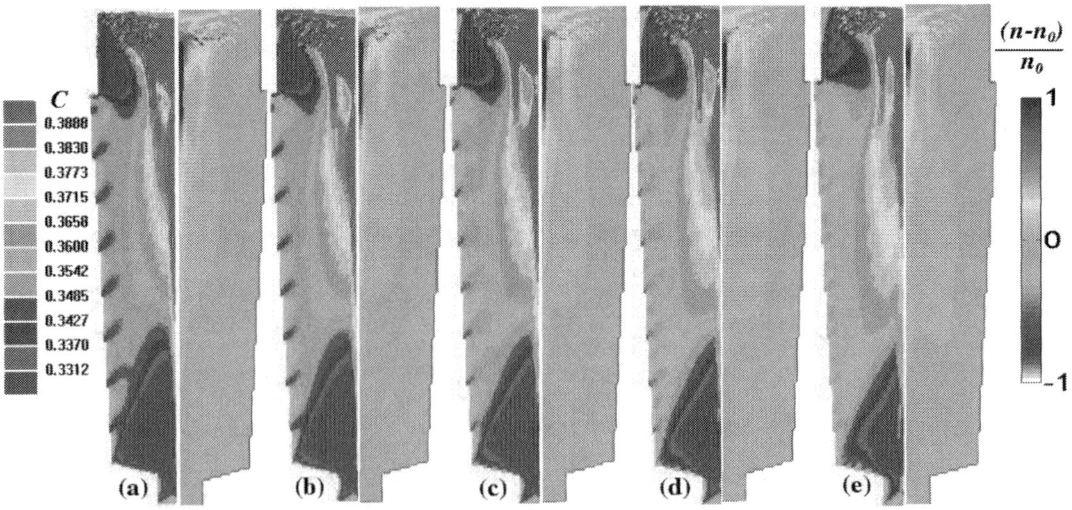

Figure 4. Final carbon (left) and inclusion (right) distribution maps with different original inclusion volume fractions n_0. Here, $k_p = 0.2$ and (a) $n_0 = 0.01$, (b) $n_0 = 0.008$, (c) $n_0 = 0.006$, (d) $n_0 = 0.004$, (e) $n_0 = 0.002$.

Besides the global macrosegregation the solute segregation along the centerline is more serious than other zones and usually is used to characterize the composition segregation extent of a steel ingot. Figure 6 shows that the severity of positive and negative segregations along the centerline become larger when k_p decreases or n_0 increases. This is closely connected to stronger convection induced by the more segregated inclusions in the mushy zone (figures 3 and 4). The experimental measurements

IOP Conf. Series: Materials Science and Engineering **84** (2015) 012098 doi:10.1088/1757-899X/84/1/012098

of carbon concentration [16] and the simulation result without inclusions are also presented in figure 6(b). The predicted carbon content at the hot top with a model not considering the effect of inclusions underestimated the experiment result, whereas the current model with consideration of inclusions can overshoot the experimental data, even though the original fraction of inclusions is as low as 0.002. This is plausible that in actual production of the experimental ingot, the inclusion content may be far below this level due to the refining techniques used before the pouring process, in which case the carbon distribution will become more uniform because of weaker convection in the mushy zone rather than being such overestimated as shown in figure 6(b). Therefore, we can postulate that taking into account the contribution of inclusions on the buoyancy force, a better agreement can be achieved with experiment at the top positive segregation zone when appropriate k_p and n_0 are chosen.

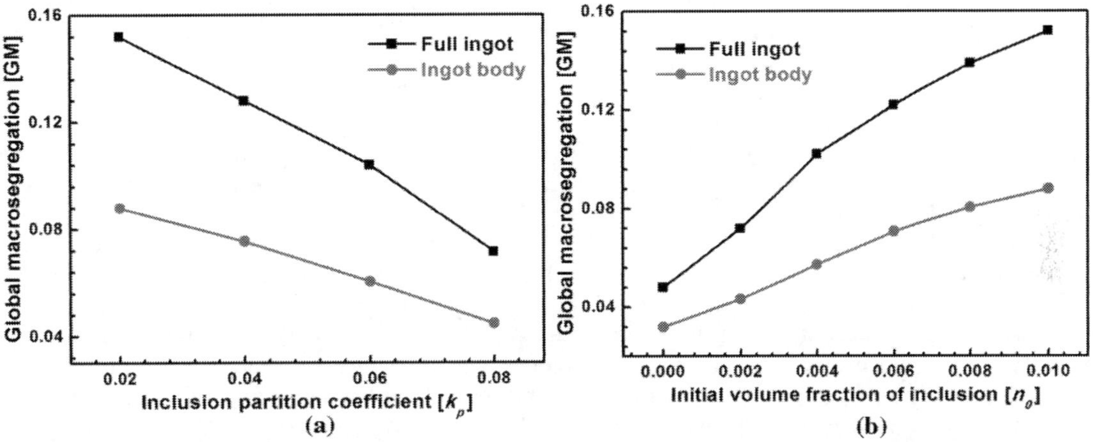

Figure 5. The global extent of macrosegregation (*GM*) in the ingot: (a) varied inclusion partition coefficients k_p with $n_0 = 0.01$; (b) varied original inclusion volume fractions n_0 with $k_p = 0.2$.

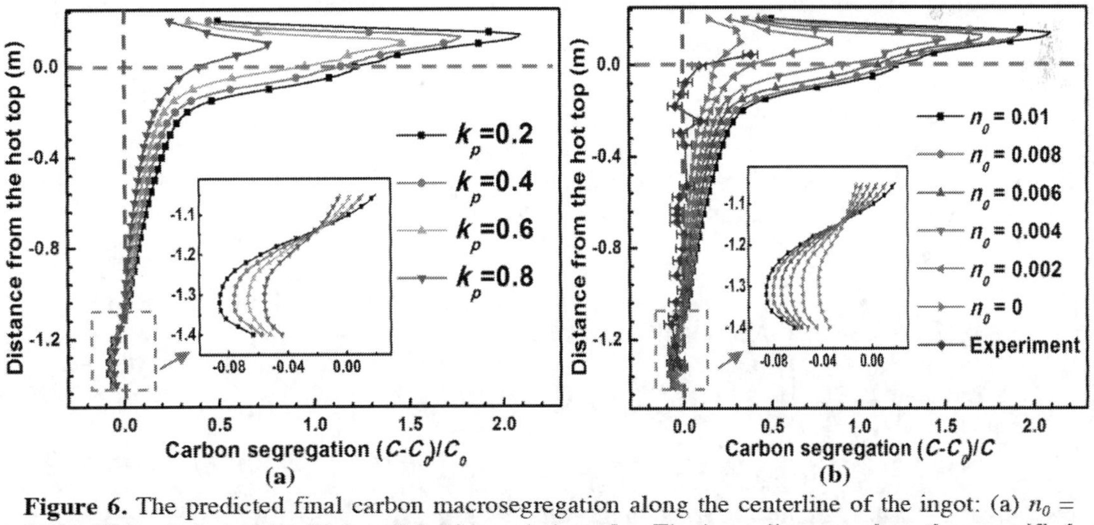

Figure 6. The predicted final carbon macrosegregation along the centerline of the ingot: (a) $n_0 = 0.01$ with variation of k_p; (b) $k_p = 0.2$ with variation of n_0.The inset diagrams show the magnified part of curves at the bottom part of the ingot.

4. Conclusions

A new mathematical model of macrosegregation including the effect of inclusions on the buoyancy force has been presented. The inclusion induced flow and the subsequent solute segregation during solidification were investigated in detail through numerically solving the governing equations. In terms of simulations, especially the effect of two important parameters (inclusion capturing probability, k_p and original inclusion volume fraction, n_0) on the flow field during solidification were firstly examined and then the model was applied to predict carbon macrosegregation pattern with different k_p and n_0 in a 3.3 ton steel ingot. The analysis and discussion on the model and application allowed for the following main conclusions. Firstly, large n_0 and small k_p both favor the upward flow in the mushy zone during the stage that solute-inclusion buoyancy force dominates the melt convection, and the average velocity of interdendritic melt can reach as high as 0.0422 cm/s even though it is free from the influence of solutes. Secondly, the model successfully predicts the carbon macrosegregation pattern, including the positive segregation at the hot top, the cone-shaped negative segregation at the bottom and A-type segregation in the riser of the steel ingot. The simulation results also demonstrate that large n_0 and small k_p will make the global macrosegregation larger and the segregation along the centerline severer. Compared with experiment results, the model with the effect of inclusions enhances the melt convection strength and results in higher ratio of segregation along the centerline at the hot top of the ingot, which is always underestimated by previous macrosegregation models, so that better agreement can be achieved between simulation and experiment when appropriate k_p and n_0 are chosen rather than the current arbitrary values used.

However, the present explorative model should be viewed as a preliminary understanding of the influence of inclusion (or impurity) on the liquid flow during solidification of steel ingot and the subsequent macrosegregation. Clearly, the practical solidification process is much more complicate than present simulations and many other important aspects should be considered in the elementary models, such as, the effect of sedimentation of free equiaxed crystals, the accurate description of the interaction between dispersed inclusions and continuous phase, mutual interactions between inclusions, different kinds of inclusions (manganese oxide, sulfide, etc.) and crystal growth.

Acknowledge

This work is supported by National Natural Science Foundation for Young Scientists of China (Grant No. 51401223) and National Natural Science Foundation of China (Grant No. 51271184).

References

[1] Schneider M C and Beckermann C 1995 *Metall. Mater. Trans.* **A26** 2373-88
[2] Wu M and Ludwig A 2009 *Acta Mater.* **57** 5621-31
[3] Založnik M and Combeau H 2010 *Comp. Mater. Sci.* **48** 1-10
[4] Carozzani T, Digonnet H, Bellet M and Gandin C A 2012 *MCWASP XIII*, Schladming, Austria
[5] Yuan L and Lee P D 2012 *Acta Mater.* **60** 4917-26
[6] Gouttebroze S, Bellet M and Combeau H 2007 *C. R. Mecanique* **335** 269-79
[7] Sawada T, Oikawa K and Anzai K 2013 *Tetsu-to-Hagane* **99** 135-40
[8] Kumar A, Walker M J, Sundarraj S and Dutta P 2011 *Metall. Mater. Trans.* **B42** 783-99
[9] Guillemot G, Gandin C A and Bellet M 2007 *J. Cryst. Growth* **303** 58-68
[10] Schneider M C, Gu J P, Beckermann C, Boettinger W J and Kattner U R 1997 *Metall. Mater. Trans.* **A42** 1517-31
[11] Flemings M C 2000 *ISIJ Int.* **40** 833-41
[12] Wang C Y and Beckerman C 1996 *Metall. Mater. Trans.* **A27** 2754-64
[13] Wu M and Ludwig A 2006 *Metall. Mater. Trans.* **A37** 1613-31
[14] Badillo A and Beckermann C 2006 *Acta Mater.* **54** 2015-26
[15] Li D, Chen X-Q, Fu P, Ma X, Liu H, Chen Y, Cao Y, Luan Y and Li Y 2014 *Nat.Commun.* **5:5572** doi: 10.1038/ncomms6572
[16] Combeau H, Založnik M, Hans S and Richy P E 2009 *Metall. Mater. Trans.* **B40** 289-304
[17] Shibata H, Poirier D R and Emi T 1998 *ISIJ Int.* **38** 339-47
[18] Kiesslin R 1968 *JOM.* **20** A20

Modelling of stress fields during LFEM DC casting of aluminium billets by a meshless method

B Mavrič[1] and B Šarler[1,2]

[1] Institute of Metals and Technology, Lepi pot 11, SI-1000 Ljubljana, Slovenia
[2] University of Nova Gorica, Vipavska 13, SI-5000 Nova Gorica, Slovenia

bostjan.mavric@imt.si, bozidar.sarler@ung.si

Abstract. Direct Chill (DC) casting of aluminium alloys is a widely established technology for efficient production of aluminium billets and slabs. The procedure is being further improved by the application of Low Frequency Electromagnetic Field (LFEM) in the area of the mold. Novel LFEM DC processing technique affects many different phenomena which occur during solidification, one of them being the stresses and deformations present in the billet. These quantities can have a significant effect on the quality of the cast piece, since they impact porosity, hot-tearing and cold cracking. In this contribution a novel local radial basis function collocation method (LRBFCM) is successfully applied to the problem of stress field calculation during the stationary state of DC casting of aluminium alloys. The formulation of the method is presented in detail, followed by the presentation of the tackled physical problem. The model describes the deformations of linearly elastic, inhomogeneous isotropic solid with a given temperature field. The temperature profile is calculated using the in-house developed heat and mass transfer model. The effects of low frequency EM casting process parameters on the vertical, circumferential and radial stress and on the deformation of billet surface are presented. The application of the LFEM appears to decrease the amplitudes of the tensile stress occurring in the billet.

1. Introduction

The stress fields during the DC casting process have important influence on the quality of the obtained billet. They can cause large deformations which influence the heat transfer and can lead to residual stress. Unwanted hot tearing and cracking [1] can occur if certain conditions are met [2], causing whole billet to be scrapped. Numerical models of the casting process, which allow us to develop better understanding of the stress fields during DC casting, are therefore of great importance for producing high quality aluminium billets.

Mathematical modelling of casting processes has been of great interest for many years and is reasonably well established [2,3]. Its developments mainly rely on the ever more sophisticated physical modelling, fast development of computer hardware and development of efficient numerical methods, especially the finite element method (FEM). The use of this method can be traced from the first modelling attempts [4,5] up to the sophisticated contemporary achievements [1,6]. In parallel, many new numerical methods have been developed, which aim to address various deficiencies of FEM. One class of the promising new methods are the meshless methods, where the space discretisation (meshing) is much simplified by removing the need for time consuming polygonization of the computational domain and circumventing the mesh distortion problems. The absence of the need for polygonization means that the nodes can be removed or added as necessary to achieve the highest possible computational efficiency at demanded precision and also that the complex geometries can be described with ease.

Local Radial Basis Function Collocation Method (LRBFCM) [7] is applied, as an alternative to FEM, to the problem of stress formation during Low-Frequency Electromagnetic Direct Chill (LFEM DC) casting in this paper. The considered alternative method interpolates the unknown solution over the localized subsets of nodes by using radial basis functions (RBF). The RBF interpolant is then used to estimate the differential operators present in the strong form of the elastostatic governing equation. This method has been already successfully applied to diffusion [7], solidification [8] and fluid flow problems, both with and without the presence of external electromagnetic fields [9]. Previous applications of the proposed method to thermoelasticity are rare and mainly focused on simple academic solid mechanics problems [10,11]. Present paper deals with application of the method in modelling of the industrial LFEM DC casting of cylindrical and axisymmetric aluminium billets and complements our previous fluid mechanics developments [12–14].

2. Physical model

Although in literature many examples of sophisticated thermomechanical physical models of DC casting can be found [4,15,16], the model used in this article is a simple one, allowing us to focus on the methodology. A linear thermoelastic model, incorporating the stress equilibrium, Hooke's law for inhomogeneous isotropic body and linear isotropic thermal expansion is considered. Since the DC casting is assumed to be a stationary process, boundary value problem is considered. The stationary governing equation, stated in terms of the deformation field $\mathbf{u}$, is given as

$$G\nabla^2\mathbf{u}+(G+\lambda)\nabla\nabla\cdot\mathbf{u}+\nabla\lambda\nabla\cdot\mathbf{u}+\nabla G\left(\nabla\mathbf{u}+(\nabla\mathbf{u})^T\right)=\nabla\left(\beta\left(T-T_{ref}\right)\right)-\mathbf{f}\,. \tag{1}$$

In the governing equation G stands for shear modulus, λ for Lamé parameter and $\mathbf{f}$ for the body force. The coupling with the temperature field is described by the coefficient β defined as $\beta=(3\lambda+2G)\alpha$, where α is the coefficient of linear thermal expansion. T_{ref} stands for reference temperature at which the thermal expansion is considered to be zero. All the material properties are allowed to vary over the computational domain in continuous manner.

For the description of the problem setting, the displacement, symmetry and traction boundary conditions are needed. The deformation of the top part of the billet is restricted by the mold. The boundary conditions in this area are nonlinear and are therefore satisfied by iteration.

In the computational domain, the material undergoes solidification and therefore significantly changes the elastic properties. The temperature dependence of alloy properties can be obtained from JMatPro database for each alloy considered [17].

3. Method formulation

3.1. Domain decomposition

In the present work, only regular geometries are considered, which allows for a regular node arrangement on the computational domain and its boundary. For each interior node l, a domain of influence is determined by a predefined constant number ${}_lN$ of nearby nodes. In this work nearest neighbor search is used. By the choice of the domain of influence, a mapping $s(i): i \rightarrow l$ from subdomain node index i, to the global index l is defined.

3.2. Construction of the interpolant

Over each domain of influence defined in the previous step, the RBF interpolant is constructed. We are using multiquadric RBFs, defined as

$$\Phi_j(\mathbf{r}) = \sqrt{\left(\frac{\epsilon}{h_l}\right)^2 |\mathbf{r}-\mathbf{r}_j|^2 + 1}\,.$$

Here ϵ is the shape parameter and $\mathbf{r}_j$ is the position of the node on which the basis function is centred. The scale of the basis function is given by the subdomain size h_l, defined as

$$h_l = \sqrt{\frac{\sum_{i=1}^{{}_lN} |\mathbf{r}_{m_l(i)} - \mathbf{r}_l|^2}{{}_lN-1}}\,.$$

The interpolation basis is expanded by constant and linear augmentation monomials. The interpolant is formally given as

$$u_\xi(\mathbf{r}) = \sum_{i=1}^{{}_lN} {}_l\alpha_{i,\xi}\Phi_{s(i)}(\mathbf{r}) + \sum_{i={}_lN+1}^{{}_lN+{}_aN} {}_l\alpha_{i,\xi} p_i(\mathbf{r}-\mathbf{r}_l) = \sum_{i=1}^{{}_lN+{}_aN} {}_l\alpha_{i,\xi}\psi_i(\mathbf{r}), \tag{2}$$

where ${}_aN$ is the number of monomials used and has for two-dimensional problems value 3. The expansion coefficients ${}_l\alpha_{i,\xi}$ are determined by collocation.

In this step also the boundary conditions are taken into account. The boundary conditions are assumed to be linear and given by $B^j_{\xi\zeta}u_\zeta(\mathbf{r}_j) = b^j_\xi$. In case an interior point has a member of influence domain that lies on the boundary, the interpolation equation for this node is replaced by the appropriate boundary condition. The system of equations obtained in this manner can be compactly written in matrix form as

$$\sum_{i,\zeta} {}_lA_{ji,\xi\zeta}\,{}_l\alpha_{i,\zeta} = {}_l\gamma_{j,\xi}\,,$$

with the right-hand side vector ${}_l\gamma$ and the collocation matrix ${}_lA$, given by

$${}_lA_{ji,\xi\chi} = \begin{cases} \Psi_i(\mathbf{r}_{ls(j)})\delta_{\xi\chi} & \text{if } \mathbf{r}_{ls(j)} \in \Omega \\ B_{\xi\chi}(\mathbf{r}_{ls(j)})\Psi_i(\mathbf{r}_{ls(j)}) & \text{if } \mathbf{r}_{ls(j)} \in \Gamma \\ \mathrm{p}_j(\mathbf{r}_{ls(i)}) & \text{if } j > {}_lN \end{cases}, \qquad {}_l\gamma_{j,\xi} = \begin{cases} \mathrm{u}_\xi(\mathbf{r}_{ls(j)}) & \text{if } \mathbf{r}_{ls(j)} \in \Omega \\ b_\chi(\mathbf{r}_{ls(j)}) & \text{if } \mathbf{r}_{ls(j)} \in \Gamma \\ 0 & \text{if } j > {}_lN \end{cases}.$$

3.3. *Discretization of the governing equation*

The interpolation of the field given in node points can be used to estimate differential operators. Since the expansion coefficients are assumed to be constant, the differential operators act only on the basis functions. This fact can be used to discretize partial differential equations. Let us assume a governing equation is given by $D_{\xi\zeta}u_\zeta = g_\xi$. By replacing the unknown solution $\mathbf{u}$ with the interpolation given by equation (2), the governing equation at every interior node can be stated in terms of the interpolation coefficients. The interpolation coefficients are further replaced by the components of vector ${}_l\boldsymbol{\gamma}$, thus expressing the governing equation at every interior node by the unknown solution values and given boundary conditions for nodes that belong to the influence domain, centred on the considered node. The resulting governing equation is for each interior node l stated as

$$\sum_{k,\chi} \Upsilon^\Omega_{ls(k),\chi} u_{ls(k),\chi} \sum_{i,\zeta} {}_lA^{-1}_{ik,\zeta\chi} D_{\xi\zeta}\psi_i(\mathbf{r}_l) = g_{l,\xi} - \sum_{k,\chi} \Upsilon^\Gamma_{ls(k),\chi} b_{ls(k),\chi} \sum_{i,\zeta} {}_lA^{-1}_{ik,\zeta\chi} D_{\xi\zeta}\psi_i(\mathbf{r}_l) \tag{3}$$

In this expression the boundary and the domain indicators Υ^Γ_j and Υ^Ω_j are used to achieve efficient notation. The indicators evaluate to one, if the corresponding point j belongs to the set under consideration and to zero otherwise.

The set of linear equations for the unknown solution values stated in (3) is sparse and can be solved efficiently by specialized solvers. The study of numerical performance and detailed statement of the method can be found in our recent publications [18,19].

4. Application to DC casting

The briefly described thermomechanics solver forms a part of the multiphysics and multiscale DC modelling framework, that is being developed in our group [13]. All the models are formulated by meshless methods. Currently, the modelling capabilities include heat and fluid flow, electromagnetic field and species transport. A special Point Automata (PA) meshless technique has been developed for simulation of the microstructure evolution [20]. The LFEM DC casting arrangement for 140 mm billet diameter is modelled from the top of the mould to 80 cm below the mould.

The computational domain is determined from the results of the heat and mass transport model. The computational domain and boundary conditions are given in Figure 1. Only the area below the highest point reached by the liquidus isotherm is considered. On the top surface, the metalostatic pressure is applied. The top part of the domain is constrained by the mould, while the rest of the outer surface is assumed to be free. In the centre, the symmetry boundary conditions are applied, and on the bottom the vertical deformation is set to zero, while zero traction is assumed in radial direction.

The computational domain consists of 10^5 discretization points arranged in nodes of a rectangular grid. The shape parameter was set to $2 \cdot 10^{-2}$ with the domains of influence consisting of ten nodes.

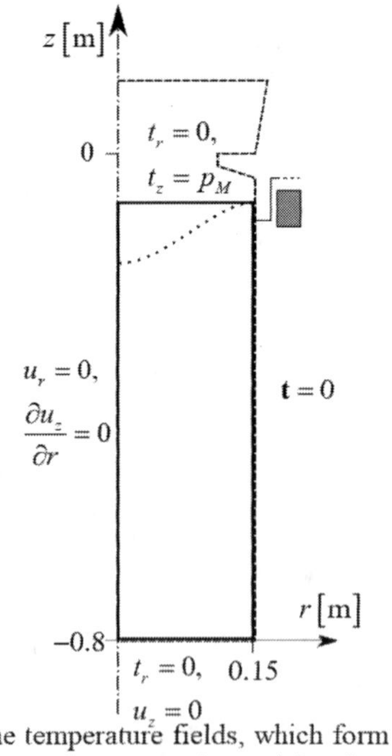

Figure 1 Scheme of geometry with indicated computational domains and boundary conditions for the thermoelastic model. The fluid flow computational domain is given by the dashed line, the liquidus isotherm by the dotted line. The solid line represents the computational domain of thermoelasticity. The position of the electromagnetic coil is indicated by the gray-filled rectangle.

The temperature fields, which form the driving term of the governing equation (1) are obtained by the heat and mas transfer model and are shown in Figure 2 for different casting speeds.

Results obtained for different casting speeds in the absence of the electromagnetic field are shown in Figures 5, 6, and 7. In general features, the structure of the stress fields remains the same. Compressive state forms in the centre of the billet and tensile state occours in the circumferential direction near the surface. The increase in casting speed only stretches those features along the casting direction due to the change of the sump depth.

More illustrative is the comparison of the surface deformations given in Figure 3. Since the transient zone in which the temperature drops from liquidus temperature to the ambient temperature is the thinnest in the case of the smaller casting speed, also the surface shrinks the fastest. The differences in surface deformation between the two faster speeds are small; however the rate of shrinkage does not monotonously decrease with increasing speed, since the two curves cross each other. The images illustrating the influence of the casting speed on the radial stress, illustrate large stresses that are present in the contact between the solidifying shell and the mould surface, implying the need for lubrication of the area if high quality surface is required.

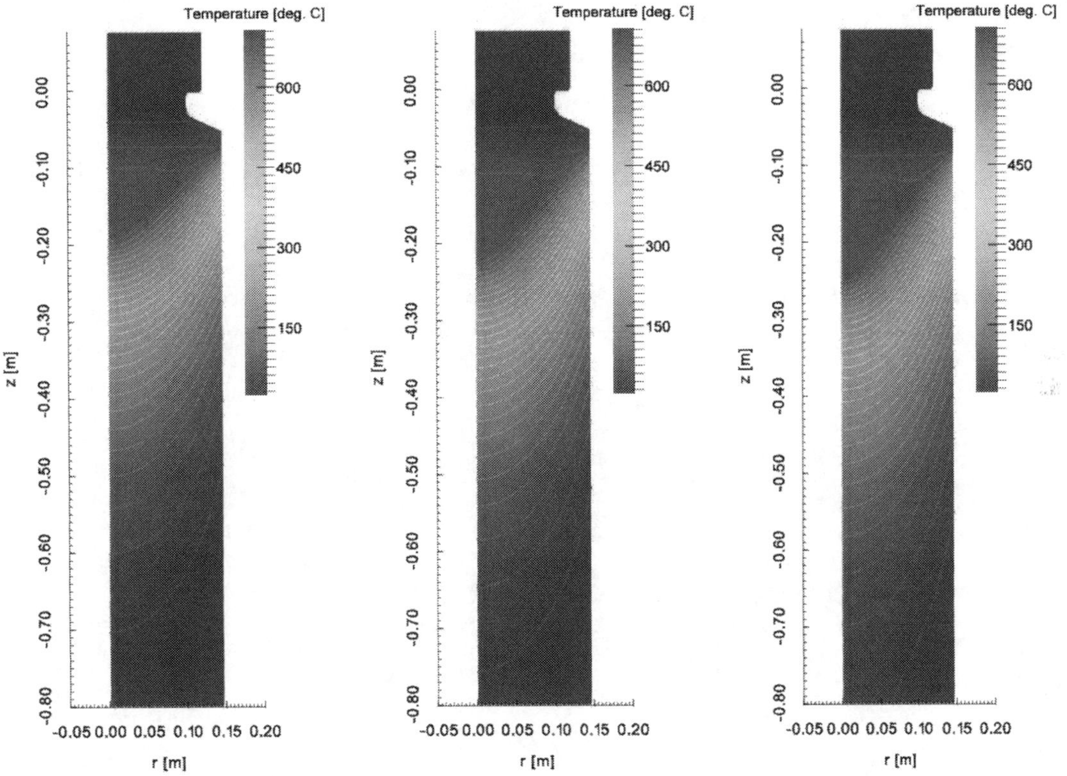

Figure 2 The temperature field at different casting speeds. From left to right: 90, 80 and 70 mm/min

The effect of electromagnetic stirring on the stress state is shown in Figures 9 and 8. The small current amplitudes, displayed in this work do not have significant impact on the stress state in the billet. The changes are small but their effect remains the same if the current amplitude and frequency are varied. The application of the electromagnetic fields results in the increase of all three stress components in the centre of the billet. This increase reduces the magnitude of radial and vertical stress components and increases the value of the circumferential stress in the billet centre. Near the mould contact, the differences of all stress components are negative. This reduces the magnitudes of all three stress components, thus reducing the possibility of cracking or other stress-related defects.

IOP Conf. Series: Materials Science and Engineering **84** (2015) 012099 doi:10.1088/1757-899X/84/1/012099

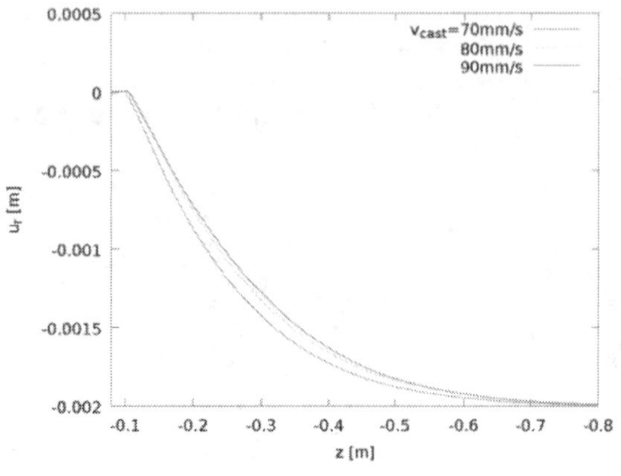

Figure 3 The deformation of the surface of the billet at different casting speeds.

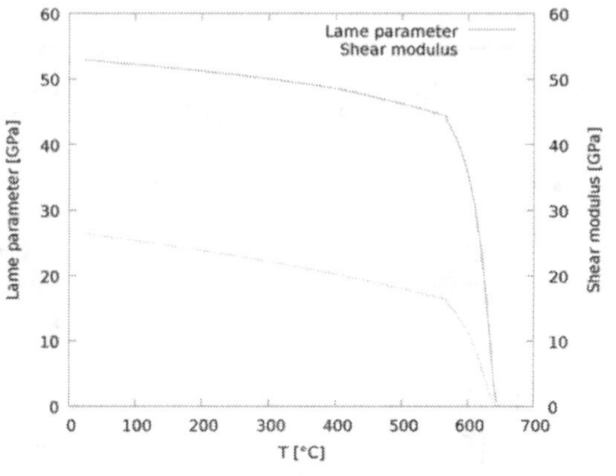

Figure 4 The dependence of the material properties on the temperature for the alloy AA6082 as given by the JMatPro database.

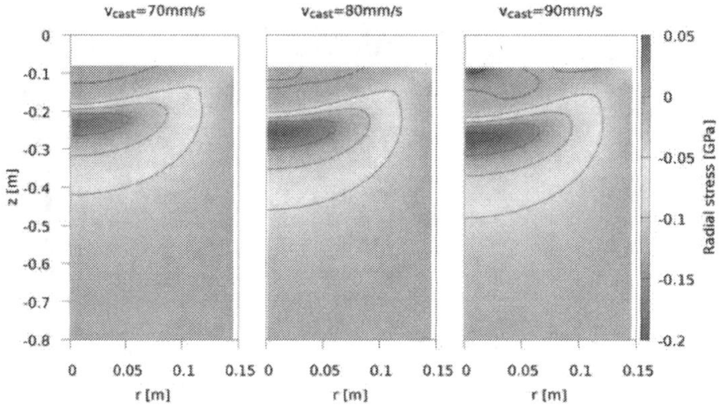

Figure 5 Radial stress at different casting speeds.

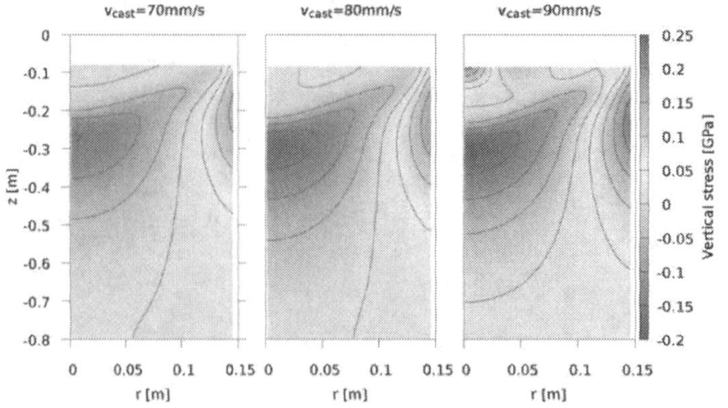

Figure 6 Vertical stress at different casting speeds.

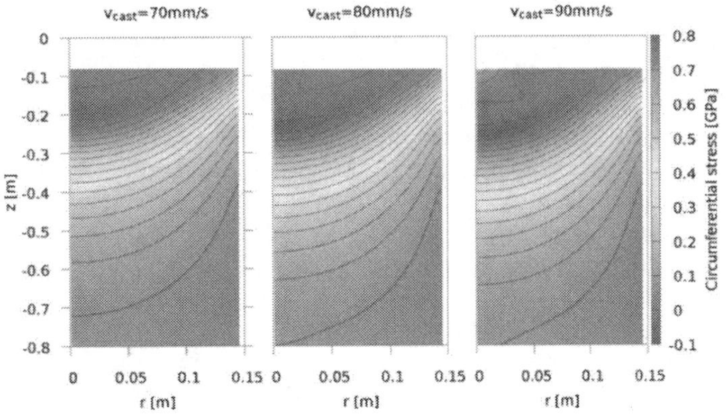

Figure 7 Circumferential stress at different casting speeds.

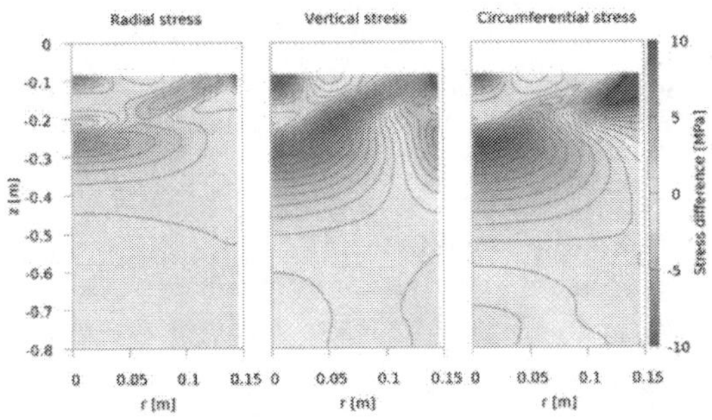

Figure 8 The difference between the stresses in the billet when the electromagnetic field is applied and when it is not. The driving current has amplitude 5A and frequency 20Hz.

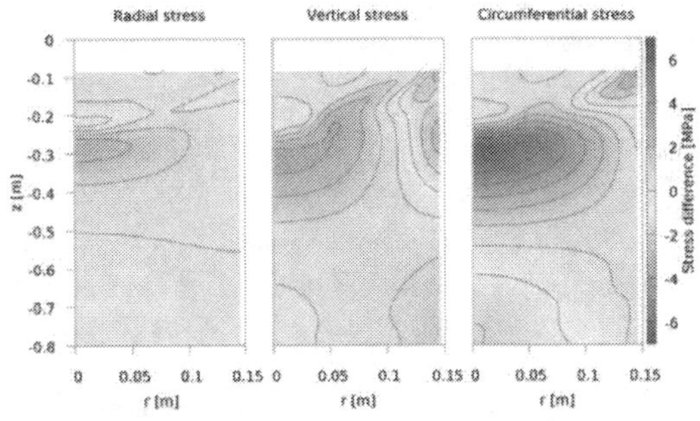

Figure 9 The difference between the stresses in the billet when the electromagnetic field is applied and when not. The driving current has amplitude 5A and frequency 10Hz.

5. Conclusions

A thermoelastic model of LFEMC DC casting of aluminium cylindrical billets, employing the meshless LRBFCM solution procedure has been developed. This method has been most probably for the first time applied to an industry related engineering problem.

The model calculates the stress field for a given thermal profile, which is calculated by a heat and mass transfer model that uses the equivalent meshless numerical method. The resulting stress fields are presented in this paper, together with the demonstration of the effect of the driving current parameters on the stress state. The application of the electromagnetic stirring to the casting process reduces the amplitude of the circumferential and radial stress, thus possibly reducing the cracking probability. The increase in the casting speed stretches the stress field along the casting direction.

The elaborated numerical method is straightforward to implement, can be easily extended to three spatial dimensions, and allows for simple applications to the cases when the material properties continuously vary. Because of the meshless nature of the LRBFCM, many potential improvements can be devised. The method can easily incorporate node refinement and can be used to solve problems defined on more complicated domains.

We next plan to extend the model with a more sophisticated material law, which will incorporate viscoplastic deformation of the mushy zone and plastic deformation of the solidified billet in the future.

6. Acknowledgements
This paper is supported by the Slovenian Funding Agency through grants L2-6775, P2-0379 and Young Researcher Programme.

References

[1] Lalpoor M, Eskin D G, Ruvalcaba D, Fjær H G, Ten Cate A, Ontijt N and Katgerman L 2011 Cold cracking in DC-cast high strength aluminum alloy ingots: An intrinsic problem intensified by casting process parameters *Mater. Sci. Eng. A* **528** 2831–42

[2] Eskin D G 2008 *Physical Metallurgy of Direct Chill Casting of Aluminum Alloys* (Taylor & Francis)

[3] Grandfield J 2013 *Direct-chill casting of light alloys: science and technology* (Hoboken, New Jersey: TMS-Wiley, a John Wiley & Sons, Inc. Publication)

[4] Fjaer H G and Mo A 1990 ALSPEN-A mathematical model for thermal stresses in direct chill casting of aluminum billets *Metall. Trans. B* **21** 1049–61

[5] Drezet J M, Rappaz M and Krähenbühl Y 1995 *Thermomechanical effects during direct chill and electromagnetic casting of aluminum alloys. Part II: numerical simulation* (TMS Publ.)

[6] Kumar P P, Nallathambi A K, Specht E and Bertram A 2011 Mechanical Behavior of Mushy Zone in DC casting using a Viscoplastic Material Model *Tech. Mech.* **32** 342–57

[7] Šarler B and Vertnik R 2006 Meshfree explicit local radial basis function collocation method for diffusion problems *Comput. Math. Appl.* **51** 1269–82

[8] Kosec G and Šarler B 2014 Simulation of macrosegregation with mesosegregates in binary metallic casts by a meshless method *Eng. Anal. Bound. Elem.* **45** 36–44

[9] Mramor K, Vertnik R and Šarler B 2013 Simulation of natural convection influenced by magnetic field with explicit local radial basis function collocation method *CMES Comput. Model. Eng. Sci.* **92** 327–52

[10] Simonenko S, Bayona V and Kindelan M 2014 Optimal shape parameter for the solution of elastostatic problems with the RBF method *J. Eng. Math.* **85** 115–29

[11] Lee C K, Liu X and Fan S C 2003 Local multiquadric approximation for solving boundary value problems *Comput. Mech.* **30** 396–409

[12] Vertnik R and Šarler B 2009 Solution of incompressible turbulent flow by a mesh-free method *Comput. Model. Eng. Sci. CMES* **44** 65

[13] Košnik N, Vertnik R and Šarler B 2014 Simulation of low frequency electromagnetic DC casting *Mater. Sci. Forum* **790-791** 390–5

[14] Mramor K, Vertnik R and Šarler B 2013 Low and intermediate Re solution of lid driven cavity problem by local radial basis function collocation method *CMES Comput. Model. Eng. Sci.* **92** 327–52

[15] M'Hamdi M, Mo A and Fjaer H G 2006 TearSim: A two-phase model addressing hot tearing formation during aluminum direct chill casting *Metall. Mater. Trans. A* **37** 3069–83

[16] Nallathambi A K, Bertram, Albrecht, Specht, Eckehard, Tyagi M, Specht E and Bertram A 2011 Thermomechanical simulation of direct chill casting *Trans. Indian Inst. Met.* **64** 13–9

[17] Saunders N, Guo U K Z, Li X, Miodownik A P and Schillé J-P 2003 Using JMatPro to model materials properties and behavior *JOM* **55** 60–5

[18] Mavrič B and Šarler B 2014 A collocation meshless method for linear thermoelasticity in 2D ThermaComp20014 (Bled) pp 279–82

[19] Mavrič B and Šarler B Local radial basis function collocation method for linear thermoelasticity in two dimensions *Int. J. Numer. Methods Heat Fluid Flow* (In review)

[20] Lorbiecka A Z and Šarler B 2010 Simulation of dendritic growth with different orientation by using the point automata method *Comput. Mater. Contin. CMC* **18** 69

IOP Conf. Series: Materials Science and Engineering **84** (2015) 012100 doi:10.1088/1757-899X/84/1/012100

Process-scale modelling of microstructure in direct chill casting of aluminium alloys

M Bedel, L Heyvaert, M Založnik, H Combeau, D Daloz and G Lesoult
Institut Jean Lamour, CNRS – Université de Lorraine, F-54011 Nancy CEDEX, France

E-mail: marie.bedel@univ-lorraine.fr

Abstract. The mechanical properties of an alloy being related to its microstructure, the understanding of the mechanisms responsible for the grain structure formation in direct chill casting is crucial. However, the grain size prediction by modelling is difficult since a variety of multi-scale coupled phenomena have to be considered. Nucleation and growth of the grains are interrelated, and the macroscopic transport phenomena such as the motion of grains and inoculant particles with the flow impact the nucleation-gowth competition. Thus we propose to study the grain size distribution of a 5182 alloy industrial scale slab of 510 mm thickness, both non-inoculated and inoculated with Al-3Ti-1B, for which experimental grain size measurements are available. We use a volume-averaged two-phase multi-scale model that describes nucleation from inoculant particles and grain growth, fully coupled with macroscopic transport phenomena: fluid flow induced by natural convection and solidification shrinkage, heat, mass and solute mass transport, grains and inoculant particles motion. We analyze the effect of liquid and grain motion as the effect of grain morphology on microstructure formation and we show in which extent those phenomena are responsible for the grain size distribution observed experimentally. The effect of the refiner level is also studied.

1. Introduction

Grain refiners are commonly used in DC casting of aluminium alloys in order to refine the microstructure. Not only the average grain size is reduced but the size distribution is homogenized [1][2]. When refiners are added in the melt, nucleation, i.e. grain initiation, is supposed to be athermal and heterogeneous. Grains form on inoculant particles instantaneously as soon as a critical undercooling, which depends on their size, is reached. The size distribution of inoculant particles was determined for industrially used refiners such as Al-Ti-C and Al-Ti-B [3]. Coupled with a growth model, it enabled to predict the competition between nucleation and growth of grains at the microscopic scale and thus the final grain size when using these refiners [4]. Indeed, the heat release induced by grain growth tends to reduce the local undercooling and thus to prevent the activation of further inoculant particles. This type of model can explain the decrease of the grain size with an increase of the cooling rate due to the competition between nucleation and growth. But what happens in macroscopic systems, where convection phenomena cannot be neglected anymore?

In most of the macroscopic solidification models, grains are supposed to nucleate instantaneously at a unique temperature. Moreover, the transport of grains is either strongly simplified or neglected [5][6]. The transport of inoculant particles has never been considered to our best knowledge. This can explain why only few models propose a comparison with experiments in terms of grain size

IOP Conf. Series: Materials Science and Engineering **84** (2015) 012100 doi:10.1088/1757-899X/84/1/012100

distribution. We can name the work of Håkonsen et al., where different nucleation laws were tested [7] but where both the motion of inoculant particles and equiaxed grains were neglected. Indeed, these transport phenomena impact the local competition between nucleation and growth of grains and thus the final grain size distribution.

Moreover, the inoculant particles and grains transport can alter the grain morphology locally. In a closed system, when the melt is rich in inoculant particles, the resulting grain density is high enough to result in a globular grain morphology. This hypothesis is made in most of macroscopic solidification models. However, when grains and inoculant particles are free to move, they can leave the nucleation area and thus their growth and their morphology is strongly affected by their motion. Thus in our work, we propose to study the impact of combined transport phenomena on the final grain size distribution, taking into account their morphology.

2. Model description

The description of the volume-averaged two-phase multiscale model we use is given in [8], so we only present its main features and the extensions which were made since. At the macroscopic scale, the model accounts for heat, mass, solute mass, grains and inoculant particles transport coupled with flow driven by natural convection (modelled through the Boussinesq assumption) and solidification shrinkage (supposed to be constant). Two flow regimes are distinguished: when the grain fraction (which is different from the solid fraction when grains are dendritic) is locally larger than the imposed packing limit g_{pack}, the grains are supposed to be blocked and to move at the casting velocity while the liquid flow through the porous solid matrix is modelled considering the drag interactions by a Darcy term. When the grain fraction is smaller than g_{pack}, the solid phase transport is modelled through a solid viscosity in a pseudo fluid model [9]. In this way, the solid phase inertia and internal stress are taken into account.

Nucleation is supposed to be athermal and heterogeneous, which means that a grain forms instantaneously on an inoculant particle of diameter d when the undercooling reaches a critical value ($\Delta T(d)=4\Gamma/d$, where Γ is the Gibbs-Thomson coefficient). The inoculant-particle size distribution is approximated in this work by an exponential law: $N(d)=N_0/d_0 \exp(-d/d_0)$, whose parameters are estimated in [3] for the Al-5Ti-1B refiner (as no such parameters are known for the Al-3Ti-1B refiner). At the macroscopic scale, the inoculant particles are supposed to be transported with the liquid, which is modelled by solving separately the transport equation of several inoculant particles classes. To do so, the size distribution law is discretized into classes, as shown in Figure 1.

Phase change is supposed to be controlled by solute diffusion in both liquid and solid phases at the grain scale. Local thermal equilibrium and thermodynamic equilibrium are supposed at the solid-liquid interface and analytical expressions are used for diffusion lengths [10][11]. When not stated otherwise, the growing solid structures are assumed to have a spherical shape, which can correspond to globular grains, but a dendritic model has been also developed. Indeed, the transport and the packing of the grains depend on their morphology. Moreover, the grain morphology impacts the growth kinetics and thus the competition between nucleation and growth of the grains. We suppose that the grains envelopes are octahedrons and the tips are considered to be paraboloidal [12].

3. Experimental results and case modeling

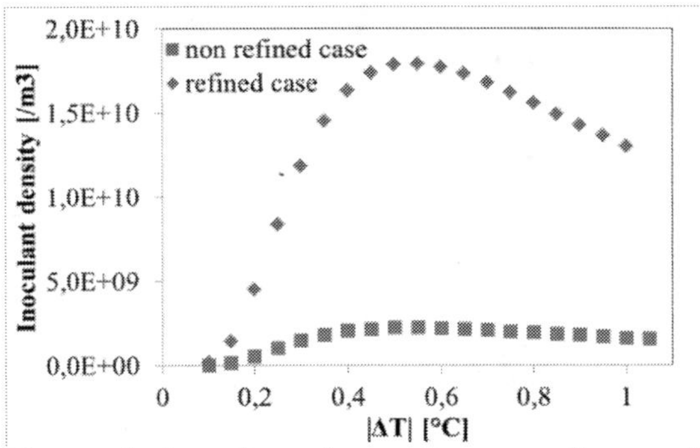

Figure 1. Inoculant classes corresponding to 1 kg/t of Al-5Ti-1B refiner in red and to 0.125 kg/t of Al-5Ti-1B refiner in red [3].

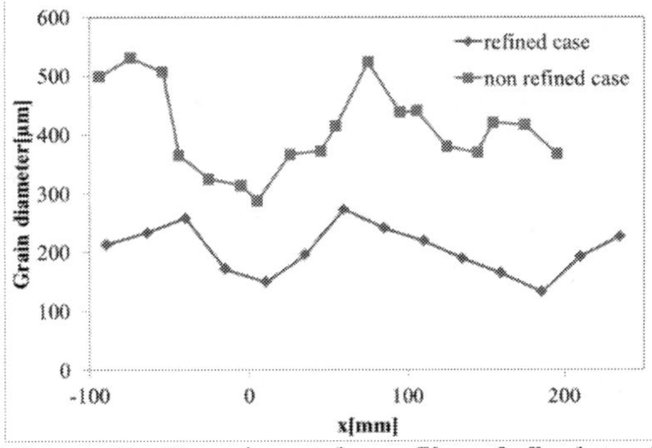

Figure 2. Experimental profile of final grain diameter, as a function of ingot thickness for the refined case in red and for the non-refined case in blue [2].

We modeled the solidification of a 5182 alloy industrial scale slab of 510 mm thickness and 1897 mm width, solidified at v_c= 1mm/s, without refinement and then with the addition of 1 kg/t of Al-3Ti-1B, casted in the frame of the BRITE EURAM project EMPACT [2] [13] [14]. Image analysis on optical microscopy images from samples taken along the ingot thickness enabled to obtain the average grain diameter profile for the non-refined and for the refined cases, as shown in Figure 2 [14].

Table 1. Thermo-physical data of Al-4.06wt.%Mg [5].

Property	Unit	Value
Density	kg m^{-3}	2 400.0
Specific heat	J kg^{-1} K^{-1}	1 107.0
Thermal conductivity, solid	W m^{-1} K^{-1}	109.0
Thermal conductivity, liquid	W m^{-1} K^{-1}	100.0
Diffusion coefficient, solid	m^2 s^{-1}	1.8 x 10^{-12}
Diffusion coefficient, liquid	m^2 s^{-1}	7.7 x 10^{-9}
Latent heat of pure Al	J kg^{-1}	3.725 x 10^5
Liquid dynamic viscosity	Pa s	1.2 x 10^{-3}
Thermal expansion coefficient	K^{-1}	1.245 x 10^{-4}
Solutal expansion coefficient	wt.%$^{-1}$	4.0 x 10^{-3}
Pure Al melting temperature	°C	660.5
Eutectic temperature	°C	450.0
Liquidus slope	°C wt.%$^{-1}$	-5.831
Partition coefficient	-	0.485
Gibbs-Thomson coefficient	K.m	1.9x10^{-7}

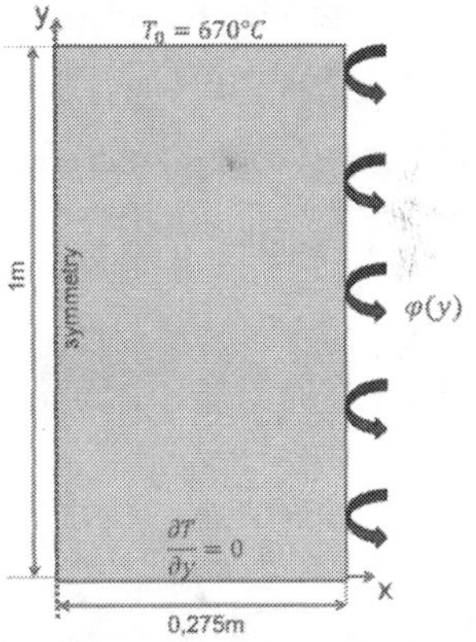

Figure 3. Model geometry and boundary conditions.

We model the 5182 alloy by a Al-4.06wt.%Mg alloy, whose thermo-physical properties are given in Table 1 [5]. The refiner size distribution for 1 kg/t of Al-5Ti-1B is taken from [3] and is discretized by considering constant undercooling intervals as shown in Figure 1. The non-refined part of the ingot is modeled by considering a modified refiner size distribution, as the nucleation is considered to occur only in refiner particles in the model. According to the experimental grain size profiles (Figure 2), the average grain size is almost doubled when the refiner is suppressed, which corresponds to a grain density divided by 8. Thus we use the same refiner size distribution for the non-refined case but for 0.125 kg/t addition of Al-5Ti-1B (Figure 1).

The ingot width being much larger than its thickness, we consider a two-dimensional system 1m high (y axis) and 275 mm thick (x axis) as represented in Figure 3. By assuming an axial symmetry in x=0, we have to model only half of the ingot thickness. Although the melt is injected through a combo-bag in the experiment, we suppose that the melt is injected through the whole ingot thickness. The entry temperature of the melt is T_0=670°, as measured by a thermocouple at the melt surface. The surface heat extraction rate $\phi(y)$ has been determined by an inverse method as a function of the height [15].

Table 2. hypotheses made for the case modeling (v_p: velocity of the product, v_l: liquid velocity).

	Liquid motion	Inoculant particle velocity	Grain motion	Dendritic morphology	Nucleation law
Case 0	No	v_p	No	No	Refined case
Case 1	Yes	v_l	No	No	Refined case
Case 2	Yes	v_l	Yes	No	Refined case
Case 3	Yes	v_l	Yes	Yes	Refined case
Case 4	Yes	v_l	Yes	Yes	Non-refined case

All the calculations are made with a time step $\Delta t=10^{-3}$s and a regular mesh composed of 51x200 square cells. In order to study the impact of macroscopic transport phenomena and grains morphology on the microstructure prediction, we study 5 cases of increasing complexity. In Case 0, no convective transport phenomenon is considered. In Case 1, liquid motion is taken into account through natural convection and solidification shrinkage and the inoculant particles are transported with the liquid phase. Then in Case 2, the grain motion is added and the packing grain fraction is fixed to g_{pack}=0.3. In Case 3, we add the effect of grain morphology by considering a paraboloidal tip shape for calculating the tip velocity of the stems of the grains. Finally, in Case 4, the model hypotheses are kept identical, but the nucleation law is changed in order to better fit the non-refined case.

4. Results and discussion

4.1. Impact of liquid motion

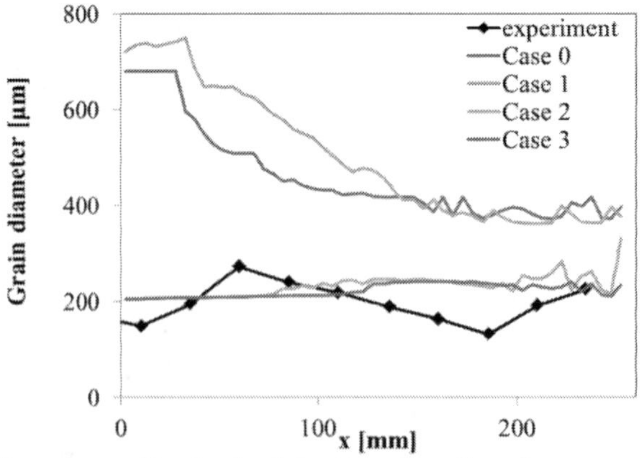

Figure 4. Grain diameter profile obtained for Case 0 to Case 3, compared to the experiment.

Even when all the convective transport phenomena are neglected, the grain size prediction is not uniform in the ingot. We can see this for Case 0 in the grain diameter profile given in Figure 4. Indeed, the heat being extracted from the ingot sides, the cooling rate and thus the undercooling are larger closer to the ingot surface. This can be seen in Figure 5-a, which shows the profile of the maximal undercooling. Thus due to the higher undercooling close to the surface, more inoculant classes can be activated and the nucleation rate is larger (Figure 5-b), which explains the smaller grain size in this region. Indeed, Figure 5-b shows the nucleation rate along the ingot thickness, which corresponds to the integral of the nucleation events per second at a given thickness in the ingot.

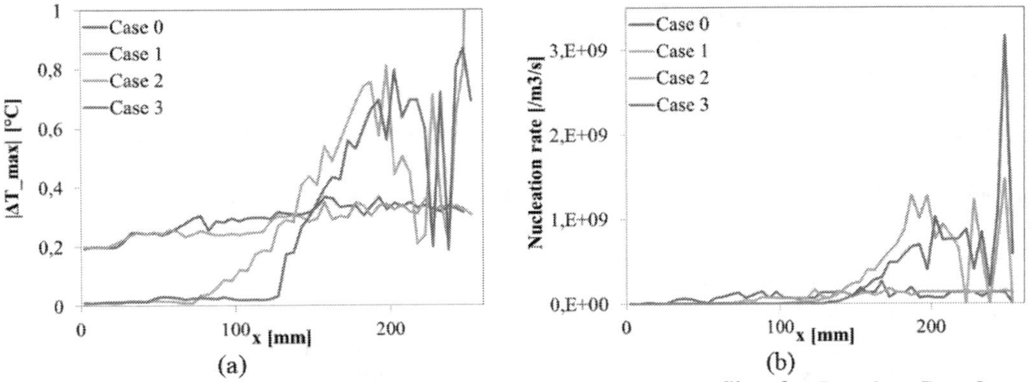

Figure 5. (a) maximal undercooling and (b) nucleation rate profiles, for Case 0 to Case 3.

When the liquid motion induced by natural convection and solidification shrinkage is added (Case 1), a clockwise flow loop appears. The cooling of the liquid close to the surface (which tends to induce a clockwise loop) more than compensates the liquid enrichment in solute induced by solidification (which would tend to reverse the flow direction). Thus the liquid cooled close to the surface is carried towards the centre of the ingot, favouring a uniform temperature distribution in the centre. However, the maximal undercooling is only slightly modified compared to Case 0 (Figure 5-a), but the final grain size tends to increase in the centre of the ingot (Figure 4). Indeed, as the inoculant particles are considered to move with the liquid, they pass by the ingot surface before reaching the centre. The inoculant particles classes activated close to the surface do not reach the centre and the competition between nucleation and growth is altered: fewer big particles reach the centre of the ingot, and with an identical maximal undercooling, fewer grains are formed. Thus we observe that the liquid and most of all the inoculant particles motion tends to amplify the grain size heterogeneities. However, the grain diameter profile is still far from fitting the experimental profile (Figure 4-a).

4.2. Impact of solid motion

When the effect of grain motion is taken into account (Case 2), we notice that the predicted grain size profile is strongly modified and is much closer to the experimental observations (Figure 4). The grain motion impacts the microstructure formation through two phenomena:

- Firstly, the grain motion in itself induces a redistribution of the grains in the ingot. In the studied case, grains are heavier than the liquid, therefore they settle in the centre of the ingot. Thus the grain motion tends to counter the thermal effect by increasing the grain density in the ingot centre.
- Secondly, the grain motion strongly affects the nucleation localization and intensity. Without grain motion, nucleation occurs in the whole ingot thickness, and starts all along the liquidus line. The undercooling only slightly increases from the centre to the surface of the ingot (Figure 5-a), and the nucleation rate increases accordingly (from 10^6 $m^{-3}.s^{-1}$ at the centre to 1.510^8 $m^{-3}.s^{-1}$ at the surface) as shown in Figure 5-b. When grain motion is added, grains continuously leave the nucleation zone (close to the surface) after being formed, maintaining a high undercooling in this zone (>1°C compared to 0,35°C in Case 1, as shown in Figure 5-a). Thus more inoculant particles can be activated. On the opposite, grains are continuously brought in the ingot centre, reducing the undercooling (<0.1°C) enough to prevent any nucleation in this area.

As the packing fraction of the grains is not known a priori, three values of packing fractions were tested, from 0.3 to 0.5, but we noticed that this parameter has a very weak influence on the grain size prediction. Thus the grain motion tends to refine and to homogenize the microstructure, and the predicted average final grain size is close to the one observed experimentally. However, the grain size variations along the ingot thickness observed experimentally are not found by the model, which can be the sign of a missing physical phenomenon in the model. Notably, in our model the packing is described by a criterion on the local grain fraction. In the regions where the packing front is inclined the blocking of the grains is certainly not well described by such a simple criterion.

4.3. Impact of grain morphology

In the cases presented until now, grains were supposed to be globular, which was supported by experimental observations of the final grain morphology. However, this observation cannot give information about the morphology of the grains during their growth. The morphology is known to impact the growth velocity and thus the coupling between nucleation and growth. Moreover, the transport of the grains is affected by the shape of the grains. This is why we study in Case 3 the effect of grain morphology through a paraboloidal tip model.

We observe that when the grain morphology is taken into account (Case 3), the grain size profile is hardly changed compared to Case 2 where grains are supposed to be globular (Figure 4). Indeed, even if the morphology is dendritic at first in Case 3, the grains globularize before reaching the packing fraction. This can be seen in Figure 7-a where the map of internal solid fraction $g_i = g_s/g_{env}$ is shown for Case 3. More precisely, the dendritisation favors solidification in the nucleation zone, but it also amplifies the grains sedimentation. In our case, these two phenomena compensate and the equilibrium between nucleation and growth of the grains in the nucleation zone is hardly changed so the finale grain size is only slightly changed.

4.4. Impact of the grain refiner level

The nucleation model we use is adapted to inoculated castings, as the nucleation law directly depends on the size distribution of the refiner particles. Thus in order to model the non-refined case, we propose to adapt this size distribution in order to fit the final average grain size obtained experimentally. An average final grain size that is two times larger corresponds to a 8 times smaller grain density, therefore the density of each inoculant class is divided by 8 in Case 4.

As expected, the grain morphology is more dendritic in the mushy zone in Case 4 compared to Case 3, as can be seen in Figure 7. The grain dendritisation does not directly impact the packing in the nucleation zone since the morphology remains globular at the packing fraction in both cases. However, the nucleation law change highly impacts the nucleation-growth coupling in this zone. The inoculant-particle density being smaller in Case 4, the undercooling becomes higher in order to activate nuclei classes with a higher critical undercooling (Figure 6). However, this increase of undercooling is not high enough to reach the same nucleation rate as in Case 3 (Figure 8). The density of formed grains is reduced, which is the expected effect when the refiner level is reduced. This can be seen in Figure 9, where the grain diameter profile is given in Case 3 and Case 4, and is compared to the experimental profiles obtained respectively for the refined and the non-refined cases.

By reducing the refiner level, the predicted average final grain size estimation fits the non-refined case measurement. This result was not obvious because the reduction of the global nuclei population could affect the nucleation-growth competition in a nonlinear way. Thus the model not only enables to predict the average grain size without any adjustment in the refined case, but by adjusting the refiner level, it enables to predict the final grain size in the non-refined case. However, the size variations along the ingot thickness observed experimentally are not correctly predicted. Moreover, although the model predicts a globular morphology in the center of the ingot for both Case 3 and Case 4, it is more difficult to characterize the microstructure close to the surface. The reason is that the grains globularize extremely rapidly around the packing fraction (from a very dendritic to a globular morphology in 5 mm, in the sub-surface region). Although the fully dendritic morphology observed experimentally in the non-refined case is not predicted by the model, the maps in Figure 7 show a more dendritic morphology around the packing fraction close to the surface in Case 4 compared to Case 3. Thus the influence of grain refinement on the grain morphology is qualitatively well captured.

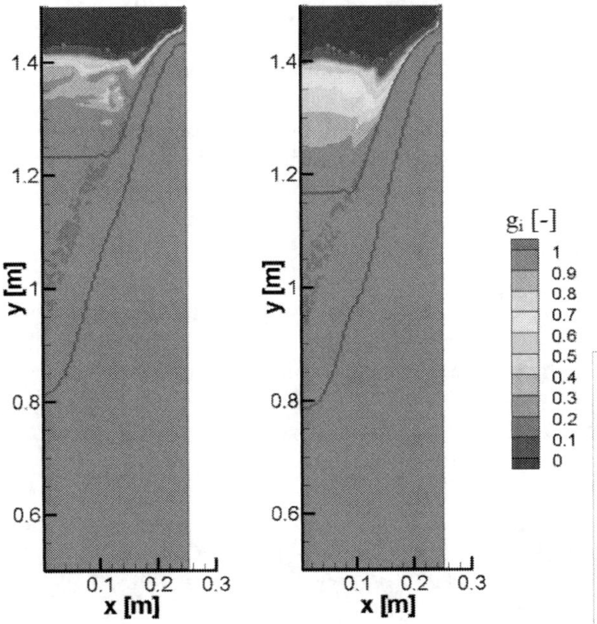

(a) Case 3 (b) Case 4

Figure 7. Internal fraction map in the mushy zone and liquidus, packing and solidus lines in red for (a) Case 3 and (b) Case 4.

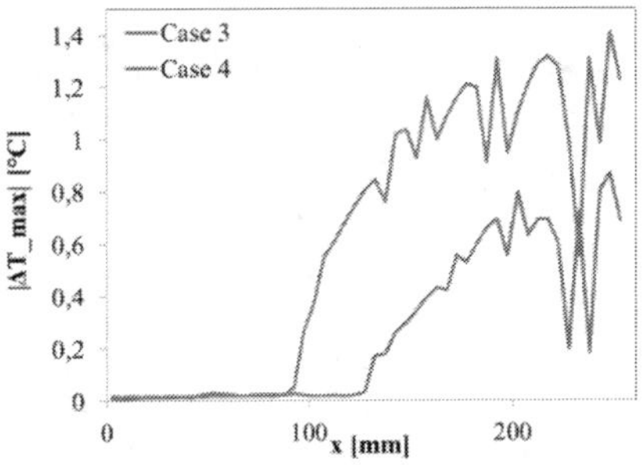

Figure 6. Maximal undercooling profiles, for Case 3 and Case 4.

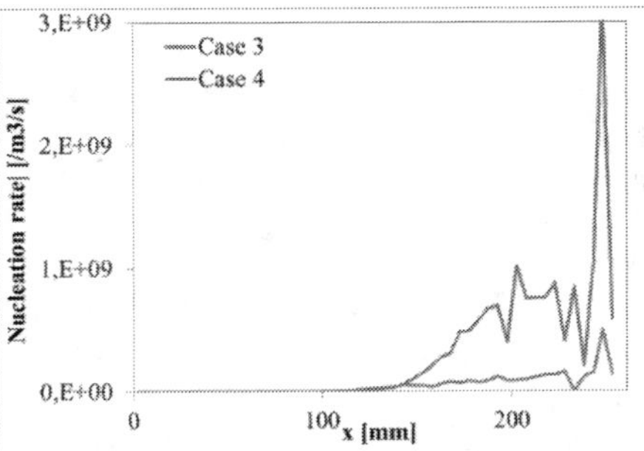

Figure 8. Nucleation rate profiles along the ingot thickness, for Case 3 and Case 4.

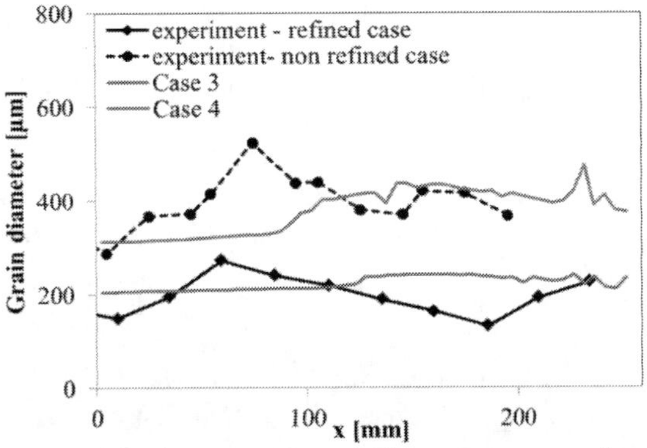

Figure 9. Grain diameter profile for Case 3 and Case 4 compared to the experimental profiles, respectively with and without refiner.

5. Summary and conclusions

The microstructure formation has been studied in a 5182-alloy industrial-scale slab of 510 mm thickness, cast without grain refiner or inoculated with 1 kg/t of Al-3Ti-1B, for which experimental grain size measurements were available. With the presented model we were able to study for the first time the coupling between nucleation and growth of the grains in presence of liquid convection and grains and inoculant particles transport at the macroscopic scale. The comparison between the experimental grain size profile and the model results enables us to analyze the impact of each phenomenon on the grain formation.

The flow induced by natural convection and solidification shrinkage slightly amplifies the grain size heterogeneities. On the contrary, grain motion reduces them and highly reduces the average grain size. We show that grain motion is a key phenomenon for the understanding of the final microstructure size. A correct modeling of these phenomena enables us to predict an average final grain size close to the experimental observations. However, the grain size evolution across the ingot thickness is not accurately predicted. To model the non-grain-refined casting, which can present dendritic grains, we further investigated the influence of the description of grain morphology on the model prediction. The same conclusions than for the inoculated ingot can be drawn: the average grain size can be predicted but not the observed variations of the grain size across the ingot thickness.

We demonstrated that our model can predict the average final grain size in an industrial size casting. A detailed modelling of both microscopic phenomena of nucleation and growth of grains and of macroscopic phenomena of convection and grains and inoculant particles was required for that. Moreover, we could show that the influence of the grain-refiner level can be reproduced. However, further phenomena should be considered to explain the grain size distribution in the ingot. Principally, further work on the description of grain packing, a fundamental phenomenon in the transport of equiaxed grains, is required.

References

[1] Nadella R, Eskin D G and Katgerman L 2008 *Metall. Mater. Trans.* A **39** 450–61
[2] Joly A, Grün G-U, Daloz D, Combeau H and Lesoult G 2000 *Mater. Sci. Forum* **329-330** 111–20
[3] Tronche A 2000 (PhD thesis, University of Cambridge, UK)
[4] Greer A L, Bunn A M, Tronche A, Evans P V and Bristow D J 2000 *Acta Mater.* **48** 2823–35
[5] Jalanti T 2000 (PhD thesis)
[6] Vreeman C J, Schloz J D and Krane M J M 2002 J *Heat Transfer* **124** 947-53
[7] Håkonsen A, Mortensen D, Benum S and Vatne H E 1999 *Light Metals* ed Eckert C E (Warrendale, USA: TMS) pp 821–7
[8] Založnik M and Combeau H 2010 *Comput. Mater. Sci.* **48** 1–10
[9] Heyvaert L, Bedel M, Založnik M and Combeau H 2014 *International Conference on Advances in Solidification Processes 4*, Old Windsor, 8-11 June 2014
[10] Bedel M, Založnik M, Kumar A, Combeau H, Jarry P and Waz E 2012 *IOP Conf. Ser. Mater. Sci. Eng.* **27** 012070
[11] Bedel M 2014 (PhD thesis, Université de Lorraine, France)
[12] Nielsen Ø, Mo A, Appolaire B and Combeau H 2001 *Metall. Mater. Trans.* A **32** 2049–60
[13] Lesoult G, Albert V, Appolaire B, Combeau H, Daloz D, Joly A, Stomp C, Grün G-U and Jarry P 2001 *Sci. Technol. Adv. Mater.* **2** 285–91
[14] Daloz D, Combeau H, Joly A, Lesoult G, Grün G-U, Jarry P and Commet B 2002 Materiaux 2002 (Tours, France)
[15] Drezet J-M, Rappaz M, Grün G-U and Gremaud M 2000 *Metall. Mater. Trans.* A **31** 1627–34
[16] Combeau H, Založnik M, Hans S and Richy P E 2009 *Metall. Mater. Trans.* B **40** 289-304

Numerical simulation of convection and inclusion distribution during solidification in a heavy steel ingot

Rui LIN and Houfa SHEN

Key Laboratory for Advanced Materials Processing Technology, Ministry of Education; School of Materials Science and Engineering, Tsinghua University, Beijing 100084, China

E-mail: shen@tsinghua.edu.cn

Abstract. Inclusions content in the steel ingot is an important index for homogeneity, and it becomes more serious for heavy steel ingots which are used for major equipment. However, knowledge about the formation of inclusion in steel ingot is limited, and modeling of inclusion distribution is still challenging, so it is of great significance to research the behavior of inclusion. In this paper, fluid flow during solidification is numerically simulated based on the equilibrium equations of mass, momentum and energy, and then inclusion distribution is modeled according to the Lagrangian Stokes trajectory method. The Results show that the inclusion distribution in the steel ingot is influenced by the flow pattern which is affected by the solidification pattern. Therefore, inclusion distribution could be controlled by the solidification front with the optimization of heat transfer condition such as the hot top design of steel ingot for the high quality steel production.

1. Introduction

Ingot casting is still commonly used for high alloying steel grades, e.g. tool steel. Non-metallic inclusion is a very serious defect because it may lead to problems that require expensive casting repairs or rejection. The mechanical properties of steel are controlled to a large degree by the volume fraction of inclusion, which acts as stress raisers. Similar pronounced property degradation caused by inclusion is observed in tests that reflect slow, rapid, or cyclic strain rates, such as creep, impact and fatigue testing [1]. Sometimes, inclusion causes voids which will induce cracks if larger than a critical value [2]. Large exogenous inclusion, which originates mainly from reoxidation of the molten steel, slag entrapment and lining erosion, also cause inferior surface appearance, poor polishability, reduced resistance to corrosion, and in severe cases, slag lines and laminations [3]. Larger inclusion has a more negative effect on the fatigue life than smaller ones [4]. So with the increasing demands on steel cleanness, the study of inclusion formation and removal in the casting process has drawn more and more attention of the researchers and engineers [5-8]. Zhang and Thomas [9] presented methods to measure and detect inclusion in steel, and the causes of exogenous inclusion, the transport as well as entrapment of inclusion during fluid flow have been studied in detail. Ragnarsson et al. [10] have investigated the impact of flow pattern on inclusion removal in ingot during mold bottom filling with experimental and numerical modeling. It is well known that many negative effects are generated during solidification. However, further understanding about the distribution and removal of inclusion during the solidification after filling are comparatively limited.

The present work attempts to depict the effects of fluid flow on the movement of inclusion during the solidification of ingot by using the computational fluid dynamics method.

2. Model formulation

2.1. Governing Equations

The investigated ingot weighs 36 t with an average diameter of about 1.4 m and height of about 3 m. Figure 1 schematically shows the ingot system with sliced meshes. The inclusion's trajectory coupled with heat transfer and fluid flow during whole solidification process after top filling is modeled with assumptions: (1) solidification starts after mold filling with an initial uniform temperature; (2) inclusions are uniformly distributed in the molten melt; and (3) convection is driven by thermal buoyancy and the residual fluid flow is not considered after filling.

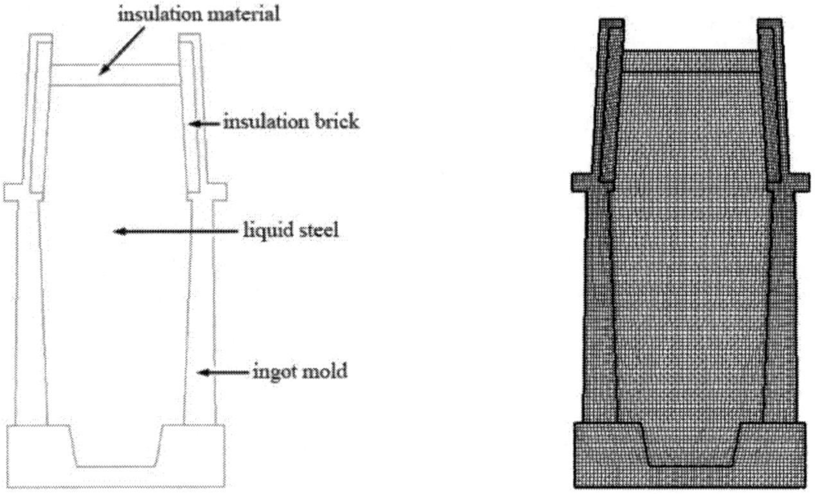

(a) (b)

Figure 1. Schematic of ingot system (a) with sliced mesh (b).

A two-dimensional axial symmetrical heat transfer and turbulence linked model is applied to describe the unsteady transport phenomena of molten steel with solidification. Mass, momentum and energy conservation equations are as follows (k-ε equations turbulence model is omitted):

$$\frac{\partial \rho}{\partial t}+\nabla \cdot(\rho \vec{v})=0 \tag{1}$$

$$\frac{\partial}{\partial t}(\rho \vec{v})+\nabla \cdot(\rho \vec{v}\vec{v})=-\nabla P+\nabla \cdot\left(\bar{\bar{\tau}}\right)+\overrightarrow{F_K} \tag{2}$$

$$\frac{\partial \rho}{\partial t}(\rho H)+\nabla \cdot(\rho \vec{v} H)=\nabla \cdot(k\nabla T)+S \tag{3}$$

where ρ is density, $\vec{v}$ is fluid velocity; P is static pressure; $\bar{\bar{\tau}}$ is stress tensor; $\overrightarrow{F_K}$ is interaction momentum per unit mass; H is enthalpy, and S is source term, which is related to thermal expansion coefficient β.

During solidification, inclusion distribution is modeled according to the Lagrangian Stokes trajectory method. Force equilibrium equation of inclusions in liquid steel is as follows:

$$\frac{\partial u_P}{\partial t} = F_D(u - u_P) + \frac{g(\rho_P - \rho)}{\rho_P} + F \tag{4}$$

where F_D is fluid resistance by unit mass of inclusion particles; u is the time average velocity; u_p is the speed of inclusion; ρ_p and ρ are the density of inclusion particle and liquid steel, respectively; g is acceleration of gravity; F is an additional acceleration term. $F_D(u-u_P)$ is the drag force per unit particle mass and is given by:

$$F_D = \frac{18\mu}{\rho_P d_P^2} \frac{C_D \mathrm{Re}}{24} \tag{5}$$

where μ is the molecular viscosity of the fluid; d_P is the particle diameter. Re is the relative Reynolds number, which is defined as

$$\mathrm{Re} = \frac{\rho d_P |u_P - u|}{\mu} \tag{6}$$

And C_D is the drag coefficient for the smooth spherical particles:

$$C_D = a_1 + \frac{a_2}{\mathrm{Re}} + \frac{a_3}{\mathrm{Re}^2} \tag{7}$$

For the formula (7), a_1, a_2, and a_3 are all constants [11]. Table 1 lists parameters used in calculation.

Table 1. Parameters used in calculation.

	Steel (ingot)	Mold	Insulation material	Insulating brick
ρ (kg·m^{-3})	7000	7300	100	2900
k (W·m^{-1}K $^{-1}$)	41	7.9	0.04	1.65
Cp (J·kg^{-1}K $^{-1}$)	785	460	1200	1200
μ (kg·m^{-1}s $^{-1}$)	0.0056	/	/	/
β (1/K)	0.0002	/	/	/
initial temperature (K)	1830	500	1830	1000

2.2. Solution Procedures
FLUENT14.0 code is employed to solve the set of conservation equations along with initial and boundary conditions. The QUICK scheme [12] is utilized for all the space derivatives, and the pressure-velocity coupling is solved using the PISO algorithm [13]. In order to avoid the well-known problems due to checker-board oscillations, the Rhie-Chow algorithm [12] is applied. An adaptive time-stepping is used, which provides automatic control of the time-step (the maximum time step allowed is $\Delta t < 6\times10^{-3}$ s), and at least 80 iterations are needed within each time-step to satisfy a strict convergence limit (all the scaled variable residuals are smaller than 5×10^{-4}).

3. Results and discussion
The computational results about temperature field, fluid flow, solid fraction and inclusion distribution at times of 2 min, 26 min, 2 h and 10 h are shown in Figure 2 to Figure 5. Because of heat release, temperature decreases from the mold centerline to the mold outside (Figures 2(a) ~ 5(a)).

At the early solidification stage of time 2 min, liquid steel begins to flow and velocity increases gradually owning to the temperature gradient. It is seen that, solid fraction is very small and circulation forms in front of the solidifying shell. The liquid steel flows downwards along the mold side wall and upwards within the central region as indicated in Figure 2. At the same time, inclusion begins to move with liquid.

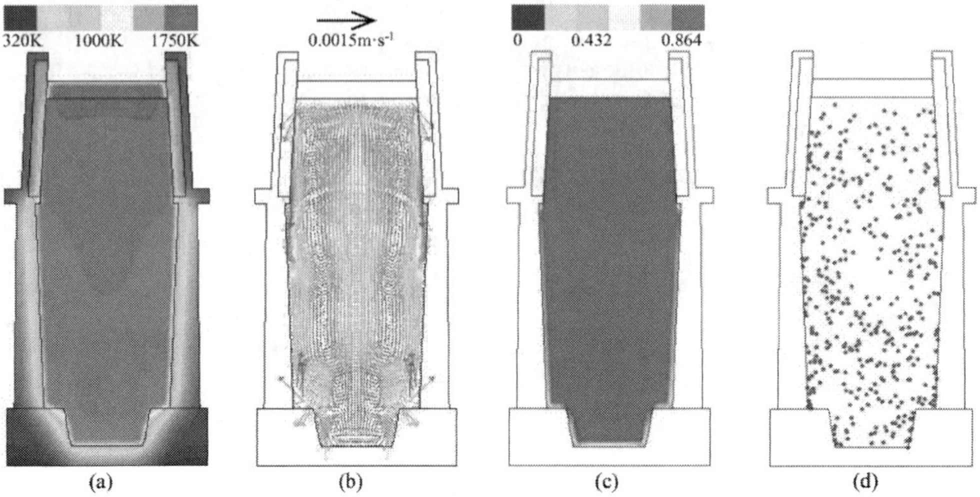

Figure 2. Calculated temperature distribution (a), fluid flow (b), solid fraction (c) and inclusion distribution (d) at the time of 2 min.

At the time of 26 min, the whole liquid steel temperature decreases further, and the solid shell becomes apparent near the mold inside wall. The temperature gradient in front of solid shell becomes large and melt flow is strong. Most of inclusion in the liquid steel flows with melt and some are caught by the solid part as shown in Figure 3. At the same time, the two circumfluences beside the centerline become violent. With the increasing time, the solid shell thickness increases and more and more inclusions are caught by the solid shell.

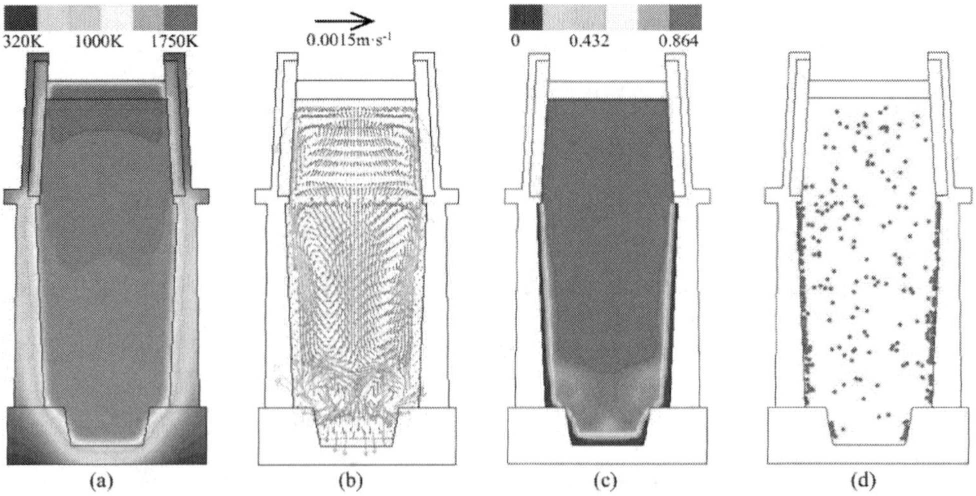

Figure 3. Calculated temperature distribution (a), fluid flow (b), solid fraction (c) and inclusion distribution (d) at the time of 26 min.

At the time of 3 h, temperature gradient in front of solid shell becomes small because more latent heat is released. The solidification rate decreases with the increasing solidifying shell, and the melt flow thereby becomes gradually weak as shown in Figure 4. Therefore, inclusions in the liquid pool have enough time to float up to the top surface of melt instead of being captured by freezing steel.

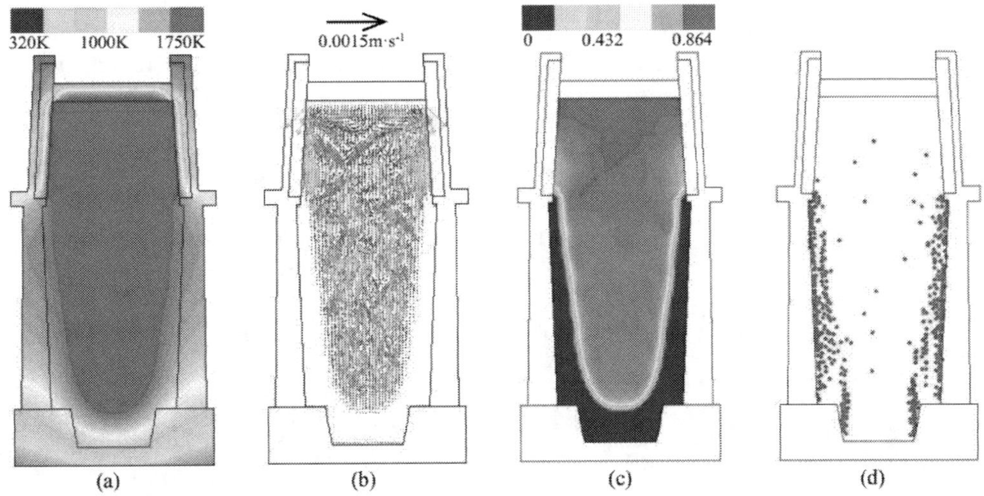

Figure 4. Calculated temperature distribution (a), fluid flow (b), solid fraction (c) and inclusion distribution (d) at the time of 2 h.

At the time of 10 h, the solid shell changes the shape from U-type to V-type and it is more open to the riser. Therefore, fluid flows directly upwards, and inclusions are concentrated in the riser or the flux at the ingot top surface as indicated in Figure 5. It is estimated that the fluid flow and inclusion entrapment are related to the shape of sold shell according to Figure 4 and Figure 5.

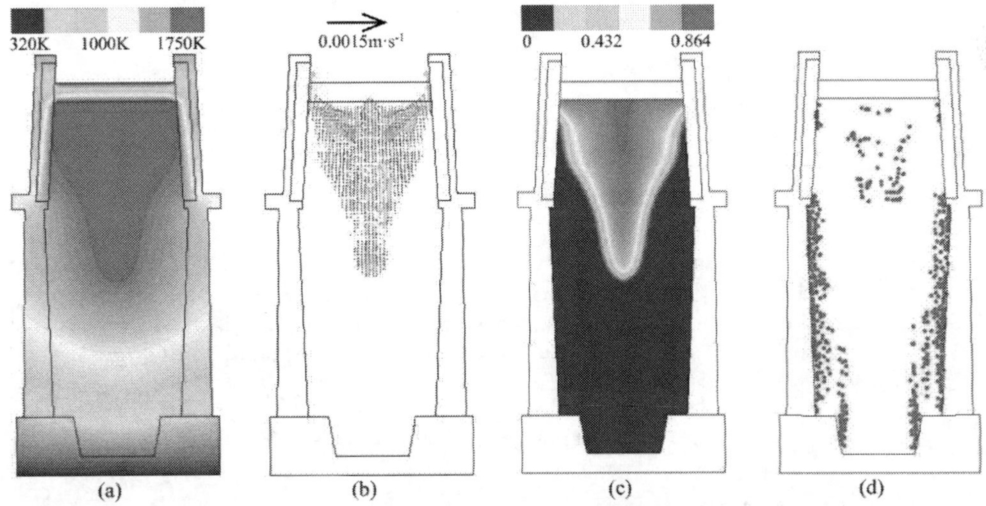

Figure 5. Calculated temperature distribution (a), fluid flow (b), solid fraction (c) and inclusion distribution (d) at the time of 10 h.

In order to research the effect of solid shell shape on the inclusion removal rate, two cooling conditions are proposed with different riser insulation materials. Case 1 is traditional riser with heat transfer coefficient of 60 W/(m^2·K) and case 2 is improved strengthen insulation riser with heat transfer coefficient of 10 W/(m^2·K). Inclusion removal rate is obtained after numerical computation for two cases.

Figure 6 compares the solid fraction of two cases. It is found that the solid shell shape is more like V-type and is more open to the riser at the late stage of solidification for case 1 compared to case 2.

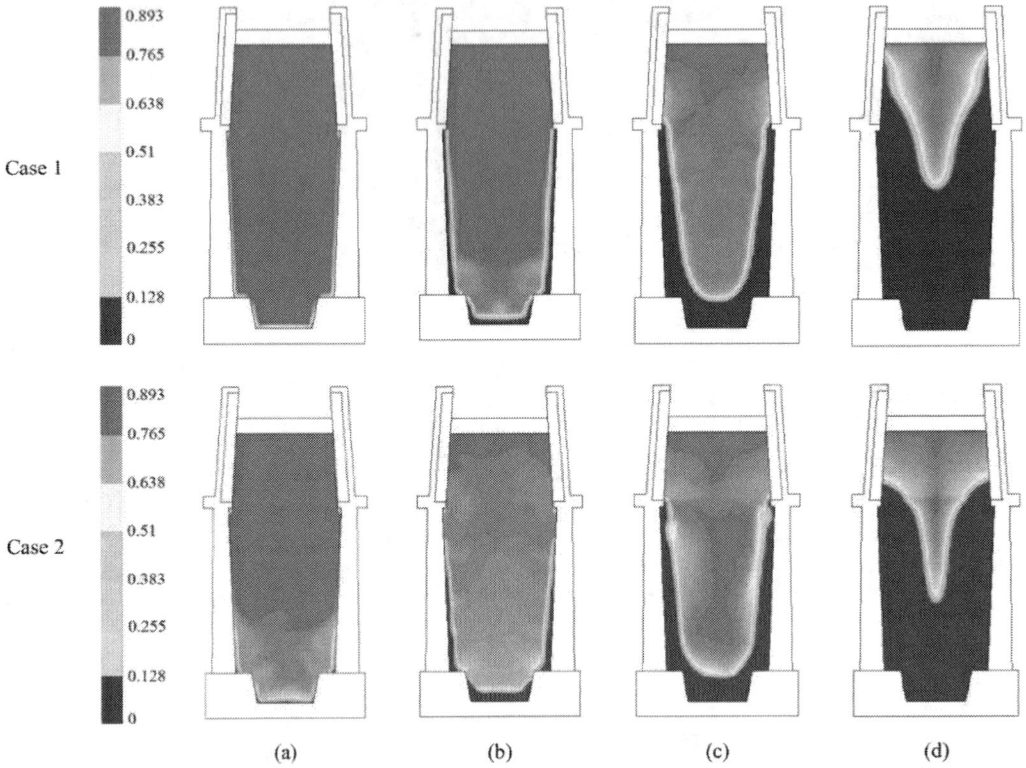

Figure 6. Calculated solid fraction at time of 2 min (a), 26 min (b), 2 h (c) and 10 h (d) with traditional riser (case 1) and improved insulated riser (case 2).

Figure 7 compares the inclusion distribution of two cases. It is found that fewer inclusions are remained in the ingot body and more inclusions are concentrated in the riser or are captured in the flux for case 2 compared to case 1. It should be mentioned that the original inclusions number for the calculation is same for two cases and the inclusions in the flux is not shown in Figure 7, also not in Figure 2 ~ Figure 5.

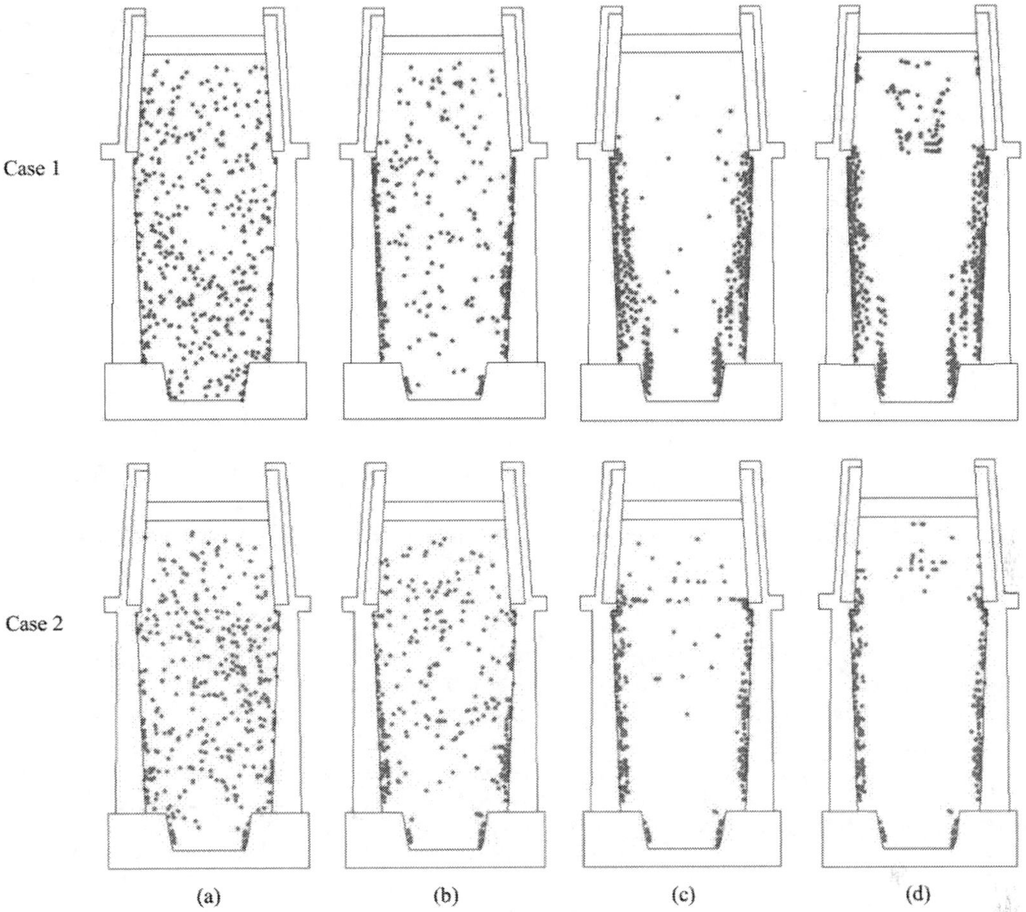

Figure 7. Calculated inclusion distribution at time of 2 min (a), 26 min (b), 2 h (c) and 10 h (d) with traditional riser (case 1) and improved insulated riser (case 2).

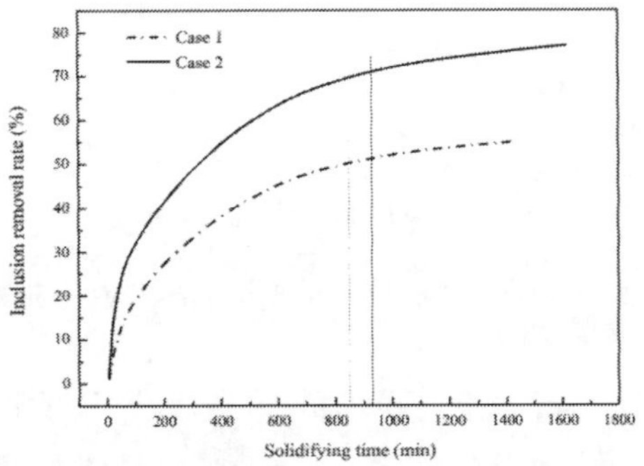

Figure 8. Inclusion removal rate at different cooling conditions.

In order to investigate the effect of cooling conditions on the inclusion distribution in the ingot, the removal rate is applied, which means the percentage of inclusion entrapped in the flux to the initial amount. Figure 8 indicates the inclusion removal rate with solidifying time for two cases. In general, inclusion removal rate increases especially at the early solidification stage for both cases. For a certain time, inclusion removal rate for case 2 is larger than that for case 1 and the difference between them becomes large with time. Because the heat transfer of riser is weaker for case 2 compared to that for case 1, the fully solidification time (i.e., 27 h) is longer than that of case 1 (i.e., 24 h). Along the ingot centerline, the solidification time at the neck of riser for case 2 is 14 h, while that for case 1 is 15 h. It can be found that the inclusion removal rate at the final solidification stage for case 2 and case 1 is 75% and 55%, respectively. When solidification starts in the riser, the inclusion removal rate for case 2 and case 1 is 70% and 50%, respectively. Therefore, the inclusion removal rate can be increased by optimization of the riser of ingot.

This is preliminary work coupling solidification and inclusion by use of CFD, further details of inclusion's character should be awaited with a precise model.

4. Conclusions

According to the numerical simulation of solidification, fluid flow and inclusion distribution in steel ingot, conclusions are summarized as following:

1. Inclusion is easy to be entrapped at the early solidification stage because the solidification is quick and the fluid flow is against the solidifying front.
2. Inclusion is easy to be removed at the late solidification stage because the solidification is slow and the fluid in the liquid pool tends to flow upwards into the riser.
3. Inclusion removal rate increases with solidifying time.
4. Inclusion distribution and removal rate are influenced by the shape of solidification front. Inclusion is easy to be removed with V-type solid shell compared to that with U-type. The inclusion distribution can be improved with control of the heat transfer of ingot.

Acknowledgments

This work was financially supported by the National Basic Research Program of China (No. 2011CB012900) and the National Science and Technology Major Project of the Ministry of Science and Technology of China (2012ZX04012011).

References

[1] Trojan P K 1988 *ASM Handbook,* vol. 15, *Casting* (Metals Park, OH: ASM International) pp 88-97
[2] Baker T J and Charles J A 1971 *Effect of Second-Phase Particles on the Mechanical Properties of Steel*, ed. May M J (Scarborough, UK: The Iron and Steel Institute) pp 88-94
[3] Lunner S E 1970 *Int. Conf. on Production and Application of Clean Steels* (London: The Iron and Steel Institute) pp 124-136
[4] Fuchs E and Jonsson P 2000 *High Temp. Mater. Proc.* **19** 333-343
[5] Tripathi N and Du S 2004 *Scand. J. Metall.* **33** 362-369
[6] Szekely J 1969 *Trans. Metall. Soc.* AIME **245** 341-343
[7] Sumitomo K, Hashio M, Kishida T and Kawami K 1985 *Iron Steel Eng.* **62** 54-61
[8] Freborg A M 1992 *Proc. 50th Conf. on 'Electric furnace'*, Vol. 50 (Atlanta: ISS-AIME) pp 235-243
[9] Zhang L Z and Thomas B G 2006 *Metallurgical and Materials Transactions B* **37**, 733-760
[10] Ragnarsson L, Ek MEliasson A and Du S 2010 *Ironmaking and Steelmaking* **37** 347-352
[11] Zhang D Y, Wen L Y and Shu Y Y 2014 *Metallurgical Reaction Engineering Conference* 18th 411-415
[12] Leonard B P 1979 *Computer Methods Appl. Mech. Eng.* **19** 59-98
[13] VanDoormal J P and Raithby G D 1984 *Num. Heat Transfer* **7** 147-163

Two-phase modelling of equiaxed crystal sedimentation and thermomechanic stress development in the sedimented packed bed

A Ludwig[1,*], A Vakhrushev[2], T Holzmann[1], M Wu[1,2] and A Kharicha[1]

[1] Chair for Modeling and Simulation of Metallurgical Processes, University of Leoben, Austria
[2] Christian-Doppler Laboratory for Advanced Process Simulation of Solidification & Melting, University of Leoben, Austria

[*] ludwig@unileoben.ac.at

Abstract. During many industrial solidification processes equiaxed crystals form, grow and move. When those crystals are small they are carried by the melt, whereas when getting larger they sediment. As long as the volume fraction of crystals is below the packing limit, they are able to move relatively free. Crystals being backed in a so called packed bed form a semi-solid slurry, which may behave like a visco-plastic material. In addition, cooling-induced density increase of both, liquid and solid phases might lead to shrinkage of the whole casting domain. So deformation happens and gaps between casting and mold occur. In the present work, a two-phase Eulerian-Eulerian volume-averaging model for describing the motion of equiaxed crystals in the melt is combined with a similar two-phase model for describing the dynamic of the packed bed. As constitutive equation for the solid skeleton in the packed bed Norton-Hoff law is applied. Shrinkage induced by density changes in the liquid or the solid phase is explicitly taken into account and handled by remeshing the calculation domain accordantly.

1. Introduction

In many practical solidification processes the equiaxed crystals that form first move with the melt when they are small and then sink down and sediment when they are getting larger. Specially, in big steel ingots or in grain refined Al-alloy slabs, large regions usually form by growth and sedimentation of equiaxed crystals. While the melt cools and crystals occur, liquid and solid densities changes with temperature. In addition, often there exist a density change between solid and liquid. For shape and ingot casting or even continuous casting, temperature and phase changed-induced density changes leads to massive shape deformations including the formation of shrinkage cavities and/or microporosity. In last decades, numerical simulation of nucleation, growth, motion and sedimentation of equiaxed crystals have made huge progress especially by means of multi-phase volume averaging methods [1-3]. Also, the simulation of deformation and stresses caused by uneven cooling from solidus to room temperature is a standard procedure today. However, the simulation of flow related phenomena on one side combined with stress and deformation which evolve already during solidification deep inside the mushy zone on the other side is still missing. This paper is supposed to do first steps to close this gap.

In the chemical industry a stack of small objects in a vessel is called packed bed. Much knowledge about permeability of dendritic mushy regions are gained from flow experiments with packed beds. Also the rheological behavior of packed beds made of porous metallic materials saturated with liquid at high temperatures were subject of investigations. Nguyen et al. found for Sn-Pb alloys in the semi-solid state that two functions are needed to describe the corresponding viscoplastic potential [4]. Bellet et al. [5,6] used this viscoplastic potential for the volume-averaged version of the effective macroscopic stress tensor of the solid phase in the mushy zone of a continuous casting of steel. In the present paper, we suggest to combine this approach with a two-phase Eulerian-Eulerian model for motion and sedimentation of equiaxed crystals and a method to deform the mesh in order to account for temperature-induced density changes.

2. Model description

2.1 Basic considerations

Based on former work of the authors [1-3], a two-phase Eulerian-Eulerian model for simulating formation and sedimentation of equiaxed crystals has been adopted and combined with the two-phase thermomechanic model suggested by Bellet et al [5-6]. The following main considerations were made:

Two phases are defined: the liquid phase (ℓ) and the solid (s). The corresponding phase fractions, g_ℓ and g_s, add up to one. A predefined number density of already existing grains was considered. The implementation of nucleation and growth of equiaxed crystals is under construction and will be topic of a subsequent paper. Mass-, momentum and enthalpy conservation is solved for both phases by considering (i) above the packing limit a submerged object drag law [6] and an effective solid viscosity which ensures the proper mixture behavior (Eq. (6),(8),(9)), and (ii) below the packing limit a Darcy term drag law with a Kozeny-Carman type permeability and a viscoplastic materials law for the packed bed (Eq. (6),(10)). Note that following the arguments in [6-7], the pressure equilibrium in the liquid phase is almost instantaneous. Subsequently, the interfacial pressures [8] in both phases equal the intrinsic average value of liquid pressure, i.e. its microscopic value.

We also allow that the density of the liquid and solid may vary with temperature. Thus, we have considered corresponding thermal strain rates according to [9]. For the packed bed regime, we use a volume-averaged Norton-Hoff law. As stated in [10], for metallic alloys at elevated temperature the constitution of the solid phase can be characterized by the Norton-Hoff law. According to [5-6] volume-averaging of this law gives a compressible viscoplastic constitutive model which tends to the incompressible viscoplastic Norton–Hoff model as solidification completes (Eq. (10)). As viscoplastic consistency, K_v, and strain rate sensitivity coefficient, m, we temperature-averaged the information given in [11] for the low carbon steel (with a composition in mass pct. of 0.06C, 0.1Ni, 0.13Mn, 0.1Si, 0.08Cu, 0.035Al, 0.015P, and 0.012S) and got $m = 0.139$. In the current studies we were not satisfied with the K_v value based on [11], predicting rather low effective viscosity at the viscoplastic regime. To derive the value of K_v the assumption of the equality of viscous stress prefactors in Newtonian regime (Eq. (9)) and in viscoplastic one (Eq. (10)) at the coherence limit was made. By ignoring the strain rate influence (assuming $m = 1$) at the transition point, one can derive $2\mu_s = 3K_v/(g_s A)$. From this algebraic calculation, we got $K_v = 7668$ at $g_s = g_{s,pack} \approx 0.637$. Please note that the transition point is somehow arbitrary. With a transition at $g_s = 0.55$ K_v would have been much smaller and so the apparent viscosity of the viscoplastic semi-solid mush. It will be subject of future work to adapt the transition point to a suitable experimental measured K_v.

Table 1. List of equations solved on a grid with moving boundaries. Symbols are defined in the Nomenclature at the end of this paper.

mass conservation

$$\frac{\partial}{\partial t}(g_\ell \rho_\ell)+\nabla\cdot(g_\ell \rho_\ell \mathbf{v}_\ell)=0 \tag{1}$$

$$\frac{\partial}{\partial t}(g_s \rho_s)+\nabla\cdot(g_s \rho_s \mathbf{v}_s)=0 \tag{2}$$

momentum conservation

$$\frac{\partial}{\partial t}(g_\ell \rho_\ell \boldsymbol{v}_\ell)+\nabla\cdot(g_\ell \rho_\ell \boldsymbol{v}_\ell \boldsymbol{v}_\ell)=\nabla\cdot g_\ell \boldsymbol{\sigma}_\ell + g_\ell \rho_\ell \boldsymbol{g} - K_{\ell s}(\boldsymbol{v}_\ell - \boldsymbol{v}_s) \tag{3}$$

$$\frac{\partial}{\partial t}(g_s \rho_s \boldsymbol{v}_s)+\nabla\cdot(g_s \rho_s \boldsymbol{v}_s \boldsymbol{v}_s)=\nabla\cdot g_s \boldsymbol{\sigma}_s + g_s \rho_s \boldsymbol{g} + K_{\ell s}(\boldsymbol{v}_\ell - \boldsymbol{v}_s) \tag{4}$$

$$\begin{aligned} K_{\ell s} &= 18 g_\ell^2 \frac{\mu_\ell g_s C_\varepsilon}{d^2} \quad \text{for } g_s < g_{s,pack} \quad \text{with } C_\varepsilon = 10\frac{g_s}{g_\ell^3} \\ &= g_\ell^2 \frac{\mu_l}{K} \quad \text{for } g_s \geq g_{s,pack} \quad \text{with } K = d\Big|^2_{g_s=g_{s,pack}} \frac{1}{180}\frac{g_\ell^3}{g_s^2}. \end{aligned} \tag{6}$$

$$\boldsymbol{\sigma}_\ell = -p_\ell \boldsymbol{I} + \boldsymbol{\tau}_\ell,\ \boldsymbol{\tau}_\ell = 2\mu_\ell\left(\dot{\boldsymbol{\varepsilon}}_\ell - \tfrac{1}{3}\mathrm{tr}(\dot{\boldsymbol{\varepsilon}}_\ell)\boldsymbol{I}\right),\ \dot{\boldsymbol{\varepsilon}}_\ell = \tfrac{1}{2}(\nabla\mathbf{v}_\ell + (\nabla\mathbf{v}_\ell)^T),\ \mathrm{tr}(\dot{\boldsymbol{\varepsilon}}_\ell) = \nabla\cdot\boldsymbol{v}_\ell \tag{7}$$

for $g_s < g_{s,pack}$:

$$\boldsymbol{\sigma}_s = -p_\ell \boldsymbol{I} + \boldsymbol{\tau}_s,\ \boldsymbol{\tau}_s = 2\mu_s\left(\dot{\boldsymbol{\varepsilon}}_s - \tfrac{1}{3}\mathrm{tr}(\dot{\boldsymbol{\varepsilon}}_s)\boldsymbol{I}\right),\ \dot{\boldsymbol{\varepsilon}}_s = \tfrac{1}{2}(\nabla\mathbf{v}_s + (\nabla\mathbf{v}_s)^T),\ \mathrm{tr}(\dot{\boldsymbol{\varepsilon}}_s) = \nabla\cdot\boldsymbol{v}_s \tag{8}$$

$$\text{with } \mu_s = \frac{\mu_\ell}{g_s}\left(\left(1-\frac{g_s}{g_{s,pack}}\right)^{-2.5 g_{s,pack}} - (1-g_s)\right) \tag{9}$$

for $g_s \geq g_{s,pack}$:

$$g_s \boldsymbol{\sigma}_s = 3K_v\left(\sqrt{3}\dot{\varepsilon}^s_{eq}\right)^{m-1}\left(\frac{1}{A}\dot{\boldsymbol{\varepsilon}}_s - \left(\frac{1}{A}-\frac{1}{3B}\right)\tfrac{1}{3}\mathrm{tr}(\dot{\boldsymbol{\varepsilon}}_s)\mathbf{I}\right) \tag{10}$$

$$\text{with } {\dot{\varepsilon}^s_{eq}}{}^2 = \frac{1}{A}\sum_{i,j}\dot{\varepsilon}_{s,ij}\dot{\varepsilon}_{s,ij} - \left(\frac{3}{A}-\frac{1}{B}\right)\left(\tfrac{1}{3}\mathrm{tr}(\dot{\boldsymbol{\varepsilon}}_s)\right)^2 \text{ and defining } \eta_s^{app} = \frac{3K_v}{g_s \rho_s A}\left(\sqrt{3}\dot{\varepsilon}^s_{eq}\right)^{m-1}$$

enthalpy conservation

$$\frac{\partial}{\partial t}(g_\ell \rho_\ell h_\ell)+\nabla\cdot(g_\ell \rho_\ell \mathbf{v}_\ell h_\ell)=\nabla\cdot(g_\ell k_\ell \nabla T_\ell)+H^*(T_\ell - T_s) \tag{11}$$

$$\frac{\partial}{\partial t}(g_s \rho_s h_s)+\nabla\cdot(g_s \rho_s \mathbf{v}_s h_s)=\nabla\cdot(g_s k_s \nabla T_s)-H^*(T_\ell - T_s) \tag{12}$$

$$h_\ell = \int_{T_{ref}}^{T_\ell} c_{p(\ell)} dT + h_\ell^{ref} \text{ and } h_s = \int_{T_{ref}}^{T_s} c_{p(s)} dT + h_s^{ref} \text{ and } H^* = 10^8 \text{ W/m}^3\text{/K} \tag{13}$$

grain number density conservation and average grain diameter

$$\frac{\partial}{\partial t}(n_s)+\nabla\cdot(n_s \mathbf{v}_s)=0 \tag{14}$$

$$d = 2\cdot\sqrt[3]{\frac{3}{4\pi}\frac{g_s}{n_s}} \tag{15}$$

The two rheological functions $A(g_s)$ and $B(g_s)$ used in Eq. (10) are valid for an equiaxed dendritic network (backed bed) and were taken from [4] and reprinted in Table 2. They satisfy $A(1)=3/2$ and $B(1)=0$, imposing in that way the incompressibility constraint by means of a penalty factor, the multiplier of $\dot{\varepsilon}_s^{th}$ in Eq. (10), which becomes infinite. Thus, the second term of Eq. (10) vanishes and the equations turns into the regular Norton-Hoff law. Numerically, the penalty factor is made arbitrarily high but finite by replacing $B(1)$ by $B(0.999)$. As coherency limit we have chosen $g_{s,cohe}=g_{s,pack}$. All equations solved are gathered in Table 1.

2.2. Numerical implementation

The numerical results are obtained by performing the two phase simulation using the in-house multiphase solver developed in the open-source CFD package OpenFOAM®. The heat transfer is included to track the domain shrinkage due to the temperature dependent density change applying the numerical grid motion / smoothing technique.

Table 2. Rheological function $A(g_s)$ and $B(g_s)$ according to Table 1 in [4]

$$A(g_s)=\frac{3}{2\cdot g_s^{6.47}} \quad \text{for } g_s > g_{s,cohe} \tag{16}$$

$$B(g_s)=0.009\cdot\left[\frac{1}{g_s^{6.94}}-1\right] \quad \text{for } g_s > g_{s,cohe} \tag{17}$$

The mesh motion algorithm is based on the information of the mass fluxes at the domain boundaries required to compensate the volume loss. Therefor a compressible mixture sub-model is applied with the opened domain boundaries as opposed to the closed one for the multiphase model. The boundary displacements are calculated by interpolating the velocity at the face centers to the boundary points. After the boundary points are moved, the internal mesh is smoothed by solving a Laplace equation for the cell centers to maintain the mesh quality [12-14]. This procedure is done for each time step and the coupling between mesh motion and flow calculation is repeated until the zero boundary fluxes are reached.

The presented shrinkage model implementation consists of the following steps:

1. The solution of the multiphase model is obtained for the current iteration.
2. Updated density distribution is calculated.
3. Boundary mass fluxes are calculated from the additional compressible mixture model to compensate the density change, provided from the settling and the heat transfer simulation. Gravity is excluded from the calculations to prevent the liquid / solid mixture run-out.
4. The mesh is moved along the boundaries and sequentially smoothed using a Laplace equation with the provided mesh diffusion coefficient: the bigger coefficient is the more the deformation of the boundaries is smoothed and redistributed through the all mesh elements. Mesh motion fluxes are then included into the convective terms of the multiphase model.
5. The complete multiphase model is recalculated for the next iteration on the modified mesh.
6. Mesh motion / solution updates are continuously repeated until the shrinkage is compensated.
7. The solution is advanced for the next time step.

IOP Conf. Series: Materials Science and Engineering **84** (2015) 012102 doi:10.1088/1757-899X/84/1/012102

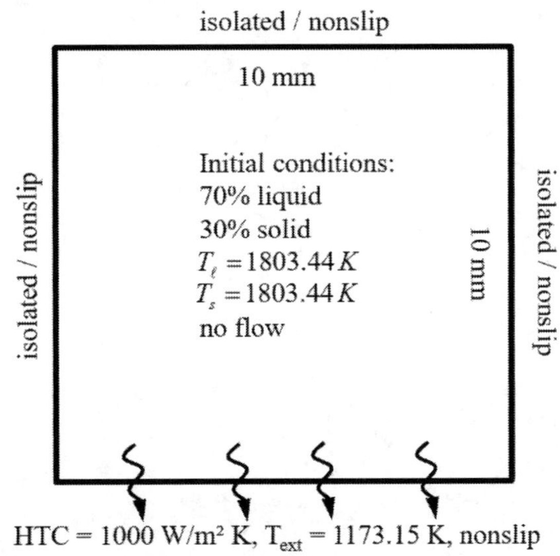

Fig. 1 Domain with the boundary and initial conditions

The simulation domain is shown in Fig. 1 and represents a square cavity (each side is 10 mm), which is isolated from all sides apart of the bottom one. Bottom wall is cooled down with the reference cooling temperature 1173.15 K and the heat transfer coefficient of 1000 W/m²K.

The simulations is started with the uniform distribution of the 70% liquid and 30% solid in the whole domain and the initial temperature being uniform for both phases. All walls of the cavity are nonslip and initially we start with a no-flow situation.

Table 3. Material properties used in the test case

	T_{ref}, (K)	c_p, (J/kg/K)	k, (W/m K)	ρ, (kg/m³)	α, (1/K)	d, (m)
liquid	1803.44	820.6	35.846	7021.95	-	-
solid	1803.44	735.9	35.771	7291	$63 \cdot 10^{-6}$	10^{-4}

3. Results and discussion

Results of the simulation are shown in Fig. 2, where the evolution of the solid fraction distribution is presented. At the very beginning, both phases obey the Newtonian liquid behavior, and the phase separation interface is quite uniform (see Fig. 2a). Later on the high concentration of the solid above the packing limit (marked with a black line at Fig. 2b) is reached at the bottom of the domain; the apparent (effective) viscosity of the solid phase rapidly increases following the visco-plastic rheology described above. The formation of the coherent solid structures with freckles in between (Fig. 2c) providing the passages for the lighter liquid can be seen. During their development the high solid content zones tend to collapse and to merge together, entrapping some liquid pockets.

The density change for the solid phase is based on the temperature distribution and the compressibility factor α (see Table 3). One can follow the deformation history of the cavity with time in Fig. 2a-f. It can be noted that (i) the total domain volume is reducing; and (ii) the bottom of the cavity has shrunk more due to the higher concentration of the solid phase, which is the only phase compressing in the presented studies. Please note that all deformations are multiplied by a factor of 2 to make them noticeable on the figures.

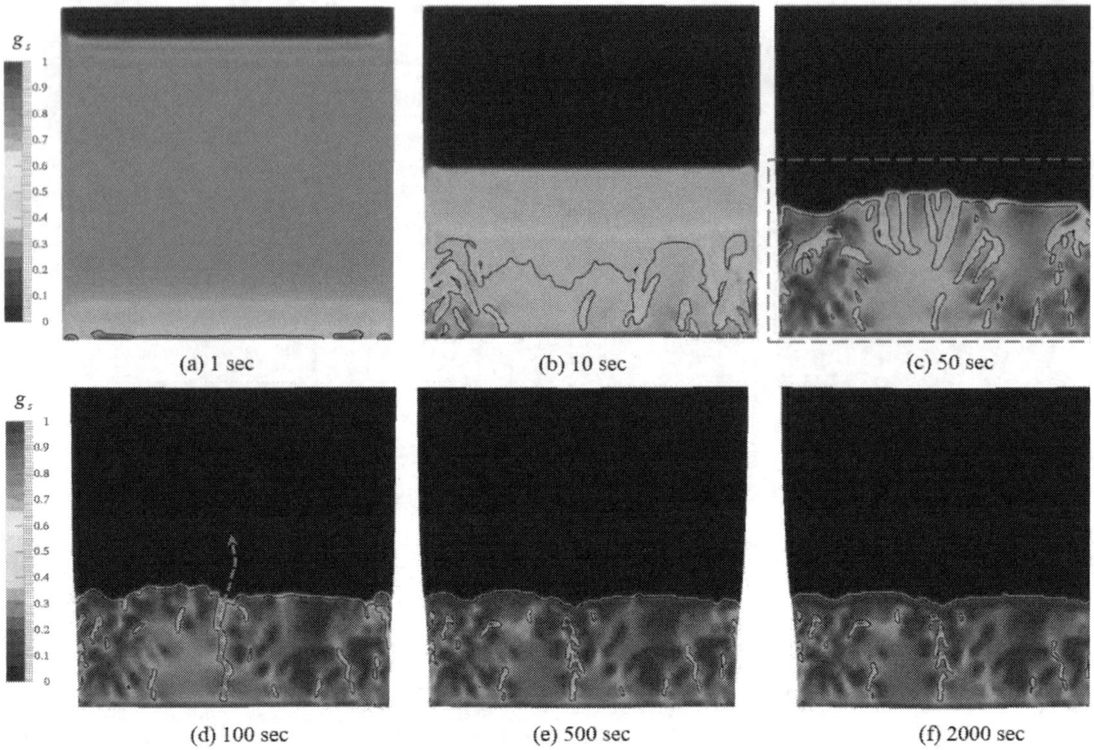

Fig. 2 Evolution of the solid fraction field. The dotted rectangle in (c) is analyzed further in Fig. 3. The arrow in (d) indicated the inner granular melt flow.

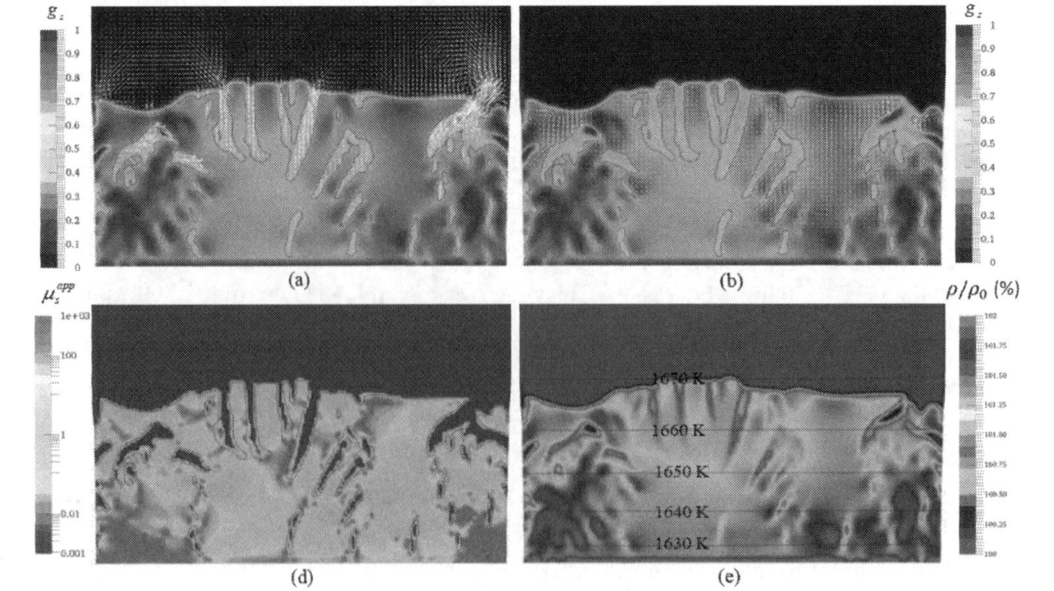

Fig. 3. Results at time $t = 50$ sec: solid fraction with (a) liquid velocity and (b) solid velocity vectors; (c) apparent kinematic viscosity of the solid μ_s^{app} distribution; (d) shrinkage rate: ratio of the calculated mixture density $\rho(T)$ related to the initial isothermal density ρ_0; isotherms show the mixture temperature distribution.

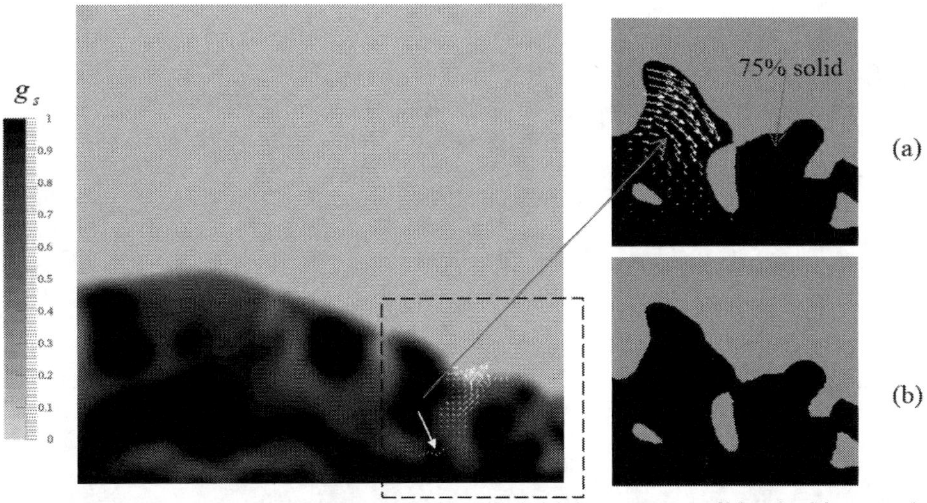

Fig. 4. Collapse of a single solid phase structure: (a) initial moment; solid velocities are shown as vectors; (b) collapsed structure is merged with the more stable one near the cavity wall. The velocity vectors in the large picture shown the liquid velocity escaping from the mush.

With time the initially more or less uniform phase interface becomes quite distorted due to the formation of stable solid (sub-)structures. More other, some zones with higher liquid content (here called liquid pockets) are entrapped between the solid structures.

After the solid phase finally settles at the bottom of the domain, the rest of the liquid floats up through the channels (similar to one marked with a red arrow in Fig. 2d) before they are closed with the coherent solid zones (see Fig. 2e). The stresses occurring in the coherent solid structures are so strong, that they resist any pressure from the uprising liquid, and the rest of it remains entrapped and the final shape of the solid fraction field remains almost unchanged until the system cools down, as shown in Fig. 2e-f.

As it was already mentioned, the heat transfer is simulated together with the solid phase sedimentation causing local density change in the solid phase and the shrinkage of the domain as a consequence of that. Fig. 3 shows the details of the solid / liquid phases interaction at $t = 50$sec at the bottom part of the simulation domain, marked with the red dashed rectangle in Fig. 2c. One can see that liquid tends to pass through the freckles / channels in the coherent solid structures, as shown with the vector field in Fig. 3a. Correspondently, the solid phase sinks down and its motion is dominantly downwards (as one can see from Fig. 3b). However, due to the high viscoplastic stresses which appear above the packing limit, the solid velocities are much smaller if compared to the liquid ones (vector scaling is the same in Fig. 3a-b). The distribution of the apparent (effective) viscosity μ_s^{app} of the solid is plotted in Fig. 3d: its maximum value is 6÷9 orders higher than in the liquid, causing a strong resistance to the strain in the dense solid regions. Additionally the shrinkage ratio is presented in Fig. 3e corresponding to the increased density (due to the thermal compression in the domain) to the initial density relation.

Finally, an example of the coherent behavior in the highly packed solid is illustrated as an avalanche / collapsing of the single bound structure (see Fig. 4). The results are taken from an additional simulation not included in the current study. So, if we follow the motion of the marked object, its starts to fall down in the direction of a white arrow in Fig. 4 and to rotate simultaneously, after the small bridge (marked with the red dotted line) collapses under its weight and the liquid escapes from a freckle formed in between coherent solids. In Fig. 4a solid velocities are shown as a vector field indicating that the bound solid structure moves as a single rotating object, until it impacts by its tip and merges with another more stable one close to the cavity wall.

4. Summary
In order to model both, motion and sedimentation of equiaxed crystals together with stress and deformation in the backed bed, a two-phase Eulerian-Eulerian volume-averaging model is suggested which uses different constitutive models for the stress-strain rate relation of the solid phase below and above the packing limit. In addition, shrinkage induced deformation of the whole domain is accounted for by an adequate remeshing model. As result, the formation of a backed bed of sedimenting equiaxed crystals is models together with the melt flow induced by the sedimentation process. As the solid skeleton is modelled by a visco-plastic flow model, also deformation of and stress evolution in the backed bed is predicted. The results show that the actual dynamic of sedimentation and visco-plastic and density-related deformation might be quite complex as semi-solid channels where liquid can flow more easily might exists in between coherent solid area. The actual dynamic of this process was found to depend highly on the assumed material parameter. As an example, the formation of a continuous crystal pill up and a corresponding initiation of a crystal avalanche / collapsing was shown. The model is the first step towards describing the formation of the solidifying shell together with the occurring gap between mold and metal during thin slap casting of steel.

Acknowledgement
This work was financially supported by the FWF Austrian Science Fund (P22614-N22), FFG Bridge Early Stage (No. 3893791), and the Austrian Federal Ministry of Economy, Family and Youth and the National Foundation for Research, Technology and Development within the framework of the Christian Doppler Laboratory for Advanced Process Simulation of Solidification and Melting.

Nomenclature

A,B	rheological function taken from [4]	m	strain rate sensitivity coefficient [-]
$c_{p(\ell)},c_{p(s)}$	ℓ/s specific heat [J/kg/K]	n_s	grain number density [-]
C_ε	settling ratio [-]	p_ℓ	intrinsic pressure in the liquid [kg/s^2/m]
d	vol. averaged grain diameter [m]	$\mathbf{v}_\ell,\mathbf{v}_s$	ℓ/s vol. averaged velocity [m/s]
$\mathbf{g}$	gravity acceleration [m/s^2]	T_ℓ,T_s	ℓ/s vol. averaged temperature [K]
g_ℓ,g_s	ℓ/s volume fraction [-]	T_{ref}	reference temperature [K]
$g_{s,pack}$	packing limit [-]	ρ_ℓ,ρ_s	ℓ/s densities [kg/m^3]
$g_{s,cohe}$	coherency limit [-]	$\boldsymbol{\tau}_\ell,\boldsymbol{\tau}_s$	ℓ/s vol. av. dev. stress tensor [kg/s^2/m]
h_ℓ,h_s	ℓ/s volume averaged enthalpy [J]	μ_ℓ,μ_s	ℓ/s viscosity [kg/m/s/]
h_ℓ^{ref},h_s^{ref}	ℓ/s enthalpy at T_{ref} [J]	μ_s^{app}	apparent kinematic viscosity of the solid
$\mathbf{I}$	identity tensor [-]	α	compressibility [1/K]
K	much permeability [m^2]		
K_v	viscoplastic consistency [kg s^{m-2}/m]		
k_ℓ,k_s	ℓ/s eff. heat conductivity [J/m^2/K]		

References
[1] A Ludwig and M Wu 2002 *Metall. Mater. Trans. A* **33**, 3673-83
[2] A Ludwig and M Wu 2005 *Mater. Sci. Eng. A* **413-414** 109-14
[3] M Wu and A Ludwig 2006 *Metall. Mater. Trans. A* **37** 1613-31
[4] TG Nguyen, D Favier, and M Suery 1994 *Int. J. Plast.* **10** 663–93
[5] M Bellet 2007 in 9th Int. Conf. Numer. Methods Ind. Form. Process., J.M.A. Cesar de Sa and A.D. Santos, eds., American Institute of Physics, Porto, Portugal, 1369-74

[6] V Fachinotti, S Le Corre, N Triolet, M Bobadilla and M Bellet 2006 *Int. J. Numer. Methods Eng.* **67** 1341-84
[7] CY Wang, S Ahuja, C Beckermann, and HC De Groh III 1995 *Metall. Mater. Trans. B* **26** 111-19
[8] J Ni and C Beckermann 1991 *Metall. Trans. B* **22** 349-61
[9] M Bellet and BG Thomas 2008 in ASM Handbook, Vol. 15 Cast., ASM International, 449-61
[10] M Rappaz, M Bellet and M Deville 1998 Numerical Modelling in Materials Science and Engineering, Springer-Verlag, Berlin, Heidelberg, New York
[11] PF Kozlowski BG Thomas JA Azzi and H Wang 1992 *Metall. Mater. Trans. A* **23** 903-18
[12] H Jasak, and Ž Tuković 2007 *Trans. FAMENA* **30** 1-18
[13] H Jasak and Ž Tuković 2010 V Eur. Conf. on Computational Fluid Dynamics 1-6
[14] T Marić J Höpken K Mooney 2014 *The OpenFOAM Technology Primer* **1** 379-391

AUTHOR INDEX

AUTHOR INDEX

AUTHOR INDEX